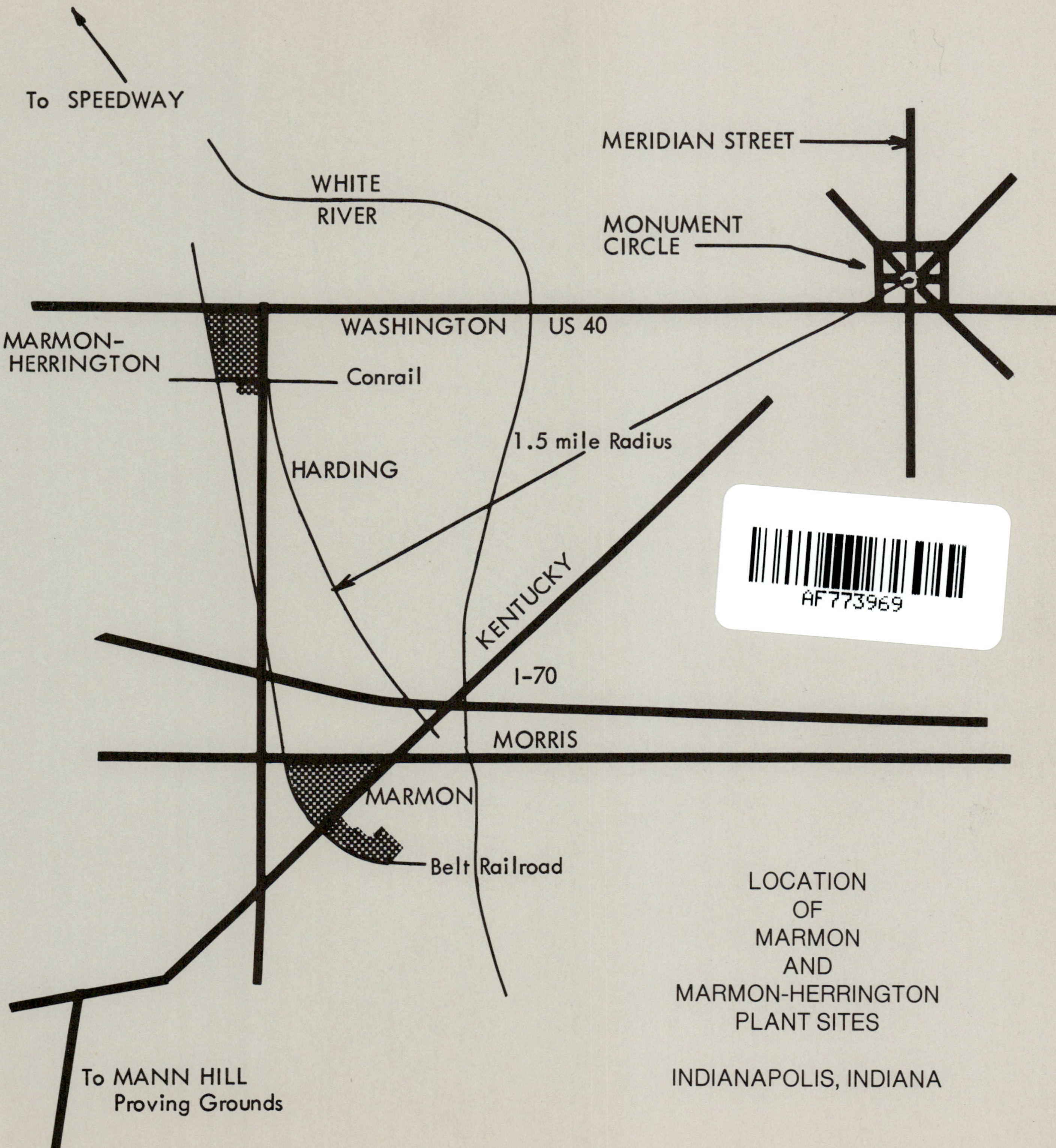


LOCATION
OF
MARMON
AND
MARMON-HERRINGTON
PLANT SITES

INDIANAPOLIS, INDIANA

# THE MARMON HERITAGE

George Philip Hanley
Stacey Pankiw Hanley

Awarded the prestigious

THOMAS McKEAN MEMORIAL CUP AWARD

for significant research
in automotive history

by the
ANTIQUE AUTOMOBILE CLUB OF AMERICA

Awarded the International

NICHOLAS JOSEPH CUGNOT AWARD

for the

OUTSTANDING BOOK OF THE YEAR

by the

Society of Automotive Historians

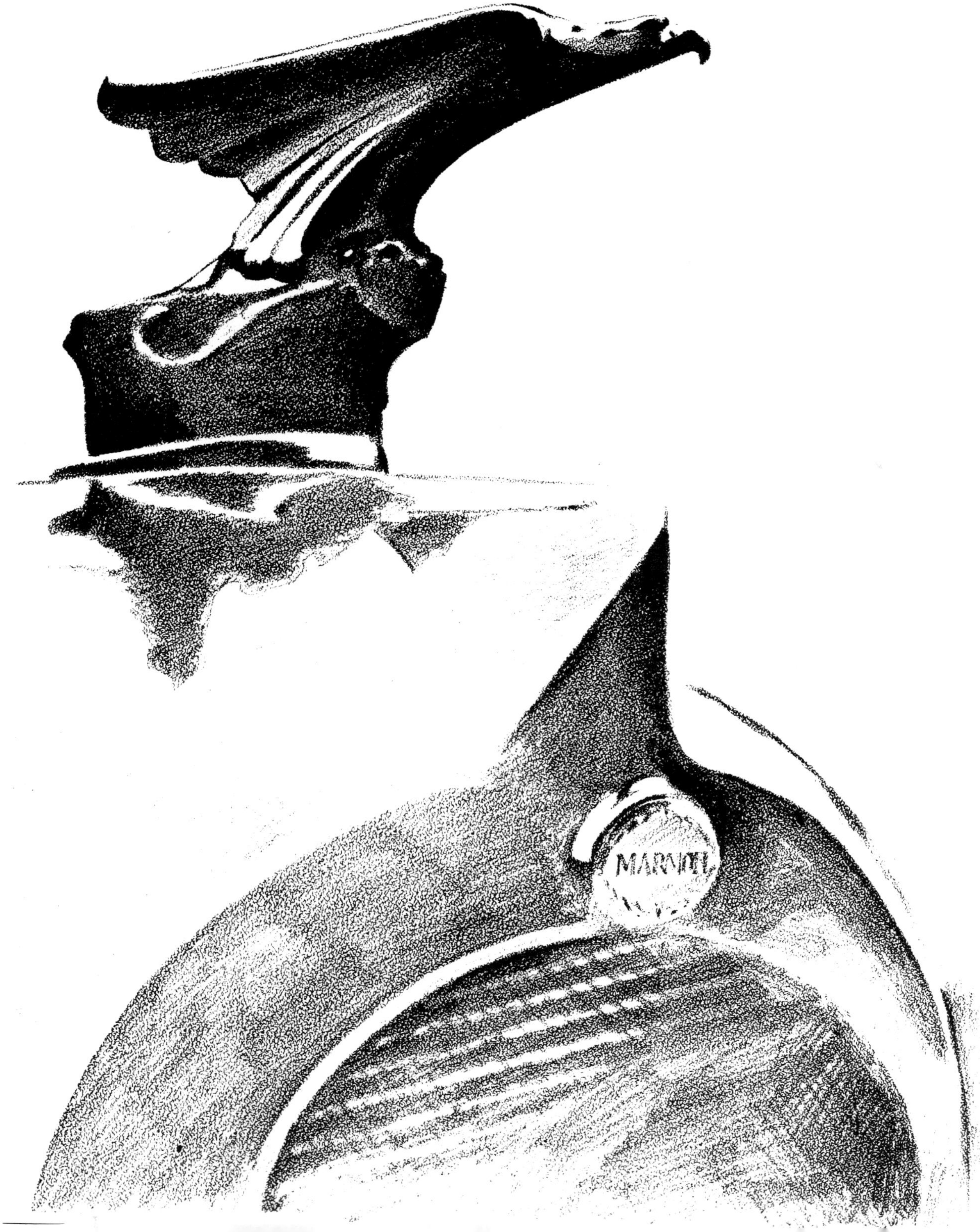

# THE MARMON HERITAGE

MORE THAN 125 YEARS OF
AMERICAN PRODUCTION
OF
WORLD RENOWNED PRODUCTS

MARMON
*
MARMON-HERRINGTON
*
MARMON-HERRINGTON
ALL-WHEEL-DRIVE FORD

BY

GEORGE PHILIP HANLEY
STACEY PANKIW HANLEY

For information on this publication contact: **Doyle Hyk Publishing Company**
**P.O. Box 1021**
**Rochester, Michigan 48308-1021**

1st Edition, Second Printing. 1990



**ISBN** 0-9615817-0-0

Library of Congress Cataloging in Publication Data

85-73082

PRINTED IN U.S.A.
TAYLOR PUBLISHING COMPANY
DALLAS, TEXAS

The authors dedicate this book to their children

JOHN

WILLIAM

MARY

# PREFACE

This work was undertaken for a number of significant reasons and both the content and organization have been influenced correspondingly.

The Marmon Heritage was primarily written because it needed to be . . . the influence of the heritage is overwhelming, yet the lack of knowledge of its origins and significance is staggering. It was written by this author because circumstances over a large segment of his life had peculiarly suited him to the task. The combined need for the subject matter to be set down and the authors' presumed credentials to undertake its accomplishment imposed an obligation that the manuscript be undertaken.

The Marmon Heritage is the comprehensive story of one very significant automobile manufacturer. It is the combined stories of the people who established and built it, of how they financed it, of the products they conceived, of how they marketed them, and of the trials and travails of economic and consumer vacillations, of depression and war . . . all appropriately told in terms of the technology leader, "The Marmon Organization," as it performed at Nordyke & Marmon, at Marmon Motor Car Company, at Marmon-Herrington.

Aside from the technology leadership aspect, the Marmon car is particularly well suited to a study of the progressive development of automotive styling because the thirty year period of manufacture can be divided into five periods during each of which a relatively stable basic chassis design was offered. These are designated "Eras" because they were also characterized by distinct management teams and faced different market circumstances.

The treatment of the subject is comprehensive. For that very reason it is unlikely that a complementary book would be undertaken by a subsequent author. That presumption imposes an obligation to the author and perhaps a small burden on the more casual reader. Information to be found only in the more obscure sites would therefore very likely be 'lost' for all practical purposes were it not included in The Marmon Heritage.

One segment of information is of great interest only to two small groups: those who restore Marmon and Marmon-Herrington vehicles and strive for true authenticity, and those who by hobby or profession are interested in the development of the art of motor vehicle finishing. The author has sought to accommodate such needs by segregation of special interest information, eg, the sub-chapter titled 'Art & Color' located in the latter part of the Marmon Car treatment.

Similarly, non-product aspects of the business are addressed in depth but isolated in individual chapters. The financial history of a firm begun only 75 years after the Declaration of Independence, which survived to spawn a descendant firm ranked in the prestigious FORTUNE 200 roster, warrants recording as does the identification of the crucial management teams responsible.

The chapters on the Marmon and Marmon-Herrington manufacturing facilities, while of specialized interest, are probably of much broader appeal. Significant in Industrial Archaeology is a site currently occupied by two highly viable industries, which includes the Nordyke & Marmon site of 1876 and the expansion site and structures added during the first World War and immediately thereafter. Both it and the Marmon-Herrington site . . . the Birthplace of the Arsenal of Democracy . . . warrant recognition on the National Roster of Historic Places.

Throughout his progress through the book, the reader will want to bear in mind the then contemporary size not only of the auto industry but of the national population, both startlingly small by todays standard, and of the great size of the dollar bill now grossly shrunken, much more so in purchasing power than in physical dimension.

## ABOUT THE AUTHORS

George and Stacey Hanley have shared an interest in the history of the automobile and of the Marmon in particular throughout their married life. They restored and own a prize-winning Model 34 Marmon and were organizers of the Henry M. Leland Chapter of the Society of Automotive Historians. Both are members of the Antique Automobile Club of America and of the Marmon Club. George's interest in the Marmon story began when his father purchased a Model 34. The evident superiority in handling and engineering detail of that car, reinforced by his brother's ownership of a Marmon Sixteen, was a major determinant in his career choice. George is a graduate of the University of Detroit and a former faculty member; Stacey is a graduate of Oakland University. He is a registered professional engineer and retired engineering executive of General Motors. In 1988 he was honored by being elected a fellow of the Society of Automotive Engineers.

Publication of the book marks the end of a project born as a joint hobby which matured into a projected book nearly a decade ago and quickened into an ever more compelling pace after George's retirement in 1981.

A work of the depth and complexity of The Marmon Heritage would be impossible without the dedicated assistance and encouragement of my wife Stacey. Beginning with library research, she became more and more involved as she typed several iterations of the manuscript, eventually drawing on her art training to undertake the design of the book and to personally layout the hundreds of pages of closely coordinated copy and supporting illustrations.

# ACKNOWLEDGMENTS

This book benefitted immeasureably from the files and staff assistance of the Motor Vehicle Manufacturers Association of the United States, Inc. — recipient of the first James Bradley Award for Literature Preservation of the Society of Automotive Historians. Particular thanks are due Ms. Bernice Huffman of the Public Relations Staff and Mr. James Wren and his Patent Library Staff.

The staff and files of the National Automotive History Collection of the Detroit Public Library were of significant assistance. The help and encouragement of the late NAHC / DPL Curator James Bradley and the late George Risley as well as the current staff is acknowledged.

Mr. Robert C. Wallace and Mr. Harold Blank contributed both by extensive interviews and by the loan of historic documents and photographs.

The late Jerome Felts shared his experiences both with the author and with the Marmon Club through his authorship of memoirs of his Marmon employment years.

Ms. Dorothy M. Ross, President of the Automotive Hall of Fame at Midland, Michigan, made available factory release photographs from the AHF Library.

The author is indebted to Ms. Betty J. Starns, Assistant Archivist for Eli Lilly and Co. for furnishing site plans and structure identifications on the former Marmon site, and to Mr. Lloyd A. Brown, a Canadian member of the Society of Automotive Historians, and an authority on Canadian Armed Forces vehicles for his contributuion in that regard. Mr. Fred M. Young, Chairman of the Young Radiator Co., provided critical documents relating to the origins of the Marmon-Herrington Co.

Mr. Marion (Slim) Swinford arranged access for the author to what remained of the pre-merger Marmon-Herrington files at Marmon Transmotive, and provided numerous significant photographs from the post-war period.

Mr. John McAlpine, retired from a career in automotive advertising, reviewed the sub-chapter on advertising and provided valuable comments.

The contributions of other Marmon historians who in the past have served to sustain interest in the marque and importantly to preserve significant facts uniquely available to the individual author must not be overlooked; the author acknowleges gratefully their dedicated research and its contribution to this work. They are:

Dr. Charles W. Bishop
John R. Bond
Griffith Borgeson
Napoleon Boz
Barney Geoff
Marion George
William R . Gibson
Maurice Hendrey
Eugene Janderquist
John Lamm
Edmund L. Robinson
W. Frank Robinson, Jr.
David H. Ross
Clifford Sklarek
William S. Snyder
Frank Taylor
Charles L. Betts
James Quinlan

# ACKNOWLEDGEMENTS, (continued)

The existence of a thriving and enthusiastic Marmon Club today is primarily the consequence of the pioneering efforts and dedication of three people. The Club was established in February 1970 and initially nurtured by Russell M. Stadt of Grand Rapids, Michigan. When personal commitments forced Mr. Stadt to place the club on an inactive basis, Philip and Leona Belote of Dunedin, Florida reactivated it and laid the groundwork for the current club organization. Bruce R. Williams, Editor and Publisher of MARMON NEWS, is responsible for developing a strong espirt d'corp among the more than 200 members. The club officers in the 1980s have been Randall "Duke" Marston, President, Ballard Crooker, Jr., Vice-president, Ben Biedron, Treasurer, and Bruce S. Williams, Secretary. Bruce S. is a son of G.M. Williams who was President of Marmon. The first edition of *The Marmon News* was dated April 1, 1970; *Marmon News* bowed January 2, 1978.

A brief history of the Marmon car was written by Clifford Sklarek for Antique Motor News & Atlantic Auto Advertiser, January 1972 issue. It is recognized herein because Mr. Sklarek found his first employment, in St. Louis, Missouri working as a chauffeur for Joe Mathes. Mr. Mathes not only owned the Mathes Iron Works in that city . . . he owned a 1906 Marmon V 4 touring car! Mr. Sklarek acknowledges that the opportunity to drive the Marmon was so thrilling that the job could have been paid for with "just thanks," but he did receive $25.00 per week which the recipient noted was a princely sum at that time. He subsequently owned a Model 32 roadster in the primitive lumber country of northern Michigan.

The book has benefitted significantly from the reporting of Hugh Dolnar. Mr. Dolnar was a member of the editorial staff of Cycle and Automotive Trade Journal in the first decade of the century. His scientific and technical comprehension and perceptiveness resulted in his recording of significant automotive design and engineering details that would otherwise have been unobserved and undocumented.

# THE MARMON HERITAGE

## BOOK ONE — THE MARMON Story

## BOOK TWO — The Story of MARMON-HERRINGTON

# REFERENCE IDENTIFICATION

References are identified by either a letter or number in parenthesis inserted in the text. Those identified by a letter provide supplemental information which some readers may find useful; the source or explanatory statement will be found at the bottom of the page. References identified by a number provide source information; the source will be found in the List of References under the particular chapter and the specific reference number.

The source name, if a periodical, will be typically a 1 to 3 digit letter code, followed by a 6 digit day, month and year, and ending with the page number (s) preceded by the letter p. MA for March, MY for May, JU for June and JY for July. The periodical code identifications appear on page 15 and on page 608.

Photographs, illustration and charts are numbered collectively and consecutively for each chapter. For example, Figure 7-19 would be the 19th in Chapter 7.

# ABBREVIATION CODE FOR SOURCE REFERENCE LIST

| | |
|---|---|
| AA | Antique Automobile/Antique Automobile Club of America |
| AB | Autobody |
| AE | Automotive Engineering/SAE (Society of Automotive Engineers, Inc.) |
| AI | Automotive Industries |
| AN | Automotive News |
| AQ | Automotive Quarterly |
| AQP | Automotive Quarterly Publications |
| AT | Automotive Topics |
| ATJ | Automotive Trade Journal |
| CATJ | Cycle and Automotive Trade Journal |
| C & P | Cars and Parts |
| CC | Classic Car |
| FDSF | Ford Dealer and Service Field |
| HA | Horseless Age |
| Indpls | Indianapolis . . . |
| MA | Motor Age |
| MBF | Marmon Book of Facts published annually 1920-29, Education Div, Sales Promotion Dept. |
| MTR | Motor (US) |
| MW | Motor World |
| NAHC | National Automotive History Collection/Detroit Public Library |
| R & T | Road and Track |
| SEP | Saturday Evening Post |
| SIA | Special Interest Autos |

Infrequently referenced sources are fully identified at specific reference location.
Certain complex references are referred at specific chapter reference to a prior chapter reference.

3120 N Y

# BOOK ONE
# THE STORY OF MARMON

CHAPTER ONE

## "ONE OF THE WORLD'S GREATEST ENGINEERING HOUSES"

### PART 1 — The Prestigious Firm

A brief review of the fifty year history of the Nordyke & Marmon firm, which preceded the involvement in the automobile industry, will introduce the reader to this remarkable enterprise. It is not the intent of the author to document a non-automotive product history. Rather an understanding of the stature and technological prowess of "One of the World's Greatest Engineering Houses" will be helpful in comprehending the resources which were available for the concept and development of a unique automotive product at the dawn of the twentieth century.

The Indianapolis Herald for Jan. 7, 1888[1] reporting on the various enterprises and public works in the city and suburbs, described the suburban Nordyke & Marmon Co. facilities, in West Indianapolis, as being of immense proportions . . . an enterprise the largest of its kind in the country and increasing annually.

The reporter continued: *"The plant altogether occupies an area extending over eleven acres, upon which are erected a number of buildings, each of which is devoted to a specific department. The general divisions are: the woodworking shops, the iron working department, the mill-stone and portable department and the foundry, in addition to a number of minor subdivisons. The machinery in operation constitutes the most complete equipment of this character, much of which has been specially constructed to fill the requirements of this establishment. Their methods of mill construction are the result of years of close application and a thorough and complete knowledge of the wants of the milling industry. The company are manufacturers of what is known as the Patent Nordyke & Marmon Roller Mill in addition to a large quantity of grain elevator machinery. The reputation which their mills have achieved throughout the country furnishes a far stronger comment as regards their superlative merits, than any remarks that we might presume to offer on the subject. The operations of the works afford constant employment to a force of about four hundred mechanics. As an indication of the extent of the Company's operations, we will state the fact that the yearly sales approximate from six to seven hundred thousand dollars in value, fifty thousand of the amount last year being for foreign trade."*

This "World Renouned Manufactory of Flour Milling Machines"[2] produced, marketed and optionally erected flouring mills, rice mills and starch works.

#### *Innovative Products*

The National Centennial Year of 1876 was also the Silver Anniversary of the Nordyke & Marmon Co. The Bicentennial Year of 1976 would be the 125th Anniversary of the founding and the 73rd year of continuous production of Marmon (and/or Marmon — Herrington) badge vehicles in the USA.

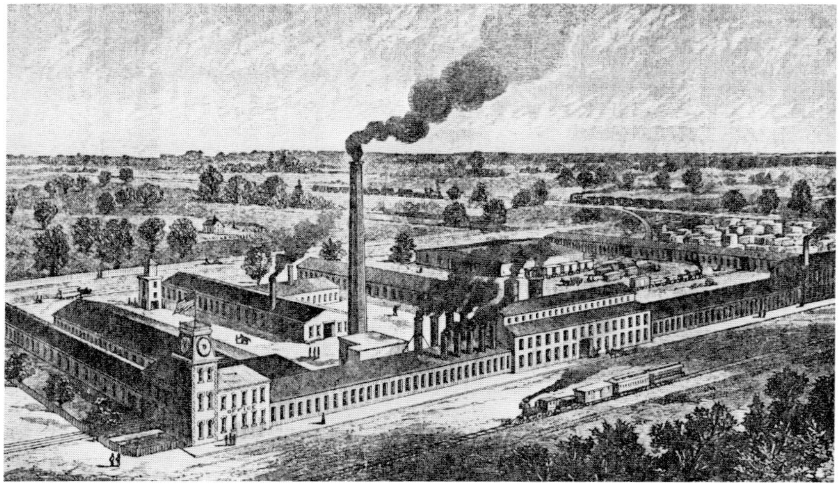

Fig. 1-1 Nordyke & Marmon Shops in suburban Indianapolis in 1892.

The Smithsonian Institution as their contribution to the national celebration of the bicentennial of the United States, recreated in their old exhibition-hall the Philadelphia Centennial Exposition of 1876. It was a remarkable accomplishment with many of the original exhibits located and refurbished, and others recreated from original drawings. For example a locomotive manufactured in 1876 was located in Mexico, restored by its manufacturer and displayed in place of the specific unit shown at Philadelphia. Other exhibits were contrived to be representative of the era and its products.

The visitor was instantly transported back through a century of time and dramatic development. The official State of Michigan exhibit was redisplayed . . . a small table with labeled mineral specimens upon it, two adjoining walls bearing maps of the state with railraod trackage and mineral deposits respectively shown.

A machinery dealership was displayed, adorned with signs truly representative of the 1870s. Conspicuous in the forefront of the establishment was a Nordyke & Marmon Corn Mill, restored to what a car collector would describe as a 100 point condition. The mill was restored and loaned by Mr. Robert L. Johnson, founder of Whistles in the Woods Museum of Early Technology in Rossville, Georgia.

Reference to Nordyke & Marmon catalogs preserved in the Indiana State Library shows it to be an example of:

Nordyke's Under-Runner Geared Mills
Awarded Highest-Prizes at Cincinnati, St. Louis, New Orleans
and Indianapolis Fairs and Exhibitions.

A most appropriate candidate for the recreated exposition. The prize-winning product exhibits surprising technology. The 'mortised' bevel ring gear incorporated individual hard-wood involute teeth, presumbly to allow grease-free operation in the dusty environment; dust removal systems were an optional extra, however. Other technical features are best described by reproducing the original text: *"For grinding wheat, corn seed, etc. The runner is balanced upon a steel point, or cock-head, projecting into a steel seat, and, in factory, is put into running balance*

Fig. 1-2 N & M restored corn mill in the 1976 Smithsonian recreation of the Philadelphia Centennial Exhibition of 1876.

*up to the full grinding speed. The driving irons in this mill are the self-adjusting kind, so long manufactured by us. No mill is perfect without them; with them they are more desirable for wheat griding than any other under-runner mill. The bed-stone is stationary in the upper part of the husk frame, and is turned over on heavy hinges whenever necessary to sharpen the buhrs. Four stiff rubber springs around bolts are provided above the bed-stone frame, holding it down while grinding, for the important duty of allowing said bed-stone to yield in case any hard substance enters between the stones. In addition we provide a metallic bush arranged to oil collar of spindle while mill is running. The curb is made of pine staves, and banded with neat iron bands. The runner has an iron and stone back to prevent any inclination to wear out of balance. The husk is made heavy and very strong. Does not require an experienced miller to successfully operate it."*

The example appears to be the smallest of three sizes: The 26 inch model which turned 440 revolutions per minute and had a capacity of 15 to 25 bushels. Weighing 1650 pounds and priced at $245, it cost but 15 cents per pound!

Production quantities of this corn mill are not known but the 1893 catalog attested to over 8000 Nordyke & Marmon reel-type flour dressing machines having been sold. It also illustrated an early and long term success: The Plantation Mill . . . a Vertical Grinding Mill for processing corn or corn and oats into family meal and (livestock) feed, and wheat into graham flour. This product had already been manufactured for twenty years and the original illustration was used, consequently bearing the pre 1876 "Nordyke, Marmon & Co." identification. The grinding surfaces were stated to be "genuine French buhr stone cemented into iron cases because that is the only substance that will last a life-time. The factory-dressed stones will grind 8000 to 10000 bushels before requiring redressing according to instructions sent with each mill."

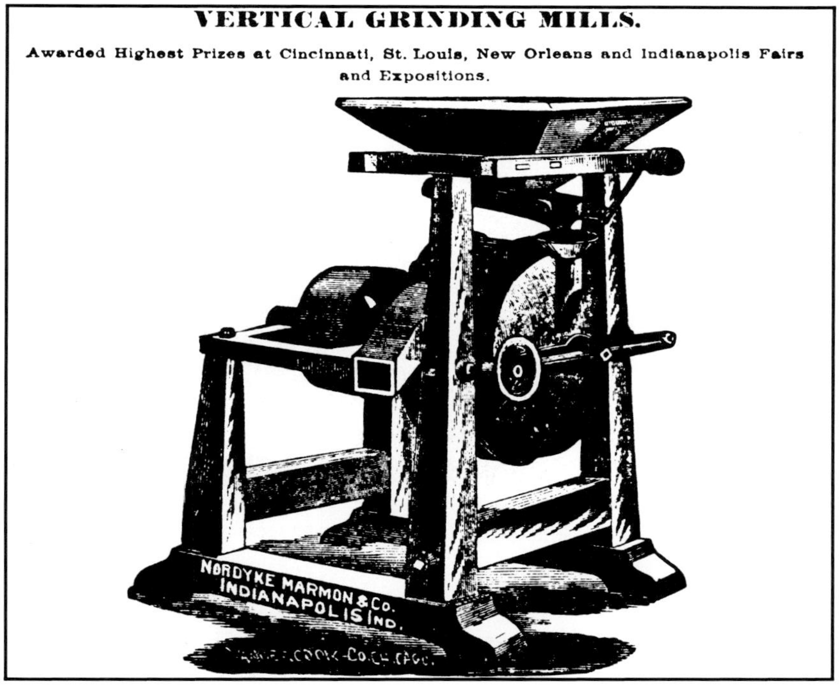


Fig. 1-3 N & M Vertical Mill from 1892 Catalog.

Also manufactured were "Iron Hurst" machinery support frames required for erection of the complete flouring mill.

The N & M Co. catalog listed in addition to hundreds of individual machines a variety of complete flouring mill packages offered complete with free architectural plans. Should the customer wish, a millwright supervisor could be provided to direct on-site erection or the complete mill would be constructed, erected and handed over on a 'turn — key' basis.

An example of such a complete package is the two story, 20 by 24 foot, frame building with an 8 foot basement, 10 foot high grinding floor and a 14 foot high 'bolting' floor equipped, including steam drive engine and a boiler, as a 30 barrel short system flouring mill which would produce 30 barrels of flour from 135-140 bushels of wheat. Price including erection (but not building construction which varied with locality) was about $4500.

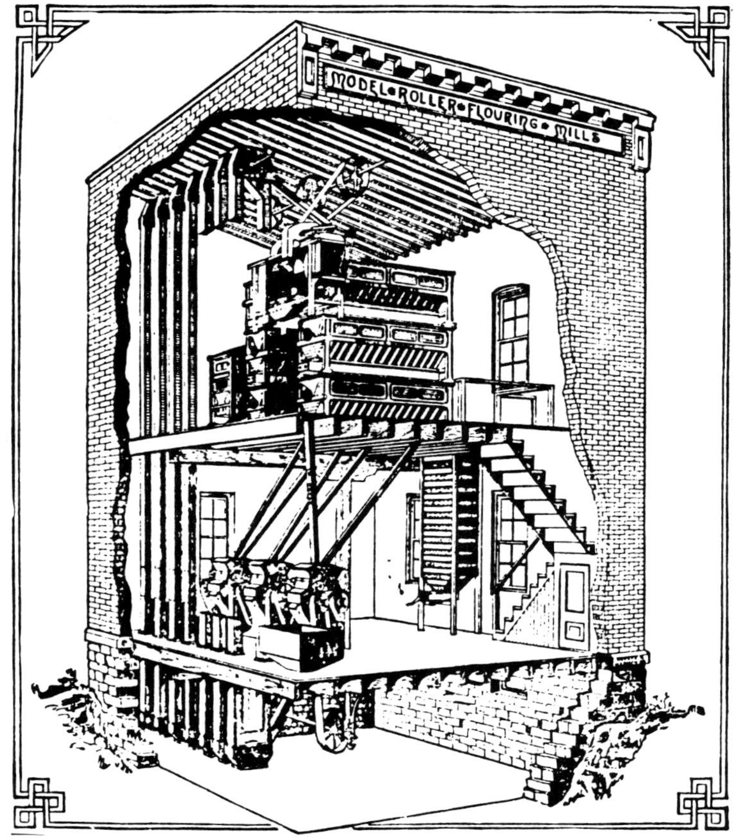


Fig. 1-4 N & M 30 Barrel Flouring Mill, availáble on a turnkey basis and machinery engineered to fit in a railroad car.

The equipment included in the above mill package provides an insight-into the variety of Nordyke & Marmon products for the flouring industry. It included: 3 Double N & M Roller Mills, 3 new style 7 foot scalping Bolts, 4 slow-speed 7 foot round reel Flour Dressers, an automatic-feeding Midlings Purifier with aspirating attachment, one automatic — feeding combined grain Separator and Scourer, the 7 elevators (enclosed basement to attic belt drive column assemblies) complete with pulleys, shafting, belts and belt tighteners, as well as a 5 x 5 x 6 foot wheat bin and spill-proof hand packing device for sacks.

The complete outfit was packaged in a single railroad car for logistic and economic reasons.

Steam engines and boilers were procured from the Atlas Engine Company of Indianapolis and ranged from a 10 horsepower unit at $545.80 to a 40 horsepower unit weighing 15,100 pounds for $1414.75. The latter had a 12 inch bore by 14 inch stroke double-acting single cylinder rated 40 horsepower at 170 rpm.

Also marketed was the Hercules line of planetary geared belt drives powered by up to four horses (oats fueled) with a 151 rpm output pulley shaft.

But the story of the firm's expansion in the Nineteenth Century would not be complete without addressing the origin of the Indianapolis Power and Light Company.

## *The Jenney Electric Venture*

In 1885, the officers (and sole owners) of Nordyke & Marmon Co. diversified their product lines by establishing a separate firm named the Jenney Electric Company. The *Marmon Book of Facts*[3] provided the following insight into this development: *"In 1885, early indeed in the history of electrical development, the officers of the Nordyke & Marmon Company learned that there was an electrical genius at Fort Wayne, Indiana, by the name of Charles D. Jenney. An investigation proved that he was the inventor of an arc light of unusual merit, an appliance that gave much promise in those days when cities were lighted principally by gas. Satisfactory arrangements having been made with the inventor, a stock company was formed, known as the Jenney Electric Company, with Mr. D.W. Marmon, president; Mr. A.H. Nordyke, vice-president; Mr. A.K. Hollowell, treasurer; Mr. Brainard Rorison, secretary, and Mr. Charles Jenney, chief electrician. The business was carried on within the Nordyke & Marmon Company as a separate company. Complete electric lighting plants, dynamos and various electrical apparatuses were manufactured. Lighting plants were installed by the company in many cities, including Indianapolis; Topeka, Kansas; Toledo, Ohio; Waco, Texas; San Diego, California; St. Louis, Missouri and Bay City, Michigan. This adjunct to the business of Nordyke & Marmon Company was sold a few years later to the General Electric company, for two reasons: First, the General Electric Company was consolidating as many of the large electrical companies as possible, and therefore offered the Jenney Electric Company an attractive price for its patents and business: second, the milling machinery industry was steadily growing greater, taxing the facilities of the plant to such an extent that it was decided to devote all of the organization to this branch. No little credit is due the Jenney Electric Company for its contribution to the development of electrical appliances. Its progressive step in the electrical field is in conformity with the rapid progress that has always kept the Nordyke & Marmon Company ahead in the mill building and motor car industries."*

A more precise accounting of the involvement of the Marmon family in the electric utility business is found in 'Electrifying Indianapolis, Part 1 1881-1927', the history of the Indianapolis Power & Light Company (IPALCO) and its predecessor companies. The following account is taken from that document. Reprinted by permission of IPALCO.

The Jenney Electric Company was organized March 23, 1885 to manufacture dynamo-electric machines, lamps, motors and electric lighting plants. They installed the first electric street light in West Indianapolis at the Kentucky-Morris intersection in June 1885 and erected a system of street lights on towers and suspension wires throughout that community. The Jenney Company used a so-called 'short' arc lamp of about 18 amperes while the Indianapolis Brush Electric Light & Power Company (founded 1881) employed a 'long' arc of 9.6 amperes, and the relative merits were much debated. When the manufacturing business of the Jenney Electric Company was purchased in 1889 by the Thomson-Houston Company of Boston, the street lighting in West Indianapolis was continued by the Nordyke & Marmon Co.

## *Marmon-Perry Company*

In 1886, Mr. Marmon had hired, as a 'utility man' at $75 per week, Charles C. Perry. The 28 year old Mr. Perry was a former railroad telegrapher in Richmond, Indiana. He quickly learned the electrical business and in 1888 entered a partnership with Daniel W. Marmon under the name Marmon-Perry Light Company. The partnership, presumably financed entirely by Mr. Marmon, provided the first incandescent lighting service in Indianapolis. A 35 kilowatt Jenney electric generator was installed at the rear of the Sentinal Printing Co. at 117 W. Market Street; steam from the printing company's boiler powered an Atlas engine connected by flat belt to the generator. That generator is today a museum piece at Purdue University, the gift of its original operator, Edward Hutchens. Electrical service (to the Park Theater) began on November 29, 1888 with 741 incandescent lamps of 16 candlepower each.

The following year Marmon-Perry constructed the city's first central station for incandescent lighting and steam heating. At the rear of 121 Monument Circle, two Atlas boilers, a McIntosh & Seymour engine and two Thomson-Houston (former Jenney) 35 kilowatt, 110 volt generators were installed. Underground service to the Grand Opera House was accomplished by running the cable in vitrified tiles and routing the four inch steam pipe along side through the basements of buildings and under streets.

When a second theater was wired, the lights were much brighter than expected; the only peculiarity in the equipment was a non-metallic interim pulley on the generator. Even Thomas Edison journeyed to Indianapolis to seek an explanation of

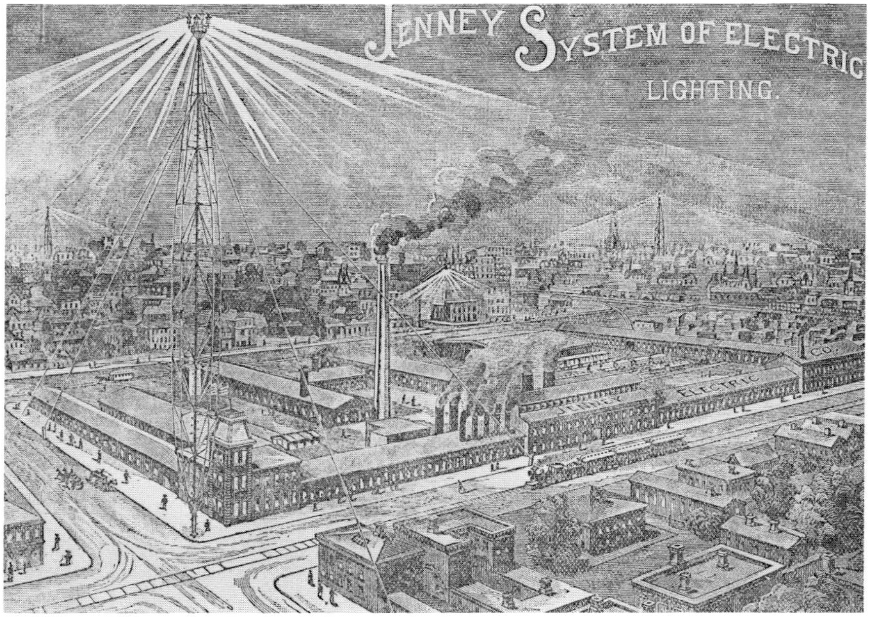

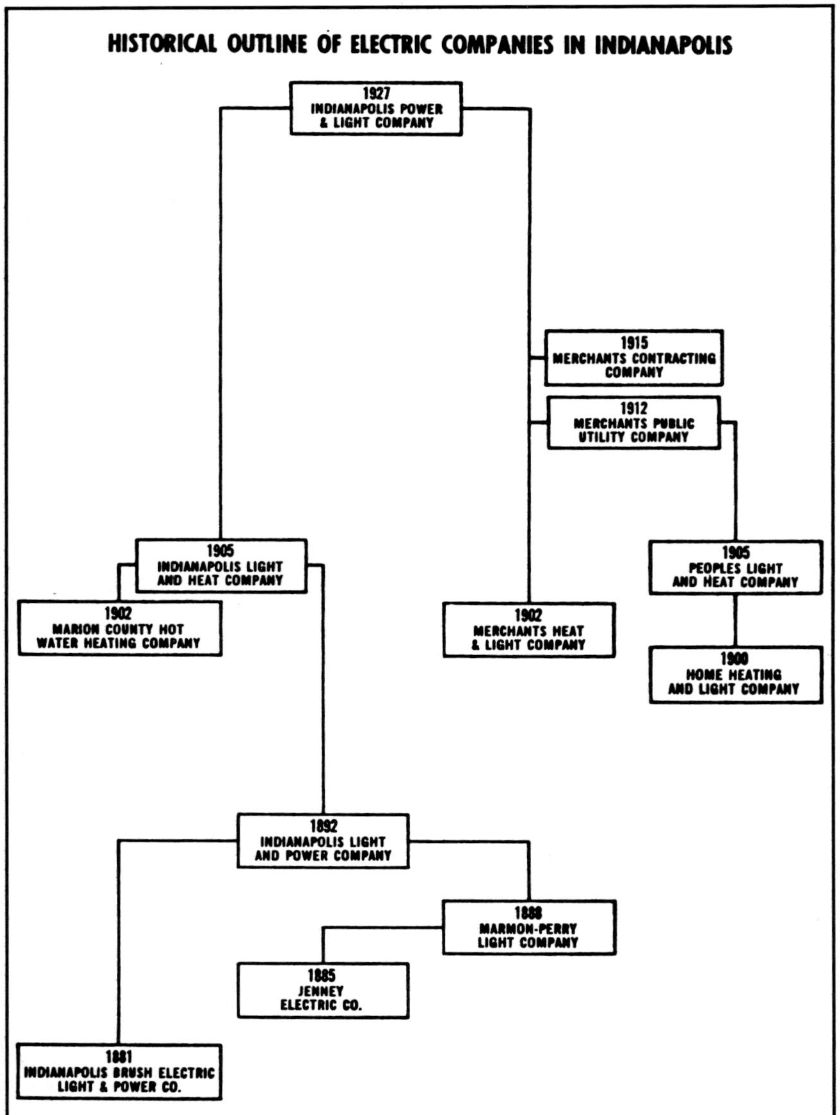


the supposed electrical influence of pulley material; the real reason was found eventually to be a higher coefficient of friction of the non-metallic pulley — less belt slippage meant higher speed of the direct-current generator!

On May 20, 1892 Marmon-Perry merged with the Indianapolis Brush Electric Light & Power Company and became the Indianapolis Light & Power Co., capitalized at $600,000. The new firm took over the operation of the West Indianapolis street lighting from Nordyke & Marmon Co. The officers of the new firm were: Daniel W. Marmon, president; John Caven (former president of Indpls. Brush E L & P) vice-president; and Charles C. Perry, secretary and treasurer.

Max R. Hymans *'Handbook of Indianapolis'* published in 1897 reported: ''Indianapolis now has twice as many street lights as London, England and three times as many as Cleveland where they were invented (Indpl., 824, W. Indpls., 57, Haughville, 30 and Mount Jackson 7). Few plants in the world equal the capacity of Indianapolis. Its underground conduits will accommodate the city for a century'' So much for prophesy.

The Marion County (IN) Hot Water Heating Co. was incorporated on June 7, 1902; D.W. Marmon and C.C. Perry were directors. The charter included electric power generation and the first turbo-generators in Indianapolis were installed in the Mill Street station of the new company — two 2500 kilowatt Curtiss units procured from General Electric Co. — which entered service in 1904. The very first Curtiss steam turbo-generator which had been placed in service in February 1903 at Newport, Rhode Island, is preserved in an Indianapolis generating station today as a public service.

In 1905 Marion County Hot Water Heating Co. was merged into Indianapolis Power & Light. The name was changed to Indianapolis Light & Heat Co., with capitalization of $1 million in common stock and $5 million in first mortgage bonds. The officers were: Daniel W. Marmon, president; Charles C. Perry, vice-president and treasurer, and Thomas A. Wynne, secretary (originally with Jenney Electric).

When Daniel Marmon died in 1909, his son Walter took his director's seat and became secretary; Charles Perry became president and Thomas Wynne moved up to vice-president and treasurer. On the deaths of Perry and Wynne in 1924, Walter Marmon became president and Norman A. Perry (son of C.C.) became secretary. When today's Indianapolis Power and Light Co. was formed by the merger of Merchants Heat & Light Co. with Indianapolis Light and Heat Co., Walter Marmon became chairman of the board of directors and served until his death. The Marmon-Perry partnership of 1888 had grown into the successor firm of eleven other utility companies operating in the city at various times from 1881 to 1927.

The merger of Indianapolis Light & Heat Co. and Merchants Heat & Light Co. was accomplished by the organization of IPALCO by Utilities Power & Light Corporation, a holding company established for the purpose. In 1940 all common stock of IPALCO formerly held by Utilities Power & Light Corp. was sold to the public. The company was the first large electric utility in the United States to have its ownership transferred from a holding company to thousands of individual investors.

## *The Proving Ground Mills*

### THE MOFFITT MILL

Although the company relocated to Indianapolis in 1876, D.W. Marmon had deep roots in Richmond, Indiana, and in 1890 he repurchased a mill originally built by his Grandfather Charles Moffitt. Moffitt, after emigrating to Richmond from North Carolina in 1811, had built his first mill on the site in 1815. A permanent structure followed and survived under several owners until destroyed by fire in 1883. A $6000 brick building was erected and equipped with what was then the latest (stone) milling machinery but a financial depression occurred and the business failed, closing during the following year. D.W. Marmon, upon purchasing the property and mill, removed all the machinery for making flour exclusively and replaced it with a full roller — milling system . . . 'The Patented Nordyke & Marmon Roller Mill' . . . which produced a range of foods such as pearl meal, cream meal, cerolene, etc. with whole wheat flour a specialty. In the winter of 1897-98 66 carloads, of 750 bushels of meal each, were shipped overseas to Liverpool.[4]

THE NOBLESVILLE MILLING COMPANY

The Nordyke and Marmon Company did architectural work in support of their milling machine business. Such activity was still carried on in the early 1920s. This probably accounts for the photographs of foreign flouring mill buildings which often characterized foreign language advertisements for the firm's milling machinery.

The flouring mill built in Noblesville, Indiana, in the early 1890s for the Noblesville Milling Company was constructed at the direction of Daniel W. Marmon. While most people associate the fame of the Nordyke and Marmon firm with its highly regarded automobiles, the firm was singularly known internationally, before and concurrently with the motor car business, as an outstanding builder of flour milling machinery. Its reputation in the milling field was unequaled internationally for quality of product and integrity of management. All new inventions and product improvements from the Indianapolis plant were taken to the Noblesville Mill to be evaluated in actual operations. In consequence Nordyke & Marmon was regarded not only as the world's largest manufacturer of milling machines but as the builder of the finest of the genre.

Expert millers came to Indianapolis from the great wheat regions of the world — the United States, China, Japan and many others. They were promptly taken to Noblesville, some ten miles north of Indianapolis, to see the latest flour making machinery perform. Because of this, the mill came to be known as "The Model Mill."[5] After all, Peter The Great, of Russia, studied milling at the Nordyke mill in Holland in 1697.

Daniel Marmon was actively associated with the mill until his death in 1909. Following his death, Mrs. Marmon disposed of the mill property.

In 1949, when the Indianapolis News[5] published an article on which this recounting is based, the 'Noblesville Mill' was being operated by the Doughnut Corp. of America. All the previous milling machinery had been removed and new equipment installed at a cost of one million dollars. The plant, which does not make doughnuts, has a capacity of 1800 barrels of mixed bread, doughnut and cake flours per day.

Although the milling machine business and the name "Nordyke" were sold to Allis Chalmers in 1925-6, the name "Nordyke" still appeared in the "1969 McRaes Business Listing." On Jan. 1, 1977 Allis-Chalmers transferred Mr. George Jennrich from the Milling Machine Group to their Continuous Casting Group. According to George Derrig of A-C, Mr. Jennrich was the only remaining employee familiar with the Nordyke acquistion. Mr. Derrig stated in a letter to the author on Jan. 31, 1977 that the Nordyke name was gradually phased out over the years but A-C still had many of the old patterns and continued to take orders for Nordyke parts. He said that "Allis Chalmers designs had superseded many of the Nordyke designs."

The manufacturing plant will be described in Chapter Eleven. There the story of the establishment of the firm in Richmond, Indiana, and its relocation to Indianapolis will be told, and the Twentieth Century expansion described. Corresponding management-organization and financial histories are presented in Chapters Twelve and Thirteen. And importantly, the reader will more likely be interested in such matters after exploring the Five Eras of the Marmon Car.

Fig. 1-6 N & M designed and directed construction of this mill at Daisi, Japan.

## PART 2 — The Nordyke and Marmon Families

### *The Nordyke Family*

The Marmon Book of Facts, published throughout the 1920s, began with the following report on the genealogy of the Nordyke family: *"In 1697 Peter the Great, inspired by the ambitious hope that the progress of Europe could be transplanted to Asia, visited the industries of Holland and England. While his barbarous courtiers amused themselves with ribald entertainment, the great Peter himself worked in flour mills, printing plants, and paper mills, traveling incognito as "Baas" or "Master Peter." Afterward, this constructive and unusual activity on the part of the medieval Russian ruler did more to modernize the East than had been accomplished in the centuries of Russian history before this first Romanov came to the throne. In Holland, Peter learned the art of grinding grist from Stephen Henry Nordyke, whose descendants, combining their skill with the genius of another family, the Marmons, established in the new world an industry that has come to be world-renowned. Stephan Nordyke's three sons, with patriarchal names, to wit, Abraham, Isaac and Jacob, joined the pioneer adventurers long before the Revolutionary War and came to America. They settled in Pennsylvania near the site of the present city of Philadelphia. Two of them engaged in shipbuilding, ultimately launching a vessel which was christened, appropriately enough, "The Two Brothers." Abraham married and settled in Virginia. Micajah, the fourth son of the fifth child of the ten children of Abraham, was born in Stafford County, Virginia in 1771, and at the age of four went to Georgia with his parents. The family later moved to Tennessee, where Micajah was married in April, 1793, to Charity Ellis, of Martinsburg, Virginia. In 1806, he emigrated with his young family to Clinton County, Ohio, and bought three hundred acres of heavily timbered land on which the city of New Vienna was afterward located. The production of food in this early part of the nineteenth century was a problem with which to be reckoned, and it was limited by the facilities for the preparation of grain suitable for consumption. Micajah Nordyke devoted much of his energy to this problem, and finally constructed a mill for griding flour. As a result of these efforts, he is credited with building the first flour mill west of the Allegheny Mountains. Eight sons and four daughters were born to Micajah and Charity Nordyke, all of whom were reared in the purity and simplicity of the Quaker religion. Ellis, the fifth child, was born in a log house on the original family farm in Clinton County, Ohio, in 1807. Educational advantages in the common schools were limited, yet he so applied himself as to acquire a liberal education. In early boyhood he taught school in his native county. When yet a child, he was accustomed to go with his father, and sometimes alone, to the old mill propelled by horse power, to wait for his grist to be ground. Even then it occurred to him that something better could be devised for the manufacture of flour and meal. Ellis later was employed by his father to care for the machinery of the grist mill. Here was his opportunity, and he soon acquired the rudiments of a mechanical education and laid the broad foundation for his success as a manufacturer. By 1830, he had become a skillful workman. There was abundant water power for milling purposes at Richmond, Indiana, not far from his home, and thither he went and hired himself to work as a millwright. His employer, Abel Thornberry, counted thirty shillings into his palm every Saturday night. His hours of labor were from sunrise to sunset; there was no clamor for an eight-hour law with an extra charge for overtime. The young Nordyke did not go to Richmond for an idle purpose. While his hands were busily engaged, his mind was ferreting into the mechanics of milling machinery and the application of power. By careful study, combined with natural ability, he was soon able to adapt new appliances to the processes of grinding. In January 1837, Ellis Nordyke was married to Catherine Haynes, of Sunbury, Ohio. Their first child, Addison Haynes Nordyke, was born in Richmond, May 5, 1838. From early childhood, he was interested in his father's experimental work with milling appliances and while attending the city schools he helped in the little shop where he and his father founded the E. and A.H. Nordyke firm in 1851. Ellis died in 1871."*

The growth of that firm into Nordyke & Marmon Co. is described in Chapter Eleven, "Home of the Marmon Car," Chapter Twelve,' "The Marmon Organization" and in Chapter Thirteen, "Financial History."

### *The Marmon Family*

The following concerning the Marmon family in the mid-nineteenth century is also taken from the Marmon Book of Facts. *"The Marmons came to Indiana in 1846 in the person of Dr. James W. Marmon, M.D. In that year he moved his family to Richmond from Logan County, Ohio (60 Miles NE). He was the son of Martin Marmon (1777-1843) and grandson of Peter Marmon (1737-1802). His wife was the daughter of Charles and Elizabeth Cox Moffitt; her grandfather, Charles Moffitt of Scotch ancestry, had emigrated from North Carolina in 1811 and her maternal grandfather Jeremiah Cox had 'originally entered the land' upon which Richmond was later built. They had married in 1840 and a son Daniel was born in 1844. It was probable that no man in the early history of Wayne County (Richmond) Indiana was held in higher esteem than Dr. Marmon for his heroic professional dedication.*

Fig. 1-7 Daniel W. Marmon 1844-1909

*During the great epidemic of cholera which swept the nation in the late 1840s, he was the only physician in his community willing to attend the victims and it cost him his life in 1849. His wife survived him by only one month; Daniel W. Marmon was orphaned at five years of age. His care and education was assumed by an uncle, Eli Stubbs, who operated a saw mill and manufactured furniture and later pumps in West Elkton, Ohio (30 miles SE of Richmond). Daniel attended and graduated from Earlham College in Richmond, Indiana. He married Elizabeth Carpenter and they had two sons, Walter Carpenter Marmon born August 26, 1872 and Howard Carpenter Marmon born May 24, 1876 and a daughter Caroline. Both boys were born in Richmond, Indiana, where the firm of E. and A.H. Nordyke had undertaken the manufacture and erection of flouring mills in 1851. Daniel Marmon's association with Ellis and Addison Nordyke in the Nordyke & Marmon firm is recounted in*

*Ellis and Addison Nordyke in the Nordyke & Marmon firm is recounted in Chapter Twelve, 'The Marmon Organization,' and Chapter Thirteen, 'Financial History.'*

The Marmon production car was born much as was the production Lincoln Continental. Each was initially created to satisfy the demands of a family heir, and subsequently entered production in response to an unanticipated market clamor. This is the first of many parallels between Marmon and Ford extending over many years. This circumstance most likely was responsible for the Marmon automobile being offered as a product of the Nordyke and Marmon Co. The founding, developing and maturing of the firm was the work of the Nordykes and Daniel Marmon. Although the firm name of Nordyke & Marmon Co. persisted until 1924-1926, A.H. Nordyke had died prior to the undertaking of the automobile venture. Although Daniel Marmon would live until 1909 and almost certainly had the final word on the decision to enter the automobile business, the development of the Marmon marque into an internationally recognized "Foremost Fine Car" was the work of Daniel's sons. The "Marmon" badge could well have read "Marmon Brothers" rather than Marmon. The "Graham Brothers" did just that, but eventually ended up with a "Graham" badge. Marmon foresight was good as evidenced by the dropping of the "t" from the French family name of Marmont soon after coming to the United States in the early nineteenth century. Louis Chevrolet retained the "t" in his name and universally correct pronunciation of Chevrolet was not achieved until the advent of radio . . . many people sold and drove "Chev roll ets" in the twenties. The "Marmon" badge stood for a firm directed by Walter Carpenter Marmon and a product whose engineering was directed by Howard Carpenter Marmon. The "Marmon-Herrington" badge of the latter-day products reflected the fact that Colonel Herrington was initially responsible for the engineering guidance.

The Marmon "Book of Facts" corporate history, in the 1922 edition, had this favorable comment on the two Marmon brothers: *"Both of these men, older employees say, have the qualities of their father that have made the name "Marmon" on the company's products synonomous with that of "Sterling" on silver (items)."*

Therefore, prior to proceeding with the history of the automotive product, the reader will be introduced to the Marmon brothers . . . Walter, the executive, and Howard the innovative engineer. Likewise, Book Two will begin with the introduction of Colonel Herrington.

## *Walter Carpenter Marmon*

The official Marmon history publication *"Book of Facts,"* published by the Education Division of the Marmon Sales Department in May 1929, provided the following resume of Walter Marmon's career up to that time: *"In 1909, Walter C. Marmon, who is now chairman of the board of directors, succeeded his father as president of the company and retained the office until 1924. Mr. Walter Marmon was born in Richmond, Indiana, August 25, 1872. Following his early education in the Richmond schools, Mr. Marmon entered Earlham College and two years later pursued his studies at the Massachusetts Institute of Technology at Boston, from which he was graduated. His summer vacations were spent in the factory and also in the mill managed by his uncle, Mr. Charles Carpenter, in Richmond. From early boyhood he was profoundly interested in athletics and clean sports. Later, he participated in bicycle races, one of the most popular sports of the time, and was so successful that he took part in national contests. Stored away with the treasures of his earlier life is a score of trophies that he won in these contests. In 1897, he began his association with the Nordyke & Marmon Company, later guiding the company through one of its great periods of development.*

On the occasion of the twenty-fifth anniversary of the MIT Class of 1895 and Walter Marmon's graduation with the degree of Bachelor of Science, he provided the following information for inclusion in the *Alumni Register* (1920, P 88):

President of Nordyke & Marmon Co.[6]
President of Noblesville Milling Co.
Secretary of Indianapolis Light and Heat Co.
Married Mar. 29, 1898 to Annie B (urroughs) Hall in Philadelphia
Children: Franklin Hall of MIT
Elizabeth C. at Smith College, Northampton, MA
Dorothy B. at Miss Capen's School, Northampton, MA
Amusements: "attending to my job"
Politics: Republican
Weight and Height: 175 pounds; 5 feet 9 1/4 inches

*"I have been connected with Nordyke & Marmon Co. since leaving Tech. My only connection with war activities was that of civilian as the head of our organization. Our company was selected soon after the outbreak of the war as one of the companies to build aviation motors and we were given the first order for aviation motors given to any motor car concern. We built a special plant for the production of these motors and built two distinct types of motors, one the Hall-Scott training motor and the other the Liberty Motor. In October, 1918, the Bureau of Aircraft Production offered a pennant for production competition among the five concerns building Liberty Motors, to be awarded to that concern which exceeded its assigned quota by the largest percentage. Our company won this pennant for the month of October, and as the Armistice was declared the following month, was given permanent possession of it. My Tech training was of undoubted value to me in helping to appreciate the various problems that have come up in an institution such as ours. I have not been in very close touch with '95 men since leaving Tech as it has been largely a question with me of sticking upon the job here and watching our plant grow which now covers 45 acres of ground instead of the 8 1/2 acres it covered when I joined the organization."*

Two years later, W.A.P. John interviewed the Presidents of the major automobile manufacturers and his recounting of his interview with Walter Marmon in *Motor Magazine* for November, 1922 gives a remarkable insight into this great man and the caliber of the Nordyke and Marmon management organization: *"But let me tell you here, at the logical place, what preceded the interview proper. As is my wont in arranging these interviews, I wrote to the director of publicity of the Nordyke & Marmon Company asking him if he would not be good enough to ascertain whether Mr. Marmon planned to be in his office the following Friday. And if that were true, whether he would not make an appointment for me. By return mail came a letter inquiring which Mr. Marmon I desired to see. Frankly, I didn't know . . . because I didn't know how many Mr. Marmons there were nor which one was the president of the company. So I dropped into the local Marmon agency, introduced myself to the salesman on the floor, explained my quandary and asked his advice. "Are you restricted," he inquired, "to interviewing one man?" "Why not exactly restricted," I answered, "but so far, I have only interviewed the presidents of the leading American automobile companies. Why?" "Because," came the prompt answer, "there are two Marmons down at Indianapolis. And although Walter Marmon is president of the company and chief executive of its both divisions, his younger brother Howard is the automobile genius and concerned wholly with the automobile division." "Two divisions of the company?" I asked. "What two divisions?" "Why don't you know," he demanded with a look that was almost pitying, "that Nordyke & Marmon are also one of the oldest, largest, and best known manufacturers of milling machinery in the world and that they have built and equipped some of the biggest mills in America, Europe, South America and the Orient?" I had to confess my ignorance. That same night I telegraphed Indianapolis that it was Walter Marmon I would like to see; two days later a letter informed me that the interview had been arranged. When I reached the office of Nordyke & Marmon I was introduced to one of the executives and we chatted while Mr. Marmon finished a conference. "The thing that surprises me," said the forementioned executive cheerfully, "is that Mr. Marmon even consented to see you. I have known dozens of instances that men have asked to see him on similar missions and he asked to be excused." "You don't mean," I asked, hoping for a negative answer, "That he is . . . er . . . a poor talker?" "Well, no. He isn't exactly a poor talker. But he is more given to action than he is to conversation. What I meant was that Mr. Marmon is one of those quiet unassuming men who are happiest out of the limelight just as some men are happiest in it." Which wasn't very reassuring news. When the group in Mr. Marmon's office broke up I was taken over and introduced. With a crisp smile that created cheerful little wrinkles about his eyes he apologized for keeping me waiting and asked me to take a chair across the big double desk at which he works. Like those adjacent, his office was utterly devoid of sumptuousness. No pictures. No hangings. No cushiony rugs. Just three eight foot plateglass and wood partitions. With a sort of half quizzical look he leaned back in his chair while I explained the general idea behind this series and reason for my presence. When I finished he chuckled. "Pshaw, if you want to get at the personality behind the Marmon car you want to talk to my brother Howard and the other men in the automobile division. They deserve all the credit." "But I want the personality behind the whole institution." "Oh . . . you want the men who have been here so many many years, you want what they've done and what they have accomplished. Well . . ." And the conversation that ensued contained as fine a story of sturdy American accomplishment as I have ever heard; one that goes back to a decade before the Civil War; the story of an institution that has lived, and prospered through almost 75 years (as of 1922) with its three great wars and its financial crisis; the story of an institution whose 71 years have "mellowed its business sense but not tamed its eagerness for progress." Both sons of Daniel W. Marmon after graduation from college were taken into the firm. Neither stepped into a soft berth. Mr. Marmon laughingly explained that for the first year he received no salary at all and that after he had by dint of hard work demonstrated to his father and his father's associates that he merited a place on the company payroll, he received a salary hardly large enough to live upon. His younger brother Howard . . . now vice-president in charge of automotive activities . . . went through a similar experience. During my visit at the Nordyke & Marmon plant I met an unusually high percentage of men who, I assumed from their conversation and bearing, were college graduates. This being a situation quite different from what I have encountered elsewhere in the industry, I mentioned it to Mr. Marmon wondering whether it were accidental or intentional. "Oh, it's intentional," he explained, "although we don't of course, make a college education one of the requirements for a position in the organization. But most of the executives are college men . . . father was an Earlham graduate, you know . . . and taking it all in all, we found that the college trained man after he has absorbed the hard knocks, turns out to be more valuable to us. He's less of an individualist and more of a co-operator. And co-operation is the fundamental of success in any large institution. We have an executive school here and I believe that every man in the particular class going through it this time is a college man." Several men I have interviewed have in no uncertain terms pronounced the "Practical" man superior in every way to the college graduate. And I was anxious for Mr. Marmon to express himself on that point. So I said, "Then the college graduate is the best material for executives?" "Not necessarily. Most young men do not realize how long and hard and slow is the road to success. They want to leap into important positions almost over night. College men are unusually so. They're impatient. They think they've outgrown their jobs before they have half mastered every phase and detail of them. They are sort of tugging at the leash all the time. But if their impatience can be curbed without dulling their enthusiasm . . . they are exceptionally fine executive timber; for every man is bettered by a good education and the best practical men would be bettered by the right kind of a college education." But I have written much without telling you about Walter C. Marmon himself . . . although if you have read between the lines, you have doubtlessly sensed the sort of man he is. A prominent banker of Indianapolis said to me: "He is, I believe absolutely the most honest man I ever met. Having been raised in a home where the Christian virtues were pre-eminent and having been very very close to his wonderful father, old Daniel W., he is so square that he actually leans over backwards. He is a man of simple tastes. He doesn't play golf; he doesn't ride horses; his home and social life are touched with modesty and serenity. How much he gives to charity no one knows because he never discusses his benefactions. But I know that it must be a sizable sum because he is impulsively generous to worthy causes. His executives literally love him . . . and it is interesting to know that the youngest executive has been there seven years and the oldest has been a member of the organization for 35 years . . . and yet he leads them a vigorous pace of hard conscientious work. For he appears at his office early and remains there until quitting time. No long vacations and cruises for him. He was raised to the doctrine of doing an honest day's work when he is on the job. And he's always on the job." Of this phase I spoke to one of the company's executives and his comment was characteristic. "That's true," he said. "We do love him. I would go through hell and hot water for him. And so would the rest of us." It is honestly a joy to work with a man so broad and so considerate and so absolutely and so beautifully schooled in the fundamentals of running a large and successful business. He sits at the helm steering a straight course, looking to us to keep our part of the ship in proper order. He never interferes with his lieutenants. He gives us our work, lets us severely alone, offers counsel and suggestions only when it is requested or asked and gives us ample credit for a good job well done. In executive meetings he guides the discussion without dominating and when decisions are being reached his opinion is only one of six and of no greater weight than that of any other man." Late in the afternoon I dropped in at the office of another Indianapolis citizen whose fame is international and who has known Mr. Marmon for many years. We were chatting aimlessly and I said: "You know Mr. Marmon. You have known him for a number of years. And so you're eminently qualified to answer the question I am going to ask. And I'm going to make it hard by requiring you to confine your answer to a single sentence. To what do you attribute the successful life and career of Walter C. Marmon?" My friend thought intently. "One sentence? Well . . . I can tell you in less than ten words: Absolute integrity, hard work, devotion to duty . . . that's seven words, isn't it?" Which seven word opinion, if you have met the man of whom they were spoken and listened to his associates, becomes more and more accurate as you turn it over in your mind."*

Following his move from the Presidency to the newly created position of Chairman of the Board of Nordyke & Marmon Co. in May 1924, Walter Marmon was elected President of the Indianapolis Power and Light company,[7] a position held by his father Daniel W. Marmon, from its founding until his death in 1909.

The *(MIT) Technology Review* for Nov. 1940 reported the death of Walter Marmon on Aug. 29, 1940, after a three month illness at his home in Brendonwood (Indianapolis), Indiana. It provides additional insight into his career: *". . . He was Chairman of the Board of Marmon-Herrington Co. and formerly held the Chairmanship of the Indianapolis Power and Light Co. He found time to contribute greatly to the betterment of his city as general chairman of the Community Fund in 1928 and 1934. Marmon had taken an active part in development of the Boy scout program in Indianapolis and was one of the leaders responsible for its excellent status in nation-wide scouting circles today. He was former president of the Indianapolis Scout Council. He was a member of the Woodstock and the Columbia clubs, the Indianapolis chapter of the Citizens' Historical Association, and the Second Presbyterian Church. Further testifying to his versatility was the fact that he owned an eighty-acre Marion County poultry farm. We all knew him as a most genial fellow, athletic and democratic. We shall miss him."*

His wife, Annie Hall Marmon, a Philadelphia native, died on Feb. 25, 1957 in St. Petersburg, Florida at age 85. The Walter Marmons' two daughters survived their parents: Dorothy . . . Mrs. Philip L. Boyd of Riverside, CA and Elizabeth . . . Mrs. Frank J. Hoke of Indianapolis.

## *Howard Carpenter Marmon*

Howard Marmon's career up to 1929 was recapped in the official history 'Marmon Book of Facts,[8] as follows: *"Col. Howard C. Marmon, Vice-President in charge of engineering, brother of Walter C. Marmon, chairman of the Board of Directors, was born in Indianapolis, May 24, 1876. He, too, received his early college education at Earlham, later studying at the University of California, at Berkeley. Since early boyhood his primary interests have been concerned with mechanics. Engines particularly fascinated him. He is said to have spent many hours watching locomotives that brought trains in and out of the city. An incident that illustrated the boy's early enthusiasm for engines was related by his uncle, Mr. Charles Carpenter. One day, while but a small boy, he desired to visit the factory, a plan that was contrary to his father's wishes at that time. Concealing himself behind the seat of his father's carriage, he would have been able to reach the factory undetected if he had not heard a locomotive whistle. Unable to resist the temptation to look at the engine, he emerged from his hiding place and was discovered by his father. Later, this boy was to become one of the greatest of all automotive engineers, pioneering in the development of the motor car, continually evolving advanced departures from conventional construction. The military record of Col. Marmon is intimately woven with the history of the first World War. Not long after the United States entered the war a committee was appointed to select aircraft equipment in Europe for American manufacture. An important post in this committee called for the services of an expert on engines . . . someone upon whose knowledge and judgment tremendous sums of money would be spent and, far more important, valuable time and effort that might easily mean the difference between success and failure in a great program of the war. The government sought for this post the most outstanding automotive engineer in the United States. The nation was searched for the most eligible man. In the end, Col. Howard Marmon was called upon to serve his country in that hour of need. Again his services were requested, with those of other prominent engineers, in designing the great wartime creation, the Liberty Motor. Marmon, as an institution, can well be proud of Col. Howard Marmon, the master engineer and the man. The fire of his genius and the kindly strength of his personality pervade the organization. As Vice-President in charge of engineering, his daily interest remains in his greatest achievement and life work, the Marmon car."*

A press release datelined Washington, D.C. June 15, 1917 reported that Howard Marmon had come to that city to locate for an idefinite time so he might give his sole attention to assisting Major Henry Souther, head of the Aircraft Production Board in the capacity of Aircraft Engineering Engineer. He was to become a Major in the Signal Corps (pre Army Air Corps) with his division one of four departments under Major Souther.

Fig. 1-8 Howard Carpenter Marmon 1876-1943

Howard Marmon's technical interests were not limited to the automobile. He was not sent to engineering college to come back to Indianapolis and start an automobile branch of the business, but rather to prepare him for leadership of "one of the worlds greatest engineering houses" dominant in the flour milling machinery field. This broader aspect of Howard's ingenuity is evidenced by US Patents on milling machinery granted to H.C. Marmon. The full range of interest and perception of product potential is revealed by the "Capehart event."

Joe Felts[9] was working on the drawing board at Nordyke & Marmon about 1922 and an Emil Nyman worked alongside. An Indianapolis manufacturer of popcorn machines and phonograph record changers, Holcomb — Hoke Co., employed a salesman by the name of Capehart. One day Howard Marmon brought Mr. Capehart into the drafting room and informed Mr. Nyman that he was to take his direction from Mr. Capehart.

A layout of a coin-operated selective record player began to take shape in the Nordyke & Marmon drafting room. Mr. Capehart departed both the Nordyke & Marmon engineering department and his employer to go to work for Wurlitzer and took draftsman Nyman with him. Mr. Capehart subsequently founded the well known firm bearing his name in Fort Wayne, Indiana and successfully campaigned for the United States Senate.

Years later, George Freers,[9] who was then Chief Draftsman, would work briefly for Holcomb — Hoke Co. during his transition from Marmon to Marmon — Herrington.

W.A.P. John, in *Motor* Magazine, wrote[10] of Walter Marmon . . "His executives literally love him . . .". But George Freers, who was the senior engineer under Howard Marmon, so respected . . . admired . . . his boss that he named his son for him: Howard Freers, a major Ford engineering executive in the 1980s.

Joe Felts, who was hired at Marmon by George Freers and who hired Howard Freers at Lincoln — Mercury, had this to say about Howard Marmon: "He was about 5′9″ tall. He was a master engineer, thought of as a genius. The kindly strength of his personality was felt by those who worked for him."[11]

Bruce S. Williams, son of Marmon President G.M. Williams (1924-33) recalls his mother telling the Williams children that Mr. (Howard) Marmon, who was invited to dinner, was shy.

When Howard Marmon died in Fort Lauderdale, Florida, on April 4, 1943 at age 66, the Society of Automotive Engineers Journal ran a full page photograph and full page obituary. This provides additional detail on his career and constitutes a professional endorsement:

. . . Howard Marmon is credited with development of the first side-entrance automobile body, the first engine with hollow crankshaft carrying oil direct to bearings, and the first all-aluminum car. He contributed to the engineering development of vacuum control of engine oil pressure, double cross rear spring suspension, and 12- and 16-cycle engines. For years, Howard Marmon developed experimental cars, incorporating what then were regarded as radical designs, but which time proved were sound engineering. Until ill health forced him to abandon the project, Mr. Marmon was developing a "whaleback" design which unitized body and chasis.

He served in World War I as Army Air Corps officer in charge of the aircraft engine experimental station at McCook Field, Dayton, and made engineering contributions to the development of the famous "Liberty" aircraft engine. He received an honorable discharge from the U.S. Army at the end of the war as lieutenant-colonel.

Mr. Marmon's engineering achievements and his personal popularity were recognized by his election as SAE president in 1913 . . . three years after he had become a member . . . and his election the same year as the only American honorary member of the British Institute of Automobile Engineers.

Mr. Marmon had lived on his 5000 acre estate at Pineola, N.C. since 1923. After Nordyke & Marmon Co., Indianapolis, Ind. . . . of which he had been vice-president and chief engineer for 25 years . . . went out of business in 1936, Mr. Marmon devoted his time to developing on his estate one of the world's largest evergreen tree nurseries, and to his interests in mining and milling operations yielding kaolin, feldspar, and mica. It was his custom to spend winters in Florida.

It should be added that Mr. Marmon raised rhododendron and had some 250,000 trout in a hatchery at Pineola.

Howard and Martha Foster Marmon had one daughter, Carol, who married a Russian prince, a Knight of Malta, Nicholas Tchkotoua and resided in Lausanne, Switzerland and later Mexico City.

On April 6, 1960 George Freers in responding to a letter from Mr. Phil Hockett wrote: "With reference to Mr. Marmon's family . . . Mr. (Howard) Marmon was married twice and had a daughter by the first marriage. The daughter and her mother lived in Europe or some foreign country. After Mr. Marmon and his first wife were divorced, he then married an Indianapolis woman and there were no children by that marriage. When Mr. Marmon died he was a very wealthy man and all of his wealth was not inherited."

The first Mrs. Marmon died at age 78 on April 25, 1959 in her home in New York City. Private services and burial were in Indianapolis Crown Hill Cemetery.

### *Caroline Marmon Fesler*

Caroline Marmon, sister of Howard and Walter, married James Fesler and was described by the Society Editor of the Indianapolis News as "one of Indianapolis great women." This assesstment was made in a discussion of the four leading Indianapolis families who had founded major firms which appeared in the Oct. 11, 1965 edition of the News.

Following her father's death in 1909 she, like her brothers, received a gift share of the mother's inheritance in the amount of one million dollars. The Feslers lived graciously but unpretentiously on North Pennsylvania Street in a home now occupied by the Robert Greenleafs . . . Mrs. Greenleaf is Mrs. Fesler's niece.

Mrs. Fesler was a knowledgeable student and lover of music; she was a close friend of Arturo Toscanini of the New York Philharmonic Orchestra. She was also a student and patron of art, particularly of the John Herron Museum of Art. When an initial purchase of "modern" art was debated, Mrs. Fesler argued in favor but was opposed by Booth Tarkington although they were good friends. Mrs. Fesler offered to contribute the $500 deficiency in funds needed to purchase a Picasso, but the board was dissuaded by Tarkington's laughter. Mrs. Felser retaliated by purchashing the Picasso for $3,500 and donating it to the museuam. In 1965 it was estimated to be worth $100,000.

The News article concluded by noting that there were no Marmon men left in Indianapolis but that Mrs. Fred J. Hoke is the granddaughter of Daniel Marmon and Mrs. Robert Greenleaf is a great-granddaughter.

### *Franklin Hall Marmon*

There was a parade in downtown Indianapolis in March 1914 for an unidentified occasion. It was led by a procession of Marmon vehicles. Preceding, in order, the Marmon '32' Winner of the first 500 Mile Race, the first Marmon car of 1902, and an early production Marmon, was the lead vehicle . . . Franklin Hall Marmon driving the car he built at age 14 and named "Marmonette."

The son of Walter Marmon, (Franklin) Hall Marmon was the heir apparent for the business — the Edsel Ford of Indianapolis. Like his father, he was educated at Massachusetts Institute of Technology, graduating with a commendable record in the Class of 1923. He married a girl from Athel, Massachusetts who bore him a child, Anne Amelia.

Hall was placed in charge of the Marmon Test Department, to begin his practical education.

In October of 1924 he drove a Marmon Test car west with the intent of climbing Pikes Peak in Colorado to evaluate a brake system designed by his Uncle Howard. However an early snow storm closed the road for the winter before his arrival. He was accompanied by another engineer, Harold L. Fisher.

As they neared Indianapolis on the Saturday night of October 11, 1924, with Mr. Marmon driving, the car skidded out of control on loose gravel and overturned near Avon, Indiana. Hall's chest and skull were crushed by the steering wheel and death was apparently instantaneous. Mr. Fisher climbed from the wreck and was driven to town by a following motorist, who believed that blinding lights from a passing car were responsible.

F.H. Marmon was a 2nd Lieutenant in the Officers Reserve Corps, and a member of SAE and ASME. Walter Marmon served as president of both the Indianapolis and the Central Indiana Boy Scout Councils; He donated a swimming pool at the Indianapolis Boy Scout Reservation in memory of his son.

Hall's daughter Anne was only five months old at his untimely death. During 1979, the *Indianapolis News* ran a series of feature articles consisting of individual interviews with prominent Indianapolis women. The series included one with Mrs. Robert (Anne Marmon) Greenleaf. Therein[12] she observed that all the Marmon men were noted as fast drivers, including her father. She spoke of her father's accident and said, "I think Uncle Howard, a rough, gruff person, felt guilty forever."

### *Marmon Ancestry and Crest*

Beginning in 1927, the traditional Marmon radiator "pyramid" and badge were augmented by a Marmon Coat-of-Arms. It appeared on the frame of the instrument cluster of the Model 75; it graced the pyramids that crowned the headlights of the Big Eight. Cadillac had the Coat-of-Arms of Antione de le Mothe Cadillac who founded Detroit (Michigan) in 1702 and showed it liberally on their automobiles. Competitive pressures and the French ancestry of the Marmon family made a comparable Marmon adornment logical. Beginning in 1927 the "Foremost Fine Car" and the "Standard of the World" both bore French coats-of-arms. Subsequent editions of the Marmon "Book of Facts" incorporated a Chapter entitled "The Marmon Crest," which addressed the Marmon Family genealogy.

The Marmon family traced its ancestry directly to the Duke of Reguse, Auguste Frederic Louis Viesse De Marmont whose coat-of-arms was the same as used in the twentieth century, and whose father Nicholas Edma Viesse De Marmont reportedly first used the name of Marmont.

About Nicholas Edma Viesse De Marmont, the Book of Facts states: "He was lord of Sainte-Columbe, a captain in the service of Marchal de Richelieu and was the father of the renowned Auguste Frederic Louis De Marmont. The mother, following marriage in 1769, was Clotilde Helene Victoria Chappron."

About Auguste Frederic, it continues: "He was born at Chatillon-sur-Seine, on July 20, 1774. He entered the military service in 1789, which began one of the most illustrious military careers ever followed by any Frenchman. Through a series of between fifteen and twenty notable military expeditions and campaigns, he gradually rose and became marchal of France; one of the major-generals of the Royal Guard; won the cross of the Royal Order of the Legion of Honor; was commander of the Royal and Military Order of Saint Louis; won the grand cross of the Royal Order of Wurtemburg; was commander of the Royal Order of the Iron Cross; and member of the Royal Academy of Sciences and of the Institute."

The Viesse ancestry, pre-Marmont, was traced directly to the great-great-great-grandfather of Nicholas Edma Viesse and less specifically to the second century when the name was first used. The account continues: "In the beginning of the third century, the house of Viesse was divided into many branches, which were established in Germany and France. From that time they flourished and produced men distinguished in the gown and the clegy. The name of the second branch which became fixed in Germany, underwent several modifications brought about by the country's language; but this is not true of the original branch, whose name and arms have remained intact."

"Upon approaching modern times, the titles of the branch located in Bourgogne established regularly a continuous lineage from Alexis Viesse, a lieutenant in the Vielille Marmine regiment. He was born 1504 and died 1590. He gave an endowment to the Latrecey church; this fact being entered in the church property registers. He was the father of Helion Viesse, lord of Latrecey, who added to the family's

Fig. 1-9 Marmon (t) family coat-of-arms.

arms. He placed on the dexter (right) side of the arms, a gold Lorraine cross on a blue ground; also on the sinister (left) side a flaming sword moving from a cloud all of silver on a red or carnation ground, leaving at the chief sinister point (the upper left hand corner) the silver lion. It is not known if the flaming sword is the result of concession or of family conventions."

The Marmon (t) coat of arms is described: "In the first and fourth quarters, fesses of silver and gules (gules denoting, by parallel lines of red, flame or fire and indicating valor and knightly prowess) : the silver standing for purity and virtue. The second quarter is gold, with a gules standard charged with a silver cross. In the third quarter the dexter side is blue with a gold Lorraine cross; while the carnation dexter is gules, with the flaming sword moving from a cloud all in silver; with the chief point green, charged with a silver lion. The helmet at the top of the shield is emblematic of the family's descent from nobility."

Some twenty years after production of such great American marques as Marmon and Duesenberg had come to a sad halt, they were widely thought to have been imports from Europe — French in the case of Marmon and German in the case of Duesenberg.

Extensive publicity in recent years has pretty well established the Indiana, USA, origin of the Duesenberg. The problem has persisted in the case of the Marmon. A relatively recent example appeared in the 1979-2 issue of the respected *Car Graphic,* a high caliber Japanese Language publication which includes English translations of article titles for western world subscribers. Each issue includes an automotive history article; in subject issue "The French Racing Cars at Indy" was

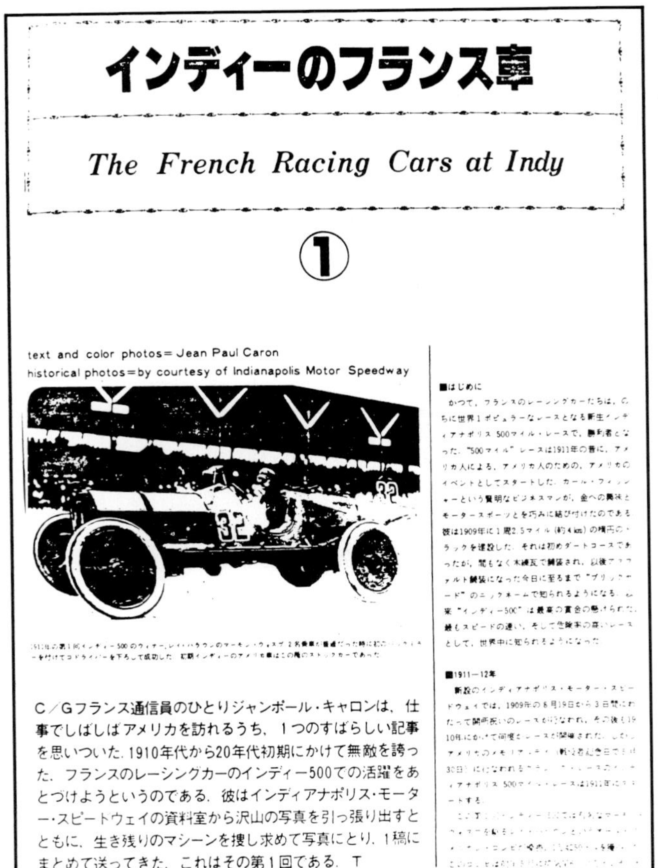
インディーのフランス車

The French Racing Cars at Indy

①

text and color photos= Jean Paul Caron
historical photos=by courtesy of Indianapolis Motor Speedway

C／Gフランス通信員のひとりジャンポール・キャロンは、仕事でしばしばアメリカを訪れるうち、１つのすばらしい記事を思いついた. 1910年代から20年代初期にかけて無敵を誇った、フランスのレーシングカーのインディー500での活躍をあとづけようというのである. 彼はインディアナポリス・モーター・スピードウェイの資料室から沢山の写真を引っ張り出すとともに、生き残りのマシーンを捜し求めて写真にとり、1稿にまとめて送ってきた. これはその第1回である. T

Fig. 1-10 *Car Graphic* (Japan) reports on the historic Marmon win at Indy.

headed by the familiar picture of Harroun winning at Indianapolis with the most famous of all Marmons on its homegrounds!

The non-anglicized original family name was perpetuated in the name of a town in Marshall County, Indiana. According to the *Marmon Book of Facts*[3], the town name was changed in 1851 from "Culver to Marmont" in honor of a French military general. Because the name was missing from contemporary atlases, the Northern Indiana Historical Society was asked by the author to identify the location and current name: According to *"Indiana Place Names"*[13], today's Culver, Indiana, was laid out in 1844 and was "Uniontown" until 1851 when it was renamed "Marmont" for unidentified reasons. The spelling deteriorated into Marmount and in 1895 it was again renamed, this time for the founder of Culver Military Academy — Henry Harrison Culver. Interestingly the first redesignation occurred in the very year that Ellis Nordyke was founding the parent organization of the Marmon firm and just two years after Dr. James Marmon's untimely death in the cholera epidemic; perhaps he was then known as Dr. Marmont.

## PART 3 — The Marmon Car

### *Introduction*

The first motor car to bear the proud name of Marmon came to the marketplace in a unique manner. It was not the product of an established builder of horse drawn conveyances seeking to avoid losing a market segment to those willing to experiment with horseless propulsion. (That product would have reflected the skilled tradesmen, — woodworkers and blacksmiths — whose skill produced the American stagecoach and international carriages). Neither was it the product of an entrepreneur recognizing on the one hand the prototype product born some ten years earlier in Europe and on the other the potential untapped market on this side of the Atlantic.

Rather it was one of an elite group of motor cars which came from the plants of established machinery builders or satellite plants. These vehicles were characterized generally by superior mechanical durability and in some cases by highly innovative mechanical design.

About the same time that press releases announced that Nordyke & Marmon Co. would enter the automobile business with a very unusual motor car, the first Nordyke & Marmon advertisement in an automotive trade magazine appeared. It was a full page addressing the stature of the firm, its experience and reputation worldwide. The new product was stated to be designed and manufactured to be without peer on either side of the Atlantic. *An introductory brochure had this to say of the offering: "There is a place for it; a place for it because it embodies new ideas, new and important features which make it better than existing motor cars and something better is needed, in fact demanded."*

An insight into the philosophy of the firm upon embarking in the automobile business can be had from the following excerpt from the 1906 catalog #630: *"The Marmon car is built in accordance with ideas, and of construction, original with us and since early in 1902 we have been engaged in **developing** and **perfecting** the car. Manufacturing, as we do, the car practically complete from the raw materials, we aim to build up the business on its merits and thus secure a reputation in keeping with our reputation in other (product) lines. Not only have we set a high standard of workmanship which will be closely adhered to, but we have introduced many new features in which we have great faith and which we firmly believe will make the Marmon Car far-famed." (Author's Emphasis)*

The extent of in-house manufacture was surprising: *"We manufacture the Marmon car, all of it from the ground up, excepting such items as the wheels, tires, lamps, ignition apparatus, etc. We make the pressed steel frames, fenders, hoods, etc., the axles, driving mechanism, transmissions and other numerous parts. We make the aluminum castings for bodies and machinery parts; brass bronze and iron castings; do all machine work and gear cutting except cutting the bevel gears. We make our air cooled motor entirely, and in this we believe we have attained greater perfection than yet reached in any other air cooled motor. We build, trim and finish the bodies, assemble and finish the cars complete ready for use."*

Only a firm able to confidently describe itself as "One of the world's greatest engineering houses" could afford the luxury of the cautious entry which Nordyke & Marmon Company made into the automobile market. They began with a prototype vehicle of such advanced concept that no less prestigious a critic than Charles E. Duryea[14] would remark in *MOTOR* magazine for January 1918: *"The Marmon of 1902 looks more like its present day descendants than do most of the veterans."* That was not surprising, considering the statement in the 1906 catalog[15] but applicable to all Era One Marmons: *"They are finely proportioned and of a pleasing design and style, which is more likely to be permanent than that of many other cars designed to cater to passing fads and fanciers."*

Nor should it be surprising after rereading the very first full-page advertisement, which addressed the product rather than the manufacturer, and which appeared in December 1904, and had reference to the 1905 Model: *". . . our 1905 Model. We might, with ample justification, call it our 1910 Model, for authorities have pronounced it five years ahead of the times.* ***Parts are made interchangeable*** *and can be carried conveniently for such emergencies as collision or other accident." (Author's Emphasis)*

The prototype was followed by a full year effort (1903) to produce a second generation prototype, only seven of which were produced in 1904 and sold to selected customers for field evaluation. Only after these cars performed above expectations was the press informed in Nov. 1904 that Nordyke & Marmon would offer a car in the coming season.

Cycle and Trade Journal subsequently commented on the pedigree of the Marmon Car:[16] *"This house, with more than half a century of manufacturing experience, in the course of which it has known no superior in the (product) lines it has undertaken, guarantees the materials and workmanship of the Marmon Car as first class in every respect."*

The "Marmon" was marketed throughout Era One under the conservative slogan "A Mechanical Masterpiece." It was just that. Perhaps then it was not so surprising after all when a perfect score in the 1906 Glidden Tour was followed by victory after victory in just about every automotive competition and was climaxed by victory in the first Indianapolis 500 mile race in 1911; nor when a model 34 set a coast-to-coast record averaging 30 miles per hour in 1916; nor when a 1931 SIXTEEN claimed the Stevens 24 Hour Trophy.

The stature of the firm that established "The Marmon Heritage" was well defined in the 1922 model Marmon brochure: *"Long before the motor car came into being, the Nordyke and Marmon Company was an institution devoted to the manufacture of flour mill machinery and had established a worldwide reputation for manufacturing integrity and business vision. Marmon flour mill machinery had long been going into every corner of the world, carrying the message of a work well done. Today (1922) Nordyke & Marmon Company stands as a vision realized. Its mills are to be found in every country where wheat is raised and flour made. Its motor cars represent the highest degree of mechanical excellence yet conceived. Its men and women, backed by the ideals of seventy-one years of unquestioned business and mechanical integrity, are engaged in the task of fulfilling these ideals. This is the Nordyke & Marmon Company of today — an institution with a vision."*

In support of the above claim, consider the assessment of British novelist Anthony Gibbs in his *"A Passion for Cars."*[17] Having accompanied his father on a trip to the United States in 1922, Gibbs was thrilled by American locomotives and motor cars . . . he wrote of it in 1974: *Long high bonnets (hoods) and low stubby seating matched the American automobile to the impressiveness of the American locomotive. Does anybody remember the names? the Apperson Beauty Eight; the Cole Aero-Eight; the Paige; the Templar;* **that very good car the Marmon;** *the Pierce-Arrow with ugly froglamps on its wings (fenders); the Packard Twin Six. Only the Cadillac is still with us, sliding nobly, as it did then, through a shoal of Fords, Chevrolets and Dodges. (Author's Emphasis)*

Beginning in 1914, Gibbs' ownership of a long string of thoroughbred cars was remarkable both for quantity and perceptive assessment of both mechanical personality and styling significance. His opinion of the Marmon given above reflects the international image achieved by that firm by 1922 because Anthony Gibbs had never owned that marque although he or his mother owned several US marques.

Perhaps the ultimate accolade was the election of the Marmon Models 74 and 75 to full "Classic" status by the *Classic Car Club of America.* Such recognition is mandatorily limited to cars built in 1925 or later. Maurice Hendry, in his "Housing 200 Horses"[18], has commented on the remarkable circumstance of the last iteration of a basic car model, originally placed in production for the 1916 model year, having achieved Classic status more than ten years later without having undergone a change in basic styling concept or basic chassis configuration.

The author presumes, or at least strongly suspects, that the reader will have developed, from the above insight into the ambitions and plans of the infant automobile arm of prestigious Nordyke & Marmon and the international recognition that was subsequently achieved, a curiosity to explore with him the matter in detail and learn the full significance of The Marmon Heritage to the world of transportation.

## *The Five Eras of the Marmon Car*

The Marmon cars do not lend themselves, for example, to the series designations that characterize the Packards; such reflect a progressive advancement and refinement process, and some common design philosophy. However, the thirty years of Marmon passenger car production can be classified into five eras, each quite distinct. The cars of the first, third and fifth were characterized by totally unique design and engineering, were created without regard to convention or pre-conceived market preferences, were a symphony of unique ingenious components cleverly intergrated into a most harmonious and pleasing whole. Quite differently, the second and fourth eras were marked by vehicles soundly engineered and reliably manufactured in apparent conformity to other contemporary offerings and presumed public preference.

The FIRST ERA Marmons were perhaps the most unique motor cars ever to travel the North American roads. They were "System Engineered," i.e., the driver, the car, the road and the tire interface were all significant considerations in the design. The reward was a perfect score in the Glidden Tours of 1906, 7, 8 and innumerable non-stop road race victories. These Marmons (1) had a unique crankcase/crankshaft interface shared today only with the engine which powers many of today's Indianapolis race cars and two of the world's most advanced diesel engines, and (2) pioneered modern pressure lubrication via a drilled crankshaft and connecting rods.

The subtle superiority of the more conventional SECOND ERA vehicles was evidenced by stellar performances on the racing circuit culminating in the famous First 500 Mile victory at Indianapolis in 1911. (Marmon returned to the bricks in 1920 and 1928 as the Official Pace Car and to competition during the FOURTH ERA in the 1928 running of the 500).

The THIRD ERA was Marmon's greatest. A single basic model, equally innovative in technology and styling was progressively refined over thirteen model years. It was a vehicle concept so original that anyone familiar with the 1916 model would immediately recognize the 1927 model accredited full "Classic" rank by the Classic Car Club of America!

The FOURTH ERA was one of growth under the Marmon Motor Car Co. marque, although sales never brought production to projected levels. The plans of G.M. Williams, who left a general managership in General Motors to assume the Marmon helm, were frustrated by the economic crisis of 1929 and the market stagnation which followed. It was also characterized by a pioneering demonstration of the fallacy of exploitation of a respected exclusive marque in an attempt to achieve rapid acceptance in the mid-range volume market, followed by tardy attention to retention of the forsaken status niche.

The FIFTH ERA, the briefest, was marked by crowning engineering and design accomplishment and alas by concurrent deterioration of market penetration and the once-strong dealer organization. Howard Marmon returned to active direction of engineering and combined his technical genius with the styling mastery of Walter Darwin Teague to create the Marmon Sixteen masterpiece. Creditors forced receivership in May, 1933 after months of reorganization efforts following the National Bank Holiday proclaimed by President Roosevelt. Despite the pleas and efforts of Harry Miller and Preston Tucker to preserve the integrity of the facilities for planned future resumption of production, as the American Automobile Corporation, the court ordered disposal and dispersal of the assets in 1934.

Prior to the demise of Marmon Motor Co., Walter Marmon had organized in March, 1931 an independent firm, sharing the Marmon manufacturing complex, for the manufacture of military vehicles and off-road vehicles to the designs of Arthur W. S. Herrington and the new products bore their hyphenated names: MARMON-HERRINGTON.

The remarkable Marmon-Herrington story in war and peace is told in Book Two of The Marmon Heritage.

In the following chapters each ERA will be addressed in a consistent format. The several car models of a particular ERA will be discussed with emphasis on styling and innovation supported by general technical specifications. That will be followed by separate detailed treatments of the engines of that ERA and of the Chasses of that ERA. The latter will be sub-titled for the various components or systems. For the reader with an interest in the products but who may be either unskilled or disinterested in technical detail, the successive reading of the CARS OF ERA X sections will provide a cohesive account and perhaps encourage exploration of the more detailed information.

For the contemporary engineer and/or designer, as well as for the serious automotive historian, this assessment by an experienced Professional Engineer and engine/power train designer combines historic press releases and reportorial assessments with information freshly gleaned from analysis of engineering-drawing illustrations, and viewed from a modern perspective.

The reader will note that the time frame of a particular ERA may be different for the product than for manufacturing, management or financial discussions. This reflects the offset of conception, planning, development and tooling in advance of production.

### *Engine Specifications 1904-1934*

A chart of Marmon Engine Specifications appears at the end of this section (Part 3 of Chapter One). It tabulates data on the twelve basic engines and variations thereof which powered the Five Eras of Marmon car production. Locating the chart here makes for convenient referral from time to time while progressing through the several Eras, and the racing chapter "International Champion" as well.

**THE MARMON**

**"The Easiest Riding Car In The World"**

| ERA | Engine Type | Engine | Cyl. | Bore | Stroke | CID | SAE HP | BHP Rated-Max | HP/CID | Cyl. Head | Car Model | Year | Compr. Ratio |
|---|---|---|---|---|---|---|---|---|---|---|---|---|---|
| 1 | Air-cooled OHV-90° | 1a | V 4 | 4 | 4 | 201 | 25.6 | 21 | 0.105 | Integral | Prototype | 1902-03 | |
| | | " | " | " | " | " | " | " | " | " | A, B | 1904-05 | |
| | | 1a spl. | V 6 | " | " | 302 | 38.4 | 31 | " | " | Prototype | 1905 | |
| | | 1b | V 4 | 4-1/4 | " | 227 | 28.9 | 20-24 | 0.09-0.11 | " | C, C-7 | 1906-07 | |
| | | 2a | " | 4-5/8 | 4-1/2 | 302 | 33.7 | 26-30 | 0.09-0.11 | " | D | 1906 | |
| | | 2b | " | 5 | " | 353 | 40.0 | 35-40 | 0.10-0.11 | " | F | 1907 | |
| | | 2b spl. | V 8 | " | " | 707 | 80.0 | 75 | 0.11 | " | V | 1907 | |
| | | 2c | V 4 | " | " | 353 | 40.0 | 35-40 | 0.10-0.11 | Separate | G: H opt'l | 1908; 1909 | |
| 2 | H2O Cooled T-Head Dual Block | 3a | L 4 | 5 | 5 | 393 | 40.0 | 45-50 | 0.12-0.13 | Integral | H opt'l; 45 | 1908; 1909 | |
| | | 3b | " | 5-3/8 | " | 454 | 46.2 | 50-60 | 0.11-0.13 | " | 50 | 1909 | |
| | | 4a | " | 4-1/2 | 4-1/2 | 286 | 32.4 | 40 | 0.14 | " | 32 | 1909 | |
| | | 4b | " | " | 5 | 318 | " | 40-60@2100 | 0.13-0.19 | " | " | 1910-12 | 3.15:1 |
| | | 4c | " | " | " | " | " | " " " | " " | " | " | 1913 LH, 1914 | |
| | | 4d | " | 4 | " | 251 | 25.6 | | | " | " Com'l | 1913-14 | |
| | | 4e | " | 4-3/4 | 7 | 495 | 36.1 | | | | " Race Spl. | 1911-14 | 3.8:1 |
| | T-Head Triple Block | 5a | L 6 | 4-1/2 | 5 | 477 | 48.6 | 60-90+@2100 | 0.13-0.19 | " | Indy 500 | 1911 | |
| | | 5b | " | " | 6 | 573 | " | | | " | 48 | 1913-14 | |
| | L-Head Dual Block* | 6a | " | 4-1/4 | 5-1/2 | 447 | 43.4 | 70 | 0.16 | " | 41 | 1914 | |
| | | 6b | " | " | " | " | " | " | " | " | " | 1915-16 | |
| 3 | OHV AL Uniblock | 7 | " | 3-3/4 | 5-1/8 | 339 | 33.8 | 74 2450 | 0.22 | Separate | 34A | 1916-19 | |
| | OHV CI | 8a | " | " | " | " | " | 84 | 0.25 | Separate | 34B/34C | 1920-24 | |
| | | 8b | " | " | " | " | " | " | " | Intake in head | D74/E75 | 1925-6/27-8 | 4.5:1 |
| 4 | OHV Uniblock | 9a | L 8 | 2-3/4 | 4 | 190 | 24.2 | 64 3300 | 0.34 | Separate | L | 1927 | 5.27:1 |
| | | 9b | " | 2-15/16 | " | 217 | 27.6 | 86 3400 | 0.40 | " | 78 | 1928-29 | 5.5:1 |
| | L-Head Uni-block | 10a | " | 2-3/4 | 4-1/4 | 201 | 24.2 | 72 3200 | 0.36 | " | 68/R/MR | 1928/29/30 | 5.5:1 |
| | | 10b | " | 2-13/16 | " | 211 | 25.3 | 76/84 32/3400 | 0.36/0.40 | " | 68/69/70 | 1929/30/31 | 5.25:1 |
| | | 11a | " | 3-3/16 | 4-3/4 | 303 | 32.5 | 107 3250 | 0.35 | " | 79 | 1930 | " |
| | | 11b | " | 3-3/4 | " | 315 | 33.8 | 125 3400 | 0.40 | " | B8/88/8-125 | 1930/31/32 | 5.5:1 |
| 5 | OHV AL Uni-block | 12a | V 16 | 3-1/8 | 4 | 491 | 62.5 | 200 3400 | 0.41 | " AL | Sixteen | 1931-33 | 6.0:1 |
| | Fork/blade rod | 12b | V 12 | " | " | 368 | 46.9 | 160 3400 | 0.43 | " " | Prototype | 1934 | 6.0:1 |

* A bore of 4-1/8 would give a 41 SAE HP rating; evidently enlarged prior start of production.

CID - Cubic Inch Displacement
OHV - Overhead Valve

Fig. 1-11 **Marmon Engine Specifications and Model Usage.** The twelve basic engines and variations which powered the Five Eras of the Marmon Car.

## IN MEMORIAM

A Marmon Memorial Collection of Paintings by Dutch and French Masters was given to the Art Association of Indianapolis in memory of Daniel W. and Elizabeth C. Marmon. The collection has been at the John Herron Art Museum since 1948. It includes Van Gogh's 'Landscape at Saint Remy', Cezanne's 'House in Provence' and George Seurat's 'Port of Gravelines' all from the late nineteenth century. Oldest is Corneille de Lyon's 'Man With a Glove' from the mid sixteenth century while four are from the seventeenth — by Van Ruisdael (The Cascade), Hobbema (The Water Mill), Kolf (Still Life with Brown Jar) and Cuyp (The Valkhof at Nymengen).

The Administration Building at Earlham College, Richmond, Indiana is named Carpenter Hall in recognition of a significant contribution for its construction about 1925 by Mrs. Elizabeth Carpenter Marmon, whose parents, Walter and Susan Carpenter, were respectively Superintendent of the Management Committee (President) and Matron of the then boarding school when she married Daniel W. Marmon. 'Marmon House', located across the street from the Campus, was the residence of the President for about nine years, was sold and reacquired as a student residence which it remains today.

NORDYKE & MARMON COMPANY

ESTABLISHED 1851

INDIANAPOLIS, IND., U.S.A.

CHAPTER TWO

# THE MARMONS OF ERA ONE

*"A Mechanical Masterpiece"*
*1901-1908*

The Marmons of Era One firmly established the marque's reputation for excellence in engineering and superior performance in competition. These were the alphabetic sequence Marmon models; in subsequent Eras, alphabetic identification was used internally and was evidenced in part number prefixes but the cars were marketed under other model designations.

Era One is the story of an innovative concept of a superior motor car and its progression from concept demonstration through the prototype/field evaluation phase and a succession of progressive design refinements to full maturement. This first Era, and only that Era, would lend itself to the "Series" treatment, for it is characterized by promotion and progressive development of a single vehicle concept and basic product.

There are three phases to Era One: The concept demonstration vehicle built and operated successfully in 1901-02; The prototype Model A developed in 1903-04, of which seven were sold in 1904 without public offering; and the successive Models B, C and D, C-7 and F, G and H marketed for four models years (they called them "seasons" then) 1905 thru 1908.

Models A and B were singular model year offerings; Models C and D, C-7 and F, G and H were companion offerings. Evidently a Model E had been planned but a decision was made to continue the Model C for 1907; the C-7 designation identified the extended wheelbase for 1907.

The advent of the first Marmon car was recorded for posterity, after the fact, by W.A.P. John in his article for the November, 1922 issue of *MOTOR.* He was reporting an interview — one of a series with presidents of major automobile manufacturers — with Walter C. Marmon. In the interest of fidelity, that portion of the interview is repeated here: *"Having exhibited from his boyhood days an amazing mechanical ingenuity and an insatiable curiosity concerning all sorts of machinery, Howard, the younger brother, after finishing his engineering studies at the University of Cali-*

Fig. 2-1 First Marmon on an Indiana road (1922 Catalog.)

*fornia, naturally gravitated to the engineering department of Nordyke & Marmon. There he experimented with various mechanical appliances and spent long hours studying the crude horseless carriages that were being produced and sold at the time. Finally he purchased one for his own use; and as Mr. Walter Marmon laughingly explained, "it was a terrible piece of machinery. It did everything but run consistently." Whereupon the youthful genius decided that he could build a better motor car himself. And he did, in 1902."*

The manufacture of seven Model A Marmons in 1904 is confirmed by a letter[1] which the Nordyke & Marmon treasurer, A.R. Heiskell, wrote to *The Automobile* on Feb. 18, 1916 in response to their inquiry. He reported that the Marmon car was 'placed on the market' in 1904 and that some seven were built and sold in that year.

## PART 1 — The Cars

Prototypes, Models A Through H, 45 and 50

### THE FIRST MARMON 1901-1902

A 1915 Nordyke and Marmon brochure, 'Marmon (Model 41) Oiling Features' provides confirmation of the date of construction and type of engine in the First Marmon. *"In 1901 Howard Marmon built his first automobile and the motive power was a V-type, two cylinder, water-cooled motor."*

The first car which Howard Marmon designed was constructed in the Nordyke & Marmon shops. The parts were machined under the personal supervision of R.E. (Bob) Roberts who had joined the firm in 1880. He is also credited with assisting in the development of the car, and was a passenger on the first (100 mile) trip to Kokomo, Indiana and return. The other passengers were Harry More and George Nichols. The road testing may have begun in 1902 and been the basis for the 1902 date generally attributed to the car.[2]

That first car is most remembered for its most conspicuous feature — a side entrance tonneau with hinged rear doors. Marmon is generally credited with this innovation.[3] Because the author has seen pictures of Oldsmobile prototype vehicles with a similar feature[4] about the same time, it must be concluded that the Marmon distinction was based on its production use in 1904.

Very significant was the placement of the engine under a forward hood, the choice of a Vee configuration and an aluminum crankcase for the engine, use of a bevel gear axle shaft-driven by the engine, adoption of a single-plate clutch and selective-shift three-speed and reverse transmission. Spark and throttle controls were brought to the top of the steering wheel through a hollow shaft. The frame was made of pressed steel construction. The body construction was cast aluminum. The tonneau was a one-piece casting and included an integral bustle trunk; access was by a door hinged at the bottom and opening to a horizontal position.

All that is known of the engine is that it was a V-2 with aluminum crankcase and was water cooled. If it had the same (4 inch) bore and stroke of the first V-4 engines, it would have been low-powered but not unreasonably so by 1902 standards.

Inspection of surviving photographs leads to the conclusion that the front springs were semi-elliptic leaf and the rear were full-elliptic. It may have incorporated a dual-frame chassis but the springs appear to all be attached to the same frame. If so, any dual frame would have avoided the use of universal joints (chains were not used) but would have had no advantage so far as tire life was concerned.

THE FIRST MARMON EVER BUILT

Fig. 2-2 From the 1928 catalog '77 Years of Craftsmanship'

### THE SECOND MARMON 1903

What the author calls "The Second Marmon" was the car which Charles Duryea illustrated and described in his 1918 article with the comment: "The Marmon of 1902 looks more like its present day descendants than do most of the veterans."

It is not the First Marmon of 1902; rather it has all the characteristics of the First Era commercial products. Because it is not a 1904 Model A Marmon, it is most likely the prototype built in the winter of 1903-04 which incorporated an entirely new chassis concept.

The unique feature of the chassis was the result of concern with the bane of the early motorist — all pre-Marmon and later non-Marmon owners — unreliable tire performance. Howard Marmon made a pioneer application of what today would be called "System Engineering," the consideration of the automobile as a system with the driver at one end and the road surface at the other. Nordyke and Marmon were not, and did not intend to be, in the tire business. Study of tire failure patterns, including punctures, led Marmon to the conclusion that 35 mile per hour travel over the roads of the early century required up to twelve inches of wheel travel.

He also observed that, in the operation of a conventionally sprung car, one tire could have left the road surface causing the opposite tire to carry a double load; should that tire encounter a bump at that time, the double load would be multiplied by the number of Gs consequent the vertical acceleration. Should a bump be encountered while the spring was already highly deflected consequent a preceding bump, extreme tire loads would occur and failure of the tire casing would result.

Fig. 2-3 Charles Duryea's comment referred to this Marmon car.

Another key determination by Marmon was that chassis deflections were the origin of the majority of engine, transmission and axle (power train) bearing problems and failures.

Howard Marmon's highly innovative solution was to design two separate suspension systems: one to accommodate road undulations (progressive deviations from a "plane surface") and one to accommodate ruts and bumps in that surface.

This was the basis for what would become the patented double three-point suspension used on all First Era Marmons. The concept utilized a main frame and a propulsion frame.

The Second Marmon differed from the Model A in the body: (1) the rear door and adjoining body differed in that the body was an integral unit (with front and rear seats) whereas the Model A was reconfigured just to the degree necessary to accommodate a detachable tonneau; and (2) a cast cowl, similar to the 1902 car, was used in place of the flat bulkhead which characterized the Model A. The cast cowl interchanged with the bulkhead and did not affect the cast seat riser/floorboard support which extended to the bulkhead line on either car.

Confirmation of "Second Marmon" is found in an article "Bodies Cast From Aluminum" which appeared in *The Motor World* for May 1907.[3] *They (Nordyke & Marmon) have been using it (aluminum) at least in parts of their bodies since 1903. During that year, for the second car built as a matter of fact, a cast seat was made which, it may be noted, is yet in good condition and use. Previously the company had been casting crank and gear cases.*

The fenders illustrated are most certainly a later rework and it is probable that the windshield was also added in later years.

It is a documented fact[5] that the V-4 air cooled engine was initially operated on the dynamometer in "the winter of 1903-04"; it could be concluded therefore that the car was concurrently constructed.

It must have been extensively demonstrated early in 1904 because the first Model A was delivered no later than May of that year. Alternatively the car could have been, and almost certainly was, built and operable in 1903 with a subsequent engine retrofit incorporating the pressure lubrication invention.

## MODEL A
## 1904

'Horseless Age' for December 24, 1904 carried an announcement that the Nordyke & Marmon Co. would enter the automotive field for the 1905 season with a very unusual motor car. This announcement signified successful completion of a trial production and field test in owners' hands.

This first automobile product of Nordyke and Marmon featured a V-4 air cooled engine, a two speed and reverse planetary type transmission, shaft drive (without universal joints) to a bevel-geared rear axle fitted with internal shoe brakes. The engine was mounted forward and longitudinal in the chassis, inclined toward the differential and enclosed under a rearward-hinged hood having an expanded metal grille.

Uniquely, it incorporated two patented features: A full pressure lubrication system for the engine with recirculated oil and a dual-frame chassis in which the complete power system was assembled to a separate frame and isolated from the frame which carried the body.

Only seven[6] Model A Marmons were produced in 1904. These had prototype Model B chasses and constituted a field test in consumer hands before a full-flown public marketing. Surprisingly, the owners of five of these cars are known; it is reasonable to assume that the other two were members of the Marmon family, and probably the Marmon brothers.

During the 1905 model year, a brochure "Good Words About the Marmon" was distributed. It contained letters of endorsement from five people who had the privilege of purchasing a 1904 Marmon. The letters provide information on the perfor-

Fig. 2-4 The 1904 Model A (from the 1905 Model B Catalog rear cover.)

mance and economy of the car, as well as a surprising insight into the usage patterns. Then, as now, mileage accumulation at about 1000 miles per month was not uncommon with these cars in 1904. A January 1905 ad stated that "Marmon Cars have been driven 8000 miles without repair."

The first Marmon owners and the delivery times were: George Van Camp, Van Camp Packing Co., Indianapolis; George M. Pangborn, Indianapolis, Insurance Banker May 15 approx.; John Perrin, Pres. The American National Bank of Indianapolis, July; E.C. Andrews, Secy, Security Mills, Nashville, TN Aug. 4; F.O. Watts, Pres. First National Bank, Nashville, TN Aug.

George Van Camp reported his car so quiet that some people believed it to be an electric. Presumably Van Camp had been the first to take delivery; although, unlike most of the others, he did not give an in-service date, he had driven 10,000 miles by Nov. 29th 1904. Even in eight months, that would be representative of car usage in the 1970s. He commented: "The Marmon rides smoother, runs faster than most people like to go and is remarkably free from petty annoyance common to many machines."

Mr. Pangborn took a 600 mile trip in six days through southern Indiana and northern Kentucky, with a party of four and luggage. Average speed was 16 miles per hour and fuel economy was 16 miles per gallon; more than 50 miles per quart of oil. He averaged 22 miles per hour on the return between Richmond, Indiana and Indianapolis (77 miles in 3.25 hours). No tire problems were encountered and no work of any kind was required on the car. He reported 5000 miles of daily usage except for the replacement of a broken spring leaf; he was driving 700 miles per month.

Mr. Perrin drove 6500 miles in 5 months (1300 miles/month) and experienced only one puncture of his G & J 3.5 inch tires. Mr. Perrin's Marmon was equipped with a Goddard top.

Mr. Andrews reported his Marmon was capable of 42 miles per hour on a level road. He had driven 4000 miles in 4 months (1000 per month) and experienced 167 miles per quart of oil. He noted that the car raised less dust and had less of a dust problem to rear seat passengers than any car with which he or his friends had experience.

Mr. Watts simply expressed a high degree of satisfaction with his car.

It is interesting that Nordyke & Marmon were evidently aware that a valid operator assessment of a prototype product can only be obtained by sale to an owner/operator.

The illustration of the Model A Marmon appeared on the rear cover of a 1905 brochure.

It is assumed that the description of the Model B which follows is equally applicable to the Second Marmon and the Model A. Comparable documentation is not available for the earlier cars.

## MODEL B
## 1905

Nordyke & Marmon were very serious about their entry into the automobile business. Full page ads in the trade press sought to acquaint the new customer with the stature of the manufacturer with nary a word about the car details. These appeared in issues of December, 1904 and were followed immediately by full page ads illustrating the "The Marmon Motor Car" — the Model B.

The new car was carefully engineered . . . it was modestly promoted as "A Mechanical Masterpiece." Economy of fuel and tires would become its hallmark.

Fig. 2-5 The Marmon Motor Car — Model B, 1905

The extensive use of aluminum was remarkable. As a consequence the Model B weighed only 2000 pounds despite the dual frame system. Its use was not an alternative; the design was predicated on aluminum with full consideration of its shortcomings (to the degree then recognized) as well as its advantages. Most significant was the barrel engine crankcase configuration cast in aluminum. The aluminum transmission case and differential housing were also noteworthy. Surprising was the use of precision thin wall (.156) castings, of aluminum-nickel alloy, to form the entire body exterior "skin" with the exception of the sheet metal back of the front seat. Running "boards" and floor "boards" were also aluminum castings. The hood and fenders were formed from sheet aluminum, but this evidently proved troublesome and they were changed to steel two years later.

The engine was an air-cooled V-4 of equal bore and stroke . . . 4 inch bore, 4 inch stroke . . . displacing 201 cubic inches and rated 25.6 SAE horsepower. Maximum brake horsepower at the SAE conditions was 21. Individual, finned cylinders with integral heads and push-rod overhead valves were mounted on an aluminum crankcase.

The crankcase was barrel type with the crankshaft inserted from the rear into bulkheads extending the full depth of the crankcase.

The engine had the patented first ancestor of the modern lubrication system. It featured a gear type lube oil pump, drilled crankshaft and connecting rod passages, a pressure gauge on the instrument panel and a sump from which oil was recirculated through the engine bearings. In the early century, engines used their lubricant only once, with delivery drop by drop through separate sight devices on the dash. Some marques surviving today did not adopt the Marmon system until the mid 1950s and one was still using a piston type oil pump in 1929.

Fig. 2-6 Engine model C-7 of 1907 — representative of pre-1908 engines.

The chassis had a 90 inch wheelbase with standard tread of 56.5 inches. It incorporated the "patented double three-point suspension" construction involving separate sub-chassis for the propulsion system and for the body-hood-fenders. The suspension system is discussed in great detail under "Chasses of Era One."

*The Cycle and Automobile Trade Journal* reporter, Hugh Dolner, had a personal demonstration with Howard Marmon at the wheel. His reaction provides additional insight into this Marmon "peculiarity": *"Save for a good hard bottom, untouched prairie would give a far better driving surface than that of the average Indianapolis dirt street surface of today (1905). The Marmon can actually be driven over these fearfully variegated Indianapolis road surfaces without the slightest discomfort to passengers at a 15 to 25 mile (per hour) rate. So long as only one front or one rear wheel goes up or down at the same time, the car merely sways gently and comfortably no matter how deep the rut or how high the bump may be. The car does not mind any twist in road surface at all, and no one-wheel lift or drop on either side can bounce the car body. To avoid the mitigating action of Marmon's suspension, a pretty deep trench or high bump straight across the roadway must be had, and in about thirty miles of Marmon car riding this was found but once. Where the impediment was the same on both sides of the road, a slight swerve in steering enabled Mr. Marmon to always lift or drop one front wheel a little in advance of the other, so that the car body felt the level variation but slightly. This double three-point peculiarity of the Marmon car enables it to go over extremely bad roads much more rapidly than is agreeable with any other suspension system and also greatly lessens tire load and the propulsive effort demanded as the body lift in front is only half that given a four-point suspended car body."* Note that it was already recognized that a rough ride resulted in energy expenditure which was reflected in a fuel economy penalty.

The unusual cast aluminum body construction avoided the cracked surfaces and consequent chipped paint of the traditional coach-builder construction which had exterior wood panel surfaces. The second Marmon car built in 1903 incorporated a cast aluminum seat. The Model B body consisted of aluminum castings assembled together and to hardwood sills and the wood support structure for the front seat. The only exterior wood was the side panels of the front seat. The dash, front seat, tonneau seat, and tonneau doors were each one-piece castings 0.156 thick. The tonneau base consisted of three single castings bolted together and forming a tool storage compartment. The aluminum surface was reported to have required little finishing from the as-cast condition prior to painting and was credited with taking a superior appearing and durable finish.

The body provided many conveniences. There were two shallow tool drawers on the front of the front seat riser. Both tool and rear compartment doors were fitted with latches and Yale locks. The cast aluminum body construction facilitated a flush rear compartment accessed by a door in the rear of the tonneau. Beneath it, provision was made for storage of a spare tire casing within the frame, slid in from the rear. The front seat was divided, the rear was not. The car was a four passenger but could accommodate three slender passengers in the rear.

The front floor was covered with a heavy rubber mat and the rear with Wilton carpeting. Rubber matting covered the running boards. The car was upholstered in dark green leather. The body was finished in Brewster Green and the "running gear" in Carmine.

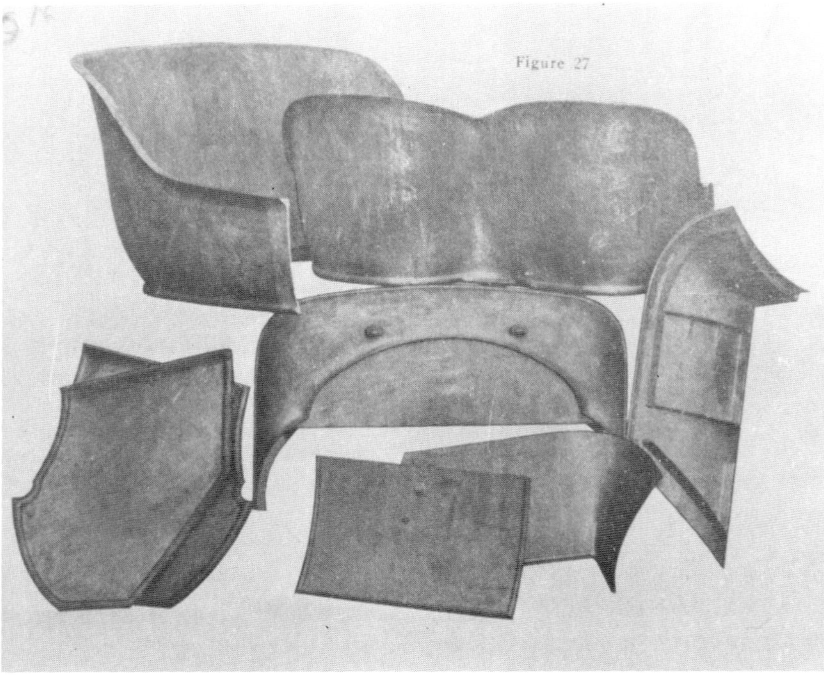


Fig. 2-7 The Marmon cast aluminum body sections — Model B of 1906.

Fig. 2-8 The Marmon cast aluminum body sections — Model F of 1908.

The 12 gallon (+2 reserve) fuel tank and the ignition batteries were under the front seat cushions. The four coils for the jump-spark and the oil pressure gauge were on the dash bulkhead.

The cast aluminum cowl supported the rear-hinged hood which was secured with a pair of latches below the expanded-metal grille. The hood had a new feature not found in the Model A; the rear half of the hood widened at the sides to provide a pair of air scoops with their own expanded-metal grilles. The purpose of these was to augment air delivery to the second pair of cylinders to compensate for the higher-temperature air coming off the forward pair of cylinders.

The Models A and B did not incorporate the manual transmission and single-plate clutch of the First Marmon. Instead they featured a two-speed and reverse planetary transmission without a main clutch. It made driving, and particularly learning-to-drive, easier and less distracting because it was capable of starting and operating from a crawl to full speed by forward and then backward (for high) movement of a single lever. The left floor pedal provided reverse. Throttle and spark-advance levers were conveniently placed **above** the three spoke steering wheel on the inclined column.

The car was unusually quiet and vibration-free. The barrel crankcase eliminated a major source of engine noise and the dual frame isolated the entire power train from the body sub-chassis.

A splash pan completely closed the gap between the running board and frame for dust control. This was a new addition on the Model B. Outwardly flared front fenders also replaced the non-flared ones on the Model A; possibly for deflection of tire puncturing items, the fender was carried to halfway between running board and ground.

Entry into the automobile business is not an easy task, as Leland, Kaiser, Bricklin, DeLorean and others would attest — even Ford Motor Co. with the Edsel fiasco. It was not necessarily easy in 1904 either. Although Nordyke & Marmon even had international marketing in mind when they embraced the automotive product, they came to market without a dealer organization. In consequence, and despite considerable advertising, sales of the Model B totaled only 25 units during the model (and calendar) year.

The Model B was priced at $2500.

## MODEL C
## 1906

Despite the disappointing sales of the Model B, the product development program moved rapidly ahead.

Even in 1906, the press referred to the Marmon arrangement of "oblique cylinders."[7] The term V-engine was not yet part of the language. The use of "circular flanges" for cylinder cooling on the Marmon engine was also viewed as a unique feature. Knox, for example, used radial studs threaded into the cylinder — the "porcupine engine."

The Model C was essentially a Model B with increased power. The bore was increased 0.25 inch, from 4.0 to 4.25; this increased the SAE rating to 28.9 and the brake horsepower to 20 at SAE conditions, to 24 maximum.

It also had modified fenders. The new front fenders terminated just below the running board, probably because the extension had proved vulnerable to damage. The new rear fenders extended slightly farther aft. Both fenders and hood were formed from sheet aluminum.

A bulb horn was mounted on the cowl below the right hand cowl lamp; the bulb was on the cowl rim.

Both head and cowl lamps were of more pleasing shape. The headlamps in particular might be called "streamlined"; they were relatively small in diameter, with length more than 1.5 times the diameter and free of step-changes in diameter. The lantern type cowl lamps had hemispherical ends, matching the headlights.

Fig. 2-9 Models C and D of 1906 (upper left and right), Models F and G of 1908 (lower left and right)

The car weight increased 150 pounds to 2150. A runabout body style was planned in addition to the touring car which was priced at $2500.

## MODEL D
## 1906

The Model D was a new companion car to the Model C. It featured an enlarged tonneau adequate for three passengers and a larger V-4 engine.

The body was new from the cowl back. The tonneau was eight inches wider (50 versus 42) and provided three inches additional leg room. The tonneau door was carried down all the way to the frame and the riser for the front seat was flared to the front-seat width and a flush alignment with the tonneau door.

A larger V-4 engine was fitted. It had a 4.625 bore and 4.50 stroke (versus 4.25 and 4.00 for the Model C), and displacement of 302 cubic inches. It was fully 50 percent larger than the Model B engine, and 33 percent larger than the Model C although the Model D at 2420 pounds, weighed only 270 pounds (12.5%) more than the Model C. The new engine was rated 33.7 SAE horsepower, developed 26 brake horsepower at SAE conditions and 30 maximum (versus 24 for the Model C).

The head and cowl lamps were different from those used on the Model C (and B as well). The headlamps were relatively large in diameter, drum shaped with a convex rear face. The cowl lamps were the lantern type, different from the Model B but failing to match the headlamps.

The Model D was priced at $3000.

## GLIDDEN TOUR VICTORY
## 1906

A Marmon Model D (Contestant No. 66) won the 1906 Glidden Tour with a perfect score; two nail punctures were repaired but there was no casing damage and four minor adjustments were made. The Glidden Tour is addressed in more detail in the chapter on Marmons in competition. It is brought up here because of available data on travel performance during and subsequent to the Tour.[8]

The 1153 mile route ended July 28 at Mount Washington, New Hampshire. The Marmon covered the distance at an average speed of 21.5 mph, consequently arriving 20 hours and 29 minutes ahead of "Par time" set before the race. Less than 5 gallons (20 quarts) of oil were required for engine, transmission and axle. There undoubtedly was appreciable oil loss from the gear boxes consequent to state-of-the-art seal technology, but if all the consumption were charged to the engine it would have used a quart every 57.5 miles.

After two days, the car was driven to Boston (July 31) covering 183 miles non-stop in 7.5 hours to average 24.5 mph.

On September 8 the car with five adult occupants departed Boston for Macedonia, Ohio. Four trouble-free days were required to reach Macedonia. After visiting there, four additional passengers (children) and the five adults made the return trip to Boston — 715 miles in three and a half days. The initial leg, Macedonia to Buffalo, New York required 8 hours and 40 minutes to cover the 197 miles; an average of 22 mph!

The Models C and D were identified as Models '35' and '26' in the *Marmon Book of Facts,* 1929 Edition. Evidently there had been the intention to market them using a horsepower value; the '26' matches the brake horsepower of the 'D' at SAE conditions whereas the comparable C value would be 20, not 35. Similar model designations do not appear elsewhere and should be discounted.

## MODEL C-7
## 1907

The Model C was carried over for a second year as the companion offering to the entirely new Model F. The wheelbase was however extended six inches to 96; the front axle and springs were relocated *forward* correspondingly. The change would shift engine weight toward the rear wheels, improving traction.

Among new features introduced on the Model C-7 was a new steering gear designed to reduce steering effort. The illustration released to the press exemplifies the sophistication of the Nordyke & Marmon organization.

The release of the Model C-7 was an alternative to release of a Model E which evidently had progressed sufficiently that the "E" designation became a void in the alphabetic list of Era One Models. The decision may have been influenced by the radical change in transmission type which characterized the Model F. The change was made in response to public demand for the clutch/manual transmission system. However the Marmon planetary system was well regarded by owners and there may have been concern that some customers might be lost if it were retired.

The C-7 tonneau was similar to that introduced on the Model D. The door extended downward to the top of frame; there was no deck extension of the front seat structure under the tonneau as on the A, B, and C. The tonneau with rear doors could still be removed if desired but there would be no finish deck revealed.

The head lamps were bell-shaped — the lens and rims were larger than the lamp housing.

The C-7 weighed 2250 pounds.

## MODEL F
## 1907

The Model F was a totally new Marmon automobile on a 102 inch wheelbase, but with one exception — the transmission — it continued the maturity process that characterized the First Era.

The car was noticeably lower, particularly so for the body. The hood contour was no longer rounded but was rather like many radiator shapes of the time. Otherwise there was no new styling approach except that the flared front fenders had been abandoned in favor of a question-mark shape, and the front axle was placed forward as on the C-7. A step was added outside the rear door. It shielded clothing from the nose of the rear spring and facilitated egress from the tonneau. It would characterize all subsequent First Era Marmons.

The Model F marked the entire elimination of wood, the body being assembled together using brackets integral with the castings. Upholstery attachment was by means of wood tacking strips bolted to bosses on the castings.

There was no provision for tonneau removal. The forward door jams for the tonneau doors were integrally cast extensions on the front seat. For the first time, the doors were rectangular with a corner cut-out for fender clearance, and all exterior body surfaces were cast aluminum. The all-aluminum body was a Marmon exclusive.[3]

Like the First Era Marmons, the hood and grille assembly pivoted on the cowl. New on the Model F was provision for routine inspection and lube oil replenishment. On each side of the hood, in the upper sloping surface, was an access door. It is interesting to note that just such access doors were added to GMC trucks fitted with rear-hinged fiberglass hood/fender/grill units beginning with 1982 production.

The new grille extended downward covering the front of the frame, not unlike the Mercedes-Benz hood-grille configurations of the 1960s and later. This gave the grille a more nearly square and handsome shape which, like its predecessors was outlined with brass trim. The grille bore the Marmon badge in brass script in a central location. Because of the gear reduction system, the hand crank was still below the grille.

The change in hood profile was a pioneer application of a styling philosophy which General Motors Design (Styling) Staff has used most notably to protect the consumer's investment in the prior year product and to ease the dealer's problems in 'moving' any left-over inventory. A water-cooled engine option was being readied for the 1908 Marmon and a hood unique to air-cooling was no longer appropriate, at least on a model completely new for 1907.

Cowl oil lamps were shared with the C-7 companion model, but the Model F had yet another iteration of gas lamp configuration . . . they were bell-shaped! And, they had bail handles as well.

Tires were 32X4.

The Model F engine was an overbored variation of the larger engine first applied to the Model D the previous year. This made it considerably oversquare: 5 inch bore (4.625 on the D) and 4.5 inch stroke displacing 353 cubic inches. It was rated 32.4 SAE horsepower, produced 35 at the same conditions and 40 maximum.

Like the first Marmon of 1902, the Model F had a single-plate cork-faced wet clutch and a selective-shift three speed and reverse transmission, and all were built into a unit power plant package with the engine. However unitization was accomplished only when both units had been assembled to the sub-rails. Note that the clutch and transmission shared a common oil sump.

The left pedal, which had been the reverse/emergency brake on Marmons with planetary transmissions, became the clutch pedal . . . Separate emergency brakes were added at the rear wheels and a brake lever with ratchet sector was mounted outboard of the gear shift lever.

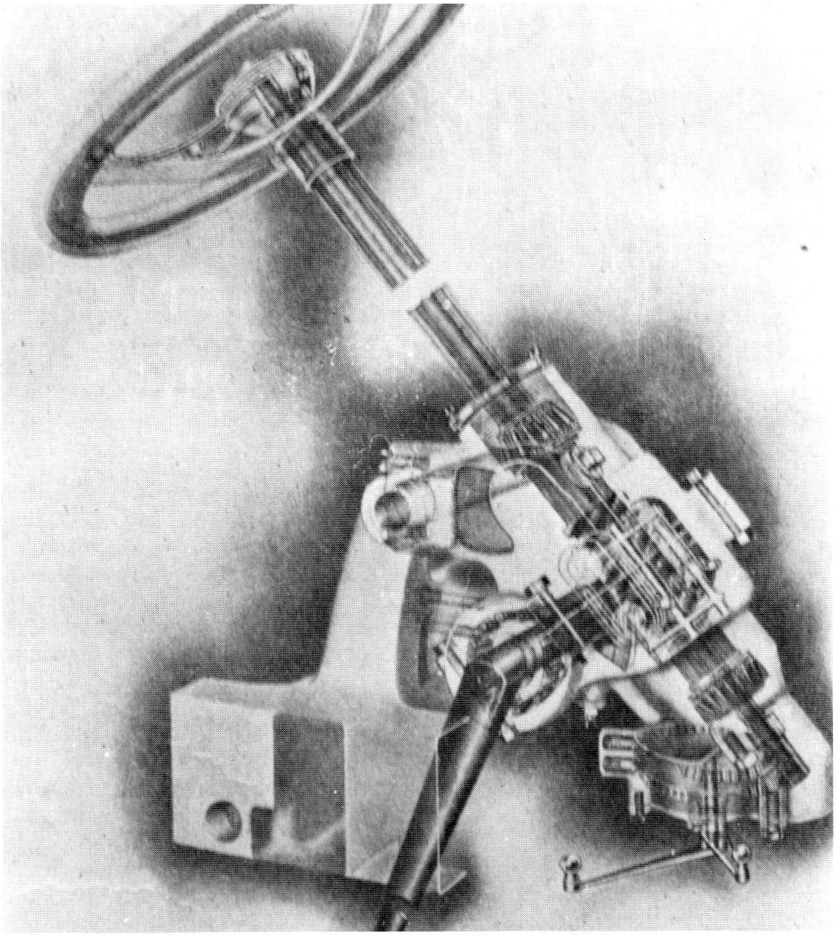

Fig. 2-10 Phantom photo-illustration of 1907 steering gear also shows sophisticated graphic capabilities

The transmission had two interesting and worthwhile features: (1) the interlock between shift rails incorporated on additional interlock with the clutch release sleeve and (2) selection of high (direct) gear disengaged the countershaft drive gear. The former prevented gear shifting and gear damage due to failure of the operator to depress the clutch pedal and the latter improved fuel economy by reducing parasitic friction losses.

The 102 inch wheelbase of the new chassis was responsible for the reduced body height . . . lower seats require extended leg-room. Tread was standard. Tires were 34X4 front and 34X4½ rear. Four shock absorbers were standard equipment.

Both body and running gear were finished in Brewster Green.

The Model F weighed 2850 pounds including lamps, mats, horn, tools, batteries and fuel. The touring was priced at $3500. A cape top was a $150 extra. A limousine and landaulet were advertised at $4500 and $4750 respectively.

Fig. 2-11 Selective transmission for Model F — Note rotating shift levers for remote linkage.

A 1907 Marmon was driven in low gear at 6 mph (879 engine RPM) for 8 continuous hours. At no time was there any visible oil smoke.

## MODEL V "Eight Cylinder" 1907

The first word of a Marmon Eight was a press release reported by Motor Age, in the issue of October 11, 1906. The car was to have an engine with 5 inch bore and 4.25 stroke and to be on a 120 inch wheelbase. Sales of 300 units in 1907 were forecast. It was acknowledged that the first prototype was not yet ready. Subsequently it was reported to have a 4.5X4.5 engine on a 128 inch wheelbase.

The "Eight Cylinder" model was not mentioned when the 1907 model offerings were announced in November 1906, but prototypes were ready in time for the December shows. Although advertised as the "Eight Cylinder" Model, it was designated Model 'V' for service purposes;[10] never however was the engine referred to as a V-8.

The car was demonstrated curbside at the Grand Central Palace Auto Show in New York in December, 1906 and was displayed at the Chicago Auto Show early in 1907.

Nordyke & Marmon was then a member of the American Motor Car Manufacturers Association, as distinct from the Association of (Selden) Licensed Automobile Manufacturers, sponsor of the Madison Square Garden Show in January.

Throughout the 1907 model year, Nordyke & Marmon ads illustrated the Model F and listed:

Model F $3500.
Model C-7 $2500.
Eight Cylinder $5000.

Fig. 2-12 1907 V-8 with executives aboard. John Conde photo.

Some ads asked the question:

"What More Can You Buy At Any Price?"

The "Eight Cylinder" was evidently an elongated Model F. It weighed 3650 pounds, 800 more than the 'F.' The wheelbase was 128 inches. An extended tonneau made a seven passenger capacity optional.

The 90 degree V-8 engine used eight Model F cylinders assembled on a double-length crankcase, crankshaft and camshaft assembly. It was accordingly an overhead valve engine of 5 inch bore and 4.5 inch stroke, displacing 707 cubic inches, with an SAE rating of 64.8 horsepower and developing 75 brake horsepower. A single carburetor on the right side of the engine fed a central intake manifold; exhaust manifolds were outboard the cylinders. A single central camshaft actuated the push-rod overhead valves and drove a central vertical shaft mounting the distributor on its upper end and driving the gear type lube oil pump from the bottom end. A remarkably perceptive and prophetic creation indeed!

The author has never located any information whatsoever on the sales, or lack thereof, of the Model V. One thing is certain — Howard Marmon had met the large sixes of e.g., Welch more than head on.

Maurice Hendry[9] reported the V-8 to have been designated the M-37, to have had an engine bore and stroke of 4X4.5 developing 65 bhp; and to have had a 3.0:1 axle ratio allowing high gear operation from 2.5 to 60 mph. Fuel economy was stated by Hendry to have been 12 miles per gallon. The *Marmon Book of Facts,* 1929 Edition, refers to the Models C and D as Models 35 and 26 respectively, but such designation has not been seen elsewhere by this author.

In November of November 1907, when two four cylinder models were introduced there was no mention of an 'Eight Cylinder' model for 1908.

Fig. 2-13 Model G Landaulet — the first enclosed Marmon.

## MODEL G & H
## 1908

In November of 1907 two Marmon lines were announced — two V-4 lines — the "Eight Cylinder" was missing. The new models were designated Models G and H.

## MODEL G

The Model G was a continuation of the Model F chassis with improved cylinder construction. Upon that chassis was mounted a new body described as "straight-lined and handsome." It evidenced a new styling sensitivity.

The front seat side panel, at a fixed height, was carried full length and around the rear of the tonneau. The lower line was carried across the near-square rear door as a molding; elsewhere the seat risers were recessed. Upon this "panel-band" were placed nearly-identical front and rear seat backs which tapered to zero height as they joined the "panel-band." Consequently the rear seat upholstery roll-over did not extend forward to the door opening as was traditional but stopped at a length identical with the front seat upholstery. The cast aluminum body construction permitted an ornamental door jam to protrude from the lower body around the lower half of the door.

The extension of the "panel-band" at the rear moved the rear seat rearward, allowing both front and rear seats to be lower without sacrificing leg room. The belt line formed by the bottom of the panel-band was at the same elevation as the breakline on the hood. This styling development had been anticipated the prior year by placement of an integral belt molding on the otherwise smooth body surface of the Model F.

A foot accelerator was a new feature. It allowed momentary increases in power without disturbing the pre-set throttle position on the steering column. Its continuous use was not then anticipated.

The Model G touring was priced at $3000.

The lower body on the 96 inch wheelbase G was necessary to have an acceptable companion to the new Model H which featured a new application of the double three-point suspension principle which was responsible for a chassis able to masquerade as a conventional one.

Fig. 2-14 Model H with executives aboard.

## MODEL H
## 1908

*Cycle and Automotive Trade Journal* commented upon the news that Marmon would offer a water cooled product as follows: *"The announcement that Nordyke & Marmon Co. was putting-out a water-cooled touring car for 1908, in addition to its air-cooled cars, was received by the public generally without surprise, it being*

Fig. 2-15 Model H Roadster — Unchanged wheelbase but repositioned axles.

*apparently understood as a business step in the right-direction. The Marmon air-cooled cars are well known for their reliability and the fact that the company has finally put a water-cooled car on the market is looked upon simply as a bid for the trade of those who prefer the water cooling system. The Marmon Air-Cooled Motor for 1908, with its improved detachable cylinder heads was described in the November Journal. The success attained with the new motor has made the company stronger than ever in its belief in the air-cooled motor and it proposes to continue the manufacture of air-cooled cars to supply its increasing demand for this type of car."*

The Model 'H' water-cooled Marmon was noticeably lower than contemporary and predecessor air-cooled Marmons, and casual observation did not reveal any peculiarities of construction. Despite appearances it did feature the exclusive Marmon double three-point suspension.

The propulsion sub-frame was shortened and pivoted at both extremities on the main frame; it became an engine/transmission sub frame. This was accomplished by introduction of a universal joint to accommodate axle movement relative the sub-frame. Actually a pair of joints on an open drive shaft were used, the axle being positioned by a torque arm.

The Model H was offered in two wheelbases and three body styles: Five-passenger touring and two/four passenger roadster on 114 inch wheelbase chasses, and a seven-passenger touring on a 118 inch chassis. The body construction was described as typical Marmon: *"made entirely of the Nordyke & Marmon Company's perfected aluminum castings mounted over a substantial hardwood frame. This construction makes a body which is practically indestructible and not affected by any climatic conditions. It is claimed it takes and holds a beautiful finish, is not easily marred and does not shrink, swell or dent."*[1]

The front floor was cast aluminum, in sections, and the rear floor was carpeted with Wilton velvet. The front seat was divided. Compartments under the rear seat accommodated the battery and tools. The fuel tank was arranged to have a two-gallon reserve and was fitted with a "supply indicator."

A separate catalog (830-B) for the Model H Roadster stated: "The chassis of the Model H Roadster is the same as that of the Model H Touring Car, except that the front axle is placed further forward ahead of the radiator and the rear axle is brought up closer to the seat, maintaining the same wheelbase, 114 inches (and presumably the same weight distribution).

The rumble seats and box are removable so that the roadster may be used to carry two and a trunk, or three or four passengers. When arranged for three passengers, one rumble seat can be placed in the center of the rumble box, making, in all, four finished combinations — a two passenger car with platform in rear, with or without box, a three passenger car or a four passenger car."

The longer hood length was required to house an optional water-cooled in-line four cylinder engine of 5 inch bore and stroke. It displaced 393 cubic inches, was rated 40 SAE horsepower but developed 45 with a maximum of 50 horsepower. The optional engine was "to be offered to those who prefer that type." The same option was stated to be available to purchasers of the Model G.

An overbored version of the engine was used for the 7 passenger style. The bore was increased to 5.375, the stroke was unchanged at 5.0. Displacement was 453 cubic inches, SAE horsepower rating was 46.2 but the engine produced 50 and had a maximum rating of 60 horsepower.

For the Model H, Marmon departed again from the semi-unit powerplant concept adopted for the Model F. The engine and transmission were independently mounted on the sub-frame, with the clutch and flywheel exposed.

The engine was a four-cylinder in-line design with T-head valve configuration (intake valves and camshaft on one side, exhaust on the other). Cylinders were cast in pairs with integral heads and port plugs to close the access openings for valve assembly. It incorporated the exclusive patented Marmon full-pressure lubrication system and the traditional Marmon barrel crankcase construction.

The clutch was a wet multiple-disc design with cork-insert facings. The transmission was a 3 speed and reverse unit of selective type, with clutch release interlock and no engaged gears when in direct drive.

Suspension was the usual Marmon with four full-elliptic springs but with the front springs underslung (ie, below the axle).

Both the G and H front and rear fenders were described as "enclosed." This meant that for the first time they incorporated a 'skirt' or dust shield between the fender rim and the frame. The Marmon was then dust shielded for the full length of the car.

Fig. 2-16 The Marmon 50 seven-passenger touring.

The roadster and five-passenger touring were priced at $3500. The seven-passenger at $3750.

"Cambric" covered electrical wiring was adopted in 1908. It was a product of Pirelli Cable Co.

## MODELS 45 & 50
## 1909

The Model H was continued into the 1909 model year as roadster and seven-passenger touring companion offerings to the new Model 32. They were designated separate models — Models 45 and 50. Note the use just for these two models, of the developed horsepower at SAE conditions, rather than the calculated power, as a model designator.

The five-passenger touring was not offered for 1909, having been replaced by the new Model '32' at a much lower price — $2400 versus $3500. The seven-passenger Model 50 was priced at $3750, the same as the comparable Model H of the prior year. There were no styling changes.

Because the Models 45 and 50 lacked innovation, were transition models between Marmon Eras, and also because they were progressively less vigorously promoted throughout the year, the author has elected to include them with the Model H here in Part 1 (Car) and Part 4 (Chassis) of Era One. Conversely, because the engines of the three models represented a major departure from Era One practice, they are discussed in the Era *Two* (Chapter Three, Part 2) Engine section.

### *The ERA of Annual Model Change*

Who conceived the annual model change concept for the automobile industry? It wasn't Alfred P. Sloan, Jr., frequently credited, for Marmon practiced it throughout the First Era.

The Model A of 1904, after some seven units and a field evaluation was followed annually by the Model B, by the C and D, by the C-7 (stretch) and F, and by the G and H in 1908. All represented significant advances in the state of the art in engines, transmissions, brakes and steering gears, as well as in body construction and styling. Some were reported licensed as recently as 1929.[2]

But by the end of Era One, Howard Marmon had determined the design elements of a totally new car which would become International Racing Champion on and off the tracks and require only progressive refinement. That single chassis, with some wheelbase elongation, would survive for six years and as the sole offering for four.

Fig. 2-17 Cast aluminum construction allowed an integral storage box with 'tailgate' door.

## PART 2 — The Engines

Models A, B, C, D, C-7, F, G
1904-1908

The Model H engine and its derivatives reflect Era Two design philosophy and will therefore be discussed in Chapter Three — Part 2.

The design of the engines which powered the Marmons of Era One would have to be described as 'inspired' had they not been the product of a mature, progressive and innovative firm which called itself "one of the world's greatest engineering houses" before undertaking the design of its first motor car.

Howard Marmon saw the packaging advantages of the V-form engine in general, but believed that configuration to be particularly well suited to uniform cooling of multiple air-cooled cylinders.

All the Marmon production engines of Era One were 90° Vee form and except for a short lived V-8 in 1907 were all V-4s. They all incorporated a patented lubrication system now universally used and an innovative barrel crankcase construction increasingly recognized today as a superior concept. The design underwent a progressive development throughout the ERA.

Howard Marmon's first air-cooled engine was as spectacular a success as was Charles Kettering's first and only air-cooled engine[10] a disaster. Each Marmon engine was run for 2 hours at full power in a test cell with ambient temperature controlled to 110° F.

The Marmon Vee engines were all of what has become classic general configuration . . . Push-rod operated overhead intake and exhaust valves actuated by tappets from a central camshaft gear-driven from the crankshaft at the front; a single central intake manifold, and outboard exhaust manifolds with dual pipes from each bank combining at, or just ahead, of the muffler.

By 1907 the 'distributor' was relocated in the Vee to the upper end of the vertical shaft by which the camshaft drove the lube-oil pump in the pan sump — the classic configuration. It had been driven from the camshaft nose, like early Ford V-8s.

Until 1908, the individual cylinders had integral heads. In that year a separate-head design was introduced which was remarkably similar to the cylinders used on air-cooled radial aircraft engines of the 1930s and 1940s and contemporary horizontal-opposed general aviation engines.

The stellar feature was a patented lubrication system utilizing drilled crankshaft and connecting rod passages fed by an engine driven gear-type pump. Every element of today's universally adopted lubrication system, including the pressure gauge on the instrument board, was included. Patent 810,959 was filed on February 13, 1905 and issued January 30, 1906, and was undoubtedly Howard Marmon's most significant contribution and his prime claim to automotive immortality. In the National Automotive History Collection of the Detroit Public Library are original letters in which the Editor of The Chilton Co., publishers, asks Howard Marmon for specifics on his invention and is advised (October 10, 1924) that the concept was conceived and developed in the winter of 1903-04.

Further evidence of then-contemporary terminology is found in the language of the patent. Crankshaft journals were 'wrist pins of the crankshaft.' Connecting rods were 'pitmen' or 'pitman arms' and the big-end with cap was the 'wrist-pin' end. The invention was described as being illustrated in connection with a gas (oline) -engine structure but was not limited thereto.

*"The object of my invention is to provide means by which the several bearings of a shaft or a crank-shaft, the wrist-pins thereof, and both ends of the pitmen, car-*

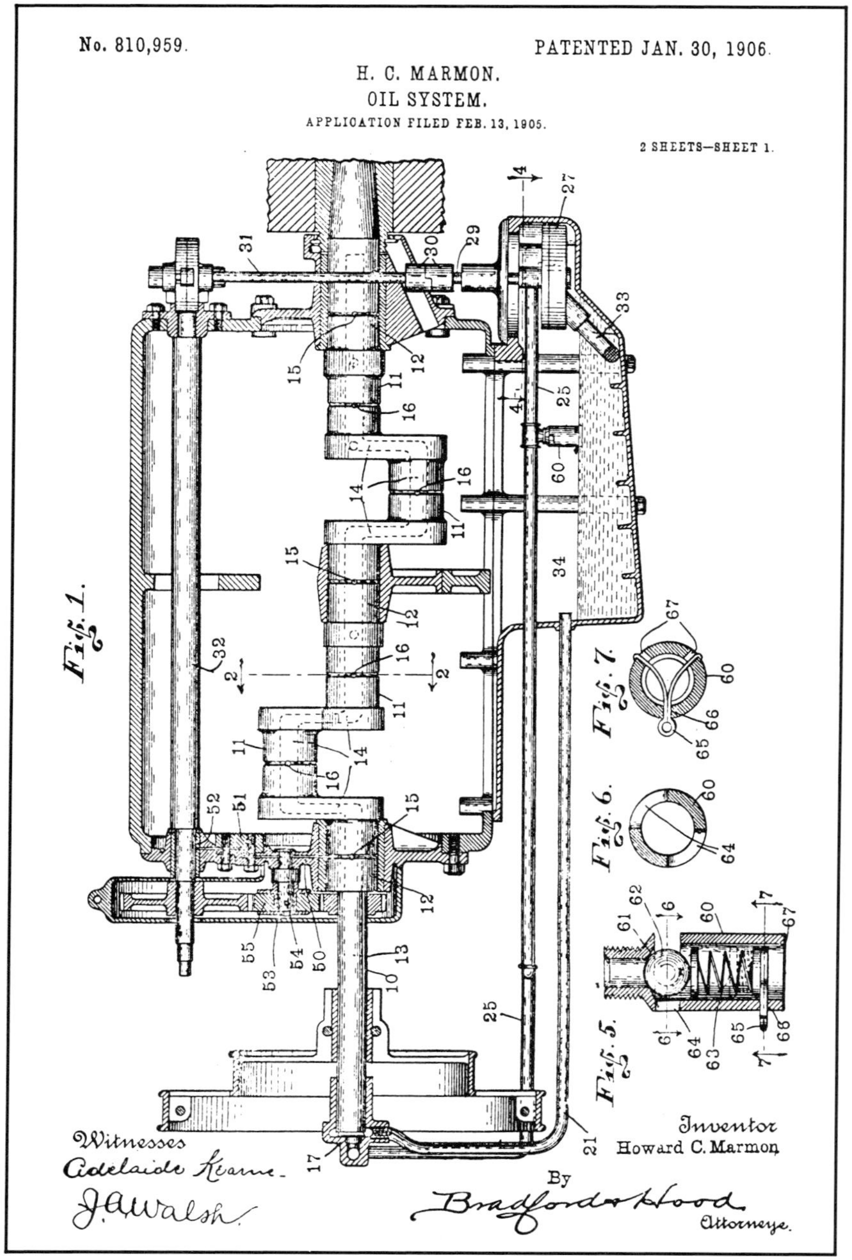


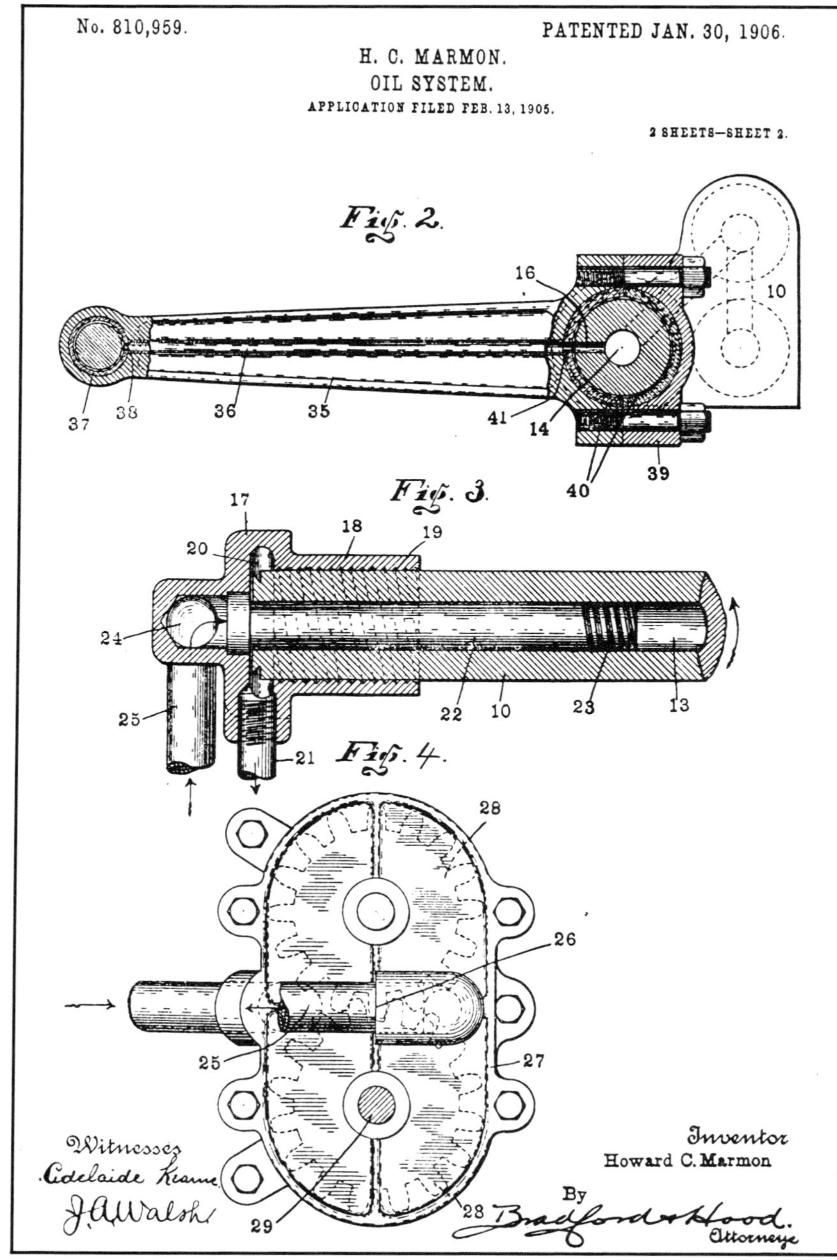


Fig. 2-18 The H.C. Marmon patent on a full pressure lubrication system.

*ried thereby, and other bearings associated therewith may be supplied with oil under pressure, the arrangement being such that a single oiling mechanism will serve to supply all of the points requiring lubrication with an adequate supply of lubricant. To this end the oil is introduced into the end of an axial bore formed in the crank-shaft; and a further object of my invention is to provide a simple means through which the introduction may be made and of such character that no packing is required to prevent leakage and all leakage is returned to the pump. The accompanying drawings illustrate my invention."*

Because of the patented pressure lubrication system, the engine could advantageously, and uniquely, be installed in the chassis in an inclined attitude, "aimed" at the rear axle pinion. This is evidenced by the contemporary-modern configuration of the oil pan . . . the bottom being inclined relative to the crankshaft but parallel to the ground and the sump restricted primarily to the rear half of the engine to provide suspension clearance without the typical drop-center axle configuration of competitive cars.

The barrel crankcase construction consisted of a crankcase in which the inter-cylinder bulkheads were carried all the way to the bottom of the crankcase and the crankshaft, with its intermediate bearing (s) in place was installed from the rear of the crankcase.

The bulkheads were bored round to a diameter sufficient to clear the crankshaft during assembly. These round bores were free of "stress risers" . . . a sharp corner can multiply imposed stresses several times. The "Achilles Heel" of the conventional construction is the notch required at the main bearing caps to locate the cap with respect to the block. Some engines have deep crankcase walls to stiffen the block but many others terminate the block at the crankshaft center line because the wall extensions are compromised by the crankshaft bore and main cap milling.

The basic problem results from the fact that the crankshaft is a dynamic part . . . it experiences forces which deflect it from its static shape . . . and the deflections are transmitted through the close-fitting bearings to the bulkhead. The bulkhead flexes under reversed stress conditions and tends to fail in severe service by fatigue cracking. The problem is escalated when aluminum is used.

Howard Marmon was concerned with fuel economy and sought maximum use of aluminum as early as 1904. The barrel crankcase allowed the use of aluminum without concern for fatigue failures.

The intermediate carrier consisted of a bearing cap assembled to a member having more than 180° of arc length. After crankshaft installation, the intermediate bearing carrier (diaphragm) was positioned axially and held against rotation by a bolt-dowel threaded into the center bulkhead from the bottom of the engine, as far off-center as permitted by the pan rail.

Very significantly, the carrier was then clamped in place by a pair of transverse bolts, assembled from outside the block and passing just forward and just rearward of the carrier, at the elevation of the bottom of the carrier bore in the bulkhead. This was a clever design . . . that location was equivalent to where the screw would be placed if the crankcase were split like a clamp, and importantly it placed the thin sections of the bulkheads in compression.

Transverse bolting is used today in the Teledyne Continental air-cooled diesel engines which power the US and NATO M-60 battle tanks. These engines employ semi-conventional bearing cap design, having the crankcase bulkhead carried down to below the bearing cap. The bearing cap then fits tightly in the milled or broached slot in the bulkhead. Two (one above the other) transverse bolts pass through the bearing cap, placing the crankcase in compression and avoiding movement of the cap relative to the bulkhead.

The front main bearing was a 360 degree bearing in a crankcase front cover, although it could have been placed directly in the block.

Because the rear main bearing of the Marmon engine was flanged to take crankshaft end thrust, the rear carrier was split horizontally and, like the intermediate carrier, was assembled to the shaft before insertion into the crankcase.

It should be noted that a feature of the Marmon tunnel crankcase construction was the ability to replace main bearings without having to repour babbitt in the cylinder block and remachine the bearing bore. As late as 1954, some prominent US automobiles still required the engine block to be removed from the engine and sent to a machine shop in the event of a bearing failure . . . quite a traumatic experience should malassembly cause such a failure on a recently purchased car. No Marmon was ever sold without replaceable main bearings; prior to the Model 34B some fitting was required.

Another feature of the Marmon tunnel crankcase construction was a quieter engine than attainable with conventional construction. In 1915 *Automobile* (July 1) had this to say: *"When the crankshaft and split intermediate bearing carriers are assembled and installed as a unit and the carriers are bolted in place, the (resultant) structure adds enormously to the stiffness of the whole motor. The crankcase has become a tube with a series of stiff partitions — a tube reinforced at seven places. It is claimed that the construction is lighter than any other which gives equal rigidity."*

In November, 1981, in *Automotive Engineering,* journal of the Society of Automotive Engineers, appeared an article[11] headed:

BEARING VIBRATION IS MAJOR
CONTRIBUTOR TO NOISE

> Crankshaft deflection during combustion causes vibration of an engine's main bearings, which, in turn, causes the crankcase skirt to vibrate.

It was a review of the formal SAE Paper 810772 by Masao Ishihama and two associates at the Nissan Motor Co., Ltd. who described the opening/closing action of the skirt of the (conventional) crankcase as the *dominant noise* in the engine tested.

## 1904 MODEL A
## 1905 MODEL B
## 1906 MODEL C
## 1907 MODEL C-7

The general layout of the Marmon V-4 was of modern V-engine configuration. It had individual cylinders due to direct air-cooling (waterless) just as does the cur-

**THE MARMON**

**"A Mechanical Masterpiece"**

rent US M-60 Battle Tank engine manufactured by Teledyne Continental Motors.

The Model B engine had a bore and stroke of 4.0 inches. It displaced 201 cubic inches; had an SAE horsepower of 25.6 and a brake horsepower under SAE conditions of 21. The compression ratio used resulted in 45 pounds per square inch (gauge) pressure, the maximum feasible without detonation.

For the Model C and C-7 the bore was increased 0.25 inches to a 4.25 bore (and 4.0 stroke). Displacement became 227 cubic inches, SAE rating became 28.9 and brake horsepower was raised to 20 at SAE conditions and 24 maximum.

The pan had an extension aft of the cylinder block, on the top surface of which was attached the mounting flange of the lube oil pump which was suspended in the sump. This extension was necessary because the pump drive was external to the engine.

An overflow pipe in the pan provided a limit on the oil level below the rotating machinery. The pump inlet was in a downward extension of the pump body. The outlet was directed through a pipe and bulkhead fittings to an external pipe terminating at the nose of the crankshaft. Oil was conducted into the crankshaft through a gland covered in the Patent . . . a long screw thread conducted leakage to a drain annulus from which a second pipe ran back to the sump.

For 1906, an inlet pipe was added to the oil pump, which terminated in a screen, to sustain suction to a lower sump level. A pressure relief valve was also added to the discharge pipe within the sump. A sight-level gage was added on the outside of the pan and replaced the overflow tube.

Oil was conducted, through drilled passages, for the full length of the crankshaft with drilled holes in main and rod journals feeding those bearings. The connecting rods were machined all over and fitted with an oil tube which conducted pressure oil to the piston pin bushing.

The crankshaft was a four-throw type (90° spacing) with three main bearings and was without balance weights. The main journals were 1.75 dia.; the front journal was 2.25 long and the other two were 2.50. The cylinder bank offset was 3.125. Rod journals were 1.75 dia. and 2.125 long. The rear main was "sealed" by a large drain relief behind the bearing and a connecting angular drilled hole to the crankcase above the oil level. The shaft mounted a flywheel on a tapered end, with key orientation and retaining nut. The 18 inch three-spoke flywheel incorporated a steel hub with a tapered-bore to fit the crankshaft and having three clutch teeth on the end to transmit power to the transmission and permit easy engagement of the transmission shaft during vehicle assembly.

No positive type oil seals were used in the engine. Slingers and adequate drain passages were used instead.

For 1906, balance masses were added to the engine to counteract unbalanced couples. A weight mass was added to the crankshaft between the first and second throws and the flywheel was cast with an integral unbalance. A curved steel bar was bolted to the inside of the rim of the three-spoke flywheel to achieve the desired specific unbalance.

The camshaft was driven by a front-end gear train consisting of gears keyed to the crankshaft and camshaft (press fit retention) and an intermediate idler bushed on a stub shaft pressed into the block. For 1906 the idler was eliminated by resort to larger diameter gears. The 1907 Model F camshaft gear was made of fiber with a bronze hub for quieter operation; this model had a lower camshaft position. The camshaft was carried on two bronze bushings which were pressed into flanged carriers.

The gear train enclosure was a cast case, split on engine center line, to permit assembly over the gear and around the camshaft and was bolted to the front main bearing carrier. The camshaft extended out the rear and had an integral spiral gear to drive the oil pump. Cams were profiled for simple harmonic motion of the valve train. They were individually hardened and assembled to the camshaft by pin-index and press-fit retention.

Engine starting was accomplished by a hand crank attached to a gear which incorporated a ratchet mechanism and was spring released from engagement with a gear on the crankshaft nose, between the gear case and the pulley for fan drive. There were two reasons for this arrangement: (1) the end of the crankshaft was utilized for oil entry through a gland, much like a hose reel connection, and was not available; and (2) the gear provided a reduction ratio making cranking easier, and feasible for a woman driver.

The individual air-cooled cylinders with eighteen 1.25 inch radial fins were flange mounted to the crankcase. They had integral heads and were threaded to receive permanent installation of elbow ports for the overhead intake and exhaust valves. Both valves were spring-returned and push rod actuated through rocker arms pivoted on the head.

The integral cooling fins appear quite sophisticated. The reason they do is evident in the description of their manufacture found in Chapter Eleven Part 2.

The cylinders were cooled by ram air assisted by a fan driven at 2.5 times crankshaft speed by a flat belt. The higher temperature air reaching the second set of cylinders was compensated by two scoops on the side of the hood at mid-length.

For 1906, the valves were installed in cages with integral elbows and rocker-arm brackets. These cages were then bolted to the head. Improved valve cooling was achieved. Individual rocker-shafts allowed optimum valve locations.

The pistons were straight, ie without taper, and had four rings, three above the pin and one below. The rings were described as "eccentric, 45 degree lap cut." By 1907 the ring complement was six rings of 0.125 inch width, unusually narrow for the time. Manufacture of the pistons is described in Chapter Eleven Part 2.

The connecting rods had aluminum babbitt bushings in the small end and half-bearings of the same material in the big end; the brass rod cap provided the other half-bearing surface.

Valves were one-piece forgings of nickel steel.

Factory pictures of Marmon engines in Era One fail to show the carburetor . . . only the end of the intake riser. The 1905 Model B used "the new Holley Vaporizer."[1] The 1906 illustration is supported by the comment that the carburetor is not shown because of new features not yet protected. It was elsewhere described as a compensating type. It was an air-valve type.

The intake manifold was a brazed assembly of copper tubing and fittings resembling modern copper plumbing. The riser extended from the carburetor, which was mounted at crankshaft level forward of the left cylinders, to a point central to the four cylinders from which fore and aft runners extended to 'T' fittings feeding the individual cylinder runners.

For 1905, the manifold ends were flared and attached by flare-nuts to the threaded ends of the elbows which were permanently threaded into the cylinders. (The exhaust elbows had welded diamond flanges). When the elbows were eliminated for 1906, the manifold incorporated brazed two-hole (diamond) flanges with slotted machine-screw attachment.

The carburetor was gravity fed from a 14 gallon tank in the body, through a heavy brass pipe along the frame to a flexible metallic tube to the carburetor.

The spark plugs were located on the inboard side of the cylinders. Ignition was by the jump-spark method. A quadruple bank of vibrator coils was located in a

Fig. 2-19 Model D engine.

Fig. 2-20 Model D engine on the test stand. Note bolted-on valve cages and valve spring retainer washers formed from square blanks, like modern car wheel discs, to minimize weight and offal (waste steel after blanking from strip). Note crankcase vent.

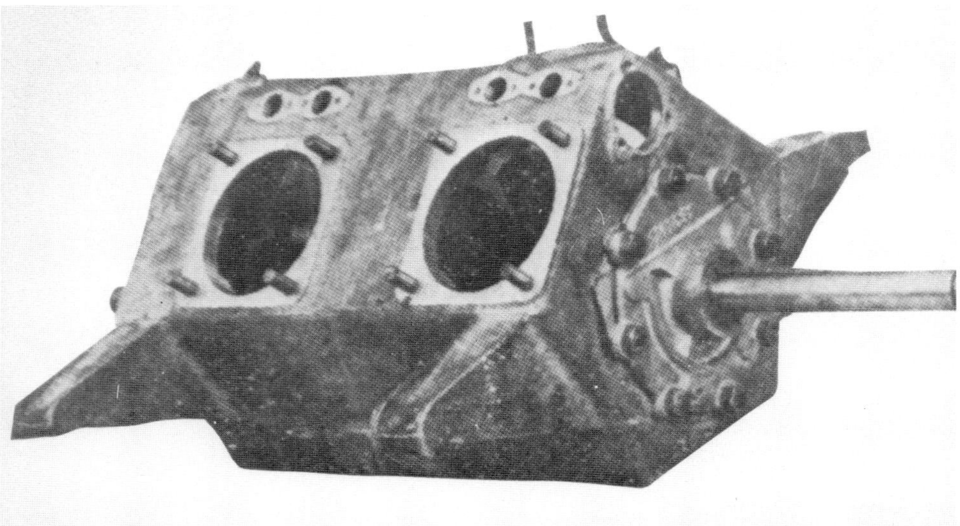

Marmon 1906 Aluminum Motor Base.

Fig. 2-21 "Oil pool base with force pump"

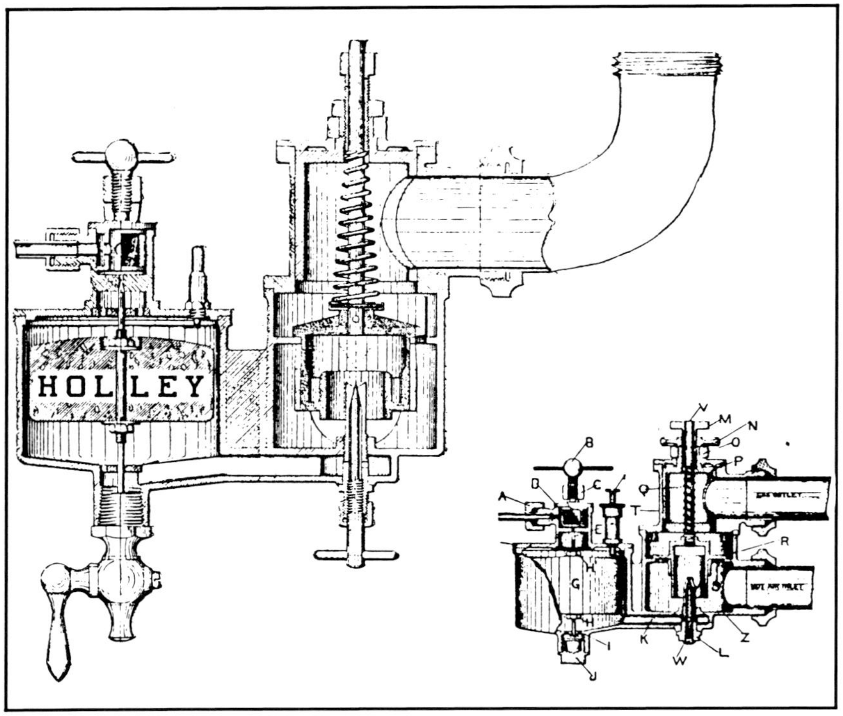


Fig. 2-22 Holley Carburetor was an air valve type — attachment with 'sink-trap' plumbing was typical.

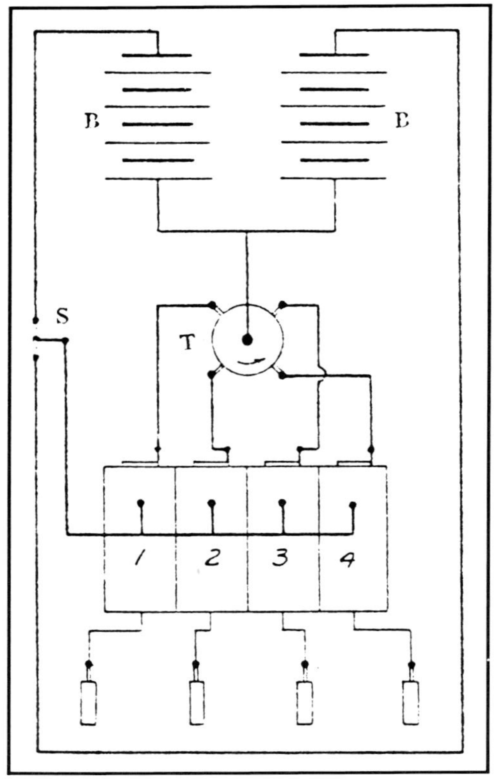


Fig. 2-23 Typical Era One ignition schematic. Marmon used one wet battery and twelve dry cells in two parallel rows.

metal box secured to the dashboard (engine bulkhead). Separate high voltage leads joined each coil to the respective spark plug. Primary current was supplied to the individual coil at the appropriate time by the 'timer' (a low-tension distributor). The basic source of ignition current was a storage battery. In an emergency, the ignition could be switched to a bank of twelve dry cells connected in two parallel rows. The spark advance and throttle controls were above the steering wheel. This accessibility was important because both spark and throttle required frequent adjustment; there was no foot accelerator and no automatic advance of timing with increased speed. Also it was probable that each fresh tank of fuel required compensating adjustment of spark timing.

A separate exhaust pipe from each cylinder led to one of two functional (but commonly housed) mufflers. The connections were so made that the mufflers received alternate firing pulses. Each cylinder was claimed to have the use of a complete muffler for a full engine revolution. Each of the two mufflers was a radial flow type with the exhaust entering one end of a central tubular plenum. Exhaust exited through perforations near the center of the plenum wall but was forced to flow toward both ends because the next concentric and larger tube had perforations only near the ends. The third concentric tube was similar to the first, forcing a flow from both ends toward the center before exiting radially into the outlet plenum which was a single chamber semi-circular at the ends and parallel walled between. This wall of the plenum was made of a sandwich formed with two sheets

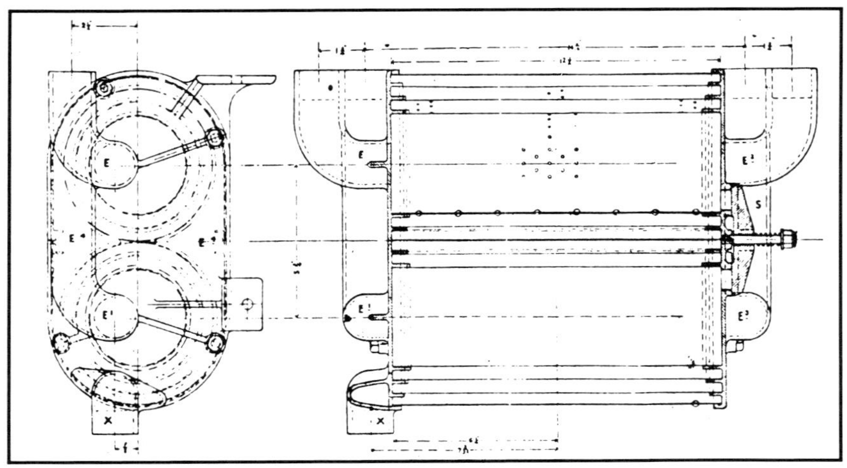

Fig. 2-24 1904-1906 Muffler Cross-Section — Owners claimed it quiet as an electric.

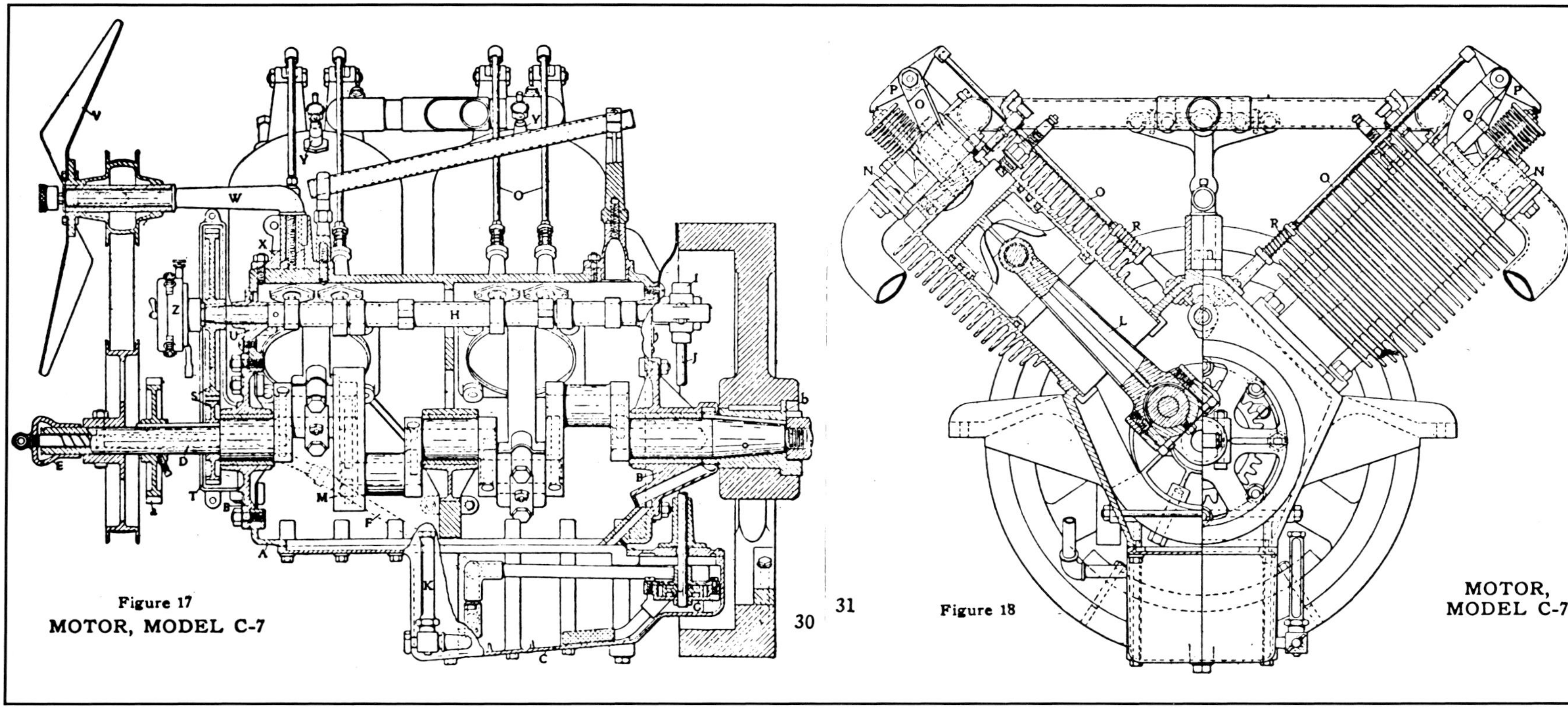


Fig. 2-25 Transverse & Longitudinal Sections of Engine — Model C-7, 1907

of steel interspaced with sheet asbestos. A pair of aluminum end castings clamped the assembly together and manifolded the gases.

This muffler was installed transversely between the flywheel and transmission. It was suspended from the propulsion frame side rails.

In the event of an engine malfunction or inept starting procedures, damage to the muffler was averted by a pressure relief valve designed into the unit.

During 1905, the prototype car, built in 1902, was reengined with an experimental V-6. That engine is presumed to have had an extended length crankcase mounting six of the Model B cylinders. A unique camshaft and crankshaft would have been required, the latter certainly having two intermediate bearing bulkheads. Otherwise, the engine would have conformed to the preceding description of the Model B. On that assumption, the engine data would be as follows:

Displacement — 302 cubic inches
SAE rated — 38.4 horsepower
Brake at SAE conditions — 31.3 horsepower

The 1902 Marmon prototype, with the V-6 conversion was occasionally used for promotional purposes or in parades and was known to be in the factory at the time of bankruptcy. It might be presumed that Howard Marmon was not satisfied with the vibratory characteristics; otherwise he would not have marketed an eight cylinder version for the 1907 season. An expedited engineering and tooling program would have to have been initiated soon after initial operation of the V-6 to meet the deadline of having a V-8 demonstrator outside the New York Auto Show in January, 1907. There had to be a compelling reason for the offering of two cars with a 100% difference in engine size. Could it have been considered essential to be counted among products such as the Welch in order to sustain prestige both in competition and the market place?

One thing is reasonably certain — the V-6 was intended to satisfy the power requirements of the larger Model D companion car for 1906. The V-4 engine actually released for that Model carried an enlarged bore and extended stroke which combined to give exactly the same displacement as the V-6 using the Model C cylinders!

## 1906 MODEL D

The Model D had a wider tonneau and weighed an additional 150 pounds. It would appear that a significant portion of that increase was consequent to the increased engine size rather than the basis for that increase in size.

Fig. 2-26 Experimental V-6. Note dual transverse legs to manifolds feeding three cylinders each.

The Model D was a big-brother to the Model C. Bore and stroke were increased .375 and .50 respectively relative the Model C (.625 and .50 relative the predecessor Model B). A 4.625 bore and 4.50 stroke gave a 302 CID, an SAE rating of 33.7 and corresponding brake horsepower of 26, and 30 maximum brake horsepower. This was 25 percent more power than the Model C (which car was only 150 pounds lighter).

Crankcase modifications, eg. height of camshaft above crankshaft, necessary for the Model D stroke were likely common with the Model C and related to the redesign of the front gear train for 1906.

## 1907 MODEL F V-4

The Model F engine was a second generation engine and, with exception of a major cylinder improvement the following year, represented the full maturity of the Era One air-cooled engine development. This is true in particular because there was also a companion V-8, concurrently developed and introduced shortly after the Model F and because both engines incorporated advancements that made especially the V-8 even more prophetic of universal practice five decades later.

It was a larger engine of the same general design philosophy . . . 5 inch bore and 4.5 inch stroke; the Model B had begun as a 4X4 engine. This engine confirmed Howard Marmon's continued commitment to the "square" (equal bore and stroke) engine and went further — initial release of a new engine with oversquare dimensions. Displacement was 353 cubic inches, SAE rating was 32.4 horsepower, brake horsepower was 35 at SAE conditions and 40 maximum.

The five principle design changes were: (1) the addition of an intermediate bearing for the camshaft, flange mounted like the end bearings; (2) adoption of a non-metallic drive for the camshaft; (3) relocation of the forward crankshaft-axis balance mass from the crank cheek (between #1 and #2 cylinders) to a more effective location as an integral part of the fan drive pulley; (4) addition of a flange to the intermediate bearing diaphragm and bolt attachment to the bulkhead; (5) relocation of the oil pump drive to within the engine and relocation of the distributor to an upward extension of the oil pump drive shaft.

The additional camshaft bearing minimized shaft deflections and was particularly desirable to maintain precise meshing of the oil pump drive gear formerly adjacent (and outboard) the rear camshaft bearing.

The camshaft gear was made of bronze and carried dual "tires" of fibre material on which the teeth were cut. These were individually piloted on the bronze "wheel," and were clamped to a central flange protruding from the rim, by machine screws and nuts. The purpose was reduced gear noise.

The combination of balance masses in the crankshaft pulley and in the flywheel of the Model F is an efficient use of balance mass that came into widespread use in the 1960s.

The addition of a flange and bolt attachment to the bulkhead was the full maturement of the barrel crankcase configuration . . . the one used on the Miller/Meyer-Drake/Offenhouser racing engines (as well as the winning Marmon in the first Indianapolis 500 Mile race). The Italian V-M diesel engines marketed in the United States by AVCO, and the Japanese Kubota industrial diesels also marketed here in the 1980s use the earlier Marmon unflanged concept.

Integration of the timing gear cover with the front bearing diaphragm eliminated the oil leakage point at the front of the camshaft. It necessitated a (vertially) split diaphragm (similar to the intermediate) because the case was split to extend behind the camshaft gear.

All air-cooled Marmon engines pumped oil thru an external pipe to a fitting, floating on the nose of the crankshaft and restrained from rotation by the rigid pipe. The flexible hose to the dash gauge now originated at this fitting thereby confirming that the indicated pressure was indeed that existing in the crankshaft, uninfluenced by any possible plumbing leak en route.

An oil filler tube was pressed into the crankcase where a pipe plug filler had previously been located. The adjacent oil level sight glass was continued.

## 1908 MODEL G

This was the last stage in the remarkable development of the Marmon air-cooled V-type engine. Even as it was announced in September of 1907[1], imminent offering of an optional water-cooled engine was revealed.

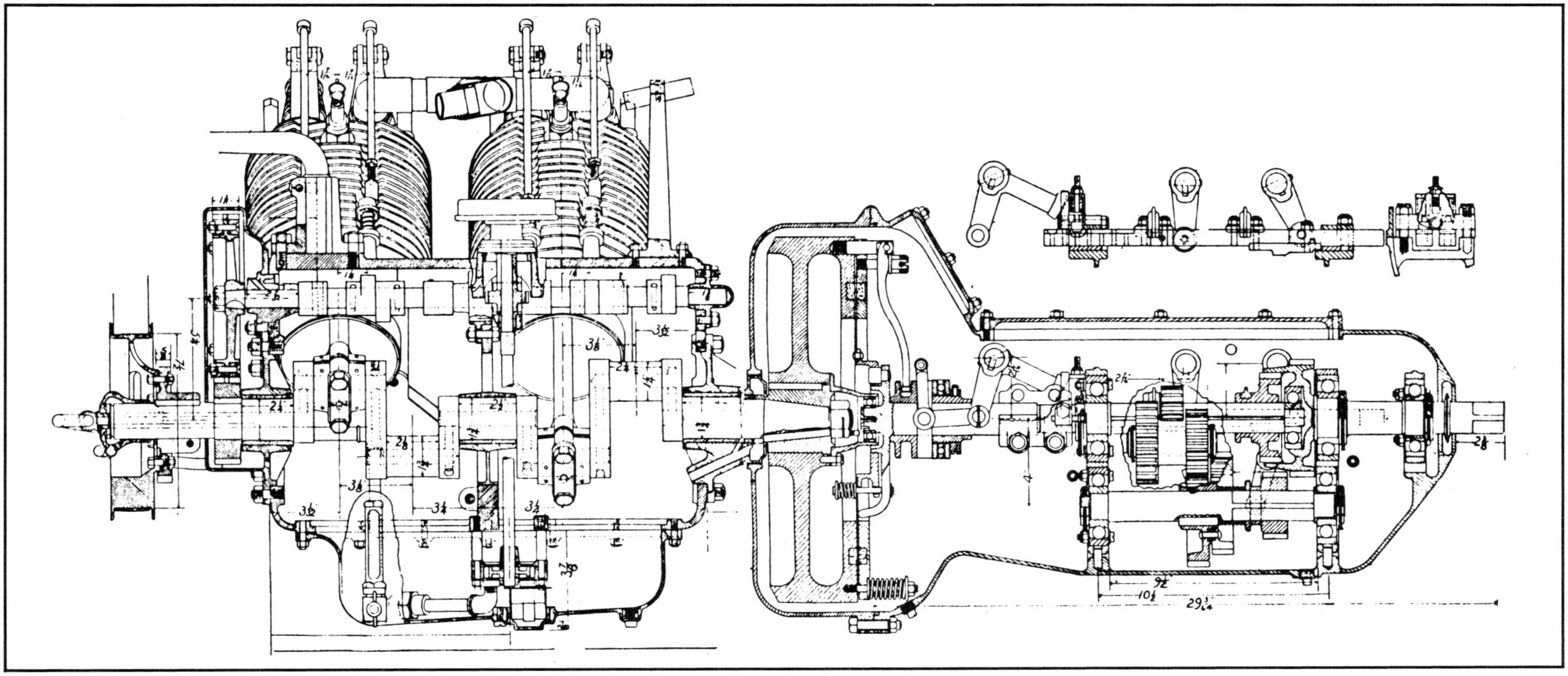

Fig. 2-27 The advanced and innovative unit powerplant of the Model F, 1907.

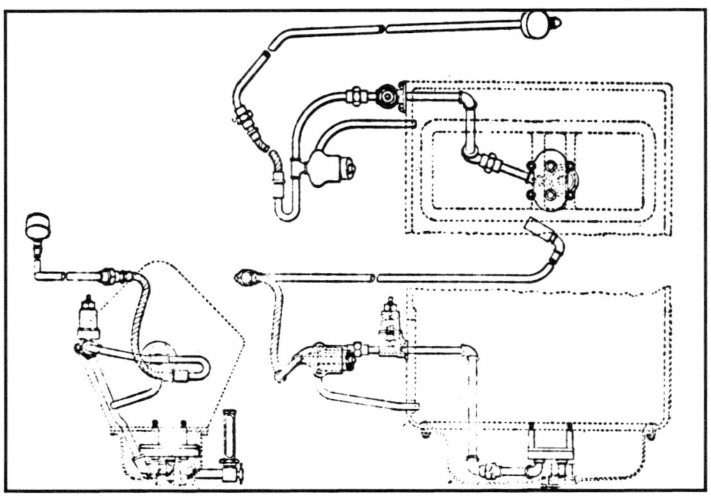

Fig. 2-28 The patented pressure lubrication system. Model F 1907

The alternative engine was reported necessary to place Marmon agents in a position to meet any demand which might arise and was not a featured innovation. The V-type air-cooled motor was to continue to be held out as the chief attraction to the buyer.

The ad announcing the 1908 air-cooled and water-cooled models had this to say: *"We offer, with increased confidence, the latest improved Marmon Air-Cooled Motor, which for many years has proven beyond question its thorough efficiency under all conditions of hard service. One of the new features of this remarkable motor which will appeal to the expert is the Marmon One-Piece Detachable Cylinder Head, removed or replaced in a moment, dispensing with valve cages entirely, reducing engine (valve) noise to a minimum, and making the valves, pistons and cylinder walls instantly accessible."*

Moving the oil pump drive within the block avoided the use of dry sand cores in casting the oil pan and made it possible to have a screened oil inlet located above an access cover for cleaning convenience. The prior design had such access prevented by the flywheel rim. Also, the rear camshaft bearing carrier could be made with a closed bore eliminating a leakage point.

The relocated oil pump drive made feasible the transfer of the distributor from the nose of the camshaft to the top center of the crankcase where it was more accessible for servicing. This subsequently became near universal practice . . . distributor on the top end of the oil pump drive shaft.

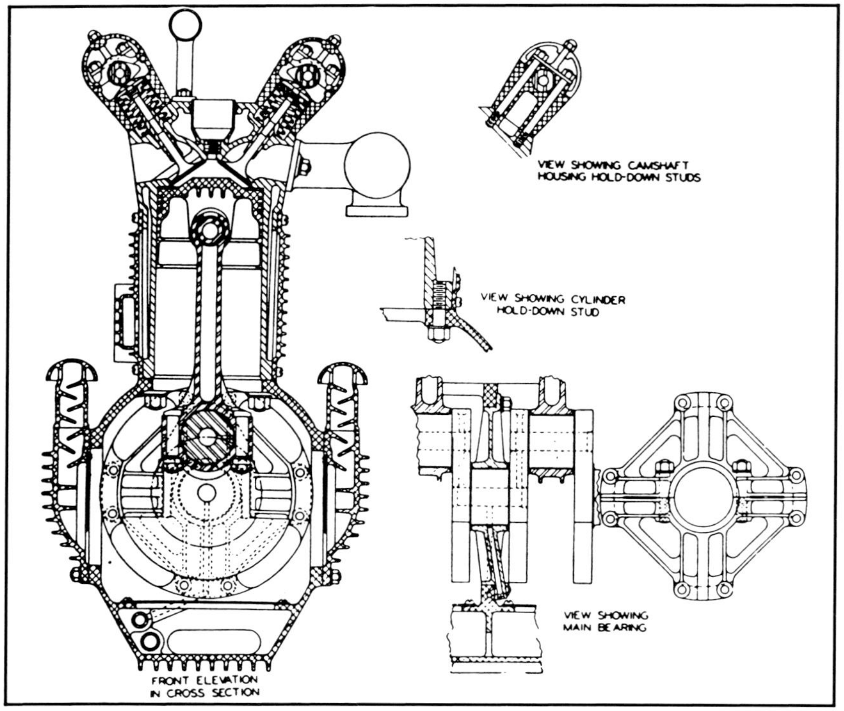


Fig. 2-30 Modern heavy-duty engines using barrel crankcase construction. Offenhouser (above) and VM (below).

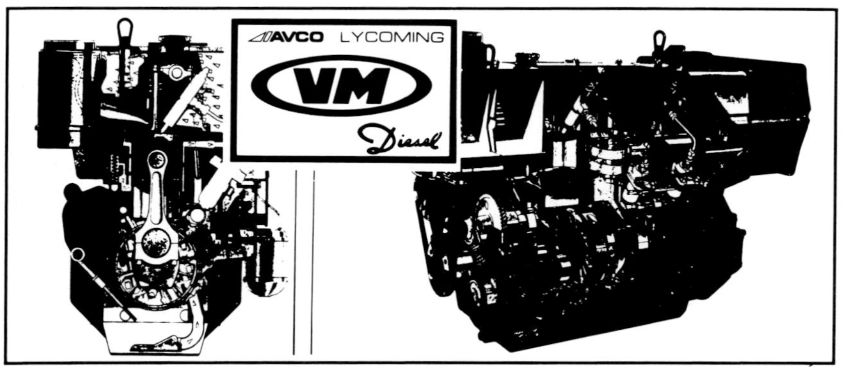


A reporter for *The Motor World* commented: *"In its latest guise, this machine embodies a most original form of cylinder construction, whereby the removable heads are made to serve the purpose of the two valve cages, being almost as readily removed as the ordinary type of independent cage, thus reducing the number of packed (gasketed) joints to a minimum of one and besides simplifying the mechanism, affording ready access to the interior of the cylinder upon occasion. The head which is flanged (finned) externally to assist the radiating function in the usual way, is assembled by means of four studs which run the whole length of the cylinder and take hold of (are threaded into) the crankcase. The removal of four nuts and release of the gas (intake and exhaust) and electrical (spark plug) connections, thus permitting the head, or even the cylinder itself, to be removed at once. The joint in the cylinder wall comes just above the upper end of the piston travel and is protected by a copper asbestos (ring) gasket. Removable valve stem guides are set in the head, and the rocker arm support of the valve gear is bolted securely in place on the head casting." (Author's parenthetic insertions)*

Exhaust elbows were flange attached, whereas the inlet elbows were threaded into the head and configured for attachment of the intake manifold by (plumbing) unions. The use of threaded parts was a reversion to the system used in the earlier Marmon cylinders before the detachable valve cages were adopted to expedite valve service.

Formerly the cylinders had a base flange and were retained by nuts on short studs. For the Model G, the studs continued upward through drilled holes in the cooling fins and through the cylinder heads. The single set of nuts on the four studs retained both the head to the cylinder and the cylinder to the crankcase.

The new cylinder head was configured to provide an air space between the intake and exhaust ports and also had radiating fins between them in addition to circular external fins, slightly smaller in diameter and spaced identically with those on the cylinder itself.

More even fuel mixture distribution was achieved by incorporation of a 90 degree elbow where the intake "riser" joined the distribution manifold. The mixture consequently entered the central manifold vertically and had to turn 90° to enter the two "Y" runners feeding each bank. The prior entrance through a 'T' fitting presumably biased the distribution.

The illustration of the Model G engine must be carefully observed to spot the head/cylinder hold-down nuts; the joint between head and cylinder is concealed by the continuous complement of cooling fins. This cylinder design, combined with the restyled crankcase (probably introduced for the Model F) and the new fan and intake detailing resulted in a remarkably handsome engine appearing to be decades ahead of its time. After all, wasn't it?

## 1907 MODEL V

*The Motor World* reporter, who covered the Grand Central Palace New York Auto Show in December 1906, noted that there was great interest in multi-cylindered cars and in particular in the Marmon eight: "The peculiar construction

Fig. 2-29 Model G engine cylinder — with detachable heads.

Fig. 2-31 Model G Engine, 1908

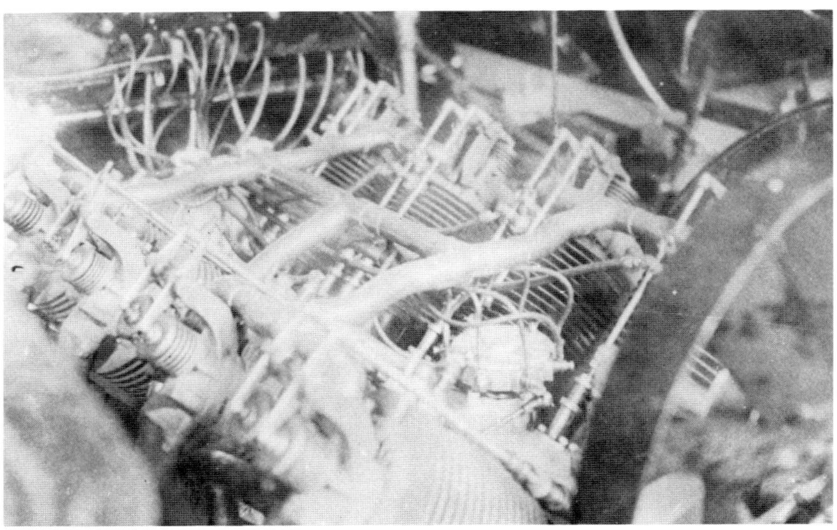

Fig. 2-32 The V-8 engine in Chassis NAHC photo.

of the Marmon engine lends itself readily to this duplication, so that the eight cylinder motor is no longer than the average."

The Model 'V' engine was an overhead valve V-8 engine of 5 inch bore and 4.5 inch stroke, displacing 707 cubic inches. It had an SAE rating of 64.8 horsepower and developed 75 brake horsepower.[1]

It required a new crankcase, camshaft and crankshaft. Considering the utilization of dry sand cores from the V-4 in casting the crankcase, it probably had three intermediate bearings or five total.

Principle difference from the V-4 was in the intake manifold and carburetor location. The carburetor was located on the right hand side of the engine, mid-length. The intake riser passed between the cylinders to join a 'T' leading to forward and aft cross legs. Heated intake air was drawn from cooling air off the number two cylinder.

There is no known information on sales of the Model V Marmon. Its availability was noted in advertisements during most of 1907, and its demise went unnoticed. Some seven years later, during a meeting of the Society of Automotive Engineers reported in the *SAE Transactions* for 1915, Howard Marmon spoke from the audience to disagree with Alanson Brush who had stated that additional cylinders were needed when a certain piston diameter was reached. Mr. Marmon's response provides some insight into the short life of the Marmon V-8 offering and the subsequent Marmon concentration on the line-six for more than a decade following that SAE meeting.

Mr. Marmon's remarks: *"I have placed the satisfactory limit of cylinder dimensions considerably higher than Mr. Brush has; 3½" to 4" bore is quite a feasible dimension to use. I put the upper proper speed limit somewhat below 3,000. It seems to me that the number of cylinders is purely a matter of the cubic capacity needed to do the work. In dealing with cars that carry five and seven passengers, a six-cylinder motor up to those limits of 3½" to 4" makes a pretty good performer. Personally I have not been "sold" with the eight or the twelve. I have never seen them do things that a six will not do, except possibly some of the stunts. Starting a car on the high gear with the starter is a common multiple-cylinder demonstration. But a good six-cylinder motor will do that and do it on the magneto. I feel rather confident that if the motor is well built, a blindfolded person will not know wheth-*

Fig. 2-33 Model V — the OHV V-8 Engine. Brigham Collection Photo

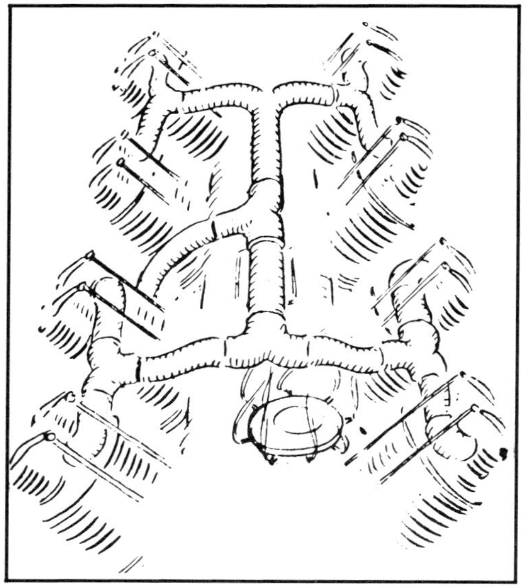

Fig. 2-34 Intake manifolding of the V-8.

Fig. 2-35 Model B Chassis

*er it is a six, an eight or a twelve, unless he catches the rhythm of the multiple-cylinder type. There are a good many explosions to take care of even in a six. The muffling proposition is rather difficult. The (human) nervous organization will feel vibrations set up in a car by frequent impacts. As the number of impulses increases the high-frequency period becomes more difficult to handle. The same thing is true of the multiplication of the moving parts of the motor. I agree with Mr. Brush that the valve-in-head motor has some thermal advantages that appear more in the cooling, possibly, than in actual consumption. I am convinced that the valve-in-head motor is much easier to cool than the L- or T-head. It has, due to easier and smoother ingress to the inlet system, a certain volumetric efficiency that is difficult to get in either the T- or L-head."*

Vibration would continue to characterize the 90° V-8 until the combination of a two-plane crankshaft and a clever redistribution of crankshaft balance weights achieved inherent balance in the V-63 Cadillac for 1922. However the author personally found the successor 314 Cadillac engine to have objectionable high frequency vibration.

(A 200 hp V-8 had appeared in a Darracq race car in 1904. It was a push-rod overhead valve design but was relatively primitive, designwise).

## PART 3 — The Chasses

### Models A Through G
### 1904-1908

There were two basic chasses used in Era One: The original concept which did not require universal joints and the subsequent variation which did require them but allowed the Model H to be significantly lower in height.

The chasses for the Models A through G were derivatives of the 1903 prototype chassis. There were 3 wheelbase variations, Models A through D having a 90 inch wheelbase and 56.5 inch tread. The C-7 was lengthened to 96 inches. The 102 inch F and G chassis was designed for a 2850 pound automobile.

Fig. 2-36 The Model C-7 Chassis Cat. 730

The prototype (1901) and the Model F, as well as the air cooled versions of the Models G and H, had single-plate, cork-faced clutches and three speed and reverse selective-shift conventional (lay-shaft) transmissions with unique features. The intervening models all had two-speed and reverse planetary transmissions using bronze-on-steel clutching and without a main clutch ahead of the transmission.

The rear axle and brakes underwent a progressive development through the Era. The brakes were all internal-expanding type using bronze shoes on iron drums. An exception, the Model H, marketed into 1909 as Models 45 and 50, had a new generation of brakes.

The chassis of the Model A is believed to be identical with the Model B. Specific confirmation is lacking. Likewise the 1903 prototype chassis is believed to be identical except as noted above and tire/wheel sizes.

This general discussion is followed by detailed treatments of the chassis components of Models A, B, C, D, C-7, F, and G. The chassis components of Models H, 45 and 50 are discussed in Part 4 of this Chapter.

## SUSPENSION SYSTEM (FRAME & SPRINGS)

Because the Era One Marmon suspension system had two compliance modes . . . non-planar road surface and roughness of surface . . . it is not practical to separate discussion of the Frame Structure which provided the first compliance and the Spring Suspension which provided the second.

The novel and unique dual-frame "double three point suspension" of these cars was covered by a US Patent, 783599 issued February 28, 1905 following application filing on March 24, 1904. It was issued in the names of J. Warrington and H.C. Marmon. Jesse Warrington was a Nordyke & Marmon engineer and/or designer to whom other patents on flour milling machinery had issued . . . eg, one on a gyratory sifter the following August (796478 Aug. 08, 1905). In fact, the converse was true . . . Patent 1443629 issued Jan. 30, 1923 to Howard Marmon for a "Heated Roll for a Flaking Mill". Anyway, Jesse and Howard shared the honors for the chassis configuration of the Era One Marmons. Patent coverage was also obtained for Canada and several foreign countries; a reflection of Nordyke & Marmon's international perspective.

To describe the operation of the system, the author will first quote the exhortation on the last page of a four page advertisement announcing the 1905 Marmon:

DON'T DO A THING
until you know what it means and
are familiar with
MARMON DOUBLE THREE-POINT SUSPENSION

and then reproduce in its entirety the lenghtly and well illustrated explanatory article which Cycle and Automobile Trade Journal felt the development warranted in October 1905, even after having been on the market for more than a year:

## MARMON PATENTED DOUBLE THREE POINT SUSPENSION

The Marmon Patented Double Three Point Suspension is a radical departure from the common construction of motor car running gears, which has not heretofore been fully treated in these columns. The accompanying cuts show it to be a simple and practical construction of plainly apparent merit. It is the old principle of the three legged stool which will rest solid where a four legged stool will not. The importance of the principle of three point suspension, applied to motor car construction, has long been realized and many attempts have been made to employ it, the Marmon method having proved an eminently successful one. The manufacturers of the Marmon car, Nordyke & Marmon Co., Indianapolis, Ind., on which this method of suspension is employed, have furnished us with the following description of it.

*"To realize the necessity of a change from common practices one has only to analyze what takes place when traveling over rough roads in a car of the construction at present prevailing. Strong construction, good springs and pneumatic tires alone have made it possible to build an automobile which would withstand, fairly well, hard usage on the road and these do not prevent cars from receiving a contin-*

Fig. 2-37 Note position of fan hub as propulsion chassis rolls

Fig. 2-38 Plan view of Model B chassis.

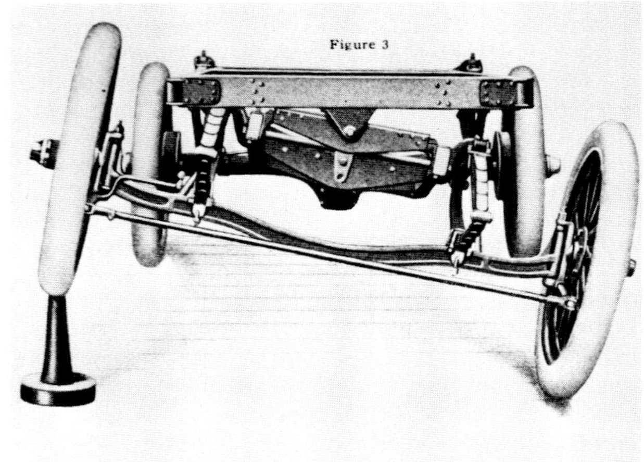


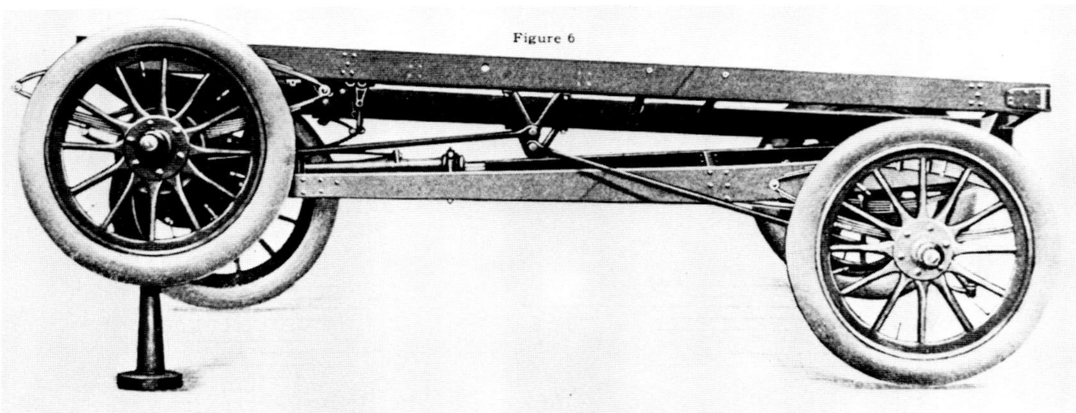

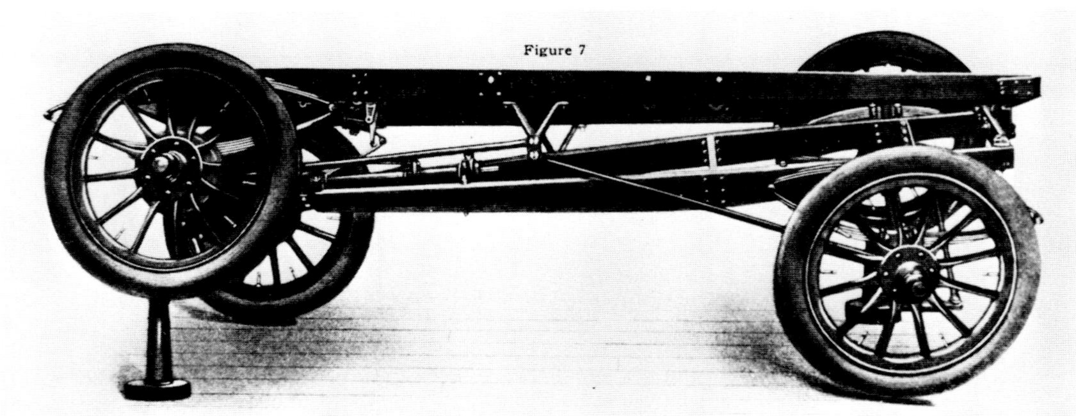


Fig. 2-39 Marmon Double Three Point Suspension

Photo 1 Right front wheel raised: Upper Left.

Photo 2 Right rear wheel raised: Middle Left

Photo 3 Left front and right rear wheels raised: Lower Left

Photo 4 Front view — Right front wheel raised: Upper Right.

Photo 5 Front view — Right rear wheel raised: Above Right

All positions without any spring deflection.

*uous and ruinous twisting and racking. The employment of mechanical principles, notably the cardan joint, cannot be avoided with the customary rigid four point suspension. It is a well known fact that very severe binding and twisting strains are transmitted through the frames, mechanism, bodies, wheels and tires of cars which have one frame carrying the body and power plant mounted rigidly on all four springs. The springs and tires, which can only afford partial relief, are called upon to perform a greater duty than should be expected of them. This is easily demonstrated by lifting any one wheel of a car so constructed. If, for instance, a front wheel is raised a few inches the opposite front wheel will begin to rise and raising still more a rear wheel will rise, thus showing that strains are not only transmitted through the frame and working parts, but that excessive weight is thrown on one or two wheels and tires."*

*"The Marmon Patented Double Three Point Suspension overcomes all this, permits the use of a straight shaft from engine to bevel gears and results in easier riding qualities than obtainable with any other construction. That this suspension is thoroughly practical, durable and a pronounced success in every way, has been proven conclusively by three season's hard usage. Better results than have been obtained with this construction could not be desired. The principle of the Marmon double three point suspension is made very clear by the accompanying illustration: "Fig. 2-37 shows the front of the car with right wheel raised — note position of fan hub in grille.*

*The engine and transmission have been omitted purposely to avoid obstructing the view of any portion of the suspension system. Fig. 2-38, plan view shows clearly all six of the pivotal points of suspension. The rectangular frame which carries the body only is mounted on the two rear springs with transverse pivotal supports and is supported in front with one longitudinal pivotal connection in the middle of the lower frame. The lower triangular frame which carries the power plant is mounted on the two front springs with transverse pivotal supports and the rear end is supported in two bearings on the sleeve or quill which contains the driving or propeller shaft. These two bearings being concentric with the driving shaft are in effect the same as one longitudinal pivotal support. These bearings permit a movement around the sleeve or a movement of the sleeve within the bearings. The driving shaft is shown protruding forward from the sleeve which is rigidly fixed to and is a part of the rear axle construction. The cross member at the front end of the triangular frame to which the upper frame is pivoted, is also pivoted at each end in bronze bearings affording a slight movement which in connection with the central pivot, permits of any angular change between the two frames. The strains occasioned by the tendency of the outer rear axle structure to rotate, when driving and setting the brakes, are taken by the sleeve projecting forward from the rear axle, it being held firmly within the two rear bearings of the lower frame. Thus the driving shaft is relieved from these strains as are also the rear springs."*

*"The springs form no part of the suspension system except as supporting members and they are relieved from driving strains by means of radius rods extending from the rear to the front axle as shown clearly in Photo 1 of Fig. 2-39.*

*"The engine and transmission on the lower frame slope rearwardly with the frame, the driving shaft being straight and rigidly clamped to the propeller shaft, thus doing away with the necessity of flexible couplings or cardan joints. No matter in what position the two frames may be in relation to each other, the perfect alignment of the driving shaft is maintained without any binding or twisting strains upon it or any other part of the driving mechanism. The illustrations showing the wheels raised not only demonstrate this fact but they show also that each of the four wheels maintain its proportion of the weight and that all four wheels have a firm*

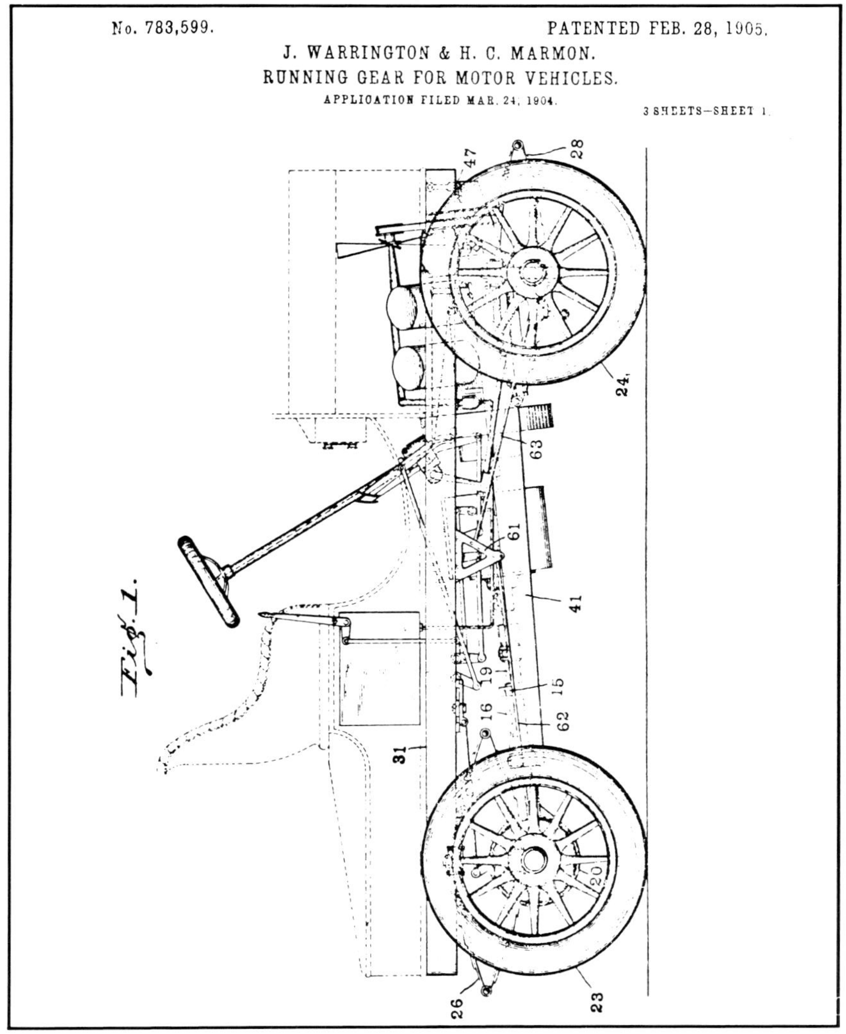


Fig. 2-40 The Warrington-Marmon patent on the Era One chassis configuration.

*footing at all times."*

*"Photo 4 of Fig.2-39, is a front view showing one front wheel raised 12 inches. Notice that the lower frame is tilted but that the body frame transversely is level. Fig. 4 shows the relative position of the two frames in a side view with one front wheel raised. Note that the upper frame is raised vertically but not tilted. It is raised vertically one-half as much as it would be in any other car. When meeting an obstruction the springs of course, absorb some of the shock, but by reason of the pivotal points of the frame suspension the wheels are permitted to ride over the obstruction with a freedom which relieves, to a remarkable degree the tires, the*

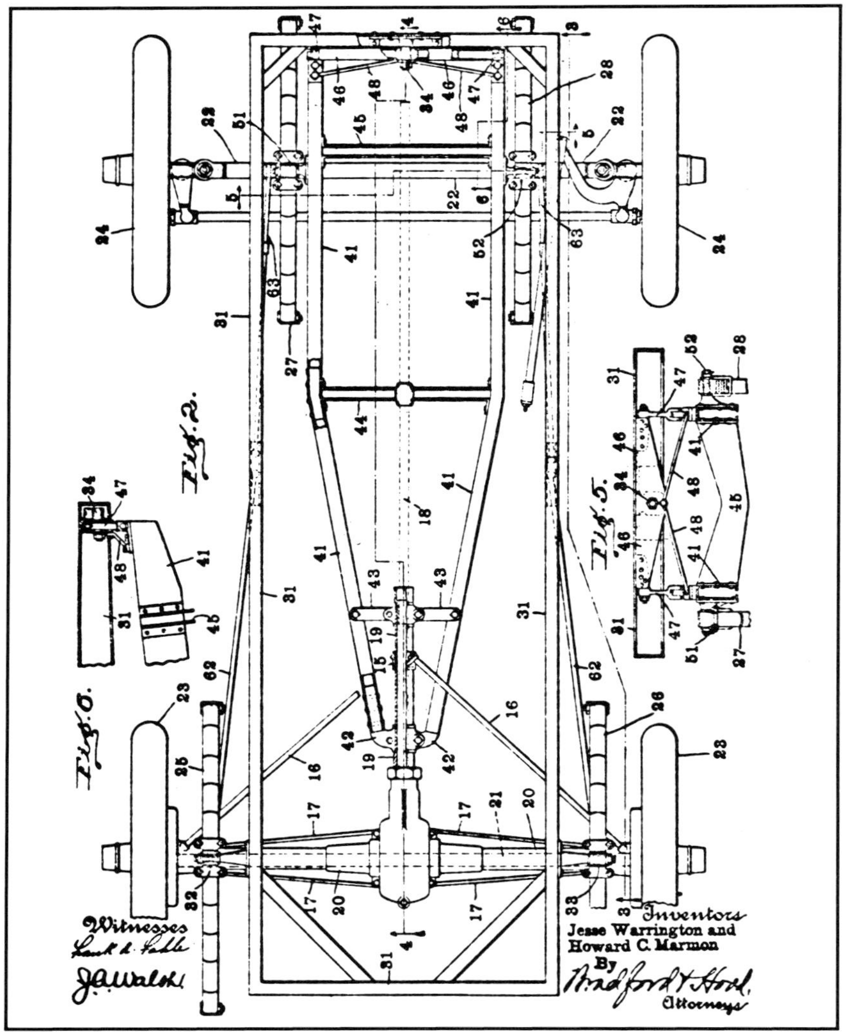


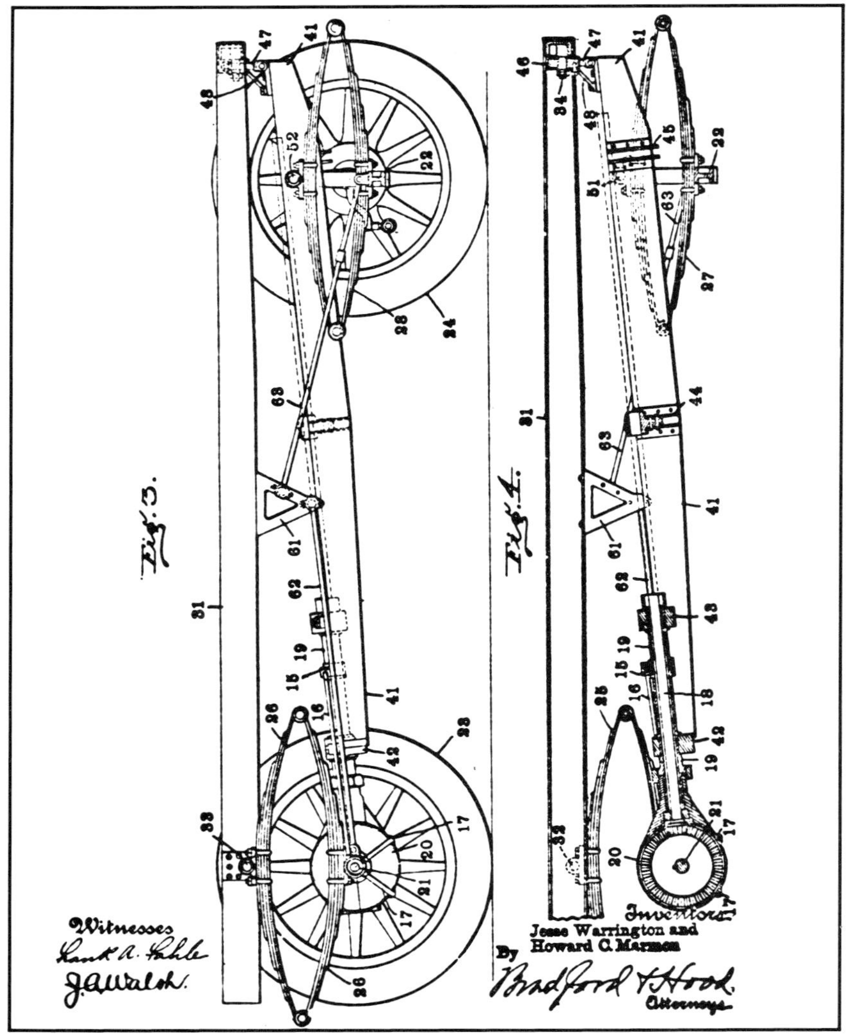


*machine and its occupants from the effects of the shock. Thus unusually easy riding qualities are obtained in addition to the many advantages of this system of suspension. The wheel base here illustrated is 90 inches and the car rides as easy as would a rigid four point suspension car of 180 inches wheel base."*

*"Fig. 5 is a front view with one rear wheel raised 12 inches. Notice that the lower frame, transversely, is level, the conditions being simply reversed from those shown in Figs. 3 and 4.*

*"Fig. 6, side view with rear wheel raised, shows that the rear of the lower frame is raised vertically, the pivots on the front springs permitting this movement without binding strains.*

*"Fig 7 is a remarkable demonstration of Marmon double three point suspension which is impossible in any other system. It is an exaggeration of road conditions which are being constantly met in small or large degree. One rear wheel is raised 12 inches and the opposite front wheel is also raised 12 inches. Both frames are tilted, each to the same extent as shown in the front views Figs. 4 and 6. The other two wheels rest solidly on the ground and as can be readily appreciated, the power plant and driving mechanism is still in perfect alignment with the rear axle drive. The power plant and transmission being rigidly fixed to the lower frame are of course tilted as the lower frame is tilted. In effect the power plant and rear axle drive are practically a unit, being maintained in perfect alignment under all road conditions.*

The Marmon frame was made of chrome nickel steel. The lower or propulsion frame was a wood filled steel tube construction accomplished by placing an inverted 'U' steel pressing over an I section hickory wood bar and then closing the steel to rectangular tubing shape. The main frame was of channel section pressed steel filled with wood from the front to the intermediate cross-members. The lower frame was readily separable from the main frame or the axle by removal of bearing caps. The thin steel from which the frames were made was held "by a long former in a special fixture and formed to channel section (of varying depth) by hand-hammering at present (Oct. 1905)."[15]

The illustration shows the transmission, clutch, accelerator (new), brake and brake equalizer linkages, as well as the castings which supported the transverse tube used to mount the steering gear (on the Model C-7 of 1907).

The dual frame chassis posed a major challenge in engineering of the control linkages between the driver (on the body frame) and the engine, clutch, transmission and brakes (on the power train frame). The steering linkage had to accommodate not only the usual frame to axle motion but the additional frame to frame motion.

All four springs on the Models A through D were single-hinged full-elliptic type. The front were 36 inches long, 2 inches wide and had five leaves. The rear were 38 long, 2 wide with seven leaves.

Fig. 2-41 The power train sub-chassis of the Model F is representative of all Era One Marmons except the Models H, 45 and 50. Only the Models F and G had 'unit' power plants.

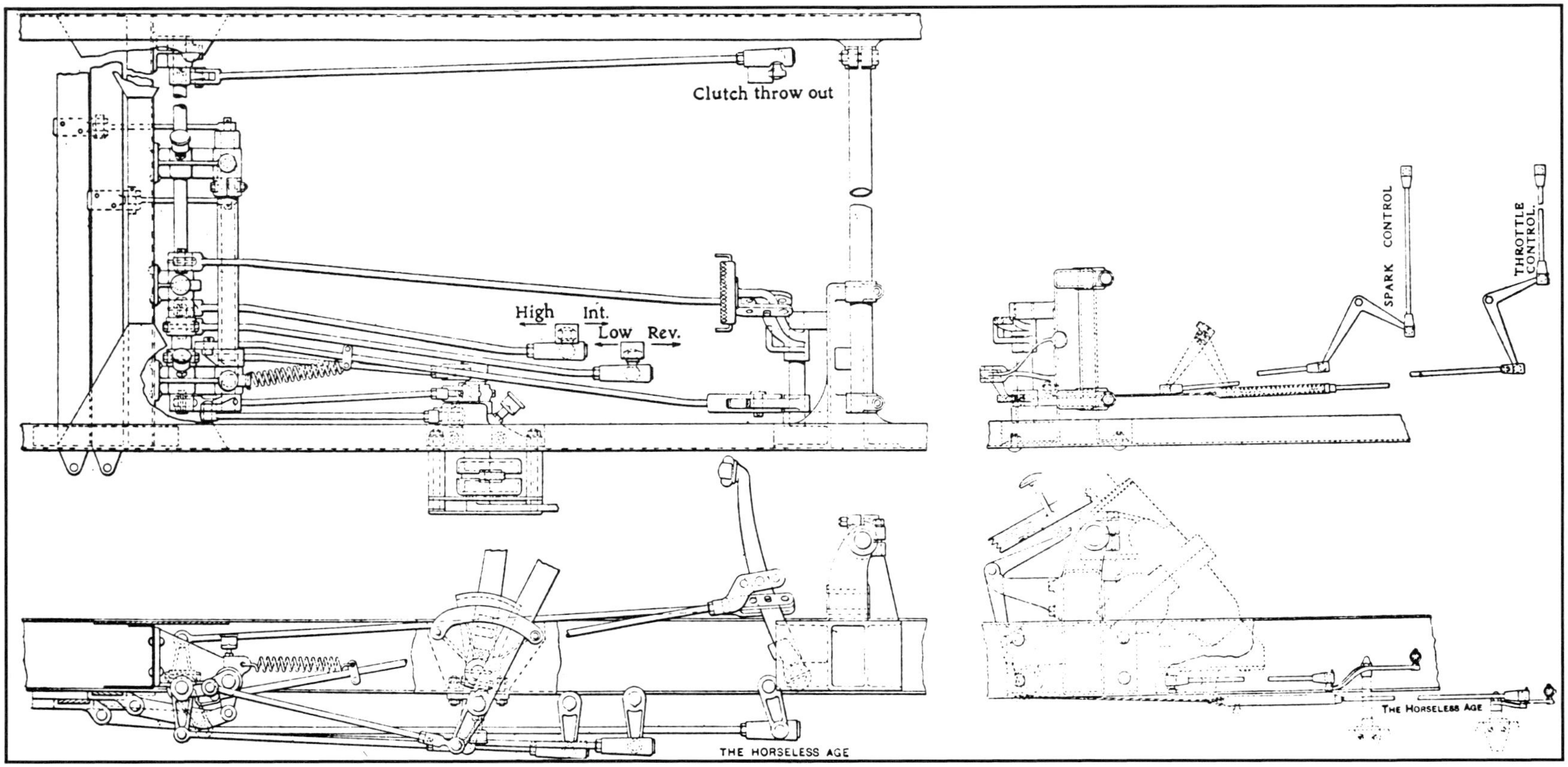


Fig. 2-42 Control system was an engineering challenge — inputs on one frame and outputs on the other. 1906.

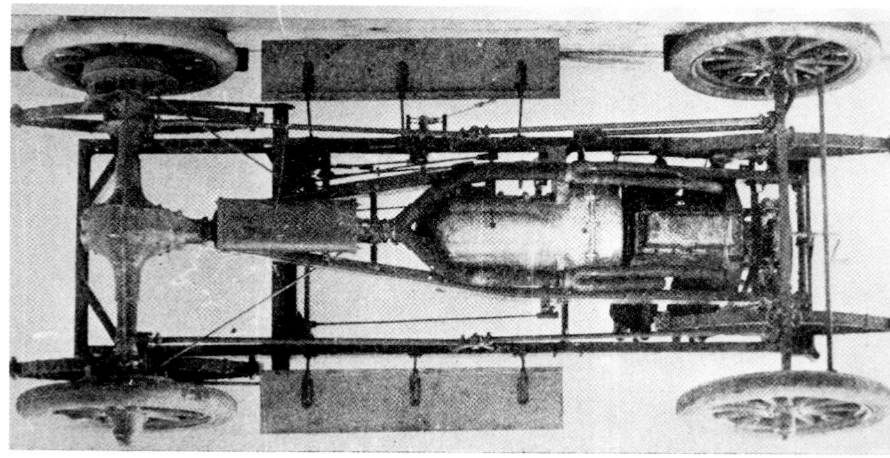

Fig. 2-43 Model F chassis — Note the exhaust system in the right photo.

The Models F and G had scroll end, double shackled, full elliptics which were 42 long with five 2 inch leaves in front and 44 long with six 2 inch leaves at the rear. The front axle was overslung.

The Models H, 45 and 50 had scroll end, double shackled, full elliptic springs used with an underslung front axle.

*Other Dual Frame Vehicles — Then and Now*

There was another car having a dual frame chassis which was a contemporary of the First Era Marmons. It was the Craig Toledo introduced in 1906.[16] The concept has reappeared several times in very recent years. Most notable because of the high degree of similarity is the Ryder Truck of the early 1970s. The several others have all utilized the forward-rearward pivoted chassis division to maintain weight distribution on all wheels despite terrain configurations; some have been rear drive and in the unsprung all-terrain (except highway) category. An exception is the four-wheel drive, independently sprung mini-truck recently announced in Great Britain. Brief descriptions follow.

## CRAIG-TOLEDO CAR 1906

Another dual frame concept had appeared in 1906.[1] The Craig-Toledo car had a chassis which at first inspection appeared similar to the Marmon. It did have a separate frame for a power train having shaft drive without universal joints; like the Marmon, this sub-chassis could be removed for mechanical service and did provide vibration isolation. However the main frame mounted all four springs in the conventional manner and therefore lacked the ride quality and tire durability features of the Marmon. The first Marmon of 1902 may have been similar.

## RYDER TRUCK/HOBBENSEIFKEN 'PAYMASTER' 1970

Another vehicle incorporating the dual frame construction concept came on the automotive scene seven decades after the first such vehicle took to the Indiana roads.

A vehicle every bit as peculiar and provocative as its distant predecessor was the Ryder "Paymaster" Truck. It was designed by Dean Hobbenseifken of Lyons, OR to whom US Patent 3732942 was issued May 15, 1973 on an application dated Dec. 16, 1970.

The initial prototype carried the badge "Paymaster". Of ten truck-tractor units ordered from Hendrickson Mfg. Co. by the Ryder System of Miami, Florida, USA, two were completed before the major recession which followed the OPEC oil embargo. These prototypes, intended for field testing, carried "Ryder" badges. Consequent to the economic crisis, plans for manufacture were abandoned.

Just as Consolidated Freightways had sought a superior truck-tractor for their inter-city freight operations and developed an advanced vehicle concept ultimately produced by a subsidiary as the "Freightliner" truck, Ryder was dissatisfied with the stereotyped vehicles offered on the domestic market and embraced the Hobbenseifken concept.

Fig. 2-44 Ryder Truck reached multiple prototype stage in 1973.

The 'Ryder' truck offered lower wind resistance and simplified maintenance. The former was attained by a slant-front cab, mounted forward of the front wheels, which was "two man" width.

The reason for detailed discussion of the 'Ryder' herein is the manner in which improved serviceability was attained. Like the Marmons of Era One, the 'Ryder' had two frames, one of which was built into a propulsion sub-chassis. This mounted a unit power plant (Detroit Diesel V-8 and manual transmission) and was rigidly attached to the rear axle while being coupled to a forward crossmember of the main frame with a ball joint. Like the Marmons, the propulsion system and rear brakes could be removed as a unit from the vehicle for service; the rear tires/wheels facilitating removal and handling. Importantly the passenger (driver) compartment was similarly isolated from the vibrations originating in the propulsion system; in the Ryder, the unit power plant was rubber-mounted in the propulsion frame and a U-jointed drive shaft was consequently required.

The main frame of the Ryder carried the front axle and its suspension *as well as the rear springs.* Upon it was mounted the cab and the fifth-wheel attachment for

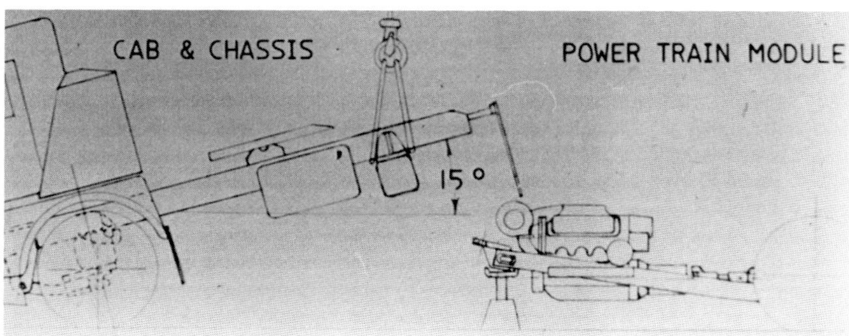


Fig. 2-45 Separation of Power train sub-chassis of Ryder truck.

the trailer. The Marmon front axle was suspended from the propulsion frame to achieve their "double three-point suspension"; this was not an objective of the Ryder truck concept. Rather the Ryder main chassis resembled a truck "Glider Kit" such as is used to expedite rebuilding of a vehicle damaged in a collision. With a 'Ryder', transfer of power-train components to the Glider would not be necessary. The propulsion sub-chassis would simply be transferred intact from the damaged truck to the replacement Glider.

## CHICO, COOT, CROCO AND RTV-1 VEHICLES

The Marmon Double Three Point Suspension system provided very superior tractive performance. This is endorsed by the claims made in 1982 for an advanced vehicle developed by the renowned German helicopter manufacturer, Messerschmidt-Bolkow-Blohm, and manufactured in the Republic of Ireland for distribution primarily in third world countries. The Chico (Spanish for Workhorse) is articulated, a concept first used in German military tractors in 1917 *and said to*

Fig. 2-46 CHICO vehicle.

*give good off-road capability without four-wheel drive.*[17] In the Chico, the passenger compartment (2 man cab) follows road undulations and the underfloor-powered cargo compartment does likewise by means of a central tubular chassis element supported by roller bearings; both front and rear axles are suspended on springs uninfluenced by terrain undulations and fully available for reaction to surface bumps and ruts. Obviously the Chico is a direct application of the engineering concept of the patented Marmon system, but utilizing a different configuration, while possibly retaining the isolation of power train noise/vibration.

The Chico is actually a two-wheel drive version of COOT, an articulated but unsprung all-terrain vehicle, acquired by and briefly marketed by Marmon-Herrington in the 1970s and discussed in Chapter Three, Part 2, of Book Two.

The CROCO (CRoss COuntry) vehicle, reported by the same source in March

1984, is another German vehicle. It is similar to the COOT but Wankle powered. Like the COOT, it is unsprung.

In November 1984, the same publication (AE Vol 92, No. 11 p83) described yet another horizontally articulated chassis vehicle, identified only as the RTV-1 by a small British firm, All wheels are driven and independently sprung using Triumph Herald swing axles. An Austin Mini engine-transmission in a below-floor, longitudinal alignment in the rear (longer) chassis powers the RTV-1. An overhung two-man closed cab is mounted on the forward (shorter) chassis. The center pivot is a five inch diameter roller bearing; the forward prop shaft passes through the bearing axis. Like the CROCO, RTV-1 weighs 2000 pounds and is rated 500 pounds load capacity on the flatbed loadfloor.

## STEERING SYSTEM (AXLE AND STEERING GEAR)

For 1905, the front axle had the steering knuckles reversed from the usual arrangement, the yokes being on the stub axles . . . then also known as the "Mercedes type", then and now as the 'Reversed Elliot'. A bronze thrust washer carried the car weight. The axle H-section was 2.0 deep and 1.5 inches wide; the ribs and web were 0.5 inches thick. The wheels ran on Hess Bright tapered roller bearings. By 1907, the web was .44 inches and the ribs .38, and the spring seats were integral. The 1907 axle is presumed to be a one piece forging.

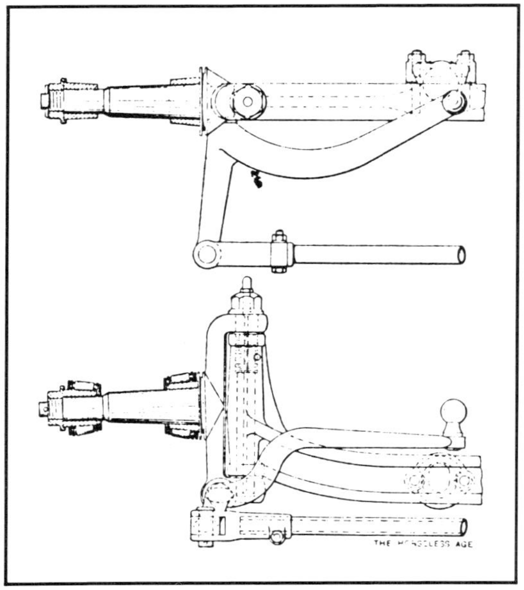

Fig. 2-47 Front axle detail — 1907

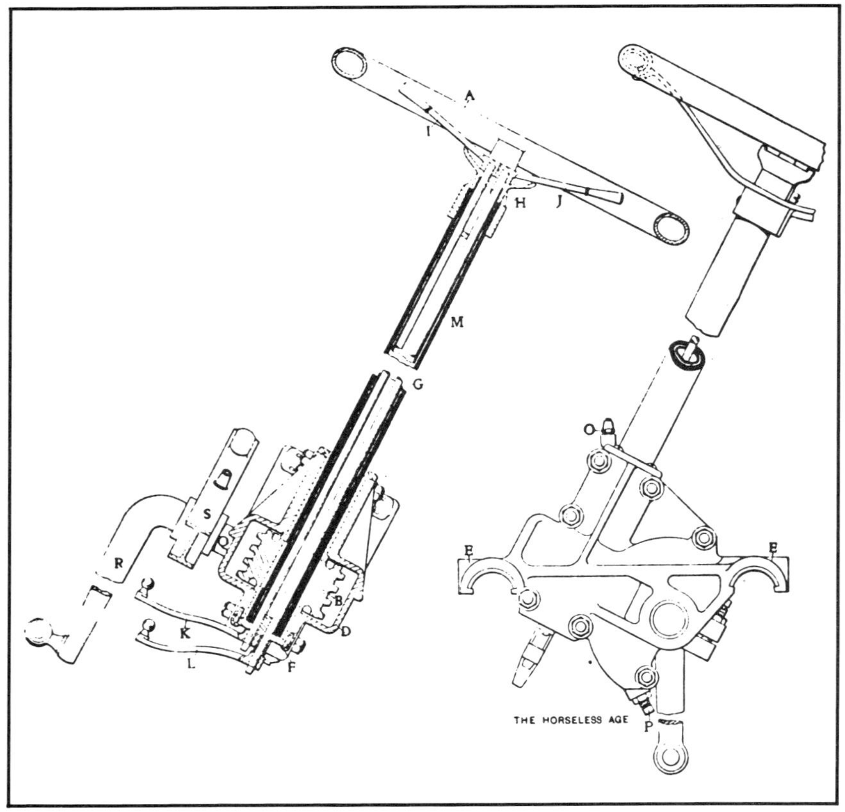


Fig. 2-48 *"The 1906 Marmon steering post, steering action and control rockers, hand levers. Globe-end and wrist bellcranks and globe-end rods are used from the two permanently placed rockers enclosed in the revolvable steering column, rockers supported by stationary inner tube. The rockers pass through a split and pinching-screw adjusted bronze block, inside the tube, screw adjustable from without, so as to regulate retaining friction to suit. These stationary rockers require no compensating devices, and as they are enclosed in the steering post the device is smooth outside. The hand levers work in a double stop-sector on top, no ratchets. The steer is by a quadruple-thread worm, applied to the steering sleeve, engaging a bronze divided worm-gear sector, parts screw adjusted on each other to take out lost motion. The steering arm outboard bearing S well supports the hard-worked and vitally important member R. The whole steering and spark and throttle control is a noteworthy example of elegant and thoroughly good machine design and construction."* (Note that the steering tube is **not** enclosed).

The axles on all Era One Marmons, front and rear, were positioned by longitudinal radius rods having ball and socket fittings at the axles and clevis attachment to a member extending downward from the frame midships. The drop-forged ends were brazed in place in the non-adjustable rods.

The 1906 and presumed 1905 steering gear was of the quadruple-thread worm and split-segment type, with the latter adjustable for gear wear. Internal stops for extremes of travel were externally adjustable. An integral outboard bearing support was provided for the output, or sector shaft. The hollow-steering shaft was *not* enclosed, precluding any support above the gear case.

Gear cases split on the shaft center-line were accepted domestic practice in the early century. The penalty of assembly, after face machining and before boring, and subsequent disassembly prior to assembly of the mechanism was evidently preferable to the setting of dry sand cores in a one-piece case.

The gear housing was split on the center line of the column or input-shaft, each half-housing incorporating a pair of trunnions by means of which attachment was made to a pair of transverse tubular cross-members elevated above the frame side rails. Plain bearings were used throughout.

Within the steering-tube was a second tube anchored at the bottom within which were two side-by-side shafts for the throttle and spark-advance control. They were guided at the top end by a split clamping-block whose pinching-screw was, surprisingly, adjustable from the outside to control the motion-resistance of the levers after an operating position had been selected.

*The Cycle and Automobile Trade Journal*[15] reporter described it thus: The whole steering and spark and throttle control is a noteworthy example of elegant and thoroughly good machine design and construction.

A new steering gear introduced for 1907 by Marmon was noted by the press to have several noteworthy features incorporated. Principle among them was the adoption of Timken taper roller bearings for both the input-shaft and the pitman-arm shaft and a method of installation which made feasible the selection of a steering column inclination which suited the individual purchaser.

Although the already familiar screw and nut principle was utilized, a steel screw operated in a bronze nut having a hard babbitt lining on the working surfaces. The bronze nut incorporated two integral horizontal journal extensions which actuated a pair of steel links, nominally parallel with the input-shaft. These links were joined pivotally to a yoke on the output, or pitman-arm, shaft. Twisting of the links was avoided by a "wing" extension of the output-shaft which engaged a slot in the nut. Except for the location of the anti-twist tang, this would be the basic configuaration of all Marmon-built steering gears until in-house gears were permanently abandoned during the 1924 model run.

Both sets of bearings were adjustable by nuts, at the top of the gear case on the stationary column (which was brass tubing) and at the inboard end of the output-shaft.

The 1907 Marmon steering gear housing was split perpendicular to the input-shaft on the centerline of the out-put shaft. The upper-half incorporated a pair of widely-spaced trunnions by which the gear assembly was attached to the chassis. The chassis incorporated a single transverse tubular cross-member, higher than the side-rails, for this purpose. Prior models had two parallel rails and a four "trunnion" attachment at a fixed inclination, although the 1902 prototype had an adjustable column.

Special consideration was given to placement of the pivot for the column such that the adjustment did not interfere with the steering, throttle or spark-advance linkages. All were fitted with ball-socket attachments and the hand controls were gear-driven to operate about a vertical axis. Concentric shafts, within the hollow input-shaft, rotated pinions engaging bevel gear sectors supported on a vertical axis within the lower-half of the housing. This half-housing required dry-sand coring and incorporated spring-loaded retention brakes acting on the periphery of the sectors.

All Era One Marmons had a handsome three-spoke steering wheel (after all,

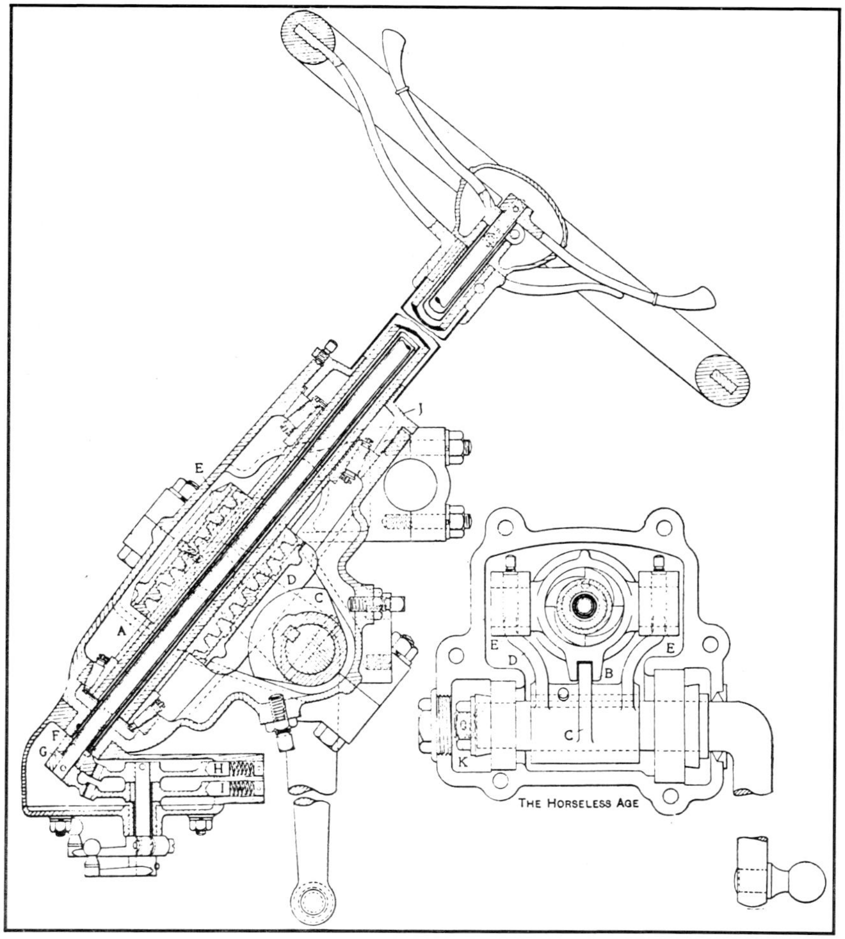


Fig. 2-49 There was rapid progress in steering gears: The 1907 redesign.

three points are all that are required to determine a plane!) with either a cherry or black walnut rim given a natural wood polish. The hub and spoke member was a brass casting.

## WHEELS AND TIRES

From the earliest advertisements Nordyke & Marmon claimed that they used only the finest wood artillery wheels available. Later they were specifically identified as the name of the vendor became more generally recognized. They were made to a patented design of spoke by the Schwarz Wheel Company of Philadelphia, Pennsylvania, and were used on most if not all Marmons with artillery wheels throughout the first four ERAS. The Schwarz Patent Spoke had "the tenons grooved and mortised for interlocking"; hence the inner end of the spoke which contacted the hub was as wide (in the plane of rotation) as the spoke itself. The

Fig. 2-50 Details of the wheel construction favored by Marmon as late as the Third Era.

spokes were assembled to the felloe in a Belleville or conical shape and then forced into a flat assembly using a press.

Schwarz claimed "the spokes can never loosen". The author has restored two 1924 Marmons and the wheels were evidently strong as new in 1980; The spokes were unable to rock about the hub bolts as would be the case with the usual tapered nose spoke.

Tires were 32 x 4 on the Models C and C-7, 34 x 4 front and 4.5 rear on Models D, F and G. All were clincher type, requiring inflation after mounting like bicycles tires and were constructed of layered fabric.

## CLUTCH
## 1907-08

Howard Marmon had used the single-plate clutch and three-speed selective transmission in his first prototype car of 1902 but for his first production car had elected to switch to a planetary gear box with dual-function clutches which engaged the gear for direct drive *and* initiated vehicle motion. He had found the latter to be more practical and free of driver training problems at a time when practically every driver was a neophyte, and it made driving more attractive to women as well.

When expressions of public preference for the clutch and transmission system were evident, Marmon reverted to his original prototype configuration but gave careful and clever attention to means to lesson the problems of mastering the starting, shifting and stopping of a car so equipped.

His new manual transmission incorporated the now-familiar interlock between the two shift levers but importantly also incorporated a balk on movement of either rocker lever unless the clutch release collar was in the released position; the collar motion was conveyed to the interlock package by an extension arm. Inability to move the shift lever alerted the driver to his omission of the clutch release action preceding gear selection.

The first Marmon had what is now known as a "single plate" clutch; a single driven plate but two driving elements . . . the flywheel face and clutch pressure-plate face . . . for a total of three. In the early years of the industry the expression "multiple-disc" was used to distinguish that type, regardless of size or number of discs. The objective of that general classification was distinguishment from several other types . . . cone, band etc., long since vanished from the automotive scene. Later "multiple-disc" was used to describe a category of clutch in which a number of small hubless discs are used on both the driving and driven sides, as distinct from "single-plate" clutches in which one, or occasionally two, large driven discs with splined hubs are driven by heavy driving members with high heat absorption capacity; this is the universal vehicle clutch today.

This problem of interpretation of nomenclature is international. It was first brought to my attention by Dr. Luigi Giovanni consultant to the firm VALEO, SpA of Turin, Italy, which published his history of (European) clutch design *"Clutches Through the Years"* in 1978.

The problem of decelerating below the engine minimum speed before releasing the clutch was averted by an unusual innovative arrangement. The engagement was accomplished by the left pedal in the usual manner and the pedal was normally utilized during gear shifts but when the brake pedal was depressed, it carried the clutch pedal down also. It is not documented anywhere, but it would appear that "hillholder" action would also be obtained . . . simultaneous clutch engagement and brake release when starting on a grade to avoid roll-back.

The clutch was a single-plate type with cork friction surfaces. Cork inserts were placed in drilled holes in the flywheel and pressure plate; the driven plate was an unfaced disc for minimum inertia during gear shifts. Coil springs were direct acting on the pressure plate and release was accomplished by means of a pull type lever system. Many industrial clutches still use the pull concept. Cork inserting of clutch surfaces was a patented process controlled by the National Brake and Clutch Co. of Boston, Massachusetts. They listed Nordyke & Marmon as one of their licensees in a 1909 advertisement. A spring system was provided for release of the driven plate from the flywheel subsequent to retraction of the pressure plate. Still used in 2 plate clutches for trucks today.

The clutch and transmission are illustrated in the section "Transmission — Layshaft Type"

## TRANSMISSION — PLANETARY TYPE 1904-07

The planetary type transmission in the Models A, B, C, D, and C-7 also incorporated the function of vehicle clutch. Like the Welch car, these Marmons were started in motion from rest by the engagement of the low-gear reaction-band in the transmission and direct-drive was obtained by release of the low-band with concurrent engagement of a compact and novel single plate clutch within the transmission. The single control-lever was moved forward to engage low and backward to the 'high' position, at the prevailing throttle setting.

Reverse was selected and motion initiated by depression of the left of the two floor pedals. This was similar to the center-pedal function of the Ford Model T.

The transmission was described in the February 1905 issue of The Horseless Age;[1] it was stated to be "of original design". Details were amplified in the Nov. 21 1906 issue which described the 1906 model cars.

The housing was made of thin-wall aluminum castings. The main case was open at the top and closed by an aluminum access cover retained by four bolts. The rear of the case had a large opening for assembly puposes, also closed by a cover; the front had a small diameter opening without obvious purpose, closed by a cover which incorporated the oil "seal" drainage groove.

There was no boring of the case itself, which surprisingly incorporated no bearings! N & M Catalog 730 stated: "It has no bearings and none are required for the driving shaft passing through". Support was provided by couplings to the flywheel and extended axle pinion shaft.

The single planetary gear set provided both low forward and reverse. It was of the spur gear type, and without any internal toothed ring gears.

The planetary drum was formed of two facing castings, carrying three planet shafts and joined on the periphery by a steel band forming the brake surface. The drum was supported by bushings in the two castings. The planets were triple gears; the smaller two pitch diameters were integral and the largest was pressed on between them. All gears in the transmission were of nickel steel, and 8 diamentral pitch.

The input shaft carried a clutch driving member and terminated in a sun gear and adjacent journal. The larger output sun gear and its integral shaft piloted on that journal. When the drum was held stationary by the bronze band, the planet axes were fixed and the planets rotated on their fixed axes driving the output at reduced speed.

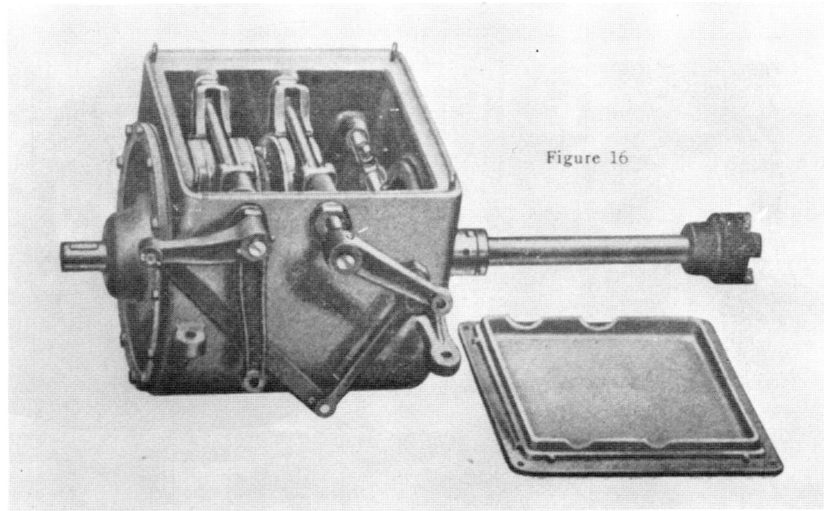

Figure 16

The shift to 'high' was accomplished by concurrent release of the low-band and engagement of the direct "single-plate" clutch. That engagement drove the third sun gear, as well as the input sun, at engine speed thereby locking the planetary gear set and effecting direct drive.

Reverse was accomplished by engagement of the second band which held the driven side of the direct clutch and its attached sun gear. Thereupon the engine-driven sun gear, in turning the planet gear cluster forced the cluster to roll around the locked sun gear, (thereby rotating the drum) and turning the output sun-gear in a reverse direction. The reverse action results from the relative diameters of the three suns . . . the driving sun being smaller, and the driven sun larger, than the locked sun. This same concept provided low and reverse in the Ford Model T.

All the friction elements in the transmission were unusual, innovative and technically sophisticated. All were bronze on iron, running in oil. All were controlled by levers on parallel shafts, making inter-connection precise and durable. The description of the direct clutch and the planetary bands and their controls is repeated from *The Horseless Age,*[18] because it is alphabetically correlated with the illustration:

> The high speed clutch consists of a phosphor bronze double cone adapted to engage with a double female cone of cast iron; it is operated by the usual sliding cone B and clutch dogs C, which press against a four armed centre of plate steel bolted to the clutch cone. A light spring, consisting of two bulging sheet metal plates riveted together at their outer edge, effects the disengagement of the clutch when the sliding cone is withdrawn.
>
> The manner of holding the low gear and reverse drums, D and E respectively, from rotation is of interest. These drums are surrounded each by a split ribbed bronze ring F, the ends of which are provided with lugs, one of which, G, is slotted, and the other, H, drilled. These rings are supported at their centre by set screws passing up through bosses in the bottom of the gear case.

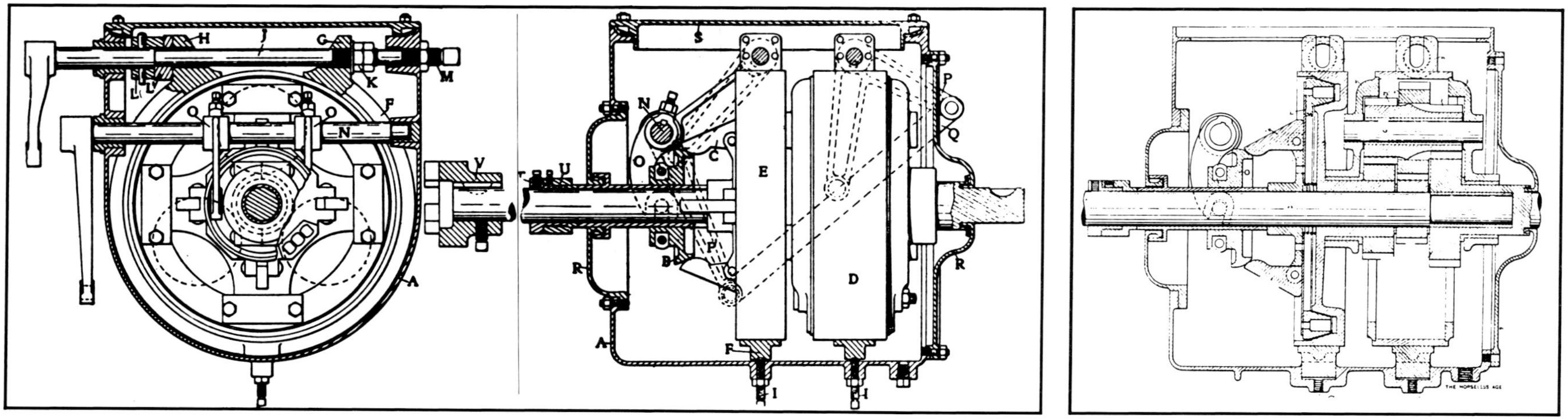

Fig. 2-52 Marmon planetary "change speed gear" of 1907. Section of 1906 unit provides added detail.

Through the lugs at the ends of the ring passes a shaft J, which is provided with a washer, nut and lock nut K just outside the slotted lug, and a pinned-on spiral cam L just outside the drilled lug. To the outer face of the drilled lug is secured a second spiral cam L', and it is obvious that when the shaft J is turned in one direction and the two spiral cams are moved relatively to one another, the bronze ring F is contracted upon the drum, and the latter is thereby held from rotation. When the shaft J is turned in the opposite direction, the bronze ring F, owing to its natural spring; expands and releases the drum. A set screw M is provided by which the ring may be adjusted so that when released it does not bear on any part of the drum.

The high and low speeds are operated by a single lever, and the shaft J by which the low speed brake is applied and the shaft N carrying the high speed clutch fork O are interconnected by means of two lever arms PP and a link Q in such a manner that when the high speed clutch is engaged the low speed brake is released, and vice versa. The reverse is operated by means of a pedal. These lever arms PP and their link connection Q are located outside the gear case on the right hand side. The gear case is cast of aluminum and consists of one main casting A with circular end plates RR and a full sized cover plate S. The cover plate is provided with a sort of stepped joint which is practically oil tight. Adjusting means for the high speed clutch are provided outside the casing at the forward end. They consist of two threaded collars, one of which, T, is fastened to the driving shaft with a set screw, and the other, U, can be turned on or off the former and thereby give the conical clutch more or less play.

The clutch in the Marmon transmission was a significant development in the field of clutch design particularly as applied to clutches used for vehicle motion initiation.

The clutch was a toggle-linkage over-center type widely used in machine tools today. The pressure on the friction surface results from the compression of the linkage, rather than from a true spring load. Howard Marmon sought a design with a soft engagement, an essential characteristic as any overlap of the direct and low clutches would result in excessive wear and objectionable vibration. The solution was to load the bronze ring friction element through a spring-steel cross-shaped plate; the plate was .125 thick and the arms 2.0 inches wide. Hugh Dolnar described it in Cycle and Automobile Trade Journal and reported: *"(It) produces a soft partial engagement, making a most satisfactory slipping drive. The great area of high-speed clutch engaging surface permits an infinite range of slipping speeds, so that the driver can pick his way through a crowded street on the high gear and go as slowly as he pleases, and at the same time be all ready for a rapid movement as occasion permits or safety demands."*

Adequate torque capacity at a 9.25 inch diameter was obtained by the V-shaped friction surfaces. The V-angle was established by trial and was the smallest angle that would release. Separation of the surfaces was assisted by a coil spring until 1906 when a pair of Belleville-type washers, riveted together at their periphery was adopted and provided considerably more force in the available space.

During 1906, a public promotional demonstration of the capability of the air-cooling system was made. Marmon was evidently having difficulty with public acceptance of air-cooling; engineering of a water-cooled optional engine must have been underway by mid 1906 for it was announced little more than a year later.

The demonstration consisted of fifty miles of non-stop operation in low gear on a summer day. In reality, it was a demonstration of transmission gear durability because low-gear usage on level terrain is conducive to improved cooling whether direct (air-cooling) or indirect (water cooling) heat rejection systems are used. Hence this event is recorded here in the transmission discussion.

## TRANSMISSION — LAYSHAFT TYPE 1907-08

The model F transmission was an industry milestone. The aluminum case included the 'bell housing' flare to flywheel diameter, and incorporated two transverse rocker shafts for shift fork movement.

The gear and shafting layout was basically that which persists to this day on rear drive cars; gears were of course spur type with sliding engagement and an internal/external tooth coupling for direct drive such as persisted into the 1930s.

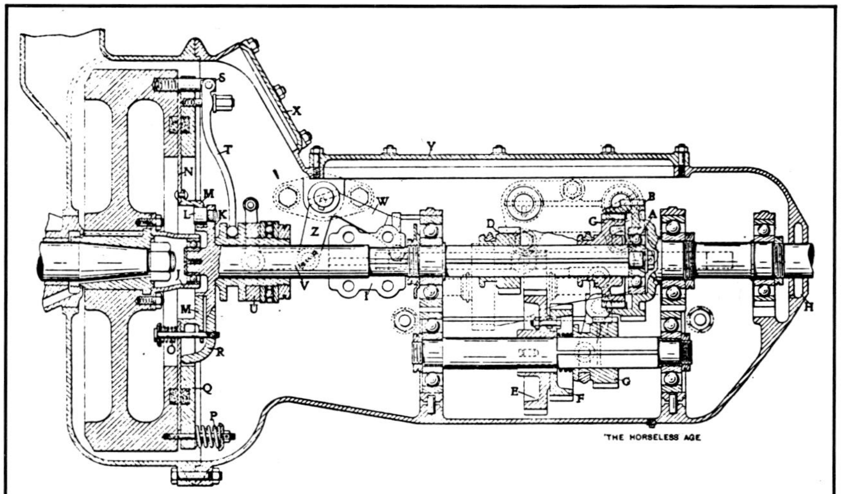

FIG. 1.—LONGITUDINAL SECTION THROUGH CLUTCH AND CHANGE GEAR.

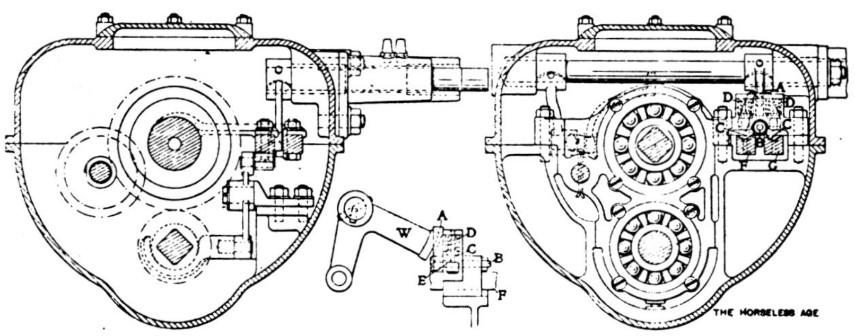

FIG. 2.—SECTIONS OF CHANGE SPEED GEAR.

Fig. 2-53 Model F Transmission.

However, the entire configuration was reversed from subsequent industry practice; in consequence, an additional bearing was requried to provide the support of the output shaft normally accomplished by the pilot bearing in the flywheel.

There were two novel and commendable features unique to the Marmon design. The interlock mechanism between the shift rocker shafts was interconnected with the clutch release internal linkage to prevent inadvertent shifting without prior release of the clutch. Also a third rocker shaft, mounted internally, moved in response to selection of direct drive and disengaged the drive gear on the countershaft in the interest of improved efficiency by reduction of transmission spin losses. The gear was fork released and spring-biased for reengagement; this assured engagement in the event of a tooth abutment on the countershaft.

Ball bearings were used throughout except for the reverse idler. They were retained by the overlapping heads of four screws threaded into the case on each bearing face. 'Splines' were square shaft type.

The case was split horizontally on the mainshaft centerline. However the bearings and the reverse idler shaft were assembled in separate bulkheads which had been straddle-milled; these bulkhead sub-assemblies were then flange mounted to the top of the lower-half case and were transversely positioned in slots milled longitudinally in the case. Each sub-assembly also engaged a dowel pin at the bottom to prevent deflection.

The Model F transmission was a truly remarkable engineering accomplishment for its time. It was an appropriate companion to the Marmon engine and an accomplishment of equal significance.

## REAR AXLE

The Marmon axle was of the three-quarter floating type, most familiar on the 1932-48 Fords. In that type axle, the weight of the vehicle is supported by the wheels and axle housing only; the axle shaft transmits torque to the wheel and is free of bending loads which would reverse with each revolution and could lead to fatigue failure; the wheel however is held in position by the shaft.

Bevel reduction gearing was used with ratios of 60 to 16 (3.75:1) and 66 to 16 (4.125:1) available. The differential was the spur gear type. Gears and shafts were of nickel steel, and were supported by anti-friction bearings throughout. Axle torque reaction was taken by a torque-tube surrounding the drive shaft and the assembly was stiffened with diagonal "hound" rods. (Properly aligned, the vehicle would not "run like a dog"). The axle was positioned by spherical ended radius rods. A central aluminum housing, split horizontally, carried the pinion bearing. The drive shaft carried the pinion on a taper-mount and was supported by a ball-bearing at the pinion and a Hyatt roller-bearing at the other end.

Novel methods were used to manufacture the axle housings . . . steel tubes whose aluminum bellmouth ends sandwiched the central housing to enclose the differential gearing. The manufacture is described in Chapter Eleven Part 2.

Hugh Dolnar of *Cycle and Automotive Trade Journal* commented on the truss rod: *One 9/16th steel truss rod is placed directly under the rear axle. This truss is believed to be needless (by Marmon), but is placed in deference to popular requirement.*

Mr. Dolnar was refering to the type of truss rod appearing currently on 4 x 4 sporting vehicle axles and ranking high as a status symbol. "Needless" or not, the attachment hardware on the Marmon reflected their high standards of detail design. A split bronze trunnion was clamped to the tube for suspension spring attachment.

The axle shaft was installed from the bell-mouth end of the housing against a ball thrust-bearing adjacent a long Hyatt roller bearing (like 1932-48 Ford rear wheel bearings) and was threaded on the outer end where a three-jaw fitting was keyed and nut-retained on a taper *after* the wheel was placed on the Hyatt roller bearing which supported it.

Careful examination of the pinion/ring-gear engagement area reveals an unusual and intriguing configuration . . . a taper-faced washer is clamped against the ends of the pinion teeth by the retaining nut and against the teeth of the ring gear which were in mesh. The underlying principle of this design would probably have been lost had not Hugh Dolnar been so dedicated to understanding and documenting significant technical detail as the automobile design art/science developed.

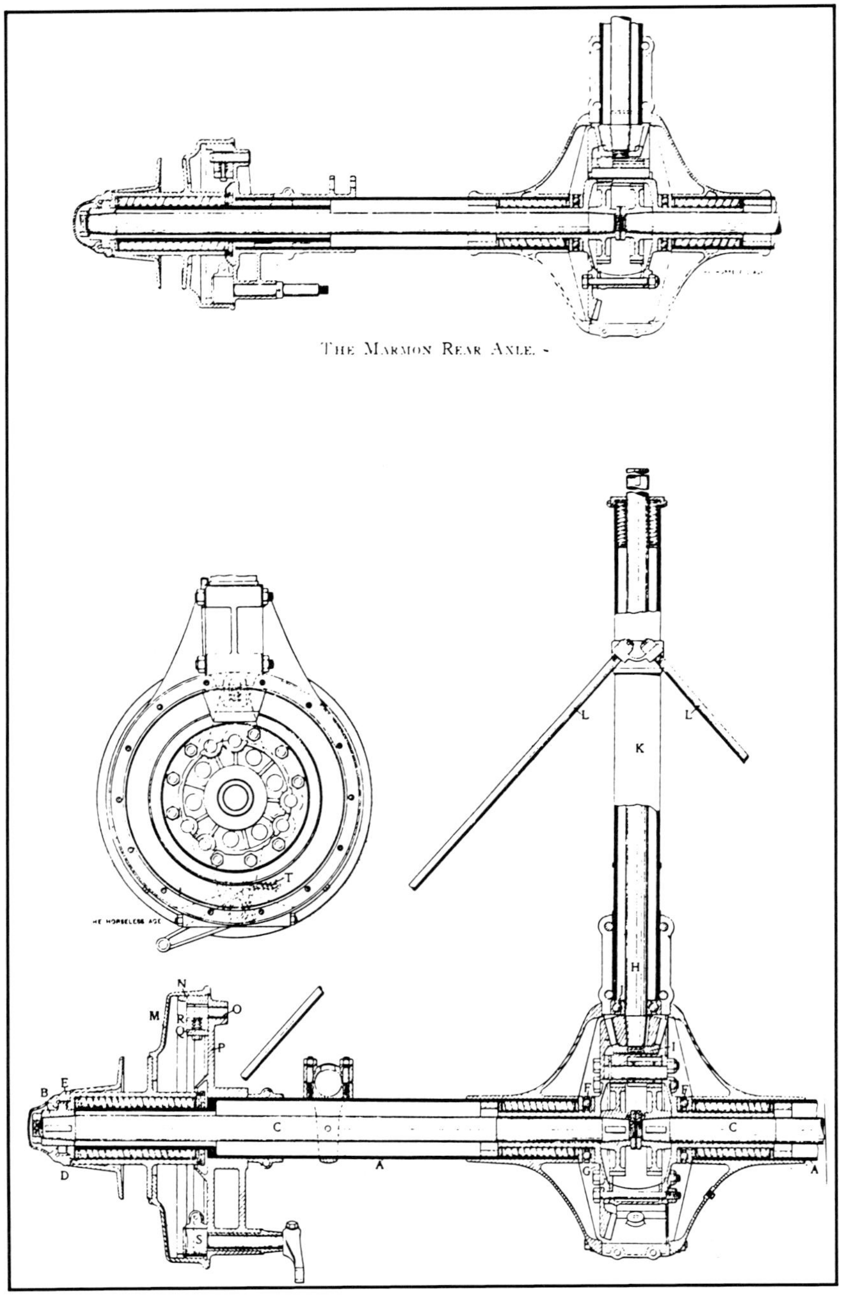


Fig. 2-54 There was rapid advancement in axle design. 1905 Model B (upper) and 1906 Model C/D (lower).

He had this to say: *The most remarkable kinetic feature of this construction is, perhaps, the manner in which (Howard) Marmon takes care of the bevel gear and change-gear (transmission) clutch thrusts. Instead of simply applying costly and adjustment-demanding bearings at the nearest points, Marmon potently surveys his whole field, and finds a hitherto undiscovered possibility of rolling surface resistance, without balls or ball-races . . . almost without sliding friction. The axle is an original design by Marmon and embodies several unique stress transmissions and resistances of extreme simplicity and elegance. The bevel-pinion resistance to forward disengagement movement is a peculiarly flanged end-washer, nut-retained, which bears against the inner ends of the bevel gear teeth, the contact being very nearly rolling.*

Because of this thrust reaction device, only a radial ball bearing adjacent the pinion and a Hyatt long-roller bearing at the transmission end of the torque tube were required,. Experience with the steel/aluminum joint and the bevel gear "rolling" thrust washer was so satisfactory that the design continued in production through 1907 and both were incorporated in a brand new heavier axle for the Model F in 1907.

The Model F axle differed in several areas where more advanced concepts were introduced. Adoption of ball wheel bearings . . . one large and one small with the same bore . . . made the axle a full-floating configuration. The wheel was now supported *and retained* by the axle housing.

The axle shafts were upset with flanges at the inner ends and the differential "side" gears were rings riveted thereto. Because the differential was of the spur-gear type, no thrust control was required on the shafts; a dowel pin spaced the shafts apart at the center. At the outer end, the axles continued the design used previously: A collar was retained on the square end of the shaft by a nut and lock-wire ring. That collar had jaws on the periphery which engaged the wheel hub. The threaded hub cap provided the outward positioning of the shaft.

The long spiral Hyatt roller bearings were completely eliminated from the entire car by 1907 (Ford used them in the rear axle through 1948). Ball bearings were placed directly adjacent the differential and outboard of them, in the same bore, were the ball thrust bearings. The rear axle housing tube was converted to a single piece formed by swaging. It replaced two tubes joined by a bushing. It was a remarkable accomplishment for the times.

## BRAKES

The approach taken in regards to brake design throughout Howard Marmon's career was well put by Hugh Dolnar of Cycle and Automotive Trade Journal:[20] "Marmon does not trifle with the retardation problem."

The brakes on the Era One Marmons were a patented construction. The first Marmon of 1902 had rear band brakes about twelve inches in diameter. Inspection of photographs shows the braking surface was a steel ring rather than drum and was attached by lugs or ears to the wooden spokes. The second Marmon and the Model A cars had drums estimated to be eight inches in diameter with an internal mechanism. The Model B for 1905 had an internal shoe brake with gray iron drums measuring 10 inch diameter and a 1.05 inch face. The reaction pin was diametrically opposite the double-faced actuating cam. The unlined single shoe was machined from a phosphor bronze casting with a rectangular opening for the reaction pin at its mid-point, and with both extremities engaging the cam faces (effectively a pair of shoes). Within the opening was a spring forcing the shoe out of drum contact; a radial screw in the shoe limited this action and thereby centered the

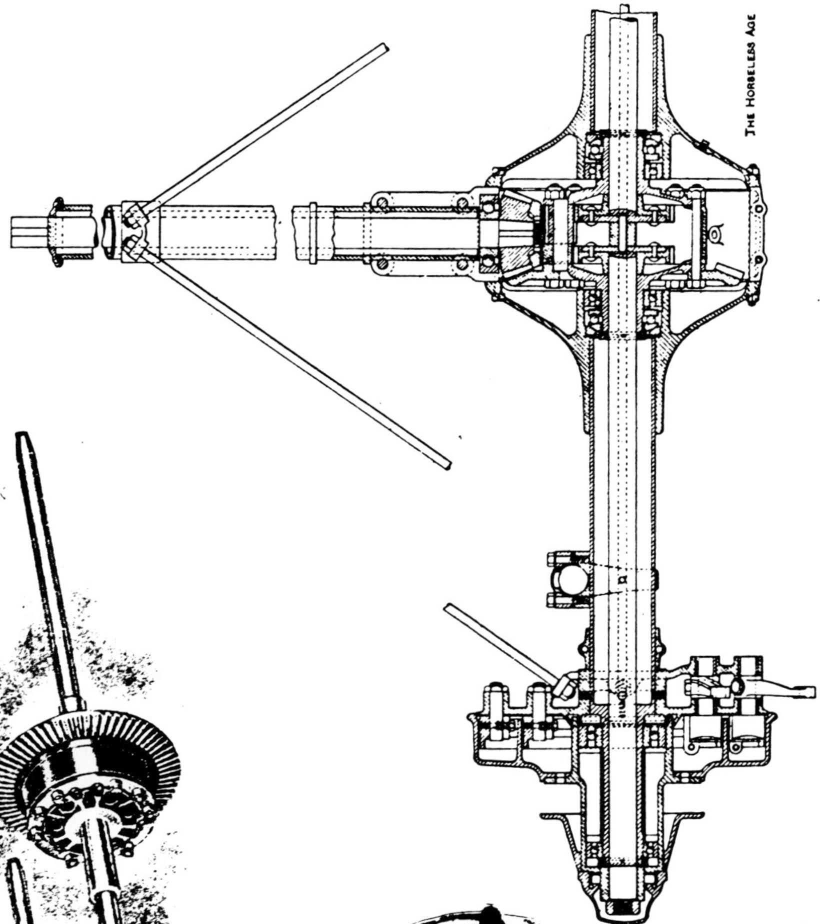


Fig. 4.—Bevel Gear Drive and Rear Axle. Model C-7.

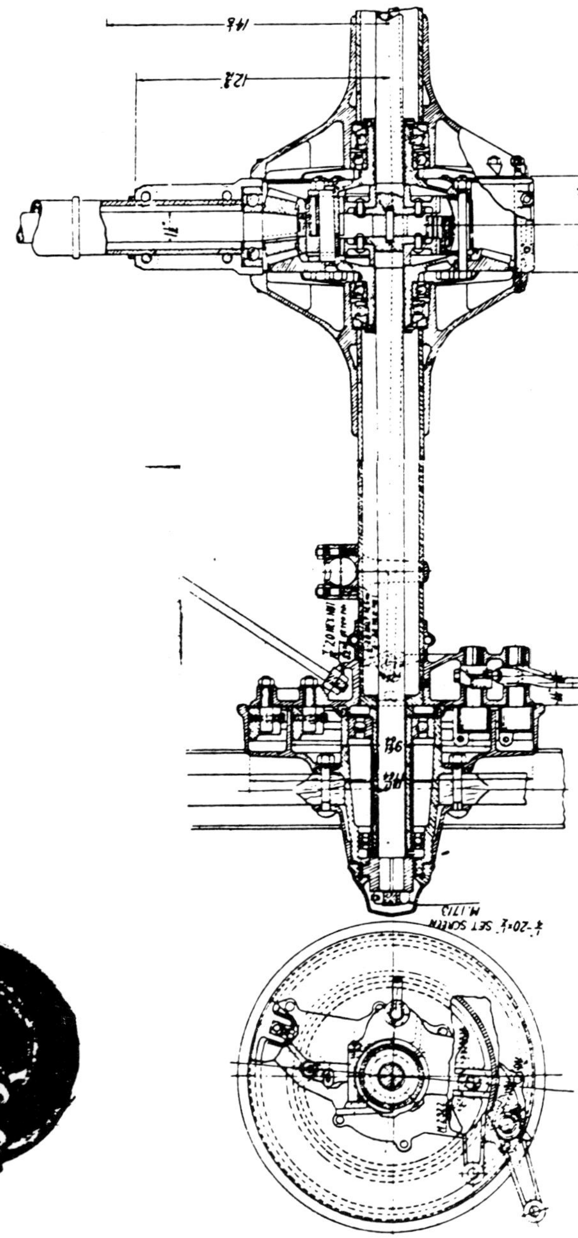

Marmon Model F, Bevel-gear Housing.

Marmon Model F, Rear Axle and Rear Wheel and Brake Construction.

Fig. 2-55 Model C axle and torque tube assembly.

Fig. 2-56 Model F axle with frame pivot.

shoe in the drum. A spring maintained cam-shoe contact by joining the ends of the shoe. All of the above construction details were covered by US Patent 793206 issued to Howard C. Marmon and assigned to Nordyke & Marmon Co. on June 27, 1905 following application on June 4, 1904. This was a bi-directional, non-self energizing brake and was, to say the least, precedent setting.

The reverse band on the planetary transmission, actuated by the left of the two pedals was the emergency brake.

For 1906, the brake size was increased to 14.0 inches diameter and 1.5 inches width. This evidently was required partly due to increased car weight, a much higher power-to-weight ratio on the model D, and partly consequent to more experience with actual consumer driving practice . . . typically 10000 miles per year based on owner surveys. The reverse transmission-band continued to serve as the emergency system, with foot actuation.

But the 1907 Model F had a hand brake lever and ratchet sector beside the gear lever. It actuated a secondary brake system identical with the primary system and concentric with it. The cast iron drum was increased to 14.5 diameter and incorporated an additional internal braking surface upon which the smaller secondary bronze brake shoe acted. All other brake parts were used in duplicate. This arrangement brings to mind the modern Chevrolet Corvette configuration of a drum parking brake surface within the rear service brake disc.

## PART 4 — Chasses of the Model H, 45 and 50 1908-1909

The Model H chassis laid claim to the traditional and unique Marmon double-frame construction and associated title "Easiest Riding Car in the World" but incorporated what Marmon called "A Special Application of the Concept".

The Warrington-Marmon Running Gear patent was amended as Reissue 12609 on February 12, 1907 upon application of May 5, 1906. Additional claims defined the variation subsequently adopted for the Model H . . . both pivots for the motor/transmission frame were located on the platform frame and drive was transmitted through a shaft having universal joints at both ends.

It remained unconventional and unique in that the engine and transmission were mounted on, and aligned by, a platform or sub-frame which was independent of deflections consequent to road surface. This sub-frame, like its predecessors, carried the front suspension and tipped to accommodate road undulations without spring deflections. Similarly it was pivoted to the center of the front crossmember of the main frame which carried the body, associated sheet metal and the rear suspension including the rear axle.

The former design pivoted the rear of the sub-frame on the nose of the differential housing, coaxially with the pinion shaft. This avoided the use of universal joints and was desirable considering the state of the art in universal joint design at the start of Era One. At least one surviving marque still used rubber/fabric laminated rings for universals in 1929. That Era One design was only feasible because the Marmon pressure lubrication system allowed the engine to be inclined in the chassis.

The Model H design accepted the use of universal joints. The engine and its sub-frame continued to be aimed at the axle pinion, but only in the design load condition. Excursions of the axle above and below the mean position were compensated for by the U-Joint propellor shaft. The result was a significantly lower top-of-main-frame and a correspondingly lower automobile . . . an important competitive consideration.

### FRAME

The general nature of the frame has been addressed in the preceding general discussion of the chassis. The features which contributed to a low chassis were a front crossmember on the main frame with a dropped center section at the pivot location and a pair of forgings which raised the rear spring perches appreciably above the frame. The pivot for the rear of the sub-frame was a forging suspended from the central crossmember. This was a large trunnion because it was coaxial with the drive shaft.

Clutch and brake pedals were attached to tubes which rotated on a tubular "crossmember". The transmission shift links were moved by levers attached to concentric tubes surrounding another tubular cross member. Because shift selection was done outside the right frame rail, an outboard support was provided for the transverse tube.

### SUSPENSION

Suspension was by full elliptic springs at all four positions. They were of the scroll end type, shackled together at both ends. The front axles were positioned by dual radius rods with spherical ends; the rear axle by a rigid torque arm which was anchored to the main frame cross member, but was cushioned *vertically between*

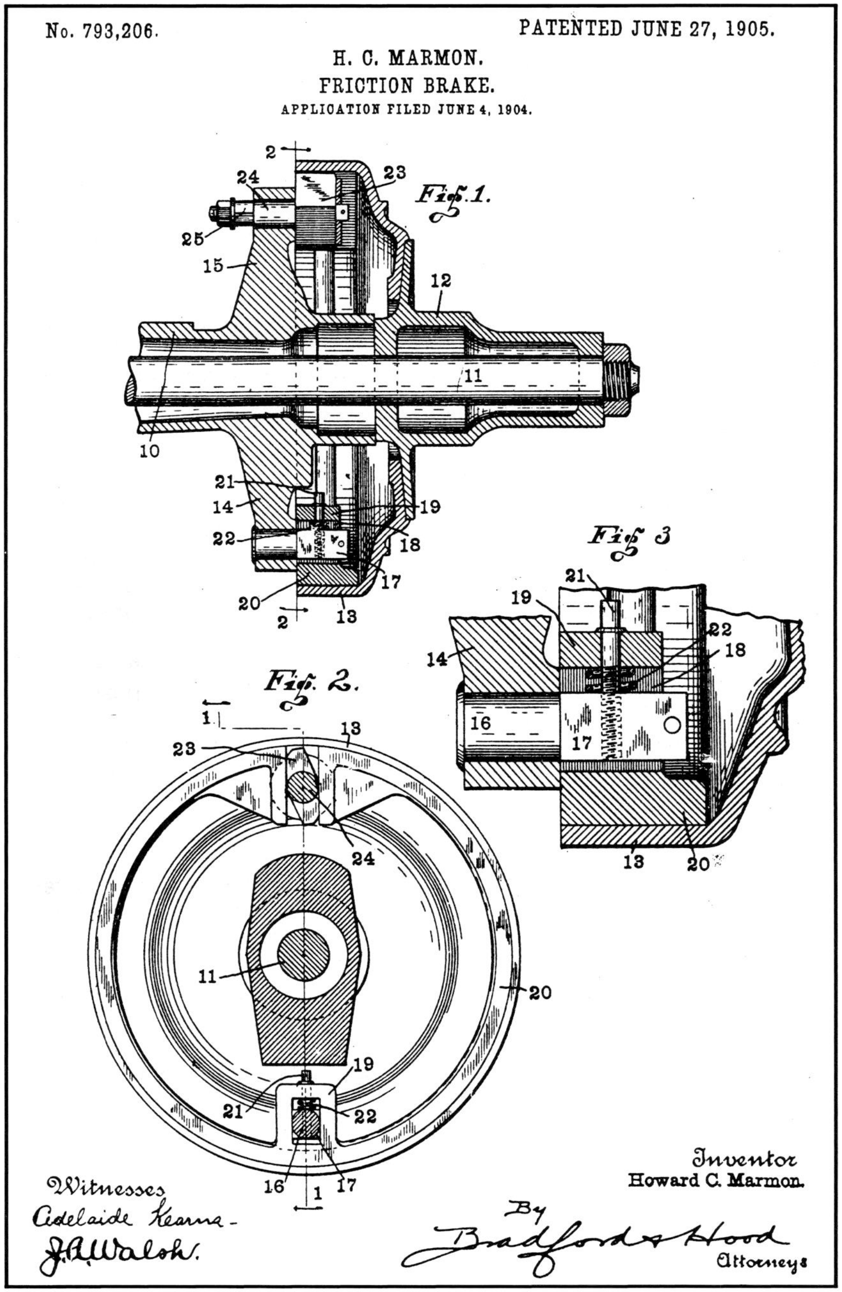


Fig. 2-57 The H.C. Marmon patent on an internal shoe brake.

Fig. 2-58 The sub-assembly of the Model H engine, transmission and independent frame which mounted the front suspension.

*springs to enhance clutch engagement.* The reporter for Cycle and Trade Journal commented that the combination of Marmon frame and springs "made even the Indiana roads seem fit for travel." The front lower springs were perched under, rather than over the front axle, ie, were "underslung". Shock absorbers were used at all four positions.

Actually, the "suspension" of the Era One Marmons was a combination of the dual-frame system for non-planar road surfaces and the springs for the road irregularities.

## WHEELS AND TIRES

Tires for the Model H were 34 x 4 front and 34 x 4½ rear. It is interesting to note that different front (26 inch) and rear (25 inch) wheel sizes were required to achieve identical outside diamter. The Model 50, being a 7-passenger car, was fitted with 36 x 4½ and 36 x 5 front and rear tires respectively. All were mounted on quick demountable rims. Wheels were twelve-spoke with cast steel hubs and brass hub caps.

## STEERING SYSTEM

"The front axle of these models was an I-section forging with integral spring seats on the lower side. Each spring was retained by four U-clips. A notable feature of these axles was that the width of the lower flange was greater than the upper, proceeding from the steering pivot to the spring seat. This strengthened the axle some hundred percent at a point where it was most liable to bend if an obstacle were struck by the wheel. Between the spring seats the beam was of uniform section. The steering knuckles were of the Lemoine type with the weight of the car being carried upon Hess-Bright ball thrust bearings."[19]

The same screw and nut Marmon steering gear introduced for 1907 models was used, but a change was made in the manner by which the selected position of the spark and throttle levers was maintained. The press described the design as inge-

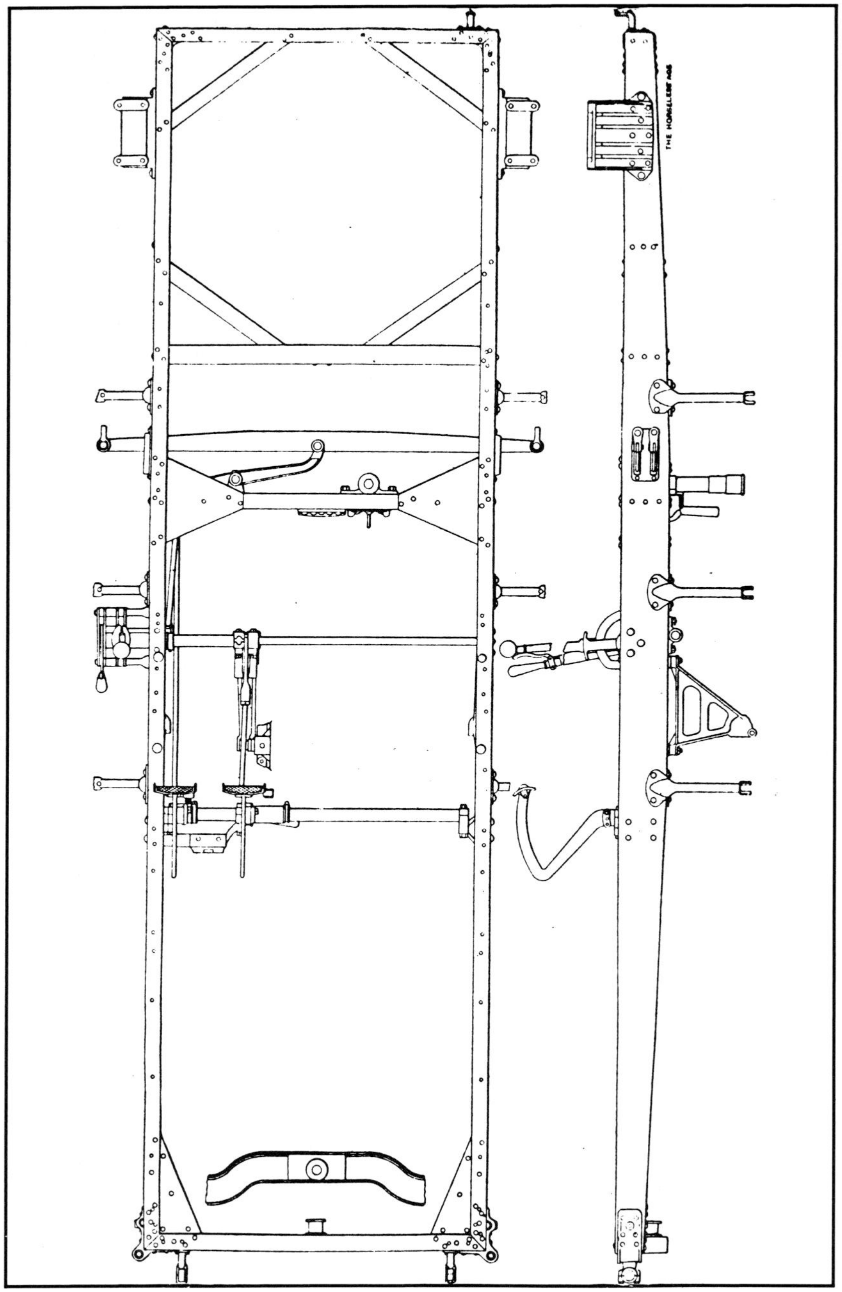


Fig. 2-59 Model H main frame showing trunnions for engine/transmission sub-frame.

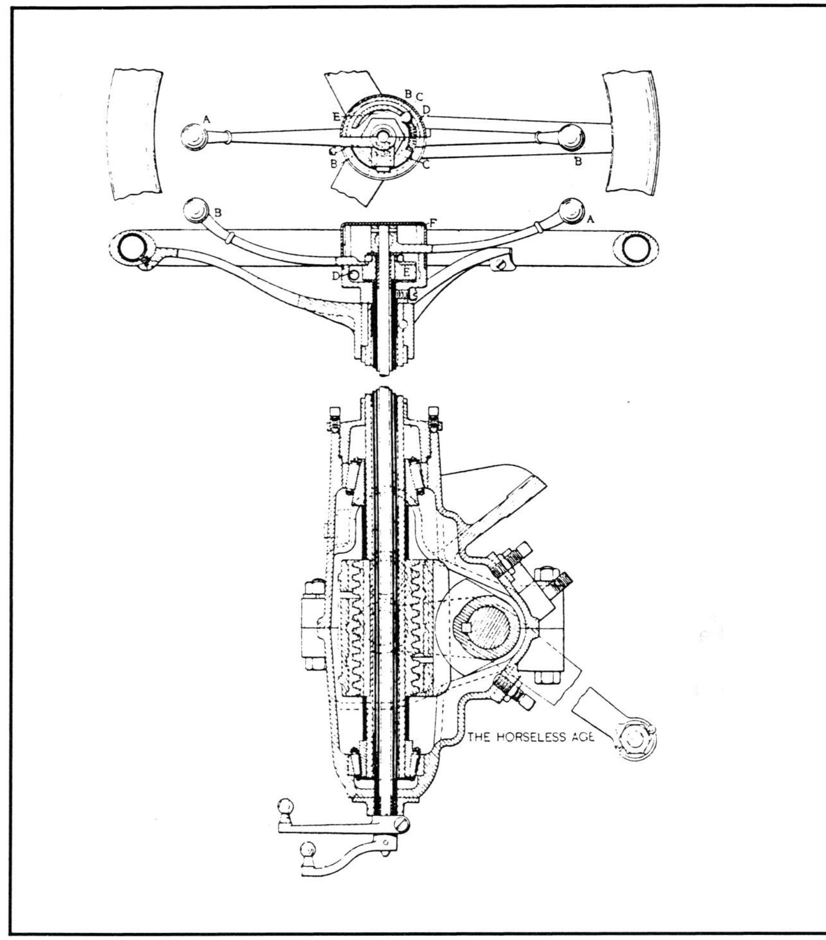


Fig. 2-60 Steering Gear — Model H

nious; the problem was a challenging one which evidently frustrated Marmon engineers for some years.

The sophistication of the 1908 "solution" warrants its description here. That published in *Horseless Age* (Dec. 5, 1907, P 8 94-5) reads like it was copied from the *Patent Gazette* of the US Patent Office. The author elects to repeat it verbatim. *The control "shafts move inside a stationary tube on the upper end of which a housing is fastened. Friction action against this housing locks the levers from being turned from the bottom. Lever A is fast on its shaft, and is held from moving from the position in which it is set by a cork plug pressed against housing F by a spring. On lever B a positive locking device is needed. Fast on its tubular shaft is cam E. Lever B has a large collar on the underside of its hub, shown partly in section in the plan view. In this collar are two notches, in one of which are balls C and $C^1$ and spring $D^1$, and in the other the end of cam E, which is narrower than the rest. This* *tail serves as a key to connect lever B loosely to its shaft. As shown, B is being turned in the direction indicated by the arrow. Ball $C^1$ is prevented from wedging between cam E and housing F by the collar on the hub of B. If the shaft is placed in any position, it cannot be moved from the lower end, for cam E will pinch (balls) C or $C^1$ between itself and the housing almost as soon as it begins to turn. The clearance at G is just enough greater than that between the cam and balls to permit this action".*

## CLUTCH

The Model H clutch, also used in the Models 45 and 50, was a wet multiple-disc type. The clutch casing which bolted to the flywheel had four keys which drove ten bronze rings. These bronze driving discs were studded closely with cork inserts finished on both ends and drove, by friction, nine steel plates splined (the *Cycle and Automotive Trade Journal* reporter used the term "feathered"!) on the driven hub. The mechanism operated in oil and was described by the CATJ reporter as "being exceptionally smooth in engagement and free from slip when fully engaged." He continued: "A series of coil springs engage the discs and small springs aid in separating the discs when the clutch is disengaged. The clutch is released when the left pedal is pressing directly on a plunger in the forward end of the transmission working in connection with a sliding sleeve; the clutch engages as the pressure on the pedal is relieved."[20]

Note that the action of the pedal is described in the technically correct terms of force change rather than pedal movement (a secondary consequence); this is evidence of the sophistication of the engineering staff who undoubtedly edited the release.

The cork faced driven disc was a characteristic of the Hudson automobile up until the merger with Nash. That was a wet single plate design. It was very smooth but suffered from requiring a high spring load due to the wet friction coefficient. Acceptable pedal force was achieved by a very long pedal travel, but the action was very smooth. A two plate design would have made it a superior design but economics, both first cost and rebuild, precluded it.

The author has been a clutch designer in his career, but has never driven a car with a dry single or double plate clutch that even approached the "feel" of a dry multiple disc clutch. Dry multiple disc clutches lacked heat absorbtion capacity and failed under extreme traffic conditions characterised by multiple starts and *idling in gear.* The wet multiple-disc cork-faced clutch must have been a real joy to operate and would have been durable due to the oil heat sink.

The release linkage was that familiar today including a ball thrust bearing, which did not become universal until the 1950s. It also included a small bronze brake drum which was automatically engaged by a bronze shoe when the clutch pedal was fully depressed. The function was described as "to overcome clutch spinning while gear changes are being made and therefore protects the gears and makes for noiseless changes".[20]

## TRANSMISSION

The transmission for the model H, 45, 50 was the same as that introduced the prior year but the housing did not incorporate a flywheel enclosure. The use of a multi-disc clutch was not compatible with the configuration of the flywheel enclosure which was predicated on the single plate clutch used on the Models F and G. The interlock preventing gear shifting without prior clutch release was continued. A ball ended gear shift lever was introduced.

"FIFTY" AXLE DETAILS

Fig. 2-61 The original banjo housing concept — Model H Rear axle.

## AXLE

The Model H rear-axle was a new design concept. It was subsequently used on the derivative 1909 Models 45 and 50. On all prior Marmons, the rear axle was integrated into the dual-frame suspension system; on the new version of that system it was not, and drive was through a shaft with universals at both ends. Torque reaction was taken through a torque arm.

The new axle was of banjo configuration but continued the Marmon practice of tubular extensions from a cast housing. The entire design is remarkably ageless; it differs from modern full floating medium truck axles primarily in the installation of the gear carrier through the *rear* opening. This detail evidently anticipated the trans-axle design of the Model 32.

## BRAKING SYSTEM

The Model H and its derivative models had redudant braking systems . . . two systems identical except for pedal and lever actuation and the common deep drums upon which they acted.

The Models F and G had used cast brake drums with double "flanges" to accommodate concentric internal expanding brakes . . . the larger for service braking and the smaller for emergency (parking) use. They employed bronze friction shoes on iron drums.

The Model H introduced side-by-side internal expanding shoes, faced with a non-metallic compound and acting on 14 (114 inch wheelbase) or 17 inch (118

Fig. 2-62 The 1909 brakes: Models 45/50 left (cover off), Model 32 (wheel and drum removed)

wb) diameter pressed steel drums wide enough to accommodate the pair of 2 inch wide shoes. The brake operating levers, at the inboard ends of the concentric shafts to which the 'S' cams were brazed, were maintained in proper leverage position by what the CATJ reporter called an "elaborate method". Adjusting worms rotated segment wheels adjacent each of the four levers. He conceded however that the adjustment design "was representative of the refined design employed throughout. The entire (chassis) mechanism is replete with ingenious design and refinement of detail."

The brake was effectively a two rigid shoe design, but actually employed a single cast shoe which flexed at a machined slot where it engaged a square reaction pin. The shoes were "reinforced at the slot with special spring steel strips."[20] These were installed in the same manner as the lining and were riveted to the shoe in the gap between the two lining segments. This is visible in the plan view of the 1912 brake. This design avoided the uneven braking encountered later with the Lockheed hydraulic two-shoe brakes (as well as innumerable mechanical brake designs) consequent to the problem of lubricating the shoe pivot pins. It had the same frictionless-pivot action as the Bendix two-shoe brake but was not self energizing because the reaction was taken at the pivot location.

Significantly the brake shoe carried rollers on the ends where engagement with the brake cam took place . . . the very construction employed today in heavy duty truck brakes.

It should be noted that in 1908 brake lining was called "brake belting"; The Model H featured brake shoes lined with camel's hair fabric.[23]

The brake linkage included a pair of large forged levers, one above the other, acting in horizontal planes, which connected with the center of stamped steel channel equalizers. These equalizer beams extended through openings in the frame side rails and connected outboard with the rods running aft to the brake levers. The principal friction points were those on the horizontal lever and therefore not influential in brake balance; however the beams rested on the bottoms of the slots in the frame rail and friction there could offset all the advantages of the linkage system.

CHAPTER THREE

# THE MARMONS OF ERA TWO

## 1909-1916

The Marmons of Era Two are primarily the Models 32, 48 and 41, in order of introduction. Models 45 and 50 were the second stage of a transition model and were gone from the production line by the end of the first year of Era Two.

In this work, Era Two is largely the story of the Models 32, 48 and 41 . . . the Models that made the name "Marmon" universally known and respected in the automotive world.

Automotive styling was developing rapidly. The tulip body architecture of Era One was transitional from the horse drawn period to the motor-propelled period. The Marmons of Era Two evidence the progressive development of the 'bathtub' period of styling. The Model 32, because the basic chassis was produced from 1909 through 1914, is an excellent 'vehicle' for study of the styling progression through those years. Styling through-out Era Two was primarily component styling as opposed to the total-vehicle styling approach that began with the next Marmon Era.

Again for this Era, a general discussion of the Era Two Marmons will be followed by engine and chassis sections which provide detail information that may not be of particular interest to many readers but which is important, not only for the sake of collective documentation and preservation, but for its value to the student of automotive history and to the contemporary automotive engineer.

The first five years of Era Two were characterized by spectacularly consistent Marmon victories wherever the Checkered Flag was waved. THE EVENT of Era Two was the culmination of the victory march with the victory at Indianapolis in the first 500 Mile Memorial Day event in 1911.

Another characteristic of Era Two was Nordyke & Marmon's repeated effort to offer a single car line that would satisfy a wide range of customers. The use of a six cylinder version of the Model 32 in the Indy 500 whetted the public appetite for a "Marmon Six" performance automobile and the Model 48 was marketed in response. But by the last year of the Era, a single car line again was offered . . . a concept that would prevail through Era Three.

Because the Marmon racing story encompasses more than Era Two, and because of its stature, the complete racing history is addressed separately . . . Chapter Eight "Marmon — International Champion".

Era Two began with an advance press release in 'Automobile' for September 24, 1908, which revealed that "Nordyke & Marmon Co. were preparing a light-touring car of approximately 35 horsepower to compete in the 1909 market."

The Marmons of Era Two differed radically from all those of the First Era. Discounting the short lived transition models, the Era Two cars dispensed with the unique patented double-frame construction which provided a deflection-free body platform, justified the slogan "The Easiest Riding Car in the World" and claims of exceptional tire life; claims supported by owner surveys and non-stop victories in over-the road competition.

But the roads of 1908 were perhaps better than those of 1902 and almost certainly there had been some progress in tire design. Anyway, Marmon sought the market pulse in 1909 and subsequent years with new chasses which appeared to be of conventional semi-elliptic front suspension and either full or three-quarter elliptic rear suspension.

Experience would confirm that the new chasses were as well suited to the roads of 1908-1916 as their predecessors had been to the turn-of-the-century roads for, soon after introduction, the first of the Era Two Marmons began to amass an enviable record in competition. For an example of continued reputation for superior tire performance: A Marmon 32 won permanent possession of the *Minneapolis Journal* Trophy by winning the fourth annual tour of the Minnesota State Automobile Association on August 16, 1912 with a perfect score (zero demerits) for each of the seven days required to travel to Winnipeg, Manitoba from Minneapolis and return. *The Marmon had no tire trouble whereas the second place finisher had two blowouts and nine punctures, all on rear tires; the fifth place finisher had three blowouts and eight punctures and all other competitors had some tire problems.*

Evidently there was more sophisticated engineering in the suspension of the Era Two Marmons than was apparent by inspection. Marmon was continuing to apply "System Analysis" to the automobile, ie, considering the entire system from operator to road and vice versa. In 1902 that analysis dictated invention of the double-frame chassis. In 1911 it dictated the target speed at which Marmon would run the bricks to victory at Indianapolis.

Ray Harroun took the leading Marmon into the pits at 450 miles of 75 mph average speed over the brick track surface . . . it was the fourth change of the left rear tire but he roared back onto the track on the other three Goodrich Silvertown tires with which he had started the race and never stopped again until he was in Victory Lane.

*And* Marmon continued to advertise "The Easiest Riding Car in the World" as late in Era Two as 1914. And customer correspondence confirmed that claim.

### PART I — The Cars — Models 32, 48, 41

The models 45 and 50 were descendants of the 1908 Model H which was a transition vehicle. They were powered by an engine which was representative of Era Two and which was, in the more obvious respects, totally different from the engines of Era One. On the other hand, they retained the dual-frame concept of the Era One Marmons although the particular configuration was transitional . . . functionally it retained the dual frame characteristics of an excellent ride and a deflection-free bedplate for body mounting, but appearance-wise the dual construction was not evident and the car conformed to more familiar norms.

The management of the transition in design philosophy was remarkable. It ranks with the stellar performances of General Motors Styling (now Design) Staff in transitioning from a current to a totally new car without seriously impacting the trade-in value or prestige of the former. In other words, the annual or bi-annual buyer was protected and his important loyalty preserved. This policy should not be overlooked in assessing the large share of the domestic market which GM has attained.

The author has therefore elected to discuss the Models 45 and 50 in general and their chassis details in particular with the Model H in Parts 1 and 4 of Chapter Two (Era One). The engine (s) of the Model H will be discussed in detail with the Model 45 and 50 engines in Part 2 of (this) Chapter Three.

In the Cycle and Trade Journal's coverage of the 1909 Marmon line, the Models 45 and 50 were given equal treatment in text and illustrations with the Model 32 and the latter was addressed last.[1] But initial Marmon advertisements appearing in January 1909 illustrated only the Model 32 but promoted also a Model 50 7 passenger touring. A "footnote" related the availability of a special Model 45 roadster. The '32' was priced at $2400., the '50' at $3750. Another source identified the '45' price at $3450.

February 1909 advertisements made no mention of the '45' and only a "footnote" made reference to a "Suburb Fifty" available in a 7 passenger tooling.

The 1909 Marmon brochure acquired by MVMA at the New York Auto Show stated: *"The Marmon presents for 1909 two chasses, the Marmon "Fifty" rated conservatively at 50-60 horsepower and the Marmon "Thirty-two" giving 32-40 horsepower on a like conservative rating."*

The Model '32' would not accommodate 7 passengers and represented a major change in design philosophy. Perhaps Marmon were concerned about acceptance of the '32' by traditional Marmon owners . . . a circumstance surprisingly like that in which GM found itself with the 1982 mid-size products (when 1981 A bodies became 1982 G and 1982 A bodies were radical new offerings).

## MODEL 32
## 1909-1914

First hint of a new Marmon model was a "sneak preview" of a prototype engine to power "a new light 35 horsepower car for 1909" which appeared in "Automobile" for September 24, 1908. The new engine appeared to be a new design and not a derivative of the Model H introduced the prior year.

The Model 32 was announced in December of 1908.[2] The press release heralding its advent stated: *"The new '32' is the result of an effort to produce a car of highest grade and most (publicly) approved design which should be moderate in weight, power and in initial and upkeep cost".*

The Marmon 32 and 50 were first displayed at the New York Show of the "National Carriage Dealers Protective Association."

A significant advantage in abandoning the dual-frame concept was the reduced vehicle height and the corresponding lower center of gravity. That Marmon was sensitive to this consideration is evident from the fact that the Model '32' was noticeably lower than its contemporaries, a fact particularly evident with the top 'up'.

A Marmon "light car" meant just that. A very serious engineering program was directed at achievement of a minimum-weight packaging of four to five adults in as simple a vehicle as could provide top speeds in excess of 60 mph and sustain the Marmon reputations for the "Easiest Riding Car in the World" and for remarkable reliability of both car and tires.

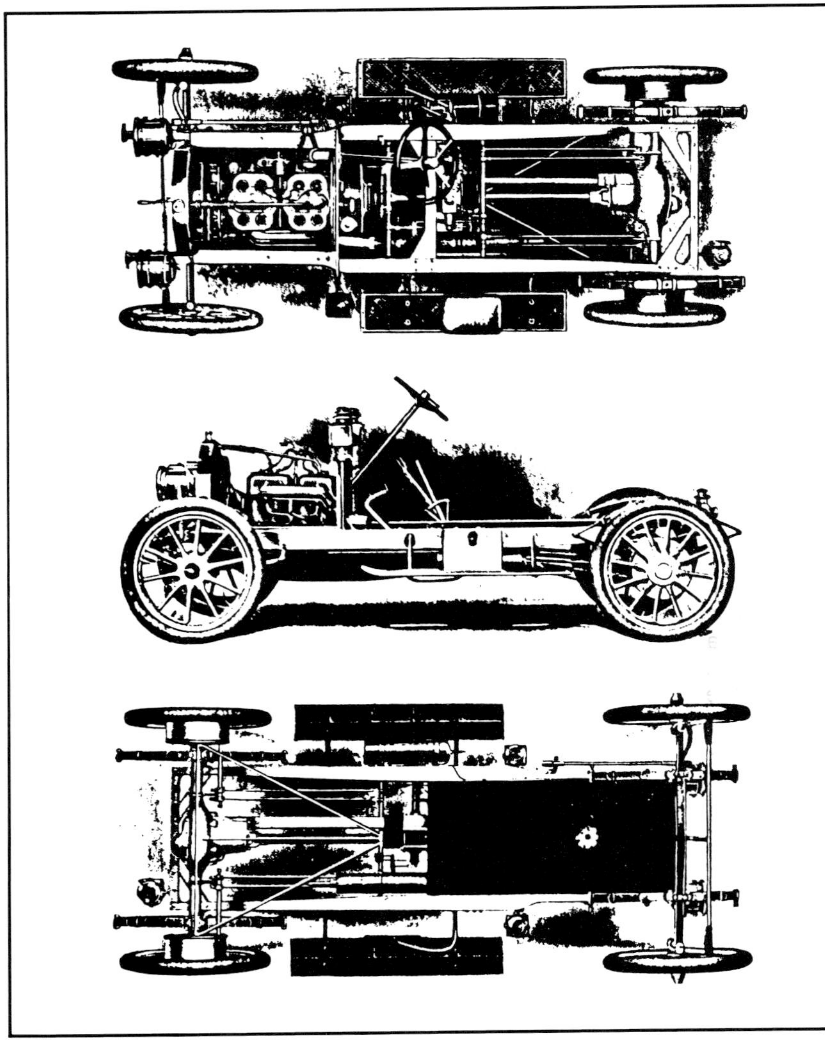

Fig. 3-1 The light-weight simple Model 32 Chassis — 1911 version.

Advertisements described the '32' philosophy: *"The Marmon 'Thirty-two' is the rational, logical car. Moderate in size and capacity; light in weight with an abundance of power; exceptionally well proportioned and well balanced; made to answer every requirement of city use and touring with the greatest comfort and economy. The easy riding of the Marmon; its quiet, smooth-running, enjoyable operation; its wonderful ease on tires and economy of fuel, oil and upkeep; these are the results of the superiority of Marmon design, Marmon materials and Marmon workmanship."*[3] *"The world produces very few cars, so well built of such fine materials, that they can stand speed around 70 and 80 miles per hour for more than a few miles. The Marmon ('32') stands it for 100 miles, 200 miles or more whenever called upon, as has been demonstrated repeatedly: 60 mile per hour non-stop runs for 225 miles at Indianapolis, 100 miles on a circular track at New Orleans and 190 miles in the Vanderbilt at Wheatley Hills. The reason for this amazing stability, this remarkably consistent reliability lies in the quality of the material and in the minute care with which every essential part of the car is manufactured — not merely assembled . . . by one of the world's oldest and most successful manufacturing institutions of America."*[4]

In announcing the 1910 edition of the 'Thirty-two', Marmon stressed a hidden feature which could have contributed to the remarkable success of this car on the competitive circuit: *"Perfect fitting interchangeable parts assured by carefully prepared tools and jigs."*

The Model 32 as introduced for 1909 would certainly fit today's classification of a "Compact" car. Interestingly, it underwent some of the growth metamorphoses that characterized the compacts introduced by the Big Three[5] in the 1960s for the domestic market. It was unique, however, in the commitment beginning with the 1910 model year to a single product line, a line which would qualify as a downsizing except that the Marmons of Era One were all conservative in size.

Increases in wheelbase the second and third model years may have been influenced by problems in mounting custom bodies; the short wheelbase of the '32' was unusual in a luxury automobile. More likely, the short wheelbase had made the 7 passenger closed bodies somewhat cramped. Had Marmon not anticipated the requirement to offer closed bodies? Or, had Marmon planned only to offer 5 passenger closed bodies? Regardless, all of the increases were in the passenger compartment and the riding qualities could only have been enhanced by the longer wheelbase. The tread increased from 56 inches to the standard wagon tread of 56.5 as tire sizes increased.

The commitment to a single product line was amplified when Marmon announced in 1912 that annual model changes were to be abandoned. The combination resulted in the 1913 catalog number 1331 being entitled "The Marmon" without reference to model or model year. In fact various publications, which traditionally carried the model year as a prefix to the two-digit code number for the character of the publication, were continued through at least 1915.

During the lifespan of the Model 32, Marmon followed a consistent practice of Announcement Advertisements which identified the product changes in a more or less uniform format. These will be quoted herein, each accompanied by additional discussion to follow the maturing and refining of this milestone automobile. Direct quotation of the press and of the manufacturer . . . speaking through advertising and interview . . . preserves the flavor of the period and greatly assists in attaining a proper perspective from which to view the various progressive developments. It is particularly important that styling development be seen from a forward-looking perspective rather than from a modern or hindsight one.

Whereas the Models G, H, 45, and 50 had button type accelerator pedals, the Model 32 introduced a knurled roller set in the floor.

The '32' was powered by a water-cooled engine with four cylinders of 4.5 inch bore and 4.5 inch stroke, displacing 286 cubic inches and rated 31.8 SAE horsepower. Valve arrangement was T-head. The unique Marmon full pressure lubrication system was incorporated as was the barrel crankcase configuration with the crankshaft inserted from the rear.

The drive train was torque tube with transaxle (but the term had not yet been coined). A cone clutch was fitted. The transmission was selective shift with three forward and one reverse ratios.

Fig. 3-2 The 1910 engine showing the novel intake manifolding.

The fuel tank was under the front seat, except on the roadster. It was fitted with a float type gauge, and held 17 gallons. The engine sump held 6 quarts of oil.

Over the production period of the '32' the styling advanced notably. The hood and radiator was characteristic of the entire series. The illustration shows the driver's view of the final evolvement.

The Model 32 Marmons had a full-floating rear axle construction with flanged axle shafts bolted to the wheel exterior. Special shallow hubs were designed for the rear only. Full page advertisements directed attention to this unique appearance

Fig. 3-3 Model 32 transaxle

Fig. 3-4 Driver's view of the 1913 Model '32'

Fig. 3-5 Driver's station of the 1910 Model 32 owned by R.C. Blanding.

Fig. 3-6 Progressive development of the Model 32 touring style: 1910 (above); 1911 (upper right); 1913 (lower right).

and the underlying technology: *"Note the flat hub of the Marmon. It is the mark of the strongest, simplest and most accessible rear axle system on any automobile. The Marmon rear system, including the transmission, is one of the superior details of Marmon design and construction. The facts are readily understood if you will but investigate these features."*

The integral flange, full-floating axle shaft concept later achieved universal application in truck axles, using the same bolted attachment to the wheel hub as did

Marmon. The improved version, using crowned gear teeth on the flange periphery, now used on General Motors truck axles had appeared on the Cadillac car . . . the Model 314 of 1926.

As the '32' grew in wheelbase over its first three model years, so did the cowl undergo a metamorphosis. Somewhat surprisingly the 1909 introductory model carried a flat slab cowl, presumably of wood . . . excepting the 1904 Model A (7 units), the first Marmon to not have a cast aluminum cowl of some depth.

For 1911, the bodies were described as "of perfected cast aluminum construction with sheet metal seat backs."

Other than the cowl, the 1912 body was "sheet aluminum assembled over cast aluminum framing".

The floor boards and running boards were of cast aluminum construction. This permitted the forward end of the running boards to incorporate the blend radius and the fender was therefore straight to its end.

When introduced, the '32' was offered in three open body styles; five-passenger touring, four-passenger 'Suburban' and a two-passenger roadster priced at $2400.[5] Also available on order were a limousine at $3400 and a landaulet at $3500 and a coupe at $3000.

The Roadster was a particularly attractive sports car. One reporter called it "Just about the 'classiest' proposition that has appeared in motordom this year. And its performance is just as impressive." A single seat was contoured into a pair of buckets and sat upon a matching riser. Behind the seat ws a shallow deck which supported a vertical cylindrical fuel tank about 24 inches in diameter with a convex upper surface. Three stacked spare tires (on rims) surrounded the fuel tank and enclosed nearly all of the cylindrical portion of the tank. The "running board" was

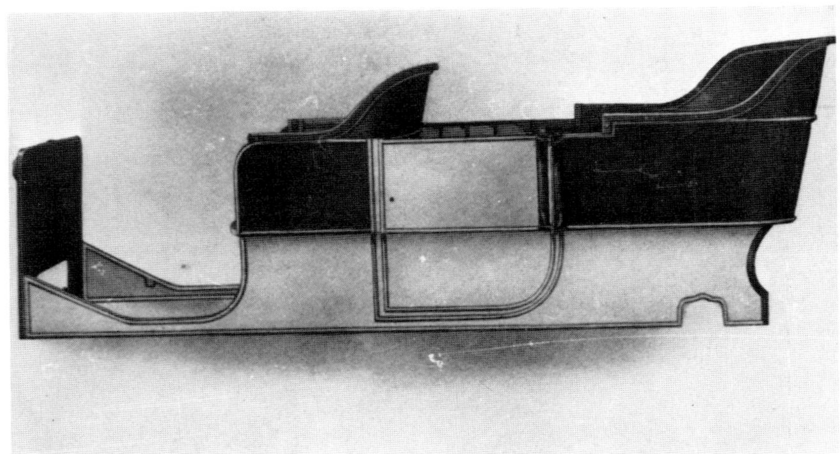

Fig. 3-7 The 1910 '32' featured a composite body structure.

only about 8 inches long with both front and rear fenders sloping uniformly to the "board" step. The fenders had peculiar splash skirts which continued down to the running board height.

The "Suburban" was a four-passenger version of the touring. The claimed "impression of speed coupled with stability" was based essentially on the narrowed, shorter tonneau and lower seats. The body was designed so that the tonneau could be removed when the extra passenger capacity was not required.

*Standard equipment* at introduction was: two seven-inch gas lamps, gas generator or Prest-O-Lite 'E' gas tank (optional); two dash mounted square *oil* lamps and one tail light; horn with flexible tube and bulb; complete set of tools, two oilers, jack, tire repair kit and pump. The following year the Prest-O-Lite tank was a 'B' type and the oiler quantity was down to one.

On November 19, 1909, little more than a year after the introduction of the Mod-

Fig. 3-8 The Model 32 roadsters were pioneer muscle cars, featuring pistons fitted with 0.002 additional clearance. Top to Bottom: 1909, 1910 and 1912 roadsters, and the newly designated speedster for 1913.

el 32, the Nordyke & Marmon Company wrote to purchasers of the car requesting letters of comment on its performance. Eighty-six of these were reprinted in a booklet entitled:

EXPERIENCE and OPINION
Being in type the expression of customers
Concerning THE MARMON

All were on professional or business letter heads and some twenty of the letters provided total mileage data and eighteen allow calculation of miles per tire puncture. Nine provided the time span of ownership as well as mileage from which monthly mileage accumulation can be calculated. Four reported the fuel economy experienced.

These owners reported an average mileage acquisition of 5400 miles during which they experienced flat tires (from punctures only) at 4400 mile intervals. Monthly mileage accumulation was at rates still representative in the 1970s. . . . averaging 1150 per month! (The range was 750 to 1500)

Fuel economy varied from 12 to 17 miles per gallon. One owner noted that his fuel price was 16.5 cents per gallon. Lubricating oil consumption varied from 75 to 87 miles per quart but the sample size for that parameter was small and the lube additions were total quantity added to both engine and transmission.

None reported having tested the top speed of their cars.

One owner reported that his "Diamond" brand tires were in "good" condition at 8500 miles. The owners frequently referred to tires as "shoes" and to the inflation air as "wind". Another had driven 4700 miles and had yet to experience a tire or mechanical problem. Another reported replacing two of the original set of "Michelins" between 5200 and 6500 miles.

A St. Louis, Missouri owner reported that his car *had been recalled* by the manufacturer and the dealer had substituted certain updated parts!

Another owner from the same city reported a puncture-free trip from St. Louis

Fig. 3-9-1 1910 Model 32 Suburban for 4 passengers.

Fig. 3-9-2 1910 Touring for 5 passengers.

to Chicago, a 362 mile run, of which 210 was covered the first day, averaging 14.2 miles per gallon.

Many reported passenger surprise at the quietness of the car, one noting that conversation with rear seat passengers could be carried on in a normal voice and another referring to the "absolute quiet" of the engine.

The letters ranged geographically from Auburn and Bangor, Maine to Los Angeles, Portland, Oregon, and Yakima, Washington. One was from Fort Worth, Texas, with the largest concentrations in Indiana, Missouri and New England in that order.

## 1910

In September 1909 the following ad appeared:

**Announcing the "THIRTY-TWO" for 1910.**

The Marmon "Thirty-two" for 1909 won its prestige as one of the best designed and best made cars in the world. Automobile engineers and manufacturers generally — our friendly rivals — recognized its thorough excellence and gave it generous praise. Nearly perfect as the car was, we have succeeded in improving it in some particulars — a little longer wheelbase, increased length of engine stroke, improved front axle, improved body construction and a slight change in radiator design. The general design and construction remain the same.

After only one year of production, the wheelbase was extended to 116 (from 112) inches. There must have been compelling reasons for this action because the consequence was a whole new line of bodies. The new open bodies were of more traditional Marmon construction practice: "Sheetmetal construction" (presumably on wood framing) was replaced by "All metal . . . a cast aluminum base with cast aluminum doors and door frames" and sheet steel seat backs. The longer wheelbase and the change in body construction brought the weight of the touring car to 2300 pounds.

The new Suburban body did not feature the removable tonneau concept of the 1909 model. It continued to feature significantly lower seating and body side height. The top was designed for folding and was adorned with an external landau bar system; the touring top appears to have been intended for permanent use. The inclusion of the optional tops in press release photographs was a new approach indicative of changing consumer preferences.

A striking advance in the 1910 version was the redesign of the fenders. The visor lip and associated rounded forward end (in plan view) of the front fenders was eliminated. The rear fenders, which had trailed off horizontally from about the 135 degree point, were continued around at a radius from the hub. Initial pictures of the 1909 car showed the horizontal trail off from the 90° position and the limousine displayed at the New York show had them.

Two side oil lamps were added equipment. They were rectangular in shape and mounted on the "cowl" board or engine bulkhead.

The suburban illustrated was fitted with a Globe cape top, distinguished by the absence of a forward bow anchorage on the front seat and the presence of a mechanical linkage allowing it to "fold like an umbrella" as one reporter described it. Globe tops were manufactured under license from a German patent. Apparently a low-slung version was not available, as a Suburban top should be lower than the touring top.

## 1911

It was August 1910 when the following appeared in the trade press:

**1911 ANNOUNCEMENT 1911**

The history of the Marmon has been a story of progressive development and improvement. The 1911 models will, we believe, prove a revelation to our friends, notwithstanding they have come to expect in the Marmon the very best that can be produced in high-class motor cars. The wheel-base has been lengthened to 120 inches, adding not only to the beauty of the car but to its roominess and easy-riding qualities. A complete line of closed-front bodies is offered while the open-front body is furnished to those who prefer that type.

The new bodies were slab-sided. The five-passenger touring body carried the width of the rear seat forward to the front passenger area. This placed the door outside of the shift and brake levers. Also the horn bulb and the tube were inside with the tube emerging from below the door and running forward to the horn which was mounted above the frame and just forward of the cowl. Because the Suburban body was narrower, the front compartment was also narrower and the shift and brake levers were outboard of the door. The horn bulb however was inside. The Suburban retained the former unique front fenders . . . much longer and with shorter running boards, but extending forward horizontally and visored.

The Suburban body was appreciably lower than the touring, and the seats were correspondingly lower. The touring body doors were the same height as the flat topped cowl panel and the hood center hinge; the Suburban cowl 'board' had rounded corners, a rounded cast aluminum cover, and was twice the length of the open-topped touring cowl.

The rear seats on both touring and Suburban were in the same position. Due to the lower cushion height, the front seat was considerably more rearward on the Suburban (hence the long cowl) and the steering column was longer and more severely raked. The sporting appearance was enhanced by additional cowl length achieved by shorter front doors. For 1911, the Suburban appearance matched the earlier claim of the "impression of speed coupled with safety."

The new roadster continued to share the elongated front fenders of the Suburban and had similar (reversed) rear fenders, but the latter rolled downward behind the wheel. It also shared the covered cowl and front doors of the Suburban. The seat riser extended to the rear as a deck and mounted a transverse well to accommodate a spare tire and rim attached to the fuel tank carrier. Last was a rectangular trunk fitted with a dust cover. The well alternately accommodated the third passenger's feet when the single rumble seat option was specified.

The bodies for the limousine and town car were unchanged except for the addition of optional half-height doors on the drivers compartment.

A design refinement which attracted much public and media attention at the auto shows was the relocation of the horn (under the hood) and of the operating bulb (on the column).

The horn was frame mounted under the hood directly behind the radiator core. A rigid tube extended from the bulb base down the steering column to near the horn where a short flexible tubing connection was made. The bulb was fitted to the end of a tube extending at 90° degrees from an elbow-fitting on the steering tube just below the wheel rim in the location of the column-mounted gear shift lever on today's cars.

This horn change was a typical product improvement early in the 1911 model year which showed up at the 1912 shows.

The body styles offered for 1911 were: 4-passenger Suburban and 5-passenger touring — $2750; roadster — $2750; limousine — $4000 and landaulet — $4100.

*Additional equipment* was standard for 1911; In addition to the two 8 inch gas lamps and Prest-O-Lite tank, dash and tail lamps were fitted. Equipment included the bulb horn, a coat rail and foot rest in the rear compartment; jack and tire pump

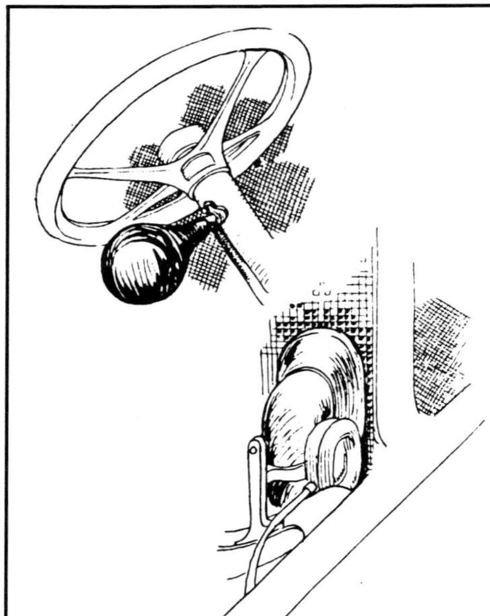

Marmon Horn and Bulb Locations

and complete set of tools.

All models had blue-black bodies with white striping and olive running gear with blue-black striping. Exposed metal work was polished brass.

For the first time, the illustrations of the touring included a top but in the folded position and covered with a boot.

## 1912

*Cycle and Automobile Trade Journal* edition of August 1911 carried the following Nordyke & Marmon advertisement:

**THE MARMON "THIRTY°TWO"**
ANNOUNCEMENT

The Marmon "THIRTY-TWO" made its appearance in the autumn of 1908 and was immediately recognized as being far in advance of contemporary designs. Since that time, this model has been continued with such alterations in body design as were in keeping with the styles and requirements of the times. Increases in size and power have been made with well developed improvements and refinements in mechanical details. Needless to say, we have taken every advantage of the valuable experience we have obtained in racing and other practical tests to make the car better.

The ensuing production of this model, which is still well in advance of competitive designs, does not differ materially from the Marmon "THRITY-TWO" marketed during the past twelve months.

The clutch and brake pedal assembly has been simplified and the steering column has ratchets for the spark and throttle controls instead of fric-

Fig. 3-10 1910 Model 32 chassis

tion controls as formerly used. The rear system is unchanged except that a differential of our own manufacture is now used. The size of the brakes is increased from 14 to 16 inches. The 5-passenger touring body has a cast aluminum hood over the dash similar to that used in the 4-passenger body.

A two-passenger Speedway Style Speedster was offered in addition to the two-passenger roadster with rumble seat. It shared with the roadster and Suburban the elongated front fender. The rear fender interfaced with a storage box on the aft running board. A side skirt full length of the fender tapered to end at the board. The seat was frame-mounted and followed by transverse round fuel and oil tanks and a pair of tires and rims. The low seat was quite far aft requiring a severe rake to the steering column. A tapered racing cowl shrouded the drivers feet and control pedals.

A new standard equipment item on the touring car was a spare rim carrier on the right running board adjacent the driver's door. The two arms of the carrier did not interfere with the door but installation of the non-standard spare tire and rim rendered the door useless. Such use was anyway compromised by the shift and brake levers.

The forward section of the limousine and landaulet bodies was made similar to the 1911-12 open bodies, with the wide cast cowl and full height doors. An optional filler panel with window at the sides of the seat offered the driver additional weather protection.

The 1912 prices were unchanged from 1911.

The Marmons at the Garden and all subsequent production sported a new radiator ornament. In commemoration of the Indianapolis victory the previous Memorial Day, the radiator filler-cap bore a "Baby Wasp" . . . a miniature reproduction of the winning "Wasp". The model was cast in one piece, had a 4 inch wheelbase and showed the clean lines and low-seated driver of the original.

## 1913

"When your family equipage is a Marmon, neither your good taste nor your judgement of automotive value can be questioned."[6]

In introducing the MARMON SIX, the announcement advertisement in July 1912 was in a two page format, each page presenting and addressing a seperate model and being a complete self-standing ad on its own. The right hand page was entitled "Announcing the Marmon Six"; the left page was entitled:

**Announcing a New Series of The Marmon "Thirty-Two"**

and continued:

> *"The features which in the past have built up so great a name for the superiority of Marmon design, materials and construction are not materially changed in this new series of the Marmon "Thirty-Two". The motor remains unchanged except for the addition of an electric lighting dynamo. The clutch pedal has an adjustment and increased leverage of the clutch throwout. The brakes, which have already received the highest praise from experts, are even further improved by a combination facing of asbestos fabric lining and composition metal in alternate segments on the brake shoes. An improved simple brake adjustment has been added which gives another point of Marmon superiority in its extreme accessibility and effectiveness. This is a new and distinctive feature brought out with this new series and the Six."*

Suprisingly, the illustration showed a left-hand drive "Thirty-Two" but no mention of the change from right-hand drive was made. Evidently the change had occurred during the 1912 model run. Both the SIX and '32' sported the "Baby Wasp" radiator cap.

Also unmentioned in the ad text, but equally significant was the incorporation of a patented engine mount system which utilized resilient bushings at three points positioned on a longitudinally inclined axis. This development found on all 1913-1916 Era Two Marmons is fully descriped in Part 2 of this Chapter in the section addressing the Model 32 engine for 1913.

In addition to the several new mechanical adjustment conveniences, the major change was in standard equipment. The illustration of the touring showed the "Special Marmon ventilating windshield" and showed a top boot extended to cover most of the length of exposed bows.

The windshield was mounted on the cowl cover casting and wind force was resisted by wires extending from the radiator base region to the top of the lower windshield panel. In the illustration, the upper panel was folded downward inside the lower fixed panel. This was probably the open car position when the rear seat was unoccupied, and perhaps useful when the top was up in warm dry weather. It was described as a "Pantasote (material) Cape Top with matching boot". Rear seat cushions were extra deep to give added luxury and comfort.

The speedometer and clock were lighted by a spot beam from a light source on the seat riser.

*Other equipment* was: electric lighting system with storage battery; electric head lights and combination oil and electric dash and tail lamps, *self-starter,* shock absorbers on rear axle, spare demountable rim and single tire carrier, coat rail, foot rest, jack and pump, tire repair outfit and complete tool equipment. Lamp bulbs used the Ediswan candelabra base.

The 1913 open style prices reflected additional standard equipment:

| | |
|---|---|
| Roadster; single rumble or trunk optional with top and full equipment | $2900 |
| Touring and Suburban (4-pass) with top and full equipment | $3000 |
| Speedster *with* shock absorbers both front and rear, double tire carrier and extra rim but *without* dash lamps, top or windshield | $2850 |
| Limousine | $4000 |
| Landaulet | $4100 |

## 1913½

A December 1912 advertisement showed a restyled Model 32 only five months after the announcement of the New Series '32' (concurrently with introduction of the '48'). That ad stated: *"The graceful and beautiful lines of the new bodies distinguish the* ***latest series*** *Marmon '32' at a glance. The many refinements in equipment and appointments add to the comfort, luxury and ease of operation".*

Marmon had announced abandonment of yearly models earlier. Evidently when inventory of the prior body style was exhausted and/or when the additional body tooling required to adapt the '48' body to the '32' was available, the "latest series" was placed in production.

A companion ad for the Model 48 showed the new windshield common with the '32' but in addition a new tool or battery box integrated into the front fenders. The '32'however had new front fenders with a pleasing reverse curvature which brought the fender tangent to the running board, similar to the initial Model 48 fender.

The body styling was entirely new. A close resemblance to the Model 48 styling was achieved. The bodies appear identical back to the rear doors which required a cut-off corner to the fender on the '32'. The touneau appears peculiar to the '32' consequent to different passenger capacities, but is similar to the '48' in appearance. For the first time, the exterior door handles were omitted.

A single panel windshield was adopted (also for the '48'). It was a handsome design with rounded corners. It was supported by pillars which reached the midpoint where a pivot attachment allowed the glass to be rotated. The pillars themselves were pivotally attached to the cowl and the combined pivotal action permitted the windshield to be folded down on top of the cowl.

The upper sloping surface of the cowl was flat and this contributed to the pleasing appearance of the lowered windshield. The cast aluminum cowl construction allowed the windshield to be several inches closer to the driver than the front edge of the doors.

*Equipment* included starting and lighting system; tailor made top with dust boot and side curtains; built-in windshield, black and nickel finished; speedometer and clock combination; electric horn with button on steering wheel; electric light to illu-

Marmon Limousine

Fig. 3-11 Progressive development of the limousine style. 1911, 1913 (landaulet) and 1914.

minate entire dash; shock absorbers front and rear; single tire carrier *at rear* and extra demountable rim; coat rail; foot rest; full assortment of tools and an *electric extension lamp.* The cowl lamps were electric and shaped similarly to the headlights.

Prices were unchanged from the New Series (1913) models introduced in July 1912.

Later a new roadster style was introduced. It was an early 1914 style offering. This new roadster body was a very significant development. The Marmon catalog described it:

> The Marmon "32" Roadster is designed to afford the utmost convenience in a sensible, logical car for city use as well as a snappy, practical car for road purposes. The lines of this type are distinctive and every effort has been extended to afford all the advantages of roadster body designs without including the many inconveniences that have heretofore marked so many bodies of this type. It is distinctive in appearance especially because of the great sweeping line of its body. The driver's seat is set well back from the dash, and the entire body to the rear of the seat is decked over to provide ample space for gasoline and oil tanks as well as a large compartment for storage.

The larger tonneau on the Model 48 also accommodated a disappearing rumble seat. The dual rear spares evident on this model marked the start of a trend that would become a status symbol in later years. Significant also was the concurrent illustration of a style in both top 'up' and top 'down' configurations; the top 'up' mode was becoming the usual mode and appearance in both was important to the customer.

## 1914

Trade periodicals from July 1913[7] carried the following ad:

> **ANNOUNCEMENT**
>
> For the coming season the Marmon "Thirty-Two" offers all the latest ideas in equipment, style and appointment.
>
> Years of satisfactory service have proved its economy in tires, fuel and upkeep . . . plus smooth, delightful operation and durability.
>
> The features of exclusive Marmon design, the scientifically selected materials and conscientious expert workmanship attain their highest development in this car.
>
> With center control and every feature for safe and easy operation, making a car that a woman may drive with assurance . . . a car that may be trusted in any emergency.
>
> *Bodies* . . . The individual design and finish of the Marmon "Thirty-Two" bodies add a distinctive tone of beauty and refinement. Distinguished in appearance they indicate the high character of every detail of the car.
>
> The running boards are clean, with metal boxes for batteries and tools set in the front fenders.

The illustrated touring car in the ad showed the full length running boards and tool/battery boxes integrated into the front fenders which had appeared on the Model '48' in December 1912. The electric horn was mounted atop the frame inboard of the left box.

The front fenders had a detail that was a miniscule beginning of the skirted fenders that widely appeared in 1933. Its purpose however was more likely a gusset function where the fender had a break line as it interfaced the tool/battery box on the running board. Previously the battery was carried in a metal box on the running board between the front and rear doors; earlier the Prest-O-Lite tank had been in that location. The cast aluminum running boards were extended full length between the wheels to support the integrated boxes. This change occurred very early in 1913½ production.

For the first time, the illustrated cars carried spare tires. Prices remained the same as had applied since mid 1913 (calendar year).

## 1915

The Model 41 introduced in February 1914 shared dual-billing with the Model 48 beginning with the 1915 models. The time of the Model 32 had ended . . . luxury cars demanded six cylinder power. The Cadillac was already a V-8 and would not offer an L-4 again until 1981.

## Summary

Over the production life of the Model '32' the wheelbase grew from 112 to 120 inches, the weight of the touring car grew from 2100 to 2600 pounds. The price of

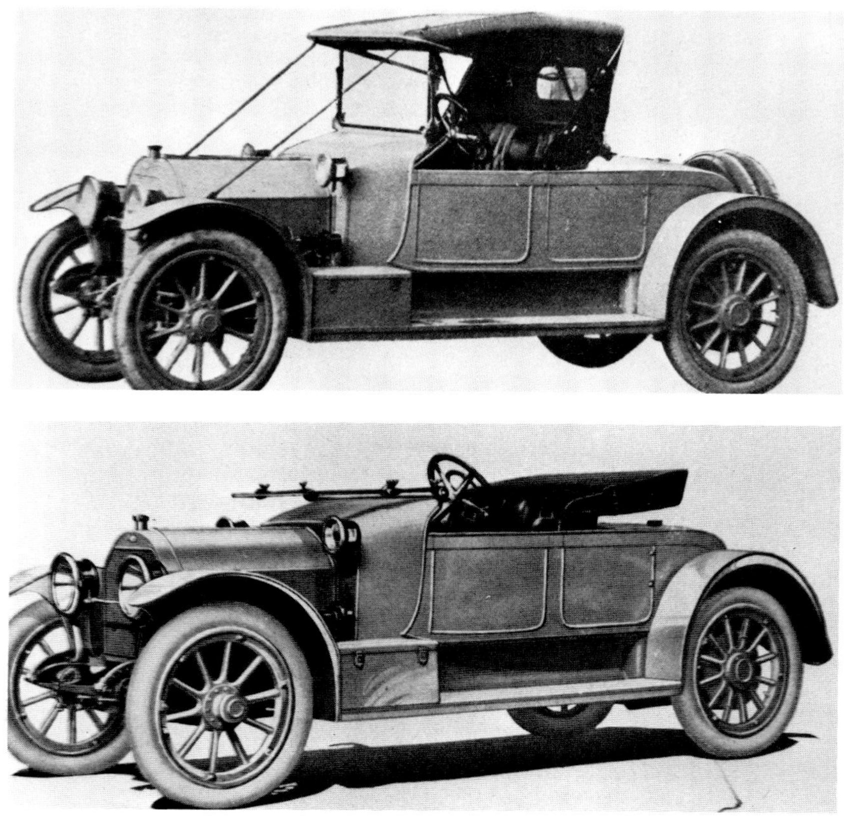

Fig. 3-12 The 1914 Model 32 roadster with locked compartment at rear for baggage.

Fig. 3-13 Model 32 progressive change — 1911 left and 1913 right. Note change in lights, in badge on radiator, in fenders and the disappearance of the starting crank.

the touring car advanced from $2400 to $3000 as additional standard equipment was included in the base price; the closed bodies advanced from $3400./3500. to $4000./4100.

The Model 32 achieved truly remarkable durability both in competition events and in regular service. The stellar success as a racing machine is told in Chapter Eight — "International Champion". But a three page ad was run in Motor World in July 1914, to proclaim "a greater Marmon record than the winning of the first 500 miles Indianapolis Sweepstakes or any of its other world-famed victories in contests — certainly a greater mechanical achievement."

Mr. Horace F. Wood, President of the transfer company in Indianapolis bearing his name, had written a letter to Nordyke & Marmon reproduced on the first page of the ad and referenced therein as "one of the greatest documents of automotive history".

Mr. Wood's fleet of 18 Model 32 chauffeured for-hire limousines included nine that had exceeded 100,000 miles and his satisfaction was such that he would not accept 75 percent of first cost for any of them, including car Number 1 at 200,000 miles, Number 2 at 250,000 and Number 4 at 150,000 miles. Car Number 1 had been driven an average 42000 miles per year for 4.75 years. It had been in service for 1700 of the 1735 calendar days of ownership — a remarkable 98 percent availabilty seven days a week summer and winter! Driver Al Fisher (135,000 of 200,000) said "I'd put that old car against any new car. Another thing about it — it's easy on tires; one set carried it almost two years through all kinds of roads and weather." Car Number 2 had been driven 53000 miles per year. Driver Dick Riley (last 80,000 of 250,000) commented "I've taken that old Marmon over the worst roads in Indiana, Illinois and Ohio, in Summer and in Winter, and she's never given me a moments trouble" Cars 1-9 were 1909 and 1910 models purchased near or after the end of the model years. Cars 10-13 were 1911 or 1912 models purchased in 1912 and 1913; Number 14 was a 1910 purchased obviously used in 1913. Numbers 15-18 were recent purchases and included a (self-drive ?) roadster and a funeral car.

The ad concluded with: "There's a big and important message here for every man who buys or sells a car".

### *The Marmon Truck* *1913-1915*

Nordyke & Marmon Co. entered the commercial vehicle market for the 1913 model year. The product was a derivative of the Model '32' and it will therefore be described here and its engine subsequently discussed in detail along with the '32' engine in Part 2 of this chapter.

First word came in September 1912 when "The Motor Truck" reported:

"The Easiest Riding Car In The World"

Marmon Car No. 1 driven over 200,000 miles

# 450,000 Miles In Daily Service

Marmon Car No. 2 driven over 250,000 miles

## It will pay you to read this letter carefully

It forms one of the greatest documents of automobile history. It is part of the uncompleted story of two cars that have covered a distance equal to 18 times around the world—and are still giving satisfactory service.

Read every word of this letter—

The Horace F. Wood Transfer Company
INDIANAPOLIS, INDIANA.

July 8, 1914.

Nordyke & Marmon Co.,
Indianapolis, Ind.

Gentlemen:

In 1909 I began using Marmon motor cars exclusively in my automobile livery service. I was influenced to adopt the Marmon as my standard equipment, after experimenting with various other makes, being convinced that for the lowest upkeep cost, I could get more and better service from the Marmon. At the present time I have a fleet of eighteen Marmon cars, which are in service 24 hours per day and 30 days per month. I believe we have not lost a day a month with them, on an average.

Car number one has been driven over two hundred thousand miles and car number two has been run over two hundred and fifty thousand miles.

The rest of these cars have gone from twenty five thousand to one hundred thousand miles each and all are in good condition today.

It is commonly stated that the depreciation on a motor car is 50 per cent the first year. I have not a single car in my fleet I would sell for 75 per cent of the first cost as I believe they are as good as when I purchased them. All are working and none are for sale.

Respectfully,

The Horace F. Wood Transfer Co.,
By Horace F. Wood
Pres.

Sixty Years of Successful Manufacturing

"The Easiest Riding Car In The World"

PART OF THE HORACE F. WOOD EQUIPMENT OF MARMONS

# Here are the records

Dick Riley

—has piloted the old Marmon No. 2 over 80,000 of the more than 250,000 miles it has covered. Riley has taken a number of parties on long trips appraising farm lands and knows about every good and bad road in Indiana, Ohio and Illinois. He says: "I've taken that old Marmon over the worst roads in three states, in Summer and in Winter, but she's never given me a minute's trouble."

No. 1 1909 Marmon purchased in September, 1909. Operated continuously since that time. Records show over 1700 days of service and a mileage in excess of **200,000 miles.**

No. 2 1909 Marmon purchased in September, 1909. Used in long road tours in summer and operated continuously in winter livery service. Records July 1st, 1914, establish a record of over **250,000 miles** in service for this car.

No. 3 1909 Marmon, purchased in March, 1910. Established a mileage record of over **100,000 miles.**

No. 4 1909 Marmon, purchased in March, 1910. Operated continuously since that time. Mileage on July 1st, 1914, over **150,000 miles.**

No. 5 1910 Marmon, purchased in August, 1910, and operated in daily service since that time. Mileage in excess of **100,000 miles.**

No. 6 1910 Marmon, purchased in April, 1911, to take the place of car of another make which failed to make good. Has covered over **100,000 miles** in service.

No. 7 1910 Marmon, purchased in September, 1911, and now entering its second hundred thousand miles of service.

250,000 miles in daily service is to our minds a greater Marmon record than the winning of the first 500 mile International Sweepstakes by the Marmon or any of its other world-famed victories in contests—certainly it's a greater mechanical achievement.

***It is the sort of record that means actual, cash value to you in the car you buy or sell.***

Sixty Years of Successful Manufacturing

Fig. 3-14 Marmon considered the performance of their cars in the Horace F. Wood fleet to be a greater record than their Indy 500 victory. The third page of the ad reported five other similar cars with 25,000 to 100,000 miles.

## MARMON TRUCK ANNOUNCED

Well Known Manufacturer of Pleasure Cars Adds
Light Delivery Wagon to Its Line

Coming somewhat in the nature of a surprise to the industry, the Nordyke & Marmon Company announces the production of a light delivery car with carrying capacity of 1200-1500 pounds. It is stated that the concern has been engaged in experimental work for some time past, and that several of these machines have been used, with satisfactory results, about the Marmon plant for several months.

The motor is the same as that utilized in the 32 horsepower pleasure car, except that the stroke has been reduced from 4.5 to 4.0 inches. Naturally the entire assembly has been strengthened throughout.

One or two *new ideas* in truck construction are evidenced, *among them being the policy of fitting dual pneumatic tires to the rear wheels."* (Author's emphasis)

The Marmon truck was introduced in November and was described and illustrated in *"Commercial Vehicle* (US) and *"The Horseless Age"* (for Nov. and Nov. 27 respectively).

The change in engine stroke reduced the displacement from 318 to 251 cubic inches and the SAE horsepower rating from '32' to 25.6. The vertical drive downward from the camshaft to the lube oil pump was extended upward to mount an engine maximum-speed governor of Marmon design; setting was for 20 miles per hour, typical of trucks at that time.

The reference to a strenghened drive train likely indicates that the Model 48 trans-axle assembly introduced the prior month was used. Although the assemblies used on the '32' and '48' pleasure cars appeared identical and had identical brake dimensions, the introductory brochure for the latter made a strong statement to the effect that the drive train was strengthened as necessary for use behind the larger engine.

The truck chassis appears to have differed from the car only in a frame reinforced with gussets and heavier gage cross members, in the stiffness and arch of the suspension springs and in the wheel and tire equipment.

Fig. 3-15 The Marmon Truck.

Fig. 3-16 Rear chassis shows pioneer dual pneumatic tire application.

It is noteworthy that Nordyke & Marmon, who would only be in the commercial vehicle business for three years, introduced the use of dual pneumatic tires to the industry. That advance in truck design appropriately came from a firm whose engineers viewed the vehicle and the road as a "system" and who consequently had been concerned with tire application engineering from their very first prototype vehicle. Also to be kept in mind was the fact that from the Marmon perspective, both manufacturer and owner, the pneumatic tire was a much more satisfactory item than it appeared to the remainder of the industry and public. Solid rubber rear tires were however available on order.

Vehicle specifications, prior to the advent of the Marmon truck, made reference to dual tires. These were however dual solid rubber tires pressed on a wide felloe, in lieu of single wide solid tire. It is significant that no less an authority than "The Motor Truck" considered the Marmon use of dual pneumatics to be a "a new idea in truck construction".

The six tires were the same size as the pleasure car tires . . . 34x4. The rear springs were 38x2, probably shorter than the 40x2 of the pleasure car only consequent to the increased arch height of the commercial application.

The wheels were wooden artillery type, the front having 10 spokes of 1.75 inch diameter and the single rear wheels (which mounted the dual tires and rims) having 12 spokes of 2.0 inch diameter. The front tread, like the car, was 56.5 and the rear tread was 65 inches.

Two standard bodies were offered . . . van delivery and open box. The fuel filler, for the underseat 10 gallon fuel tank, was extended out the right hand side of the seat panel (cab). Although the car had front doors (a delete option beginning in

1911) the truck did not. Standard equipment was the same as for the car, including the spare rim carrier and lighting equipment.

The Marmon delivery car was furnished with dash, front seat and floor boards. The outside body builder completed the vehicle.

The combination of a stamped-steel banjo rear axle housing with full-floating flanged axle-shafts, of dual pneumatic tires, and of 4 inch wide brake drums over internal brakes with S-cam actuation of roller-ended rigid brake shoes, was indeed prophetic in 1912!

The Marmon truck had a right hand driver position and was introduced just after the pleasure car had been switched to left hand drive; Any left-over engine assemblies could have been conveniently used in the truck.

There is no discussion of the Marmon ¾ ton truck in the Model 32 chassis section of Part 3 because of the pleasure car components other than discussed above. The truck engine, however, is discussed in the Model 32 Engine section of Part 2.

*Model 48 — The Marmon Six*
*1913-1915*

The first announcement alert from Nordyke & Marmon came in Horseless Age for June 26, 1912. It noted that two years of intensive development preceded production release. Considering the demonstration of durability in the Indy 500 victory and the high degree of Model '32' componetry in the Model '48', it is more than likely that the introduction was delayed by the decision to switch to left hand drive for the 1913 model year and the consequent requirement for rearrangement of engine auxiliary equipment.

Initial advertising referred to the new offering as "The Marmon Six". It was not until late 1913 that the car was promoted as the Model 48. Introductory ads appeared in issues of July, 1912 with companion, facing full page ads for a New Series Model 32.

New bodies were described: *"The (Model 48) bodies with their graceful, symmetrical lines, their deep and luxurious upholstering and the many refinements, added to the strength of the Marmon perfected cast aluminum construction are the best examples of the art of body design and building."*

The seat backs were sheet metal.

The car, and particulary the body, was exceptionally low, ie. of course, with the top down. The cowl sloped upward from the terminus of the hood hinge, but was curved whereas the hood continued the traditional Model 32 radiator profile back to the cowl. The belt line at the top of the door curved upward and across the cowl; an integral hood over the dash continued the cowl aft over the forward edge of the front doors, placing the windshield closer to the driver than would otherwise have been possible. The long low hood, long cowl and low body sides combined with the long (148 inch) wheelbase to make a very racy automobile.

The front doors were quite low; the top-of-body line stepped upward just behind the door and again just behind the rear door. In this way the necessary elevation of the rear seat, consequent to the frame kick-up, was accommodated somewhat inconspicously while keeping the hood and cowl low to minimize frontal area.

Fig. 3-17 The production Marmon Six was more conservative — it arrived 6 months after the announcement and 12 months after the Indy win.

The four-passenger 'Suburban' touring shared the cowl and low front doors with the five and seven passgenger bodies. However that same height was continued all the way to the point where the tonneau rose to pass behind the rear seat. The central (inter-door) section was extended, at expense of rear door width, to accommodate the rearward relocation of the front seat associated with the much lower seat height. The tonneau was "close-coupled", allowing for inclined mounting of the two spare tire/rim units. It should be noted that this was not an inclined mounting of a dual set of spares but rather two side-by-side inclined spares. This was a very sleek automobile.

The low tonneau on this model was achieved by compromising the rear seat height by the amount of frame kick-up . . . a practice that was followed in later four-passenger Speedsters as well.

Initial ads showed no hood louvers, but by December seven large vertical louvers were evident, spaced as if another six could be accommodated later. At introduction the hood was retained by a single transverse strap at the midpoint; Two "footmen loops" atop the hood positioned the strap. Conventional latches were also present. Further attention was directed to its Speedway ancestry by the model Marmon race car surmounting the radiator cap. Neither strap nor cap was evident in the December 1912 ads.

The fenders were the type with an angle-iron frame. The front fenders were an 'S' shape, the two opposite curvatures meeting without a straight section. The rear followed the wheel radius for a full 180° and incorporated a 90° radius juncture with the running board. Associated with the latter was an enclosure of the forward spring shackle. A new feature was a lamp, actuated by opening the tonneau door, and located on the bottom of the door sill where it overhung the splash pan.

At announcement time, full-height windshield pillars pivoted at the cowl and supported upper and lower framed glass panels, each pivoted near its upper edge. By December, the pivoted pillars were three-quarter height and carried a single framed glass panel. The latter enchanced the car's appearance and the view from the front seat.

The big appearance change by year-end was a new fender treatment, surprisingly soon after announcement. It could be conjectured that actual production began with the "second generation" fenders. The 'new' rear fender incorporated an edge angle which was specially formed to be of varying height (greatest at wheel center-line) for styling considerations.

Fig. 3-18 The SIX front compartment.

The 'new' front fenders were an original concept. The cast aluminum running boards were extended to the front wheels and supported storage lockers on both sides. The fenders were still of the 'angle-iron' edging type but as they approached the juncture with the top of the lockers they acquired a mini skirt . . . a portent of the big swing to skirted fenders in 1933.

For the 1913 and 1914 model years, the left locker was described as a "tool and battery box", but for 1915 it was identified as a "tool and dry cell (battery) box".

Fig. 3-19 The 1914 Model 48 roadster came in two options — 3 or 4 passengers — and both were unusual and clever designs.

Fig. 3-20 For 1914, four-passenger touring, quite low appearing, was offered in addition to five and seven passenger versions.

Fig. 3-21 The 1913 Marmon Six limousine for seven.

THE MARMON

"The Easiest Riding Car in the World"

Fig. 3-22 The Model '48' was completely restyled for 1915. Note that appearance was equally important then with the top up or down.

For the first two years, the right locker was fitted with three identical cases. For the third year, two of the cases were identical to those previously used and the third was deeper; the locker height was correspondingly increased. The attractiveness of the locker-fender integration was compromised by this change.

Semi-drum shaped electric cowl lamps were fitted matching the headlamps.

Seven body types were offered. The five and seven passenger touring, four-passenger Suburban and two-passenger roadster were priced at $5000. The limousine (7-pass) was priced at $6250, the Berline Limousine (7 pass) at $6450 and the Landaulet (7 pass) at $6350.

*EQUIPMENT* included: electric starting and lighting with storage battery, cape top, integrated windshield, speedometer and clock with electric light to illuminate entire dash, electric emergency lamp with long extension cord, electric horn, power tire pump, shock absorbers front and rear, tire carrier and extra rim, coat rail, foot rest and assortment of tools, oiler, jack and tire repair kit. The power tire pump was a Ten Ecyke brand located to the left of the flywheel and operated by friction drive from that member. Once the inflation hose was connected, depressing a button engaged the friction drive; disconnecting the hose automatically disengaged the pump from the flywheel.

A new three-passenger roadster was placed on the market in May 1913. It was a completely new cast aluminum body styled to have a "snappy" appearance. The cowl was lower than the other models. A "coaching" molding was cast on the cowl, similar to that shape on the 1928-29 Ford Model As. The **four** doors were of uniform, and low height; the *door hinges were concealed* and there were no exterior door handles. The driver's seat was set well back from the dash and the steering column was severely raked.

The body appeared from the side to be a four-passanger with close-coupled seating. Behind the seat the body radiused down to the frame height. Acutally there was only a single central seat with storage lockers on each side. When not in use the seat folded up into the body completing a closed rear deck; the rear doors had keylocks for security of the luggage area. The top was conveniently removable and without the top the car was a very handsome two-passenger sports car.

## 1914 MODEL 48

There were major changes in the electric system for 1914. The former NorthEast double-wire, 6 volt lighting system and 12 volt starting motor-generator unit were replaced with a JESCO single-wire, 6 volt lighting system and 6 volt starting motor-generator unit. Door switch actuated tonneau lighting was introduced.

## 1915 MODEL 48

For 1915, the '48' was extensively restyled from the radiator as far aft as the tonneau. The hood tapered, rising and widening to mate with a cowl of identical contour. The rear doors were somewhat higher and the front doors much higher to provide a uniform height. The doors were increased in width for easier entry and the running-board lockers were increased in height for additional capacity. Additional louvers were added, to the hood and to the engine dust pans.

The new cowl achieved the larger front doors by mounting the windshield forward of the wider doors, very much closer to the hood.

Different headlights were fitted and the cowl lamps were discontinued.

## MODEL 48 CHASSIS

The '48' chassis used semi-elliptic springs with conventional shackling at the front. Unlike the '32' (full elliptic) the rear springs were three-quarter elliptic. The "quarter" spring was clamped to a perch on the frame above the axle and extended aft. The "half" spring was shacked to the frame forward and to the scroll end of the "quarter" spring at the rear; shackles were the inverted type.

The principle innovation in the '48' was the "Center-Point" steering spindle design which contributed to convenient handling of the 145 inch wheelbase automobile. In addition Marmon claimed it be significantly safer due to greatly enhanced driver potential for retaining control in an emergency. The steering pivot (king pin) was within the plane of the wheel; this necessitated a large diameter hub structure within the spokes, similar but larger than that required at the rear to accommodate the full-floating rear axle.

## MODEL 48 ENGINE

The '48' was powered by a six cylinder version of the Model 32 (4½x5) but with the stroke increased to six inches (4½x6). Three blocks, rather than two, were mounted on a new aluminum crankcase of traditional Marmon barrel construction. For the first time, a main bearing was provided between each cylinder — seven in all.

The engine displaced 572 cubic inches, had an SAE rating of 48.8 horsepower and 80 brake horsepower.

A siamesed intake mainfold fed a pair of 'T' fittings between the three blocks. A Harroun 2.0 inch carburetor was used in 1913-14 and a Zenith 1.75 inch Type H7-A in 1915.

**Both main and connecting rod bearings were die cast on bronze backs and were removable for service.**

A NorthEast 12 volt starter-generator was used. Ignition was by Bosch magneto

Fig. 3-23 1915 version of the Model 48 front compartment.

Fig. 3-24 1914 Model 41 7-passenger touring

or battery with common distributor and dual spark plugs. A two-bladed (aircraft) propellor-type cast fan was driven by a link-belt. A Powell multi-section muffler was fitted.

A dry multiple-disc clutch drove a strengthened Model 32 torque-tube and transaxle assembly. A surprising range of axle ratios was available ranging from a fast 3.0:1 to 3.77:1 undoubtedly used on the (seven-passenger) closed bodies. In between were 3.2, 3.4. and 3.47:1.

The Model 48 touring weighed 3600 pounds.

## *Model 41*
## *1914-1916*

Marmon had concluded that the day of the large four-cylinder engine was ending and prepared a replacement automobile for the 1914 model year. The Model 32 chassis had been in production since 1908 and when the 1914 Model 48 was announced in December 1913 there was no reference to the Model 32 which however continued unchanged.

In February 1914, a new six-cylinder car was introduced . . . the Model 41. It was priced only slightly higher than the '32' and was midway in size between the '32' and '48' and was obviously a replacement for the '32'. Marmon would concentrate on two somewhat similar automobiles with six-cylinder powerplants.

The Model designation . . . Marmon Era Two practice . . . was based on the SAE rating of the new engine, 41 horsepower. This compared with 32 and 48 for the predecessor and companion offerings respectively. The new chassis had a 132 inch wheelbase, compared similarly to 120 and 144 inches. A 138 inch chassis was subsequently released for limousine and coachbuilder styles.

The chassis was similar to Model 48 but the 132 inch wheelbase did not warrant the center-plane steering system and a midships transmission was adopted rather than the transaxle location. An open driveshaft was used with torque reaction taken by a parallel torque arm.

The engine was a 4.25 bore by 5.5 stroke unit displacing 468 cubic inches.

Fig. 3-25 Model 41 instruments and controls; 1914.

Fig. 3-26 Model 41 instruments and controls; 1915.

Fig. 3-27 A 1914 Marmon Model 41 Speedster in 'high-classic' company at the Indianapolis Motor Speedway Museum. Indy Speedway photo.

Fig. 3-28 After only one year the '41' underwent a metamorphosis! The change in styling is very evident.

The new bodies were "substantial" sheet steel over hard-wood framing. The fenders were steel and similar to the '48' except for a 90 degree juncture of the rear fender with the running board.

Styling was generally similar to the '48', including the distinctive battery/tool box integrated with the left front fender and the fitted luggage compartment on the right.

Body styles included a roadster and five and seven passenger tourings.

For the first time, the cowl was fitted with a separate mahogany dash panel (previously instruments were mounted on the engine bulkhead or 'firewall'). The individual instruments were set in, not on, that panel in a pleasing array, and were illuminated by a spot beam from a light source on the face of the seat riser at car centerline. The hand air-pump for the fuel supply was also mounted on the riser face at the left edge. The central electrical switch, adjacent clock (left) and speedometer (right) and smaller fuel (left) and oil (right) pressure gages were arranged on an arc. The straight lower edge of the panel was interrupted at center for an extension mounting the carburetor choke button. Above it appeared the word "MARMON".

The standardization on Bosch electrical equipment for 1915 combined with the enlargement of the cowl to precipitate a new generation panel and array.

An innovative integral tail light, license holder/frame, license lamp and mounting bracket was adopted.

A speedster model was introduced during the 1914 model year. The chassis was characterized by elongated front fenders and shortened running boards. The fenders extended directly forward from wheel centerline with heavy 'visors' at the front end. Running-board lockers were smaller and integrated into the rear, rather than front, fenders.

The body had an extended cowl without windshield, miniscule front doors, about 6 or 8 inches high, and a single seat contoured into two buckets.

Behind the seat was a large round fuel tank with twin racing-type fillers followed successively by a tank-shaped storage compartment and tandem spare tires/rims.

An excellent example of this model is exhibited at the Indianapolis Speedwary Museum. It has a retangular trunk and a folding single-panel windshield which appear to have been a factory installation.

The Model 41 engine was Marmon's first L-head design. Two three-cylinder blocks with integral heads were mounted on an aluminum crankcase which housed a seven-bearing crankshaft. Tunnel crankcase design was continued, as was the concept of integral intake manifolding for temperature control. A dual ignition system fired a single set of plugs.

A combination starter-generator unit was used. It was procured from NorthEast Electric Co.

## 1915 MODEL 41

The entirely new Model 41 had been announced in December 1913. Very surprisingly (especially if you were a recent purchaser), *Automotive Trade Journal* for August 1914 announced an extensively restyled Model 41. In fact, totally new exterior metal was evident.

The object of this major change was to achieve "more beautiful body lines by smoothing body lines to the latest dictates of streamline fashion"; To conform to the streamline form of the body, fenders were "given a different shaped camber or crown than before"

At model introduction, only a roadster and 5 passenger touring were mentioned. By 1915, the following open styles were available: speedster, 4, 5, and 7 passenger tourings. All were priced at $3250 except the 7 passenger at $3350. The 7-passenger limousine was priced at $4750. and the berline-limousine at $5350. Limousines and berlines were on a 138 inch wheelbase.[8] A chassis, without dash and rear fenders, painted "in lead", was available to custom bodybuilders at $3300. Bodies and running gear were finished in blue black with a cream stripe. Wheels were cream with black stripes.

Starting at the front of the automobile:

The head lights were ellipsoid in shape with flush rims, they replaced a flared drum shape where the rim was much larger in dia than the rest of the light. Both had nickle rims.

The new radiator was oval shaped at the top, blending smoothly into the sides and had an inconspicuous filler neck; the former radiator shape, traditional for Marmon but not exclusive by any measure, had a rounded top center sloping flat to the side panel hinge and a long filler neck. A larger radiator core was achieved but without distinction.

The hood sloped upward and widened to match the contour of the cowl, whereas the prior year hood had maintained the radiator shape for its full length. This resulted in a flat face on the cowl much like the modern school bus body-to-cowl interface, it was a long term carry-over from the concept of body building by coach building firms and chasses by automobile manufacturers . . . the modern circumstances in school bus construction.

The front fenders were question-mark shape but quite short. The integrated lockers were gone; the higher seats provided additional under-seat storage.

The body was convex surfaced with flush doors and was devoid of any moldings or interruptions in surface. The sides were considerably higher . . . several inches . . . reportedly to "conform to the latest dictates of streamline fashion". The prior year body was slab-sided with integral moldings outlining the doors.

The top was of the "one-man" (to lower or erect) type with body attachment only at the tonneau pivots and the windshield. The prior year top was traditional with the front half supported off fittings on the front seat section, independently of the windshield, and retained by cables from the forward corners of the top to the cowl lamp brackets. Consequently the new body had pillars anchored in the body framework to support both windshield sections and the top.

It would appear that there had been a change in styling philosophy between the two versions of the Model 41. The 1914 version appeared to have been styled without a top with the result that with the top down the low body was attractive, but when the top was raised the car appeared high and ungainly. Conversely the 1915 version appears to have been styled with the top raised; the height of the doors was about equal to the gap between the top of the doors and the bottom of the top. The changed philosophy of design resulted from a change in consumer habits: the top was initially raised only in inclement weather and the motorists dressed appropriately; later the top was lowered only infrequently and still later (the 1920s) fixed tops were popular because of superior durability and weather resistance . . . steps in the progressive shift to the closed automobile of the 1930s.

Having reached the rear of the automobile in this discussion . . . the 1915 spare tire was inclined outward at the top; the 1914 tire had been vertical. This was certainly a negative move. The author does not believe that there was ever anything more ugly done to an automobile than an outward/upward slope to the spare —

not even the Hupmobile radiator neck! That sloping spare was as far from handsome dual side mounts as one could get!

The removal of the lockers made evident the double-drop frame configuration of the Model 41. The new full length splash panel between body and hood and the running board curved upward under the hood; a triangular filler panel was used under the straight-bottom hood. This splash panel was outboard of the body to enclose the forward spring-shackle without the separate box enclosure used the prior year.

A Rostand windshield was adopted in 1915. This was a multi-position type, including ability to be folded in half or to have both halves angled for ventilation.

A small casting closed the opening in the cowl where the windshield pillars entered the cowl for attachment to the framing. This same casting supported a novel parking lamp . . . a closed horizontal cylinder, with a vertical slot on the front face, which covered a glass cylinder and bulb. It was quite inconspicuous and a close examination of photographs is necessary to detect it.

Marmon had learned, even before they entered the automobile business, that direct (ie, in-house) control of all aspects of product manufacture was the keystone of their quality control program and consequent worldwide reputation for superior product and warranty support. They continued this policy, producing even iron and aluminum castings and purchasing only tires, major specialty products like wheels, bearings, lamps, carburetors and ignition components. Nordyke & Marmon was responsible for the dominant mechanical content of the product.

With the rapid introduction of electric starting and lighting, the electrical system content of the total product was greatly enlarged and Marmon found the number of vendors proliferating with a corresponding increase in electrical system problems.

The solution, at least for 1915, was adoption of a complete electrical system from a single vendor . . . Robert Bosch.

The Model '41' incorporated a Bosch-Elba battery (but Willard in 1916), and Bosch generator, (independent) starting motor, generator regulator, switches, wiring-harness, lamps and bulbs. The horn was the exception. The system was 12 volt, Bosch claiming to have solved the bulb filament problem.

The New Series Model 41 was introduced as "the only American car having a Bosch starting and lighting system, designed and guranteed, down to the smallest connection, as the finest electrical equipment ever made for an automobile."[10]

## 1916 MODEL 41

A new body style was introduced in July 1915 concurrently with a new series of Model 41 cars subsequently officially identified as a 1916 model. This was a Three-Passenger Club Roadster with a novel seating arrangement and unique concept of top.

The description of the rear passenger accommodation could well be used for the rear passenger provisions of some recent sports car offerings: The *Automobile Trade Journal* reporter described it thus: *The rear seat provides a snug comfortable **place** for the extra passenger. The new roadster affords a design which takes care of three passengers very comfortably without the objectionable feature of having the third person **far away** from the front in a place that is not easy of access or that is not covered by a top (the rumble seat). There is an aisleway between the two front seats which allows easy passage to the broad and comfortable third seat. (The passengers legs remained in that aisleway). Author's emphasis and parenthetic additions.*

Fig. 3-29 The 3-passenger roadster was an unusual design. The rounded rear surface and consequent curved bows of the top.

A convex rear deck began behind the two front seats and curved downward to the frame following the third seat. It incorporated luggage compartments on each side of the third seat with top-door access and a rear lid hinged low with a latch handle near the top; All three had key locks.

A new feature for Marmon was the use of natural wood top bows with nickle plated attaching hardware on this model. Unlike usual top practice, the rearmost bow was curved throughout its length and the top interfaced with the body deck along a near semi-circular attachment molding. Evidently the rear window glass must have been curved as well.

The three touring bodies (4, 5, and 7 passenger) were noticeably changed in appearance. The change was primarily due to a new method of upholstery finish which they shared with the new 3-passenger club roadster: the upholstery ended flush with the top of the body and was not visible on the exterior. However the ATJ reporter stated: *"the body is thoroughly redesigned and is of **sheet aluminum** construction. It is very substantially built and is finished, upholstered and painted in the Marmon shops. The cowl is raised and widened and the width of the front seat is increased. The pedals are set forward slightly so as to give additional leg room for the driver. Instead of bringing the side of the body back from the cowl face with a gradual curve, a sharper angle is employed, allowing greater width and more room in the front of the car."*

The new sheet aluminum bodies featured what the manufacture's press release called "The Greatest Forward Step in the Upholsterer's Art — the new straight pipe upholstery which gives a new and added luxury to these distinctive new designs."[9] The new upholstery concept restricted the leather trim to the interior of the car. The door aluminum sills rolled over the top of the doors. The upholstery was fastened flush on the inside so that none of it was visible from the side view of the car. The binding was concealed by a molding just inside of the rolled top of the side of the body. The period of overlapping upholstery was ended. All body hardware was new and was carefully and harmoniously designed. A minor change to the windshield made the upper glass overlap the lower for the purpose of making the joint raintight.

Fig. 3-30 Plan view of the 1916 (1915½) Model 41 shows the new 'Straight-pipe' upholstery.

Axle ratios were: Speedster 3.06:1; Club Roadster 3.50:1; other models 3.77:1.

The performance of the car was demonstrated by driving the car under AAA auspices on the Indianapolis track at a speed of 62.89 miles per hour with the **top up** on a stock 1914 five-passenger touring car.

Besides the factory sponsored demonstration of the Model 41's performance on the Indianapolis Speedway another demonstration, monitored by the Technical Committee of the ACA (Automobile Club of America), was sponsored by the Marmon New York Company... a wholly owned subsidiary of Nordyke & Marmon Co.

Car number 115218, a Model 6-41 five-passenger touring, certified to be a stock model in every respect, was tested in May 1915 in New York City and awarded Technical Committee Test Certificate No. 18. Fuel economy was measured using a one gallon ACA test tank.

The transmission was sealed in the high gear position and the car was driven from the Marmon Branch on Broadway to Battery Park, returning on Fifth Avenue. This run of 1 hour, 25 minutes required 75 full stops and high gear starts. The clutch performed without fault or difficulty for the driver and a fuel economy of **12.2 mpg.** was obtained. Subsequently the car was driven on Ocean Parkway in Brooklyn at an average speed of 17.1 miles per hour. An economy of **16.1 miles per gallon** was demonstrated.

The Model 41 averaged 55 miles per hour for 1030 miles on the Speedway. The last lap was turned at 63.3 mph with straightway speeds over 70 mph. The car accelerated in **high gear** from a **standing start** to 72 miles per hour in 400 yards (120 feet less than a quarter mile).

Over-the-highway testing took the prototype car to Seattle, San Francisco, and Globe Arizona.

Specification tables in trade publications do not show a Model 41 subsequent to announcement of the one-model '34' program in December 1915.

### PART 2 — The Engines

Of course, the obvious differences in the Era Two engines from their predecessors were water-cooling, in-line cylinder arrangements, and side valves... instead of air-cooling, 90° Vee cylinders and overhead valves.

At least in initial versions of some engines, the "square" engine, or equal bore and stroke, was retained from Era One, Importantly, the Marmon unique features of full-pressure lubrication and tunnel crankcase construction were retained.

The exclusive Marmon lubrication system differed in the manner of introducing oil into the crankshaft. The Era One engines had an external oil line feeding the crankshaft nose through a packing gland; the Era Two engines fed all the main bearings individually through a brazed tubular manifold with fittings threaded into the caps. Crankshaft drilling was restricted to supply passages feeding rod journals adjacent to a main journal. With this change, the Marmon system became the full prototype of todays universally used lubrication system.

At the close of Era Two, Nordyke & Marmon were still using the designation of left and right sides of the engine, and of the car, based on observation when standing in front of the car. That orientation persists today in aircraft engines and in the European motor industry.

## MODEL H, 45 and 50 ENGINES

The Model H engine introduced for 1908 as a companion to the air-cooled engines in the last year of Era One, was also the power unit for the Model 45 of 1909. An enlarged displacement version powered the companion Model 50.

This basic engine as introduced in December of 1908 was a four cylinder in-line type with a 'T' — head valve configuration... side valves arranged on both sides of the cylinder. "Cross-flow" was achieved with this configuration and separate camshafts actuated the intake and exhaust valves. Two identical cast iron blocks, each with a pair of cylinders and integral heads, were mounted on an aluminum crankcase and topped with aluminum castings which closed a large opening in each water jacket and together formed an outlet manifold. Presumably this opening provided access for better inspection of water passages, assuring that no blockages with core-sand went undetected.

The exhaust manifold was a two-piece casting with a slip-joint to provide for expansion. Each had two legs, providing an exhaust port from each cylinder.

The intake manifold system was unusual and prophetic, evidently based on sound engineering studies. The manifolding was cast internally in the block; a tee-fitting between the blocks connected the inlet from the carburetor to the two blocks. The fitting included provision for adjusting its length to fill the inter-block space. The inlet leg was very short with a high-mounted Schebler carburetor.

A most unusual aspect of this design, which would characterize all Marmon engines until near the end of Era Two, was that the fuel-air mixture for the first and fourth cylinders was inducted over the closed intake valves of the second and third cylinders. A similar situation applied to the Model 48 six cylinder engine.

The high carburetor location necessitated a power fuel-delivery system despite the tank being located under the front seat. A symmetric front gear train was housed half in the crankcase and drove two accessary shafts outboard the crankcase and the camshafts. The high carburetor location permitted a magneto to be mounted and driven on the right (intake) side and a large centrifugal water pump

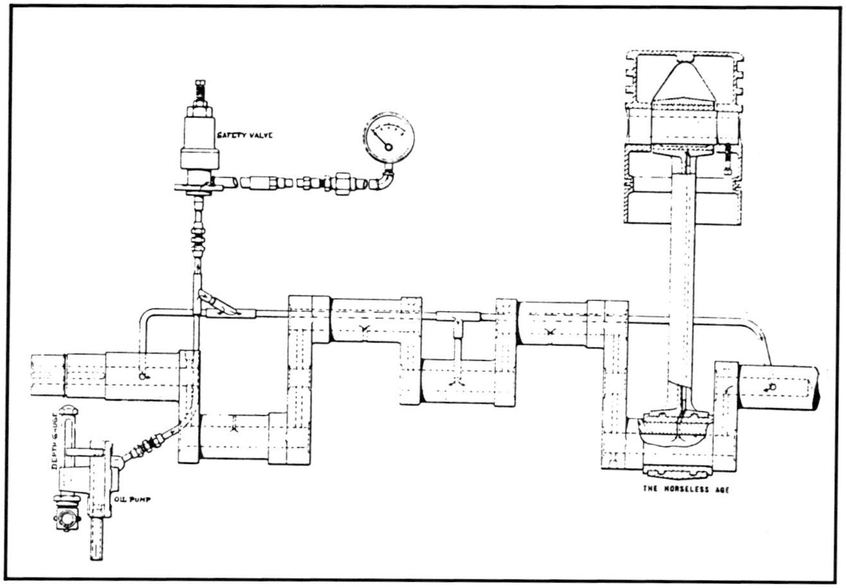


Fig. 3-31 Model H crankshaft oiling system — the first with oil introduced via a main bearing cap.

on the left side.

The camshafts incorporated an intergral flange for bolted attachment of the drive gears. Cams, which provided close to uniformly accelerated lift motion were attached to the shaft by keys and pins. Each shaft was supported by three long bronze bearings in the crankcase where it benefited from splash lubrication.

From the rear end of the intake camshaft, a roller timer was driven by spiral gears, and from the rear of the exhaust cam a small cam-driven air pump supplied pressure to the fuel tank.

Valve actuation was accomplished through a yoked pivot arm and short push rod (or solid tappet). This system was fundamentally similar to that which characterized the Packard Eights right to the end of production: External castings attached to the crankcase supported pivot shafts which carried levers yoked to receive a cam-following roller and having an upper contact to lift the valve tappet. The Marmon Model H, 45, 50 system differed in the method of attachment . . . a single retaining yoke clamped the units for two cylinders to the crankcase . . . and means for clearance adjustment by shims rather than threaded adjustments with lock nuts.

"Upon removal of a retaining yoke, the followers may be withdrawn from the case and wear in the valve operating mechanism taken up by the insertion of hardened steel shim washers under the head of a piece the stem of which fits into a counterbore in the plain push-rod noted above."[1] Freedom from loosening in service was the claimed advantage.

The cylinder blocks had "port-plugs" over the valves to permit manufacture, and assembly of the valves, since the head was integral . . . a construction providing uninterrupted (by the block and head decks of separate head construction) water cooling of the combustion chamber walls. In recent years British Leyland introduced a diesel truck engine with integral head (but overhead valves) and the BMW Steyr diesel engine furnished to Ford Motor Co. for US passenger cars has a similar construction.

"Over each valve opening seat, a ground-in disc is held by a threaded ring screwing into the cylinder casting. Each of the ground-seat discs carries a spark plug"[1]

Two independent jump-spark ignition systems were fitted: one by storage battery, timer and quadruple coil; and the other a Bosch high tension magneto driven from the intake camshaft in the same way as the water pump was driven from the exhaust camshaft. However only one spark plug per cylinder, located over the intake valve, was used, on the 45 and 50; two had been used on the Model H. The battery dependant ignition system of Era One Marmons was relegated to starting duty. By 1912, 85 percent of domestic car models would be equipped with both magneto and battery ignition systems.

The Model H and Model 45 (1909) engine was of five inch bore and five inch stroke displacing 393 CID and having a 'SAE' Formula horsepower rating of 40, but producing 45 brake horsepower at the SAE piston speed. The Model 50 engine had the bore enlarged to 5.375 and was rated 50 SAE and 60 brake horsepower. Maximum brake horsepowers were 50 and 60.

The additional power of the Model 50 required a stronger crankshaft . . . the crank webs were increased from .875 x 2.375 to 2.875 except the webs between cylinders 1-2 and 3-4 which were thickened as well, from 1/0 x 2.375 to 1.375 x 2.875. The crankshafts were both of the three main-bearing type and were cut from billets and carefully machined all over to secure balance. The main journals were 2.0 inches in diameter and 3.125 long, except the rear which was 4.5 inches. The connecting rod journals were 1.875 diameter by 3.0 long.

Valves were 2.29 clear dia with 45 degree seats on all versions. The hollow piston pins were 1.25 inches diameter and connecting rods 13.5 long. The rods were forged and machined all over (to secure equal weights) and fitted with oil tubes paralleling the I-section. The big end was bushed with poured-in-place Parson's white bearing bronze. The upper ends were bushed with phosphor bronze bearings.

The pistons (7.5 inches long) exceeded the stroke in length and this feature combined with the long connecting rod length "to minimize wear and secure a long life."[1] The crankshaft, camshafts, and wrist pins were all heat-treated, hardened and ground.

Initially three compression rings were used and there were four oil grooves below the pin. The rings were Marmon three-piece rings — two thin rings backed up by a "bull ring."

Later three compression rings and two oil rings, below the pin, were used. The compression rings were Marmon three-piece rings.

The barrel-type crankcase was cast open on the bottom and closed by a large aluminum oil "pit" of 12 quart capacity. The case itself was cast of an aluminum alloy in the "home" foundry. It included 4 tapered arms with rectangular sections for engine support in addition to the gear train enclosure at the front. The radiator fan support bracket was attached to the top front of the case.

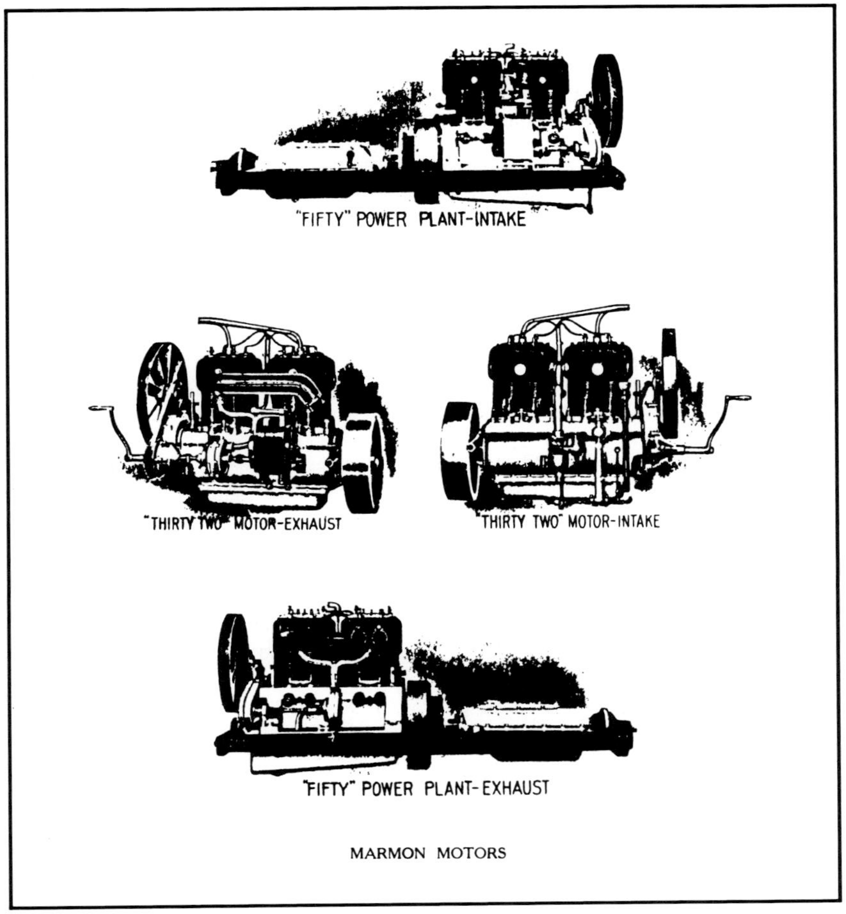


Fig. 3-32 Engines for Models 32 and 50, 1909.

## MODEL 32 ENGINE

The "Thirty-two" engine was a smaller companion to the '45' and its derivative, the '50'. It followed the design philosophy established the previous year . . . four cylinders in-line with 'T'-head valve arrangement, blocks cast in cylinder pairs with integral intake manifold, 'square' bore and stroke, barrel type crankcase with full pressure lubrication to main, rod and wrist pin bearings, and dual (magneto/coil) ignition systems with a single set of plugs.

The novel pivoted cam follower concept with shim adjustment of valve clearances introduced on the '45' was conspicuously absent. Perhaps it was too sophisticated for the service industry of the time; more probably it was inconsistent with the simplicity and light weight philosophy of the Model 32 vehicle concept. Instead roller-ended cam followers (tappets) with axial bolt and locknut adjustment means were adopted. Square bolt heads and nuts were initially used but were superceded by hex shapes. Roller alignment (to prevent the roller axis turning 90 degrees and sliding) was maintained (1) by a rectangular lower end of the tappet riding in a slot in the retaining bushing and (2) by a double-ended forked retention clamp which engaged straddle-milled seats on a pair of adjacent bushings. The forked clamp and seat configuration was identical to that used for injector retention and indexing on the Detroit Diesel Series 71, 53 and 92 diesel engines beginning in 1939. All Era Two and Era Three Marmons used roller cam followers. Evidently Marmon had identified the appreciable friction loss associated with the flat follower (which has all the elements of a brake device). When Pontiac adopted roller-followers for the 1985 Tech IV version of the GM 2.5 liter four-cylinder engine, they claimed a 5 percent reduction in total engine friction losses.

That same simple and light philosophy was also probably responsible for the single accessory axis (on the left side below the exhaust manifold); a double water pump packing was tolerated to gain a drive aft to the magneto. The relocation of the latter also permitted lowering the carburetor and making possible the use of gravity fuel feed. A Schebler carburetor was used.

The Model 32 fan drive pulley was mounted on the front of the accessory drive shaft; the 45-50 on the crankshaft.

The Marmon 32 Sales Brochure had this to say about the valves and springs: *"The valves have a superior method of holding the springs which insures the dura-*

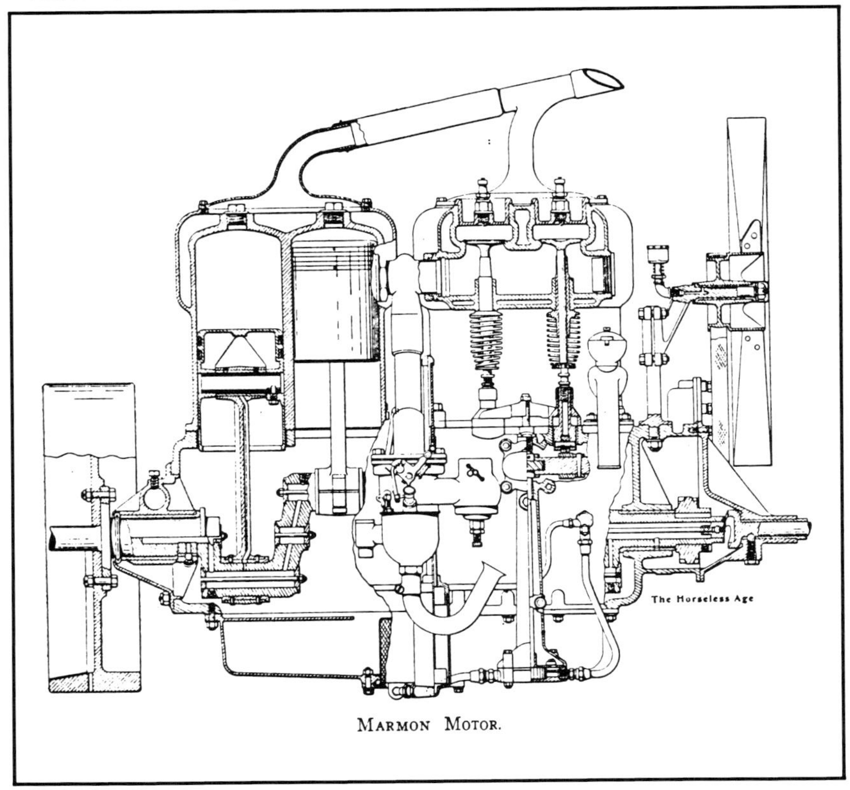


Fig. 3-33 Model 32 engine — 1910

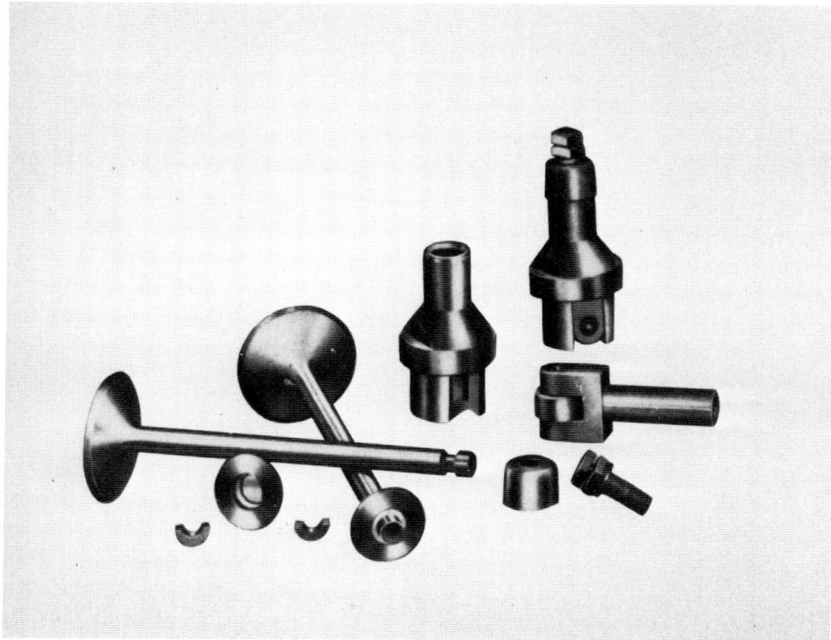

Fig. 3-34 Model 32 valve train components. Note use of roller tappets and modern split valve keys.

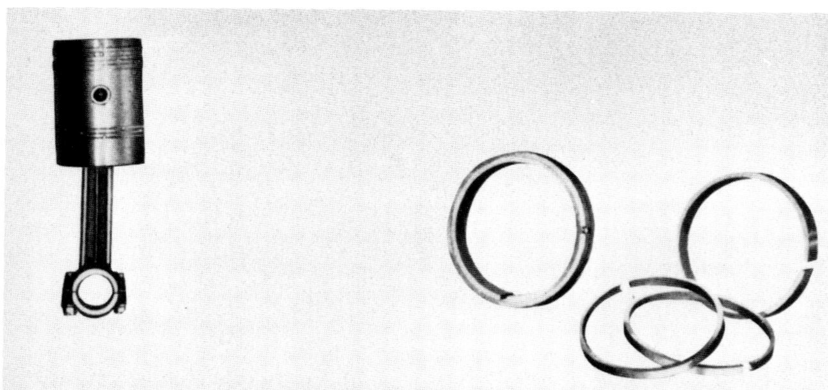

Fig. 3-35 Model 32 piston rings and location on piston.

*bility and facilitates the quick removal and replacement of valves and springs." This superior method was evidently just that . . . it referred to the innovative use of split, tapered keys which engaged grooves in the valve stem . . . for the method quickly became universal. Prior practice was retention by a pin in a drilled hole in the valve stem, with consequent severe weakening of the stem.*

The "Thirty-two" had a bore and stroke of 4.5 by 4.5 inches giving a displacement of 286 cubic inches and an SAE taxable horsepower rating of 32.4. Actual developed horsepower was 40 under the same conditions. Clear valve diameter was 2 inches.

The water jackets were shorter but deeper than typical practice as exemplified by the '45' and '50'. Presumably they were restricted to the length of bore exposed to the combustion gases, the remaining bore being air cooled (unfinned) consequent to the exposed valve spring arrangement on both sides.

The engine was described in press releases as being comparatively light in weight but exceptionally strong throughout and built especially with a view to accessibility of all its parts.

Engine block and pistons were of the same special gray iron. The pistons and bores were ground to size. The cam follower rollers and journal pins were hardened and ground; the rollers were bushed with white brass. The camshaft was flanged for gear attachment by bolts and was fitted with hardened cams individually attached with pin retention.

Interestingly, after some sixty years of camshafts with integral arms, the 1982 Toyota introduced a shaft with individual cams formed of powdered metal and furnace brazed into place.

The fan and bearing assembly was supported by a vertical steel tube attached to the forward crankcase.

Northworthy was a road draft-tube extending downward from an oil-separator atop the forward right corner of the crankcase. Oil sump capacity was five quarts.

The oil pump and pump drive was a separate assembly, flange attached to the crankcase at the camshaft level and extending downward to the pumping gears and adjacent the lowest part of the pan. External lines connected the pump to an inlet sump in the pan and to the block passage. An easily accessed screen in the pan provided protection to pump and engine.

A two-legged water outlet manifold of brazed tubing construction with flange attachment to the top of the two blocks was used at start of production; sometime later and presumably prior to introduction of the 1910 models, the '45' concept of water outlet was adopted . . . two cast water-outlet manifolds closing large openings in the top of the blocks, and joined by a hose.

## 1910

Engine improvements for the 1910 model year were evident in the illustrations accompanying the introduction. A stamped transverse member bolted to the rear of the crankcase, forward of the open flywheel, replaced the tubular member of the prior year (or of early production). Similary a cast bracket for the fan replaced the tubular part. The water-outlet manifold change was described above.

The Schebler carburetor now incorporated a throttle-body water jacket which was fed by a tube from the forward cylinder block.

For 1910, the stroke was increased to 5 inches from 4.5 and the displacement to 318 CID. The bore/stroke ratio changed from unity to 0.9:1. Engine drawings were identified as Type 6210. Presumably the last digits refer to the model year.

During 1910, a Marmon '32' race car was remarkably successful in both road and track events all across the country. Joe Dawson was the victorious driver of the two-man car which qualified as a stock Marmon 32 chassis under 1910 requirements of the American Automobile Association (AAA). Marmon press releases, (for 1911) stated "AAA Contest Classifications 15A, 4B and 4C" for the production car.

Fig. 3-36 The 1910 Model 32 engine.

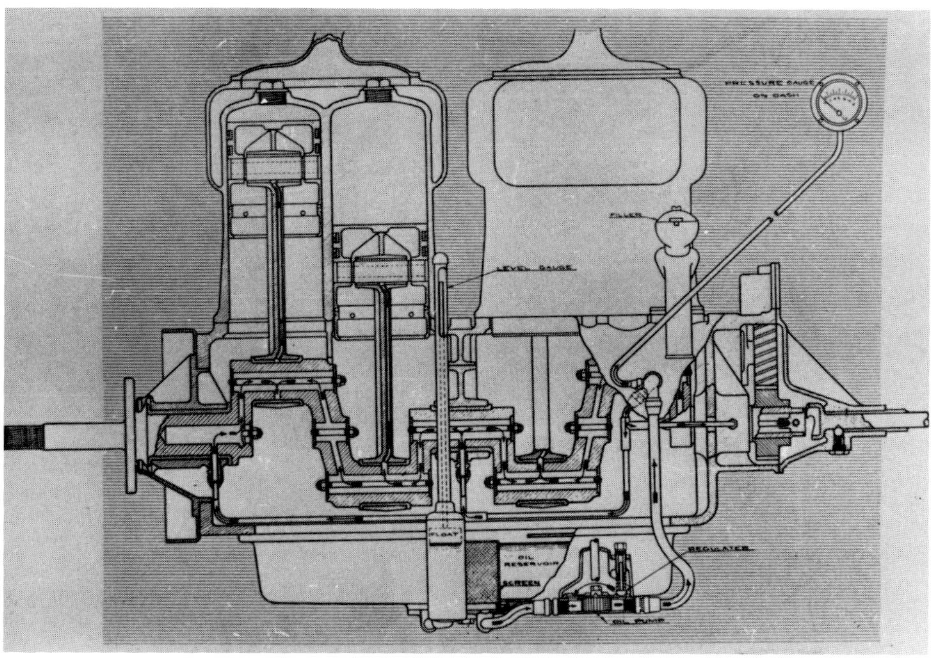


Fig. 3-37 The Model 32 lubrication and main bearing systems were typically and uniquely Marmon.

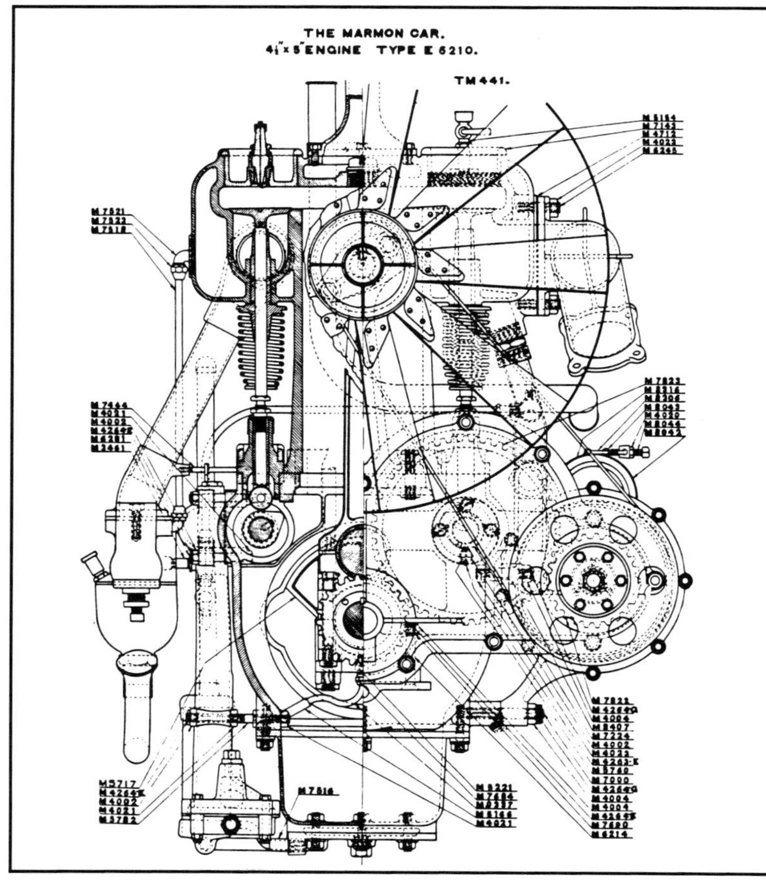


Fig. 3-38 The 1910 Model 32 engine.

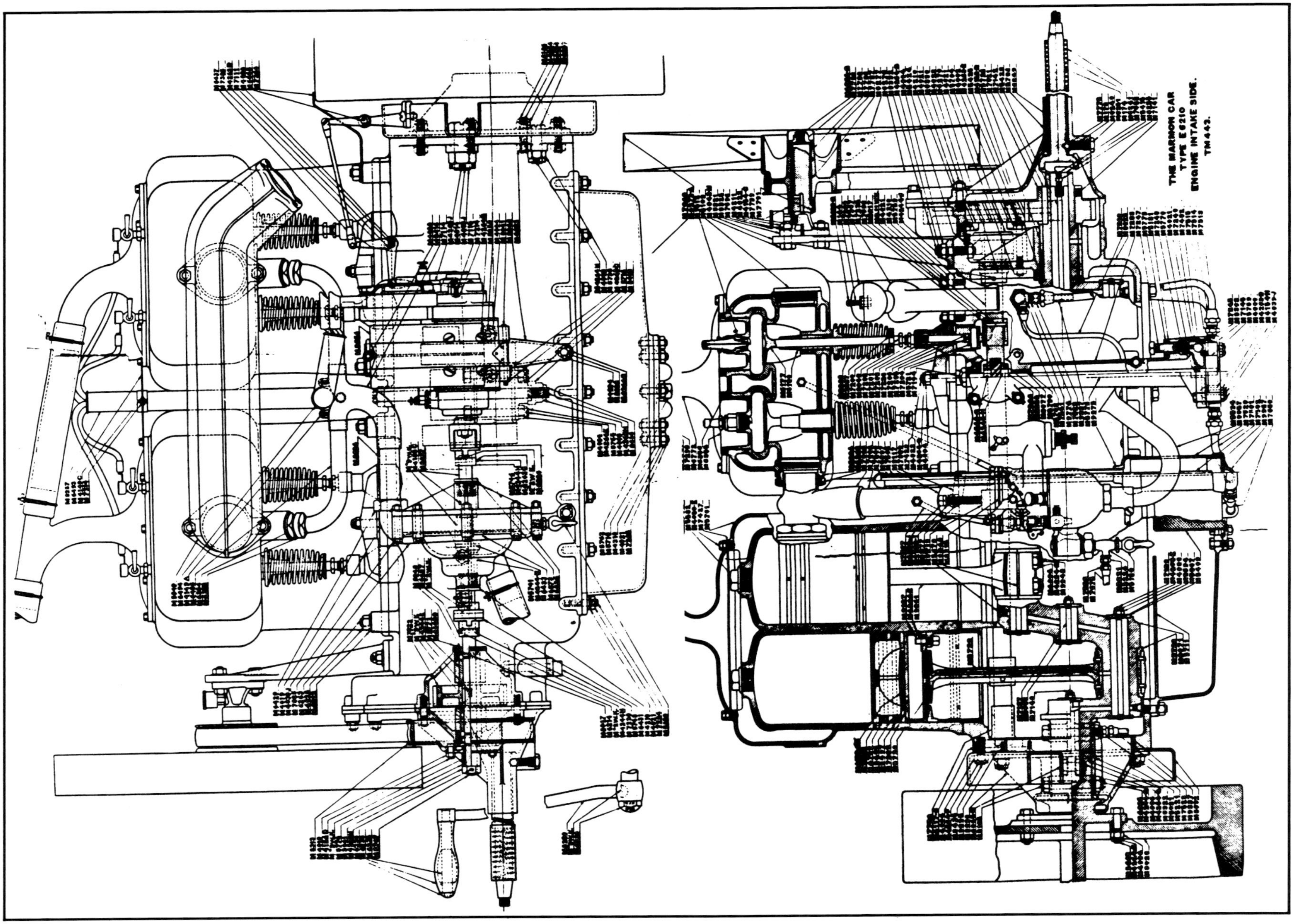


Fig. 3-39 The 1910 Model 32 engine.

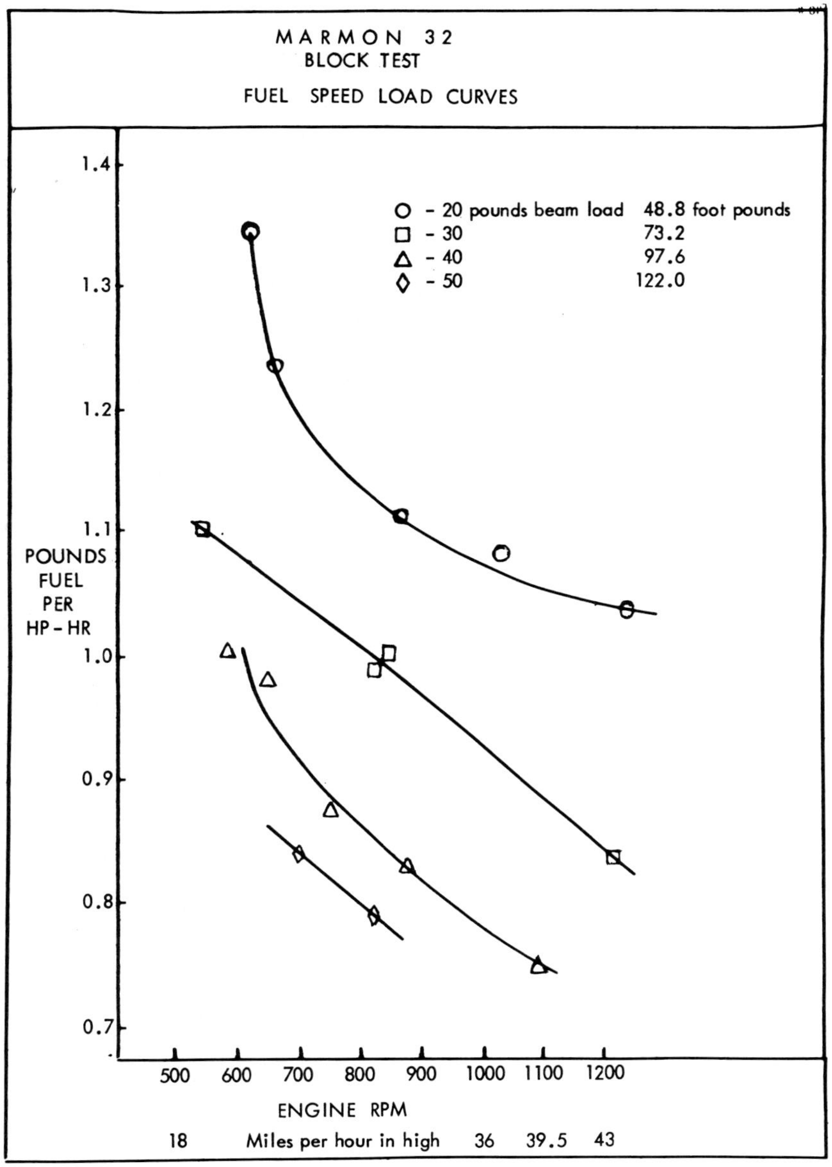


Fig. 3-40-1 Plate XII of the Taylor et al thesis, Purdue University, 1911. Brake specific fuel consumption (BSFC) versus RPM for several loads. Original reported gallons per hour.

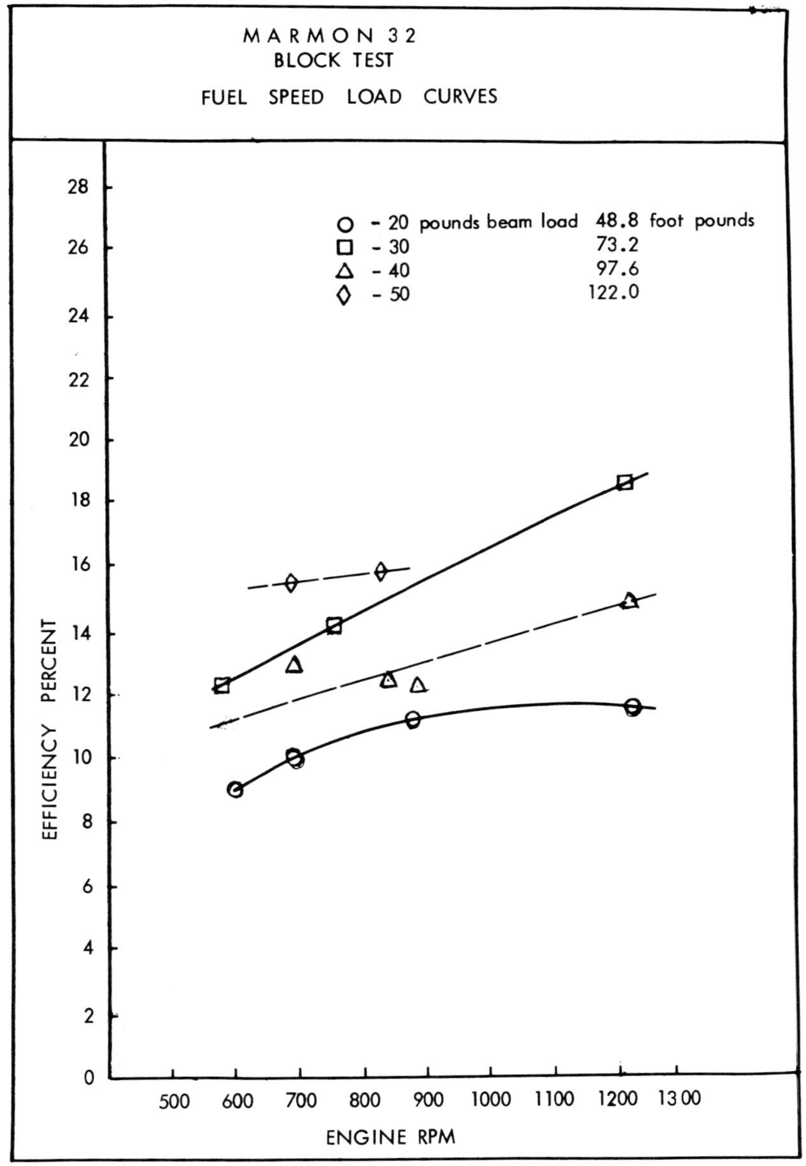


Fig. 3-40-2 Plate XIII of the Taylor et al thesis, Purdue University, 1911. Engine efficiency versus RPM for several loads.

Fig. 3-41 Chassis dynamometer test at Purdue.

Fig. 3-42 Engine dynamometer test at Purdue.

This particular race car was the principal in a Thesis for the Bachelor Degree in Mechanical Engineering at Purdue University. Howard Taylor and three classmates (Roscoe Charles Hoffman, Guy Alwyn Wainwright and Thomas Sharpe Wilson) of the Class of 1911 sought to determine the efficiency of power transmission from the engine to the road, in a project supported by Nordyke and Marmon who provided the vehicle and Joe Dawson to operate it. A literature search at the time found no precedent for the tests except a somewhat similar test of an electric vehicle. The thesis is discussed in greater detail under the Marmon 32 Chassis section of Part 3. However, in the process of determining the chassis power loss, Mr. Taylor and associates produced a remarkably comprehensive record of the performance and efficiency of the Thirty-Two engine whose track record established it as being in the forefront of contemporary engine design. It is unlikely that such a complete set of data exists today for any engine of the period. It is fortunate indeed that it was representative of the highest technical accomplishment of the time, a fact confirmed by its unparalleled success in racing during 1910 and the victory of a six cylinder version the following year in the first Indy 500.

Plates XII, XIII, and combined Plates VI and IX are replotted from the Taylor et al thesis because the originals could not be satisfactorily reproduced. The data in Table III of the thesis has been plotted to provide a typical horsepower (engine full throttle and wheel) vs rpm presentation. The thesis did not provide such a plot. Plates XII and XIII provide plots of (1) engine brake specific fuel consumption and (2) engine efficiency both versus engine RPM at several arbitrary loads. Plates VI/IX provide similar plots for the complete vehicle of (1) miles per gallon and (2) efficiency, both versus vehicle speed at several arbitrary draw-bar loads. In the engine test, 'load' is force on a scale at a fixed leverarm (ie, a measure of torque); in the vehicle test, 'load' is the force on a scale restraining the vehicle against motion. Certain terms then used require clarification today: "Block Test" refers to operation of the engine while loaded by a dynamometer and with engine cooling provided by a temperature controlled system feeding water to the engine without the usual radiator. "Plant Test" refers to operation of the engine in the vehicle with loading (resistance) being provided by a chasis dynamometer driven by the rear wheels.

Some readers may be unfamiliar with dynamometer measurement of power and with the units which are used to express engine efficiency. It is therefore appropriate to provide some helpful interpretation herein. The Purdue tests of 1911 utilized a primitive form of dynamometer called a "Prony Brake". It is, however, easier to explain the principal using such a device. For the subject tests, advantage was taken of the smooth surface on the outside of the flywheel: it was encircled by a brake band fitted with oak wood blocks as "lining". When tightened on the running flywheel, the band would seek to rotate therewith, but was restrained by a member acting at 90° on a "Platform" scale. The scale load in 'pounds' at balance, multiplied by the length in 'feet' of the lever arm gives, as product, 'foot pounds' torque (or twist) of the engine flywheel available to propel the vehicle.

With the engine running "no load" at a desired rpm, if the engine throttle is opened at the same time that the brake band is tightened (at an appropriate rate) the engine rpm (speed) can be maintained and torque values measured at various throttle openings all the way to 'wide-open' position.

Horsepower is then calculated . . . it is the product of torque and corresponding rpm (and a certain numerical constant value).

Engine efficiency is expressed in "specific fuel consumption": "specific mean-

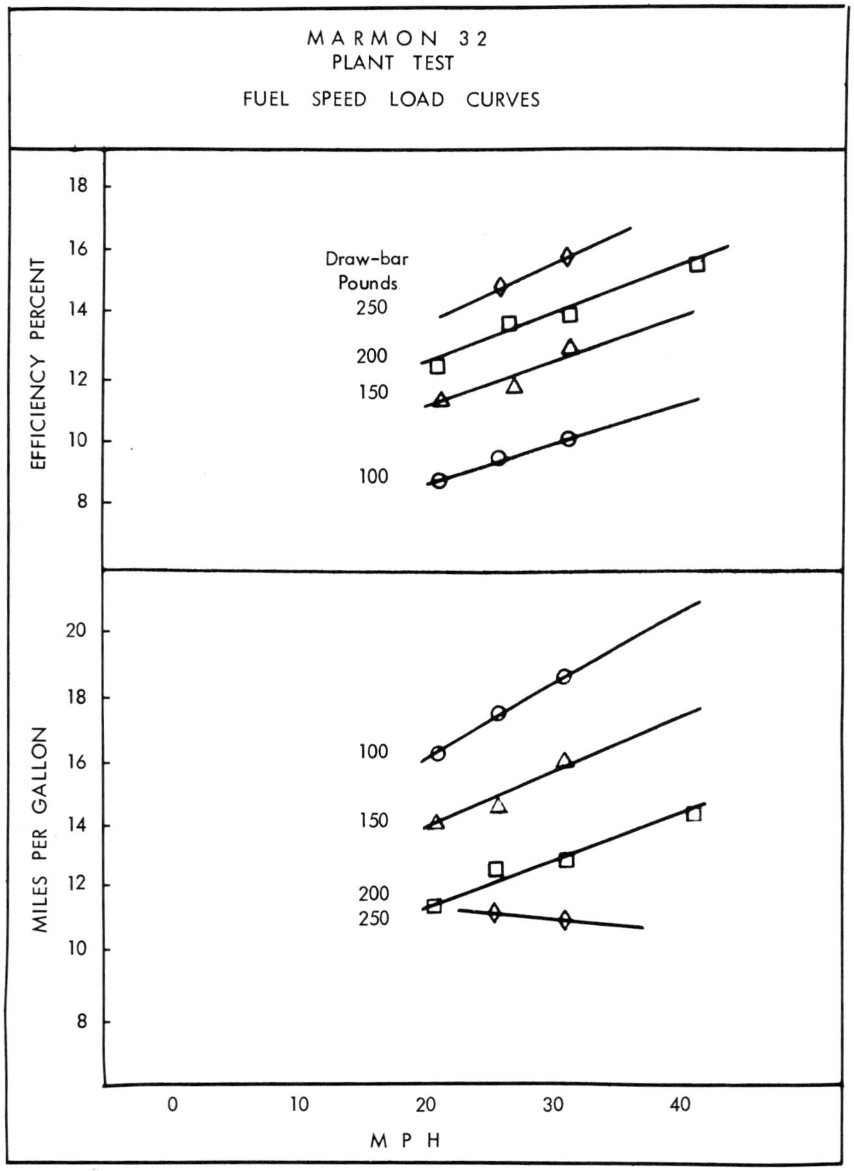


Fig. 3-43 Plates VI and IX of the Taylor et al thesis combined. Draw bar efficiency and fuel economy of Marmon Model 32.

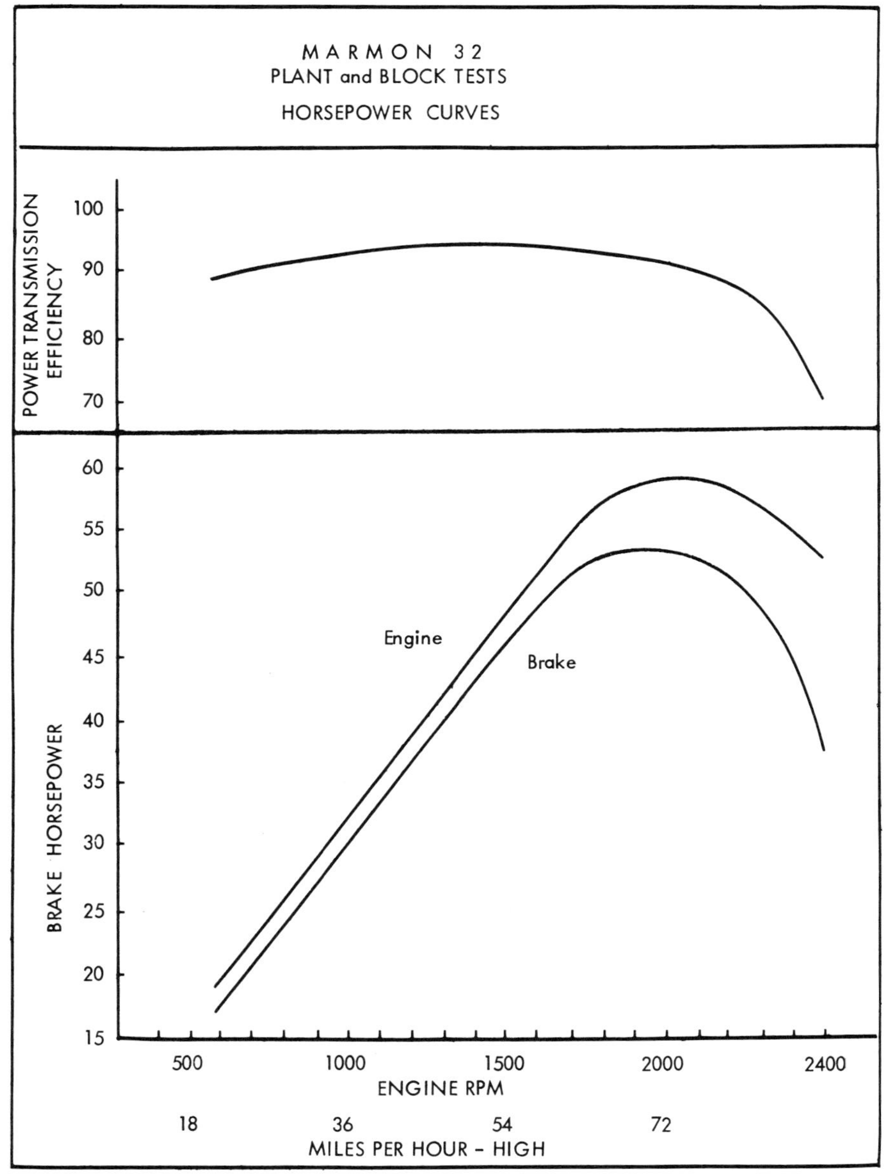


Fig. 3-44 Data from Table 3 of the Taylor et al thesis, Purdue 1911. Engine and chassis dynamometer horsepower versus engine RPM (lower) and Power transmission efficiency in high gear versus engine RPM and miles per hour.

ing the fuel used to produce one horsepower for one hour, thereby allowing comparison of engines of various sizes and powers. If the power is measured at the flywheel, efficiency is expressed as "brake specific fuel consumption" or bsfc. Note the continued use today of the term "brake" despite more than a half century of 'dynamometer' substitution.

Because the energy present in a pound of fuel is a fixed value, and the volume of a pound varies with temperature (due to expansion), the engineer today expresses bsfc in "pounds fuel per horsepower-hour."

For the Taylor thesis, the Chemistry Department at Purdue measured the specific gravity of the gasoline used in the tests, 0.735, and the measured gallons were converted to pounds for use in expressing "efficiency". The author has converted the scale on Plate XII from gallons per horsepower-hour to pounds per horsepower-hour using the reported specific gravity, .735, and 62.355 pounds of water per cubic foot.

The Purdue tests determined a maximum engine efficiency of 18.43 percent, at 1214 rpm and 97.6 foot pounds (40 pounds at 29.28/12 feet), 22.5 horsepower. The corresponding brake specific fuel consumption (bsfc) was 0.80 pounds per horsepower hour and the corresponding high gear speed was 40 mph.

The thesis reported the working cylinder dimensions as: displacement 79.4 cubic inches; clearance space 37.0 cubic inches and a resultant 'percent clearance' of 46.6. No mention of 'compression ratio' is made. From the above dimensions, the compression ratio can be calculated to have been 3.15:1.

## SAE HORSEPOWER vs BRAKE HORSEPOWER

There is much misunderstanding of the meaning of 'SAE' and 'Brake' horsepower, and numerical values of the former are often stated, without the SAE prefix, as "horsepower". Therefore a word of explanation is in order here.

The term 'brake horsepower' refers to actually measured power availabe at the flywheel as measured by a dynamometer or 'prony brake'. **Maximum** brake horsepower is usually inferred and the speed at which the value was observed is concurrently stated.

When automobiles were first taxed for road use, some states based the tax on vehicle weight, while others adopted engine size as the parameter. But some vehicles had large displacement engines that ran slower than others with smaller displacements and equitable taxing could not be based on displacement. To provide uniformity, a taxable horsepower formula was adopted by the Association of Licensed Automobile Manufacturers (ALAM) which subsequently became the Motor Vehicle Manufacturers Association (of the US) (MVMA); taxable horsepower was therefore also reported as ALAM horsepower and MVMA Horsepower. The ALAM formula was published by SAE in keeping with SAE's function of documenting industry standards, giving rise to the contemporary term "SAE Horsepower". Manufacturers today furnish annually, to tax authorities, the SAE horsepower of all current model vehicle-engine options.

In his "75 years of SAE", William (Bill) Sherman states that the taxable horsepower formula "originated with the Royal Automobile Club of England about 1900. It was adopted in this country by ALAM; published by SAE for the convenience of engineers. It falsely became known as the SAE Horsepower Rating".

The taxable horsepower formula is:

$$\frac{D^2 \quad N}{2\ .\ 5} = \text{SAE HP}$$

where D is cylinder bore diameter
N is number of cylinders
and piston speed is 1000 feet per minute.

Compression ratio and breathing efficiency are ignored. A 1982 four cylinder engine with a given bore and stroke would have the same taxable horsepower as a 1900 engine of the same bore, stroke and cylinder number!

To illustrate the peculiarities and unrepresentativeness of 'SAE' Horsepower and confirm the current usage, despite SAE not calling it by that name, the "1975 Chevrolet Truck Licensing Data" release by the Engineering Product Information Department of the Chevrolet Engineering Center is partially reproduced.

Note that the "SAE Horsepower" is idential (36.03) for both the 250 and 292 CID engines (same bore, different strokes) and similarly identical (57.80) for the 427 and 454 (Heavy Duty) CID engines. The two smaller engines are both rated at 115 SAE net brake horsepower at 2000 rpm, at 200 and 215 foot pounds max torque. The two larger engines are rated 210 and 230 SAE net brake horsepower at 3800 rpm, at 350 and 360 foot pounds max torque.

ENGINE DATA

GASOLINE ENGINES

| | | | | | | | | | |
|---|---|---|---|---|---|---|---|---|---|
| No. of Cylinders | 4 | 4 | 6 | 6 | 8 | 8 | 8 | 8 | 8 |
| Displacement (Cu. In.) | 110.8 | 140 | 250 | 292 | 350 | 366* | 400 | 454 | 427* |
| SAE Horsepower | 17.53 | 19.60 | 36.03 | 36.03 | 51.20 | 49.60 | 54.5 | 57.80 | 57.80 |
| Bore | 3.31 | 3.50 | 3.875 | 3.875 | 4.000 | 3.937 | 4.125 | 4.251 | 4.25 |
| Stroke | 3.23 | 3.625 | 3.530 | 4.125 | 3.480 | 3.760 | 3.75 | 4.00 | 3.760 |

DIESEL ENGINES

| | | | | | | | |
|---|---|---|---|---|---|---|---|
| Engine Series | 6V-53* | 6-71* | 6V-71* | 8V-71* | 12V-71* | NH230, NHC250, NHC270, NTC270E NTC350, NTC335* | V-903* |
| No. of Cylinders | 6 | 6 | 6 | 8 | 12 | 6 | 8 |
| Displacement (Cu.In.) | 318.4 | 425.6 | 425.6 | 567.5 | 852 | 855 | 903 |
| SAE Horsepower | 36.04 | 43.35 | 43.35 | 57.80 | 86.68 | 72.6 | 96.8 |
| Bore | 3.875 | 4.25 | 4.25 | 4.25 | 4.25 | 5.50 | 5.50 |
| Stroke | 4.50 | 5.0 | 5.0 | 5.0 | 5.0 | 6.0 | 4.75 |

Fig. 3-45 Data on SAE horsepower ratings still furnished to state registration authorities by the manufacturers.

The Purdue thesis reported that the Marmon '32' which had a taxable horsepower of 32.4 actually developed 38.3 horsepower at 1200 rpm (the 1000 foot per minute piston speed on which the taxable horsepower is predicted). Marmon rated the Model at 32-40 horsepower meaning 32 taxable, 40 brake, but the Marmon engine achieved 58.75 horsepower maximum at 2060 rpm.

The thesis authors were surprised by the Marmon's exceeding, by 18%, the taxable value (at the 1000 feet per minute piston speed). They stated "This is remarkable in the light of the fact that the highest power ever obtained from a car on the Purdue University testing plant **at any speed** was 88% (of the taxable value) and that a similar powered car tested in 1910 developed only 76.7 per cent".

The engine, as installed in the car, would (on level road in high gear) reach 2060 rpm at a car speed of 73.5 miles per hour . . . a speed feasible with 58.75 horsepower. The car would travel 47 miles per hour at the piston speed corresponding to taxable horsepower; Marmon apparently concluded that brake horsepower corresponding to taxable horsepower conditions was an appropriate power level to advertise. Hence the 32-40 rating by Marmon in 1911 would have been stated 32-60 horsepower by typical practice of the 1940-1970 period.

Marmon made just such a shift in policy when the Model 48 was introduced: It was rated 48-80 reflecting 48 taxable horsepower and 80 maximum measured on the engine dynamometer. However a projection of the Purdue test data would indicate ability to demonstrate at least 85 horsepower at 2100 rpm. The difference can be explained . . . the Purdue tests used the racing exhaust manifold, with dual legs extending several feet from the valve ports of the two blocks, to assure no reflected pressure waves, before juncture into a single pipe and the system incorporated no muffler.

## 1912

Detail refinements were incorporated in the engines offered for 1912.[10] Some were of considerable significance, however.

The valve lift became .406 and the clear diameter was increased to 2.25 from 2.00. The release stated: *"To add to the life of the valve gear and make it as noiseless as possible, the length of the bushing upon which the roller followers rotate has been increased nearly 50 percent."*

An important change was the adoption of the (imported) Bosch double ignition system with two entirely separate sets of magneto spark plugs, one set placed as formerly over the inlet valves and the other added over the exhaust valves. All plugs were threaded into "port plugs" which closed the openings in the block through which the valves were inserted when integral head construction was used. With the Bosch system, ignition (and flame front development) took place simultaneously at two points in the combustion chamber.

The 1913 Marmon brochure subsequently noted that "Marmon was the first (domestic) car on which this dual (spark plug) system was used!.

International conferences sponsored by the Champion Spark Plug Company in the 1970's confirmed that such dual initiation of combustion resulted in some 5 percent improvement in fuel economy. However some reduction in knock-limited horsepower resulted. For 1982 production, Toyota Motor Company of Japan adopted the dual plug concept for emission control and fuel economy but incorporated automatic control for single plug operation under certain load conditions.

Four detail improvements were made to the exclusive Marmon lubrication system. They are particularly noteworthy in that they confirm the extent of development engineering then being conducted in the Marmon plant laboratories in support of the product.

In an effort to reduce oil consumption without compromising lubrication, baffle plates were introduced beneath each cylinder. These were composed of two pressed steel stampings which fitted in the opening in the top of the crankcase beneath each cylinder bore. A slot in their central portion admitted the connecting rod. The baffles were hemispherical in form so that oil dropping from the pistons ran outward and downward into the crankcase through a small hole near the edge of the baffle plate.

The oil passages in the crankshaft had formerly been drilled all in the same plane. They were changed to be at 120 degrees to one another to minimize fluctuation in oil pressure during rotation. This had been reflected in vibration of the oil gauge 'hand' on the dashboard.

The oil inlet strainer was redesigned by raising both the screen and the suction inlet. This produced a "waste oil sump" below which level no oil could be drawn into the circulatory system. Any sediment or waste could be removed by access through the hand hole cover to which the suction line was attached.

The routing of the lubricant in the pressure oil system was changed to provide direct delivery of oil from the pump to the two front camshaft bearings and hence through the hollow camshafts to the other camshaft bearings.

Carburetor and intake manifold size was increased and an exclusive feature incorporated in the carburetor.

A Schebler 1.5 diameter Model L carburetor replaced the 1.25 predecessor. Speedster engines went to 1.75 diameter and also carried an oversize intake manifold, heated-air duct and "hot-air pocket" on the exhaust manifold in addition to assembly of unique cams to the camshaft.[11]

The carburetor was of the air-valve type. Adjustment was by change in tension of the air-valve spring. Special provision was incorporated to permit adjustment by a dashboard control. This was accomplished by attachment of a small rack to the air valve spring in place of the usual nut; A rod from the dash rotated a small pinion which engaged the rack.[10]

The engine hand-starting crank was improved by the adoption of a four jaw engagement which allowed the operator to select the most advantageous position . . . an improvement in line with Marmon's consistent encouragement of woman drivers.[12]

Fig. 3-46 The truck version of the Model 32 engine incorporated an integrated maximum speed governor with enclosed linkage to the manifold.

Greater accessibility of the valve tappets was obtained by locating the water distribution manifold higher and having the entry to the cylinder blocks horizontal instead of at an angle of 45%.

## 1913 MARMON LIGHT TRUCK ENGINE

Just as the Marmon truck for 1913-14 was a modified Model 32 chassis, so was the engine derived from the Model 32 engine.

The stroke was increased to 5 inches from 4.5 and the bore was reduced to 4 from 4.5 to provide increased low speed torque appropriate to the commercial service. This resulted in a bore/stroke ratio of 0.8:1 versus the 0.9 of the Model 32 (1.00 in 1909), and a piston displacement of 251 cubic inches reduced from 318. 'SAE' horsepower was reduced to 25.6 from 32.4.

The truck engine featured a neat integral engine-speed governor. The fully enclosed flyball governor was flange mounted to an upward extension of the oil-pump drive assembly and interfaced at the rear with an intake manifold having a governor-controlled throttle plate above the carburetor flange. The linkage to the governor throttle was completely enclosed to discourage tampering.

The extended oil pump drive shaft was fitted with annular ball bearings to control thrust from the flyweights against the spring. The flyweights moved a worm gear in mesh with a small gear on a horizontal shaft which through universal joints operated the throttle valve in the intake manifold.

## 1913 PASSENGER CAR ENGINE

The adoption of electric starting and particulary of left-hand drive required the reconfiguration of the engine including a mirror image reversal of the basic power unit. This latter change was facilitated by the 1910 change which separated the front gear case from the crankcase. A mirror image gear case was required, as was a similar relocation of mounting pads on the crankcase for the several auxiliaries. The cylinder blocks were reversible, but a new exhaust manifold was required.

Integration of a motor-generator unit into an engine already in production posed a major challenge to the engine designer. The '32' crankcase had a separate stamped steel crossmember at the rear to provide mounting to the frame. This facilitated the task. The member was converted to a casting and fitted with a cover thereby forming a chain case and bearing support for the outboard sprocket of the chain. The crankshaft sprocket was attached by a thin flange sandwiched between the flywheel and crankshaft flange.

The NorthEast starter/generator unit consisted of the motor/generator, clamped by a steel strap to a cast base which incorporated a gear train. The complete unit, as purchased, was bolted to a machined pad extending from the crankcase casting. The layshaft-type double reduction gear set drove the sprocket, at a 40:1 reduction in starting mode, through an over-running clutch of the ramp-roller type using six spring/shoe roller sets. Upon starting of the engine the sprocket, which was continously coupled by the chain to the crankshaft, then over-ran the reduction gear train and, in so doing, engaged a second (inner, coplanar, concentric) over-running clutch of the same type, having four element sets, and through it drove directly the motor/generator in the generating mode. A spring-biased cone clutch in the gear train limited torque resistance to expected cranking values. The rating of the starter was a 12 foot-pounds cranking torque on the sprocket at 46 amperes. The charging rate in the generator mode was 7 amperes. The current was regulated by reverse series field and a vibrating regulator.

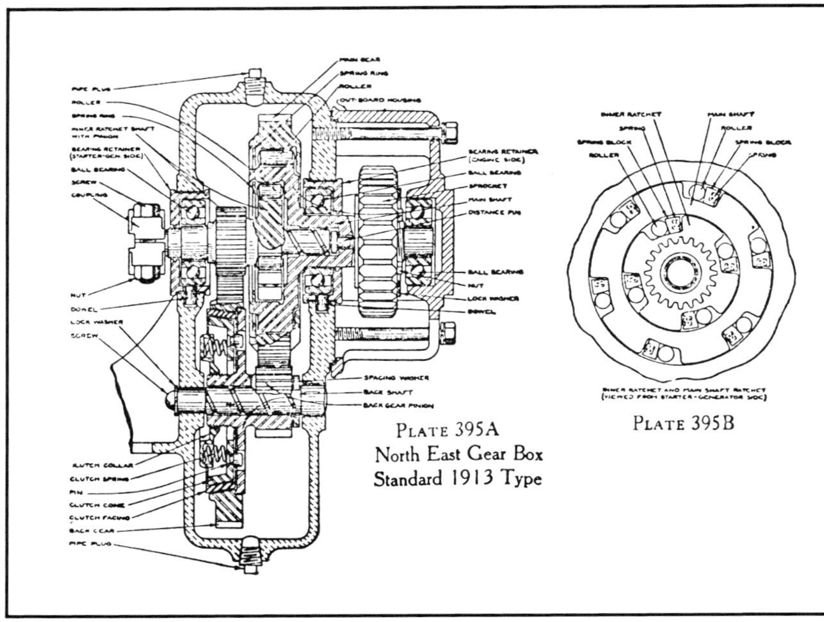


Fig. 3-47 North East starter-generator gear box which permitted single shaft connection to the engine.

Edward A. Halbeib, who with his brother founded NorthEast Electric Co., "had a vision of combining the power of electricity with the power of the automobile motor to elimate hand-cranking, the kerosene lamps and the rubber-bulbed horn". The starting and lighting system he developed was first applied to the Galt car (Galt Motor Co., Galt, Ontario) in 1911; the electric horn was developed in 1912. The Dodge application of the NorthEast system followed Marmon by one year; Dodge used NorthEast products until 1929 when Chrysler acquired Dodge and General Motors acquired NorthEast, merging it into Delco-Remy.[13] The DELCO system appeared on the Cadillac during 1912. It, too, was a motor-generator type but used separate drives (to and from the engine) for each mode. By placing teeth on the flywheel, the reduction gear train was not required. The Kettering system drove the generator through a ratchet-type clutch, from the front gear train. The ignition switch also by-passed the generator cut-out relay (a non-driven generator will motor backwards if connected to the battery) causing the generator to motor slowly, with a very audible ratchet sound, turning the direct driven starter pinion. When the starter pedal was depressed, the pinion was first slid into engagement, facilitated by the slow rotation. Additional travel of the pedal cut out the generator windings and cut in the heavy starter windings whereupon cranking began.

In the 1960s the motor-generator was tried again — on the fleet of some 50 Chrysler gas turbine cars; there were still problems associated with brush durability and concurrent satisfactory commutation over the range of ambient temperatures and the dual function requirement.

The new rear engine-crossmember had a bore at each extremity, parallel to the crankshaft. These carried a clevis-pin engaging a vertical clevis on the frame rail.

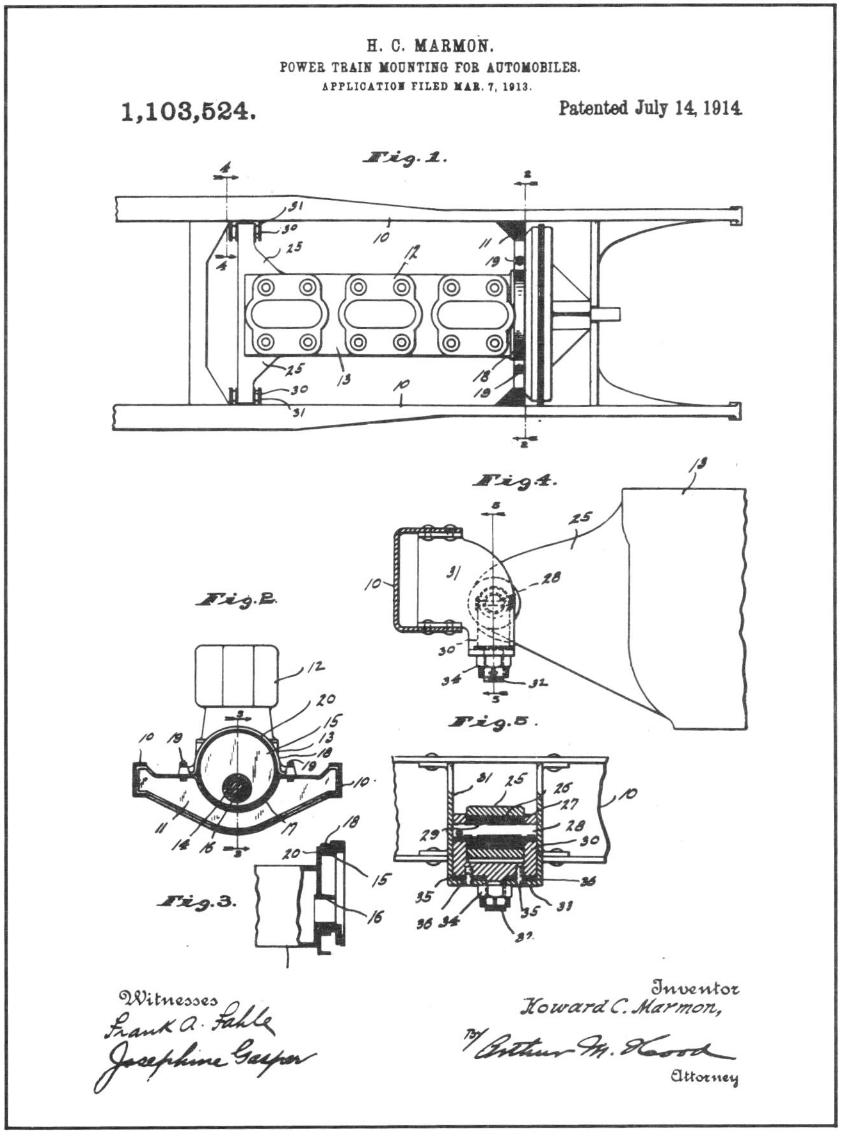


Fig. 3-48 The H.C. Marmon patent for an engine mount along the axis of minimum moment of inertia of the power train.

Fig. 3-49 Model 32 engine — 1913

Fig. 3-50 Model 32 engine — 1913.

Fig. 3-51 Model 32 engine — 1915.

Fig. 3-52 The SIX engine was much more than a stretched '32' — Note the enclosed valve tappets and springs, and the unique intake mainfolding of the '32' adapted to a SIX.

It was essentially a universal joint attachment between engine and frame.

This new engine mount system was shared with the new model 48 and would be the Marmon system for the rest of Era Two. It was covered by U.S. Patent 1,103,524 issued July 14, 1914 to H.C. Marmon following an application filed March 7, 1913.

The patent covered three aspects of engine mounting. The first was the mechanical linkage on the rear mounts to compensate for frame flexure, the second was the use of resilient (eg, rubber) bushings in a three-point engine mount to isolate vibration and facilitate accommodation of flexure. The third was indicative of the advanced understanding of the scientific aspects of motor vehicle engineering possessed by Howard Marmon and his associates at Nordyke & Marmon. In this pioneer resilient engine mount patent, the positioning of the mounts was along an axis inclined upward from the rear of the crankshaft . . . the round front mount centered appreciably above the crankshaft . . . to minimize engine motion on the resilient mounts due to forces of engine origin.

This anticipated both the "Plymouth Floating Power" (1932) with the high front trunnion above the chain-case and the modern practice of two angular mounts whose center-lines, if projected, would meet at that location and function accordingly.

The carburetor was relocated at the level of the crankshaft centerline; A probable cost reduction because gravity feed could replace the air-pressure fuel-delivery system with its engine-driven and manual-emergency pumps. This lowering required a forward displacement of the carburetor to clear the starter/generator. Parallel inclined ducts carried heated air to the top of the carburetor and fuel-mixture from the top of the carburetor. A control valve was incorporated in the cast elbow which joined the flexible duct to the carburetor.

## 1914-15

A Zenith carburetor replaced the Schebler and the carburetor location reverted to the close-hauled intake elbow and high mounted carburetor location of 1909. The flexible metal duct from the "Hot-air Pocket" on the exhaust manifold was reduced to about a quarter of its former length . . . perhaps the motivation for the change was heat loss over the long length, most of it directly exposed to fan air discharge.

## MODEL 48 ENGINE
## 1913

The '48' engine in the winning car at Indianapolis had an extended '32' crankcase, crankshaft and camshafts to accommodate three rather than two '32' cylinder blocks. It retained the 4.5 X 5 bore/stroke of the '32' and had 477 CID. It is relatively easy to cast an extended case using the same sand cores for the inside of the standard engine; a new pattern would be required for the exterior.

It was two years after the victory before the Model 48 was announced. The production engine design reflected considerable refinement from the Model 32 design which was announced four years earlier.

It also represented an additional step away from the square, or unity bore/stroke ratio of the earlier Marmons. It retained the 4.5 bore of the '32' but had a 6 inch stroke. Because stroke was ignored in the taxable horsepower formula, the relative displacement of the '48' engine is greater (180%) then would be indicated by SAE horsepowers (150%).

Marmon "Forty Eight" Motor, left side, showing the electric lighting and startin motor-generator, the carburetor, the hinged rear motor support and the multiple disc dry-plate clutch at the extreme right.

Right side of the Marmon "Forty Eight" Motor. The centrifugal water pump, the oil pump, air pump and magneto.

Fig. 3-53 The Model '48' engine in 1914 dress.

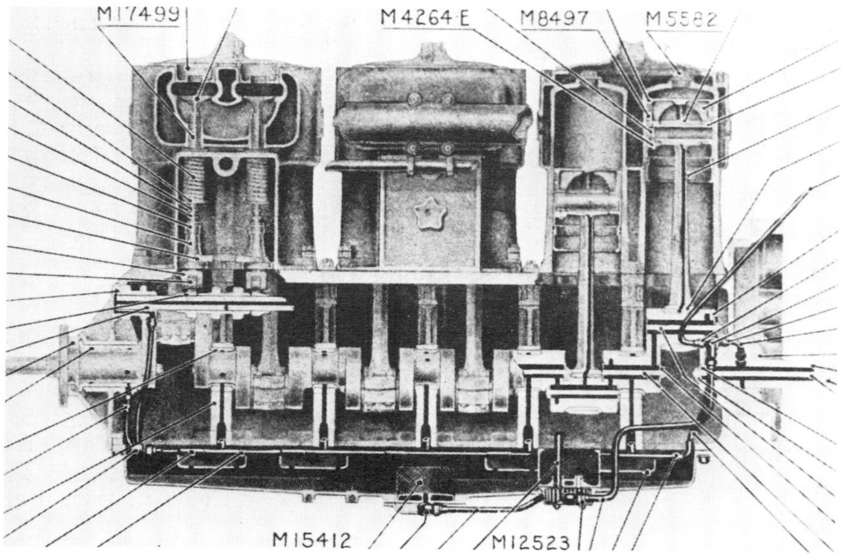


Fig. 3-54 The Model 48 engine was Marmon's first with a full complement of main bearings.

The obvious advance was enclosure of the valve springs and tappets, with access through doors each retained by a hand-wheel; one door per two-cylinder block.

The general arrangement of the '48' engine was similar to the reconfigured '32' (1913-14) as updated for left-hand driving and electric starting and lighting.

The central cylinder block interfered with application of the model '32' inter-block cross-over duct for heated intake air. Consequently the flexible metal duct was run under the oilpan. Air flow was downward through a "stove" tee surrounding the exhaust manifold, down the duct which hugged the block, and, after looping below the engine, through an elbow into the carburetor. The carburetor was a Harroun 2 inch size. The riser separated into a two branch intake manifold similar to the single Model 32 manifold.

A new and noteworthy improvement incorporated in the '48' engine was the use of intermediate main bearings between every cylinder, rather than a single intermediate as used on the '32'. On the Model 48, seven main bearings were used, five of them assembled to the crankshaft by bolting conventional main bearing caps to semi-circular members, prior to assembly of the crankshaft into the engine. When the nose of the crankshaft had entered the bearing in the front wall of the crankcase, the five intermediate bearing carriers would have entered the large diameter bores in the several bulkheads; the rear cover plate would then close the opening through which the crankshaft had passed and would support the rear main bearing. The intermediate bearing assemblies were retained axially and rotationally by a pair of main bearing studs, which were inserted vertically from below, through openings in the bulkheads, into the bearing caps and upward through drilled holes in the upper half of the bearing carriers (diaphragms) and the crankcase. Nuts were then applied to both ends of the studs and tightened. On the Model '32', the studs emerged from the top of the crankcase between the two cylinder blocks; a similar situation prevailed for the second and fourth intermediate bearings of the Model '48'. However, the first, third and fifth intermediates were located at the midpoints of the three cylinder blocks; for these positions, the bearing studs also extended through the lower flange of the blocks with the nuts being concealed by the covers over the valve springs and tappets.

Also new and noteworthy was the extension of the pressure lubrication system to include the camshaft bearings. The lube oil pump was driven by a vertical shaft from the camshaft via a skew gear set at the #2 cylinder location. The pump was flange mounted to the bottom face of the cast aluminum oil pan. External plumbing connected the pump inlet to the central screened sump area and the outlet to the rear of a brazed oil gallery located via a tee fitting at the front main, to the front camshaft bearing. The **hollow camshaft** carried lubricant to the other three camshaft bearings. Oil pressure was sensed at the entry to the rear main.

The front gear train included a drive spindle for the North East unit on the left side of the engine and gear driven by the left, or intake, camshaft. The motor/generator and associated transmission, as purchased, was mounted like the Model 32, on a machined pad extending from the forward left side of the crankcase.

Although the Model 48 installation of the North East dual function starter/generator was similar in philosophy to the Model 32, the application design was more sophisticated. The designers had to consider the logistics of service parts support for the handcranked Model 32s in operation when determining the electric unit application . . . the parts were essentially new parts not required on prior engines. Undoubtedly the Model 48 design was well advanced prior to the decision to adopt electric starting and lighting for 1913. The designers were not however restrained from extensive modification of engine parts by logistic concerns.

The '48' engine had the same three-point mounting provision of the 1913 '32'. The front support system was a saddle in the front frame cross-member on which rested a large trunnion, machined on the rear of the front gear case, which was upwardly eccentric with the crankshaft axis. A light trunnion cap and non-metallic 360 degree isolation strip completed the front mount. The rear mount provision was a cast transverse member having clevis bores parallel the crankshaft which mounted vertical clevises piloted in frame brackets. These mounts incorporated isolation bushings and essentially constituted universal joints and effectively prevented propagation of frame deflections to the engine as well as engine vibrations to the frame. They were covered by U.S. Letters Patent.

Apparently the 5 quart sump had proven adequate on the Model 32, because the Model 48 was released to production with an 8 quart capacity.

### 1914

The carburetor and engine auxiliaries were changed for the 1914 model year.

A Zenith H 7-A carburetor was adopted for both the '32' and '48'. A larger "stove" surrounded the exhaust manifold and the central leg thereof; the 1913 stove was similar to the '32' and was located between the middle and aft manifold legs. The flexible duct was of a larger diameter and entered the carburetor without the 180 degree turn previously necessary.

## MODEL 41 ENGINE
### 1914

The Model 41 engine, like the entire product, was a fresh design unencumbered by utilization of prior components or tooling. It therefore reflected truly the best design philosophy of 1914, or more properly of 1912 when the design must have been laid down to allow detailing, prototype manufacture and testing, as well as

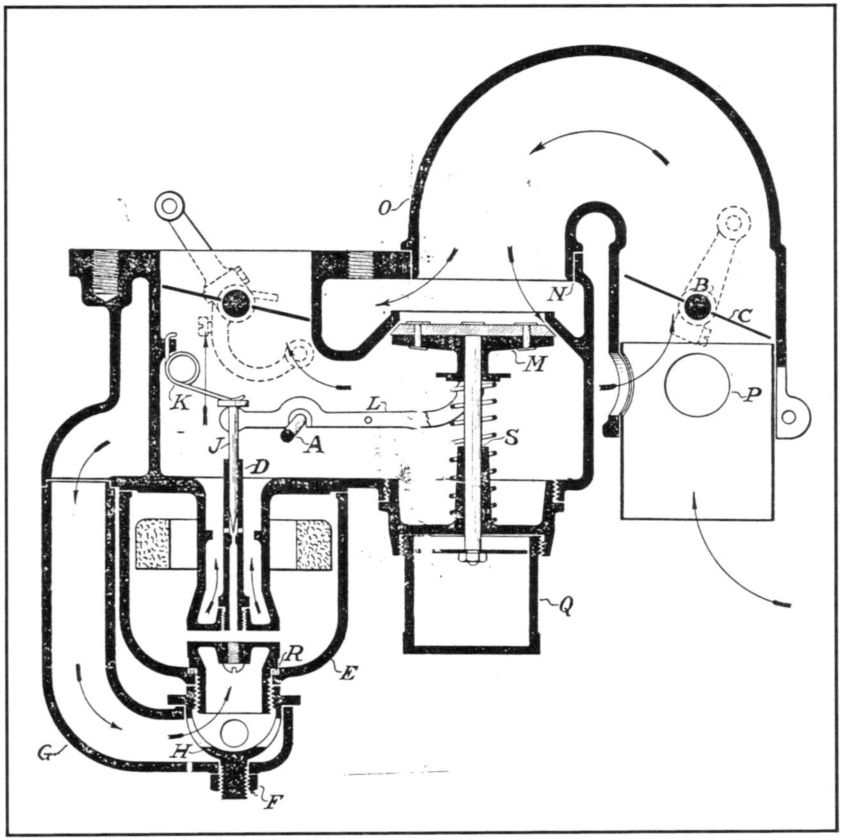


Fig. 3-55 Harroun carburetor used on 1913 Marmon engines.

On the Model 32 the intake valves opened 4 degrees before top center and the exhaust closed 13 degrees after top center, for 17 degrees overlap.

On the Models 48 and 41, the intake valves opened 10° after top center and the exhaust valves closed 14 degrees after top center for only 4 degrees overlap.[1]

The camshaft operated in a closed tubular tunnel section of the crankcase which was flooded with oil under pressure and the valve tappeu/valve interface was enclosed by covers. Each tappet was drilled axially with a tiny hole through which oil escaped the tunnel immediately beneath the bottom of the valve stem. Every part of the valve operating mechanism was adequately lubricated and the sound of the valve action was significantly quieted because the oil between the tappet (screw) face and the valve stem had to be forced out at the start of valve lift. This was a new development.

A crankshaft main bearing was provided between each cylinder to minimize crankshaft deflection and consequent vibration of the shaft and the crankcase and attendant noise generation. This had been introduced in the Model 48.

This full bearing compliment concept may not appear unusual to one most familiar with recent contemporary practice. To emphasize its significance, in 1948 the only domestic passenger car offering this feature was the Nash Ambassador Six with seven main bearings and — additionally . . . the only domestic offering of an

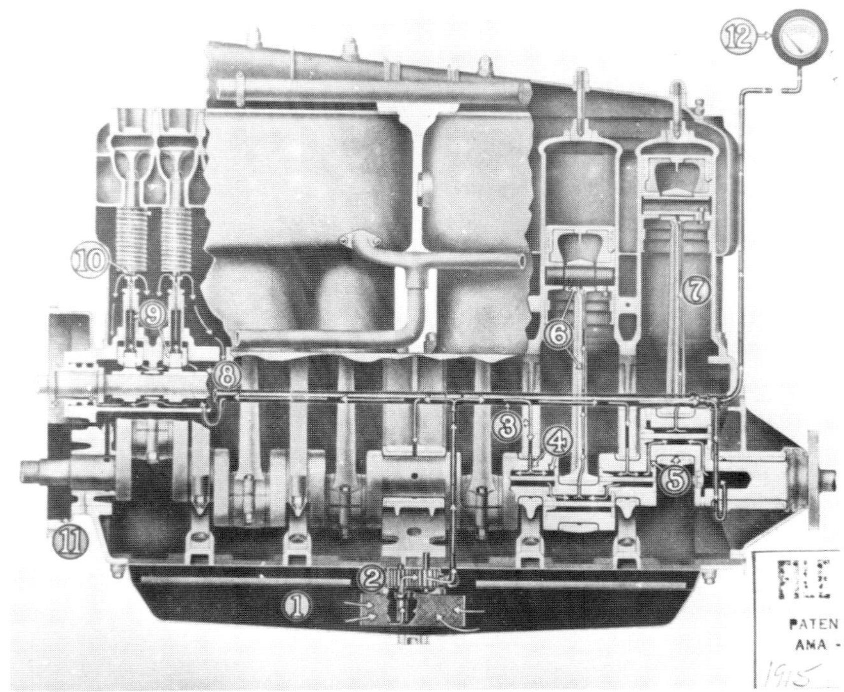


Fig. 3-56 Lubrication system of the Model 41 engine.

design and fabrication of tooling prior to the January 1914 announcement of the new Model.

The traditional barrel crankcase construction and full pressure lubrication features were continued: (1) full pressure lubrication to each main bearing and distribution, through drilled holes and reservoirs in the crankshaft, to connecting rod journals and thence through oil tubes to the upper ends of the rods to lubricate the piston (or wrist) pin; and (2) superior crankcase durability and engine rigidity (freedom from vibration) consequent to full skirt depth bulkheads at bearing locations and insertion of the crankshaft from the rear (rather than bottom) of the case into a front bearing in a bore in the front wall and closure of the rear opening by a cylindrical flanged plug incorporating the rear bearing; the split intermediate bearings being assembled in diaphragms on the crankshaft prior to insertion and being secured to the crankcase following insertion of the crankshaft.

In addition to the traditional featured constructions, two significant features, one shared with the Model 48, were incorporated in the Model 41 engine.

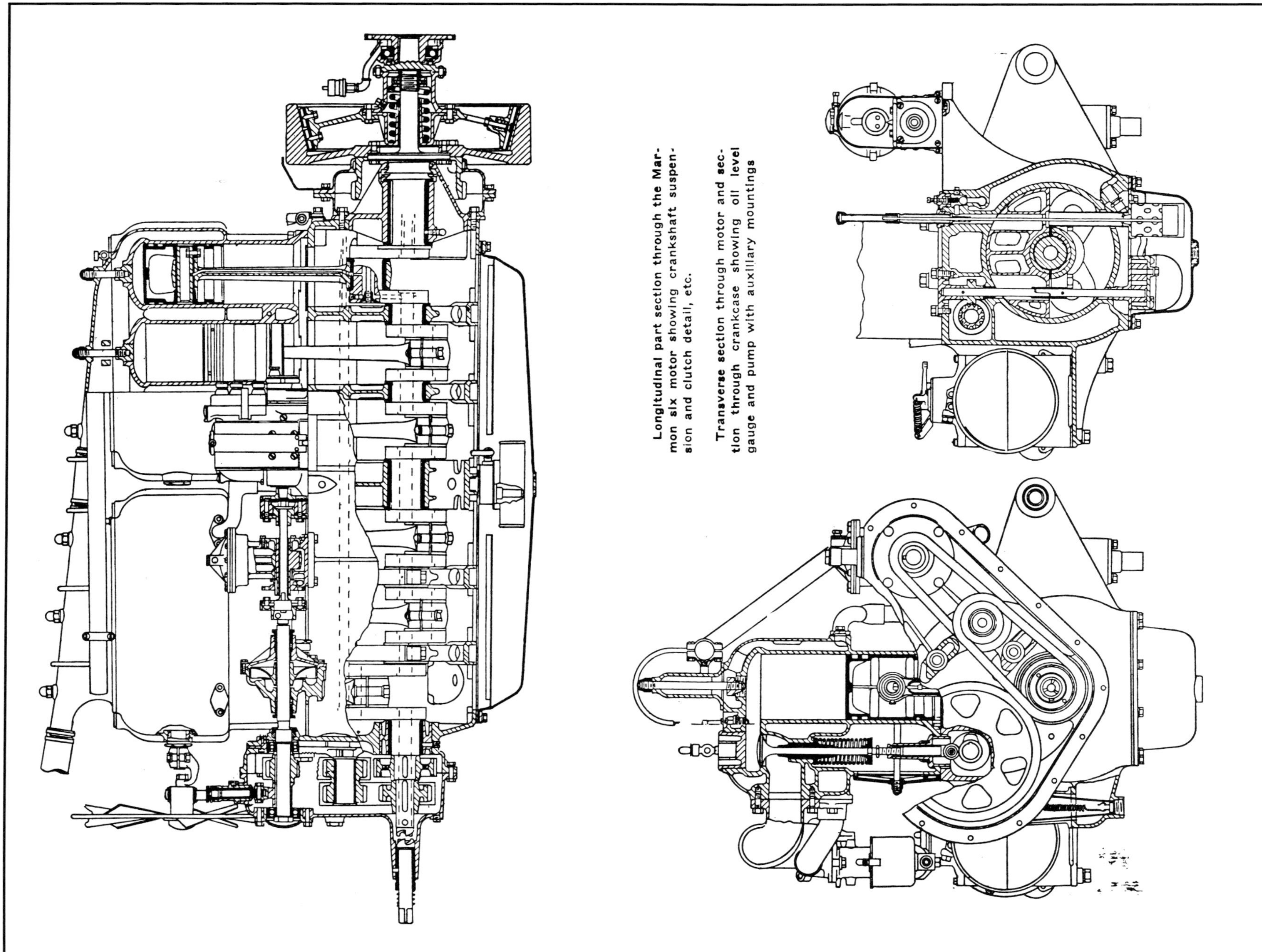

Longitudinal part section through the Marmon six motor showing crankshaft suspension and clutch detail, etc.

Transverse section through motor and section through crankcase showing oil level gauge and pump with auxiliary mountings

Fig. 3-57 The Model 41 engine was Marmon's first L-head engine, and the first designed from scratch with generator and starter.

intake manifold within the engine water jacket. The Models T and A Fords, and numerous contemporaries, had no intermediate main bearings at all. General Motors completely water-jacketed the intake manifold on the famous 215 CID aluminum Olds/Buick V-8 of 1961-3, and incorporated the intake passages in the cylinder head of the Chevrolet 250 CID six beginning in the mid 1970s. In both cases, improved fuel economy was a consideration. The "Kettering Engine" and all its industry wide descendant V-8s, beginning with the Cadillac and Olds models for 1949, had a full complement of main bearings; So also does the latest example, the all-new Cadillac 4.1 liter aluminum-block engine.

The new engine was an L-head configuration with both intake and exhaust valves operated by a single camshaft. Bore and stroke were 4.25 and 5.5 inches, giving a displacement of 447 cubic inches. The resulting taxable horsepower was 43.4. This was inconsistent with the model designation and resulted from a late decision to increase the bore from 4.125 to 4.25 (presumably for increased performance). The smaller bore agrees with a taxable horsepower of 41. The engine was rated at 70 brake horsepower maximum. The bore/stroke ratio of 0.77 was the **smallest** yet used by Marmon.

The Model 41 had two identical cylinder blocks of cast iron with three cylinder bores each. It is presumed that the use of two blocks made service handling easier, replacement of damaged blocks less costly, and possibly production tooling less costly. The Model 48 had used three blocks, but that undoubtedly was a consequence of commonality with the four cylinder Model 32.

Unlike prior engines of Era Two, the '41' had an 'L-head', rather than 'T-head' valve arrangement. Economics certainly favored the L-head but it imposed restrictions on air-breathing capacity in the forms of limited valve diameter and heating of the inlet air by the adjacent exhaust passage. Howard Marmon may have decided that the valve size penalty could be overcome by the increased engine length associated with the additional main bearing between each pair of cylinders. Valve diameter was 2.187 and lift was .375.

The L-head configuration was to become the basic design of the vast majority of domestic passenger car engines until the rush to overhead valve V-eights began with the 1949 models of Cadillac and Oldsmobile. There were however, the exceptions of Chevrolet, Buick and Nash Ambassador.

The '41' blocks had one exhaust port from each cylinder but only one central intake port. A two-legged intake manifold fed these ports. Intake air was not heated prior to entry to the carburetor nor during conduct to the cylinder block ports. The remainder of the intake passage was temperature controlled by the engine coolant. A one-piece exhaust manifold of cast design was used.

Although retaining dual ignition systems, the 41 engine reverted to the pre 1914 Marmon practice of a single set of spark plugs. Unlike all prior Era Two engines, the '41' plugs screwed into the cylinder block near the center of the offset combustion chamber formed above the bore and the two valves. Prior engines had the plugs threaded into port-plugs above the valves. In July 1914 when announcing the New Series Model 41, the manufacturer claimed it to be "the only American car with a 'Center-Fire', L-head, six cylinder engine."[14]

For 1914, or perhaps during the 1913 production run, the Models '32' and '48' had been changed to dual spark plugs, adding plugs over the exhaust valves to those over the intake valves. All were threaded into port-plugs. Probably that was a practical solution, on an existing engine, to the determination that a central plug location was preferable, eg, to promote a faster burning of the charge. A long standing objection to dual plugs **with dual ignition sources** was the probability that one set would become fouled and useless during an interval when the driver elected to operate only one system.

The barrel crankcase of the Model '41' was similar to the '48' in frequency of main bearings and method of intermediate diaphragm retention. The principal improvement was the avoidance of the oil distribution manifold, after only one year of experience with that construction on the '48'. (Such manifolds were still in use on some marques, still surviving today, into the 1930s). A steel tube was cast in-place in the Model 41 aluminum crankcase and was intercepted by transverse drilled passages. The front main bearing was poured in a replacable carrier sleeve retained in the aluminum case by a threaded dowel; previous Marmon engines had the bearing in an iron front cover which was bolted in place.

The aluminum crankcase included the rear portion of the enclosure for the front geartrain and the large machined trunnion for the front mount on the rear face of that enclosure. The front "gear" train was unusual. In the absence of the second camshaft drive gear (of a T-head configuration), the drive to the accessory axis was by chain from a sprocket immediately forward of the camshaft gear train. In the interest of minimizing lash in the drive to the ignition distributor on the magneto, a pair of chains were used, with an adjustable location of the intermediate idler-sprocket shaft which could be shifted to correct for stretch and/or wear in the two chains. The idler offset placed the second chain in the plane of the camshaft gear set. The accessory drive sprocket was carried on ball bearings and drove a water pump mounted on the crankcase. The magneto was located just aft of the engine center and was driven by a shaft, with flexible couplings, from the rear of the water pump. This shaft incorporated a cam driven diaphragm-type pump.

The starter/generator unit was coupled with the crankshaft by chain at the rear,

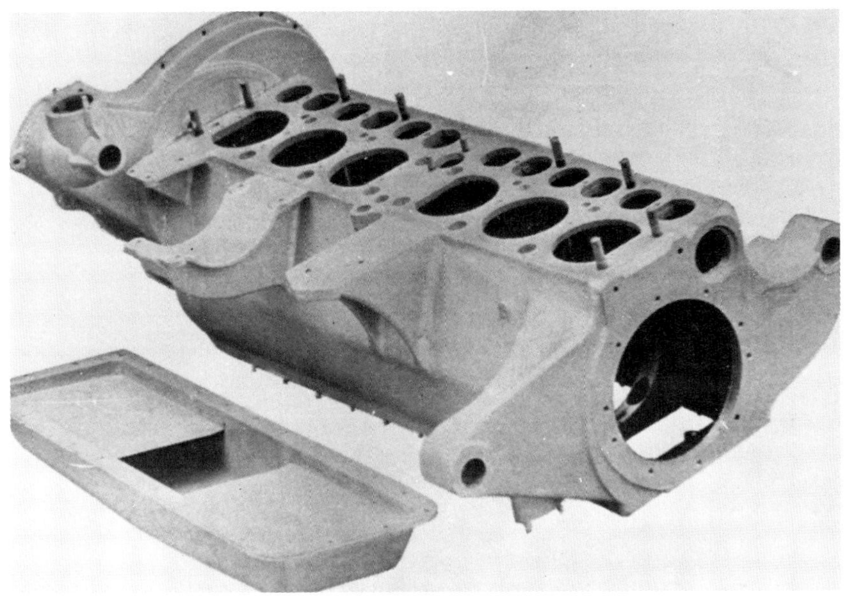

Fig. 3-58 Aluminum crankcase and pan — Model 41

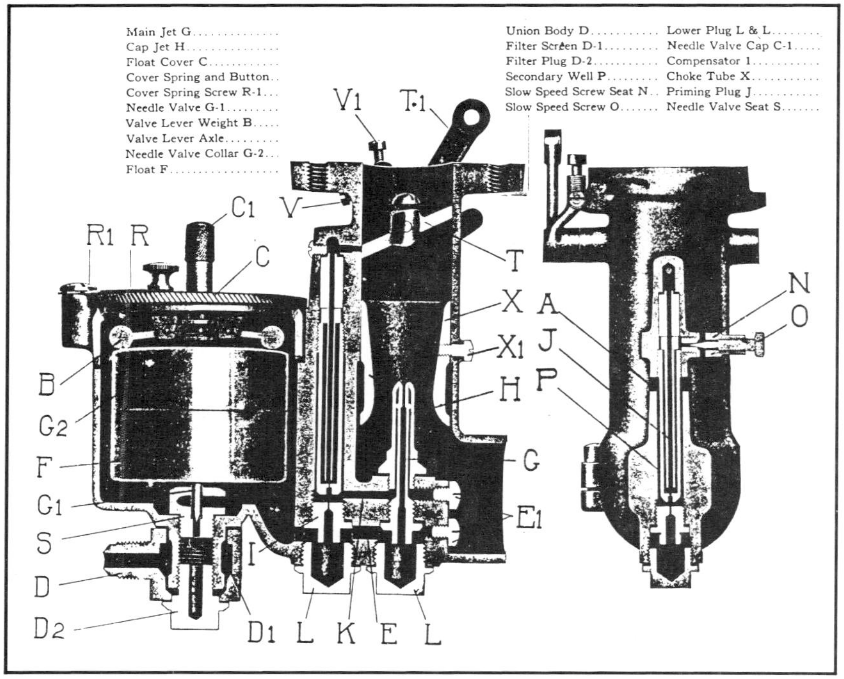


Fig. 3-59 Zenith carburetor adopted for the 1914 Models 32 and 48.

the location of the Model '32' system. A probable prime consideration was the width of the complete engine associated with the forward location; this could be tolerated on the expensive Model '48' because the center-plane steering system allowed a wider frame. It is likely that experience with the front chain on the Model '48' influenced both the rear location of the starter/generator and the adjustable dual-chain drive for the accessory axis on the Model 41. Drives from the front end of the crankshaft can be very active torsionally. In Marmon Era Two, the crankshaft front vibration damper had not yet been invented nor had means to measure, either qualitively or quantitively, the vibratory activity of the crank; that would have to wait for the Marmon innovations of Era Three.

The North East starter/generator unit and its combination saddle-mount and two speed transmission was the identical assembly used on the Model 48.

Extensions cast on the crankcase provided horizontal mounting pads for the diaphragm pump, the magneto and, on the right hand side, the generator/starter. A drive flange for a chain sprocket was sandwiched between the crankshaft flange and flywheel. A closed case for the generator/starter drive chain was bolted to the rear of the crankcase. This construction was similar to that already in use on the Model 32. Because this was a totally new engine, and possibly in anticipation of a subsequently adopted electrical system, the engine rear mount legs were cast integrally with the crankcase; the '32' mounted on the starter/generator chain case.

The carburetor location directly above the starter/generator forced the intake manifold to be negligable in length below the 90° turns to the nearly horizontal forward and rearward legs. Because the carburetor throttle shaft was transverse, this configuration undoubtedly resulted in maldistribution of the mixture at part throttle conditions. The carburetor was a Zenith Type L-6 (1.5 dia.).

Some limited metallurgical information is available and it is recorded herein as indicative of the best practices of the mid-teens.

Connecting rods had a length of 12.75 inches, were forged from 0.35-0.40 carbon steel and heat treated.

The camshaft had **integral cams** and was machined from hot rolled steel bar of 0.30-0.40 carbon content and heat treated.

The crankshaft was forged from an open-hearth steel billet of 0.38-0.40 carbon and heat treated.

The crankshaft had seven main journals. The front one was 2 diameter, 2.56 long; the intermediates were 2.125 diameter and 1.137 long except the center one which was 3.562. The rear journal was 2.25 diameter and 4.495 long.

Bearing technology was then such that individual bearings had to be hand-scraped to final size. The Marmon design of full 360 degree front bearings with intermediate and rear bearings pre-assembled in split diaphragms greatly facilitated achievement of precision bearing fits. Each individual bearing could be fitted and clearance confirmed prior to insertion of the shaft through the rear end of the tunnel crankcase. But the tunnel crankcase also greatly increased resistance of the crankcase to fatigue failure at the point of bearing cap attachment in conventional engines.

## 1915

The keystone of Marmon quality was the fact that Nordyke & Marmon, like Mack truck, produced nearly all of the product including castings and forgings in their own shops. A necessary exception was the electric equipment; had Nordyke & Marmon not sold their Jenny Electric venture to General Electric prior to the start of the automotive era, that capability too might have existed in-house.

Marmon had used a Bosch magneto. The advent of electric starting saw adoption of a North East starter/generator. The associated electric lighting and controls proliferated both vendors and responsibility for quality control and for field performance. After three years of North East usage, a change was made for the 1915 model year.

The 1915 Marmon Model 41 was the first car to be announced with the full Bosch system. The magneto, generator and starter (three separate units) and all switches, wiring harnesses, lamps and bulbs were procured from a single source . . . Bosch. The battery was a 12 volt, 60 amp-hour Bosch-Elba, positive grounded, and the entire system was 12 volt; the Bosch firm had claimed (prematurely by decades) to have solved the problem of failure of fragile lamp filaments by adoption of a special tungsten alloy combined with a mushroom filament configuration. (When voltage is doubled, the same energy is transmitted at half the current flow. Therefore the filament resistance must be doubled by halving the cross-section area of the filament. The diameter must be reduced to one fourth diameter to achieve this area reduction.) A fuse was located in the starter circuit. The Bosch-Rushmore generator was rated 8-10 amperes at or above 20 mph. A **single-wire** distribution system was adopted. headlights used 12-16 volt, 24 candle-

Fig. 3-60 1915 Model 41 engine with Bosch generator and starter.

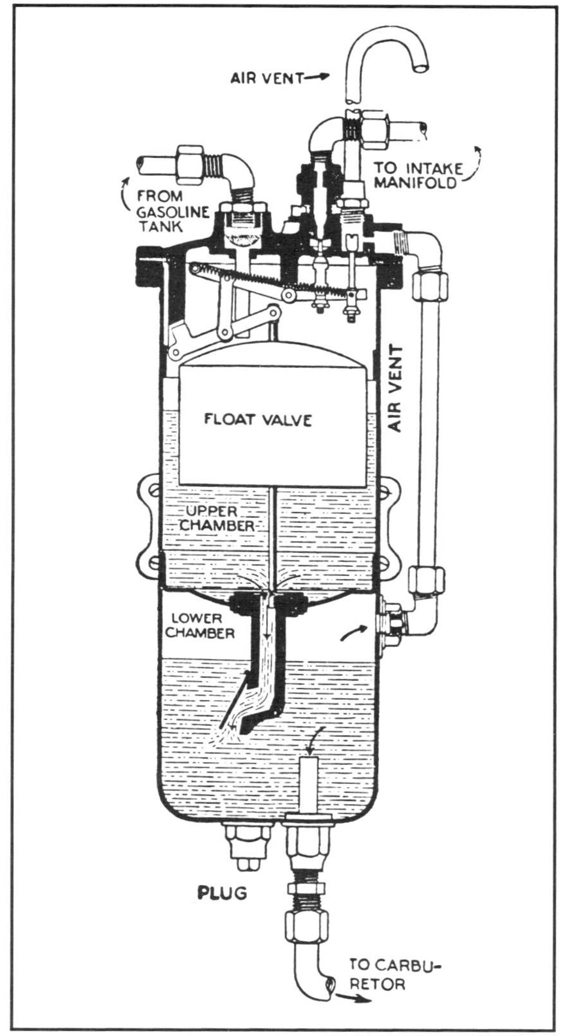


Fig. 3-61 Vacuum gasoline feed device adopted for Model 41. In operation, as the upper chamber fills, the float rises and finally permits the two springs to move the valve lever up, closing the valve on the intake line and opening one to the outside air. The gasoline then flows to the lower chamber. When the float is lowered the valves reverse and the upper chamber refills.

power bulbs. All bulbs were single contact and a 20 ampere fuse protected the wiring system. Generator output was controlled by a carbon-pile regulator mounted on the dash in the control box with voltmeter, **and** armmeter and switch.

The Marmon catalog provided the following arguments in support of the 12 volt system (over the alternative 12/6 volt): "This E.M.F. (ElectroMotive Force) had many advantages over the 6 volt circuit, such as increased lamp and cable efficiency, more certain generator and starting motor commutation, better current regulation and improved battery service conditions."

A ring gear was added to the periphery of the flywheel and the starter unit was strap-mounted to a saddle machined in the right-rear mounting leg of the crankcase. This location and drive was to become the universal practice by WWII. The starter pinion was attached to the armature which was free to move axially in the housing but was biased in the non-engaged direction by a spring. Engagement resulted from electro-magnetic thrust when the motor was energized; when the engine started, the current decreased and the spring force disengaged the gear. The switch was floor-mounted, heel operated.

To accomplish engagement of the gear teeth, the operating switch of the Rushmore starter was so designed that when in the inactive position it almost shortcircuited the armature winding through a limiting resistance grid; thus when the switch was closed, a very heavy series field was built up. The strength of the resulting magnetic field was such that the armature and its attached pinion were pulled into the housing thereby engaging the flywheel gear. Because of the partial short, the armature turned very slowly and engagement occurred without abutting or clashing. Continued travel of the foot-operated switch plunger jumpered the grid resistance and allowed the armature to be in full series connection with the field windings and cranking to begin. Because the motor was series wound, when the engine started the removal of the load resulted in a 'run-away' of the armature; the field was consequently weakened and the axial spring overcame the magnetic thrust, disengaging the pinion. During the return stroke of the switch plunger, removal of the grid resistance jumper was averted by a simple override spring or clip which lifted the switch arm until it had passed the contact.

The starter was capable of spinning the Model 41 engine at 120 to 150 rpm under average conditions.

In the event that hand cranking was necessary, a switch activated by insertion of the crank switched the ignition from magneto to battery. Upon attaining a start, automatic reversion to magneto took place. This feature was not carried over to 1916 possibly because of negative perspective . . . ie, the need to ever handcrank.

Both magneto and single-coil battery ignition utilized the magneto integrated distributor and a single set of spark plugs.

The 1914 engine had required an unusually long rear main bearing consequent to the extreme overhang of the flywheel required for the enclosed chain sprocket associated with the combination starter-generator. With adoption of separate electrical units, and relocation of the starter interface to the rim of the flywheel, the rear main and crankshaft were shortened to a more conventional configuration.

Disposal of the starter and generator functions to individual units independently mounted made feasible a much improved intake manifold. The former T-shaped manifold was replaced by a Y-shape and the negligable riser length of the former was appreciably extended.

A Stromberg Type H-3 carburetor was adopted, replacing the Zenith. Both choke and air mixture adjustment were controlled from the dash. Fuel feed was by a vacuum system, replacing the former pressure system.

The Marmon '41' Speedster in the Brooks-Stevens Museum at Mequon, Wisconsin has the 1915 engine configuration. However the inlet manifold, although of the new 'Y' configuration, incorporates two integral exhaust-jacket 'muffs' on the two manifold legs, joined by a cored passage . . . the manifold between the muffs has a figure-eight cross-section. Plumbing, similar to fuel-line hardware, connected the forward muff to the number two exhaust leg and the rear muff to the manifold just ahead of the outlet flange. This particular car has an unusual serial number . . . S/N C/C005 on the N & M plate. It apparently left the factory as a chassis forwarded to a custom body builder, and no model year reference is evident. It also has the Zenith carburetor of the pre-1915 engines.

The water pump was relocated forward of the drive chain with consequent avoidance of the dual packing necessary when the drive to the magneto (and generator) passed through the pump. Single alignment couplings connected the drive spindle with the generator and the magneto, the generator in principle, occupying the former water pump location.

A Kellogg tire-inflation pump was driven by exposed gearing from the forward generator coupling. The generator was attached to the vertical crankcase surface by two split clamping rings, bolt connected.

There were some detail mechanical changes to the engine. For manufacturing convenience, the front gear/chain enclosure was made separate from the crankcase. This obviously had advantages as the gear enclosure projected above the machined upper surface of the crankcase and similarly obstructed the finishing of the side mounting pads. However the acknowledged reason was to improve the machining operation on the large front trunnion, upwardly eccentric to the crankshaft axis, which was between the gear casing and the front cylinder block. The finished surface was recessed between the gear casing and the crankcase below the crankshaft but extended above the crankcase at the top . . . a difficult production operation.

The pistons were altered to have a length of 5.455 inches and all three piston rings were above the piston pin. The lower ring was an oil control ring; nine .125 holes were drilled in the piston around the circumference of the groove, downward at a 45 degree angle. Marmon 41 pistons and rings were licensed under U.S. Patents 908569 and 908570 issued in 1909 to an F.P. Howe outside the Nordyke & Marmon organization. The former covered the use of a sharp shoulder on the piston to act as an oil scraper, an adjacent relief connecting with drilled holes for drainage; the latter covered the use of an oil control ring with a sharp lower corner but with the upper portion of the ring length tapered inward to avoid scraping action on the upward stroke and to spread the lubricant over the wall.

## 1916

A by-pass valve was added to the pressure lubricating system in addition to the pressure regulating valve. The setting of the latter was increased and the dash gauge scale was expanded from 12 to 60 psi. The higher oil pressure was only carried when the throttle was opened and was in some proportion thereto. This action resulted from interconnection of the throttle and the oil by-pass valve. Thereby excess smoking at idle and excessive consumption under cruise conditions were avoided. Greatly improved lubrication and durability at power was achieved. Some form of oil pressure modulation would become a continuing Marmon characteristic.

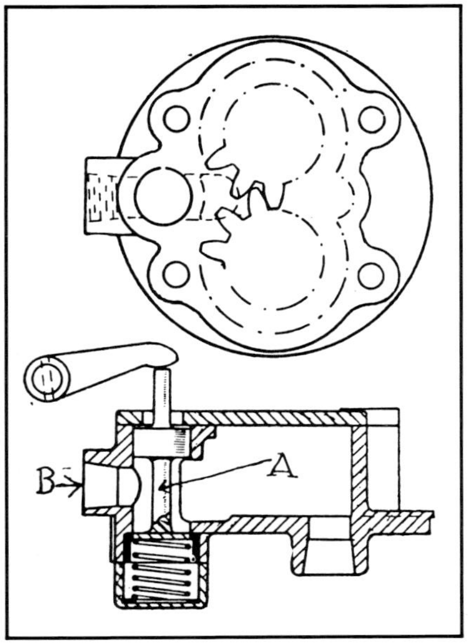


Fig. 3-62 Marmon Model 41 oil pump with throttle which adjusted oil pressure in accordance with carburetor throttle opening.

The water-jacketed version of the carburetor was adopted . . . the Stromberg Type HA-3.

## PART 3 — The Chasses

### MODEL 32 CHASSIS 1909-1914

The introductory brochure for the new Model 32 said of the chassis: *"The chassis of the Marmon "Thirty-two" is an evolution of the highest degree of motor car engineering ability. It is thoroughly rational in power, weight, and carrying capacity, and consistent in price: that is to say, a car of similar power, weight and carrying capacity cannot be designed to weigh less and still maintain a proper factor of safety; at the same time, any greater weight in a car of this size would be entirely unnecessary and a detriment, rather than an advantage. The simplicity of the design is striking, and the extreme accessibility of all units and parts is manifest upon the most casual examination. No car has ever been designed with more thought bestowed upon these qualities. Special attention has been paid to the protection of all essential moving parts from dust and mud. The neat, trim and well proportioned appearance of the chassis cannot help but appeal strongly to the novice, as well as the expert. The facilities for adjusting, cleaning and care-taking are apparent and unique as compared to the general run of motor cars."*

The *Motor*[15] reporter (in reporting the 1910 model) commented, "Looking upon the chassis from above, one is impressed by the simplicity and directness of action of the several parts" and directed the observers attention to the enclosure of the gear shifting rods.

#### FRAME

The '32' frame was of pressed steel construction. It was characterized by the complete absence of castings; pressed steel was used for the brackets which supported the floor boards and at other points. The side rails were parallel but were 4 inches closer together foreward from the rear of the engine; this assisted in achieving a narrow turning radius. At the point of offset, the flanges were wide but progressively narrowed the remainder of the length to the rear. The rail depth decreased slightly in the rear to a 3 inch kick-up at the rear axle. The frame terminated not far beyond the rear axle and spring upper trannions.

The frame of course was lengthened as the wheelbase grew from 112 to 116 in 1909 and to 120 in 1911. The 1910 semi-chassis is illustrated as the mature frame constructon.

**For 1910,** the frame was also modified and subsequent years in a manner that reflected the continuous search for detail refinements. Forward of the midships cross-member, the lower flange of the frame rails was turned downward at its edge to provide an attachment for securing the mud-pan under the forward chassis by means of bolts accessible from the skdes of the car; Evidently the frame flange was tapped. Additional stiffness of the frame structure was also claimed, and perhaps necessary with the lengthened wheelbase.

**For 1911,** the headlight support system became primary with the front fenders being supported thereby. Cast malleable iron brackets, riveted to the side of the frame rail but inboard of the fender apron carried the headlamps. A transverse tie bar located in the same transverse plane contributed to the rigidity of the structure. Fender vibration was significantly reduced.

The rear spring mount forgings were redesigned from a pad configuration to a trunnion and were extended inboard to one third the frame width, following the gusset channels to an inboard anchorage on the rear cross member.

The forward (engine front support) and central frame cross members had integral gussets. An additional frame bracket was provided to allow shortening the clutch control shaft which had been full width of the frame; improved clutch control resulted.

#### SUSPENSION

Suspension of the Model 32 was a radical departure for Marmon . . . the first model to bear the marque above an apparently conventional suspension. Front springs were semi-elliptic, the rear were full elliptic. The former were 35 inches long and 2 inches wide with conventional shackling at the rear of the spring. The front axle was located only 15 inches from the front spring eye; an arrangement which increased the effective wheelbase and placed the front axle directly beneath the radiator without compromising riding quality. Marmon designers considered the axle/radiator alignment important to the appearance of the car. The rear springs were 40 inches long and 2 inches wide; they were of scroll-end construction, ie, the upper and lower semi-elliptics were joined by shackles which compensated for the slight forward and backward movement of the axle as it moved about its fulcrum where the torque-tube was pivoted to the central frame crossmember. The lower spring was pivoted on the axle housing, a trunnion journal being machined on the housing concentric with the axle shaft centerline; the upper spring was clamped to a forged bracket on the frame rail.

Marmon designers, responsible for "The Easiest Riding Car in the World," were very concerned about forces of unknown magnitude which would result from dis-

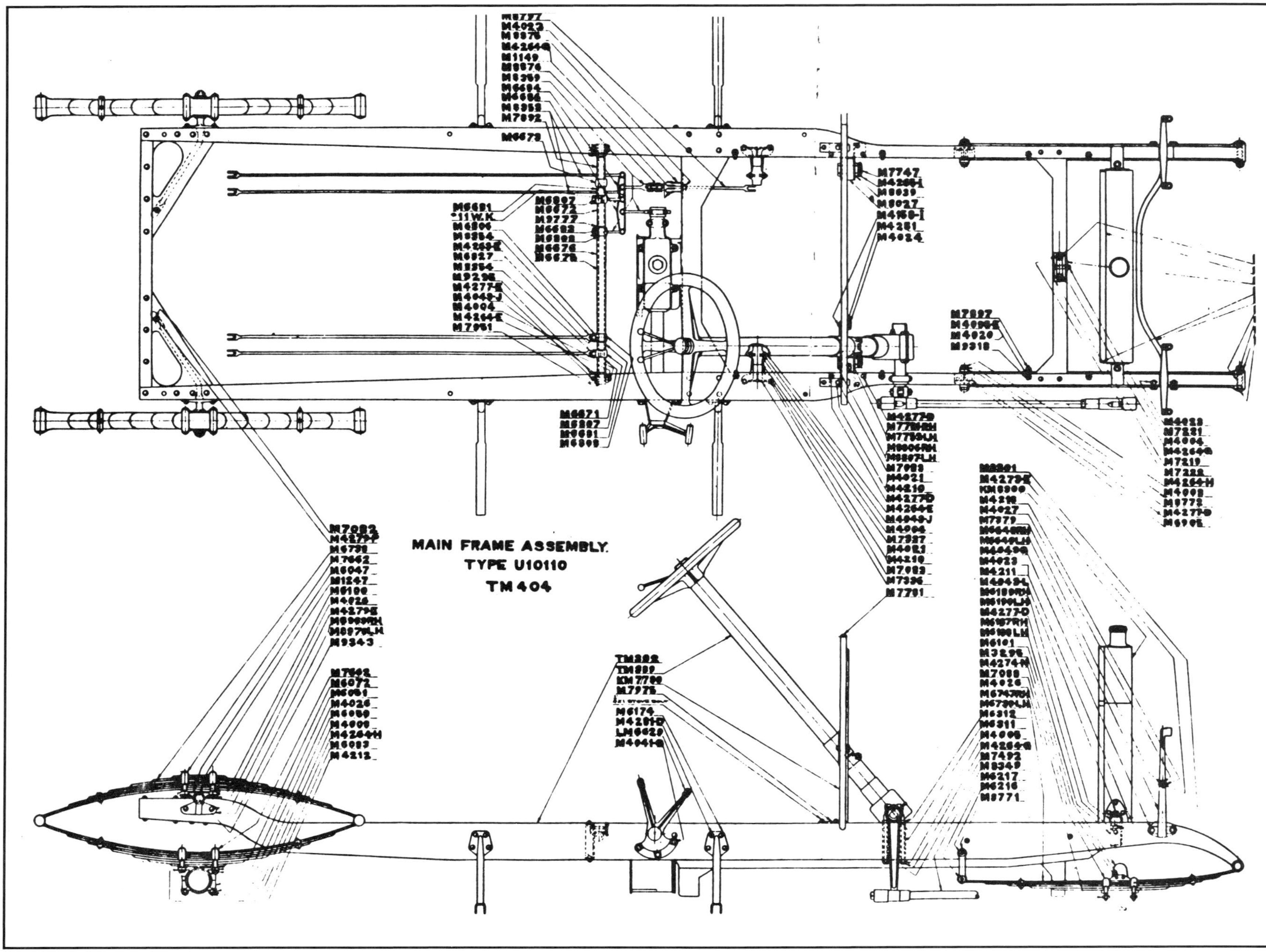


Fig. 3-63 Engineering drawing of 1910 Model 32 main frame assembly.

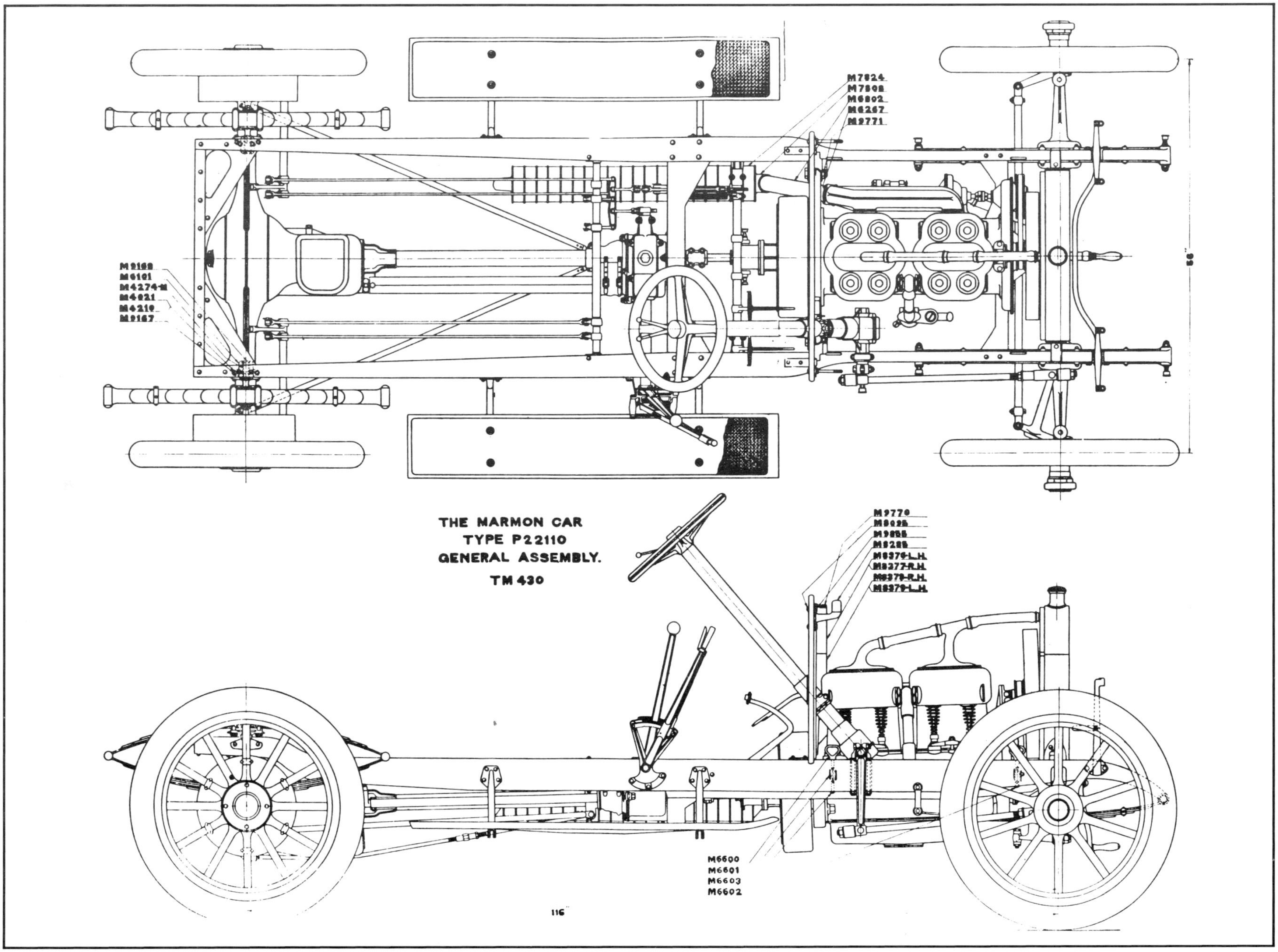


Fig. 3-64 Engineering drawing of 1910 Model 32 chassis.

Fig. 3-65 Rear chassis of Model 32 (1909) with shift rods and brakes exposed

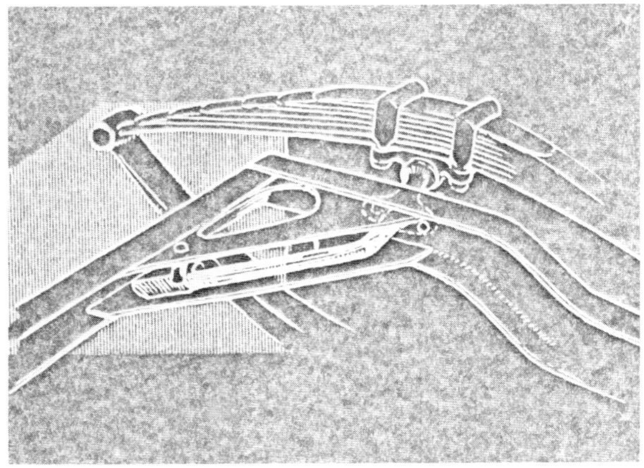

Fig. 3-66 Model 32 rear spring pivot with extension for broad-based anchorage.

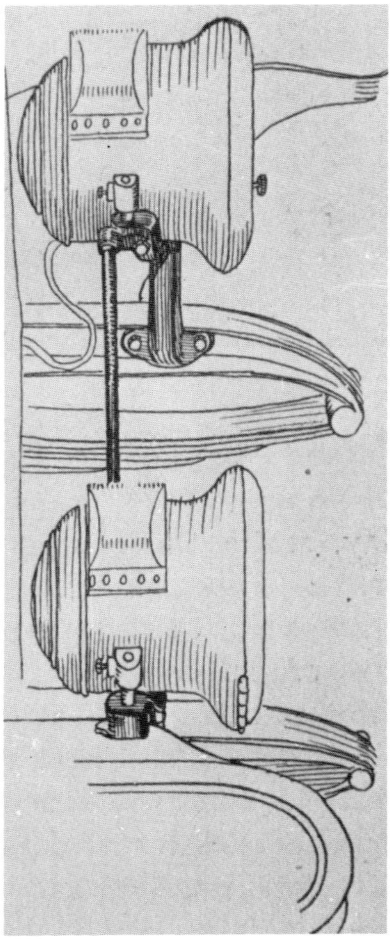

Fig. 3-67 Rugged headlight mounting of Model 32

tortions of linkages and could compromise predicted durability and ride performance. An example, one of many in various Marmon models, is the use of the scroll end rear spring and pivoted lower spring pad on the axle.

**For 1912,** a reinforced front spring shackle was adopted. The one piece shackle forging incorporated transverse elements above and below the spring pivot axis.

The rear spring attachment to the frame was changed from a rigid clamping to a pivoted attachment to a trunnion on the frame. The use of both upper and lower trunnions eliminated the need for the scroll end upper springs and the associated shackles. The upper and lower springs were both semi-elliptic and were joined with a single pivot bolt at each end.

## WHEELS AND TIRES

Tires were "plain" tread 34 X 4 (26 inch wheel) front and rear; for 1912 they were mounted on Marsh demountable rims and for 1913 on demountable, quick-detachable rims. The latter offered two methods of tire repair: off the car by removal of the tire/rim assembly before repair or substitution of a spare assembly or alternatively by tire removal from the mounted rim, after removing the T-bolt and acorn nut lock on the split outer ring which had a tongue and groove engagement with the rim. In 1911 32 X 4 (24 inch wheel) tires were optional; by mid 1913 tire size was 33 X 4.5 (24 inch wheel). Spokes were made slightly heavier at the start of 1912 production.

Front wheels turned on imported Hess-Bright tapered roller bearings.

The Model 32 used Schwarz wheels as did nearly all Marmon cars ever built with wood artillery wheels. The wheel used a patented and superior construction which is described in the "Tire and Wheel" section of the "First Era Chasses", Chapter Two, Part 3.

## THE STEERING SYSTEM

The '32' front axle was basically similar to the Model H. It was a new unit however as the extension of the integral spring perch outward to the spindle was not evident. The steering spindle (king pin) bushings were installed in the yoke which was forged integrally with the wheel spindle. The steering spindle bolt was threaded

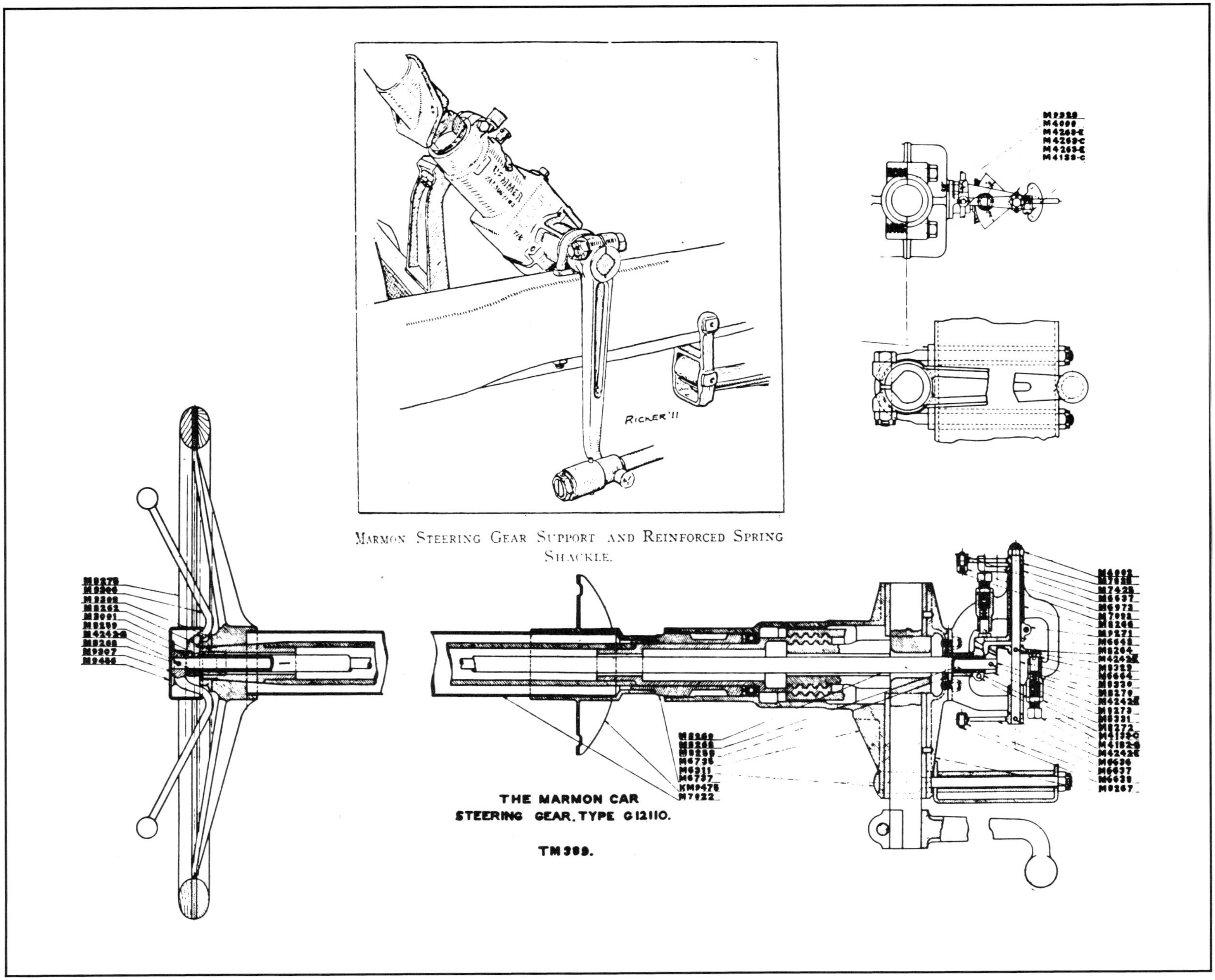


Fig. 3-68 The 1912 Model 32 steering gear with Gemmer gear box.

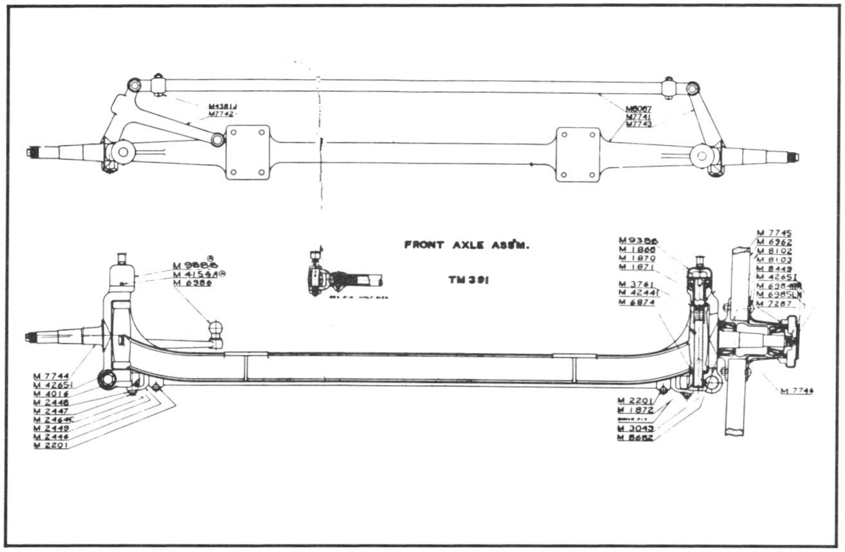


Fig. 3-69 Model 32 front axle assembly and speedometer gear support.

into the axle end and carried the weight of the car on its head through a ball thrust bearing which was enclosed with a dust cover.

The tie-rod end yokes were forgings threaded into the tie rod and were locked in adjustment by circular split clamps. They were bronze bushed with provision for lubrication and fitted with hardened pivot pins. In keeping with the light weight objective of the Model, and to allow an interchangeable spindle, a single forging incorporated both reach-roe and tie-rod arms. It would have been feasible and probably economic to saw off the reach-rod arm for the opposite side part. Beginning with the 1910 model, the tie rod was located behind the front axle to protect it from damage.

The high content of in-house manufactured components was reduced with the introduction of the Model 32 . . . a Gemmer steering gear was specified. Uniquely, the gear box incorporated both the trunnion mount concentric with the output shaft and a mounting pad on the opposite side. These combined with the flanged-clamp, for firewall support of the tube, to give a rigid three point mounting conservative of weight. It did mark another departure . . . late Era One Marmons all featured steering gears with the column rake-angle adjustable for the driver. The '32' column was fixed in position, perhaps in keeping with the light weight and simplicity objectives. In the Third Era, Marmon would again design and manufacture steering gears.

With the adoption of a vendor steering gear, the complex Marmon position-retention device for the throttle and brake levers on the steering column was replaced by a friction retention system but evidently was unsatisfactory for a ratchet retention system was adopted for 1910 production. Apparently another friction system was used in 1911 because another reversion from friction to ratchet was reported for the 1912 model year.

## POWER TRAIN SYSTEM

The power transmission system of the '32' was a complete departure from previous Marmon practice. A complete change was of course consistent with abandonment of the exclusive dual-frame suspension. But, typically Marmon, the new power train was innovative both in concept and in manufacturing process. Inherent "maintainability" was a continuing characteristic of the new model. A trade reporter commented "Great interest centers about the rear axle components of the Marmon."

The transmission location on the Model 32 was surprising. Marmon had begun Era One with a midships transmission, and had progressed to the unit powerplant. To adopt a transaxle configuration would seem to ignore the impact of greatly increased unsprung weight on riding quality; for a manufacturer claiming his Era One product to be "The Easiest Riding Car in the World" it was unexpected.

But Marmon had set out to build a very light, very efficient automobile that was very serviceable. Their engineers had integrated the transmission into the rear axle with an increase in weight of only 18 pounds over the basic axle. The penalty was limited to only 18 unsprung pounds. The benefit was a chassis with only 18 pounds chargeable to the transmission! The result of the total engineering effort was a 1900 pound chassis.

The innovative drive train that was to survive for 6 model years and for 500 miles to victory lane at Indianapolis began with a breakthrough in axle housing design. The banjo housing was produced by welding a forward and aft half banjo together using an "autogenous" welding process[15] to join the projecting flanges formed on the axle halves. Cast tubular-like fittings welded to the housing ends supported the wheel bearings and the brakes and provided a machined journal surface on which the lower spring attachment was free to rotate. Because the axle motion pivoted around the ball joint at the engine end of the torque tube, the instantaneous center of rotation of the axle was different from that of the spring assembly, which deflected nearly vertically. The trunnion attachment avoided introduction of secondary

Fig. 3-70 The press was impressed by the speedometer drive installation.

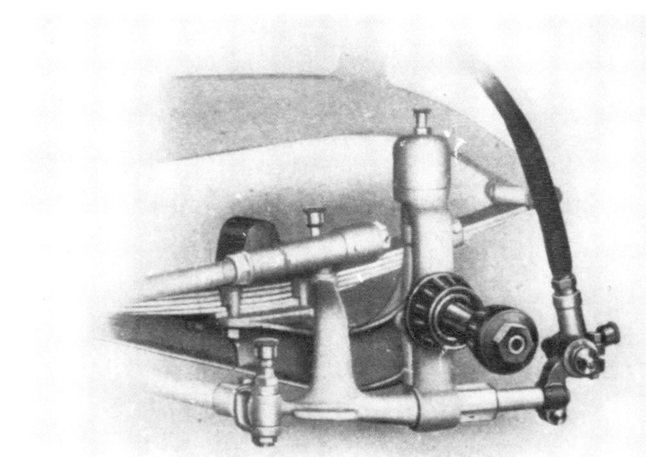

loads into the axle and possible distortion of predicted ride performance.

Beneath the banjo, and extending upwards and outwards to clevises welded to the bottom of the housing outboard of the trunnions and inboard!of the brakes, was a truss rod about .4 inch diameter. Such reinforcing devices have been appearing recently on axles of contemporary 4 X 4s equipped with outsized wheel and tire assemblies for off-road competition. In the author's youth, he recalls them only as a Buick characteristic in the early 1920s.

## CLUTCH

When *Horseless Age* ran "sneak preview" photos of the Marmon light car engine in September of 1908,[16] the illustrated clutch was a multiple disc type . . . probably the Model H clutch. But when the '32' was announced it had a cone clutch . . . another instance of the drive for simplicity and light weight.

The cone clutch was of the pull (to release) type. Cone clutches did not require a lever system to multiply force to overcome the loading spring (s) as is necessary in single flat-plate clutches because the wedging action of the cone provided the force multiplication.

The Model 32 cone clutch was faced with Thermoid material rather than the traditional leather. It was another example of refined design with great attention to detail. It marked the first appearance of a significant development in the design of dry passenger car clutches ultimately universally adopted: progressive engagement was accomplished by the forced compliance of a deformed member carrying the facing material. Midway between each pair of facing rivets, the designers placed eight spring-biased radial plungers which arched the facing locally.

A groove was machined at the inner end of the cone finished surface on the flywheel. It was intersected by a radial drilled hole to permit escape of any surplus oil before it could move across the friction surface.

A trunnion type universal-joint was housed in the end of the central hub behind the spring. The translatory motion of the clutch cone during release was accommodated by axial movement of the two hub grooves over the roller drive trunnions of the U-joint. It was a type of joint such as Chrysler products used for many years; used in pairs on the drive shaft, they eliminated the need for a splined joint in the shaft as is required with the usual fixed-cross joint construction. They are used frequently in transverse engine front drive applications for the same reason. There was no seal where the shaft exited the clutch. Lubrication of the bushing, on which the cone rotated and translated, was by grease cup on the exposed rear surface of the cone. The clutch turned upon an integral extension of the crankshaft which was threaded for loading and adjusting the spring.

Note that the flywheel was relatively light for that time, and symmetrical.

### 1911

The clutch was completely redesigned for the 1911 model year, although the crankshaft nose and flywheel were unchanged. The redesign extended to the front of the torque tube and marked the introduction of sophisticated sealing devices for lubricant retention and dirt exclusion. The plunger-type U-joint was relocated to the forward end of the torque-tube, replacing the fixed cross type joint.

The cone member now incorporated a large four trunnion cross in the plane of the cone and surrounding the clutch spring. The output member was sealed by machined spherical surfaces, i.e. a ball-joint construction with four axial springs in the cone member loading the seal face. A ball thrust bearing replaced the thrust washers previously used.

The central housing was packed with grease. However, the prototypes, and possibly early production cars, had a construction in which the clutch, when released, uncovered an oil hole in the crankshaft nose which was part of the engine pressure lubrication system.

This clutch was unusual in accomplishing release by translating the output power shaft, the motion being allowed by the trunnion type joint at the torque tube ball.

Also adopted for 1911 was a marine type clutch release bearing: A split collar was assembled around a flange on the drive shaft.

Alanson P. Brush either conceived this design while serving as a consultant to Nordyke & Marmon or brought it to Marmon, because on June 1, 1914 he filed application for letters-patent and Patent 842, 118 issued November 2, 1915. That patent covered both the release by translation of the power shaft and the integral universal joint.

### 1913

A simpler concept of integral universal coupling was adopted. A redesign of the extension on the crankshaft was necessary to mount a pair of cylindrical roller-bearings which were packed with grease at assembly. The flex-ring coupling adopted required no lubricant. Two ears on the release/power shaft were maneuvered through slots in an extension of the driven-cone and bolted to a flex-ring. The extension also had two lugs on its periphery, at 90° to the slots for attachment to the ring which appears to have been of Thermoid material which would provide some noise and vibration isolation. This was a very clever design.

## TRANSAXLE

The *Cycle and Automobile Trade Journal* reporter went to great detail to explain the now-familiar enclosed torque-tube drive system, evidently because it represented new and innovative construction in the fall of 1908. It is appropriate to use his descriptive text herein: *"Power is transmitted from the clutch to the propellor (drive) shaft through a universal joint of the yoke type enclosed within a spherical forward end of the combination torque and propelling shaft tube. The spherical forward end of the tube, the rear end of which is flanged and bolted to the change gear housing, provides a universal connection with the rear axle. Rods extend from just back of the sphere to the outer ends of the axle tubes and maintain proper relationship between axle and enclosing tube."* These rods were later referred to as "hound rods."

For the description of the novel, and challenging to the reporter, differential construction, that which appeared in *Automobile* (Sept. 8, 1910) will be used here. The language would be an asset to any political aspirant: *"In this it will be apparent that a unit plan of assembling is brought into play. The differential gears of the bevel type are shown in a housing to which is attached the large beveled gear, and conical roller bearings are mounted in place on the two ends of the differential housing. To the left of this assembly will be found the supporting grid, which is also contrived to accommodate the bearing back of the bevel pinion on the end of the main shaft of the sliding-gear system. The lay shaft of the system is also in view, and through the good office of the supporting grid, considering the methods of design, the entire aggregation may be assembled independent of the axle housing, after which they may be thrust into place with the assurance that they will be in proper alignment and so related to each other that the intended functions will be performed*

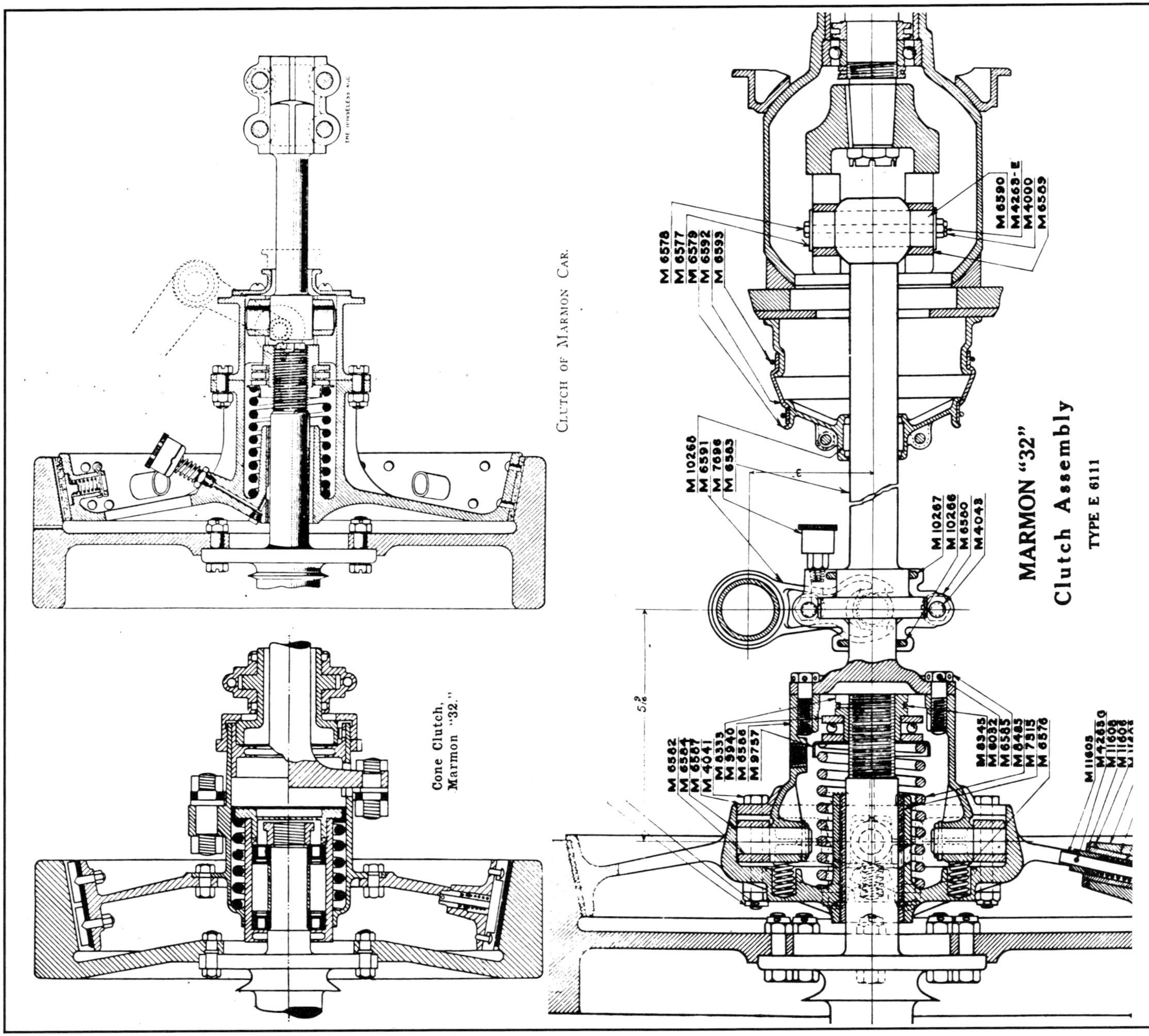


Fig. 3-71 Model 32 clutches — 1909 (upper left), 1913 (lower left), and 1911 (right).

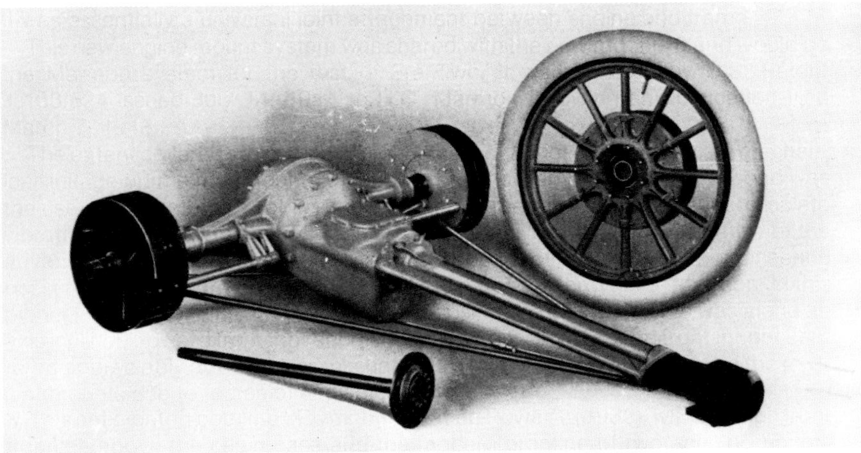

Fig. 3-72 1912 Model 32 drive train

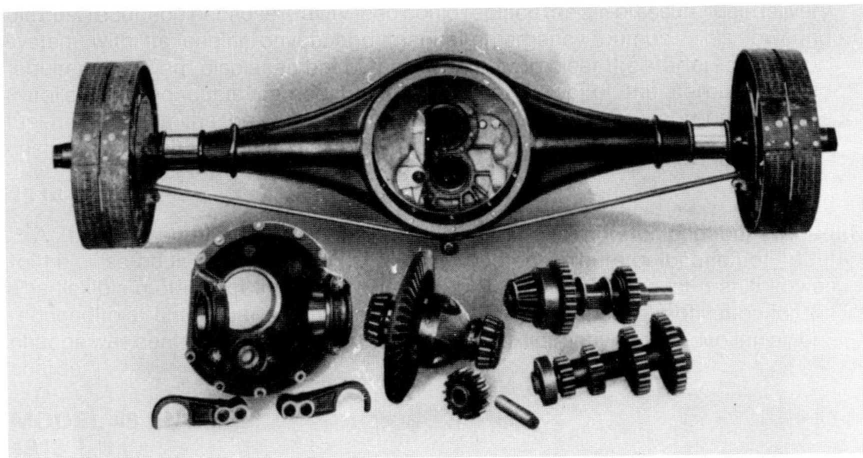

Fig. 3-73 Rear view of banjo axle with differential and transmission removed through access cover — 1909.

*noiselessly and well."*

The 1909 design incorporated the pinion and pinion shaft in the carrier, making the carrier a contemporary configuration. It was however installed from the rear. The forward banjo flange had two concentric bolt circles: One to attach the transmission from the front and a second for attaching the carrier assembly from the rear after it had first passed through the rear (cover) flange opening.

Subsequently and by start of 1911 production the pinion was made integral with the transmission output shaft. All gears were heat treated.

The rear openimg of the banjo axle housing was closed by a cast aluminum cover which, when removed, permitted the removal of the differential/axle ring gear unit (without disturbing the adjustable bearings controlling gear mesh) and subsequent removal of the change gear main and lay shaft assemblies. This could be accomplished without jacking of the car or disturbance of any of the other parts, except the partial withdrawal of the floating axle shafts. Factory mechanics could remove all axle and transmission components in **15** minutes.

Since this was the axle construction used in the Marmon which won the first Indianapolis 500 Mile Race, it should be noted that today's Indy race cars use an axle gear construction which allows ratio changes by helical gear-set substitution after removing the differential cover. Ratios are changed between qualification and race day and for different tracks.

The axle was of "full-floating" type. In that class of axle, the wheels turn upon bearings supported by the axle housing and are fixed in position on the housing. The road impacts are transmitted directly to the housing structure. The axle shafts are installed and bolted to the wheels by a flanged end of the shaft; they can be removed or replaced without jacking the car. Importantly the axle shafts are free from bending loads which are very detrimental because they reverse bend the axle every revolution and result in breakage by fatigue failure. Single row ball wheel bearings were used with vertical stability of the wheel obtained from the rigid assembly of the axle shaft flange to the wheel.

If the reader is not familiar with axle constructions, the contemporary car axle shaft carries the weight of the car and impact loads from the wheel to the housing, whereas all trucks use the full-floating design. The Fords of the flat-head V-8/transverse-spring era used a three-quarter floating axle design. The wheel/axle bearing was between the brake drum and axle housing . . . loads were carried from the wheel to the housing, but the wheel and drum were attached to the axle conventionally.

The Marmon "32" torque tube was paralleled by a housing which enclosed the gear-shifting rods. This feature was added to the prototype at start of model production. The shift lever was outboard the frame on the right hand side and operated a frame-mounted lever box in line with the shift control tube. An extension of the housing enclosed the shift-shaft all the way to the lever so that the entire shift selection and engagement mechanism was enclosed and oil tight. Beginning in 1913 (left hand drive) the shift lever (and brake) was incorporated in this central housing, which was a part of the same unit which mounted the ball-joint for the torque tube and through which propulsion and braking forces were transmitted to/from the car.

Marmon originated the transaxle concept, but not the term. Factory engineering drawings referred to the combined unit as the "transmission assembly." The press reidentified the pictures as "Marmon transmission and rear axle."

The unit, by any name, was extensively redesigned between the original 1909 version and the 1912 design. Changes included:

(1): The attachment of the central truss rod (which provided vertical rigidity to the very slender torque tube) to the banjo housing was redesigned to fasten to both sides of the axle weld.

(2): The unusual outboard pinion bearing was repositioned behind the pinion in a now conventional location. It had been in the location of the secondary bearing of a conventional design with straddle-mounted pinion. The pinion was also made integral with the shaft.

(3): The new differential carrier was supported on tapered roller bearings in the modern manner, but outwardly loaded by adjustment nuts. Previously the carrier was supported on a pair of ball bearings with the gear thrust taken by a single ball bearing.

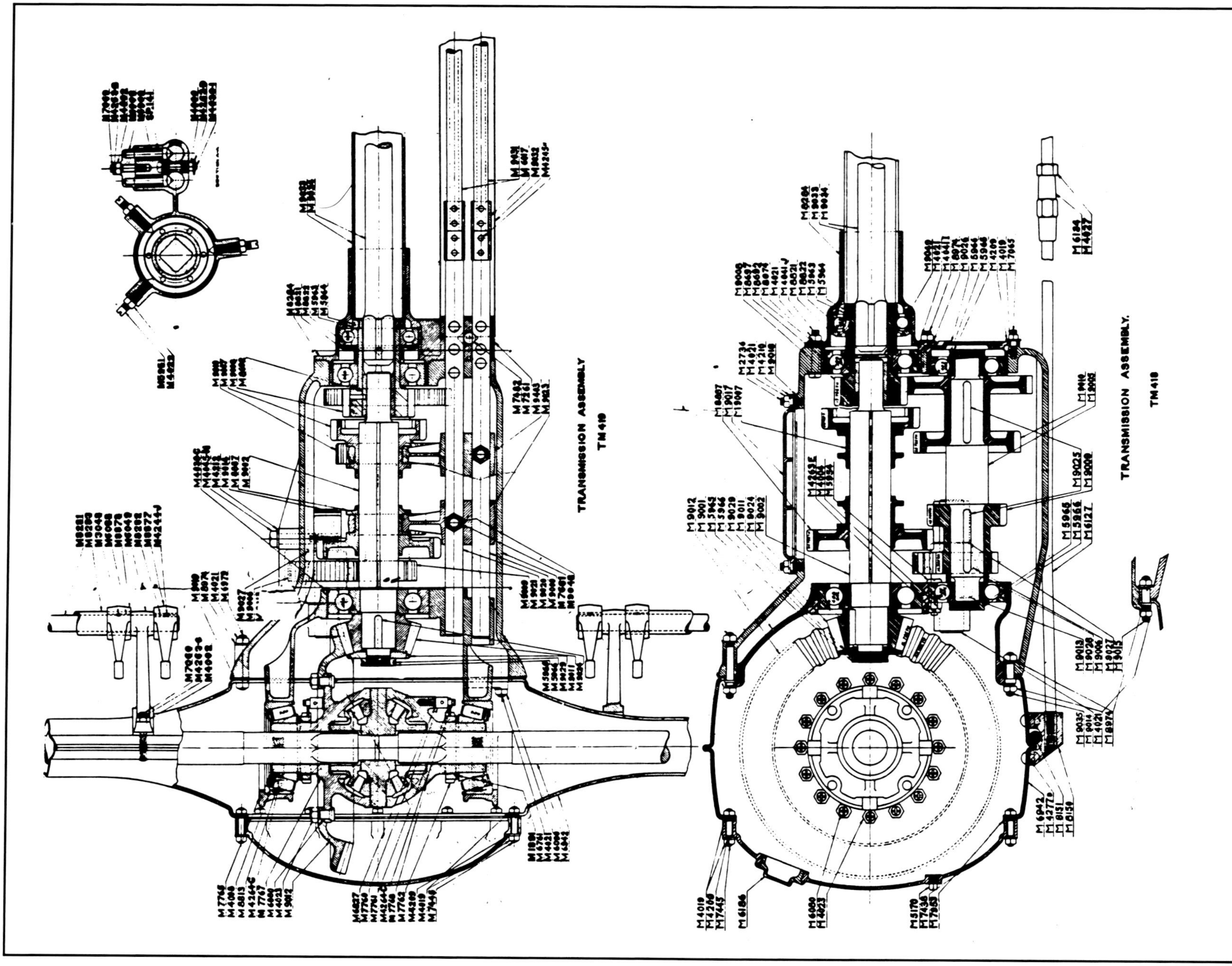


Fig. 3-74 Transaxle for Model 32, 1910.

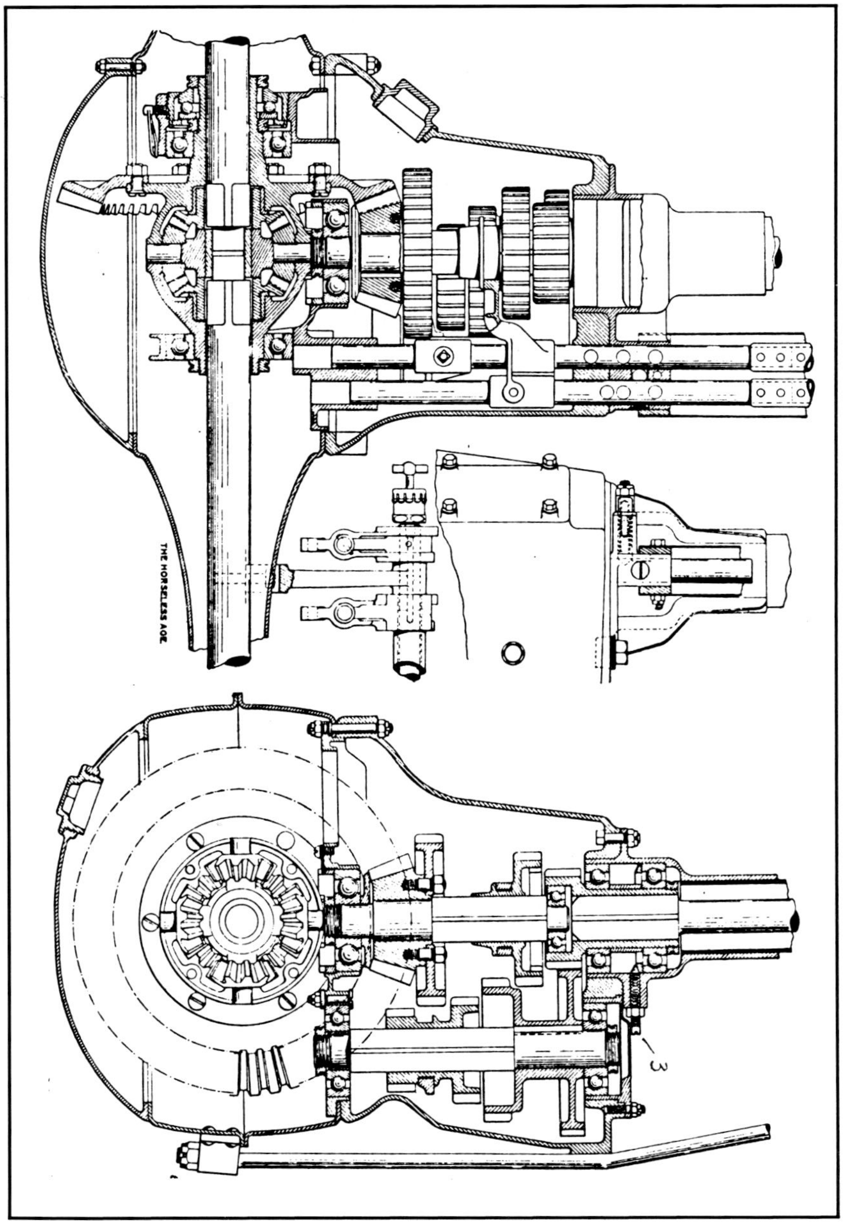


Fig. 3-75-1 Transaxle for Model 32, 1909

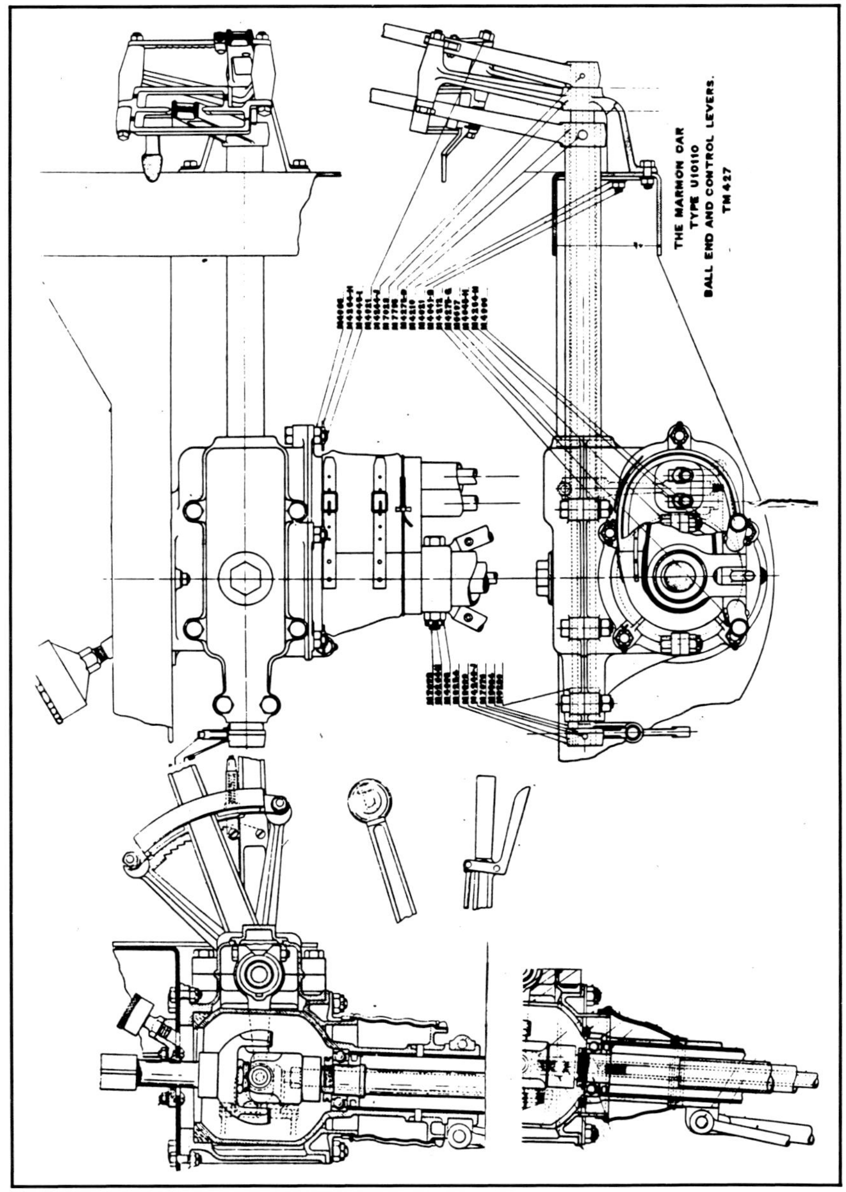


Fig. 3-76 Transmission controls, 1912, with 1909 detail in lower right.

(4): Because in the original design the low-reverse gear on the mainshaft was bolted to the axle pinion, selection of low or reverse was made by a sliding cluster on the layshaft. Second-high selection was made on the mainshaft in the usual manner in both designs. The new design used a rigid layshaft, of three element construction.

(5): The ring gear was attached with 16 bolts and nuts, instead of the original 6, for noise control. The new differential was identified as being of Nordyke & Marmon manufacture; evidently the original was not. The axle ratio was 3.375:1 (54/16 teeth).

(6): The original design boasted of very light weight consequent to the integration of the two units. The redesign was much heavier because it was essentially a common enclosure of two units; however the significant feature of access to both through the rear banjo cover was retained. For competition purposes, either transmission or axle ratios could be changed without jacking the car and only by through the banjo cover access. Not only was the Model 32 weight increasing annually, but the upcoming requirements for the Model 48 (and a current requirement in competition) and the commercial version of the Model 32 dictated a much more rugged package.

(7): Axle shaft diameter was increased 0.125 to 1.562 diameter. The entire flange was forged integrally on the shaft; previously the upset flange was riveted to the larger wheel flange. A special flat-faced rear hub cap was adopted and became a distinguishing characteristic of all subsequent Marmons of Era Two

Subsequently the design was modified to provide "a neat means to adjust the pinion with reference to the differential bevel: The pinion shaft is carried in a ring which threads into the housing. The outer face of this ring is serrated and meshes with a worm, the upper end of which extends through the differential housing. By turning this worm the adjustment can be effected from the outside."[12] This outside access to the pinion adjustment was uniquely required because of the Marmon concept of transmission access by removal of the differential/ring gear assembly and incorporation of the pinion on the transmission output shaft.

## TRANSMISSION

The '32' transmission element of the transaxle was a 3-speed and reverse selective shift-unit which by 1910 was of today's universal configuration. It was of course unsynchronized and spur-geared. Both main and lay shafts ran on cylindrical roller bearings of the type used currently, with short rollers packaged with inner and outer races (As opposed to the Hyatt type special rollers and loose races).

## BRAKING SYSTEM

The '32' brakes closely resembled those introduced the prior year on the Model H. They featured side-by-side parking and service brake shoes, of internal-expanding shoe type, acting on a 14 inch steel drum. Shoes were 2 inch wide faced with an asbestos material identified as "thermoid."

Noteworthy was the absence of the self-guided floating equalizer beams introduced the prior year on the Model H. That system was replaced by a transverse shaft system. In keeping with the minimum weight philosophy, the brakes were

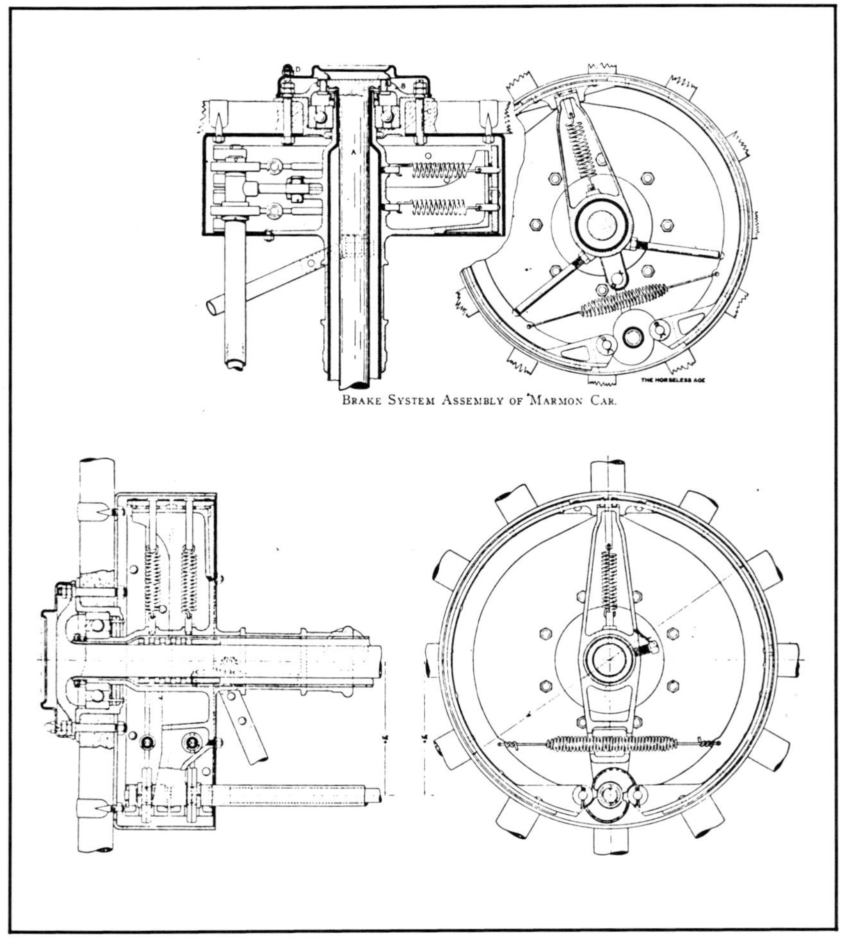


Fig. 3-77 Model 32 brake development. The upper illustration shows the 1909 configuration of Model 32 brakes. The lower shows the 1912 and final configuration. Note that the original concept of the brake cam location being radially fixed but tangentially free (between shoe-rollers) was abandoned in favor of a fixed position; the wheel bearing exterior seal had matured from a felt ring in a groove to a leather laminate stack riveted in place; the entire axle flange had become integral with the shaft; and that the axle gear lubricant had become independently sealed by a felt washer/labyrinth seal system so satisfactory that it would survive the switch to an oil lubricant in 1916 and be retained until 1928. The seal consisted of ten felt rings, five of which contacted the housing and five alternately which contacted the shaft. There was appreciable clearance with the opposite member, forming reservoirs.

Fig. 3-78 Model 32 14 inch brakes for 1909.

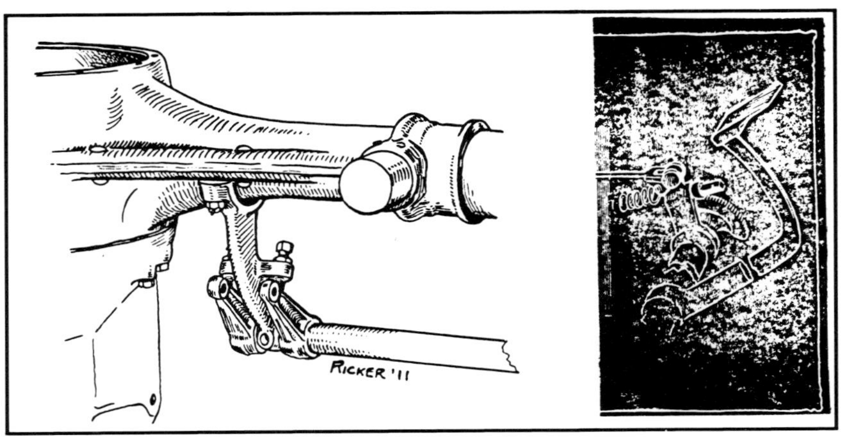


Fig. 3-79 Brake adjustment and equalization, 1910 and 1912.

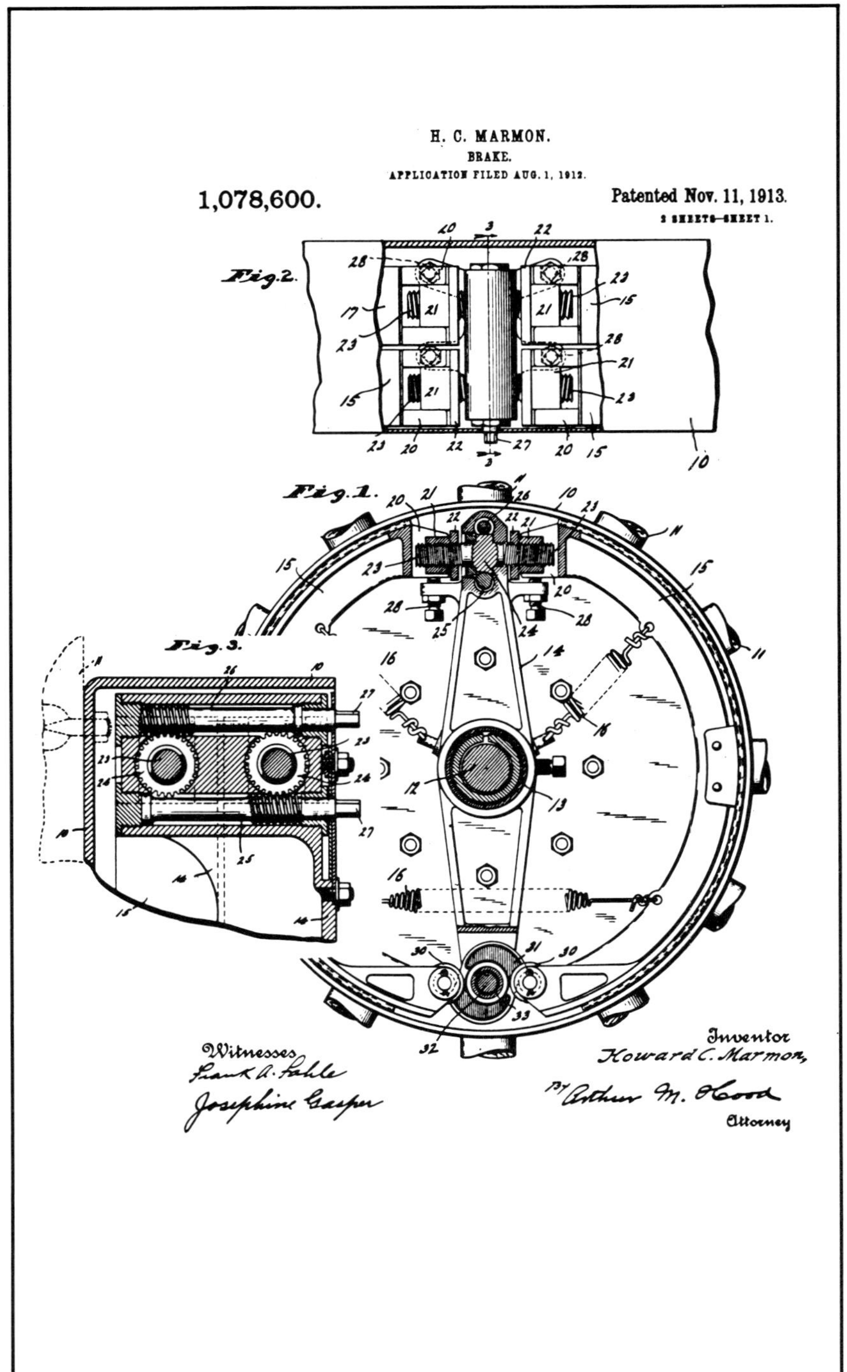


Fig. 3-80 The H.C. Marmon patent for the 1913 Model 32 brake adjustment.

reoriented from the Model H . . . the cross shafts were forward of the axle tubes rather than behind, to reduce the rod lengths, but access for adjustment was compromised.

The introductory brochure described the brakes as "dust-proof, very effective and durable; easy of access."

The more detailed discussion in Part 4 of Chapter Two under "Chasses of Model H, 45 and 50" should be consulted for a full description.

Following the wheelbase growth of 1910 and 1911 and availability of limousine bodies the brakes were increased for 1912 from 14 to 16 inches diameter, to 403 square inches of braking area.

The enlargement of the brakes required retooling of the outer axle sleeve structure which supported the brakes and wheel bearings and provided the spring trunnion. The new design weighed less than the original. The brake related structure was limited to opposite arms which carried the brake cam and the reaction block at their extremeties. Unlike the prior design, the remainder of the 360° was enclosed by a sheet metal dust shield.

For 1910, a single screw-and-segment brake adjustment device was located at the linkage lever on the pedal shaft. It was more accessible (under removable floor board) than the former (four similar devices all on the rear axle) and more economical. For 1912, the levers on the brake cam shafts returned against adjustable stop screws to facilitate equalization.

Brake adjustment continued to receive study and for 1913 a milestone advance was made. Provision was introduced to expand the brake shoes at the anchor point (180° from cam) to compensate for lining wear. The cam throw was also made adjustable to return the linkage to normal leverage while compensating for wear. Together, the two adjustments moved the shoes apart in a parallel motion to achieve more uniform lining wear and extend relining intervals.

The adjustment concept required two independent shoes. Each shoe was pivoted to a block which was joined to the reaction block by a right and left-handed screw having a gear at its center. The gear was rotated by a worm which extended through the dust shield and terminated in a square head for wrenching. The adjustment mechanism was covered by US Patent 1,078,600 issued to Howard C. Marmon on November 11, 1913 following application on August 1, 1912.

A change in brake materials was also made for 1913 and was prophetic of later use. Segments of asbestos fabric alternated with segments of "composition metal" named "Edkin Metal" on the brake shoes. Marmon claimed the two materials were appropriate to superior operation under wet and dry conditions and may have reflected Marmon's experience with fully metallic brakes in Era One. Chevrolet adopted a similar alternation of materials for the Corvette application about 1956.

## MISCELLANEOUS

The '32' radiator was attached to the frame side rails by means of a saddle and cap arrangement which retained the trunnions extending from the sides of the radiator but allowed limited rotation. Because of the latter limitation, a clever design used a single bolt to retain the cap by having it pass through a clearance hole in the trunnion. This protected the radiator from distortion due to frame deflection.

The radiator was a 'Mayo' honeycomb type.

## *Transmission Gears — Efficiency Test of a Marmon 32 Automobile — June 1911 (and horsepower test of the engine)*

Howard G. Taylor of Indianapolis has preserved a remarkable document, *"Tests to Determine the Efficiency of the Transmission Gear* (Drive train) *of a Marmon* (32) *Automobile."* It was the thesis project be shared jointly with R.C. Hoffman, G.A. Wainwright, and T.S. Wilson as candidates for the B S Degree in Mechanical Engineering at Purdue University **in 1911.**

The project was sponsored by Nordyke & Marmon Co. who furnished Joe Dawson's Marmon racer and Joe himself during actual testing. Howard Taylor recalls that Dawson commuted by interurban from Indianapolis whenever the car was to be operated. The reward for the students' efforts was first a fast ride, one at a time, through the streets of Lafayette in the famous racer with Dawson at the wheel; Mr. Taylor, at 93 in 1983, recounted his experience as if it were a recent event. That was followed by box seats at the finish line for the running of the first Indy 500 . . . with Harroun winning in a Marmon Six and Dawson fifth.

The test was a combined engine and chassis dynamometer test of the car using both an engine dynamometer (prony brake) and a chassis rolls dynamometer.

This was accomplished by taking advantage of the smooth exterior surface of the flywheel which was similar to a flat-belt pulley. A prony brake band (steel band with wood friction blocks) was wrapped around the flywheel, with the engine in the chassis; the torque arm extended outboard of the chassis and acted on a platform scale where the observed loads were used to calculate engine torque.

For those not familiar with the procedure, the engine throttle is opened while the brake band is tightened, at a rate to hold the desired engine for a given data point measurement, until the throttle is wide open.

The car, and the engine dynamometer equipment were mounted on the chassis dynmometer in the Purdue University Mechanical Engineering Laboratory.

The objective was measurement of the efficiency of the drive train . . . transmission and axle, wheel bearings and tires. This required determination of the power coming from the engine and the power being input to the road (chassis rolls) under the identical engine conditions (speed, load). The necessary engine test data at both full load and road load provided a valuable historic record of the state-of-the-art because the test vehicle had a significant record of victories during the prior year and the winning car at Indianapolis the following summer differed primarily only in the increased number of cylinders. The results of the engine tests are reported in Chapter Three Part 2 but the conclusions of the authors on the complete test are presented here.

In the "Conclusions" of their 1911 Purdue Thesis, the authors stated: *"The enormous amount of horsepower developed by this (Marmon 32) motor in excess of many others of nearly the same cylinder dimensions is rather difficult to account for. The following points in the design are worthy of note as possible factors in this high power production: Large valves with fillets of extremely small curvature, roomy manifolds and large carburetor. The high speeds are made possible by the thorough oil circulation of the lubricating system and the fact that the pistons are ground to .015 inch smaller diameter than the cylinder instead of the usual..003 to .005 inches."* The engine was rated 32.4 horsepower by the Association of

*Licensed Automobile Manufacturers formula (subsequently referred to as SAE Horsepower) which meant the engine should be expected to produce 32 horsepower at* **1000 rpm** *(without test; only a function of engine dimensions). The engine tested produced 38.3 or 118% of the formula power. The authors remarked: '"This is remarkable in the light of the fact that the highest power ever obtained from a car (engine) at the Purdue University testing plant,* **at any speed,** *was 88% of formula rating, and that a similarly powered car tested in 1910 (at 1000 rpm) developed only 76.7%." "If the speed requirement of the rating be disregarded, the Marmon developed 163% at the rear wheels and 180% at the (flywheel) block." "The maximum power obtained in the block (engine dyno) test was 58.4 horsepower at 2165 rpm, corresponding to a car speed of 77.26 mph." "The highest power obtained from the rear wheels was 52.93 horsepower at 70.9 mph when the engine was rotating at a speed of 1984 rpm." "The maximum speed at which the engine was run was 2912 rpm." which would correspond to 104 mph (in high gear)." "The per cent of power lost in the transmission, rear axle and rear wheels (including tires) was almost constant for ordinary speeds. It decreased slowly from 10.5 per cent at 20 mph to 5.2 per cent at 50 or 60 mph where it is the least. Above this speed, the friction loss rises rapidly and becomes 30.2 per cent*

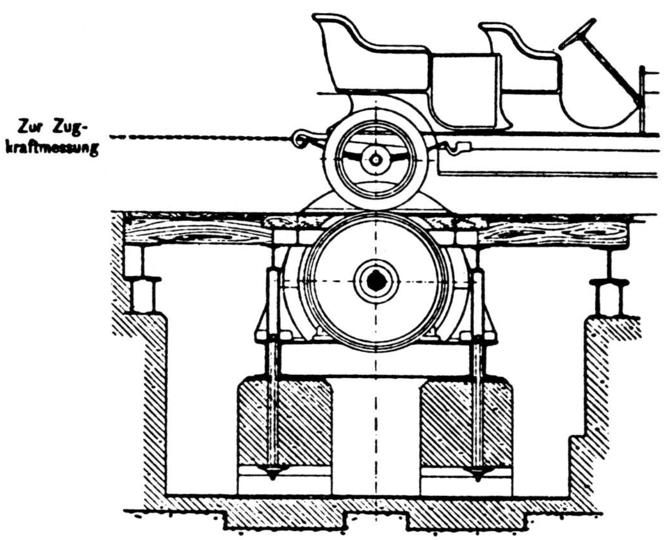


Bild 1. Seitenansicht des Trommelprüfstandes im Riedlerschen Laboratorium für Kraftfahrzeuge an der Technischen Hochschule Berlin, 1908. Verstellbare Spurweite, mechanische Bremsen und 2 elektrische Gleichstrom-Bremsgeneratoren, die auch als Motoren zum Leistungsaufnahme-Antrieb betrieben werden konnten, 120 PS, 150 km/h

Fig. 3-81 Chassis dynamometer at Berlin Technical Highschool.

*at a speed of 89 mph . . . The maximum power tests on intermediate gear . . . indicate that the amount of power absorbed by the transmission gears proper (excluding axle and tires) is less than can be determined within the limits of accuracy of the test."*

Direct projections of this test data to the Marmon winner of the first '500' on May 30, 1911 is reasonable because the two cars differed essentially only in the 50% increase in engine displacement accomplished by the use of three cylinder blocks of two-cylinders each instead of two used on a Model 32 and the correspondingly longer crankcase, crankshaft and camshafts. It can be assumed that the 1911 Indy winner had at least 88 (1.5 times 58.4) horsepower at 2165 rpm which, **if** the axle ratios were identical, and tire sizes the same, would be 77 mph.

The actual racing exhaust manifold was used by inverting it to direct the exhaust outdoors. An 8 foot diameter fan was used in front of the vehicle . . . like contemporary EPA emission test procedures. . . to provide radiator air flow. A separate heat exchanger was used during the engine testing over the full load range just as engines are dynamometer tested today.

The thesis program began in January with the construction of the platform to support the vehicle and rolls. The entire program required three months and 'was the talk of the campus, attracting many visitors.' The Wainwright family lived in Indianapolis and Guy's father was a friend of the Marmon brothers. He was instrumental in the decision to undertake the project.

At that time, according to Mr. Taylor, Purdue offered three fields of mechanical engineering: Gas Engines, Heating and Ventilating, and Railroad. Lab equipment included an "Otto" engine manufactured in Germany.

Although not known in Lafayette, a chassis dynamometer had been constructed at the Technical University of Berlin in 1908 which had both friction and alternate electric power absorption from the rolls. It had capacity for 120 horsepower and 90 miles per hour. In 1910 it had been used to test the 75 horsepower Adler race wagon.[17] However, no prior test program of the magnitude of the 1911 Purdue tests is known to the author.

## MODEL 48 CHASSIS 1913-1915

### FRAME

The '48' frame was kicked-up at the rear and extended beyond the spring perch to support the rear tire carrier.

Unlike the '32' frame, the '48' did not narrow at the forward end. The forward placement of the starter/generator required a wider forward chassis and the increased heat rejection of the engine . . . more than a 50% increase . . . dictated a wider radiator. And, the unique center-point-steering axle design allowed adequate wheel angle for steering without interference with the wider frame.

Surprisingly, the frame rails were very shallow and of nearly uniform depth between springs. Strength was attained by a truss rod, bolted to the frame at mid-engine and at the start of the rear kick-up. Two "hold-offs" were flange-attached to the lower frame flange, about forward and aft of the front seat location. Each engaged the truss rod with a tubular section which was slightly angled to provide a uniform curvature to the truss rod.

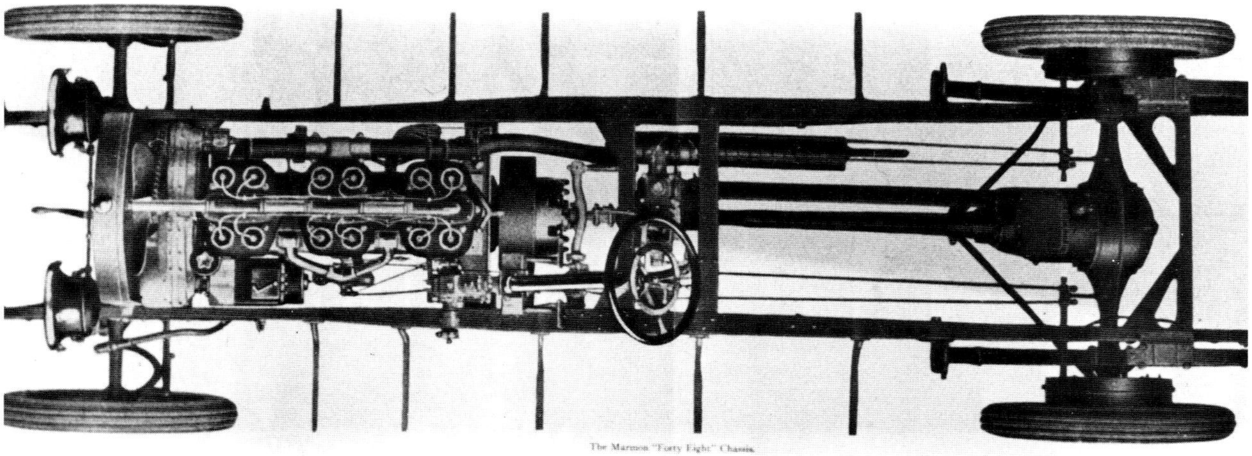

Fig. 3-82 The Model 48 Chassis featured a truss side rail configuration and different springing from the '32' — half elliptical in front and three-quarter in rear.

Fig. 3-83 Model 48 front axle with center-point steering.

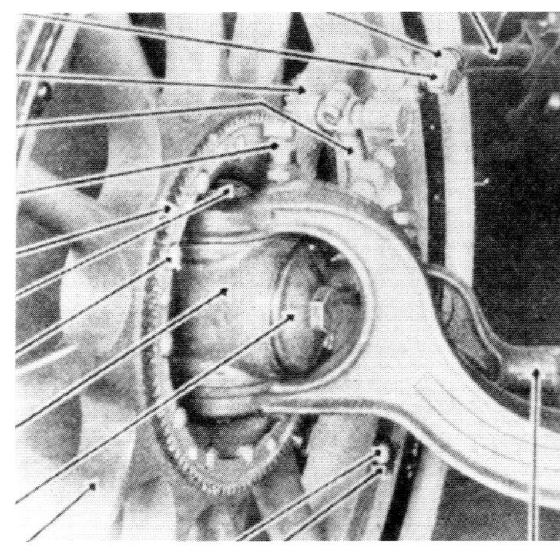

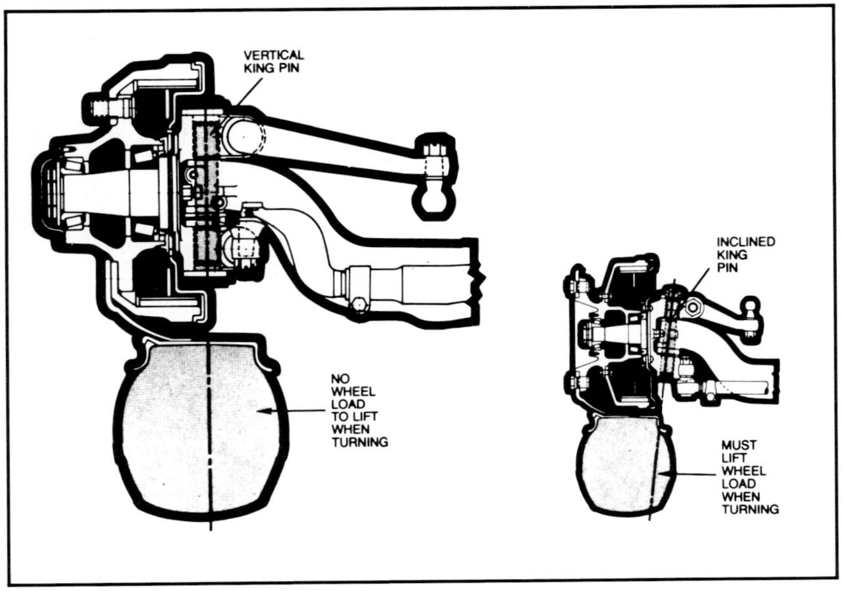


Fig. 3-84 Rockwell center-plane steer axle for heavy duty trucks and comparison with typical 1970s truck practice. Rockwell release in the SAE *Automotive Engineering.*

## SUSPENSION

The Model 48 suspension differed from the Model 32 in the adoption of three-quarter rather than full elliptic rear springs. The forward upper quarter of the full elliptic spring had always encroached on passenger access to the rear seat and seriously complicated efforts to hide chassis elements from view.

## WHEELS AND TIRES

Tire sizes were 36 X 4.5 front and 37 X 5 rear, all on 27 inch wheels. The fabric tires required 80 psi in the front and 85 in the rear to prevent the fabric chafing; it was imperative that the tire stand round under the car load. Rims were detachable and quick-demountable.

## STEERING SYSTEM (Front Axle and Gear)

Because the Model 48 utilized basically the same drive train . . . engine-clutch to tire as the Model 32 . . . it also had the characteristic Marmon "flat" rear hub, indicative of the integral-flanged full-floating drive axle. But the Model 48 also had an equally distinctive and equally original front hub configuration.

The announcement for the "Marmon Six" described this new feature: *"The front axle is another new and distinctive feature first introduced with this model. The vertical spindle is placed directly in the center line of the wheel, making the surest, safest and easiest steering mechanism ever devised. It permits this car with a 145-inch wheel base to be turned in a forty-five foot street."*

It failed to specifically mention that impact of a wheel with a rut would not produce a steering input to the car, as occurs with conventional configuration. The Model brochure did stress ability to maintain control during and after a blowout.

Were the Marmon claims for this new development valid? Were they exaggerated? Certainly not, for the verdict has come in. The jury? No less than the National Highway Transportation Safety Administration of the United States Department of Transportation! The verdict: *"The vertical kingpin design rated highest of the commercially available front-end arrangements (for trucks) tested in a DOT sponsored study which included conventional axles with both manual and hydraulic power steering, spring loaded (centering) devices, and air-assist power steering."*

The DOT report was referring to tests reported in 1978 subsequent to introduction in 1977 by Rockwell International Automotive Operations, of their Model FF-971 Center-Point steering axle for Class VIII trucks (the heaviest inter-city highway trucks). The Rockwell description of this axle design and comparison with the conventional system provides a better understanding of the new feature of the 1913 Marmon Model 48: *"The vertical kingpin design is contrasted with the inclined kingpin design (right). The Center-Point design requires less steering effort because the vertical kingpin acts simply as a pivot point, eliminating any lifting of the (vehicle) front-end load that necessarily occurs when turning an axle using an inclined kingpin."*

Coleman (Truck) Company designed and manufactures a front wheel drive axle for heavy duty vehicles which incorporates the center-point concept.

A Gemmer steering gear, with special mounting provisions for Marmon, was used. The gear box had a second trunnion journal on the side opposite the usual trunnion which was concentric with the output shaft. This permitted a light-weight but very rigid gear mount. The pitman arm had a universal joint connection to the drag-link.

## CLUTCH

The '48' clutch was a multiple disc type with nine steel driving discs and eight driven discs faced with asbestos on a wire mesh foundation. The disc pack was loaded by a cluster of springs surrounding it and reacting against the bottom of deep drawn cups inserted in holes drilled in the flywheel.

A light weight aluminum casting enclosed the mechanism and was supported on the same long studs, attached to the flywheel, which both piloted and drove the drive discs.

The driven discs were locally splined on an output-hub which was carried, on a pair of cylindrical roller bearings, on an extension of the crankshaft. Grease lubricant was retained by felt seals, and provision was made for replenishment. Two large studs attached to the hub drove a **laminated steel** flexible coupling. A corresponding pair at 90 degrees transmitted output torque to a yoke on the output shaft to the transmission.

## BRAKES

The Model 48 initially used the Model 32 brakes. For 1914 the brakes on the '48' were increased to 17 from 16 inch diameter and were fitted with the same double worm screw design of adjustment adopted for the '32'. The '32' brakes continued with 16 inch drums.

The second adjustment feature allowed rotation of the brake cam to compensate for wear while retaining the rod and lever apply linkage in its optimum configuration to maximize leverage.

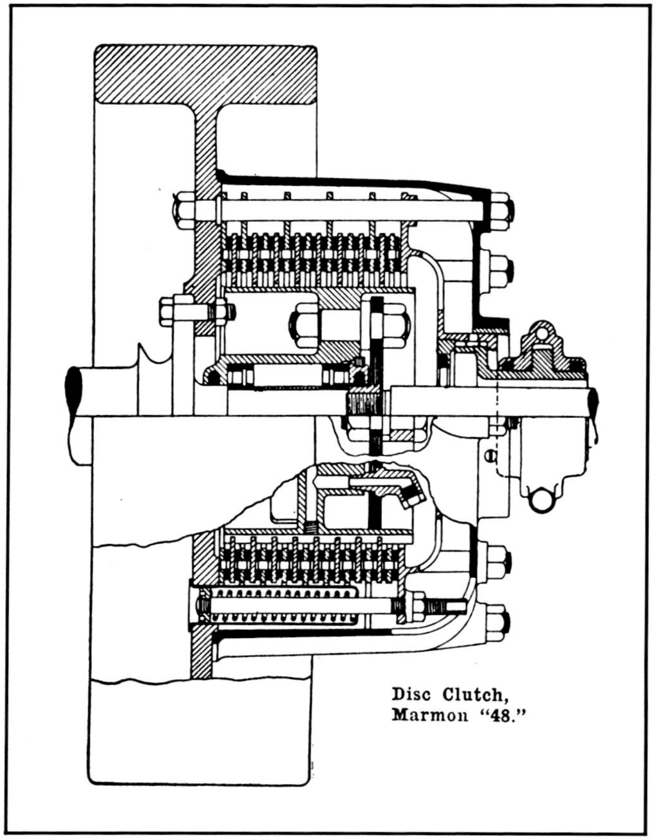


Fig. 3-85 Model 48 multiple-disc clutch.

Fig. 3-86 Model 48 transaxle and brakes. Note sealing rib on drum and split cover which engaged it.

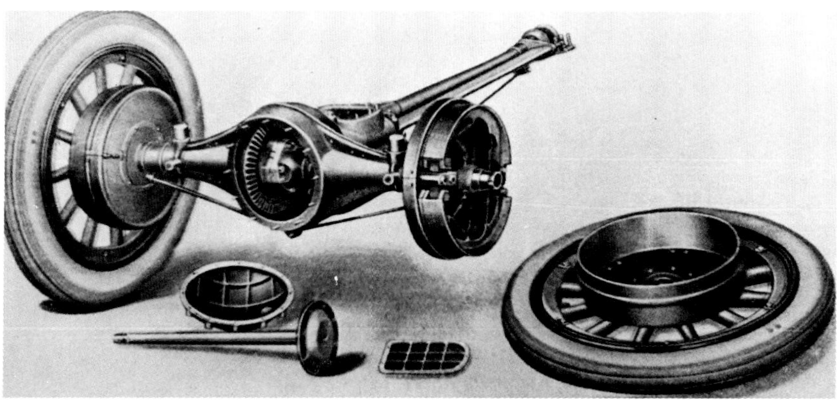

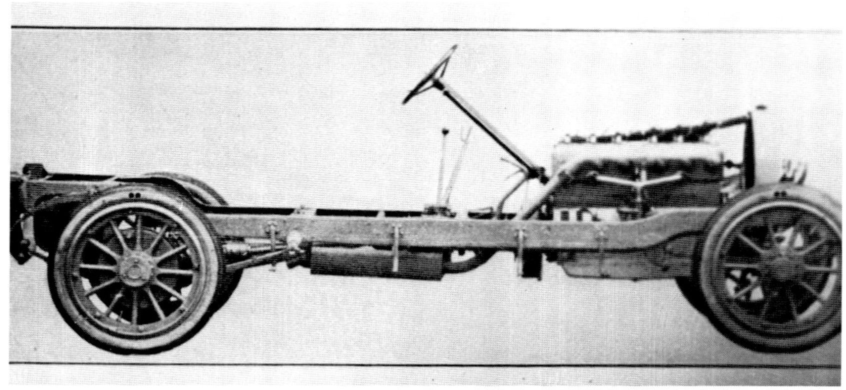

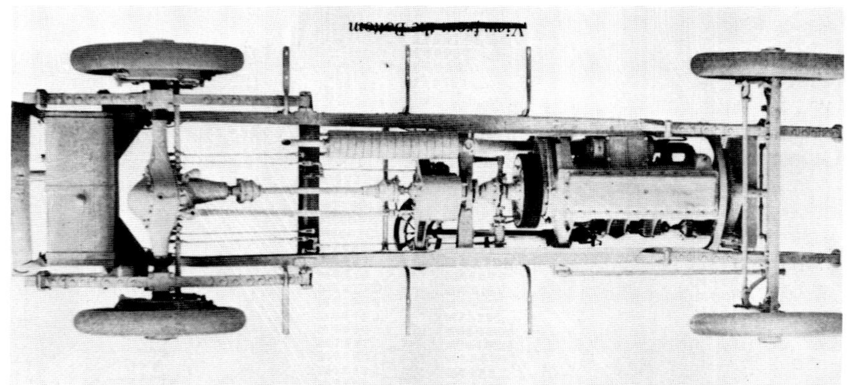

Fig. 3-87 Chassis of the Model 41. Note double drop frame.

## MODEL 41 CHASSIS 1914-1916

The Model 41 chassis was intended to be the ultimate survivor of the three models offered in 1914-15. As such it reflected the rapidly changing state-of-the-art as assessed by Howard Marmon, his staff and consultants.

Its stature was evident when *'The Automobile'* (July 1, 1915) reported that the 1916 version of the Model 41 differed but little from the 1915 model year. The article however was sub-titled "A Study of the Engineering Principles in the Big Six." (Oddly, the same magazine (June 18, 1914) reported "Marmon Continues **Little** Six (for 1915)"; a matter of relevancy — the Model 48 was big brother in 1915 but was absent from marketing plans for 1916).

Two years after its introduction, the reporter for *"The Automobile"* stated: "The chassis as a whole is one of the most original in thought that America is producing today."

### FRAME

The Model '41' was the first Marmon to have a double-drop frame. The forward kick-up was just aft the rear shackle of the front spring. The rear kick-up was extended beyond the spring perch (for the 3/4 elliptic springs) over the rear axle to support the fuel tank and an unusual transverse tire well integral with the frame. The 6″ rails were straight in plain view and the flanges nearly constant in width. There were four cross-members in addition to the wheel well. The front engine cross member was formed into a saddle perch for the patented engine mount system. Wheelbase was 132.5 inches but an extended base of 138 was available for custom limousine and Landaulet bodies.

## SUSPENSION

Suspension was basically similar to the '48': semi-elliptic at the front, 39 inches long and 2 inches wide; three-quarter elliptic at the rear, 57 inches long and 2.5 inches wide, interconnected by scroll-end and inverted shackle.

Unlike the '48', the front springs had inverted shackles, and the rear half-elliptic elements were pivoted at the forward end. This was somewhat surprising because the '41' axle was positioned by a torque-arm paralleling the open two-joint drive shaft.

The Model 41 was the first Marmon to use a single torque arm to transmit the rotational reaction of the axle housing to the frame. Such a construction allows the use of two universal joints on the drive shaft, with one canceling the irregular rotation characteristic of the other, whereas a torque tube configuration allows only a single joint. It also permits engineering the rear springs for ride and handling without compromise for torque reactions as necessary with Hotchkiss drive. The most recent domestic appearance of the torque arm concept was the "fix" incorporated in the Chevrolet Vega and derivatives when all other efforts to control rear wheel hop on full acceleration were inefeective.

The description of the "Rear System" and "Torque Bar" from the Marmon 41 catalog warrants repetition here as it describes another advantage over torque-tube drive . . . the cushioning of sudden torque reactions:

*Rear System:* The rear axle housing construction follows the Marmon precedent having two pressed steel sections welded together. It performs the important function of supporting the load on the rear wheels, carrying the differential members (bevel gears, differential and axle shaft) in a bath of grease, and also carries the brakes.

*Torque Bar:* As power from the motor tends to make the rear wheels revolve, the axle housing tends to revolve in the opposite direction. The long torque arm takes care of this torque or tendency to revolve. It does not take any driving or side sway stress; as these are efficiently cared for by the springs. A long pin or swivel at the rear end joins, the torque bar to the axle housing, rigidly for vertical motion or rotary motion, but flexible sideways. The ball which terminates the forward end of the torque bar is carried in a vertical tube which contains cups above and below held against the ball by powerful coil springs. This flexible universal joint automatically cushions the stresses of sudden axle torque. The vertical tube is itself mounted on a cross member of the frame which may revolve slightly to prevent strains.

## WHEELS AND TIRES

Wheels were all artillery type; the front with ten spokes, the rear with twelve and three clips to the brake drum for torque transfer.

Tire size was 35 X 5 for 1914-15 and 36 X 4.5 for 1916. Corresponding wheel sizes were 25 and 27 inches. Goodrich Silvertown Tires were standard equipment.

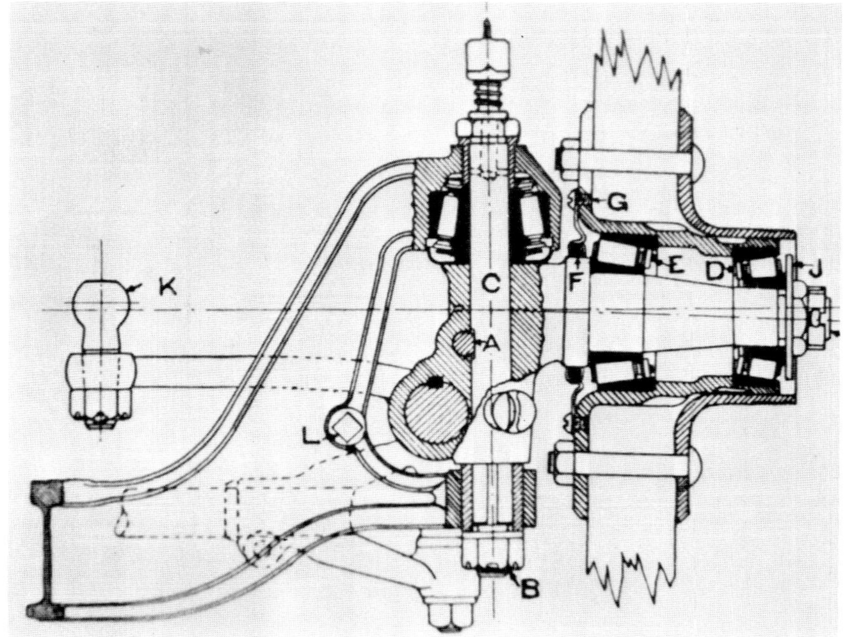


Fig. 3-88 Left steering knuckle and wheel bearings — Model 41

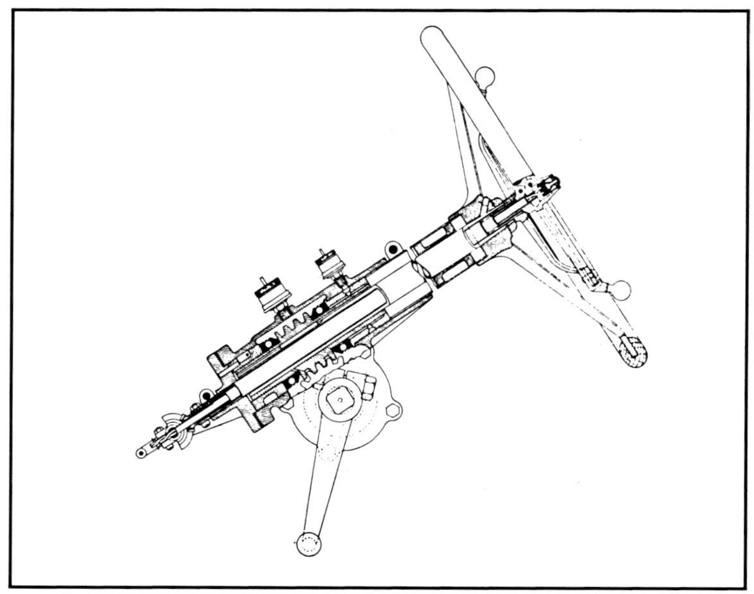

Fig. 3-89 Steering gear and wheel — Model 41

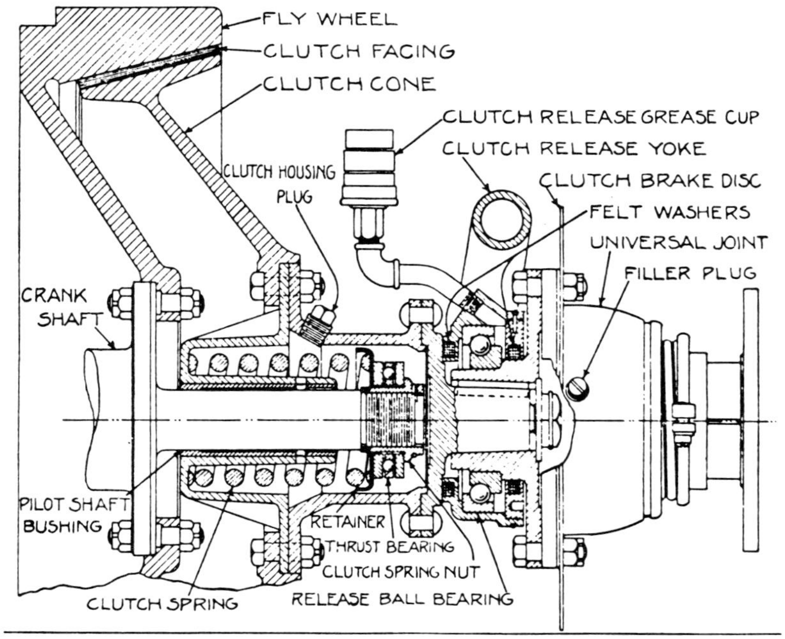


Figure 13. Marmon "41" Clutch

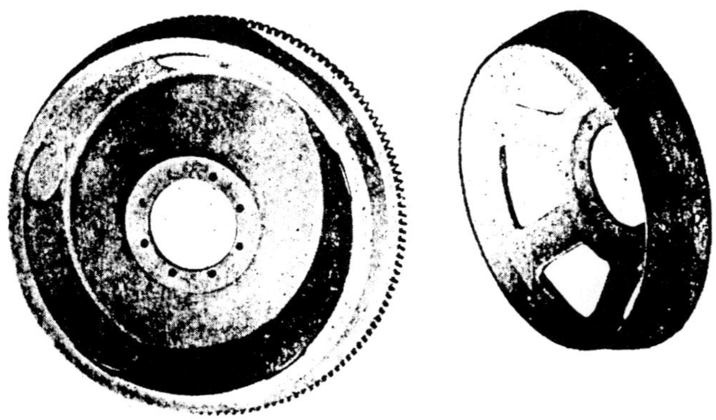

**Marmon clutch with cushion springs set in flywheel instead of beneath fabric facing of cone**

Fig. 3-90 Model 41 clutch.

## STEERING SYSTEM (Axle and Gear)

The front axle was an Elliot type, unlike prior Marmons and highly conventional then. The axle beam, wheel spindles and steering arms were all heat-treated forgings. The upper bearing was a tapered roller type to transfer the car weight to the steering knuckle. Wheel bearings were opposed tapered rollers of slightly different size. Tie rod ends were clevis type. Adjustable stop screws in the axle beam limited steering motion.

The steering gear was a worm and sector type with plain bearings for radial loads and ball thrust bearings. Large grease cups were fitted to lubricate the upper bearing and to replenish the gear box lubricant.

## CLUTCH

For 1915, the mechanism for cushioning clutch engagement was redesigngd. Five "pistons", backed up by springs, were incorporated in the flywheel on the inner conical surface; formerly these devices had been on the outside of the driven cone, beneath the leather facing. The latter was much less challenging to manufacturing engineers; special machinery was necessary for the new location.

The change was triggered by adoption of a fabric made of woven asbestos and wire, which was a heavier and stiffer material unsuited to interrupted riveting and flexure.

## TRANSMISSION

The transmission was independently mounted to the frame aft of the engine and, for the first time, Marmon integrated the shift and hand-brake in the transmission package.

Mounting of the transmission was almost identical to the patented engine mounting. Transverse, channel-section arms extended from each side of the aluminum case to nearly the width between frame flanges; in their outer ends were parallel horizontal bores for clevis-pin attachment to sheet metal clevises riveted to the frame side rails. Presumably non-metallic isolators lined the bores as described in the patent. The third mount was an isolated trunnion concentric with the input which was clamped upward, against a saddle formed on the lower flange of the cross-member, by a band clamp.

Internally the three-speed and reverse box was generally of typical Marmon practice. It was procured from an outside source, but for 1915 was brought in-house and some design improvements were incorporated in the process of retooling. The direct drive engagement was made by male and female gear coupling, replacing a dog-clutch (face-type) coupling. Also the gear-selector cross shaft was mounted on ball-bearings to facilitate easy shifting. The reporter for *The Automobile* (July 1, 1914 issue) commented: *"It is to be doubted whether there are any cars which have an easier handling gear-shift and certainly very few indeed have one that is even nearly as good."*

The transmission gears were machined from 3.5 percent nickel steel amd were case hardened. Each individual gear tooth was tested with a Scleroscope Instrument for acceptable hardness. The shafts were 0.45-0.55 carbon, chrome oil-tempered steel.

Gear ratios were: Low 3.36:1, Second 1.63:1 and direct; reverse was 4.33:1.

The output was an open propellor shaft with Spicer universal joints at both ends and spline provision for length variation. The shaft was made of 0.3-0.4 carbon hot-rolled steel, heat treated.

The open speedometer drive gear, traditionally mounted on the inside of the

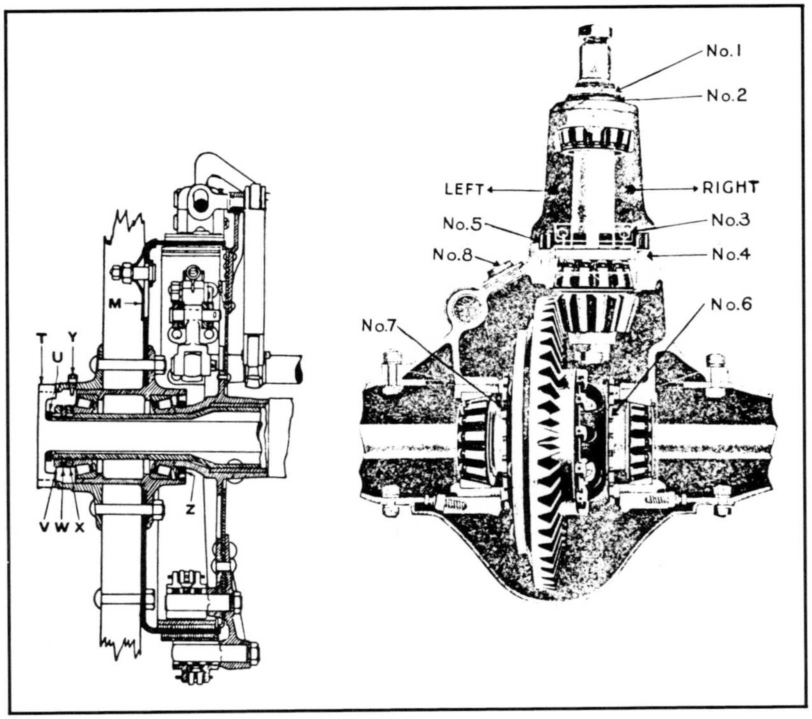


Fig. 3-91 Model 41 axle gears and full-floating wheel drive (hub cap removed)
*"The pinion may be moved back into deeper mesh by turning the adjustment nut about the pinion bearing to the left, or forward by turning to the right. To do this, loosen nut No. 1, then nut No. 2. Remove cover No. 3 and loosen clamp bolt No. 4. Turn adjusting nut No. 5 one slot."*

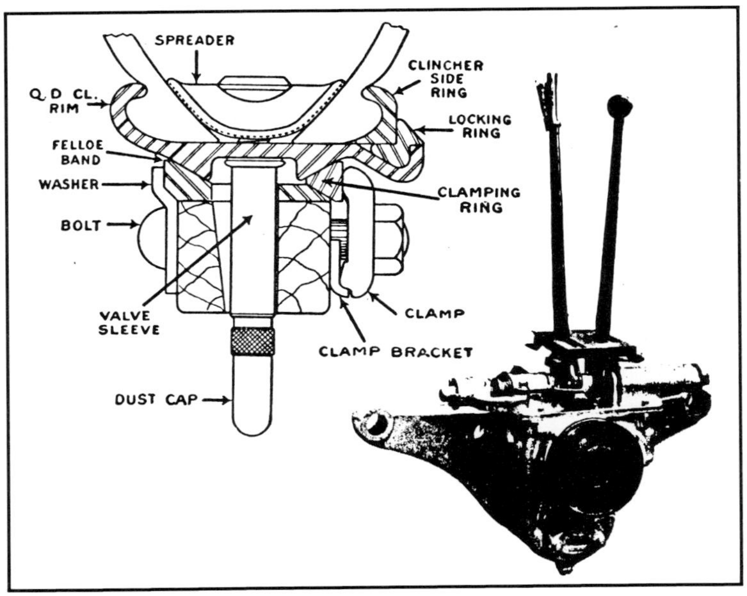


Fig. 3-92 Model 41 demountable rim for clincher type tire.

right front wheel, was mounted on the transmission output flange.

Alignment variations between engine/clutch and transmission were compensated by a Spicer U-joint forward of the transmission and a crowned tooth coupling at the clutch.

## REAR AXLE

The spiral bevel rear axle gears were carried on roller bearings and the assembly was flange-mounted to a pressed-steel axle housing. Ratios of 3.08, 3.50, 3.77 and 4.08:1 were available and were offered on the speedster, 3-4 passenger, 5-7 passenger and for special application respectively.

Howard Marmon's invention, the banjo axle housing, first appeared on the Model H car and immediately thereafter on the transaxle of the Model 32. However the Model H gear carrier anticipated the transaxle concept and accordingly was designed for insertion thorugh the rear opening with attachment to the inside of the forward flange. The Model 41 axle was the first with the exact configuration that became the near universal configuration of banjo axle. In the Marmon Book of Facts, 1929 Edition, reference was made to "Howard Marmon's invention of the banjo axle." However the quotation in the Model 41 Catalog under "Suspension" refers to the "Marmon precedent"; evidently there was no patent coverage. The use of *spiral bevel gearing* was claimed in a Nordyke & Marmon ad which ran in Motor Print for July 15, 1915 (p33).

## BRAKES

Brake design departed from the Marmon tradition of unique design with exclusively internal brakes and semi-rigid shoes. The '41' service brakes were of the external contracting type with toggle actuating linkage on the band. They were 17 inches in diameter, and 2.5 inches wide and faced with Thermoid material.

The inside of the pressed steel drum was 16.625 diameter and the 2.5 inch wide parking-brake band was toggle expanded. For 1915 the internal brake was cam-actuated and the linkage was revised to increase leverage.[18]

The change in brake philosophy was major. No Marmon had ever had external brakes because of their self-destruct characteristic if they operated with insufficient running clearance and because of their exposure to the elements. It is presumed that the change was made because it allowed a wider stance of the rear springs for cornering and allowed a wider rear seat between wheel housings.

## FUEL SYSTEM

The '41' had a 22 gallon tank at the rear. For 1915 a Taylor fuel level gauge was mounted midway on the rear vertical face of the tank.

Also for 1915, fuel delivery by vacuum "device" replaced the air pressure delivery system which utilized an engine driven diaphragm-type air pump to pressurize the fuel tank; A hand-pump on the seat riser was required to pressurize the tank if the dash gauge showed loss of pressure, eg. during refueling.

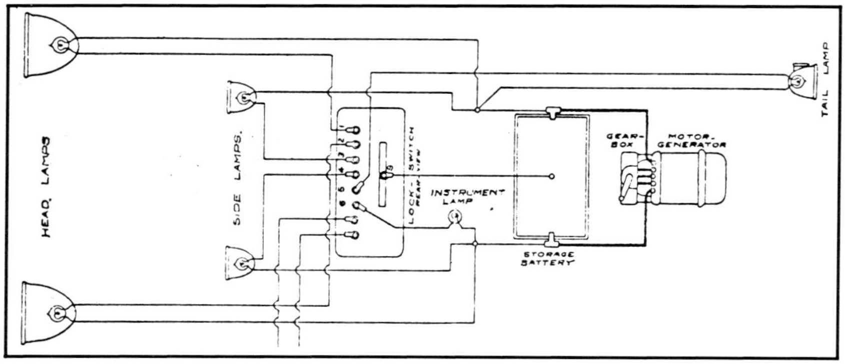


Fig. 3-93 Diagram of the electric system used on 1914 Marmon cars.

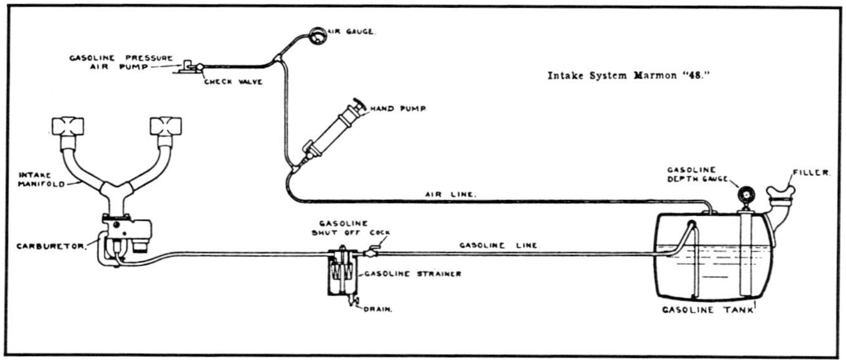


Fig. 3-94 Fuel system used on Model 48 (1913 shown) and Model 32.

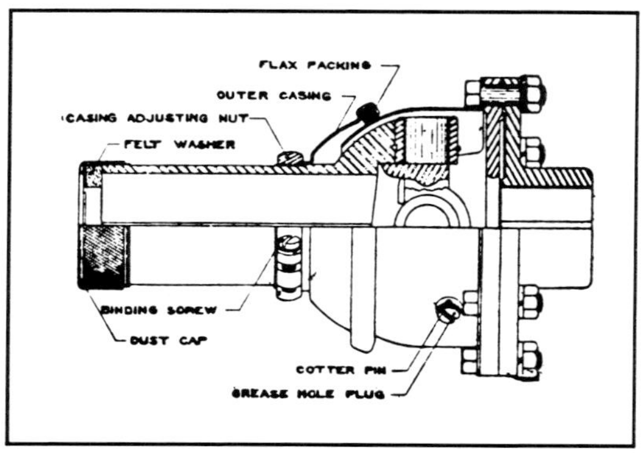


Fig. 3-95 Spicer universal joint.

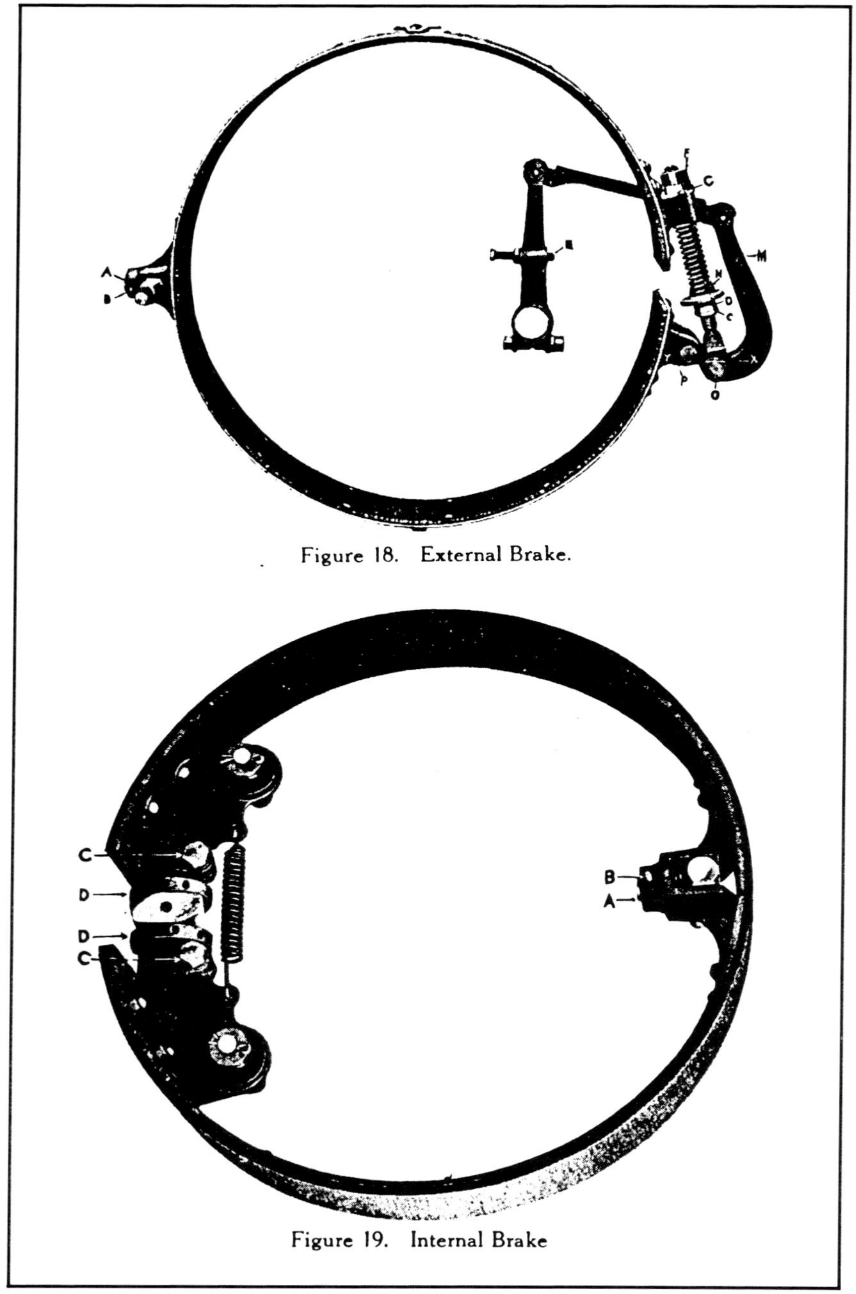


Figure 18. External Brake.

Figure 19. Internal Brake

Fig. 3-96 Model 41 service (ext.) and parking (int.) brakes.

CHAPTER FOUR

# THE MARMONS OF ERA THREE

*"The Foremost Fine Car"*

*1916-1927*

## PART 1 — The Cars — Model 34A, 34B, 34C — Sub-Era 1916-1924

The advent of the THIRD ERA stands as a major milestone in the development of the automobile. In fact the introduction of the Model 34 Marmon ranks as probably the greatest single step-change in domestic motor car design and additionally marked the strong beginning of the automotive styling age. Although a survey of advertising language, including much documented in preceding chapters of this work, would indicate a much earlier date, the prior efforts appear directed to the styling of component elements of the vehicle.

That the Model 34 advent was a milestone of yet unequalled stature was confirmed some half a century later when the fourth and fifth iterations of the Model 34, . . . the 1925-26 Model D-74 and 1927-28 Model E-75 . . . both readily recognizable by a twentieth century Rip Van Winkle who dozed off in 1916 . . . were granted "Full Classic" status by the Classic Car Club of America which limits candidates to 1925 or later models. This remarkable tribute was first brought to public attention by Maurice Hendrey, author of *'Cadillac — The Complete History'*[1] and *'Marmon — a quest for perfection'*,[2] in his *'Housing Two Hundred Horses'*.[3]

The coming automotive milestone was intimated in a series of six major multi-page advertisements which ran on a weekly basis in weekly periodicals beginning in late November 1915. The Marmon name did not appear until the last six page ad; only an unfamiliar Marmon emblem background emblazoned "Made in USA" was shown. The ads were attributed to and signed by the management of the periodical who offered therein to reply to inquiries about the new product. All the ads were on special hard-surface paper and built up from 2 to 4 to a final 6 page introduction.

The ads included pictures of Marmon '34' test cars at Dog Canyon on the Santa Fe Trail, at Borderland Trail, Douglas and Lordsbury, and Starvation Peak, La Bajada Hill and Colmor, New Mexico, San Carlos and Roosevelt, Arizona, as one traversed the rugged road from Phoenix, Arizona to El Paso, Texas which ranged from sea level to 11,000 feet altitude. Testing extended from summer throughout the winter on that route.

The introductory series of advertisements was followed by what Marmon called a "brief" . . . a 45 page brochure detailing the new features of each vehicle component and the scientific reasons for their adoption. That outstanding and unique document was reprinted by Floyd Clymer in 1951 when any antique automotive publication was rare and remained a popular item for a quarter century . . . a tribute to both its contents and advanced format.

On the cover was a picture of a Model 34 touring with a lady driver. The Executive Committee direction to the Engineering Department had specified a series of objectives for the Model 34 design and "drivable by a woman in the winter" was a high priority objective. (In consequence the steering spindles (king pins) operated in an **oil** bath and the transmission and rear axle lubricant was oil; castor oil was the steering gear lubricant) . Other directions were: "Car weight, ready for the road, should be kept as low as safety permits; At speed, the car should handle with ease and certainty and: Cost is of secondary importance[1]." Maurice Hendrey's[2] comment on these directions warrants repetition: "Thus a virtually free hand was given to create a high efficiency connoisseur's machine. It was an engineer's dream". In 1961 no less an authority than John R. Bond called the Marmon 34 the "product of Howard Marmon's nimble brain, the ambition of many and the chattel of few."[4]

The two pages of the 'Brief' which address the selection of wire wheels as standard equipment for the Model 34 are reproduced because they express the design philosophy behind 'The Scientifically Constructed Motor Car' and illustrate the remarkable effort to share that philosophy with the consumer.

Actual work on the '34' had begun in November 1912 and the motor had run in October 1914. The technical announcement came on Dec. 23, 1915: "New Marmon is Unique Car"; Amen. Discontinuance of all other models and rearrangement of the plant for single model production was reported. Deliveries to dealers began immediately after the announcement.

Howard Marmon had carefully monitored European automotive developments right from the turn of the century. He had recognized domestic innovation outside the Nordyke & Marmon organization and Marmons of the FIRST and SECOND ERAS incorporated under license a number of significant patented features.

At the very beginning of the Model 34 project, Marmon sought the services of the respected Alanson P. Brush as a consultant and both warrant equal credit for the masterpiece which resulted. Brush maintained his own design staff in residence at Nordyke & Marmon; it included Owen M. Nacker of later Cadillac V-16 fame. Brush was issued patents on the frame concept, on the clutch and the rear axle/differential design. He was assigned a half-interest in Nacker's patent on the phase I rear transverse spring, although the final configuration incorporating variable rate was covered by a H.C. Marmon patent.

The application for patent on a chassis — frame for motor vehicles was made by Brush on Oct. 21, 1913 and patent, 1,120,900 issued on Dec. 15, 1914. The new concept was discussed before a meeting of the Detroit Section of the Society of Automotive Engineers[5] in May 1915. Mr. Brush explained that in the conventional construction the running boards, their supporting members and the splash pans add to the frame weight while making no contribution to its strength. His design

# Wire Wheels on the Marmon 34

## Wheels

*Wire Wheels Standard Equipment*

Two points were considerably emphasized in the executive committee directions to the engineering department for the design of the Marmon 34. These were: First—car weight, ready for the road, should be kept as low as safety permitted; Second—at speed the car should handle with ease and certainty. Inasmuch as an extremely low list was not insisted upon, these two requirements inevitably led to the selection of wire wheels for standard equipment instead of wood wheels.

*Weight Saving—105 Pounds*

The net saving in weight of the Marmon 34, equipped ready for the road with five wire wheels, as against the same car with wood wheels and a spare rim is 95 pounds. The extra tire carrier for the demountable rim weighs about 10 pounds more than the carrier for the spare wheel, so the actual difference on the cars, as turned out, is 105 pounds. So much for the weight saving.

*Rim Weight Reduced*

At 60 M. P. H. the road wheels are revolving at the rate of 600 turns per minute. The wood wheel with its demountable rim and felloe band concentrates the major part of the weight at the outer diameter of the wheel, which, of course, is moving at a speed nearly as great as the car. With the wire wheel, this rim weight is reduced more than half.

*Gyroscopic Action*

You are familiar with the gyroscope, and know how, even in the little toy ones only a few inches in diameter, when once they are set spinning, it is very difficult to move them so as to change the plane of their rotation. Of course, an automobile wheel when rotating is a gyroscope, and as they bounce over the road, or are moved with the steering apparatus, their plane of rotation is changed, to which changes the gyroscopic action offers a very considerable resistance.

*Why Wire Wheels are Best*

It would be possible to build a wheel so heavy at the rim that it could easily break the axle ends off from this gyroscopic action at high speeds. To make the car safer and easier to handle, this gyroscopic action should be kept as low as possible, which means that the rim weight should be kept to the lowest possible point. This inevitably leads to the selection of wire wheels, as you have noticed that all modern racing cars that develop any speed invariably use, it being practically impossible to handle the cars at the speeds attained with the heavier wood wheel equipment with demountable rims.

*The Relation Between Sprung and Unsprung Weight*

In designing the springs of a car, the ease with which the car will ride, and the accuracy with which it will steer, will depend not upon the total weight of the car, but upon the ratio between the weight below the springs, commonly called unsprung weight, and the weight above the springs. To make this plain, imagine the body and chassis to weigh only as much as the axles, and the axles to weigh as much as body and chassis. In going over a bump in the road, the wheels as they reach the top of the bump will not be compelled to fall down on the far side,

## Wheels

# *Wire Wheels on the Marmon 34*

*How it Affects Comfort and Handling*

as there is not sufficient weight above the springs for the springs to react against, and as a result, the wheels will continue to rise in the air until stopped by gravity, and of course, carry the body with them.

However, if the body is heavy in comparison with the axles, the springs will be compressed as the bump is gone over, and the body being considerably in excess of the weight of the axle, will quickly return the axle to the ground with a comparatively small movement of the body. Keeping the wheels in contact with the ground, by keeping the ratio of the unsprung and the sprung weight as high as possible, not only results in very easy riding, but very much more accurate and safe steering, as the wheels can do a very much better job of guiding the car when in contact with the road than when in the air.

*Decreasing Unsprung Weight*

The front axle of the Marmon 34, complete with the wire wheels and tires, weighs 205 pounds. Under the same conditions, with wood wheels and demountable rims, this weight is 265 pounds, an increase of about 30 percent, this 30 percent increase in the unsprung weight makes such a difference in the handling of the car that it has to be tried to be realized.

*Wire Wheel Stronger*

The wire wheel is very much stronger than the wood wheel, and is very much more difficult to dish when the car skids against some object. The resiliency of the wire wheel and its better action in cooling the tire, results in longer tire life. When tire replacements are necessary, it is vastly easier to change the entire wheel than to change the demountable rims. In weight, strength, accuracy, ability to perform its functions properly, and with the ease with which the tire changes may be made, and its immunity from shrinkage and cracking from weather conditions and climatic changes, the wire wheel is superior to the wood wheel.

The only point in which the wood wheel is superior to the wire one lies in the fact that the wood wheel is somewhat easier to clean.

*Steel will Eventually Supplant Wood*

Point by point, wood has been eliminated from automobile construction. Steel has replaced it for frame members, and metals have entirely replaced it in body construction except for tacking strips and floor boards. Cheapness has been the main reason for maintaining it in the construction of wheels, one of the most vital points about the car where the safety of the passengers is concerned. It is inevitable that, for this purpose the safer and more desirable material — steel — will eventually entirely supplant wood as a constructive material.

incorporated those elements into the load-bearing structure. He did not state the resultant weight-savings but did remark that for a car with a wheelbase of 11 feet, 3 inches (135 inches . . . the model 34 would be 136) a frame constructed of 0.10 inch steel had been found amply strong.

The concept dispensed with the lower flange of the frame siderail for the length of the running board and substituted a running board attached by a riveted flange and having an outer flange doubled over on itself for reinforcement.

The rear "kick-up" was accomplished by a sandwich overlap channel construction reinforced at the juncture by the upward-turned running board. At the time of the SAE presentation, the side rails abruptly spread aft of the front wheels; the torque-box of the transition was reinforced by the upward turned 'board'. The transition was necessary to a uniform width running board, but was eliminated in

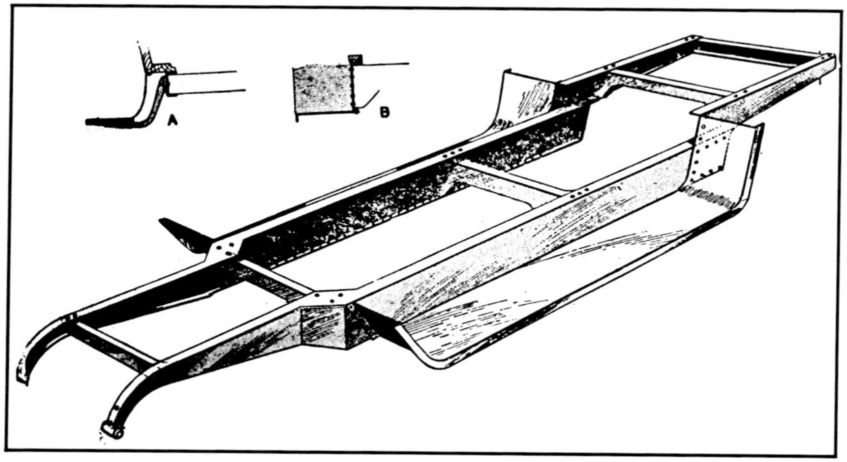


Fig. 4-1 Views showing details of the novel frame structure.

the final design by curving the rails to exactly follow the body and hood plan shape. Nordyke & Marmon could engineer into their cars features requiring break-throughs in manufacturing technology because of their large experienced staff. The variable width running board was a remarkable product in 1916.

The Introductory Brochure for the Model 34 stressed that the location of the fuel tank in the cowl was selected in the interest of consistent vehicle handling; The usual rear location would result in a variable weight at a very unfavorable lever arm.

The '34' was a handling machine unequaled in its time. The balanced front to rear loading and the near equal loading of both rear wheels during roll combined to invite aggressive maneuvering of the automobile. The wisdom of the Model 34 fuel tank location is confirmed today by the widespread use of cockpit mounted brake (front/rear) biasing valves used in road racing cars to compensate for diminishing fuel weight.[6]

The 50-50 front-rear weight distribution of the Model 34 is confirmed by the design philosophy behind the Porsche 944 . . . for many authorities, a yardstock by which its contemporaries are measured. In a Porsche ad run in 1982, the following statement appeared above the signature of Dr. F. Porsche. *"So we formed Porsche . . . to build Porsches. To our own standards. Without compromise. The 944 has the Porsche Transaxle design proven at Le Mans. It produces a nearly-perfect 50-50 front to rear weight distribution for balanced braking and improved cornering."*

The '34' was destined to become the personal transportation of the Henry Fords (prior to the Lincoln acquisition), the Warren Hardings, the Vernon Castles, the Franklin Roosevelts (secretary of the Navy Era), Glenn L. Martin, F.B. Rentschler (Pres., Wright Aeronautical Corp.), the F. Scott Fitzgeralds and several European heads of state.

Zelda (Mrs. Scott) Fitzgerald unknowingly foretold the future restoration and treasuring of Marmon 34s in recognition of the mechanical and styling arts they represented, and their exchange under the exclusive hammer of Soothby, Park-Bernet Galleries. Describing the expenditure of the proceeds of Scott's first book-rights sale, Zelda wrote: *"Lustily splashing their dreams in the dark pool of gratification, the $50,000 bought a cardboard doll house for Bonnie,* ***a second-hand Marmon*** *(34),* ***a Picasso etching,*** *white satin dress, a dress as green as fresh wet paint, two white knickerbacker suits exactly alike, a broker's suit and two first-class tickets to Europe (on the Aquitania)."*[7] *(Authors' emphasis)*

Leon Ruth was a friend of Zelda's studying at Columbia University, and the Fitzgeralds sought his advice on their first car purchase. He later recalled the event: *"As I remember it, we went down to the Battery (NYC) and it was a choice between a new sedan and a second-hand Marmon sports coupe (Cloverleaf roadster). Of course they couldn't resist the Marmon; they bought it."*[7]

Much has been said and written for decades about the American love affair with the automobile. With owners of the Marmon 34, it was a personal relationship. A Marmon advertisement[8] boasted of it: *"If there is such a thing as a hairy fisted man falling in love with a machine, that thing can happen with respect to the Marmon. Strong men have fallen in love with good ships, good horses, good battalions and good cannon. Why should they feel any less strongly toward a good automobile?"*

The ad writer addressed the personality of the Marman 34: *"The first big difference between a Marmon and other truly fine cars is its mechanical disposition . . . its mechanical manners . . . its attitude toward its job. You can't get it fussed up or red in the face. It has a great big mechanical reserve and can afford to be good natured. And the Marmon is probably the only fine car in the world with the same kind of uncanny, mysterious road adhesiveness . . . with automatic self-balance. The steering wheel of the Marmon is the quickest thing there is to take a hint, and, at the same time, the most steadfast."*

This personal relationship with individual Marmon 34s is supported by a letter[9] which the Marmon factory received from an enthusiast in Chile, South America, extending congratulations on the launching of the SIXTEEN. Frederico L. Claude, of Vina del Mar, Chile wrote on Feb. 28, 1931: *"I have long known Marmon cars, stcrting my acquaintance in 1917 with a Model 34, two of which are still in my possession, a seven seater touring car which still runs gallantly on the farm, and a "clover-leaf" roadster which occupies an honored position among my old carriages. Both of these cars have served me far too well to ever be got rid of. They have become old friends".*

Almost from birth in December 1915, the '34' went to war. The American Defense Society (now American Defense Preparedness Assoc.) sponsored a coast to coast record run in the summer of 1916 to demonstrate to army officials the potential for rapid and reliable motor transport which the radical new car offered. The trip from New York to San Francisco was made under AAA surveil-

Fig. 4-2 The departure from Columbus Circle, New York City, beginning the record transcontinental run to San Francisco.

Fig. 4-3 The Marmon 34 in service of the French Army Staff.

lance in five days, eighteen and one-half hours at an average speed of 25.1 mph (34 mph New York to Cleveland). All AAA seals were intact on arrival except for tires and crankcase fill. Remember there were no by-pass routes in those days and the route traversed major cities as well as wagon tracked prairie.

This effort led to experimental service of the '34' with troops during the Mexican border problems in 1916. After 5044 miles (157 miles per day) of on and off-road service, the sealed car was driven 2831 miles to Washington, D.C. under AAA surveillance. During patrol service, the Marmon demonstrated 12.6 mpg of fuel and 360 miles per quart of oil; on the Marfa, Texas-Washington, D.C. run it demonstrated 13.26 mpg of fuel and 325 miles per quart of oil. Drivers on the Marfa-Washington run were Joe Dawson and H.H. Brooks (Sales Manager in 1927).

The French vehicle industry was disrupted by the carnage of the World War and French authorities sought advice from the US Army on selection of an American car for volume procurement as an official vehicle for French army and government use. The US Army experience with the Marmon led to procurement of a fleet of several hundred Marmon 34s for the French Army General Staff. This was a signal honor to Marmon for the family was of French origin under the spelling Marmont. The letter T had been deleted to anglicize the spelling and force correct pronunciation.

The author has in his collection a Marmon color brochure showing President Poincare of the French Republic in a seven passenger Marmon open touring passing through a ruined village en route to dedicate the Battle of the Marne monument and, on another page, showing joyous newly liberated villagers surrounding a Marmon staff car carrying French officers.

Liberty engine manufacture shared the Indianapolis facility with the passenger car manufacture during 1917 and 1918. Nordyke & Marmon, Leland-Lincoln, Packard and General Motors joined in the production of the Liberty V12 aircraft engine crucial to the US war effort. Marmon however was awarded the permanent designation "Champion Liberty Engine Builder" in a gala ceremony in Monument Square, Indianapolis on November 16, 1918.

Nordyke and Marmon was the only domestic manufacturer to have contracts for the manufacture of two distinct aircraft engines . . . the Liberty V-12 and the Hall-Scott line-four. The Henry Ford Museum displays an example of the Nordyke & Marmon built Hall-Scott engine. More detail is included in "Marmon and Aviation" in Chapter Fourteen, "Marmon Musings".

The reporter for *The Automobile,* Dec. 23, 1915 was fairly confident of his prediction of events to come in consequence of the Marmon 34, but underestimated the time for that widespread reaction by three score years! He wrote: *Just as the Cadillac V-8 ushered in an epoch, so, it is probable, has Marmon, in which lightweight construction has been studied with such exceptional care. Of course that car is not by any means the lightest seven-passenger now built, but is far lighter than the average car of the class it represents. It is from 20 to 30% lighter according to actual figures. A cut of between one fifth and one-third in weight is no small accomplishment and it is difficult to see how it could have been done without the special frame and aluminum engine. Whether this is so or not, there is little doubt but that Marmon will spur on others to attempt the same result. Light weight is the greatest factor in reducing the cost of upkeep and so is of immense importance to the owner. Except in isolated cases the automotive engineer still has a tendency to disregard the weight of his production, but once his attention is forced upon it, we may expect advances as rapid as we have seen before with other engineering problems.*

But for the Marmon class of automobile, the consumer was little interested whether his car got 11-14 mpg or the usual 8-10, or whether it went 40% farther on a set of tires. In 1961 General Motors engineers made a valiant effort to sense evidence of public interest in light weight fuel efficient cars (with the 215 CID V8 aluminum engine, Olds F85 Cutlass and Buick Special.) but it was met with disdain

Fig. 4-4 The styling breakthrough of 1916 — first example of styling the car as an entity rather than component styling. The basic chassis, hood and fenders would be recognized as full classic in the 1925-28 editions.

and clamor for more power, more performance, more pizazz.

But the energy emergency of 1973 got the public's attention and the legislators' attention and, as *The Automobile* reporter prophesied, "once the engineer's attention was forced upon it" advances were as rapid as had ever been seen.

The Model 34 basic engine, chassis and styling were to persist for thirteen years. There would be five model variations A through E. The latter two would use the rated brake horsepower as the model designation although the SAE rating remained unchanged; a marketing ploy.

The models 34A 1916-19, 34B 1920-24, 34C 1923-24, D74 1925-26, and E75 1927-28 will be individually discussed here in Part 1. The engines and the chasses will then be discussed in much greater detail respectively in Parts 2 and 3.

*Model 34A*
*1916-1919*

The Scientifically Constructed Light Weight Car

## STYLING

The Model 34 at its introduction had an exterior surface free of discontinuities and sharp edges and unmarred by handles or hinges except for the hood latches and lift handles. Fenders and running boards flowed smoothly into each other and the hood was devoid of louvers. The handsome rounded radiator shell and full crown, smooth edged fenders were, e.g. a full decade ahead of Cadillac. The doors were flush and had rounded corners and the body sides continued across the rear edge of the front seat in a rounded hint of later dual-cowl custom configurations. A scattering of shallow rivet heads was evident on the exposed frame surface below the doors. The frame of the lower windshield glass exactly followed the cowl contour between free-standing pillars having roots within the cowl skin.

The driver's zone received equally careful attention. The combination ammeter/ignition switch unit was centered on the dash beneath an extended flood-light. The other instruments were symmetrically arrayed, large 8-day clock to the left and speedometer to the right, oil pressure and fuel level above and nearer to center. Shift and hand brake levers were offset to the left bringing them adjacent the driver's right hand on the steering wheel rim; The accelerator pedal was consequently **between** the clutch and brake. The transmission could be locked in neutral by a Yale lock in the shift knob. The mahogany wheel rim was carried above the hub on an aluminum spider; this 'sank' the spark and throttle control ring and levers nearly flush with the rim. Floorboards were linoleum covered and edged with bright metal.

At announcement, only a seven passenger touring body was shown. That body style weighed 1100 pounds less than comparable conventional automobiles . . . 3540 pounds ready for the road. The performance was consequently equally impressive: acceleration from 10 to 50 miles per hour in less than 18 seconds on a hard level road, a top speed of 65 to 70 miles per hour.

*The Automobile* reported:[10] "The light weight has been secured by means of highly scientific design in combination with careful choice of material. There is much aluminum employed and much high tensile steel in thin section. The new car is new from stem to stern, from the motor to the rear axle no opportunity has been missed where there was a chance to increase efficiency or reduce weight."

The Marmon Model 34 construction "featured 'unification' making the body and chassis practically one unit". The integration of the running boards and splash aprons into the frame has already been discussed. Marmon combined this frame construction with a patented "Compound Cantilever Rear Spring" which effectively gave a three point attachment of frame to running gear. This resulted in a distortion-free mounting surface for the body, reminiscent of the Era One Marmons, and allowed Marmon to dispense with the usual full-length body sill.

The open bodies consisted of three modules: cowl, with front doors; front seat with rear doors; and tonneau. Polished aluminum step-plates were attached directly to the top flange of the frame. The floor boards were supported at the step

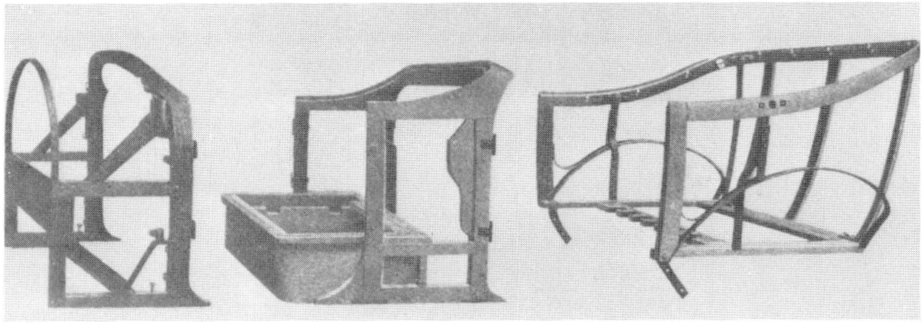

Fig. 4-5 The three-piece open body construction made possible by the Model 34 frame and suspension system.

Fig. 4-6 The closed body version of the sill-less open body concept.

Fig. 4-7 Every control of the 34A was in easy reach.

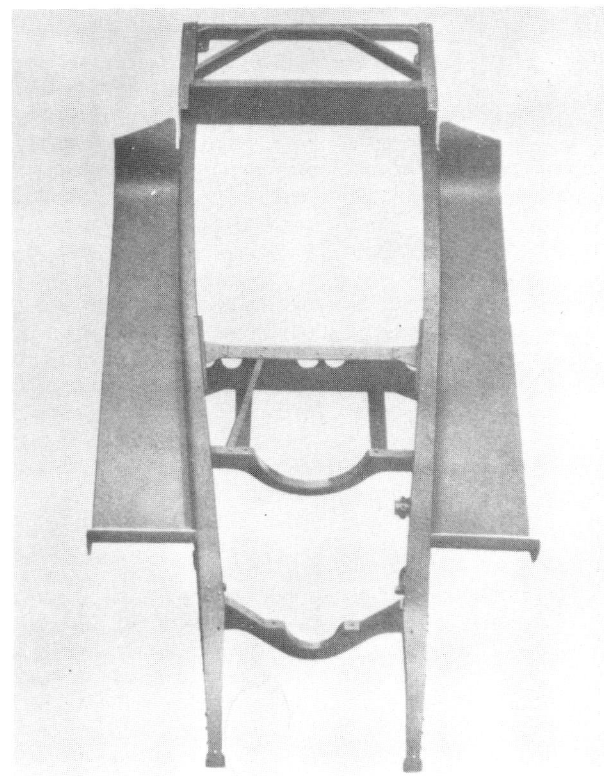

Fig. 4-8 The unusual Model 34A frame

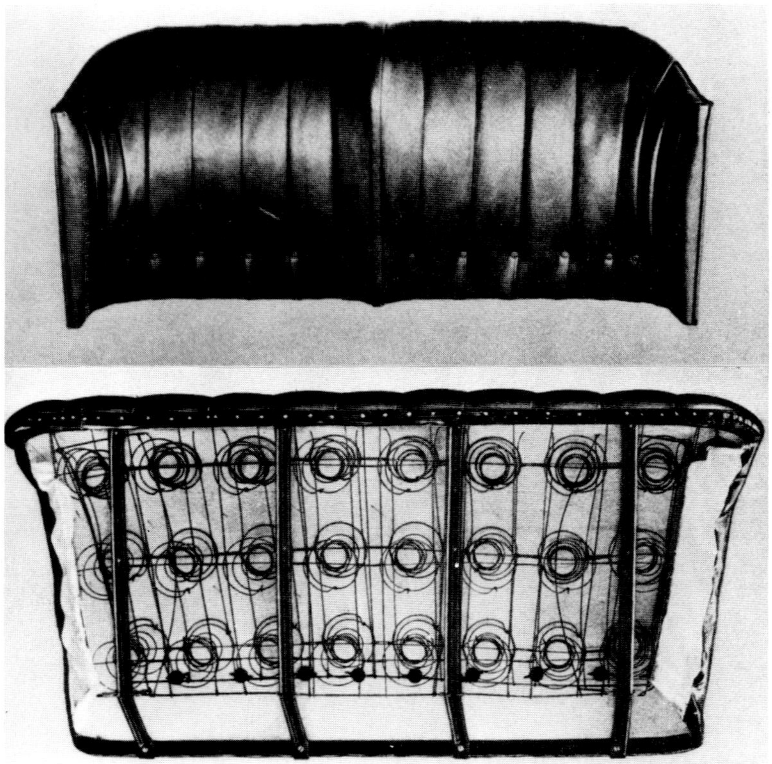

Fig. 4-9 *"Removable upholstery adds refinement and convenience in a new way. The upholstery is made on a form ready to fasten in the car."*

plate height by Z shaped supports spot-welded to the bottom of the flange.

Closed bodies were, of course, single units but lacked sills beneath the doors; attachment to the frame provided completion of the door "frames".

Elimination of the body sill beneath the doors reduced the body height correspondingly . . . some 2.5 inches typically. Because of this feature, Model 34 Marmons were noticeably lower and handsomer than their contemporaries, the floor boards being only 24.5 inches from the ground.

The hood, fenders and splash pans were aluminum. The body was aluminum skin on an ash wood frame so constructed that all loads were transferred to steel reinforcement members without being reacted through a wood structure. In addition to the novelty of sectional body construction, there was significant innovation, decades ahead of its time, in the method of upholstering. Seat cushions were built on light weight metal frames complete and ready to install in the car and readily removable in event of body or upholstery damage. Panels and cushions were fuliy interchangeable and the Introductory Brochure suggested ordering an extra set of different color.

The following tabulation shows the surprisingly light weight of the three-section sill-less Marmon 7-passenger touring body and remarkably light weight of the aluminum fenders. Both of course were direct results of the unique frame made under the Brush patent.

The weight of the Marmon 7 passenger touring car body, minus gasoline tank was:

| | |
|---|---|
| Front section | 72 pounds |
| Middle section | 105 pounds |
| Rear section | 110 pounds |
| Two front doors | 43 pounds |
| Two rear doors | 45 pounds |
| Tonneau floor | 33 pounds |
| Spare seats (two) | 49 pounds |
| Front seat cushions (two) | 24 pounds |
| Rear seat cushion | 29½ pounds |
| Body | 510½ pounds |
| Fenders (4) | 36½ pounds |
| Total | 547 pounds |

A Marmon salesman's booklet provides the following specifics on the body tooling and construction: *"In conventional construction with bodies of unit design, any injury requires a replacement of the complete body instead of merely the injured portion, as is possible in the Marmon. To insure this interchangeability of sections in body construction, complete assembling jigs and dummy frames have been provided. Each of the three sections of the touring car body are built up in cast iron assembly jigs, which not only insure that the units are interchangeable, but also act as gauges for the parts going into the body framework. If any of the mill work, that is, the wooden part of the body, which is shaped out exactly to size on special machines, does not fit, it is at once noticed and rejected. The same is true of all the iron braces that are provided between the wooden portion of the body frame and the main frame of the chassis. In examining the rear unit, it will be noted that none of the top body irons depend upon the wooden frame or the sheet metal work for support. This main top iron is bolted to the steel wheel house arch, to the rear door pillar and to the brace which is attached to the main frame just in front of the rear seat heel board. Hence any strains in this iron are not transmitted to the body, but are thrown at once upon the steel frame of the car. The same is also true of the rear top irons. These are not attached to the wood frame of the unit for support, but are continued down the back to the main frame of the car so as to relieve the entire unit from strain. The front seat unit only has to carry the passenger load and provide a carrying space for the storage battery. It therefore does not have to take the top strains and only has to be braced crosswise and diagonally so as to be rigid. There is no metal in its construction save the aluminum sheets which form the sides of the body. The front section is also very rigidly braced so as to carry the fuel tank and windshield. A single sheet of aluminum is used for the back unit of the body, and the same is true of the cowl. Hence there are no joints which might cause trouble in service. The aluminum sheets are formed over special dies so that they fit the framework without straining."*

Pictures of a second style , the 4-passenger roadster of 'cloverleaf' type, were released in late January.[11]

Cost was indeed a greater consideration by the time the 34 was introduced than earlier when the Board of Directors considered it a secondary consideration. The 34 project had begun about the time that the Model 48 was being readied for

Fig. 4-10, Fig. 4-11 The cloverleaf roadster style came in both three and four passenger configurations. Note the lid on the integral rear stowage compartment, and the compartments beside the rear seat of the three-seater.

Fig. 4-12 Zelda (Mrs. F. Scott) with the Fitzgerald Marmon 34A

Fig. 4-13 Special order victoria top on the popular Model 34A — 1916 cloverleaf roadster for four.

announcement. Roy Milner told the author that, when he and Chief Engineer Jeffers arrived at Marmon Jan. 15, 1914, the factory was saddled with a large inventory of new Model 48 cars which had proved difficult to move because of the $5000 price. A sales promotion specialist was employed and a major promotion (could it have been rebates?) undertaken to move them.

Fortunately the Marmon product and manufacturing engineers were equally innovative and Nordyke & Marmon escaped the escalated vendor costs for unfamiliar hardware that would have handicapped any other manufacturer introducing such a radical product as the Model 34. The 7 passenger touring car was priced, at introduction, at $2750 and both the 5 passenger touring and the Club Roadster at $2700. The wisdom of the Board was evident, their decision vindicated.

The later announcement of closed bodies commented: "Nowhere are Art and Engineering more ideally mated than in the production of Marmon Closed Cars". The unique construction (no sills below doors) resulted in closed bodies only **81 inches** high without sacrifice of usual headroom. To reconcile the concept with practicality of handling the body before unitization, the sedan bodies had only a center door on each side; the driver and companion seats were buckets reached through the narrow center aisle between them . . . the reverse of the 'cloverleaf' entry arrangement.

Because of the rounded cowl, the windshield pillar line of the closed body extended downward about six inches below the lower window line. Closed bodies also had concealed hinges but, of necessity, exterior door handles.

Open car bodies were Marmon built. The limousine bodies were by New Haven or Thompson priced at $4400 or by Holbrook at $5250. The landaulet and Town car bodies were by New Haven and Holbrook and were priced $100 over the limousines. The convertible sedan was built by Thompson and was priced at $3950.

The limousines, landaulets and town cars had identical dimensions for the rear compartment and featured reversible jump-seats.

The "Convertible Sedan" was of the "Springfield Type:" (originated by Springfield Body . . .) . Entry was through center doors to front bucket seats, the driver's being adjustable fore and aft. The door window was raised and lowered by a 'Perfect' brand patented regulator and the window and its frame could be stored within the door. The two side windows were removable with their frames; Provision was made for secure rattle-free storage behind the rear seat cushion. The converted body had a true "hard-top" appearance.

Regular equipment included roll curtains, a carpeted foot roll, interior lights, interior heater, a lady's vanity case and a gentleman's smoking set. The vanity case contained a mirror, address book and pencil, card case, (smelling) salts bottle, and powder puff case. Engine exhaust provided indirect heat through a grill in the heelboard of the rear compartment.

The convertible sedan was made of steel over ash wood framing; the others of aluminum over ash. The convertible interior was done in a neutral color 'Craveneted' cloth of pleated type over spring cushions and white curled hair. Exteriors were all two color, with black uppers on all closed cars, above the molding line. Brewster green, Dark Maroon, or Coach Blue Deep were optional.

Rear doors were equipped with double latches and Yale locks. Rotary buffers near the top and bottom of the rear door jambs eliminated noise in closing and kept the doors tight.

The frame was formed from 1025 carbon steel, allowing a thin gauge and was assembled with hot riveting; spring hangers were heat-treated forgings attached with hot riveting.

Fig. 4-14 The Thompson-built 'hardtop' sedan.

## ENGINE AND CHASSIS

The Automobile reported: "The engine is the first automobile motor to be placed in regular production with an aluminum cylinder casting . . . into the aluminum cylinder block, thin (white) iron sleeves are pressed after machining is completed, in order to provide a working surface for the pistons, which are also aluminum alloy made by the permanent mold process. Cast in unit with the cylinders is the upper part (half) of the crankcase." The engine was developed with an aluminum cylinder head but difficulty in producing sound castings at the required production rates forced the use of iron heads at start of production. The reporter continued: "It is hoped uo be able to make this also in aluminum with iron valve seats when the supply of castings becomes a little more plentiful"; there is no evidence that the aluminum head ever was released for production.

The engine was a push-rod overhead valve six cylinder 'enbloc' design of 3.75 inch bore and 5.125 stroke with a piston displacement of 339.7 cubic inches. The SAE rating was 33.75 (hence the 34 Model designation) and the engine developed 70 horsepower at 2500 rpm. It was the first Marmon engine to have a conventional crankcase configuration . . . split on the shaft centerline and the crankshaft installed from below with the bearing-half caps. The shaft was unusually massive for control of torsional vibration and had four main bearings, as did the gear-driven camshaft.

In addition to the cylinder block, the rocker cover, lower crankcase, oil pan, water jacket side closure, water pump housing, intake manifold, fan pulley **and the pushrods** (for control of valve clearance) were of aluminum. The water pump impeller was crankshaft mounted; the oil pump directly driven by the camshaft.

The patented cone clutch with exclusive progressive engagement was continued but featured an aluminum cone member.

The transmission was a three-speed and reverse sliding-gear type in an aluminum case mounted on the forward end of the torque-tube and utilizing the drive shaft as mainshaft. The differential was a patented design by Alanson Brush and was housed in an aluminum casting. Torque tube and axle halves were formed of drawn steel tubes; the latter had stamped steel brake-supporting flanges attached by welding.

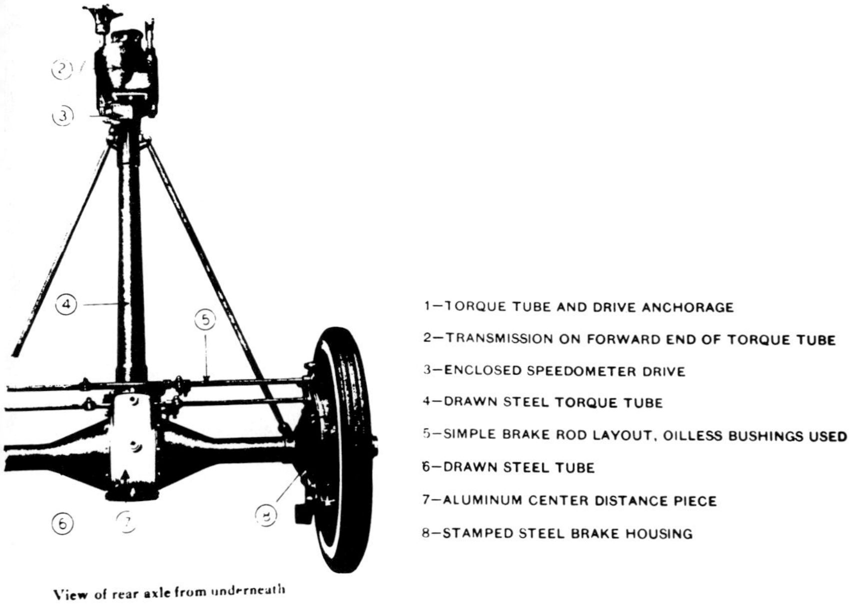


Fig. 4-16 Drive train of the Model 34A

Fig. 4-17 Lubrication and cooling systems of the Model 34A

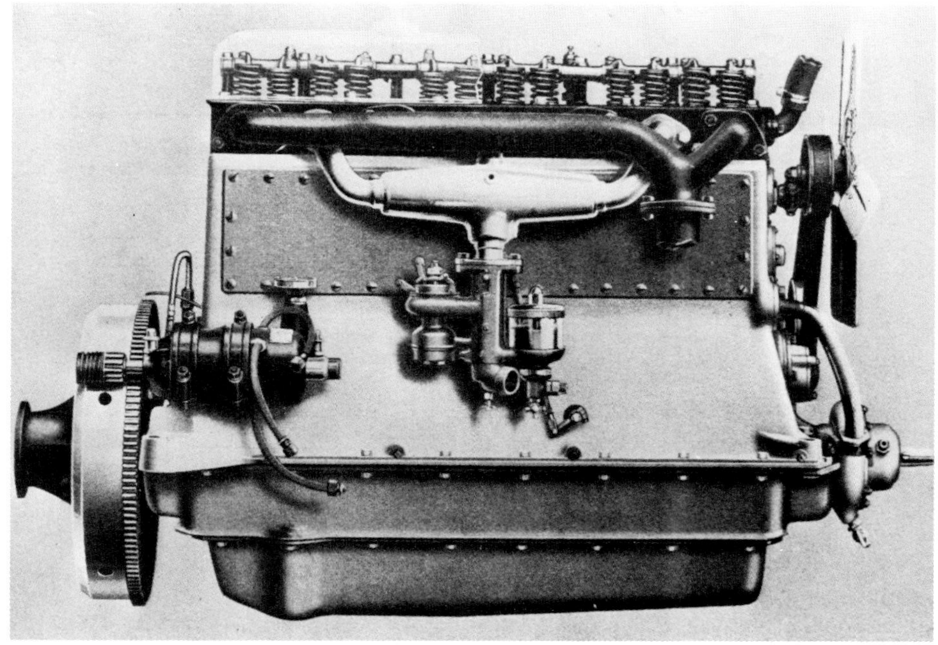

Fig. 4-18 Model 34A aluminum, wet-sleeve engine.

Service brakes were external band type, 2 inches wide acting on 17.5 inch diameter drums. Emergency or parking brakes were internal band type on the same drums.

Front suspension was semi-elliptic with reversed rear shackles.

Marmon had adopted the Bosch electrical system for the 1914 models.

Favorable experience over the two year period warranted its adoption for the Model 34A.

## MODEL 34A
## 1917

The principle mechanical change was the adoption of a thermostatically controlled cooling system as standard equipment. This introduction was particularly significant because the system was of the "full-flow" type wherein the thermostatic element actuates a valve which reroutes the outlet water flow back to the pump inlet without passing through the radiator. This type system became an industry standard for heavy duty diesel truck and bus engines in the 1950s. Such a system was appropriate to the Model 34A because passenger car load factors at highway speeds with a 70 horsepower engine were not unlike those of a modern highway truck.

The upper corners of the windshield were squared, having been round previously. This signified the marriage of windshield and top; the windshield had begun as an accessory independent of a free standing top and the rounded upper corners had been perpetuated when the top became supported by the windshield.

## COLONEL HOWARD MARMON, US ARMY SIGNAL CORPS.

On June 5, 1917 *The Automobile* announced that Howard Marmon had gone to Washington, D.C. for an indefinite period. He was to be commissioned a Major in the Signal Corps. and would assist Major Henry Souther, head of the Aircraft Production Board as Aircraft Engineering Engineer.

During 1917 when a Commission was sent to Europe to study and select airplanes for the US military, Marmon was placed in charge of the technical group. Lt. Col. Marmon supervised construction of McCook Field, Dayton, Ohio, and became the first commanding officer of that prestigious installation. While at McCook, he directed the development and testing of the Liberty V12 aero engine. [2] In his absence, engineering responsibility was carried entirely by C.E. Jeffers, Chief Engineer.

## MODEL 34A
## 1918

The appearance of both open and closed bodies was appreciably improved for 1918 by a 10 degree inclination of the windshield. On closed bodies, this was accomplished by resort to a wedge shaped pillar. The market for closed bodies had become important enough to provoke a higher cowl and corresponding increased rise of the hood from the radiator; this eliminated the pillar extension below the belt line. The doors on the open bodies were correspondingly higher, as was the tonneau. The latter had a reduced curvature at the rear by virtue of intersecting the chassis farther aft. The semi-divided front seat was changed to a plain double seat. The windshield pillars were extended downward inside the cowl to the frame channels.

Fig. 4-19 Model 34A sedan; 1918 two-door (upper) Model 34A sedan; 1919 four-do (lower). Applegate and Applegate Photo

The center entrance constraint evidently was not too well accepted; everyone is not that athletic. In June of 1918 Marmon announced a new 4-door sedan: "Front doors have been added to the Marmon sedan to afford easy access to the front seats without passing through them." Evidently the individually adjustable bucket seats were retained.

The cowl was redesigned to accommodate two side triangular windows converging toward the narrower lower windshield. This effected a smoother transition

Fig. 4-20 The under-hood refueling in process. The impressive engine was exposed and usually attracted an audience.

from the rounded cowl to the front door.

A magnetic type fuel level gauge on the instrument board was adopted in conjunction with the cowl mounted fuel tank previously used. This allowed sealing of the tank for fume control except for the vent line which was carried down the cowl face to below the body. The odometer reset knob was relocated from the speedometer face to below the dash.

Hayes wire wheels were continued but changed to 23 inch to accommodate 32 X 4½ tires on open models and 33 X 5 on closed bodies. Tire size had been 34 X 4 on 24 inch wheels.

An automatic tonneau light was adopted, actuated by opening of the right rear door, as well as by manual switch. A new electrical switch was fitted on the instrument board, with capability of being locked in any position, one of which was "Magneto on". A key lock was also added to the ball on the gear shift lever to lock it in the neutral position. The single key locked all locks including the new tool compartment in the left front door where tools were held in pockets of the compartment which swung downward to a horizontal position maintained by chains.

The availability, upon special order, of closed bodies by Rubay was announced. A limousine was priced at $5750 and town car at $5700. For other custom body applications, a bodyless chassis was priced at $3200.

By mid 1918 owners' experience with superior fuel and tire economy consequent the "high-efficiency" engine and the light weight construction led to rather surprising relative economy claims in Marmon ads: 50% to 75% more mileage from fuel and 40 to 50% more mileage from tires. The slogan had become "Stabilized Design — Advanced Engineering". Later the fuel mileage claim changed to "11 to 14 miles per gallon".

A button release for the parking break was introduced.

The new Bijur starting motor was actuated by a starter pedal forward of the shift lever. Previously an electric foot switch was located convenient to the driver's heel. The generator was also procured from Bijur, but the Bosch magneto was retained.

MARMON

The Following Men Connected with Mechanical Manufacturing Interests have Purchased MARMON 34 Cars of the New Series

COL. H. W. ALDEN, Vice-Pres. and Chief Eng.
*Timken-Detroit Axle Co., Detroit*

F. A. ALDRICH, Secretary and Treasurer
*Dort Motor Car Co., Flint, Mich.*

E. E. ALLYNE, President
*Aluminum Castings Co., Cleveland*

FRED L. BAKER, President
*Los Angeles Shipping & Dry Dock Co., Los Angeles*

M. M. BAKER, General Manager
*Holt Manufacturing Co., Peoria, Ill.*

C. F. BLUE, JR., Vice-President
*Carbon Steel Co., Pittsburgh*

D. S. BROOKS, President
*Spacke Machine Co., Indianapolis*

NEWTON CHAPIN, Secretary-Treasurer
*Interstate Pipe Co., Carnegie, Pa.*

HARRY H. GRAY, President
*United Engineering Co , San Francisco*

COL. E. J. HALL, Vice-President
*Hall-Scott Motor Co. (Co-Designer Liberty Airplane Motor), Berkeley, Calif.*

JAMER R. HARRIS, Vice-President
*Busch-Sulzer Bros., Diesel Engine Co., St. Louis*

NORVAL HAWKINS, Capitalist
*(Formerly Sales Manager Ford Motor Co.), Detroit*

EDWARD HEINA, Vice-President
*Cleveland Metal Products Co., Cleveland*

HERMAN HOLLERITH, Formerly Gen. Mgr.
*Tabulating Machine Co., Washington*

L. G. HOOKER, Assistant Treasurer
*General Aluminum and Brass Co., Detroit*

E. S. HOOPES, Director
*Union Drawn Steel Co., Beaver, Pa.*

H. A. JACKSON, President
*Chicago Pneumatic Tool Co. (Pneumatic Tools and Motor Trucks), Chicago*

C. F. KETTERING, President
*Dayton Engineering Laboratories Co., Dayton, Ohio*

L. H. LA CHANCE, President
*Stewart Manufacturing Co., Chicago*

JOHN B. SEMPLE, President
*John B. Semple Co. (Ordnance Mfrs.), Pittsburgh*

GEORGE E. SMITH, President
*Royal Typewriter Co., New York City*

LON R. SMITH, Director of Sales
*Midwest Engine Co., Indianapolis*

S. B. STEVENS, Capitalist and Manufacturer
*Well-known amateur racing driver, Rome, N. Y.*

R. A. STRANAHAN, President
*Champion Spark Plug Co., Toledo*

G. A. STROM
*Strom Ball Bearing Co., Chicago*

MARMON

JAMES R. CHRISTY, Manager
*Alameda Bethlehem Shipbuilding Corporation Alameda, Calif.*

J. R. CLARK, President
*American La France Fire Engine Co., Elmira, N. Y.*

EARL COOPER
*Formerly Automobile Racing Champion Pacific Coast, San Francisco*

COL. E. A. DEEDS, President
*Domestic Engineering Co. (formerly Chief Equipment Division Bureau of Aircraft Production), Dayton*

FRANK DISSTON, President
*Henry Disston & Sons, Philadelphia*

CLARE . H. DRAPER
*Draper Corporation (Woolen and Cotton Machinery), Hopedale, Mass.*

J. B. ECCLESTON
*Brandenburg Bros. & Eccleston (Motor Car Parts Manufacturers), Detroit*

VAL A. FYNN, Consulting Engineer
*Wagner Electric Co., St. Louis*

C. P. GOSS, General Superintendent
*Scovill Mfg Co., Waterbury, Conn.*

B. A. GRAMM, Vice-President and Designer
*Gramm-Bernstein Truck Co., Lima, Ohio*

ALBERT C. LANG, Vice-President
*Crucible Steel Castings Co., Milwaukee*

E. F. LUNKEN, President
*The Lunkenheimer Co., Cincinnati*

E. H. LUNKEN, Vice-President
*The Lunkenheimer Co., Cincinnati*

ALBERT H. MOHR, President
*John Mohr & Sons (Iron Works), Chicago*

W. J. O'NEILL, Vice-President
*Stromberg Motor Devices Co., Chicago*

O. C. PANTALL, Vice-President
*Midwest Engine Co., Indianapolis*

MASON PHELPS, President
*Phoell Mfg. Co. (Screw Machine Products), Chicago*

J. H. POWELL, Vice-President
*Bethlehem Steel Corp., Bethlehem Shipbuilding Corp., Bethlehem, Pa*

A. F. ROCKWELL, Chairman Board of Directors
*Marlin-Rockwell Corporation, New York City*

FRED SCHAEFER, Secretary
*The Lunkenheimer Co., Cincinnati*

LELAND SCOTT, Secretary
*Hall-Scott Motor Co., Berkeley, Calif.*

H. M. SWETLAND, President
*United Publishers Corp. (Publishers of automobile and technical magazines), New York City*

FRANK A. URWAN, President
*U. & J. Carburetor Co., Chicago*

SAMUEL VAUCLAIN, President
*Baldwin Locomotive Works, Philadelphia*

GUY W. VAUGHAN, Vice-Pres. and Gen. Mgr.
*Van Blerck Motor Co., Monroe, Mich.*

F. L. WALTON, Sales Manager
*Union Switch & Signal Co. (during war, Mfrs. La Rhone Aviation Motors), Swissvale, Pa.*

In 1918 Marmon published a sales booklet titled "Deciding Factors in Buying a Motor Car". It included five letters expressing satisfaction with the Model 34 — one of them was from Edsel B. Ford:

Again, at the 1921 shows, Marmon distributed a booklet 'Man Who Appreciate Manufacturing Excellence'. Particularly worthy of note are the names of Col. Alden of Timken, Col. Hall and Mr. Scott of Hall-Scott Motor, and C.F. Kettering of Delco. For the record and possible interest, the full list is reproduced.

## MODEL 34B
## 1920-1924

THE FOREMOST FINE CAR

In November 1919 a series of unidentified advertisements began in weekly magazines, similar in concept to those which had preceded the introduction of the Marmon '34' in November 1915. Four successive two-page unidentified ads printed on heavy stock were followed by a four-page spread identifying the manufacturer and announcing a "New Series Marmon 34 with High Efficiency Motor."

The Nordyke & Marmon Company's knowledge of metallurgy and manufacturing processes had advanced so remarkably, as a consequence of their experience in building both Hall-Scott and Liberty engines during 1918, that Howard Marmon concluded the '34A' engine to be obsolete. Dimensional tolerances of one-quarter of one thousandth of an inch (.00025) had been achieved in wartime production and would be employed in the New Series 34 engine.

Dimensional control would be maintained by three levels of gaging: workman, inspector and master gages with twice weekly reference to the masters. Incoming material stocks would be subjected to chemical and physical testing. Critical parts would be subject to 100% testing. Engine crankshafts would be statically and dynamically balanced.

Each new production engine would be run by electric power for one hour following assembly. This would be followed by a three hour run under its own power. A complete tear-down, inspection and reassembly would be followed by an additional hour of dynamometer running, finishing the test sequence at wide open throttle. No other known production engine worldwide was subjected to a comparable test program prior installation in the chassis.

Performance of the New Series cars was stated to have been witnessed by the press as follows:

| **Axle Ratio** | | **3.75:1** | **4:1** |
|---|---|---|---|
| 10 to 50 mph, high gear | | 18.8 sec | 15.8 sec |
| Half mile avg. | | 69.75 mph | 65.7 mph |
| Hill climb from | 5 mph | | |
| 1000 feet | 8% grade | 24 mph | 27 mph |
| 1775 | 6% | 40 | 43 |
| Hill climb from | 10 mph | | |
| 1000 | 8% | 26 | 30 |
| 1775 | 6% | 43 | 47 |

Operation was demonstrated for 300 yards at a walking pace. One reporter commented: "The car is delightful to handle. Not only does the engine possess that elasticity and smoothness which characterizes the first half dozen cars of the world but every control operates with small effort. On a rough road it is easy to steer with one hand, and then with but a light grip on the wheel. The clutch needs only a little pressure to disengage it and the service brake has leverage enough to pull the car up smartly without more force on the pedal than the average woman could easily supply. Both gearshift and emergency brake levers are positioned so that they can be grasped without moving the body.

"Practically the entire car would be constructed in the Marmon plant from raw material to finished product. The new car would exhibit "extraordinary smoothness of running, a freedom from vibration and an immunity from trouble and wear."

"The engines would be built from new tooling installed in the machine shop . . . 600 feet X 600 feet . . . constructed for military production. Bodies would be built in a brand new shop five stories high and 600 feet long. Vehicle assembly would take place in a brand new final assembly building 800 feet in length. Supporting wood processing and woodworking shops were being constructed. When and if capacity production of 600 cars per month were achieved, 5000 employees would be required."

The *Marmon News* described the strenuous test program conducted to confirm the design of the Model 34B: *"C.E. Jeffers, chief engineer of the Nordyke & Marmon Company, followed a unique course in developing the early experimental models of the new series Marmon 34. Believing in the "Ride-'em-cowboy" methods of the West, he evolved a world series contest that stretched over months. He developed and built two models and gave them into the hands of two hardened graduates of the testing crew, who have the guaranteed ability to break any car ever built. They had instructions to "bust" them if they could, to ride them and drive them and to give the new creations the beating of their lives. One of these gentlemen was Joe Dawson, winner of the 1912 500-mile Indianapolis Motor Sweepstakes Race, and a veteran road racer. The other was Bruce Keen, who has ruled the final assembly and testing of Marmon's these many years. With the cars in the hands of these two men, Jeffers was sure he would learn of any weakness "muy pronto," as the men of the cow country put it. More than this, with two such men contesting to find the worst and the best of the cars, one was sure of results. From the tabulated findings of months of tests so continued, Jeffers developed this car, which has so remarkable a performance record, and he gives no little credit to Joe Dawson and Bruce Keen."*

### 1920

The Model 34B closed bodies appeared at first glance unchanged from their immediate predecessors. The same could be said of the chassis exterior as well. However it quickly became evident that the "greenhouse" (body immediately surrounding the windows) had been lowered several inches. This was accomplished by resort to a curved lower windshield edge, following the shape of the hood. Some 1916 **custom** sedan bodies had a similar treatment. The sedan however now had four doors. No mention was made of the demise of the handsome four door with sloping windshield and triangular windows at its sides introduced in mid 1918.

A new body style was the close-coupled four-passenger four-door touring; the body ended at about centerline of the rear wheels and made a very handsome car. Like the five and seven passenger tourings, it featured traditional Model 34 body configuration and aluminum skin.

The radiator shell and core were reproportioned to allow increased core surface by having the core wider. This requirement was probably related to increased heat rejection from the iron block engine. The shell sides were parallel; previously the shell was tapered upward.

Unusual round parking lights were attached to the bottom of the headlights by a coupling nut somewhat like a pipe union. A very neat detail design.

Fig. 4-21 Although lower than its contemporaries, the 1920-21 sedan had deep windows and tapered windshield pillars.

Fig. 4-22 The Model 34B four-passenger close-coupled touring.

A crowning touch to the 34B masterpiece was the indirectly-lighted cluster of matched instruments under a single glass. Marmon identified it as a "new feature quite likely to be copied". The announcement advertisement for this Marmon innovation warrants reproduction here.

The Model 34B engine substituted an aluminum split crankcase and two cast-iron cylinder blocks for the aluminum block/wet steel liner construction of the Model 34A. Delco coil ignition replaced the magneto. A multiple-disc dry clutch replaced the cone clutch of the 34A.

"By actual test I have found that I can get more mileage from gasoline, oil and tires in my seven-passenger Marmon than I could from a good Roadster of another make which I used for years before purchasing a Marmon."

(*Signed*) M. M. LEFTWICH,
*M. M. Leftwich Company, Contracting Engineers,*
*Nashville, Tenn.*

"The Marmon 34 has a remarkable spring suspension and I do not believe that I have ever ridden in a car which handles as easily as this car does over the rough country roads. The motor is very quiet and free from vibration of every sort."

(*Signed*) EDSEL B. FORD,
*Vice-President Ford Motor Company,*
*Detroit, Mich.*

"The user gets excellent economy—ease of control—superior riding qualities at any speed up to sixty miles an hour—and as for appearance—there is nothing on the market—in my opinion—that compares with it."

(*Signed*) LON R. SMITH

Fig. 4-23 Exerpt from a Marmon publication of fan mail.

The Model 34B engine was the first automotive engine to incorporate precision insert-type main bearings and to dispense with shims which Marmon had determined to be detrimental to bearing performance. Both main-bearing inserts and connecting rods were serviced with interchangeable parts; diamond-tipped tools for boring the crankcase and machining the inserts and rods were used. Newly assembled engines turned freely; previous industry practice resulted in tight engines requiring special break-in practice. It was not unusual as late as post WWII for a certain high volume domestic car to experience a run-down battery within one or two days of delivery consequent to excessive engine friction.

The chassis of the 34B differed from the 34A in detail only, with the redesign of the steering pivots being most significant. Of course, the engine and clutch assembly were entirely new.

## MODEL 34B
## 1921

Louvers were added to the hood for the 1921 model year, presumably because experience dictated it.

Fig. 4-24 The 1920-24 milestone instrumentation.

Fig. 4-25 Front passenger's view — Model 34B closed styles 1922-24

In early 1920, "newspapers were debating whether or not a standard fully equipped touring car could cover a **standing** (start) Indianapolis lap at an average of a mile a minute. Ever ready to meet a challenge, Howard Marmon sent a stock 34 to the speedway on March 13th, 1920 and it accomplished the feat in 2 minutes, 22.4 seconds, equal to **63.2 mph.**"[2]

Subsequently the Marmon 34B was selected to be the pace car for the 1920 500 Mile Race. Barney Oldfield was named the driver and a special Marmon car was prepared for the event. It was a 2-passenger automobile with what the author calls a "beeback" rather than a boat-tail . . . an appropriate design for a Marmon 'Wasp' Speedster. Frontal area was reduced by a slightly lower radiator and a hood which rose only slightly to the cowl. Straight and parallel nickel plated pillars supported a relatively narrow and low windshield; the lower panel seated in a groove in the cowl skin. The cowl was elongated compared to the regular Marmons which had to accommodate rear seat passengers and the seating and general layout did not have to contend with a rear cushion resting on the kick-up of the frame. Consequently the driver was considerably aft of the usual placement, requiring a severe rake to the steering column and a deeply-dished steering wheel. The spare wheel rested on top of the sloping bustle. The top was designed for easy removal, being snapped rather than tacked to the body; it was only 70.75 inches high and was characterized by diamond-shaped beveled-glass single side and multiple-rear windows essential to protection from breakage during stowage. The car was styled to be particularly handsome in the topless configuration.

It was indeed a handsome automobile that appeared at the head of the pack on Memorial Day to await the starter's flag. Aided by the special 3.0:1 axle (3.75:1 standard) Oldfield brought the pack to 80 mph before pulling off at the end of the pace lap.

Barney Oldfield acquired the Marmon Wasp Speedster and used it extensively in promoting a personal business venture — the Oldfield Tire. He crossed the continent eight times before switching to a Marmon coupe[2] following an accident while the car was stolen.

Who was responsible for the design and construction of the aluminum pace car body is not positively known to the author. It may well have been been W.E. Pierce of the Hume Body Co. of Rochester, New York, for the car created great interest and the Hume firm was contracted to supply production bodies beginning with the 1921 model year; they would continue to the end of the Model 34 production. The public offering of the pace-car speedster was announced on August 20, 1920. It carried a 0-100 mph speedometer and 3.0:1 rear axle as standard equipment. Standard Speedster colors for 1921 were Marmon Racing Yellow on the body, hood and wheels with everything else black. An exception — red hub centers.

A variety of seating arrangements is evident on surviving 2-passenger speedsters. Presumably the earlier cars had two individual fixed bucket seats, later ones had a single seat formed into individual seating either with a free standing back or built into the body. Even later a full bench seat was fitted; the true sequence, if one existed, is not certain.

The beetail provided some storage space which was accessed through what later were called golf-bag doors on either side. These doors had concealed hinges and showed only a key-lock: the key became the handle. Some speedsters surviving today have an additional button type handle; this identifies a different usage of the space. The "door" was not that; it pulled straight out to the edge of the car like a drawer, whereupon a folded back rest could be positioned. The seats could then each accommodate one small lean passenger facing forward with his (her) feet

Fig. 4-26 Barney Oldfield in the 1920 Indy pace car on race day.

on the running board. Presumably the intended passengers were children; I have never seen the seats in use. It is remarkable that both the seat belt and mandatory-use legislation were not both concurrently born in 1921!

Marmon was riding high as they brought their 1921 models to the New York show; they proceded to "steal" the show not just with their 1921 models but with their "model Models". "Because lack of space would not permit showing its entire line, Nordyke & Marmon Co. has built miniature models of its eight body types and is displaying them at the New York show, together with some of the original cars. Built to one-quarter scale, they are 46 inches long."[12] These model cars attracted much attention and some survive today under most interesting circumstances. They are the subject of a full discussion in Part 1 of Chapter Ten: "Advertising the Marmon."

The New Series 34B was priced surprisingly higher than the 34 A had been at its introduction. The immediate post-war period was marked by a highly cyclic economy. For 1921, after a 7.5 percent raise, the 4 and 7-passenger touring cars and the roadster were priced at $5000. The speedster was $5300. Closed body prices were: 4-passenger coupe $6150; 7-passenger sedan $6600; Limousine and Town Car $6800.

Interestingly, **dual bodies** were also offered and priced, one 'open' for summer and one 'closed' for winter. Prices were based on both bodies being specified on the original order for concurrent shipment: chassis with coupe and any open body $6700; Sedan and any open body $7150; Limousine and any open body $7350. Closed bodies, without chassis, were priced as follows: Coupe $1700, Sedan $2150, Limousine and Town Car $2350.

The dual body solution to all-weather motoring was evidently an alternative to the alternate color open-car upholstering option offered on the 34A; reversion to traditional upholstery construction for the 34B precluded continuance of that option.

For the custom body builders, the Marmon 34B was offered without body for $4450, or with cowl included for $4525. The latter weighed 3125 pounds.

Closed body styles averaged about 4050 pounds. The Speedster weighed 3420, the 4-passenger touring 3460 and the 7-passenger touring 3700 pounds.

Hood locks were available for $10.00 installed. Rear view mirrors either on fenders or windshield were a $7.50 option.

In 1921 the domestic automotive industry was in a market difficulty not unlike that which beleaguered the industry in the 1980-1982 period. The nation was experiencing financial repercussions from the first World War and prices, including automobiles, were volatile. The Marmon 34, because of the consistency of its design, could be said to have a near constant 'value' to the consumer and therefore provides a yardstick to measure the price excursions during the earlier part of the Third Era.

The 7-passenger touring at the time of introduction of the Model 34A in 1915 was priced at $2750. The Model 34B bowed in 1919 at $4650 and by 1921 was up to $5000 as an across the board increase of 7.5 percent brought the total increase to a full **80 percent.**

Fig. 4-27 1921 Victoria

Fig. 4-28 Interior of the four-passenger coupe. Note floor exhaust heater.

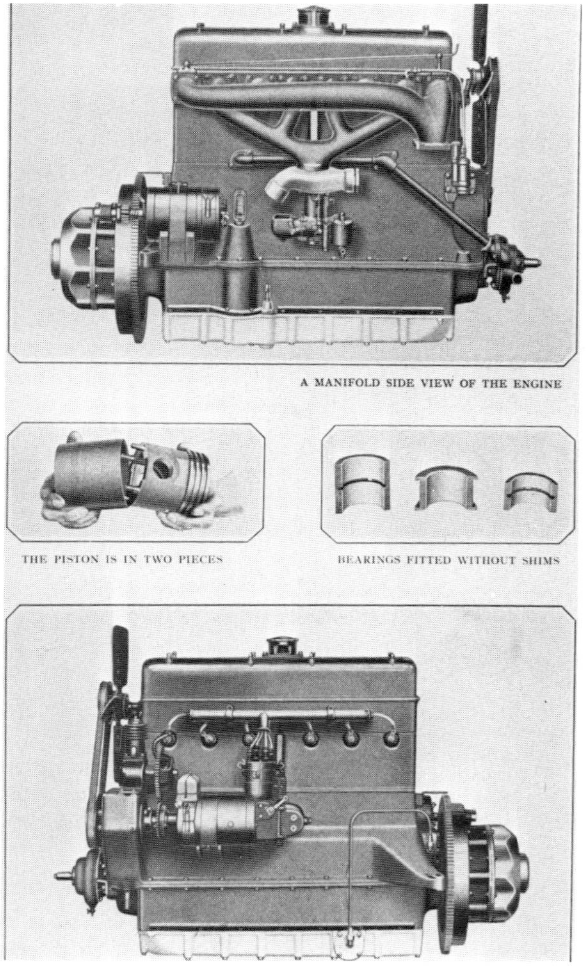

Fig. 4-29 The Model 34B engine was a remarkably clean design with great attention to detail — note the water pump location on the nose of the crankshaft.

Fig. 4-30 Door compartments stored tools and side-curtains. Note the height of the seat riser and the steering column notched into the dash of the 1923 seven-passenger phaeton body.

The industry was in the doldrums in 1921 when a newsworthy press release came from Nordyke & Marmon. Automobile sales had been poor because car prices had been maintained in the face of significantly declining personal incomes . . . maintained because the bulging inventories of unsold cars and unassembled components had been produced or procured at the previous high rates. *Automotive Industries*[13] reported the development under the heading:

**Marmon Price Reductions from $1015 to $1400**
Look for Business Stimulation

In announcing a 20 percent across the board price reduction effective immediately — May 2, 1921, Vice-President Moskovics said: "*We believe the spirit of our reduction is in keeping with the demands of the American public and in accord with the spirit of the times. We do not*

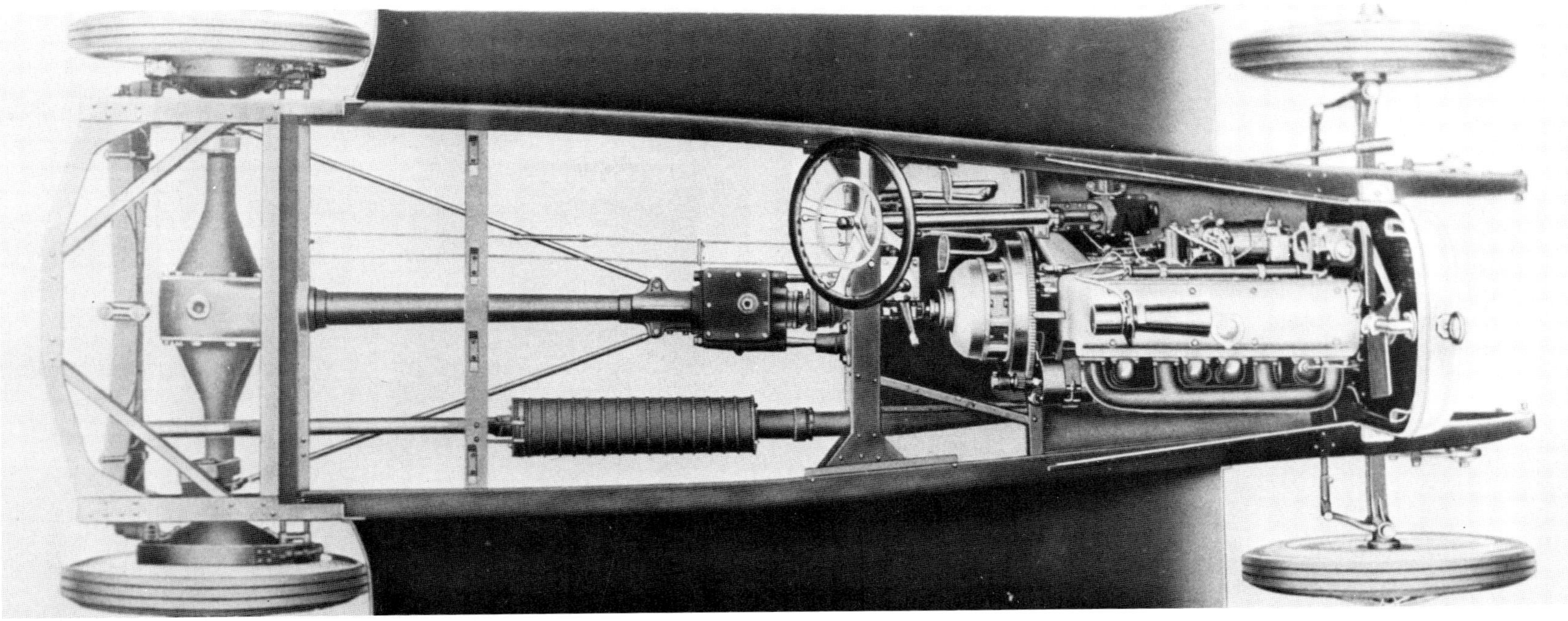

Fig. 4-31 Plan view of the unique Model 34 (34B shown) chassis shows the integral running boards and the general clean and clever design.

Fig. 4-32 The 1920 seven-passenger touring was light and sporting in appearance. Note resemblance to the 1916 Model 34A.

*believe that any company can sit tight on inventories purchased during the period of high prices and (at the same time) assist in starting the prosperity of the country. Present inventories are worth only their reproduction value. New (Marmon) prices are based on material and labor costs now or in the immediate future."*

One week later, the same magazine reported[14] a development probably related to the market stalemate. The Marmon Automobile Company of New York came under the direction of officers of Nordyke and Marmon with the treasurer, A.R. Heiskell becoming president of the New York firm which would remain a separate entity . . . not become a factory branch.

## 1922

Three specially equipped open style cars were "brought-out" in September 1921. They were 'DeLuxe' models of the 4 and 7 passenger tourings and the 2 passenger speedster and were "specially equipped so that a purchaser desirous of the usual list of extras (options) and the other special order equipment need not wait while his special order is filled but can get his car at once".[15]

"All three models carry the following special finish and equipment: Special color paint, special fabric top material, special leather upholstery, nickeled radiator bar-cap (dogbone), nickeled deluxe Motometer, nickeled hood (key) locks, nickeled barrel (drum) type headlights and tail light, nickeled Hartford front bumper, spotlight (on pillar) with integral "mirrorscope" (convex mirror!), "Cozy" (brand) windshield wings, (glass) sun visor on windshield, vacuum operated windshield cleaner (wiper), rear view mirror on windshield, cigar lighter on instrument board and two spare wheels, tires and tire covers."

"On the Four-Passenger Special there is a trunk rack and trunk, with nickel bars on the body. Two tires are mounted on each side in special wire wheel **carriers which are sunk into the running board.**"

The Seven-Passenger Special also included dual rear-mounted spare wheels.

Fig. 4-33 The 1922 sedan was Marmon-built. It was lower than its predecessor while retaining the glass area.

Fig. 4-35 This 1922 four-passenger version of the very popular Speedster. Standard trunk and sidemounts.

Fig. 4-34 The 1922 close-coupled four-door, four-passenger responded to the combined demand for a sporty style and for a closed car.

Fig. 4-36 This 1922 seven-passenger touring shows the new standard woodspoke wheels and new drum type headlamps.

"For the Speedster Special, the rear compartment is lined and fitted with an electric light. Instead of having the extra wheel and tire on the back (deck) as in the standard Speedster, there is a bracket and two extra tires (and wheels) mounted on the rear of the frame.

These Deluxe Special cars were painted in one of four color combinations unique to each style. These were used on the pilot production of four cars of each model pending receipt of dealer orders. Very specific information is found in 'Art and Color', Chapter 9 Part 2.

In a concession to practicability, the Speedster single spare wheel was subsequently relocated to an upright-rear mount. Wrestling the wheel off and on to the near-vertical spindle on the deck must have required athletic prowess or a pit crew. The flat relief was removed from the deck. At least one Speedster was delivered with a one-man rumble seat.

Four all new closed car bodies were displayed at the New York Palace Show. They were Marmon-built in the brand new 5 story body plant; previous Marmon closed car bodies were procured from an outside source. The completion in mid-1921 "of the immense and completely equipped body shop of Nordyke and Marmon Company" made the change possible.

All were characterized by metal sun visors over severely sloping windshields, which narrowed towards the bottom and had large fixed triangular windows at the sides separated by narrow pillars, and by a coaching motif in body side treatment (somewhat similar to the early Ford Model A cowl side molding). Coachwork was described by the press as being on a par with the Marmon chassis. The new windshield was recognized as contributing to both appearance and superior vision.

Open bodies were continued essentially unchanged.

Following its surprise showing at the Chicago Auto Show[16], Marmon made an exciting announcement: A 4-passenger Speedster, companion to the very popular 2-passenger Speedster, had entered production in March 1922. This was, in 1922, an event comparable to the announcement in 1982 of a four-passenger Corvette. It was identical to the 2 passenger, except for a shorter cowl, from bumper to back of the front doors. Top bows were natural wood with nickel plated slats (hardware). A special frame cross-member at the forward edge of the rear seat allowed the rear seat cushion to be below the top of the rear frame kick-up, rather than resting on top of it. A trunk with fitted luggage rested on a fixed rack behind the body. Side mounted spare wheels were standard, the tires resting on aluminum pedestals on the running boards, and strapped to brackets extending from the cowl.

The 4 passenger speedster was upholstered in unpleated Spanish leather with a single central interruption in the smooth surfaces.

The 1922 Brochure 2230 continued: *"The ideals of Marmon design and Marmon workmanship have an unrestricted development in these nine cars. That same perfection attained in the engine is found in the coach-work of the open and closed models. "Marmon made from first to last, these nine cars are the seven-passenger Touring, four-passenger Touring, Roadster, two-passenger Speedster, four-passenger Speedster, seven-passenger Sedan, four-passenger Sedan, seven-passenger Suburban and (Victoria) Coupe. The two custom-built bodies are the Limousine and the Town Car."*

The inclusion of the two Speedster models in the list of totally Marmon-made bodies is the result of the Marmon acquisition of the Hume firm by mid 1921. Hugo Pfau in his *'Custom Car Era'*[17] reported that Nordyke & Marmon had purchased the Hume Body Co. of Boston as a means to acquiring the services of W.E. Pierce whose creations on Marmon chassis were especially admired by Howard Marmon.[18] The appointment of Pierce as Marmon's Chief Body Engineer and Designer in 1924 apparently coincided with the closing down of the Hume operation at the end of thg Model 34 production run.

Drum type nickel-plated headlights were adopted on May 1, 1922. They were an outside-sourced item, but had specially located electrical plugs for the Marmon application because Marmon recognized the convex mirror property of the rear headlight surface and did not want to distort the reflection of the car. Also effective May 1, 1922 Marmon prices were reduced and wood wheels made standard equipment, the former standard wire wheel equipment being priced at $135. Previously wood wheels were optional at $50 additional charge, including the spare rim carrier. Also the Hartford shock absorbers and MotoMeter (temperature gauge) became extra equipment items priced at $45 and $10 respectively. Hartford (Single bar) bumpers could be ordered for $18.00 installed, front **or** rear. However, during the model year, optional Hartford single bar bumpers were superceded by optional Bi-Flex bumpers which had a central double bar arrangement . . . somewhat mouthlike. A spot light was priced at $7.50 or $10 if equipped with Mirroscope (Convex mirror). A Victoria Top (over rear seat only) was available in leather or burbank finish for $550.00 including curtains. Casing (tire) covers were offered in black enamel duck for either wire or wood wheels for $9.50 (32 X 4.5) and $10.00 (33 X 5). Special colors were $2 additional.

## MODELS 34B and 34C
## 1923

All prior year body styles were carried over. Although the four and seven passenger tourings were continued, new seven and four passenger phaetons were announced on Aug. 19, 1922[18] and in April, 1923.[19]

The new phaetons were so different as to warrant a Model 34C designation. They were built on a chassis with reduced frame kick-up (and less suspension travel) which was fitted with the lower radiator shell previously used on the Hume bodied speedsters. The hood was similar to the speedsters, having only a slight rise to the cowl, and was slightly longer due to a forward repositioning of the radiator. The frame change permitted a lower body with an adequate rear seat whereas the four-passenger speedster rear seat had been compromised by the high frame kick-up and low top.

The additional length of the 7 passenger body was minimized for styling considerations by not providing the former flush storage of the rear jump seats and by a special front seat structure which traded-off front seat leg-room for a higher front seat located closer to the cowl. This necessitated a reduced rake of the steering column and a notch, for the column, in the instrument "board" which was black-lacquered sheet steel on all Model 34C cars.

The reduced height and enhanced styling of the Model 34C relative the 34B 7-passenger touring is clearly evident in the sales-promotion overlay of the two models which appeared on August 31, 1922.[19] A four-passenger style followed in April 1923.

The new bodies were steel-skinned and characterized by rectangular doors with square corners. The doors came down to the top-of-frame level, concealing the threshold plates. The thresholds were made from aluminum bar stock whereas on the other models they were castings which were visible, flush below the doors. The top-of-door line was unbroken at the front seat and rose slightly as the front door neared the cowl. The styling of the sides was influenced by requirements of a new option . . . the convertible phaeton. Actually there were two alternate options: the "Panorama Top Phaeton" and the "Convertible Phaeton", available on both the four and seven passenger models.

The "Marmon Panorama Top" which was non-folding, and whose standard covering was waterproof "Pantasote" material, similar to Model A Ford closed car top covering, provided rear seat passengers with a view unobstructed by the usual forward-facing bows required for a collapsible structure. The "Panorama Top" had continuous framing full length and the sail panels were reinforced with sheet steel for rigidity; the top was fully removable, with disconnection at the usual two windshield pillar and two rear "pivot" points. The interior was bow-free; a cloth curved headlining similar to modern closed cars was used. The four-passenger phaeton was priced at $3185; Model 34 prices which had risen to 180 percent of 1916 by 1921 had declined to 115 percent.

There was another reason for this top construction . . . the "Convertible Phaeton" concurrently announced. It was identical to the above phaeton except for threaded inserts on the inside of the doors and body sides which accommodated attachment of glass windows mounted in leather covered frames. They were essentially the traditional side curtain attached with the usual snap fittings but having the rod support replaced by a pair of nickel plated cast arms capable of posi-

Fig. 3-37 The Convertible Phaeton reflected changing consumer attitudes. Weather protection was by rigid leather panels with glass windows; when removed, it became a Panorama phaeton.

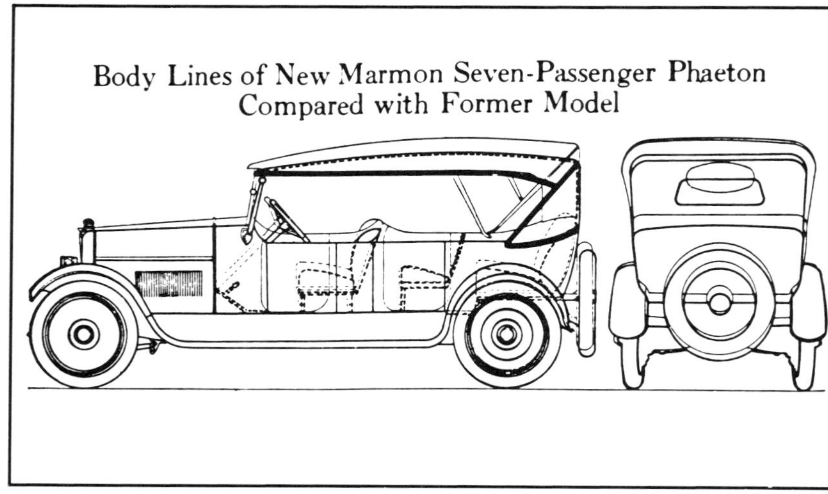


Fig. 4-38 1923-24 Model 34C seven passenger phaeton superimposed on the 1922 Model 34B seven passenger touring.

tioning and supporting the glass pane; the arms incorporated hand grips to facilitate handling of the loose panels. The enclosures were individually fitted to the particular car at the factory and were claimed "to be rattle and rumble proof". The Convertible Phaeton was priced $165 above the regular phaeton with conventional side curtains; the option price included a regular side curtain set for summer emergency use.

The initial brochure identified the 7-passenger as a 1922 model. However official Nordyke & Marmon serial number documents furnished to state registration authorities confirm that no phaetons carried 1922 serials; the 4 and 7 passenger were coded styles '5' and '8' ('5' had identified the 5 passenger touring prior to 1920 and had been inactive because the 4 passenger touring was coded '4').

In April 1923 a new 4-passenger Speedster was announced with deliveries to begin in May. It was based on the Model 34C chassis and on the 4-passenger version of the 7-passenger phaeton introduced the previous summer. The Hume-bodied 34B 4-passenger speedster was discontinued; The style prefix to the serial number changed from '1' to '9'.

A trunk with three pieces of fitted luggage and a cloth cover was standard on the Speedster, as were polished aluminum protective bars on the body behind the trunk. A single right hand spare tire and rim was carried on a cast aluminum arm bolted to the frame; The tire was suspended in a polished aluminum well in the running board.

The folding (or collapsible) top was fitted with natural wood top bows and nickel plated hardware. Soon after introduction of the 4-passenger, a polished radiator shell, dual-bar nickel bumpers and glass wind wings on the windshield became standard on both 2 and 4 passenger speedsters. The side mounted tire could be ordered on the phaetons and a second one for the speedster. The phaeton side mount option included a trunk rack which could be rotated to a vertical position when not in use. The Speedster was priced at $3435, the Panarama Top Phaeton at $3185 and the Convertible Phaeton at $3350.

The new body was characterized by square cornered doors extending to the top-of-frame. The speedster windshield was new but similar to the straight-vertical pillared one on the Hume speedsters; it had nickel plated pillars and fittings. The phaeton windshield fitted in the same cowl sockets but had painted pillars which curved first outward and then upward; that windshield was wider to better protect rear seat passengers and higher to allow thicker seat cushions, in particular the extra thick front seat required to shift leg room to the rear compartment for the jump seats of a seven-passenger car. Both windshields had a pivoted upper glass and a fixed lower glass. Glass wind-wings were standard on the Speedster windshield. Three different steering-column angles were provided to accommodate the three bodies; the steep column on the 7-passenger phaeton required a notch in the instrument panel.

The four-passenger bodies continued the traditional Marmon 34 foot-rest/tool box integrated into the rear of the front seat. All metal in the new bodies was steel. A Delco ignition/light switch was adopted; it included an on-off dimmer switch for the instrument lights and a resistor for headlight 'dimming'.

Although the engine underwent external and some internal modifications, it could **not** be identified as a Model **34C** because of use in both the 34C and continued 34B styles; owner's manuals and parts books refer to "late 1923, 1924" engines. Another factor was that the engine changes were not all concurrent, having been brought into production during 1923 as they became available and parts stocks were exhausted.

Fig. 4-39 The four-passenger phaeton on the Model 34C chassis. It featured the Panorama top, a non-folding, headlined design with unobstructed rear seat vision.

Fig. 4-40 The Model 34C four-passenger Sport Speedster which Frank L. Hambly took to Europe to develop foreign dealers. Note the 20 inch wheels which became optional during the 1924 model year.

## MODEL 34B AND 34C
## 1924

Just prior to start of 1924 model production in September 1923 Marmon announced optional front brakes. They were a self energizing, internal type of original Marmon design. In January 1924 the availability of balloon tire equipment was announced. Their availability was initially limited to the wire wheel option but was soon extended to wood wheels with Firestone lock-ring rims.

The cars at the New York show, with the exception of the featured all-weather phaeton, were equipped with optional brakes and tires. The balloon tires were 7.3 X 20 and appeared somewhat like 'General Jumbo' tires of a later period. Soon thereafter 6.2 X 21 tires and wheels, wood and wire, were released.

A screw operated cowl-top ventilator door was added to all styles in addition to the lever-operated cowl-side vent doors.

For the first time, the 2-passenger Speedster had side mounts and wood wheels. Both it and the companion 4-passenger were merchandised as 'Sport Speedsters.' Windshield wings again became an option on the Speedsters, and the trunk on the 4-passenger sedan became an option. During the model year, Watson single-acting shock absorbers replaced the Hartford shocks.

A cigarette lighter became standard equipment and was located centrally in the blank area to the right of the cluster. In had been an option during 1922. It was of the extension cord type.

Model 34 production ended October 31, 1924.

A large inventory of Model 34 cars which existed at the time of introduction of the Model 74 was successfully reduced during the latter part of 1924. New management controls were introduced which were to achieve an inventory revolving rate of 10.3 per year (35.5 day supply) in 1925.

The plant was closed for about two months during which essentially all equipment in the assembly operations was relocated for more efficient operation with minimum inventory.

The end of Model 34 production also marked the end of body manufacture by Marmon. Model 74 bodies were built by the American Body Co. of Buffalo, New York.

The 1916 Model 34A five passenger touring weighed 3215 pounds. (dry) . The 1924 Model 34C four passenger phaeton weighed 3861, (dry) an increase of 61 pounds over the 1923 car.

## ARMSTRONG-SIDDELEY CLONE
## 1920

A veiled but significant accolade was given to the Marmon 34 engine and chassis design when the Armstrong-Siddeley car was unveiled in Great Britain in the fall of 1919. The (Sir W.G.) Armstrong, Whitworth and Co., Ltd. and the Siddeley-Deasy Motor Car Co., Ltd. of Coventry combined to form Armstrong-Siddeley Motors, Ltd.

A large new facility was prepared for a relatively high production rate of a large luxury automobile representative of the state-of-the-art. The increased sales volume was to result from offering this new automobile in the combined market of the two predecessor products. The production design of the Armstrong-Siddeley chassis clearly discloses that the "Model" for the design was the American Marmon car, Model 34A introduced in 1916. It is possible and probable that Alanson Brush was also a consultant on the British car.

The frame incorporated the same integral running-board concept of the Brush patents including the radially upswept forward and aft ends of the board. The entire Model 34 drive assembly was adopted — transmission, torque tube, rear axle including the offset ball joint — so completely that it would be difficult to identify an assembly as to origin. The design copy included the aluminum central housing, the drawn steel tube axle housings, the single ball-wheel bearings and the splined and tapered axle shaft to hub interface.

The engine incorporated extensive aluminum usage but used two iron cylinder blocks . . . a "feature" which appeared on the Marmon 34B the following year . . . and two aluminum cylinder heads. Also uniquely, the push-rods were enclosed in tubes extending from the crankcase to the cylinder heads, similar to contemporary air-cooled engine practice. Two silent chains drove the camshaft and accessory shaft whereas the Marmon used a 3 gear train. The Armstrong-Siddeley engine incorporated the Marmon design of hollow crank pins and utilization of the shaft as the oil gallery. The car followed the Marmon lead in having the fuel tank in the cowl and in having a radiator shell separate from the core, with the latter removable. Maurice Hendrey provided this insight into the origin of the Armstrong-Siddeley[2]: *"In England, an almost cloak and dagger atmosphere prevailed at Armstrong-Siddeley. In 1918, harboring the idea of building a "car of the future", J.D. Siddeley, head of the British firm, imported a Marmon 34 for study. Key men on the Armstrong-Siddeley technical staff were released from aero engine work and given the job of evaluating the Marmon design as a model of the post war Armstrong-Siddeley. The Marmon had been imported in great secrecy, and the studies were made on Siddeley's own estate, even to the extent of setting up a drafting office in the billiard room of his house. The result was the highly successful Armstrong-Siddeley 30, which faithfully followed Howard's ideas throughout."*

### CHANGE AT THE HELM

A management upheaval . . . the first in Nordyke & Marmon's seventy year history . . . occurred in May 1924. It had significant short term effects and dramatic long term effects. When it occurred, the Model 74 was being readied for start of production in just four months; it was too late for any major changes. But the manufacturing process and the sales promotion activities received immediate critical attention. The impact on the engineering department would necessarily be a delayed appearance, evident in the products of the FOURTH ERA beginning in the 1927 model year.

Meanwhile the change in management and in its philosophy warrants a Sub-Era classification of the Models 74 and 75.The two sub-eras of the THIRD ERA will be separated by an introduction of the reader to Mr. G.M. Williams.

## PART 2 — Introducing G.M. Williams

### THE (NEW) MARMON MOTOR CAR CO.

George Montague Williams was born in 1892 in Nanaimo, Vancouver Island, British Columbia, Canada. That city was then the departure port for those seeking their fortunes in the gold rush up the Fraser River. Through it passed the prospectors and miners, aglow with gold fever, en route from San Francisco. Monte's father owned a bank in Nanaimo and employed a teller who succumbed to the allure and headed up the Fraser; he would become the famous Canadian poet, Robert W. Service.[1]

The following account of Monte's youth is from the book *"Boot Straps"* by Tom M. Girdler, then Chairman of the Board of the Republic Steel Co.[1]: "When Monte was twelve, coming East for the first time in a sleeping car, he stayed up through the night to see what lay beyond those mountains that seemed an insurmountable

Fig. 4-41 George Montague Williams 1892-1965

barrier between his world and another one beyond. It is significant, too, that many Horatio Alger books were in his baggage. Even between semesters, in technical schools and college, (he graduated from the Sault Sainte Marie Institute of Technology in Michigan) he was accompanying parties making hydrographic surveys and exploring up in the Hudson's Bay country. Monte was fearful that all the frontiers would be erased before he got to see them, never supposing that in his business career he would be concerned incessantly with bringing our civilization to even more exciting frontiers".

"Most of that exploratory work was concerned with finding water power sites far up in the wilderness of Quebec and Ontario. Later on as a junior engineer for the H.E. Talbot Company of Dayton, Ohio, he helped to construct dams and power houses on those sites that he had discovered for the government. Much of that part of his life was spent up on the Michichipicoten River. At 23 years of age he was chief engineer of the Talbot Company in 1915 when it acquired the Dayton Metal Products Company and through an association with Orville Wright there was then formed the Wright Field Company to train students of flying; an outgrowth of this was the Dayton-Wright Airplane Company. Monte, starting as assistant general manager, became general manager. During the first world war, this company produced the most planes (3670 planes, 1500 of them before the first delivery by any other manufacturer) [1] of any company in the United States. Later on it became a division of General Motors, concerned with experimental body work and engine developments and Monte, as its general manager, continued until 1923.

Then a group of bankers hired him to run the Wire Wheel Corporation of America, at Buffalo, which was in bad shape. The popularity of the disc and artillery wheels for automobiles had just about ruined the wire wheel business. A few weeks after he took hold, Monte, in a Pullman berth between Buffalo and Detroit, was worrying about the almost hopeless task of restoring life to a broken-down company. Before snapping out the light he had been reading in the Journal of the Society of Automotive Engineers an account of a new development, balloon tires. Suddenly it occurred to him that here was an opportunity. These new tires would require a new wheel. At Cleveland, in the middle of the night, Monte left the train so as to be in Akron at the start of the business day. He spent 48 hours there interviewing engineers and production men of three or four tire companies. He learned to his delight and astonishment that none of them had gone more than vaguely into the matter of suitable wheels for the new tires. Here was an unstaked claim in the field of enterprise! He learned Firestone was going to announce balloon tires in a month with a big advertisement in the Saturday Evening Post.

After an exchange of telephone calls with the chief engineer of his plant he decided all that was needed was to give the old wire wheel a new rim and new spokes. The hub would be exactly the same. They could be in production in a week. The next day he spent with an advertising agency man in Buffalo, with the result that Monte's announcement of a wheel for balloon tires ran in the same issue of the Saturday Evening Post that contained the announcement of Firestone. Almost at once the old wire wheel business came back to life. They were swamped with orders and began shipping five or more carloads of new wheels a day. Manufacturers of disc and artillery wheels had been caught flatfooted. Wire wheels recaptured their popularity. With that one feat Monte justified the faith of those bankers in his ability to run a business. It is an inadequate measure of the value of what he had done to report that the stock of the company went, in a few weeks' time, from 55 cents a share to about $10 a share."

Because of his reputation for sound handling of firms suffering economic distress, a group of bankers in 1934 invited Williams to run the century old Russell Manufacturing Company (Reorganized) of Middleboro, Connecticut.

Tom Girdler's account continued: "Originally (this firm) had a fleet of sailing vessels engaged in trade with the Orient and first engaged in manufacturing to make cargoes for its ships. In the Civil War days, it was making hoops for hoop skirts. By the time Monte took it over (in January 1934) it was really a group of manufacturing enterprises (a conglomerate today) that might well have been called Russell Industries. Under Monte Williams this concern in a few years expanded from a gross of a couple of million dollars to more than $20 million".

By the time that WW11 threatened, Williams was Chairman of the Board and President of Russell who were then the world's largest manufacturer of the tape on which slide fasteners — zippers — are mounted. The stringent requirements were recounted by Tom Girdler: "It is the only textile product made to the exact tolerances of metal (parts). This tape is woven to a tolerance of 1/2000 of an inch in thickness. From the selection of the cotton in the field to the point where the tape is dyed and finished they have to take meticulous steps in a bewildering succession of manufacturing processes to have their product of a precise thickness."

When the US became involved in WW11 the separate boards of directors of Republic Steel and Russell Manufacturing voted to grant their chairmen wartime leaves of absence. Tom Girdler became chairman and chief executive officer of Vultee-Aircraft (later Consolidated-Vultee) and G.M. Williams became his vice-chairman and assistant chief executive officer . . . hand picked by Girdler because of his WW1 aircraft building experience and superior management talents demonstrated in diverse precision manufacturing industries.

Now that the reader is no longer a stranger to G.M. Williams, the recounting of the Fourth Era can begin.

In March 1924, Monte Williams came to Nordyke & Marmon Co. Tom Girdler recounted: " (His) skill as an industrial corporation executive was enlisted by his former chief (at Dayton-Wright Division of General Motors) Alfred Sloan to help a (Sloan's) personal friend Walter Marmon (they were classmates at MIT). Nordyke & Marmon Co. was in a snarl and three New York banks had taken charge."

Following purchase of a substantial block of stock in the privately held firm,[1] G.M. Williams was named President of Nordyke & Marmon Co.

Williams was the architect of the separation of the traditional milling business and the automobile business. The latter became the publicly held Marmon Motor Car Co. in September 1926. Full details of "going public" and the consequences are provided in Chapter Thirteen, 'Financial History'.

Mr. Williams elected to prophesy at the opening of the 1929 Chicago Auto Show: "The internal combustion engine will be obsoleted in this century . . . cars will be electrical, (driven) with power received directly from centrally located transmitter stations. *Automotive News* repeated the prediction in the September 23, 1974 edition; forty-five years had passed and twenty-six of the century remained.

## PART 3 — The Cars of the Models 74-75 Sub-era 1925-1928

During May 1924, there had been a drastic change in Marmon management. G.M. Williams, who had been general manager of a General Motors division, had been brought into Marmon and named president succeeding Walter Marmon who took the newly created position of Chairman of the Board of Directors. It was unquestionably a turning point for Marmon Motor Car Co. 'The Foremost Fine Car' was to become 'The New Marmon — It's a Great Automobile.' Williams mission was to accelerate sales to fully utilize the full production capacity; despite profitability, the post WWI sales had been at a steady pace with no promise of growth to the installed capacity. He was obviously the corporate spokesman, and in full charge of the business, including engineering. Walter Marmon transferred his enthusiasm to the former Marmon subsidiary, Indianapolis Power and Light Co., where he would become president and later Chairman of the Board. Howard Marmon, while retaining his title of Vice-President of Engineering, became progressively less evident as his involvement in his Pineola, North Carolina estate grew. Chief Engineer Jeffers, who deserved a great deal of the credit for the Model 34B consequent the absense of Howard Marmon beginning in June 1917 to serve as Aircraft Engineering Engineer of the Aircraft Production Board, had left Marmon in January 1924 to enter private business.

The Model 74 was the first new model introduction following the change at the helm. Significant changes in product philosophy were evident. The plant had been closed for two months, for a major realignment of the car assembly process, following the end of Model 34 production.

It is therefore appropriate to separate the Third Era into two Sub-Eras, reflective of different managements. The marketplace was also changing: government experiments had proved that phaeton models had less wind resistance (i.e., a lower drag coefficient) with the top and wind-shield **up** than when they were down! This second Sub-Era saw the open car shift to the minority segment.

### MODEL 74
### 1925

Introductory ads for the Series 74 proclaimed:

THE NEW MARMON SEVEN-PASSENGER SEDAN DE LUXE

*The New*
*MARMON*
*It's a **Great** Automobile*

New policies, new strength and a new plan have produced a New Marmon. The aspect of our boulevards will be freshened by its grace of line. It will lend tone to the country clubs of America and to fashion's fondest rendezvous. In every detail it reveals the hand of a great designer, close to the likes of people of exquisite and cultivated taste. In the fine, sweet-running mechanism of the New Marmon you will find the touch of time. Years of unfaltering concentration on the Marmon chassis and the six-cylinder Marmon engine have given the New Marmon its great, inherent goodness. Among all the axioms there is none truer than, "Practice Makes Perfect." Certainly there is no substitute for experience. And herein lies the secret of Marmon's sure-footed mastery of the road and of Marmon's confidence-inspiring dependability. Upon the inherent goodness of the Marmon chassis, the greatest custom tailors of motordom have lavished unstinted charm and beauty in ten different body styles.

In the mid-twenties Marmon was riding the heights of acceptance and prosperity. A dealer remanufacturing program was nationally advertised, took advantage of the durability and maintainability inherent in the design of the Model 34, and brought "new" Marmon ownership to a less affluent market segment. In financial circles, Marmon Motor Co. was considered among the very strongest of the automobile firms; profit on sales was a favorable 10 percent, $375 per car.

On the other hand, Marmon sales had never achieved the levels of Packard and Cadillac. A major overhaul of the Model 34 was undertaken for the 1925 model year. Everything that could be done to convey newness to the Marmon line was done. It was advertised consistently throughout the 1925 model year as 'The New Marmon.' Buttons shaped like the Marmon badge and carrying the slogan "It's a Great Automobile' were distributed at the shows; the slogan appeared in every ad. The car was the fourth edition of the Model 34 and the new parts released for it carried a D prefix signifying first usage on the 34D. The search for newness led to substituting the engine brake horsepower (74) for the SAE horsepower (34) as the Model designation: officially it became the D 74.

The frame was extended at the rear to mount both a relocated fuel tank and bumpers. The integral running board was continued but camouflaged: The front fender extended downward over the frame several inches and the rear end of the boards no longer curled upwards but rather squarely abutted the conventional rear fender. Behind the fender was a triangular brace completing the integration of the board. The thickness of the side rail was increased slightly to permit attachment of the cross-member to the top flange of the rail and, at the bottom, to the rivets joining the rail and "board.' The steering gear attachment bracket was riveted to the top flange only. This eliminated the visible rivets on the side of the rails and completed the camouflage. The body sill no longer followed the frame contour, being overhung to allow a wider rear seat.

A number of "74" improvements had made their appearance in late 1923 and 1924 models 34's. Optional internal self-energizing front brakes (1924), standard diameter Delco generator/distributor and starter (late 1923), conventional bronze-bushed spring shackles, water pump driven from the head end of the generator drive (late 1923), and conventional rocker-shaft (1924).

The model 74 sales promotion highlighted the offering of closed bodies at only $130 over the open bodies. This really constituted a price reduction and corresponding reductions in manufacturing cost were necessary. In the Model 34 it was evident that engineering was the dominant determinant in vehicle specifications; not so on the Model 74.

The connecting rods and crankshaft of the "74" were machined only at the bearings whereas their predecessors had been machined all over to reduce weight, the standard practice on aircraft and racing engines. The "instrument face" engine oil level indicator was replaced with an exposed ball-end rod (late 1924). Certain chassis fittings became massive iron castings rather than slender forgings of heat-treated alloy steel.

The "74" open body styling was a big improvement over the immediate predecessor "34." Efforts to reduce the height of the "34" and the styling compromise necessitated by conversion to steel body panels had lost some of the symphony of line that had characterized earlier open-bodied 34s. The '74s' were unmistakably Marmons, and appeared much larger and heavier than any 34. Concessions were made to automotive fashion — terraced fender surfaces were adopted, as were belt moldings and exposed door hinges.

The Marmon body plant was idled and all Model 74 bodies were procured from outside sources. All **open and deluxe closed** "bodies are built by the American Body Company of Buffalo, New York in accordance with designs of Hume, of Bos-

THE NEW MARMON SEDAN

ton, working in conjunction with Marmon and American Body designers. The New Marmon (Standard 5 passenger) Sedan is built by the C.R. Wilson Body Company, of Detroit."

The Marmon contract was celebrated by American Body with a two page ad and the description of the 74 styling philosophy therein warrants repeating: *"There has been no effort to establish a new "school of design" but to the best of the accepted things has been added a note of originality that sets them apart from all other cars. The student of color will pronounce their harmony perfect. The student of contour and design will find their silhouettes faultless. The woman of limitless ease will be quite content to go from the luxury of her home to the luxury of her Marmon."*

There were three open styles: Roadster with rumble seat, 5-passenger Phaeton and 7-passenger Touring Car. The Phaeton continued the Marmon Panorama Top concept. Five deluxe closed body styles included: Coupe, 5-passenger Sedan and Sedan-Limousine, 7-passenger Sedan and Sedan-Limousine.

All deluxe closed bodies incorporated the Marmon windshield configuration but with a more inclined center section and correspondingly larger triangular side windows. The visor was now die formed to blend into the curvature of the roof and with the front pillar; it was integrated rather than added. The (2-passenger) coupe had a padded leather top aft of the door which was fitted with ornamental landau bars.

The deluxe sedans had a nearly horizontal roof line whereas the standard sedans had curved roof lines rising to about the rear quarter pillar and then falling slightly.

The new bodies were wider, particularly at the rear seat. They no longer followed the frame side rail contour as they had on all Model 34s. This change further obscured the traditional Marmon chassis construction but allowed competitive interior dimensions. It also marked the end of the sill-less body construction and the sectional open bodies. A full length belt molding began at the nickel plated radiator shell and was paralleled by another below the windows.

Headlights were more massive and incorporated the Marmon pyramid at the top center. Matching drum lights were mounted on the cowl. A handsome oval tail light incorporated a central 'tail' light lens beneath the Marmon badge, with adjacent 'stop' and 'back-up' lens. It was supported by a bracket from the frame.

The fuel tank relocation both utilized the available space created by the extension of the frame to accommodate standard-equipment bumpers and made the car more conventional. The fuel gauge was still feasible on the dash cluster by virtue of a mechanical coupling device: a ribbon of steel with holes for balls moved back and forth in a steel tube to couple the instrument needle to the tank float.

The author has always considered the Model 34 styling to be dynamic . . . the car wanting to be in motion . . . and the Model 74 styling to be static . . . handsome in the driveway. This perception may be due to the massive motif generally adopted but most likely results from the dominant wheels and tires. The large section balloon tires, 6.2 x 21 on open cars 7.3 x 20 on closed cars, combined with round spokes (previously elliptical) and an increased tread which brought the tires close to the outside width of the fenders, to cause the wheel areas to dominate.

The rear tire carrier was designed to accommodate either one or two spares. Cars were shown and illustrated with the dual spares. This further contributed to the wheel/tire emphasis.

The increased weight consequent the several economies and conventionality of body construction, including wood wheels and balloon tires added appreciably to the weight of the car. Published weights are suspect.

The accelerator pedal was relocated from between the clutch and brake pedals, to the conventional location. This necessitated an offset in the shift lever to the right; it was no longer feasible to have the key-lock in the lever ball. The lock was relocated to the base of the lever.

A line of standard bodies was anticipated at Model 74 introduction as evidenced by the deluxe designation of all closed styles.

By December 1924, a standard seven-passenger sedan and a (standard) four-door, Brougham-coupe had joined the model line. Ads stressed the fact that "the Standard bodies are mounted on the same chassis and engine which go into the New Marmon deluxe models."

The Standard and open bodies featured rearward-opening front doors, forward-opening rear doors, with consequent adjacent door handles. The De Luxe bodies had forward opening doors at both front and rear locations.

In February a Standard seven-passenger sedan was available. Surprisingly, an ad six weeks later featured the Standard seven-passenger sedan but illustrated a different body!

The 'new' body appeared longer and had a flatter roof line identical to the brougham; the five-passenger and initial seven-passenger bodies had a rising roof line particularly evident over the front door. The longer appearance resulted at least partially from a longer side window and narrower rear corner 'pillar'; rear corners of the window were rounded. The two doors abutted each other, like the Brougham-coupe: the earlier bodies had a raised section of the pillar flush with and between, the doors with a consequent massive appearing composite upright member.

Interior door handles, like the brougham, were visible through the windows because they had a vertical shaft through the window sill; the earlier Standard and the De Luxe bodies had handles on the inside door face.

The close coupled 4-door, 4-window, Brougham-coupe (called Brougham in 1926) featured leatherlike rear quarters with both opera windows and landau bars. Standard equipment included a rear trunk, detached from the body and

Fig. 4-42 The Model 74 seven-passenger sedan as advertised on Feb. 7 (lower) and on March 21 (upper) 1925.

Fig. 4-43 Club Touring introduced mid 1925.

The NEW MARMON
"It's a Great Automobile"

Fig. 4-44 Model D74 Four passenger Speedster, companion to the two passenger.

between the rear half of the fenders. This style proved very popular among the sedans.

A companion Victoria Coupe was announced near the end of the model run. Only 50 carried 1925 serial numbers but it was destined to be a top seller. It had a high degree of commonality with the two-passenger coupe which would change to the standard windshield for 1926. Seating was typical victoria . . . a fixed bucket seat for the driver, a smaller front passenger seat which folded under the dash and presumably was to be used only when a fourth passenger was aboard, a rear seat for two which extended partly behind the driver. The rear luggage compartment extended forward to the drivers seat, to the left of the rear seat. This provided sufficient fore and aft length in the trunk for stowage of golf clubs. Above the trunk extension was a shallow compartment accessed from above by rear seat passengers. The rear quarter exterior was the same as the brougham and similar to the coupe. This brought the body style selection to twelve. The Victoria was procured from the Rollins firm.

In April 1925 a Club Phaeton and subsequently a Club Touring were announced.

They were distinguished by cowl lights supported by brackets extending from a nickel-plated cowl band at the hood interface, and by nickeled rear view mirrors on the front fenders.

Following an increase, on July 31, 1925, in open car prices to the same price as the sedan and brougham-coupe, Marmon ads proclaimed no extra charge for a closed car. In this promotion, Marmon followed the lead of Essex.

All Model 74 Marmons had dual rear view mirrors as standard equipment. Two side-by-side mirrors on the windshield header improved the driver's vision and gave the passenger a rear view. The mirrors differed . . . how the author doesn't know . . . because one had square corners and one had round corners, presumably as an assembly identification guide. Judges at antique car shows often criticize this really authentic variance.

It is much to Marmon's credit that they perceived dual mirrors as essential to safe driving so many decades before their widespread adoption. Open models carried mirrors on both front fenders.

The instrument cluster was similar to the Model 34B. However the instruments were white with black numerals and the frame carried the Marmon pyramid at the top center. A rotary carburetor mixture-control replaced the pull-type. The steering wheel was entirely of wood and the levers for throttle and spark were very short. The Model D-74 substituted a 'PREST-AIR' bottle for the previously used engine driven Tire pump.

Model 74 prices during the 1925 model year were:

| | |
|---|---|
| Roadster, 5-passenger Phaeton, 7-passenger Touring | $3165 |
| Increased on July 31, 1925 to | $3295 |
| Standard Sedan and Brougham-Coupe | $3295 |
| Coupe De Luxe | $3455 |
| Sedan De Luxe 5-passenger | $3900 |
| Victoria Coupe | $3937 |
| 'Club' versions of the Phaeton and Touring | $3465 |
| 7-passenger bodies were $75 higher. | |

## MODEL 74
## 1926

During the first year of the Model 74 there was a conspicuous void in the Marmon model selections. The flagship Speedsters were missing. In their place was a relatively "Plain Jane" Roadster with rumble seat and rear mounted spares. The forward half of the roadster was common with the phaeton and touring bodies. The situation, which was alienating some customers and deglamourizing the marque, was corrected in July 1925 with announcement of a new Speedster. Deliveries were to begin in September as a 1926 Model. Coachwork was by C.R. Wilson Body Company; American Body had built the roadster.

The Speedster with rumble seat was in the Marmon tradition of extended cowl, had a severely raked windshield, and was a proper successor to the Bee-tailed Model 34. The removable speedster top featured mahogany natural finish exposed bows with nickel-plated slat irons. The rear deck was fitted with nickel finishing bars. The front seat cushion was only 10 inches from the floor at the front, 6 inches at the rear. Rear axle ratio was 3.73:1.

Duco color options included: two-tone Newport Gray or Hollywood Green Gray, both with hand crushed full grain tooled leather in Green Gray color; two-tone Miami Taupe with Black leather; all with an American Vermilion Dark stripe on a black molding. Despite this offering, only about 100 of which were sold, the production of open cars fell from 375 units (9% of sales) in 1925 to only 150 units (4.5% of sales) in the 1926 model year.

The Victoria Coupe which had been introduced quite late in the prior model year (only 50 units) was the stellar sales performer for 1926. It accounted for no less than 43 percent of all 1926 production, 1341 units out of 3140. But despite this success, production was down from a high of 4160 in 1925. The decline in sales could very well have been attributable to the introduction of the Cadillac 314, the first highly styled Cadillac which replaced the rather stodgy Model 63. Whoever at Marmon was responsible for abandoning the smooth crowned fenders, used since 1916, for the Model 74 should have been severely embarrassed when Harley Earl incorporated smooth crowned fenders on the 1926 Cadillac; the ill-advised paneled fenders of the Model 74 would only survive for two model years.

The Victoria coupe and the (standard) two-passenger coupe with rumble seat were advertised as the 'Outstanding Coupe Designs of the (1926) Year.' The former achieved remarkable popularity for a Marmon style, displacing the four door sedan as the leader, much as the Coupe de Ville became the popular Cadillac in the 1960s. Both were built by Rollins, the latter replacing the deluxe coupe built by American Body the prior year.

A Marmon ad for the coupes stated: *(They) are produced by one of the leading custom builders, yet they are grouped among the New Marmon Standard Closed Cars. Combining intimate charm and extraordinary grace of line with a surprising degree of roominess, they have won instant popularity as the personal cars of men and women.*

The custom character of these styles was accented by the leather rear quarters with landau bar trim as well as the opera window on the Victoria style. The peculiar seating arrangement which characterized the popular Victoria style is illustrated in the dimensioned drawing. Not evident is the forward extension of the trunk luggage compartment forward beneath the inside-accessed compartment to permit longitudinal stowage of a golf bag.

The Rollins bodies were built using No. 18 USS gauge steel over an ash and maple frame which was hot glued and screwed together at mortised and tenoned joints which were reinforced with steel brackets. "Extremely heavy" steel was used for the firewall which was coated with thick felt and had a heat insulating board bolted in place.

The 1926 Model D-74 had fenders which were 1.25 inches wider than used the prior year. A probably unexpected consequence was the shipment of cars with a demounted left rear fender when the rail carrier was a narrow-door box car. (Bulletin D-49)

Two of the three major changes incorporated in 1926 models at no increase in cost were engine improvements: "Double-fire" ignition and a "three-way" oil purifier system.

"In the mid 1920s, Marmon and other engineers were still confronted with three sources of constant difficulty . . . crankcase oil dilution, spark-knock and power deficiencies due to insufficient heat to sustain the expansion of the charge (i.e., combustion at lighter loads. Efforts were initially directed at the improvement of combustion (by dual spark plug primarily for the purpose of reducing crankcase oil dilution at lighter loads."

The third major improvement for 1926 was the adoption of the Bowen "sure-shot" system of chassis self-lubrication. Oil, rather than grease, was distributed to 19 points on the chassis simultaneously by a single stroke of a foot-operated pump whose pedal was accessible from the drivers seated position.

A Grolan hydraulic gasoline gauge replaced the mechanical type used the prior year.

Marmon 74 sales established significant gains. Sales of the Model 34 in May 1924 had been exceeded 52% for May 1925 and that again exceeded by nearly 20% in May 1926.[1]

A hint of impending developments at Marmon was given when *Automobile Topics* reported on June 2, 1926, that Judge C.C. Simons had granted the Guardian Trust Co. of Detroit, receiver of the bankrupt Murray Body Corporation, permission to establish a new Murray Body plant in Indianapolis. The Trust Company was authorized to use $200,000 of the Corporations funds to form an Indiana Corporation to operate the Marmon Motor Car Co. body plant. The plant had been idle for two years following discontinuance of body building by Marmon and substitute procurement from American Body Co. of Buffalo. The report noted that the plant was practically new, partially equipped and well suited to the purpose.

More insight on Marmon plans was revealed when *Automobile Topics* reported on July 23, 1926 that the Murray Body-Marmon Body Deal had been consummated. The contract was reported to be valued at $15,000,000 and included lease of Marmon Plant 3 and commitment to build all Marmon bodies for three years (1927, 28, 29). Work was to begin on the project immediately with production of about 75 bodies per day planned. Two thousand workmen were to be employed in the 400,000 square foot facility.

Charles H. Widman, vice-president of Murray, was quoted by the AT reporter: "Our Corporation realizes the importance of the Marmon contract, and in face of the greatly increased volume in which Marmon cars are being produced and of the prospective Marmon program, we (Murray) feel warranted in making a substantial investment here, and in establishing a separate Indianapolis plant."

Widman continued: "The inspection of bodies at Marmon is so unusually stringent that it is not practical to handle such a large project at long range. Under this method, Marmon will get the finest bodies that it is possible to build and the necessary delay involved in shipments by rail and otherwise will be eliminated."

Presumably Marmon was dissatisfied with either the quality of the American Bodies or with their timely delivery to Indianapolis. The author reworked a Model 74 (American Body) touring door to use on a Model 34 (Marmon body) and was shocked at the wood construction — more like the work of an unqualified amateur restorer.

The Murray contract was presumed to indicate that further developments were in definite perspective and the AT reporter quoted Marmon president, G.M. Williams as saying: *"Marmon engineers and executives have recently spent much time abroad making an extensive study of the engineering phases of the European motor cars which have found favor with the Americans. It seems that the solution is a combination of the best elements of European body design with tried American engineering practice. Unquestionably this country is realizing the necessity for the small, beautiful, high-powered type car that will go far toward at least temporarily relieving our traffic congestion and solving the parking problem Beauty, speed, safety and power in miniature seem to be the next call in prospect for the American manufacturer."*

Williams pronouncement is as appropriate today as it was in 1926, or at least it would be if "foreign" or "Japanese" were substituted for "European."

The AT account concluded: "Should such a car be developed by Marmon, Williams added it would be purely a supplementary car, for the Company feels that the present type of large car will always be in demand by a certain class of buyers."

Prices were unchanged at the start of 1926 production in mid-September 1925 with seven of twelve models offered at a fixed price of $3295: 2/4 passenger speedster, 5 passenger phaeton, 7 passenger touring, 5 passenger sedan, 5 passenger brougham, 2 passenger Coupe and 4 passenger victoria. Other offerings were: 7 passenger standard sedan $3370, 5 and 7 passenger deluxe sedans $3775-3880, 5 and 7 passenger deluxe sedan-limousines $3900-3975. Marmon front wheel brakes were a $100 option. The other available options were said to exceed those of any other manufacturer. Speedster equipment included windshield wings and front fender mirrors. Standard closed bodies had nearly vertical windshields, the deluxe bodies had sloping windshields with fixed triangular corner windows.

Some 350 sales representatives gathered at the N & M plant January 4, 5 and 6, 1926 to celebrate the firm's diamond jubilee, hear plans for the coming year and travel by special train to the National Show in New York. They were shown a new series of cars to be known as the Marmon "Special" line, consisting of 5 passenger brougham, 5 passenger sedan and 7 passenger sedan. This new line was to be in addition to the Standard and De Luxe lines. However registration documents identify only the Standard and De Luxe models.

On May 1, 1926 a 4 passenger speedster was introduced. The review by *Automobile Topics* said it was characterized by an air of jauntiness and distinction and was decidedly appealing. It listed at $3490 and was said to deserve a place among the finest of the "swagger" cars on the market. Illustrative of this stature were the extended cowl, the sharply sloping windshield, the natural wood top bows, and the top so-constructed that when folded down it lay almost flat. The car was built close-coupled, leaving space for a trunk between the body and rear mounted spare. The latter item reflected a change in styling philosophy . . . the side mounted spares having been a trademark of the sporty car models. Although a trunk rack was standard, and body bars, no truck was furnished. Fender mounted rear-view mirrors were standard, solving the problem of where to mount what had been spare-tire mirrors the prior year. The steering wheel and door garnish moldings were walnut wood.

This model was to become Barney Oldfield's third Marmon. Barney however would elect to order the standard two-wheel brakes . . . he drove with the accelerator pedal, not the brake pedal.

The period 1925-26 was characterized by mergers among the body builders and several firms ceasing business. In 1925 Towson, J.C. Murray and Charles Widman became Murray Body Corp. and were joined by C.R. Wilson in 1926. Also that year, LeBaron was acquired by Briggs Mfg. Co. and Marmon closed down its Hume subsidiary in Boston. Naturally there was a lot of jockeying for position, both among companies and personnel. American Body was desperate enough to warrant Marmon idling their relatively new body plant in favor of outsourcing from American. Unfortunately American's superintendent, Raymond L. Birge promptly shifted to the 'new' Le Baron operation of Briggs. Anticipated Marmon (and other) volume was not realized and American was out of business sometime in 1926.[1] The demise was preceded by deteriorating quality, for which they had been well regarded, and by failure to maintain shipping schedules . . . often interacting problems.

It is likely that Locke and Co. took over the building of Marmon deluxe bodies during the Model 74 production rather than at start of Series 75 production as publicized.

The following serial numbers for individual body styles are the starting numbers

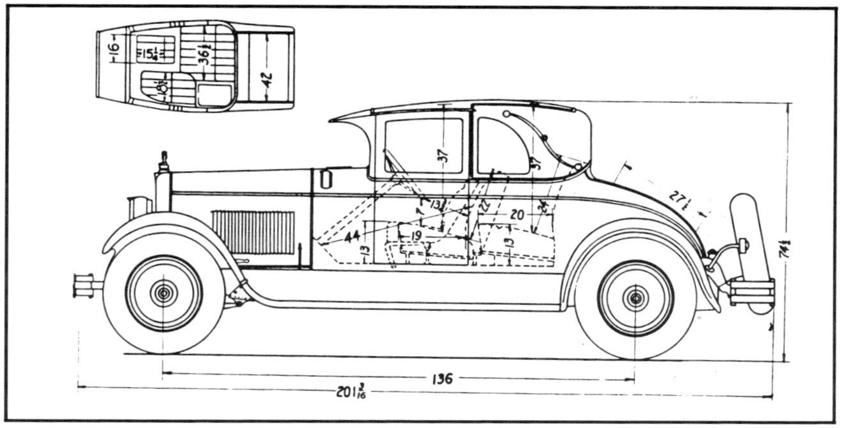

Fig. 4-45 Body arrangement and interior dimensions of the victoria style.

for 1926 production: D2-5222, D5-5104, D8-5172, D16-5342, D17-5253, D18-5223, D19-5037, D20-5886, D21-6109, D22-5874, D23-5052, D24-5001.

## MODEL E-75

The Model 75 was the fifth and final version of the Model 34 which had bowed in 1916. Except for a much lessened interest in light weight construction (it had proved irrelevant in the competitive marketplace) it was also the ultimate version having benefitted from successive technical advances and evidencing a fair number of fresh developments as well.

A Marmon advertisement commented on the Series 75:

> It is the easiest riding car in the world today.
> It is an unusually dependable and long-lived car.
> It is without question the most precisely built — among fine cars.
> It is the honest product of a house of 75 years standing.

Note the clue to the origin of the '75' designation. It wasn't just one digit added to the '74' . . . it recognized the 75th anniversary of the Marmon firm!

The Model 75 was very similar in appearance to the predecessor 74. Initial picture releases showed the paneled fenders of the 74 but the accompanying text noted the readoption of the full crown configuration for all models except the price-leader town-coupe. The retention on that model was probably a solution to the inventory exhaustion problem which resulted when it was realized Cadillac had gone to full crown fenders on the all-new 314 Series.

Fig. 4-46 Not misplaced from 'Foreign Marketing' — Series 74 Deluxe Sedan with actor Mitchell Lewis.

Fig. 4-47 Instruments of Series 75 standard closed styles. Model 74 differed only in having a drum type fuel gauge.

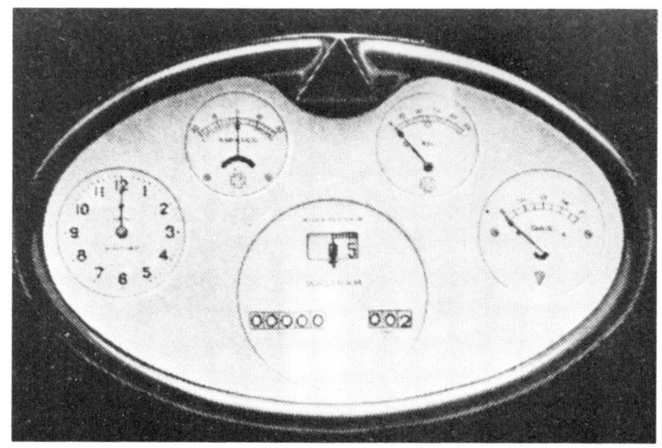

Interior comfort was enhanced by adoption of molded sponge rubber arm rests.

Six new body types were added to the line, four of them custom-built by Locke & Co. These were a touring speedster for seven passengers, 5 and 7 passenger sedans and a 7 passenger sedan-limousine. The two others were a 5 passenger

Fig. 4-48 Series 75 display at 1927 New York Show — Locke-bodied four passenger Speedster in foreground.

Fig.4-49 The Series 75 version of the popular two-passenger Speec

town coupe and a collapsible coupe-roadster.

The 5 and 7 passenger standard sedans, the brougham, victoria and 2 passenger coupe were continued, as were the three open types: 2/4 passenger speedster, 4 passenger speedster and 5 passenger phaeton.

The town-coupe was a close coupled two-door sedan with space for a trunk. It was priced at $3195, the lowest ever for a closed Marmon.

The Locke custom bodies replaced the De Luxe styles manufactured by the American Body Company. There had been serious delays in body deliveries from American; the new alignment would assure adequate capacity and quality appropriate to the top of the line styles.

The Locke bodies were aluminum skinned. There was an elliptical panel embossed on the doors between the two moldings. The lower molding stepped closer to the upper just beyond the rear door. The upper one extended across the cowl directly in front of the windshield.

Locke was not identified in ads or press releases which simply said "A new series of custom built motor cars by distinguished body builders . . . accurately reflecting the best feminine taste in America." A single ad referred to custom-built motor cars with bodies by Locke . . . and other distinguished body builders.

Certain standard bodies for the Series 75 were made by Murray Body Corporation consequent to the mid 1926 merger into Murray of the Model 74 supplier, C.R. Wilson Co.

Whereas the Model 74 had been touted as 'The New Marmon,' the Model 75 was sometimes referred to as the 'Greater New Marmon' but more often just as 'The New Marmon.'

The Custom-built touring speedster for seven had side-mount spare tires and a second fixed cowl across the rear of the front seat. The single full length belt molding became double on the rear doors only and had a French canne 'sham cane' filler panel. A trunk was contoured to the rear body curvature and fitted between, and flush with, the regular rear bumperettes.

A custom-built 4-passenger speedster was subsequently added to the line. This model was both displayed and advertised with dual searchlights mounted on stancions.

Later in the year, Marmon Jewel-inspired colors were featured in ads promoting the 'Series 75 MARMON.'

The Marmon Jewel Colors were said to be, "in every sense of the word, coleurs naturelle, derived directly from jewel stones where nature has achieved her most unusual color harmonies while working in a hard substance. Marmon has again taken the initiative in the creation of characterful color schemes for the motor car by turning for inspiration to jewel stones and their matrices, where luminous brilliancy is combined with more subdued tonings to produce perfect color symphonies."

Buffalo wire wheels were illustrated and the custom-built sedan and custom-built 4-door speedsters were shown with side mounted wire wheels. The sedan had a folding trunk rack with bumperettes.

Unlike the aluminum well placed in the running board on side-mounted Model 34s, the Model 74 and 75 spare tire/wheel fitted between a pair of integral 'chocks' above the board surface. The tire was strapped to a plate supported off the cowl, and locked to the running board.

All Series 75 bodies were seasoned clear ash or oak framed with double mortised and tenoned joints, glued and screwed, and reinforced with metal angle braces.

Closed car roofs had closely spaced transverse ribs; on custom (Locke) bodies they were set in mortised joints, glued and screwed. The ribs were covered with heavy wadding over which heavy wire mesh was placed followed by thick Kersey felt. Covering was high grade top material; on custom bodies it was bright black high grade roofing material.

The external skin on the speedster was #14 gauge aluminum. The custom bodies were #20 gauge aluminum. On all aluminum bodies, joints were welded and soldered prior to assembly on the framing. The other closed bodies were skinned with #18 USS gauge steel.

A heavy insulating board was mounted on the engine side of the dash (firewall or bulkhead).

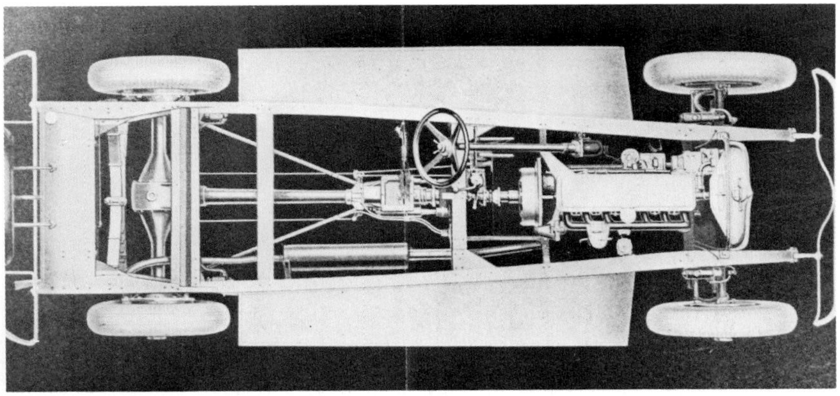

Fig. 4-50 Model E-75 chassis

Custom-built bodies were fitted with Ternstedt's overhead door checks, Atwood rubber bumpers set in nickel plated housings and Ternstedt's 'ideal' dove tails. Other bodies were fitted with Ternstedt's rubber bumpers and rubber insulated dove tails set in nickel plated metal housings. All closed bodies used Ternstedt remote-control type latches with double-check nickel plated bronze strikers. The speedster doors were equipped with Atwood bumpers and Ternstedt's 'ideal' dove tail, all nickel plated. Door locks and catches were by English & Mersick. Hinges were by Soss of Detroit; they were individual units but of piano type construction.

Front compartment floors were covered with metal-bound corrugated linoleum. Rear floors were done in highest grade loom carpet; custom bodies had rubber matting under carpet of extremely high quality. The victoria common floor was high grade loom carpet.

All 1927 Marmon seat cushions used Nachman springs. Closed standard and custom-built bodies used pleated padding of the piped type, using 6 inch pipes filled with thick cotton batts over a thick layer of curled hair, thoroughly stitched in place. Cushions were of tufted pillow type with three rows of buttons to make them form fitting; Custom-built bodies had two rows on the seat and one on the cushion.

A number of chassis changes were made to "secure a super-silent chassis of extreme durability and ease of control. Increased silence and smoothness were obtained by changes in the valve mechanism, by the addition of a Lanchester type damper to the front of the crankshaft, by mounting the rear of the engine in Belflex rubberized fabric shackles while the front remaining spring cushioned, by introducing a rubberized fabric coupling in the clutch driven member to prevent synchronization of vibrations between the engine and drive system, by the use of patented rubber bushings in the steering column to prevent small vibrations being telegraphed up the column; and by balancing tire valves by placing the splice opposite the valve."

Marmon advertisements modestly proclaimed the Series 75 as being "developed to a point very closely approaching mechanical perfection."

Fig. 4-51 Model E-75 engine. The final version of the engine introduced on the Model 34B for 1920. The crankshaft damper was a new feature.

The Fedco number plate—the basis of the Fedco system of theft prevention and detection. The Fedco-Marmon number plate similar to the above, bearing the serial number, is attached to the instrument board as illustrated below.

This photograph shows the location of the Fedco number plate on Marmon cars. It is permanently attached to the instrument board where it may be easily inspected

This shows what happens if a thief tries to remove the Fedco number plate. Two holes are left in the plate, mutilating the plate beyond repair

In prying off a Fedco plate, parts of the plate adhere to the plugs by which it is fastened to the instrument board thus destroying the plate

MARMON MOTOR CAR COMPANY

INDIANAPOLIS, U.S.A.

July 7, 1928.

FEDCO SYSTEM NUMBER CARD

*The Fedco Number of your* MARMON *is* **Y1SL61**

PROTECTED UNDER THE
FEDCO SYSTEM
OF THEFT PREVENTION AND DETECTION
*In case of theft, immediately notify the police, your insurance company and*
THE WM. J. BURNS INTERNATIONAL DETECTIVE AGENCY, Inc.
[Select nearest office from list on reverse of this card]
KEEP THIS CARD IN YOUR WALLET

Mr. Henery Youngs,
Mexico, N. Y.

Dear Mr. Youngs:

Attached is a card bearing the serial number of your new Marmon.

The single plate clutch application was noteworthy because the driven member was piloted on an extension of the crankshaft (roller bearings) and coupled to the transmission via a rubberized fabric torsional isolator. The transmission shift-rail interlock was redesigned for more positive action.

A superior type of low-friction steering gear box had been invented by an engineer named Marles and had been promptly introduced, under license, to the domestic trade by the Gemmer Co. Marmon was quick to adopt this worm and roller gear for the Model 75. Basically the same design is used in todays heavy duty trucks, and is manufactured by Ross Gear Division of TRW who acquired Gemmer.

Balloon tires had posed difficulties for car manufacturers because of uncontrollable cyclic wheel movements which developed after some usage. This was called "shimmy" and, although it did not prevent directional control, it produced violent vibrations; once initiated, a full stop was required to terminate the phenomena.

Marmon, as a pioneer in balloon tire usage, had studied the problem. Adoption on all four wheels of the new Bendix three-shoe internal brake was coordinated with other front suspension redesign and innovation to eliminate the causes of shimmy.

The front spring shackles were transferred to the forward end of the spring to allow the axle movement about the fixed spring eye to approximate the movement of the steering reach rod end; steering inputs from spring deflection were greatly reduced.

The Bowen system of chassis lubrication was continued.

A steering stablizer was fitted between the front axle and the tie rod. Hysteresis necessary to damp out steering oscillations was provided by a novel application of the Hartford friction "shock absorber" mechanism. When the 1961 Pontiac Tempest experienced shimmy problems in the field, General Motors engineers retrofitted essentially the same device.

The rear engine mounting was isolated from the frame by a non-metallic Belflex tension member. Rivets reappeared on the outside of the frame — necessary to attach the hanger for the rear engine mounts.

Two filament headlight bulbs were adopted; the alternate filament provided a **depressed beam** of comparable intensity. Previously, when opposed traffic was encountered, the single filament bulb was reduced in intensity, i.e., **dimmed,** by switching a resistance into the circuit. A foot switch on the toeboard selected the desired beam. In 1927 "dim your lights" became "depress your beam"; who then would have thought that fifty-five years and two generations of motorists later, the expression "dim your lights" would still be the usual one in North America!

The Series 75 styles were more than ten percent heavier than the last Model 34.

Manufacture of a Series 75 Marmon involved 15,859 separate and distinct operations in the Marmon plant: 13193 production operations and 1946 inspection operations.

Shipping weights for the Series 75 were significantly heavier than the Model 34s with their scientific light weight construction. They were:

| | |
|---|---|
| Speedster | 4251 pounds |
| Victoria | 4346 pounds |
| Brougham | 4525 pounds |
| Sedan — Custom 7 Pass. | 4678 pounds |

The FEDCO SYSTEM of theft discouragement was adopted for 1927 Marmon production. Throughout the Third Era Marmon had taken defensive measures against car theft; presumably the car was an attractive target.

*The Custom-Built Sedan*
FOR SEVEN PASSENGERS

*The Standard seven-Passenger Sedan*

The Victoria for four Passengers

The two-Passenger Coupe

The Brougham for five Passengers

The Standard five-Passenger Sedan

Fig. 4-52 The last of the Third Era — The 1928 Series 75 styles

## MODEL 75
## 1928

In 1927 the Little Marmon 8 had been promoted as the 'Companion to the Large Marmon'; the 1928 advertising promoted the Marmon Eight-in-Line Models without reference to the Large Marmon . . . the Series 75. There was however a 1928 edition of the Series 75: Color catalogs portraying each body style in unique colors were available at the New York and Chicago Shows. No changes from 1927 were claimed except possibly colors but inspection of the illustrations reveals a wide point band (about .75 inch) in a dark color directly below the full length belt molding. They also were predominantly illustrated equipped with Buffalo wire wheels.

The 1928 Series 75 was offered in three model categories:

1. Special Custom Built Models
   Built in accordance with individual taste and design. Town cars, Limousines, Sedans, Cabriolets and Speedsters. Bodies by Locke, Holbrook, Judkins, Willoughby and others. *Priced from $7,000 to $10,000.*

| | 2. Regular Custom Built | 3. Standard |
|---|---|---|
| 7 p Sedan | $4,500 | 3,965 |
| 7 p Limousine | $4,600 | . . . |
| 5 p Sedan | $4,375 | 3,885 |
| 5 p Brougham | . . . | 3,885 |
| 2 p Coupe Roadster | . . . | 3,885 |
| 2 p Coupe | . . . | 3,800 |
| 4 p Victoria | . . . | 3,800 |
| 5 p Town Coupe (2dr.) | . . . | 3,485 |
| 2 p Speedster | . . . | 3,800 |
| 4 p Speedster | $3,800 | . . . |
| 7 p Speedster | $3,885 | . . . |

Prices included full equipment (bumpers, shock absorbers, motometer with lock-bar cap, automatic windshield cleaner, rear view mirrors, cigar lighter/lamp, fuel gauge on dash and 2 Presto air bottles.)

The regular custom built — **Series 75 cost twice the price of the Series 78** with 'A' equipment-group option.

Third Era car production was approximately 43,000 units in thirteen model years. Model 34 production was approximately 32,000 in 9 years; model 74-75 production about 11,000 in 3 years.

### *ERA THREE DURABILITY*

The 1928 series 75 was also offered in the export market although the nature of that market would be expected to effect a rapid switch to the Marmon Eights. A brochure produced by the United Kingdom distributor of Marmons, Pass & Joyce, reported a remarkable endurance test of a 1928 Marmon 75 which was a fitting climax to the Third Era. The test was conducted on public roads under the official observation of the Automobile Club of France in the spring of 1928 and provides documentation of the durability, and economy of lubricant and tires of the Third Era Marmons.

The car was a Series 75 two-passenger coupe with right-hand drive and was taken from the London, England showrooms of Pass & Joyce without special preparation in any way. It was equipped throughout the test with North British Rapson tyres, K.L.G. spark plugs, BP (British Petroleum) petrol and Wakefield's Castrol Oil.

"The Marmon was driven continuously . . . day and night . . . in all kinds of weather without cleaning or other attention except during the few moments when it was brought to rest for re-fueling and changing drivers." The test continued until 50,000 miles had been covered; the trial had required nine weeks.

It had averaged in excess of 33 miles per hour. . . 800 miles per day. . . despite a mishap which the brochure described: *"After covering 11,000 miles the car was 'ditched' in avoiding a flock of sheep, necessitating the replacement of a leaf spring, a shock absorber and a wheel rim. The only other spares fitted throughout the test were four headlight bulbs, one valve spring and one speedometer cable." "The Rapson covers (tyres) were changed only after 20,000 and 40,000 miles, the original inner tubes running the whole 50,000 miles. Only two sets of plugs (25,000 mile interval) were used.* ***The Engine oil consumption showed the wonderful average of over*** *4,500 miles per gallon (1100 miles per UK quart. . .* ***900 miles per US quart****)."*

It was appropriate that the Third Era — the Era of one basic Marmon model. . . which began with the spectacular New York to San Francisco dash of the Model 34A in the summer of 1916, 3476 miles in 5 days, 18.5 hours (25.1 miles per hour). . . should close with the Automobile Club of France durability run which broke all existing distance, speed and endurance records for 20,000 to 50,000 miles.

The ultimate documented Third Era durability experience is that of a 1924 Marmon Speedster named 'Riley.'

In the summer of 1971, Leo Gephart was offering for sale, at his Englewood, Ohio establishment, a Marmon 1924 Model 34 two-passenger speedster. This car appeared to be representative of an accident-free automobile with less than excessive mileage. The appearance masked the unique history of this particular car built in April 1924 and purchased by New York Mayor Jimmy Walker. This information and much more was revealed by an accompanying diary authored by the third owner which covered the span 1932-1936. It was stipulated therein that the car's name was "Riley" and that no subsequent owner should ever presume to address it otherwise.

Ownership had changed as follows:

Built — April 11, 1924
Sold — April 16, 1924
Resold — Nov. 1931 for $300.
Resold — April 1932 to C.G. Smith P.O. Box 203 Redondo Beach, CA

The log recorded the following mileage points:

| | |
|---|---|
| April 1932: | 65,000 Purchased by Mr. C.G. Smith |
| | 291,585 Overhaul; first time out of chassis |
| Sept. 27, 1944: | 294, 600 |
| Sept. 25, 1946: | 304, 540 |

The car reflected the change in the Marmon serial numbering practice which occurred on April 11, 1924. The engine number was 1538A and the car carried Hume body number 1224HU 933. Although the car was a true 34B (unlike the other Marmon open models of 1924 which were 34C) it featured the top-of-cowl vent, Delco ignition switch and optional 20″ wood spoke wheel characteristic of other 1924 models. The front axle I-beam reflected modifications for four-wheel brake usage but was not machined for the brake cross-shaft attachment.

Surprisingly the diary revealed that the car had originally been fitted with dual side-mount rims like the Marmon bodied 4 passenger Sport Speedster of the same

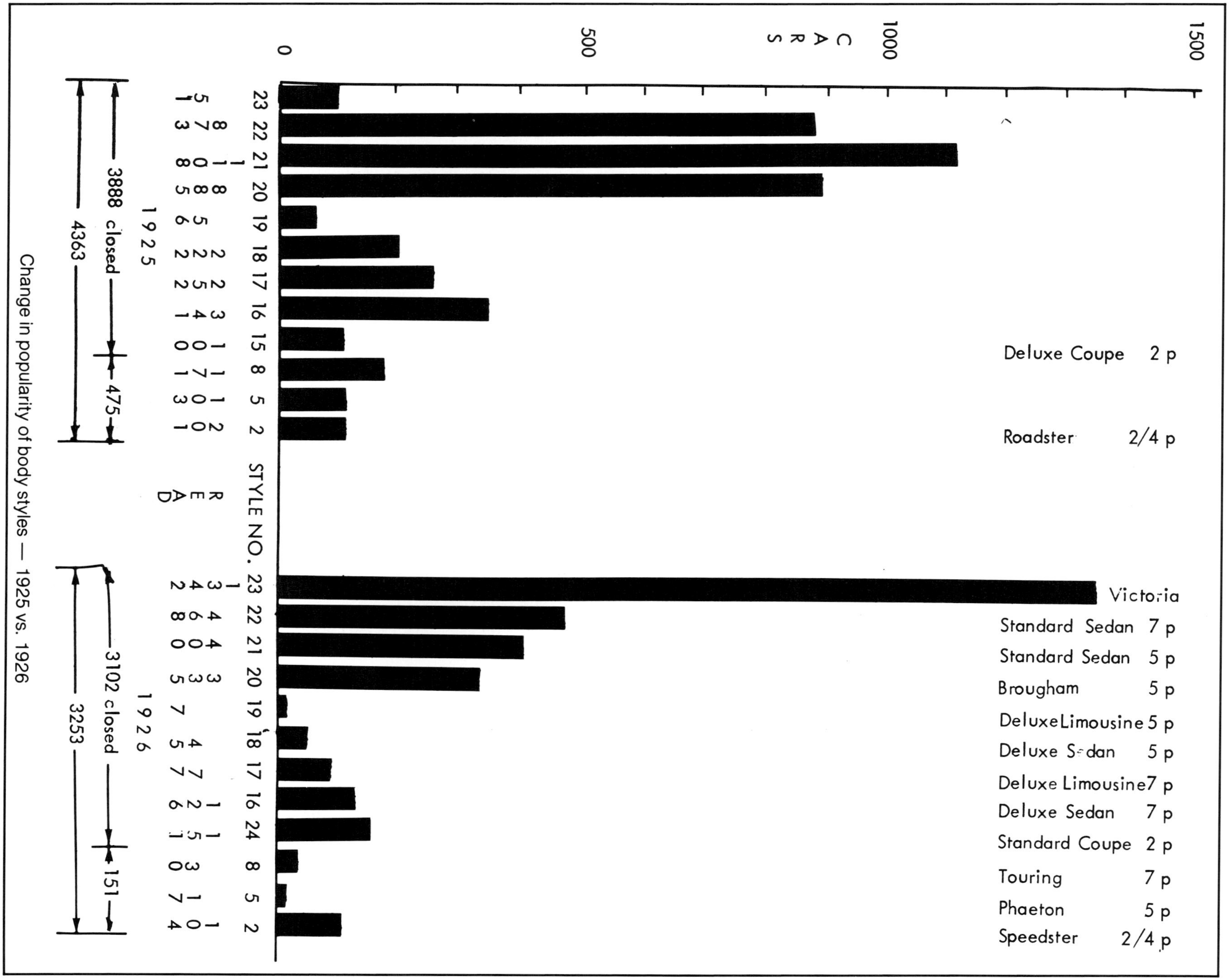


Change in popularity of body styles — 1925 vs. 1926

year. The car as observed was equipped with a standard Marmon rear mounted spare tire/rim, and no evidence of side mount attachment remained. The diary recounted removal of the **right hand** (therefore dual) spare tire bracket on July 23, 1942. This coincided with the government order that only one spare tire per car could be legally possessed; additional spares had to be turned in to the government. An under-car inspection revealed a partial remanufacture of the running boards, removing any prior provision for the tire wells. The car carried California plates and had recently been used in the movie "The Comic" starring Dick Van Dyke. The asking price was $15,000.

Not all Marmon 34s took as many years as 'Riley' to travel a third of a million miles . . . or more. The Marmon House Organ *'Men of Marmon'* for May 1931 reported a garage owner in Ohio then owned a Model 34 which had been driven 430,000 miles . . . he wrote the factory requesting information on a possible retrofit of the cone clutch in his Model 34A with a 34B multiple-disc clutch.

For just plain longevity, take the 1917 Model 34A Cloverleaf Roadster purchased in original condition in 1982 by Dr. Leo Parnagian of Romeo, Michigan. After freeing a stuck tappet, this sixty-five year young Marmon with an aluminum block was ready for the highway.

The original owner of a 1918 Model 34A wrote to the factory to report that in four years he had driven the car 54437 miles with practically 100 percent availability and exceedingly small maintenance expense. The car had not been overhauled and 'runs just as well as when it was delivered!'

Another owner of the same year Marmon reported traveling 90,000 miles in four years of ownership and experiencing tire life of 14-16,000 miles and fuel economy in the 14-15 mile per gallon range.

President Warren G. Harding (1921-1923) owned a 1917 Marmon 34A. It is recorded that the Harding's traveled at least twice to California in that car. After his death, (in office) the car was sold to an unidentified US Senator, and some years later to a US Army Captain stationed in New York. The fourth owner eventually traded it in at a dealership in Monticello, Indiana. It was 1929 and the odometer had recorded 125,000 miles.

In 1968, a 1916 Marmon 34A was reported[1] to be the oldest car still in use in Poland. It had been acquired by the American Charity Mission in 1919. In 1921 it had been used by Silisian insurgents to transport ammunition in a campaign against Germany.

In the 1970s, a Marmon 34B was advertised in the United States[2] as being offered in Poland for export sale. It was reported to have been a presidential car in the Polish Republic era.

### *Commentary on Transition to Closed Car Bodies*

The popularity of closed body styles at Marmon increased dramatically one year after the introduction of the Model 34. It is likely that they were not available until well into the first model year (1916) and that there were supply shortages from the outside sources. The closed proportion increased steadily: 5 percent in 1916, 18 percent in 1917 and averaging 25 percent for the four years of the model 34A. The lightweight Marmon — built bodies increased the proportion to 42 percent in 1923 whereas the level had held at 25 percent the prior two years. When the price premium for the closed styles was reduced by Marmon to $130 for the 1925 model year and then to zero (by raising open style prices $130), the closed styles increased spectacularly to 90 percent.

Early closed body automobiles suffered greatly in performance and handling characteristics.

With rare exceptions, all cars had very low specific power, i.e., horsepower per pound of car even in the open configuration. Consequently the increased performance of a two seater due to the lighter weight was appreciable, particularly if enhanced by a monacle windshield or better yet by substitution of goggles for the windshield.

The aerodynamics of a Second Era closed bodies was about comparable to a modern Class VI city delivery van. Due to the huge frontal area, and the consequent air load, the rear axle ratio had to be adjusted, trading off top speed for more lower speed torque. Performance was degraded greatly, and perhaps fortunately for the high center of gravity threatened upset in response to any quick steering maneuver.

The higher than usual specific power of the Marmons in any of the Eras minimized the penality of the closed body, particularly in the Third Era with sill-less construction and later aluminum skinning. Nevertheless the open car symbolized performance motoring and the closed car stodgy transportation for older folk.

People who lived their teen years in the first third of the century acquired, consciously or unconsciously, the above perspective and the love affair with the automobile was directed to the **open** automobile. When in later life they became collectors or restorers of antique cars, they were biased to the open examples. Often a very low mileage closed car in very fine original shape became a 'parts car' to provide mechanical components for a high mileage, badly deteriorated, open car of the same make and model.

Subsequent generations who find the performance levels of all early cars, open or closed, to be untenable are consequently free of this early bias. A 1957 Chevrolet convertible may be highly desirable but the overwhelming majority of restored 1957 Chevrolets are closed styles. Early closed cars are therefore so rare that each one restored is a significant historic preservation as well as a hobby project. Closed Marmons of the Second and Third Eras were particularly well styled and often handsome motor cars and are so scarce today that any example warrants a serious restoration regardless of the condition as found. Most surviving closed Marmons thru Era Three are sole examples of the particular body; only multiple survivor group is the standard series 5 passenger sedan of the Marmon 74 which includes 13 examples. There are no pre-1922 closed body survivors known to the author.

Although 25 percent of the model 34A production was closed cars, none survive. Fourteen percent of surviving Model 34B/Cs are closed cars, less than half the percentage as built; however for the 1924 model year, the survivors are in the production proportion . . . 42%. The Marmon percentage of closed styles surpassed the industry average of 30 percent in 1923 and 35 percent in 1924.

## PART 4 — The Engines of Era Three 1916-1928

Although the Marmons of Era Three are addressed in Parts 1 and 3 of this chapter individually for the five versions, there was really one fundamental engine used throughout the Era, despite a near total redesign for the Model 34B and some major improvements for the Model 74. Accordingly a general discussion of common features will be followed by sections on the 34A engine and the 34B, D-74, E-75 engine.

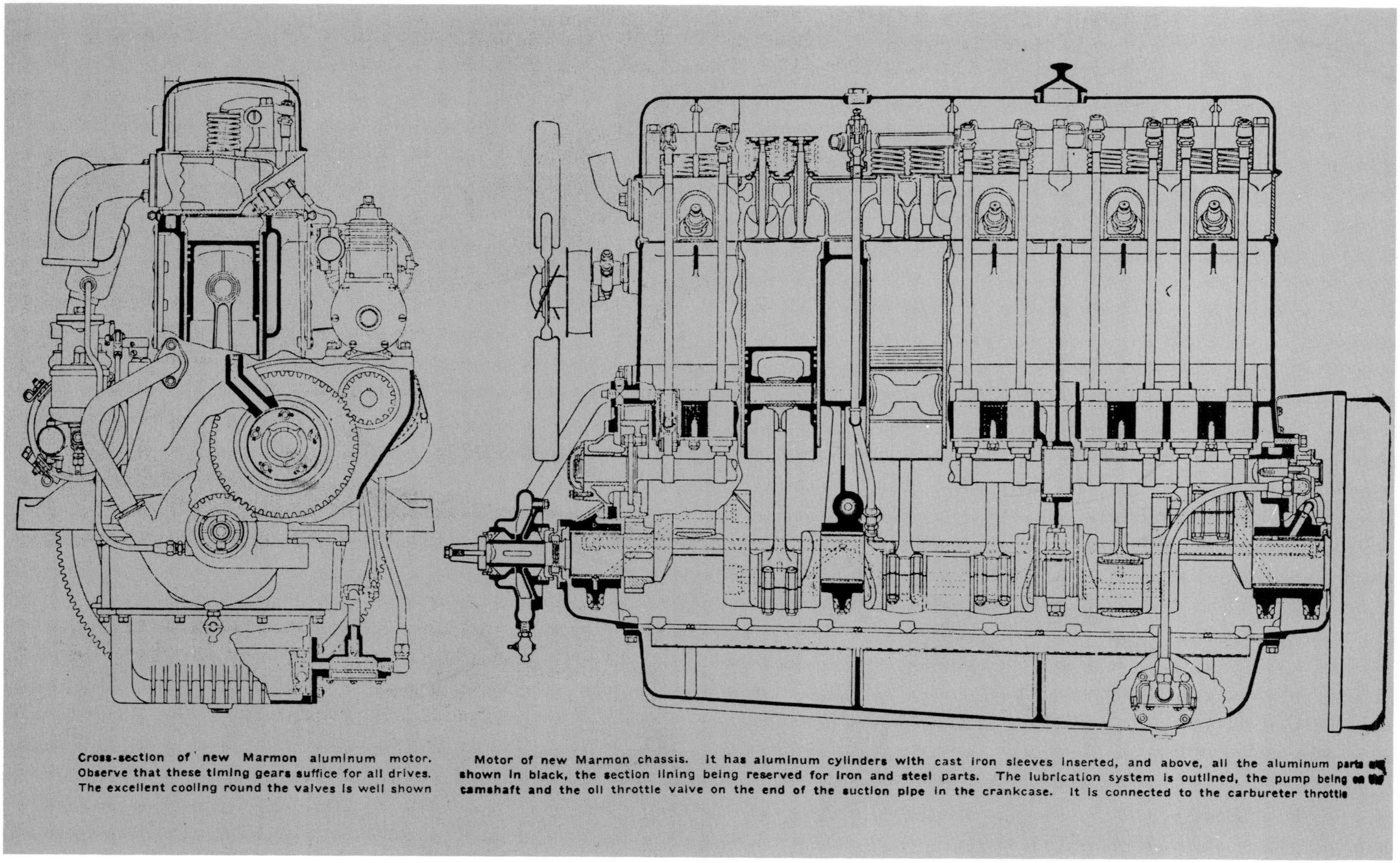


Fig. 4-53 Engineering drawings of the Model 34A engine.

That fundamental engine was an in-line water-cooled six with push-rod operated overhead valves. With a 3.75 inch bore and 5.125 inch stroke, it displaced 339.7 cubic inches. It had an SAE rating of 33.75 and developed 74 brake horsepower at 2450 rpm. The car model designation was derived from the horsepower rating. The 34A, 34B and 34C used the rounded SAE rating in the tradition of all Marmon models of Era Two; the D and E versions, the D 74 and E 75 used the brake horsepower rating.

The introductory brochure of 1916 entitled *"MARMON 34,"* reprinted by Floyd Clymer in 1951 . . . fortunately, for I have never seen an original copy . . ., discussed and illustrated engine features and the philosophy responsible for their adoption. Because many of these features represented new state-of-the-art subsequently widely adopted, appropriate portions of the text will be used verbatim. Because some of them were continued througiout the Era whereas others were not, the former will be addressed in this introductory section leaving the latter to the Model 34A section.

The design sought the lightest feasible weight and the lowest achievable levels of noise and vibration. The Marmons of Eras One and Two used barrel crankcase/crankshaft construction for both noise and vibration control as well as durability of the aluminum case. Howard Marmon sought to retain those qualities in a minimum weight engine by use of a massive crankshaft, fully machined (all-over as were the connecting rods). The Clymer reprint said:

"The crankshaft is put in standing and running balance, and with the flywheel bolted to the crankshaft, the two together are then put in running balance. Each crankshaft bearing is of nickel babbitt with phosphor bronze back, the upper half fitting into the aluminum crankcase, the lower half fitting into a cap which is bolted to the casting."

"The valves are located side by side in the (single) cast iron cylinder head. They are all the same size, 1.937 diameter, and are very light, having .375 stems only 5 inches long and weighing but 5.5 ounces. Two valve springs are used, one 1.75 diameter and the other 1.0, one within the other. The use of the two springs prevents synchronism of the spring and actuating mechanism at certain speeds and so prevents tossing of the valves at the higher speeds. If a single spring were used it would have to be nearly twice as stiff as the two smaller springs combined."

"The water-jacketed intake manifold is quite short, facilitating carburetion; it has three connections to the cylinder head, each supplying two cylinders."

"The exhaust manifold, made of steel tubing, has four ports and points to the front of the motor rather than the rear, as is ordinary practice. The primary value of this placement lies in the cooling of the exhaust gases, due to the close proximity to the fan. Lowering the temperatures in the exhaust pipe, close to its connection to the cylinders, allows a freer passage of burned gases. Another advantage is in having a relatively cool pipe under the car and away from the floor boards."

"The camshaft and cams are integral (and the drive gear flange mounted). The cams are of the hollow ground type (requiring small dia finish grinding wheels) which gives a quicker opening, allows the valves to remain open longer and a very rapid closing. In designing, the uniform acceleration principal has been followed to secure gentle opening and closing. The lifting surface of the cam is concave rather than flat as in conventional design (i.e., concave parallel the shaft axis)." Valve lift was .437 inches.

The features retained for the entire Era have been discussed. The construction of the engines used in specific models will now be addressed in detail.

## MODEL 34A
## 1916-1919

The Model 34A engine was **the** first aluminum block engine and was a pioneer of the 'en bloc' integration of crankcase and cylinders in a single casting.

The design of the engine, particularly the block and cylinder head, was far ahead of its time and reflected clever sophisticated design engineering with hardware that today appears contemporary. The timing gear housing was integral with the cylinder block and did not require a full cover. The oil galleries to the main bearings were formed by cast-in-place tubes in the block. The cylinder head was very advanced as to rocker-arm design and spark plug location.

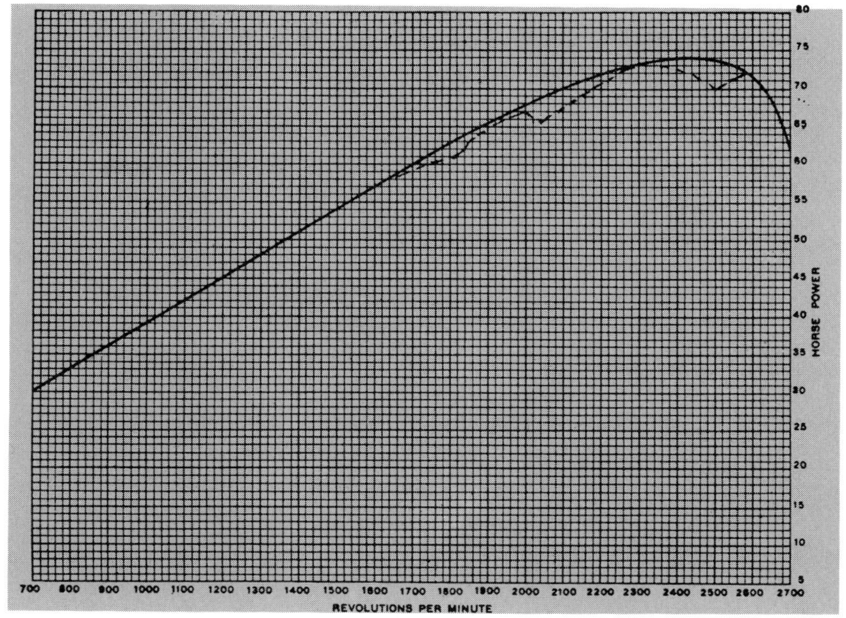


Fig. 4-54 Dynamometer performance of the Model 34A engine (34 SAE Horsepower).

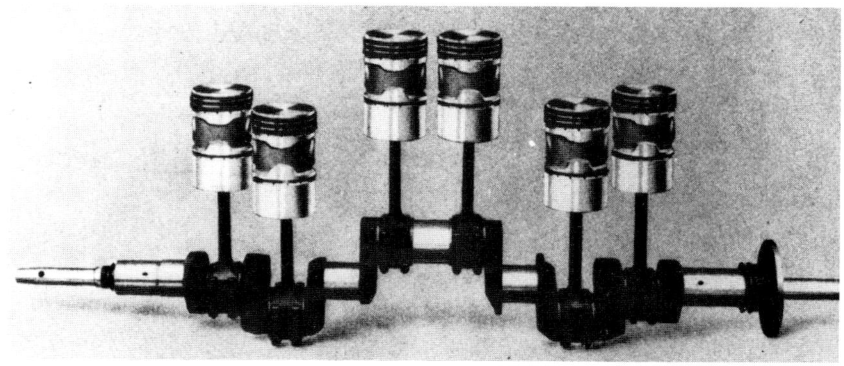

Fig. 4-55 The crankshaft and aluminum alloy pistons of the Model 34A engine.

The engine was developed with an aluminum cylinder head but was released with an iron head consequent to inability to produce sound castings at production rates. The 1982 V-6 Oldsmobile used an aluminum block with iron heads.

The exact language of the Clymer reprint *MARMON 34* brochure will be used whenever possible to describe this innovative masterpiece:

"Instead of using cast-iron cylinders bolted to a crankcase, the cylinders are integral with the crankcase and are made of a single aluminum casting which extends to the center line of the crankshaft. This large aluminum casting is the main structural member of the motor. It is interesting to note the advantage gained by forming the cylinders and crankcase in one casting. The design of this member, which supports the main crankshaft bearings, is of extreme importance in eliminating vibration and wear. The Marmon 34 single-piece crankcase is of extraordinary depth, as compared with conventional (separate crankcase) design, and results in a motor of exceptionally sturdy construction."

"The cylinders are machined accurately and hard white cast-iron sleeves (are) dropped into the aluminum bores. This can easily be done, as steam is turned into the aluminum water jacket while the iron sleeves are being inserted. Collars (shoulders) at the top of the sleeves hold them up while the cylinder head holds them down firmly in the aluminum water jacket. These inserts always fit tight, because the cast iron being exposed to the flame of combustion expands more rapidly than the aluminum, which is thoroughly water cooled."

"The crankshaft runs on four main bearings, the front being 2.75 inches long by 2.125 diameter, the center two being 2.25 by 2.25 dia and the rear 3.75 long by 2.25 diameter."

The connecting rod bearings were bronze backed nickel babbitt with a bronze bushing at the pin which was lubricated by splash to drilled holes in the piston boss.

"The overhead-valve rocker arms receive their lubrication through the hollow shaft on which they are mounted. The oil is forced through a lead into the shaft at (which) point the oil overflows (from the pressure regulator) into oil wells immediately above the valve tappet guides. The rollers, pins, bottom ends of push rods and guides are submerged in oil." The push rods were steel-ball tipped aluminum rods. The adjustment screws were drilled to lube the upper push rod ends."

"An aluminum cap (rocker cover) fastens over the top of the motor enclosing the valve mechanism and giving the motor an extremely clean appearance. This cap (cover) gives a most logical means for disposing of the crankcase gases and smoke which are disagreeable to the passengers. The smoke and gases are drawn from the cover through a pipe to the air intake of the carburetor and burned." The oil expand filler was on top of the rocker cover."

"The (four) camshaft bearings and the timing gears are lubricated through leads from the oil wells above the tappet guides."

"The pistons are of cast aluminum alloy. Four rings are used on each piston, three above .187 wide and one oil ring below .125 wide. Below the oil ring and the lower main (compression) ring is a circular space with holes leading to the interior of the piston through which excess oil runs to the crankcase." The skirt extended a considerable distance below the oil ring. The skirt was fully relieved (360°) in the region of the piston pin.

Third Era Marmons Models 34A and 34B had a fabricated exhaust manifold. Four 1982-83 models of Ford 4, Pontiac (A car), Chevrolet, Corvette V8, and Buick V6 are newly equipped with fabricated stainless steel manifolds. According to the American Iron and Steel Institute.[32] *"Lower back pressure in a fabricated tubular manifold increases horsepower, sometimes as much as 15 percent."*

An unusual design detail was the placement of the timing gears **behind** the crankshaft front main and camshaft front bearings. The latter bearing was in a cover over the cam-gear attaching-bolts; the gear having been placed in position prior to insertion of the camshaft. A fundamental consideration in the engine design was the achievement of a quiet precise set of timing gears. The crankshaft was massive to control torsional vibrations of the shaft and the front timing drive gear. Presumably the unusual placement of the gear sought to place it at a point of minimum run-out during engine operation and thereby permit reduced gear backlash; however the relationship of this gear placement to the feasibility of the integral gear housing concept must not be overlooked.

In several respects the engine was a surprising departure from traditional Marmon features: The crankcase was conventional, i.e., with the crankshaft inserted from below into half-bearings in the block transverse bulkheads and retention by bearing caps, unlike the barrel crankcase of all preceding Marmons. The Model 41 had a main bearing between every cylinder; the new engine reverted to the Model 48 complement of four, not seven bearings. The patented traditional pressure oil supply to the wrist pins was foregone in favor of splash lubrication. The engine was "long stroke" whereas all prior Marmons had been near "square" in initial versions.

Obviously weight considerations were major factors in these changes of philosophy. Although the crankcase appeared conventional, the use of a massive crankshaft was not conventional.

The gear type oil pump was driven directly from the end of the camshaft to save the weight of drive gears and shafting. This application proved that submersion of the pump was not necessary and was copied by Henry Ford for his flathead V-8.

The inlet to the oil suction line incorporated a check valve, considered unnecessary today. Gearotor type oil pumps on the nose of the crankshaft are becomming popular on new passenger car applications. Influenced by the Marmon 34, the author designed the pioneer application of this type into the Detroit Diesel truck and bus engines in 1956.

The oil pressure was regulated in proportion to engine load by a throttle restriction of the pump inlet which was interconnected with the engine throttle.

The water pump housing was aluminum. The bronze impeller incorporated a long sleeve which was key driven from the supporting crankshaft and interfaced on the outside with the two packing glands in the aluminum body. The water contacting surfaces were entirely of bronze or aluminum. System capacity was 6 gallons.

The complete Bosch 12 volt electrical system adopted for the Model 41 had proven so satisfactory that it was basically retained for the Model 34A. The Bosch magneto was driven from the rear of the Bosch-Rushmore generator which was driven from the accessory drive gear at 1.5 times engine speed. Both voltage and amperage controls for the shunt-wound generator were packaged directly above and engine mounted. The generator provided 10 amperes at speeds of 25 or more mph. Headlight bulbs were increased in candlepower from 24 to 30 and featured flat filament bulbs claimed "to diffuse the light to the sides of the road in an extraordinary manner."[40] Additional 6 candlepower bulbs were incorporated in the headlights.

The starter motor weighed 30 pounds and was capable of turning the engine at 135-145 rpm. Engagement continued to be by electromagnetic displacement of the armature in an axial direction with disengagement resulting from counter currents set up in the starter when the engine started. However the rigidly attached pinion was replaced by a new concept. "A very unique method of avoiding clashing of gears in the engaging of the starter has been worked out by the Marmon engineers. It was found that sometimes the pinion teeth would be directly in line with the gear teeth; consequently engagement could not be accomplished without considerable noise and clashing of gears. This has been overcome by putting a pin into the projecting nut which holds the pinion on the shaft, with a spring behind it.

This pin enters into a spiral slot in the hub of the starter pinion. The spring is between the head of the nut and the head of the spiral, its function being to keep the spiral in forward position at all times. This is so mounted that when the switch is closed and the teeth of the flywheel and the pinion strike each other instead of engaging, the additional amount of draw-in pull of the starter shaft revolves the pinion by means of this spiral groove and pin. The amount of motion is equal to the circular pitch of the gear, consequently there is always enough to give one complete tooth turn. The pinion will always engage without noise or clashing."[40]

The carburetor was a Stromberg Model 0-3, of the air valve type. It featured a glass float chamber and access, through removal of an acorn nut, to the float level adjustment. The float could be depressed by the same access to clear a seat obstruction by flooding action.

## MODEL 34A
## 1917

A full-flow by-pass thermostatically-controlled cooling system was introduced — decades ahead of its widespread use on heavy duty diesel truck engines. Because of the significance of this pioneer application, it is appropriate to repeat the description of the system in the language of the Marmon 1917 brochure: *"A feature of the Marmon 34 for 1917 is a device for controlling he temperature of the water circulating system . . . an important factor if maximum performance and economy are to be achieved. The thermostat used is a simple device which automatically keeps the temperature of the water between 125 and 175 degrees Fahrenheit at all times when the engine is running. Within a cylindrical case, mounted just back of the radiator upper inlet, is an expansion unit to which is attached two poppet valves so placed that they alternately open and close two ports. When the engine is cold, the unit is contracted so that the water in the radiator is forced, through a by-pass, directly down into the suction side of the pump. Consequently no water circulates through the radiator and the water in the engine heats up rapidly. When the water becomes heated, however, the unit expands, the by-pass chamber is automatically closed, the port leading into the radiator is opened and water flows into the radiator from the engine. The little unit . . . known as a "Sylphon" unit . . . always does the right thing automatically, regulates the circulating system so that the water warms up very rapidly when the engine is started and sees that it is kept at the proper temperature for the best results. This "Sylphon" unit is practically everlasting, as tests have shown that it will expand and contract more than 8,000,000 times without injury." Similar performance by production bellows was evidently not experienced . . . the 1918 brochure illustrated the 1916 cooling system (conventional non-thermostatic) without mention that it had been different, at least in early 1917.*

Blocking type thermostats appeared in the late 1920s (in 1927 on the Little Marmon) but full-flow by-pass systems were not seen again until they were adopted for heavy duty diesel truck engines in the early 1950s. Thermostatically controlled shutters were adopted by Cadillac for 1926 and by Marmon (for the Big Eight) in 1930. They persisted on the Cadillac through 1948, and on all top-of-the-line Marmons until the end of production. The physical configuration of the Model 34A system has a remarkable resemblance to the full-flow by-pass system adopted for the Detroit Diesel Model 6-71 truck engine in the early 1950s and still in production.

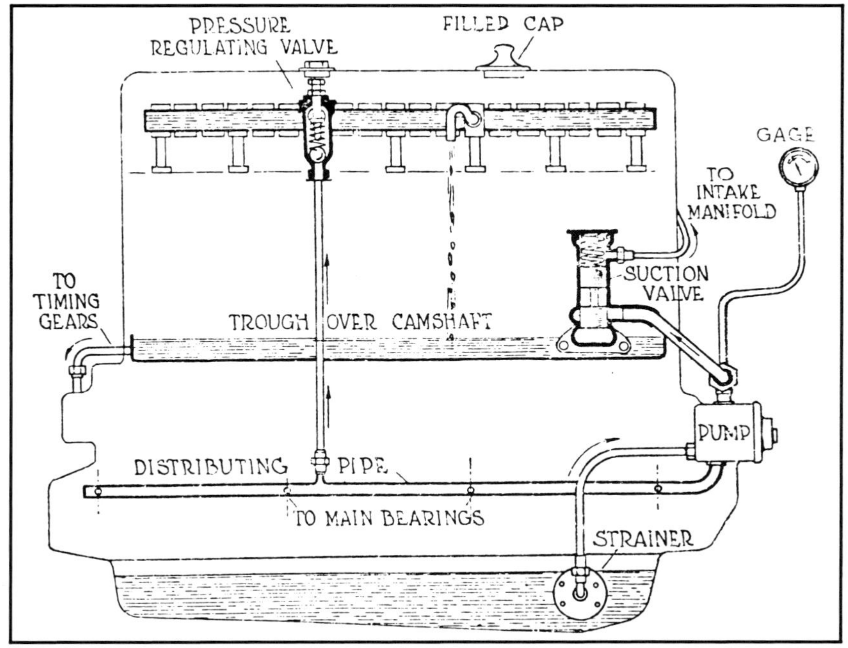


Fig. 4-56 Lubrication system of the Model 34A engine.

MARMON

*It's a Great Automobile*

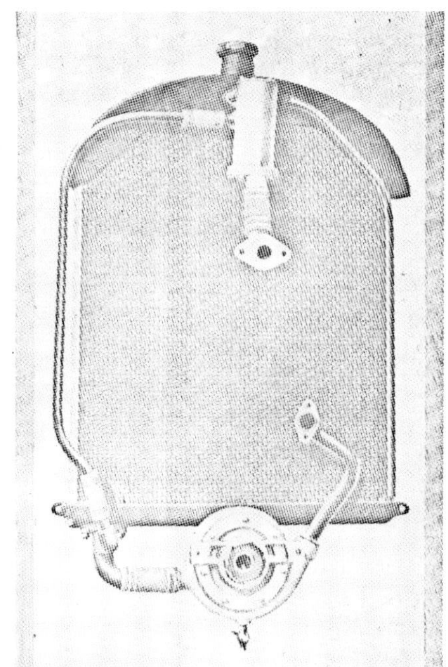

Fig. 4-57 Ahead of its time — The Model 34A thermostatic full-flow by-pass cooling system for 1917.

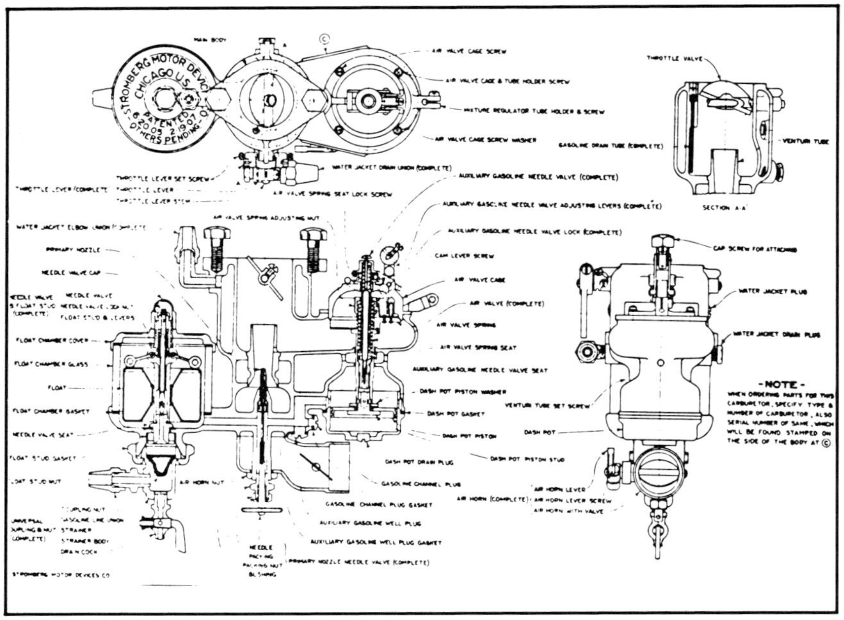

Fig. 4-58 The Stromberg 'H' Carburetor used on the Model 34.

## MODEL 34A
## 1918

The appearance of the engine was enhanced by the elimination of the two large aluminum plates required to close the valve push-rod cavity on one side and the water jacket on the other. At time of introduction these openings were claimed beneficial to quality by providing superior locating of the casting for cylinder boring; either experience had shown it unnecessary or the foundry had learned how to core and clean the casting without them.

During the 1917 model year a vacuum operated valve was incorporated in the lubrication system. The pressure system was dumped by a spool valve at a vacuum level indicative of idle operation. Because the valve was above camshaft height the pump then maintained pressure on main and rod bearings at a pressure head less than the height of the engine. This was intended to prevent a tendency to carbon formation and oil waste.

A novel Marmon fan hub design replaced the ball bearing design procured from a vendor. It utilized plain bearings pressure lubricated from an integral rotating reservoir. The rotating oil impacted a stationary tube at the periphery of the reservoir, which led to oil galleries in the fan shaft.

The thermostatic water temperature control system was changed to a special-order option.

With exception of the Bosch magneto ignition, the electrical system was changed to a Bijur system during 1917 production. The generator provided from 25 maximum amperes down to 5 under fully charged conditions. The starter featured the new Bijur screw-engagement drive subsequently known as Bendix Drive, with an axially fixed armature. The new system was 6 volt and therefore a 120 amp-hour battery was used. The positive terminal was grounded but the generator was capable of charging with either polarity. There were unspecified problems with the Bijur starter because a Bosch starter was subsequently retrofitted after the end of Model 34A production.

## MODEL 34B
## INTRODUCTION

Soon after the introduction of the Model 34, Nordyke & Marmon became deeply involved in the war effort and consequently were particulary aware of the extremely rapid growth in the technology base which characterized the first (and the second) World War. The construction and tooling of a new manufacturing facility for two independent designs of aircraft engines and series production of those engines at degrees of dimensional precision previously considered impossible made Marmon aware that new standards of perfection in motor car engine manufacture were dawning.

Some kind of problems had been encountered with the Model 34A integrated aluminum block and crankcase engine. Some authors cite field durability problems possibly associated with galvanic corrosion. P.M. Heldt wrote, long after the engine was out of production, that keeping the cylinder sleeves aligned in the aluminum block was a continual problem in service. Maurice Hendry had this to say:

2 *Heldt's revelation had been missed by latter day historians but is confirmed by a confidential letter from Howard Marmon to David Fergusson, chief engineer of Pierce-Arrow.*[33] *Fergusson, interested in the 34, had written to Moskovics and Marmon telling them about his own experiments and problems with light alloy engines at Pierce-Arrow. Four days later, on Dec. 5, 1916, Howard replied in a letter that explains what went wrong with the 34A:*

*The Howard Marmon letter:*

*"In a general way you have had exactly the same experience that we have had," he wrote Fergusson. "Where the aluminum surrounds the full length of the sleeve we have had more or less trouble keeping the sleeve round, but find that the condition is materially bettered by violently boiling the crankcase casting for two or three hours before machining. This preliminary boiling also assists in keeping the crankshaft bearings in line. Otherwise they come out quite considerably as soon as the motor is used." Marmon then mentioned his experiments with wet liner engines in which there was only a short bearing at top and bottom of the liner. Under these conditions, with a well seasoned barrel, the factory had had no trouble with liners getting more than .0005 out of round. "This is a construction which we will eventually go to entirely," Howard continued hopefully, unaware that this lay a full fifteen years...and sixteen cylinders...ahead. Howard also admitted to trouble with aluminum pistons, which required so great a clearance that ring damage could occur, but this matter he had well in hand. He then made the interesting revelation that sealing of the castings againt water leakage was carried out by filling the entire water jacket with baking enamel, putting a slight pressure on it to drive it into the porous spots, then draining the block and baking in the enameling oven. This process is usually credited to Hispano-Suiza but Marmon may well have anticipated them."*

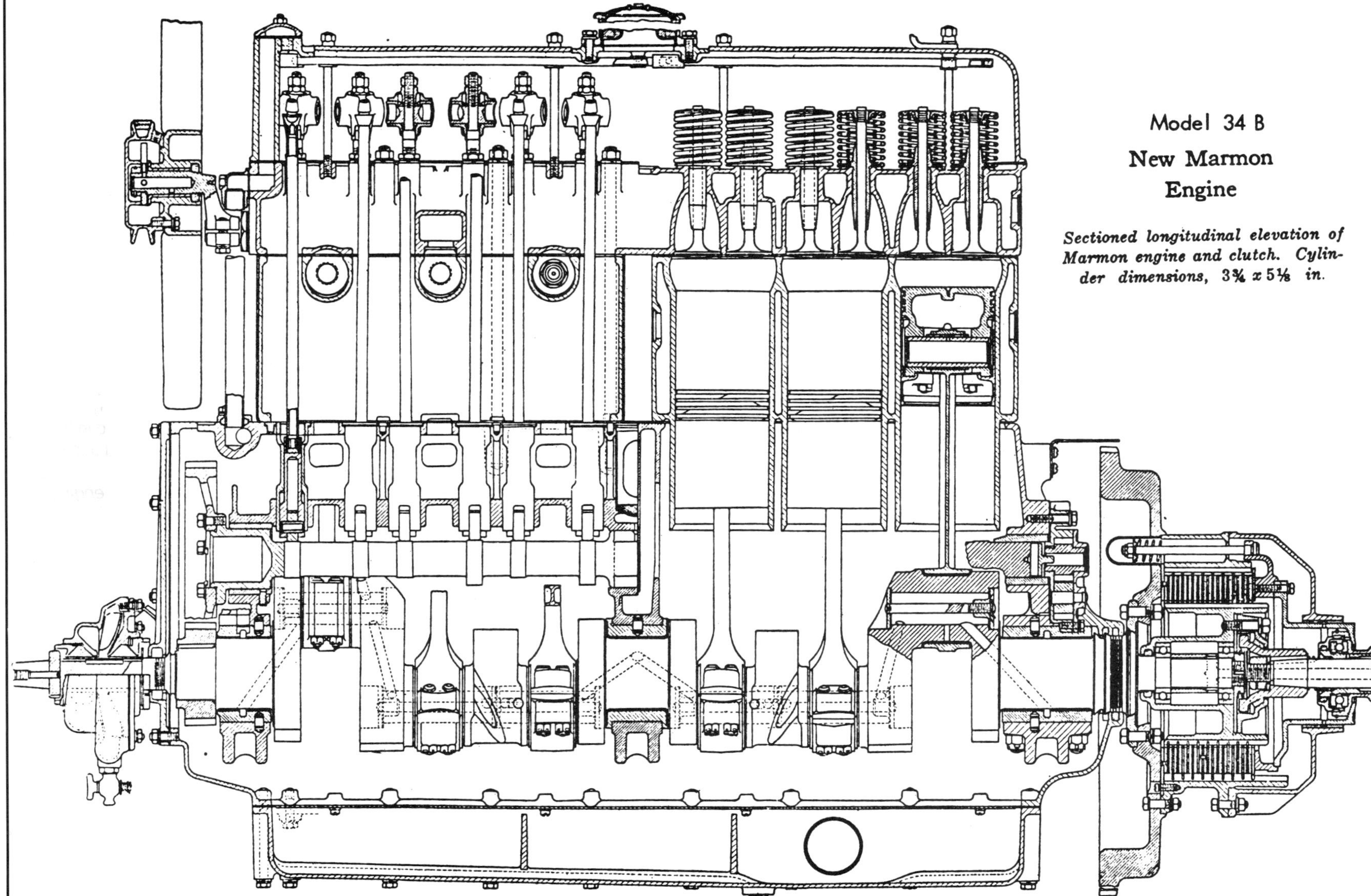

Model 34 B
New Marmon
Engine

*Sectioned longitudinal elevation of Marmon engine and clutch. Cylinder dimensions, 3¾ x 5⅛ in.*

New Marmon Engine

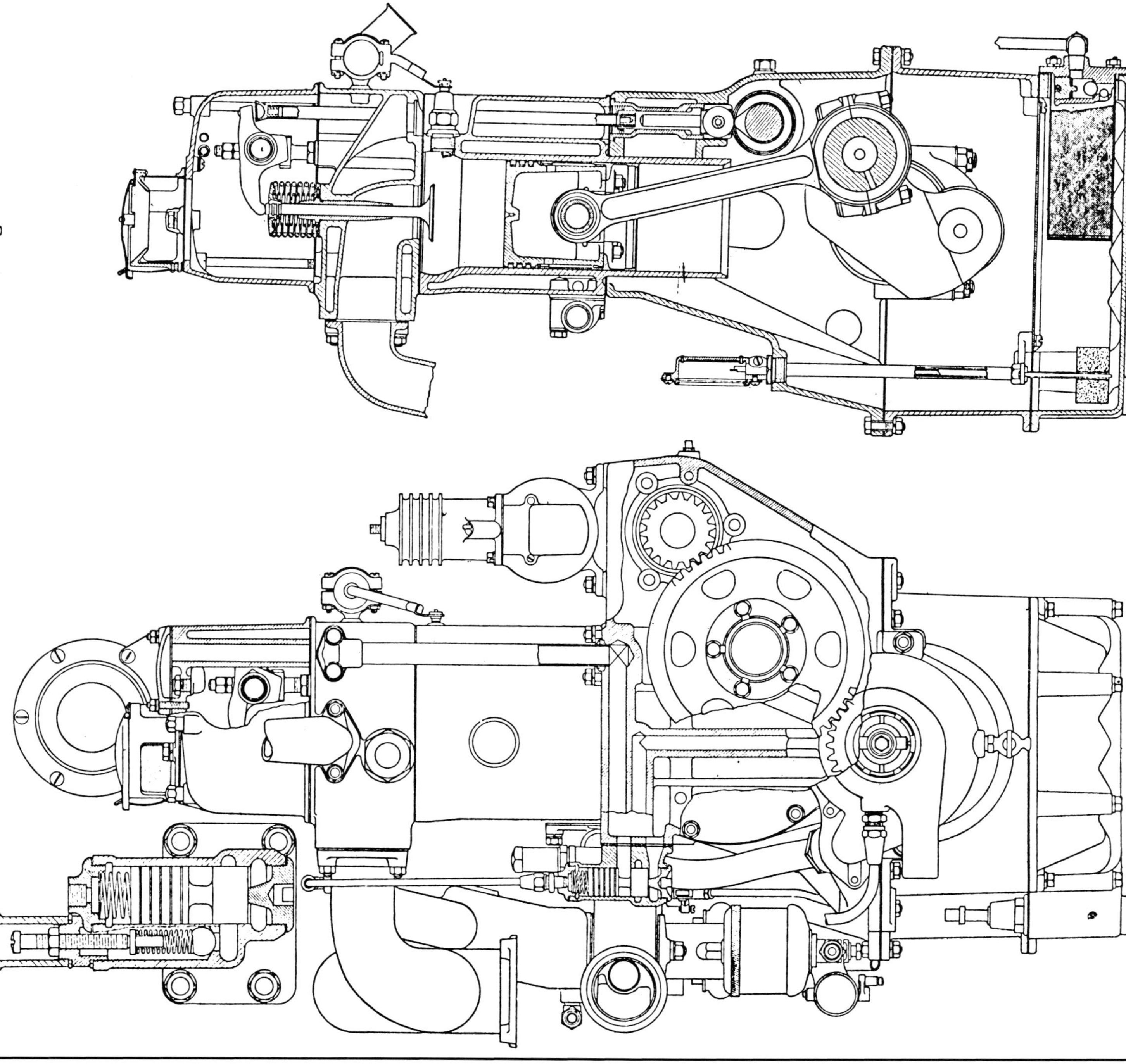

Cross section through crankcase, cylinder and head

Front elevation, showing details of oiling system

Today . . . automotive aluminum castings are specified to be "Condition T51," a modern day variation of Marmon's solution to the dimensional stability problem by boiling the crankcase.

The author is inclined to believe that the problems with the 34A were no different than those with the General Motors Buick/Oldsmobile 215 CID Aluminum V-8 of 1961-63, presently powering the Rover automobile from the UK . . . an excessive rate of unacceptable rough or finished castings. Marmon never reached a production rate with the planned aluminum cylinder head which would allow its incorporation in the engine; perhaps the block persisted in high rejection rates. In the case of the GM 215 engine, the block casting problems were overcome but not before a commitment had been made to go to iron. A friend recently purchased an unrestored 1917 Marmon 34A and is thrilled with its handling and performance after freeing a sticky valve tappet!

Detroit Diesel Division of GM produced 6-71 diesels in the 1950s with optional aluminum cylinder blocks. These were dry liner engines and it was necessary to have a steel washer under the liner flange to avoid brinnelling of the counterbore in the cylinder block, only when aluminum was used.

Howard Marmon was absent, serving as Aircraft Engineering Engineer reporting to the head of the Aircraft Production Board and the engineering of the Marmon automobile was largely left to Chief Engineer Jeffers who undoubtedly deserves a fair share of the credit for the Model 34 B engine.

## MODEL 34B
## 1920

The totally new Model 34B engine did retain the 3.75 bore and 5.125 stroke of the predecessor Model 34A. It was again an overhead valve engine but with the pistons running in integral iron cylinder bores.

The 34B engine, operating on the dynamometer **without** exhaust manifold or pipes, produced 82 brake horsepower using the standard Stromberg H carburetor with 1½ inch air valve. Maximum torque was 222 foots pounds at 1450 rpm.

The 34B engine "saved three inches under the hood" due to the election to use 3 main bearings rather than the 4 on the Model 34A. Full 360° cooling of the cylinders was retained, however.

The integral block/crankcase design was abandoned in favor of an aluminum crankcase and two cast-iron cylinder blocks. The single iron head concept was continued but the entire engine was given a detailed redesign.

Having lost the crankcase rigidity of the integral block/crankcase, the already massive crankshaft was strengthened and the new crankcase was carried up over the cylinder bores to the top of the piston at bottom center position. This limited the water jacket to the length of the expanded combustion chamber and left the cylinder walls protruding, for half their length, into the crankcase; but it did maximize the vertical regidity of the case.

The neat appearance of the 34A was retained by assembly of the 34B blocks over 16 studs screwed into the crankcase, followed by assembly of the cylinder head theron, with the nuts being hidden by the rocker cover.

The new crankshaft had three 2.75 diameter main journals which were 2.25, 2.50 and 3.75 inches long, front to rear respectively. This gave 23.375 square inches of projected area. The connecting rod journals were 2.56 diameter. The interchangeable bearing shells were 0.5 inch thick. Nickel babbitt .06 inch thick was 'die-cast' in the cast iron shells. A warning label on the rocker cover alerted the mechanic to the precision bearings and the prohibition of the usual adjustment of

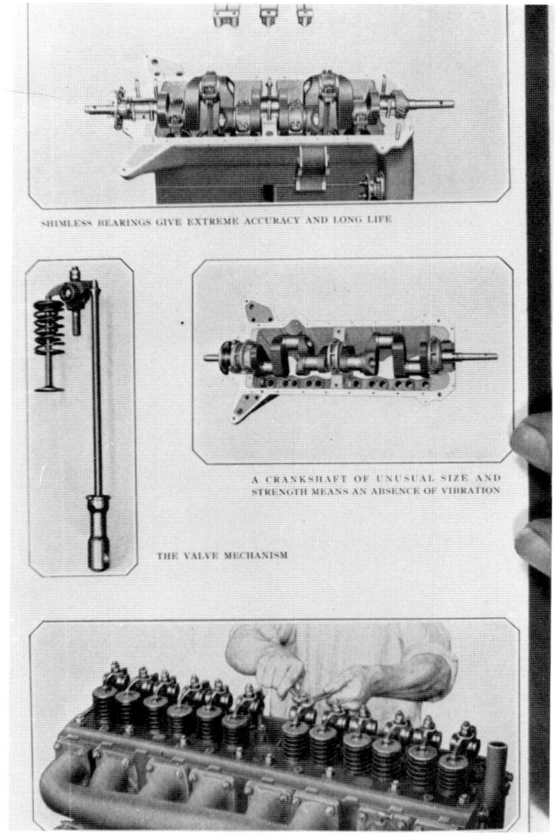


Fig. 4-60 Details of the Model 34B engine.

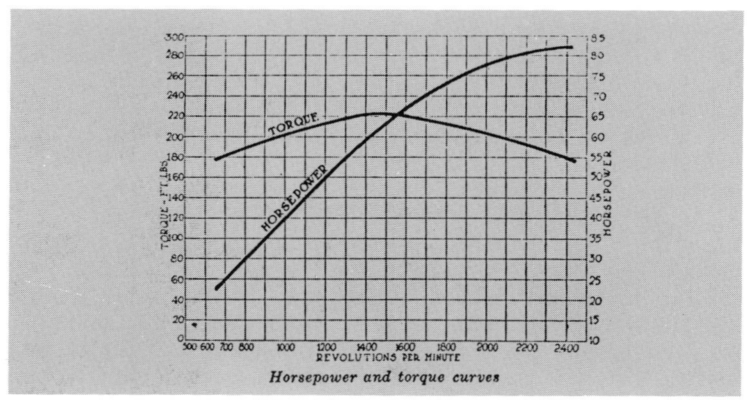


Fig. 4-59 Dynamometer performance of Model 34B.

bearings by filing metal off the cap face. Bearing caps were retained by studs 8.50 inches long.

The Marmon 34B brochure had this to say about the new crankshaft and its bearings: *On account of the design and size of the crankshaft used in the Marmon 34 (B) it runs at all speeds without vibration. A shaft of this size is twice as rigid as a shaft 0.5 inch less in (journal) diameter (the 34 A size). It is of greater diameter than the shaft used in the 400 horsepower Liberty airplane motor. The webs of the shaft are so formed that they give a perfectly balanced shaft, without the use of counterweights. The bearings with a 0.06 inch facing of anti-friction metal, are backed with cast iron. Since cast iron has practically the same rate of expansion as that of the crankshaft, uniform bearing clearance is maintained through all ranges of temperature and engine speed. Cast iron backed bearings will not contract and pinch the crankshaft, as bronze would do, when the engine is cold; yet they provide a snug fit at the temperatures of high speed. These bearings, as well as connecting rod bearings, are fitted without shims, and for this reason afford many advantages over other designs. An unbroken surface is presented to the crankshaft bearing surfaces. This is an inestimable aid to proper lubrication and long-life bearings. When oil is forced into these bearings under pressure it forms a continuous film between the shaft and bearing; there are no shim notches to wipe or destroy the film and it remains a permanent protection. A special bearing finishing machine eliminates hand scraping and insures perfect fits and interchangeability."*

The manufacture of the crankcase was described: *The design and construction of the aluminum crankcase is such that utmost rigidity is obtained. In addition to the usual methods of finishing, Marmon had added a series of hand cleaning and lapping processes. After being machined, the case is hand-scraped, put on a lapping plate and rubbed down to a true surface, so as to give an accurate reference plane for the locating of the case in subsequent boring operations. Ordinary machining of crankcase surfaces can not be made more accurate then .008 inch, even under the best conditions, whereas by lapping the surfaces can be made absolutely true. This allows for perfect alignment of studs for holding the bearings and consequently the perfect placement of bearings themselves. The backs of the bearings are ground, so as to make a perfect fit in the seats, which are cut with diamond-pointed tools.*

Main bearing caps were positioned by sleeve dowels with a ground finish. The Marmon 34B crankshaft system was practically indestructible. A log exists for one car which reports the first disassembly of the crankshaft beaings at 305,000 miles.

The problem of fit experienced with the aluminum pistons in the 34A was solved by a two-piece two-metal piston design which was the single all-time popular Marmon detail feature. There was hardly an engine textbook anywhere for decades without an illustration and description. The Marmon 34B brochure described this development: *The head and central structure for holding the wrist pin is of aluminum; the skirt of cast iron. This combination gives the advantages of aluminum, its lightness and freedom from carbon; while the cast iron skirt, expanding uniformly (unhampered by irregular sections) with the cylinder walls, fits snugly, prevents piston slaps, and acts as an oil-wiping unit ring and keeps oil out of the combustion chamber. The wrist pin fitted in the aluminum part (exclusive of the skirt) cannot come in contact with or score the cylinder walls because the cast iron skirt surrounds it. The total length of the piston is 4.69 inches. Of this 3.5 inches is cast iron skirt. The rest of the length consists of the piston rings and their supporting portion of aluminum. Three rings are fitted in the .187 grooves. The cast iron skirt surrounds the aluminum part of the piston just below the piston rings to the bottom. It is held as light in weight as possible and the surfaces where the parts are bolted together are ground and lapped true so that the assembled portions will be in line. The aluminum may expand and contract without affecting the cast iron skirt. The hardened and ground pins are 3.25 long and 1.187 in diameter.*

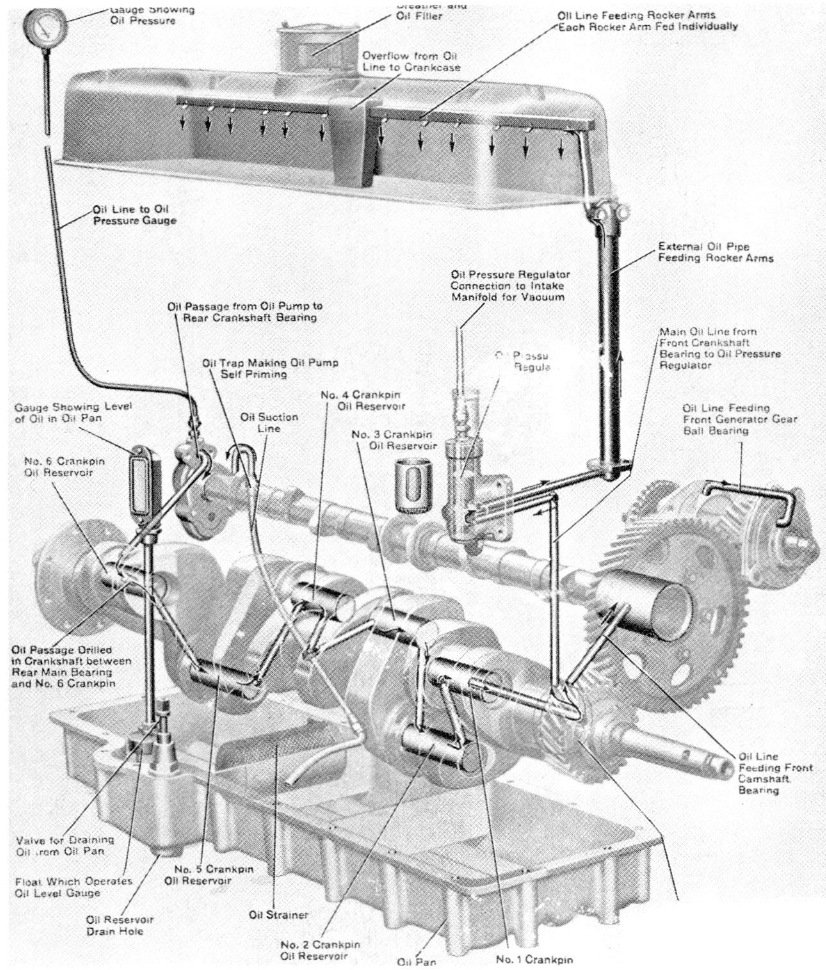


Fig. 4-61 Model 34B lubrication system.

Max Thoms, an engine designer specializing in pistons and rings, was probably responsible for the two-piece piston used on the Model 34B which received wide acclaim.

The piston had a .06 inch annular gap beneath the head which functioned to scrape oil from the cylinder wall into a duct feeding the wrist pin bushing.

Piston and connecting rod assemblies were held to plus or minus 0.5 ounce by machining an 8 pound 12 ounce forging all-over to a finished weight, less bushing

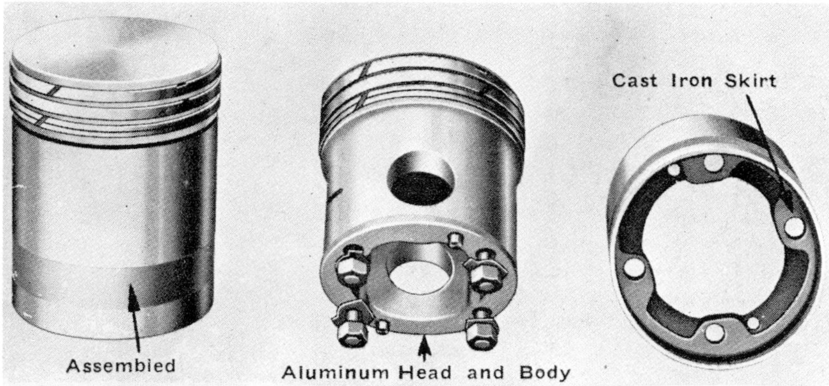


Fig. 4-62 The famous two-piece piston — possible the most widely reproduced component illustration of the period.

and 4 bolts, of 3 pounds 14 ounces. The rods were 10.125 long, center to center, with a uniform H — section of 1 inch by .625 inch with a .125 wall. They were forged from nickel alloy steel. The caps were retained by four .312 diameter chrome nickel steel bolts.

Valves were slightly larger, 2.0 diameter instead of 1.9375.

The valve mechanism was very light. Components of the valve train and their individual weight were: Push rod with button — 4 oz.; Valve — 5.5 oz; Rocker arm and screw — 7.5 oz.; tappet with roller and pin — 5.75 oz.; valve spring retaining washer and key — 1.75 oz. Total weight was 24.5 oz. The valves were 0.375 diameter, five inches long. Lift was 0.375 inch. Dual valve springs of 65 and 45 pounds load were used. The large valves precluded water between combustion chambers — on the engine centerline.

The means of rocker arm support and lash adjustment was new; the 34 A had a rocker arm shaft and adjustment screws in the rocker arm over the push rods: *"The valve mechanism is quite a departure from conventional design. It permits valve adjustment — while the engine is running; by placing the point of clearance take-up at the fulcrum of the rocker arm instead of at the valve stem, the entire mechanism is lightened, the inertia of the parts reduced, and quiet operation and long life result."*

When the industry went to overhead-valves and V-8s beginning in 1949, the Cadillac and Oldsmobile had conventional rocker-shaft construction. But when the Chevrolet and Pontiac appeared in 1955 they had stud mounted rocker arms . . . stamped steel arms with a spherical seat and a hemispherical "washer" below the adjustment nut. The Marmon innovation had not been completely overlooked. Charles Chayne had become General Motors Vice-President in charge of the corporate Engineering Staff; Chayne was a graduate of MIT and the Nordyke & Marmon Executive Training Program in the Model 34 era. His last patent was a hydraulic lash adjuster integrated into the stud-mounted rocker arm: the patent illustrates application of the invention to a spherical seated arm with oil supply through the stud (Pontiac) and with oil supply through a hollow push rod (Chevrolet), but it also shows application to a cylindrical-bushed, stud-mounted rocker arm (Marmon 34 B)! Later the Cleveland-built Ford 350-C engine appeared with cylindrical-bushed stud-mounted arms (but with usual hydraulic lifter tappets).

The rocker cover incorporated an oil distribution tube which drip fed the rocker arm bushings via funnels formed in the arms.

A number of features were continued, but redesigned, from the 34 A: roller tappets installed from below, concave cams integral with shaft and flange, aluminum water pump on the crankshaft nose.

The timing gears were relocated to a conventional outboard location; they were helical and 1.75 wide. The camshaft gear had a fiber rim on a metal hub, and was bolted to a 3.5 dia. flange integral with the camshaft.

Valve timing was as follows:

| | | | |
|---|---|---|---|
| Intake | open | 19° | after top center |
| | close | 35° | after bottom center |
| | duration | 196° | |
| Exhaust | open | 45° | before bottom center |
| | close | 12° | after top center |
| | duration | 237° | |

The camshaft was 1.375 diameter with three instead of four bearings. The front bearing was 2.37 dia and 1.812 long, the middle was 1.375 dia and 1.875 long, the rear 1.375 dia and 1.50 long. The front end was bored out for weight reduction. The middle bearing was innovative. It was in two halves assembled together with two screws, in counterbores so that the round exterior was not violated. The shaft with the bearing in place was then inserted in the engine and the bearing rotated by hand until a bolt dowel from the outside of the crankcase could enter the dowel hole. This exact concept was adopted in 1938 for the very first Detroit Diesel Series 71 engine and has been used ever since; it differs only in retention of the two halves by means of a pair of wire snap rings in appropriate grooves. The end bearings were pressure lubricated by drilled passages from the main bearings; the middle bearing by splash only. Crankcase capacity was slightly less than the 34A . . .

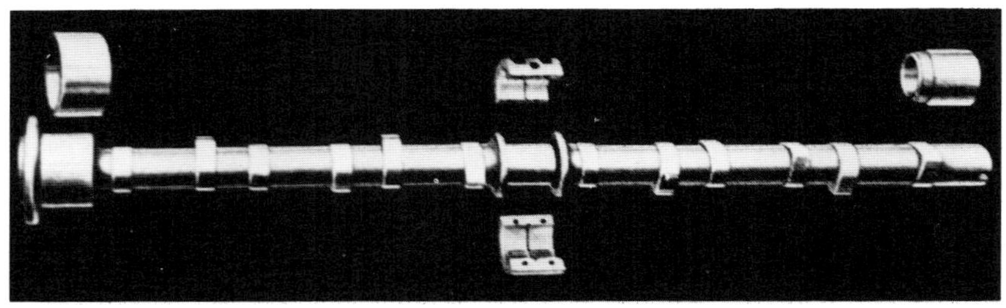

Fig. 4-63 Model 34, 74, 75 camshaft and bearings.

twelve quarts.

The 34A engine incorporated a cast-in oil gallery in the crankcase. The 34B relocated the oil pressure regulator to the front of the engine and pumped the full oil pump delivery through .375 diameter passages for the full length of the crankshaft. A large circumferential groove, .25 wide and .31 deep, in the rear and front main bearing shells conducted the oil into and out of the crankshaft. This large oil flow assisted in cooling the shaft and bearings.

The quest for optimum means of oil pressure modulation was ended with incorporation of a vacuum-modulated oil pressure regulator. A separate spring-loaded

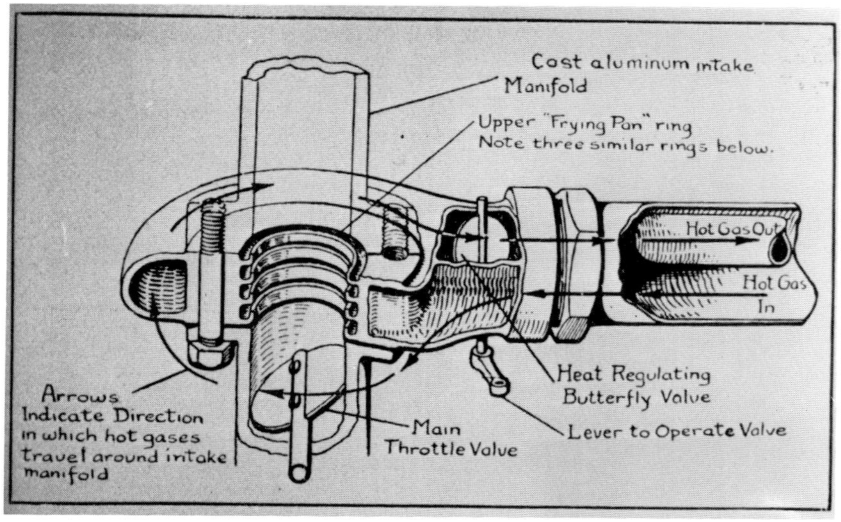


Fig. 4-64 Carburetor heating system to assure vaporization.

ball valve limited pressure to 30 psi. Operation of this valve has puzzled Marmon restorers; The original description[35] of its function is repeated here:

## VACUUM REGULATOR

The vacuum regulator in construction is a two-headed piston operating in a cylinder whose ends are closed by caps. The upper cap is provided with a connection to the intake manifold — in this manner the vacuum of the intake manifold is communicated to the top end of the cylinder and tends to suck the piston upward. This tendency of the vacuum to suck the piston upward is, of course, resisted by gravity, but in addition a coiled spring has been provided.

The bottom end of the automatic valve cylinder is closed by a cap with a projection so that the piston valve will be held, when in its lowermost position, at a definite distance from the bottom of the valve cylinder. The automatic valve piston is made from a single steel bar, but is in appearance a double piston. The upper piston is connected to the bottom piston by a 5/16 inch stem and the space between the two is 7/16 of an inch. The lower and upper pistons are both of the same diameter, namely, 1 inch, while their lengths, top and bottom are respectively 11/16 and 5/8 inches. The upper piston is hollow to carry the return spring and has five grooves turned on its outside to hold oil and thus keep it packed.

The action of the automatic piston valve is very simple to follow. There is an annular groove turned in the automatic valve cylinder. This communicates with the pressure line of the oil system. When the vacuum in the intake manifold is low, as under full load or full throttle opening, the valve is in its lower most position and the annular port is covered by the lower portion of the piston valve. Hence no oil can escape past the piston valve and the maximum pressure of oil is obtained upon the bearings, subject, however, to such relief as is obtained through the ball safety valve located adjacent to the automatic valve. As the vacuum in the intake manifold increases the suction increases on the piston valve and it is raised until the suction is balanced by the spring pressure. In this position it may have been raised enough to uncover the lower edge of the annular opening. When this happens oil escapes or is by-passed from the pressure line of the lubricating system, and this overflow is conducted to a pipe line conveying it to the overhead valve operating mechanism. Thus, for full throttle opening, i.e. when the engine is developing maximum power and doing its greatest work, the vacuum is low, the spring holds the piston down — the by-passage is closed and the oil is being forced to the main and connecting rod bearings under high pressure. When the throttle is closed the vacuum is high, the piston is sucked upward, the by-passage is open and the oil passes through the crankshaft with little resistance, the oil is being forced to the main and connecting rod bearings under low pressure. In other words, the vacuum regulator so adjusts the oil supply that it is furnished in direct proportion to the needs of the engine.

In contrast to the ordinary pressure system, this method does not supply excessive lubrication with the engine running light at high speed, neither does it fail to supply adequate lubrication when running under load at low speed.

These are both recognized faults of the conventional system; the former resulting in fouled plugs and a smoky engine, the latter in a heated motor and burned-out bearings. The Marmon vacuum control has corrected both deficiencies.

A DELCO (Dayton Electrical Laboratories CO.) electrical system was adopted. The system was 6 volt with positive ground. The generator, distributor, coil, condensor and cut-out were packaged in a single unit which was retained in a saddle mount on the crankcase by a dowel and steel strap, and was driven via an Oldham type coupling. It was a handsome design with the distributor housed in the generator end plate and with both the ignition coil and the condenser flange mounted thereon. The generator was third-brush regulated with four poles and a maximum of 15 amperes output.

The wiring was protected by two circuit breakers housed in a single enclosure on the cowl face. One protected the hom, tonneau and trouble-light wiring and opened the circuit until corrective action was taken. The other reduced the current to the main lighting system by vibratory interruption.

The distributor used two sets of breaker points wired in parallel. Both were opened simultaneously by the 6 lobe cam driven at half engine speed. The dual points were claimed to provide a greater factor of reliability and to improve ignition.

The starting motor was of four pole, series wound type and was equipped with Bendix Drive. The pinion was mounted on a threaded sleeve and was driven through a torsion spring. When the motor started to rotate, the inertial resistance of the pinion caused it to be screwed into mesh with the flywheel gear (integral with the flywheel). Upon engine start-up, the pinion was overdriven and screwed out of engagement. A toeboard switch energized the starter motor.

Overseas orders were fitted, only upon request, with a magneto similar to the Model 34A. It was driven from the rear of the generator and was mounted on a bracket attached to the three bosses on the crankcase which were left unfinished on domestic engines.

Marmon assessment of the relative merits of the two systems was stated in the brochure for overseas dealers: "We strongly recommend however, that the standard Marmon ignition of the Delco type be used instead, as our experience has taught us that the Delco system of ignition is superior to the magneto." Superior or not, it is probable that some domestic cars were ordered with the magneto by parties aware of its restricted availability and unconvinced of the merits of the Delco ignition.

 

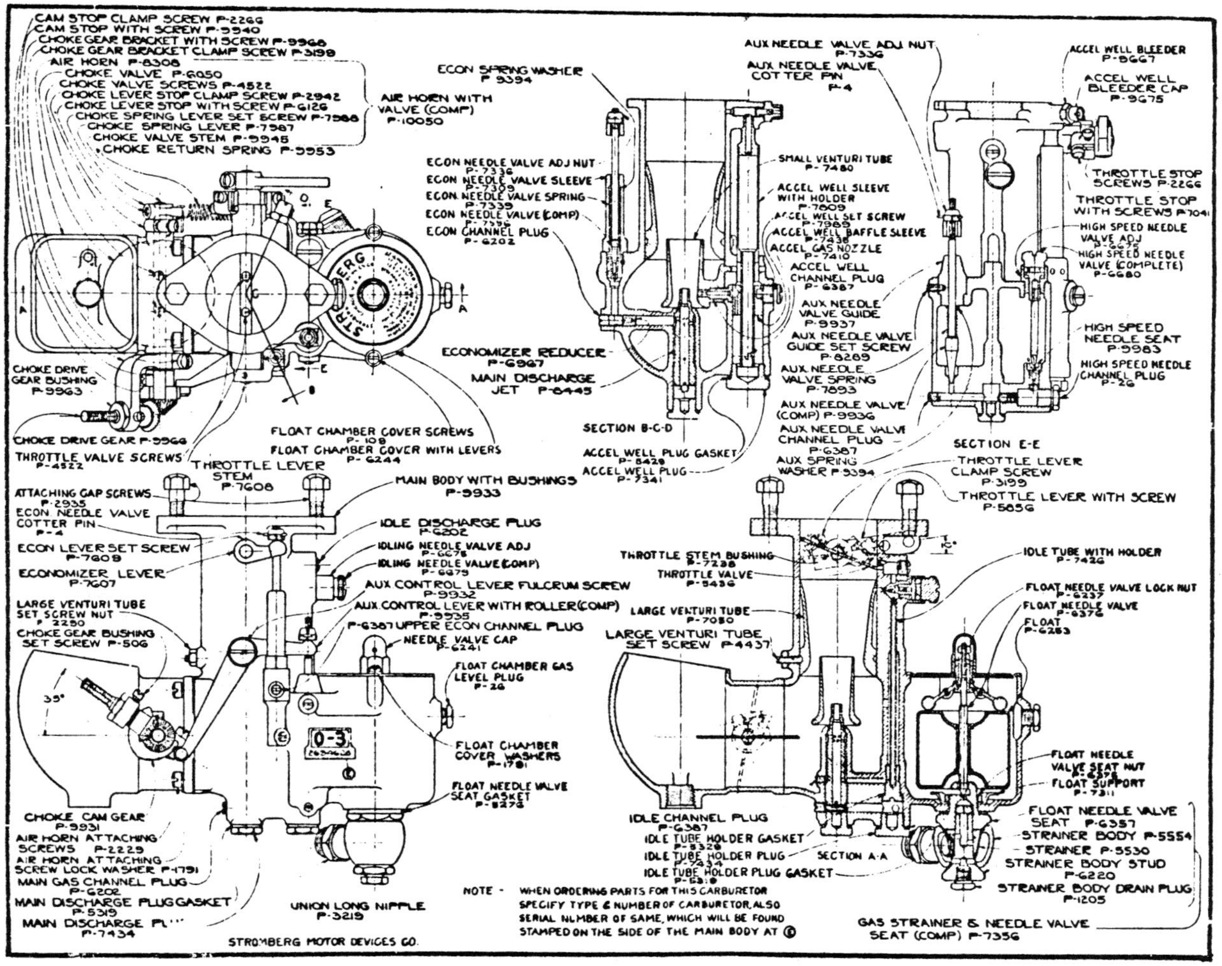


Fig. 4-65 The Stromberg, 0-3 carburetor

By 1920 the force feed lubrication system invented by Howard Marmon in 1903 had spread to one out of three cars made in the USA. That was also the year when the US motor vehicle population equaled the combined horse and mule population . . . 27 million of each.

Each Model 34B engine was block tested for seven hours, torn down and reassembled. The completed chassis was driven 50 miles and the finished car 5 miles. Details of the test procedures are reported in Chapter Eleven Part 2.

## MODEL 34B
## 1921

The integral water jacket was eliminated from the intake manifold. The mixture heater (hot spot) was redesigned. In place of the coaxial to/from exhaust connector, the exhaust return was by flexible metal pipe rearward from the heater and downward to a juncture with the exhaust pipe.

The cast horizontal 'tee', sandwiched between the manifold and exhaust pipe, for feeding the mixture heater was redesigned to an integral 90 degree elbow and inverted vertical 'tee'. This presumably was done to direct the main exhaust flow toward the heater before turning sharply downward.

The Stromberg carburetor was changed from a Model H to a Model 0-2 and later to a Model 0-3.

The accessory drive axis was made adjustable in June 1921 to accomplish more precise gear lash and ignition timing.

Tappet roller shafts were increased from .437 to .500 diameter. Steel fan blades replaced aluminum blades.

## MODEL 34B
## 1922

At engine 5970 a change was made in the oil feed to the rocker arms. Previously the distributor tube was in the rocker cover; the redesign used a box section tube supported on a pair of studs and fed through a tube projecting upward from the head. The lube system was therefore unaffected by rocker cover removal.

## MODEL 34B, (34C)
## 1923

A number of engine changes were included in the "Sixteen Improvements" announced in the mid-1923 model year.

After 6500 Model 34B and 34C cars, another redesign of the mixture heater went into production. Parallel tubes, one inch in diameter, carried exhaust to and from the heater with flow controlled by a one inch throttle valve in the heater inlet. The tubes were welded to a curved plate which was clamped to the exhaust pipe. The pipe was decreased in size from the former 3.0 inch diameter to 2.5 and a new exhaust manifold was required. The tube ends projected into the exhaust stream where each was formed into a box-like end with the inlet pointed upward and the outled downward; subsequently a die formed downward reflector was spotwelded to the curved plate.

It is worthy of note that a nearly identical system was used on the Chrysler V-8 engines which featured ram induction from a pair of four-barrel carburetors. The high-flow cylinder heads did not allow for the usual exhaust cross-over passage for mixture heating. Instead the manifolds were joined by a cross-over pipe at the rear with a pair of tubes running forward to the heater and having the same configuration at the ends as the Marmon.

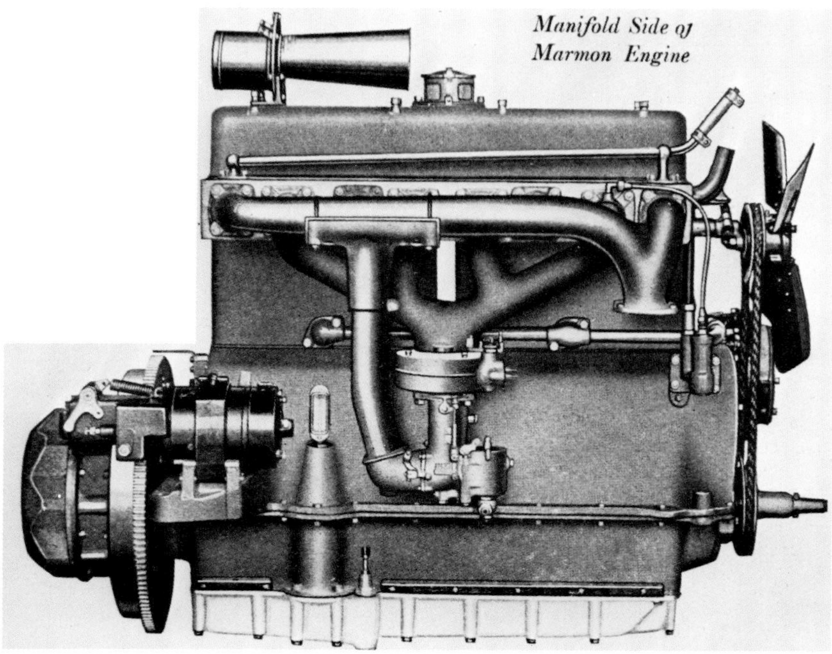


Fig. 4-66 The engine appearance after the numerous changes progressively made during 1923. The final configuration of the engine under the 34 B designation.

The new mixture heater was, as previously, an aluminum casting with an offset mixture duct and trough for catching liquid particles. It differed in being designed to accommodate an insulating jacket. A pair of spun steel covers enclosed a 0.4 inch thick insulation blanket comprised of two circular parts and a circumferential strip.

Intake air was ducted to the carburetor inlet from a 'stove' on the exhaust manifold regardless of ambient temperatures. There were concerns at that time with reduced volatility of available gasolines, consequent to a perceived onset of exhaustion of supply. The duct tube was bent so that a 180° rotation uncovered the carburetor intake when more volatile fuels were used in hot weather.

A change in camshaft contours was made to improve valve action and reduce noise. The new shaft required different valve springs. The new outer spring had a load at 1.69 length reduced from 75 to 62 pounds. The new inner spring had a load, at 2.09 length, reduced from 35 to 30 pounds. The exact timing of this changepoint is not certain.

New and more compact DELCO electrical components were adopted during 1923 model production. The compactness was consequent to a change to two-pole machines. Crankcase saddles were revised accordingly; cast iron shims were temporarily used. Maximum charging rate was increased to 18 amperes. The generator end plate mounted a distributor assembly, pivotable for ignition manual advance. The ignition coil was mounted separately on the engine block and the condenser was internally mounted in the distributor. A resistance coil was mounted externally on the distributor for the purpose of limiting battery drain should the ignition be left 'on' when the engine was not running, and to "provide more nearly uniform spark at different engine speeds." Single breaker points were used in the new distributor. The new coil had a surprising resemblance to the coil configuration which because common in the early 1980s.

The new starter was a two pole, series wound motor which incorporated a "Marmon ingeniously designed over-running clutch drive"[36] and integral switch replacing the Bendix drive. A floor pedal and linkage replaced the floor switch and harness. The significance of this development warrants repetition of the description provided in the owner's Information Book Number 2434: "A mechanically operated pinion shift is employed. The starter switch is incorporated in the rear housing of the starting motor. When the pedal is depressed, the pinion meshes with the flywheel teeth before the switch contacts close, causing the motor armature to revolve. The adjustment of the length of the switch plunger should be such that the distance from the pinion to the over-running clutch face should be 0.125 to 0.187 inches when the electrical circuit is completed. When the driver releases the pedal, a stiff spring returns the pinion to the out-of-mesh position. To provide for the few occasions when the pinion fails to mesh due to the teeth butting squarely, a stiff compression spring steel cup housing (shell) attached to the pinion is mounted on the splined shaft between the pinion and the operating lever. When the pinion does not mesh, the control lever moves on, compressing the spring until the motor contacts close. Then with the first slightest motion of the pinion shaft, the pinion moves enough to clear the flywheel teeth and the compressed spring instantly forces it into mesh before the motor can turn fast enough to cause clashing or stripping of the gear teeth.

The starting motor pinion shaft is provided with an over-running clutch between it and the armature shaft, so that the moment the engine starts running on its own power, the pinion shaft over-runs and prevents the armature from being driven by the engine at dangerous speeds. The pedal should be released in the shortest reasonable time to conserve the lubricant in the over-running clutch. There is also a brake at the inner end of the splined shaft which causes the shaft and motor armature to come to rest almost instantly when pressure is removed from the pedal. This feature makes it unnecessary to wait several seconds for the starter pinion to come to rest before making another attempt to start an engine which has started and then stopped.

The intake manifold tap for vacuum to the oil pressure regulator was relocated to the center of the manifold from the forward leg, probably due to the advent of optional vacuum wipers.

The water pump was relocated from the nose of the crankshaft to a forwardly extended accessory drive shaft. The fan drive pulley was interchanged with the pump, and the fan bracket and hub were redesigned to place the fan forward of the belt rather than behind.

The new pulley was a slip fit on the crankshaft with a bar-key drive having a surprising amount of side play in the keyway. In the event of inadequate lubricant in the fan hub, the slip-stick characteristic of impending seisure produced a hammer-like noise at the pulley key, alerting the driver to the need of immediate shutdown and correction.

There had been a persistent problem with leakage of the crankshaft mounted pump. During production of 6500 cars there had been four different pumps used. The first and fourth were essentially identical and were characterized by large triangular packings sealing on the outside diameter of the impeller hub extensions. The second design doweled the packing rings to the adjustment nuts and they acted on slightly tapered hub surfaces. The third design used semi-face seal packings also doweled to the adjustment nuts. It was ultimately determined that the problem had its origin in deflections of the massive crankshaft on which the impeller was mounted. Relative movement with respect to the pump body which was mounted rigidly to the crankcase resulted in packing failure. The solution was ultra-simple and so effective that the original seal design was satisfactory. Coil springs were placed on the studs mounting the pump and the nuts tightened to a controlled spring length. The pump then followed the miniscule orbital motions of the crankshaft nose without damage to the packing! It is probable that the relocation of the pump had been committed to production before the problem solution was identified.

## MODEL 34B, (34C) 1924

Another engine change came unheralded during the 1924 model run. The stud-mounted rocker arms were replaced by conventional shaft-mounted arms. Several modifications to reduce side wear on the arm guides had not proven successful; a sound similar to excessive tappet clearance had resulted. The new shaft provided pressure lubrication to the rocker arms; pressure was limited by a discharge orifice and baffle at the end of the shaft.

The oil pressure gage was replumbed to sense oil pressure at the regulator body instead of pump discharge pressure. The new measurement point was **after** the main and rod bearing galleries (in the crankshaft).

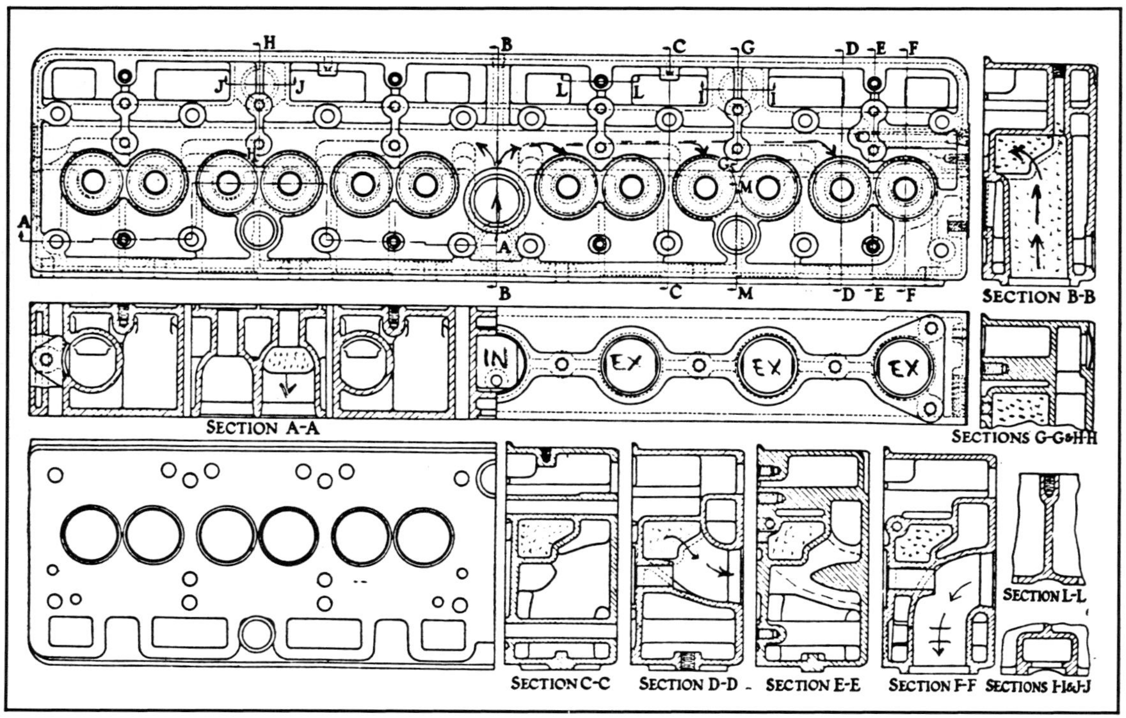


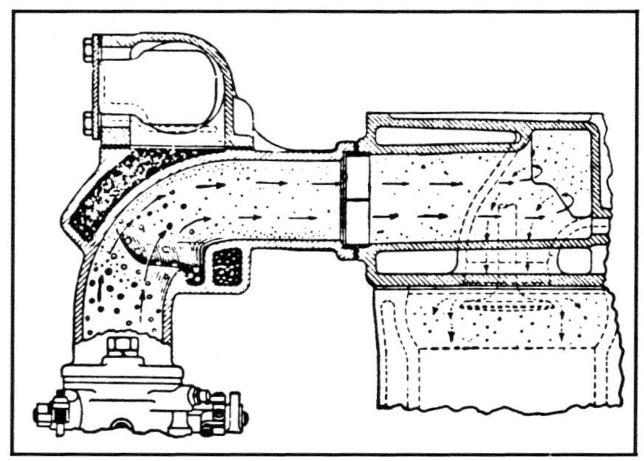

Fig. 4-67 Engineering drawing of the cross-flow cylinder head with individual exhaust ports and integral intake manifold.

Fig. 4-68 The integration of the intake manifold simplified the exterior of the Model D-74 engine. Note the relocated and redesigned water pump and fan drive pulley introduced mid 1923.

## MODEL D-74
## 1925

An entirely new cylinder head and manifolding were introduced. The intake manifold was integrated into the cylinder head for the purpose of achieving more rapid warm-up with consequent improved fuel economy and of avoiding excessive mixture temperatures in high load and high ambient temperature circumstances with consequent improved volumetric efficiency and power.

Test data published by Chrysler Corporation in the 1940s confirmed that normal fuel economy was not achieved until conventional intake manifold temperature had reached normal operating temperatures. The tests revealed that this could require up to 20 miles in winter weather despite normal oil and coolant temperatures being achieved within five miles.

The Models 32 and 48 had featured integral intake manifolds; the Model 41 had adopted a water jacketed carburetor. The D-74 engine applied this traditional Marmon feature to the model 34 engine but in a new configuration appropriate to overhead valve engines with detachable cylinder heads. Similar designs would later be adopted by Nash for the Ambassador Six (1940) and for the Willys Four (1950).

Chevrolet adopted an integral manifold for the 250 CID engine in the early 1970s. The General Motors 215 CID aluminum V-8 of 1961-63 would feature a full-flow water-jacketed intake manifold and the practice would become widespread during the EPA regulatory era.

The carburetor fed the cylinder head through an abbreviated aluminum intake manifold which was simply an elbow. It was jacketed for exhaust heating and interfaced with the exhaust manifold on the top. A trough on the inboard side of the passage collected liquid fuel droplets for vaporization.

The inlet elbow fed a transverse passage (Sec. B-B) which took advantage of the wider cylinder spacing over the center main bearing to convey the mixture to the distribution manifold located within, but on the opposite side of the head to the exhaust ports. The new head was, to use a current term, a 'cross-flow' design. The close proximity of the intake duct to the individual valve ports (Sec. D-D) made the rail type manifold feasible from an equality of distribution perspective.

The cross flow arrangement made individual exhaust ports feasible. The new six-leg manifold was an iron casting of smaller cross-section than the previously used tubular design. Both head and manifold had machined counterbores for alignment sleeves.

The Stromberg 0-3 carburetor was continued. Due to relocation of the fuel tank to the rear of the chassis, a Stewart-Warner vacuum feed system was adopted.

## MODEL D-74
## 1926

Two major improvements were made in the engine for the 1926 model year: Dual ignition was adopted and an oil purifier system was added.

### DUAL IGNITION

The Champion Spark Plug Company sponsors at intervals a world conference of automotive engineers concerned with ignition and combustion. At the conference held at Geneva, Switzerland in 1971, dual ignition by diametrically opposed spark plugs generating two opposing flame fronts was acknowledged to offer a 5 percent fuel economy advantage. Mr. R.D. Miller of Champion Spark Plug Co. (USA) stated: ". . . with dual ignition systems using two plugs, you can run leaner and run where a single spark plug would produce missing. Dual ignition can produce a drivable motor car with lower carbon monoxide than single ignition". Because of this recently expressed technical position, it is appropriate to describe the adoption to the '74' of such a system in the words of the Nordyke and Marmon press release of September 1925: *"In the mid 1920s, Marmon and other engineers were still confronted with three sources of constant difficulty . . . crankcase oil dilution, spark-knock and power deficiencies due to insufficient heat to sustain the expansion of the charge (i.e., combustion) at lighter loads. Efforts were initially directed at the improvement of combustion primarily for the purpose of reducing crankcase oil dilution at lighter loads. Attention was directed to improving the manifolding of both inlet and exhaust gases . . . the result was the new Model 74 cylinder head with crossflow from an integral water-cooled/heated intake to six individual exhaust ports. Initially the dual opposed ignition was incorporated to provide means whereby the fuel could be burned more quickly and completely and thereby diminish crankcase dilution. However the actual result was reported to be quickly evident to the driver . . . improved acceleration and remarkable lugging ability."*

With the car operating at a steady speed of 25 mph, cutting off the ignition to one set of plugs resulted in a reduction in steady speed to 19 mph . . . a fair measure of the improved economy at light loads.[37]

The Marmon patented oil purifier was a subsequent development which "practically eliminated" crankcase dilution. The dual ignition was however released for production because of the improved drivability experienced.[37]

This was the first application of the old idea of two point ignition to the overhead valve engine. A single twelve wire Delco distributor was used, with twin coils con-

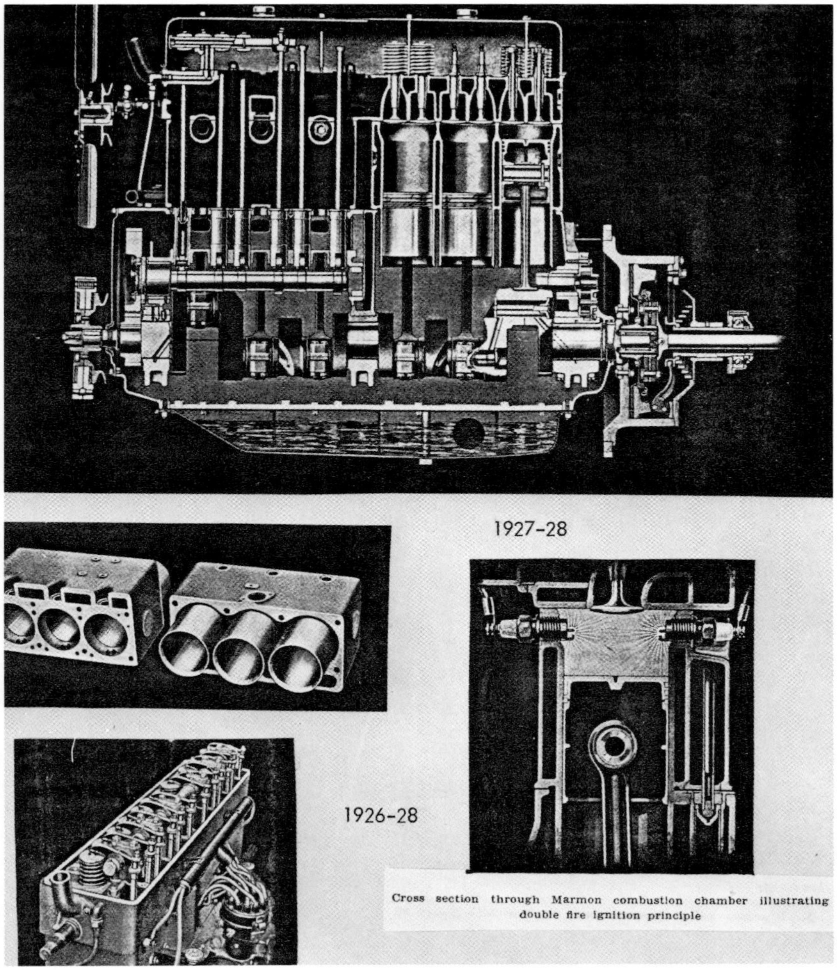


Fig. 4-69 Longitudinal section of Model E-75 engine. The modified crankshaft mounts a vibration damper at the front and a new single-plate clutch at the rear; otherwise representative of the Model D-74.

nected, one centrally to the cap and one offset. A single cam operated two sets of breakers. Marmon described the Delco distributor as remarkable for its compactness and workmanlike design, including interchangeability with the former 6 wire unit. The leads to the additional plugs were routed directly from the distributor through the gap between the forward and aft cylinder blocks, maintaining the traditional neat appearance of the engine.

A reputed disadvantage of dual ignition of opposing flame fronts was a tendency to detonation at higher power levels, requiring some sacrifice in rated power to achieve the efficiency advantage.

The Japanese Nissan firm began using opposed dual plugs on its NAPS-Z

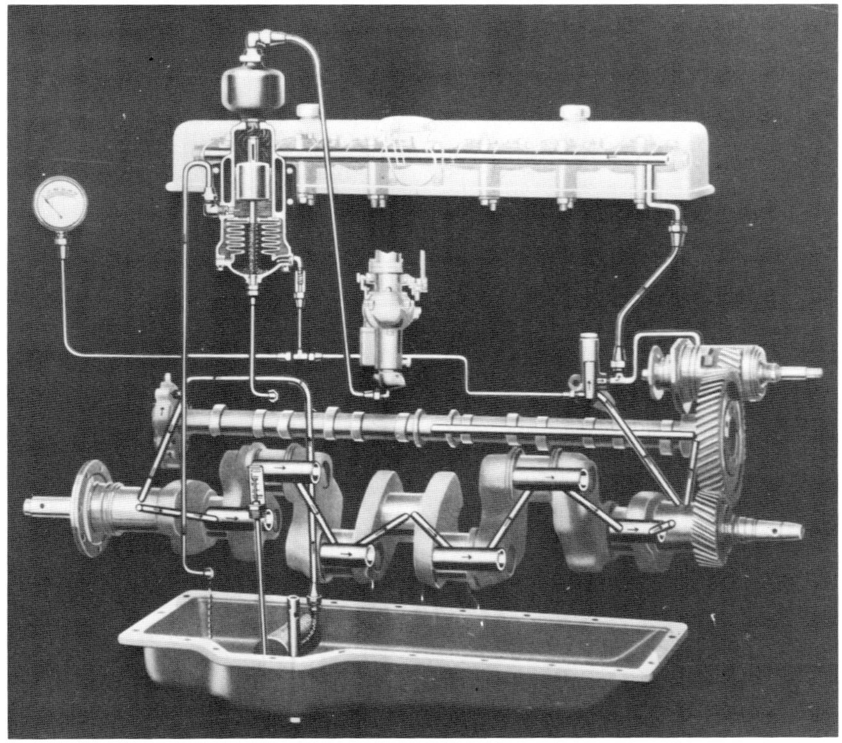

Fig. 4-70 The lubrication system of the E-75 with pressure oil to the middle camshaft bearing and the oil rectifier plumbing.

engine for Datsun cars sold in California during the 1980-81 model years as a means of offsetting the fuel economy penalty normally resulting from the more stringent California emission control regulations. Beginning with the 1982 model, the second set of plugs is electronically switched off at high loads thereby avoiding any power limitation from dual plug use.

## OIL PURIFIER

The "Marmon" Three-Way Oil Purifier was actually the Wall Oil Rectifier, a development of the Rectifier Mfg. Co. of Chicago[38] adopted exclusively by Marmon for one year. It was a device which combined filtering with distillation.

Automobile Topics[37] described the function of the oil purifier: *"The three-way oil purifier, which is an exclusive development, removes by filtration all ponderable solids, eliminates water and the more volatile constituents, which are capable of being burned in the cylinders and finally restores the clarified lubricant to the crankcase. It does all this automatically, and is so arranged that should its operation for any reason cease, the flow of lubricant to the bearings would be in no way impaired, but the oiling system proper would function as well as it has ever done. Its operation depends upon the principal that water, which is a by-product of combustion and therefore always present in the exhaust, can be driven off only at temperatures at or above its boiling point, and also that filtration is more successfully*

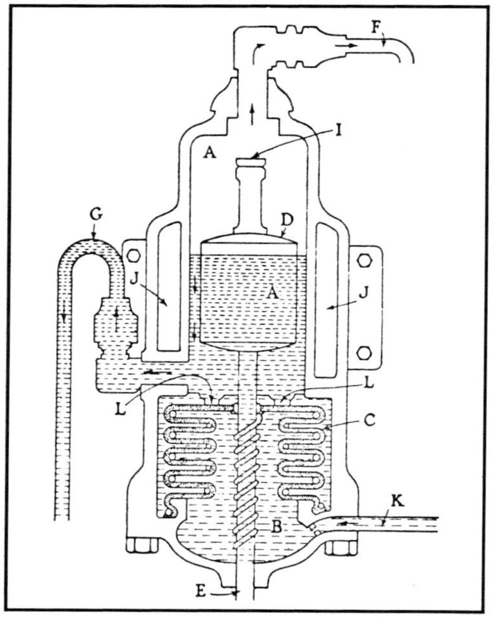


Fig. 4-71 Oil purifier device adopted for the Series 75.

*conducted when the oil is heated, rather than cold, and therefore, so to say, more fluid. The device is attached to the exhaust manifold and operates continuously and automatically. A considerably higher temperature than prevails in the crankcase, together with a partial vacuum in the appliance combine to assist both filtration and distillation of the oil. The purified oil is restored to the crankcase intermittently whenever the distillation chamber becomes filled. The distillation chamber "A" in the diagram is fed from the supply chamber "B" through the asbestos filtering sack "C", which frees it of all sizable particles of foreign matter, especially solid carbon and abrasives. This lower chamber is fed by constant flow through the pipe "K" from the oil pressure regulator on the engine. Exhaust gas from cylinders No. 5 and 6 circulates through the jacket 'J' maintaining a relatively high temperature in the distillation chamber at all times. This causes the more volatile constituents of the oil to be evaporated, the vapors passing off at the top and reentering the engine through the pipe "F", which is connected with the carburetor. Filtering and distillation are thus promoted by the suction in the manifold. Less volatile constituents in the oil accumulate in the distillation chamber until it is filled level with the reverse bend in the outlet pipe "G". The float "D" simultaneously rises and at the end of its travel automatically opens a valve "I", which admits to the upper chamber air from the crankcase, which arrives through the tube "E". In this way the vacuum in the chamber is broken, whereupon the tube "G" becomes a syphon, drawing off the entire contents of the chamber. As the chamber is thus emptied the float falls closing the air-admission valve and restoring the vacuum, permitting distillation to be resumed. This completes the cycle of the device. Once in every 2,500 miles it is recommended that the filter sack be removed and cleaned in gasoline or renewed. There is provision for winter and summer regulation of the exhaust heat."*

## MODEL E-75
## 1927-28

Changes to the engine included adoption of an air cleaner and addition of two vents in the rocker cover (instead of one) to reduce corosion of the valve mechanism. Valve springs and cam contours were modified for improved follow and reduced noise.

The Marmon full-pressure lubrication system was improved for 1927 and incorporated some additional innovation. The camshaft was drilled for the purpose of conducting oil from the front camshaft bearing (pressure fed from the front main bearing) to the middle-bearing previously only supplied by splash oil. This change made every bearing in the engine pressure lubricated. Because the connecting rods were no longer machined over their entirety, the oil tube was replaced by a rifle drilled hole in the modified H section forging.

An air cushion chamber was added to the pressure oil system. The purpose of this was described as prevention of excessive pressure in the oil pump when the engine was first started in cold weather. When the engine was first started there would be a sluggish flow of oil from the pump through the crankshaft and via the front main bearing to the air cushion chamber, and oil will rise in the chamber compressing the air. "A certain time will then elapse during which oil can start through the oil relief valve and begin to flow through the other lines to the overhead rocker shaft and accessory drive shaft."

The front pulley on the crankshaft incorporated a split flywheel torsional vibration damper. . . a pioneer application. Marmon called it a "modulator". The damper was designed to slip at 35 to 40 pounds pull on a belt wrapped around it.

Light weight cast iron pistons were adopted. Oil sump capacity was 10 quarts.

A Schebler Model S Carburetor replaced the Stromberg 0-3. The Schebler was an air-valve type with an accleration pump whereas the Stromberg was a plain-tube type with an acceleration well. Both were 1.5 diameter single throat designs.

The vacuum modulated oil pressure regulation feature was eliminated.

A standard DELCO-REMY starting motor, including pinion drive and externally mounted switch actuated by the engagement lever, was adopted. It was retained in the crankcase saddle by a rugged cast semi-hoop attached with a pair of bolts.

## PART 5 — The Chasses of Era Three
## Models 34, 74, 75
## 1916-1928

The Marmons of ERA THREE were basically built upon the same chassis. The changes which characterized the several models were evolutionary rather than revolutionary. Changes in tire equipment, the advent of four wheel braking and remote fuel level sensing, and improved road surfacing all precipitated local redesigns of the chassis. There was a continual effort to enhance the isolation of the driver and passengers from road and engine vibrations.

Although the five versions of the same basic vehicle that was the product of the Third Era are addressed individually in Parts 1 and 3 of this chapter, the progressive development of the chassis throughout the Era warrants a single package treatment.

Therefore the chassis components for the entire Era will each be addressed fully, eg., the Era Three frame will be fully discussed as used in the Model 34A, followed immediately by the changes for the several other models.

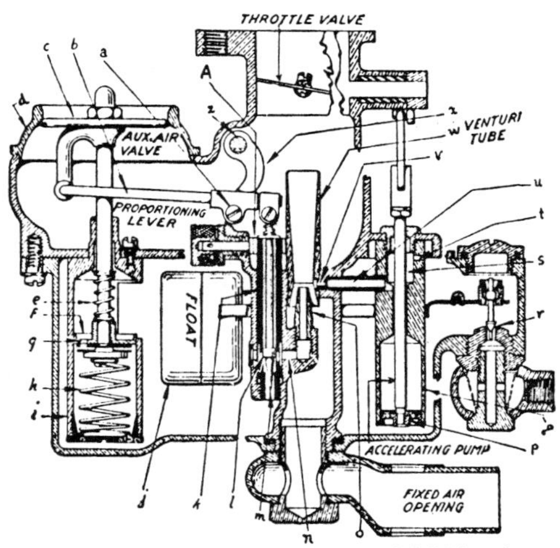


Fig. 1. Sectional view of Schebler model "S" carburetor, showing the relation of the various parts and the adjusting members. In order to illustrate plainly the various parts inside the carburetor, it has been necessary to show them somewhat distorted.

**Action of accelerating pump:** When the throttle is opened, the piston (**p**) of the accelerating pump is raised, lifting gasoline into the upper chamber (**s**). From here, the gasoline flows through the passageway (**u**) to the venturi (**w**), where it mixes with the air entering the carburetor.

No gasoline at all will pass through (**u**) except immediately following an opening of the throttle; so that the accelerating pump acts only during periods of acceleration.

**The capacity of the pump** is large compared to the amount of gasoline used for accelerating. This is for a definite reason. When the throttle is opened just a few degrees from the idling position at low speed, the engine will be running under full load; and a full load charge of accelerating gasoline will be needed. The accelerating pump must therefore be able to deliver the full load charge of gasoline with just a few degrees of motion.

**When a very rich mixture is needed,** as in starting a cold engine, the dash-control lever (**F**), (Fig. 3) is pulled down by means of the wire (**E**). This will rotate the fulcrum lever (**x**) (Fig. 1) about the shaft (**z**) toward the air valve (**c**). This will give the gasoline needle an initial lift plus an increased percentage of the motion of the air valve.

The velocity of the air through the venturi (**w**) is ample to pick up and atomize whatever gasoline may pass through the nozzle (**o**), so that a mixture many times as rich as normal may be effectively supplied to the manifold of the engine, and starting may be made correspondingly easy.

**As the engine speeds up,** the effect of the initial lift of the needle will diminish, so that the percent of excess gasoline will decrease but not disappear at high speed. This corresponds to the requirements of an engine, as a greater percentage of excess gasoline is required to run a cold engine slowly than fast.

Fig. 4-72 The Schebler Model S air-valve carburetor adopted for the D-74.

Fig. 4-73 Plan view of Series 75 chassis.

The 1926 Model D-74 and the 1927-28 Model E-75 featured a self-lubricated chassis. Because that system involves several chassis systems, it will be discussed following the several chassis components and systems.

## FRAME

### MODEL 34A
### 1916-1919

The Model 34 frame closely followed the Brush patent discussed earlier. It differed primarily in having wood floor boards; the patent envisaged a stamped steel floorpan with a drive shaft tunnel.

The side rail was a channel section for the length of the front spring attachments. The lower flange then rapidly decreased to zero and the angle section, 2x10 deep, continued to about the leading edge of the rear brakes. The angle was changed to a strong Z section by the attachment, by hot riveting, of the running board inner flange.

The running board had an outer flange 2.5 inches deep doubled over on itself for 2.0 inches for additional strength and to serve as a sideguard member in event of accident. The inner flange of the board, and the side rail were curved in the plan

Fig. 4-74 Top and bottom views of the Model 34A chassis. Note the generally smooth lower surface and the bottom enclosure of the forward half.

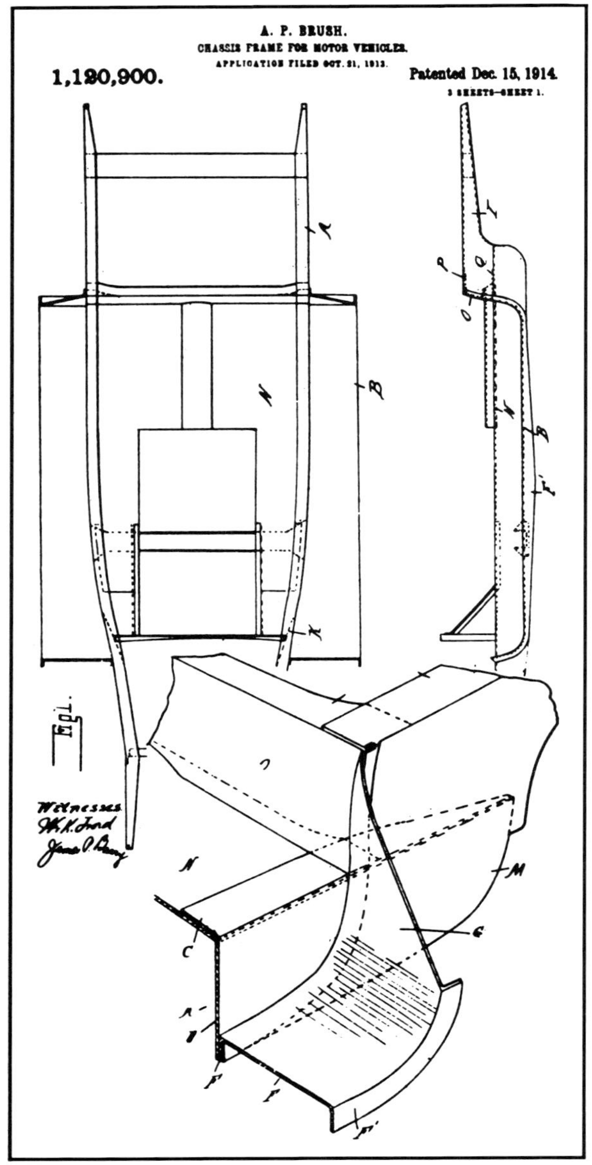


Fig. 4-75 The A.P. Brush patent on which the frame design was based.

view to follow the body contour and eliminate the splash pans normally used for that purpose with conventional frames. In consequence the rivets for attaching cross members and mounting the steering gear bracket were evident on the car exterior; the heads were very shallow and they were external evidence of the unseen novelty within.

The running boards curved upwards at each end, a full 90° at the rear and less at the front, continuing to the elevation of the top of siderail. The manufacture of these 'boards' was a great challenge in 1916 . . . actually in 1914 when the design commitment was made. The double outer flange and the single inner flange continued around the rolled-up ends and the inner flange was curved . . . the width of the member varying from 16 inches to 11 inches, and its finished length about 8 feet. Manufacture was further complicated by the specification of SAE 1025 steel.

The rear construction used a double-deck configuration in lieu of a kick-up and the inner portion of the running board continued up, with reversed curvature, to reinforce the joint. The rear siderail member began at the rear spring hangers as a channel section; the lower flange curved downward ahead of the axle location but to the rear of the uplifted running-board, while the upper continued horizontal to the top of the 'board'. The result was a horizontal channel section of about 8 inches depth at the 'kick-up' area.

The side rails were formed from .094 stock but were reinforced by lamination at areas of higher stress; these were spot-welded as well as being gas welded at the common edges. In some local areas there were two additional layers, each of unique extent. At each body attaching point, a stamping extended from the body bolt to a running-board rivet.

## MODEL 34B
## 1920-1924

The 34B frame differed principally in changes associated with the shorter engine and in having the rear terminal cross-member bowed outward to facilitate spare wheel mounting. The 4-passenger Speedster had a special cross-member.

## MODEL 34C
## 1923-1924

The Model 34C chassis combined the low radiator of the 34B two-passenger speedster and a similar hood with a rear frame and suspension construction necessary to accommodate a rear seat of reasonable depth.

Roads were continuing to improve and it was concluded that the frame "kick-up" could be reduced 2 inches to 4.25 inches. This was accomplished by repositioning the rear section of the side member before riveting, and trimming off the 2 leg protrusion below the side rail which resulted. The upper spring of the transverse spring set was reduced in loaded height by 2 inches. Concurrently the rebound leather straps were omitted.

## MODEL 74
## 1925-1926

The rear section of the side rails was extended to provide a rear bumper attachment and to mount the relocated fuel tank. The forgings at the forward end of the rails were eliminated by forming the full 'horn' integral and adding a welded reinforcement at the spring pivot.

The running boards were changed to eliminate the rounding into the rear fender. A triangular frame out-rigger behind the conventionally shaped rear fender retained and concealed the function of the former round-up. Model 74 fenders were made of steel and did not require the 'dry' joint between aluminum and steel that was a feature of the rounded-up running boards.

The gauge of the frame was increased to eliminate the local laminations required previously and to strengthen the top flange for the relocated attachment of the cross-members necessitated by the decision to eliminate rivet heads on the exterior frame surface. Massive iron castings replaced delicate spring hangers forged from alloy steels and heat treated. Cold rivets were adopted and necessitated thick flanges and spot-faced surfaces on fittings. The engine rear cross member and the torque-tube support cross-member were joined by two parrallel tapered-channel members to increase rigidity.

## MODEL 75
## 1927-28

The forward frame rails were modified to provide for reversal of the spring attachments. A forging was again used at the terminus of the frame horn; it provided for shackle attachment and for direct mounting of the front bumper. With the rear modification effective on the '74', both bumpers were directly attached.

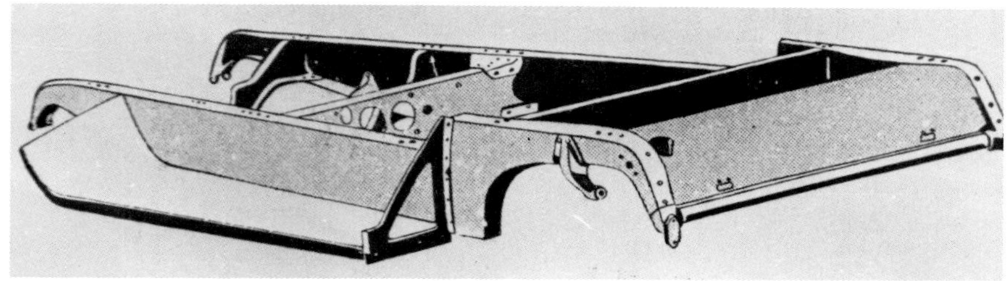

Fig. 4-76 The extended rear and modified running board termination that characerized the D-74 and E-75 frames.

### SUSPENSION

## MODEL 34A; 34B
## 1916-19; 1920-22

The feature which gave the Era Three Marmons freedom from body flexure, superior tractive ability, a remarkable ride for the time, and excellent "tranquil" handling was the rear suspension. This was the patented "Marmon Compound Cantilever" cross suspension. It consisted of two semi-elliptical springs mounted transversely and clamped together at their opposed centers.

The front springs were parallel semi-elliptic with inverted rear shackles. Two rebound leaves were placed above the main leaf which was 39.06 long by 2 inches wide.

Bushings for front and rear springs were dustproof, silent and required no lubrication. Each bushing consisted of steel inner and outer sections with a Thermoid non-metallic bearing strip between. The inner casing of the bushing carried two keys which engaged keyways in the springbolt.

The rear suspension was by "compound cantilever cross-suspension type" located at the rear extremity of the frame behind the rear axle. The opposed semi-elliptic springs were assembled in an 'X' configuration, clamped together at their midpoints. The upper spring was 39.25 inches long and the lower half was 45.0 inches long; both were 2.5 wide. Both springs wre pivoted on the right side and shackled on the left, resulting in essentially vertical movement of the body and providing freedom from sidesway either from side deflection of conventional parallel springs or from deflection-induced motion.

During development of the Model 34 vehicle concept, the rear suspension was a design by Owen Nacker to whom US Patent 1,126,832 was issued on Feb. 2, 1915 following application on June 15, 1914. A half-interest was assigned to his employer, Marmon consultant Alan Brush. In the Nacker design, the central portion of the two springs were clamped together, in the manner in which the single transverse spring of the Ford transverse suspension was clamped. This rigid attachment results in the Ford spring being effectively two quarter-elliptic springs and the Nacker concept four quarter-elliptic springs joined at their common anchorage.

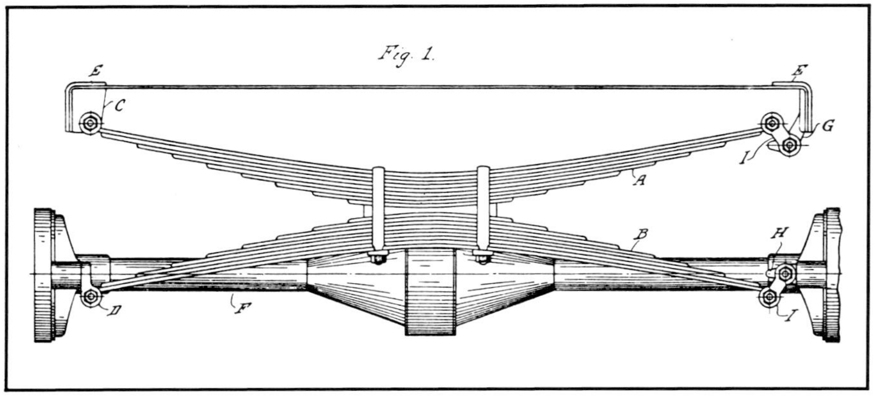


Fig. 4-77 The Nacker patent that influenced the final suspension concept.

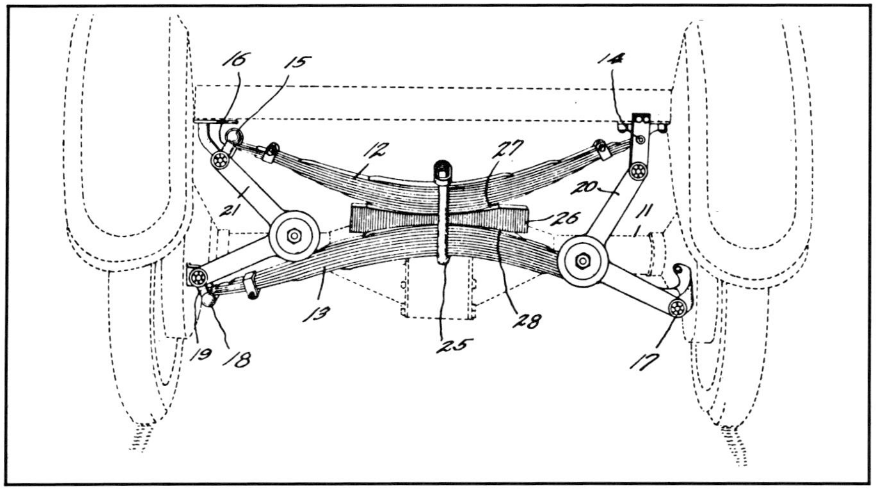

Fig. 4-78 Howard Marmon's patented rear suspension with single U-bolt and cam block for variable rate.

The rear spring also had a desirable response to roll inputs, becoming very rapidly stiffer. This allowed the car to be rolled slightly by a front wheel encountering a bump thereby reducing the harshness; it also allowed the rear axle to pivot when a rear wheel encountered a bump and thereby to effectively cut in half the bump height. Upon entering a sharp turn at speed, the roll stiffness rapidly increased and no unusual body roll was evident.

The Nacker patent rear suspension, and the Brush patent frame with integral running boards subsequently appeared together on the 1920 Saxon; Mr. Brush was obviously a consultant on that car, too.[39] The Saxon runnng boards curled up

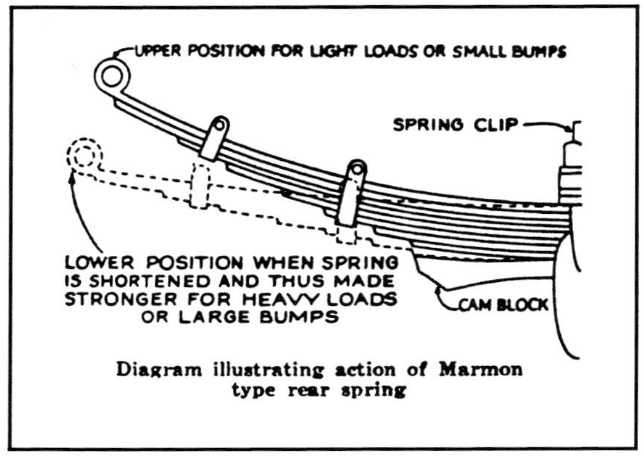


Fig. 4-79 Diagram illustrating action of Marmon type rear spring.

at both ends like the Marmon 34; this signature of the Brush frame was also evident on the Lexington and the Yellow Cab.

The final design released for production used a single large U-bolt clamp at the midpoint. This utilized the full length of the spring, achieving more effective spring length without additional weight. It also achieved a three point suspension with effectively a deflection-free frame, reminiscent of the double three-point suspension of the dual-frame Marmons of Era One. The upper and lower springs were separated by a maple wood cam which served to provide a soft flexible spring for normal conditions, but an effectively shorter stiffer spring when deep ruts were encountered . . . a variable rate spring.

In addition to the upper and lower springs being of different length, the number of leaves in each were different to avoid any resonance of the spring system.

This variable rate, single clamp configuration was covered by US Patent 1,182,192 issued to H.C. Marmon on May 9, 1916 following application on Nov. 8, 1915.

The Model 34 Introductory Brochure commented: *"This method of rear spring suspension results in a reduction of weight and less unsprung weight; its location and peculiar method of mounting results in less violent spring action over rough roads; does away with sidesway which means that the car handles with marvelous ease on the road, and effects a great saving in the wear of the rear tires. The riding qualities of the Marmon 34 are different and better than any car we have ever produced. No other car can compare with it."*

The reference to tire economy makes note of the important characteristic of the suspension . . . that the weight of the rear of the car was approximately divided between the rear wheels regardless of the terrain encountered. Slippage of tires due to unequal loading was avoided and superior tire mileage resulted. Importantly, uniform tractive ability was maintained over a broad range of terrains and the car could navigate in mud up to the running boards.

The Model 34 spring leaves were forged with tapered thickness at the ends, which were also rounded semi-elliptically in the plan view. The spring surfaces were polished for uniform action and grooved for graphite solid lubricant.

Hartford friction-type adjustable (individually) shock absorbers were standard equipment at all four wheels. They were fitted with universal-action attachments to avoid any unknown forces consequent to distortion when a single spring was impacted.

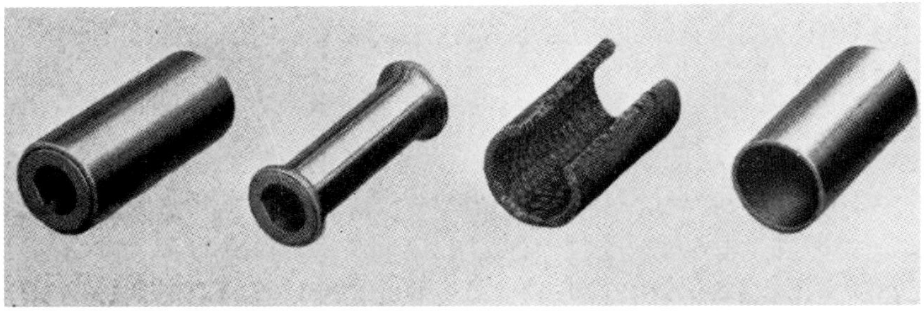

Fig. 4-80 Oilless bushings used on Model 34, 1916-22.

Leather rebound-limiting straps were fitted below the rear axle.

The Nacker patent rear spring and the unusual steering spindle design also appeared on the 1916 Overland; A.P. Brush probably was also a consultant to Overland. The Nacker spring also was used on the Moline-Knight.

For those unfamiliar with rebound straps, they also acted as passenger launchers . . . at the strap limit, the body **and the top bows** ceased to rise (or did so at a greatly reduced rate) but the rear seat passengers continued upward alone. The purpose of the device was to prevent bending of the main spring leaf upward away from the other leaves. Rebound clips are an alternative, forcing the leaves below the main leaf to follow on rebound.

Each spring shackle consisted of a pair of journal and side-plate units . . . the bolt was welded to the side-plate. This design maintained squareness of wear on the bushings and thereby sustained car handling characteristics in the worn mode.

The rebound leaves on all four springs were discontinued in 1922.

Conventional bronze bushings and Alemite-type grease fittings were substituted about 1921.

## MODEL 34C
## 1923-24

Rear axle jounce travel was reduced by 2 inches only on the 34C and the upper half of the rear spring had a corresponding reduction in loaded height. In mid 1924 Watson shock absorbers replaced the Hartfords; they were single-acting inertial-friction type with strap, rather than link, attachment to the axles.

## MODEL D-74
## 1925

Compromise in technical excellence first appeared at Marmon in the D-74 and was particularly evident in the uniform-thickness spring leaves used throughout the Model 75 car; the corners were chamfered. The leaves were pierced and spring center bolts used; the pierced hole weakens the leaf at that location. All Model 34 leaves were half-sheared, forming a nesting dimple and dome on each leaf. The leaves however were polished for reduced friction.

As the 1980s dawned there was a rediscovery that good leaf spring design includes tapered, rounded end leaves. GMC and other builders of heavy duty trucks adopted them following agitation of the National Highway Traffic Safety Administration by the Teamsters Union over the Union's concern about heavy truck riding characteristics.

However there was no evidence of compromise in the new spring shackle design adopted. Separate bolts and side-plates were used, but one side plate was threaded. The spring eyes were straddle-ground to make them square and uniform. Bronze washers were assembled on each side of the leaf eye and the bolts tightened to the desired clearance for lubrication; then a lock nut was tightened against the threaded side plate. Occasional readjustment would maintain the configuration integrity and minimize entry of dirt; the washers could ultimately be readily replaced.

The front frame 'horn' was extended to include the clevis for the front spring. The channel section was closed to a box locally by a welded plate. Previously the 'horn' had stopped just short of the clevis area and a steel forging riveted to the channel web incorporated the spring eye clevis. The redesign included a higher positioning of the spring eye, for unknown reasons but possibly from experience in fitting balloon-tire options to Model 34s the prior year, or due to practicality of forming the integral 'horn'.

Lovejoy hydraulic single-acting shock absorbers were adopted. An aftermarket kit specifically for the Marmon 34 B/C had been nationally advertised during 1924.

## MODEL D-74
## 1926

The one-shot chassis lubrication system, adopted as standard equipment, required changes in the spring shackles to provide, and maintain alignment of, lubricant passages. The spring bolts were welded to one side-plate; the other was no longer threaded. The adjustment-for-wear feature was retained by the use of two lock-nuts.

## MODEL E75
## 1927-28

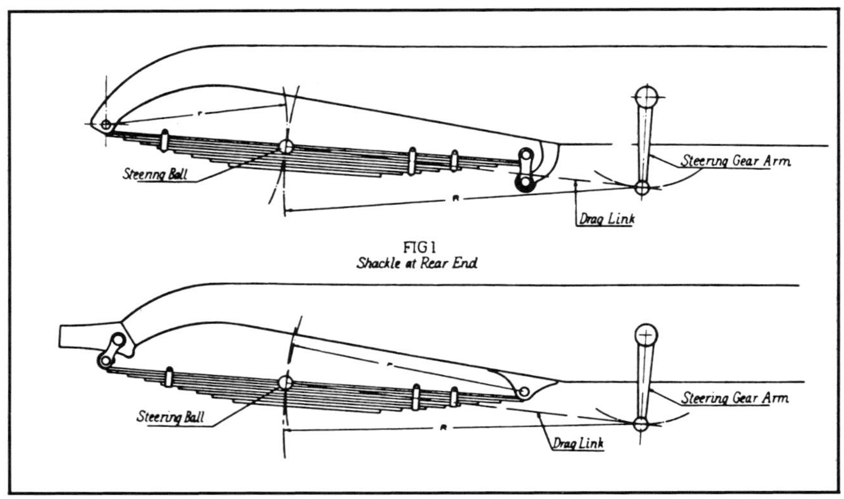


Fig. 4-81 Advantage of forward shackling.

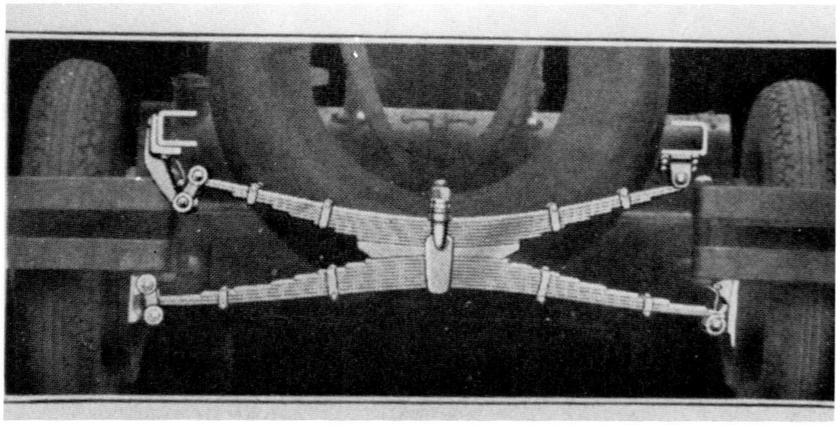

Fig. 4-82 Series 75 rear spring.

Both front and rear springs were re engineered for the Series 75 but retained the traditional arrangement of 1916. Concurrently with extension of the front frame to incorporate bumper attachment, the shackling of the front springs was moved to the front and conventional, rather than inverted, shackles adopted. This change was dictated by the necessity to avoid suspension feedback to steering which could provide periodic steering forces and provoke onset of 'shimmy' . . . uncontrolled oscillation between left and right steer modes. The closer the spring eye pivot is located to the reach rod (drag link) pivot on the steering (pitman) arm, the less steering input results from spring deflections.

The rear transverse spring set was redesigned to appreciably reduce the height of the roll center. The 34 and 74 rear springs were highly arched; the 75 upper and lower springs were quite flat. The adoption of Bendix brakes on all four wheels required a redesign of the rear spring attachment to the axle for the first time since the 1916 introduction. Spring material was identified as 'silichrome manganese steel'.

Both front and rear (upper and lower) spring lengths remained unchanged despite the redesign. The Model 75 had 9 front spring leaves and a 395 pound per inch rate.

The rear suspension configuration of the 1980-81 Corvette bears a surprising similarity in overall layout to the Marmons of Era Three. It however is functionally different . . . effectively two quarter elliptic springs, just like the front suspension of the Chevrolets of the early 1920s; the upper beam is rigid.

WHEELS AND TIRES

## MODEL 34A
## 1916-

Houk demountable wire wheels were standard equipment; wood wheels were available at extra cost. One spare wire wheel was included. These wheels were driven by dowels on the hub, retained by an octagon hub-nut, and removed with a special wrench. A split ring was removed for tire repair.

'Wire wheels were adopted after three years of testing. Strength, lightness, resiliency and effect on tires were carefully noted and it was found that the wire wheels were superior in all of these points and effected a considerable reduction in weight.'[40]

Tires were Goodrich Silvertown 34 x 4½ on 25 inch wheels. Inflation pressure was 65 psi. Type C (Clincher) Cord tires were fitted despite higher cost than fabric tires. The cord tire differed in that the main body of the tire consisted of two plies of cords, each extending from bead to bead diagonally across the tread, one on top of and at right angles to the other:

'The cords had a strength of 250 pounds each, being tightly twisted and bound together and thoroughly 'lubricated' with rubber'. This construction eliminated the working and wearing of the fabric tire due to the cross weave, and permitted the tire to conform more readily to the shape of the road. They were described as immensely durable because of the strength of the cords and the pure gum rubber binding them together, and as representing the closest approach to ideal tire efficiency.

The Houk wire wheels used on the Model 34A during 1916 and 1917 were the triple spoke type and were driven by six dowel pins on the hub which engaged corresponding holes in the wheel hub or core. The wheel was clamped between a conical surface on the hub and a reverse one on the threaded cap. The latter had an octagonal wrenching surface and a corresponding special wrench. Initially the design was claimed to be self-tightening with any displacement of the wheel from the true axis being claimed to result in hub rotation. Therefore the hubs on the RIGHT side of the car had LEFT handed threads, and vice versa.

## MODEL 34A
## 1918

Tire size was changed to 32 x 4½ on open cars and 33 x 5 on closed cars. Tires and rims were straight sided rather than clincher type. Hayes Wire Wheels were adopted. They had the drive dowels riveted to the wheel core effecting a neater appearance. They featured a locking device to prevent loosening of the hubcap. A perforated ring inside the outer end of the core was engaged by a pawl fitted rotatably in a slot in the octagon cap and spring biased in the locked direction. When in the locked position, the pawl protruded from one of the octagon faces. Application of the wheel-change wrench depressed the pawl, allowing rotation of the hub and wheel removal. If the tire-changer was unfamiliar with the device and it was not in his (her) line of sight, he could remove the wheel and replace it without checking for reappearance of the pawl to confirm locking. Abuse and contamination led to stuck pawls; use of end wrenches on the octagon resulted in sheared pawls, as attested by many car restorers. Correct procedure required reapplication of the wrench and a slight indexing to encourage reappearance of the pawl. F. Scott Fitzgerald's experiences with lost wheels on his Marmon 34A were well publicized. A pawl-lock wheel fell off my father's 34B when the engine was started the morning following a 300 mile trip at 60 miles per hour.

## MODEL 34B
## 1920-1921

Tire size continued at 32 x 4½ on 23 inch wheels. Marmon tire equipment was stated to be 'oversized' . . . those measured 34 inches in diameter. On closed body cars 33 x 5 tires were used. Tires and wheel rims were Type SS (Straight Sided).

Optional wood artillery wheels were by Schwarz with wooden fellow construction. Split rims were demounted upon removal of 6 nuts and lugs, and a split

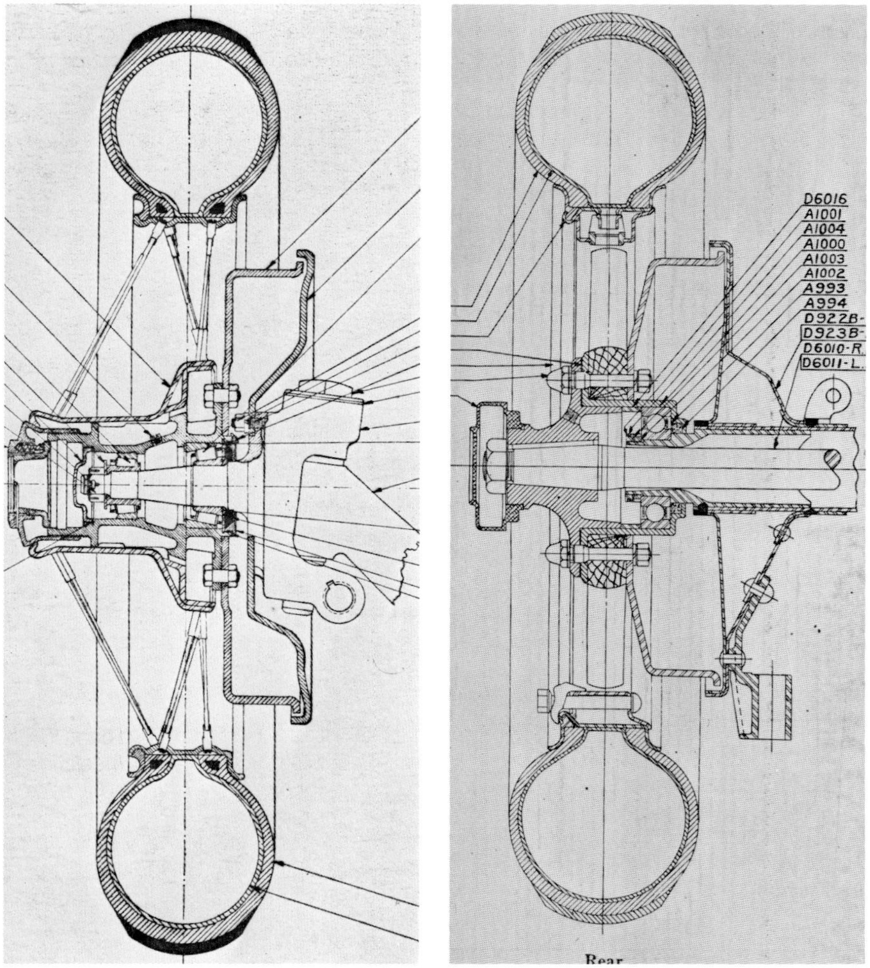


Fig. 4-83 Series 75 front wire and rear wood wheels.

wedge-section ring.

## MODEL 34B
## 1922-1923

On May 1, 1922 wooden artillery wheels became standard; wire wheels were optional at $135.

Buffalo wire wheels were the optional wheels. They were driven by convolutions on the hub and on the inner end of the wheel core and were retained by an octagon hub which was locked against loosening by a spring-loaded splined center section of the cap. This center-section was withdrawn by a toggle lever on the octagon wrench before rotation. Marmon specified the MS series rather than the usual HC series.

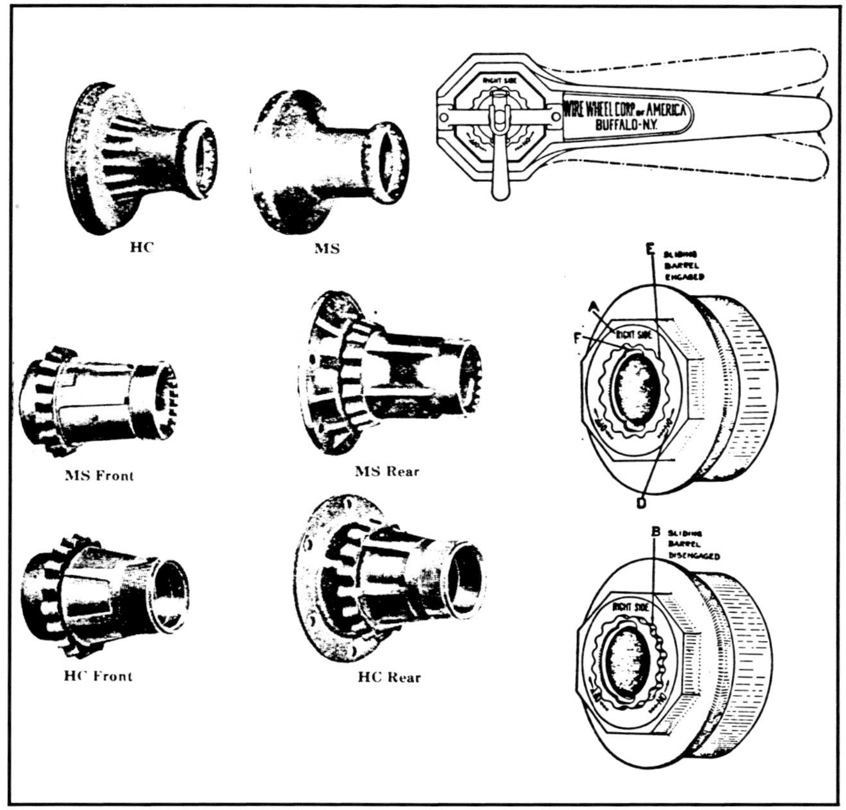


Fig. 4-84 Model 34B wire wheel hubs and cores.

The MS Series had a greater distance between conical surfaces which strengthened the spoke structure against transverse loads and featured a two piece center hub or core on the wheel which concealed the drive convolutions.

The Schwarz wood wheels had steel felloes replacing the sectional wood construction and steel band 'tire'.

## MODEL 34B, 34C
## 1924

Twenty inch wire and wood artillery wheels become optional; tire size was 6.00 x 20. Balloon tires were also available; they were mounted on 21 inch wheels. Balloon tire sizes were 6.20 x 21 on open bodies and 7.30 x 21 on closed bodies, with 45 psi inflation pressure.

## MODEL D-74
## 1925-1926

Open styles used 6.20 x 20 balloon tires, closed styles 7.30 x 20. Round spoked wood wheels were standard equipment. Wire wheels were Buffalo with the new large hub cap and were priced at $100 extra. The wood wheels were fitted with

Firestone rims having a split snap ring for tire removal.

## MODEL E-75
## 1927-1928

All Series 75 cars were equipped with 33 x 6.75 (20 inch wheel) tires, eliminating the several sizes previously required.

Over the lifespan of the Model 34 and its four variations, 34B, 34C, D-74 and E-75 the tire size grew from 4 inches (34 x 4) to 7.20 (7.20 x 20).

STEERING SYSTEM (AXLE AND GEAR)

## MODEL 34A
## 1916-1919

The complete steering system for the Model 34A reflected the design philosophy — drivable by a woman in the winter. This constraint established the lubricants to be used and the containment of those lubricants then influenced the mechanical design.

Oil was selected as the steering knuckle (king-pin) lubricant and castor oil for the steering gear box.

The front axle was of typical I-beam section but uniquely featured double-spindle knuckles which were one-piece forgings of chrome vanadium steel. The wheel end of the spindle was conventional with dual tapered roller bearings. The other end of the L-shaped (approximately 90°) knuckles fitted into phosphor-bronze bushings which entered downward into bored holes in the axle. This type of construction is known as LeMoine type.

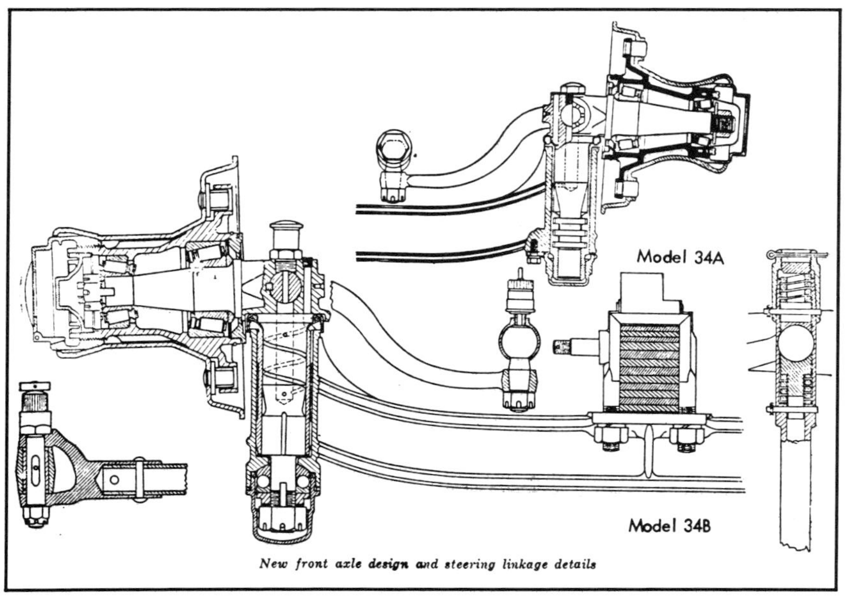


Fig. 4-85 Model 34A and 34B Steering spindles.

The knuckle bushings were of the type commonly used for marine drive shaft thrust and support loads. The knuckle was machined with three protruding rings which provided three thrust surfaces for support of the car weight. The mating phosphor-bronze bushing was split to permit assembly. This construction demands extreme precision in manufacture to ascertain that the load is indeed shared by the three thrust surfaces. The radial loads were taken by the straight sections above (larger) and below (smaller) the three thrust rings.

The bushing halves were flanged at the top and retained at the bottom by a special acorn nut which was locked into position by a single bolt and special washer.

The construction provided an oil reservoir which was filled with medium-weight machine oil. At long intervals, replenishment was through a hex-headed plug in the top of the knuckle. Lubricant retainment was by means of a copper gasket in the acorn nut and a felt washer in a steel retainer at the top and **above the reservoir.**

Marmon described the "axle steering spindle design as making for great durability, strength and ease of operation." The author considers the choice of the word "great" as a particularly wise selection . . . it was just that. A prominent domestic automotive editor with early training in Europe was quoted in the Introductory Brochure:[40] "It is the easiest handling car of American manufacture that I have ever driven", followed a full days journey in a Model 34A.

The steering gear was of the irreversible worm and worm wheel segment type.

## MODEL 34B
## 1920-22

The novel front axle construction was retained but the marine-type spindle bearing was replaced with a bronze bushing for radial loads and a ball thrust bearing for carrying the car weight. The latter was located at the bottom of the spindle and fitted into a spherical bottomed counterbore in the axle. A threaded cap with copper gasket closed the opening. The L-shaped double spindle now even more resembled a typical wheel spindle with the bearing and "hub cap" closure.

An oil "well" (about 1 inch in diameter and length) replaced the nex filler plug; the 'well' was probably more likely to receive service attention by unfamiliar service attendants. An air vent hole was drilled parallel the oil hole and was closed by a pipe plug with slotted 'head'; previously the vent hole was closed by an extension of a washer below the filler plug.

The one-piece bronze bushing was flanged at the top, had a single-diameter bore and outside diameter. The OD was relieved to provide reservoir volume; a circular groove in the bore received oil from a transverse hole in the spindle which communicated with the oil filler. A radial hole connected groove and relief. The inside diameter at the upper end was spiral grooved to enhance lubrication.

Uhe steering pivot was not inclined except due to the 4 degree caster which was spring induced. Wheel spindles were inclined 2° to the ground (camber) and toe-in was .312. This was representative of all Third Era cars except the Series 75.

The axle I-beam section was 2.5 inches high and 2.0 inches wide with .312 web and flanges. Material was 0.40 percent carbon, chrome nickel steel, as were the spindles.

A new Marmon built steering gear was used on the 34B. It used anti-friction bearings throughout: a needle-like roller bearing at the steering wheel end of the tube and tapered roller bearings above and below the worm in the aluminum gear case; a pair of single-row ball bearings carried the output shaft.

The gear was the worm and nut type. The torque reaction on the nut was taken by engagement of a tang with a slot in the case cover. Nut motion was relayed to a yoke on the output shaft via a pair of links paralleling the worm.

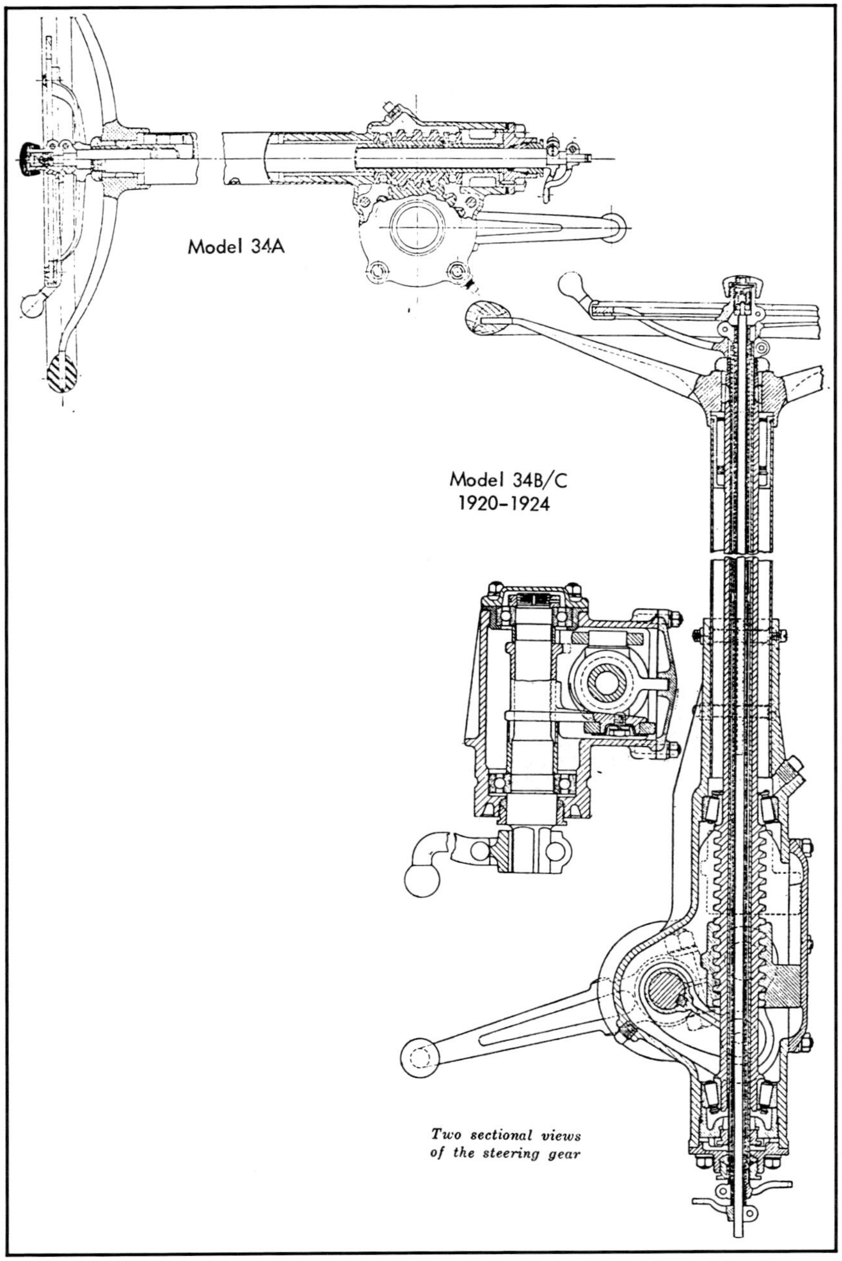


Fig. 4-86 The Marmon designed steering gear used on Model 34s until sometime in 1924 when the Gemmer gear was substituted.

The steering input shaft was tubular to permit passage of a stationary tube carrying the control sector ring and having within it throttle and spark shafts, one solid and one tubular. The control levers had friction retention by brake pads against the sector ring.

## MODEL 34B
## 1923

The oil reservoirs on the steering spindles were replaced by Alemite-type grease fittings, probably concurrent with the change to greased bushings in the suspension.

## MODEL 34B, 34C; D-74
## 1924; 1925-26

The front axle was strengthened when front wheels brakes were offered as optional equipment in September 1923. The steering spindle of the knuckle was enlarged in diameter and length. The flanged bushing was short and served only as the upper radial bearing; a tapered roller bearing (midway in size between the two wheel bearings) at the bottom of the spindle took both lower radial loads and (car weight) thrust loads.

The I-beam section was enlarged and a pair of bosses added to the web. These were drilled on cars with front brakes to mount the brake cross-shaft. When the brakes were fitted, the usual steering arms were omitted and the tie rod was connected to curved ball-studs protruding from the brake backing plates.

Always concerned with minimizing unsprung weight, Marmon made the brake spider serve as the steering arm by extending the brake shoe reaction pivot to include the steering ball. The design was both clever in concept and very neat in appearance.

Late production of 34B, C models carried the Gemmer steering gear that would be used on the Model 74. It used bronze bushings at the three locations on the worm shaft; a pair of thrust ball-bearings were located above and below the worm which was welded to a steel tube steering shaft. The housing was cast iron and the worm nut acted directly on the output yoke. Spark and throttle controls were similar to the prior design.

## MODEL D-74
## 1925-26

A new drag-link ball-socket construction was introduced which eliminated springs and had a seal on the ball-stud. A three part socket fitted in the enlarged end of the tube and was clamped in place by a threaded plug. Clearance was determined by the precision manufacture of the ball and socket components. The previous design with an adjustable spring load on the ball seat was considered a contributor to shimmy encountered with balloon tires.

A ball-end tie rod replaced the clevis construction used previously. It was of conventional construction.

## MODEL E-75
## 1927-28

The steering system on the Series 75 was entirely new. The combination of experience with balloon tires and the adoption of the new Bendix four wheel brakes as standard equipment required a full redesign of the front axle and the invention and licensing of the design made available a superior steering gear.

Because the new Bendix brake had rigid shoes essentially for the full 360° of the drum, it was not compatible with the unique Marmon steering knuckle used since 1916. The new front axle was of the reversed Elliott type but was a unique Marmon design. The I-beam section transitioned to a round section outboard of the spring perches.

The upper yoke bushing was flanged at the lower end for rebound loads; car weight was carried to a ball thrust bearing, above the upper bushing, by a nut threaded onto the upper end of the king pin, which was pinned in the axle beam. This ball thrust bearing had a spherical face which seated in the yoke; ball loading was uniform regardless of deflections. Closure was by threaded cap and copper asbestos ring gasket above the thrust bearing.

The lower yoke bushing was also flanged, but the flange was finished to a conical shape to position and load a non-metallic seal ring against the axle beam. At its lower end, the bushing extended beyond the king-pin and yoke, and was threaded on the inside to receive a threaded cap and copper asbestos ring gasket.

Marmon had abandoned the distinctive LeMoine steering knuckle after eleven years but the 'conventional' design adopted was upon close inspection just as unique and superior as its predecessor. It represented a degree of sophistication in knuckle design that would stand alone until the demands of heavy truck operators forced comparable designs in the 1970s.

The king pin was inclined 6.5 degrees in the transverse plane. The wheel spindle was at a 2 degree angle to the ground (2 degree camber). Caster of 4 degrees was spring induced.

A new tie-rod ball and socket design was adopted. Spring loading was retained but the spring acted at 90 degrees to the steering forces; the split line of the bushing was therefore on the loaded side of the ball. Undamped oscillation of tie rod ends was eliminated.

A special Hartford friction-type shock absorber was fitted between the axle and tie-rod to introduce hysteresis (energy absorption) as a deterent to shimmy.

P.M. Heldt reported this development:[42] *"In cooperation with B.G. Roos, chief engineer of the Nordyke & Marmon Co., Edward V. Hartford, Inc., Jersey City, N.J. has developed a device for preventing shimmy at all car speeds. This device, known as a steering damper, works in the same principle and is similar in construction to the well-known Hartford shock absorber, which has been used for controlling spring vibration for many years. It is the contention of the Hartford concern that shimmy, wheel wobble, galloping and steering wheel throw, are simply different forms of periodic vibration of the front end, and that they could be controlled by the same means as other forms of periodic vibration, such as chassis vibration on the springs, and torsional vibration of the crankshaft... that is, by friction. The first application of the new damper was made to the Marmon car, and on that the device proved so successful that the Marmon Co. decided to furnish it as standard equipment on its (Series 75) cars for 1927. The frictional element of the shimmy damper as fitted to Marmon cars corresponds in dimensions to the smallest size of Hartford shock absorber. This latter device, as is well known, consists of two sets of friction disks which are being pressed together by a spider type of spring under a pressure which can be closely adjusted by means of a nut. The same as the shock absorber, the steering damper has two arms, each rigidly connected with one set of friction disks, so that these disks must rotate with it. In the Marmon application, one of these arms is clamped between the front spring and the spring pad on the front axle. No special clamping devices are required, the arm being stamped with a pad of the same width as the spring and of a length somewhat greater than that of the spring pad on the front axle in the fore and aft direction. At the center of the pad there is a hole to accommodate the spring bolt. The pad is merely placed between the spring and the axle and the tightening of the nuts on the spring clips then secures it rigidly in place. In this particular case the friction disks are located directly over the front axle, and the other arm extends toward the rear. From its end a link with universal connections connects to the steering tie rod. In the Marmon car the connector at the end of the tie rod is ordinarily provided with a clamp bolt, and this clamp bolt has been replaced by a ball stud which takes one end of the link."*

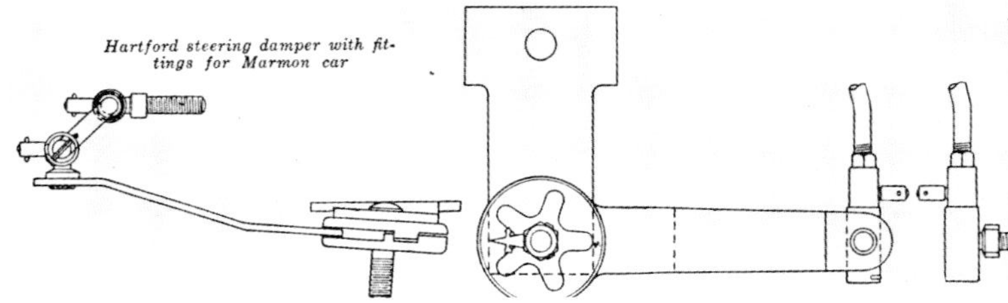


Fig. 4-87 Series 75 steering system with gear detail.

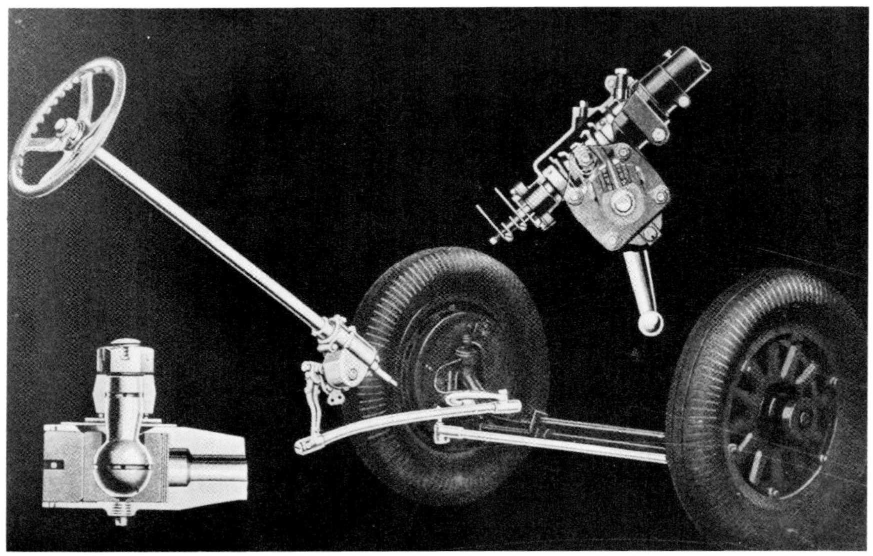

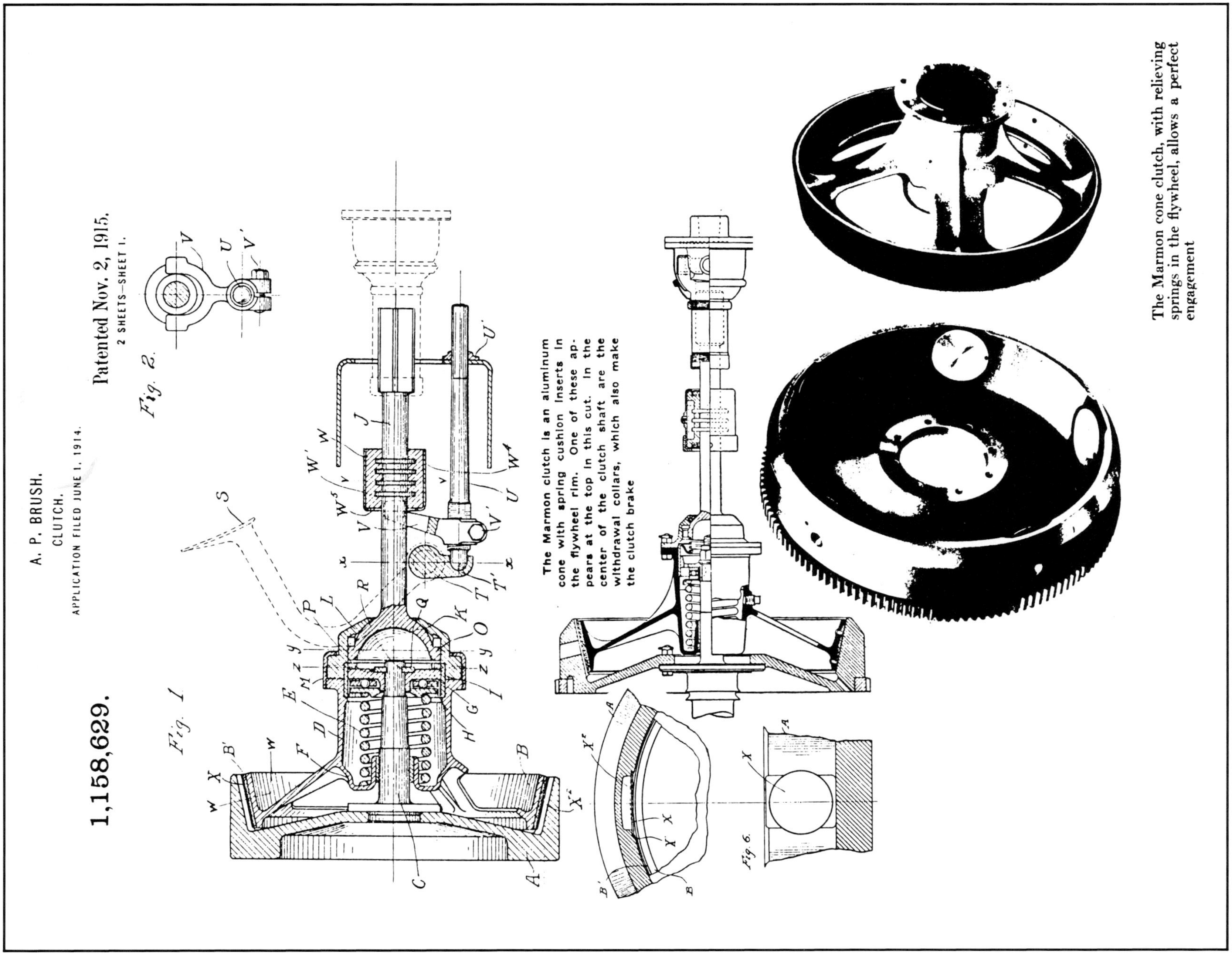


Fig. 4-88 The Model 34A cone clutch featured progressive engagement by virtue of the Brush-patented flywheel inserts.

Henry Marles invented a low-friction steering gear and assigned his patent rights (subsequently issued as 1847721, March 1, 1932 and 1862692, June 14, 1932) to Gemmer. It used a roller on anti-friction bearings as the 'tooth' engaged by the worm thread. Essentially the same design gear is presently manufactured by the successor company — Ross-Gear Division of TRW Inc. and is used extensively in heavy duty trucks. Unique to the Marmon application was isolation of the steering shaft from the tube by the use of rubber-backed bushings. These rubber elements had circular grooves (somewhat like an Acme or square tooth thread in section) so that they were under a combinatin of compression and deflection after assembly. They were made under a Marmon patent.

## CLUTCH

Quite unlike the other major components which underwent a progressive development over the five versions of the Model 34, three distinct types of clutches were used.

The first Marmon and the Models F and G used single plate clutches but Marmon elected a cone clutch for the Model 32, then adopted a multiple-disc clutch for the Model 48 and returned to the cone type for the Model 41. Cone clutches were very popular in the teens and Marmon had carried the type to a high degree of refinement including a progressive engagement.

It was therefore not surprising when the Model 34A appeared with a cone clutch. It used a four spoke aluminum cone faced with Thermoid type material. The flywheel had five spring steel inserts, on the inside conical surface of the flywheel. Marmon described[40] its action: *"The center of surface of each disc is above the surface of the inner (flywheel) cone. As the clutch engages, the discs act as relieving springs and as the clutch pushes the cone into complete engagement, the discs yield gradually so that the whole face of the cone engages the entire surface in the flywheel."*

The clutch release mechanism was unusual and novel; the Introductory Brochure modestly stated: *"The clutch throw-out collar is a beautiful piece of mechanism . . . extremely logical in design. The function of the throw-out collar is to release the clutch while the gears are being shifted. In the Marmon 34 it is located on the drive shaft between the clutch and transmission. Again we see the marine type thrust bearing being used. There are three ring collars cut on the steel shaft. These fit into a white die-cast bronze bushing, made to a free fit and with ample lubricating facilities. This bushing is covered by a spun tube steel sleeve on which a collar of larger diameter is secured. A sheet of wood fibre, one-quarter inch thick, is riveted to the sleeve. A shoe attached to the clutch pedal works aginst the wood fibre and the adjustment is so perfect that the slightest pressure of the shoe against the fibre stops the spinning of the collar. When the clutch is engaged, the shaft either runs free in the loose bushing, or carries the bushing with it, without friction loss, there being one-eighth inch clearance between the shoe and the fibre. The gears shift smoothly. The forward end of the drive shaft connecting with the clutch cone has a universal action in oil and the rear splined end is carried in a universal joint immediately forward of the transmission."*

The Model 34A clutch, release mechanism, provision for misalignment and simplified engagement cushioning device were all covered by US Patent 1,158,629 issued on Nov. 2, 1915 to Alanson P. Brush of Detroit, Michigan following application on June 1, 1914. The patent also covered the Model 34B provision for misalignment.

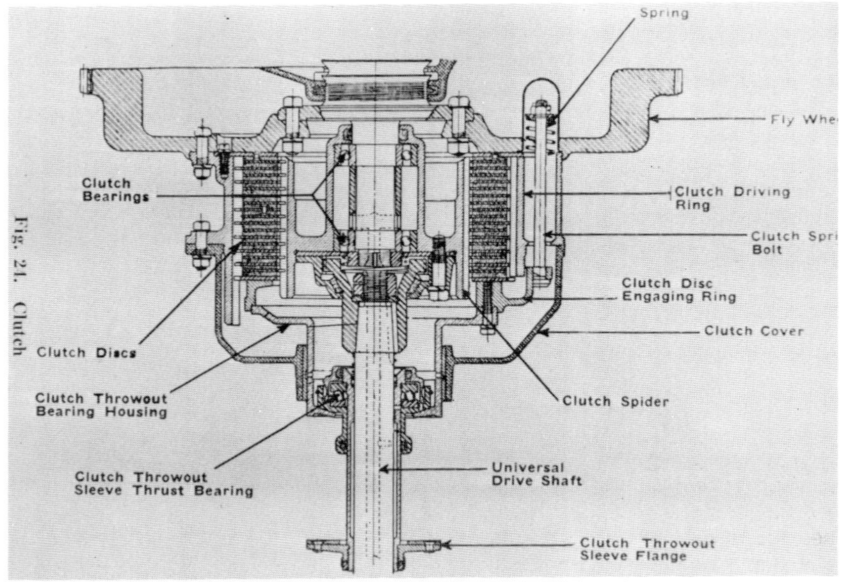


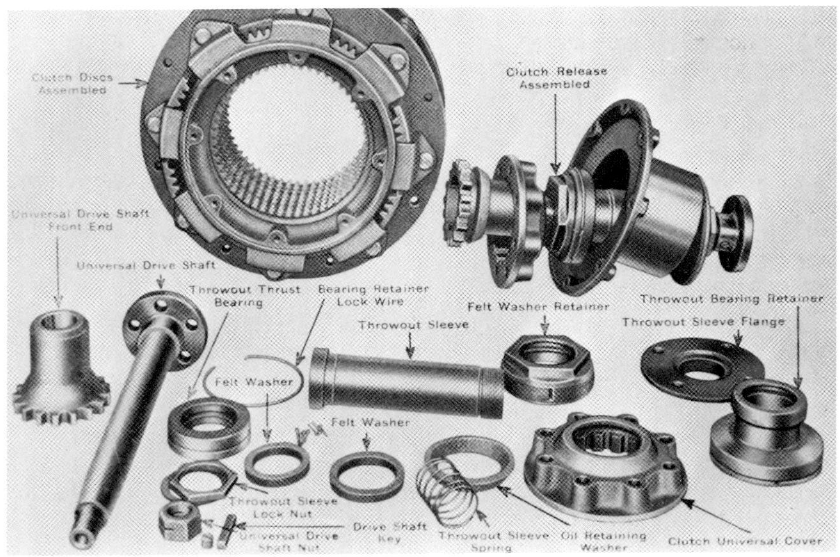


Fig. 4-89 A multiple disc dry-type clutch was selected for the Model 34B. It was cleverly redesigned, presumably for more economic manufacture, for the 1924 model year.

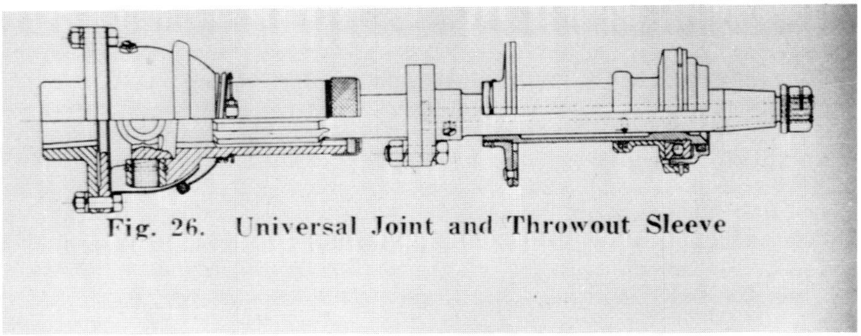


Fig. 4-90 Universal joint and throw-out sleeve.

Significant coverages were: (1) the dual purpose use of the output shaft . . . rotation for power transmission and translation for release . . . made practical by the marine type thrust bearing; (2) the use of a crowned tooth coupling and hemi-spherical enclosure to tolerate small degrees of misalignment (between the clutch and the U-joint on the forward end of the torque tube); (3) The use of flat round spring steel discs in the flywheel to accomplish progressive engagement of the clutch . . . the discs, in flywheel counter bores, were in a "chord of an arch of the surface" when free and were forced to a curved shape at full engagement. (The curved-surface pistons with biasing springs were no longer necessary).

## MODEL 34A
## 1918

The clutch release mechanism was modified to incorporate a felt washer soaked with oil.

## MODEL 34B (AND 34C)
## 1920-EARLY 1924

A shift was made to the dry multiple-disc type of clutch with pull release, and a throw-out mechanism redesigned but similar in concept to the 34A. The 34B clutch had three times the 111.9 square inches of facing area of the 34A. The Marmon Owners Information Book[42] described it well: *"The multiple disc clutch with which the Marmon 34 is equipped insures a gradual, smooth engagement which is at the same time very positive when the clutch becomes fully engaged. There are two sets of steel discs in the clutch, one set of ten being the driving discs and the other set of nine the driven discs. Eight of the driving discs are faced on each side and the two end discs on one side so that the steel surfaces of the two sets of discs do not come in contact with each other. The driving discs are carried by the driving ring which is bolted to the flywheel while the driven discs are carried on the clutch spider to the rear end of which is fastened the drive collar. When the clutch is engaged the two sets of disc are held in close contact by eight equally spaced helical or coiled springs that are carried in the clutch driving ring. The power of the engine, when the clutch is engaged, drives the driving discs which in turn by friction drive the driven discs and these discs in turn transmit the power through the drive collar to the clutch shaft, universal joint, and the transmission. When the clutch pedal is depressed the spring pressure which holds the two sets of discs together is released, allowing the driving discs to run free of the driven discs. This releases the connection between the engine and the transmission."*

The driven hub was carried by a pair of ball bearings on a spindle extension of the crankshaft beyond the flywheel mounting flange; the hub assembly incorporated the crowned tooth gear coupling which served a universal-joint function.

The marine thrust bearing concept was eliminated from the clutch throw-out mechanism for the 34B, just as it was eliminated from the steering knuckles. A ball thrust bearing was used instead, and the release was accomplished by a sleeve concentric with and surrounding the output shaft instead of by dual purpose use of the shaft itself.

Because it was essential to protect the dry friction surfaces from lubricant contamination, elaborate means were used to replenish the lubricant for the ball bearings on the crankshaft end, the toothed universal coupling and the spring loaded spherical closure, as well as the floating bronze throw-out sleeve.

By removal of a plug in the drive shaft, grease could be forced through the shaft into the toothed coupling and then through drilled passages in the crankshaft spindle.

Grease could be added to the throwout sleeve by removal of a plug in the throw-out bearing nose and rotating it to align with a hole in the sleeve.

Based on the authors observation of several Model 34Bs, release sleeve wear was a problem; rattling of the sleeve, with the clutch engaged, resulted. There was no reservoir below the shaft outside diameter and unless the space between bearing surfaces of the sleeve was filled with grease, centrifugal force would hold the grease off the shaft.

The serviceability of the clutch was remarkable and evidenced a high degree of design refinement. The author has removed and replaced the entire clutch assembly, piece-by-piece, while standing beside the car and leaning in through the open front door after removing the floor boards!

Clutch removal was feasible because of two unique design features (1) the torque-tube ball (for transmitting propulsion and braking forces to the frame) was non concentric with the universal joint as used on other torque tube drive cars, some as late as the 1950s; and (2) the shaft transmitting torque from the clutch to the universal joint was separable by virtue of a bolted-flange coupling at its midpoint.

When the coupling was separated, the rear half of the shaft could be pushed into the U-joint spline far enough to allow the clutch cover casting to be threaded through the gap and removed.

The coupling was an example of the precision manufacture claimed for the the Model 34B. The flanges on the two half-shafts were located and retained by 5 ground-finish body-fit bolts, without any male/female pilot. Precision was such that random indexing on reassembly presented no problem. Just such a construction is used to join the two piece crankshafts used in the Detroit Diesel 16 V-71 and both 12 and 16 V-149 diesel engines.

## MODEL 34B (AND 34C); D-74
## LATE 1924; 1925-26

The multiple-disc clutch was totally redesigned. Eight driving discs were faced on both sides; only one of the nine driven discs was faced . . . the rearmost disc and only on one side. The flywheel was faced, eliminating the single faced disc against the flywheel on the previous design.

The drive housing was changed from a closed aluminum casting to a clever stamped steel open member. Limited heat capacity was!the 'Achilles Heel' of dry

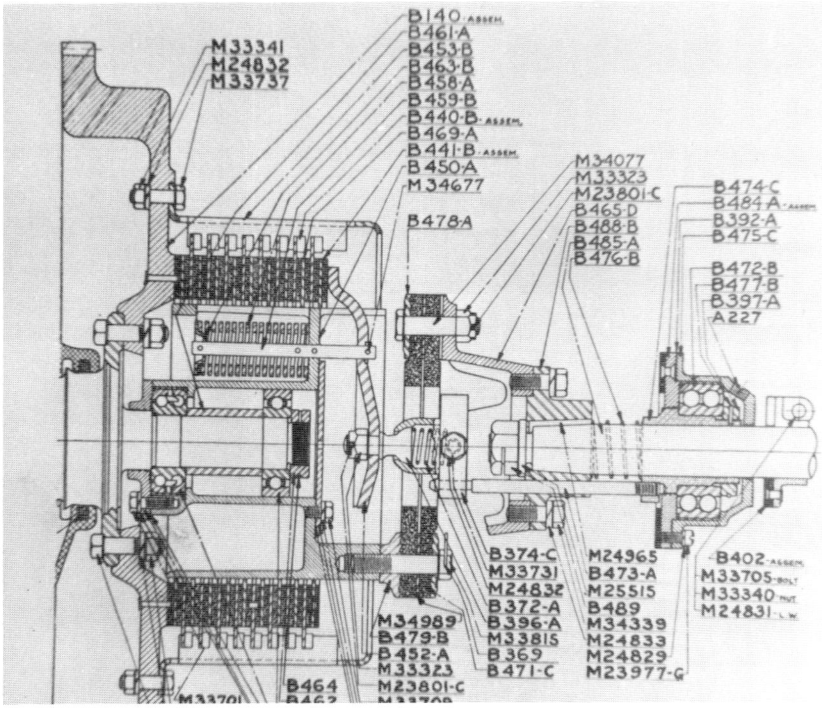


multiple-disc clutches; the open construction would improve heat rejection significantly. The driving discs had drive ears which were folded over 90 degrees to form drive tabs; this increased contact surface assured satisfactory durability despite a limited number of drive contacts. The tabs contacted the sides of drive channels formed by piercing and folding the housing. The splined output member was carried on the crankshaft spindle using a double-row and a single-row ball bearing; it drove a non-metallic coupling disc (actually a sandwich of two discs) which not only served the universal function but also isolated the transmission from engine torsional vibrations.

The throw-out bearing was a double-row ball type. The rear face of the bearing housing engaged a friction-faced member, at the end of release travel, to provide a clutch-brake function helpful to clashless gear-shifting. The braking action was between the transmission input shaft and the semi-stationary throw-out bearing housing.

"Ordinary yellow heavy grease" was prescribed for the throw-out bearing every 1000 miles.

Removal of the flexible discs permitted the drive yokes to nest and allowed removal of the clutch without disturbing the transmission. The output shaft was one-piece, the two-piece shaft no longer being required for clutch removal.

## MODEL D-75

## 1927-28

The 75 had a single-plate dry clutch with 11.83 x 7.25 dia facings. It was of the pull release type with single volute spring as still commonly used on heavy trucks and equipment. The unique part of the clutch system was the driven member. Instead of the splined hub engaging the transmission or output shaft, the Marmon splined hub engaged a hub riding on a pair of ball bearings supported by an extension of the crankshaft. The hub had three arms at its rear engaging a pair of non-metallic discs, one forward and one aft thereof. The output shaft had three similar forged arms which engaged bolts in the discs midway between those on the hub. The isolation of the transmission concept introduced in mid 1924 model year had been successfully extended to the single plate clutch. The clutch cover plate assembly was purchasd from Borg & Beck; while this action might appear the result of G.M. William's cost reduction efforts, it reflected the industry transition to single-plate out-sourced clutches.

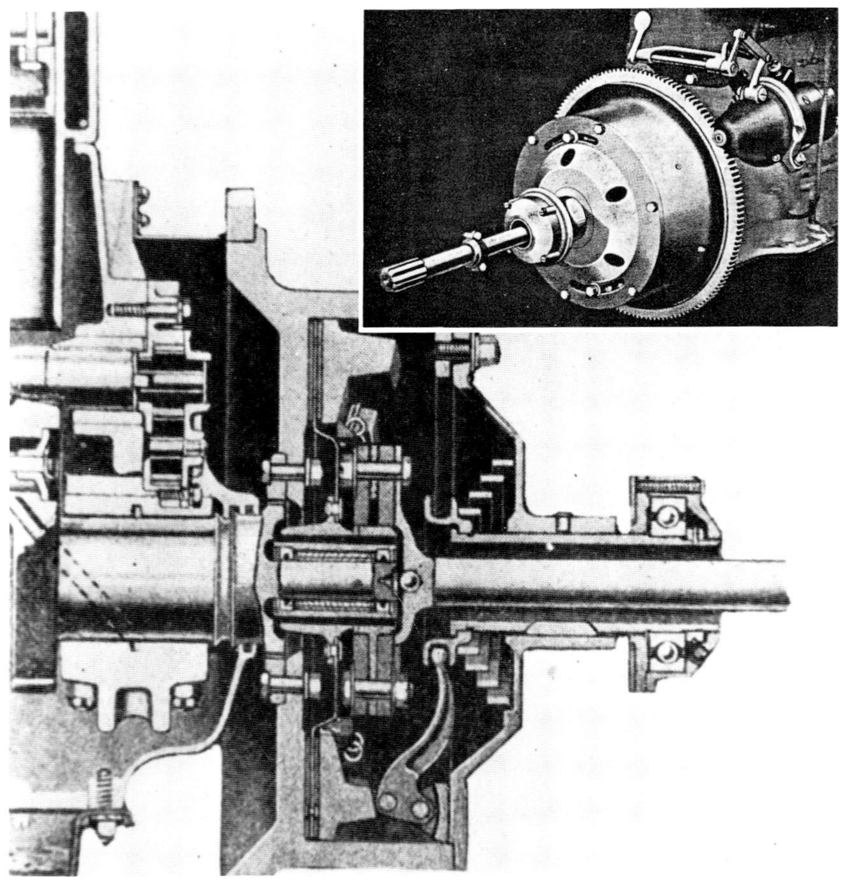

Fig. 4-92 For the Model 75, a single plate clutch was adopted. It was a special application of a vendor design.

DRIVELINE AND AXLE

## MODEL 34A, 34B, (AND 34C) 1916-24

The transaxle (axle mounted transmission) concept, embraced by Marmon beginning with the '32 and continued in the '48', although not on the most recent '41', was retained on the Model 34, at least in principle. However the transmission was relocated to the forward end of the torque tube to reduce unspring mass and reduce the length, complexity and mass of the shift linkage. The advantage of subjecting the Spicer universal-joint only to engine toque was retained (a 2.5:1 low gear ratio would impose a load 250 percent of engine torque).

A clever design avoided the serious access difficulties normally associated with torque-tube drives. For example, the last users of torque-tube drive . . . Buick, Chevrolet and Nash/Rambler all encased the U-joint within a concentric ball-joint through which the car was propelled and rear-wheel braked. The torque-tube drive was claimed to apply propulsion and braking forces at or near the vehicle center of gravity with consequent superior handling. Access to the clutch or transmission required removal of the rear axle/torque-tube assembly . . . an operation involving disconnection of brake lines, shock absorbers and suspension springs.

The Marmon 34 design placed the ball joint beside, rather than around, the U-joint. This offset resulted in a misalignment of the U-joint with respect to the engine/clutch axis whenever the front and rear wheels were not on parallel surfaces. Because the offset was only about four inches compared to the 56 inch tread, the misalignment was small.

But the advantage was great: The U-joint could be removed, by separation of the joint from the transmission flange and separation of a flange joint on the clutch output shaft. Then either the clutch, or transmission internal parts, could be serviced while the car remained on its wheels.

The Brush patented alignment coupling incorporated in the clutch tolerated the error introduced by the offset ball joint, which used only a 1.5 inch ball socketed to a cross member.

Like the Model 34 frame, the rear axle and torque-tube were massive in appearance yet light in weight. The half-axle housings were deep drawn from sheet steel and reinforced: at the inner ends by circular rings brazed in place, at the outer ends by welding to a heavy-wall tubular part which was machined to mount the wheel bearing. Brake mechanisms were supported and reacted by backing plates formed of two opposing thin wall, slightly conical discs joined at the periphery by rivets and welded to the axle tube at their centers. The torque tube was a tapered (swaged) tube welded to cast steel end flanges, the forward one forming the rear wall of the transmission (thru 1924). The rear end carried two large bearings, one a thrust bearing and the other an annular bearing immediately behind the pinion.

The center unit of the axle was an aluminum casting. The exterior axle configuration appeared similar to that found eleven years later on the Ford Model A. Internally it was an unusual construction specifically covered by US Patent 1,168,245 issued Jan. 11, 1916 to A.P. Brush following application on Feb. 1, 1915.

The cast center section was only wide enough to accommodate the pinion shaft ball bearings. It featured a support for a double row ball bearing, located directly in the plane of the pinion shaft, but positioning the ring-gear/differential gear assembly.

The differential gear set was consequently located on the opposite or outboard side of the ring gear; it included an extension journal carried in a wide Hyatt (spiral

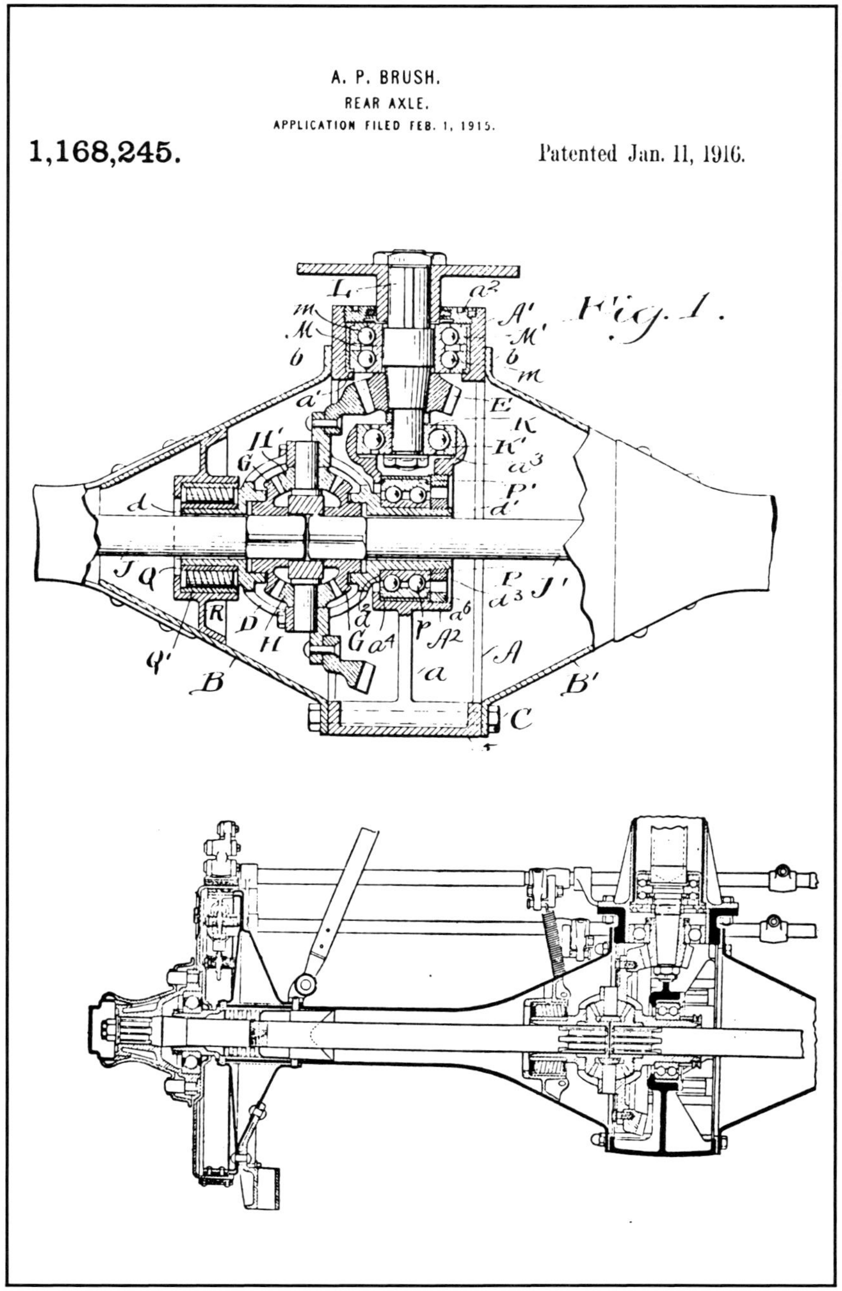


Fig. 4-93 The differences between the A.P. Brush patent concept and the production Marmon axle can be seen in the illustration.

roller type) roller bearing. The left half-housing had a pilot provision, for this outrigger bearing, welded in place.

A feature of this rear axle system was the fact that **medium weight oil instead of grease** was used for lubrication "without any danger of leaking either at the joints or out through the axle end." The axle shafts had flat-top threads at the location where the axle passed through a stack of felt washers . . . of constant ID, but alternating full and relieved ODs.

The axle was of the three-quarter floating type with spiral bevel axle gears. The wheels turned on a large single-row ball bearing in the plane of the tire. The axle shafts were large diameter forgings having 6 tapered splines at the outer end and 6 straight splines at the inner end. A forged hub on the outer spline bolted to the wheel, both for torque transmission and to maintain the wheel square with the shaft.

The patented center housing was claimed to be significantly stiffer than conventional overhung bearing-cap designs. "The mounting of the differential is so accurately made and so firmly and rigidly supported that no adjustment is needed and none is provided. Shims behind the pinion bearing in the torque-tube established the factory gear lash setting.

The following rear axle and overall ratios were used:

| | **34 A** 7 p open all closed | **34 B, C** 7p open all closed | **34 A, B, C** 2/4 p open | **34B** 2 p Speedster | **34 B, C** 4 p Speedster |
|---|---|---|---|---|---|
| Direct (axle) | 4.0:1 | 4.1:1 | 3.75:1 | 3.0:1 | 3.58:1 |
| Second 1.63:1 | 6.52:1 | 6.7:1 | 6.1:1 | 4.9:1 | 5.85:1 |
| Low (3.36:1) | 13.44:1 | 13.8:1 | 12.6:1 | 10.1:1 | 12.05:1 |
| Reverse (4.33:1) | 17.32:1 | 17.7:1 | 16.2:1 | 13.0:1 | 15.5:1 |
| Axle Teeth | 44/11 | 45/11 | 45/12 | | |

## MODEL D-74
## 1925-26

The following rear axle and overall ratios were used:

| | **2 p Closed 4-7 p Open** | **Closed Cars** | **2 p Speedster** |
|---|---|---|---|
| Direct (axle) | 4.10:1 | 4.50:1 | 3.73:1 |
| Second (1.76:1) | 7.12:1 | 7.92:1 | 6.1:1 |
| Low (3.36:1) | 13.78:1 | 15.12:1 | 12.6:1 |
| Reverse (4.33:1) | 17.75:1 | 19.39:1 | 16.2.1 |
| Axle Teeth | 41/10 | 45/10 | 41/11 |

Ratios of 3.0:1 and 4.77:1 were also released.

The original design for the 34A and 34B used an overhung pinion supported by a single row ball bearing. Thrust was taken by a bi-directional ball thrust bearing forward of the single row bearing.

The pinion was mounted on a tapered end of the drive shaft and mounted the ball bearing, requiring extreme precision for run-out control of the pitch diameter.

The model 74 had a redesigned pinion support system. A double-row ball bearing capable of thrust control was adopted; the original Brush patent had illustrated a double-row bearing with separate outer races and adjustment means. The design was strengthened for the heavier car by integrating the pinion and shaft, and by providing a second (single-row) bearing near the aft end of the torque tube.

The aluminum transmission case was extended to incorporate the rear bearing and the speedometer drive gears; the torque-tube was correspondingly shortened. Simplified manufacture of the torque-tube could have been a consideration. Second gear ratio was increased from 1.63 to 1.76:1.

The pinion shaft was attached to the solid drive shaft (and transmission main shaft) by a splined coupling, replacing the welded joint. The splined sleeve coupling and the shafts were drilled thru after assembly and pins were installed and riveted over for axial retention. These couplings experienced considerable wear. Note the parallel with the 1934 Ford V-8 construction and the loose pin problem which became a 'lathe' and cut off the tube! During the production period of the D-74 the forward half of the coupling became a weldment to eliminate one degree of freedom.

## MODEL E-75
## 1927-28

The axle outer end construction was changed consistent with the new Bendix internal brake requirements. Axle shaft diameters were increased to 1.5 diameter; Material was a 0.40 percent carbon nickel steel.

TRANSMISSION

## MODEL 34A
## 1916-17

The transmission was a conventional three-speed and reverse gear box with sliding spur gears and selective shift (ie, from neutral directly to any gear). The gears were made of a special heat-treated steel and **each** gear **tooth** was inspected for hardness using the Scleroscope instrument. The countershaft was hollow and stationary in the aluminum case; the gear cluster rotating on cartridge roller bearings. The input and output shafts had Hyatt roller bearings and bronze thrust washers. In lieu of splines, the gears slid on a square shaft and the output shaft engaged the drive shaft also with a square end.

**The lubricant was medium weight cylinder oil** instead of grease. This was consistent with the near total absense of grease in the entire car and made operation practical for a woman in the winter.

Speedometer drive gears were located in the end housing of the torque-tube.

Transmission ratios were: direct, 1.63:1 second; 3.36:1 low; and 4.32:1 reverse.

The gear selector lever and companion hand brake lever were packaged together in an ingenious manner and were located slightly to the left of car center line, bringing the shift lever beneath the steering wheel rim and the longer brake lever adjacent to the rim. The offset also placed the lever in line with the shift rails in the transmission which were one above the other on the left **side** of the transmission opposite to the shank of the torque ball.

The shift rails operated in bronye bushings in the transmission and extended forward through felt seals. The rails terminated in offset arms which brought the two linkage clevises side by side in line with the ball joint on the opposite side.

Each shift rail was fitted with an individual spring-biased detent pin.

The hand brake lever pivot shaft carried the hand lever on the right and the brake linkage lever on the left end, beyond the shift mechanism. On that shaft, the two shift levers pivoted independently, and were connected to the shift linkage;

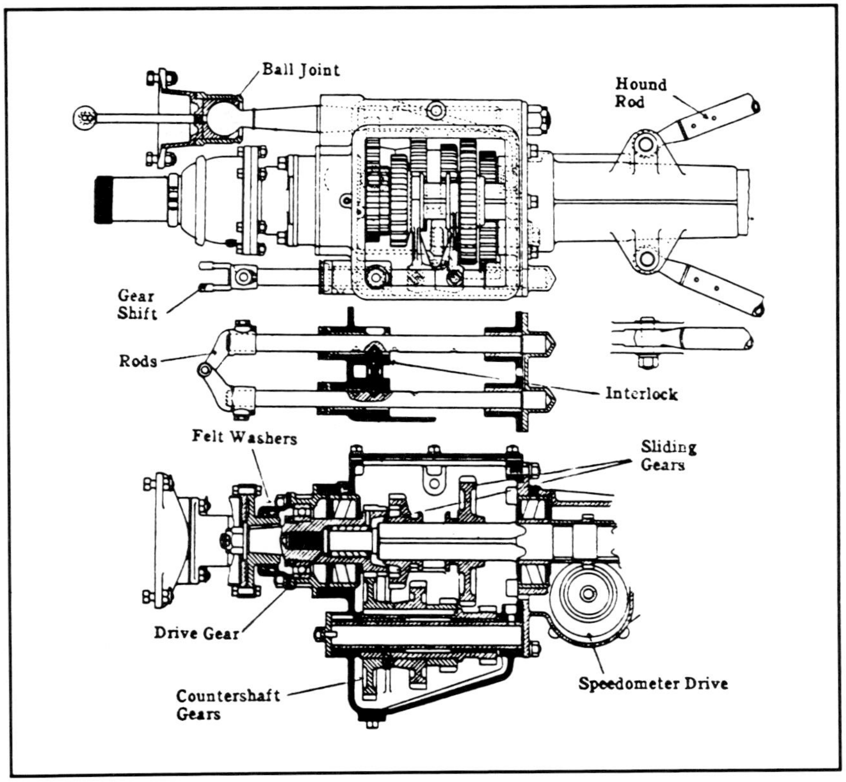


Fig. 4-94 The Model 34B transmission.

between them pivoted a sleeve carrying the shift yoke for "rail" selection. Shift lever movements were short and positive providing for rapid precise shifts . . . The Hurst of its day.

## MODEL 34A
## 1918

A key lock provision in the shift lever ball was introduced. Multiple rotations of the key withdrew a pin operating in the hollow gear lever which acted as a dowel between the lever and its pivot shaft; the lever if then moved did so independently of the "secured" shift mechanism. Biasing springs returned the lever to the neutral position, assuring alignment of the dowel and hole for 'unlocking', which was actually a re-engaging.

## MODEL 34B
## 1920

The transmission main shaft was welded directly to the drive shaft tube. Previously, a splined (square) end was welded to the tube and it engaged the mating provision on the mainshaft. In both designs, the transmission rear bearing was in the torque tube end housing. Grease lubricant was specified.

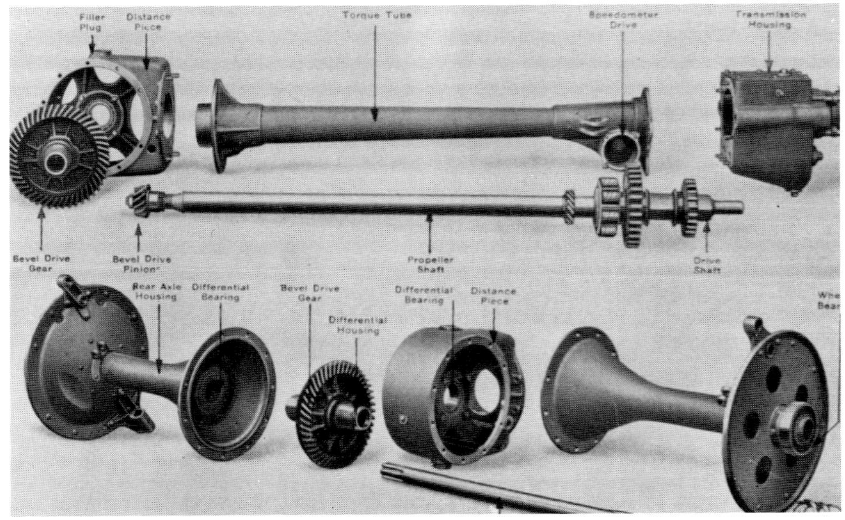


Fig. 4-95 Exploded components of the Model 34 drive train. The massive appearing parts were quite light due to thin walled deep drawn technology.

## MODEL E-75
## 1927-28

The forward end of the torque-tube was fitted with provisions for a short brake cross-shaft, dual rods being used to apply the rear Bendix brakes. The drive shaft was a 2 inch diameter tube with a 0.235 wall. Material was a .40% Carbon nickel steel.

### BRAKES

Three brake systems were used during the Third Era. The first was a very conventional external-contracting band service brake with internal-expanding band emergency/parking brake, both acting on the same rear drums. The third replaced the first as standard equipment; it was a four wheel internal shoe system with self energizing action. The second was an optional four wheel system in which self-energizing internal band front brakes were added to the first system.

## MODEL 34A, 34B/C, D-74
## 1916-1926

The Model 34 was introduced with a very conventional braking system as regards the type. It was however a very well engineered system of Marmon caliber.

Both service and emergency/parking brakes were Thermoid faced band type acting on the outside (17 inch diameter) and inside (16.7 diameter) of drums on the rear wheels. The drums were stamped steel. Service brakes were 2.0 inches

Fig. 4-96 1924-26 front optional brakes. Lower right is Model 34/74 rear brake.

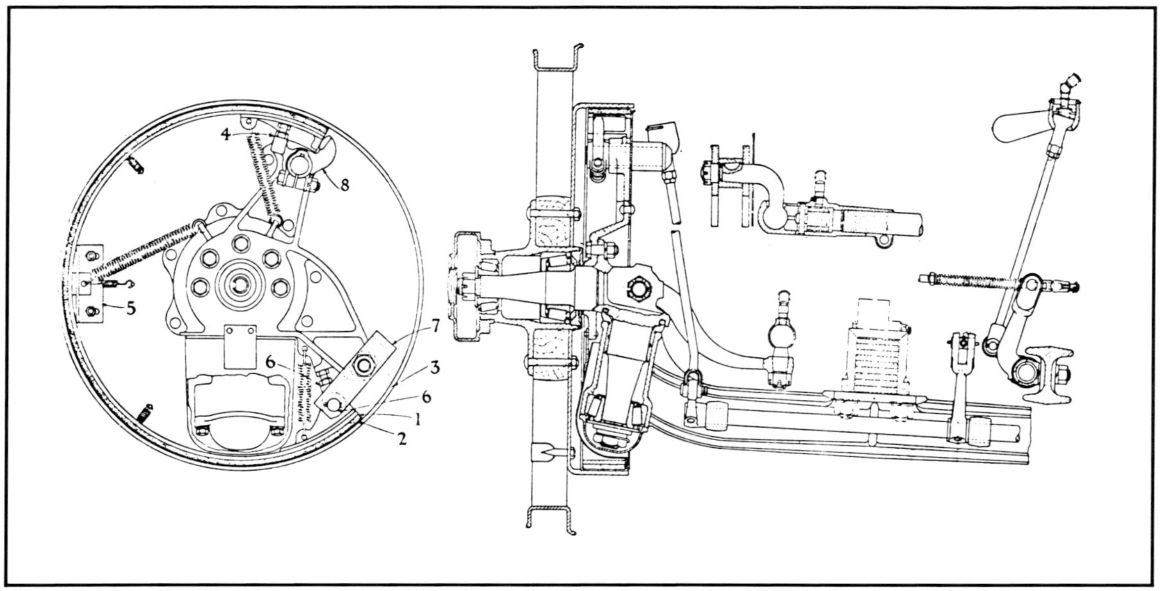


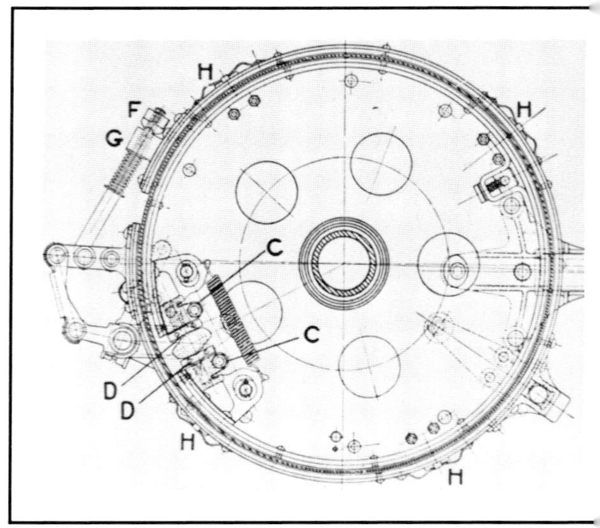

wide while the emergency/parking brakes were 1.5 inches wide; the former had 106.5 square inches of facing on each wheel, and the latter had 71.8 square inches.

Both brakes were actuated by parallel cross shafts, mounted on the axle which were in two basic sections but actually three for assembly considerations to allow precise positioning of the two ends. The rod from the pedal or lever was attached to a lever on one half-shaft; the lever on the other half-shaft was clamped to the main lever to provide relative adjustment.

The service cross-shaft carried levers at the ends for toggle band tightening. The parking cross-shaft carried cams for band expansion. Both internal and external bands reacted on square shafts attached to the backing plates 180 degrees from the actuating shafts. The external band clearance was controlled by a spring-biased adjustment nut (band length control) and by four adjustable arms hooked under leaf springs on the band (band contour control). The internal shoes had forged end sections which incorporated threaded capstans bearing on the cams. Loosening transverse locking bolts permitted rotation of the capstans by progressive insertion of a rod; both of these were accessed, and the clearance measured, through an inspection port in the drum.

## MODEL 34 B, 34 C; D-74
## 1924; 1925-26; optional

Four wheel brakes became optional on the Marmon 34 with start of 1924 production in September 1923. An extra charge of $125 list was made. The new brakes were connected to the standard rear wheel linkage which was unchanged except for addition of an arm on the brake pedal necessary to attach the front linkage. They were an internal expanding band type with forward self-energizing action. The Marmon front brake design was intended to provide four wheel brake actuation without an increase in either pedal travel or required pedal force.

The significance of this development warrants repetition of the description of the brake function from the press release published[43] in Sept. 1923: *The front wheel brake mechanism is shown in the left view, the left front brake having the drum removed. The shoe assembly rests against three supports, 4, 5, and 6, being drawn inward against these supports by the tension of three springs. Each support is adjustable and can be used in regulating the clearance between the brake band and the drum. The lower end of the band assembly is hinged to the link 7 at 9. This link is also hinged at its upper end where it is secured to the support. At the upper end the brake lever is free to move circumferentially, being actuated by the lever arm, 8. This lever arm is keyed on a shaft which extends through the wall of the support and dust shield, where it forms on the outer end a crank arm for connecting the control linkage. When the brake pedal is depressed it causes the lever arm, 8, to force the brake drum outward into contact with the inside of the brake drum. The initial braking effort is caused by the friction due to direct pressure of the shoe against the drum. Further application of the brake by the driver causes the band to exert a greater pressure against the drum, first to greater direct force, but, second, greatly augmented by the tendency of the band to rotate with the drum and press outward against it thus securing the powerful wrapping effect. The brake control linkage is shown in the illustration. The lever arm 8 right terminates externally in the crank arm. This arm is operated by a cross shaft 4, which lies within the rear channel of the front axle I-beam and is supported by it in self-lubricating bearings. The ends of the cross shaft are connected to the crank arms on the left and right brakes through levers and links. The upper end of the links terminate in a ball and socket joint entirely enclosed and provided with a lubrication fitting. The ball end is threaded and provided with a lock nut 12, serving as an adjustment for equalizing the left and right front brakes. The position of the ball joint is in alignment with the axis of the steering knuckle so that steering the car does not cause any relative motion in the brake mechanism and the application of the brakes is not affected. Regulation of the relative braking effect of the front and rear wheel brakes is secured by turning a wing nut, which serves to increase or shorten the effective length of the brake rod.*

The similarity to the Huck brake used on Chevrolets in the 1930s should be noted. The Huck brake was not self-energizing however because of opposing sets of shoes and links.

The issue of four versus two wheel brakes was a controversial one with some major manufacturers advertising that they would never offer four wheel brakes. The Marmon option won only limited acceptance and very few surviving cars from the option years incorporate the four wheel brakes. Extensive searches by restorers seeking to convert their cars have generally been unsuccessful in locating front axle assemblies with brakes. Barney Oldfield's fourth Marmon was a Model 74 four-passenger Speedster . . . with two-wheel brakes; but after all, Barney was known to favor the accelerator, not the brake pedal.

### THE KINEMATICS OF VEHICLE BRAKING

A discussion of the kinematics of vehicle braking may assist the reader in understanding the issues of four versus two wheel braking:

Maximum deceleration is achieved only when the braking action at each wheel on a vehicle is distributed exactly in proportion to the portion of the vehicle dynamic weight carried by that wheel. The Marmon 34 had about a 50/50 front/rear static weight distribution and consequently could have no more than half of the ideal braking when equipped with two-wheel brakes. Many cars had less than 50 percent static load on the rear wheels and braking was correspondingly further compromised.

However, the action of braking causes a transfer of car dynamic weight to the front wheels. This is evident from the posture of a decelerating vehicle . . . the increased front loading shortens the height of the front springs and correspondingly the rear spring height increases consequent the reduced load. This action results from the simple fact that the inertia force being reduced is centered collectively at the height of the center of gravity . . . some 18 inches or more above the road . . . whereas the braking force opposing the inertia force is applied at the road surface. Consequently a Marmon 34 with static 50/50 weight distribution could have a 60/40 dynamic distribution during hard braking and could decelerate only at 40 percent the rate of a car with ideal braking.

Because of the limited braking effort available before the rear wheels locked, brake pedal forces were low on two-wheel braked cars. A Marmon 34 **if equipped for ideal braking** could have required two and one-half times (250%) the pedal force of the production car when stopping 250 percent faster. Alternatively the stroke of the pedal could be increased correspondingly; actually a combination of both was the usual solution. Marmon however elected to utilize the car energy to provide some or all of the required additional pedal force.

There were two other important considerations which limited the degree to which ideal braking could be incorporated. They both concerned the effect of front braking on steering characteristics.

The first consideration was a compromise in turning capability. A car is steered from a straight ahead direction by a side force generated at the tire/road contact when the wheel rotates in a plane other than the car direction; Cornering is limited to the friction force limit associated with the weight on the wheel. Any portion of the friction force used for deceleration (in the direction of car motion) reduces the force available for cornering. For example, when brakes are locked a car slides straight ahead regardless of steering wheel position. When most roads were gravel surfaced, the concern with any compromise in cornering ability was understandable.

The second consideration was a disconcerting 'feel' of braking on the steering wheel. Until the 1934 advent of independent front suspension, in which the wheels were positioned by structural members other than the spring, the amount of front braking allowable was significantly limited. This resulted from the rotation of the front axle during braking consequent to the braking reaction torque being taken by the spring; the rotation changed the wheel alignment and therefore the castering effect which provided inherent centering of the steering.

## MODEL E-75 BRAKES

With the introduction of the Model 75, Marmon was out of the brake engineering and manufacturing business. The move could have been seen as reflective of the new philosophy brought in by Mr. Williams had it not been so perceptive an assessment of the future of the new Bendix self-energizing internal shoe brakes.

For three years Marmon had offered a Marmon self-energizing internal shoe brake as an option on the front wheels but with relatively low market penetration; the external band rear service brake introduced for 1916 was continued through 1926. It is a credit to Marmon that, when the decision was made to adopt four-wheel brakes as standard, a full new brake system was incorporated even though it meant major changes to the rear axle.

The Marmon 75 had Bendix three-shoe brakes on all four wheels with mechanical linkage actuatiom. The drums were 15.625 diameter and provided 360 square inches of braking surface. The following year the Bendix brake was adopted for the front axle of the 1928 Oldsmobile while retaining the external rear brakes through 1929; a half-breed system similar to the Marmon option first offered in 1924.

The Bendix engineers would solve the pivot friction problem with the two-shoe version and the actuation friction problem with hydraulic application; the brake became the universal passenger car brake of the 1950s and 1960s and still persists on the rear wheels in the disc brake era.

Because Era Four was characterized by supplier designed componentry and Era Three at Marmon was not, a more detailed discussiom of Bendix brakes is found in the Chassis part of Chapter Five.

### CHASSIS SELF-LUBRICATION

## MODEL D-74, 1926
## MODEL E75, 1927-28

The third major improvement for 1926 was the adoption of the Bowen "sure-shot" system of self-lubrication. Oil, rather than grease was distributed to 19 points on the chassis simultaneously by a single stroke of the foot-operated pump. The means by which the lubricant was metered and delivered was ingenious and warrants repetition here of the Marmon description of the system function. *"The pump is of the displacement type located beneath a reservoir containing a*

Fig. 4-97 The Model 75 application of the Bendix three-shoe self-energizing brake. Rear shown.

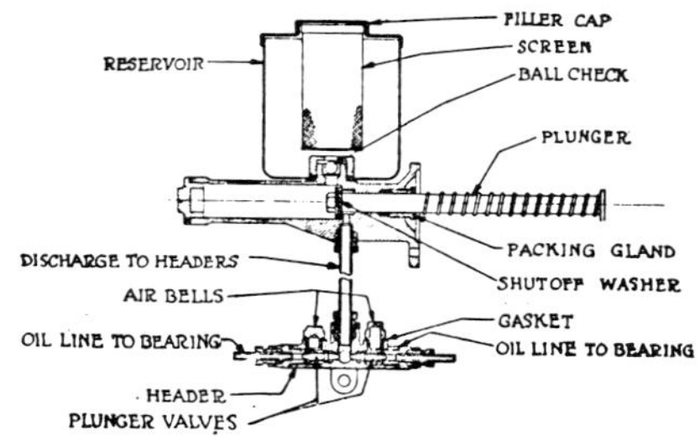


*Sectional view of automatic chassis lubricator*

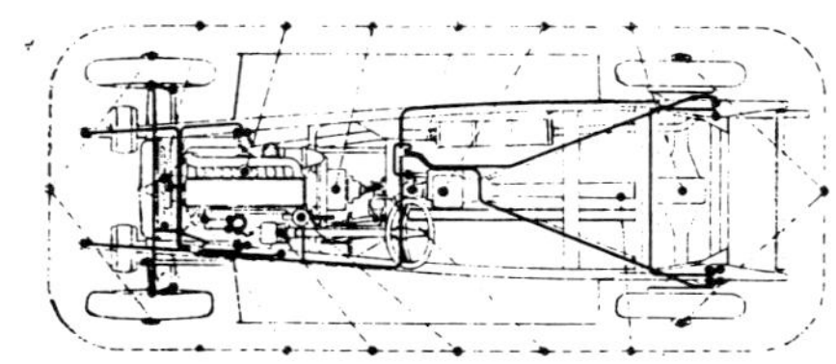

***Plan of chassis, showing oil leads***

Fig. 4-98 Schematic shows manner of distribution of oil throughout the Model 75 chassis from the manual pump adjacent the driver's foot.

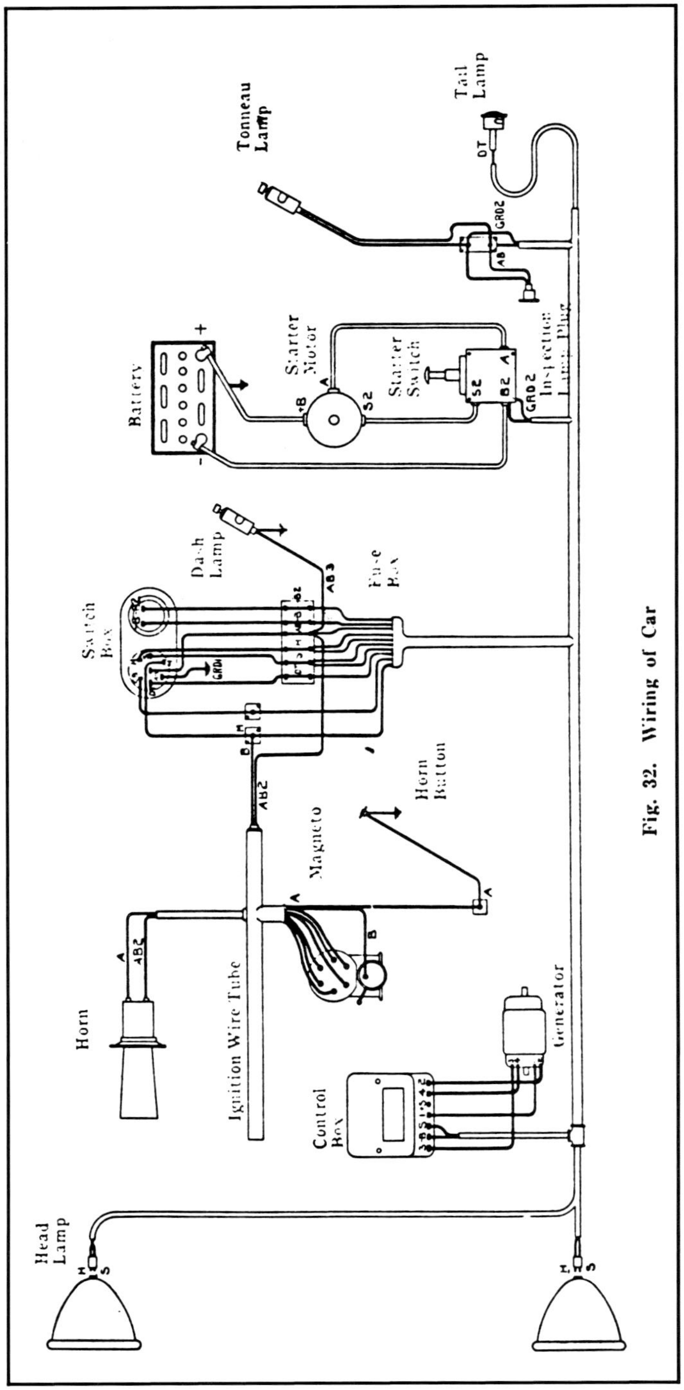


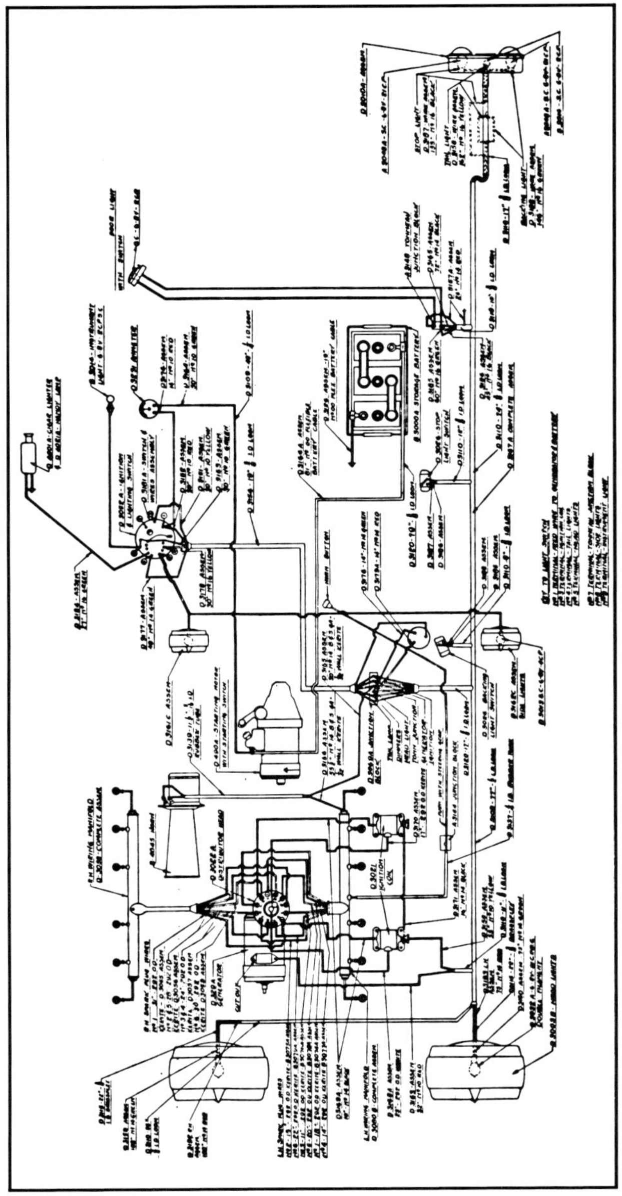

Fig. 4-99 Increasing complexity of electrical systems is evident from the diagram of the Model 34A (1916 upper) and the E75 (1927 lower).

*screen. The plunger is returned to its normal position by spring pressure. When the plunger is pushed home, oil is forced to three "headers," which distribute and measure the oil for the various bearings. Each header contains a number of units, depending on the number of bearings which are to be served. Each of these metering units embodies a tiny double-ended plunger valve, normally held in one position by very light spring pressure. This serves to close the outlet from the header to the chamber in which the plunger valve is located. As pressure is applied to the oil behind the little plunger it is unseated, its travel causing it to seat on the opposite end, thus sealing the line which runs to the bearing. With the outlet from the unit thus closed, oil flows into a little air-bell directly above, comqressing the air which it contains. Upon the size of this air-bell depends the quantity of oil charged into the unit, and also the pressure built up behind it. When the main plunger is permitted to return to its normal position the pressure behind the headers immediately drops. The pressure in the little air bells then causes the little double-ended plunger valves to return to their former position, closing the inlet from the supply line and at the same time opening the outlets to the bearings. The pressure in the air-bells then forces the oil into the distributing lines and so feeds the bearings. All spring shackle bearings, all front axle bearings except those in the wheels, the two steering cross-rod bearings and the two drag link bearings are lubricated in this way. Naturally the system can be operated at any time, regardless whether the car is at rest or in motion".*

CHAPTER FIVE

# THE MARMONS OF ERA FOUR

*"UNFULFILLED AMBITION"*
*1927-1932*

The Fourth Era is symbolized by the Marmon Lightning Flash. The Lightning Flash first appeared exclusively in Marmon ads in July 1926 which revealed for the first time that Marmon Motor Car Co. would build a small car. The Badge emblazoned with the Flash was used only in advertising; it never appeared on a production car.

During the 1928 Auto Show in New York, several Marmons exhibited at the Belmont and Commodore Hotels carried chromium plated lightning flashes on their radiator cores. They became standard equipment on 1929 model Marmons.

The idea became very popular . . . so much so that aftermarket firms began to market similar trim items. The factory received numerous letters from owners of Roosevelt cars and dealers demanding that the Roosevelt radiator core be similarly emblazoned. Why not? The factory conceded and it was added.[1]

When radiator shutters were adopted on senior models late in the Era the lightning flash appeared on hub caps and bumper trim plates.

Marmon Motor Car Co. received a copyright on the Lightning Flash prior to October 1929.[1]

When the Marmon body plant, which had been idled for two years due to outsourcing of bodies, was leased to Murray Body Corp. in December 1925, the announcement carried a hint of a different message, noting that further developments at Marmon were in definite perspective.[2] G.M. Williams was quoted as saying: *"Marmon engineers and executives have recently spent much time abroad making an investigative study of the engineering phases of European motor cars which have found favor with the Americans. It seems that the solution is a combination of the best elements of European body design with tried American engineering practice. Unquestionably this Country is realizing the necessity of a small, beautiful, high-powered type of car that will go far toward at least temporarily relieving our traffic conjestion and solving the parking problem."*

Mr. Williams neglected to mention the auto-related air quality problem . . . the New York Academy of Medicine had undertaken in 1923 an investigation to determine how automobile exhaust gases affected the health, life, liberty and pursuit of happiness of the citizens of New York.[3]

Mr. Williams continued: *"Should such a car be developed by Marmon it would be purely a supplementary car, for the Company feels the present type of large car will always be in demand by a certain class of buyes."*

During the decade of the twenties, a relatively small number of automobile manufacturers were producing a progressively larger proportion of the total. It was evident that the smaller firms producing "assembled cars" from purchased major components would not survive. The larger firms of this category were typically purchased by major firms to achieve rapid expansion of production capacity. It was also probable that a smaller firm such as Marmon, which engineered and manufactured both major components and the finished vehicle, would have difficulty surviving. Factors making survival difficult were the limited extent of dealer service outlets and the rapid expansion of the travel area of the typical car as roads improved. No minor consideration to management was the attraction of a segment of the huge financial "pie" associated with the multi-fold expansion of the automotive market for small and medium size cars.

Marmon management decided to expand market coverage for 1927 by the introduction of a "Compact" car, "The Little Marmon", to be offered in addition to the Model 75. (Locomobile had made a similar move the previous year: The Locomobile Junior 8.)

The announcement stated that three years earlier the question was asked "Why is it that none of our fine car manufacturers have built a small fine car? The market is here . . . if the car is really distinctive and outstanding in workmanship and performance". The answer to that question was the "Little Marmon".

'MARMON IS READY TO BUILD ITS LIGHT CAR' headed press reports[4] of a news release in late October 1926. It revealed that Plant 3 had been revamped and especially equipped for the newcomer and that it would be turned out in its entirety in the one plant consequent to the Murray Body Corporation operation of the Plant 3 body facility. Capacity was to be 24,000 cars per year (200 per day).

"The new car was characterized by G.M. Williams as not being a 'small' car in the popular conception of the term but rather a 'big car compacted'. It was said to be an outgrowth of an idea which took root with Williams some seven years earlier (1919) when he was making a survey of European countries. The car which was so popular there, he believed, could be adapted to the American market were certain changes and refinements made. It was...his belief that the American public wanted more riding comfort, more distinguished lines and appointments, and more power, speed and stamina than found in the majority of small cars in Europe."

"The idea remained in Williams' mind and when he joined the Marmon company in 1923 it was incorporated into the general scheme on which Howard Marmon had been working for a number of years. Marmon had built according to his conception, various (study) models of small cars and one after another they had been discarded. Williams ideas were added and **more test cars were built.** One after another they were studied until finally the forthcoming Little Marmon was evolved."

### *The Mysterious Marmon Race Car*

A fully restored Marmon dirt-track racer was shown at the 1981 Hoosier Auto Show and Swap meet held at the Indianapolis Speedway. It incorporated features both interesting and puzzling to the historian.

The engine was a small overhead-valve six which obviously was a relative of the Model 34 B but exhibited unique features. A **single** six-cylinder block was mounted on an aluminum crankcase which had a side extension housing the water pump. Spark plugs were in the block, the intake manifold was in the head like the Model 74. Engine and transmission were in a unit configuration like Fourth Era Marmons.

The chassis was characterized by rubber shock insulators at spring attachments as used on the fourth Era Marmons.

The third owner and restorer reported the car had been built as a convertible coupe. Following an accident it had been reconstructed as a race car by the second owner who understood that the car was built in 1924. It evidently was the last of Howard Marmon's small concept cars and the prototype Little Marmon. Introduced in the fall of 1926, after a claimed two years of test development, there would have been prototype Little Marmons built in 1924.

It can be conjectured that the Little Marmon was planned to have a six-cylinder engine like the "full-size" Marmon. Howard Marmon had expressed a position on the adequacy of six cylinders at an SAE meeting in 1915; he obviously thought otherwise when the first sketches were made for the SIXTEEN. There is reason to believe that he had not changed his mind in 1924, particularly for the case of a "Little" companion to a "Large" car when the latter had a six.

Joe Felts recalled Al Leamy, Marmon designer, working on design of a small six cylinder engine as part of a joint program with a now unidentified European manufacturer; the project was not carried to maturity.

But G.M. Williams was a man in a hurry. He lacked the time necessary for development of one of Howard Marmon's several candidate designs.

Joe Felts recalls that when he returned to Marmon in 1926 he was put to work by George Freers making tracings (engineering drawings) of original Continental Motors Corporation drawings on Marmon Motor Car Co. drawing forms.

Without question the Little Marmon engine was manufactured in the Marmon plant. Layouts and pictures of the engine manufacturing and assembly facilities exist. Evidently the time for development of the combustion chamber and the durability development of the moving elements was saved by purchase of rights to supposedly proven engine components.

When the call 'Gentlemen Start Your Engines' was heard at the Speedway on May 30, 1927, thirty-three straight eight engines roared into mechanical life. It certainly appeared that the era of the straight-eight had arrived . . . G.M. Williams was convinced and for Marmon Motor Car Co. the Fourth Era would indeed be the Straight-Eight Era.

## PART 1 — The Cars "Eight Firing in Line"

Marmon made certain that the arrival of the Little Marmon was an event of significance. It was publicly presented for the first time at the New York Automobile Show on Jan. 8, 1927 accompanied by the largest display of cars ever assembled by Marmon. A total of forty cars was shown on the floor at Grand Central Palace, at the Commodore and Belmont Hotels and at the showrooms of the Marmon Automobile Co. of New York.[5]

Marmon claimed the Little Marmon to be "a small car with the distinction and individuality which heretofore have been associated with only the largest, most luxurious and most expensive automobiles".[5] In support of this claim, a very handsome Little Marmon town car with body by Locke was exhibited at the Belmont Hotel show. The press was impressed at the amount of comfort space provided in such a small automobile. It is pictured in Chapter Nine Part 1.

Fig. 5-1 LeBaron four-door brougham was a regular premium offering.

Fig. 5-2 The two passenger Speedster performed as superbly as it looked.

The chassis was remarkably low because the severe double drop in the frame took maximum advantage of the hypoid rear axle. As a result the open body styles were correspondingly low, racy and attractive; An appearance appropriate to the spectacular performance of the car.

Unfortunately Marmon sought to provide in the sedan bodies the interior dimensions of a large car and used the low chassis to that end, sacrificing reduced overall height otherwise feasible. This combined with the packaging problems of the narrow tread to result in styling compromises not unlike those evidenced by the short-lived Hudson Jet of 1954 . . . the appearance of a larger car 'compacted' to smaller dimensions . . . just as described by Mr. Williams in the initial press release.

The front compartment was contemporary and not distinctive except in the use of inlaid wood trim on the upper dash and wainscoating below the window sills and in the grade of cloth trim in a car of that price range. The instrument cluster reflected possible influence of the 1926 Cadillac 314 cluster. The top-pivoted, swing-out, windshield required ugly slotted support arms on both pillars; it lacked the neat appearance and convenient one-handed operation of the Fisher Body system already on some General Motors car lines.

The Little Marmon was offered in both standard and custom body styles. The most attractive Little Marmon closed body was the Custom Victoria . . . actually a close-coupled two-door sedan with permanent separate trunk. It lacked the integral sun visor that characterized, probably detrimentally, the sedans and the Series 75 as well. It was such a pleasant design that it would survive the Little Marmon and grace the Marmon eights as late as 1931.

Body styles were:
Coupe, 2 door, 2 window sedan, $1795; Coupe roadster, $1995
2/4 pass Speedster, 4-pass Speedster, 4 door, 3 window sedan, $1895
Custom styles extended the range up to $4175.
4 door, 2 window sedan Victoria
4 door, 3 window sedan Cabriolet

The custom four-door bodies had a five passenger rating, with exception of the Cabriolet (or Brougham) which was a custom version of the standard sedan. All standard sedan bodies were rated at four passengers. The Cabriolet had a leather covered rear quarter with landau bar. The two custom sedans had an unusual belt moulding which curved upwards on the cowl and then ran aft higher than the hood moulding.

In mid-April, the addition of a 4-door brougham was announced (The Series 75 brougham was very popular) at a price of $1895. That brought the model line to 11, including 4 custom built bodies.

The Little Marmon departed from the novel frame and suspension that had characterized every Marmon since 1915. It is probable that improved road conditions dictated the suspension change. The suspension was however novel in (1) the frame attachment of the 4 semi-elliptic springs with rubber isolators exactly as used today on all domestically sourced Mack Trucks in lieu of spring shackles, and (2) the long length (22″ less than the wheelbase) of the springs. Hotchkiss drive via metal (Cardon-type) U-joints replaced the torque tube drive. The entire chassis was unusually low, as a result of double-drop side rails and a pioneer application of a hypoid (2½″ vertically offset pinion) gear set in the rear axle which was otherwise of conventional banjo construction. Wheelbase was 116 inches. Tread was 54 inches.

Although the integral running board concept was abandoned, the otherwise conventional "boards" were of all-steel construction and mounted to provide side-impact protection. Four wheel Bendix 3-shoe self-energizing brakes (12 x 2) as adopted also for the model '75' were fitted and actuated by mechanical linkage of notably short cross-shaft configuration for rigidity.

The sedan weighed 3092 pounds, the Custom 3-window sedan at 3172 was heaviest; the 4-passenger, 4-door Speedster at 2977 was lightest.

Fuel capacity was 12 gallons.

The engine was a new OHV straight eight of 2¾ inch bore and 4 inch stroke, displacing 190 cubic inches. With a compression ratio of 5.27:1 it was SAE rated at 24.2 horsepower and developed 64 brake horsepower @ 3200 rpm.

The engine developed .33 horsepower per cubic inch of displacement resulting in "more power and wider range of performance than any other engine in the world, outside of aviation and the speedway". The Little Marmon was also advertised as having "Fastest Acceleration Ever Known" and getting 18-22 mpg with negligable oil consumption."[6]

Fig. 5-3 The size and performance of the Speedster made it a favorite of young ladies.

The superior engine performance was credited to large breathing capacity, scientific valve mechanism design and a "Marmon invention . . . Equi-distant dual downdraft intake manifolding". The manifold was fed from a single updraft carburetor and established the need for carburetor manufacturers to design downdraft carburetors. The engine also featured a dynamically balanced crankshaft carried on five main bearings and fitted with a front vibration damper.

The engine was unit mounted (i.e., integral with clutch and transmission). A cam and lever steering gear was fitted. The axle ratio was 4.7:1; 5:1 on closed cars. Tire size was 29 x 5.25.

The Little Marmon offered a compact car with superior rather than typical inferi-

Fig. 5-4 Impressive appearing Model L engine.

or acceleration, speed and handling. Public enthusiasm was noteworthy but not necessarily reflected in production totals. The traditional radiator shell made it unmistakably a Marmon. The radiator cap was topped with a newly sculped "Eagle at rest" . . . "Sinewy Steeled Muscle under feathered wing" unique to the Little Marmon which incidently carried no other advertised model designation. Specification data listed the model as "L". The meaning of "Little" is evident from the advertisement showing the "Little" and "75" phaetons side by side.

It was soon evident that the enthusiasm of the expanding dealer organization was not shared by the public. Traditional Marmon owners crowded showrooms at introduction time but did not represent the market segment at which the Little Marmon was targeted. Production settled down to a rate of about 100 cars per day. By July 1927, magazine ads promoted the "Marmon Eight" without any reference

Fig. 5-5 Little Marmon Chassis.

to "Little", placing emphasis on performance. Frantic efforts were underway in Indianapolis to implement styling changes departing from the "compact" appearance: normal fender side-overhang, larger headlights, wider radiator. And after only one year, a new body of pleasing lines and devoid of resemblence to the Series 75 bodies.

## *Models (T) 68 and (N) 78*
## *1928-29*

The "Little Marmon" lasted only one year . . . 1927. In December 1927 Marmon announced a Series 78, a completely new eight-cylinder car on a 120 inch wheel base and priced the same as the Little Marmon . . . just under $2000. The 'compacted' look was gone consequent to adoption of standard tread and substantial side overhang. Also absent was the integral visor; the new bodies were much more pleasing in appearance. The series 78 however was not the successor to the Little Marmon.

Fig. 5-6 Series 75 with new little brother.

At the New York Auto Show in January, 1928, a companion Series 68 was introduced. Slightly shorter than the Little Marmon at a 114 inch wheelbase, it was priced surprisingly lower . . . under $1500, and also had a standard tread.

For engineering development and technical accounting for procurement and for service logistics, the Fourth and Fifth Era Marmons were designated by letter model designations. Many parts and component assemblies were common to more than one concurrent or successive model and retained the letter model prefix initially used. Merchandising model designations were often changed in response to external market requirements rather than internal needs; hence service parts were all procured via the letter model route.

## SERIES 78
## 1928

The Series 78 was an upgraded and enlarged automobile intended to replace the 'Large Marmon' Series 75 which persisted into 1928. Although it certainly was not, sizewise nor pricewise, in the same market segment as the Series 75 had been, every effort was made to build it in the best Marmon tradition in hope of converting the traditional large car buyers while attracting a vastly larger market. Prices ranged from $1895 to $1995.

The press was impressed. One reporter[7] wrote: *"The interiors are more interesting than the exteriors. Inside, the closed cars look like $4000 automobiles. An unusually fine job of harmonizing the interior and exterior colors has been done. Saddle cushions with form-fitting (seat) backs are upholstered in a warm tan broadcloth with a darker brown figure on the background. Bronze hardware in a tan tone for harmony lends to the great richness of the interior; a similar finish is used on dash and control levers. Window moldings and belt panel are solid walnut. The ladies vanity and the smoking set are covered in tan leather. A decorative inlay on the belt panels and cowl panel is a nice touch."*

Exteriors were finished in the 'Jewel Effects' color schemes introduced on the Series 75 the prior year. Mica Schist Gray, Emerald Green and various combinations of blues, grays and browns were available. Many full color advertisements were run to communicate the beauty of these automobiles to the consumer.

Although the Series 78 (sedan) was 317 pounds heavier than the Little Marmon, the Bendix 12 x 1.75 (12 x 2 drum) brakes were retained.

The radiator was larger for both appearance and increased cooling considerations.

Headlights were unusually large and a new type of parabolic depressed-beam type. Cowl lamps were mounted on a 'cowl surcingle' which also mounted the brackets for side mount spares. The headlamps were supported by vertical nickel-plated standards as well as the usual cross-brace. The Marmon 'pyramid' was die formed on the leading edge of the front fenders.

The instrument cluster was a panel of etched bronze situated unusually far forward. It was indirectly lighted.

Fig. 5-7 The Model 78 benefitted from a full length belt molding and long hood and was always shown in six wire wheel trim.

Fig. 5-8 Driver's compartment of (first series) Model 78.

Weights and Prices of the Model 78 styles.

Available body styles:

| | Shipping Weight pounds | Price |
|---|---|---|
| Sedan, 4 door, 5-passenger | 3104 | $1895 |
| Coupe with rumble seat, 2/4 passengers | 3097 | $1895 |
| Victoria, 4-passenger | 3206 | $1995 |
| Speedster with rumble seat, 2/4 passengers | 3007 | $1895 |
| Speedster, 4 door, 4 passenger | 3052 | $1965 |
| Coupe roadster (collapsible) 2/4 passengers | 2987 | $1995 |

The Series 78 was powered by the Little Marmon engine with the bore enlarged. It had a 2.937 inch bore and 4 inch stroke, displaced 216.8 cubic inches. The SAE rating was 27.6 horsepower and the developed horsepower rating was 86. Speeds in excess of 70 miles per hour and acceleration from 10 to 50 miles per hour in 15 seconds were claimed. The 'Lightning Performance' of the Little Marmon had been retained.

Fuel was supplied by a Stewart Vacuum (tank) system from a 14 gallon tank, via an Alemite fuel filter.

## SECOND SERIES 78
## 1929

Surprisingly Marmon announced in mid August, 1928 that the '68' and '78' had undergone styling changes. Evidently the frantic effort to eliminate "Little" from the Marmon Eight had delayed introduction of the '68' and '78' until January 1928 and the mid-August 1929 restyling was a return to the usual Marmon model-year.

The hood louvers were horizontal rather than vertical and smaller headlights combined with a larger, higher radiator shell to enhance appearance. The standards were eliminated and the lights mounted higher and more separated. Beneath each light was an 'M' ornament. The Lightning Flash on the radiator core was continued. The double-concealed hood hinge, used since 1916, was replaced with a chromium piano hinge.

A major improvement was the fitting of new bumpers. The first Series had dual-bar bumpers which appeared appropriate to a very small car, were quite flat and had two grooves which, paint-filled or not, dulled the appearance of the nickel plated bumpers. The new bumpers were of specially rolled oval faced bumper stock and were chromium plated, as was all bright work on the Second Series. A center ornament carried the Lightning Flash.

A new instrument panel was featured. All instruments were in line on an etched silver panel and all were full circular shape. Heavy black numerals on white background and adequately wide pointers made the gages much more legible than the First Series where the engine monitoring instruments were curved and deep set, revealing only the tip of the pointers. The clock was electric, the speedmeter read to 90 mph; ammeter, oil pressure and engine temperature gages completed the set.

The ignition switch, carburetor control and instrument light switch were included in the cluster.

A celotex panel was placed "on the dashboard" to reduce heat transmission. Side mount spare hardware was revised for increased rigidity and a 'theft-proof' flush lock provided.

The Marmon badge on the radiator became gold-plated.

A new coupe body replaced the former and had styling consistent with the sedan. At start of production there was no four-door Speedster or touring body nor a collapsible coupe.

Marmon advertisements during 1929 carried the slogan "Great Cars Since 1902" officially establishing that year as the reference year for future anniversaries. The Marmon Club, an association of historians and restorers interested in the Marmon marque, met at the Indianapolis Speedway in September 1982 to mark the eightieth anniversary of the Marmon car.

The Marmon cars at the 1929 Auto Shows were the same as those extensively enhanced in styling in Aug. 1928 and identified as "Second Series". There were a number of interior refinements, added comfort features and new combinations of Marmon Jewel colors. Adjustable front seats and foam rubber rear arm rests were new.

On the Series 78 only, the genuine walnut wainscoating beneath the windows was extended to the third side windows; Nachman-type cushion springs were adopted for both front and rear seats.

Subsequent to the shows, two entirely new body styles were introduced. They were a 'collapsible' coupe and a 'touring speedster'. The latter was planned with a unique seating capacity for six, which was achieved by resort to a curved body side; Marmon called it a 'canoe shape'. These were handsome cars with tops that were engineered to fold and store in a six inch boot. Some data indicates that only 17 touring speedsters were built; at least one Series 78 example survives.

Second Series Model 78 sedan, coupe and roadster were priced at $1965; the (custom) victoria!was higher at $2065.

Fig. 5-9 The-Second Series — an early introduction of the 1929 version.

## SERIES 68
## 1928

The Series 68 differed from its larger companion Series in wheelbase and body but lacked the compacted appearance of the Little Marmon. The 114 inch wheelbase made it 2 inches shorter than its predecessor, and 6 inches shorter than the Series 78. While handsomely styled, the apparent difference in length was greater than the actual due to use of 'coach' styling which separated the '68' body proper from the smooth cowl and hood and accentuated the difference. The Series 78 had full length belt moldings and the hood length was exaggerated by an overlap of the cowl.

Body interior space was comparable to the Series 78 but appointments were less luxurious. Whereas the senior series had Ternstedt body hardware, the Series 68 used Butler fittings in a silver finish.

Fig. 5-10 Instrument panel of the New Series (1928½) Model 78.

Fig. 5-11 The Model 68 was typically shown in standard form.

The 1928 '68' was powered by a new L-head straight eight engine having the same 2.75 bore as the overhead valve Little Marmon (and probably the same cylinder spacing). Stroke was longer at 4.25 versus 4.0, Displacement was 201.2 compared to 190, the SAE rating remained at 24.2 but the horsepower was 72 @ 3200 compared to 64 @ 3300.

Body styles included a sedan and 2-passenger coupe priced at $1395, a victoria at $1450 and a rumble seat coupe with leather-like roof at $1495. The sedan weighed 2897 pounds (shipping weight).

## SECOND SERIES 68
## 1929

Both the 68 and 78 experienced design refinements in August 1928, some seven months after announcement of the Series 68. The '68' shared most changes with the '78' . . . the horizontal hood louvers, larger and higher radiator, new and differently placed headlights with 'M' ornaments beneath, attractive new oval-faced bumper bars joined at center with a plate bearing the Marmon Lightning Flash, a piano-type plated hood hinge, and chromium rather than nickel plating.

The changes were particularly helpful to the Series 68 as they muted the stubby hood appearance of the First Series. In addition, the '68' acquired the Lightning Flash on the radiator core for the first time and the side mount option was extended to the Series 68. The '68' also acquired a new instrument cluster but one unique to the series . . . lacking the clock but otherwise similar to the '78'.

The performance of the car was increased by enlarging the bore of the engine .062 to 2.812 and the displacement to 211 cubic inches. SAE horsepower became 25.3 and rated horsepower was 76 at 3200, up from 72. A new volute type water pump was adopted and, with the larger radiator, the cooling system capacity increased to five gallons.

Subsequent to the shows in January 1929, the '68' shared with the '78' two new bodies: a 'collapsible coupe' and a 'touring speedster'. Common bodies meant common styling and these two '68' styles therefore had full length belt moldings and were distinguishable only by the hood length.

Second Series prices were higher by about $100: Sedan and rumble seat coupe $1465, Victoria $1520 and roadster with rumble seat $1565.

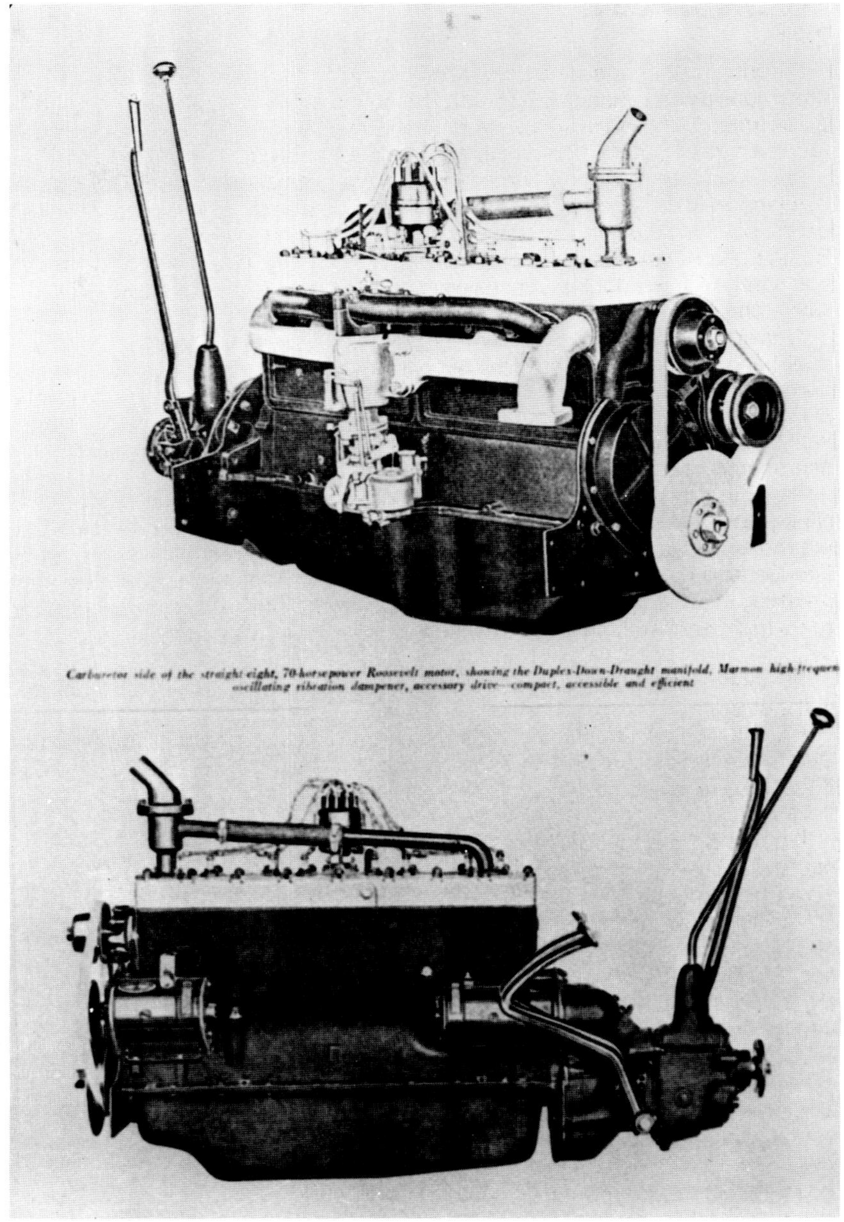


Fig. 5-12 The Model 68 engine rated 72 horsepower at 3200 RPM.

Fig. 5-13 The Second Series '68' collapsible coupe.

In April 1927 Marmon had reported the addition of 300 dealers and distributors in the first quarter of 1927 and projected a production beyond the then current 100 per day. But even the mid-sized '68' and '78' team did not provide a solution to the evasive attainment of large sales volume. In announcing, on Dec. 22, 1928, the imminent introduction of the Roosevelt, President G.M. Williams acknowledged that production in 1928 had been a steady 100 per day since introduction... no better than the 1927 rate of the Little Marmon, and now the production of the Series 75 had ended as well.

*The Roosevelt Venture*
*"World's First Eight Under $1000"*

## MODEL (S) ROOSEVELT 1929

The debut of the Marmon-built Roosevelt was heralded in advance when Marmon introduced their refined Models '68' and '78' at the 1929 New York Show. Details other than a $1000 price tag were withheld. In a pre-show press conference reported on Dec. 22, 1928, G.M. Williams was quoted as follows: *"In his statement regarding the new car, Mr. Williams said its design was the result of a long period of experimental work by the Marmon Engineering Staff headed by Col. Howard Marmon, Vice-President in charge of Engineering, and T.J. Lytle, Jr. Chief Engineer. He said most factory equipment required (for production at 200/day) was already installed.*

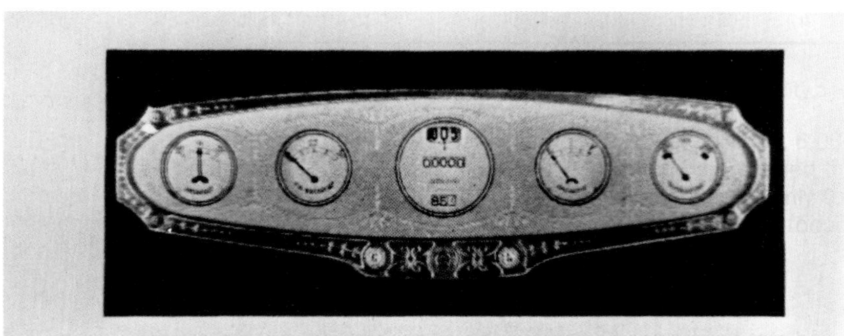

Fig. 5-14 Second Series 68 instrument cluster.

The new car was not seen at the 1929 auto shows. It was shown to the press and trade only in the Palm Room of the Belmont Hotel during the New York Shows. The first unit came off the conveyor line in Indianapolis on February 20, 1929. Public announcement followed in March.

The new Roosevelt had the distinction of being the first eight-cylinder car in the world to sell for less than $1000 . . . the four-door, five-passenger sedan listed at $995. This was a distinction of sufficient caliber to warrant listing today in the Guinness Book of Records. (Marmon's other Guinness listing is for factory installed radio, also in 1929.)

Marmon didn't do anything 'half-way'. The Marmon Book of Facts, 1929 Edition, included a Part IV Technical Section which that year addressed the new Roo-

sevelt product offering in great technical detail. It didn't stop there! In fact it rather began with a dissertation on the pronunciation of the new product's name:

"There are (or were in 1929) several pronunciations of the name Roosevelt. All Marmon salesmen should be in accord on this point; so in the following is an authority for the correct pronunciation.

In a letter received from the Roosevelt Memorial Foundation of New York City the following quotation is given from an article which appeared in the New York Sun for May 25, 1903, written by Robert B. Roosevelt, the uncle of Theodore Roosevelt: *"In English when we try to distinguish the long from the short 'o' we get into trouble. In Dutch they do not. The double 'o' is simply a long 'o'. The word 'Ross'; means rose and is pronounced identically the same way under all circumstances and in all combinations. So the first syllable of the President's name is 'Rose' pure and simple. "But the following 'e' like the short German 'e' or like the silent French 'e' when read in poetry, is slightly aspirated. An English analogy is the word 'the'.*

*It is not pronounced at all as it is spelt, not like 'thee', but with a sort of 'th' and a breach stopped by the tongue on the teeth. So the name is 'Rose- (uh) -velt'."*

Other body styles besides the sedan were a victoria (close-coupled two-door with separate trunk), coupe and collapsible coupe both with rumble seat, and all at slightly higher prices. They were mounted on a 112.75 inch wheelbase chassis of standard tread, and measured 172 inches overall. This made the Roosevelt only 1.25 shorter in wheelbase and only 2 inches shorter overall.

Trade editors described the new offering as characteristically Marmon in appearance but with a high narrow radiator shell bearing the cameo portrait of President Teddy Roosevelt at the top and a large letter 'R' on the crank hole cover. The letter 'R' also was evident on the bumper center plate and on the octagon hubs of the Hayes wire wheels.

The Roosevelt was a handsome automobile, viewed from any vantage point, particularly in the optional side mount configuration. It was the credible work of Count Alexis De Sakhnoffsky, styling consultant to Hayes Body and Marmon, who gave it a lithe athletic look. Reportedly he proposed, and promoted unsuccessfully, the placement of the cameo portrait on the radiator core, pendant to a necklace from the shell.

Other than the shorter wheelbase, there was really very little compromise from the Marmon 68. The engine was the 1928 Model 68 (prior to the 1929 increase in cylinder bore) and carried the Marmon "oscillating modulator" (crankshaft vibration damper) and themostatic cooling system. Although the chassis was unique to the Roosevelt, it featured, like the other Marmons, side-guard steel running boards, a 6 x 2 inch channel frame, cam and lever steering gear, elastic-drive clutch and "single-button" control on the steering wheel for starting, lights and horn.

The Roosevelt however carried self-adjusting metal spring shackles of a new design — the side plates were clamped against tapered shackle pins by a center bolt and belleville washer unit which compensated for wear. Single-acting Gabriel hydraulic shock absorbers were fitted but priced separately. (The Marmon 68 and 78 had single-acting Lovejoy shock absorbers also priced separately).

Brakes were Bendix 'duo-servo' but 11 inches in diameter whereas the '68' and '78' had 12 inch drums. Raybestos molded linings (146 square inches) were used whereas the '68' and '78' had woven asbestos linings (224 sq. in.) with 50 percent more area. The Roosevelt was only 200 pounds lighter. A 50/50 brake effort distribution was used on all three car lines.

The Roosevelt had a leaner appearance than the Model 68. The smaller tire size and the unique radiator and hood, headlamps and support bridge were contributing factors. Parking lights were attached to the cowl just ahead of the windshield. Tire size was 5.00-19 (versus 5.50-19 on the '68') and sedan weight was 2695 pounds. (versus 2896 for the '68'). Dual vent doors in the top of the cowl avoided disturbance of the raised hood and cowl center section molding.

To avert any suspicion that the Roosevelt represented compromised quality or durability, an unusual test run was made on the Indianapolis Speedway Track in June 1929. A Roosevelt sedan, selected at random by AAA representatives, who subsequently officially observed the test, was driven 13,457 miles without stopping. Refueling and remanning was accomplished by use of a companion "Chase car". Average speed for the entire test run was 30.5 miles per hour. The test was initiated to demonstrate economy of operation but could better be classed as an endurance run.

Following completion of the non-stop automotive endurance record of 440 hours and 30 minutes, the engine and driving mechanism were immediately disassembled in conformance with the rules of the American Automobile Association. The purpose of the inspection was to determine that the test car was in every respect a true stock automobile. The examination team consisted of Louis Chevrolet, veteran racing engineer, Charles Merz, AAA representative, and S.E. Rowe, chassis engineer at Stutz Motor Car Co. (later to join Marmon).

Fig. 5-15 The Roosevelt was well styled with great attention to detail.

Subsequently "every part was subjected to careful scrutiny by Thomas J. Litle Jr. and members of his engineering staff. They found the engine to be in perfect working order and examination of the pistons valves and other engine mechanism failed to disclose appreciable wear or defects. Axles, wheels, steering gear and transmission were found in original condition."[8]

The end of the record breaking run was forced by a cloudburst and violent windstorm which descended on the Speedway and made driving practically impossible.

Despite the constant running over the eighteen year old brick surface, no sign of fatique crystallization of metal parts was evident.

Another pre-production Roosevelt embarked on a 30 day trip of 18,224 miles (608 per day) which took it to all (then) 48 state capitols. The trip was reported completely trouble free.

Two initial production cars were driven cross-country at a rate exceeding 400 miles per day.

The 1929 Pikes Peak Hill Climb Championship was won by William P. Bentrup driving a Roosevelt. It was first in its class reaching the summit in 26 minutes, 56 seconds.

The Roosevelt promotion was extended to the radio media. In late October of 1929 a series of thirteen Monday evening programs under Marmon sponsorship was announced. A biography of Theodore Roosevelt was to be presented in thirteen incident portrayals.

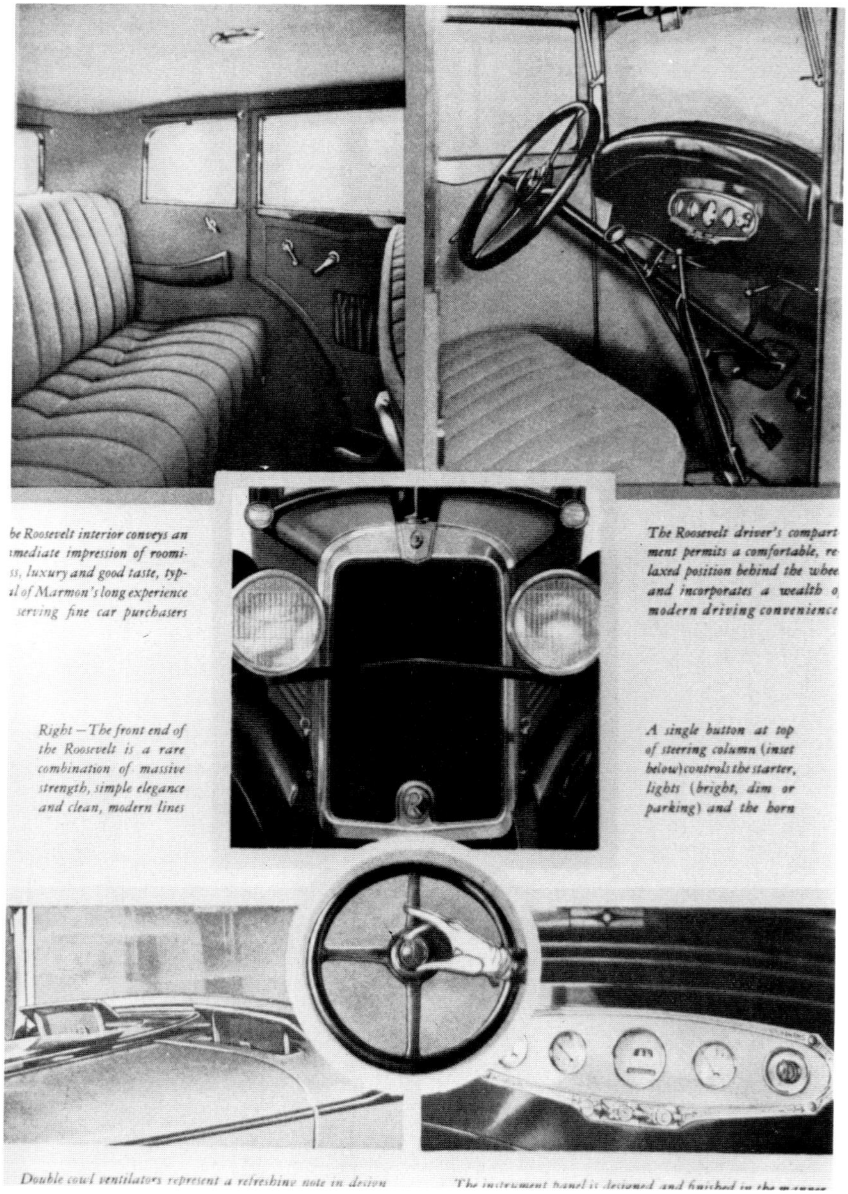


Fig. 5-16 Upper Left: Roominess, luxury and good taste.
Upper Right: Doors had 'walnutized' metal trim.
Center: Shell and hood had raised center section.
Lower Left: which widened between twin vent doors.
Lower Center: Starter, lights and horn on a single knob.
Lower Right: Ignition key replaced '68' temperature gauge.

## MODEL (S) MARMON-BUILT ROOSEVELT 1929 3/4

In Oct. 1929, design refinements were evident. An enlarged bumper plate carried "Roosevelt 8" and the radiator core was emblazoned with the Marmon Lightning Flash. Headlights were larger. Significantly, all advertising extolled the "Marmon-Built Roosevelt".

When Marmon and Roosevelt dealers met at the factory December 9-11, 1929, the Jubilee celebration was highlighted by a Roosevelt driveaway. The dealers were addressed by Walter Dorwin Teague who was introduced as styling consultant to Marmon Motor Car Co. This was the first evidence of Mr. Teague's association with Marmon.

Note that another official baseline for anniversaries had been established . . . 1904 (start of production) whereas 1902 (first car) was the basis for the jubilee celebration in 1927. Much like Gilbert & Sillivan's "Let the punishment fit the crime" — "Let the base date fit current needs".

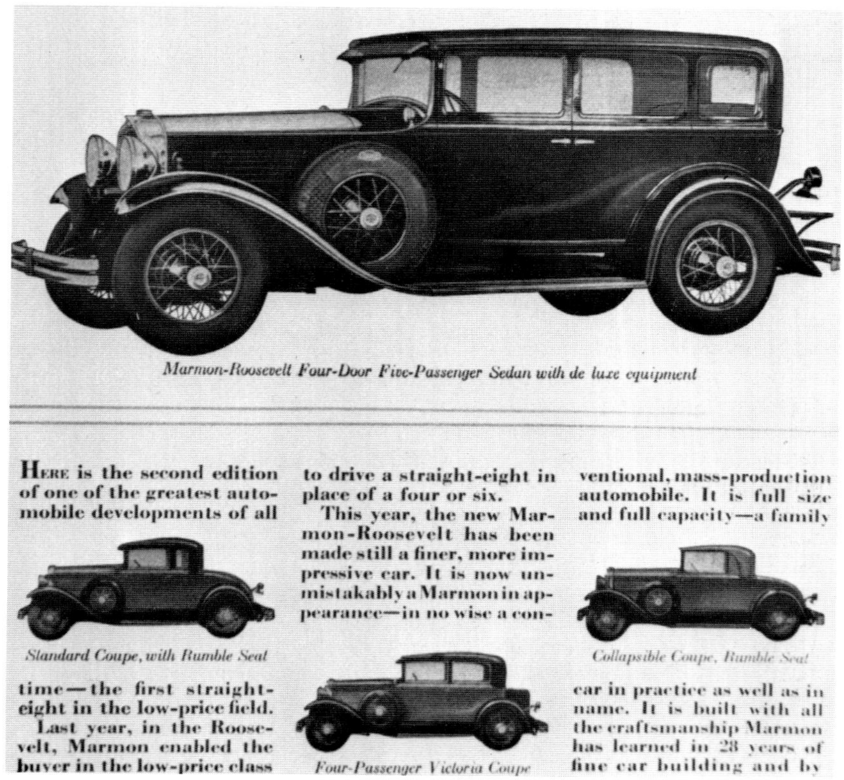


Fig. 5-17 The 1930 MARMON-Roosevelt.

## BODY CONSTRUCTION AND TRIM: ROOSEVELT, 68, 78

The Roosevelt body and fenders were identical with the 1928 Model 68 except for the cowl and the extension of the non-metallic roofing down to the gutter over

Fig. 5-18 Roosevelt being refueled during non-stop record run.

the doors. The new cowl extended the belt molding forward for continuation on the hood and allowed the greater engine intrusion which resulted from the shortened wheelbase. Like the Model 78 body, the cowl extended under the hood several inches so that a longer hood would contribute to perceived length, despite the shortened wheelbase and retained interior dimensions.

The Roosevelt weighed about 200 pounds less than the Marmon 68. This resulted partly from smaller brakes but largely from advanced body technology applied to the forward part of the basic '68' body. That body was of traditional composite wood/steel construction but featured a continuous steel skin from one rear door pillar around the rear to the opposite door pillar; it was formed by flash welding several panels.

The cowl sub assembly, including windshield pillars, was a unit steel structure. The sides and top, including the new distinctive integral moldings, was formed from one piece of steel; it was electrically welded to the bulkhead (or firewall) stamping and the pillars similarly attached.

In addition, the forward pillar of the front door was converted to steel. The front door was therefore supported on two steel members hinged together. In addition to superior strength and durability, the very slender combined corner post enhanced vision. This was markedly superior to the contemporary Fisher Body construction used in low and medium price cars.

The center post on the sedan was wood encased in a one piece steel casing. Each door panel was formed from a single steel stamping extending the full height of the door.

A review of the seat construction used in these three car lines, in the $1000, $1500 and $2000 markets, will serve to document practices then used.

The Roosevelt seat cushions were described as 'davenport type'; the '68' were described as 'comfort type' and the '78' as 'saddle type'.

The Roosevelt springs were covered with a steel mesh beneath the 'thick padding of the upholstery'. The '68' used 5-inch pipes filled with thick cotton batts over

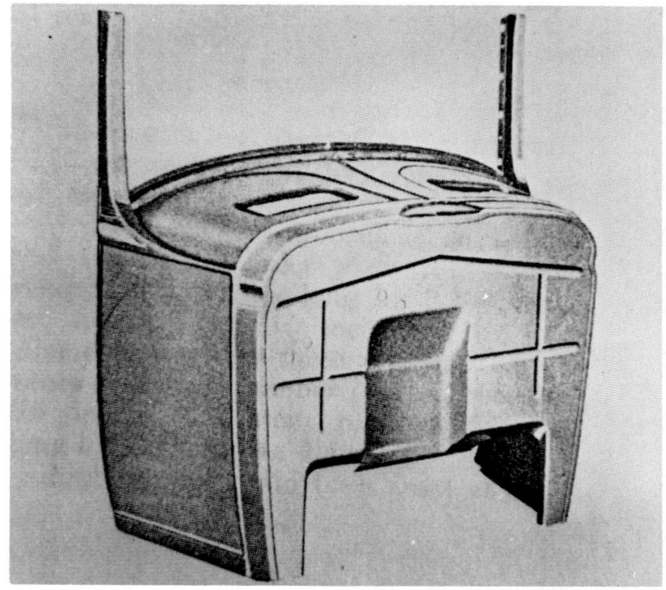

Fig. 5-19 The welded steel cowl structure of the Roosevelt body.

Fig. 5-20 The composite construction of the Roosevelt body.

Fig. 5-21 The Big Eight was large, handsome, distinctly Marmon with an agile look.

a thick padding, thoroughly stitched in place. The '78' was similar but with a thick layer of curled hair beneath the cotton batts.

The Roosevelt upholstery cloth was pleated Broadcloth with the stitching retaining the padding. Buttons were retained by rustproof chain rather than conventional twine. The '68' was upholstered in mohair, the '78' in broadcloth; the latter had two rows of buttons on the seat, one on the cushion. All three had sponge rubber filled rear-seat arm-rests.

Interior lighting was by dome light; on the Roosevelt the switch was on the fixture, on the '68' on the pillar, and on the '78' both on the pillar and automatically upon rear door opening; the '78' also had 'torchiere' reading lamps in each rear quarter panel on the same circuit.

The '68' and '78' had tufted elastic pockets on the rear doors and a cowl pocket 'on the right hand side underneath the cowl for small hand tools'. The Roosevelt instead had tufted elastic pockets on the left rear and on the right front door.

All had vertical, top-hinged windshields with nickel support quadrants.

The Roosevelt had two hinges per door, the others three, but all three had Atwood rubber bumpers, integral rubber insulated dovetails and heavy bronze double-check striker plates. The Roosevelt had concealed metal door checks, the others exposed rubber check straps.

Hardware on the Marmons was by Ternstedt; of special design. Roosevelt doors were specially treated to prevent rumble.

The full 6 inch chassis differential length of the '78' over the '68' appeared in the hood panel length. However the cowl projected under the hood an additional 2.25 inches for additional front compartment leg room and thicker cushions.

**Roosevelt Market Comparison on Sedan Models**

A comparison of competitive cars[9] as to price, horsepower and specific horsepower (hp per cubic inch) shows the significance of the Roosevelt.

| Name | List Price | Cubic Inch Displacement | Horse-power Developed | Horse-power per Cu. In. |
|---|---|---|---|---|
| ROOSEVELT | $995.00 | 201.9 | 70 | .347 |
| Chevrolet | 675.00 | 194.0 | 46 | .237 |
| Chrysler 65 | 1,145.00 | 195.6 | 65 | .332 |
| Oldsmobile | 1,025.00 | 197.5 | 55 | .278 |
| Pontiac 6-29 | 845.00 | 200.0 | 58 | .290 |
| Dodge 6 | 995.00 | 208.0 | 58 | .279 |
| Hupmobile A-6 | 1,395.00 | 211.5 | 67 | .317 |
| Moon 6-72 | 1,495.00 | 214.7 | 61 | .284 |
| Peerless 6-61 | 1,195.00 | 214.7 | 62 | .289 |
| Reo Mate | 1,395.00 | 214.7 | 65 | .303 |
| Jordan 6 RE | 1,395.00 | 214.7 | 66 | .307 |
| Nash Spec. 6 | 1,345.00 | 224.0 | 65 | .290 |
| Graham-Paige 615 | 1,195.00 | 224.0 | 76 | .339 |
| Gardner 120 | 1,595.00 | 225.7 | 65 | .288 |
| Oakland | 1,245.00 | 228.0 | 68 | .299 |
| Buick 116 | 1,320.00 | 239.1 | 74 | .310 |
| Studebaker Dict. | 1,295.00 | 242.0 | 67 | .277 |
| Studebaker Com. | 1,495.00 | 248.3 | 74 | .298 |
| Stearns-Knt. 68-80 | 2,495.00 | 255.0 | 70 | .274 |

## MODEL '(S)' MARMON-ROOSEVELT 1930, FIRST SERIES 1931

Roosevelt production had been 25089 cars in 1929. After the auto shows, a new edition of the Marmon-Built Roosevelt was announced . . . the 1930 Marmon-Roosevelt with traditional Marmon radiator shell!

The registration weight of the Marmon-Roosevelt was 128 pounds more than the Roosevelt, evidencing increased commonality of components. The 2823 pound weight was only 73 less than the '68' but retained the distinct body and frame of the Roosevelt. Fuel (14 gallons) and spare tire added 126 pounds for a total of 2959.

The 1930 Marmon-Roosevelt was the only car in its price class to be offered with a full one-year warranty. Also, every 1930 Marmon-Roosevelt was subjected to a 50 mile road test before acceptance for shipping. Both were maneuvers obviously intended to offset the widespread rumors of engine durability problems with the 1929 offering; a high price to pay for the 1929 quality problems.

The sales department announced on Jan. 13, 1931 that all Marmon-Roosevelts scheduled for build prior to model discontinuance had been sold. This situation served to expedite the start of production of the 1931 (Second Series) version — The Marmon 70.

The Roosevelt marque was abandoned completely at the end of the 1930 calendar year. Cars built after August were designated 'First Series' 1931. Little more than a year later, political activity began which would make Franklin Roosevelt the democratic candidate and the next President of the United States. What appeal would have accompanied a Roosevelt shell emblazoned with overlapping cameos of Teddy and Franklin!

## MODEL R

The export version of the Marmon-Roosevelt was advertised abroad as the Marmon Model R. It differed from the domestic version in using the Model Q engine (adopted for the Model S during 1930 model run). A running change was also made to adopt the oval Marmon badge (as used on the Model N) in place of the Marmon-Roosevelt badge.

Roosevelt sales (1929) were less than 20,000 sedans, 2500 coupes, 1500 victorias and 1000 convertibles. This limitation is established by the starting serial numbers for the Marmon-Roosevelt. The total would then be less than **24,500** units. Marmon registration information establishes production at 3309 sedans, 652 coupes, 171 victorias and 211 convertible coupes, for a total of **4343** cars. Total production of the Model S . . . Roosevelt and Marmon-Roosevelt . . . then would appear to be something less than 30,000 units.

However, Marmon cars built after August 1930 were identified as 'First Series 1931' with true 1931 models coming in January 1931. Production was very low during the 'First Series 1931 duration. Assuming 4000 units, total Model S production would be about 34,000 units in two years.

In an effort to counter widespread talk of serious engine durability problems which were drastically curtailing sales (while the economy was still robust), G.M. Williams announced that Marmon straight-eight cars purchased after May 1, 1930 carried a 12 month, 12,000 mile warranty instead of the usual industry 90 day warranty. The press release noted the manufacture of 65000 Marmon eights by approximately April 30, 1930.[10]

## MODEL (H) BIG EIGHT
## 1930, FIRST SERIES 1931

During 1928 and 1929 Marmon had ignored their traditional market segment. Admittedly tabular industry data does list a **1928** Model E-75 and a quality full-color brochure for that model was distributed at the 1928 New York Show. But there was no mention of the offering in national advertising and the sole difference evident was an additional black paint band beneath the belt molding . . . readily added to left over 1927 models. So for all practical purposes the Marmon 78 was the luxury offering from Indianapolis for those two years. The unheralded demise of the most successful of all Marmons found the traditional Marmon purchaser abandoned. With $3000 to $4000 to spend, he found only $2000 models, however well finished, in his Marmon dealer's showroom; he presumably then sought a match of product and pocketbook elsewhere.

In October 1929 an announcement was made that Marmon was to make "A reentry into the higher-priced field." The new car was to be called the Marmon Big Eight and president G.M. Williams acknowledged the recent void in the traditional market segment when he said: "It will serve to re-establish Marmon in the fine car field in which it has been noted in the past and in which it has a large clientele of owners." He directed attention to the growth of the Marmon distributor and dealer organization to five times its size at the time Marmon produced large cars exclusively.

Introduced in December 1929, the Marmon Big Eight was a magnificent automobile. It was large with a look of both agility and dignity. It was unmistakingly a car in the best Marmon tradition yet was superbly styled in the best contemporary flair.

Advertisements proclaimed: "The Big Eight is immediately Marmon, but unmistakably Marmon of this new day, with a charm and unusualness which sets it apart from all other cars" and "The Big Eight preserves and carries forward all Marmon fine car traditions". It was 'Big': 202 inches long on a 136 inch wheelbase and 6 inches more on the 7-passenger styles.

Noted automotive authority J.B. Nethercutt[11], in describing his restored Marmon Big Eight seven-passenger sedan, had this to say about the car and its "quality very evident to the eye":

"It's Cadillac all over again and maybe even a bit more so." Mr. Nethercutt's comment supports the initial advertisement text: "the last note in body architecture and fitments — INSIDE... roominess and luxurious detail beyond any previous fine car fitments".

Unusually large headlights each topped with the Marmon pyramid and coat-of-arms combined with vertical radiator shutters (thermostaticly controlled) and matching parking lamps on the fenders to present a distinguished frontal appearance. This was complimented by ventilator doors in the hood side panels and standard dual side mounts . . . standard wood artillery or optional Dayton wire wheels.

The bodies were characterized by sweeping lines enhanced by long windows, slightly curved at top with curved reveal moldings at bottom. The doors likewise were slightly curved at the bottom. Beneath each door, on the spalsh pan, was a rubber kick-pad edged in chromium molding.

The chromium hood piano-hinge continued over the cowl as a molding widening to six inches at the windshield.

With side mounts standard equipment, a supporting structure between the mounts was integrated in the body structure.

Innovations were safety plate glass in all windows and dual adjustable interior sun visors. Exterior door handles were 'theft-proof' with free wheeling action when locked and the detent was overcome.

On seven passenger models, when raised together, the two auxiliary seats formed a full-width continuous seat and were fully cushioned.

All doors had shirred pockets. Instead of a robe rail, two spring-filled silk ropes were fitted. The outer anchor of each rope had an assist extension.

The limousine had a clock sunk in the walnut header bar over the partition glass, and an umbrella-in-holder located in the chauffeur's compartment. The telephone had a transmitter in the right rear door pocket and a receiver in the center of the roof in the front compartment.

Upholstery was broadcloth with the headlining a lighter tone. There were four upholstery tones to harmonize with the four exterior color options. Two carpet covered hassocks were supplied rather than a foot rail; they were covered with corrugated carpet on the bottom to retain position once placed. Nachman upholstery springs were used... each in a burlap bag. One and a half pounds of genuine curled hair was used in each cushion and seat back. The same construction was used in the folding arm rests found in the rear seats and the coupe front seat.

The headlinings had a central panel of lighter tone cloth set off from the rest of the liner by a corded colored binding which was a nice finishing touch.

There were ash trays on the left and right dash, and in the ends of both rear arm rests. A vanity case was set in the rear of the front seat except in the limousine, where it and a matching smoking case were set in an enlargement of the walnut panel below the windows.

Carved piano-finish black walnut woodwork was used throughout all body styles, and included the header bar over the windshield.

The deeply recessed instrument 'board' and cluster concept used in previous Era Four cars was abandoned. Contributing to the pleasant interior appearance was the grouping of the spark, throttle, manifold-heat and carburetor choke buttons together with the ignition lock all at the center of the instrument board and the integration of horn (push), starter (pull) and headlight (rotate) controls in a single knob on the steering wheel hub.

The ignition lock was located directly in the end of a metal-sheathed 'theft-proof' coil. A pass-about type **cigar** lighter was fitted at the right of the instrument 'board'.

The front seats (fore and aft), steering column (angle), clutch and brake pedal (length) were all adjustable. The accelerator pedal was a rubber covered treadle type.

One automatic windshield wiper was standard with the second optional. Dual cowl-top ventilators were used.

The body side of the metal bulkhead was covered successively with layers of asbestos (heat), felt (noise) and finish material.

There were six closed body styles: two-passenger coupe, five-passenger four-door sedan $3260, seven-passenger sedan and limousine $3860; four-passenger, four-door (close-coupled) club-sedan and a brougham version with fabric covered rear quarter . . . both with trunks, at $3310. On the brougham, the tire covers (full with wood wheels) and trunk cover matched the rear quarter fabric.

The two-passenger coupe was a particularly handsome automobile, having quarter windows unlike the blind quarters of other contemporaries.

The Big Eight was powered by an all new large streight-eight "of Marmon design and manufacture". It was an L-head type . . . a surprising departure from company

Fig. 5-22 *"Marmon Big Eight powerplant with new Double Dome Combustion chamber, duplex carburetor, air cleaner, fuel pump and all modern engineering advancement."*

Fig. 5-23 Model 8-79 chassis; representative of the Big Eight.

tradition for the L-head engine in the Series 68 and the Roosevelt had been assumed to be a compromise for cost considerations on the lowest priced series. It had a bore and stroke of 3.25 by 4.75, displaced 315 cubic inches, and was rated 33.8 SAE forumla horsepower and developed 125 horsepower at 3400 rpm. That was 0.4 horsepower per cubic inch of displacement and a new Marmon 'double-dome' combustion chamber was responsible. While most manufacturers followed the lead and guidance of Ricardo in Great Britain, Marmon researched and developed their own intake and combustion systems. The Big Eight had the Marmon dual downdraft intake manifold concept of the '78' but was equipped with a dual updraft Schebler carburetor and an AC diaphragm fuel pump driven by a cam on the camshaft.

The Big Eight chassis had a 136 inch wheelbase, 202 inch overall length and had a widened tread (wide-track) of 58 front and 59 rear. It continued the Marmon practice of using rubber spring insulators and carried the new Bendix two-shoe self-energizing brake system with cable-in-conduit actuation.

A Warner four speed transmission was standard. It featured a silent third gear achieved with straight gear teeth. Hotchkiss drive by a tubular shaft was used.

A 22 gallon fuel tank was used.

By 1930 a wide variety of options was available. Surprisingly many of them were factory installed on all cars resulting in advertised prices which were unrealistically low.

The Marmon dealer in New York City listed the following additions[12] to the base price of the Big Eight:

| | | | |
|---|---|---|---|
| Dual wipers | $20. | Wire wheels (6) | $ 185. |
| Front bumper | 25. | Trunk | 50. |
| Rear bumber | 25. | Trunk with cases | 75. |
| Shock Absorbers | 65. | Trunk cover | 18. |
| Rear view mirror | 3.50 | Heater — hot air | 17.50 |
| Fender lamps | 15. | Heater — water | 45. |
| Spare tire (2), covers | 85. | Motovox horn | 10.50 |
| Spare rims (2) | 10. | Mirrorscopes (2) | 20. |
| Fender wells | 75. | Spotlight | 35. |
| Spring covers | 20. | Pilot Ray Lamp | 45. |
| Shutters (automatic) | 50. | Spl Robes | 75. |
| Folding arm rest | 25. | Spl Fatigue Cushions | 15. |
| Sun visors (2) | 15. | Opera Boots | 75. |
| Ash trays | 10. | Tire Chains | 8.50 |
| Cigar lighter | 5.50 | | |
| Electric Clock | 22. | | |
| Built in at Factory: | $471.00 | Dealer installed | Variable |

While the advertised price of $3300 may have helped attract new buyers, the actual price of about $3800 was not evident in the advertising and that could have further deterred the interest of the traditional Marmon clientel.

The Big Eight 5-passenger 4-door sedan weighed 4363 pounds without fuel (22 gallons) and two spare tires which added 215 pounds for a total of 4578. Optional wire wheel equipment added another 100 pounds.

## MODELS (Q) 8-69 and (W) 8-79 1930; FIRST SERIES 1931

In January 1930, replacements for the '68 and '78 were introduced at the New York Show. They reflected the new Marmon 'bigger is better' philosophy...the inverse of that adopted in 1927. The new cars were not only restyled; they were, with the exception of the '69' engine, new automobiles.

The 8-69 and 8-79 bore an unmistakeable family resemblance to the Big Eight announced the previous month. They did have the radiator shutters and pyramided headlights and fender lamps, but the hood side panels were almost fully utilized for multiple long horizontal louvers. On the 8-69, the shutters were manually operated and the headlights on both were smaller.

Fig. 5-24 The EIGHT-79 four door brougham.

Fig. 5-25 The 8-69 standard sedan — a much larger and longer car than its predecessor.

A single window reveal extended around the entire set of window openings instead of the previous practice of individual reveals. With the Marmon practice of using contrasting shades of paint on the reveals, this enhanced the apparent length of the new bodies. The full length belt molding enlarged over the cowl establishing a curved line carried upward to the roof in a slightly, and pleasantly, concave front pillar. Across the doors the molding was quite broad and then gracefully diminished as it swept around the rear quarter. The bottom of the body, where it met the splash pan, was curved downward at the center, in what the designers call a 'wach sill'.

Fig. 5-26 The 8-79 sedan was good looking but properly lacked the charisma of the top-of-the-line Big Eight.

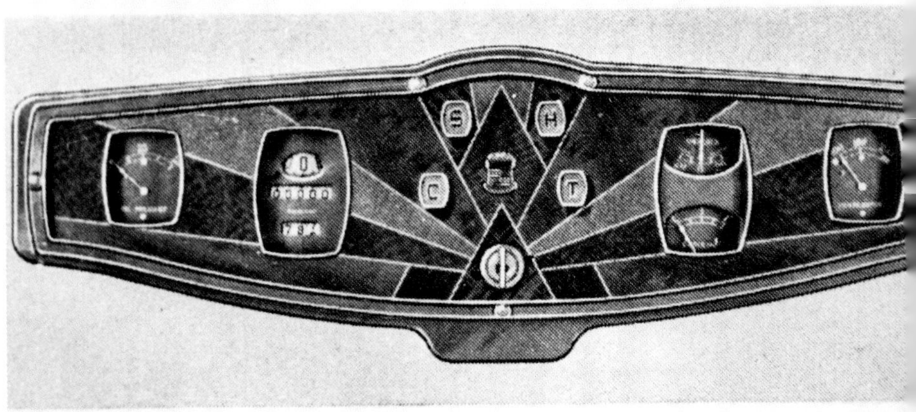


Fig. 5-27 Instrument cluster for Model 8-79.

Windshields were top pivoted and monocontrolled. Whereas the Big Eight had full shatter-proof glass, the 8-79 had it only in the windshield and the 8-69 had none at all.

Both broadcloth and mohair upholstery was used in the 8-69 but only a high-grade and durable broadcloth was used in the 8-79. Curtains were furnished on rear quarter and rear windows. For the greatest possible comfort, the rear seat was hollowed out at the shoulder line. Steering column angle and front seat position were adjustable.

A feature of the 8-79 only was a steel running 'board' with a molded rubber covering and durable scuff-resistant chromium plated moldings.

The bodies were of composite wood and steel construction with all steel panels flash welded to hide visible joints. Hardwood, free from knots and treated with a preservative was used throughout. All wood joints were hot glued, screwed and tenoned. Permanent joints were painted in special oil for enduring silence. Steel angle bracing reinforced all critical points and graphite webbing was placed between all reinforcements. Closely spaced wood top bows were covered with felt strips and woven galvanized wire mesh before installation of the padding and top material.

The two cars used chasses identical except for the wheelbases, of 119 and 125 inches, and the brake size. The 8-79 had 14 inch diameter drums, the 8-69 had 13 inch; both were 2 inches wide. The brake systems were otherwise the same Bendix two-shoe self-energizing type used on the Big Eight. On the 8-79 the Warner four-speed transmission, with silent internal-geared third, was optional.

Overall length of the 8-79 was 191 inches . . . eleven less than the Big Eight but ten inches longer than the '78' it replaced. The 8-69 was 179 inches long . . . twelve inches shorter than the 8-79 but four inches longer than the '68' it replaced.

The same body styles as the standard wheelbase Big Eight were offered on the 8-69 and 8-79 plus a convertible (**not** 'collapsible') coupe. They were two-passenger coupe and convertible coupe, both with rumble seat; five-pasenger sedan; four-passenger club sedan and brougham, both with trunks.

The 8-79 5-passenger 4-door sedan weighed 4028 pounds without fuel (22 gallons) spare tire which added 180 pounds for a total of 4208. The 8-69 5-passenger 4-door sedan weighed 3103 pounds without fuel (22 gallons) and spare tire which added 126 pounds for a total of 3229.

The '69' was powered by an improved version of the '68' engine. The patented double-dome combustion chamber introduced on the Big Eight was also applied to the 8-69 engine. The compression ratio was consequently raised to 5.5:1 and horse power to 84. The cylinder block was revised to incorporate full (cylinder) length water jackets.

The 8-79 was powered by an underbored version of the new big Eight engine. Decreasing the bore one sixteenth of an inch to 3.187 diameter reduced the displacement to 303 cubic inches and the SAE rating to 32.5 (from 33.8). A single updraft carburetor was fitted and the horsepower rating was 107 at 3250 rpm (versus 125 for the big Eight).

All four Marmon lines for 1930 carried L-head engines . . . a full reversal of prior Marmon policy. From an historical perspective, it may well have been a prudent and appropriate policy. The L-head engine valve trains were much quieter, and increasingly so with higher mileages. The technically uninformed could thereby be led to purchase decisions to the detriment of the manufacturer dedicated to the more sophisticated type. After all, the all-time ultimate in silent internal combustion engines was the flat head Cadillac introduced in 1936 and built through 1948. The GM and subsequent industry swing to overhead valve V-8s beginning in 1949 took place only after fuel technology and combustion chamber research made feasible compression ratios impractical in L-head engines.

In March 1930 an open 'touring' body was added to the three product lines. It provided a comparison of relative price ranges of the three Marmon lines for 1930; 8-69, $1610; 8-79, $2020; Big Eight, $3120. All three were characterized by natural wood top bows, nickel plated top irons and the use of 'Teal' top fabric. The folded top assembly fitted into a boot only six inches high!. The common 69-79 body was built by Hayes but the Big Eight came from LeBaron.[1]

Fig. 5-28 Big Eight seven-passenger touring with LeBaron body.

Fig. 5-29 Grand National Sportsman's Coupe by Locke & Co.

In late summer, a 'new Custom Series' was announced. It included two new styles: a four-door, four window close-coupled brougham in steel-wood composite construction, by Weymann Body Co. and a two-door, five-passenger convertible by Locke & Co. The former was characterized by a semi-integral metal trunk in body finish and a fuel filler accessed through a hinged door on the side of the trunk. The brougham carried the 'Longchamps' name and the convertible was named the 'Grand National Sportsman's Coupe'. Locke had designed the style originally to specifications laid down by their customer, Mahmoud Sabit Bey of Cairo, Egypt for mounting on a Marmon Big Eight chassis. The specs included a 40 inch front door for easy access to both front and rear seats. Marmon management was impressed by the car and commissioned Locke to produce the body for Marmon assembly. Each of these 'Custom Series' styles was featured in full page color advertisements and was identified by a regular production body code number.

## FIRST SERIES 1931

The Models '70' and '88' were delayed and introduced at the New York Show.

Fig. 5-31 Hayes advertisement confirms commonality of Marmon and Peerless bodies.

The Marmon-Roosevelt, Models 69, 79 and Big Eight built after the usual Marmon model change point in September were designated officially as First Series 1931 Models. Marmon ads in the fall featured 'the New Custom Series'.

## SHARED BODIES

The bodies and fenders of the 1930 Models 69, 79 and Big Eight were not entirely unique to Marmon, being shared with three models of another make. These Marmon models were designed in the studios of the Hayes Body Corporation by "staff designer Alexis de Sakhoffsky (Alsac) who had previously been Art Director of Van Den Plas and for five successive years had been the winner of the grands prix in the Monte Carlo Concours de' Elegance competition. Sakhnoffsky had left his native Russia at eighteen and studied two years at the Electromechanical Institute at Brussels, Belgium. Eighteen months of designing women's clothes preceded his five year stint with Carossiere Van den Plas, S.A. of Brussels where he was art director in charge of line and color research. He came to the USA in October 1928.

Fig. 5-32 The only factory-release picture of the Model 70 line.

An advertisement of Hayes in the *Michigan Manufacturer and Financial Record* for January 18,1930 showed a line drawing of the Marmon '8-79' and noted that "Already, such discerning manufacturers as Marmon and Peerless have adopted these Hayes...Alsac creations which features "Through Reveals"...window reveals which ignored the inter-window pillars...and a curved bottom sill configuration. Hayes, of course, manufactured the bodies and sheet metal and finish upholstered them as well, based on observation of a Peerless 155 and Marmon Big Eight preserved in mint condition.

The Marmon body plant, following its lease to Murray Body Corp. for the 1927-28-29 period was reported in Oct. 1928 to have been acquired ay the Hayes Body Corp. and to already be in production under their control. Probably the Peerless, and other, bodies were built and trimmed there.

At first glance the Peerless and Marmon appear similar. Actually they shared basic dies and assembly fixtures but secondary dies imparted more originality into the bodies. The original and more artistic treatment with rather massive moldings was the Marmon; the Peerless had uniform width belt molding and an odd panel above the molding and below the window.

## MODEL (U) 70

The Marmon '70' was announced to sell in the $1000 field . . . the Roosevelt/Marmon Roosevelt market segment. It was indeed, despite a new sedan body, a third year of the basic Roosevelt automobile.

The Model 70 bumpers carried a lightning flash on the central bumper plate and the upper bar was curved upward in the central area, rather lip-like.

Inside visors replaced the external one and the dual cowl-top ventilators of the Roosevelt were present.

Wire wheels, an option, were bolt attached and concealed by a large snap-in hubcap.

Production did not begin until February 1931.

The press described the new appearance of the car as both distinctive and unusual in its price class. The radiator in addition to sporting the traditional Marmon pyramid and badge, was higher and brought the hood to a horizontal line. Ventilator doors replaced louvers in the hood side panels.

The model '69' engine was used and had been adopted for Marmon-Roosevelt production earlier. It provided 84 horsepower with the larger bore and incorporated full length water jackets. Both Marmon eights for 1931 had 0.4 horsepower per cubic inch due to the patented 'double-dome' combustion chamber and 5.5:1 compression ratios.

The Marmon 70 was list-priced as follows, plus $110 for wire wheels and fender wells:

| | |
|---|---|
| Five Passenger sedan | $1128 |
| Two Passenger coupe | $1083 |
| Four Passenger victoria | $1128 |
| Two Passenger coupe | $1178 |

The display ad announcing the first showing of the Model 70 at the Marmon-Detroit Co. offered the base Model 70 for $950 . . . the lowest price ever on a Marmon or Marmon-built automobile.

In April 1931, availability of the Warner 'free-wheeling' device was announced for the '70' (and '88'). The 70 was the lowest priced eight cylinder car to offer the free-wheeling feature, which was of the premium type with control by a push button on top of the shift ball. In June 1931, a radio antenna was made standard equipment and three makes of radios offered in the range $140 to $170.

## MODEL (CC) 8-88
## 1931

This senior eight cylinder offering was announced as "THE NEW MARMON EIGHTY-EIGHT". The introductory brochure called it "Marmon's finest eight-cylinder achievement" and described the styling changes as achieving "Impressive yet bulkless size, (the) exterior stripped of non-essentials, making for a pleasing dignity and simplicity."

Evidently marmon marketing people did not want the SIXTEEN to have a car of the Big Eight proportions as a lower priced competitor. The luxury consumer class was to be encouraged to buy the $5000 SIXTEEN; the '88' was planned to fit below it comfortably.

The two-passenger coupe and convertible coupe as well as the five-passenger sedan and club sedan were built on a 130 inch wheelbase chassis. The radiator and engine were moved forward in the chassis about five inches; hence there was little if any compromise in interior dimensions of the styles on the 130 inch wheelbase.

Fig. 5-33 The only factory-release picture of the '88' line. Note the forward placement of the radiator relative the Big Eight predecessor.

The new bodies had straight sills and door bottoms unlike the Big Eight bodies of the prior year. The forward placement of the engine contributed to handsome rear quarter contours. On the club sedan, the trunk was attached to the body although not integrated as on the SIXTEEN. The fuel filler cap was accessed through a hinged, locked door on the side of the trunk.

Side mount spares were standard. Wood wheels were standard, wire optional. Wide single bar bumpers were adopted and for the first time dual windshield wipers were installed. Like the big Eight,, shatterproof glass was used in all windows. On four door sedans, the driver's half of the front seat was independently adjustable fore-and aft. Pedals and steering column could be adjusted one inch either way from normal.

The 130 inch chassis used the 4.45:1 axle ratio of the Big Eight, and the 6.50x19 tires, wider rear tread and 15 inch Bendix brakes of the Big Eight were retained. Overall length was 197 inches. The rear engine mounts were now "rubber relieved" to eliminate vibration.

The '88' was priced as follows, plus $75 for wire wheels.

| | |
|---|---|
| Five Passenger sedan | $2485 |
| Five Passenger club sedan | $2535 |
| Two Passenger coupe | $2465 |
| Two Passenger conv. coupe | $2585 |
| Seven Passenger limousine | $2785 |

These prices were about $250. lower than the 1930 Big Eight.

In April 1931, availability of the Warner 'free-wheeling' device was announced for the '88' (and '70'), with the sedan being repriced at $2385.

Free-wheeling consisted of an over-running clutch on the transmission output shaft and means to lock it out when reverse was selected or when the action was not desired. It permitted gear shifting, once the car was in motion, without releasing the clutch provided that the accelerator pedal was released. The shifts were effortless because only the inertia of the internal transmission parts had to be overcome by the synchronizers. But to lift your foot from the accelerator, as the car entered a curve at speed, was a traumatic experience if the free-wheel was not locked-out! The sudden silence and the absence of any evident deceleration combined to convey an airborne feeling. (Free-wheeling reappeared as an essential element of

Fig. 5-34 The Model 8-125 deluxe sedan when owned by the author in 1937.

Fig. 5-35 The handsome 8-125 convertible coupe.

automatic overdrive in the mid 1930s, but the shift to overdrive provided a solid drive at touring speeds while retaining the clutchless shift feature so long as shifting was completed before the 45 miles per hour engagement speed was reached.)

In June, a radio antenna was made standard equipment and three makes of radios offered in the range $140 to $170.

The start of production was delayed until February and possibly March 1931 due to an unsold inventory of Model 69, 79 and Big Eight cars.

The Fourth Era overlapped its successor era by two years. The SIXTEEN is described fully under the Fifth Era.

Fig. 5-36 The 8-125 instrument cluster. It was Marmon's first in front of the driver; the clock (deluxe only) was in the glove compartment door. The production cars used a T-handle in a slot to the left of the cluster rather than the rotary lever shown on the column.

## MODEL (HH) 8-125
## 1932

The body lines of the new 8-125 were described as 'quite new' by the press. It had a V-shaped radiator with a slight slant and vertical grill bars, a slanting visorless windshield, a long high hood, sweeping full-crown fenders and treatment of the front side and rear sides such that all chassis parts were concelaed from view. In all these respects, it resembled the Marmon Sixteen. Hood length was accentuated by very low seating, which approached the seat heights popular in the seventies. A large Marmon name plate with super imposed '8' was centered between two parallel chromeplated bars bridged between the headlight supports.

The bodies however were a new generation of Hayes designed bodies shared with the Franklin Olympic and the Reo. It was not another Teague work of art; but an automobile quite a bit above the rest of the crowd.

It was a spectacular performer . . . in this regard a fitting companion to the SIXTEEN. It was powered by the 125 horsepower engine introduced for the Big Eight but the engine was mounted in a chassis where the 5-passenger sedan weighed a surprising 3400 pounds. The specific weight was only 27 pounds per!horsepower, just about the same as the SIXTEEN! It not only had GTO type acceleration, it also had a wide track of 58.5 inches front and rear.

The radiator was topped by the same ornament as the SIXTEEN and bore just a hint of the Marmon pyramid. The core was concealed by vertical chromium-plated bars.

Although not as distinctive as the SIXTEEN, the front fenders also effectively concealed the shock absorbers. The SIXTEEN configuration was only feasible in aluminum; despite the light overall weight aluminum was conspicuously absent in the 8-125. A shield at the rear concealed fuel tank and spring shackles.

The head lights were carefully sculped to effectively enhance the overall impact of the front end. Were it not for the overwhelming beauty of the SIXTEEN, the 8-125 would have been the sensation of the 1932 shows.

Three body styles were available, each in standard and deluxe trim. They were: 5 passenger sedan, 2 passenger coupe and convertible coupe both with rumble seats. At least on the deluxe versions, with which the author had personal experience, the sound deadening package was equal to the SIXTEEN's. Not only was the dash or bulkhead faced with insulation, but the toeboards and floorboards as well. The convertible was one of the first to feature inward-folding headers over the side windows. They moved the windshield header back to the rear of the windows as the first step in collapsing the top, which then was compact enough to disappear below a hinged cowling. The 8-125 convertible was a handsome car indeed.

The model designation came from the 125 horsepower on a 125 inch wheelbase.

Upholstering was in whipcord or mohair edged in leather piping. All hardware was chromium finished. The front seat and steering column were adjustable.

An important new feature was a vacuum-operated automatic clutch. It was used with a three speed and reverse transmission having silent helical second and synchronized engagement of second gear and direct. A free-wheeling unit integral in the transmission was optional. It was controlled by a button on the shift lever ball, and was locked-out by a knob on the dash.

A dash control cancelled the automatic clutch function. Another, moving vertically in a slot to the left of the instrument cluster, adjusted the ride characteristic of the four shock absorbers via a linkage system. The instrument cluster matched a large glove-box door on the passenger side. A large speedometer was central and high, and bore both miles per hour and revolutions per minute (high gear only) scales on its black face; it was the first Marmon eight with a rotating-needle type. The ammeter, fuel level, oil pressure and coolant temperature gauges were arranged in slated pairs on each side of the speedometer. Central on the dash were the same ashtray and lighter as the SIXTEEN. It was in the authors opinion the best panel of the time after the SIXTEEN panel. Starter, lights and horn were on the wheel hub knob.

The steering wheel was a 3-spoke molded rubber design similar to the SIXTEEN.

Although the 8-125 carried the top-of-the-line eight engine, it also incorporated the spring shackling method of the bottom-of-the-line eights of prior years. Perhaps this was consistent with the remarkably low price of the '8-125'. It did however incorporate the kick-shackle as used on the rear of the left front spring on previous senior cars.

The '8-125' sedan with five standard wire wheels was priced at only $1395 F.O.B. (Free-On-Board) Indianapolis.

## EQUIPMENT GROUPS

A comparison of typical equipment on cars in the $1000, $1500 and $2000 brackets should be of interest. The cars were manufactured with Groups A and B installed; hence the actual price was appreciably above that advertised.

Group A in 1928 was priced at $265 (68) to $270 (78) and represented a price addition of 19.5 percent on the '68' and 14 percent on the '78', bringing the four door sedans to $1650 and $2165 and narrowing the effective price differential between the two Series. Group A included essentials such as shock absorbers, spare tire and cover, front and rear bumpers, rear view mirror and powered windshield wiper. In 1928 it even included the gasoline and temperature gauges, the electric clock and the combination tail, stop and back-up lamp!

Group B was the combined dual side mount and wire wheel 'option'; it included the folding rear trunk rack. It evidently could be deleted on special order and was made a true option on the Marmon-Roosevelt during the model run. It was priced at $115 but was held to $100 on the Roosevelt by deletion of the rack. It included two fabric-covered spare tires.

During 1928, (and 1930) a Group C was offered as an alternative to Group B. It provided dual side-mounting and rear trunk rack, but included two fabric-covered spare tires mounted on rims.

## COMMENTARY ON FOURTH ERA

Marmon faced a unique and probably unrecognized marketing problem as they transitioned from the Third to the FOURTH ERA. The very advanced styling of the 1916 Model 34 permitted the basic styling theme to be perpetuated for a decade. This, combined with the characteristic longevity of the model and the dealer remanufacture program/promotion of the early twenties to give the appearance of a much larger segment of the luxury car market than Marmon had actually attained. It is probable that, except for accident attrition, the total production of the THIRD ERA, some 35,000 units, was still on the road in 1928.

The failure to achieve the projected production rates with the early FOURTH ERA products made the Marmon eights a rare sight on the streets. Even the impressive production of 1929 (0.6% of domestic sales) was more dispersed geographically than previously consequent to the greatly expanded dealer system. Hence it appeared that the new models were not achieving even traditional Marmon penetration of the market-place. This impression and the relative rarity (on the order of one third 1977 American Motors penetration) were deterrants to whetting the public appetite for the FOURTH ERA Marmons.

In addition, in the later (Sept. 29, 1953) words of George Freers: "The Company experienced considerable trouble and criticism because the buyers thought that they were not getting Marmon quality". Mr. Freers was Vice-President of Marmon-Herrington at the time, had been an assistant chief engineer in the FOURTH ERA and Chief Engineer in the FIFTH with engine design responsibility in both eras.

Maurice Hendrey, author of Cadillac — the Complete History[16], asessed Marmon's problems during Era Four with this candid statement[17]; "At least when the Marmon brothers were running it, everybody knew what a Marmon was. But in the late Twenties the name could mean anything."

The Fourth Era was unquestionably the Williams Era. The truth of that is most evident in the Marmon ad from which the heading of Page 1 of this Chapter was exerpted. Immediately beneath MARMON but in very much smaller type appeared: "Under the direction of G.M. Williams."

The overhead valve Marmon eight engines and the Little Marmon car were designed under direction of the team of Delmar J. (Barney) Roos, Chief Engineer (mid 1925 through November 1926) and George Freers, Assistant Chief Engineer (1925-1929).

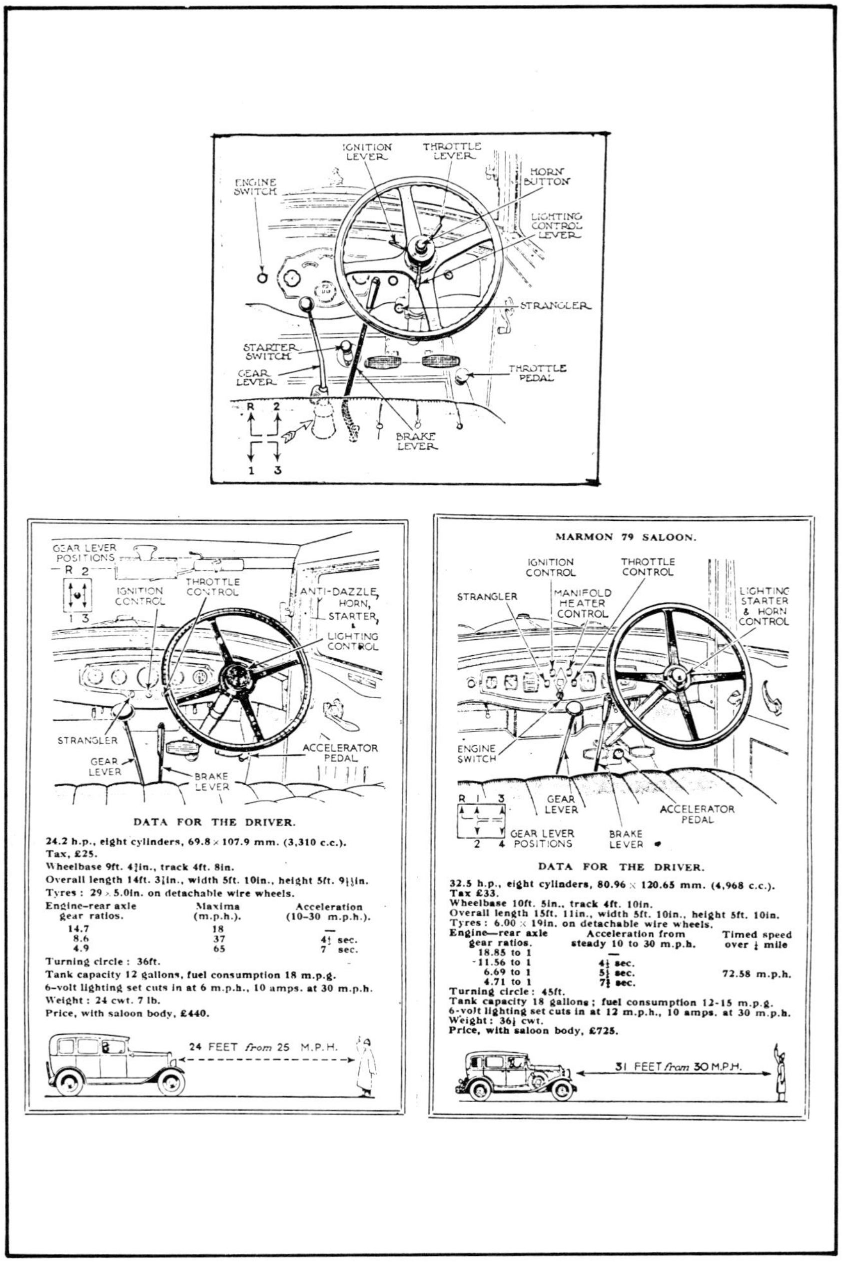


Fig. 5-37 Comparison of controls: L 1927 (top), R 1929 (left) and 8-79 (1930). *Autocar* (Br) Road Test Reports.

The flat-head Marmon eights were products of the team of T.J. Litle, Jr., Chief Engineer (Nov. 17, 1925 through Oct. 26 1929) and George Freers, Assistant Chief Engineer.

Joe Felts and Chet Blair reported to Bob Lewis and worked in the Advanced Engineering Department. In that partitioned-off secluded area, they designed the L-head series of production engines as well as several experimental engines including a 6 cylinder design. Lewis was subsequently sent to Germany to participate in a joint venture with an unidentified German firm; the project came to naught.

George Freers was promoted to Chief Engineer January 1, 1930 just as the last of the flat-head eight engines entered production. S.E. Rowe joined Marmon in February 1930 as Assistant Chief Engineer. They were responsible for the last cars of the Fourth Era — the 70, 88 and 8-125.

Although Williams had raised Marmon from 27th place in 1926 to 13th place briefly in 1929, reputed quality problems and the onset of the depression quickly terminated the growth period. Marmon had achieved 0.5% of domestic sales with registrations of 14,759 units in 1928, 0.6% (22323 units) in 1929... but only 0.45 percent (13,369 units) in 1930. A rapid decline followed with 0.35 percent (5687 units) and 0.13 percent (1365 units) in 1931 and 1932 respectively.

The styling of the eights with the exception of the standard closed bodies on the Little Marmons, had been competitive and in conformance with contemporary norms. The engineering of the cars had generally been superior particularly in the areas of noise and vibration suppression and in the achievement of high specific powers by advanced combustion and valve gear technology. And the pricing policy was surprisingly liberal, and probably too much so. But, unlike the Third Era, the technical novelty and superiority were inconspicuous, and the overall personality blended into the crowd of competitive eights.

Fourth Era Production was approximately 62,500 units produced in six model years.

1932 saw the end of the eights and Era Four; for 1933 the SIXTEEN was the sole offering. And its price was beyond the reach of practically everyone in 1933.

The durability problems with the Roosevelt engine are discussed and documented in Part Two of this Chapter. The influence on the fortunes of the company of the Roosevelt Venture is analyzed in Chapter Thirteen.

## PART 2 — Engines

There were three basic engines used in the Marmons of Era Four. They were all straight eights. The first was a small overhead valve eight for the Model L and an overbored version for the Model 78 used in 1927-29. It was the largest Marmon engine in 1929. The second was a flat or L-head eight using the Model L cylinder bore in the 1928 Model 68, the 1929 Roosevelt and 1930 Marmon-Roosevelt. A slightly overbored version powered the Second Series Model 68 (1929), and successively the Models 69 and 70 the following two years. Third was the large L-head eight introduced for the Big Eight in 1930 and used in the 8-88 and 8-125 the following two years. An underbored derivative engine powered the 8-79 in 1930.

The following detailed discussion of the Engines of Era Four will address each of these three basic engines, and their derivatives, in the order of their initial introduction as indicated above.

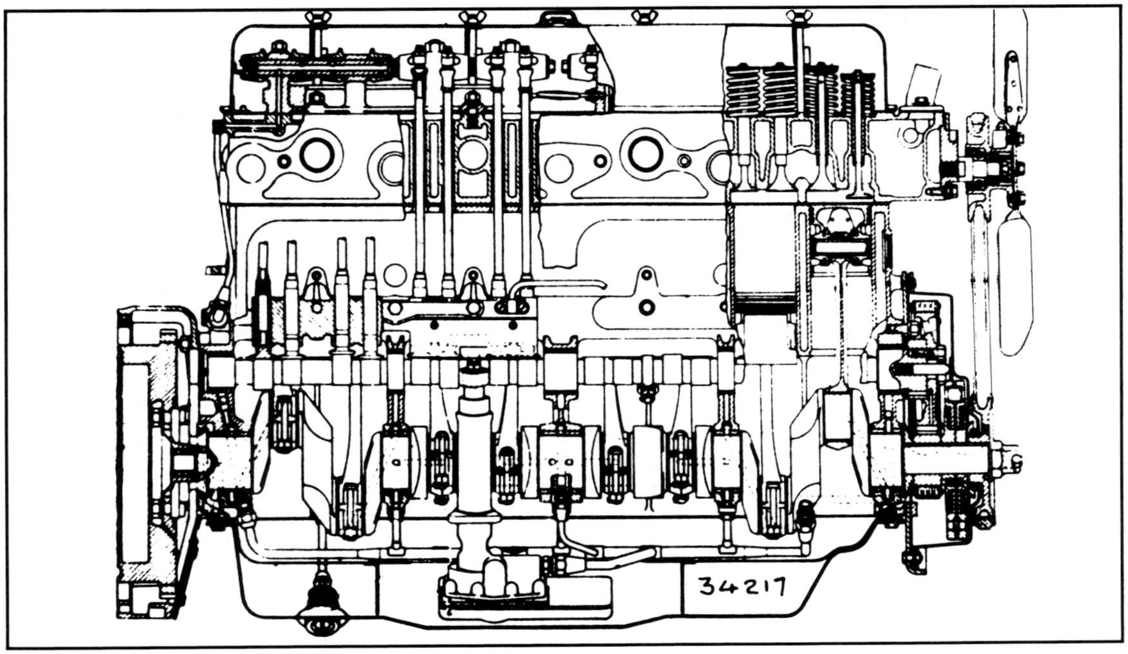


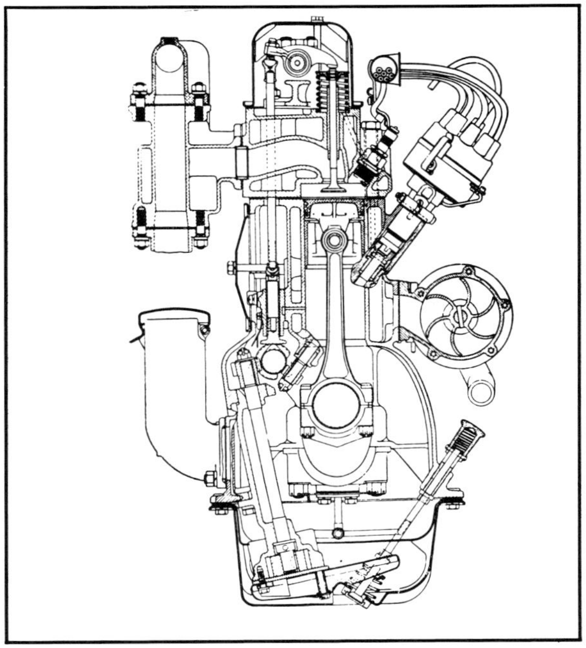

Fig. 5-38 Engineering drawings of the Model L engine.

## MODEL L
## 1927

The Little Marmon engine was an overhead valve straight eight. It had a 2.75 inch bore and 4.0 inch stroke. Displacement was 190 cubic inches; This made it the smallest eight on the domestic market at the time of introduction. It had an SAE horsepower rating of 24.2 and developed 64 horsepower at 3300 rpm.

The cylinder head and the cylinder block, as well as the flywheel housing were gray iron castings. The timing chain/damper housing, oil pan and rocker cover were steel stampings. The flywheel housing incorporated the rear engine mount attachment; a steel stamping sandwiched between block and timing cover provided the front mounts.

The cylinder block had five main bearing supports and extended downward 2.5 inches below crank centerline where it was flanged externally for oil pan attachment and internally for additional rigidity. It featured full length water jackets a full decade before they gained major acceptance. The combustion chamber was cylindrical but in the head, unlike the '34' design, where it was in the block, but like the '34' the chamber was larger in diameter than the bore and this precluded a water passage between chambers. It however permitted larger valves than would otherwise be feasible. It should not be overlooked that the placement of the combustion chamber entirely within the head left open the possibility of building an L-head engine on the same block tooling.

The crankshaft was a stiff five-bearing design weighing 66 pounds. Main journals were all 2.375 inches in diameter. Starting from the front, the lengths were: 1.53, 1.03, 1.78, 1.03 and 1.47 inches. The main bearings were babbitt faced and bronze-backed; they were interchangeable for service. The **front** bearing was flanged for thrust control. On the front of the shaft was mounted a torsional vibration damper, of the Lanchester type; it was adjacent to the timing gear and **within** the timing gear cover.

Crankshaft oil sealing was accomplished at the front by what was then conventional practice . . . a felt seal running on the hub of the crankshaft pulley behind a conical slinger. The design of the rear seal was unusual . . . a split packing in a counterbore which bore against a curved slinger surface on the crankshaft at approximately a 45 degree angle.

The connecting rods were steel forgings weighing 28 ounces complete with cap. Center to center distance was 7.75 inches. The big end bearing was babbitt poured in the rod and was 2.125 diameter. The bearing was 1.25 long but somewhat surprisingly was offset severely. This was the penalty of having full 360 degree water jacketing.

The piston pin was of full-floating (free to rotate in both rod and piston) type and was 0.735 diameter. There was a bronze bushing in the rod eye but the aluminum piston provided integral bushings. Snap-rings retained the pin. Holes drilled in the piston bosses and in the top of the rod big-end provided lubrication by spray. This was the first Marmon engine ever to not have pressure feed to the piston pin. But about April 1927, at engine S/N 4291, the connecting rods were revised to incorporate rifle drilled holes for wrist pin lubrication.

The aluminum alloy pistons were of invar strut type but were cast and machined in the Marmon plant. They were fitted with 0.001 skirt clearance. Length was 3.25

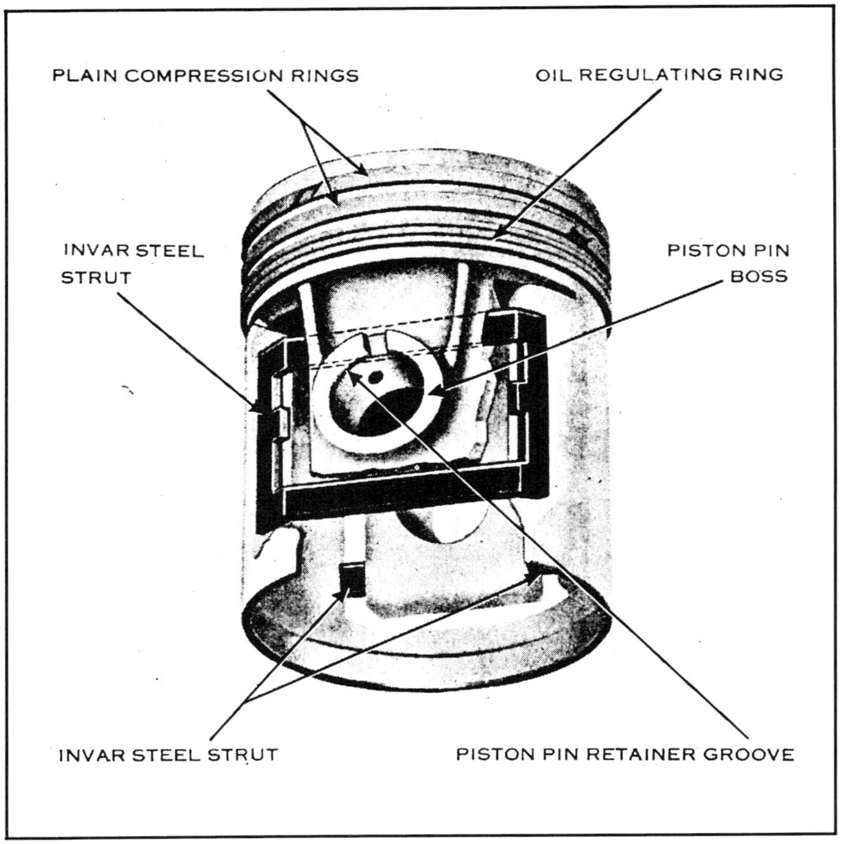


Fig. 5-39 The piston used in the Model L engine.

Fig. 5-40 Model L power plant.

inches with the pin 1.937 from the top of the head. Two 0.125 wide compression rings and a single 0.187 wide oil control ring were all located above the pin.

The crankcase permitted withdrawal of the piston and rod asemblies past the crankshaft.[18] A desirable service feature when rod bearing replacement meant rod and piston removal.

The camshaft was carried on five bronze-backed, babbitt-lined bearings. From front to rear the sizes were: 1.875 dia. x 1.312 long; 1.812 dia. x 0.75; 1.75 dia. x 1.50; 1.687 dia. x 0.75; and 1.437 dia. x 1.187. The shaft had an integral flange for gear attachment and two integral spiral gears for oil pump and distributor drive. The front bearing had a single flange (necessary for assembly + for rearward thrust; a spring-biased plunger, inserted in the center of the shaft, resisted forward thrust against a plate secured to the timing gear cover.

The camshaft was driven by an innovative timing chain without the traditional third adjustable sprocket to compensate for wear. It was Marmon's first use of a timing chain since Era Two; it was 1.25 inches wide and of silent-chain type.

The cast iron cylinder head had drilled holes for the push rods, four siamesed inlet ports and five exhaust ports, of which three were siamesed. The spark plugs

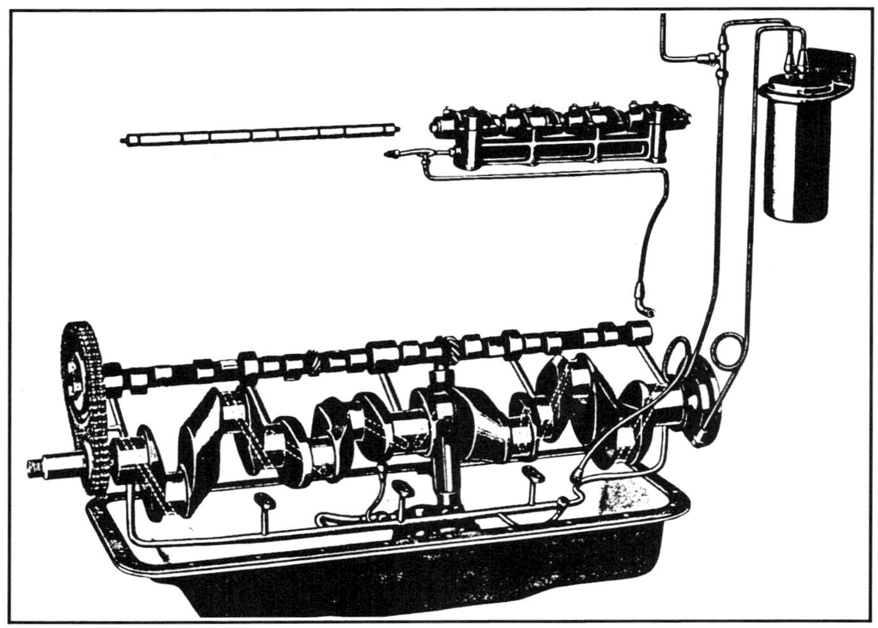

Fig. 5-41 Model N engine oil circuit.

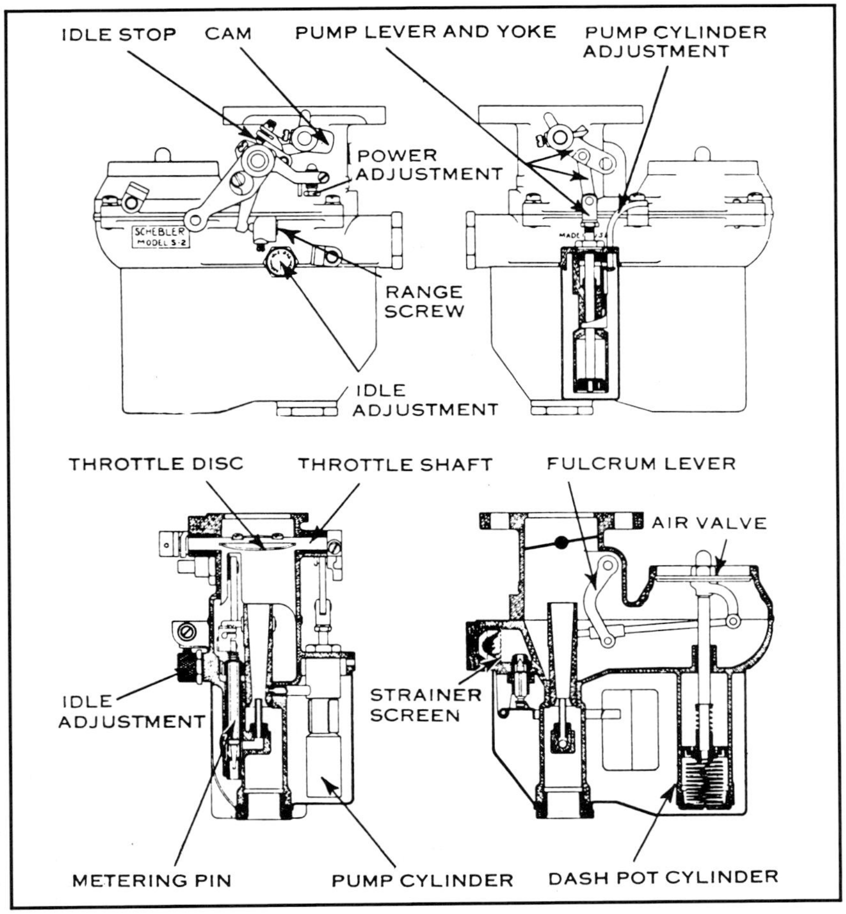


Fig. 5-42 Schebler Model S-2 carburetor used on the Model L engine. The air valve was manually closed for starting. Lean adjustment was clockwise.

were threaded into the cylinder head and together with the distributor location and the full length water jacketing gave the engine an ageless appearance.

The valves operated in pressed-in guides. The tulip inlet valves were mde of chrome-nickel steel and were 1.187 diameter. The exhaust valves were made of silchrome and were 1.0 diameter. Lift was 0.64 inches. The inlet valve timing was: open 6 degrees before top center and close 40 degrees after bottom center. Exhaust timing was: open 40 degrees before bottom center and close at 6 degrees after top center with 12 degrees overlap.

The firing order was 1, 6, 2, 5, 8, 3, 7, 4. This was chosen for manifolding considerations and required a two-plane (at 90 degrees) crankshaft. The head was extended forward of the block and rocker cover, to mount the water outlet elbow on top and provide a pilot for the eccentric fan hub required to adjust the belt tension.

There were two identical rocker shaft bracket clusters, each with four support bosses, located between the pairs of valves, on cylinder centerlines. The two rocker shafts were tapped at their ends for retention of the end arms. The hollow shafts served as an oil gallery for rocker arm lubrication. The push rods were fitted with hardened end pieces.

A flanged steel stamping formed the rear enclosure of the timing chain and served both as engine front mounting legs and a frame cross-member. The flywheel housing was an aluminum casting incorporating the rear engine legs and serving also as a frame cross-member. A separate 'bell' housing enclosed the clutch and mounted the transmission.

The Model L engine featured an innovative downdraft intake manifold with identical runners to every cylinder. From the Shebler Model S-2 1.25 inch single-throat carburetor, the mixture moved vertically through the exhaust manifold hot-spot to enter the aluminum intake manifold central 'tee' where the mixture was directed through horizontal runners to the front and rear sets of cylinders. These runners ended in 90 degree downward elbows feeding 'tees' and equal runners to the four siamesed inlet ports. It was a 'downdraft' manifold so that fuel which condensed in the manifold (after the heater) was carried **with** the mixture and distributed evenly to the cylinders. During development of the Model L engine Donat A. Gauthier (interview JY76) consulted with Marmon Staff consultant Mr. Turyman and Chief Engineer Roos and his assistant George Freers on carburetion of the engine to power the two prototypes of the Little Marmon. For a while, a two-stage Ball & Ball carburetor was seriously considered.

The lube oil pump was in the sump driven by a vertical shaft from the camshaft. It fed the rear main bearing by oil-line to the block adjacent. A brazed manifold with threaded fittings was fed **from** the rear main cap and fed to the three central main bearings and to the crankcase via drilled passages to the front main; this allowed a pan configuration giving needed chassis clearance. An external oil line fed the cylinder head.

There was concern with possible piston skirt scuffing during engine start-up. The Model L engine incorporated a novel means for cylinder wall lubrication during the brief interval before oil viscosity decreased sufficiently to provide spray from the crankshaft journals.

Directly above the rear main bearing (fed directly from the pump) on the top of the crankcase was mounted a valve body. Normally oil passed through to an external line feeding the cylinder head. When the starter pedal was depressed, a cam shifted a spool-valve directing oil also to an internal passage feeding through drilled holes directly into the cylinder bores. A groove in the tappet carriers formed the distribution passage; a jumper line joined the two carriers.

The front accessory drive was innovative. A new and more durable V-belt made a three pulley drive of fan and generator plus water pump feasible. The generator, in typical Marmon fashion, was strap mounted to a saddle on the engine front closure/mount plate. It drove the centrifugal water pump via a hose type coupling. The fan axis was used to control belt tension by means of an eccentric pilot in the front of the cylinder head.

The three pulley fan drive and accessory drive introduced on the 1927 Model L was adopted by sixteen manufacturers by 1929.

Oil sump capacity was six quarts. Oil was drained by pressing on the upper end of a rod projecting from the crankcase, and accessible under the hood, which carried on its lower extremity a spring-biased face seal on the **outside** of the pan.

The cooling system capacity was five gallons. A cellular type radiator was used.

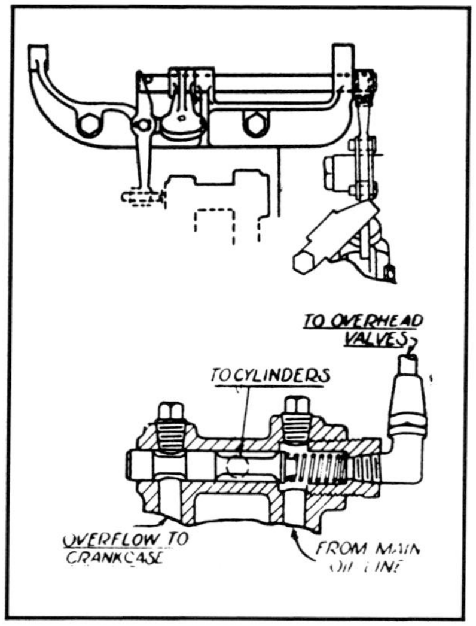


Fig. 5-43 Anti-scuff valve

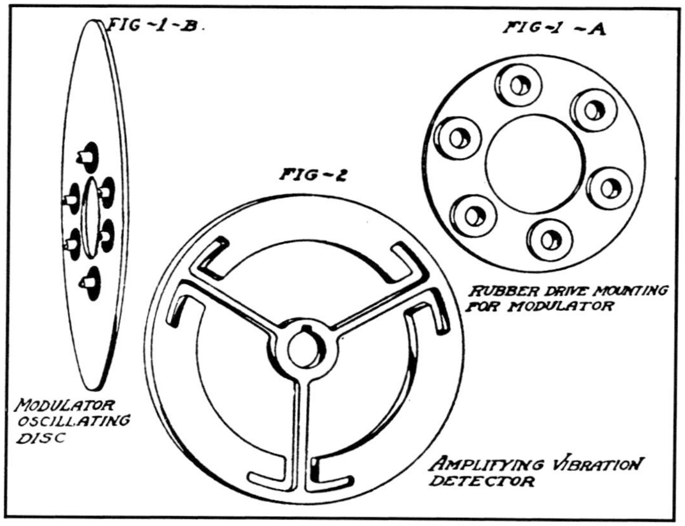


Fig. 5-44 Circular tuning fork

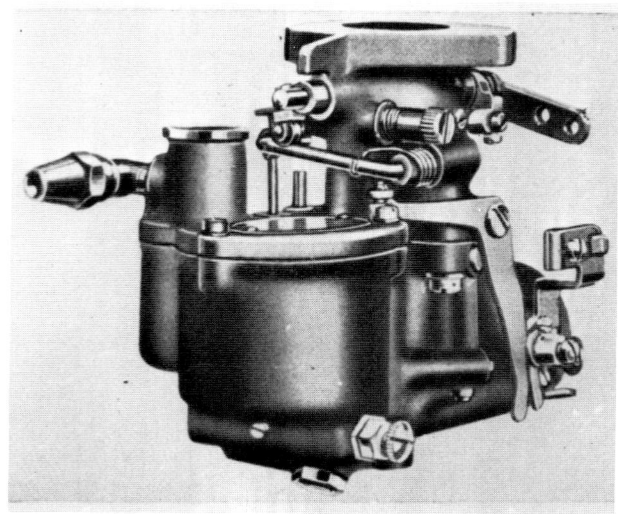

Fig. 5-45 Stromberg UX-2 Carburetor.

The 17 inch fan was carried on a pair of ball bearings (two sizes).

A blocking type thermostat was installed in the two-piece water outlet hose.

Both the Series 75 and the Little Marmon engines featured crankshaft vibration dampers of the split flywheel type. Similar dampers appeared on some other 1927 models. But the 1928 Models 68 and 78 featured a patented design of what the inventor called a 'High-Frequency Oscillating Modulator'. There is no evidence that a patent was ever issued to Mr. Litle. It was described as a new form of torsional damper consisting of a relatively light metal disc mounted on a special rubber support from the crankshaft. Mr. Litle explained "The period of vibration of the crankshaft is comparatively low because of its large mass. The period of vibration of the damper disc is so high that there are a great number of oscillations of the disc to one of the shaft. The great restoring force of the rubber will dominate over the inertia of the disc". Because the damper disc oscillated at multiples — harmonics — of the crankshaft speed, the device became known as a harmonic damper.

An interesting device was used to establish the existence of numerous periods of crankshaft torsional vibration. It was a triple tuning fork cut from a circular steel disc and mounted on the crankshaft nose. Different thicknesses of disc were alternately installed. When the car so equipped was started in motion in second gear, e.g., three recurring audible notes were heard; in high gear at certain speeds five distinct tones were heard. Some of these notes represented vibration periods not previously known to exist in crankshafts. All of them were effectively damped out by the new device which would be a feature of all subsequent Marmon-built automobile models. The lighter damper allowed more rapid engine (and car) acceleration.

## MODEL 78
## 1928

The Model L cylinder bore was increased .187 to 2.937 and the revised cylinder gasket was released for service on the Model L.

The enlarged cylinder increased the displacement to 216.8 cubic inches, raised the SAE rating to 27.6 horsepower and, combined with a new intake manifold concept, developed 86 brake horsepower at 3400 rpm. The compression ratio was increased from 5.27 to 5.50:1 and was claimed to be the highest ratio in use on any automobile at that time. Knock-free operation was claimed.

The oil fill location was moved from the top of the rocker cover and became an aluminum casting on the right forward part of the crankcase. It had a hinged cover with snap latch. A by-pass type oil filter was mounted on the firewall.

The increased loading, due to the bore enlargement, required detail block and crankshaft changes to accommodate longer main bearings. The Model 78 bearing sizes were, from front to rear: 1.531 (no change), 1.125 (+.094), 1,875 (+.094), 1.125 (+.094) and 1.875 (+.406).

Valves were increased in size, even more than the increase in bore. Surprisingly both valves were increased equally in diameter by 0.265 inch to 1.452 intake and 1.265 exhaust. The exhaust valve material was silchrome for the head and tungsten steel for the welded-on stem. Tappet clearance was stated as 0.008 hot and valve spring load at 120 pounds; presumably these figures applied also to the 1927 engine.

Lynite split-skirt aluminum pistons were adopted. They weighed 13 ounces (versus 11) and were 3.187 inches (versus 3.25) long. Presumably they were procured from an outside source. The top compression ring and the oil ring were identified as sourced from Perfect Circle, whereas the second compression ring was sourced from McQuay-Norris. The piston pin was unchanged.

Piston pins were pressure lubricated, that feature had been a running change addition to the Model L engine.[19]

The Chevrolet LT-1 engine option, available in Corvettes only and prior to 1973, included, in addition to a forged crankshaft and connecting rods, a double roller timing chain.

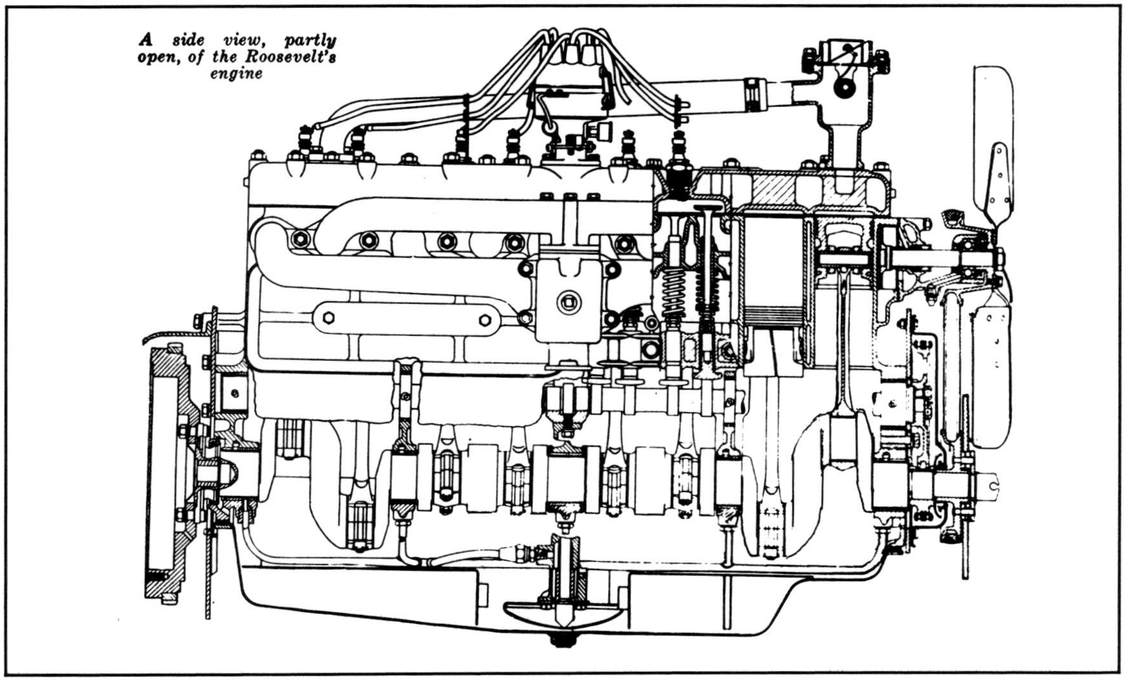


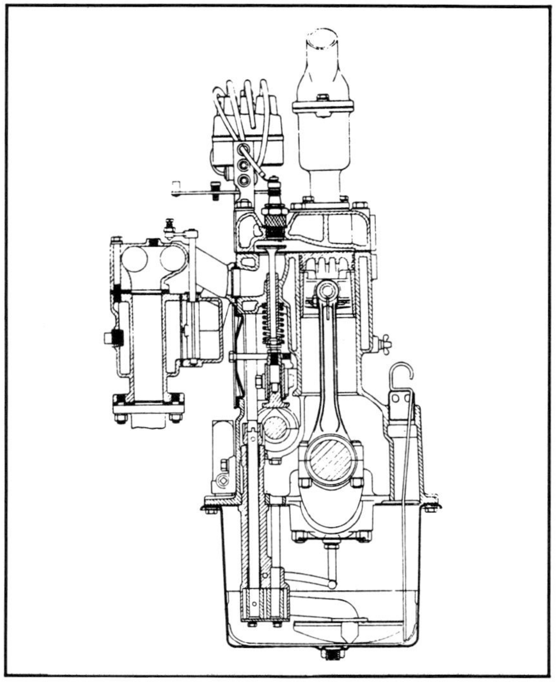

Fig. 5-46 Engineering drawings of the Marmon Model T engine used on the '68' and Roosevelt cars.

The totally new (in 1981) 1.8 and 2.0 liter four cylinder Chevrolet gasoline engines (for the GM J-Car) incorporated the second usage of double-roller timing chain by that Division. In the SAE Paper 820111 describing the new engine and the election to use a chain driven camshaft with push rod over-head valves, in preference to a belt-driven overhead camshaft, the superior durability of the roller chain over "silent" type chains in development tests was credited for the specification of roller chain.

All Marmons of the Fourth Era . . . except the Little Marmon . . . featured the pioneer usage of double-roller timing chains (two parallel roller links joined by three plates and engaging two side-by-side sprockets). Competitive engines (as well as the Marmon 16) used a chain driven generator as an adjustment device for chain stretch; the fan was typically driven by a two-pulley belt using, eg., an adjustable pulley sheave on the water pump/fan.

The drive replaced the 'silent-chain' adopted the prior year. It was developed by Marmon engineers in collaboration with the Diamond Chain Co. of Indianapolis (employer of Howard Taylor who tested the Marmon race car at Purdue 1911).[20] It was claimed that the roller chain, consequent lighter weight and smaller sprockets, reduced the centrifugal forces on the links and associated wear on the pins. In the Marmon 78, there were only 4.5 links between the dual-sprockets which practically eliminated chain whip; the chain length was only 24 inches (64 links).

Recognized by the press as an outstanding feature was a new intake manifold concept; it was a further development of the downdraft manifold introduced the prior year. It was a duplex downdraft (draught in those days) configuration, being in reality two manifolds within one aluminum casting. One manifold fed cylinders 1, 2 and 7, 8; the other fed 3, 4 and 5, 6. Successive intake strokes were from alternate manifolds (firing order 1, 6, 2, 5, 8, 3, 7, 4).

This was the manifold concept which Ford would adopt for his V-8 in 1934 and which would become a world standard for V-8s into the Seventies.

On the Marmon 78, the duplex manifold was fed from a single-barrel carburetor with the heated riser (section of the exhaust manifold) transitioning from single to dual passages. Marmon was evidently ready for the dual downdraft carburetor . . . they certainly had an application.

There was considerable design activity in the area of fuel vaporization and distribution in the late 1920s, which led to the advent of the (dual) downdraft carburetor.

Paul Bastien, of the Marmon design staff, was granted patent 1,785,108 on Dec. 16, 1930. It covered an intake manifold arranged horizontally in an overhead-valve type cylinder head. It covered an associated firing order and basically was a merger of the Marmon dual downdraft manifold concept (in plan view only) with the integral manifold concept of the Marmon 75.

One week later Earle S. McPherson, assignor to Hupp Motor Car Co., was granted 1,785,869 for an identical plan layout but using an external intake manifold.

Neither included the downdraft runner concept nor the attention to equality of runner effective length evident in the 1928-1931 Marmon 'dual downdraft' manifolds.

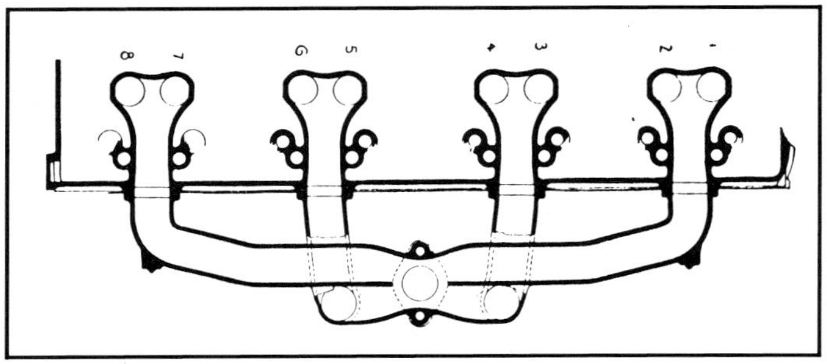

Fig. 5-47 The Model T engine used a dual downdraft manifold but a single barrel updraft carburetor.

A 1.25 diameter Stromberg UX-2 carburetor replaced the Model S Schebler. Fuel feed was by Stewart vacuum system from a 14 gallon rear tank. An alemite gasoline filter and a Wall oil filter (by-pass type on the firewall) were used.

A thermostat was added to the cooling system outlet.

Delco-Remy electrical units were used. The distributor was of "semi-automatic" type with 20 degrees of manual and 25 degrees of automatic advance. A Bendix pinion drive was used on the starter with mechanical linkage to a starter-mounted switch.

The author in 1937 reworked a Marmon duplex manifold to mount a dual downdraft carburetor. A flange was welded atop the manifold and two cylindrical slugs were welded-in to close the original entry but leave shallow basins to evaporate any liquid fuel. The operation was very successful.

The Model 78 engine weighed 715 pounds, of which the crankshaft was 66 pounds.

## MODEL 78
## 1929

Only changes in the engine were detail changes for increased turbulence in the combustion chamber and addition of an insulator to reduce heat transfer to the carburetor. Both changes were related to the increased volatility of commercial gasolines. A new oil seal (they were then appropriately called 'oil retainers') was adopted for the front chain cover.

## MODEL 68; R, MR (THRU S-25544)
## 1928; 1929-30

The new engine for the smallest Marmon, no longer the Little Marmon, was a side-valve or L-head type. It was certainly appropriate considering the $1395. base price for the Model 68 and the widespread popularity of flat-head sixes and eights.

Although generally conventional in appearance and detail, it did evidence some significant features for its time. The water outlet manifold on the top of the cylinder head incorporated a blocking type thermostat. The intake manifold, like the Model 78, was the new Marmon dual downdraft type fed from a single updraft carburator. Most significantly, it featured a prototype accessory configuration that would become universal.

Roosevelt specifications for this engine stated '70 plus horsepower'. It had a 2.75 inch bore and 4.25 inch stroke. Displacement was 201.9 cubic inches. SAE formula horsepower was 24.2 and rated power was 72 horsepower at 3200 rpm with 5.25:1 compression ratio for operation on anti-knock fuel.

The crankshaft was mounted in five bronze-backed, babbitt-lined bearings of 2.125 diameter. Bearing lengths from front to rear were 1.75, 1.06, 1.75, 1.06, 1.75 inches. Crankpins were 2.0 diameter. The new Marmon rubber-hub harmonic damper was mounted on the nose of the crankshaft.

The connecting rod center to center length was 8.5 inches. The rod was clamped to the piston pin . . . never before seen on a Marmon engine. The big end bearing was poured, not spun, babbitt.

Full pressure lubrication except to the piston pins was used. Oil pressure was 40 psi and a spray nozzle lubricated the Diamond (brand) double-roller timing chain which had 62 links of .375 pitch.

The camshaft ran on five bearings. Diameters, front to rear, were 2.136, 2.116, 2.125, 2.076 and 1.625. Lengths were 1.125, 0.5, 2.0, 0.5, 1.125. Cam lift was .327 inches.

Forged steel intake valves were 1.469 diameter, and had grooved stem ends.

Exhaust valves were Silchrome steel, 1.344 diameter. Spring pressure was 70 pounds and tappet clearance was 0.006 to 0.008 inch.

The pistons, like all the other engine parts, were of Marmon manufacture; the pistons were both cast and machined in-house. They were 3.25 inches long of split-skirt aluminum construction. The top compression ring and oil ring were procured from Perfect Circle, the second compression ring from McQuay-Norris.

A Stromberg Model UX-2 carburetor was used, as on the 78.

The '68' engine did not have full length water jackets like the '78'. Specifications for the engines cleverly de-emphasized this discrepancy by describing the '78' as having cooling over the full length of the cylinder and the '68' as having cooling over the full length of the piston travel (the stroke).

The Marmon-Diamond double-roller camshaft drive chain was used as on the Model 78. A flanged steel stamping, lacking the generator saddle of the '78', formed the rear of the chain enclosure and served as front mount and frame cross-member. Unlike the '78', another flanged stamping was bolted to the rear of the cylinder block to provide rear mounts and serve as a cross member. The 'bell' housing casting included a flywheel enclosure.

The three pulley belt accessory drive concept originated on the Little Marmon was further developed for the Model 68. The water pump carried the fan blades and was mounted on the front of the block. The generator was pivoted on the U-bracket bolted to the side of the crankcase and served as the adjustment means for belt tension.

A Fedders (brand) 'honeycomb' radiator was used and cooling system capacity was 4.5 gallons.

The Model 68 engine weighed 568 pounds of which the crankshaft was 60 pounds.

It is almost a certainty that during the design engineering phase of the Model 68 engine, that engine would have been planned for the 1928 **Little Marmon.** Due to lead time considerations this design work had to appreciably precede the introduction of the 1927 Little Marmon. Redesignation as a Model 68 did not occur until marketing problems became evident early in the 1927 model year. Engineering used alphabetic model designations in the Fourth Era and the Little Marmon was the Model L, the '78' was the Model N because Model M had been assigned to a car which did not reach production until Era Five. The '68' should have been the Model O but instead was identified as Model T. The author finds the deviation from alphabetic sequence most interesting . . . the election to reach down the alphabet may have symbolized Monte Williams' expectation that it was destined to be a Ford-like success. Marmon had produced in prior years the Models A through H, the 32, 48, 41, 34, 74 and 75; the letters I, J and K had been by-passed. One could conjecture that the 32/48, 41, and 34A through E constituted three basic models corresponding to the unassigned letters, which might have been in-house designations prior to announcement.

There are several strong supports for the argument that the Model '68' engine was based on purchased technology to expedite the start of production.

Joe Felts[21] recalls copying Continental Motors Corp. drawings on to Marmon Motor Car Co. letterhead drawing forms specifically for the Little Marmon. The author is convinced that the Marmon Model T drawings had a similar origin.

At introduction, Marmon ads for the Series 68 (Model T) noted, "The engine is entirely of Marmon manufacture." Later when the large L-head engine was introduced for the 1930 model year the ads noted: "**Designed** and manufactured by Marmon."

Continental Motors Corporation had introduced an overhead valve straight eight engine in 1925. It was their first straight engine, and the only one they would ever build with that valve arrangement, despite a number of subsequent models of larger displacement. It was a **2.81** bore by 4.00 stroke engine displacing 198.2 cubic inches and designated the Model 8P. It was offered for industrial applications at a 57 horsepower rating and powered the Locomobile Junior 8 for model years 1925, 26, and 27[22]. Barney Roos had been Chief Engineer at Locomobile before coming to Marmon and, with Harry Miller as a consultant, had designed an overhead-valve straight eight of around 200 cubic inches displacement for the new Locomobile Junior model. It was mounted in the prototype chassis but the firms finances would not support the production tooling expense and a stock Continental of the same type and size was substituted[23]. This information is strong support for Joe Felt's argument that the Model L Marmon engine (**2.75**x4.00, 190CID) originated as a Continental basic design and was underbored 0.06 inch from the stock engine used by Locomobile.

On the other hand, Continental built a flat-head straight eight of 3.00 bore and 4.25 stroke displacing 240.3 cubic inches and rated 72 brake horsepower at 2700 rpm.[22]

This engine could well have been the basis for the Marmon Model T engine, which also had a 4.25 stroke, and was built in bore sizes of 2.75 (201.9 CID) and 2.812 (211.2 CID).

The Model L engine was increased to a 2.937 bore (216.8 CID) for the Model 78. It could be concluded that both Continental and Marmon cylinder block tooling had bores on a center-distance appropriate to a maximum bore of 3.0 inches and that the Model L and Model T Marmon engines were derived from the Continental Model 8P and Model 12Z respectively.

Acquisition of design rights to a supposedly fully developed set of moving parts and a mature combustion chamber configuration would allow rapid finalization of the engine exterior and immediate procurment of production tooling.

It would have been illogical for Marmon, who had introduced full length water jackets on the Little Marmon only the prior year, to immediately revert to the general industry practice of the Era on their following model.

The Model T engine and its derivatives powered the Models 68,69,70, the Roosevelt and Marmon-Roosevelt. All of these models had reportedly serious engine durability problems including extreme cam lobe wear which severely compromised performance. The resulting publicity critically damanged Marmon's market position and potential. As recently as the spring of 1982, *The Detroit Free Press* carried an automotive memorabilia item recalling the advent of the Roosevelt and "Its apparently unsolvable crankshaft and leaky roof problems."

## *A QUESTIONABLE DURABILITY*

The problem of poor durability of the Roosevelt engine was nicely documented for posterity in *Automotive Digest* Magazine for April 1934. In the January issue the magazines "Trouble Man" had requested advice from the readership for a subscriber on the resolution of a low oil pressure problem on a Roosevelt engine. The editor noted that "Trouble Man" had received an abnormally high response to that particular request: "Scads of letters reached us with most all of them stating specific cases and specific cures." There was space to print only a few of the replies but the nature of the problem was made quite clear.

The universal symptom: low oil pressure and severe bearing knocking after 8 to 10 miles at a consistent 35,000 miles on the odometer.

Solution: remove and disassemble engine, regrind main and rod journals of crankshaft, install new connecting rods and new main bearings.

Typically the owner had recently had new rings installed to reduce excessive oil consumption which grew progressively worse as the crankshaft journals wore out of round and the pistons were showered with oil.

Obviously there was a serious quality control problem at Marmon Motor Car Co., or Marmon lacked sufficient process routing detail from the source of the engine design.

The engine of one of today's most highly regarded marques suffered the same malady in 1929...the author can personally attest to that...but the marque had an established reputation and survived the storm. The Roosevelt was in its first year and the consumer turned thumbs down.

Resolution of quality control and design deficiency problems after start of production is difficult and time-consuming even in the case of purchased components. Resolution of such problems in the case of purchased technology and in-house manufacture is orders of magnitude more difficult. Communication, placement and acceptance of responsibility are difficult and controversial matters.

Engineering drawings do not provide all the information necessary to successive manufacture of parts identical in dimension, physical characteristics and durability potential. Metallurgical information is often provided in the form of referenced specification numbers. Specific manufacturing information such as the sequence of operations is recorded on routing sheets prepared by manufacturing engineers. A surprising amount of "know-how" is unrecorded and passed along by foremen over various process operations. In some cases, the concentration of information is avoided due to confidentiality concerns.

The early mileage engine problems with the Marmon 68/Roosevelt engine were compounded by the rapidly expanded dealer organization. Dealers typically were ill-prepared for large scale work on engine internals so soon after acquiring the franchise.

A careful review of announcement press releases and advertisements reveals different terminology used for the origin of the '68' and the '78' engines. In a six glossy-page announcement ad, separate pages addressed the '68' as 'Made by

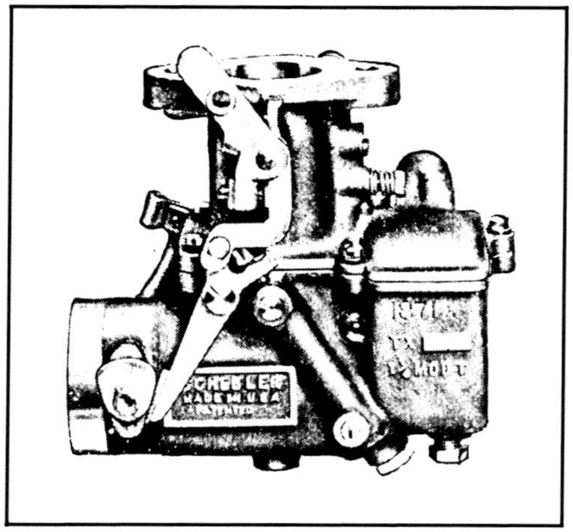

Fig. 5-48 Schebler Model T carburetor used on Marmon Model T engine.

Marmon' and the '78' as "Manufactured by Marmon". Releases stressed that every part of the '68' was **Made** in the Marmon plant. Later when the Big Eight engine was announced the releases would read "Designed and Manufactured by Marmon".

No engine cross-section drawings were released for the Model 68 engine. This was contrary to all prior and subsequent Marmon practice. If the engine internals were a purchased design, there would be no need to prepare costly sectional drawings, at least not on an expedited basis. They were available by the time of the Roosevelt announcement and were released as Roosevelt engine sections.

## MODEL 68
## 1929

The cylinder bore was increased 0.062 making the bore 2.812 (x4.25) and increasing displacement to 211.2 cubic inches. Horsepower was increased to 25.3 SAE and to 76 (from 72) at 3200 rpm. Compression ratio remained at 5.25:1 but the combustion chamber was revised for greater turbulence of the mixture charge.

Bronze-backed babbitt lined camshaft bearings of improved composition were adopted. The diameter of the main bearing studs was also increased, reportedly (and obviously) for additional safety factor.

## MODEL 69; 70; MR (STARTING S-25545)
## 1930; 1931; 1930

Adoption of a new cylinder head incorporating the Marmon patented double dome combustion chamber made an increase in compression ratio to 5.5:1 (from 5.25:1) feasible. Horsepower was increased to 84 at 3400 rpm (from 76 at 3200). For the model 70, the division wall between combustion chambers was reduced slightly and was machined to a uniform height presumably to attain a more consistent combustion chamber over all eight cylinders. A 3 blade 18 inch fan was used with a new radiator. A Schebler Model T carburetor was adopted.

## MODELS H (BIG 8), 8-88, 8-125; 8-79
## 1930-1932; 1930

Marmon's return to the prestige automobile market called for a new engine as well as a new car. The management's dedication to the straight-eight left no question as to the cylinder complement. The election of a 'flat-head' or L-head valve arrangement was possibly a surprise; but considering that Cadillac, Lincoln and Packard were all users of that type, it should have been expected as in keeping with Monty Williams' formula for success in the marketplace. In fact, the L-head would dominate the domestic market until the mid Fifties.

Basically the engine was quite conventional. It did incorporate the traditional Marmon full pressure lubrication system but like the Model 34A resorted to splash lubrication of piston pins. In a configuration very popular at the time, the oil pump/distributor drive shaft was vertical. This placed the pump in the pan and the distributor atop the cylinder head. Mushroom tappets were used. Flat heavy gage steel plates were used front and rear to extend the block for timing gear and flywheel enclosure and to serve as motor mount legs. Rubber isolation was used at the rear mounts.

But incorporation of several Marmon exclusive developments made the engine noteworthy: the camshaft was driven by a two-sprocket Diamond double-roller timing chain; the intake manifold was a dual downdraft configuration; and a new patented combustion chamber was revealed for the first time.

The engine was offered in two versions differing in bore diameter and in the number of carburetor barrels. The Big Eight was the first engine to combine a dual (updraft) carburetor with a dual downdraft manifold; the 8-79 version had a 0.06 lesser cylinder bore and retained the dual downdraft manifold but fed it through a transition section from a single-barrel carburetor.

The Model H engine had a bore and stroke of 3.25 by 4.75, a displacement of 315.2 cubic inches. The SAE rating was 33.8 horsepower and the actual developed power was 125 at 3400 rpm, with a 5.5:1 compression ratio. The 8-79 version was 3.187 by 4.75, 303.2 cubic inches. SAE and rated horsepowers were 32.5 and 107 at 3250 rpm.

The "double-dome" combustion chamber was incorporated in the whole Marmon line for 1930. The design was covered by US Patent 1,812,474 issued to George Freers on June 30, 1931. The Marmon press[1] release described the innovation: *"One of the major features of the engine, and one that is to a large degree responsible for its power and flexibility, is the new Marmon designed double-dome cylinder head (Pat. applied for).[25] This head is unique in that it contains* ***two*** *compartments or combustion chambers which greatly increase the turbulence of the explosive mixture, and gives an operating result that is comparable to the use of two spark plugs. In this head the spark plug, intake valve and exhaust valve are located in the smaller of the two chambers. As the mixture is ignited the flame spreads instantly to the larger chamber over the piston in which a high degree of turbulence has been created, and the exposive charge delivers its full power uniformly and without a trace of unused fuel"*

The compression ratio (5.5:1) was claimed to be unusually high for the recommended regular grade fuel. A full 10 percent power advantage was claimed.

The crankshaft had five main bearings, all 2.625 in diameter. Lengths were 1.844, 1.187, 2.125, 1.187 and 2.094, front to rear. The main bearings were fed oil from a manifold threaded on to the bearing caps. The shaft carried a Marmon rubber-hub vibration damper. Clearance on the bronze-backed babbitt-lined bearings was 0.0025. Thrust was taken by the middle bearing. The rod journals

Fig. 5-49 The basic engine which successively powered the Big Eight, 88 and 8-125 Models, and was underbored for the 8-79.

Fig. 5-51 Pressure lube system for 8-79 fan.

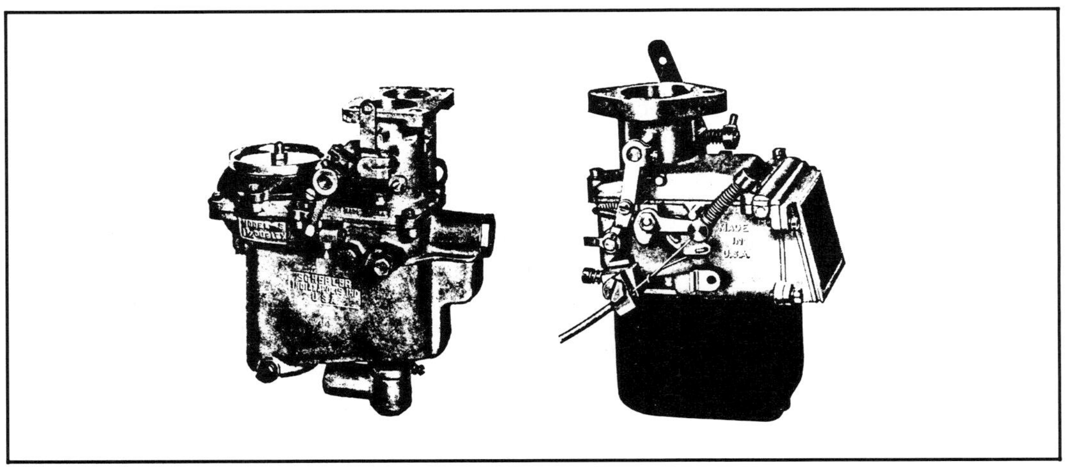

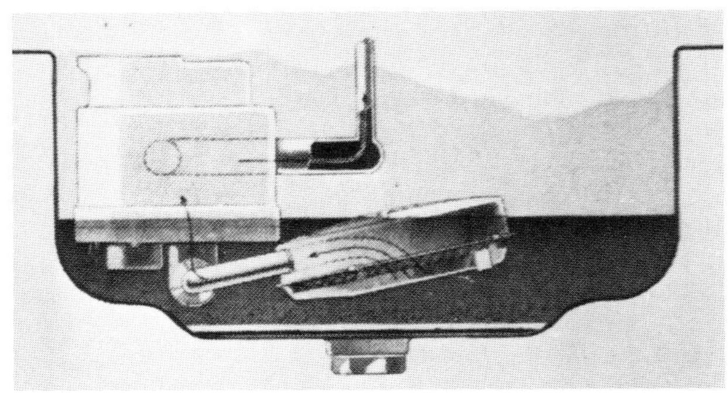

Fig. 5-52 'Float-O' oil intake.

Fig. 5-50 Big Eight (left) and 8-79 Schebler Carburetors.

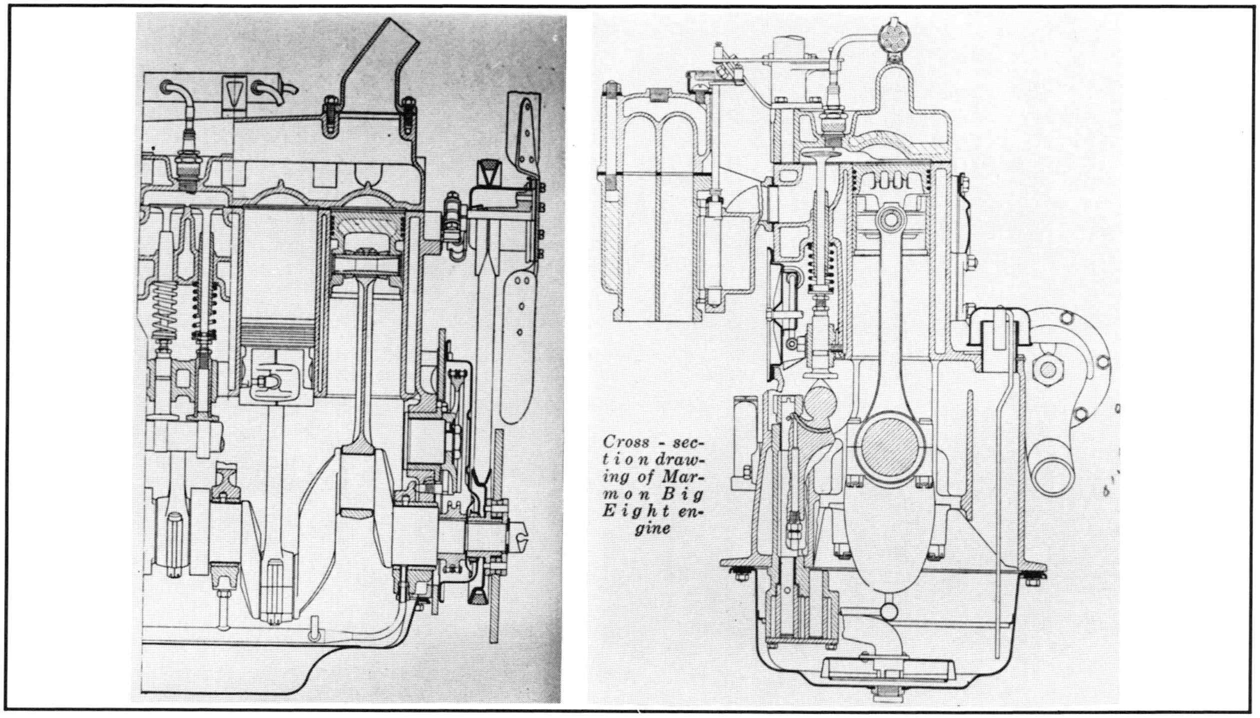


Fig. 5-53 Engineering drawings of the Big Eight engine.

were 2.375 diameter and 1.0 wide.

The camshaft had five bearings and was driven by a non-adjustable Diamond Chain Co. double roller chain 24.75 inches long. This was the same construction already proven on other Marmon models.

Connecting rods were 9.5 inches between centers and had babbitt poured in the lower end. No shims were used and clearance was fitted to 0.0015-0.00175 inch.

Four ring Permite split-skirt pistons were used. The piston was 3.812 long and the oil ring groove was drilled radially. The piston pins were full-floating 0.875 diameter and 2.812 long.

Inlet valves were Wilcox-Rich, SAE 3140 chrome nickel steel, 1.687 in diameter. Exhaust valves were Wilcox-Rich Silchrome, 1.562 in diameter. Both had .341 diameter stems and 45° seats. Operating tappet clearance was 0.007 inch. Valve spring load was 55 pounds closed and 100 pounds open. Intake valves opened at top center.

The Big Eight carburetor was a Schebler Model S Duplex air-valve with dual 1.25 throats. The 8-79 carburetor was a Schebler 1.5 single throat. An AC diaphragm fuel pump was camshaft driven and beginning with the 1931 model year a fuel filter was incorported.

The electrical system was Delco-Remy. The flange-mounted starter turned the engine at 125 rpm through a Bendix inertial engagement.

The fan hub incorporated a gear-type oil pump and oil reservoir. For 1932, 'Float-O' floating oil intake screen was adopted but the oil filter was abandoned.

Exhaust heat to the intake manifold was manually controlled from the dash.

A 2.25 diameter exhaust pipe was adapted.

The cooling system had a capacity of seven gallons.

The sump held 10 quarts of oil. Normal pressure was 40 psi. A spear-type (bayonet) oil level gage was used and an AC oil filter (by-pass type) was standard equipment, and engine mounted on the Big Eight and the 8-69 and 8-79 as well.

## PART 3 — Chasses of Era Four

There were five basic chassis designs used in Era Four. Two of them came in more than one wheelbase.

All Era Four Marmon chasses used ladder type frames, with four semi-elliptic springs. They all had straight-eight engines of Marmon manufacture, unit-mounted, with single plate clutches and various transmissions from outside sources. All were equipped with Bendix self-energizing internal shoe four-wheel brakes. And nearly all had four point rigid engine mounts making the engine structure part of the frame.

Era Four was characterized by concurrent significant developments in transmission design. It saw the 1928 advent of the Cadillac syncho-mesh transmission, the Warner four-speed with silent third (non-helical) transmission and later the Warner Gear synchronized three-speed helical-geared transmission with 'free wheeling'. The Warner Gear four speed was on the Big Eight Marmon and the Warner three-speed, with optional 'free wheeling,' was on the Models 70, 88, 8-125.

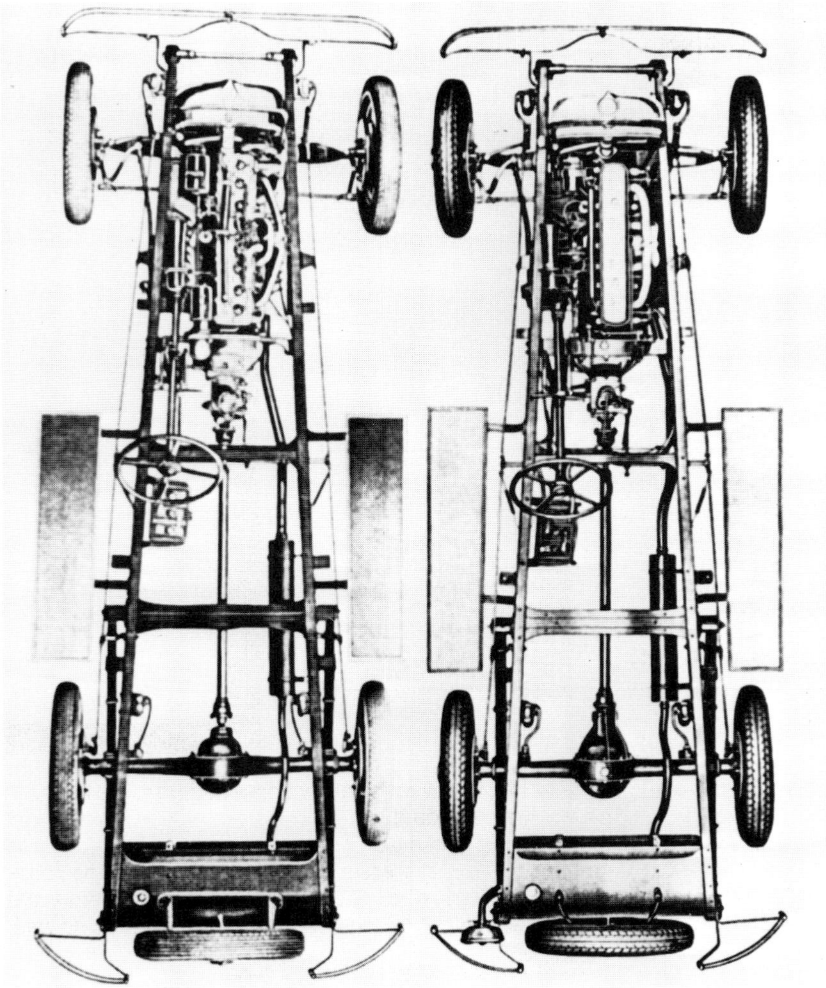

Fig. 5-54 Chassis of the Model 68 (left) and Model 78.

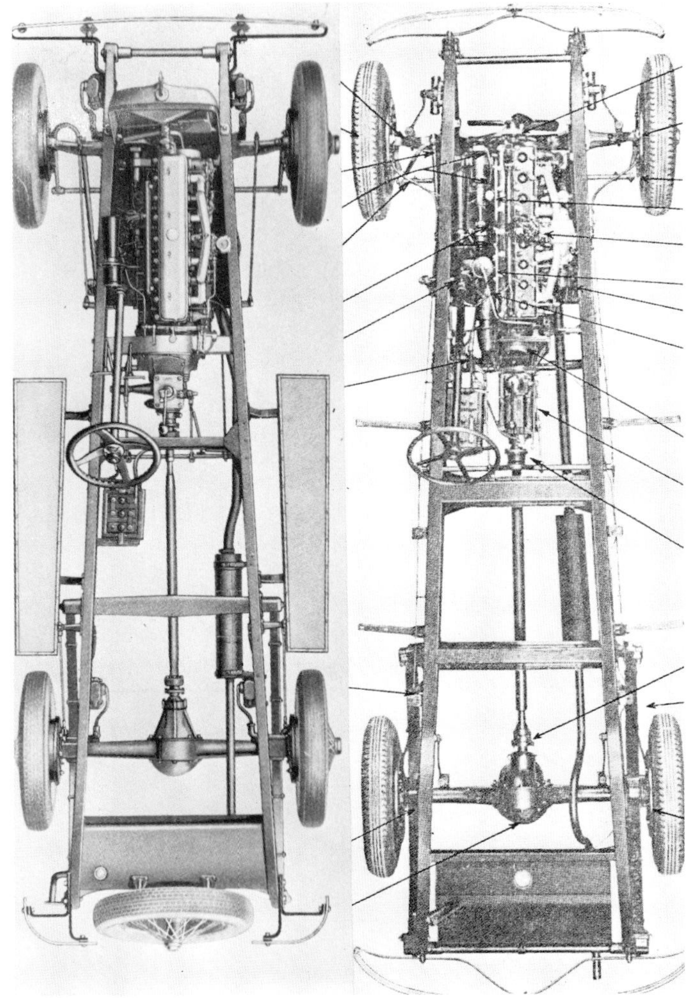

Fig. 5-55 Chassis of the Little Marmon (left) and Big Eight.

## MODEL L (LITTLE MARMON) 1927

The Little Marmon chassis was a radical change for Marmon. It was conventional in general layout but incorporated a number of features reflective of sophisticated engineering and little or no economic compromise.

It had a double-drop ladder frame which took advantage of the hypoid gear rear axle (exclusive to Packard and the little Marmon) to provide an unusually low chassis.

It had rubber spring isolators in place of conventional shackles. These eliminated the transmission of running gear vibrations to the frame and could readily be renewed. These had appeared on the Chrysler Imperial the prior year, following a 200,000 mile demonstration in a vehicle of the Yellow Cab and Mfg. Co. They were a development of A.F. Masury, President of the Rubber Shock Insulator Co. and Chief Engineer (Mack) International Motor Truck Co., and Frank Pampinella.[26]

Today they are as much a part of the Mack truck machismo as the very Bulldog itself. Bendix 3-shoe internal brakes were used with Perrot type front brake actuation.

## MODELS T (68) N (78)
## 1928-29

The models 68 and 78 had 114 and 120 inch wheelbases respectively, with 56 inch tread. The chasses differed only in the wheelbase and attachments for the unique springs and engines of the two models. Specifically, the steering gear, brakes and clutch were identical. The transmission housings were peculiar to the different clutch housing design and relative location of shift and brake lever to the transmission. The engine compartment length was 3.75 inches shorter on the '68'; the body interior accounted for the remaining 2.25 inches of the 6.0 inch wheelbase differential. Rubber shock insulators were used. Only the Model N retained the hypoid rear axle and in consequence both models were set higher off the ground than the Model L.

## MODEL Q (8-69)
## 1930 & First Series 1931

This 119 inch wheelbase chassis was similar to the Model N (78) but had a 58.5 inch tread and Bendix 2-shoe brakes with sheathed cable actuation.

## MODEL S (ROOSEVELT, MARMON-ROOSEELT), U (70)
## 1929, 1930 & FIRST SERIES 1931; 1931

This lighter 112 inch wheelbase chassis had what Marmon described as a bottle-neck frame shape with forward and main sections having parallel rails. It was a single chassis differing only in engine displacement and availability of an optional 'three speed with free-wheeling' transmission on the Model 70. Unlike most Fourth Era Marmons, metallic spring shackles were used.

## MODEL W (79), H (Big Eight); CC (8-88)
## 1930 & FIRST SERIES 1931; 1931

This heavier chassis came in three wheelbases: 125 (W), 130 (CC) and 136 (H and CC). It featured rubber shock insulators in lieu of shackles; The Model CC featured an anti-shimmy 'kick' shackle on the rear of the left front spring. The engine on Model CC was moved forward approximately five inches. Front tread was 58 inches and rear was 59.

## MODEL HH (8-125)
## 1932

The light weight of this model, despite the heavy engine, axles and brakes (from the Model H/CC) and the use of metallic shackling indicate a new frame. No illustrations of the 8-125 chassis are known to have been made.

FRAME

## MODEL L 116 INCH WHEELBASE
## 1927

The frame was a double drop design with 6x0.125 gage principal section. The side rails were bent to parallel at the radiator to provide a 36 foot turning circle despite the narrow tread.

The rails terminated at both ends in steel forgings machined to accommodate the rubber spring isolators and their retaining caps.

The frame was a rigid structure. The front forged steel members were joined by a tubular cross member which much more than doubled the resistance of the front 'horns' to bending in a corner accident. At the rear a steel plate cross-member concealed and protected the fuel tank. A massive cross-member joined the forward rear-spring attachments and there was a mid-ships cross-member just aft of the transmission which supported the brake cross-shaft.

## MODEL N (78) 120 INCH WHEELBASE
## 1928-29

Although the Model 78 double-drop frame was similar to the Little Marmon, it was a unique design. The standard tread allowed straight side rails in the plan view. Tubular front and 16 inch wide steel-plate rear cross-members were again used. The other cross members were similarly located but differently configured from the Model L. The central part of the frame was not as low as the Model L.

## MODEL 68 114 INCH WHEELBASE
## 1928-29

The '68' frame differed from the '78' only in the length of the central side rails to accommodate the six inch shorter wheelbase and in the forged ends to accommodate the narrower (1.75 vs 2.00) spring leaves.

This commonality of '68' and '78' frames was responsible for the higher top-of-frame height . . . the '68' did not use the hypoid axle of the Little Marmon and the '78'.

## MODEL Q (8-69) 119 INCH WHEELBASE
## 1930; FIRST SERIES 1931

The Model Q frame appears to have been a derivative of the Model N (78) frame with modifications in the forward half to provide stiffness in the frame formerly achieved by use of the engine structure. The central cross member was a wide hat section member dipping below the drive-shaft. Brackets on the frame interfaced with the vertical flat steel transverse member interposed between the block and bell housing of the engine.

Frame end forgings were extended to directly mount the bumper crossbar and eliminate the intermediate brackets characteristic of optional bumpers. Such a change had been made to the 34 chassis for the 1927 E-75.

The Model Q wheelbase was one inch shorter than the Model N.

## MODELS S (ROOSEVELT, MARMON-ROOSEVELT); U (70)
## 1929, 1930 & FIRST SERIES 1931; 1931

The frame for the price-leader line was a departure from Fourth Era practice in the use of metallic spring attachments.

The frame side rails were parallel for the length of both the front and the rear springs, widening rather rapidly between the springs, Marmon described it as a "bottle-neck' configuration and claimed a weight advantage because the bends in the plan view did not add weight and they made feasible lighter-weight spring attachments because bending moments were eliminated.

The Roosevelt frame was assembled by hot-riveting. It was six inches deep with 2.0 upper and lower flanges. It was stamped from 0.125 steel. Counting the engine as two cross members, the total member count was six.

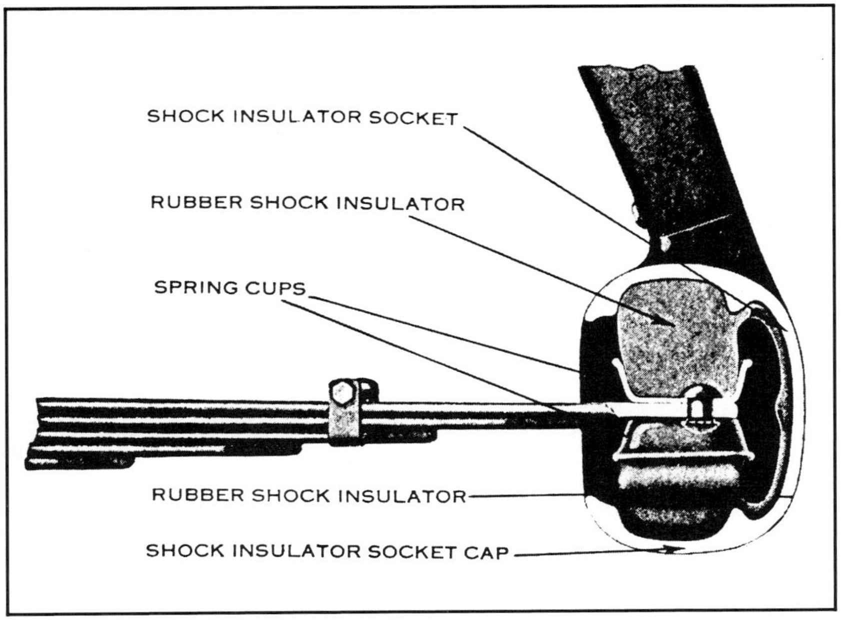


Fig. 5-56 Rubber spring insulator; 'shackle' end.

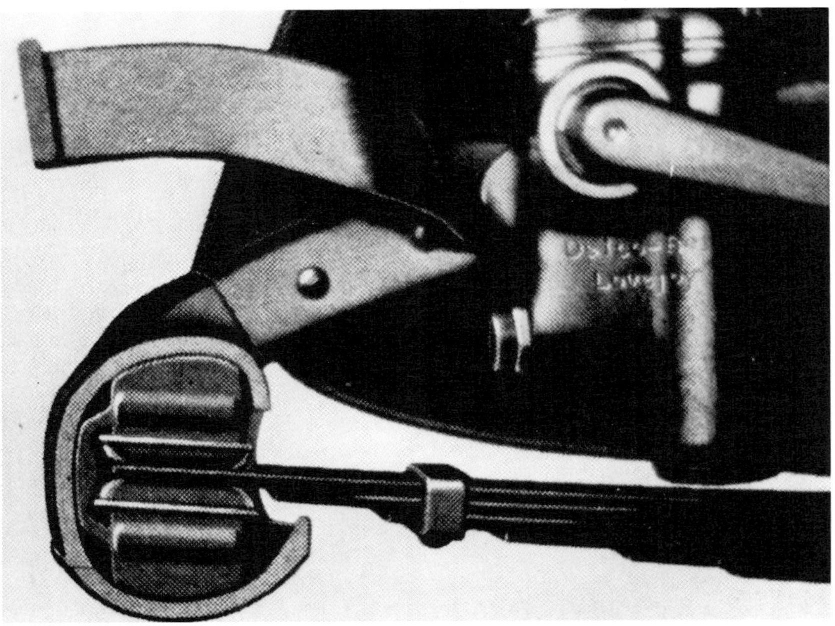

Fig. 5-57 Rubber spring insulator; Fixed end.

## MODEL H (BIG EIGHT) 136 INCH WHEELBASE 1930 & FIRST SERIES 1931

This was a more massive structure than other Fourth Era frames with five cross members and a detachable tubular cross member beneath the rear of the engine to provide rigidity for the side-mounted spare wheels. The engine rear cross member . . . a vertical flat steel plate sandwiched between block and bell housing . . . was rubber isolated from the side rails.

Frame and forgings were made to accommodate rubber isolators for wide spring leaves and to directly attach the front bumpers. Side rails were SAE 1020 steel in #9 gage (.156 thick); the channel section was 7.5 inches deep with upper and lower flanges 2.75 and 3.00 wide respectively.

## MODEL W (8-79) 125 INCH WHEELBASE 1930 & FIRST SERIES 1931

## MODEL CC (8-88) 130 AND 136 INCH WHEELBASES 1931

The Model W and CC-130 inch frames were shortened versions of the Model H frame. However both CC frames had the rear rubber insulator in the left front spring replaced by a Marmon "chassis modulator" . . . a metal shackle incorporating means to immobilize it under normal spring action. Two pairs of short coil springs acted in opposition on arms extending from the shackle. It deflected under shock or vibratory loads. Engine attachments and related chassis items were moved forward approximately five inches, and the forward engine crossmember was also rubber isolated from the side rails, on the Model CC.

## MODEL HH (8-125) 1932

This was a new frame which incorporated metallilc spring shackles. Both front springs were shckled at their forward ends. The left one was positioned at the rear by a coil spring-centered 'kick' shackle which Marmon termed a 'modulator'.

SUSPENSION

## MODEL L 1927

The Model L front springs were semi-elliptic 38 inches long and 1.75 wide. The rear springs, also semi-elliptic, were 56 inches long and 1.75 wide. Together they were 81 percent of the 116 inch wheelbase. A pair of special 'washers' was riveted to the ends of the main leaves. These mated with the configuration of a molded rubber vibration-insulator block which fitted into an appropriately configured cavity in the frame bracket and were retained by a mating cap bolted in place. If the caps were removed and the car lifted by the frame, the axles and springs would remain on the ground; new rubber 'blocks' could then be readily slipped in place and the spring attachment system would be quickly and totally restored.

Because of the narrow tread of the Little Marmon in a car having nearly normal passenger compartment dimensions, the rear springs were immediately adjacent the brakes.

Lovejoy single-acting (jounce control) hydraulic shock absorbers, with strap connection, were standard equipment at all four wheels.

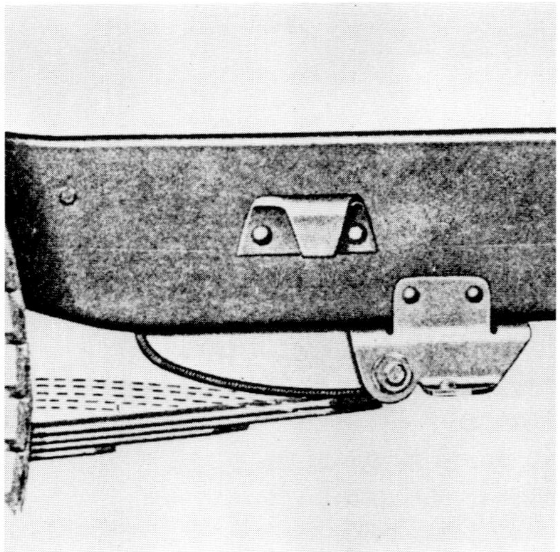

Fig. 5-58 Extra leaf gave variable rate spring.

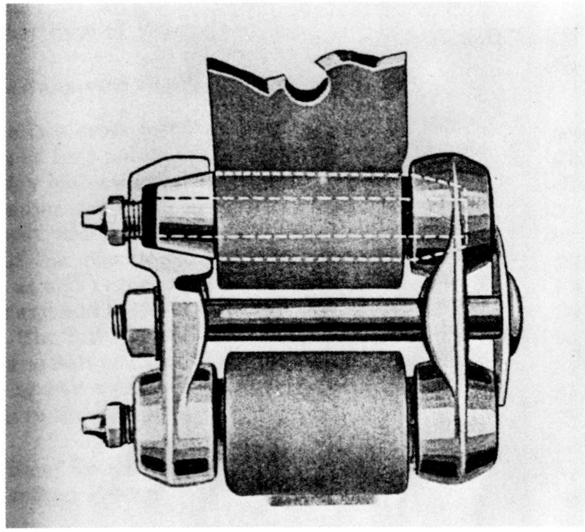

Fig. 5-59 Roosevelt spring shackle.

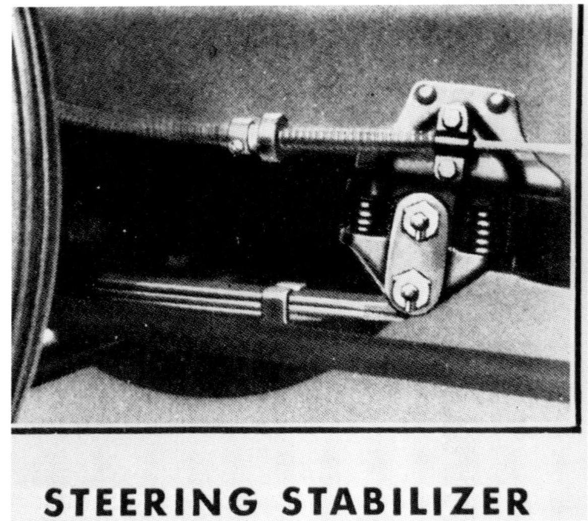


Fig. 5-60 Semi-rigid 'kick' shackle anchor for left front spring.

The front and rear insulator blocks were so configured as to anchor the front springs at the rear end and the rear springs at the forward end, with 'shackle-effect' movement allowed at the other ends.

This was accomplished by presence or absence of a void beyond the 'spring cups' and within the insulator socket.

## MODEL N(78)
## 1928-29

The springs were semi-elliptic. The front were 38.187 inches long and 2.0 wide. There were nine leaves, all made of chrome vanadium steel. The rear were 56.562 inches long and 2.0 wide. There were ten leaves with the main leaf being chrome vanadium steel and the others carbon steel.

The front springs were effectively 'shackled' at their forward ends and were inclined two degrees downward to the rear. The rear springs were effectively shackled at their rear extremeties and were inclined one degree downward to the rear.

Lovejoy single-acting shock absorbers (and bumpers) were assembled to all cars at the factory but invoiced separately at extra cost. The shock absorbers were coupled by webbing straps.

## MODEL T(68)
## 1928-29

The '68' springs differed from the '78' in width, being 1.75 wide, Lengths were essentially the same, being 38.312 and 56.375.

## MODEL S (R) (MR); U (70)
## 1929, 30 FIRST SERIES 1931; 1931

The front semi-elliptic springs were 37 inches long and 1.75 wide. The rear were 54.125 inches long and 1.75 wide. Together they were again 81 percent of the wheelbase of 112.75 inches. The springs were made of silico-manganese steel because it allowed thinner leaves than would have been feasible with carbon steel, it being claimed "more resilient and less brittle."

Marmon engineers had always taken the attachment of springs very seriously, and their solution for the Roosevelt shackles was again innovative. They sought a more durable design appropriate to a Marmon product but sought equally a low cost design: It was a "Tryon"[27] brand design, procured from a supplier firm.

The fixed, hollow shackle pins engaged the side plates on tapered surfaces. Clearance was determined, and readjusted by a single bolt, between the pins, which moved the plates against the pins. The crucial design element was a belleville-type (conical) spring washer on the center bolt which maintained a constant load on the assembly despite wear which might occur between adjustments.

The modern motorist sees the very satisfactory performance of the rubber-bushed shackles on today's cars using leaf springs. However spring shackle lubrication and dust exclusion baffled automotive designers for several decades; the typical used car exhibited severely skewed shackles on the rear springs.

A new Marmon innovation on the Roosevelt was a low cost means of achieving variable rate spring action. The lighter Roosevelt car, loaded to the same passenger capacity as the heavier Marmon models, would suffer a greater spring deflection or alternatively a stiffer ride when lightly loaded . . . unless a variable rate characteristic could be incorporated. Marmon had used a maple block as a cam to induce variable rate on the rear springs of their Third Era cars. An equivalent device for the Roosevelt was cleverly contrived . . . a special short spring leaf was placed

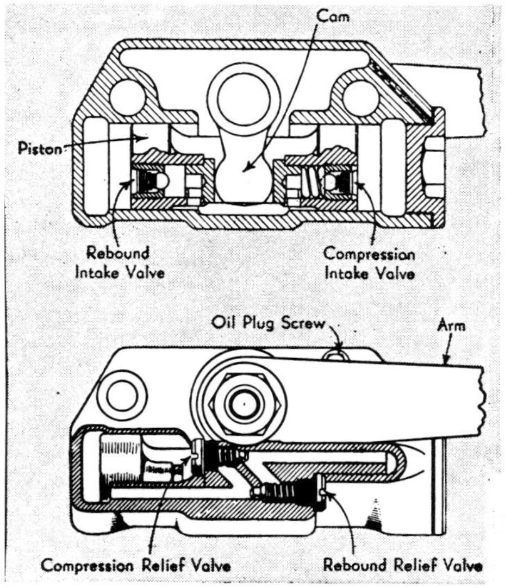


*Another type Lovejoy duodraulic shock absorber also controls car springs on both compression and rebound strokes but has only one piston*

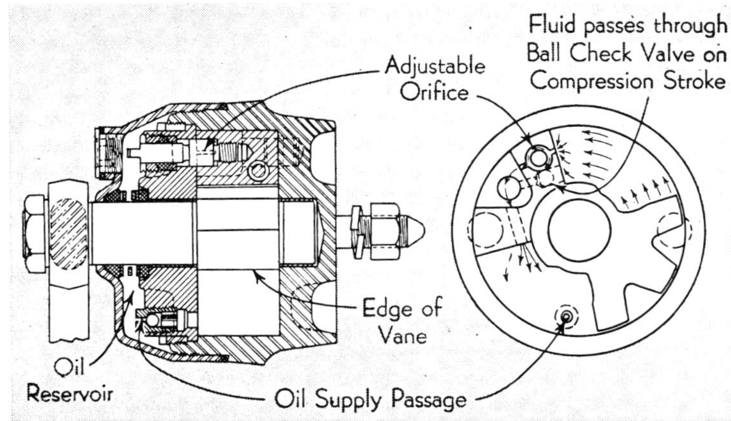


*The Gabriel shock absorbers have three vanes which force fluid through an adjustable orifice*

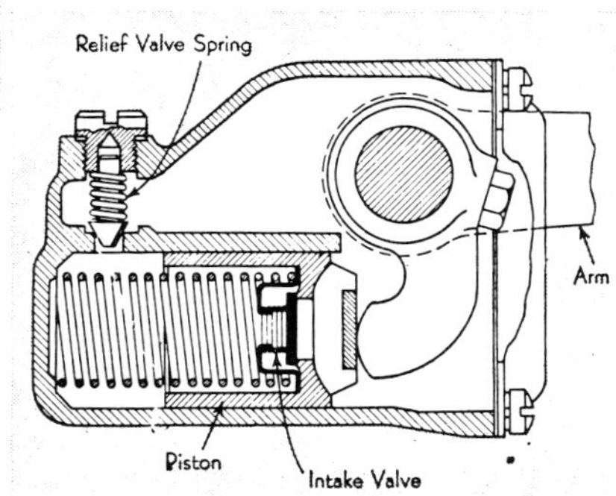


*The Lovejoy single acting shock absorber controls spring rebound*

Fig. 5-61 Era Four Shock Absorbers: Single-acting Delco-Remy Lovejoy as used on the Model 70; double-acting Lovejoy Duodraulic as used on the Models 69, 79, and Big Eight; Gabriel double-acting rotary type used on First Series 1931 Big Eight, 8-88 and 8-125 and on the SIXTEEN.

on top of the forward end of the rear spring and the two were coiled together. The special leaf however was unusually configured with a progressively increasing curvature ultimately turning 90° and contacting the bottom flange of the frame in the normal-load mode. It was therefore a fixed spring steel element of the frame and as the load-bearing leaves were deflected it acted to shorten the effective length of the main leaf and thereby stiffen it to bear the additional load with less deflection.

## MODEL Q (8-69)
## 1930

The springs were mounted in "chimney-type" rubber shock insulators wherein a larger volume of rubber was provided for cushioning as compared to the predecessor type. The springs, same width and length as the Model T, were placed as close to the axles as possible to allow a firmer support for the chassis and thus minimize sidesway, as well as reduce axle stresses since the point of weight application was closer to the wheels.[28] This could well have been a means to accommodate the increased weight and wider tread of the Model Q on the previously used basic axle and springs.

## MODEL W (79) H (BIG EIGHT); CC (8-88)
## 1930 AND FIRST SERIES 1931; 1931

The springs were mounted in rubber shock insulators. The front springs were 42 inches long and 2.25 wide; the rear were 60 inches long and 2.5 wide. Total spring length was 78 percent of wheelbase. All the leaves were silico-manganese steel.

Two-way Lovejoy hydraulic shock absorbers of horizontal cylinder type were fitted for 1930; for 1931 Gabriel Wahl rotary type were used.

The springs were metal-sheathed for lubricant retention and dust exclusion. These same spring dimensions were used on the Fifth Era SIXTEEN.

On the Model CC, the springs were mounted in rubber insulators with exception of the rear of the left front spring where attachment was made through a "chassis modulator" device. Its purpose was to halt any tendency toward oscillation (shimmy) of the steering. The device was basically a regular metallic shackle but was normally maintained in a rigid configuration by two pairs of heavy coil springs acting in opposition to each other.

### MODEL 8-125
### 1932

Metal spring shackles were used in conjunction with the same 'kick-shackle' or, 'chassis modulator,' as Marmon called it, as used previously with rubber insulator spring attachments. The 8-79 had used the Big Eight springs; the 8-125 had shorter springs: the front were 38 inches and the rear 57 inches long.

## WHEELS AND TIRES

### MODELS L, T (68)
### 1927, 1928-29

Tire size was 29 x 5.25 inches (18 inch wheel) and 4-ply tires were fitted. Wood wheels were standard. Octagon hub Hayes wire wheels were optional. No side mounts nor white walls were evident in Model L photos, and rear mounts were always single. Wire wheels and side mounts were Deluxe Model 68 equipment. For 1928 only, side mount tires and rims were available with standard wood wheels.

### MODEL N (78)
### 1928-29

Tire size was 29 x 5.50 inches (18 inch wheel). Octagon hub Hayes wire wheels with side mounts were optional. Wood wheels were 'standard' but were illustrated only in early 1928 catalogs and may not have been installed at the factory; simply a means to lower the advertised price. However side mounted tire and rims were 'available' with the wood wheels.

### MODELS S (ROOSEVELT, MARMON-ROOSEVELT), U(70)
### 1929, 1930 AND FIRST SERIES 1931, 1931

Tire size was 29 x 5.00 inches (19 inch wheel). Wood wheels were 'standard' for advertising price purposes but wire wheels (octagon hub Hayes) and side mounts were factory installed and separately priced. They were discontinued after production of 26190[29] Model S cars (i.e., wire wheels became a true option). The Model U used Bimel brand wood wheels.

During Marmon-Roosevelt production a change in the wire-wheel option occurred. Multi-bolt attached wheels were adopted with a 7.81 diameter stamped hubcap snapped in place to conceal the fasteners.

### MODEL Q (8-69), W (8-79)
### 1930 AND FIRST SERIES 1931

The '69' used 30 x 5.50 (5.50 x 19) tires; the '79' 31 x 6.00 (6.00 x 19) on 19 inch wheels. Bimel brand wood wheels were standard equipment.

In the January 1930 issue, Automotive Topics noted a new development. Marmon was balancing front tire and wheel assemblies at the factory to prevent the possibility of 'shimmy' and/or 'tramp.'

### MODEL H (BIG EIGHT); CC (8-88)
### 1930 AND FIRST SERIES 1931; 1931

Tire size was 31 x 6.50 (6.50 x 19). Dual side mounts were standard. Bimel wood wheels were standard equipment; spare tire/rim assemblies were encased by full cloth covers. Wire wheels were optional at additional cost; they were multi-bolted in place and the opening closed with a large diameter hub of stamped steel, chromium plated.

### MODEL HH (8-125)
### 1932

Tire size was 6.00 x 18. Welded spoke wire wheels with drop-center rims were used.

## STEERING SYSTEM (AXLE AND GEAR)

### MODEL L
### 1927

The Little Marmon front axle was a semi-conventional I-beam axle of reversed Elliot type. There was, however, distinctive detail design in the manner of steering

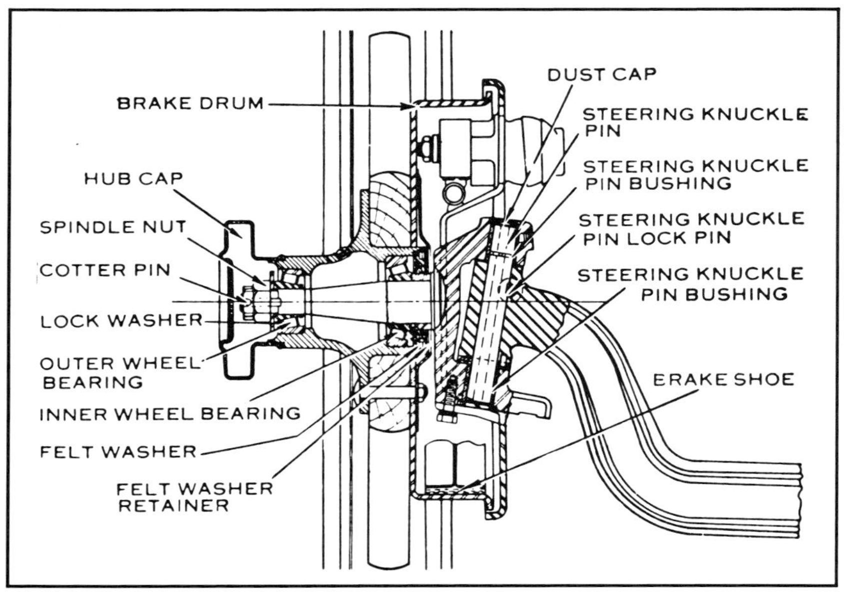


Fig. 5-61 Section through Model L steering knuckle

## Straightaway Driving (1)

**In mid-position, the general direction of the pitch of the cam-screw approaches parallel with the lever-arm. (Note large photograph on opposite page.) Road-shocks meet an abrupt check as they force the lever arm, up or down, against the flat, slow pitch of the screw. In mid-position your car almost controls itself—and turns are made with proportionate ease.**

## THE CAM

The cam, shown below, is crossed by three black dotted lines, to show the steep, fast pitch of the cam-screw at both ends (for right and left turns), as contrasted with the flat, slow pitch at the center, which is mid-, or driving-position (for straightaways). The round stud of the lever-arm slides up or down in the cam-screw as you rotate the cam by turning the steering wheel—and this lever rotates the steering arm controlling your front wheels.

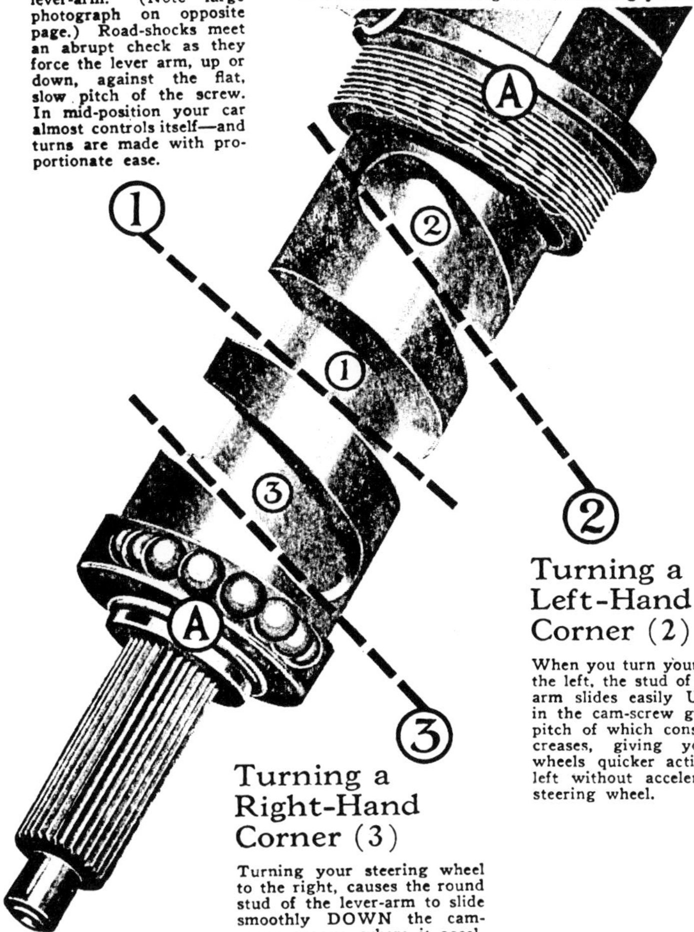


## Turning a Left-Hand Corner (2)

When you turn your wheel to the left, the stud of the lever-arm slides easily UPWARD in the cam-screw groove, the pitch of which constantly increases, giving your front wheels quicker action to the left without accelerating the steering wheel.

## Turning a Right-Hand Corner (3)

Turning your steering wheel to the right, causes the round stud of the lever-arm to slide smoothly DOWN the cam-screw groove, where it accelerates because of the increasing pitch—giving the same surer, quicker, easier turn at right-hand corners.

Note that the cam is mounted between ball-bearings (A-A), which take both thrust and radial load.

## THE LEVER

This view of the lever-arm shows the side which faces the cam, and pictures the round stud which travels up or down in the cam-screw. The unusual length of this lever-arm greatly decreases the pressure between the cam and the lever's stud—which, together with the principle of leverage, produces the easiest steering the world has ever known and greatly increases the durability of the gear.

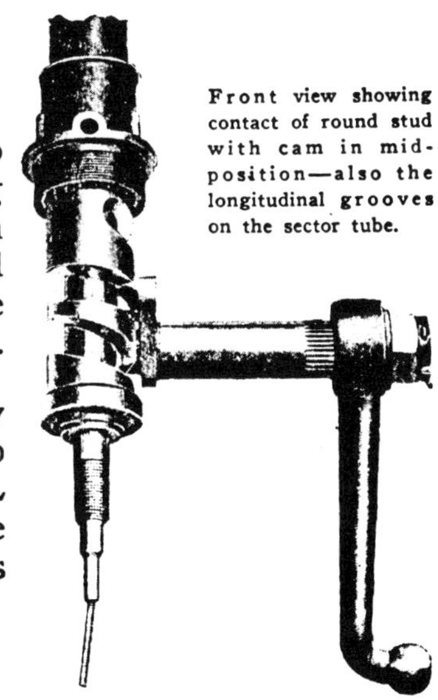

Front view showing contact of round stud with cam in mid-position—also the longitudinal grooves on the sector tube.

The variable pitch of the cam-screw permits a driver to turn corners with only a slight movement of the steering wheel—and, likewise, it stops road-shock in a new way, in all straight-ahead driving. The long lever reduces internal pressures and wear to a new minimum and helps produce the enormous power which makes Ross steering so easy.

The Ross Cam and Lever Steering Gear permits any woman to handle any car with ease—even in flat-to-the-curb parking. And it brings relief, ease and security to ALL drivers, no matter how heavy the vehicle or how rough the road. Especially recommended by leading manufacturers of balloon tires.

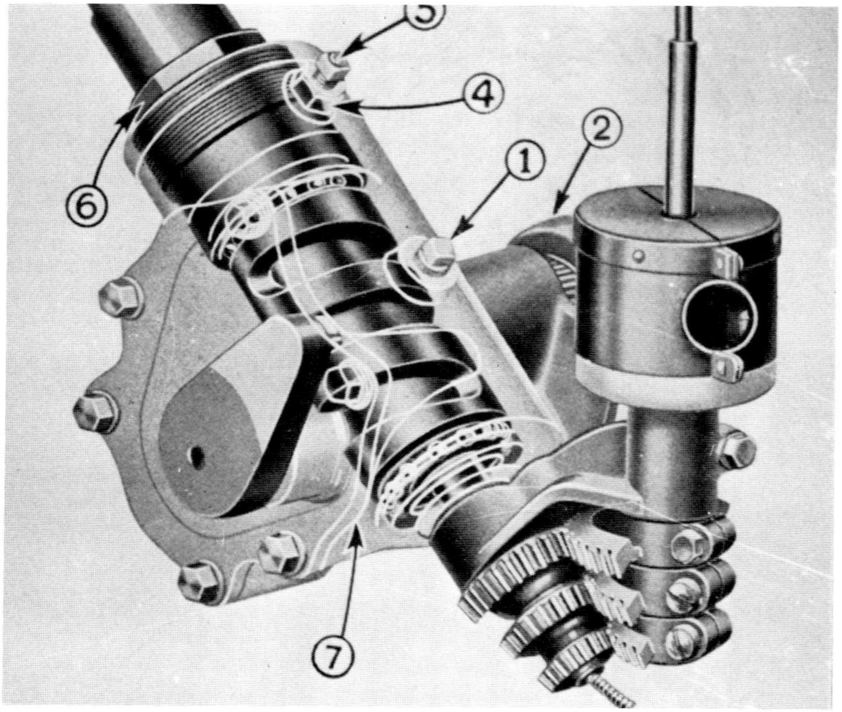


Fig. 5-62 Ross cam and lever steering gear on Model L. Note enclosure for headlight switch.

arm attachment. The arm attached, to the bottom of a flange extending from the yoke, with two bolts. The flange had an extension which closed the lower spindle bore and incorporated a third smaller bolt attachment.

Although apparently of Marmon design, the axle was procured from the Salisbury Axle Company. This axle design established the procurement source and criteria for all Fourth Era Marmon front axles: Pin inclination of 7 degrees and camber angle of 1.5 degrees. Model L toe-in was .187 inches.

The steering gear was a cam and lever type procured from Gemmer but which incorporated considerable custom application details. An external set of bevel drive gears and driven sector gears on a vertical axis operated the headlight switch, spark and throttle levers.

The steering ratio was 15.5:1 with an 18 inch wheel entirely of wood construction. Turning circle was 38 feet.

## MODELS T (68), N (78), S (ROOSEVELT)

These models shared a common axle of standard tread. It was a conventional I-beam of reversed Elliot type. In a transverse plane, the king-pins were inclined 7 degrees and the wheel (rolling) spindle was at 2 degrees to the horizontal. Caster angle was 4.5 degrees with .187 toe-in. The axle incorporated vertical journals for brake linkage mounting.

A Ross cam and lever steering gear with 15.5:1 ratio was used with an 18 inch wheel on the 68 and a 19 inch wheel on the 78. Turning circle was 40 feet. Ball thrust bearings on the cam were the only anti-friction bearings. Only adjustment was for cam thrust wear. The concentric spark and throttle shafts terminated in levers.

## MODELS Q (69), W (79), H (BIG 8), CC (88)
## 1930 AND FIRST SERIES 1931; 1931

A conventional I-Beam axle of reversed Elliot type was used. Like the Models L, 68 and 78, pin inclination was 7 degrees and the camber (wheelspindle) angle was 1.5 degrees. The caster angle was reduced to 4 degrees and toe-in increased to .25 inches.

The 'new' Ross cam and lever gear was used. For 1931, the steering shaft (which mounted the cam) was carried on double Timken tapered roller bearings . . . A construction adopted for all Marmon-built steering gears beginning in 1907.

CLUTCH

## MODEL L
## 1927

The Rockford single plate clutch was used. It was a 6-spring, push-release design with flat cast cover mounting to a counterbored flywheel. Adjustment for wear was made on the three levers. The pressure plate was driven by three lugs projecting through the cover.

The driven member was faced both sides with **woven** asbestos facings and slotted (and presumably twisted) "to provide an elastic driving engagement."[1]

A rubber insulated attachment of the disc to the hub was jointly developed by Marmon and engineers of the Rockford Clutch Company. 'East' brand stranded facings were .327 thick and the facings measured 8.875 OD and 5.875 ID. A straight-sided spline was used.

The clutch pedal was mounted on the transmission case with a relay link to the shaft in the aluminum clutch housing which carried the release yoke. This arrangement reflected the forward placement of the engine with respect to the passenger compartment which characterized early Fourth Era cars.

## MODELS T (68), N (78),
## 1928-29

A 10 inch Rockford clutch was used. Facings were 9.88 OD, 6.88 ID and 0.125 thick.

Molded facings were adopted. A 90 degree segment was molded with constant radial thickness but with thickness uniformly tapering over the angular length; four of these were riveted to the hub. Engagement was consequently progressive because the disc was twisted at full engagement because the total thickness of two opposing segments was constant.

## MODELS S (ROOSEVELT, M-R); U (70)
## 1929-30 AND FIRST SERIES 1931; 1931

A 9 inch Rockford clutch was used.

The driven member was faced with 'East' brand woven stranded asbestos facings which were 8.875 inches outside diameter and 5.875 inside. Four thin and four thick facing segments were riveted to the disc on each side with a thin segment

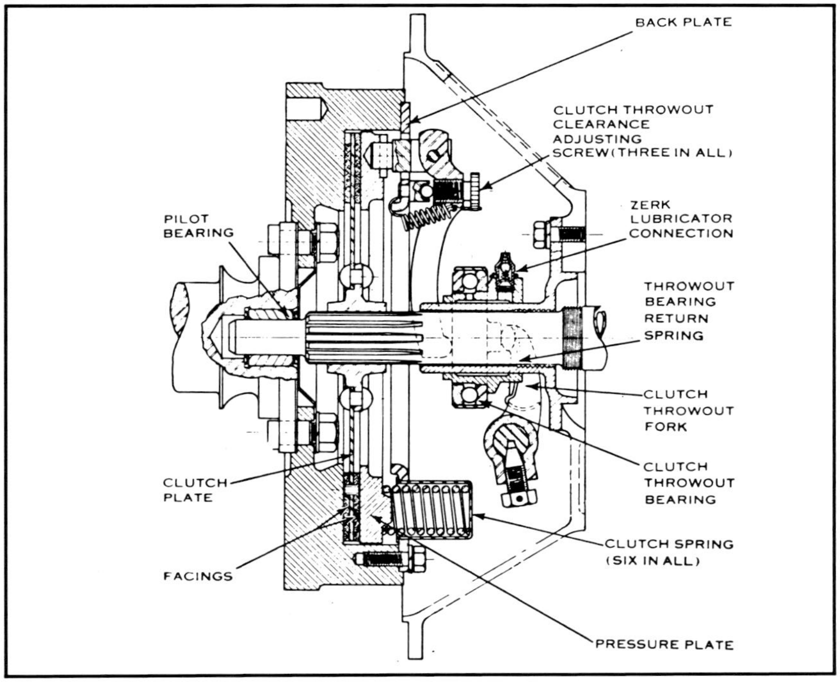


Fig. 5-63 Rockford clutch on the Model L.

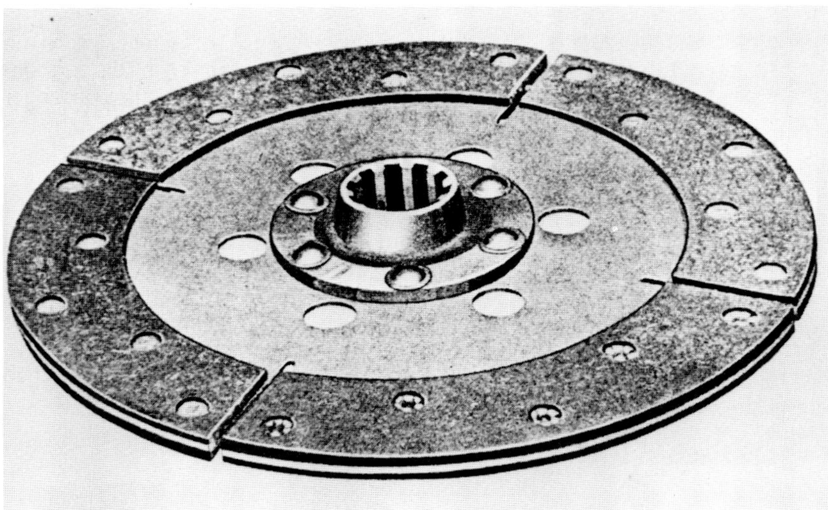

Fig. 5-64 Light-weight, rubber insulated clutch plate with alternate thick and thin facings.

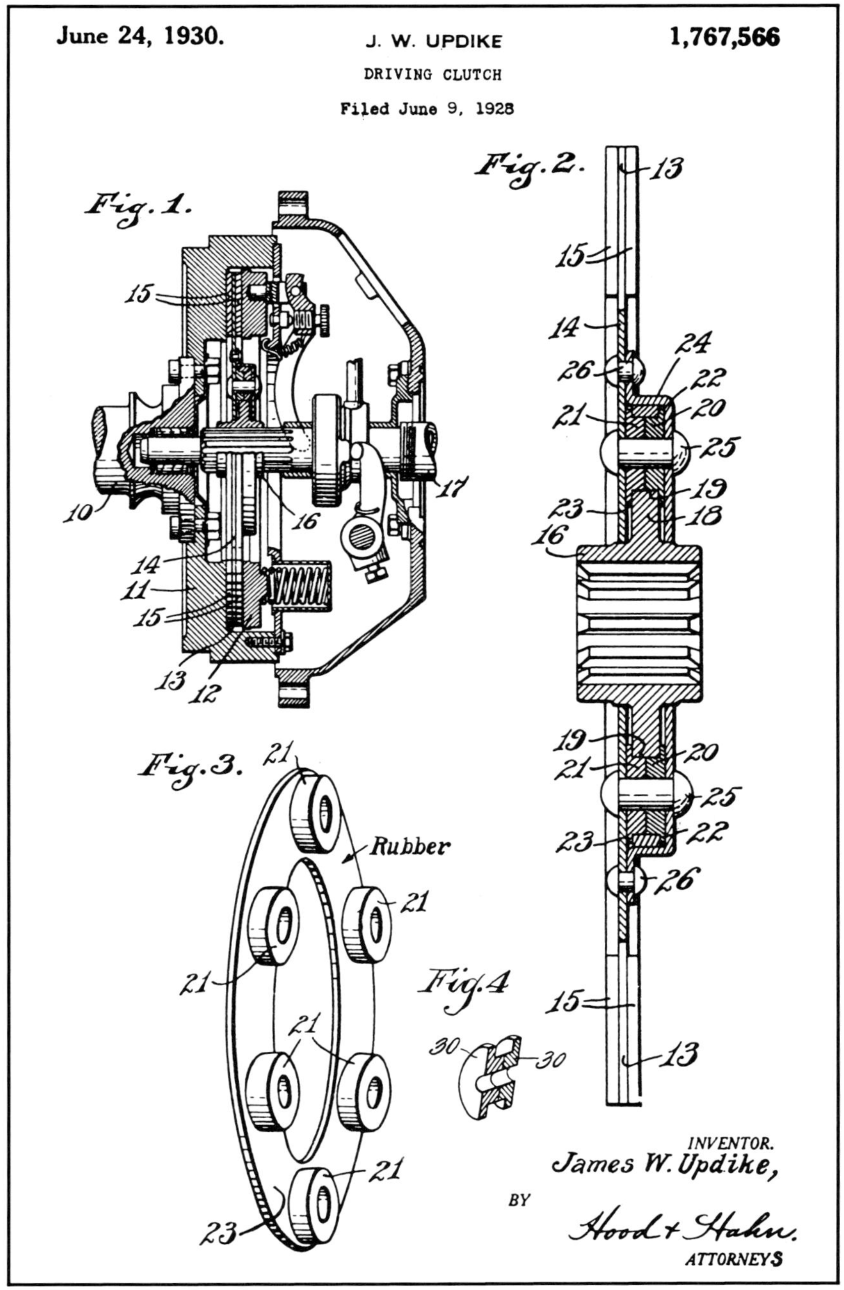


Fig. 5-65 The Updyke patent for the Marmon-Rockford clutch disc with rubber isolation.

opposite a thick one. Upon full engagement, the disc spokes were deflected in consequence and a smooth progressive engagement resulted. A straight-sided spline was used to drive the transmission. Torque was transmitted from the disc to the hub via rubber grommets which provided isolation of engine torsional vibration from the driveline gearing. It was developed and exclusively manufactured for Marmon by the Rockford Clutch Co. under US Patent 1,767,566 issued to James W. Updike of the Marmon engineering staff on June 24, 1930 following application on June 9, 1928. Included in the patent claims was the preloading of the elastomeric elements during assembly.

## MODELS Q (8-69), W (8-79), H (BIG EIGHT) 1930 AND FIRST SERIES 1931

Rockford (Russell) clutches were used. Model Q used a 10 inch clutch and 9.88 OD, 6.88 ID facings. The others used an 11 inch clutch with 11.0 OD, 5.88 ID facings. Nominal facing thickness was .125. Alternate thickness facings were used on a *Marmon-Rockford insulated disc.*

## MODEL CC (8-88), HH (8-125) 1931, 1932

These models used a 'Tower Clutch' manufactured by Rockford Clutch. It was a single-plate version of the two-plate clutch used in the SIXTEEN and featured a variable ratio lever system conducive to smooth engagement.

For 1932 a Bendix vacuum control for the clutch was optional. It provided automatic operation when a foot control button was depressed by the left foot.

TRANSMISSION

## MODEL L 1927

The transmission was a three-speed and reverse spur gear type of conventional sliding-gear, selective and interlocking design. Ratios were Direct, 1.76:1, 3.07:1 and 4.02:1 reverse.

The transmission was, for the first time in Marmon history, a purchased vendor-designed product. It was procured from the Mechanics Machine Co. of Rockford, Illinois, who had introduced it in 1924. Its general configuration differed from the Model 34 transmission of 1916-28 only in the incorporation of the conventional location of the shift lever and rails in the transmission cover and the use of ball bearings on the input and main shafts in place of Hyatt type cartridge roller bearings and bronze thrust washers.

The input shaft was piloted in a Hyatt type roller bearing in the flywheel and was supported and located by a ball bearing at the other end. The mainshaft was ball-bearing mounted: the counter shaft was bushed and ran on a fixed shaft.

## MODELS T (68), N (78) 1928-1929

The ratios were: Direct, 1.75:1, 3.0:1 and 3.87:1 reverse. Otherwise it was similar to the Model L. Nickel steel gears were used. Cluster gear bearings were bronze.

A Briggs & Stratton lock at the base of the gearshift lever permitted locking the gears in neutral.

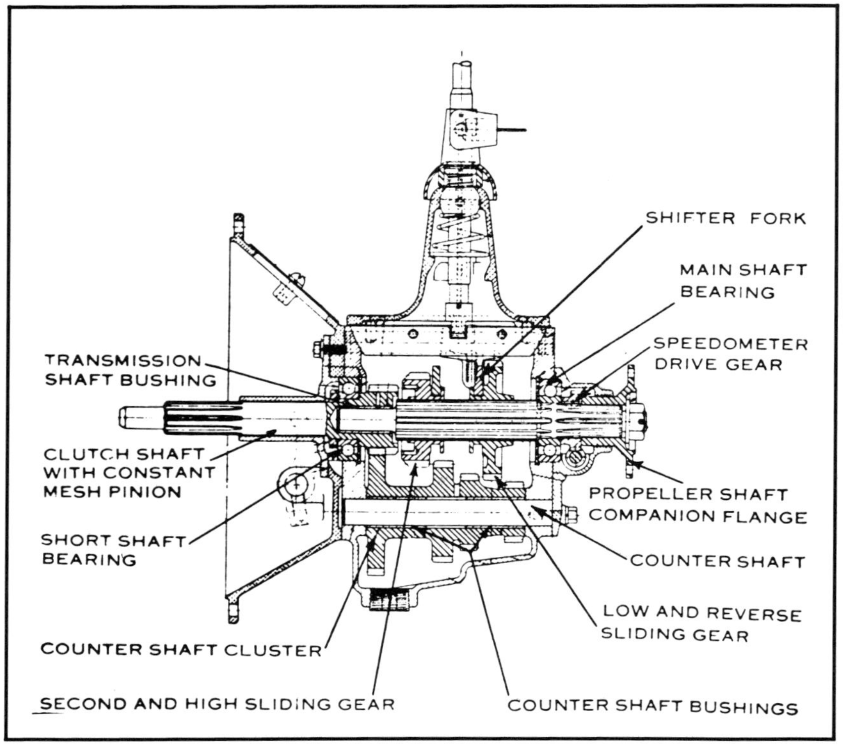


Fig. 5-66 Model L transmission.

## MODEL H (BIG EIGHT) 1930 AND FIRST SERIES 1931

The 'standard' transmission was a conventional spur gear unit procured from Detroit Gear.

Optional (for pricing purposes) was a four speed transmission procured from Wamer Gear. It incorporated a clever silent third gear achieved by use of two spur internal/external gear sets. Ratios were: Direct, 1.475:1, 2.46:1, 4.01:1 and 3.46:1 reverse. The shift pattern was standard with the addition of a spring-biased lowest gear position to the left of reverse. This was standard on 7-passenger styles.

This transmission had been pioneered on the Graham application in 1928. It featured the first gearing system to provide a silent gear drive. The progressive engagement and load-sharing by more than one tooth which subsequently characterized helical-geared transmissions was accomplished for a single ratio . . . third of four . . . by the use of two internal-external spur gear sets in series; The internal gear of the initial gear set and the external gear of the other gear set were on a shaft mounted, eccentrically to the main shaft axis, on a pair of roller bearings.

The input shaft carried on adjacent pair of gears; the first was the usual drive to the cluster (for first, second and reverse), and the second was the input to the eccentric sleeve.

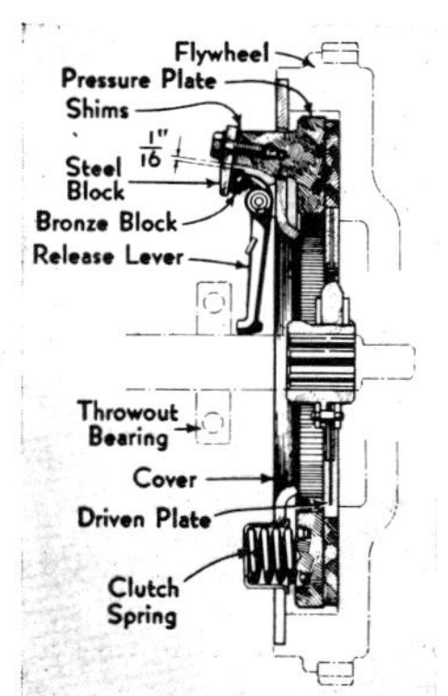


*Marmon 8 clutch uses shims to make adjustments for clutch wear*

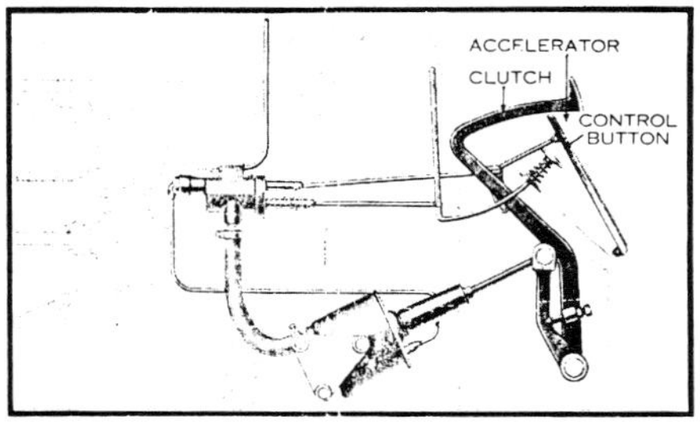


**IMPROVED AUTOMATIC CLUTCH**

The Marmon 8-125 is equipped with the most approved automatic clutch. When the driver holds his foot on the convenient button on the floor board of the car, the clutch is controlled by the action of the accelerator. When coasting on an idle throttle, the clutch automatically disengages, providing every result and advantage of advanced free wheeling.

Fig. 5-68 TOWER clutch used on the 88 and the 8-125.

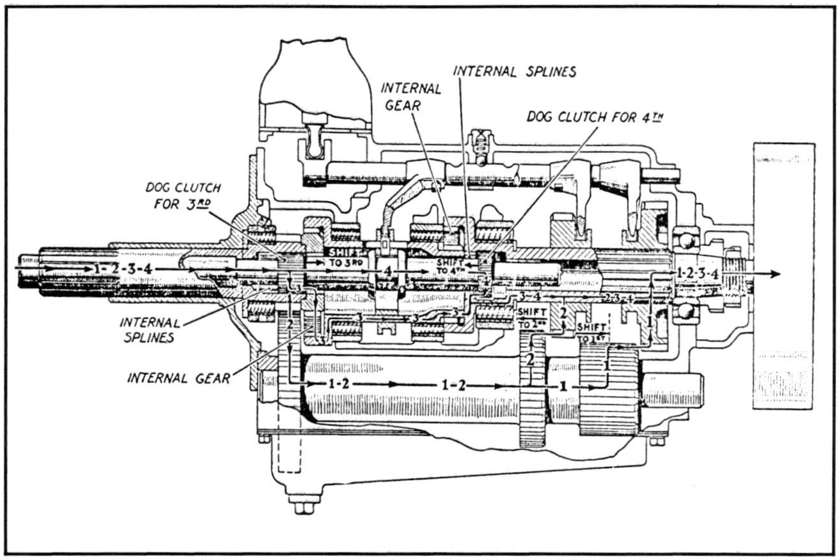


Fig. 5-69 Four speed, silent third, Warner transmission used on Marmon Big Eight and Graham-Paige cars. Brake drum shown is the G-P application.

The 'key' to the Warner Gear four-speed, silent-third, transmission was the inversion of the rotating shift elements . . . the constant mesh gears were engaged by a 'control rod' (shift-shaft) within the main shaft (rather than by a sleeve on the outside). Three slots in the tubular shaft permitted coupling of the control rod to a shift-sleeve on the outside of the shaft, positional by conventional means.

For the Graham-Paige and Marmon application, the transmission was configured to retain a conventional shift pattern for 2nd, 3rd, high and reverse; 1st gear was the 'odd' position and not anticipated to be used in normal level road operations. This feature necessitated separate sliding gears (moved by different shift rails) for 1st and 2nd gear, and a correspondingly longer case. The alternative shift pattern (used on the Dodge Senior Six, for example) engaged reverse by sliding a reverse idler cluster with the third shift rail ('odd' position). Stutz used the same gearing concept in a unique configuration which avoided the internal shift-shaft which was prone to rattle in neutral.

For the Marmon application, the ratios were: (4) 1:1; (3) 1.475:1; (2) 2.46:1; (1) 4.01:1; (R) 3.44:1.

## MODEL CC (88)

A Warner four speed transmission with synchronized third and high gears and silent helical drive in third was used.

## MODELS U (70) CC (88)
## 1931 ½ OPTION

The free-wheeling silent-second transmission released as an option on the Marmon 70 and 88 in April 1931 was basically shared with Lincoln and Pierce-Arrow. It differed from other free-wheeling applications in that the overrunning roller-type clutch was incorporated in the coupling moved by the shift rail to engage direct drive or to secure the free-running helical second gear to the shaft. Consequently the free-wheeling mode was available in both second and 'high' gears.

For the incorporation of the free-wheeling feature, **two** sets of coupling teeth . . . internal and external . . . were attached permanently to the adjacent gears. The sliding selector member was divided into inner and outer elements, separated by the overrunning clutch rollers, having respectively external and internal coupling teeth.

Depending upon whether the driver depressed the control button, as he grasped the shift lever, either the inner or outer sleeve was engaged and either direct or free-wheeling drive resulted in the gear (second or high-direct) selected.

## MODEL H H (8-125)
## 1932

The standard transmission was three-speed and reverse with synchronized second and high gear engagements and silent helical second gear. The free-wheeling transmission described for the Models 70 and 88 above was optional. Ratios were: low 2.87:1, second 1.68:1, and reverse 3.76:1. The axle ratio was 4.08:1. Detroit Gear and Machine Co. was the transmission vendor.

Bendix Power Clutch Control was also an available option; the clutch was automatically actuated in response to accelerator pedal movement when a mode select button was depressed by the left foot.

### AXLE

## MODEL L; N (78)
## 1927; 28-29

The Little Marmon featured a hypoid-geared, semi-floating axle with a banjo housing. It provided a 2.5 inch lower pinion shaft location.

The axle housing (they called it the 'bridge' then) was of pressed steel construction.

Although the axle was of three-quarter-floating type (in which the vehicle weight is borne by the outer portion of the axle **shaft**) the design was sophisticated and representative of previous Marmon practice with full-floating axles.

The axle shaft carried a pair of tapered-roller bearings on opposite sides of an integral ridge. These bearings were mounted in a flanged sleeve which bolted to the end of the axle housing, with the brake backing plate sandwiched between. Grease was retained by a leather diaphragm on the wheel side and by the traditional Marmon multiple felt washer seal in the axle housing running on the shaft.

The differential carrier was supported by bearings equally disposed on opposite sides of the pinion centerline in the usual configuration. However, reflective of the Model 34 design, a single row ball-bearing was used on the ring-gear side and a pair of opposed radial-contact ball bearings on the other side. Loosening and tightening the pair of threaded retainer rings, one on each side of the bearing set, adjusted the ring gear along its axis. The pinion was carried on a pair of widely spaced bearings: a massive single-row ball bearing adjacent the pinion and a pair of opposed angular-contact ball bearings in a flanged enclosure. Shims behind that flange provided adjustment; a cartridge leather lip-seal with spring loading provided lubricant retention.

Axle ratios available were 4.7:1 and 5.1:1 for the Model L; 4.9 for the Model 78 with 4.45:1 on 1929 open body styles.

Marmon described[9] the hypoid type axle: *'The Hypoid bevel gear has the advantage, over the alternative underslung worm, of permitting ample road clearance (8.75 inches). It has an advantage over the spiral bevel in that the area of tooth contact is considerably greater, thus adding to the life of the gear. There is a small amount of sliding action between the gear teeth, which tends to generate quietness, and any Hypoid gear that may be (initially) noisy becomes quiet after a short period of usage.'*

## MODEL 68
## 1928, 29

A spiral-bevel gear in a semi-floating banjo axle was used.

The differential carrier was mounted in the now universal manner in a pair of Timken tapered roller bearings located in opposition on opposite sides of the pinion centerline. However, the dual Timken bearing on the axle shaft at the wheel was the same as the '78' Hypoid axle construction.

Axle shafts were hot-rolled, heat-treated steel tapered from 1.156 to 1.375 inches.

The ratio was 4.9:1 (49 and 10 teeth). For 1929, open styles had a 4.45:1 ratio.

## MODELS Q (69) W (79)

A semi-floating rear axle featuring tapered roller bearings throughout was procured from the Salisbury Axle Co. All closed Model 69 cars had a 4.9:1 ratio while open body styles used a 4.45:1 ratio. Closed Model 79 cars had a 4.7:1 ratio and open styles had a 4.45:1; later production used the 4.70:1 ratio exclusively.

A 2.0 diameter steel tubular drive shaft was used with Spicer universals.

## SURPRISING RECOGNITION

In the lobby of the Spicer Axle Division of the Dana Corporation in Fort Wayne, Indiana is displayed an auto show type cut-away passenger car axle. Internal parts

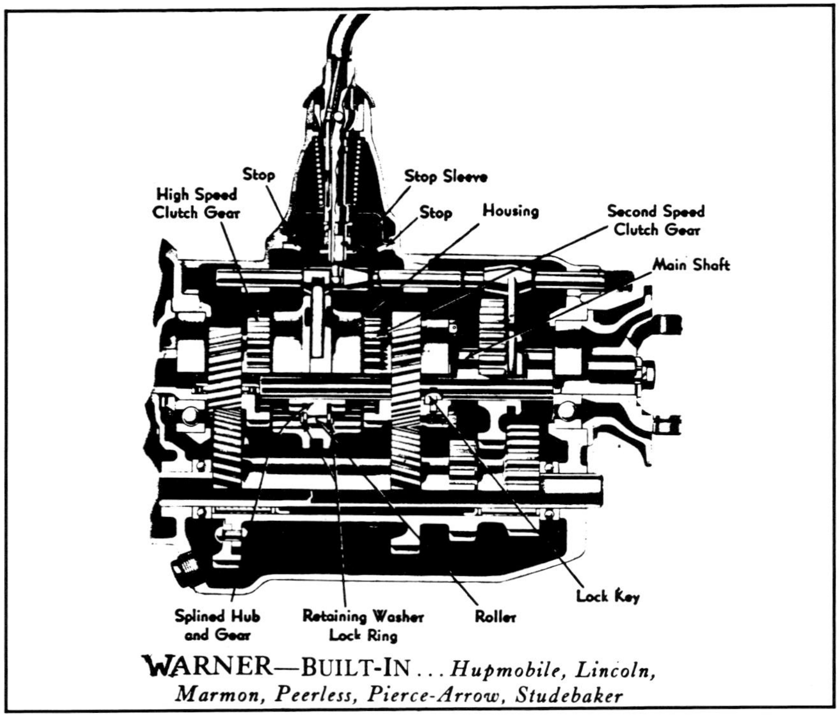


Fig. 5-70 Silent second, non-synchronized, transmission for the Models 88 and 8-125 with integrated free-wheeling in the second-high gear-type clutch.

# HOW IT WORKS

WHEN the hub (engine) is driving, the roller is forced into the narrow end of the pocket until it locks against the housing (rear wheels) and transmits power to it

When the speed of the housing exceeds that of the hub, the roller is forced into the deep end of the pocket so that the roller is loose in the housing and permits the housing to rotate free of the hub

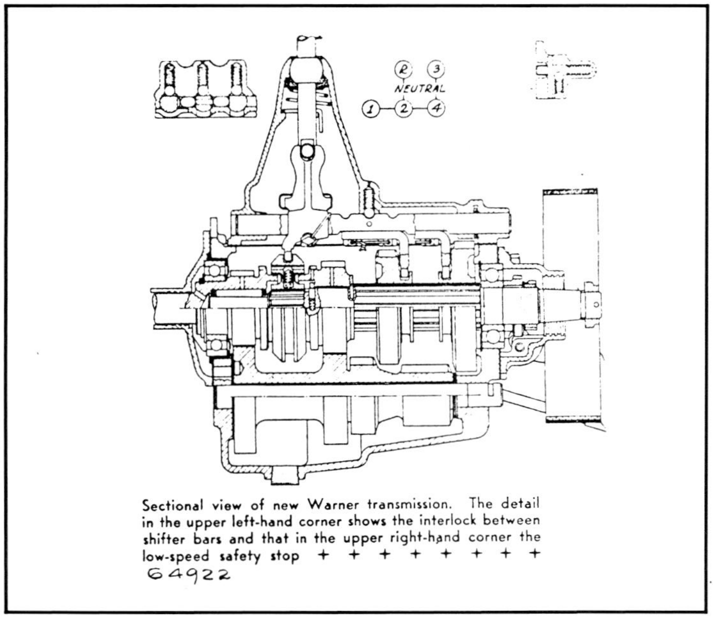


Fig. 5-71 Warner four-speed, synchronized silent third, transmission used on Marmon 88, 1931. Brake drum not applicable to Marmon application.

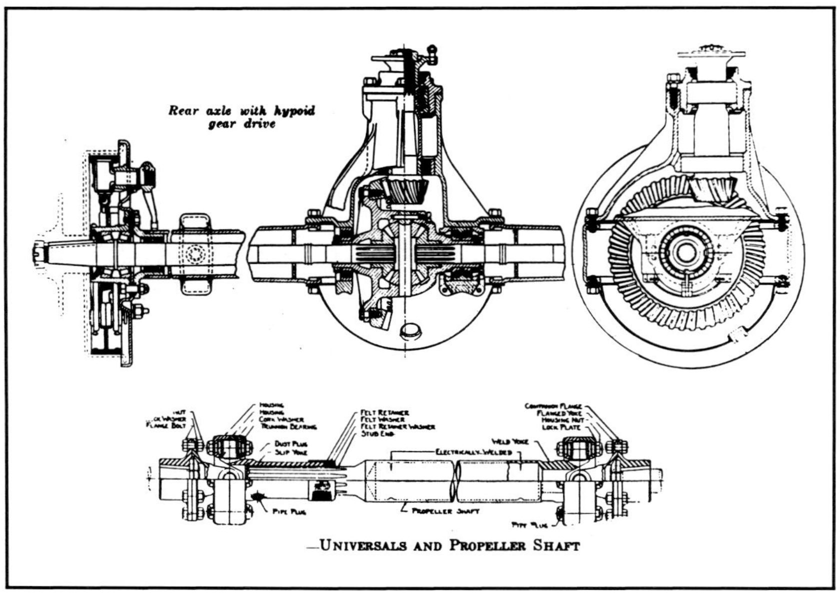


Fig-5-72 Model L hypoid rear axle.

are carefully cut-away, and each part chromium plated. The exterior is flawlessly finished in gloss black.

Unsuspecting visitors casually inspecting the axle are surprised to find Marmon identified as the customer and even more so at the following legend engraved on a brass placard and standing near the display:

> 1930 SALISBURY MODEL "J"
> Built for Marmon Corp. Model Q Motor Car
> Dedicated by Lowell Brewer
> Technical Research — Leonard W. Harry
> This axle found under a farm wagon in the year 1976 fully intact. It has been restored for the sole purpose of displaying the quality of workmanship that has made our Axle Division of the Dana Corp. what it is today.

By 'restoration' Dana meant an exhibit ready in 1978 for the 1930 auto shows! A remarkable story, indeed.

## MODEL H (BIG EIGHT) 1930, FIRST SERIES 1931

A Salisbury (brand) rear axle of semi-floating type was used. The spiral bevel pinion was straddle-mounted or ball bearings. The differential and wheels were mounted on tapered roller bearings. Gear ratios were 4.45:1 except on the long-wheelbase 7-passenger styles where a 4.81:1 axle was used.

A 3.0 diameter seamless steel tube was used in the driveshaft with Spicer metal universals at both ends.

### BRAKES

The very first Marmon had relatively large brakes. Despite that, successive increases in size followed and were accompanied by a progression of engineering advances, both metallurgic and kinematic, and usually innovative enough to be patented. The self-energizing internal front wheel brake introduced in September 1923 was the last new brake design to come from the Marmon design room.

The Bendix brake became available for application to 1927 models. Thereafter, the Bendix brake represented the state-of-the art and all subsequent Marmons were fitted with them. Marmon engineers continued to design the application systems and to develop optimum friction characteristics for specific vehicles.

The Bendix self-energizing brake used the drag induced by initial contact of one shoe to "power" the contact of the second interconnected shoe with the drum. Initially, three shoes were used: An interconnected pair for forward braking and a single shoe for reverse braking. The brake linkage on the Little Marmon was so arranged that the seizure of one brake mechanism would not prevent the application of brakes on the other wheels.

A significant feature of the Bendix concept was the bi-directional self-energizing action. This fully matured when the two-shoe design replaced the three, and identical linings were used. Some differential braking action however persists due to use of spring biasing the forward shoe against the cam in the interest of uniform action on all wheels, in the forward apply mode.

Later the cams would be replaced with hydraulic cylinders and the Bendix type brake would become the universal domestic brake of the 1960s.

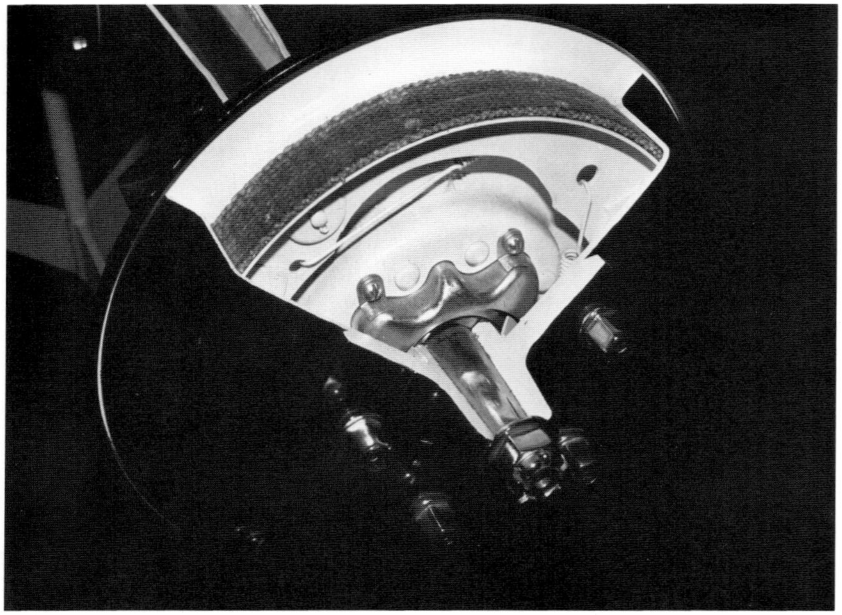

Fig. 5-73 Model Q rear axle salvaged and restored by Dana Corp.

The pivots between shoes are otherwise unretained and the force applied by the cam is reacted adjacent the cam.

The Bendix 3-shoe brakes were used in 1927 on the last of the Third Era Marmons . . . the 1927-28 Model E75 and the first of the Fourth Era Marmons . . . the 1927 Little Marmon.

The Roosevelt was among the first to use the Bendix two-shoe brake. The 1929 Oldsmobile had the Bendix three-shoe on the front wheels and the (pre 4-wheel) external band brake on the rear wheels.

## MODEL L
## 1927

The new Bendix three-shoe self-energizing internal brake was adopted for all four wheels. Drums were 12 inches diameter, 2 inches wide. These brakes served both the service and parking functions and had 224 square inches of lining.

The front brakes were operated by shafts supported between the frame and the brake, with a spherical seat at the frame and a covered universal joint at the brake located on the projected centerline of the steering spindle — the Perrot system used on the Cadillac 314 the prior year. A lever extension on the shaft was connected to a rod clevis. From a central cross-shaft between the side rails, which was linked to the pedal and the hand-lever, four rods ran inside the rail to short shafts which relayed the motion to rods outside the rail which extended to the brakes. The short shafts were supported in cast housings riveted to the frame siderails.

## MODEL T (68), N(78)
## 1928-1929

The 12 by 2 brake size of the Little Marmon was retained. Woven asbestos linings were used; 224 square inches of friction surface was provided. The operating linkage was simplified by extending the central cross-shaft through the frame and routing all rods on the outside. The short cross shafts were replaced with relay levers pivoted on bolts in the siderail.

The front brakes were actuated by an arm pivoted on a vertical journal atop the axlebeam; the arm contacted a lever on the brake actuating cam. When properly adjusted the contact point of the two arms would be directly over the projected centerline of the steering spindle, with the brake fully applied. This is called the Perrot system.

## MODEL S (ROOSEVELT, M-R) 70
## 1929, 1930, 1931

Two-shoe Bendix brakes were used. The 11 inch diameter drums were used with only 146.25 sq. inches of lining surface. Raybestos molding linings were used.

## MODEL W (8-79)
## 1930, FIRST SERIES 1931

Two shoe Bendix brakes, with drums 14 inches in diameter and 2 inches wide,

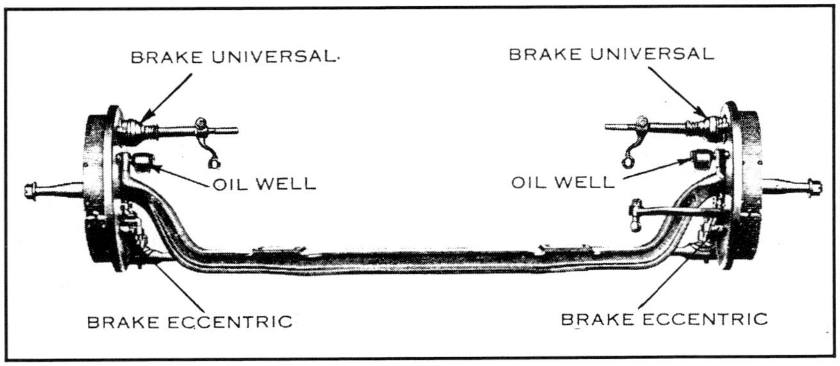


Fig. 5-74 Brake linkage: Model L (upper), Model S.

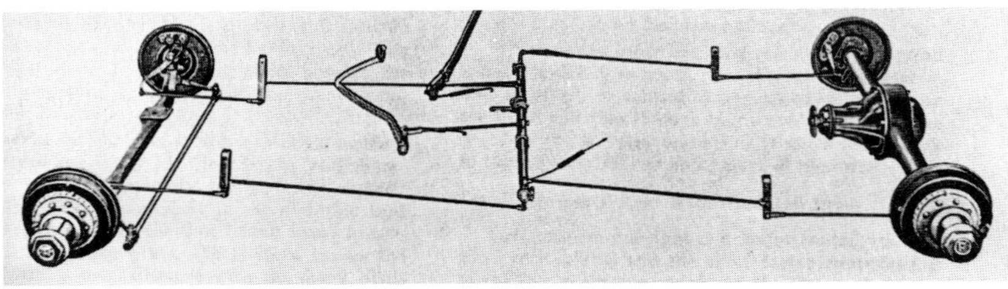

were cable operated.[30] A floating cross shaft, actuated by the pedal, operated cables running to the four wheels. Each ran along the frame side rail exterior to a point near the wheel where it entered a flexible conduit, anchored at the entry attached to the frame and at the exit to the brake backing plate. Consequently the position of the wheel had no influence on the brake function. The conduit was made of a specially designed tightly-coiled steel wire covered with fabric on the outside. This brake system was described as unusually simple and having no points to rattle or which required lubrication.

## MODEL H (BIG EIGHT); CC (8-88); HH (8-125) 1930, AND FIRST SERIES 1931; 1931; 1932

Bendix two-shoe brakes, with 15 inch diameter drums 2 inches wide, were used. Cable-in-sheath actuation was used. Beginning in 1931, sheaths were packed in grease.

For 1931 new double acting cams were adopted which obtained maximum braking with light pedal pressure. Brake effort was directed equally to front and rear wheels like all Era Four Marmons.

Service brake cable assemblies incorporated a grease fitting midway down the conduit for lubricant replenishment.

Cars for export incorporated a parking brake, of drum type, on the rear of the transmission.

# MARMONSIXTEEN PRE-SHOWING

The Marmon Sixteen, world's most advanced motor car, will be formally presented to dealers and to the public at the New York and other principal automobile shows of 1931.

The Marmon Sixteen, mechanically and exteriorly, stands as the outstanding new car, not only of 1931, but of recent years. Its 200-horsepower all-aluminum power plant was engineered by Howard C. Marmon, who devoted more than four years to its development and perfection.

Features of the Marmon Sixteen, from the mechanical viewpoint, include the highest power-weight possessed by any car to date, non-scorable case-hardened steel cylinders, simple yet efficient carburetion, manifolding and valve mechanism, and surprisingly economical fuel consumption. The performance of the car is amazing.

In appearance, the Marmon Sixteen is strikingly new and modern. It was designed by Walter Dorwin Teague who has translated the true spirit of its performance into the beauty of modern simplicity. The Marmon Sixteen has a wheelbase of 145 inches. It retails under $5,000.

TWO NEW MARMON EIGHTS

As companion cars to the Sixteen, Marmon also announces two outstanding new straight-eights, the New Marmon 70 and the New Marmon 88, in the $1,000 and $2,000 fields respectively. Distributors and dealers in open territory who are interested in learning further details of the 1931 Marmon line and the Marmon franchise are cordially invited to communicate with the Marmon factory or visit the company headquarters at the principal shows.

MARMON MOTOR CAR CO., INDIANAPOLIS

LEFT—The Marmon Sixteen Two-Passenger Coupe

CHAPTER SIX

# THE MARMON OF ERA FIVE

*"The Worlds Most Advanced Motor Car"[1]*
*1931-1933*

In "The Five Eras of the Marmon Car"[1] I titled the Fifth Era 'An American Rolls-Royce', quoting John Bell Rae.[2] I was dissuaded by Maurice Hendrey from perpetuating that association in The Marmon Heritage, for two excellent reasons: (1) Avoidance of confusion with the Springfield, Massachusetts manufacture of 'American' Rolls-Royce cars under license from the parent firm during the 1920s, and (2) recognition that Marmon never had nor deserved, **that** particular image of an extremely conservatively engineered high-quality car of high prestige ownership. In a letter to the author, Hendrey noted that Marmon engineering was as advanced as Rolls-Royce engineering was conservative and suggested that the Pierce-Arrow was more of a domestic parallel. He suggested that comparisons of the contemporary pairs — Marmon Model 34 and Rolls-Royce Silver Ghost, Marmon SIXTEEN and Rolls-Royce Phantom II — would lead to the conclusion that "Rolls-Royce has been greatly **over-rated** while Marmon has been much **under-rated.** The Marmon was more like an 'American Lanchester' (if an English parallel is sought)." I agreed with Hendrey that a title from Marmon's own contemporary literature would be more appropriate —

"THE WORLD'S MOST ADVANCED MOTOR CAR"

The preceding Four Eras were each characterized by several Marmon models. While in two of the three years of the period 1931-1933, Marmon did market two or three models, the author sees the Fifth Era as a one product . . . the ultimate product . . . Era. The presence of the other offerings during those years, the author ascribes to an overlapping of Eras, extending the Fourth Era to include 1932.

Because of the One Era — One Car approach, the format for the Fifth Era differs from the others. Following an introductory discussion, the styling, engine, chassis and chassis components will be addressed only once but fully. Because the FIFTH ERA was also the final era for the production Marmon automobile, the chapter will close (following Chassis-Chassis Components) with sections entitled "Trouble in the Marketplace" and "Demise of the Marmon Motor Car Co."

The designs that characterized the Fourth Era had failed to achieve the expanded production that was their ambitious target. Marmon had indeed advanced from twenty-seventh place in sales in 1927 to thirteenth . . . apparently half way up the ladder, but the rungs were far from evenly spaced . . . in May 1929, and the installed capacity was not fully or acceptably utilized. Additional models and extended price range coverages had simply sub-divided the volume and made profitable operations more difficult. The automobile business in Era Four was fast moving and the successive thrusts by Marmon into different markets had little chance to mature and establish a foothold.

The idea that small luxury, even at a remarkably low relative price, would compete with or supplant large luxury was short lived. The Little Marmon Town car at the Belmont Hotel Show in 1927 symbolized an unresponsive market explored for less than a year.

The 1928-29 offerings were directed at market segments unfamiliar with what Marmon prestige stood for, and were sold and serviced by a grossly expanded dealer network unlikely to be conducive to owner satisfaction. The Roosevelts fame that gained it remembrance in the Guinness Book of Records . . . first eight in the world under $1000 . . . was not so enduring as the reputation it acquired for being a trouble-ridden automobile. In its three year life, that model bore the marques: Roosevelt, Marmon-Roosevelt and Marmon 70.

The Big Eight and its companion Marmons of 1930 lacked exclusive styling, however distinctive it may have been. The Big Eight was a stop-gap measure. It re-entered the large luxury class after a two-year absence from the market, and did so with a superior but yet contemporary styling motif, and with sound but unspectacular engineering. It was simply not competitive with the twelve and sixteen cylinder models being offered by other luxury car builders although it was formidable competiton to the eight cylinder versions of those models.

Marmon management had determined that success required a new generation of the kind of car that had been the success of the Third Era; a car that would re-establish or reinforce the prestige of the Marque so essential to success of the quality car offering in any price range.

Howard Marmon, Vice-President of Engineering but residing in North Carolina, again took an active role in product design. He and George Herman Freers, Chief Engineer, created an engine masterpiece to power the new luxury car. Freers prepared a fine companion chassis. Prominent industrial stylist Walter Darwin Teague was commissioned to style the car and LeBaron was contracted to build all the 'factory' bodies. Sophisticated tooling was procured for the manufacture of the engine. An advertising promotion unprecedented even by Marmon standards was undertaken to herald the arrival of the new car.

The result was the MARMON SIXTEEN which John Bell Rae designated 'An American Rolls-Royce[2] and which J.B. Nethercutt[3] describes: *"The Marmon (Sixteen) is a miracle of American ingenuity in automotive engineering and design."*

The Fall 1968 issue of Automobile Quarterly featured a series of articles on the Marmon and on the Mille Miglia . . . the Ultimate Road Race. The issue is prefaced with a comment by the Editors: *"Marmon and Mille Miglia. What could link the two? It is simply the pursuit of the ultimate. Just as many of us who love racing consider the Mille Miglia the ultimate road race, there are those who have driven countless classic cars who consider the* ***Marmon Sixteen the ultimate touring car.*** *" (Author's Emphasis). Roy A. Schneider in his 'Sixteen Cylinder Cars'[4] stated: "Marmon Sixteen styling was a pivotal milestone in automotive history."*

## PART 1 — Birth of a Classic 1926-1931

Robert P. Lewis was an Experimental Design Engineer at Marmon Motor Car Co. in 1926. On February 23, 1967 he identified from his personal records the following timetable for the gestation of a classic Classic. Jerome Joseph (Joe) Felts confirmed the milestones and provided additional supportive facts.

In early November of 1926 Howard Marmon, Vice-President and Chief Engineer came to the office of George Freers, Assistant Chief Engineer, to discuss the planning of a sixteen cylinder engine. Mr. Freer's office and the Engine Design Group was situated at the rear of the Engineering Department behind locked doors. Marmon and Freers were joined by Lewis and a sixteen cylinder engine was discussed which could be accomplished by integration of two Model L (Little Marmon) engines. At that time, the Model L engine would have advanced to the tooling and field test phases.

The discussion then got around to the need for abnormal secrecy of the project and from it was born the "Midwest Aircraft Engine Corporation," of Indianapolis, as a cover for the project. Things moved quickly and on Armistice Day, Nov. 11, 1926, a segment of the Marmon organization relocated to the Old Trails Building at Illinois and Market Street in downtown Indianapolis. Separate organizations were established for accounting, purchasing, testing and engineering — all separate to, and divorced from, the parent organization. Frank Fithian, an assistant to G.M. Williams handled inter-site liaison.

The Engineering Group in the Old Trails Building included Bob Lewis and a group of designers supervised by Joe Felts. Work was initiated on an 'aircraft' engine designated the 'Model M,' sequentially following the Model L and symbolic of the undercover Marmon parentage. Chester A. Blair made the transverse engine section layout drawing and Jerome J. Felts made the longitudinal section layout drawing.

As the design progressed over the following months, Mister (not Colonel in the Engineering Department) Marmon made regular visits of inspection accompanied always by George Freers who came even more frequently himself.

Joe Felts recalled that quite surprisingly the overall length of the engine was controlled by the cam followers, there being 32 in a row over the single camshaft; the 45° angle between banks and followers offered negligible relief lengthwise. It was therefore concluded that followers with flats on the sides should be used; this necessitated guides, such as are used with roller-foilers, be added. At the time, that decision was a matter of concern because the rotation feature of round tappets was lost; tappet rotation enhances lubrication of the line contact between the flat tappet surface and the curved cam contour. However, experience proved the decision to be correct.[5]

Engine layouts, Joe Felts reported, initially showed "an up-draft dual carburetor and an 'octopus' intake manifold. Jack Simmons, a confidential carburetor engineer from Stromberg, brought in a newly designed so-called (then) downdraft carburetor. The layouts were redone with the Stromberg DD-X carburetor which ultimately went into production. At the time (mid 1927) this was thought to be the first time a carburetor of this type had been applied to a production type engine design."[5].

The downdraft carburetor was a crucial element in the packaging of a 45° Vee overhead-valve engine.

Joe Felts recalled that all material was procured from separate sources not in the automotive or aircraft fields. Finished parts were accumulated and shipped to a garage in Cincinnati, Ohio where the first pair of engines was assembled.

After one year of effort on the Model M, one of these engines was installed on the dynamometer and ran for the first time on November 4, 1927. Progress had been slowed by a recession during 1927 and at one time Joe Felts recalls he was the only draftsman remaining and that he himself consequently made the final engine cross-section drawings.

In early December, 1927, the Number One engine was installed in a test bed vehicle . . . test sheets identify it as W-501 which would be the Model 79 prototype. On December 15th, a test crew consisting of Howard Marmon, Hal Purdy, Bob Lewis and two test mechanics drove the car. Experience with the test bed led to a decision, on December 20, 1927, to design and build an entirely new chassis to complement the engine. But early in 1928 a 'hold' was placed on the Model M project. Joe Felts believed this was partly consequent to the high cost estimates for production tooling and partly consequent the status of the company's finances. Had the 'hold' not occurred, the Model M could have been ready for the 1930 auto shows. Instead, Cadillac was at those shows with a new sixteen cylinder car.

An article of veiled significance then, and partly of contemporary prophetic interest today, was authored by T.J. Litle, Jr. when Marmon Chief Engineer. It appeared in *MOTOR* for January 1929[6] (the big show issue) and was titled: "How many cylinders . . . 6?, 8?, 12?, 24?" (Note the omission of '16?'). It began: *"It might be said that thought is being given to 10, 12 or even 16 cylinder automobile engines in the large and high-priced cars, even though the 12 was once (more than once-author) tried and discarded as impractical in view of manufacturing obstacles that existed at the time it was introduced. If the capability to multiple bore cylinders had existed in the early years of the industry, it is doubtful that many fours would ever have been made (due to vibatory characteristics-author). The engines should have gone from one and two cylinders directly to six. It will be difficult to satisfy the most discriminatory buyers in the upper end of the $1000 to $3000 range with an eight. Even the inherent smoothness of the eight requires reasonable size cylinders; at least one very large eight cylinder automobile has been criticized on the score of roughness."*

The recent panic reversion to four cylinder engines from V-eights appears to ignore Mr. Little's assessment. However two generations separate those, who had lived with four-cylinder automobiles and happily moved from Model A Fords to the Chevrolet six (and later to the Ford **V-8),** from those who must now 'retrace the same road' and learn for themselves. General Motors who had designed an entirely new V-6 transverse engine option for their 1980 ½ 'X' cars had difficulty producing sufficient four cylinder engines to meet the public demand and disinterest in the comparably sized V-6.

When the Model M project was shut down Joe Felts noted "all the drawings were bundled together and stored in the Marmon Motor Car Company's main steel vault, for secrecy was still paramount." But periodically Joe would be directed to work on them **in the vault** and he always remembered that it was a very cold place to work. Upon learning of the impending Cadillac announcement, Marmon reactivated the Model M project and undertook design of the Model DD car.

The Cadillac V-16, like the earlier Marmon Sixteen test engine (and the Little Marmon), had a 3.0 diameter bore. Perhaps Howard Marmon considered it essential to have a larger bore and directed that the model DD have a 3.125 bore. However, there was considerable discussion by Mr. Marmon and Mr. Freers following the surprising defeat of Stutz in the famous $25,000 match race with Hispano-Suissa and that event might also have influenced a displacement increase. Joe Felts who made the drawings recalls that the change required milling flats on the liner flanges to eliminate liner-to-liner interference. He also remembers much concern about another consequence . . . an offsetting of the narrow big-end bearing on the blade connecting rod. That change evidently was not a valid cause for concern.

Dr. Charles W. Bishop, longtime automotive historian and Marmon enthusiast, has researched the surviving documentation of the origins of the MARMON SIXTEEN engine design and explored possible 'leaks' which might have triggered the design and introduction of the Cadillac V-16.[7] Dr. Bishop, to whom all Marmon restorers are indebted for his long time preservation of a cache of factory parts, examined several documents not previously available to the general public. He refers to the Marmon booklet "The Engineer's Story" mailed to prospective buyers of the SIXTEEN in which appeared the preliminary layout of the engine cross-section bearing the initials HCM and the date **11-4-26** and then reveals existence of a blueprint (copy reproduction of an engineering drawing before black-on-white copies were feasible) of an ignition distributor for a 16 cylinder engine issued by the Dayton Engineering Laboratories Company (DELCO) of Dayton, Ohio. It is a full size outline drawing for installation purposes numbered 004022 and marked 'Approved by . . . HCM." On the reverse it significantly bears the red crayon notation "63 DD," and on the face the letters, 'DD.' Those letters are the internal Marmon model designation for the SIXTEEN! It also carried a note "New style distributor with contact arm springs attached to terminal #4028 **and** the date of **Sept. 2, 1926.** The crayon notations must have been made after reactivation of the project in 1930 when the 'Model DD' designation came into usage.

Three other distributor prints are in Dr. Bishop's collection. Two bear the notation 'Approved . . . Siler 10-7-30': One is marked on the back 'XDD' and the other 'Drawing and Part No. 004084' is marked in yellow crayon DD 3036 . . . the correct Marmon part number. The third has attached an Engineering Change slip dated July 29, 1932.

From the earlier date of the distributor print, Dr. Bishop concludes that the engine sketch of 11-4-26 was not the first but simply the first that would reproduce well and establish a 1926 date, and that earlier layout drawings must have existed.

From the author's personal experience as an engine designer: (1) outlines of the 'space claims' required by the distributor and other purchased component assemblies would be required before serious planning could begin; and it is not uncommon for a design executive to put his ideas on drafting paper at home. He may seek unquestionable patent documentation, or have concerns about secrecy or rumor generation, or wish to consolidate ideas before involving subordinates. Concept drawings of the Detroit Diesel Allison Series V-71 and Series V-149 engines were made by the author to scale of only one thirty-second inch to the inch. Dr. Bishop is therefore correct in his hypothesis, but Bob Lewis' calendar of events is an equally correct recounting of events that occurred once Mr. Marmon had decided to pursue his concept.

The distributor prints bearing 1930 and later dates bore 'Delco-Remy, Anderson Indiana' titleblocks (letterheads) . During the interim, General Motors Corporation had acquired DELCO. Dr. Bishop explores the possibility that this acquisition could have inadvertently revealed to GM management that a sixteen cylinder engine was being planned or developed in Indianapolis. He also notes the departure from Marmon of designer Owen Nacker (the employee of Alanson Brush — Marmon consultant on the Model 34) to Cadillac where he designed the Cadillac V-16 engine. Maurice Hendrey[8] places the date of that event in 1927 but Joe Felts returned to the Marmon payroll on May 10, 1926, and does not recall encountering Nacker and specifically states that Nacker never worked on the SIXTEEN.

On November 10, 1930 . . . a full four years from the Model M project initiation . . . the press[9] reported the first appearance of the MARMON SIXTEEN at the Chicago Auto Salon in the Drake Hotel. This was the prototype close-coupled sedan. A fuel economy of 10 to 12 miles per gallon and a top speed of over 100 miles per hour was claimed for the 200 horsepower car!

## PART 2 — The Marmon Sixteen — Model DD

The first production SIXTEEN was shipped in February 1931 and series production was finally underway in March 1931. On April 21 it was reported that for the first time in two years, a second shift was operating in the Marmon plant.[10] Increased demand for all three model offerings was reported. 'The building of one Marmon Sixteen required the services of approximately 300 skilled workmen in 27 different industries for a period of two full working days.[10] That equates to 600 man days per car.

The eight 'standard' LeBaron bodies were: two-passenger coupe and convertible coupe; five-passenger sedan, close-coupled four-window sedan and convertible sedan both with integral trunk, victoria coupe (a second version with integral trunk was never offered) , seven-passenger sedan and limousine.

The car weighed 4800 pounds; the competition ranged up to 6000 pounds. It was 216 inches overall . . . 14 inches longer than the Big Eight with an additional twelve inches of wheelbase. It weighed in the sedan version 24 pounds per horsepower, 33 pounds per inch of wheelbase and had, for its time, phenomenal acceleration. One reporter commented "that demonstration when experienced is a thing to remember."[11] (Production cars were heavier than the prototype; the registation weight of the 5 passenger sedan was 5360 pounds . . . more than a 10 percent growth.)

The prototype had beveled edges on the two upper corners of the radiator while at most other points surfaces were merged by softening curves. That effect started a beveled corner running back along the hood which merged gradually into the curve of the hood and was completely lost before the cowl. They were smoothed before start of production; perhaps someone remembered the hood characteristic of the first Essex cars.

The caliber of interior design matched the overall treatment. Teague publically acknowledged the contribution of Helen Dryden, whom he described as one of the country's foremost authorities on style, to the design of both interior and exterior hardware. A rectangular motif was followed with the external handles being long straight bars. All the interior hardware down to the smallest knob repeated the motif.

The unidentified reporter for *Automotive Industries*[11] described the many details of the styling masterpiece so effectively that his accounting is repeated verbatim below: *"The bodies must perforce be designated as unconventional, although they do not embody any really radical features. The fuel tank at the rear is completely concealed either by a trunk or a rear shield depending upon body style, with the filler cap in the rear fender. The radiator filler is invisible, being under the hood. The usual suggestion of the presence of a frame side-rail by a ledge in the running board side aprons is eliminated by extending the doors almost down to the running board with only a short concave sheet metal splasher between the two."*
The reporter was describing the prototype car which lacked the radiator top ornament subsequently familiar.

He continued: *"The single bead which serves as a belt molding is unusual in that it extends parallel to the ground over its entire length. It serves to emphasize the long wheelbase . . . 145 inches . . . of the car. At the rear a continuous horizontal line is formed by the bottom edges of the fenders and of the trunk or fuel tank. The rear fenders serve to carry the double rear traffic signals. They have deep side-flanges and sharply reverse their curvature to join the running boards gracefully and inconspicuously. "There are two cowl-top ventilators, a decidedly sloping windshield and hinged interior visors (as introduced the prior year on the Big Eight). Inside the car, the unusual position of the instruments is notable. These are placed on a cowl-bar extension directly below the windshield, with the main part of the panel recessed back below the cowl in the usual position at either side. Instruments are all of the dial type (white on black background) with large faces and revolving hands. They include the larger speedometer and clock, under convex glass lenses (located at left and right extremities of the instrument cluster). An ash tray (drawer) which telescopes into the instrument panel (centrally below a plate bearing the legend "MARMON SIXTEEN") is provided. There is a (key-locked) compartment for gloves or small packages on each side of the instrument panel and a (leather padded) open compartment just below the instruments."*

The other four instruments were ammeter, fuel level, oil pressure and engine temperature. A double throw toggle switch, protruding from the bottom of the instrument cluster extension, allowed illumination of the full panel or only the speedometer and oil pressure gauge. Throttle, spark and manifold-heat levers, of the pivoted ball-end type, also protruded from the bottom of the extension. The speedometer was unusual in being asymmetric with the zero at 1:30 o'clock so that 120 (max) was at the 10:30 position near the bottom of the windshield to minimize eye depression for observation of the highest speed indication. Today's race car tachometers are similarly arranged.

The innovative extension of the doors downward besides the frame was cleverly utilized to enhance the interior finish. The opened door revealed to the entering passenger a chromium plated panel-facing of the body sill extension downward from floor level. With the door closed, the upholstery came down to floor level; opened, a carpeted kick-pad was revealed below floor level.

The front fender supports were attached to a special frame cross member below the radiator which provided the necessary structural rigidity without the usual tie bar between the headlights. The slightly oval headlights were supported individually from the fender brackets; contributing to the suggestion of massiveness of the large chromium-plated, V-shaped radiator shell. The prototype had a screen grille, the production cars had dual horizontal shutters.

The fenders swept down on the inside into a horizontal splash pan (provided with louvers only in the prototype). A wide flange on the front fender's forward inner edge concealed shock absorbers, steering connections, etc. from the front.

Fig. 6-1 Front compartment of the SIXTEEN. Reproduced from Motor Trend by permission.

Fig. 6-2 Rear compartment of an original 1933 sedan. W. Snyder photo.

Fig. 6-3 Marmon Sixteen Five Passenger Sedan

Fig. 6-5 Marmon Sixteen Five-Passenger Two-Door Coupe

Fig. 6-4 Marmon Sixteen Five-Passenger Close-Coupled Sedan

Fig. 5-6 Marmon Sixteen Two-Passenger Coupe

It also enhanced and distinguished the frontal appearance.

A treadle type accelerator pedal, like the prior year Big Eight, was used. A wireless, quick-heat, pass-about cigarette lighter was provided. A three-spoke rubber-covered steering wheel, like the three-spoke wheels of the First Era, could be adjusted for column angle. Horn and light controls were operated by a single button on the steering wheel hub. Both front and rear seats had folding center arm rests. The entire front seat was adjustable fore and aft by a screw mechanism. Additionally, the driver's half of the back cushion was adjusted for angle by a crank revealed when the arm rest was lowered.

*The Automotive Industries'* reported continued: "The upholstery is noted for its simplicity, these being no tufting for the pillow-type seats and backs. Heat and noise insulation for the (forward) body is provided by making the bulkhead of a layer of Formica, a layer of insulating material, the metal dash proper, and another layer of insulating material (inside the car). Insulating material has been applied also to the cowl sides while a rubber pad is placed under the carpeting. There are carpet covered hassocks in the rear." These had a corrugated rubber facing on the bottom surface for position retention.

In September 1931, a new line of custom bodies was added to the SIXTEEN's eight standard (LeBaron) bodies. According to Marmon Sales Manager George C. Tenney, the custom line consisted of no less than 32 body styles. They included five and seven-passenger town cars, all-weather type cars, various limousine arrangements and sports models. Also included were five and seven-passenger sedans with "**sunshine roofs"** and many custom features of both comfort and appearance refinement."[12]

The fact, that the two latter-day styling-milestone automobiles (after the Marmon Model 34 of 1916) — the MARMON SIXTEEN and the 1938 Cadillac 60 Special — were both offered with optional sun-roofs, is worthy of note here.

The MARMON SIXTEEN masterpiece was a marriage of the unique skills of Howard Marmon and Walter Dorwin Teague. It is fortunate and remarkable that each of them recorded this philosophy of design with respect to this specific milestone automobile in separate articles which appeared in *Automobile Topics* magazine and in separate booklets mailed to potential customers. Teague's account is repeated as it appeared in the January 24, 1931 issue: *A year ago Walter Darwin Teague, one of the best known of the modern artists who have devoted their tal-*

Fig. 6-7 Marmon Sixteen Two-Passenger Convertible Coupe

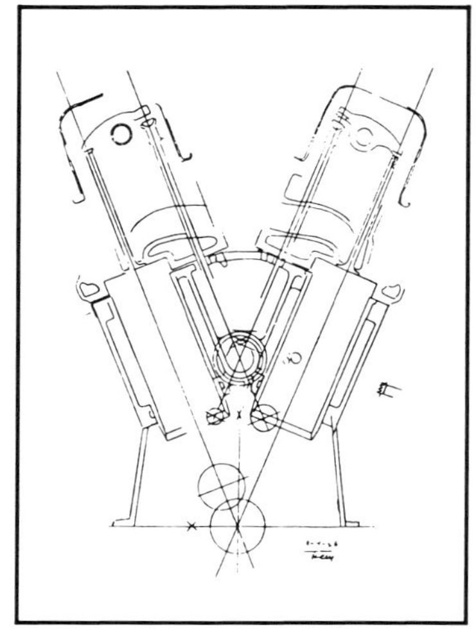

Fig. 6-8 "H.C.M., 11-4-26"

*ents to designing for industry joined the Marmon Motor Co. as designer and style consultant. In collaboration with the Marmon engineering department Mr. Teague designed the bodies and the interior details of the Marmon Sixteen. Here, writing expressly for Automobile Topics.*

## *Housing Two Hundred Horsepower*

### THE DESIGNER OF THE MARMON SIXTEEN BODIES TELLS HIS STORY

By Walter Dorwin Teague

With all the world being done over in terms of a machine age, there is no end of interesting things to be designed today. But there are two that top the list, for sheer excitement: skyscrapers and automobiles. One a problem in vertical, the other in horizontal lines and masses; one static, firmly planted, aspiring, the other mobile, eager, swift; one conditioned wholly by structural technique and solidity, the other to be worked out in terms of movement, speed, power. And both are among the most conspicuous and significant features of the modern scene.

No stork has yet left a little steel building on my doorstep, to be brought up in the way it should go. But it has been my privilege to design an automobile, unhampered and untrammeled, in every particular; and an automobile, at that, of such advanced development that nothing I might do could possibly belie its performance. I refer to the Marmon Sixteen.

Here was a challenge. Here was a body, a large body too, but small with respect to the two hundred horsepower stored within it; to be sent along the highway at any speed from three to one hundred miles an hour, but sensitive always at the most delicate control; and carrying its passengers in the utmost refinement of comfort. To attempt to give this machine a form expressive of its function, as appropriate to its mission in life, if possible, as the bodies of a race-horse or a panther are to theirs — that was a worthwhile job.

A designer, if he knows his business, attacks any problem from just this angle. He doesn't worry about making a thing beautiful, or different, or appealing. He concentrates on making it right, adapting it perfectly to the work it has to do, enabling it to function honestly, simply, candidly, with the highest degree of efficiency. He knows that if he is so fortunate as to succeed in this aim, the way will be marked out for him to make those subtle adjustments of line and proportion, to create those unexpected but delightful rhythms, which will lift his product out of manufacturing into art. It will be beautiful, distictive, economical and practical, and its rightness will be instantly appreciated.

Perhaps no one ever does fully succeed. But Nature, and the history of all beautiful and useful things have taught us that success cannot even be approximated by any other method.

Automobiles have grown beautiful as they approached this ideal of perfect adaption to their end, but they have been a long time living down their unfortunate beginnings. Airplanes and dirigibles had no ancestry and so could be designed straight-away as flying machines, and we marvel at their beauty. But automobiles at first were horseless carriages.

The carriage was a highly developed mechanism, often perfect in its own sphere, but it bequeathed to motor cars a form wholly impractical in any speed machine. And like all hereditary traits this one has been hard to outgrow. Until it has been entirely eliminated however, and automobiles designed directly as automobiles, they will not have reached their ultimate development.

It is here that an outsider entering the automotive field finds his advantage, if any. He is not burdened first of all, by weight of habit and tradition. He can brush aside, at first all details — forget about mouldings, fenders, radiators, reveals, hoods — until their turn comes. He stands a chance of being able to concentrate on the fundamentals of the problem.

Many factors in the design are pre-determined at present, of course, beyond the possibility of immediate change. Such as the position of the motor under the hood, the method of radiation, the relation of fenders to wheels and body, the location of luggage space. If we could rearrange all these elements purely in the interest of efficiency, we could go much further than at present, and we may get around to that soon. But today this scheme is fixed and roughly blocks out the form which must be modeled into one swift and powerful organism.

"Organism" is the word: no design is successful unless it is organized as one unit, poured in one mould, so that every part integrates and harmonizes with every other part as in a living thing. The parts, in fact, must have no importance except in relation to the whole.

If we are to make this moving body into one unified projectile, obviously the consideration of wind resistance will guide us in modelling every line and surface. The resistance of the air in front and the suction of the vacuum behind; with these in mind, our form takes shape inevitably, almost without effort. It is moulded by logical necessity.

Our body must be as low as the comfort of the passenger permits, so as to oppose as little surface as possible to the wind. And these surfaces we model to deflect the air rather than resist it. Our radiator is pointed, and slanted backward. We lay back the windshield, and above it is no visor, but a rolled header over which the wind may slide with least resistance. The roof has a low, flat curve, and falls away at the back in a carefully modelled slope down which the air may slide to fill the vacuum which a rapidly moving body tends to leave behind it. This curved back line is continued down the sloping trunk, of the shorter models, and carried as low as the clearance allows. The sides are similarly modelled in a long almost unbroken curve like a yacht's hull, from radiator to trunk, and the fenders front and rear are drawn out in long, streaming sweeps.

This car is to be one thing, a single unit, not an assemblage of parts; so it is impossible to set a body on a chassis and let it go at that. The chassis is like the skeleton of an animal, necessary, and determining form, but not visible. It must be possible to trace an unbroken line over our body from the outer edge of one running board up to the side across the roof and down to the outer edge of the opposite running board; from the dust shield in front of the radiator to the bottom of the trunk in the rear.

Shields below the fenders in front fill the gap between them and the wheels, deflecting air and dust and concealing all the mechanism of the axle; the rear fenders stream out to the bottom line of the trunk, and continue its base across the back in an unbroken curve. Every detail not in itself significant is suppressed or omitted. There is no radiator cap to break the line from the radiator tip to the windshield, no filler cap, no parking lights nor horns visible, no gadgets or superfluities of any kind. At front, sides and rear, no viscera can be seen. This car must not trail its guts in the public view: it must be sleek, well-groomed, "poured in one mould." I regret the necessity for license plates.

This is a projectile moving in a straight line forward. So when the larger aspects of its mass are determined, its silhouette drawn, the necessary details take their place as a matter of course in the general scheme. Every line must emphasize the direction of movement — vertical lines being eliminated or suppressed, horizontal lines being exactly and decisively horizontal. A single narrow moulding marks the length of the car, visibly straight from the radiator until it disappears around the trunk in the rear. Windows are bounded by straight lines top and bottom, and reveals are as narrow and simple as is practical. Long, straight door handles, and similar miniature handles on the louvre doors of the hood, repeat and accent this horizontal direction. Practically no other external details are in evidence.

The same principles of simplicity and unity must control the finishing of the interior. Even the biggest car interiors are still small considered as rooms, which they are. Hence they can stand no fussiness or elaboration. The ladies and gentlemen who are to ride here will expect a maximum of luxurious comfort, and a maximum too of restrained good taste. The ladies, especially, our severest critics of interiors, do not want competition with their own toilettes; they want a setting, a frame, that will enhance their personal elegance and not distract attention from it. We can employ the finest fabrics and the finest woods available, every detail of hardware and trim can be worked out to exquisite finality, but they all must take their subordinate places in a simple and unified scheme. Here again, all superfluities are to be eliminated, and there are to be no discordant notes.

Of the actual working out of these principles, until a car embodying them one day sped up Park Avenue like a sleek black cat, it is only necessary to say one thing: there is here no history of innumerable sketches, endless trials and experiments, patient elimination of many possibilities. The "trial and error method" has not entered into this car's production. From the first rough sketches the conception of this car has been clear and definite. Progressing through small scale drawings, the making of a small scale model (in which even the hardware and hub caps are exactly as they finally appear), the making of full size draughts and the building of sample bodies, there has been a logical and consistent advance, all along, toward a predetermined goal.

There have been slight changes, readjustments of proportions and dimensions, as we went along, of course, but their total effect is negligible. We had a hard time, too, to realize in metal some of the forms I had in mind, such as the fenders and the radiator, for instance. But this was due to the difficulty of expressing these subtle forms by means of drawings, and I found it necessary at last to model them in plasticene, full size, on a car, as a guide to the draughtsmen.

But my colleagues in the Marmon Company, whose confidence has surprised me as much as it has pleased me, and the skillful engineers and workmen of the LeBaron organization all have seemed to grasp the conception I had in mind, and have found ways to accomplish whatever needed to be done. The thing grew without trouble or difficulty, almost like a plant in spring. Which no doubt is why I have enjoyed the whole experience so hugely.

Few people realize the importance of reducing wind-resistance in the design of motor car bodies. This is not a vital factor so long as a car moves at low speeds, but when a speed of forty-five miles or more an hour is reached, a large part of the engine's power may be used up in overcoming the resistance of the medium through which the car is passing. This retarding force is felt in two ways: in opening a pathway through the air, and in the tendency to form a vacuum behind. The latter is the more effective by far, and increases its backward pull very rapidly as the speed rises. This suction is the same that you feel in pulling a cork out of a bottle.

In stream-lining a car body, these two factors control the design. The Marmon Sixteen has been built on yacht lines, with practically no "ogee" in the hood and cowl, but with a smooth sweep from front to stern. There are no pockets to catch the wind and create eddies, and a minimum of obstruction is offered to the passage of the air along the body top and sides.

At the back there is a noticeable rake forward so that the wind slides most easily down the top and around the rear-quarters. Where the length of the body permits, as in the close-coupled and convertible sedans and in one model of the victoria, a

Fig. 6-9 The Marmon Sixteen compared with its contemporary Cadillac offering by Walter Dorwin Teague (Sr.) in his "Design This Day".

Fig. 6-10 Walter Darwin Teague, Sr. after designing the interior of the Boeing 707.

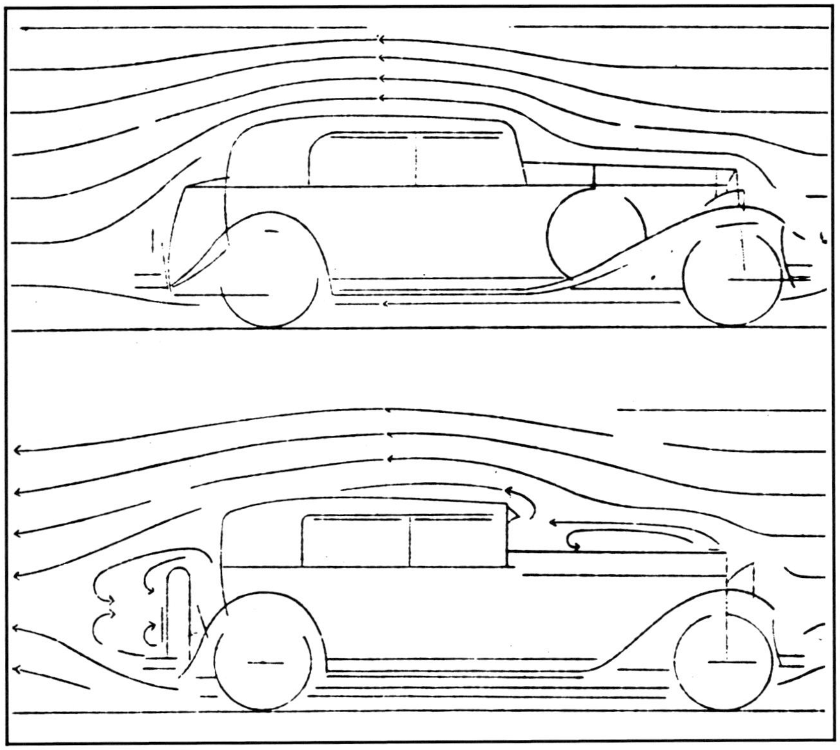

Fig. 6-11 Outlines of the Marmon Sixteen close-coupled sedan (above) and a conventional car show differences in air disturbance. Eliminating "cupping" in front and rear obviates the eddy currents which slow the car, raising dust.

trunk has been built on the rear as an integral part of the body. This serves still further to fill the vacuum-creating space at the rear. The profile of these trunks has been plotted in a perfect curve with the top.

Engineers estimate that an advantage of five miles an hour is gained at high speeds by this treatment, as compared with the conventional rear-end treatment. Its effect may also be observed by a marked lessening of the amount of dust raised by the car in motion. There is far less suction exerted, consequently less dust, more comfort and a cleaner car.

In longer models such as the five and seven-passenger sedans, the body itself has been modeled to effect a rapid filling of the air space behind. The result is not only advantageous but noticeably smart and rakish. In these models an apron has been carried down at the back to cover all the unsightly mechanism of spring-hangers, gas tank, frame and other viscera usually left exposed. This apron perfects the stream-lining as well, and aligns perfectly, as does the trunk, for instance, in the case of the shorter models, with the sweeping ends of the rear fenders.

In the coupes and one model of the victoria, the rear deck has been carried out and down to this same base line of the rear fenders. Thus in all models a perfectly clean and well organized rear end is accomplished — perhaps more completely

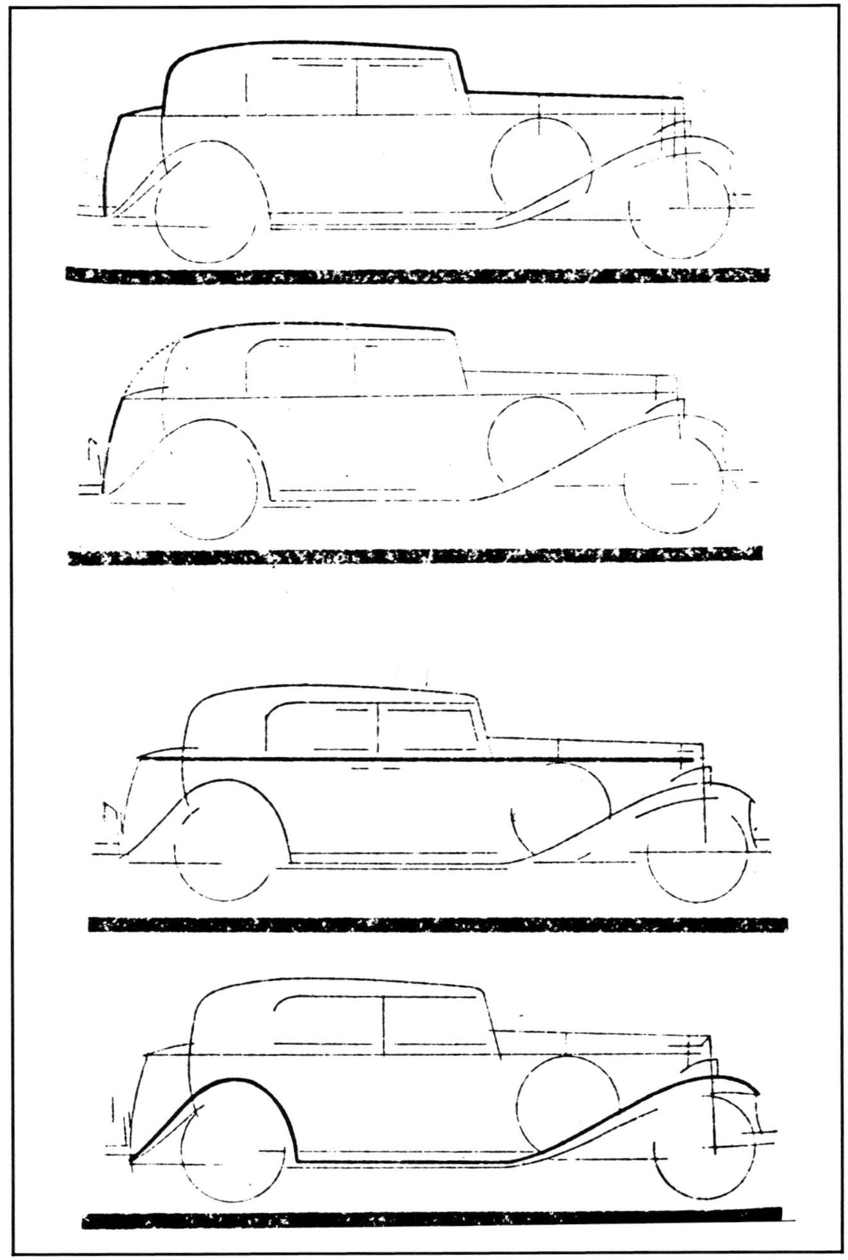

Fig. 6-12 Principles of Marmon Sixteen streamlines — (1) The horizontal base line at belt, (2) Unbroken fender line employing kindred curves, (3) Top silhouette enhances projectile effect, and (4) Trunk continues roof line.

than ever before in motor car design.

In order to avoid breaking up this effect, the gasoline filler cap has been placed in the left rear fender. Here the flat cap does not break the fender line, and yet is most convenient for filling.

A similar effect of unity is obtained by dropping the doors over the frame almost to the running board, and filling the space between with a carefully modelled apron. Thus the body and chassis are visibly one unit.

In front, screens have been placed under the front fenders to deflect air and at the same time hide all unsightly front axle mechanism. As this motor is not to be cranked by hand, the space between the frames in front can be filled completely by a metal pan, extending out to and covering the tie-bar. Thus the front-end is cleaned up as effectively as the rear.

The filler cap, being seldom used, is omitted from the radiator and placed under the hood. Thus it is possible to draw an unbroken line from radiator tip to windshield, increasing the car's rakish effect of speed. No horns, parking lights or other accessories are exposed. These details are present for use but not for vulgar display.

It has not been thought advisable to plaster this car with the maker's advertisements. We believe that in a car of this type and quality, where an effect of restrained elegance is most desirable, our reticence will be appreciated by the ladies and gentlemen who are to be its owners.

The instrument board has been planned primarily for easy reading, with large dials conveniently placed and boldly lettered. Numerals especially designed with a view to increased legibility are used on all instruments.

All hardware, interior and exterior, has been designed in a uniform style, simple, straight-lined, efficient. Here we have had the assistance of Helen Dryden, one of the country's foremost authorities on style. Her impeccable taste has insured a result beyond criticism. The door-handles are long straight bars, both practical and decorative. Corner lights and dome lights are in simple rectangular frames repeating the motif of the hardware, and this same motif is carried out even on the small handles of the cowl compartments and louvre doors.

Where wainscot panels are used, great care has been taken to select beautiful grains of rare wood such as ambayna burl walnut and, ebony. These are used in simple straight panels, wholly undecorated and therefore doubly effective. The partition in the limousine is finished with exquisite cabinet work to match the panels on the doors, and equipped with fittings of crystal and silver such as a lady of taste would choose for her personal use. The folding seats disappear into this partition and leave its surface unbroken.

* * *

Walter Dorwin Teague later discussed[13] the styling design of the MARMON SIXTEEN in comparison with its contemporaries: *Sure-fire design of the automobile of the future must be controlled by three factors: (1) an analysis of the function of the automobile and the adaptation of body design to that function in terms of pure engineering; (2) a fine sense of line and proportion . . . a special gift which a skilled designer brings to his work and provides refinement of grace and charm; and (3) a comprehensive knowledge of style trends in all related and unrelated lines of production. It may seem at first glance superfluous to ask the automotive designer to be thoroughly familiar with what the dressmakers of the Place Vendome or the Rue Montaigue are doing, the jewelers and shoemakers of the Rue de la Paix, the architects and furniture designers of the International School, the silk manufacturers of Lyons. Yet unless he knows what the preferences of tomorrow are to be in all these stylistically vital fields, how can he deal with the many details of automobile production in which style is the controlling factor? This knowledge is useful not only in planning the interior fittings, upholstery and hardware of the car; it is also essential in determining the kind of line and the general effect of the whole . . . simplicity or fussiness, severity or elaboration, significant form or ornament. To make assertions along these lines is easy enough; demonstration is far more convincing. It is for this purpose that I have assembled this series of designs showing how it has been possible for several years to predict the advancing trend of motor car design. First look at Design A. This shows one of the finest cars made in America (the Cadillac V-16) of the vintage of January 1931, and is typical of American motor car design of that time. It belongs to what we may call the Up-and-Down School, since the vertical lines are heavily accentuated and the horizontal lines of movement and power are practically lost. There is still a very strong reminiscence of the carriage tradition and no concessions have been made to the fact that this car is expected to cruise at speeds of 40 to 60 mph. Thus no effort has been made to adapt its form to the flow of air currents which use up most of the horsepower at such speeds. There is also confusion of detail at the rear and at the front. Body and chassis have not been integrated and many details of the body bear no relation to the design as a whole. "Now look at Design B. It shows a car (the MARMON SIXTEEN) of the same price class as Design A. The designs for this model were completed in March 1930 and the first (prototype) car shown in December of that year. It is a 200 horsepower car capable of speeds of 100 miles per hour. In it the first steps have been taken toward adapting the body form to air flow. Also the design has been brought into line with current style trends in other fields, and great care has been spent on the grace of line and form. The radical features of this design are evident when it is compared with Design A. Horizontal lines of speed and power are emphasized; retarding vertical lines are minimized. In it appeared together for the first time practically all the features which have become common during the three succeeding years: slanting painted radiator, simplified straight line molding treatment, slanting windshield, rolled-back header, rear end faired forward, trunk built-on, dropped (back) sides, long sweeping fenders and all irrelevant detail, front and rear, either covered or eliminated. This design was dominated by the cult of simplicity and refinement which dominated the best costume and furniture design and architecture of its period. Also, great pains were taken to develop the lines and proportions of the design to the most pleasing degree possible, with what result the observer may judge for himself. Other body styles of this series introduced features which also became common practice in succeeding years, such as the rear end treatment of the victoria and the coupe."*

Note that Teague intended the radiator shell to be finished in body color!

Evidently Walter Dorwin Teague was the first automotive stylist who had first become a successful industrial stylist, although Raymond Loewy is generally regarded as the founder of the modern school of industrial design.

Briggs Cunningham in his museum in Costa Mesa, California, protrays Teague's comparison by displaying a 1931 Marmon and Cadillac side-by-side, using convertible styles.

Now that the reader understands exactly how the MARMON SIXTEEN was designed, the author must change that conception with a most surprising revelation that occurred in 1965 but had been rumored earlier. The following letter[14] was dated December 27 of that year:

Fig. 6-13 The handsome engine room of the SIXTEEN. E.J. Robinson photo.

*Mr. Warren W. Fitzgerald*
*General Motors Styling*
*General Motors Technical Center*
*Warren, Michigan 48090*

*Dear Mr. Fitzgerald:*

*Thanks for your letter of December 21st regarding the Marmon 16. I did indeed design the bodies while in my second year at M.I.T. starting when I was nineteen.*

*At this time my father had just started in the Industrial Design business, previously he was an illustrator and specialized in advertising borders in which field he was the leading artist. His main design account was Eastman Kodak and at the time he got the Marmon account he had only one or two other accounts. I don't know how he met Colonel Marmon but they became good friends and remained so until Colonel Marmon's death in 1943. I do remember that they got to tracing back their family history and found that both Colonel Marmon and my father had ancestors who migrated from the South in covered wagons and settled in Indiana in the same year and possibly in the same expedition. Also both ancestors were millwrights.*

*The first work my father did for Colonel Marmon was to make some color suggestions for some of the models preceding the 16. These were pretty startling compared to the rather drab choices available at that time and Colonel Marmon must have liked them because he assigned my father the task of designing the 16 bodies.*

*This job was pretty exciting to me because cars had been my principle hobby since I was very young and at nineteen I already had covered lots of paper with projected designs. I had been driving since I was sixteen when I worked my way out to San Francisco on a freighter and there bought my first car (a Chevrolet) and with another boy drove it back across the United States to New York on the Lincoln Highway, much of which was then gravel and mud. My father had not been particularly interested in cars and, in fact, didn't drive a car until the Fifties when he finally got a license. (It was quite a hair-raising experience to ride with him and the rest of the family used to try tactfully to dissuade him from getting behind the wheel.)*

*So when he got the task of designing the 16, I had already put up some strong arguments for the design job, and it was agreed that I would do this. The design was done during the school year and during the summer vacation of 1930. I commuted between Boston and New York on weekends and whenever the weather was good enough I would fly on Colonial Airlines. At that time they used trimotor Fords and when you bought a ticket you had to sign a statement that you wouldn't sue Colonial Airlines if anything happened. I also used to drive back and forth. My cars then included a LaSalle touring car and an Auburn 120 convertible.*

*Colonel Marmon had the chassis and engine pretty well designed by the time I started so I had definite dimensions to work with. I did the initial concept in the form of sketches and went right into 1/10 scale drawings as there was no doubt in my mind whatsoever as to the design I wanted. No one else, including my father, made any sketches, drawings, renderings or design concepts of any kind until after the first models were built. Like many father-and-son combinations, we unfortunately tended to disagree on a lot of things but my feel for car design was obviously on the right track so there wasn't much argument on the basic concept. My father had a very good eye for lines and as the design took shape he knew it was pretty good and he kept Colonel Marmon's enthusiasm at a high pitch. All of the contact with Colonel Marmon, and later with the people making the first bodies was handled by my father.*

*The most serious argument we had about the Marmon 16 was with respect to the corner of the hood and radiator. This was during my father's "octagon" period when he used an octagonal motif in a lot of the work he was doing. This was o.k. for some things but not for others, and I maintained that the corners of the hood and radiator should be rounded. My father wanted them beveled and this became quite a battle with the temporary victory his.*

*The first designs had this beveled longitudinal edge to the hood and radiator and we had Boucher, the well-known New York model builders, build a 1/6 size scale model with this feature. This rather odd scale was chosen to fit the 600-20 model rubber tires which were used on Firestone ashtrays.*

*During 1930 I also made the full-sized lofting drawings for the car. Our offices then were at 210 Madison Avenue and we had one wall big enough to make the drawings on. I used wrapping paper which was the only paper I could seem to locate around New York which was big enough to use without a lot of patching together. These drawings including sections in all three planes similar to the line drawings of boat hulls in which I had done some work. I don't know whether anyone else has ever used this method for cars but if you know what you want it goes fairly fast and it worked very well in this case. The first mock-up was made from these full-sized drawings and comparatively little modification was needed.*

*The only part of the interior I worked on, aside from the seating dimensions and wheel position was the instrument panel. In 1930 the universal U.S. practice was to make each instrument any shape but round and combine them in various unpleasant clusters. The dial faces were usually beige or some other hard to read color. I designed round instruments, each one separate, with black faces and white numerals, located high and spread out so that each was easy to distinguish. The first hand-made car used the beveled hood edges. Colonel Marmon loaned us this car for a period of several months with, of course, a chauffeur so that my father could use it. Before production started, however, the bevel was eliminated (I don't remember how or why) and all production cars had rounded edges. The radiator cap was under the hood and there was considerable favor about leaving this off the outside of the radiator shell. Later on the temptation became too great and over my father's and my protest a little ornament was stuck on here.*

*My original design and the Boucher model were done using a modification of some special headlights which were used in several custom cars in those days. I don't remember the name of these but they were supposedly quite good and they tapered to a very small elliptical lens in front. We modified these a little to match the radiator shape. This really wasn't a very good idea and (fortunately I think) more conventional headlights were substituted for the actual cars.*

*I did several body styles for the 16 including a convertible coupe, convertible sedan and a quite nice Victoria.*

*The next year, in Boston, I built for fun a model of a more advanced design concept (photo enclosed). Colonel Marmon admired this and since he had been working on a radical twelve-cylinder design, he decided to build the prototype body to my design. I drew this up on the twelve cylinder chassis but was not in on any of the mock-up work and unfortunately the actual car didn't come up to the aesthetic standard of the first model. But it was a very advanced design for 1932 when it first appeared and it got around quite a bit — several people have told me that it had some influence on the Cord and others.*

*I have the Boucher model of the 16 in my office; we recently rebuilt this model which had gotten sort of beat up and our lab had to mold six new tires using the only one of the original tires which was still in good shape as a pattern. I also have the twelve-cylinder model at home.*

*About the time that the twelve-cylinder prototype came out, the whole Marmon operation finally came to a halt. No matter how advanced the engineering or how good the body design, 1932 was the height of the depression and another $5,000 car was not what the country was looking for.*

*You probably didn't expect such volume in reply to your letter, but this is a subject which is very close to my heart and I am quite interested that my part in it should be written up correctly. My father originally agreed that I should receive credit for the design and at first saw that I did, but as the years progressed he began to think of it more and more as his own and to omit mention of my part in it. Opportunities for this kind of automobile design become rarer as time goes on so I don't expect to get many more chances like this. Therefore I appreciate very much your allowing me to give you the above information.*

*Don't hesitate to get in touch if there is any more you need, and I wish you all success with your article.*

*Sincerely yours,*
*W. Dorwin Teague*
*New York, New York*

The Boucher model of the SIXTEEN is mentioned in the section on the Marmon Miniatures in the Third Era. It was presented by Mr. Teague, Jr. to the Cooper Union Museum in the late 1970's.

Walter Dorwin Teague, Sr. was a very versatile individual, having published in 1942 a mystery novel 'Your Can't Ignore Murder.' After designing (via Walter Jr.) the MARMON SIXTEEN in 1930, he was on the Design Board for the 1939 World's Fair and then he designed the entire interior of the Boeing 707.[15] How significant, that Marmon went to Teague for their milestone automobile and twenty-five years later Boeing went to him for their milestone jet transport!

In 1981 Maurice Hendrey would write: *'Legend endures that the first important collaboration in the industry between top engineering and a leading industrial designer had produced a modern, unified and simple design. Credit goes to Howard Marmon for recognizing the importance of a profession then in its infancy and recruiting its talents in this manner.*[16]

MARMON SIXTEEN PRICES

| Style | Jan 1931 | 1931 | Mid 1932 | 1933 | Weight pounds |
|---|---|---|---|---|---|
| 5 p sedan | $4925 | 5200 | 5700 | 4825 | 5360 |
| 5 p club sedan | 4995 | 5270 | 5800 | 4925 | 5335 |
| 2 p coupe | 4945 | 5220 | 5700 | 4825 | 5090 |
| 5 p victoria | 4995 | 5270 | 5800 | 4925 | 5150 |
| 2 p convertible | 5025 | 5300 | 5850 | 4975 | 5100 |
| 5 p conv. sedan | 5145 | 5420 | 5950 | 5075 | 5285 |
| 7 p sedan | | 5400 | 5900 | 4975 | 5440 |
| 7 p limousine | 5225 | 5500 | 6100 | 5175 | 5480 |
| — chassis | | | 4375 | 3750 | |

Table reprinted by permission of *Car Collector and Car Classics,* Copyright 1949.

Prices shown were for fully equipped cars. The 1931 "list prices" were $150 less than shown; the weight of the sedan had been reported at 4800 pounds in initial press releases. Subsequent reinforcement of the frame was probably responsible for the increased weight shown above.

Fig. 6-14 W.D. Teague, Jr. with the 16 model.

The 1933 price reductions reflected falling prices of purchased components

The full range of body styles is shown by the five factory press release photographs. Note that the radiator ornament first appeared on the Convertible Coupe.

### 1932

There were detail changes made in the 1932 models. During the 1931 model run, the shutters became chromium plated instead of painted. Although brochures showed fender lamps matching the headlights, only one car is known to have had them. A padded map compartment in the center of the lower dash was discontinued . . . probably to facilitate radio installations. A rear assist cord was added and could be slid along a track for convenience.

During 1932, a vacuum power booster was added to the braking system. Also the mini door-handles on the hood ventilator doors were changed to streamlined ones extending across the doors; the mini-handles of necessity were attached with a single screw and proved a nuisance.

### 1933

The 1933 SIXTEENS exhibited no sheet metal changes but there were metal trim revisions and new paint combinations.

Parking lights of a streamlined configuration were mounted atop the front fenders. The rear fender carried upward extensions, of progressively longer length, of the chromium tread moldings on the running boards. The horns were relocated from their nests behind the spare wheels to the front of the car and became chromium plated. A new octagonal hub cap emblem was adopted; it was a desirable enhancement.

New two-tone paint combinations in a new high-lustre finish were offered. A new opalescent finish was applied to the instrument board. Seat cushions were more luxurious, being both deeper and softer. Control of the cowl vents was made more convenient by use of control levers in place of the former knob and screw system.

'Selective Ride Control' . . . driver adjustment of the shock absorber valves while in motion . . . was adopted for the SIXTEEN following its introduction on the 1932 Model 8-125.

More rigid brake drums and a new design vacuum power booster were features.

## PRODUCTION VOLUME
## 390 AUTOMOTIVE JEWELS

Considering that the MARMON SIXTEEN is a jewel among motor cars, some assessment of the original volume produced is appropriate. Robert Mellin, an engineer with General Motors Research Laboratories, seriously and cleverly researched this subject and published his findings in the March 1968 issue of *'Cars and Parts'* magazine. The total . . . only 390 automobiles. The caliber of Mellin's study warrants its reproduction here:[17] *Model and Serial Numbers*

*The engine numbers on the Marmon V-Sixteen started at 16-501 and continued to the end of production in 1933. Although the total number produced is not positively known, it appears quite probable that a total of 390 were produced since the highest number engine for which any record has been found was 16-890. Although this car is apparently not in existence, the serial and engine numbers were obtained by Marmon V-Sixteen enthusiasts in Indianapolis where the cars were built. It is quite likely that the last ones built would be sold in Indiana when the company went out of business, and the numbers for this car were obtained from Indiana title records. This car was registered in the name of George Tenney, a former sales manager for the Marmon V-Sixteen. On the basis of this evidence it has been concluded that the total production of Marmon V-Sixteens was 390 cars from 1931 to 1933.*

*Marmon V-Sixteen car serial numbers were of the form 16-140-501 where the 140 represents the body style (5-passenger sedan) and the 501 represents the serial number of the car. These numbers started at 501 and were not repeated regardless of body style to the end of production in 1933. The 1931 models started at 501 and while the actual beginning number for 1932 models is not known, (the late) Bill Gibson, and Ed Robinson, authorities on the Marmon V-Sixteen, have arbitrarily selected car number 724 on the basis of present evidence. From the years in which the cars were titled it can only be concluded that almost all 500 series cars were 1931 models and almost all 800 series were 1932 models and significant changes were made on the 800 series or late 1932 cars. There are many cars in the 600 and 700 series titled as 1932's and many as 1931's. It would seem quite probable that 600 series cars, at least, were produced in 1931 but not all sold until 1932; therefore the 724 number probably is quite a good selection. Cars with serial numbers from 901 and up were known to be 1933 models. The highest serial number located was number 956 which was also obtained from Indiana title records and was titled in June, 1933, after the company had quit building cars. This seems to be a reasonable number for the last car which would mean a production of 56 cars as 1933 models, although these were probably built primarily in the fall of 1932.*

*At this point a comparison can be made between the total number of engines (390) and the total range of car serial numbers (456) and it is seen that there must be 66 serial numbers not used in the range of 501 to 956. If all the known serial number-engine number combinations for 1933 (900 series) are written down, it is found that there is an average lag of about 66 numbers on the engine number compared with the serial number. The departure of each individual car from this difference is quite small, so it appears that the discrepancy was previous to the 900 series cars. A check of 800 series cars shows that there is a lag on numbers of about 33 (if two cars with very large differences are excluded; therefore about 33 numbers must have been left out in the 800 series. A check of the known serial numbers shows that the highest 800 series number was 867 which was titled in October of 1932, the same month that some 900 series cars were titled. It appears possible, therefore, that the highest 1932 serial number was 867 which would account for the 33 missing numbers. A check of 500 series numbers and 600 series numbers shows relatively little difference in engine and serial numbers, while the 700 series cars seem to have no particular pattern. Of the known cars there are four in the 700 series with engine numbers of 538 or below. It appears that probably 33 of the 1931 cars that were unsold, were given new serial numbers (apparently in the 700 series) which would account for the difference in engine and serial numbers. This practice of renumbering unsold cars was not uncommon among the producers of low production cars of that era; for example, 1937 Cords which had serial numbers below 1525 were reserialized 1936 models according to statements in their engineering change notes. (The practice persisted in the truck business into the 1970s . . . author).*

From this evidence, then, it was concluded that there were 390 Marmon V-Sixteens built of which possibly 223 were built as 1931 models, 111 as 1932 models, and 56 as 1933 models. The table below sums up these conclusions.

| Year | Serial Range | Total Built | Total Sold |
|---|---|---|---|
| 1931 | 501-723 | 223 | 190 |
| 1932 | 724-867 | 111 | 144 |
| 1933 | 901-956 | 56 | 56 |
| | Total | 390 | 390 |

Another number which appears on each car is the body number which is found on a tag on the firewall. The form of this number was the same as the serial number (16-140-501) except that the last three digits represent the number in that body style, but again starting at 501. Through a study of the known body numbers for each style and their relative sequential location in terms of the corresponding serial numbers, an estimate has been made of the total production of each body style and is shown in the table below.

| No. | Style | Highest Known | Estimated Production |
|---|---|---|---|
| 140 | 5-pass. Sedan | 108 | 109 |
| 141 | 2-pass. Coupe | 18 | 19 |
| 143 | 5-pass. Victoria | 33 | 34 |
| 144 | 2-pass. Conv. Coupe | 43 | 44 |
| 145 | 5-pass. Conv. Sedan | 46 | 47 |
| 146 | 7-pass. Sedan | 62 | 63 |
| 147 | 7-pass. Limousine | 12 | 15 |
| 149 | 5-pass. Club Sedan | 52 | 56 |
| | Custom Bodied | | 3 |
| | | Total | 390 |

A remarkable percentage of SIXTEENS has survived . . . a tribute to the evident milestone caliber of the car. Mr. (Duke) Marston, 1982 President of the Marmon Club has identified over 75 surviving examples . . . about 19 percent of original production.

One SIXTEEN served a foreign legation in Washingtion, DC, until the 1970s.

Each MARMON SIXTEEN chassis was initially fitted with a test body and operated on the Indianapolis Speedway . . . the Marmon test track . . . for 210 miles with the speed increased in progressive steps to a 50 mile segment in excess of 75 miles per hour and a final 10 miles at not less than 105 miles per hour. In the dealer showroom, each car carried, strapped to the steering column, a plastic-faced leather envelope displaying a signed certificate attesting to the successful completion of the Speedway test. The reverse side noted that the car was mechanically an exact duplicate of the car holding the Stevens Trophy and all AAA stock car records for Class A closed cars for 5 to 1000 miles and for 24 hours. It further noted that the engine was a duplicate of the engine for which Howard Marmon was awarded the medal of the Metropolitan Section of the Society of Automotive Engineers.

One of these certificates, in its envelope, is in the National Automotive History Collection of the Detroit Public Library. It bears the signatures of George H. Freers as Chief Engineer and J.W. Anderson as General Works Manager and is reproduced herein.

Marmon enthusiasts are indebted to Road & Track Magazine for a latter day Road & Track Classic Test in which the dearth of media road tests in the 1930s is corrected by a hypothetic July 1932 road test report by the Road & Track technical department using a performance prediction system which had evolved and been determined accurate. It appeared in the magazine's July 1959 issue and is reproduced here just as it appeared. Note particularly the 13 percent gradeability in high gear, the 68 mile per hour quarter-mile time of 18.8 seconds, and the 70 mph shift to high. It was the author's experience that passengers were not aware of the car being in gear at 70 in second.

The SIXTEEN sedan weighed only 500 pounds more than the bare Duesenberg S J chassis.[18] Eugene Janderquist and Griffith Borgeson reported the latter's "265 horsepower rating to be an inflated figure with 207 being shown on one power curve and others not reaching 200."[19] Recently two restored MARMON SIXTEEN engines have been run on the engine dynomometer and exceeded 200 horsepower.

Fred Moskovics told E.L. Robinson that in 1932 he drove a MARMON SIXTEEN sedan, that was stock in every respect, for two laps on the Indianapolis Speedway at an electronically clocked speed of 111 miles per hour.[18] This appears consistent for a fully broken-in car considering the 105 miles per hour attained during the Speedway test of each car prior to shipment.

Dr. Bishop provides a rare commentary from one who had the experience of first hand acquaintance with both Marmon and Cadillac sixteens when they were first introduced. He saw the Cadillac at the 1930 New York Auto Show . . . he writes:[7] *"I was among those suitably impressed by the overpowering size of the cars. I was not, however, intimidated by their cumbersome progress on the road a few months later, when friends who had acquired one of the Cadillacs shared their good fortune by putting it through its paces."*

It was a full year later that he saw the *MARMON SIXTEEN* at the New York Auto Show: *"Its slender lines beguiled me, and forty-eight years later it still holds me under its spell. It was the lightest big car I have ever known . . . light in spirit, in line and (as I came to discover) in handling."*

He also provided an assessment of the knock sensitivity of the Marmon: *"The two MARMON SIXTEEN cars I had the pleasure of owning were very casual about the sort of fuel they would accept. An Esso station in Georgetown (D.C.) which sold, in addition to regular and the relatively new Ethyl gaosline, a third grade for which I had to pay 11 cents a gallon (regular was 13¢) and I never knew the engine to knock, although its compression ratio was 6 to 1 (highest in the industry) compared with 5.5 to 1 for the Cadillac which had the fault of being very sensitive to fuel (octane) ratings and requiring premium gasolines."*

In the summer of 1932 Marmon published a 5 x 7 brochure entitled 'A Distinguished Clientele'. Certainly the owners of sixteen cylinder automobiles of any make constituted an elite group . . . with combined Cadillac and Marmon production totaling less than 500 in 1932. The brochure began "If ever there was a car without a single dissatisfied owner it is the MARMON SIXTEEN" and then presented photocopied letters of commendation from a bevy of distinguished owners. Like the similar publications of the first three Marmon Eras, this Era Five reporting provides valuable information on the in-service experience, including fuel economy and the driving habits of the more adventurous owners.

The first letter was from Gar Wood . . . most familiarly remembered at the wheel of Miss America X or her predecessors and nearly dwarfed behind the massive Packard aircraft/marine engines which drove the Misses America. On land, the Gar Woods drove, not Packards, but 'his' and 'hers' MARMON SIXTEENS!

But the Woods were outdone by Frank Gannett, President of The Gannett Newspapers headquartered in Rochester, New York, who reported ownership of two other MARMON SIXTEENS besides his personal convertible coupe, and by Dr. Rost Omnes of the Netherlands who likewise owned a convertible and two others.

The letter of J.J. Holcomb, President of the Indianapolis firm of Holcomb & Hoke Mfg. Co., compares the thrill of ownership to that he experienced with his first car in 1903 and warrants reproduction in its entirely.

Eli Lilly, whose firm today occupies the Marmon site and some of the original buildings, reported 9.64 miles per gallon on a trip and commented: *"I had always felt that eight cylinders were enough for any automobile, but after driving your Marmon 16 for one day, my mind was quickly changed, and after looking at other multiple cylinder cars, was all the more impressed with the outstanding features of the Marmon 16."*

Fred Brown, the president of Tempo Brooks Inc., of New York, took his two week old MARMON SIXTEEN on a trouble-free 7000 mile tour of Europe where German police several times requested removal of the car to the rear of hotels due to street blockage by onlookers. He concluded his letter: *"On the Riviera I entered the car in Concours de Elegance and came out 'Premier Prix' "*.

A George Bell of Marion, Indiana, also took his SIXTEEN to Europe for a trouble-free 12,000 kilometer trip immediately following purchase.

The owners consistently reported eight miles per gallon in city driving and from nine to ten in the country; one owner reported 8.5mpg with seven people and luggage. Mileage acquisition rates ranged from 1000 to 1500 miles per month.

Mr. and Mrs. Jay Eitel drove their 1933 SIXTEEN Victoria home to Atherton (San Francisco) CA, after participating in the MARMON MUSTER of the Marmom Club in September 1982 (marking the eightieth anniversary of the first Marmon). They reported the events of their trip in the *Marmon* (Club) *News:*[20] They cruised at 60 to 65 mph, used 255.4 gallons for the 2496 miles for an average 9.77 miles per gallon (despite high head winds and crossing the mountains). Four quarts of oil were

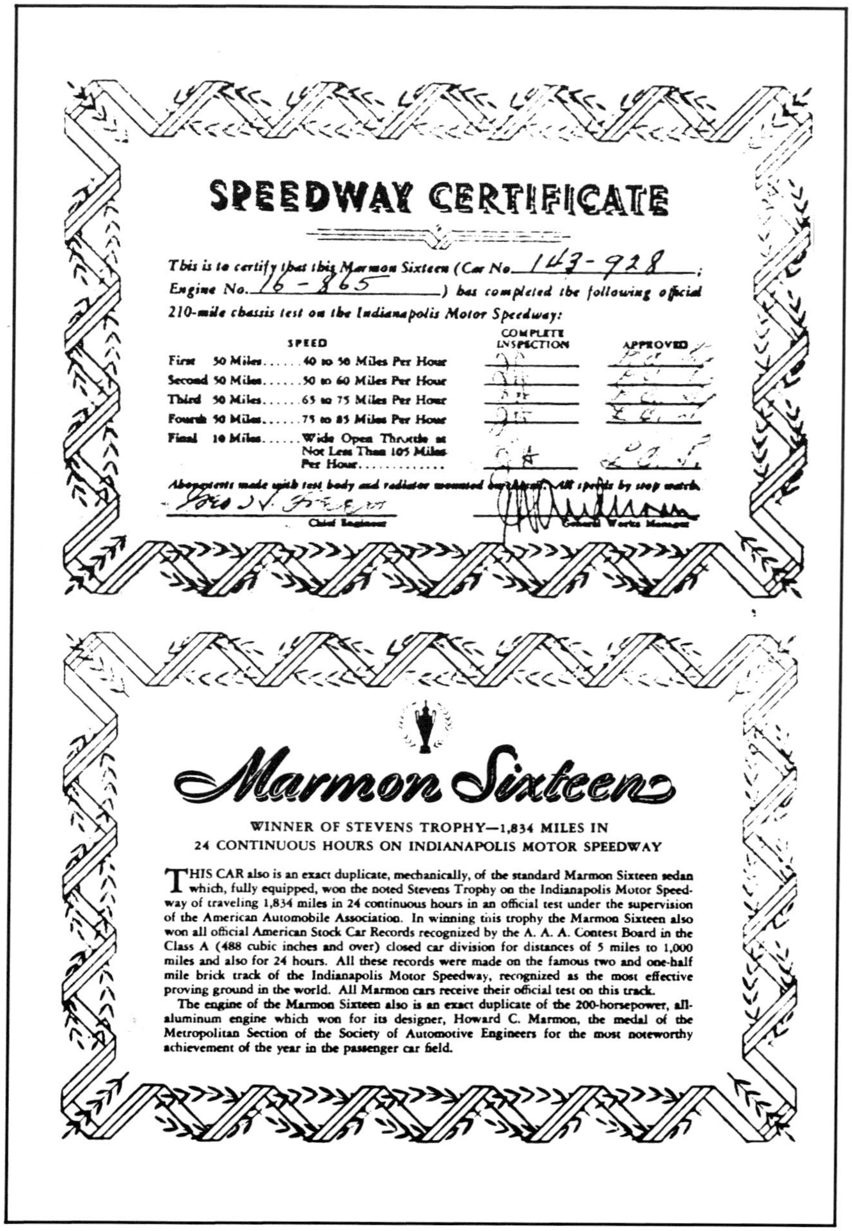

# SPEEDWAY CERTIFICATE

This is to certify that this Marmon Sixteen (Car No. 143-928; Engine No. 16-865) has completed the following official 210-mile chassis test on the Indianapolis Motor Speedway:

| SPEED | | COMPLETE INSPECTION | APPROVED |
|---|---|---|---|
| First 50 Miles | 40 to 50 Miles Per Hour | | |
| Second 50 Miles | 50 to 60 Miles Per Hour | | |
| Third 50 Miles | 65 to 75 Miles Per Hour | | |
| Fourth 50 Miles | 75 to 85 Miles Per Hour | | |
| Final 10 Miles | Wide Open Throttle at Not Less Than 105 Miles Per Hour | | |

Abonements made with test body and radiator mounted ... All speeds by stop watch.

Chief Engineer

General Works Manager

# Marmon Sixteen

WINNER OF STEVENS TROPHY—1,834 MILES IN 24 CONTINUOUS HOURS ON INDIANAPOLIS MOTOR SPEEDWAY

THIS CAR also is an exact duplicate, mechanically, of the standard Marmon Sixteen sedan which, fully equipped, won the noted Stevens Trophy on the Indianapolis Motor Speedway of traveling 1,834 miles in 24 continuous hours in an official test under the supervision of the American Automobile Association. In winning this trophy the Marmon Sixteen also won all official American Stock Car Records recognized by the A. A. A. Contest Board in the Class A (488 cubic inches and over) closed car division for distances of 5 miles to 1,000 miles and also for 24 hours. All these records were made on the famous two and one-half mile brick track of the Indianapolis Motor Speedway, recognized as the most effective proving ground in the world. All Marmon cars receive their official test on this track.

The engine of the Marmon Sixteen also is an exact duplicate of the 200-horsepower, all-aluminum engine which won for its designer, Howard C. Marmon, the medal of the Metropolitan Section of the Society of Automotive Engineers for the most noteworthy achievement of the year in the passenger car field.

Fig. 6-15 Certificate which hung from steering column.

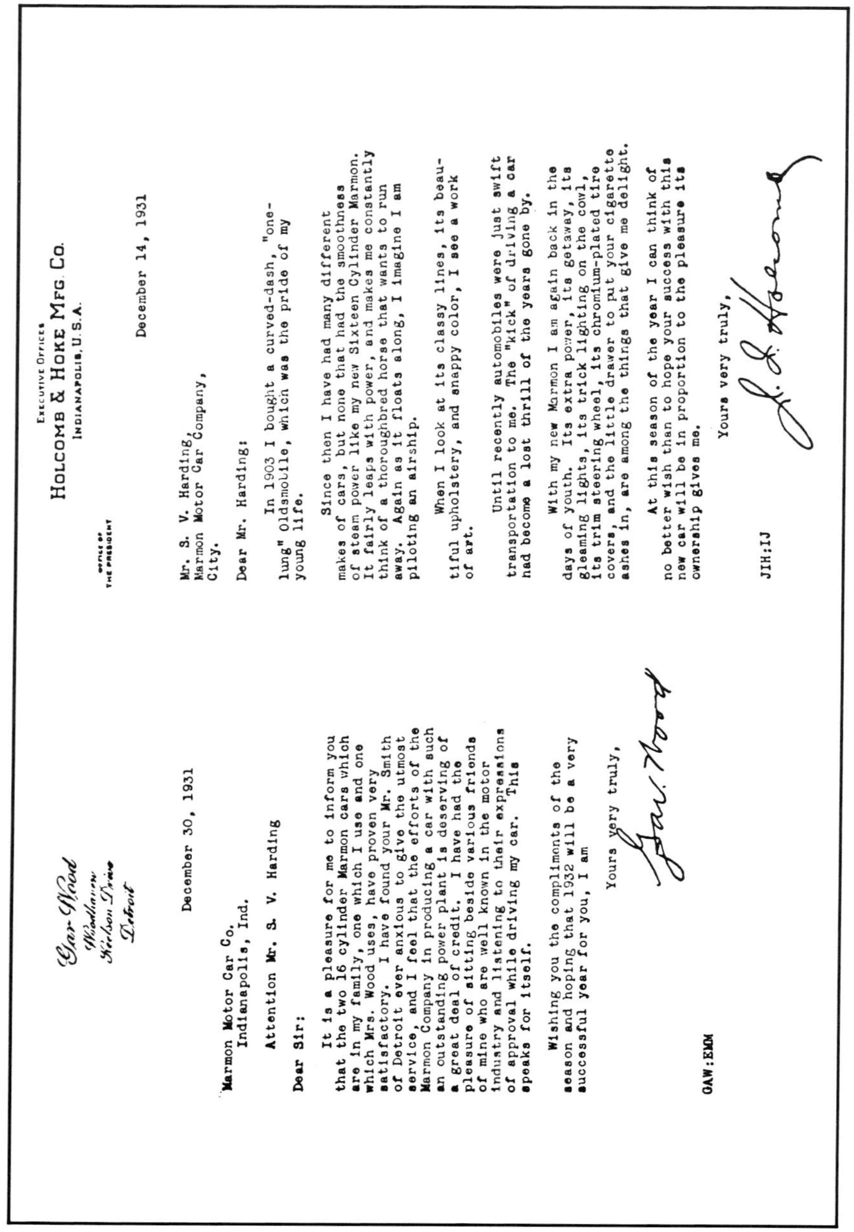

Executive Offices
HOLCOMB & HOKE MFG. CO.
INDIANAPOLIS, U.S.A.

Office of The President

December 14, 1931

Mr. S. V. Harding,
Marmon Motor Car Company,
City.

Dear Mr. Harding:

In 1903 I bought a curved-dash, "one-lung" Oldsmobile, which was the pride of my young life.

Since then I have had many different makes of cars, but none that had the smoothness of steam power like my new Sixteen Cylinder Marmon. It fairly leaps with power, and makes me constantly think of a thoroughbred horse that wants to run away. Again as it floats along, I imagine I am piloting an airship.

When I look at its classy lines, its beautiful upholstery, and snappy color, I see a work of art.

Until recently automobiles were just swift transportation to me. The "kick" of driving a car had become a lost thrill of the years gone by.

With my new Marmon I am again back in the days of youth. Its extra power, its getaway, its gleaming lights, its trick lighting on the cowl, its trim steering wheel, its chromium-plated tire covers, and the little drawer to put your cigarette ashes in, are among the things that give me delight.

At this season of the year I can think of no better wish than to hope your success with this new car will be in proportion to the pleasure its ownership gives me.

Yours very truly,

JIH:IJ

Gar Wood
Woodhaven
Keelson Drive
Detroit

December 30, 1931

Marmon Motor Car Co.
Indianapolis, Ind.

Attention Mr. S. V. Harding

Dear Sir:

It is a pleasure for me to inform you that the two 16 cylinder Marmon cars which are in my family, one which I use and one which Mrs. Wood uses, have proven very satisfactory. I have found your Mr. Smith of Detroit ever anxious to give the utmost service, and I feel that the efforts of the Marmon Company in producing a car with such an outstanding power plant is deserving of a great deal of credit. I have had the pleasure of sitting beside various friends of mine who are well known in the motor industry and listening to their expressions of approval while driving my car. This speaks for itself.

Wishing you the compliments of the season and hoping that 1932 will be a very successful year for you, I am

Yours very truly,

Gar Wood

GAW:EMM

Fig. 6-16 Endorsements by SIXTEEN owners.

# 1932 Marmon Sixteen

*Ed. Note: This is not an actual road test of the car described; in this case, no factual test data could be found. However, our technical department has evolved a predicted performance system which has proved to be accurate.*

JULY 1932. While the old-line firm of Nordyke and Marmon of Indianapolis cannot claim to be first with a 16-cylinder automobile, there is little argument over the truly outstanding engineering merit of their entry in the top-quality luxury car market. This is in line with their long-established policy, even though their reputation was more than a little tarnished by a low-priced straight-8 which proved to be one of the most miserable automobiles ever offered to a gullible and unsuspecting public (specifically, the Roosevelt).

The justly famous Marmon 34 of the early Twenties was a large, comfortable car of remarkably low weight, and the new 16 introduced only last year carries this theme to its logical conclusion. Here is a car of magnificent design and proportion which represents an almost unique wedding of engineering ideals and the stylist's art. As on the model 34, aluminum is used extensively and without regard for cost. Despite its size the 16 weighs only just over 5000 pounds, ready for the road. It is powered by one of the largest engines ever offered to the public (490.8 cubic inches and 200 brake horsepower), yet the engine weighs only 930 lb.

The net result of all this is a truly luxurious machine, yet one that is remarkably easy to drive, offering tremendous performance with incredible smoothness. We will state without equivocation that the Marmon V-16 is the finest and best all-around luxury car on the market today.

Consider, for example, the performance figures. Very few cars can even approach its 0-60 acceleration time of 14 seconds, done, incidentally, using only 1st and 2nd gears. Likewise, the top speed of 105 honest miles per hour can only be exceeded by a few cars—all but one of which are built in Europe—and none can do it so smoothly and quietly. And the 16 is not even designed to be fast, for its matching of axle ratio and power curve was designed for flexibility (105 mph requires 4100 revolutions per minute, far beyond the peak power point of the engine). It is quite possible to pop the gear lever into high at 5 mph and forget it, for this car accelerates from this speed as most cars do in 2nd, and noiselessly. Even in high gear, hills simply do not exist, as shown by a Tapley reading of 260 pounds per ton, equivalent to a grade ability of over 13%.

The 3-speed transmission is totally silent in 2nd gear. With the servo-cam synchronizers provided it is possible, though not recommended, to shift down from high to 2nd at 75 mph, with no assist by the driver. There is a faint grunt, the engine buzzes a bit, and the car slows down.

As with the transmission, all other controls are remarkably light, and despite its size the car is very easy to drive. Even when braking down hard from 100 mph, the pedal pressure required is extremely light, thanks to a combination of enormous brake drums, duo-servo shoes and a competent power booster. The ride, too, is as good as in any car we have ever tried. An anti-kick shackle on the left front spring insures accurate steering without kickback at the wheel.

Everything about this car, inside and out, shows quality; yet its appearance (by Walter Dorwin Teague) is extremely dignified and in quiet, unostentatious taste. Personally we think the Marmon should make a certain firm, which advertises "The Best Car in the World," change its tune—or introduce a new model with more cylinders. ■

## ROAD & TRACK CLASSIC TEST 29

## MARMON V-16

### SPECIFICATIONS

| | |
|---|---|
| List price (1932) | $5500 |
| Curb weight | 5300 |
| Test weight | 5600 |
| distribution, % | 48/52 |
| Dimensions, length | 210 |
| width | 72 |
| height | 68 |
| Wheelbase | 145 |
| Tread, f and r | 58 |
| Tire size | 7.00-18 |
| Brake lining area | n.a. |
| Steering, turns | 4.0 |
| turning circle, ft | 42 |
| Engine type | V-16, ohv |
| Bore & stroke | 3.13 x 4.0 |
| Displacement, cu in | 490.8 |
| cc | 8046 |
| Compression ratio | 6.00 |
| Bhp @ rpm | 200 @ 3400 |
| equivalent mph | 87.1 |
| Torque, lb-ft (est) | 390 @ 1800 |
| equivalent mph | 46.1 |

### GEAR RATIOS

| | |
|---|---|
| O d (n.a.), overall | |
| 4th (n.a.) | |
| 3rd (1.00) | 3.69 |
| 2nd (1.57) | 5.79 |
| 1st (3.08) | 11.4 |

### CALCULATED DATA

| | |
|---|---|
| Lb/hp (test wt) | 28.0 |
| Cu ft/ton mile | 120 |
| Mph/1000 rpm (3rd) | 25.6 |
| Engine revs/mile | 2340 |
| Piston travel, ft/mile | 1560 |
| Rpm @ 2500 ft/min | 3750 |
| equivalent mph | 96.0 |
| R&T wear index | 36.5 |

### PERFORMANCE

| | |
|---|---|
| Top speed (3rd), mph | 105 |
| best timed run | 107 |
| 3rd ( ) | |
| 2nd (4300) | 70 |
| 1st (4300) | 36 |

### FUEL CONSUMPTION

| | |
|---|---|
| Normal range, mpg | 8/11 |

### ACCELERATION

| | |
|---|---|
| 0-30 mph, sec | 3.5 |
| 0-40 mph | 6.0 |
| 0-50 mph | 9.4 |
| 0-60 mph | 14.0 |
| 0-70 mph | 20.1 |
| 0-80 mph | 28.6 |
| 0-90 mph | 38.5 |
| 0-100 mph | 54.0 |
| Standing ¼ mile | 18.8 |
| speed at end, mph | 68 |

### TAPLEY DATA

| | |
|---|---|
| 4th, lb/ton @ mph | @ |
| 3rd | 260 @ 45 |
| 2nd | 400 @ 35 |
| 1st | off scale |
| Total drag at 60 mph, lb. | 240 |

### SPEEDOMETER ERROR

| | |
|---|---|
| 30 mph | actual, n.a. |
| 40 mph | |
| 50 mph | |
| 60 mph | |
| 70 mph | |
| 80 mph | |
| 90 mph | |
| 100 mph | |

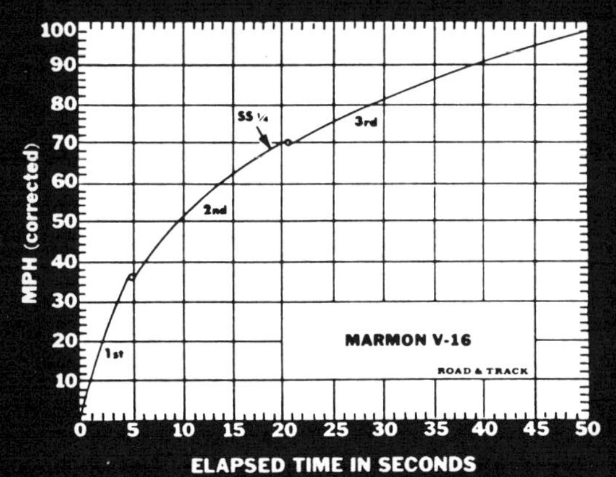


Fig. 6-17 *Road & Track* Classic Test 29 reprinted with permission of Road & Track.

used . . . mostly lost through the original rear crankshaft seal, during the five day run.

The Eitels had owned the car except for the first eleven months of its fifty year life. The engine was reringed (without removing the heads) in 1939 and again in 1982. The original Ray-Day[21] pistons and hardened cylinder liners (now tapered .020) were retained. It still had the original rod and main bearing inserts and timing chain. The camshaft was replaced in 1982 and the tappets reground, and a leaking cylinder head replaced.

In 1974 the wheels were repainted, the lock rings replated and the fender paint touched-up. Result?: "Most Elegant of Show" at the Silverado Concours and at Pebble Beach Concours!

Several brake relines and clutch facing replacements were required. The intermediate clutch pressure plate was replaced in 1982.

It was 1955 when George E. Domer, owner-restorer of classic automobiles, of Milton, Pennsylvania wrote to the Editor of Classic Car[22] about the unrestored MARMOM SIXTEEN he had acquired: *"The car has only 25,000 miles on it . . . Much has been said about acceleration, etc., but the point that I would like to emphasize is the 'sport car' feel the big brute possesses. Our Pierce-Arrow handles like a truck in comparison. That statement comes not easily for I really like our Pierce-Arrow. But the Marmon is a real delight . . . 'light on its feet'. My wife drives our MG sedan regularly and pronounces the SIXTEEN a King-Size MG! That will give you the idea. Up-hill all the way from our town (Milton, PA) to Watkins Glen, (New York) 125 miles away gave just a fraction over ten miles per gallon. It also starts easily in winter weather even after 2 or 3 weeks rest."*

The late Patron of Automotive History, William (Bill) Harrah, took his freshly and fully restored MARMOM SIXTEEN convertible coupe for a test run in 1967. The man whose Reno museum staff restored a thousand cars had this to say:[23] *"To me, the Marmom Sixteen is a very striking car. I especially like the turtleback curve on the coupe, the way it blends in with the fender line . . . and the high hood and execution of the grille design. Overall, the effect is very pleasing. Our Marmon Sixteen was acquired in 1958 and its restoration . . . which required about eight months concentrated work . . . was completed in June of 1967. It is now totally debugged and handles remarkably well for a car of its vintage. The Marmon Sixteen was an extremely well-built car; the workmanship is exemplary, and one need only look it over to realize that in the Thirties you were really getting a lot of car for your money when you bought one. The Sixteen is a big car, naturally, and very easy to get into. The leg room and positioning of controls are perfect for a man of my height . . . six feet, three inches. Actually I can't imagine anyone being uncomfortable in a Marmon. The car starts immediately, and it's extremely quiet. So much so that a very little other noise . . . like people talking nearby . . . can hide the engine idling sound. One must be careful not to start it twice! The clutch works easily, and the shifting is very simple, if you allow just a little time between second and third. First to second can be shifted as fast as one can move the lever. I drove the car around Reno and out onto Interstate 80, attracting a good deal of attention all along the way. Other motorists on the highway reacted as I expected . . . with surprise or astonishment, the latter most particularly from those I passed, and I passed quite a few. Acceleration is very good, and the small sixteen cylinders wind up extremely fast. I didn't try any 0 to 60 runs, but found acceleration better than adequate for the high-spirited Nevada traffic and the indicated top speed . . . 93 mph . . . more than sufficient to keep moving along with or ahead of the brisk traffic flow. I think with the top and windows up and a few more break-in miles on the speedometer 100 mph could be shown, which with calibration and altitude allowances would come out to a true 100+mph at sea level. In any case, the Sixteen's speed is quite satisfactory. A cruising 80 to 85 is very comfortable. The final drive ratio is 3.78 to 1 . . . but it's a long way around those 7.50x18 tires. The secondary road ride is luxurious . . . firm and quiet. One recognizes that this is a big, solid car, but there's no mistaking that it feels like a car . . . and not a truck. It handles very well, almost lithely. Cornering is remarkable for a car of its height and weight, and braking is almost too good to believe . . . perfectly equalized with a wonderful solid feel. Many cars currently produced should have such a feel. All in all, the Marmon Sixteen is a car I would love to drive from coast to coast. With its fuel consumption of eight miles a gallon, it certainly wouldn't be the most economical way to traverse the country, but it would be a lot of fun."*

## PUBLIC DEMONSTRATION AND RECOGNITION

At the very close of the Third Era, a Marmon 75 coupe under official observation of the Automobile club of France had demonstrated unprecedented durability by traveling 50,000 miles over public road in nine weeks to break all existing records for 20,000 to 50,000 miles.

During 1931 the Sixteen showed that it was also that breed of automobile by establishing several speed records, one of which was particularly noteworthy and an achievement comparable to the 1927 durability run in France. The Stevens Challenge Trophy was awarded for the longest distance covered by a closed stock automobile in a 24 hour period on the Indianapolis Speedway. It was the gift of Samuel B. Stevens who had sponsored the 1916 coast to coast run of the Marmon 34 (as well as driving part way).

Albert Bochroah, in his American Automobile Racing,[24] lists four successful runs for the Stevens Challenge Trophy:

| CAR | DATE | MILES | MPH | DRIVER |
|---|---|---|---|---|
| STUTZ | April 14, 1927 | 1642.6 | 68.4 | Anderson Rooney, Keene |
| MARMON SIXTEEN | October 24, 1931 | 1834.215 | 76.43 | Fisher, Harold Updike, Bert Cox, James Hopkins, Fred |
| CORD | July 22, 1937 | 1909.85 | 79.577 | Jenkins, Ab |
| CHRYSLER | October 30, 1955 | 2157.5 | 89.89 | Bettenhausen Taylor, O'Connor |

Keene, one of the Stutz drivers, had driven a Marmon to victory at Santa Monica, CA in 1911.

A back-up companion SIXTEEN covered 1801 trouble-free miles over the bricks at an average speed of 75.07 miles per hour. Both cars were random selections by AAA officials and made Indy 500 type pit stops for refueling and drive shift changes.

*The automotive engineering profession took official note of the advent of the Marmon Sixteen; the Metropolitan Section (New York) of the Society of Automotive Engineers awarded Howard Marmon the Moskovics Medal on January 6, 1931, for the outstanding achievement of 1930 . . . the Marmon 16 engine.*

The Moskovics Medal was the gift of Frederick E. Moskovics then recently president of Stutz and who had left Marmon in 1920. The medal was to be presented annually. The official title of the Medal is *"The Metropolitan Section, SAE,* for the individual responsible for some outstanding improvement in some branch of the

American passenger car industry."[25] The Medal Committee consisted of Col. W.G. Wall, T.J. Litle, chief-engineer Holley Carburetor and previously of Marmon, both past presidents of SAE, Myron Pope, editor of *"Automotive Topics,"* T.C. Smith of AT and T and Mr. Moskovics. At that time Mr. Moskovics was the President of Improved Products Co.

Marmon's last day in the spotlight was the ticker-tape parade to welcome Amelia Earhart upon completion of the first solo flight over the Atlantic by a woman.

Six Marmon Sixteens led the parade up Broadway, Amelia and husband Geoge Putnam riding in the first of two convertible four door sedans and driven by Walter C. Hiser, the Sales Manager of the factory branch in New York City. When the advertising Manager, Fred Benham, asked for a Sixteen fleet for the occasion, only three cars were in stock in New York; Mr. Hiser borrowed three more from their owners; aviator Clyde Pangborn, golfer Bobby Jones and orchestra leader Ozzie Nelson.

Mr. Hiser who has provided the pictures and details of the parade joined Marmon in 1917 for a one year pre-college training program which had been arranged for (Franklin) Hall Marmon. They started in the foundry, worked in every department and ended with two weeks in final test operations where each Model 34A was run at least one hundred miles on the Speedway bricks. Mr. Hiser graduated from Purdue, Hall Marmon from MIT, both in mechanical engineering.

The New York branch was located on Broadway across the street from today's Lincoln Center.

## TROUBLE IN THE MARKETPLACE

The introduction of the Cadillac 'V-16' at the 1930 New York Show and the subsequent unexpected popularity . . . which occasioned four successive increases in the planned production rate . . . sent a clear signal to Marmon management that they had unwisely elected to delay the finalization and tooling of their sixteen cylinder product.

Teague had given the SIXTEEN the same distinction and individuality that had characterized the Marmon 34 of the Third Era. A Marmon ad for the SIXTEEN stated "There is a 'Raison D'Etre' for every line and curve." Immediately identifiable and aloof from the need for change for sake of change, the SIXTEEN should have seen a glorious and prolonged Fifth Era. But such was not to be.

Marmon had a record of building winning automobiles. The SIXTEEN should have been the greatest winner of them all. It didn't lose the race for the sixteen cylinder luxury market; it would be more appropriate to say that the race had been called off before the Marmon prototype appeared before the public.

The 1930 market for 16 cylinder automobiles was a bonanza far beyond the projections of Cadillac's marketing staff. But the worsening economic climate grossly diminished that market for 1931, as total industry sales plummeted 42 percent. Cadillac sold only 750 V-16s in 1931, only 300 in 1932 and 126 in 1933 whereas they had sold 2500 in 1930.[8]

But Cadillac evidently saw trouble ahead and in mid 1931, only months after the first MARMON SIXTEEN production car, announced they would build a V-12 for the 1931 market. It was a cut-down V-16 but with the 3.0 inch bore increased to the 3.125 of the MARMON SIXTEEN. Whereas the proper bank angle for a V-16 is 45 degrees, a V-12 required a 60 degree bank angle for even firing. (Only in the late 1970s did the use of 'half-sheared' crank pins permit uniform firing of both V-6s and V-8s made from common tooling.) But the Cadillac V-12 had a 45 degree angle inherited with the tooling for the V-16 and relied on the plurality of cylinders

Fig. 6-18 Five SIXTEENS were in the New York welcome to Amelia Earhart. Walter C. Hiser drove Ms. Earhart and husband George Putnam.

# MARMON SIXTEEN

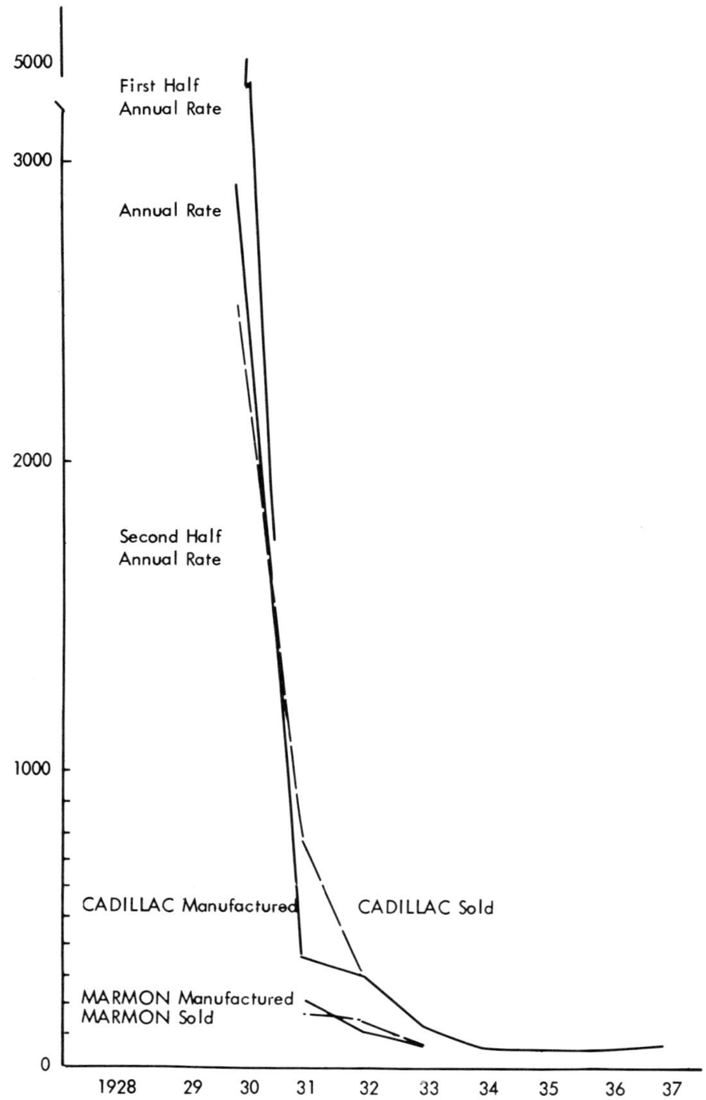


| | | 1930 | 31 | 32 | 33 | 34 | 35 | 36 | 37 |
|---|---|---|---|---|---|---|---|---|---|
| CADILLAC | MFG'D | 2887 | 364 | 300 | 124 | 56 | 50 | 50 | 56 |
| | SOLD | 2500 | 751 | " | " | " | " | " | " |
| MARMON | MFG'D | | 223 | 111 | 56 | | | | |
| | SOLD | | 190 | 144 | " | | | | |

Figure 6-19: PRODUCTION and SALES of SIXTEEN CYLINDER CARS
Annual Rate 1930 - 1937; Semi-annual 1930

to mask the uneven firing sequence. The economies of four less cylinders should not have much effect on the selling price of the car, but in Cadillac's case it did. The V-8 Cadillac had an unnecessarily long hood for competitive considerations (the long-hooded straight eights) and the V-12 fitted in the standard chassis. Cadillac could market a V-12 option in a Fisher-bodied basic (V-8) chassis and consequently sold 1984 of them in the first **six months.** To viably compete, Marmon would have had to offer the 1930 Big Eight automobile with a 12 cylinder version of the SIXTEEN. Instead Marmon was very late in bringing the SIXTEEN to market and lacked liquid financial resources to pursue any V-12 or other new Model activity.

Cadillac was producing their V-16 at a monthly rate of 500 cars in the first half of 1930; but in the second half the rate plummeted, for less than 1000 cars were sold.

Marmon did not start production until March of 1931 and sold 190 cars to Cadillac's 750 (20% of 940 total). And 1932 sales of 444 cars were divided 144 to Marmon (32%) and 300 to Cadillac. Even worse was 1933 with only 182 sixteen cylinder cars . . . 56 Marmons (31%) and 126 Cadillacs.

The total market for sixteen cylinder cars during the three model years of Marmon Sixteen 'offering' (not necessarily production) was 1566 units of which the 390 Marmons represented 25 percent. The market share was great, the market miniscule. The Cadillac V-16 survived through 1937 at typical rates of 53 per year.

The total production of sixteen cylinder cars for the **four** years . . . 1930 through 1933 was so heavily biased by the 2500 units in 1930 that Marmon's 390 units represented only 9.6 percent.

The Marmon Sixteen bodies were built by the LeBaron-Detroit Co. A number of bodies for Marmon (as well as Stutz) were built but not finished until scheduled by the customer. Because of the collapse of the market in 1931, a number of them were still in the unfinished state when the two chassis builders failed.[26]

A most appropriate epitaph for the Fifth Era is the statement of Warren W. Fitzgerald in *'Car Life Classic'* for June 1967:[27]

* MARMON V SIXTEEN *
MECHANICAL MAGNIFICENCE
IN AN ERA OF
* EVAPORATING AFFLUENCE *

## DEMISE OF THE MARMON MOTOR CAR CO.

The Fifth Era, the briefest, was marked by crowning engineering and design accomplishment, by the return of Howard Marmon to active direction of engineering, by the remarkable marriage of his technical genius with the styling mastery of the Walter Dorwin Teagues. Unfortunately it was also marked by concurrent deterioration of market penetration and of the once-strong dealer organization.

The same month that production of the SIXTEEN began, March 1931, Walter Marmon had organized first a subsidiary firm, and soon after an independent firm, sharing the Marmon complex for the manufacture of all-wheel drive military and off-road vehicles. Arthur Herrington headed the engineering activity and the new products bore their hyphenated names: MARMON-HERRINGTON. The remarkable Marmon-Herrington story in war and peace is told in Book Two of the 'Marmon Heritage'.

The precarious state of Marmon Motor Car Co. in the first quarter of 1933 would have been difficult to conceal. The gross contraction of active plant area, and painful relocation of the offices to the York Avenue address in late 1931, the long delayed start of SIXTEEN production and the similar situation with the 8-125 in 1932 and the sole offering for 1933 of a luxury entry in a national market shriveled to a hundred or so cars per year . . . all were portents of impending financial disaster.

*'Classic Cars'* a series of fine photographic essays on a limited member of the choicest motor cars in the world by Henry Rasmussen[28] included a remarkable account of W.B. Ansted's purchase of a Marmon in 1933. It warrants partial repeating here: *The Ansteds had been in the automotive industry for more than three decades. E.W. Ansted bought Ansted Spring and Axle Company in 1895. The most well known of the family's ventures was the Lexington Motor Company, which E.W. Ansted had purchased in 1913. William Ansted owned Metal Auto Products and manufactured fenders, splashpans, hoods and similar body details for Auburn, Stutz, Marmon and many other car makers. The factory was located on West Henry Street in Indianapolis, only four or five blocks north of the Marmon plant. He had been concerned about Marmon's financial situation for over a year now, and had lately become even more alarmed when he learned that they were discontinuing their eight-cylinder line and were only going to market the Sixteens. The night before, unofficial sources confirmed that Marmon would, in fact, soon receive the final blow. I should have seen it coming, he reflected, and I should not have been so generous with their credit. I have to do something about it immediately, he decided. He gave orders to have the unpaid Marmon invoices totaled; they added up to more than $70,000. He decided to go over to the Marmon people and talk to them. Just before entering the Marmon office building, he stopped for a moment in front of the showroom window. There was a shiny gray Sixteen on display, a Convertible Coupe. Very beautiful, he reflected, and then continued on to the office building and went upstairs. The situation at the Marmon Company was worse than expected. There wasn't any cash available. He was offered some tooling equipment as partial payment, but it didn't amount to much. He became extremely depressed when he realized the extent of the loss his company would have to take. But suddenly an idea shot into his mind. "How about the Convertible Coupe you have in the showroom window? I'd be willing to take the tooling and the car and write the rest off. What do you say?" "It's the last one we have, and we planned to sell it and raise some cash for the payroll."*

There was a brief pause. *"But if you agree to write everything off today, we will agree." "Agreed then! Have the car removed from the showroom. I'll sign the papers and take the car with me right now. This has got to be the world's most expensive automobile!"*

William B. Ansted was co-sponsor of cars that won at Indianapolis in 1964 and 1967. The Marmon survives in Indianapolis with only one owner since 1955 and still carries the skirts Ansted added to the fenders and the multi-louvered hood-side panels he also added.

The Fifth Era ended for all practical purposes on Monday of the first full week of May 1933. Eaton Spring and Axle Co. filed a petition asking that Marmon Motor Car Co. be declared a bankrupt consequent to failure to pay an $1100 claim by Eaton, and that a receiver be appointed to conserve assets to the benefit of creditors. Ironically, General Motors declared a regular $0.25 quarterly dividend on May 11, 1933.

The court appointed Fred Barnickol (President of Indianapolis Drop Forging Co.) and Robert D. Adams (attorney) as the receivers for the Marmon Motor Car Co. This action followed months of negotiations with creditors and efforts at reorganization following President Roosevelt's proclamation of a National Bank "Holiday".

Walter C. Marmon, Chairman of the Board of Marmon-Herrington Co., explained to the press that the legal action did not affect the truck operations in any way because the two firms had been completely independent for more than a year.

In January 1934, purchase of the product name and certain other assets were reported affected by the American Automobile Corporation of Indianapolis, a national investment bankers' syndicate having been formed by A.D. Sterner and S.N. Wheeler. Harry A. Miller of race car fame was to be an officer. Capitalization of $500,000 was to be sought for resumption of MARMON SIXTEEN production, and later for introduction of a Miller car, in part of the Marmon properties (Plant 2).[29]

Despite the pleas of Preston Tucker and Harry Miller that the court preserve the integrity of the Marmon facilities for planned resumption of production as the American Automobile Corporation, the court ordered disposal and dispersal of assets in 1934.

Details of the financial collapse of the Marmon Motor Car Co. are to be found in Chapter Thirteen 'Financial History' where the full record of private and later public financing is reviewed. Additional detail on the American Automobile Corporation is also provided in that chapter.

## RETROSPECT AND FANTASY

For the first twenty-five years Marmon product and market strategy was above criticism. The strong financial position of the company reflected that success. But circumstances changed drastically in the five years before the National Bank Holiday in 1933 delivered the financial coup de grace. In previous chapters relating the history of the product offerings, attention was directed to the vacillation in market segment concentration and the failure to expand sales volume subsequent to model proliferation. It should be interesting in retrospect to seek to identify the critical errors in strategy that set the stage for the final demise.

Neither the challenge of the Packard Twin-Six or the short lived Haynes V-twelve had been critical. But the Packard decision to switch to a straight-eight was significant. The 1923 Packard combined the elongated silhouette with pleasing and distinctive styling; The styling was refined progressively over the following ten years establishing universal prestigious recognition.

Cadillac styling made a major advance with the introduction of the 1926 models. The addition of young Harley Earl to that styling team for design of the 1927 La Salle gave Cadillac a quantum advance over the 1925 offering.

Styling was certainly not the best forte of the Leland Lincoln era. When Henry Ford acquired Lincoln, the characteristic Lincoln lines were significantly refined and were given a major assist by extensive fitting of handsome bodies by several custom builders.

Yet Marmon achieved their best sales volume in the face of these developments by their three principal competitors for the domestic fine car market. I believe that this circumstance obscured the true trends. As a consquence Marmon developed and tooled a small straight-eight of 190 CID for 1927 introduction when they should have devoted the effort to an engine twice that size to be a successor to the 339 CID six in the full size car. They finally did just that . . . the SIXTEEN began as a two-bank derivative of the Little Marmon . . . but most unfortunately they shelved

Fig. 6-20 Hanley-built SIXTEEN with dual-ignition, Stromberg EE carburetor. 'Quad'fire apparatus for Dover, Ohio, circa 1941.

Fig. 6-21 Hanley fire pumper as delivered to Conneaught, Ohio. Built on a specially manufactured incomplete GMC base, it was powered by a (Marmon) sixteen engine driving a Hanley two-stage centrifugal fire pump.

Fig. 6-22 Delivery photograph of Hanley triple combination pumper for Marion, Ohio; Chief T.J. McFarland and apparatus manufacturer Keenan Hanley who also styled the vehicle.

the engine after an in-vehicle demonstration in **1927.** An increase in wheel base would have been a minor addition to the extensive list of redesigns that characterized the 1927 '75' chassis. And most importantly, an extended hood and silhouette would have continued Marmon as a viable competitor in the luxury car field.

The period 1927-1932 was characterized by the most rapid transition the industry has ever known. The Marmon Lightning Flash was perhaps symbolic of the lightning rapidity of automotive change which saw the progression from the last Model 'T,' through the love affair with the 'A,' to the birth of the Ford V8 all in the short space of four model years.

That same period saw the first and the last of the Marmon eights. It spanned the era between the last of the big six Marmons to the sole survival of the magnificent "Marmon Sixteen".

And survive the Marmon Sixteen did. But not the way that might have been envisaged when the first production cars rolled out. For the 1933 Sixteen was to be the last production model passenger car of the Marque. However . . .

Race cars powered by Marmon Sixteens started in the Indy 500 in 1933, 1934 and 1936.

In 1937 it powered a pioneer race car concept. The SIXTEEN was midships mounted and formed the car structure between the forward-cockpit structure and the rear transaxle. The car barely missed being ready in time for the 1937 race and a rules change for the following year made the "pint sized" Marmon racing pistons much too large.

Lee Oldfield had created a beautiful car and it was most unfortunate that it was not seen in competition on the bricks.

The story of the SIXTEEN in the Indy 500 is told in Chapter Eight, 'International Champion'.

In 1940 design and manufacturing rights to the engine were acquired by the Hanley Engineering Service of Prospect, Ohio, a manufacturer of fire pumps and fire apparatus utilizing commercial chassis and proprietory pumps. Marmon Sixteen engines were manufactured during 1941 at Prospect in limited numbers and mounted in modified GMC Chassis, driving 1000 gallon per minute pumps. Production was terminated following the Pearl Harbor raid to concentrate on production of water-jet propelled fire tugs pioneered by the firm. A Hanley water-jet boat is in the Smithsonian Collection and a Hanley 16 cylinder triple combination pumper truck, formerly in the Harrah collection, is in the Ponderosa Museum Collection.

The Hanley-built Sixteen engines had dual ignition. Twin distributors on a gear case were substituted for the single automotive distributor and a pair of 10 mm spark plugs in a V configuration replaced the automotive plugs.

The pair of Hanley firetrucks built for Marion, Ohio — a triple combination (pumper) and a quad combination (pumper and ladder truck) — carried distinctive front ends styled by Keenan Hanley whose earlier (1931) design of the Prospect (Fire Engine Co.) Type B Master Fire Fighter was judged "Classic" by the editors of the *'World Encyclopedia of Commercial Vehicles'*.

The Los Angeles Daily News was quoted in a December 14, 1946 news service release: "Another hot job was a 16 cylinder Marmon which Tony Capanna spent three years building. It develops more than 200 horsepower and 6000 revolutions per minute. It holds the (California Timing Association) Dry Lake record in the unlimited class . . . 145 miles per hour; (137.24 for the two way run). Capanna has worked out a water injection system to improve carburetion." The engine design was fifteen years old, the only basic modification was a Harmon camshaft.

The Sixteen won significant recognition of its true stature in the automotive world. It was personified in the words of Fred Horsley, editor of a Trend Book (No. 107, 1953) entitled, "Dream Cars":

"A rival to Dusenberg on the American road in 1931-32 was the 200 hp Marmon 16. Styled by industrial designer Walter Dorwin Teague, the Marmon had a **clean, forthright beauty** without the dash and glamour of the Big D. Under the hood where the Marmon's 45 degree **V-16 dwelt,** was where the car really sparkled. The 930 pound eight liter engine was aluminum and steel; parts gleamed with chrome plate. Top speed for the 5000 pound Marmon was about 105 mph. The car's advanced 200 hp aluminum engine and 105 mph speed would look good even today (1953) among Detroit's covered wagons of 200 plus horses".

"The **handsome** 16-cylinder aluminum power plant of the 1932 Marmom weighed only 930 pounds".

"Although the Dusenberg had unquestionably the leading glamour appeal, the Marmon V-16 had a **much superior engine** and a more satisfactory chassis and performance, according to some enthusiasts".

Again "Although the Marmon didn't have the 'oomph' of the Dussie, its **impressive figure** justified comparison. This (1932 V 16 sedan) is as **fleet as it looks** and **even more mechanically** precise". (Author's emphasis)

When the Perfect Circle Piston Ring Co., sought an aluminum engine on which to develop and demonstrate a patented cylinder liner construction for use in future die-cast aluminum engines, they found the only candidate was a Marmon Sixteen. One bank was readily modified to prototype die-cast configuration. The engine was then operated with one original bank of liners and the experimental components were evaluated on the other for comparison. Upon completion of development, the engine was given a "Show" finish and mounted in the lobby of the Perfect Circle engineering building for use in promotion of markets for the new product. A photograph of a P-C executive and several automotive editors admiring the engine was published in the trade press in 1959 when the engine design was 28 years old.

Fig. 6-23 Perfect Circle Co. die-cast demonstrator.

## PART 3 — The Engine

Howard Marmon apparently considered the problems to be resolved which had resulted in abandonment of the aluminum integral block-crankcase and wet cylinder sleeve configuration pioneered by Marmon on the 34A (1916-1919).

The Marmon Sixteen was a 490.8 CID engine with 6.0 compression ratio, developing 200 brake horsepower at 3400 rpm (.42 horsepower per CID). Bore and stroke was 3.125 x 4.00. A one-piece aluminum block carried wet cylinder liners in banks at 45°, providing even firing impulses (comparable to a 90° V-8) and possessing a narrow width essential to packaging an OHV vee engine under a traditional hood.

Howard Marmon discussed the advantages of the widespread use of aluminum in the SIXTEEN ENGINE in an article in *Automotive Topics*[30] and provided a tabulation of aluminum parts, their individual weights and the weights of iron counterparts:

How Aluminum Alloy Lowered the Weight to Horsepower Ratio of the NEW MARMON SIXTEEN

| Part | Weight in Aluminum (Pounds) | Equivalent in Cast Iron or Steel (Pounds) | Increase (Pounds) | H.P. Wt. Ratio Al. | H.P. Wt. Ratio C.I. | Increase Pounds per Horsepower |
|---|---|---|---|---|---|---|
| Complete Engine Weight | 930 | 1274.27 | 344.27 | 4.65 | 6.37 | 1.72 |
| Cylinder Block and Crankcase | 110.23 | 260 | 149.67 | 4.65 | 5.398 | .748 |
| Cylinder Heads and Covers | 80 | 173 | 93 | 4.65 | 5.115 | .465 |
| Oil Pan | 19 | 44 | 25 | 4.65 | 4.775 | .125 |
| Fly Wheel Housing | 18.5 | 40 | 21.5 | 4.65 | 4.757 | .107 |
| Intake Manifold | 13 | 28 | 15 | 4.65 | 4.725 | .075 |
| Pistons | 12 | 26 | 14 | 4.65 | 4.72 | .07 |
| Front Motor Support | 10.8 | 25.27 | 14.4 | 4.65 | 4.722 | .072 |
| Water Pump Housing | 5.75 | 12.45 | 6.7 | 4.65 | 4.683 | .033 |
| Chain Housing Cover | 4 | 9 | 5 | 4.65 | 4.675 | .025 |
| Weight Totals | 273.38 | 617.65 | 344.27 | | | |

In the scale are shown the various parts of the Marmon Sixteen engine which are of aluminum alloy, their equivalent weight in cast iron or steel, and the increase in the weight to horsepower ratio if the heavier metals were used.

Aluminum cylinder heads of cross-flow type were used, allowing superior breathing and placing the dual exhaust manifolds on the outside of the engine. But this required a central intake manifold and the pioneer application of a dual downdraft carburetor. (Multiple single-barrel downdraft curburetors had appeared at the Speedway in 1929). The intake manifold, rocker covers, pan, chain case and flywheel housing were all of cast aluminum resulting in an engine weight of only 930 pounds (4.65 pounds per horsepower).

The polished rocker covers had an inboard sharp corner and were retained by four knurled knobs. The intake manifold was covered fore and aft the carburetor by polished aluminum covers bridging the gap between the rocker covers and secured by four knobs identical to, and aligned with, the rocker cover knobs. The result was a smooth polished upper surface of the engine broken only by the central carburetor and the forward-mounted distributor.

Fig. 6-24 The compact even-firing 45 degree SIXTEEN engine.

The intake manifold was a single aluminum casting and consisted of two parallel "dual downdraft manifolds", one for each bank, and an integral transverse passage for exhaust heating of the manifold floor, below the carburetor throats. Because the SIXTEEN firing sequence was that of two eights, there was no requirement for mixture feed to opposite banks from the same carburetor barrel, as used on all V-8 engines beginning with the 1934 Ford (except for a few hi-output plenum chamber manifolds).

A manually operated valve in the right hand exhaust manifold closed off the four forward cylinders and forced the gases through a cylinder head passage to the intake cross-over and then to the left exhaust manifold. Exhaust manifolds were finished in black porcelain.

The Stromberg DDR-3 dual downdraft carburetor featured parallel throttle shafts for the two barrels, with geared synchronization. One gear was rigidly attached; the other was coupled via a worm and wheel to allow adjustment for concurrent closing of the two blades.

The 45° Vee angle made the engine high and in consequence this pioneer downdraft vee-engine application had a very short vertical section of intake manifold above the horizontal runners much the same as todays 90° Vee-eights.

The parallel axis arrangement avoided any tendency for the blades to bias mixture mass flow towards one end of the engine, or to bias the fuel/air mixture ratio towards richer on one end than on the other. An AC dry air cleaner behind the carburetor was coupled to it by an aluminum elbow.

The procedure for adjusting the carburetor involved running the engine on one bank (as a straight eight) at a time. After achieving correct adjustment of the idle (air) screw for a given bank, the engine was shut down for two minutes and then briefly cranked, without ignition, to purge the muffler on the unfired bank.

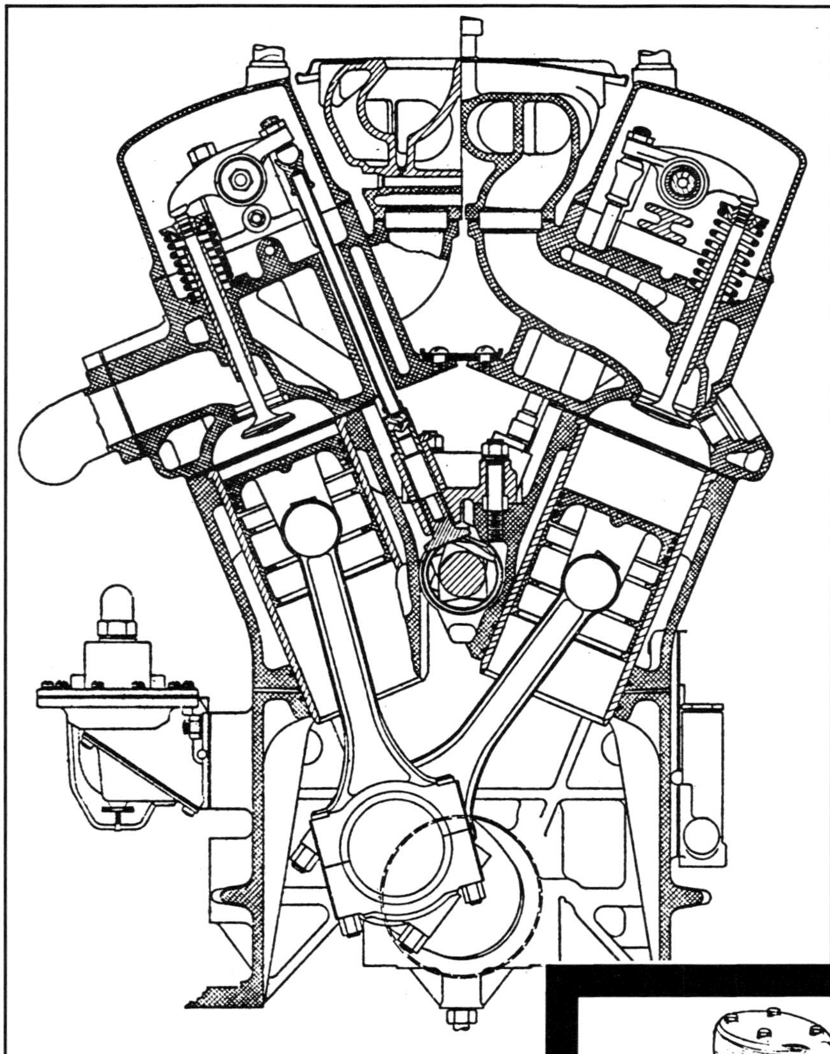

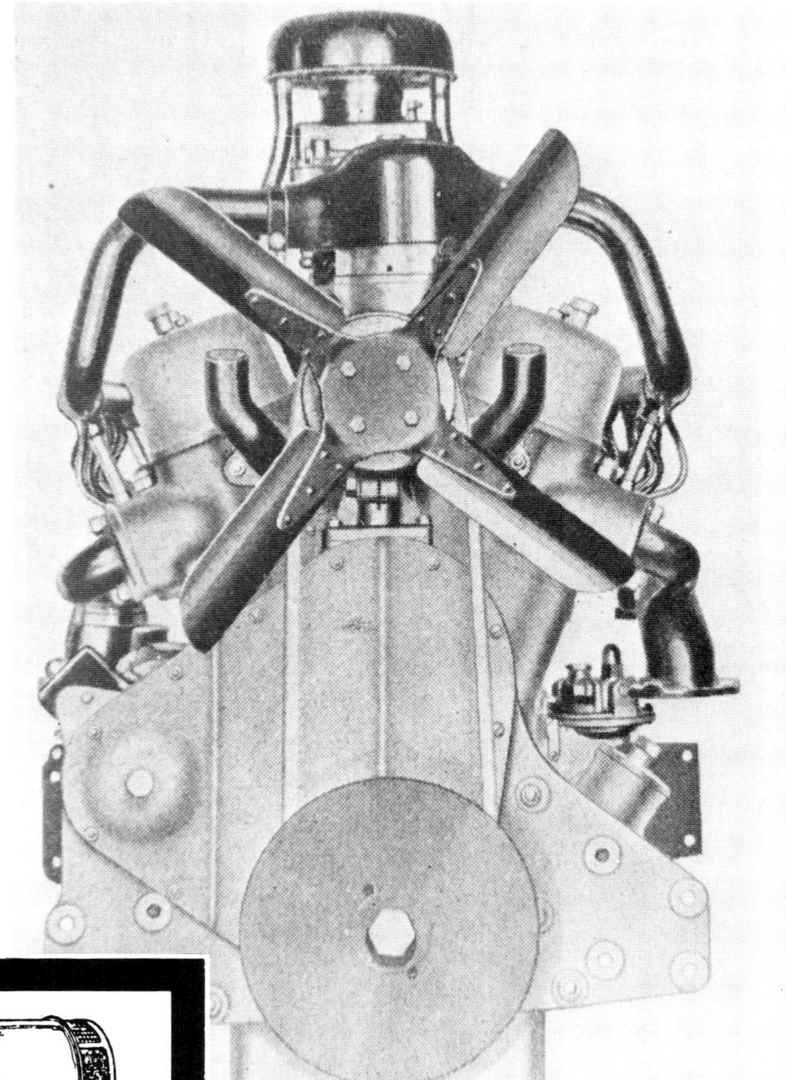

Fig. 6-25 Front view (right), transverse section (left) and detail of carburetor area of the SIXTEEN.

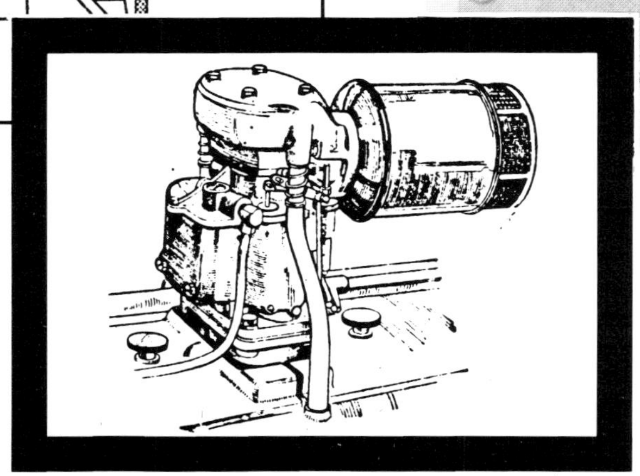

The Steel Cylinder Liners

Fig. 6-26 Details of the SIXTEEN engine construction.

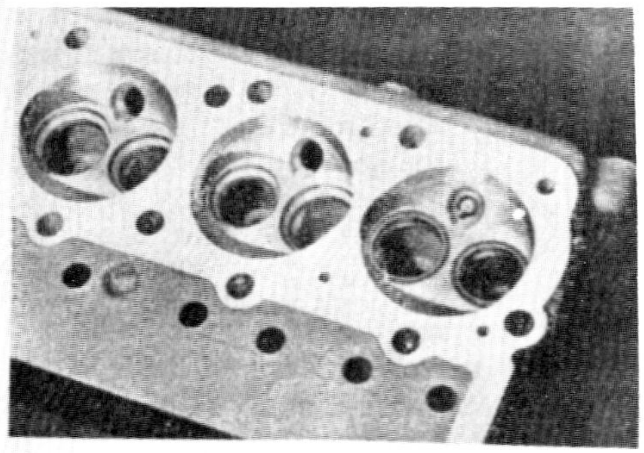

Aluminum Bronze Valve Seats

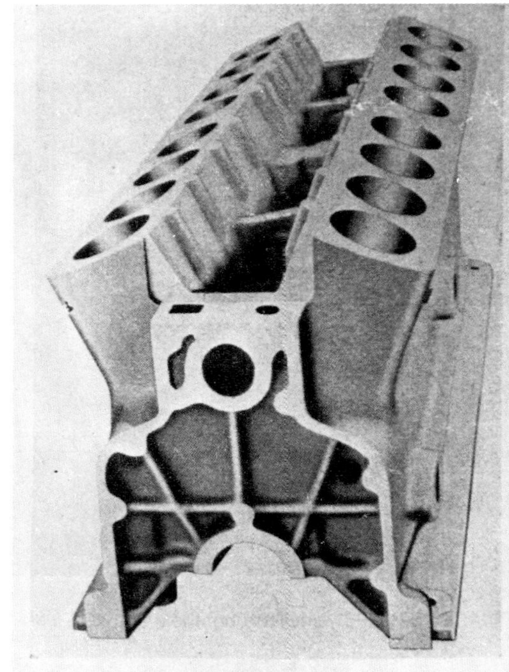

The Aluminum Cylinder Block

After both banks had been mixture adjusted, the ignition was alternately interrupted and any change in rpm noted. Correction was then made by rotation of the adjustable throttle plate to eliminate the differential rpm. The idle system ceased to function as 1000 rpm was approached.

Dual ignition coils mounted in the cowl combined with dual breaker points to provide independent circuits for each bank. An exception to this duality was the single distributor cap which, like the distributor, was larger in diameter than prior contemporary practice for a 16 cylinder unit and unique to the MARMON SIXTEEN. It was located directly behind the fan, driven by a vertical shaft from the central camshaft. Spark plug wires were run through chromium plated conduits which connected with a plated dome-shaped cover over the distributor. Ignition units were Delco-Remy.

The engine would operate "normally" with ignition to only one bank. The normal idle was so smooth that a coin could be stood on edge on the engine upper surfaces; and the same was true when running on one bank. This characteristic warranted careful monitoring of the ignition system; reasonable acceleration on eight cylinders could leave one-bank running undetected until realization resulted from the extreme rate of fuel reserve depletion.

With eight power strokes each revolution, there was not much flywheel requirement and a very light wheel was fitted in the interest of low inertia to enhance acceleration. The light flywheel combined with the general inadequacy of available 6 volt battery-starting motor systems to make the typical start a series of cranking lunges as the starter struggled against the individual cylinder resistances combined with the integrated resistance of the remaining cylinders. The throttle was usually opened wide to lower cranking resistance pending initial firing. This resulted in a breath-taking "explosion" into action as the open throttle combined with the very low inertia. The Delco-Remy starting motor extended rearward beside the transmission. The pinion was mechanically engaged before cranking.

The valve gear utilized very compact rocker arms, on conventional rocker shafts, and with mechanical adjustment. Careful selection of materials for push rods and rocker pedestals combined with rapid heating by oil flooding to provide an absolute minimum of mechanical noise. The rocker arm tips were drilled for tappet lubrication.

Inlet valves were 1.96 inches in diameter; exhaust valves were 1.375 diameter. Both had 0.309 stems. Lash was set at 0.008 hot. Valve lift was 0.64 inch.

Intake valves opened 6 degrees before top center and closed 40 degrees after bottom center, or 226 degrees later. Exhaust valves opened 40 degrees before bottom center and closed 6 degrees after top center, or 226 degrees later. Valve overlap, ie, the period after the intake valve has opened and before the exhaust valve has closed, was 12 degrees.

The one piece cylinder block was cast from 'Lynite', a copper-iron-silicon alloy of aluminum which was heat treated to a tensile strength of 30000 to 35000 pounds per square inch. A closer grained 'Lynite' alloy was used for the cylinder head castings. It contained a somewhat smaller percentage of copper and an additional small percentage of magnesium. The block and head alloys had identical coefficients of expansion. Valve seats were separate rings (inserts) of aluminum bronze having high heat resistance which were shrunk in place. Combustion chambers were fully machined for a uniform 6.0:1 ratio.

There was no cylinder bank offset because a fork and blade rod design was used.

Up-draft crankcase ventilation was featured. A current of air entered through a breather on the side of the crankcase and passed upward through the crankcase to the camshaft alley and up to the valve rocker covers and finally through vertical tubes from the covers to the bottom of the aircleaner elbow.

Full pressure lubrication was supplied to every rotating bearing including the rocker arms which were oiled through the tubular shafts on which they were mounted. The connecting rods were fork and blade type. The bearing shell clamped in the two arms of the forked rod was lubricated from the drilled hole in the crankshaft. The blade rods from the opposite cylinder bank moved on the outside of the bronze bearing shell, receiving oil through 5 holes in the shell. The large gear type oil pump was mounted on the front main bearing cap and was chain driven from the crankshaft . . . a construction used on the first generation of Detroit Diesel Series 71 engines in the 1940s. Crankcase capacity was 10 quarts of oil.

The cylinder bore liners in which the aluminum pistons operated were case-hardened and ground steel tubing, positioned axially by an integral top flange. Except for the top and bottom pilot bores, the outer surface of the liners was exposed, ie, "wet", to the coolant. Sealing at the upper end was by the cylinder head gasket, and at the lower end was by two special rubber rings, each recessed in a groove in the block. The ring cross sectional area was greater than that of the groove and assembly placed the rubber under a considerable compressive stress. Consequently during the initial running-in of the engine, the combination of mechanical and thermal stress resulted in vulcanizing the rings into a permanent seal. Between the rings was a relief intersected by a drilled hole from the exterior. Should water leak down, or lube oil up, it would be evident on the exterior of the aluminum block.

The hardened cylinder liners necessitated use of piston rings comparable to those used in aircraft engines. An ill-advised overhaul with typical automotive rings was promptly followed by another. The piston ring surfaces were left with a satin finish (visible tool marks) to expedite an otherwise prolonged break-in period. The upper compression ring was a Pedrick, 0.125 wide; the second ring was a McQuay-Norris 0.125 wide combination oil wiper and compression ring. The third was a Perfect Circle 0.187 oil control ring.

The pistons in odd and even numbered cylinders were different due to the connecting rod offsets adjacent to the main bearings. Pistons weighed 12 ounces.

Superb quality control eliminated random tappet adjustment and very high (100,000) mileages were attained before a general adjustment was indicated.

The manufacture of the cylinder liners and other manufacturing detail is reported in Chapter Eleven Part 2.

The generator and water pump were driven in tandem by the triple-roller Diamond timing chain. The generator axis was adjustable to correct for chain stretch and the error introduced was tolerated by the hose-type drive-coupling to the water pump. Stretch and compensatory adjustment was minimized by the use of the highly durable roller chain.

Engine mounts were of the rubber in tension type and were located at the four corners of the engine.

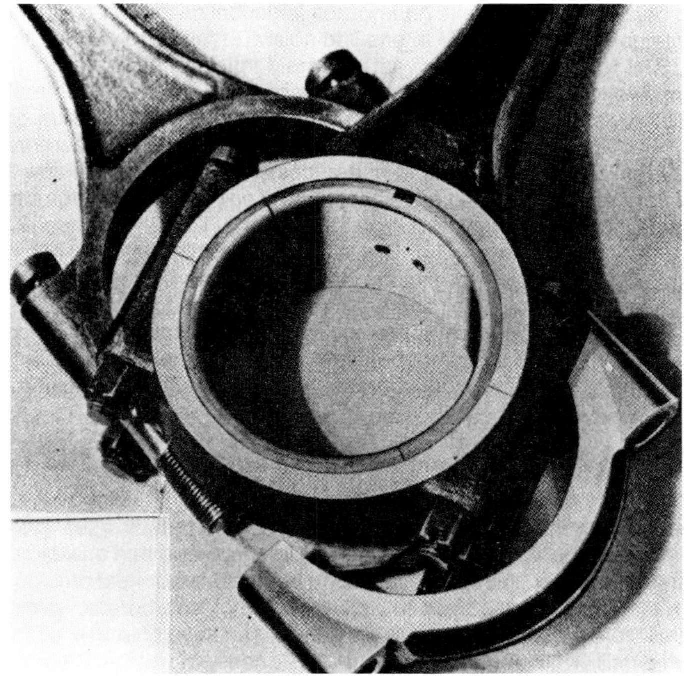

Fig. 6-27 The SIXTEEN used fork and blade connecting rods to avoid the extra length of offset banks for side by side rods.

The water pump impeller was double faced and straddle mounted on ball bearings. The water inlet passage forked to feed the dual pump which discharged equal flows to the two banks using a passage cored in the block to reach the left bank.

Dual 2.25 inch diameter exhaust systems were used, but with a connecting pipe forward of the mufflers. Each tail pipe terminated in a flare end fitting about .50 high and 6 inches wide.

The cooling system used a cellular radiator fitted with two sets of automatic shutters. The thermostat was on the rear face of the tank with a link passing through a tube in the tank to operate the shutters. (The author had several frustrating experiences with flat-head Cadillac shutters in the 1938-40 period when the 'stat' was between the core and grille). The top and upper sides of the radiator were fitted with a shroud to enhance air flow. A 20 inch 4 bladed fan on ball bearings was driven by V-belt from the crankshaft pulley. Coolant capacity was eight gallons.

## PART 4 — The Chassis Components

Because of the single car and chassis of the Fifth Era, the chassis is discussed in general terms in the overall review of the automobile and a general discussion here would be redundant.

Fig. 6-28 The SIXTEEN chassis — a transmission to match the engine in sophistication; torque reaction by torque-arm and a rubber isolated spring suspension.

## FRAME

The frame was severely dropped to take advantage of the hypoid axle, but was not dropped to the front. There were seven cross-members on the frame, three of which were tubular. One of these was under the oil pan just ahead of the dash to strengthen the frame at the point where the two spare wheels were supported. This was particularly important because there was no tie bar between the fender supports as then universally used. Like the Model 34, the frame featured reinforcing cuffs-welded into the center section giving the frame locally a stock thickness of 0.25 inches.

The side rails contributed to "an amazingly rigid frame with a comparatively narrow web . . . it is 7.5 inches deep at maximum . . . but with very wide flanges, (which) give a substantial foundation. The top flange is 3 inches wide and the bottom flange 3.25 inches."

## SUSPENSION

The SIXTEEN rode superbly on four semi-elliptic springs. They were attached to the frame with rubber shock insulators except for the kick-shackle at the rear of the left front spring, which was of usual Marmon design. The front springs were 42 inches long and 2.25 wide, the rear were 60 inches long and 2.50 wide. These were the same dimensions as used on the 8-79 and Big Eight, except for the width on the 8-79.

The author believes this was the last passenger car to be designed with the rubber insulator system which persists to this day on the domestic Mack truck line which is highly regarded worldwide.

The rear axle torque was reacted by a torque-arm paralleling the open drive shaft . . . a construction used by Cadillac in the 1920s and most recently seen in the 'fixed' Chevrolet Vega rear suspension.

The rear semi-elliptic springs were attached to the axle via saddles to which the springs were clamped. The saddles were free to rotate on the axle housing and lubrication fittings were provided. This relative motion compensated for the axle rotation about the midships anchorage of the torque arm and avoided the otherwise consequent twisting of the spring; better ride control resulted. During the First Era, a design change had been made to incorporate such a mounting.

Watson rotary double-acting hydraulic shock absorbers were fitted.

For the 1933 Model year, 'Selective Ride Control' was adopted as standard equipment. It had appeared on the 1932 8-125 and consisted of a control linkage extending from a dash lever to the four shock absorbers. Yet in 1982 Nissan world introduced a turbocharged 'Skyline' model for the Japanese home market and claim that the featured shock absorbers which could be adjusted from the driver's seat while the car was in motion was a world first . . . fifty years after the Marmon application.[31]

## WHEELS AND TIRES

Tires were 6-ply 7.00 x 18, bias construction mounted on drop-center rims. Six wire wheels, with adjustable spokes, were standard. Also metal tire covers were standard . . . a chromium plated 'tread cover' and a body-color side doughnut. Eight **bolts** were used to mount the wheels. They were concealed by snap-in hub caps. Chrome-plated split rim-sides were removed for tire replacement.

## STEERING (FRONT AXLE AND GEAR)

The front axle was a conventional reversed Elliot type procured from Salisbury. It was fitted with a hydraulic stabilizer acting between the tire rod and axle beam.

The steering gear was a Ross cam and lever design with the steering input shaft and cam on tapered-roller bearings adjacent the cam. The column was chromium plated.

An 18.5 inch slender Bakelite steering wheel was used. It had, like the First Era Marmons, three spokes.

## CLUTCH

The Russell "Tower" clutch was placed on the market in June 1930 by Russell Manufacturing Company of Middletown, Connecticut, but was manufactured by Spicer Manufacturing Company of Toledo, Ohio, with exception of the driven plate assembly.

It incorporated a number of innovations and its adoption for the Marmon Sixteen was appropriate and in the best Marmon tradition. It advanced the state of the art in both friction reduction and damping of torsional vibration . . . areas in which Marmon had pioneered. Engine acceleration was enhanced by the low inertia of the complete assembly.

A unique feature of the clutch was a lever system which, through the use of a variable ratio leverage, introduced a mechanical retardation of the engagement. This was accomplished by means of the ball and shoe end of the release levers,

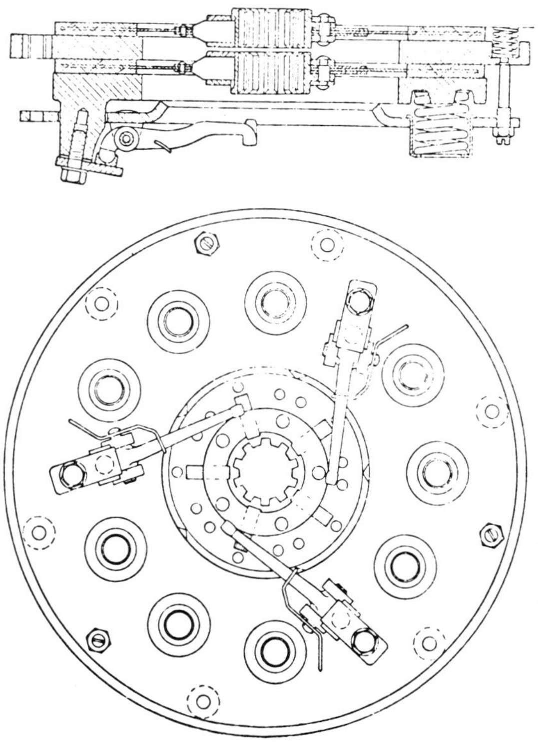

*Side view and cross-section of the Tower clutch*

Fig. 6-29 Two-plate TOWER clutch with variable ratio to facilitate shiftless driving if desired.

which moved along an inclined plane. The result was avoidance of grabbing or premature clutch engagement. This was particularly desirable on the Sixteen should the owner elect to operate exclusively in high gear, or simply wish to avoid the very low first gear ratio . . . evidently not intended for routine starts. Mr. Eitel's father did indeed use the no-shift procedure with the Eitel's Sixteen.

The outer end of each clutch lever incorporated a hardened and ground ball on which a spherical seated slider block was mounted and retained. The block was made of a self-lubricating alloy called "Sabeco" which also had a wear characteristic which compensated for clutch facing wear, thereby maintaining clutch pedal adjustment. The fulcrum pins were hollow and filled with a wicking which was saturated with oil under high pressure.

The driven disc featured five separate steel spokes which were molded into the rivetless facings. After grinding to 0.002 inches of parallelism, they were assembled into the hub/damper. Only 0.012 inches release movement was required, resulting in low pedal pressure.

The spoke/disc assembly drove the hub through five sets of multi-leaf springs; inter-leaf friction provided the damping. The design concentrated the damper mass at a very minimal radius and resulted in low inertia of the complete clutch despite the requirement for two dampers.

Facings were 9.75 O.D. and 6.12 I.D. The center pressure plate had ample clearance on the studs so that there was no danger of binding when hot; rattle was prevented by a small spring-backed ball in the plate which pressed against each of the three studs.

## TRANSMISSION

The transmission was appropriate to the SIXTEEN. It was silent in all gears, unbelievably quiet in second. Although a 3-speed and reverse unit, it was especially developed for the MARMON SIXTEEN by Muncie Gear Division of General Motors Corp. but was subsequently adopted by Stutz. John Bond of Road and Track fame recalls that his father, an engineer at Muncie, was assigned to the Marmon project. *"With cam-type synchro-mesh, it was a beautiful design with a very unusual arrangement of cone clutches for the synchronizing duties. Briefly, the high gear clutch was as normal, but the medium speed clutch was on the countershaft. This gave some space-saving for wider gears, a shorter span between bearings, weight savings, etc."*[32]

The mainshaft and countershaft were similar but reversed constructions. Each consisted of an inner and outer shaft extending nearly full length; a cageless roller-set at the nose and a straight roller bearing at the other end made the concentric shafts a rigid structure between the large double-row ball bearings which supported the assembly.[33]

Both main and countershaft assemblies included one synchronizer unit. These differed, due to one being below the lubricant level and one being above it. Shift rails were on the right side, which was closed by an access cover.

Note that **two** GM divisions were involved in the SIXTEEN development . . . Delco-Remy and Muncie. The shift lever pivot was located on the forward right corner of the transmission. Ratios of 1.57 and 3.09 gave overall reductions of 3.69, 5.79 and 11.40.

## REAR AXLE

The rear axle was a Salisbury product with a banjo housing mounting a hypoid gear set with 3.69: 1 ratio. Combined with the 7.00 x 18 tires, an N/V of 2485 rpm/mile resulted. It was semi-floating type and the second time Marmon had predicated a chassis design on a hypoid axle; the first was the Little Marmon. Axle torque reaction was by torque arm paralleling the open two-joint drive shaft.

The axle was of the three-quarter floating type. Like the Little Marmon, the outer ends of the axle shafts were carried on an opposed-pair of tapered roller bearings.

## BRAKES

Brakes were Bendix Duo-Serve two-shoe self-energizing type 16 inches in diameter and 2.5 inches wide. Marmon had always matched his brakes to the performance of his cars, the size and sophistication of the Era One brakes being noteworthy. The 355 square inches of lining acting on an 8 inch radius and actuated (1932-33 models) with the combined assistance of the self-energizing action and a Bendix vacuum power booster provided braking matched to the performance capabilities of the SIXTEEN.

The brakes were actuated by a cable system with sheathed sections providing for steering and suspension motions. The sheaths were factory lubricated and sealed.

For the 1933 Model year, there were two brake system improvements. A new type of vacuum brake booster was adopted and a new type of brake drum offering extreme rigidity was used . . . the 'Kelsifuse' drum with an iron braking surface in a ribbed steel drum.

The brakes left very little to be desired so far as vehicle deceleration and control were concerned.[23]

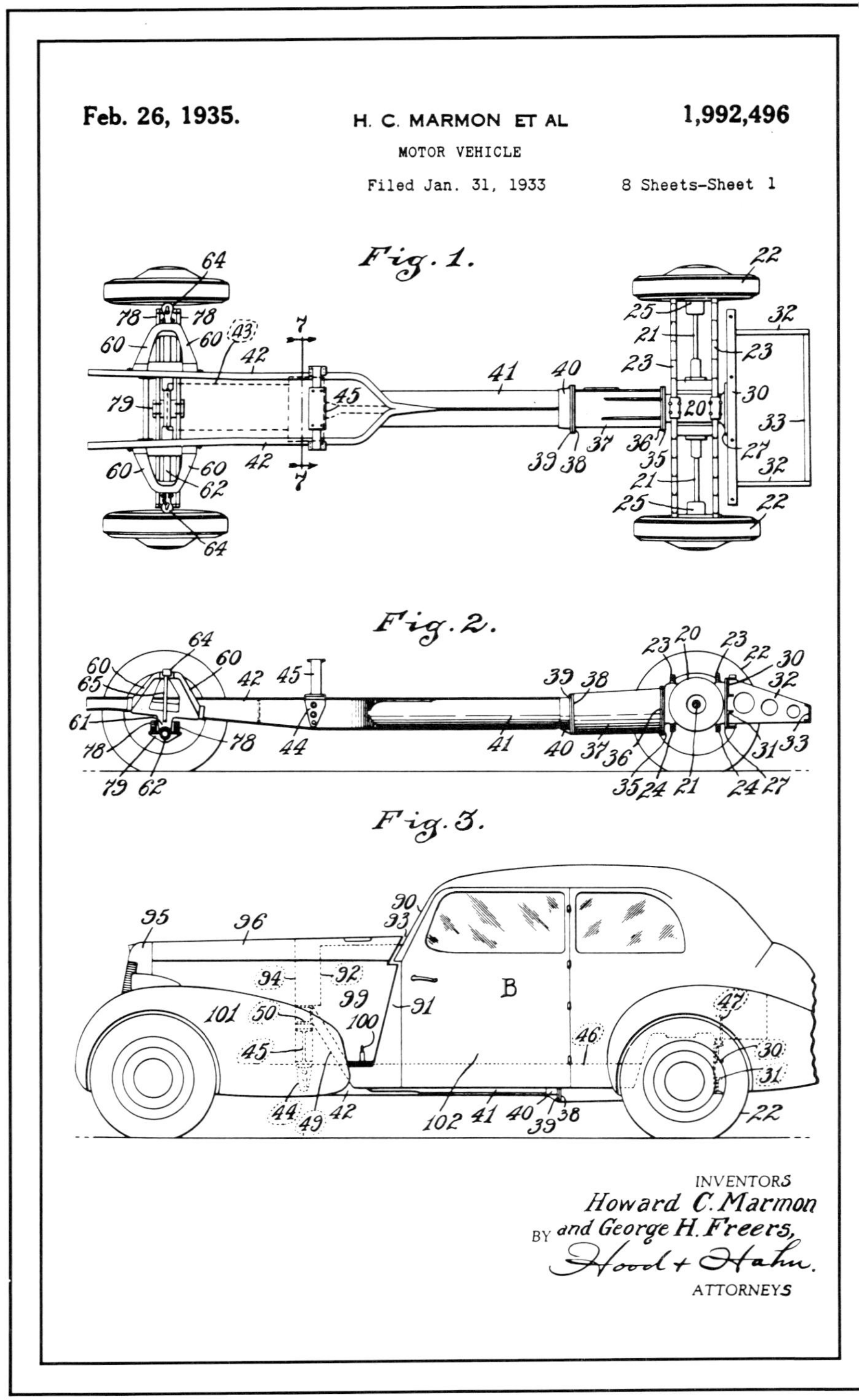

CHAPTER SEVEN

# THE HCM — AN AUTOMOTIVE MONA LISA

First it was Ranson Olds who, having sold his interest in the Company bearing a derivative of his surname, found himself again in the automobile business and in need of an appropriate name. His initials R.E.O. were pronounceable and so the REO was born; it was 1904. Next came the unpronounceable EMF, the 1909-1913 product of the firm which bore the composite initials of its founders: Byron Everitt, William Metzger and Walter Flanders. Only known pronunciations: 'ElectroMotive Force' and 'Every Morning Fixit', the latter's popularity being a marketing deterrant.

Harry C. Stutz, having disposed of his Stutz interests and still savoring the auto business, lent his initials, also unpronounceable, to the flashy H.C.S. car line beginning in 1920.

Howard C. Marmon joined that illustrious fraternity in 1932, when the Marmon Motor Car Co. found itself severely strapped financially and devoid of resources for the development of a new product line desperately needed. Strongly dedicated to what had been the family business and inheritance, Howard committed himself and his personal fortune to the creation of the H.C.M. prototype. He, of course, fully expected that the initialed hub caps were an interim matter, to be replaced with the Marmon logo when the storm had been weathered and he had been compensated for his timely action.

Despite 1929 having been Marmon's best sales year ever . . . in unit volume . . . "MARMON . . . Under the direction of G.M. Williams"[1] was in serious trouble by 1930.

In March of 1931, just as the first production SIXTEEN was being assembled, Walter Marmon had overseen the commitment of Marmon Motor Car Co. to a stock subscription in the original incorporation of a new Marmon subsidiary firm . . . The Marmon-Herrington Co., which would manufacture four wheel drive trucks in the Marmon plant. Only months later, the infant firm would have to be refinanced and reorganized as an independent firm consequent to the failure of Marmon Motor Car Co. to honor its stock subscription commitment and the subsequent resignation of the Marmon officers and directors as the sole officers of the new firm other than Herrington and Moskovics, the promoters.

Joe Felts recalls:[2] *"With the stock market crash of 1929, a general gloom set in. Marmon Motor Car Co. had its problems as so many other car companies did. We had already had two ten percent salary cuts when in the spring of 1931, George Freers had to brace himself and tell all engineering personnal except the supervisors and myself that they were laid off. The supervisors all went back to work on the drawing board and I went to work on a special assignment which turned out to be the last dream car of Howard C. Marmon. With the emotional goodbyes over, the lucky personnel went back to work wondering what would happen next. George Freers came to my drawing board with an exceptionally detailed, quite accurate sketch made by Howard Marmon at the Columbia Club where he stayed during Indianapolis visits from his Pineola, N.C. estate. It was on an 8 1/2 x 11 piece of paper. It could be termed a radical or advanced design of a car . . . My instructions from George Freers were to make a preliminary design layout of the overall chassis and then break it down into engine, transmission . . . axle, front and rear suspension, frame, etc."*

On August 10, 1931 Joe Felts was transferred to Howard Marmon's personal payroll. Subsequently "a decision was made to build one prototype vehicle and several draftsmen were hired and Walter Dorwin Teague (and son) was retained as styling consultant." George McMath, an engineer, was also transferred.

Mr. Marmon's faith in Marmon Motor Car Co. was such that he was confident it would weather the storm and that he would eventually recover his investment which would then provide the critically needed new product ready for production release.

The exact personal cost to Howard Marmon for the HCM project is not documented. A figure of $50,000 has been quoted by some historians. Gold increased in value about twenty times in the fifty years since 1934 . . . from $20 to $400 per ounce; using that measure of inflation, the $50,000 would be $1 million today.

Only by gross contraction of the active office and production facilities, exclusively to the Plant 2 site south of Kentucky Avenue, was survival accomplished into 1932. Also during 1931 there was a disastrous contraction of the dealer organization which seriously imperiled ability to market the limited 1932 offerings. Even the factory managed Marmon-Detroit Company closed in 1931. And following the introductory ads, there was negligable media promotion to assist the struggling dealers, the majority of whom were entrepreneurs who had undertaken the Marmon franchise in the prior three or four years and were ill-prepared for any siege, however short.

Just how desperate things had become in Indianapolis is evidenced by correspondence between a consumer and the Marmon factory[3] during the month following Felt's new assignment.

Mr. A.L. Dyke of St. Louis (Dyke's Encyclopedia) wrote to the Marmon offices on September 26, 1931 requesting a copy of a technical book on the SIXTEEN; he was promptly advised by a Mr. H.A. Fletcher that there was none available. Again he wrote in June 1932 seeking a manual for the 8-125. In response, a Mr. H.O. Osborn sent a set of ten component sub-manuals on the SIXTEEN (typewritten, not typeset) and an 8-88 manual to use for the 8-125.

Later, on April 4, 1933 Joe Felts and the several other designers working on Howard Marmon's payroll moved downtown to the Old Trails Building to avoid possible interference from the receivers appointed for the Marmon Motor Car Co. But they soon found it feasible to move back into the plant.

## THE HCM SPECIAL

The sketch which Howard Marmon had made and passed down to Joe Felts envisioned an advanced motor car powered by a V-12 engine and having a wheelbase of 134 inches. Tread was eventually set at 58.125 front and 60 inches rear. It would be the first American car with, not only independent front suspension, but with all wheels independently sprung. It would also be the first American car with

Fig. 7-1 The HCM — Last of the Marmon cars.

a tubular backbone chassis. It would be a car in the best tradition of Marmon performance and handling and would sustain the old Marmon slogan "Easiest Riding Car in the World." The latter would result from Howard Marmons expertise in ride engineering combined with a remarkably low ratio of car weight to unsprung weight.

Sixteen years into the twentieth century, Marmon had made a quantum advance in the engineering and styling of the motor car with the Model 34 introduction; seventeen years later the HCM was a milestone of comparable magnitude in every way.

The chassis design and associated body mounting concept was patented. US Patent 1,992,496 for a "Motor Vehicle" was issued Feb. 26, 1935 to George Freers and Howard Marmon with assignment by Freers, of his half interest, to his employer Howard Marmon. Application had been filed January 31, 1933.

In conceiving the HCM, Howard Marmon almost certainly had in mind his first car, for in some respects the HCM utilized the same principles. The first car had 'double three-point suspension' . . . a propulsion frame separate from the frame supporting the passenger body; an objective of the HCM design was again separation of the body from the propulsion chassis but this time by a rubber-isolated three point body mounting.

It is interesting that the language of the Marmon ads announcing entry into the automobile business in 1904 was repeated, perhaps inadvertently but certainly without reference, in the sales brochure prepared by Marmon and Freers in seeking a buyer for the HCM design following the bankruptcy: "this new type of twelve cylinder motor car was designed, carried through a long period of experimentation (i.e., developed) **and finally perfected."** It is that last step that seems to have been omitted from the non-Marmon and contemporary engineering programs. There is much talk about 'R & D' but who ever hears anything about 'R, D and P'!

When it became apparent that none of the reorganization attempts were going to result in resumption of manufacture by the Marmon Motor Car Co., Howard Marmon undertook the recovery of his reported $50,000 (1933 dollars) investment in the HCM Special by seeking an 'outside' purchaser of the design. A typewritten promotional brochure, and drawings illustrating the application of the HCM concept to a high volume light car with a V-8 engine, were prepared. Freers, often accompanied by Marmon, traveled from one automobile manufacturer to another, driving and demonstrating the HCM. The author met in 1983 a retired engineer who recalled that the only Marmon he ever rode in was on a manufacturer test track . . . a 113 mile per hour demonstration of the HCM.

The trip was a great disappointment, not only because it failed to identify any prospective buyer for the design but particularly because of the manner in which Howard Marmon was reportedly received by executives of some of the other films . . . much like some inventor from outside the industry with an ill-advised brain-child.

Fig. 7-2 The sleek high-technology V-12 prototype chassis.

Fig. 7-3 HCM chassis as originally built with transverse leaf springs.

Having covered the industry circuit, the car was driven to Howard's estate at Pineola, Avery County, North Carolina and permanently garaged there.

Upon Mr. Marmon's death, April 6, 1943, James Melton . . . the opera star and pioneer auto collector . . . sought to purchase the HCM but was refused by Mrs. Marmon. Instead she followed her late husband's wish and presented the car to Fred Moskovics who at that time was with the New York office of the A.O. Smith Co., a Milwaukee, Wisconsin manufacturer of automobile frames. The gift was acknowledged in a letter from Moskovics to George Freers at Marmon-Herrington dated August 11, 1950, in which Moskovics reported he had given the car to his friend, John Floyd, who was Executive Vice-President of A.O. Smith and located in Milwaukee; Floyd had a son Alan who could assist in restoration of the car. In the mid 1950s the car was placed on loan to the Brook Stevens Museum in Mequon, Wisconsin with a signboard attesting to its ownership and loaner-status. Alan inherited the HCM from his father and circa 1970 presented it as a gift to the Brooks Stevens Museum. The car is presently insured for a magnificent sum; after all there is only one HCM and there are lots of Duesenbergs!

It is a sad but most appropriate final memorial to the "Foremost Fine Car" and the engineering genius responsible.

A detailed description of the car follows: Chassis, Body, Engine, Performance, and Styling are addressed in detail in that order.

The loneliest and eventually most bitter man in the world is he who has built a better mousetrap and subsequently finds not customers but creditors beating the proverbial path to his door. And if the harried and homeless creator looks for solace in the thought that his name will go down in history he might well study the case of a man who built a succession of eminently worthy mousetraps and now rates not a line in Webster's Biographical Dictionary . . . Howard Marmon.

With those words, Eugene Jaderquist and historian Griffith Borgeson (La Tour d'Aigues, France) began their article 'not enough people bought Marmons';[4] The year was 1954. Since then matters have gotten worse, not better.

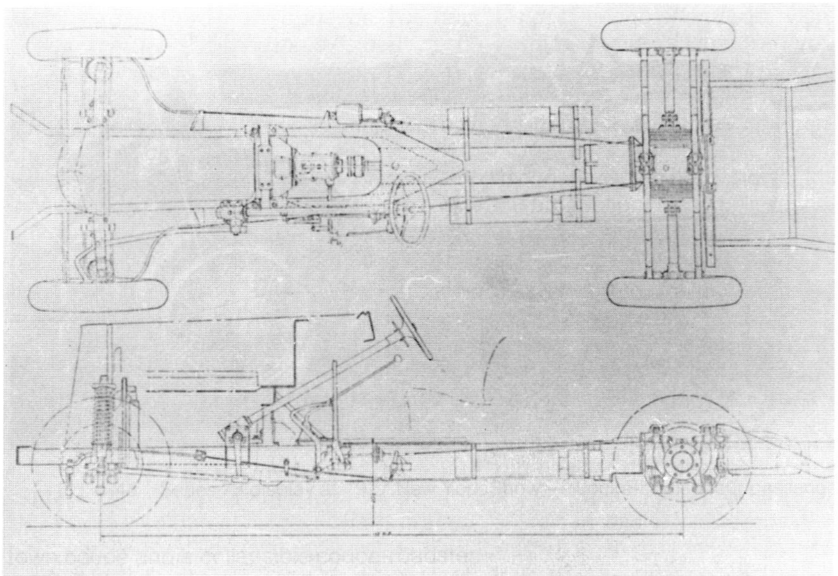

Fig. 7-4 The HCM chassis in final form.

In 1939, what is now The Automotive Hall of Fame, Inc. was established by the Automotive Old Timers, Inc. A magnificent building was erected at Midland, Michigan where (beginning in 1969) the greats of the automobile manufacturing and sales industry are enshrined, with a small uniform-size display recording and illustrating their accomplishments and contributions. By 1983 no less than sixty-seven men had been nominated and elected to recognition in the Hall of Fame . . . but not Howard (or Walter) Marmon!

Yet in 1981 when Professor David E. Lewis[5] polled (unofficially) the membership of the Society of Automotive Historians as to the people who had most influenced the industry from its inception to that date, Howard Marmon ranked thirty-first.

The name of Howard Marmon was placed in nomination by the author and by Joe Felts in 1980 and 1981 but he is evidently so little known to the present electorate that ultimate recognition by the Hall of Fame is unlikely.

## CHASSIS

Howard Marmon's sketch combined the tubular backbone chassis concept with the sliding pillar form of independent front suspension. In Europe, the Czechoslovakian Tatra and the Austrian Austro-Daimler featured tubular backbone chassis much discussed in American automotive publications.

The slipper type suspension is one in which the steering spindle wheel spindle assembly is free to move vertically along a steel shaft rigidly supported at the extremeties by outriggers of the frame structure; a coil spring surrounding the shaft supports the vehicle load through the upper outrigger. This concept was pioneered by Lancia in 1921, was later adopted by Morgan and by Nash for their very successful Nash 600 introduced in 1941 and built through 1948. Had the author ridden only in a Morgan and not in a Nash 600, he would have thought Marmon foolish to have embraced the concept. Marmon certainly had driven a Lancia at some time earlier. Frank Robinson, Jr.[6] suspects that Howard Marmon may have been present at a December 1930 meeting of the Indiana Section of the Society of Automotive Engineers at which the SAE Journal reported the showing of a film on a six-wheeled Tatra car. Regardless, right from the turn of the century, Marmon had monitored the progress of the European industry. Joe Felts did not recall the Tatra ever being discussed in his presence.

The Howard Marmon frame concept involved a semi-conventional forward frame. The parallel rails were only 15 inches apart . . . the minimum that would allow for the engine . . . and therefore were splayed outward, forward of the radiator location, for bumper attachment. A favorable turning circle diameter was gained thereby. The rails were a box section construction of lighter than the usual gage of frame steel. Behind the engine, the rails converged, following the constraints of clutch and transmission housings to be welded to the forward end of an 8.5 inch diameter steel tube through which the drive shaft passed.

At its rear extremity, the tube was flanged and bolted to the transaxle in the initial prototype configuration and to the differential in the final prototype.

An arch shaped cross-member forging spanned the flywheel housing to provide the single forward body mount. Brackets on the outboard side of the left rail mounted the steering gear and the clutch/brake pedal cluster.

A separate frame assembly was bolted to the rear of the differential. It was primarily a transverse structure providing the two rear body mounting pads and having a pair of widely-spaced tapering-height rails extending rearward for bumper attachment and fuel tank support.

The front suspension of the chassis layout which developed on Joe Felts' drafting table used a side-by-side pair of transverse leaf springs, one ahead and one behind the transverse frame cross-member which supported the left and right hand outriggers. The narrow (15 inch) forward frame greatly facilitated accomplishment of a rigid triangular structure, for suspension support, which was comprised of the upper and lower outrigger arms and the vertical shaft on which the steering spindle moved in response to road irregularities (vertical road force inputs).

It was immediately evident, on the first boulevard test ride of the prototype chassis, that the transverse leaf springs, despite the shackled attachment, imposed unacceptable side forces on the slider mechanism, essentially disabling spring action on surfaces with minor irregularities. Subsequent to Felt's departure, Chester Blair accomplished the design work necessary to modify the prototype for substitution of a coil spring surrounding, and eccentric with, the guide shaft; side forces and ride harshness were eliminated. No picture of the suspension with coil springs is available; however the concept is illustrated by the design proposed for a V-8 light car.

The shock absorbers appear to have been furnished by Delco Division of General Motors and appear to be the Delco shocks which served as the upper control arms in the parallel-arm independent suspension introduced on most GM cars in 1934. On the HCM, the shock absorbers were mounted on the upper support arm.

In the HCM prototype, the upper outrigger was a channel-section stamping fitted with a forged steel spring seat and guide shaft anchor. The lower was a transverse steel tube carried on saddles welded to downward extensions of the upper outriggers below the side rail attachments. The tube was bent downward at the outer ends to increase central road clearance and was fitted with forged ends to receive the vertical guide shafts. A 'production' design layout showed lower outrig-

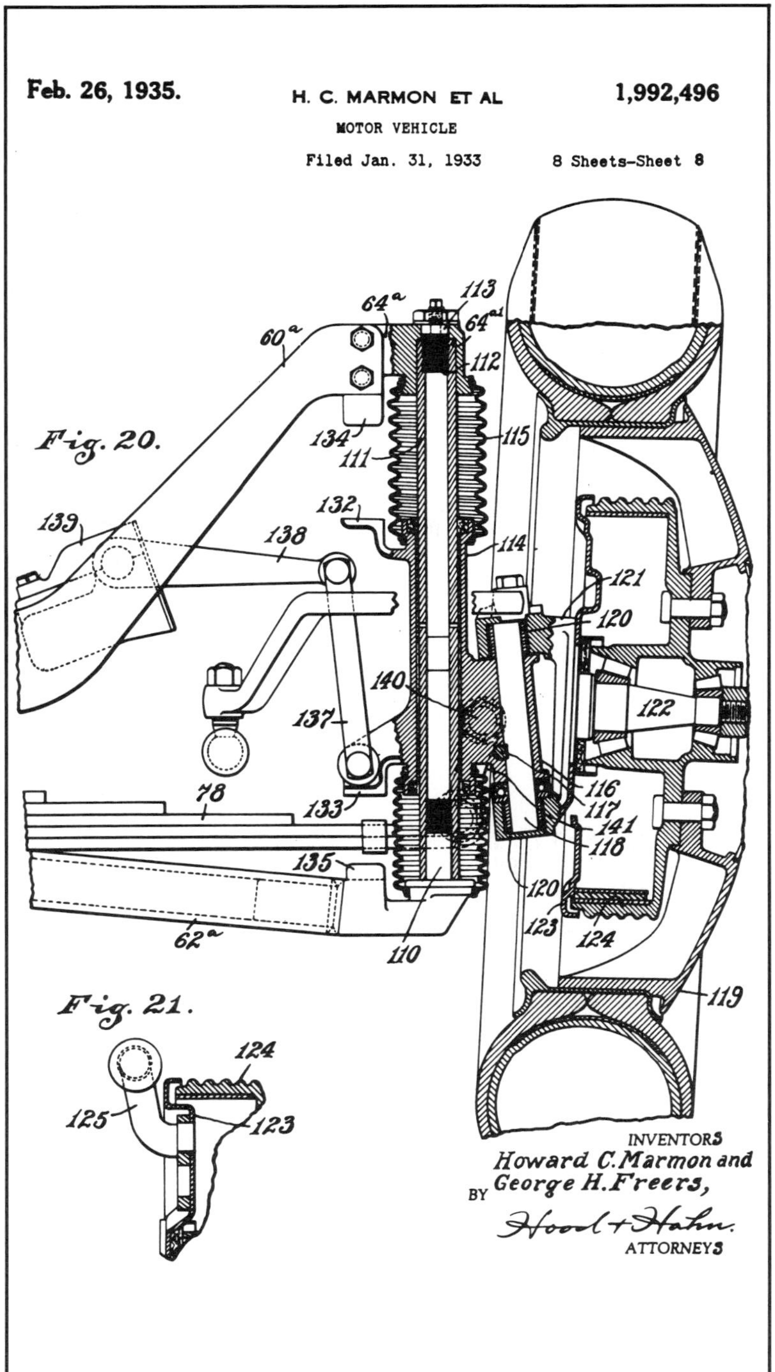


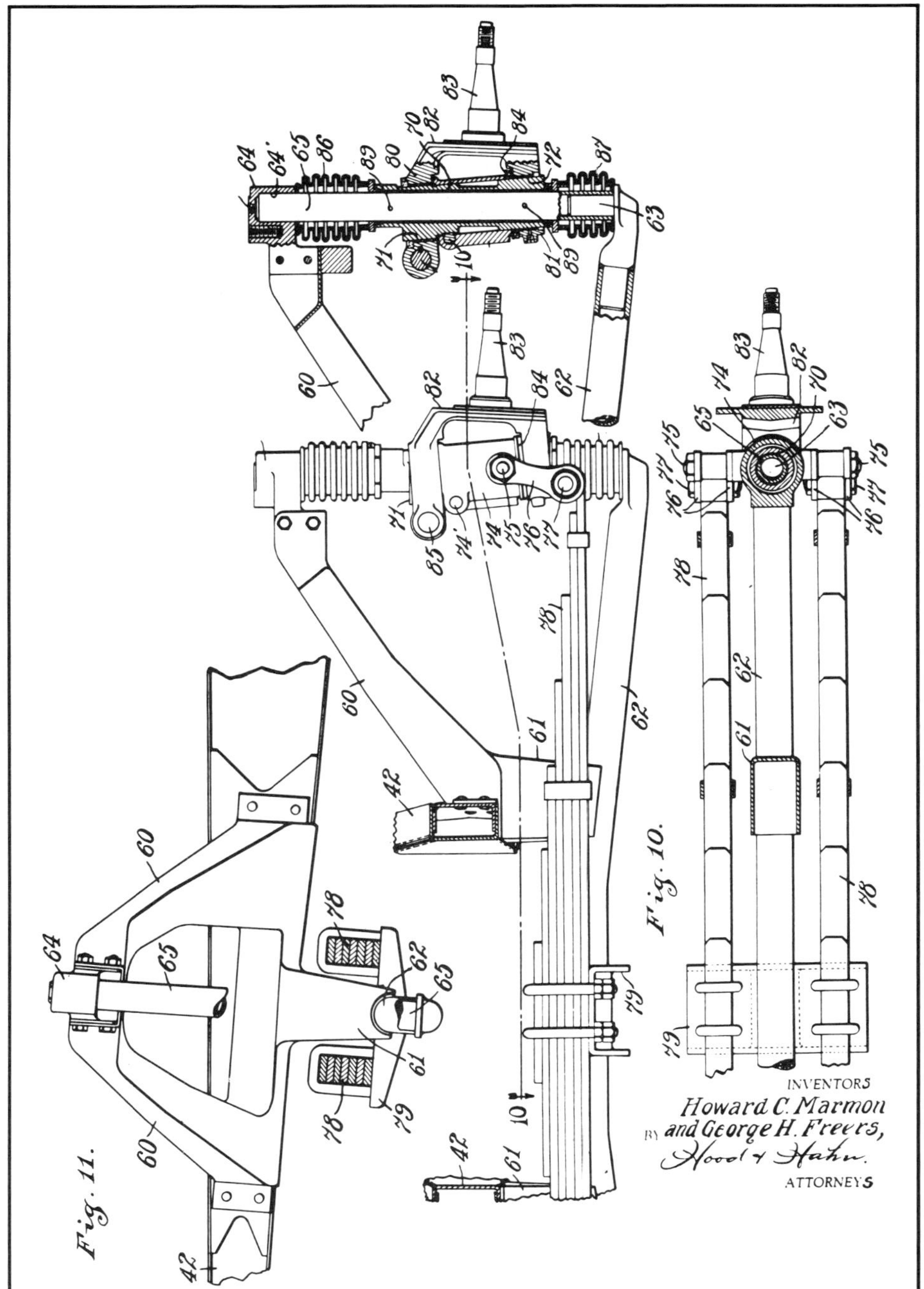


Fig. 7-5 The H.C. Marmon/G.H. Freers patent front suspension.

Fig. 7-6 Original transverse leaf spring suspension of the HCM. Note the fuel pump location on the V-12 engine.

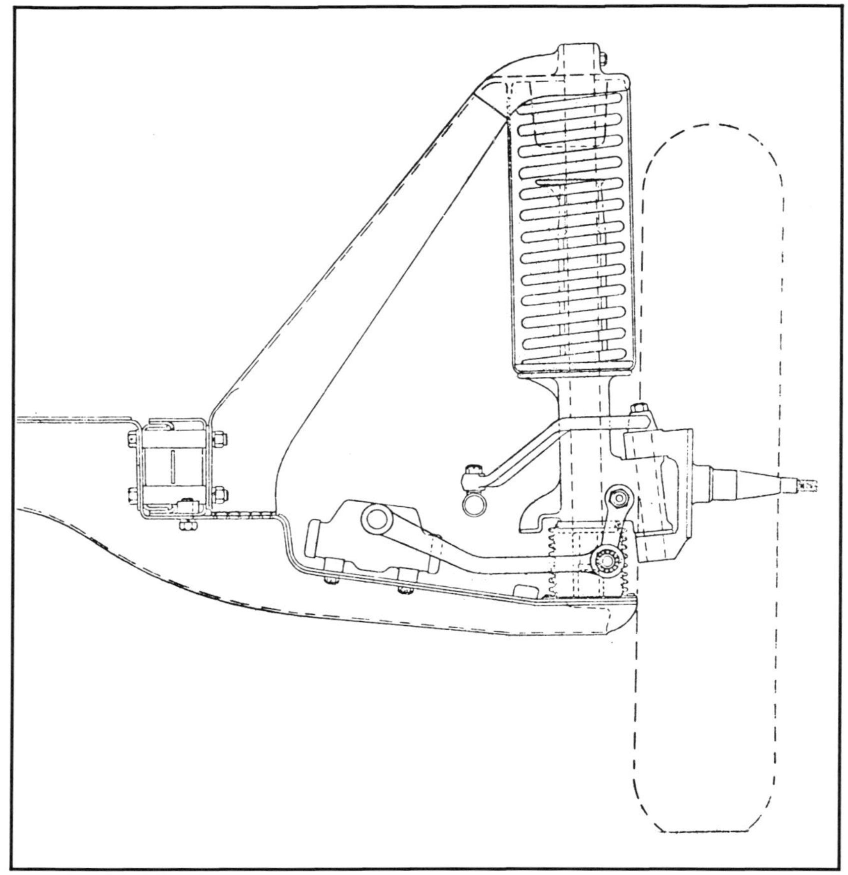

Fig. 7-7 Proposal for a high volume application of the HCM suspension concept.

Fig. 7-8 Rear suspension; springs acted as radius rods.

gers formed from a single transverse stamping which incorporated the front engine mount. It was an inverted 'hat' section member closed to a box section by welding a flat sheet to the flanges.

The lower arm mounted, on its upper surface, a double acting hydraulic shock absorber with parallel arms extending from both ends of the 'shock' cross-shaft. These arms were shackled to the steering spindle support forging and thereby restrained it from any steering motions about the vertical guide shaft. That forging extended upward inside the spring to give a wide 'purchase' on the guide shaft . . . about 16 inches over the bushings. Bellows, with the upper one enclosing the spring, sealed the mechanism, just as later used on the Nash 600.

The patent covered both designs in which the **suspension guide post functioned as the steering pivot** and in which it did not. The HCM used the latter design, as did the 'production design' for the V-8 proposal. The former concept is illustrated in the Patent Figure 20 which also shows the wheel/tire/brake construction. Note in Figure 21 the attachment of the tie rod to a brake structural member, as had been done on the 1924-1926 Marmon front brakes.

The central tube of the chassis was flange-attached to the differential housing. Rear suspension was by four transverse leaf springs, clamped to the upper and lower surfaces of the differential by a pair of clamps quite close to each other. Power was transmitted to the rear wheels by axle shafts having Cardon-type U-joints at their extremities. The Cottin et Desgouttes had adopted a rear suspension somewhat similar to the HCM in November 1931.[7]

Bendix two-shoe Duo-Servo brakes were used. Drums were 12 inches in diameter and 2.25 inches wide versus 16 x 2.5 on the SIXTEEN. Like the inboard Bendix brake application on the front drive 'Marmon 68 Specials' that started in the 1929 Indy 500, the HCM rear brakes were inboard on the differential housing. The objective of course was to minimize unsprung weight and for that same reason the wheels were cast in a magnesium alloy and had a smooth and flush exterior surface. They anticipated the wheels found on the luxury models of the mid 1980s. The design was accomplished by reversing the rim section . . . by placing the tire retaining ring on the inside of the wheel rim.

Initially the prototype car was assembled with a new transmission in a transaxle configuration. It included an overdrive ratio. Problems with gear noise and the shift control linkage led to substitution of the fine Muncey SIXTEEN transmission and extension of the backbone tube to the differential; There were neither funds nor manpower for transmission development.

The rear springs were clamped to the transaxle for only a very short distance. This made the majority of spring length on active spring, although the ineffective length was unsprung and provided superior roll characteristics reminiscent of the Model 34 rear suspension. The springs wre 53.375 long and 1.5 inches wide, with five leaves each and a deflection rate of 85 pounds per inch.

Marmon and Freers described the rear suspension: *"Because of the parallel action of all four springs, the wheels move in a vertical path, the tread changing only slightly. The vertical and parallel motion of the wheels, requiring less wheel house clearance permits the use of a wider rear seat cushion. The arrangement of the rear springs and their geometrical action is such that they effectively prevent any tendency toward body roll . . . even when turning a corner at the highest possible speed. Thus any form of stabilizer, with its attendant ill effects on riding qualities, is absolutely unnecessary."*

They continued; describing the overall ride and handling: *"This new car, on any road and at any safe speed, is a revelation in riding comfort and smoothness, steering qualities, and mental ease of driver and passengers. Its performance . . . from any standpoint . . . is both different and outstanding."*[8]

George Freers was Chief Engineer at Marmon-Herrington in 1952 when Roger Huntington, author of "Yesterday's LaSabre" (the HCM) in *Speed Age* (Oct. 1952) wrote to him and quoted him therein: *"Due to the front end construction and the friction introduced (when cornering) in the slipper-type of (suspension) spindle, we were able to drive this car around the Indianapolis Speedway on the* ***inside*** *of the track at 95 miles per hour without experiencing any roll. The riding qualities were far ahead of anything built at that time."*

Mr. Huntington's comparison was with the 'LaSabre' General Motors Dream Car, for which the Buick line is now named, and which is now in the collection of the Sloan Museum in Flint, Michigan.

Fig. 7-9 The composite structure of the HCM body.

## BODY

It would be presumptious for the author to reiterate when the descriptive text of the designers is available. They describe the body and its mounting: *"The body was constructed and mounted in such a manner that neither body nor chassis depends on the other for rigidity. It may be said that the body and the chassis are, in reality, two distinct units of the car." "The body contacts the chassis at three points only. One point is centrally at the dash and the other two are under the rear seat. At each the body rests on blocks of live rubber 1.25 inches thick." "Two pressed steel wheel housings form the body rear supports (being joined with a transverse channel atop the rubber mounts). Two aluminum rails extending forward from the wheel housings form the body sills and (curving upward) the floor board supports. These (two) rails are joined at the front of the body (above the toeboard) by a rigid, truss-like channel construction which forms the body front mount and serves to stiffen the entire body structure against deformation."* The roof rails curved downward as windshield pillars and were extended to also be joined to the referenced transverse channel member.

The body design was predicated on the absence of running boards . . . a radical departure . . . and benefitted accordingly in being wider than traditional design. However, 'boards' were subsequently added prior to completion.

The remainder of the body structure was of Douglas Spruce wood with aluminum skin.

The body was built from full-size drafts in the Marmon plant, and the car was assembled under the direction of Bert Updike, Marmon Development Engineer. However, even the body materials were furnished by Howard Marmon.[2]

In addition to the more obvious innovations found in the HCM, there are some subtle but highly significant ideas. These latter come to attention only when the description of the car in US Patent 1,992,496 is analyzed.

The extension of the hood to completely hide the cowl . . . the top panel extends to the windshield, the side panels to the front doors . . . is one of the immediately evident design features.

Not evident is the similarity of the hood to body relationship to that widely adopted for heavy duty trucks in the 1970s. The HCM body was mounted to the chassis on three rubber mounts thereby isolating it from chassis deflections; truck cabs are now similarly mounted. The HCM hood was supported off the chassis and was independent of the cowl over which it extended; the unit fiberglass truck hoods . . . the White Road Boss II of the 1970s for example . . . are also supported off the chassis and allow clearance over the cowl to accommodate movement of the cab on its own suspension.

The HCM hood upper panel was supported by an integral bulkhead behind the engine which extended down to the chassis arch which carried the body front mount. In Patent Figure 15, it is Part 94 attached by horizontal bolts to the arch, Part 45. Figure 16 sections the hood with the cowl in the background, whereas Figure 17 sections both hood and cowl and illustrates the gap between them.

## ENGINE

The HCM was powered by a V-12 version of the MARMON SIXTEEN engine. Both Cadillac and Marmon 16 cylinder engine were designed and tooled for production with a 45° angle between the cylinders. That angle providing even firing — once every 45°, 16 firings in 720° or two revolutions just as a 90° V-eight fires evenly — once every 90°. A V-12 properly would have a 60° bank angle but Cadillac had been successful in designing a 12 cylinder segment of their 16 for production on the same tooling. Howard Marmon did the same for the HCM; for the prototype block, a SIXTEEN casting was twice sawcut, removing the central four cylinders, and then welded back together and machined. The two rear legs of the intake manifold were cut off and the opening closed by welding. Cylinder heads were similarly shortened, then machined. A new crankshaft and camshaft, were of course required, as was a 12-terminal Delco-Remy distributor. Rocker covers, oil pan and exhaust manifolds were all cut and welded. The task was the handiwork of craftsman Charles Collier who later worked at Marmon-Herrington.[6] A 180° elbow relocated the SIXTEEN AC air cleaner beside the carburetor because of the shortened engine length.

Dimensions of the HCM TWELVE were; 3.125 bore and 4.00 stroke, a displacement of 368 cubic inches and an SAE power rating of 46.8. The compression ratio was 5.75:1 and the engine weighed 850 pounds versus 930 for the SIXTEEN engine.

The engine was first run on May 18, 1932 about one year after the start of the program. The engine was still 'stiff' and the performance with the Stromberg DDR-l carburetor carried over from the SIXTEEN was questionable. A Stromberg EE-3 carburetor was procured in several variations and full-power testing begun on June 21, 1932. With the EE-3 NE (without economizer) carburetor, more than satisfactory performance was attained.

## PERFORMANCE

On July 8, 1932 performance tests were run on the Speedway track for the purpose of establishing ignition and carburetion calibration. A Stromberg engineer, John P. Simons, participated and jointly signed the test report with Bert Updike.

The test car . . . serial W 501 following typical Marmon practice of prefixing 500 numbers to the first unit . . . was fitted with 6:00 — 19 tires (7:00 — 18 on SIXTEEN) and a 4.09:1 (45-11) axle ratio (3.69 on SIXTEEN). Together there would be a net increase in the N/V (engine revs/mile per hour) ratio of 11 percent relative the SIXTEEN. This represents a trade-off of top speed in favor of acceleration. It helps to understand Brook Steven's remark when he first rode in the HCM in the early 1950s:[6] *"This thing just took off like mad. I can remember thinking . . . My God, I never felt anything like this!"*

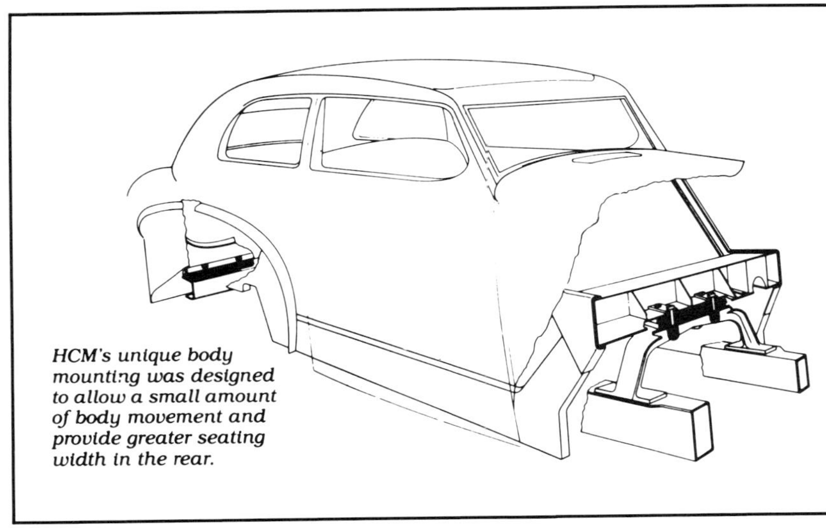


Fig. 7-10 The hardwood/aluminum composite body was mounted to the chassis on one forward and two aft rubber biscuits.

A series of North-South run pairs were made to nullify wind effects. Speedomter readings were corrected based on a calibration curve. Three two-way runs were averaged to arrive at the following times:

| MPH | Seconds |
|---|---|
| 5 to 25 | 5.95 |
| 10 to 30 | 6.33 |
| 10 to 50 | 12.77 |

Maximum speeds of 91.83 miles per hour northbound and 81.65 mph southbound (on Harding Street) were recorded.

Constant speed fuel economics were determined:

| MPH | MPG |
|---|---|
| 20 | 15.75 |
| 30 | 14.40 |
| 40 | 13.95 |
| 50 | 13.50 |
| 60 | 11.47 |
| 70 | 9.0 |

MOTOR VEHICLE

Filed Jan. 31, 1933 8 Sheets-Sheet 6

Fig. 16.

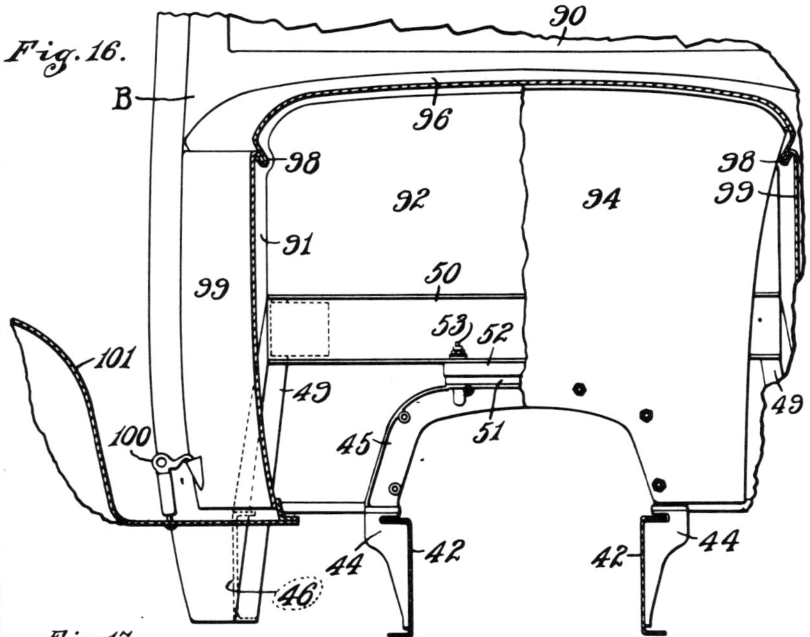


Fig. 17.

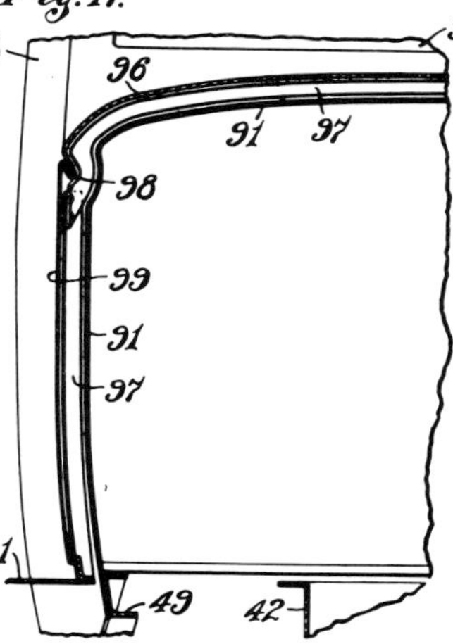


Fig. 18.

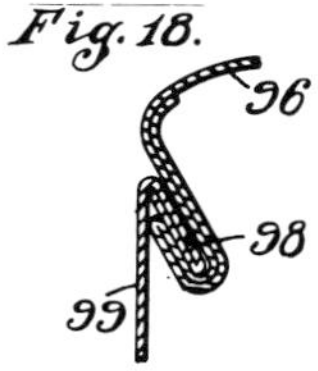


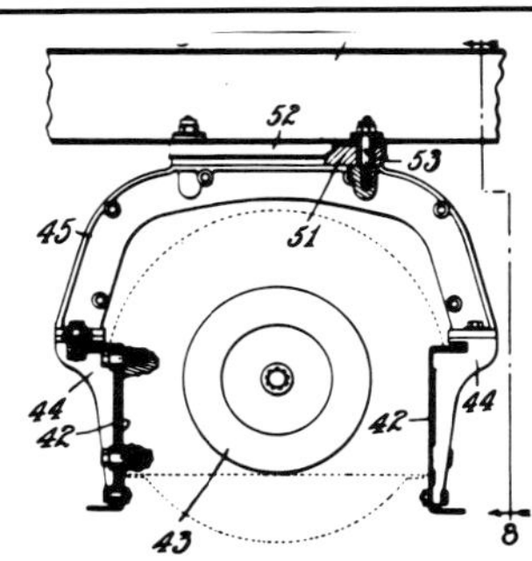


**Feb. 26, 1935.** H. C. MARMON ET AL **1,992,496**

MOTOR VEHICLE

Filed Jan. 31, 1933 8 Sheets-Sheet 5

Fig. 14.

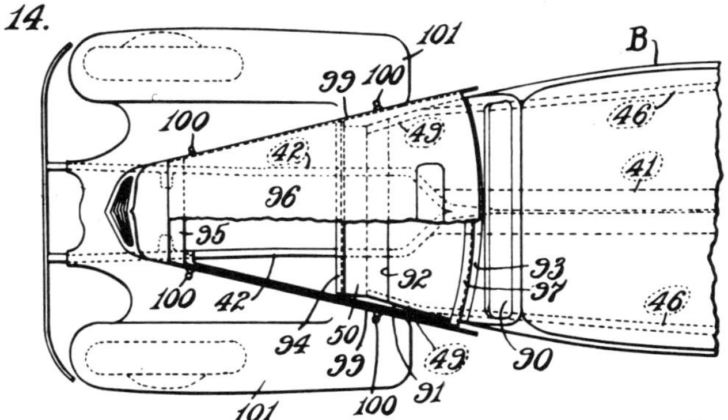


Fig. 15.

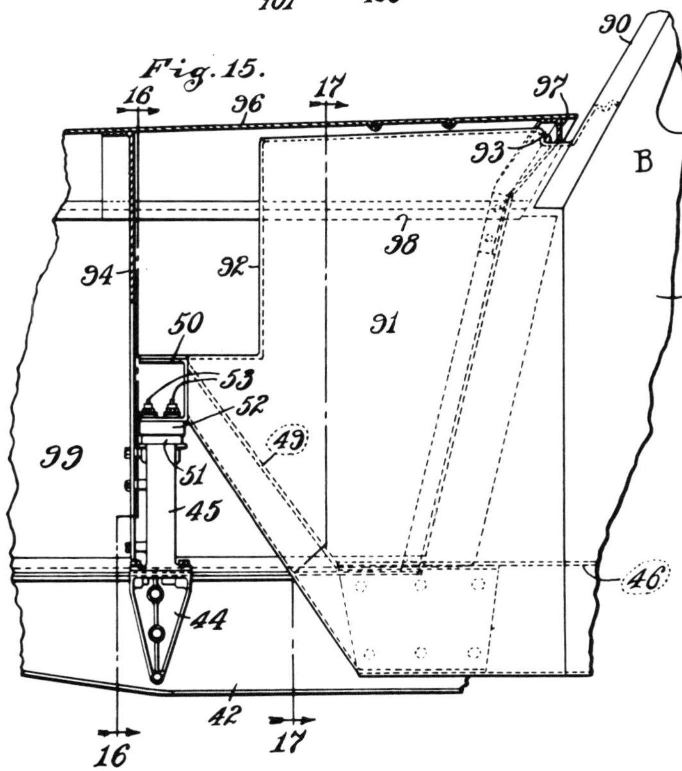

## 1933 Marmon HCM Special

**ENGINE**

| | |
|---|---|
| Type | 45 degree V-12 aluminum block in unit with crankcase, steel cylinders, four main bearings, overhead valves, pushrod. |
| Bore & stroke | 3.125 x 4.00. |
| Displacement | 368 cu. in. |
| Max. bhp @ rpm | 151 @ 3700. |
| Max. torque @ rpm | 263 @ 1200. |
| Compression ratio | 5.75:1. |
| Induction system | Dual downdraft carburetor, Stromberg EE. |
| Exhaust system | Dual, separate for each bank of cylinders. |
| Electrical system | NA. |

**CLUTCH**

| | |
|---|---|
| Type | Double dry plate |
| Disc diameter | 10 in. |
| Actuation | Mechanical by foot pedal. |

**TRANSMISSION**

| | |
|---|---|
| Type | Three speed manual, synchromesh on two higher ratios. |

**DIFFERENTIAL**

| | |
|---|---|
| Type | NA. |
| Ratio | 3.78:1. |
| Drive axles | NA. |

**STEERING**

| | |
|---|---|
| Type | Cam and lever, Ross. |
| Turns lock to lock | NA. |
| Ratio | NA. |
| Turn circle | NA. |

**BRAKES**

| | |
|---|---|
| Type | Mechanical four wheel, Bendix. |
| Drum diameter | 12 in., 2.25 in. wide. |

**CHASSIS & BODY**

| | |
|---|---|
| Frame | Steel, central tube 8.5 in. diameter. |
| Body construction | Composite. |
| Body style | Two door sedan with integral trunk. |

**SUSPENSION**

| | |
|---|---|
| Front | Independent coil, sliding pillar type. |
| Rear | Independent, four tranverse leaf springs. |
| Tires | 7:00 x 18, six ply. |
| Wheels | Disc, bolt on type, cast magnesium. |

**WEIGHTS & MEASURES**

| | |
|---|---|
| Wheelbase | 134 in. |
| Overall length | NA. |
| Overall height | NA. |
| Overall width | NA. |
| Front tread | 58.125 in. |
| Rear tread | 60.00 in. |
| Ground clearance | NA. |
| Curb weight | 4600 lb. shipping weight. |

**CAPACITIES**

| | |
|---|---|
| Crankcase | NA. |
| Cooling system | NA. |
| Fuel tank | 25 gallons, at rear of car. |

**PERFORMANCE**

| | |
|---|---|
| Acceleration: | |
| 5-25 mph | 5.95 sec. |
| 10-30 mph | 6.33 sec. |
| 10-50 mph | 12.77 sec. |
| Average top speed | Over 90 mph. |

**FUEL CONSUMPTION**

| | |
|---|---|
| Average city driving | 14.4 mpg. |
| Average open road driving | 13.5 mpg. |

### STYLING

Like several other styling efforts notably ahead of their time, the HCM'S appeal increases with the passing years. Terry Boyce of the staff of Special Interest Autos journeyed to Mequon (Milwaukee area) Wisconsin to photograph the HCM in the car collection of automotive stylist Brooks Stevens. Although he had earlier seen pictures of the car, he wrote: *"A strange thing happened to me as I began to photograph the car. I found that the HCM Special was, in its own weird way, a pretty exciting machine. Standing at the cowl and looking forward, one cannot avoid comparison with the later (1936) 810-812 Cord. There are other Cord-like details . . . the front bumpers, for instance." "The totally flat surfaces don't help (the appearance) even after Brooks Stevens explained his theory of why they are that way. The car seems to have been hewn out of a solid block of a solid substance.*

Later we kept thinking about this car. It was much larger than the Cord 810, yet it had a certain affinity to that (subsequent) design. And it had engineering and mechanical innovation just like in the 810 Cord . . . in a different manner of course, but still the excitement was there."

The author sees the side view of the car as kind of a composite of Cord from the windshield forward and Edsel Ford's first Continental personal car from the windshield back. Only on the exterior of course; no greater extremes could exist than the Henry Ford (HF1) transverse front and rear springs under the Continental and the all-independent suspension of the HCM.

The HCM would really have been the first production luxury 'personal car'; something on the caliber of the 1979 and later Cadillac Eldorado Coupe with its front-drive and fully independent suspension. Both those cars share the very large single door approach to rear seat access. On the HCM, the single large door was requested of the stylist to demonstrate the rigidity of the body structure; the doors, individually and together, operated perfectly regardless of positions in which the four wheels were placed.[8]

The HCM, like the SIXTEEN, was truly the creative work of Walter Dorwin Teague, Jr. Frank Robinson Jr. assessed young Teague's accomplishment: "It speaks volumes, both about Teague's capacity to grow as an artist and the fluidity of automotive design in the early thirties, for him to have produced two entirely distinct designs within such a short period. The SIXTEEN and the HCM are dramatically different cars, with the former a mature expression of the Classic period and the latter a bold step toward a new styling direction."

Teague's initial design and the model made from it incorporated "Woodlite" headlamps and bumpers prophetic of the 1980s. Although these were on the scale drawings furnished to Howard Marmon, those areas were redesigned at Indianapolis before fabrication. The 'Woodlite' was a patented headlamp construction sold in the aftermarket and OEM on the RUXTON; it had a very narrow width tapering to a near point at the bottom. Teague designed this headlamp half into the grill of his proposal and faired it along the hood side.

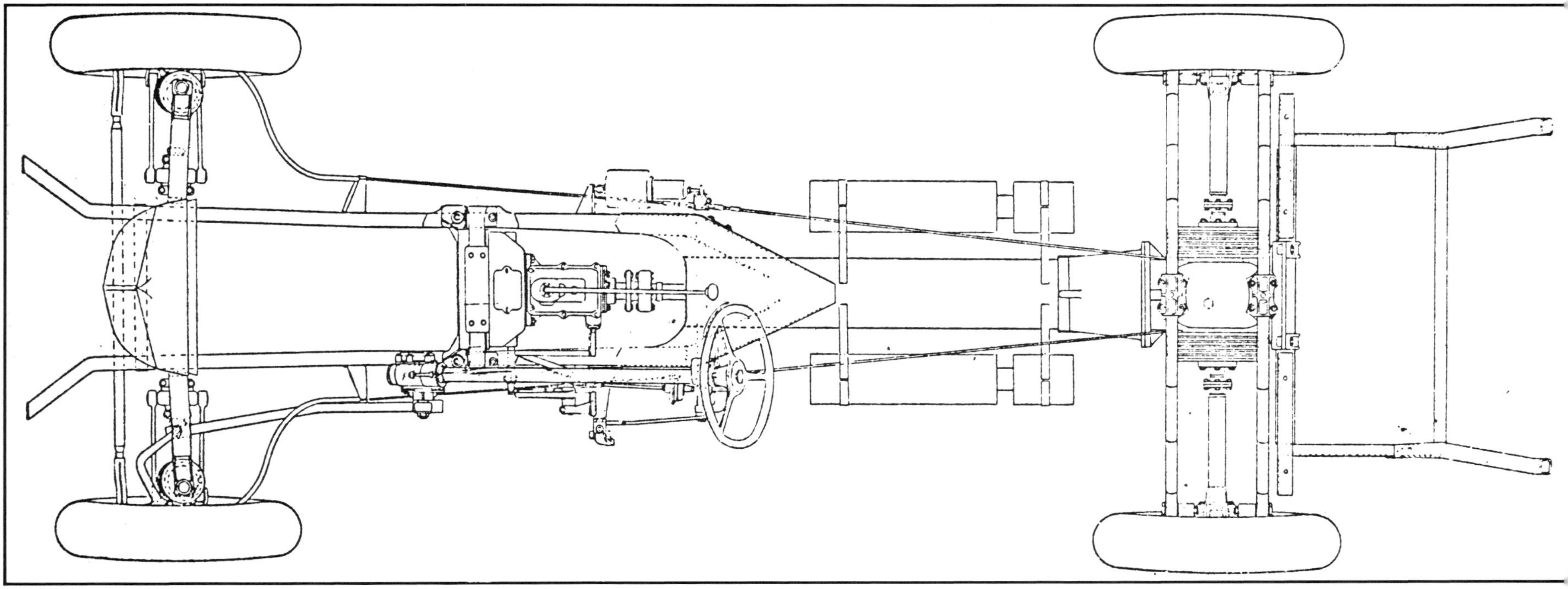

Fig. 7-11 Plan view of HCM chassis as it appeared in speed Age, Oct. 1952.

A commercial rights agreement including financial terms would have to be in hand before serious consideration could be given to predicating a prototype design on a proprietary item such as the 'Woodlite' and there would be some concern about survivability of the vending firm as well. Howard Marmon wisely evaded these problems, opting for the Pierce-Arrow like in-fender construction.

Teague proposed bumpers formed of a uniform-section bar fitted with rubber end caps and a central rubber facing. And he wrapped the end caps around the fenders. All of these ideas suddenly became very popular overseas in the late seventies and soon were found on domestic offerings.

The HCM prototype was built with the innovative bumper concept of detached bars joined by three round bars which permitted air-flow to the lower radiator area and did not obscure the lower grill aesthetically.

The hood was a three piece construction. For fluid replenishment and minor adjustments, the side panels were removed. The top panel was bolted in place as part of the structure but was removed for any serious engine work. This was somewhat the reverse of the 1938 Cadillac approach.

The vestigial running boards were an addition during construction of the prototype.

The integrated bustle trunk anticipated the stowage concept adopted on the 1936 GM 'B' bodies. The latter differed only in using a single door to enclose both the luggage area and the spare tucked below the luggage shelf. On the HCM, separate doors were elected . . . most likely because of a desire to retain the bottom-hinged, tail-gate type of luggage access like the luggage door of the SIXTEEN close-coupled sedans.

The interior was entirely done in brown leather, including the headlining. The transverse wood ribs supporting the non-metallic central roof (characteristic of all bodies before the advent of the 'Turret Top' on the 1935 Chevrolet) were exposed. This same treatment had been used on the custom Hume bodied Model 34 sedans in 1923. Was Teague aware? Window openings were metal faced but the upholstery extended right up to the window opening.

The instruments and the general dash layout was similar to that of the 1932 Marmon 8-125.

# THE MARMON

"The Easiest Riding Car In The World"

*Actual Photograph of the Winning Marmon*

## INTERNATIONAL CHAMPION

The greatest speed event mankind has ever known—the International Sweepstakes—run on the Indianapolis Speedway, May 30, was won by Ray Harroun driving a Marmon, 500 miles in 6 hours 42 minutes and 8 seconds, averaging

**74.61 Miles per Hour for 500 Miles**

In the competition were the most famous drivers in the world and forty cars—among them the finest that Europe and America have produced.

The Marmon hood was not raised during the long grind of 500 miles. Not an additional drop of water was put in the radiator. Perfect lubrication and no mechanical trouble of any kind. There were but four tire changes on the winning Marmon—three of the original tires finished the race. Joe Dawson in

**Another Marmon Took Fifth Place**

*This is a fitting climax to the brilliant racing record of the Victorious Marmon which already held the longest and most remarkable list of racing victories ever won by any make of motor car. There can be but one logical conclusion that will interest the buyer—Marmon design, Marmon materials, Marmon workmanship* **Must Be Right.**

**NORDYKE & MARMON CO., Indianapolis, Indiana**

*Established 1851* *Sixty Years of Successful Manufacturing*

CHAPTER EIGHT

# MARMON — "INTERNATIONAL CHAMPION"

The title of this chapter is derived from Marmon's victory on May 30, 1911. That was the first running of the Indy 500. It was called the "World's Championship 500 mile International Sweepstakes Race" that year. It was a truly international competition and that is the legend on the award — or rather awards . . . that Marmon won that day; two of them, one engraved "First" and one "Fifth".

That particular event which climaxed Marmon's racing era has never gathered much historical dust. There were 80,000 spectators at the Speedway that day and the race annually tends to bring to mind the past events on the bricks and later on the asphalt, but such would be but a minor disturbance of the dust.

The winning Marmon Wasp is remembered not only for its victory, but for its one-man concept and origination of the rear-view mirror that made feasible the concept. The car resides permanently in the beautiful surroundings of the Speedway Museum atop a tile platform set apart from a select few of the cars that have followed the Wasp in Victory Lane. On the fiftieth anniversary of the first 500, the original car and driver sped around the oval before the start of that year's race.

In 1970 a 'CBS (Columbia Broadcasting System) Radio Spot Sales' advertisement in Time Magazine listed, after the 17 all-time favorite movies and the eleven greatest Presidents of the United States, "The World's Greatest Motor Cars (according to Jeff Scott of CBS Radio's "Wheels" series) ":

Rolls Royce "Silver Ghost" (1920s)
Duesenberg SJ (1931) (Shown)
Mercedes-Benz 540-K (1938)
Bugatti (1932-35)
Ferrari (1950 on)
Cord Supercharger (1937)
Stutz Bearcat (1920s)
Marmon Wasp (1911)
Jaguar XK 120 (1948)
Hispano Suiza (1928)
Model T Ford (1910-27)
Cobra (1965)
Alfo Romeo (1935)
Pierce Arrow (1932)
Lincoln Continental (1941)

Note that only the Model T and Marmon Wasp predate 1920! The dates and sequence are as they appeared in the ad.

The promotion of the 1979 edition of the Chevrolet Corvette included a combination brochure-poster.[1] An 8 1/2 x 11 folder opened into a 22 x 24.5 inch multi-color poster by artist Ken Dollison. Arranged among the various detail views of the Corvette was a very nice color rendering of the Marmon Indy winner with the text: *The 'Golden Age' of auto racing began in 1911 when Number 32, a Marmon Wasp won the first Indy 500; the same year Louis Chevrolet designed and built the first Chevrolet car!*

Also shown were a Louis Joseph Chevrolet designed car typical of the 1920 and 1921 winners at Indianapolis and the 1978 Corvette Official Pace Car.

The victory at Indianapolis climaxed a remarkable and perhaps unequalled three years of factory supported racing.

## Part 1 — Era One and Two Competition

Howard Marmon had designed an engine and suspension system both unique and vastly superior to any competitor worldwide. Nordyke and Marmon Co. were masters of precision manufacturer of machinery. The early century was marked by gentleman race drivers...only the more wealthy could afford the luxury of the motor car; racing as a profession was in its infancy, symbolized by Barney Oldfield. So it was not surprising to see Howard C. Marmon listed as the driver of a Marmon entry. It was however quite surprising to see a Miss Martha Foster listed as driver of a Marmon entry on occasion; she would become Mrs. Marmon.

### FIRST ERA COMPETITION

The First Era Marmons (1904-1909) were characterized by vastly superior tire reliability consequent to the Double Three-Point Suspension Chassis. Their engines already incorporated: (1) the barrel crankcase construction that would characterize Indianapolis 500 winners from the first Marmon winner into the 1970's, and (2) the then unique and patented but now universal engine lubrication system. All this resulted in a unique capability to operate non-stop in competition over significant distances either on the track or on public roads. Their proven air cooling system was also a significant contributor to this durability performance.

Nordyke & Marmon joined the Indiana Automobile Racing Association in its founding year, 1905, and in October of that year successfully argued that a Premier factory entry be withdrawn from a race because it was a "straight eight monstrosity" bearing no relation to the Premier production car. Marmon took the position, and evidently was supported by the Association, that the objective of motor racing was to prove the dependability of products in competition.[2]

Era One competition was restricted to on-road contests. In 1906, 1908 and 1909, the only years a Marmon was entered in the Glidden Tours, the Marmon finished each time with a perfect score!

The two Model D cars entered in the 1906 Glidden Tour were driven by T.E. Schultz (the New York distributor) and F.E. Wing (the Boston distributor) . It was car number 66 driven by Frank Wing which had the perfect score; the other suffered a broken front axle and some minor problems.

A Model H was entered in several endurance contests and finished with a perfect score in each. Among these was the 1908 Glidden Tour, the two-day Indianapolis-French Lick Springs Sealed-hood Endurance run, the Chicago 1000 Mile Four-Day Sealed Endurance Contest and the Boston-Mount Washington Twenty-Four Hour Non-stop Sealed Contest. Car 21 was the Glidden victor, the other two Marmons encountering cooling problems: Number 22 overheated due to a brake drag problem and Number 23 developed a leak in a coolant pipe. The 1908 catalog 830-B stated that the Model H roadster was available with air or water cooling. It is likely that the 1908 Glidden victor was water cooled for promotional

Fig. 8-1 Marmon car finishing Glidden tour with perfect score — Bretton Woods, New Hampshire, July 28, 1906.

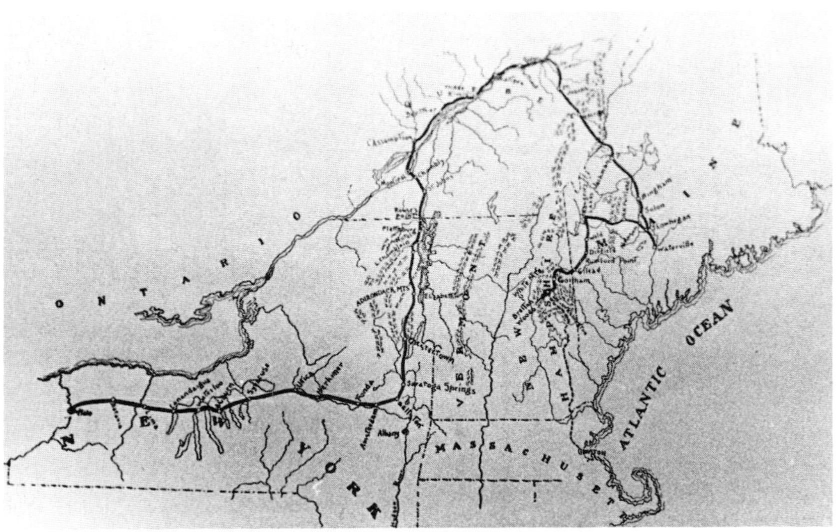


Fig. 8-2 Route of the 1906 Glidden Tour. Dotted line shows the Marmon's non-stop run to Boston after the finish.

Fig.8-3 Car "66" on the steep muddy grade near St. George, Quebec. The only car to make it without chains.

Fig. 8-4 Car "66" at Three Rivers, Quebec.

Fig. 8-5 A Mrs. Rickey checks in at starting line in Newark, New Jersey at the wheel of a Marmon C-7. Her companion and those in the Model H entry in the same event are unidentified. Note the changed appearance of the 1908 Model H from the 1907 car.

Fig. 8-6 Unidentified couple ready for a road event in a Model 32 Roadster.

reasons; Marmon had widely distributed a booklet '66' extoling the performance of the **air-cooled** Marmon in the 1906 Glidden.

## SECOND ERA COMPETITION

Marmon introduced the Model 32 for 1909 and it dominated the Second Era. It was a relatively small, cleverly simple and consequently very light automobile with apparently conventional suspension masking remarkable ride and handling characteristics. In street dress, it performed with a perfect score in the 1909 Glidden Tour. Two Marmon 32s were entered: Car number 4 entered and driven by Frank E. Wing and number 5 entered and driven by Howard C. Marmon. Number 4 had the perfect score. Number 5, renumbered 30, was then driven in the Frank A. Munsey Reliability Tour, Washington, D.C. by Ray Harroun.

It was in the late summer of 1909, when the Model 32 made its first appearance in track races, that the Marmon racing colors of yellow and black, with yellow predominating were adopted. The cars immediately were recognized as a threat in any race and were dubbed 'The Yellow Jackets'. Other competitors soon called it "The Yellow Peril"!

'The Marmon Model 32's racing career began at the dirt surfaced Indianapolis Motor Speedway Races, in August 1909. A three-car team of '32s' was fielded, piloted by Ray Harroun, . . . Stillman and Foxhall Keene. Harroun won the ten mile race at an average 71 miles per hour, Stillman finished third in the 100 mile race on the second day and in the 300 mile Wheeler-Schebler Trophy on the third day. It was actually a 235 mile race terminated after Keene hit a chuckhole and struck a pedestrian bridge hospitalizing his mechanic Jim Schiller, an event which followed an accident fatal to two spectators. Harroun was a 28 year old Marmon engineering employee destined for racing fame. The Marmon's consistent non-stop running in the long races (100 and 225 miles non-stop) amazed the racing experts. "In quick succession came its victories in the Vanderbilt, the Atlanta Races and the New Orleans Races, running 100 miles and over at a mile-a-minute speed, without a stop."[3]

In October 1909, the Marmon won the Wheatley Hills Trophy, for stock chassis 231 to 300 cubic inches, on the Vanderbilt Cup Course. Harroun drove the 189.6 mile course **non-stop** in 190 minutes, 24.4 seconds.

On November 11, 1909 the Model 32 was driven to victory by Ray Harroun "without a mechanician" in the Atlanta Automobile Association Trophy Race for stock chassis on the Southern Speedway. The Marmon ran the **120 miles non-stop** in 109 minutes, 26 seconds. The race was restricted to engines 231 to 300 cubic inch displacement and the Marmon was presumably underbored from its production 318 CID!

That same month at the New Orleans Horse Track, the Marmon, with Harroun driving, **won non/stop** the Twenty-Mile and Fifty-Mile Races and the New Orleans Prize Trophy 100 mile race, the latter two in 54 minutes (55.55 miles per hour) and 107 minutes, 14 seconds (56.23 miles per hour).

Had not a tire blow-out sent the car plunging off the track, the Marmon was also headed for victory in the New Orleans 200 mile race. The blow-out was not surprising because the car had averaged 71 mph for 164 non-stop miles.

The two-car Marmon team of Harroun and Stillman came in One-Two in the first Five Mile; finished second and third in the Twenty Mile, third and fourth in the second Five Mile Free-for-All Handicaps. Stillman would have won the Twenty Mile by 13 seconds except for a granted handicap.

Fig. 8-7 Winning the Wheatley Trophy in the 1909 Vanderbilt — 190 miles in 190 minutes.

Fig. 8-8 News of the Wheatley and Atlanta wins was quickly spread.

The Marmon box score for the 1909 racing season:[4]

Entered....................20
Won....................10
Second....................6
Third....................7

In some of the races where the Marmon did not win, opponents had been given handicaps of from thirty seconds to one minute! The "Yellow Peril" indeed! Victories had been in its own displacement class but it had competed against larger engines as well.

Following the accident experience during the August meeting, Speedway officials had arranged to have the dirt track surfaced with bricks. On December 17th, to commemorate the laying of the last brick, the Marmon team returned to the Speedway. After the ceremony, despite the sub-freezing weather, Stillman and Howard Marmon himself gave the track its official baptism. Stillman in Number 6, warmly hooded against the weather, ran eight laps of which the first two were 70 mph laps; **Howard Marmon at the wheel of Number 7** turned two laps at an average **68.6 mph** and continued for a total eight laps.

The 1910 Marmon racing season began on the West Coast at the Ascot Park dirt track in Los Angeles in March. Marmons driven by Harroun and Stone finished a close One-Two running 100 miles non-stop for a winning time of 103 minutes, 14.2 seconds or 58.25 mph. Harroun also won the Ascot 50 mile and Twenty Mile races in 50 minutes, 16.4 seconds (59.83 mph) and 19 minutes, 40 seconds (60.97) the latter being a Pacific Coast record.

Next came the inaugural racing on the new board track of the Los Angeles Motordrome. On April 8, Harroun averaged 70.5 miles per hour to win the One-Hundred Mile race for 231-300 CID engined cars in 85 minutes, 22 seconds **non-stop.** A week later the pair of Marmons running **non-stop** took first and second in the Fifty-Mile Race. Harroun averaging a surprising 75.2 mph covering the distance in 39 minutes, 53.55 seconds. The next day, the Two Hour Free-for-All for engines up to 600 CID was won by Harroun when he covered 148 miles on the boards at nearly 75 miles per hour **non-stop.**

The following day April 17th, the Marmon, which was the smallest car in the race, traveled 100 miles **non-stop** in 76 minutes 20 seconds at the then unheard of rate of 78.6 miles per hour to win the Grand Prize Trohy (and $1000 cash) in a race for cars with engines up to 600 CID. Some of the finest cars in Europe and America were competitors.

Every long distance event in the whole Motordrome series was won by a Marmon, and all non-stop.

The racing scene shifted to the East Coast to compete for the Atlanta Speedway Trophy in May . . . a 200 mile competition for stock chassis with engines from 301 to 450 CID. Harroun was the victor, averaging 65.4 miles per hour for 182 minutes, 30 seconds.

Then 'back home in Indiana . . .' and the Speedway. On May 27, Harroun won the Remy Grand Brassard and Trophy 50-Mile event for stock chassis (231 to 300 CID) again presumably driving an underbored Model 32 (318 normal CID). He averaged 70.58 mph for 42 minutes, 30 sec. non stop; and team-mate Joe Dawson was close behind, also running **non-stop.** The Remy Grand Trophy Cup was a Tiffany art piece valued at $2500 and presented to the manufacturer . . . Nordyke & Marmon Co.; the Brassard was a small silver arm band bearing a weekly stipend for the victoriious driver . . . Harroun.

The next day May 28, 1910, was the first Memorial Day Indianapolis Race . . . but it was the 200 mile competition for the Wheeler-Schebler Trophy by cars with engines up to 600 CID. "The trophy was a Tiffany art piece valued at $10,000 . . . the largest and most valuable ever offered in any contest at that time . . . and it attracted one of the greatest fields of cars and drivers America has ever known".[5] This race saw the first major appearance of the six-cylinder version of the Marmon 32, a car destined for automotive immortality. Harroun piloted his new mount to victory averaging 72 mph after making one stop to refuel, covering the distance in 166 minutes, 30 seconds and a full six miles ahead! This performance was indeed a portent of what was to be on the next Memorial Day. Press reports referred to the new Marmon racer as a 'Speedway Wasp'. Marion George[2] attributes the origin of the unusual car body design to sketches made by Harroun for Howard Marmon.

Fig. 8-9 Ray Harroun ready to climb to victory at 'Giants' Despair.

The 32 Marmon racing engines evidently used a cylinder bore which was 0.145 inch undersize on diameter. This 4.355 x 5 engine then displaced 298 cubic inches. (93.7% of the production 318) with four cylinders and 447 cubic inches with six. This placed the four in the 231 to 300 CID racing class, the six in the 300 to 451 CID class, each being only 0.66 percent below the upper limit.

On June 14, 1910 a Marmon 32 won the Giant's Despair Hill Climb at Wilkesbarre, PA. It broke the class record for 231 to 300 CID engines with a time of 1 minute, 50.2 sections. This was the only hill-climb competition entered by Marmon.

The second race for the Remy Grand Brassard and Trophy (the first was May 27) was held at the Speedway on July 2. Dawson and Harroun were awarded first and second place following disqualification of the 'winning' car. **Both ran non-stop** and Harroun was a close second while Dawson averaged 74.38 miles per hour, covering the 100 miles in 80 minutes, 40.7 seconds. Now both Harroun and Dawson had a Brassard stipend.

The Cobe Cup Race, at the time considered the "west's" greatest stock-classic for cars with engines 451 to 600 CID, was run on July 4 at the Indianapolis Speedway. The win went to twenty year old Joe Dawson in a Marmon '32'. Harroun finished third but was awarded second place in the other Marmon '32.' Dawson pitted twice for tire changes, Harroun only once. Dawson averaged 73.74 miles per hour . . .the fastest for stock-chassis cars in automobile records at that time. Dawson's time clipped 10 minutes, 28 seconds off the prior record; the prior year Vanderbilt winner came in fourth!

On August 26 at Elgin, Illinois, the Marmon team finished first and third to win the Kane County Trophy Race for the 231 to 300 CID class. It was a 170 mile road race and the winning time was 184 minutes, 45.79 seconds. In the companion Illinois Trophy race Joe Dawson finished third, covering the 203 miles in 214 minutes, 9.62 seconds (56.09 mph).

Marmon recognized the potential for recognition inherent in the spectacular string of victories and their advertising pace quickened accordingly. From time to time they featured ever enlarging trophy displays. Often the ads resembled a newspaper sporting page. Two examples are shown reporting the Cobe and Elgin victories. The illustrated car and the magazine name seem somehow incongruous with the track performances recounted herein.

Marmon was denied a Vanderbilt victory on October 1, 1910, by an unfortunate accident totally beyond their control. The Marmon 32 traveled over the course faster than any competitor, averaging 67.44 miles per hour, but a static delay of six minutes lowered the average speed to 65.08 miles over the 278.08 mile course. Dawson's time was 256 minutes, 23 seconds including the six lost minutes following collision, on the eighteenth lap, with a spectator who had run onto the course. The winning car, an ALCO driven by Harry Grant had an engine 83.5 percent larger than the Marmon and sold commercially at twice the price. Despite this David and Goliath relationship, Joe Dawson was just 23 seconds from making up those six minutes when the checkered flag was waved!

The race for the City of Atlanta trophy on November 5 provided a historic 'neck and neck' contest. It was for stock chassis cars 451 to 600 CID, so the Marmon '32' six would have been entered. The race was somewhat of a preview of the first Indy 500 for the Marmon hung close behind Mulford, driving a Lozier, for over a hundred miles only to pass on the last lap and win by four seconds for a total time of 171 minutes, 12.75 seconds or 70.125 miles per hour. It was described (in 1911) as one of the most exciting speed battles known.

Two days before, racing for the Coca-Cola Trophy at Atlanta, Dawson had finished second after leading for 98 miles when with a lead of three laps, and only two miles to go, a mishap resulted in crossing the line 1 minute, 20 seconds behind the winner, for an 87 minute, 27.38 second time.

The 1910 racing season ended at Savannah, Georgia with the Savannah Challenge Trophy race on November 11 and the Savannah Grand Prize the next day. The first was for 231 to 300 CID and Dawson's Marmon won . . . **276.8 miles** in 263 minutes, 40 seconds to average 63 miles per hour **non-stop.** This was the largest non-stop race on record at that time. The following day was an unlimited class and Marmon apparently found some large engined competition . . . should have had the **six** there . . and came in sixth, covering a credible **415 miles** in 390 minutes, 22 seconds to average 63.82 miles per hour **running on the same tires raced 276.8 miles the day before.**

The Marmon box-score for the 1910 racing season:

| | |
|---|---|
| Entered | 93 |
| Won | 25 |
| Second | 24 |
| Third | 13 |

As of August 1910 Marmon held 61 out of 70 official racing speed records, including all in the two-class range from 231 to 450 CID.[4] By the end of the season Ray Harroun was National AAA Driving Champion.

As noted in the ad, all these victories had been won by a fully stock Model 32 chassis (including engine) or, in the case of certain unlimited class records, of a fully stock Model 32 chassis modified only to accommodate a six cylinder version for the four cylinder engine . . . an engine using three instead of two of the stock cylinder blocks and an extended crankcase, crankshaft and camshaft. As early as mid-1910 these six cylinder racers were designated Marmon 48. For competition in certain classes, the stock '32' engine was underbored; for other classes it was overbored.

Two of the Marmon victories were in races for cars with engines 231 to 300 CID. Kane County Trophy, Elgin, Illinois on August 26, 1910 and the Savannah Challenge Trophy on Nov. 11, 1910. The 318 CID Model 32 engine must have been underbored to qualify, perhaps giving rise to the 4.25 bore reported incorrectly for the Indy Car No. 31. (4.25 x 5 would be 285 CID).

Fig. 8-10 Joe Dawson (Marmon) turning from La Roche into Dale Avenue in the 1910 Vanderbilt.

One picture exists showing presumably an early version of the six cylinder racing car. It had a severely pointed radiator shell and core and a kayak type cockpit. It is likely that the radiator was abandoned either to satisfy the 'stock' requirement of the chassis or because it was determined ineffective on vehicle drag resistance.

THE MARMON, the easiest riding car in the world, won the Cobe race. The great Western classic. Broke the record 10 Minutes 28 Seconds; 200 Miles. Time 163 Minutes 20 1-10 Seconds, average 73 1-2 Miles per hour. Defeated many larger cars; beat the Vanderbilt winner. But two Marmons started. Both finished in the pink of condition, Dawson first and Harroun a close third. Both were stock cars. Duplicates of the kind you buy. This completes the most remarkable string of victories ever won by any car in big competitions, including the Atlanta Speedway Trophy (200 Miles); The Wheeler & Schebler Trophy (200 Miles); The Wheatley Hills-Vanderbilt (190 Miles), and many other famous races.

NORDYKE & MARMON CO.

Indianapolis (ESTABLISHED 1851), Indiana

# THE MARMON

"The Easiest Riding Car in The World"

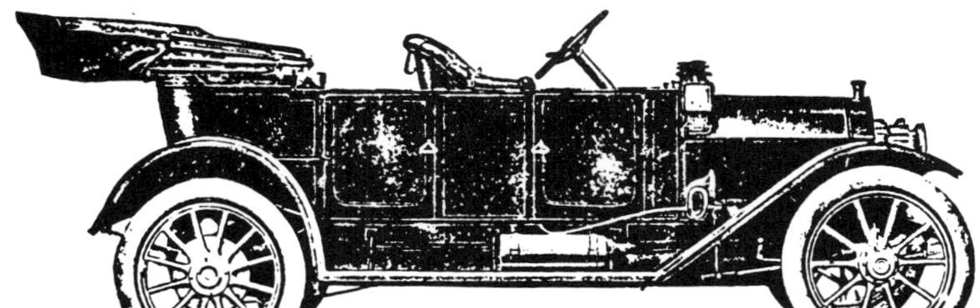

Closed Front, Five-Passenger Touring Car, $2,750

# WINS AGAIN This Time at Elgin

Following up its unparalleled string of victories which include the Wheeler & Schebler, the Cobe, the Remy, the Vanderbilt, Wheatley Hills, the Atlanta Speedway Trophies and other notable events, the victorious Marmon, champion of the 1910 season, captured Kane County Trophy in the Elgin races Friday. Another Marmon finishing a close third in the same event, and still another third in the Illinois Trophy event Friday. Three started, three finished, again demonstrating Marmon stability and consistency of performance. Dependability in the hands of the private owner is its reputation. In classy appearance and quiet smoothness of operation, the Marmon fully equals its amazing performance on road and track.

NORDYKE & MARMON CO., (Established 1851) Indianapolis, Ind.

(Sixty years of successful manufacturing)

Licensed under Selden Patent

Fig. 8-11 Spreading the news that a stock Marmon had won the Cobe race averaging 73.5 miles per hour, and the Elgin as well.

## PART 2 — Victory in the First Indy 500

The 1911 racing season began early for Marmon, for Howard Marmon, that is. *The Automobile* reported in its January 26, 1911 issue that Howard Marmon had returned to Indianapolis from the New York automobile show where he had been elected Chairman of the General Rules Committee and Vice-president of the Manufacturers' Contest Association. It continued: *"Mr. Marmon is a recognized authority in this realm and his words, owing to his long and successful experiences in all manner of automobile contests and activities, carry considerable weight with both the trade and the public." 'Particular attention is being paid by Mr. Marmon to the study of four subjects: motor car racing, the Glidden Tour, the industry in general and the rules governing contests.' "I believe that this seasons motor races will be the best ever," said Mr. Marmon. "This is not idle talk but take my own case for instance. I cannot afford to stop with a jolt, because I have gained too much of practical value. The severe races have helped the study of the construction of my cars and so with every manufacturer. I also believe that the appetite of the public is whetted to a keen taste for automobile speed battles. The 500 mile race proposed at the Indianapolis Motor Speedway promises to attract a great field and will be perhaps the greatest race ever held. Many new firms that have never entered cars before this season will do so this year. Concerning the automotive industry in general, I am an optimist. I am positive that the public demand is steadily growing. The fact that it is steady and not a sudden fad is encouragement to me. People need automobiles now the same as the telephone or the electric light in their homes. Manufacturers who last year announced that they intended building great numbers of machines (cars) found it impossible to construct them in so short a time. This weakens the rumors of critics that the country is flooded with machines. I know it is not as our agents are the men in touch with the people and we are now behind our orders for 1911. The Glidden Tour, I believe, should go through the Middle-West south and swing around into the East. I am in favor of its starting from Indianapolis. We will get as many entries from that city as any other in the country. Indianapolis is the center of the population of the United States and the famous motor speedway at that city makes an ideal camping ground with its many private garages and the city will give the Gliddenites a grand send-off. We will also rewrite the rules determining a stock car. As it now stands it is possible to interpret it more*

Fig. 8-12 Heileman (No. 32) and Joe Dawson (No. 36) at Savannah, and Dawson winning. Reprinted from Automobile Quarterly, Vol. 7 No. 2 by permission.

Fig. 8-13 Marmon racer — Joe Dawson, driver, Howard C. Marmon, designer.

# The Supreme Test of a Car

EVERY man that has driven a motor car understands that the supreme test of a car lies in the long races. The terrific strain of long sustained high speed searches out every possible weakness of construction, workmanship or design and proves the stamina of a car as no other test can.

We doubt if the motoring public fully appreciates the amazing showing made by Marmon stock chassis under this merciless test. No car in Europe or America has ever won such a remarkable list of records. Unlike the records at Brooklands, England, they were not made against time under favorable conditions, but they have been won in many important races on various tracks, under all kinds of conditions and in conflict with the fastest cars and best drivers in the world.

All these records were won by strictly stock Marmons—the kind you buy—except the free-for-all, or "regardless of class" records between 100 and 200 miles—and it is to be noted that these were not as fast as the Marmon stock car records covering the same distance. The Marmon "Wasp," with which these free-for-all records from 100 to 200 miles were made, varied from the stock chassis only in the fact that it had six cylinders instead of four.

In the 231-300 class the Marmon **holds all the records.** The longest race for the class thus far has been 120 miles and the Marmon records run from 5 to 120 miles without a break.

In the 301-450 class the Marmon holds all the records from 40 to 100 miles. We are appending all the official records from 10 miles up in the big-car class (451-600) and in the free-for-alls, where the Marmon was in competition with much larger and more expensive cars, together with the Marmon records in the 231-300 and 300-450 classes:

## OFFICIAL RACING RECORDS

### Regardless of Class.

| | | | |
|---|---|---|---|
| 10 Miles | Simplex 90 | Robertson | 6:35.62 |
| 15 Miles | Benz 120 | Oldfield | 12:34.04 |
| 20 Miles | Benz 120 | Hearne | 14:08.72 |
| 50 Miles | Fiat 90 | DePalma | 37:55.53 |
| 55 Miles | Marmon 32 | Harroun | 42:03.35 |
| 60 Miles | Marmon 32 | Harroun | 45:51.72 |
| 65 Miles | Marmon 32 | Harroun | 49:38.34 |
| 70 Miles | Marmon 32 | Harroun | 53:25.28 |
| 75 Miles | Marmon 32 | Harroun | 57:15.79 |
| 80 Miles | Marmon 32 | Harroun | 1:01:04.77 |
| 85 Miles | Marmon 32 | Harroun | 1:04:53.12 |
| 90 Miles | Marmon 32 | Harroun | 1:08:41.2 |
| 95 Miles | Marmon 32 | Harroun | 1:12:32.98 |
| 100 Miles | Marmon 32 | Harroun | 1:16:21.9 |
| 110 Miles | Marmon 48 | Harroun | 1:29:25 |
| 120 Miles | Marmon 48 | Harroun | 1:37:18 |
| 130 Miles | Marmon 48 | Harroun | 1:45:30 |
| 140 Miles | Marmon 48 | Harroun | 1:43:57 |
| 150 Miles | Marmon 48 | Harroun | 2:02:16 |
| 160 Miles | Marmon 48 | Harroun | 2:10:54 |
| 170 Miles | Marmon 48 | Harroun | 2:19:31 |
| 180 Miles | Marmon 48 | Harroun | 2:28:18 |
| 190 Miles | Marmon 48 | Harroun | 2:37:09 |
| 200 Miles | Marmon 48 | Harroun | 2:46:32 |

### 451-600 Class Stock Chassis.

| | | | |
|---|---|---|---|
| 10 Miles | Fiat 60 | Robertson | 7:47.71 |
| 20 Miles | Fiat 60 | Robertson | 15:57.41 |
| 30 Miles | Fiat 60 | Robertson | 24:53.63 |
| 40 Miles | Fiat 60 | Robertson | 33:00.00 |
| 50 Miles | Fiat 60 | Robertson | 42:02.98 |
| 60 Miles | Marmon 32 | Dawson | 50:21.35 |
| 70 Miles | Marmon 32 | Dawson | 58:22.49 |
| 80 Miles | Marmon 32 | Dawson | 1:06:21.30 |
| 90 Miles | Marmon 32 | Dawson | 1:14:22.23 |
| 100 Miles | Marmon 32 | Dawson | 1:22:15.24 |
| 110 Miles | Marmon 32 | Dawson | 1:30:08.52 |
| 120 Miles | Marmon 32 | Dawson | 1:38:16.63 |
| 130 Miles | Marmon 32 | Dawson | 1:47:42.20 |
| 140 Miles | Marmon 32 | Dawson | 1:55:40.77 |
| 150 Miles | Marmon 32 | Dawson | 2:03:36.23 |
| 160 Miles | Marmon 32 | Dawson | 2:11:31.77 |
| 170 Miles | Marmon 32 | Dawson | 2:19:25.61 |
| 180 Miles | Marmon 32 | Dawson | 2:27:17.07 |
| 190 Miles | Marmon 32 | Dawson | 2:35:17.9 |
| 200 Miles | Marmon 32 | Dawson | 2:43:20.13 |

### 301-450 Class Stock Chassis.

| | | | |
|---|---|---|---|
| 40 Miles | Marmon 32 | Dawson | 32:30.37 |
| 50 Miles | Marmon 32 | Dawson | 40:28.1 |
| 60 Miles | Marmon 32 | Dawson | 48:15.2 |
| 70 Miles | Marmon 32 | Dawson | 56:05.65 |
| 75 Miles | Marmon 32 | Dawson | 1:04:18.7 |
| 80 Miles | Marmon 32 | Dawson | 1:08:0[illegible] |
| 90 Miles | Marmon 32 | Dawson | 1:12:58.94 |
| 100 Miles | Marmon 32 | Dawson | 1:20:40.70 |

### 231-300 Class Stock Chassis.

| | | | |
|---|---|---|---|
| 5 Miles | Marmon 32 | Dawson | 4:19.36 |
| 6 Miles | Marmon 32 | Harroun | 5:40.83 |
| 10 Miles | Marmon 32 | Harroun | 8:16.8 |
| 20 Miles | Marmon 32 | Dawson | 16:23.5 |
| 25 Miles | Marmon 32 | Dawson | 20:40.75 |
| 30 Miles | Marmon 32 | Dawson | 24:52.51 |
| 35 Miles | Marmon 32 | Dawson | 29:39.85 |
| 40 Miles | Marmon 32 | Harroun | 34:25.4 |
| 45 Miles | Marmon 32 | Harroun | 38:11.6 |
| 50 Miles | Marmon 32 | Harroun | 42:03. |
| 60 Miles | Marmon 32 | Harroun | 53:53.12 |
| 70 Miles | Marmon 32 | Harroun | 1:03:01.23 |
| 75 Miles | Marmon 32 | Harroun | 1:07:31.07 |
| 80 Miles | Marmon 32 | Harroun | 1:12:08.43 |
| 90 Miles | Marmon 32 | Harroun | 1:21:12.06 |
| 100 Miles | Marmon 32 | Harroun | 1:30:08.31 |
| 110 Miles | Marmon 32 | Harroun | 1:40:14.85 |
| 120 Miles | Marmon 32 | Harroun | 1:49:26.94 |

**NORDYKE & MARMON CO. (Estab. 1851) Indianapolis, Ind.**

Licensed under Selden patent.

Fig. 8-14 Nikrent starting and finishing third at 1911 Savannah Challenge. Reprinted from Automobile Quarterly Vol. 7 No. 2 by permission.

Fig. 8-15 Joe Dawson and Cyrus Patchke in their famous mounts.

Fig. 8-16 Harroun in Marmon '32' getting the checkered flag May 30, 1911. Note fin of Dawson's '31' in foreground. Indianapolis Speedway photo.

Fig. 8-17 The 495 CID 'Model 32' racer for the 1914 500 race. (Indy Speedway photo)

*than one way. We will not lose the sense of the rule but make it more rigid in application."*

There was little news of Marmon victories on the various tracks across the country in the spring of 1911. Lou Heinemann finished second and Joe Nikrent, driving a '32' numbered 25, came in third in the 1911 Savannah Challenge Trophy but Cyrus Patscshke's car, as number 51, was forced to retire in the Grand Prize Race and, as number 12, on the tenth lap of the Vanderbilt. He had held the lead on laps nine and ten . . . the first American car to lead in a Grand Prix race. Marmons did finish second and twelfth, driven by Bob Burman and B. Brown.

Howard Marmon meanwhile was deeply involved in the Manufacturers' Contest association and related plans for the first 500 mile race anywhere. Howard however had some more personal plans with regard to the race . . . win it! What could be a more fitting climax to "Sixty Years of Successful Manufacturing" by Nordyke & Marmon.

Engineering tests were underway to determine the speed thresholds of tire failure and the efficiency of the entire power train . . . the latter a thesis effort at Purdue University discussed in the Second Era chapter.

Harroun had decided to retire as National Driving Champion, but Howard Marmon approached him in the drafting room the day after public announcement of the first 500 Mile Race. Marion George in his 'Marmon Racing — Winging It With The Wasp'[2] recounted the conversation: *When Marmon suggested he drive the Wasp "just once more", Ray demurred. "No, I'm finished with racing. You won't need me anyway. Dawson could do it just as well as I could." To this Howard replied that he was considering building another new car for Dawson and entering both him and Harroun in the race. "Think about it," he said. Then, pouring forth ideas faster than Harroun could reply, he added that he'd like Dawson's car to be a four... the same type engine they were offering to the public. "It probably would be just as fast if the piston displacement is comparable," Harroun replied, when given the chance, "but it wouldn't be as smooth, and I don't believe it would have as much chance of going the distance." Marmon, however, was insistent and asked Harroun to work out some bore and stroke combinations for a four. Within a week he had drafted out preliminary designs and specifications for a large four, and an improved six for the Wasp as well. The four would have bore/stroke dimensions of 4 1/2 by 7 giving 495 CID, while the six would remain at 4 1/2 by 5 (447 CID). Yet Harroun remained noncommittal about driving the race, offering all sorts of reasons why he shouldn't, including the fact that "five hundred miles is too far for anyone to drive at high speed." Marmon agreed, but was one step ahead of him on that point. He had a relief driver available, Cyrus Patschke, who was prepared to give Harroun a break at 200 miles. "He's had a lot of road racing experience in the East," Marmon continued. "I had a letter from him asking for an opening on the Marmon team. He helped Mulford win a twenty-four hour race at Long Island last year and shouldn't have any trouble getting acquainted with the car and track." "How about it?" Marmon asked. Harroun, finally, consented. And in mid-May the two of them took the Wasp to the track for a preliminary practice. They found that while the Wasp could lap comfortably at more than 80 mph, (93 on straightway) several rivals were considerably faster. Slightly worried, Marmon suggested a higher gear ratio. Harroun agreed, but pointed out something that Howard had failed to notice. "Almost every time someone has passed me in practice, it is back in the pits a few laps to change tires. Anyone running flat out all the way will beat himself changing tires" Heartened, the two men worked out a race strategy. They calculated that an average of 75 mph would win the race, although Harroun assured Howard that this speed could be increased if, after the initial 100 miles it proved too slow. After trying the car, Patschke confirmed their opinions.*

When a large field of 40 cars lined up in what is today a surprising five abreast array, there were two Marmon cars in the pack: Ray Harroun alone in the six cylinder Speedway Wasp, Joe Dawson in a new four-cylinder Marmon similar to that which Harroun had driven to the National Drivers Championship in 1910 but with displacement increased. Unlike the highly successful racer of the 1910 season which met AAA stock requirements, this car had a specially modified engine with stroke increased to 7 inches from the production 5 inches and was overbored to 4.75 from 4.5 inches. Displacement grew from 318 CID to 495, an increase of 56 percent. The winning Marmon had a CID only 150 percent of 318 or 477.

Exhaust valves were 2.25 inch diameter (presumed the same as the 2 inch "clear" diameter of the production engine) but the intake valves were enlarged to 2.50 diameter. Valve lift was .44 inch. The blocks had enlarged inlet passages fed, through a single port on the right side of each block, by a siamesed intake manifold.

The account of the race in *Automobile* for June 1, 1911 reported a 4.25 (undersize) bore inconsistent with the reported 7 inch stroke and 495 CID. *The Motor World* reported a 4.5 (standard) bore.

There was a remarkable similarity between the Howard Marmon/Ray Harroun racing team and the Roger Penske/Mark Donahue team of the Sixties and early Seventies. In each case the car was prepared by an engineering organization without peer in its time . . . masters of precision and quality control, students of the forces to be jousted with, and of the optimum way to marshall resources for victory. And in each case the car was managed, not just driven, on the track by a mechanical engineer.

This remarkable parallel was memorialized, perhaps inadvertently, when in 1973 Ron Burton painted matched pictures of the Marmon Speedway Wasp with Harroun at the wheel and of the Sunoco Special with Donahue at the wheel. Both drivers were living at the time.

Napoleon Boz, writing in Car Classics for August 1971, wrote of Ray Harroun, the Marmon Wasp and the First Indy 500, and recounted Harroun's strategy and his confidence of victory during the judges' long debate following the race: *"Fortunately for Howard Marmon and Harroun, the standard Model 32 was a very fast and dependable car even in stock form, so the grueling run of 200 laps over the spine-jolting Brick Yard was absorbed without problem. The speed of Harroun was an official 74.59 mph average. This was extremely fast for an American track in 1911 . . . as Ralph Mulford and several of Harroun's other rivals discovered in the course of the six hour, 42-minute race. Racing strategy was still in its infancy, but Harroun used what he felt were winning tactics, and proved his judgement to be correct. He had decided Mulford was the strongest driver he would be facing and made every attempt to stay ahead of Mulford . . . if it didn't put an undue strain on his mount. During the race, both Mulford and Harroun stayed in pretty much the same lap, and Mulford finished second. Harroun didn't allow himself to be panicked into pushing his Marmon too hard. It was meant to run, and run fast, but over-winding the engine would have spelled disaster to his hopes of winning. Prob-*ably because it was the first race of such long duration, many of the officials working in the timer's boxes and recording laps became a bit confused. About midway through the contest, the Marmon team started sending messages to the Wasp that Mulford was several laps ahead, and Harroun was to try and close the lead. To do so would have forced his car beyond the margin of safety he had allowed, and Harroun refused to follow the instructions. He was certain Mulford had never passed him and that the team probably had faulty information, so he continued his steady pace. Pit stops during the race were a breeze for the Wasp. There was never any need to lift the big car's hood during the bout as the motor functioned perfectly. *The radiator held up and no water was needed, so with four tire changes (he finished the race with three of the original tires still on the ground), and fuel stops, Harroun blasted around the oval. This was probably the extra margin of time he needed to beat Mulford's car, combined with the fact he had less weight to carry with only one driver instead of two. Fifth place during the event went to Joe Dawson and his riding mechanic also aboard a Marmon of similar design and construction as the Wasp. But the confusion over the correct lap count seemed to point toward Mulford as the winner. Howard Marmon was in a state of nerves, and blamed Harroun's failure to pass Mulford for the loss of the race. But the driver kept his cool. He knew that he had won, and had even lapped the track a few more times to make sure he had completed the entire 200 circuits. The Speedway officials worked far into the night laboriously checking each detail of the race and making cross checks of each driver's laps, times and speeds. Marmon wandered about the track all night worrying until the final announcement was made. Harroun had won. Years later, Harroun told a reporter, "I guess I was the only member of the Marmon organization that was sure who won."* Reprinted by permission *Car Collector and Car Classics*

The prize money was $16000; the value of the publicity comparable to that of a major pronouncement by Henry Ford.

No other car with more than four cylinders would win any race of Grand Prix proportions right up to the outbreak of World War I.[6]

The excitement of the race was recorded in a booklet published by Nordyke & Marmon immediately following the race. It was entitled "International Champion" and its perspective warrants its reproduction here in lieu of any contemporary effort at recounting the race details:

## "INTERNATIONAL CHAMPION"

Shortly after dawn on the morning of the race the crowds started for the Speedway . . . by eight o'clock the stands and parking spaces began to fill. Before nine o'clock over 80,000 people taxed the enormous capacity of the great course eagerly awaiting the start.

Before the race six of the cars were eliminated through withdrawals and failure to meet the requirements of seventy-five miles per hour in the preliminary trials. The remaining forty entries were sent around the track one at a time before the start of the race to familiarize the spectators with the cars and their drivers.

Among them were the flower of all motordom: the Simplex driven by Ralph DePalma; the German Mercedes driven by Spencer Wishart; Eddie Hearne in his specially built Fiat; Harry Grant, winner of the Vanderbilt, in his Alco; an Italian Fiat driven by David Bruce-Brown, winner of the grand prize at Savannah; Ralph Mulford and Teddy Tetzlaff in the Loziers; Billy Knipper and "Wild Bob" Burman, holder of short distance records in the great German Benz cars; Joe Dawson, winner of the Cobe trophy and many other important events, in his Marmon . . . and Ray Harroun, Champion of the Speedway in 1910, with his rakish Marmon "Wasp" which was to write a new chapter in the world's history of speed and endurance before the finish of the great event.

Shortly before ten o'clock, from the south end of the long line of pits, a bomb went shrieking skyward and its explosion warned the vast crowd that all was in readiness. The two score grim racing cars took their positions, five in a line across the track, eight ranks. The exhaust of almost two hundred cylinders filled the air and the power of over two thousand speeding horses awaited the starter's signal.

Another bomb exploded, and as if moved by one hand, the forty panting racers rolled forward, paced by the starter's car. Around the first turn and up the back stretch like some fiery, crawling dragon moved the cars in even lines steadily gaining in speed. Around the final turn they came, and down past the stands . . . the drivers adding speed with every yard.

A final bomb went screaming overhead and burst into a huge, starry flag . . . the starting car dropped to one side . . . the roar of the racers crackled into a hum of high speed and across the starting line leaped the pack of forty cars.

The race was on!!

Words are not made to express the excitement of that start. Here a man rose on tip-toe tearing his score card, his every muscle straining with the turn of the cars . . . there a woman clenched her hands and gasped . . . the nerves of eighty thousand people were at high tension.

And so the first lap was made. No entire minute found a car in the same mile . . . down the stretch they came, not the crowded trackful of the start, but an ever-lengthening string of speeding cars that hurled defiance at all records of time and distance.

As lap followed lap, the stubby, powerful grey Mercedes forged to the front and set the pace. The daring amateur, Spencer Wishart, turned the first twenty miles in fifteen minutes and six seconds. Closely following came the roaring, crackling, speeding racers . . . now Belcher's Knox leading . . . again Aitken's National, the Fiats and Simplex pressing close behind. Hanging steadily behind the leaders . . . the purr of their powerful engines indicating every assurance . . . came the two Marmons.

Hardly had the first lap been completed before many of the cars began to have trouble. Like wounded steeds they came limping down the course to the pits with bursted tires. Scarcely a minute passed without some one of the racers drawing up for needed repairs. The terrific pace of over seventy miles an hour was searching out every weakness of the competing cars.

At fifty miles DePalma's big, ninety horsepower Simplex was hurled into the lead . . . a length behind like the tail of some hurtling meteor flashed Bruce-Brown's Italian Fiat . . . behind them trailed the field. The two Marmons sped evenly, close behind the leaders.

Ten laps farther and the great brown Fiat leaped ahead . . . Mulford's Lozier creeping up into second place, and the steady speed of the Marmon "Wasp" began to tell . . . Harroun was fifth.

One hundred miles . . . one-fifth of the distance . . . the Fiat thundering at the lead, the Simplex one second behind and the Marmon's smooth running, remorseless, unvarying speed another second behind the leaders . . . third.

Six of the forty starters were out of the race . . . others were miles behind the leaders. Tire and engine trouble were the allies of the forces against which the speeding cars did battle.

At one hundred and fifty miles Harroun's Marmon had passed the Simplex into second place . . . the Fiat still in the lead. Dawson crowded his Marmon into sixth place. The great throng of spectators grew more excited.

At one hundred and seventy-five miles Harroun drew up at the pits and sought relief from the rigorous, nerve-racking drive. Cyrus Patschke, famed as a driver in twenty-four hour races, took the wheel.

Again the Marmon took on speed . . . again and again it flashed past the cheering stands . . . above eighty miles an hour . . . clipping off the bare second that stood between it and the great pride of Europe . . . the Fiat. Patschke was driving the race of his life . . . his one endeavor was to pass the brown, thundering blur of the leading car.

The race half over . . . a hundred times had the leaders circled the two-and-a-half-mile brick course . . . but thirty of the cars had withstood the grinding test. Three of America's Speedway records, for 150, 200 and 250 miles had fallen before their whirlwind sweep.

The crowds cheered wildly as Patschke pulled up at the pits . . . his Marmon in second place, a bare second behind the leader. Harroun slipped into the lone driver's seat, and refreshed by his hour's rest, renewed the race.

Three hundred miles . . . and the rakish yellow-bodied "Wasp" has flashed into the lead. Around the back turns it hummed . . . taking the banked curves as if moving in grooves . . . never varying . . . never faltering . . . but devouring mile after mile at relentless speed. The 300-mile American Speedway record fell before that dashing pace in two hundred and forty-one minutes and twenty-five seconds.

On to three hundred and fifty miles and the race changed with every lap. For a moment Harroun paused to change a tire . . . his third tire change of the entire race . . . all on the right rear wheel that took the brunt of the course at the turns. Mulford's white Lozier thundered into the lead for a moment. Harroun's Marmon took up the chase over a second behind . . . third came the Fiat and less than a second behind him Joe Dawson drove his Marmon "No. 31".

The crowds moved anxiously . . . the jaded nerves of the spectators, worn by the long grind, anticipated the end and took on new life.

Four hundred miles . . . Harroun's Marmon in the lead. The world's Speedway record went to the Marmon car . . . its driver had traveled faster than man ever knew for that distance. The white Lozier followed a mile behind. Then came the ninety horsepower Fiat and Dawson's Marmon in a speed battle that brought the crowds to their feet . . . two-tenths of a second separated the cars in third and fourth places.

The last hundred miles began.

But half of the cars originally entered remained in the race . . . the merciless pace of over seventy-four miles per hour shattered the world's records and gave the most terrific test of the cars to which they could be subjected.

The Marmon still led but the race was not won. At 450 miles Harroun stopped again for a new tire on the right rear wheel . . . three of the original tires had held throughout the long grind . . . seconds were precious. The big Lozier roared past and gained forty-four seconds. Then Harroun's last tire change was completed.

The Marmon "Wasp" was back in the race.

Like remorseless Fate . . . the Marmon overtook the Lozier . . . the two cars were separated by a few seconds . . . by fewer seconds every lap.

Back down the miles came the Fiat . . . Dawson's Marmon and Wishart's Mercedes followed by the eighteen remaining cars.

The cheering redoubled in the stands . . . like billows of the sea the four score thousand enthusiasts moved and shouted and roared. At an average of nearly ninety miles an hour Harroun's Marmon passed into its last few laps in the lead, having battled against all the forces of time and speed for over six hours without once stopping its engine.

The end approached.

Into the home stretch flashed the wonderful car that was breaking all records and winning the championship of the world.

The starter seized the checkered flag from the rack and darted on the track.

The Marmon "Wasp" and its intrepid driver hurtled past the waving signal. The huge packed mass of spectators went wild and amid a great shout of victory proclaiming it the International Champion, came the voice of the judges . . .

**"Marmon No. 32 wins!"**

* * *

Joe Dawson was driving Marmon Number 31 to a certain fourth (beating the 200 hp Mercedes) and probable third place (beating Fiat) when a radiator failure occurred on the last lap, after 498 trouble-free miles; Shrapnel on the track was responsible. Patschke was Dawson's alternate driver in the Indy race. They finished fifth.

This same car, or a twin, had won the City of Atlanta Trophy Race on Nov. 5, 1910 where it raced in the 451 to 600 CID class with stock chassis.

The name of Harroun is forever attached to the first Indy 500 victory. Nordyke & Marmon publications referred to the 'driver' as 'Harroun-Patschke', acknowledging that Cyrus Patschke of the Marmon Racing Team had handled the car flawlessly for 75 miles starting at the one hundred seventy-five mile point. Patschke had increased the lead on Mulford's Lozier to a full minute and cut the lead of the first place Fiat to a bare second. By the 300 mile point, the Marmon and Lozier had passed Bruce-Brown's Fiat and the pattern for the remainder of the race was set.

Harroun was called 'Bedouin' by associates due to his Arab ancestry. When he won at the Speedway, he was a 30 year old Nordyke & Marmon engineer. He promptly capitalized on his national reputation, leaving Marmon to start a carburetor manufacturing business. The 1913 Model 32 would have a Harroun carburetor as original equipment. The carburetor business was short lived, after much research into heated carburetion to accomplish the burning of a wide range of distillate fuels. By 1916 Harroun was chief engineer of Maxwell, which not surprisingly was fitted with a Harroun carburetor.[7] Harroun automobiles were built in Wayne, Michigan in the period 1917-22 by Harroun Motors Corporation.[8]

He returned to the Speedway on May 30, 1962 to again lap the track in the Marmon Wasp number 32 prior to the pace lap, to mark the fiftieth running of the 500. His death was reported in 1969 at age 89. It was indeed quite a career for a young Chicago dentist who succumbed to the excitement of the automobile.

Nordyke & Marmon sought to refute false claims of competitors concerning reputed special construction of the Marmon Wasp winner of the Indy 500. Their defense statement appeared in the N & M booklet "Marmon — International Champion and warrants repetition here in its entirety to assure fidelity of the information and because several bore/stroke dimensions are found in print: *"It has been noticeable that makers of cars which have been defeated by the "Wasp" have been very careful to refer to it as a "special" racing car, and to convey the impression that it is a large and powerful machine. It is a special car, but not in the sense that the term "special", as used, is intended to convey. It occurred to us early in the season of 1910 that so long as we were racing our small stock models we might just as well be prepared with a more powerful and faster car, which could take part in free-for-all and non-stock events. At that time there was only one way to build a car of this kind quickly, and that was to build a six-cylinder motor, using our regular production of parts in so far as they could be used. As a result the Marmon "Wasp" is, with the exception of the six-cylinder crank shaft and crank case, very much the same as our stock model, including the oiling system, since otherwise the engine is built up of the same material we use in our daily production. The frame, the rear system, the front axle and the steering gear, with the exception of the connection to the steering column, are such as are used every day in the manufacture of our regular stock models. The car was made to carry the driver only, without a mechanician, and as a consequence it was necessary to rearrange pedals, levers, controls and other small items which have no bearing on the stability of the car. The steering yoke and arms on the front axle differ from those we always use only because of the fact that the spindles are placed at the top of the knuckles in order to lower the center of gravity. The engine has six cylinders, four and one-half-inch bore by five-inch stroke, rated at forty-eight horsepower, A.L.A.M. The weight of the car in racing trim is 2,865 pounds. The wheels and wheelbase are the same and the springs are half elliptic front and full elliptic rear, as used on stock models. It will be seen that the Marmon "Wasp" is not the large and powerful "special" car it has been wrongfully represented to be, and its performance, in view of the fact that it is not so much of a special job, speaks volumes for the soundness and capability of Marmon construction, design and materials."*

The special Wasp rear axle ratio was 2.16:1.

When W.A.P. John of Motor Magazine reported on his interview of Walter Marmon in November 1922, he commented on the Marmon racing program as follows: *"Conceived by an ardent enthusiastic engineering genius, and produced by a sound substantial experienced company whose executives were keenly awake to the value of publicity, the Marmon Car, from its earliest date was a well known entrant in the various "endurance" contests that were so popular during the infancy of the industry and so influential in establishing fundamentals of design." "In 1906 the first Glidden (tour) was won by Marmon with a perfect score; in 1908 the Glidden and four or five well known "non-stop sealed-hood" contests were captured with perfect scores. In 1909 the famous Model 32 won its first race — a ten mile free for all handicap with Ray Harroun at the wheel. And until 1911 the name Marmon appeared with clocklike regularity in the newspaper columns as the winner of this and that famous race culminating with the capture of the first International 500 Mile championship on May 30, 1911. And that is exactly the point I wandered from several thousand words back. 'Why did you stop racing?' I asked. 'I'll tell you,' answered Mr. Marmon with a shrewd little smile. "There are several reasons. First, if we went on racing, we had everything to lose and nothing to gain from the sporting standpoint. We had won about everything in sight and to keep from going backward we had to keep on winning with clocklike regularity. And there were other reasons more important. There is a tremendous expense to maintaining a racing team.. and the benefits accruing from racing must be in keeping with the cost if it is to be regarded from a sound business angle. And in 1911 the name Marmon was as well known as winning races could possibly make a name. Also, the grueling test of maintaining high speeds over race tracks and roads had enabled our engineers in three years to establish what they wanted to know... the best and most efficient principals of design and construction such as oiling systems, bearings, spring suspension, character of steel alloys in certain parts and similar information that could be obtained most quickly and most accurately by designing cars for continuous high speed work."*

## PART 3 — Competition After May 30, 1911

Although the Indianapolis 500 victory marked the close of track racing for Nordyke and Marmon, they remained active in reliability demonstration events.

In August 1912, a Marmon 32 won the Badger the fourth annual tour of the Minnesota State Automobile Association... a seven day run from Minneapolis to Winnipeg, Manitoba, Canada and return. It was a combination route — blazing and durability demonstration.

The Marmon victory was spectacular and reminiscent of the Era One Glidden Tour performances. The race results are of interest because they highlight the gap that existed between Marmon technology and the rest of the industry.

Demerit Score of Cars Finishing Minnesota Tour[9]

| No. | Name / Day | 1 | 2 | 3 | 4 | 5 | 6 | 7 | Total |
|---|---|---|---|---|---|---|---|---|---|
| 1 | Marmon | 0 | 0 | 0 | 0 | 0 | 0 | 0 | 0 |
| 3 | Flanders (Stude) | 9 | 0 | 0 | 12 | 0 | 0 | 0 | 21 |
| 2 | Hupmobile | 5 | 0 | 7 | 0 | 3 | 4 | 3 | 22 |
| 11 | Reo | 0 | 37 | 3 | 7 | 3 | 15 | 0 | 65 |
| 7 | Mitchell | 0 | 0 | 3 | 0 | 0 | 191 | 3 | 197 |
| 8 | Paige-Detroit | 9 | 21 | 114 | 65 | 12 | 23 | 0 | 244 |
| 9 | Warren-Detroit | 0 | 0 | 15 | 276 | 27 | 2 | 0 | 320 |
| 5 | Cadillac | 0 | 0 | 35 | 148 | 272 | 0 | 0 | 455 |
| 4 | MacFarlan | | | | | | | | |
| 6 | Cutting | | | | | | | | |
| 10 | Staver | | | | | | | | |

Fig. 8-18 A stripped Model 34 stock touring car driven by the Buenos Aires distributor won the 200 mile Copa Provincia de Buenos Aires which was run over a 25 mile course of rural roads. Only 5 of 13 entered finished the race.

The tire experience was also enlightening, especially considering Marmon's unique tire durability reputation which was again dramatically confirmed:

Tire Failures Experienced on Badger Run[9]

| | Name | Tire Brand | Blowouts | Punctures | Total |
|---|---|---|---|---|---|
| 1 | Marmon | Goodrich | 0 | 0 | 0 |
| 3 | Flanders | Firestone (f) | 0 | 3 | 3 |
| | | Morgan and Wright (r) | 0 | 0 | 0 |
| 2 | Hupmobile | Goodyear | 2 (r) | 9 (r) | 11 |
| 11 | Reo | Goodyear | 3 | 6 | 9 |
| 7 | Mitchell | Kelly Racine | 3 | 8 | 11 |
| 8 | Paige-Detroit | Kelly Racine | 1 | 1 | 2 |
| 9 | Warren-Detroit | Kelly Racine | 1 | 2 (r) | 3 |
| 5 | Cadillac | Goodyear (f) | 0 | 0 | |
| | | Hartford (f) | 0 | 0 | |
| | | Miller (r) | 2 | 0 | 2 |
| Pace Car | Stoddard-Knight | Hartford (3) Goodrich (1) | 2 | 4 | 6 |

## THIRD ERA COMPETITION

In the Third Era, Marmon 34s (without benefit of factory support) were upholding the Marmon racing reputation, overseas as well as at home.

A stripped Marmon 34A stock touring car won the Buenos Aires Motor Classic . . . The Grand Prize Race. The Marmon was stripped down a lot more than the coast-to-coast Marmon of 1916 . . . fenders and windshield were removed and the central and tonneau body sections replaced with a "fire truck" configuration of seat riser mounting a church pew seat for two, followed by a cylindrical fuel tank.

The field consisted of twelve cars, ten of them American, but only five withstood the assault on a 25 mile course consisting largely of rough country roads. Four cars were out by the third lap of the eight lap — 200 mile race, three more were gone by the sixth lap. The victorious Marmon was driven by A.M. Magnasco, the Marmon distributor at Buenos Aires; the riding mechanic is unidentified.

The ad concluded "Here again the Marmon demonstrated that scientific engineering achieved a new degree of roadability, strength and endurance . . . in a major class car of a half-ton lighter weight."

In the 1923 running of the International Mountain Race on the 5000 foot Klausen Hill of the Swiss Alps, Marmon '34' stock cars finished first, second and third. They were followed across the line by racing cars of well known makes.[10] The winning Marmon touring car was driven by Werner Risch of Zurich.

Another Marmon 34, a 1924 Sport Speedster 4 passenger finished fifth in the 1925 Hainberg Hill Climb at Hainberg, Germany. It followed four special racing cars across the finish line. The route traversed was described as very dangerous due to a combination of hairpin turns and a heavy rain on the day of the race having made the roadbed very soft and extremely hazardous. A real tribute to the Marmon suspension and handling.

During this period Nordyke & Marmon publicized the popularity of the Model 34 among the leading race drivers. Full page ads featured texts written by Tommy Milton and by Cliff Durant, describing their travel to the 1923 Indy 500 in their personal Marmons.

A photograph was releasd to the press in late 1923 showing Tommy Milton (1923 7 p touring), Jimmy Murphy (1924 4 p phaeton), Barney Oldfield (1923 Hume — bodied Coupe), Eddie Hearne (1923 sedan) and Bennett Hill (1921 touring) standing beside their personal cars.

Subsequently (Nov. 1923) when Joe Boyer, Jr. and Cliff Durant had also made Marmon purchases, The Marmon Post declared that the Supreme Court of the Racing Profession had placed their stamp of approval on the Marmon 34 for superior engine design and safety features.

*Automotive* (Daily) *News* for December 7, 1925 carried a photograph of Lieutenant T.K. Matthews, 95th Pursuit Group, Selfridge Field, MI, winner of the Mitchell Trophy Race, with his 1924 Marmon 34C Phaeton and the military aircraft with which he achieved 161.47 miles per hour.

## FOURTH ERA COMPETITION

The straight eight had appeared at Indianapolis as the decade of the twenties began and quickly became dominant at the Speedway. The Miller-Murphy **front drive straight eight** appeared there in 1925 and finished second; Billy Arnold would be the first to win with front drive, in 1930 with a Miller straight eight.

Monte Williams had correctly sensed the resultant public enthusiasm and had made Era Four at Marmon the Straight Eight Era. There is evidence that he expected the subsequent developments at the Speedway to further influence the marketplace and planned to have Marmon ready with a front-drive straight eight . . . most likely a top-of-the-line luxury sports car.

Two somewhat parallel programs give evidence of such a Marmon development program. One was deliberately in the public view; the other known until very recently to only a few participants.

Early in 1928 there was a change in the Marmon position on factory racing support. In quick succession, Earl Cooper and Peter Kreis were added to the Marmon engineering staff. In the preceding four years, Kreis had made an enviable record for himself on the speedways of Europe and the US, having distinguished himself in the Italian Grand Prix, in Culver City and Indianapolis races.[11]

Fig. 8-19 A fully equipped Marmon 34 defeated a field of specially prepared racing cars, and set the highest speed on straightaways.

Earl P. Cooper was an accomplished race driver who was also skilled in the preparation of racing machinery. In the 1924 running of the 500 Mile Race, he started from the outside of the second row in a Studebaker Special and was in second place at the finish.

He had been enthused by the front-drive Miller that ran at Indy in 1925. Three Cooper Specials appeared on the 1927 racing scene and were entered in the Indy 500. One Cooper led for some 30 laps but engine problems resulted in a sixth place finish.

Fig.8-20 A Marmon 34B four passenger Speedster finished fifth in the Hainberg Hill Climb, Germany. Rain and hairpin curves made the course dangerous.

Fig.8-21 Jimmy Murphy, world champion driver (first from USA), with his personal 1924 Marmon 34 C four-passenger phaeton.

Two Cooper Specials had been entered in the 1927 Italian Grand Prix and were driven by Kreis and Cooper himself. Unbelievably the cars as raced at Monza had only two wheel brakes. That compounded their inappropriateness to road racing consequent to a two-speed transmission.

The race was marked by the absence of Bugatti and only a single entry from Delage and by rain. Kreis, whose car was out of the race on the first lap, later relieved Cooper but gave a lackluster performance on the wet track. Nevertheless it finished third at 77.02 mph behind Benoist's winning 90.04.[6]

In his ''American Auto Racing — An Illustrated History''[12] Arthur R. Bochrach had this to say about Earl Cooper (AAA Champion in 1913, 1915, 1917): *''A small quiet man with a reputation for being a smart driver who showed concern for his car. He enjoyed a distinguished racing career that did not end until his Marmon finished third in the 1927 Italian Grand Prix.''*

Obviously Mr. Cooper's credentials were impeccable for in 1965 he was the recipient of a ''Distinguished Service Award'' from the Automotive Hall of Fame. His citation reads: *For his role in the automotive industry as a pioneer race driver and engineer . . . three-time winner of the AAA National Driving Championship . . . designed the front-wheel drive Cooper Specials.*

It was very appropriate that his co-inductee that year was none other than Ray Harroun whose citation read: *For his victory in the **1910** Indianapolis 500 Mile Race . . . for his engineering ability used to build the Marmon ''Wasp'' and his help on the Maxwell Special of 1914 . . . an expert on carburetion.*

Cooper and Harroun were preceded as honorees in the Racing Category only by William K. Vanderbilt — 1941, Ralph DePalma — 1946, Joseph Tracey — 1948, Wilbur Shaw — 1951, and C.W. VanRanst — 1964.

Fig. 8-22 Barney Oldfield in the Marmon 34B Wasp Speedster ready to pace the 1920 500. (Indy Speedway)

Ray Harroun and Bert Dingley were among the first ten inductees, in 1952, into the Indy Speedway Hall of Fame, along with Henry Ford, Louis Chevrolet and Barney Oldfield. Earl Cooper was one of seven inducted the following year, along with Bob Burman, Tommy Milton and Eddie Rickenbacker.

Joe Felts said that Earl Cooper was a familiar figure in the Marmon drafting room during the following year, and that some changes to the Cooper cars were made in the Marmon engineering department.

Meanwhile there was an active front-drive program underway at the Marmon engineering department.

Joe Felts recalled George Freers bringing into the design room a self-employed engineer named R.C. Hoffman. Prior to this appearance, Hoffman had designed several front wheel drive (concept) cars incorporating his patented trans-axle assembly which minimized the amount that the engine had to be set back behind the front wheel center-line. The severe elongation of the L-29 Cord was the consequence of the use of conventional transmission and axle (differential) gearing. According to Felts, Howard Marmon had made Hoffman's acquaintance when visiting Purdue to present a Marmon engine and the latter was a student there. Freers assigned Felts the task of designing a trans-axle for Marmon under Hoffman's guidance. The design was completed and approved. Detail parts drawings were made, parts were ordered and some had already been received when the project was cancelled. This action occurred just prior to the decision to build the SIXTEEN.

A win, of even a place or show at the Speedway on May 30, 1929 would have firmly established Marmon Front Drive in the publics' cognizance. A win, the real objective, would have been the first win with a front drive car; that milestone did come the following year with Billy Arnold's victory. For Marmon who had been the first (rear drive) winner of the 500, being the first **front-drive** winner would have provided the ideal springboard from which to launch a front-drive straight-eight passenger car.

There were four Marmons on the brick oval when the cars lined up for the pace lap at the Speedway on May 30, 1928.

Joe Dawson of Marmon racing fame drove the Marmon 78 roadster Pace Car. The pack that he led on the pace lap included, for the first time since 1913, Marmon factory sponsored race cars. There were three identical front drive entries bearing the numbers 32, 33 and 34. The '32' and '34' recalled the famous Marmon models of the Second Era.

They were driven by Peter Kreis, Johnny Seymour (famous motorcycle racer) and Russell Snowberger.

Perennial problems with the Miller-like straight-eight engines, and not with the reportedly Marmon designed power trains, forced all three out of the race. Snowberger's No. 34 was the very first car to leave the race due to stripped supercharger drive gears. On the 72nd lap, Kreis No. 32 failed a connecting rod bearing, and the Marmon challenge ended on lap 151 when supercharger trouble sent Seymour to the pits for good. "Kreis had been a serious contender for one of the first three places."[13]

Disrespect for public astuteness and perception is a capital sin in the auto industry. Some of the major firms have practiced it rather consistently in recent years. Marmon appeared to be blatantly guily in May 1928, but may not have been, based on recently surfaced information.

*Barney Oldfield Buys His Fourth Marmon, a Four-Passenger Speedster*

# "The Old Master" Buys His Fourth Marmon

No one is a better judge of automobiles or of road performance than Barney Oldfield. No other man so habitually drives back and forth across the continent as Barney. He always drives a Marmon and recently took delivery of his fourth.

There is but one Barney Oldfield, but there are thousands of other good judges of cars who habitually drive Marmons. These repeat orders from men who know cars Make Marmon Dealers Make Money

MARMON MOTOR CAR COMPANY : INDIANAPOLIS

*"It's a Great Automobile"*

Fig. 8-23 Joe Dawson returned to the Speedway to pace the 1928 running with a Marmon 78. (Indy Speedway Photo)

To all appearances, the three Earl Cooper racing cars which Marmon sponsored in the 1928 running of the Indianapolis 500 were powered by the famous Miller 91 cubic inch straight eight and Miller front-drive De Dion-type axle.

But . . . In the May 5, 1928 issue of *Automotive Industries* appeared an article by M. Warren Baker of the AI staff which must have been basically a Marmon Motor Car Co. press release. It was profusely illustrated with photographs of familiar Miller components and a Miller transverse engine section drawing . . . the very same one appearing on Page 213 of Borgeson's 'Golden Age of the American Racing Car,[14] Chapter 21: Miller — Technical Notes'. However, all were captioned 'Marmon 68 Special' or 'Marmon Racing Engine'.

The article began by saying that the three new cars combined in their construction the racing knowledge of Earl Cooper and the engineering **ideals** (not ideas) of Col. Howard Marmon. This was followed by: *"In, appearance, the Marmon Specials closely resemble the little 91.5 cubic inch race jobs that appeared on the speedway at the last few races. The radiator shell, however, incorporates the familiar Marmon pyramid and on the front is attached the lightning flash which was shown for the first time on the new '68' and '78' lines this year."*

That statement was lily white, all truth, and wouldn't tilt anyone's halo. But not so that which followed.

Fig. 8-24 Car #33 for the 1928 Indianapolis 500, The Cooper 'Marmon 68 Special' with driver Pete Kreis. (Indy Speedway Photo)

The two-speed transmission reportedly designed by Cooper and Col. Marmon appeared to be a planetary type located between the clutch and differential gear set. Then the Miller clutch, the extremely distinctive and characteristic Miller engine trunnion (rear in this front drive application), the Miller suspension, supercharger, engine bearing details were all recited with nary a word 'Miller' anywhere. Very significant was the use of constant velocity universal joints; they were Bendix-Weiss brand and certainly a pioneer application.

Bendix brakes were claimed . . . Millers had Miller brakes . . . but weren't illustrated. A camshaft driven air pump for fuel delivery was presumably properly credited to Earl Cooper . . . "one of Cooper's own developments."

Mechanical Specifications published in *Motor Age,* (May 31-28) showed minor differences from Millers cars: 4.5 axle vs. Miller 4.58; 5.16:1 supercharger drive vs. Miller 5.0:1; 6.25:1 compression ratio versus Miller 5.5:1. Firing order and valve timing were identical.

The final 'coup d'e-tat' was an insert: ***'In Race for Engineering Reasons, Says Col. Marmon'.*** *The Company's reasons for its re-entry into the sport are explained by Col. Howard Marmon as follows: "Changes in the future design of passenger*

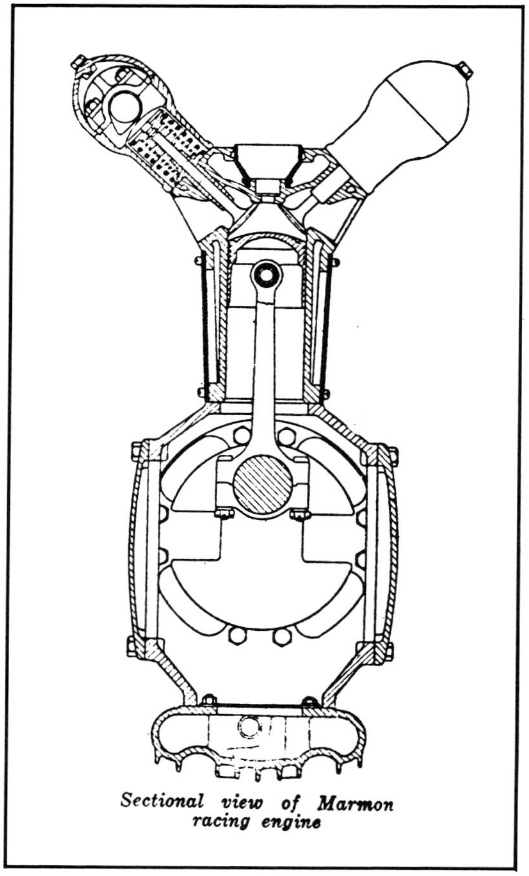

Fig. 8-25 Views of the "Marmon Racing Engine" for the Marmon 68 Special.

*automobiles are coming so fast that we decided to take some of our advanced engineering ideas to the race course to try them out. We are entering the race . . . because Marmon engineering research has developed many innovations which will revolutionize the passenger automobile as we know it today. We want to see just how near a state of perfection these innovations have progressed."*

Subtle credibility was given the release by the barrel crankcase construction described . . . every First and Second Era Marmon including the winning car in the first 500 mile race had just such a construction. But in the Third and Fourth Eras, only Miller and Duesenberg racing engines featured that superior construction.

The Cooper cars ran at Indianapolis again in 1929 but as 'Cooper Specials' without Marmon support. Fred Frame drove the ex-Marmon 3X to tenth place after having held the lead for laps 95 through 105.

## FIFTH ERA COMPETITION

One front drive Cooper car reappeared at the Speedway in 1933, this time powered by a Marmon engine . . . the classic SIXTEEN powerplant. It was equipped with four Winfield carburetors and had a displacement of only 190.8 cubic inches consequent to a 2.25 inch bore and 3.0 inch stroke. Turning the tables, it was identified as a 'Cooper 16'. A 4.5:1 axle ratio and 6.00 X 20 tires were used.

Ralph Hepburn qualified at 106.76 mph and the car carried number 23 as the Highway Parts Special. On lap 29 a connecting rod broke.

The following year Dusty Fehrnow qualified the rear drive Superior Trailer Special at 113.07 mph and ended the 1933 '500' in 24th place having been forced to withdraw on the 28th lap. The engine was identified as a Marmon SIXTEEN with a 2.916 inch bore and a 4.00 stroke, a CID of 330. Two Stromberg carburetors were fitted.

Fig. 8-26 Lee Oldfield with racer he built for Joe Thorne: During construction and at the Speedway in May 1937.

For the 1934 race, fuel consumption restrictions were imposed to slow the field. An average eleven mpg was sought by restricting the usage to 2.5 gallons for the 25 mile qualification and to 45 gallons for the 500 mile race. Mark Dees in his 'Miller Dynasty'[15] notes that Doc Williams was unsuccessful in qualifying the Cooper 16

Fig. 8-27 Twin-carbureted SIXTEEN photographed in the shops of Leo Bertelson in San Francisco circa 1949. Reprinted from *Popular Science* with permission. Copyright 1949, Popular Science.

because he found it was too thirsty to run the required distance on the allowed 2.5 gallons. Had Williams qualified, there would have been two Marmon SIXTEENs in the same Indy 500 . . . the rear drive Miller Superior Trailer Special and the front drive Cooper!

The Cooper 16 reappears in 500 race records for the 1936 running. That year Doc Williams successfully qualified and at 112.837 mph for a mid-field starting position. Displacement was reported at 201 cubic inches, inconsistent with the reportedly unchanged bore and stroke. Running as car #54, the Superior Trailer Special, Williams ended in 14th place when he ran out of fuel on the 192nd lap. Still thirsty!

Mark L. Dees in his fine 'Miller Dynasty'[15] provides some surprising facts in the 'case' of the Marmon Specials. Mr. Dees interviewed Cooper's Chief Mechanic, Reeves Dutton, while researching his book and learned that the Cooper cars were new from the ground up, including the engine and power train but were based on the Miller designs. The engine and suspension closely followed the Miller, but the transaxle, steering, brakes, constant velocity universal joints and body were original designs (as built before Marmon involvement). And most surprisingly, the cars were the work of some twenty men who had worked a year and a half under sponsorship of the Buick Motor Division of General Motors only to have that relationship severed just before race preparations. Cooper was allowed to acquire the cars and they raced at Indy with odd radiator shells only subtlely reminiscent of a 1927

Buick shell. Mr. Dees states that people who had driven both Cooper and Miller cars found the Coopers to have superior handling. He further notes that one of them in winning at Detroit Fair Grounds set an all time record on the dirt for a front drive car.

It is highly probably that Buick required a promise from Cooper that he would keep the secret of their aborted involvement as a condition to their release of the cars to him. Under this constraint he could only tell Marmon officials that the car and its engine were unique, albeit their suspicious resemblence. Based on this information, the advertising/sales promotion people presumably prepared the press release in good faith and would not have been in a position to recognize the deception that would be presumed by technically qualified readers. I also doubt, considering who were making decisions at Marmon Motor Car Co. in 1928, whether Howard Marmon saw the article before he read it in the magazine.

For the 1937 race, Lee Oldfield prepared a radical new car. It featured rear wheel drive from a midships Marmon SIXTEEN engine. It displaced 365 (of the allowed 366) cubic inches with a 2.697 inch bore and 4.00 inch stroke.

The magneto was driven from, and mounted on, the transaxle. The Stromberg DD carburetor was fully exposed above the engine cowling. Four exhaust 'stacks' protruded through each side of the cowling; the enclosure profile was limited to that of the basic engine.

The car is listed among "The 132 Most Unusual Cars That Ever (or almost) Ran at Indianapolis."[16]

Lee Oldfield had begun his career as a race driver but, after his Knox crashed into a crowd of people at the Syracuse, NY fair in 1911 with a high death toll, he

Fig. 8-28 Automotive Experts — Famous race drivers with their personal Marmon 34s: Tommy Milton 34 A Touring, Jimmy Murphy 34 C Phaeton, Barney Oldfield 34B Coupe, Eddie Hearns 34 B four-passenger sedan, and Bennett Hill 34 B Touring.

Fig. 8-29 Dick Teague's drawing of XVI. Reprinted by permission. Copyright *Car Collector and Car Classics* 1971.

became a test engineer for EMF, Amplex, Marmon and Nash and then became chief engineer at Stutz (12).

The engine drove a transaxle configured like the 1937 Cord automobile (with transmission aft of the differential and engine forward). All four wheels were independently suspended. The front suspension resembled the Packard type, with a transverse lower link positioned by a longitudinal link (clevis attached) from a ball joint anchor on the sub-frame. The central cockpit was of monocoque construction (although coolant plumbing on the exterior gave the impression of a tubular space frame). The forward sub-frame for suspension and radiator, and the unit powerplant, were self-supporting structures attached to the forward and aft cockpit bulkheads. The rear suspension was swing axle type with the wheel positioned by a wide-stanced 'A' frame and a transverse spring. The 'A' frame was attached to the engine block and transmission case by ball joints in line with the (inboard) universal joint. The leaf spring was attached to the bottom of the transmission case and to the 'A' frame trailing arms by vertical, double ball/socket, links. Consequently the car had the handling advantages of the same three-point suspension which featured the Models 34 and the current Covettes.

The car was built for Joel Thorne but most unfortunately failed to be ready in time to successfuly qualify for the 500. Severe displacement restrictions were imposed for the 1938 races. The International Racing formula 274.59 CID limit on unsupercharged engines eliminated the Oldfield-Marmon car from competition at the Speedway.

Lee Oldfield was a 1968 recipient of the Automotive Hall of Fame 'Distinguished Service Citation' in the Racing Category. His citation reads: *For a dedicated racing career as a driver, designer, mechanic and official of the Indianapolis '500' for forty-three years . . . credited with the 'Oldfield Special' powered by Marmon V-16 engine.*

The 1940s and 1950s were characterized by rampant enthusiasm for the modified cars dubbed 'hot rods'. Following the end of World War II, the dry lake beds of southern California provided a place where enthusiasts could test the effectiveness of their engine and car modifications. A fine example was the car owned and driven by Tony Capanna, who spent $2000 and several years of work to produce a car looking every bit like an Indy roadster. The roman numeral XVI on the cowl provided a hint as to what lived beneath the leather strapped hood . . . a Marmon SIXTEEN, stock except for a special Harmon camshaft and a water injection system worked-out by Capanna; it was reported to turn 6000 rpm.[17]

Cabanna set the unlimited class record of 137.24 miles per hour average for a two-way run. His one-way run of 145.39 mph was the fastest time ever clocked at any Southern California Timing Association Dry Lake meet up to that day in 1946. The event and car were commemorated in an illustration done by Dick Teague in 1947, the Richard Teague who for many years was Vice-president of styling for American Motors.

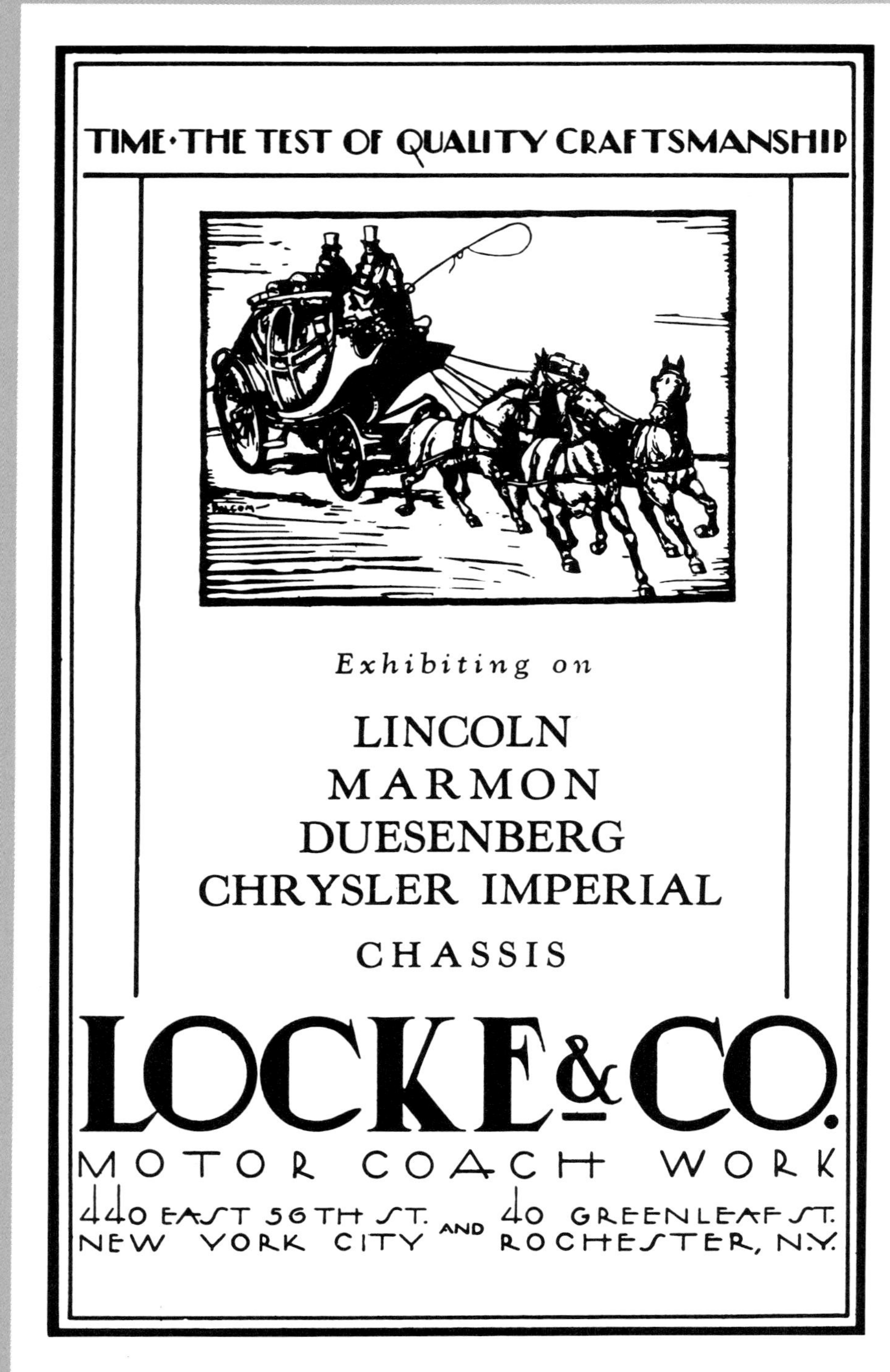


Locke ad from the souvenir catalog for the 1930 New York Salon from the W.E. Gosden Collection.

CHAPTER NINE

# THE MARMON CAR AS OBJECT D'ART

Three Marmon advertisements serve best to set the tone for addressing, in the following two parts, the coach builder creations On-Marmon chassis and the Interior and Exterior Finishing by the factory Styling Studios . . . Art and Color. The first ad appeared in June 1927, the second in June 1918 and the third in January 1927.

> Founder and Master of the new school of scientific car construction which has reduced the weight of the present day automobile, increased its strength, lengthened its life and multiplied its fuel and oil mileage.

> Nowhere are Art and Engineering more ideally mated than in the production of Marmon closed cars.

> Marmon designers have successfully sought colors most suitable to making the painting of cars into an Art. Nothing short of inspiration prompted Marmon to seek out sources where Nature works upon hard surfaces . . . the realm of precious and semi-precious stones . . . and Marmon Jewel Colors were born.

## PART 1 — "On Marmon" The Coachbuilder Creations

The title of this sub-chapter on the custom-bodied Marmon reflects the perspective of the coachbuilders, firms who in many cases had their origin in the designing and building of vehicles complete except for the horses provided by the customer. A Judkins custom limousine body mounted on, e.g., a Lincoln, did not become a Lincoln-Judkins Limousine; it became a Judkins Limousine, on Lincoln (Chassis). In fact, and particularly more so earlier in the century, the customizing often fully obscured the chassis origin.

The author is not aware of any custom bodywork of record on the First Era Marmon chasses. A limousine and landaulet by an unidentified bodybuilder were offered by the factory in 1908; a photograph of the latter appears in Chapter Two.

### SECOND ERA

The introduction of the Model 32 at the 1909 New York Auto Show was inherently an exciting display. But displayed, with the several examples of the new standard open styles, was a "Torpedo Style" touring. This was a custom creation, unlisted on the model roster, which attracted considerable press attention. The body builder, and it could have been the Nordyke & Marmon shops, is unknown to the author. It was described as being a fine example of the flush-sided design, very distinctive and representative of the latest in body design. It shows evidence of cast aluminum construction as practiced by Nordyke & Marmon.

Very significant was the display of the Torpedo Touring with the top in place and erect, because the car was a pioneer example of concurrent design of compatible body and top. It was in fact designed to be normally used with the top erect. Because the body sides were carried up to 'armpit' height, the top bows were quite short. Most unusual was the symmetrical design of the top with the upper surface sloping downward over the touneau exactly as it sloped downward forward of the windshield to form a visor several feet in length. The top of the cowl arched gracefully up from the rear of the hood and extended beneath the relatively low windshield and beyond the door opening nearly to the steering wheel rim. It was a three door design with the brake lever outboard the non-door at the driver location; the shift lever was inboard.

Fig. 9-1 Model 32 Torpedo Touring in the New York show.

Fig. 9-2 Model 32 limousine at the 1910 New York show.

'Torpedo' bodywork had been introduced by the French coachbuilder, Lamplough, in 1908 . . . it was characterized by a coherent integrated shape with unbroken horizontal lines. Marmon evidently saw its aesthetic features and sought the reaction of the American customer.

It might well be classed as a semi-enclosed design for it was a real effort to protect the passengers from the elements. The high 'freeboard' and low slung top anticipated the appearance of the Marmon-Herrington scout car Model A-7 of 1934 which served as prototype for a whole generation of scout cars worldwide.

All First Era Marmons had side-doored access to the rear set. The Second Era continued the practice at its onset but by 1911 all access was doored with optional deletion of the front doors. By 1909 an enclosed limousine, landaulet and coupe were listed styles, available only on order. Nordyke and Marmon did not manufacture closed car bodies until 1921. Chasses were shipped to the body plants or the bodies were assembled at Indianapolis in a separate facility subsequent chassis completion beginning in 1916.

Marmon had increased the wheelbase of the Model 32 for both 1910 and 1911, probably because the chassis could not accommodate closed bodies available from body vendors or constrained the ability of coachbuilders to respond to demands for 7-passenger bodies. The 1910 limousine illustrates the problem.

At the 1912 New York Show Marmon exhibited two coachbuilder bodies 'on Marmon'. They were closed bodies of usually pleasant appearance for the time; they of course benefited from the unusually low Model 32 chassis. The step transition from the individual chassis builders' unique hood profiles to the particular coachbuilder's cowl profile was still in vogue. Both were creations of the Denniston Company of Buffalo, N.Y. The coupe was identified as a "3-passenger Colonial

Fig. 9-3 Fig. Limousine by Dennison. An advanced concept.

Fig. 9-4 Colonial Couple by Dennison on the Model 32 chassis. At least one example was built without the pane dividers.

Coupe." It was in the "Coaching" style which lent itself well to fitting a variety of chassis as the rear bustle could easily be adapted to individual marque requirements. An identical body is preserved on an Oakland chassis in the Movieland Cars of the Stars in Los Angeles, Ca. It is characterized by three windows on each side, each divided into four panes, with the center window being in the door. Evidently one passenger faced rearward on the right side.

An *Automobile Topics* reporter described the limousine at the Marmon exhibit at the New York Auto Show in Madison Square Garden: *No matter what part of the Garden floor the visitor happened to be, he could easily spot the Marmon exhibit by the towering group of racing trophies. They proved a most potent attraction,*

Fig. 9-5 Model 32 landaulet, circa 1913. John Conde Collection by permission.

LIMOUSINE LUXURY

Containing a Description of New Limousine Bodies Marmon "Thirty-Two" & Marmon "Six"

**Nordyke & Marmon Company**
(Established 1851)
**Indianapolis, Indiana**

Fig. 9-6 Nordyke & Marmon catalog specifically for the limousine market.

Fig. 9-7 Marmon 32 (right) and 48 (left) limousines for 1913. Note how coachbuilder bodies were tailored to various chassis.

*but once attracted the visitor found lots of other things of interest . . . For instance, there was a novel enclosed body by Denniston, of Buffalo . . . officially termed a torpedo limousine . . . just because, no one knew why, but facetiously dubbed a "four-room apartment" which seemed to fit it a heap better, its interior looked so inviting. With its delicately finished exterior and lady-like upholstery, "bridal car" would have been equally appropriate.*

The limousine showed surprising perception of later trends in automotive styling.

Interest in enclosed bodies had increased by the following year to the extent that a separate Nordyke & Marmon catalog was available at the 1913 shows.

Fortunately when automotive historian Fred Roe encountered a custom-bodied Marmon Model 41 in 1941 (when it was 26 years old) he photographed it. It was a 1915 chassis with a roadster body by Hayes and Miller of New York. It is of interest because it shows the influence of the aircraft cockpit on the sports automobile, and evidences close attention to details and a pioneer effort in overall integration of a styling theme. Front fenders and running boards were modified and the cockpit coming (upper edge) extended rearward and downward, as a break line in the curved deck, to become tangent to the rear fender. The steering column was extended and severely raked to accommodate an extended cowl and consequent rearward displacement of the seat. The rounded top of the hood was carried through the cowl and the front doors, diminishing over the length of the door due to the downward slope of the body upper extremity, which formed a break in the quarter panel.

## THIRD ERA

Possibly because the Model 34 factory bodies were generally pleasant appearing and typically quite a bit lower than competitive cars, the custom-bodied Model 34 was typically a chauffeured town car creation.

Three other people accompanied C.E. Jeffers in the transition from Studebaker to Marmon on January 15, 1913. They were Chief Draftsman John Porter, Detailers Michael Puff and W. Roy Milner[1] They had been working in Studebaker's Detroit Plant (former E M F facility).

Fig. 9-8 Interior of Model 32 limousine

Fig. 9-9 Roadster body by Hayes & Miller on Model 41. F.D. Roe Collection.

Fig. 9-10 Rubay limousine, regularly offered, on the 1916-17 Model 34A.

Fig. 9-11 Limousine, by unidentified builder, as furnished to the French Army General Staff; 1917 Model 34A.

Fig. 9-12 Limousine by unidentified builder, Model 34A circa 1919. Note unusual attachment of windshield pillars to cowl band.

Late in 1913, when Milner was 24 years old, he was shifted from engine detailing to body design, reporting directly to Jeffers. The latter, according to Milner, was not particularly interested in body work and assigned Milner the task of liason with the custom body builders.

At that time N & M built only the open body models. Roy's responsibilities included the making of ⅛ scale drawings of suggested custom closed bodies and then performing liaison with the several builders who produced full size drawings for his approval before construction. The custom builders involved were:

| | |
|---|---|
| Kimball of Chicago | Brewster of New York City |
| Willoughby of Utica | Rubay of Cleveland |
| New Haven of Connecticut | Thompson of Pittsburg |

He liked the Rubay bodies best, but recalled that the Rubay limousines (an order of 100 by N & M) for the 34A had insufficient head room for entry and seating (by 1917 standards).

Two of his contacts, Fred Walthers of Rubay and Robert Lauer of New Haven, subsequently joined General Motors; the former was one of GM's best in Mr. Milner's opinion.

In July 1916, a two-page advertisement announced[2] "the permanent establishment of a closed body department in which limousine, landaulet, town car, and sedan bodies by such famous custom coach builders as Holbrook & Co. of New

Fig. 9-13 Rubay Town Car on 1919 Model 34 A chassis. Coachwork befitting the chassis. Applegate & Applegate Photo by permission.

Fig. 9-14 Holbrook town car on the Model 34A at the Palace Show in the late teens. Note 'fire-engine' striping favored by Holbrook.

Fig.9-15 The best of both Europe and America? New Haven body and RR-like disguise. MVMA Public Relations

York, Kimball & Co. of Chicago, New Haven Carriage Co. of Connecticut, E.J. Thompson Co. of Pittsburg and others will be regularly produced from now on."

"Formerly Marmon chassis were furnished with closed bodies only on special order. We will of course continue to supply special bodies built to order but the demand for such work, we believe, will practically be eliminated in view of the character, distinction, and exclusive style obtainable in the new designs by the above named well-known coach builders."

The ad reproduced the text of letters signed by the presidents of the several named custom body firms. All praised the styling advantages of the Marmon chassis which did not require structural strength in body sills and thereby made feasible attractive low bodies not otherwise practicable.

A most unusual car was the creation of the New Haven Carriage Co. for an unidentified customer who evidently wanted his personal transportation to represent the very best on both sides of the Atlantic. The 1916 New Haven coachwork included a custom radiator and hood patterned after the Rolls Royce. The myriad rivets were present but the R R badge was of course absent, and none was substituted. Knowledgeable motorists would have no difficulty in identifying the underlying chassis by the unique Marmon 34 fender and running-board/frame configuration and the equally unique Model 34A headlamps with their cylindrical parking lamps below.

Glenn Curtiss, who in 1910 had won a $10,000 prize for a non-stop flight from Albany to New York City and whose greatest contribution was the invention of the aileron for airplane control, was a proponent of aerodynamic styling of the motor car. In the following six years he had four custom-built bodies produced by Brunn Co. in Buffalo, N.Y., site of his firm at that time.

The first two were on Pierce-Arrow chassis. A luxurious center door sedan on the new Marmon 34A chassis in 1916 was a favorite of Mr. Curtiss . . . he even allowed himself to be photographed standing beside the car.

This was a Brunn center entrance sedan finished in a light shade including fenders and chassis with the "greenhouse" in black. It was characterized by a vertical V-windshield and a fabric-covered roof above the drip molding which extended only to the downward curve of the rear roof, like the later Model A Fords.

The Fourth Brunn body was also built in 1916 on a Cadillac chassis. It was a town car for Mrs. Curtiss but masqueraded as a Rolls Royce having a German Silver shell and headlamps exactly duplicated . . . the owners initials appearing instead of the R R on the shell.

Hermann C. Brunn (son of Hermann A. Brunn) in his article "Four Custom bodied Cars Designed for Glenn Hammond Curtiss" (Antique Automobile M-A 1977) stated "Brunn built quite a few of these Rolls radiator jobs during the teens. No complaints ever came from England...(Not so in the case of the Chevrolet Monte Carlo "Rolls" in 1976-1977).

Mr. Curtiss relocated to Garden City, L.I.N.Y. soon after acquiring the 1916 Marmon and Cadillac Town car. This was not far distant from the plant and "airdrome"

Fig. 9-16 Aviation pioneer Glenn L. Curtiss with Brunn & Co. body on Model 34 A chassis. (Herman L. Brunn)

Fig. 9-17 Dual-cowl phaeton by a Los Angeles coachbuilder on a Model 34 A chassis circa 1919. Applegate & Applegate Photo by permission.

Fig. 9-18 Exterior and interior of special Rubay body on Marmon 34B advance ordered for finishing and trimming in accord with buyer's options, 1917.

of the Wright-Aeronautical Corp. in Patterson, N.J. where the president of that builder of aircraft engines. F.B. Rentschler, not only later owned a 1922 Marmon 34B four-passenger speedster but authored a full page national advertisement for Marmon. In it, he said:

"In choosing a car for personal driving, my final selection was a Marmon Speedster. I was influenced primarily by the fact that I consider the six-cylinder Marmon engine by far the most desirable type for a motor car. In my opinion it is supreme in design and workmanship. I have carefully compared all types."

"In the selection of a sedan for Mrs. Rentschler, the deciding factors were ease of driving and control, smooth riding qualities and attractive and practical coachwork. So her car is a Marmon also."

A trend setting style appeared in December 1917 when an announcement was made that Rubay would offer special limousine and towncar bodies as well as a striking new close-coupled four passenger sedan on Marmon chasses with custom finish and trim at the buyer's option. The latter was characterized by center entry and very slender windshield pillars consequent to omission of front doors, by oval opera windows in the leather rear upper quarters, and by a general box-like roof and rear treatment.

The basic seating arrangement was three passenger with a folding right front seat stowed sideways against the bodyside, unless its use was necessary or desired. Rubay also built, for the Marmon Chicago distributor, several collapsible coupe versions of the close-coupled four passenger. Rear leg room was reduced but the seating arrangement was otherwise the same.

A handsome dual-cowl touring on the Model 34A chassis was part of the Marmon display at West Coast shows in 1919-20. The California based coachbuilder is unidentified.

In May 1920, the Marmon Chicago Co. Inc. (2430 Michigan Avenue) made available, for $10.00, to any Marmon dealer a 27 page book "Specially Constructed Bodies . . . for the Marmon Chassis by America's Master Coach Builders." It was color illustrated. The following firms were represented:

| | |
|---|---|
| C.P. Kimball | Hume Body Corp. |
| Larkin & Co. | New Haven Carriage Co. |
| Brewster & Co. | The Rubay Co. |
| Brunn & Co. | Graff Mfg. Co. |

Illustrations included a Brunn 4-passenger close-coupled touring with the tonneau rear slanted aft rather than forward, as more familiar on Brunn Lincolns in the late 1920s. That body (without installation) was priced at $2750, as was a victoria (top) touring. A Brunn sedan with disappearing upper door structures was priced at $3500. Most unusual was a limousine with the body continuously rounded from rear door to rear door in the plan view; the rear seat held only two passengers, with wide arm rests and a parcel shelf behind accommodating the curvature.[3]

About the same time the Marmon New York Co. had available four New Haven landaulet bodies with rear-quarter windows and leather aft of the windows. They were priced at $1900 each (F.O.B. New Haven) plus $75 installation charge.[4]

It was April 1921 when Marmon announced the availability of New Haven bodies on Marmon chassis . . . a suburban style and a collapsible town car fully convertible to a touring car. "The windows lowered flush, then the door-window frames and the divider window frame folded flat. The leather top then folded like any touring top." All body joints were covered with Liberty nickel castings to prevent leaking and rattling. Body upholstery was to be selected by the purchaser subject only to a $7.00 per yard limit. Body and chassis were to be finished in accord with the customer's preference. The body was priced at $2100 f.o.b. New Haven, CT (plus War Tax) net to the dealer. A five week delivery was possible assuming the chassis reached New Haven within two weeks of the order date.

Fig. 9-19 The four door sedan version of the Hume series of styles on the 1923 Model 34 B chassis.

Fig. 9-20 Victoria of the Hume body series available on Marmon 34B in 1923. The two passenger coupe followed similar lines. MVMA Public Relations photo.

Holbrook released a rendering of a cabriolet town car on the Marmon 34B chassis for 1921. Unusual was the extension of the sloping windshield downward as a molding curving forward near the frame; the shape repeated the rear body contour.

Brewster & Co. had mounted their fine coachwork exclusively on their own distinctive chassis from the beginning. When that chassis was discontinued, Brewster offered a choice of Marmon, Packard of Rolls Royce chassis; the purchase of the complete car was made from Brewster exclusively.[5] The Brewster relationship was between the Marmon New York Co. and the Brewster firm. The first showing of the joint product was at the Closed Car Show in Grand Central Palace, September 23-30, 1922.

Throughout the decade of the twenties there was an increasingly close relationship between the Hume Body Corporation of Boston and the Marmon firm. W.E. Pierce was president of the Boston firm and personally did some of their designs which proved to be stellar attractions at the Shows. Pierce evidently favored Marmon chassis for his creations.

The relationship evidently began with the pace car for the 1920 Indy 500 race which became a regular model offering by Marmon, including its Hume designed and built body, through the 1924 model year.

The Marmon Post announced in December 1920 the introduction of two new body styles: a 4-passenger sedan and a 7-passenger Suburban. Each would be built by both Hume Body Corp. and the Springfield Metal Body Co. to expedite early availability; the bodies would be identical in design, construction, paint and upholstery. All doors were rear-hinged. These styles were to be available to all Marmon dealers via the Marmon New York Co. and the F.E. Wing Co. of Boston.

Beginning in 1923, Hume Body offered what amounted to a family of special coachbuilder bodies 'On Marmon'. For two successive years, Marmon included two examples in their New York Show exhibit. These were mounted on the new Model 34 C Speedster chassis with its very low radiator and hood, as well as provisions for side mount spare wheels and fixed rear trunk. All factory closed bodies were mounted on the standard chassis. The 2-passenger Speedster was built on the model 34 B frame; the lower rear kick-up of the 34 C allowed a 4-passenger Speedster and a number of rear seated coachbuilder creations.

At the 1923 New York Show, a Hume-bodied 4 door sedan was shown. Marmon releases called it a "Special Sedan with real dignity and distinction of appearance, striking in its lowness and in the long sweep of its lines. The rear seat, 46 inches wide, is fitted with a removable arm rest. With this in place the capacity is four, but the increase to five (very slender people — author) can be made with utmost ease by removing the rest."[6]

The interior was done in what would have been termed "hardtop styling" in the 1950s. The several transverse top bows were exposed between the main two longitudinal bows; the central portion of the cloth headlining was then recessed and outlined in wood which was stained to match the American walnut panels underneath the windows.

The body exterior was characterized by restriction of the fabric roofing to the upper surface of the roof and the use of generous spherical radii on the upper rear quarters. The roof extended forward as an integral metal sun visor; beneath it was the sloping windshield already seen on some factory models.

These same features were also evident in a novel 4-passenger victoria coupe shown at the 1924 New York Show. With the exception of the windshield area, they were likewise evident in a town car built during the 1923 model year. Together they constituted a full line of custom-level bodies but were never assigned Marmon

Fig. 9-21 Holbrook proposal for a cabriolet town car on 1921 Marmon 34 B.

Fig. 9-22 The builder is unknown of this one closed body that truly matched the new standard in overall vehicle styling set by the Model 34 factory open cars. Note complete absence of joints or joint-concealing moldings. Note also the clever reveal — the only departure in the entire vehicle from smooth skin — below the rear door window which contributes continuity between the forward open and rear closed body portions. 1920 Model 34B.

body prefix numbers; they thereby qualify as true coachbuilder offerings.

The four-passenger victoria coupe featured a small folding seat for the right front passenger. But unlike other victorias where that seat was made as comfortable as feasible with the constraint that it must fold beneath the dash, the Hume seat resembled the more austere jump seats of a 7-passenger body. The unique feature was a 2/4 passenger arrangement: For the two passenger mode, one half of the rear seat rolled forward on a track, stopping in line with the matching driver's seat.

In that mode, there was no 'victoria compromise' in front seating. The jump seat came into use only when a fourth person was carried.

When Marmon announced in March 1925 the appointment of Mr. Pierce as Chief Body Engineer and Designer, it was revealed that he, at Hume Body, had designed the eleven new body styles introduced with the Model 74 chassis in November 1924.

Mr. Pierce was credited with the introduction of two-tone painting in the industry and the Marmon 'Jewel Colors' promotion.

Two 1923 examples of this Hume body line survive. They are very low mileage cars and attest to the excellant craftsmanship and fine artistry employed. Even today, their interiors are impressive. One is a 4 door sedan and the other is 2-passenger coupe both of which appeared at the 1923 Show. The coupe incorporated essentially the same features as the sedan. Based on the odometer (under 7000 miles) the survivor cannot be the example owned by Mrs. (Barney) Oldfield for she drove that car on two cross-country trips. (Barney himself made eight such journeys in his ex-pace car speedster while promoting 'Oldfield Brand' tires.) [7]

Two very unusual town cars were designed and built in 1923 by C.P. Kimball & Co., who were located at Michigan Blvd. and Pershing Road in Chicago, Illinois. Each was unusual for the consistency of styling despite the two representing opposite extremes of design philosophy.

The Kimball Town Cabriolet was built for Dr. Frank X. Walls of Chicago on a Marmon 34C 4-passenger speedster chassis. Only the tonneau and top were removed from the factory assembled 4-passenger speedster; the dual sidemount wire wheels, 32 X 4.5 whitewall tires, and small rear-mounted trunk were retained. The dealer-installed optional Bi-Flex bumpers front and rear were also retained. The headlights were the more massive, and suitably more formal yoke mounted optional headlamps included in the 1922 factory offering of fully-equipped cars including the speedster styles. The new body accommodated only two passengers and featured a false landau top with nickel sidebars; both rear quarters and roof were fabric covered. Below the beltline, the rear door and quarter were finished in sham French canne. A handsome carriage indeed!

The Kimball Town Landaulet customer was unidentified. It was described as a Victoria Six Fender Design, and was built on a modified Model 34 B chassis; The unique Marmon frame was reworked. The rear roll-up sections were removed from a second set of running 'boards' and inserted in the original 'boards' to produce a symmetric step section beneath the rear door. A bottom flange was evidently welded to the frame side member extending all the way rearward to the front of the rear door; this was necessary to replace the lost running 'board' structure. The original 'boards' were retained beneath the front doors. The front fenders were reworked and extended in a sweeping contour to the front door area. The Kimball body incorporated a new cowl and was of conventional continuous sill construction. The factory wood wheels and rear mounted spare were retained and the car remained bumperless. The cowl terminated at the high sloping windshield. The rather upright steering column position normal for 7-passenger bodies was

"Town Phæton Landaulet"
(Victoria Six Fender Design)
Mounted on Marmon Chassis

Kimball Town Cabriolet on Marmon Phaeton Chassis
Built for Dr. Frank X. Walls, Chicago, Illinois

KIMBALL MOTOR COACH WORK

Upon request special designs for any make
chassis will be submitted for your approva

C. P. KIMBALL & CO.
Michigan Blvd. and Pershing Road
CHICAGO ILLINOIS

Fig. 9-23 Two very different town cars by the same builder.

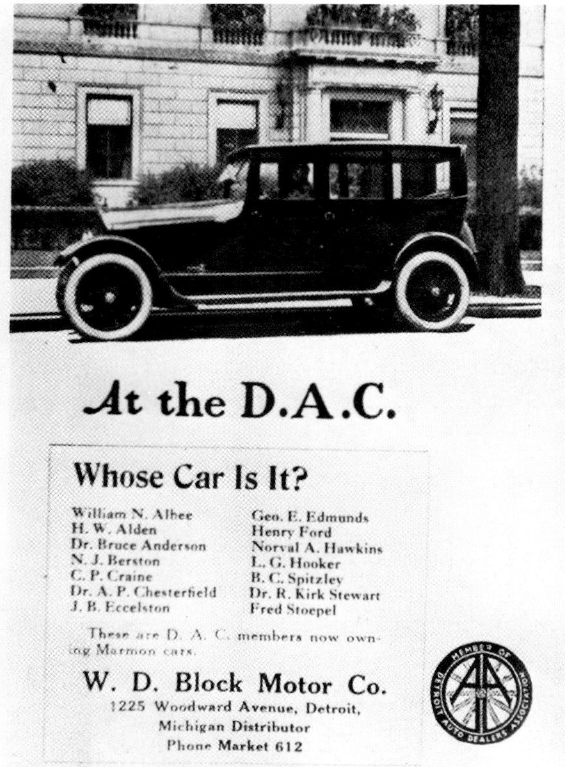

Fig. 9-24 The W.D. Block Co's. roster of distinguished members of the Detroit Athletic Club from a Garrick Theater program in 1919.

Fig. 9-25 Dealer ad offering custom bodies on Marmon 34B.

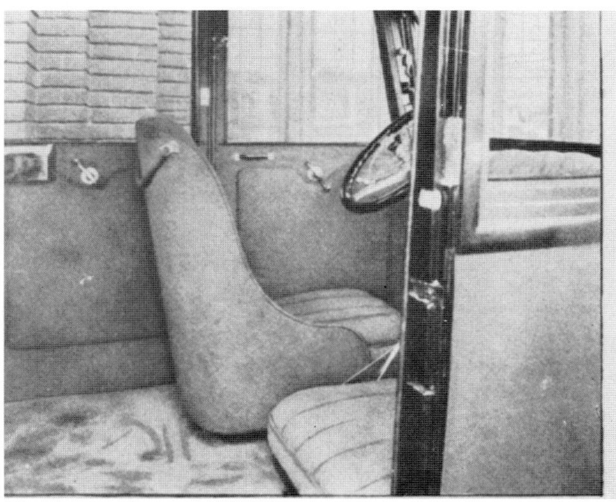

Fig. 9-26 *"This car which was built for Mr. Henry Ford of Detroit, is painted in Kimball gray-green satin finish. Its interior is upholstered in green Bedford cord. The guest or companion seat has a reversible back, and the pitch of the seat slants automatically with the shifting of the back."* Kimball body on a 1919 Marmon 34A. Note polished aluminum hood.

Fig. 9-27 The Rubay 'Moskovics II' on 34 B chassis.

Fig. 9-28 Kimball Body; short coupled sedan for five.

retained and the Kimball body is presumed to have accommodated five passengers (3 + 2 on jumpseats). Dr. Walls car had the rakish steering column and low seating of the speedster models.

C.P. Kimball & Co. were also commissioned to design a custom body on a Marmon 34 A (1919) chassis for Mr. Henry Ford of Detroit, Michigan. The driver's rigid bucket seat was accessed through the front (and only) left door. The rear seat and right-front seat were accessed via the rear (and only) right door. Both doors were hinged at the rear! The guest or companion seat had a reversible back and a mechanism which reversed the slant of the bottom cushion as the change was made; Clara could sit beside her husband or face her guests as occasion required.

The Ford Marmon "was painted in Kimball gray-green satin finish." Inspection of the photograph indicates the chassis, running gear and upper body (except the window reveals) were black, and the hood was published aluminum. The interior was upholstered in green Bedford cord.[8]

The 1920 Model 34 B was not Mr. Ford's first Marmon, but it may have been his last for his acquisition of Lincoln Motor Car Co. certainly mandated a change. Marmon's Michigan distributor, the W.D. Block Motor Co., ran an advertisement in local prestige publications, eg., the Garrick Theater Program of September 7, 1919 which showed a Marmon 34 A seven-passenger sedan in front of the Detroit Athletic club. Beneath it was the legend: At the D.A.C. . . . Whose Car Is It?

A tabulation of DAC members owning Marmons followed:

| | |
|---|---|
| William N. Albee | George E. Edmunds |
| H.W. Alden | Henry Ford |
| Dr. Bruce Anderson | Norval A. Hawkins |
| N.J. Berston | L.G. Hooker |
| C.P. Craine | B.C. Spitzley |
| Dr. A.A. Chesterfield | Dr. R. Kirk Stewart |
| J.B. Eccleston | Fred Stoepel |

This was prior to the introduction of the 1920 Model 34 B.

The W.D. Block Motor Company, ran a series of advertisements in the *Detroit Motor News* (Detroit Automobile Club — AAA) which featured original art work and messages. Their ad in the November 1919 issue featured the "Marmon Classic Foursome," a four-door, four-window sedan with opera window in leather rear quarters. The ad noted "The all-aluminum body makes the weight of this special closed car, including the tourist trunk, just equal the weight of a touring car."

The prices and other information in the Block ad are of interest. Presumably the prices, although F.O.B. factory (Indianapolis or coachbuilder city), are for fully equipped cars:

1920 MARMON PRICES

| | |
|---|---|
| Deluxe Sedan (Seven Passenger) | $6250.00 |
| Imperial Coupe (Four Passenger) | 5800.00 |
| Town Car (Four Passenger) | 6450.00 |
| Limousine (Five passenger | 6450.00 |
| Marmon Chassis equipped with Custom Bodies by these noted designers: Graff of Chicago, Rubay of Cleveland, and Erdman Guider of Detroit. | |
| Suburban Sedan (Seven Passenger) | $7175.00 |
| Classic Foursome Sedan (Four Passenger) | 6800.00 |
| "Moskovics II" Touring Sedan (Four Passenger) | 6800.00 |
| Imperial Coupe II | 5900.00 |

The ad identified three coachbuilders (Graff of Chicago, Rubay of Cleveland and Erdman Guider of Detroit) and four body styles, but failed to provide a hint at correlation. Other sources link Rubay to the Suburban and the Moskovics II (body $2675 F.O.B. Cleveland). The Rubay Classic Foursome and the Kimball five-passenger, both close-coupled, shared certain unusual styling details in addition to the rear-hinged doors. The bodies, because of the close-coupled four passenger arrangement; ended about mid-fender and were square cornered up to the beltline with a raised molding of contrasting or black color highlighting the unusual corner treatment. In consequence, the rear upper quarter had a relatively small corner radius which was also carried forward to the windshield. This allowed the

Fig. 9-29 The regularly offered Willoughly coupe on a 1921 34 B chassis. Mrs. and Mr. James Bertram of Stratford, Ontario, Canada. (Don Warren)

doors to extend upward, outboard of the roof supporting structure with only a slender roof extension above them. The arrangement resulted in the glass extending higher, to the level of the roof structure.

The Kimball version of this body accommodated five passengers (being wider over the forward part of the fender) and incorporated a rear integral luggage compartment similar to that on the Marmon cloverleaf roadsters. The spare was inclined, but the windshield was vertical. Renault-like hood louvers were used and a landau bar replaced the opera window.

## MODEL 34 COACHBUILDER STANDARD CLOSED BODIES

Nordyke & Marmon purchased their closed car bodies from outside sources during the Third Era until the opening of the new Marmon body plant.

When the 1921 Marmons were introduced about August 1920, two coupe and two sedan body styles were offered. The previous practice of prefixing all closed car bodies with an '8' prefix and serial numbering total closed cars in build sequence was replaced by specific prefixes... the open car practice of long standing. The 'purchased' (from Willoughby) coupe and (7-pass) sedan bodies were assigned '9' and '10' prefixes; "N & M Co. bodies" were assigned '15' and '16'

Fig. 9-30 The aluminum factory coupe introduced for 1922. (MVMA PR)

Fig.9-31 New Rubay limousine and town car announced in mid July 1921. Antique Automobile Club of America Library/Russell Stadt Collection.

Fig. 9-33 The 1921 Willoughby limousine style. Note the apparent use of the touring body cowl and front doors, although the latter are rehinged to open at the forward edge.

Fig. 9-34 Two-passenger N & M 34B coupe at the 1924 New York show. MVMA Public Relations Photo.

respectively. The N & M Co. Coupe weighed 3985 pounds versus the Willoughby coupe at 4875 pounds; the N & M Co. sedan weighed 4385 pounds, the Willoughby 5275 or about 900 pounds more in each case. The N & M Co. Coupe was featured in advertisements by ALCOA as confirmation that closed bodies of aluminum were feasible. The Marmon 4 door close-coupled sedan with N & M Co. body introduced for 1922 also weighed 4385 pounds.

The Rubay and Willoughby limousines were assigned '11' and '13'; their town cars '12' and '14'. Only the Rubay limousine and town car were carried in 1923-24; there were no other vendor bodies.

A 2-passenger Marmon coupe was exhibited at the 1924 auto shows. It featured a leather covered rear quarter with opera window and landau bar, two interior lockers. The cowl top vent, new for 1924, is evident, and the new optional balloon tires are conspicuous. This car and the unmodified coupe of 1922 offer an interesting assessment of the impact of tire equipment progress on car appearance: 4.5 X 23 tires versus 7.3 X 20.

## MODEL 74

The Judkins advertisement in the Automobile Salon program for 1925 included sketches of a Lincoln Berline Landaulet, a Pierce-Arrow Series 80 Coupe, and a Marmon 74 four-window Landau Sedan with leather roof and rear quarters.[3]

By the 1925 Automobile Salon, New Haven Carriage, Rubay and Kimball had retired from the coachbuilder body business. New Haven went out of business after losing the Locomobile body account; Rubay following an abortive ill-timed entry into the automobile business in 1922. Hume had been acquired and operated by Marmon but was liquidated at the end of Model 34 production.

The Robbins Body Company of Indianapolis reportedly built a number of custom bodies for Marmon (and Stutz).[3] Milspaugh and Irish was another body builder in the same city and a likely builder of bodies 'On Marmon'. By the late 1920s, Robbins had been absorbed by Briggs, and Milspaugh and Irish were out of business.

## SERIES 75
## COACHBUILDER DELUXE BODIES

The Marmon Series 75 was offered (1927-28) with both the standard line of Marmon bodies and a custom line of bodies by Locke & Co. The latter, although unique closed bodies, shared a common styling motif of Marmon design studio origin. Although characterized by a wide selection of fabrics and exterior color com-

Fig. 9-35 The regular DeLuxe Series 75, Locke bodied sedan. Applegate & Applegate photo by permission.

Fig. 9-36 The term 'Marmon Speedster' was so respected that a Series 75 seven-passenger version by Locke was released. Applegate & Applegate photo by permission.

Fig. 9-37 The De Luxe Series 75 four-passenger speedster by Locke with the running board searchlights with which it was usually shown in advertisements.

binations, they do not meet the criteria for custom bodies. The factory-mounted Series 75 Locke bodied line was similar in that respect to the Le Baron bodied MARMON SIXTEEN.

The Locke closed bodies were distinguished particularly by embossed oval monogram pads on the doors between the belt and greenhouse moldings, by the upward offset of the belt molding just aft of the rear door, and by the loop form door handles. Seven interior color and trim options were offered, with the fabrics for seats and walls being different.

The four-passenger and seven-passenger speedster bodies were Locke styled with handsome dual-cowl configurations and integral-formed body trim moldings. Both closed and open Le Baron bodies on the Series 75 were aluminum skinned.

Spot lights, supported on stanchions from the running boards, were a factory option as evidenced by their appearance in Marmon ads for the Speedsters.

## ON MARMON SERIES 75

The Willoughby Company of Utica, N.Y. built a 4-passenger sport sedan body 'On Marmon 75' in the fall of 1926[9] It was described as a distinct departure from usual design practice, having a very low belt line and correspondingly larger windows. The window increase was limited somewhat by having the reveals below the windows several times the usual width and painting them in a lighter shade. Windshield pillars were unusually slender; an English & Merrick vertical windshield was fitted and Triplex Safety Glass used throughout. It was painted Ocean Blue and

Fig. 9-38 The Locke town car on Little Marmon shown at the Commodore Hotel exhibit conveyed the expectation that class would be recognized even when 'downsized'. Reprinted by permission; "Locke & Co." Harold Emmons, *TORQUE* (MI region CCCA)

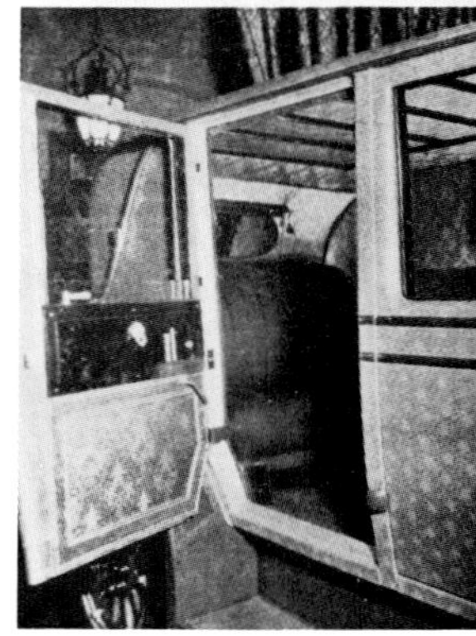

Fig. 9-39 CHILD'S Fabric Body on Marmon Series 75.

Seafog Gray with an interior of two shades of green gray.

An original (ie, unrestored) example of a very handsome landau sedan by Judkins on Marmon 75 survives. Although having the characteristic Judkins 'retreating' windshield (ie, forward leaning) this automobile has a racehorse trim and grace like the Model 34 and quite unlike the typical Series 75 closed cars. The side-mount wire wheel equipment and rear trunk of course contribute to the sporting appearance.

When the new firm of Dietrich, Inc. first appeared at the Automobile Salon in December 1926, they displayed bodies on one Lincoln, one Marmon and two Packards.

Locke created designs for the 1928 Series 75; the artwork with legend 'Coachwork Executed by Locke & Co.' survives but no examples or photographs are known. They were a landau town car and a cabriolet style 4 door, 4 window sedan with separate trunk, both with side mount wire wheels, and a formal 7-passenger Berline limousine.

Body squeaks plagued the metal skinned wood-framed bodies; it was a real quality body that was and remained squeak-free. A claimed solution was the Weymann construction in which the wood members were joined by flexible coupling elements; this novel structure was covered with a fabric skin. A competing flexible construction was the Childs system, employed by the Fabric body Corporation of Detroit, under the Kenneth L. Childs patents.

Both firms exhibited examples of their bodies on 1927 chasses at the Custom Body Salon in New York. Fabric Body Corporation displayed a four window sedan on the Marmon 75 chassis. The *Motor Age* reporter wrote: *"Appearing at a distance to be a creation of silver, this body has an outer surface which is fabricated with Meritas metallic brocade in a patterned silver gray with matching tire cover. The car carried a radio set in the instrument board, an expanded metal lath used in the body construction serving as the antenna. Fabric Body's silvery sedan and Judkin's coaching brougham, on Lincoln, reproducing the 1845 privately owned stage coach, were the two outstanding novelties of the exhibit."* The Judkin's Lincoln survives in the Harrah Collection and is well known; where is the silver Marmon?

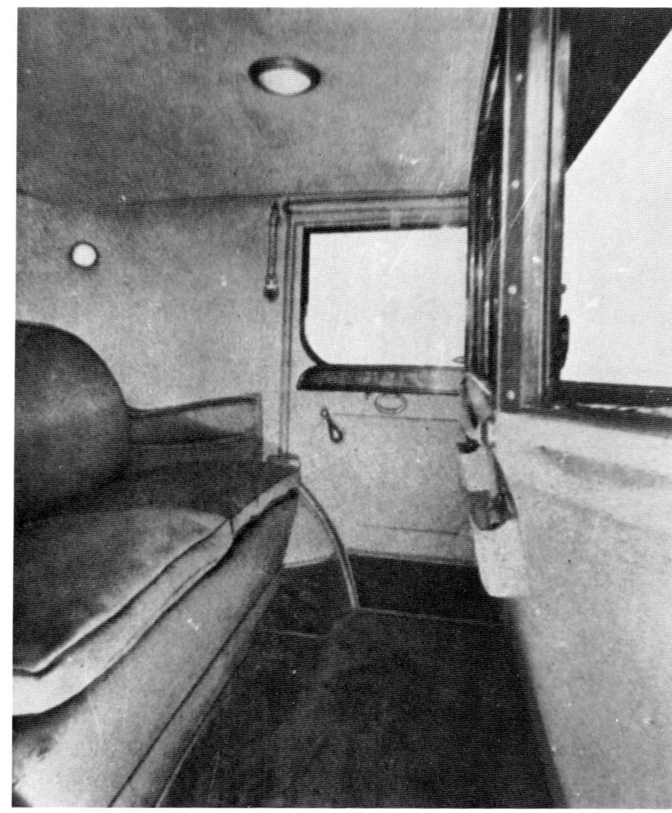

Fig. 9-40 Tasteful interior of the LeBaron town car on Little Marmon.

Fig.9-41 The Victoria made a very handsome Little Marmon. Applegate & Applegate photo.

Fig. 9-42 LeBaron four door brougham was a regular offering on Little Marmon. Applegate & Applegate photo.

Fig. 9-43 The First Series '78' town car by Locke. John Conde photo by permission.

The roster of coachbuilders exhibiting read like an industry census and warrants listing here: Brewster, Brunn, De Causse, Derham, Dietrich, Fleetwood (pre GM), Fisher, Holbrook, Judkins, Le Baron, Willoughby, Locke, Fabric Body and Weymann. Among the missing were New Haven, Hume, Rubay, and Van den Plas.

The chasses mounting them were: Cadillac, Cunningham, Duesenberg, Franklin, Isotta-Fraschini, Lincoln, Marmon, Minerva, Packard, Pierce-Arrow, Rolls-Royce, and Stutz.

## FOURTH ERA

Marmon sought to enhance the Luxury car status of the relatively low-priced Little Marmon at its (1927) introduction by the display, at the Belmont Hotel, of a Locke-bodied town car. It was a cabriolet style with leather covered roof and rear quarters, the latter being trimmed with landau bars. *Autobody* for January 1927 considered it to be so pleasing in appearance that it warranted congratulations to the body builder for such an achievement on so short a chassis (116 inch wheelbase). There were of course no jump seats. The rear seat was fitted with two attached untufted seat bottom cushions (similar to the Regency style of the 1970s). Below the window were wood panels and an embossed effect on the cloth panels below. Interior color was three tone: Light wall covering, darker seats and darkest carpet. Walls and headlining were champagne-colored broadcloth, the seats were a darker Laidlaw broadcloth.

The Little Marmon four-window custom sedan was Locke-bodied and styled. Its appearance far outshone the factory bodied jobs which the author considers mediocre. It was devoid of the heavy overhung integral visor; instead a 'cadet' visor was used with a slight forward lean to the roof above the pillars. The belt molding rose sharply to end at the front pillar; it separated a dark finish on the top of the hood from the light side panel color. The latter was also used on the window reveals of the greenhouse.

In addition to the above sedan, the Little Marmon factory-offered "Custom-Built" line included a 6 window sedan, a victoria and a cabriolet. The victoria, which was really a close coupled two-door sedan with large rear trunk between the fenders, was a very handsome car. Marmon evidently owned or acquired the design rights as this body was offered (without the custom-built notation) on the Series 68, 78, (1928-29) and the Roosevelt/Marmon Roosevelt/Marmon 70 (1929-31). It may have been Locke-styled but otherwise manufactured even in 1927.

Van den Plas built at least one example of Weymann construction on the Little Marmon chassis. It was a four-window sedan of unusual width at the rear seat. The separate luggage trunk was contoured to fit closely against the body and was finished in the same grayish-blue coated fabric. The chassis and molding were black with a silver stripe on the latter. The interior was done in a beige-rose fabric. The side mounted spares were suspended from brackets and were independent of the well-less front fenders.

The following year (1928) the Little Marmon had grown considerably into the Model 78. Again a Locke Town Car was shown, this example benefiting from the longer chassis and being much more, though pleasantly, formal with vertical slender windshield and coaching lamps. However sidemounts were used, whereas they were not on the Little Marmon.

Mr. C.J. Joyce of Pass and Joyce, Marmon distributors in England, commissioned a handsome sports body by James Young and Co. Ltd. on the Marmon 78

Fig. 9-44 Pass & Joyce was the Marmon distributor in Great Britain. This was Mr. Joyce's personal Model 78 with unidentified British coachwork.

Fig. 9-45 A four-passenger cabriolet by the firm Million-Guiet on Marmon '78' at the Paris show.

Fig. 9-46 Town Car on Big Eight for Her Royal Highness Queen Maud of Norway.

Fig. 9-47 Model 88 'Longchamps' with Weymann metal-skin body.

chassis. Very nice and very British.

Million-Guiet displayed at the 1929 Paris Salon a four passenger sport cabriolet on Marmon Model 78. Noteworthy were the chromium plating of the Marmon pyramid embossed on the forward fender edge and the white canvas piped with blue to match the two-tone blue exterior.

Weymann had established an Indianapolis plant and produced a 4-door brougham using their fabric-skin construction. The Marmon president drove a Weymann fabric-bodied Big 8 (probably a First Series 1931) until the start of SIX-

Fig. 9-48 'Grand National' four-passenger convertible by Locke.

TEEN production in the spring of 1931. His son, Bruce, recalls it being stolen; when recovered some weeks later, the odometer registered an additional 7500 miles. That Big 8 was preceded by another having an Italian coach-builder body which made the car quite distinctive.

At least one town car was built on Marmon Big Eight. It was a royal coach . . . constructed by LeBaron to the design by Walter Dorwin Teague and delivered to Queen Maud of Norway.

Fig. 9-49 The LeBaron seven-passenger touring on Big Eight. At least one has survived. Owner R.S. Ehinger.

Following an announcement by Weymann Body of their entry into the metal-skinned body business, a Weymann 5 passenger was a factory offering (body prefix 5) on the First Series 1931 Big Eight (Sept.-Dec. 1930). It weighed 4175 pounds versus the 4195 pound club sedan of conventional construction. The Marmon release identified it as derived from the 'Longchamps' model created by Weymann for el Duque de Hernani of Madrid, Spain.

The 'Longchamps' was a companion offering to the "Grand National Sportsman" model with body by Locke which was a two-door close-coupled convertible sedan characterized by very large (40 inch) doors offering convenient entry to front or rear seats and a large rear trunk integrated into the body. A small locked door on the side of the trunk gave access to the concealed fuel filler cap on both styles.

The 'Grand National' was a production release of what had been expected to be a one-off coachbuilder job 'on Marmon Big Eight' for Mahmaud Sabit Bey of Cairo, Egypt. Marmon officials admired the car and ordered it released. Radio personality "Smilin Ed" McConnell took delivery of his from the Marmon-Detroit Co.

When Marmon added phaetons to its three 'Marmon' lines in May 1930, the Big Eight offering was a 7-passenger custom body by LeBaron with a top covered in 'Teal' and fitting, folded, into a 6 inch boot.

## FIFTH ERA

The SIXTEENS all carried Le Baron bodies, with at least three exceptions. They were not custom-bodies in the usual meaning of the term. The exterior styling was the work of the Walter Dorwin Teagues and the interior the artistry of Helen Dryden. The fabrication of the massive and handsome bodies was then entrusted to the highly regarded Le Baron craftsmen.

The SIXTEEN was marketed to the custom-body clientele and the Le Baron production of bodies was an appropriate solution. Invariably, a custom-bodied car

Fig.9-50 The Waterhouse soft-top touring on Marmon Sixteen for Colonel Prentice, son-in-law of John D. Rockefeller.

Fig. 9-51 The prototype SIXTEEN — a close coupled sedan — first displayed for the news media. Note the absence of a hood ornament and the chamfer on the hood and radiator corner. Applegate & Applegate photo by permission.

was a step, style-wise and luxury-wise, above the available factory bodies. A good example of this stature is the handsome Judkins-bodied landau sedan on the Marmon Series 75 chassis which survives today. But the MARMON SIXTEEN factory styling concept was not only a quantum advance in automotive styling but the body was integrated into the overall styling and personality of the car to such a degree that stylists in the studios of the custom builders must have found the SIXTEEN chassis a most demanding challenge.

The stylists at the Waterhouse studios in Webster, Massachusetts, were one group who saw their solution mature; it was not however an alternative to the Le Baron body, rather an additional body style. A pair of Waterhouse phaeton bodies were mounted on a pair of special (1932) SIXTEEN chasses with wheelbase increased from 144 to 153 inches.

Both the Waterhouse phaetons have survived without ever having experienced a period of neglect. Both were built for Colonel and Mrs. E.P. Prentice. One had a fixed top and "performed so well and was such a delight to Mrs. Prentice (the daughter of John D. Rockefeller) that it was used in her family at their estate until 1963. Or roughly 32 years!"[10] When her brother, Winthrop Rockefeller opened his Museum of Automobiles at Petit Jean, Arkansas, the car was retired and presented to the collection. The second was identical except for a folding type top. Both were seven-passenger bodies which featured secondary instrument clusters for the rear seat passengers. The second Waterhouse phaeton was acquired and meticuously maintained by Mr. Edmund L. Robinson.

Besides the two lengthened chasses (153" for the Waterhouse phaetons) there was at least one shortened chassis delivered by Marmon to a coachbuilder. It was a 136 inch wheelbase cowl chassis shipped to Le Baron-Detroit Co. (3600 Maldrum St.) in early 1932 to mount a sedan-limousine body designed by Hugo Pfau for Mr. Michael L. Benedum of Pittsburgh, Pennsylvania. Mr. Pfau discussed this car in his 'Custom Body Era'[3] and reproduced the Le Baron-Detroit Customer Specifications form as completed for Mr. Benedum's car.

The rear folding seats were Rostand type, rear facing. The divider glass, for rear compartment privacy, was required to lower flush and to have minimal side guides and no header above. Mr. Benedum intended to do considerable driving himself and an extra pillow for the front cushion was furnished since the customer and his chauffeur differed in height only in the legs. There were minor customizings (after all, what do you do to a SIXTEEN?) such as a door on the central compartment, side-marker lens on the tail lights (NHTSA take note!) and Waycross windshield wipers. A skeleton type outside visor was fitted and had a colored glass panel.

The exterior was finished in Buckingham gray including the radiator shell and shutters. Cornice Gray was used on the window reveals, wheels and hubs except for the center octagon portion.

Interior trim was Weiss striped broadcloth, plain paneled with semi-loose cushion effect. The doors were done in matching leather and there were curtains on the front doors as well as the usual locations, plus a special flap to close the 8 X 25 rear window. A monogrammed lap robe was made of trim cloth with a plush backing to match.

A chromium plate with the initial B engraved in black was to be attached to the radiator shell.

A cowhide trunk to fit the standard Marmon rack was colored to match the car and fitted with the usual two black suitcases and one hatbox. Oddly, attachment by clamps and thumbscrews was specified.

The car was to be used for travel to and from the Oklahoma oil holdings of Mr. Benedum.

For the first showing of the SIXTEEN prototype, Marmon requested the prototype be included in the Le Baron exhibit at the 1931 Automobile Salon (attendance by invitation only) in the Hotel Commodore ballrooms in New York on the basis that the SIXTEEN qualified for the coachbuilders' show because all 'standard' bodies would be built by Le Baron. Hugo Pfau, then with Le Baron, recounted the 'Battle of the Marmon' in his *'Custom Car Era'*[3] and it is repeated here because of the styling assessment therein: *The Marmon V-16 engine was introduced here. While the bodies were built at the Le Baron-Detroit plant, they'd been designed by Walter Teague with a very angular lines and we weren't very proud of them. The Marmon was relegated to the back of the East Ballroom stand behind a Lincoln Convertible Roadster. Some of the Marmon executives complained but we convinced them that we controlled the space and had sole decision on the placement of the cars. Or so we thought. When the Salon was about to open the second day, the Marmon stood in front and the Lincoln in back. Not only that, but every door was locked and the brakes tightened so that the cars couldn't be budged. I still haven't found out who switched them during the night. At any rate, we got the cars back to their proper places before too many visitors arrived. Next morning, guess where the Marmon was! This time we got a top-level conference to settle matters for good and even some Marmon mechanics to help move the cars."*

Perhaps the displeasure of the Le Baron stylists with the Teague prototype influenced the production configuration? The sharp edged chamfers on the hood and radiator did not appear on the production car but it is almost a certainty that Marmon officials had made the decision much earlier and showed the number one prototype at the Salon due to unreadiness of the second set of prototypes shown at the public shows in January.

Another Le Baron-built body style, although no example is known to exist, would qualify as a custom body despite its non-LeBaron stylist. That was the alternate victoria body style described by Teague, Jr. as having an integral trunk built into the sloping rear deck.

Alexis de Sakhnoffsky, the Hayes Body Co. stylist, designed a custom five-passenger victoria built by Hayes on a Marmon SIXTEEN chassis. W.H. Hoagland, President of Hayes, was a member of the Marmon board of directors in 1932-33 representing the financial interests of Hayes Body Co. as a principal creditor; it is possible, if not probable, that the first example of the Hayes victoria was created for his personal use. More than one were evidently built as attested by the photograph presumably showing golf-star Bobby Jones taking delivery from Jack Hendricks, Jr., manager of the factory branch operation in New York City.

Design proposals were made by Judkins and by Murphy but are not known to have been built.

Marmon offered the SIXTEEN chassis to the coachbuilding trade for $4375.00 in 1931-32 and $3750.00 in 1933.

## PART 2 — Art and Color — Exterior and Interior Finish

Not only because Marmon was in the forefront of the development of automobile decor but also because of the importance of such information to the authentic restoration of surviving examples of the marque, the author has included this chapter with its rather detailed recounting of paint and trim details and options. Such data is also significant to the degree that it documents public preferences and perceptions. Particularly because it is gathered together in a single chapter (although

Fig. 9-52 The Custom Victoria on SIXTEEN designed by Alexis de Sakhnoffsky, the Hayes stylist/designer. MVMA photo.

Fig. 9-54 Comparison of 1917 (upper) and 1918 tourings with optional victoria top. Note belt highlight with new higher doors.

there are paint and trim references elsewhere in the text), it may be of special interest to those who today work in the industry's art and color studios.

Time has taken its toll of paint formulas . . . fire disasters and other distressing circumstances. However, even if the formulae and the code numbers for the color names used in the Marmon Eras were available, the paints could not be reproduced today because the specific pigments then used are unavailable to the paint formulator. However, paint technicians can and do precisely match old paint samples with contemporary formulations.

First Era Marmons had a Brewster Green body. Prior to 1907, the 'running gear'

Fig. 9-53 Golfer Bobby Jones with Marmon's New York manager, J. Hendricks, and one of the DeSakhnoffsky victorias.

was carmine; thereafter it was also Brewster Green. Upholstery was in dark green leather using spring edge cushion seats with spring backs.

## SECOND ERA MARMONS

The Model 32 finish varied with the model year. For 1910, the body and chassis were black with red striping; the finish was baked. The following year a blue black with gray striping was substituted and the wheels were gray with black striping. For 1912 and 1913, blue black was continued on the body and fenders, with white striping; The running gear (chassis) was olive with blue black striping. Upholstery was leather with the best grade of curled hair and with leather welts and bindings.

The New Series Model 41 announced in July 1915 featured straight piping of the leather upholstery, a claimed innovation. Special tools were required to install the padding.

## THIRD ERA MARMONS

### MODEL 34 A

The Model 34 Marmons were finished attractively both within and without and continued the reputation begun with the very first cars of the marque.

The upholstering panels and seat back assemblies on open cars were innovatively assembled individually and clipped in place to allow removal for body accident repair. The seat backs were entirely of steel construction, far ahead of their time. Surprisingly, Marmon suggested purchase of an extra set of upholstery for the alternate season . . . dark for winter, light for summer. Leather upholstering was tufted in long straight pipes.

The 1916 '34 A' Models had a standard finish of Rich Dark Blue, with fine Robin Egg Blue striping, black frame and chassis and fenders. The standard wheel color was Ivory. Two extra cost ($25) options were available: Gray with Napier Green striping, and Maroon with Red striping. Any other color was available at a premium of $50., excepting white and delicate colors, for which the extra charge was $75. A Mohair Brussels carpet was used in the rear; the front floor was metal-bound linoleum.

The 1917 '34 A' Models had some colors with which special chassis finish was standard. Coach Blue medium and Dark Maroon were standard options. With blue, the chassis and radiator shell were black; with maroon, the shell and front suspension were also in that color, but the fenders and remaining chassis were black.

The illustrated 1917 touring with optional victoria top is representative of the extra-cost option of chassis and body all in one color. It carries the vertical windshield of the 1916-17 Model 34s. The sloping windshield of the 1918 Model 34s is seen on the second illustration of the same body style. That example has the standard finish but illustrates a subtle improvement in appearance consequent to all new body panels. The two illustrations are taken from Marmon ads; the latter stressed the sloping windshield and the higher door tops and more flowing lines at cowl and seat. Note that a full length highlight results . . . a perceived belt line . . . making the car appear longer and therefore lower.

### MODEL 34 B

The 1920 34 B open cars were upholstered in long-grained, hand buffed black leather over curled white hair and Marshall springs. However the 2-passenger speedster was done in bright black leather, and the seat was divided. A Marmon release described an upholstering detail: "A listing is sewed to a foundation and then the leather is sewed to this listing, making a very flexible pad."

Closed bodies were upholstered with best quality all-wool fabric in pleasing patterns, over curled hair and Marshall springs. **Carpets, front** and **rear,** were in colors harmonizing with the fabrics.

The colors on the 34 Bs were the traditional Coach Blue Medium and Maroon on the open models and Coach Blue, Maroon and Brewster Green on the closed models. But the new 2-passenger speedster model was finished exclusively during the 1920 model year, unless special ordered, in (Wasp) yellow. All chassis and radiator shells were black, except for the maroon open cars.

A vanity case, smoking case, electric cigar lighter, dome lights were standard, as were foot lights for entrance and exit illumination which were actuated by door opening. Town Car and Limousine had a dictaphone for communication with the chauffeur.

The basic character and design modifications for 1922 highlight the sophistication of the Marmon "art and color" team — truly remarkable for the time and certainly trend-setting.

The doors of the four-passenger touring were slightly lower than those of the seven-passenger touring; this reflected the higher cushions and more erect seating necessary to accommodate seven. The top of the (cloverleaf) roadster was lower than on the tourings to offset the effect of the shorter length on apparent height.

For 1922 the seven-passenger touring was lengthened three inches, distributed one inch to the front compartment and two inches to the rear compartment, whose seat was 48 inches wide. The front seats of the tourings and speedsters were undivided.

Both open and closed bodies were of aluminum skin construction and described as "unit" construction with the chassis. "All the body joints were welded both on the inside and the outside to give the body its maximum strength and to insure against squeaks, rattles and strains coming from the hardest usage. This unusual practice assures a tight, long-wearing, non-rattling body."

Closed bodies were upholstered in a striped cloth on the seats and a plain cloth for the headlining and sides. Curtains and carpets were trimmed in a cloth matching the upholstery. A rear view mirror was standard and called a "mirrorscope."

The trunks on the 4-passenger speedster and close-coupled sedan, although of dust-proof construction were fitted with a protective covering and included two fitted cases.

Tonneau lights were actuated by door-opening.

The 1922 open models were "furnished in two color combinations — Coach Blue Medium and Maroon Deep. The standard blue finish consisted of body and hood in blue, while the frame, wheels, axles, fenders, dust deflectors, radiator shell and moulding are black; striping around the body, light azure blue. The combination used for the standard maroon finish is black frame, moulding, axles, fenders and dust deflectors with maroon body, hood, wheels, and **radiator shell;** body striping red. With the (2-passenger) speedster, two options are given: Dust-Proof Gray and Blue, Medium.

'Closed cars are offered in three options of colors — Brewster Green, Dark Maroon, Coach Blue Deep. These colors are all used in combination with black, the lighter color being used for all panels and wheels, the black being used on fenders, frame, running-gear and the upper portions of the body, above the molding line.'

'On special order, and at extra cost, cars can be finished in any color or combi-

Fig. 9-55 The 7-passenger touring style of the Model 34B Special offered to dealers in 1922 as fully equipped with all popular options and including special paint combinations.

Fig. 9-56 Interior of the sedan body built by Locke & Co. for Marmon DeLuxe Series 75. The wide tufted plaits give an air of extreme comfort while the spring-upholstered auxiliary seats offer similar comfort and ample knee room. "Locke & Co.", Harold Emmons in *TORQUE* (MI region CCCA), by permission.

nation of colors desired, if sufficient time is allowed. Prices for painting cars other than standard will be supplied on application.[11] The 7-passenger Suburban model displayed at the 1922 New York Show was black with exception of Holland Blue hood and body below the greenhouse belt. Upholstery was blue with black stripes.

Fig. 9-57 The Model 78 interior showed no compromise despite being priced in the $2000 class — inlaid wood and luxurious fabrics.

Front and rear floors were fitted with a hair carpet . . . chosen for its durability and cleanliness.

The three deluxe open cars introduced in the fall of 1921 carried "Special" finish and equipment installed at the factory to avoid the delay associated with custom orders for the most popular custom items and colors.

The Special Seven-Passenger Touring included a second rear spare tire and wheel with tandem bracket and two tire covers.

The Special Four-Passenger Touring included a rear trunk and rack with nickel rear body bars (to protect the finish from the trunk). Two spare tire and wheel units were "sunk into the running boards" and fitted with tire covers.

The Special Two-Passenger Speedster had a lined and lighted rear compartment and tandem rear mounted and covered spares (instead of the single spare on top of the bustle).

All of the Specials had special fabric top material and Spanish leather upholstery. The front of the car was dressed with a nickel plated Hartford single bar bumper, a "nickeled radiator bar cap" (dogbone) and nickeled Motometer, nickeled drum type headlights in nickeled yoke mounts including dual crossbars. The windshields were fitted with "cozy" windshield wings, a spotlight-mirror unit on the left pillar, and a glass sun visor. Nickeled Yale hood locks, a nickeled drum rear tail light, vacuum-operated windshield wiper, rear view mirror and a cigar lighter on the right hand instrument board completed the impressive list.

Very specific information on these special models is documented here because the information has been somewhat obscure. In consequence, competition judges and owners themselves have doubted authenticity of some of the Special equipment, particularly the headlights which are unique to the Specials. The degree of color and trim detail specification is also unique and justifies a wide latitude in refinishing a Marmon of that vintage, particularly because the options reflected popular special orders.

De Luxe Four-Passenger Touring Paint and Trim Options

| | |
|---|---|
| 1. Body and hood | Motor Car Red, deep; |
| Fenders and chassis | Motor Car Red, deep |
| Upholstery | Radel's Armster Gray #890 |
| 2. Body and hood | French Brown |
| Fenders and chassis | Olive Brown |
| Wheels | Straw |
| Upholstery | Radel's Spanish leather — light brown #1400 |
| 3. Body and hood | Dustproof Gray |
| Fenders and chassis | Dustproof Gray |
| Wheels | Napier Green, light |
| Upholstery | Radel's smooth green leather #115, plain |
| 4. Body and hood | Sage brush green grey, light |
| Fenders and chassis | Sage brush green grey, dark |
| Wheels | Sage brush green grey, light |
| Upholstery | Radel's Spanish leather — Silver Gray #2400 |
| Top and sunshade | Laidlaw's Gray |

De Luxe Seven-Passenger Touring

| | |
|---|---|
| 1. Body and hood | Dustproof Gray |
| Fenders and chassis | Motor Car Red, deep |
| Wheels | Motor Car Red, deep |
| Upholstery | Radel's Spanish leather — deep wine #2200 |
| 2. Body and hood | Jet Black |
| Fenders and chassis | Jet Black |
| Wheels | Orange |
| Upholstery | Pigskin Strauss #6, plain |
| 3. Body and hood | Napier Green, light |
| Fenders and chassis | Napier Green, light |
| Wheels | Hobart Red |
| Upholstery | Radel's Auto, Fancy Bright #578 |
| 4. Body and hood | Blue Devil Blue |
| Fenders and chassis | Blue Devil Blue |
| Wheels | Pearl Gray |
| Upholstery | Radel's Spanish leather — gray #2400 plain (plain means not-tufted or plaited) |

When the 4-passenger speedster was introduced in mid 1922, Dust-proof Gray Dark with Spanish leather upholstery or Coach Blue and black leather were options. The spanish leather was plain with a seam mid seat whereas the black leather was pleated. The corresponding top materials were Laidlaw no. 2, Gray and Standard Black Pantasote. The top bows were natural wood with nickel hardware. Like the 2-passenger, the steering column and shift lever were nickel plated.

The 1923 2-passenger and early 1923 4-passenger speedsters came standard only with a coach blue medium finish. The right hand spare wire wheel of the 4-passenger was changed to a left hand spare tire/rim supported off the frame in a saddle but not in contact with it. A similar change was made to the rear mounted spare of the 2-passenger. The optional wire wheels were mounted, on the 4-passenger on the same support off the frame. The 1922 wire wheel was supported vertically by the "saddle" in the running board and horizontally by strap attachment to body brackets. The speedsters were equipped with buffed aluminum radiator shells.

At the 1924 auto shows, the Marmon exhibit was particularly attractive. The display included the following:[12]

2 passenger Speedster in French Ivory (semi-dull egg shell); This was an available but special option. The 1924 Speedster in the Harrah collection is done in this color.

4 passenger Custom Coupe by Hume in Beige Brown Duco and varnish with two-tone velour interior

7 passenger Sedan in two shade green gray with taupe velour interior.

Town Car in Maxine Blue with blue velour upholstery.

Special paint colors were available but resulted in a 6 to 8 week delay. Special leather colors were $50.00; Seat covers were available for $75.00. Special blues and grays were $75 extra, ivory and yellow were $100 and delicate colors such as white, cream and purple were $125 extra. Quotations could be requested on maroons, vermilions and wines. There was no charge for special color wheels.

For 1924, the Model 34 C radiator shell was stamped steel and on the speedster was nickel plated. The plated shell and side mount (s) of the 1924 4-passenger speedster were available on special order on other open models, as evidenced by the factory photography of Jimmy Murphy's 4 passenger phaeton.

## MODEL 74

W.E. Pierce, who had been president of the Hume Body Co., joined Marmon Motor Car Co. in March 1925 with the position of Chief Body Engineer. Mr. Pierce had personally styled several cars which had been sensations at major shows. He was personally responsible for the styling, at the Hume Body, of the eleven new body styles introduced with the Model 74 in November 1924. All this was reflected in the sensational two-tone and two-color Duco finishes which were standard on the new Model 74 at its introduction in the late fall of 1924. Construction of new on-site styling studios was undertaken in 1926 for new body work and to experiment with new color schemes.

The standard closed models (with vertical windshield) were finished in two-tone gray. the lighter shade was on the top of the hood extending rearward between the full-length belt molding and the belt molding around the bottom of the 'greenhouse'; it separated the darker side panels and the greenhouse. A vermillion stripe was applied to the moldings and hood louvers.

The deluxe coupe (sloping windshield) came in two tone green gray with cream striping. The rear quarter was black leather covered and fitted with a 'landaulet' bar. Deluxe sedans and limousines were finished in Buckingham Gray with a black greenhouse and vermillion striping. Blue was an alternate to gray on the sedans, and green on the limousine; cream striping was applied. The very popular (standard) brougham sedan had gray leather rear quarters with 'landaulet' bars and opera windows.

Phaetons were offered either in two-tone gray with vermillion striping or in blue and black with cream striping. The roadster came in two-tone green gray with yellow stripes. Interiors of open cars were upholstered in hand-buffed leather.

All models carried natural wood varnished wheels, previously only offered on sport models. Interiors of closed bodies were done in Laidlaw broadcloth in harmonizing shades. The front compartment of limousines was done in hand buffed leather.

The Model 74 Speedster introduced in the mid 1925 model year had three standard options with black fenders and chassis and two-tone colors:

| | | |
|---|---|---|
| #1 | Duco Newport Gray on belt panel and top of cowl | #871 |
| | Duco Newport Gray on lower body panels | #873 |
| | Black-moldings with vermillion center stripe | |
| | Hand crushed full grain tooled leather in green gray. | |
| #2 | Duco Miami Taupe on belt panel and top of cowl | #813 |
| | Duco Miami Taupe on lower body panels | #397 |
| | Peach moldings with vermillion center stripe | |
| | Hand crushed full grain tooled leather in black | |
| #3 | Duco Hollywood Green Gray on belt panel and top cowl | #799 |
| | Duco Hollywood Green Gray on lower body panels | #662 |
| | Moldings and upholstery same as #1 option. | |

## MODEL 75

The Model 75 for 1927 continued the offering of a variety of two-tone color options, with the addition of exclusive colors on the Custom Series which featured Locke bodies.

The Locke sedans were available in single tone Lido Blue, Marmaroneck Gray, or Snipe Gray. Two-tone colors were Gray, Brown, Green and Palmetto Green. A distinctive Laidlaw Broadcloth interior was coordinated with each of the color options. Arcadia Mohair was also available but only on the Blue, Two-Tone Gray and Palmetto Green jobs.

The 'standard' Brougham was available in single-tone Bloomfield Gray and Two-tone Ascot Taupe and Naples Blue Green. A single Laidlaw Cloth or Chase 'de Salle' mohair interior was offered. The Victoria options differed only in the substitution of Black for Bloomfield Gray.

A wide range of exterior colors was offered on the 2-passenger (with rumble seat) speedster model. Single tone French Gray and Two-Tone Brown, Tan, Ascot Taupe, Green, Maroon and Newport Gray were options and all came with an interior finished with Blanchard Bros. & Lane Co. leather in a matching shade. The Custom Touring Speedster came in Lido Blue or Two-Tone Green or Tan with matching leather interior and either a Burbank or Pantasole top.

An unrestored original Locke 5 passenger sedan is (or was in 1978) in the Harrah Collection at Reno, Nevada. It was finished in black with exception of the top of the hood and inter-belt area which was dark green, as was the striping. The interior was light gray broadcloth with white pin striping. Carpets were short pile wool in a gray color. The dash panel was medium stain walnut burl; the window frames were similarly trimmed.

## FOURTH ERA MARMONS

Although priced far lower than traditionally, the cars of this Era evidenced little if any compromise in luxury and beauty.

## LITTLE MARMON — 1927

Body interiors were richly finished. Closed cars were upholstered in Laidlaw broadcloth. Walnut grain was used throughout the interior including the cowl panels while the small panel in which the instruments were grouped was burled walnut. Several models had leather covered vanity and smoking sets of special design.

Cigar lighter, windshield wiper and dual rear-view mirror were featured on all models. A dash mounted engine temperature gauge and electric fuel gauge were innovations.

## SERIES 78 — 1928

The Jewel Colors which characterized the 1927 Series 75 were continued on this new smaller senior model: Micha Schist Gray, Emerald Green and various combinations of blue, maroon, gray, and brown were offered.

Cushions were of the 'new and comfortable saddle' type, with a single row of buttons to the rear of the seat and to the top of the cushion.

Interiors were done in broadcloth or bedford cord and the instrument boards were harmonized with the cloth used.

The instruments were in an etched, bronze panel finished in dull silver and gold etched, and indirectly lighted. Walnut garnish moldings with Italian parquetry inlay of needlepoint design were used below windows and windshield.

Interior hardware was figured bronze. New was an electric clock.

For 1929, the Series 78 instrument panel was silver chased and repeated the design of the interior hardware. The 'Jewel' color promotion was ended.

The four color combinations available on the sedan style were described as representative:

| | |
|---|---|
| Option 1: | **Maroon** on body, belt moulding, visor, roof rails and reveals.<br>**Gray** upper body and belt panel.<br>**Dark Maroon** wheels<br>**Black** fenders and aprons. |
| Option 2: | **Black** upper body embellished by **green pastel** shade on belt moulding, visor, roof rails reveals, top of hood, and wheels.<br>**Lighter Pastel Green** on lower body.<br>**Black** fenders and aprons. |
| Option 3: | **Beige** on lower body.<br>**Brown** on belt panel, roof rails, visor and reveals.<br>**Green Stripe** on belt panel<br>**Green** wheels<br>**Black** fenders and apron |
| Option 4: | **Blue** on upper and lower body<br>**Green** wheels and belt panel<br>**Black** fenders and aprons |

Another catalog listed two options differing from the above:

| | |
|---|---|
| Option 1: | **Cedar** and **Tomahawk Red** with **Cedar Chassis** aluminum color wire wheels. |
| Option 2: | **Bianca Cream** and **Auto Brown, Bolero Cream** wire wheels and harmonizing chassis |

## ROOSEVELT — 1929

Door and window moldings were 'walnutized metal'.

The instrument panel had a silver etched background with a decorative bronze border.

Cushions used 'davenport' type springs covered with fine padding and layers of fine cotton batting securely stitched and tufted. Interiors were done in broadcloth of a neutral shade.

Exterior colors were unique to the Roosevelt and were very attractive, particularly the two-color combinations which took advantage of the handsome molding treatments.

## BIG EIGHT — 1930

Two-tone broadcloth upholstery was used to obtain "really distinctive appearance." The seats and seat backs were darker than the top and headlinings. Four selections of upholstery corresponded to the four exterior colors.

Cushions and seat backs were fitted with Nachman upholstery springs, as was the folding arm rest on all rear seats and the coupe front seat. The springs were mounted in individual burlap cases. One and one-half pounds of genuine curled hair was used as padding in cushions and backs.

Carved piano finish black walnut woodwork was used throughout, including a panel on the header bar over the windshield.

Prior to about 1927, phaeton tops were very bulky in the folded position and detracted from the racy appearance sought by folding the top. Even roadster tops were quite bulky; premium 2-passenger speedsters such as the Marmon Wasp and Kissel Bug featured removable tops with storage provisions. In the southwestern states where probability of rain was negligable, false top-boots became popular in the mid 20s; the top was removed and stored off-vehicle and replaced with a falsework supporting a slender top boot. When new models appeared for 1927 with tops that were designed with equal consideration to 'up' and 'down' appearance, the owners of prior model cars could resort to the false boot solution.

The 1932 Marmon 8-125 Convertible Coupe was among the first to feature a top which disappeared upon collapsing and was concealed by a hinged body panel.

The compact folding was assisted by the use of Laidlaw Burbank top material, imported from England, which was waterproof in a single closely woven layer; competing materials, e.g. Haartz cloth, were a lamination of rubber and two fabric layers.

### FIFTH ERA

The FIFTH being an Era of a single car series . . . The MARMON SIXTEEN . . . would warrant a somewhat different treatment even of Art and Color. But the SIXTEEN was so uniquely harmonious and symphonic that it should not be dissected even for discussion purposes. Hence the reader is referred to the FIFTH ERA product chapter. Comment here is restricted to noting that the SIXTEENS were typically finished in deep dark colors, broken only by a very light shade on the belt molding, even the tire-cover side-panels being of the same body color. The beauty was in the overall sculpture of the car and was not to be broken by two-tone or two-color treatments. The convertibles however were typically finished in light shades, often in brown tones to integrate the cloth top into the whole. The extreme of this concept was the Le Baron coach-builder limousine for Mr. Benadun, described in Part 1 of this Chapter, where the uniform color was extended to the radiator shell, wheels and hubs.

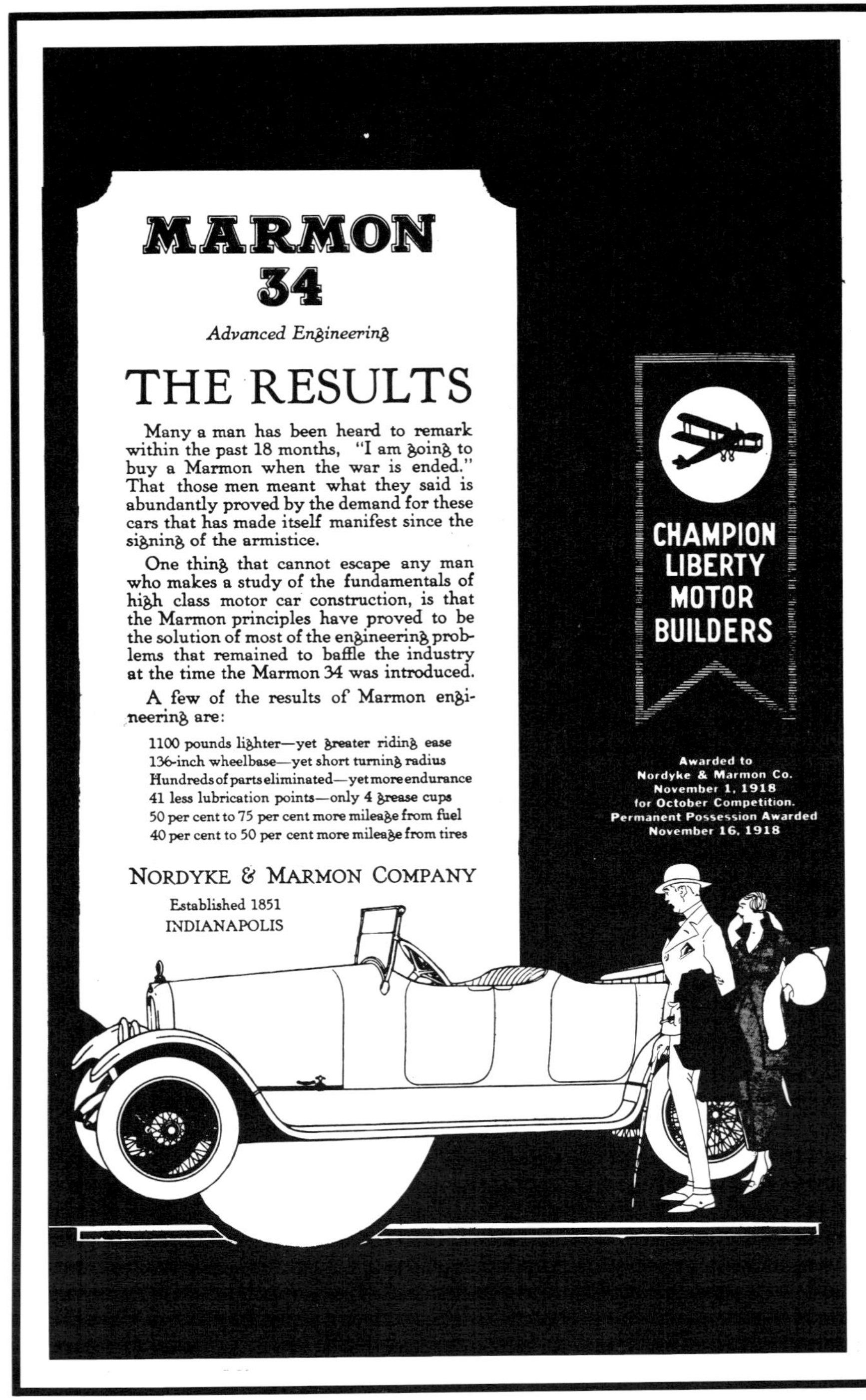

CHAPTER TEN

# MARKETING THE MARMON

The manner in which the Marmon car was merchandised is presented in four parts:

Part 1 Advertising the Marmon
Part 2 The Distribution System
Part 3 Production Volume and Market Penetration

In all three parts, the subject is discussed on an Era basis. Foreign marketing is addressed following the domestic distribution system in Part 2.

The treatment of Production Volume, for the Fourth Era, stresses the critical ratio of sales to installed capacity and the consequences of over production.

The discussion of Market Penetration seeks to convey a comprehension of the Marmon population in its time in terms of a current offering. The relative market performance of Marmon is compared to its competition.

## PART 1 — Advertising the Marmon

When a book devoted exculsively to automobile advertising history was published in the 1970s, it very appropriately carried a Marmon color ad on the hard cover. Marmon used advertising very effectively and their formats were as innovative as the product they promoted.

Marmon's advertising stature was evident in the frequency . . . almost annually . . . with which the cover of the annual show issue of MOTOR magazine portrayed some goddess-like figure holding a miniature car . . . the latest Marmon, but unidentified except by inspection.

Advertising was an important factor in the Nordyke & Marmon Co. becoming recognized as 'one of the world's greatest engineering houses' during the nineteenth century. That promotion continued with worldwide advertising of the flour milling products and complete turn-key mills after the entry into the automotive market until the sale of that product activity in 1926.

Consequently Nordyke & Marmon Co. were familiar with the press media when they entered the automotive field. They had in fact published a subscription magazine in the 1870s which carried . . . in fact, originated . . . features appealing to a broad spectrum of audience interest while accomplishing the primary objective of advertising the milling products.

### *Automotive Advertising*

One theme consistently stressed from 1904 through 1928 was the high quality level maintained by virtue of a very high in-house content in the product. Even the Little Marmon of 1927 . . . the first of the eight cylinder Fourth Era cars . . . was advertised as having an 80 percent Marmon content. Out-sourced items were typically: carburetor and electrical systems, springs, tires, roller/ball bearings and instruments, shock absorbers and later bumpers. Closed bodies were out-sourced until 1922 and in later years were built on-site by a supplier firm.

Equally stressed was the integrity and experience of the firm and the attendant abililty to engineer, develop, **perfect,** market and support the product.

The First Era advertising set a precedent . . . each of the Eras would feature a slogan; in some cases some major event resulted in a second slogan within the Era. The First Era product was appropriately described as 'A Mechanical Masterpiece' and later as 'The Easiest Riding Car in the World'. The latter slogan would persist through several Eras.

### FIRST ERA ADVERTISING 1905-1908

Introductory two page spreads in the automotive (and cycle) press were entirely devoted to the stature of the established firm making an entry into a new market. They concluded with the statement: 'A Marmon Car has been driven 8000 miles without repair.'

These ads were followed by single-column small ads which primarily provided the firm's address and stated catalog offerings while the public's interest was promoted by the extensive and enthusiastic coverage by the press in feature articles on the unique features incorporated in the cars.

Next came half and then full page ads stressing the unique features and tabulating the growing list of distributors and dealers.

Nordyke & Marmon did not engage in price competition in a selected market segment. During the First Era the Marmon was offered at a determined price for a unique product not otherwise available. A typical paragraph from a 1908 ad: *"Mechanical superiority is your clear gain in buying a Marmon, for you buy at the same time all the beauty of appearance, all the quality and all the advantage of reputation to be found in any other car at any price."*

The prospective purchaser was advised to drive any car before committing to purchase. An interesting safety consideration was also stressed in one ad: *"Do you want a car in which a woman can tour without fear of exhaustion and without injury to her health? The Marmon is the only car that fully satisfies this equipment."*

The concern for ladies comfort was stressed earlier in a 1905 ad headlined 'THE MARMON CAR' and showing front and side views of the vehicle on a white oval superimposed on a painting of a sedan chair borne by two men in colonial dress. Beneath it the following interesting 'official' definition of the unique riding qualities of First Era Marmons: *"In every other car the body is tilted in proportion to the depth of the rut or height of the obstacle encountered by the wheels, straining every part of the machine, and the occupants are suddenly jolted and thrown to one side of the other. Not so with the Marmon. It combines the exhilaration of speed with the gentle swing of mi-la-dy's sedan chair...the extreme modern luxury of motion. Rough roads are cushioned and uneven roads are leveled. DOUBLE THREE POINT SUSPENSION an exclusive Marmon feature, makes possible this marvelous flexibility. Mechanism and body are on separate frames, each suspended on three points. Freedom from binding strains prolongs the life of every part. No annoyance from either road or engine vibration."*

There followed a highly perceptive assessment of early automotive offerings: *"If you are interested in real Twentieth Century automobile building, in the original and*

# The Marmon

## A Mechanical Masterpiece

*The Marmon Car, 1905 Model. Price $2,500*

ONE GLANCE is not enough to familiarize yourself with the surprising and superior features of the Marmon Car, the machine without a weak part. Numerous ingenious devices have eliminated the annoying difficulties owners of automobiles have been meeting with for years. Our solution of the problem has been conclusively and unsparingly tested for three years—tested as a whole and tested in every part—in the factory and on the road. The 1905 Model is the finished product. We might, with ample justification, call it our 1910 Model, for authorities have pronounced it five years ahead of the times.

THREE POINT SUSPENSION—The mechanic's long recognized ideal—the object of constant and hitherto fruitless effort. Repeated failures to attain this have forced automobile builders to substitute mechanical arrangements admittedly complicated and wrong in theory. This company, famous the world over for its high grade machinery, has embodied in the Marmon Car the only successful Double Three Point Suspension System yet devised. Broad patents cover this feature. The under frame, carrying the engine and transmission, is on three point suspension. Direct drive is had by a rigid shaft from the engine to the bevel gear in a unique rear axle, moving within which and bearing no weight is the driving shaft. With flexible connection of the under frame with rear axle and location of the engine over the front springs, the jar of the rear axle is removed from engine and transmission. Cardan joints are entirely eliminated, transmission is never out of alignment, and all the energy of the engine is delivered in work to the rear wheels. Absence of vibration allows the mechanism to do its full duty without strain or interference; hence in our construction more is accomplished by 20 horse power than by 35 horse power in many other designs. On a separate frame, also hung on three point suspension, above and removed from the engine, is the body. Without any engine vibration and almost none even from rough roads, it rides like a parlor car. The mechanical beauty of this arrangement will be as apparent to one who is not an expert as it has been and is to the automobile designers who have vainly labored to effect it.

THE MOTOR is a four cylinder gasoline engine, thoroughly and perfectly air cooled. Automatic force oiling system—itself a marvel—providing copious lubrication and absolutely insuring against any oil troubles. All gears run in oil and the entire mechanism, while readily accessible at every point, is oil tight and dust proof. Marmon Cars have been driven 8,000 miles without repairs. Parts are made interchangeable, however, and can be conveniently carried for such emergencies as collision or other accident.

THE CONTROL is simple, quick and sure.

THE BODY is roomy and luxurious, with seats, dash, fenders and bonnet of single piece aluminum castings, which take and retain a high finish. Wheel base, 90 inches. Side entrance tonneau.

The Marmon is built from the ground up by this Company, which guarantees every part of it. It is made throughout of the very best material, with the most minute and precise workmanship. It is an interesting, high grade proposition for a high class trade.

# Nordyke & Marmon Company

*Factory and Executive Offices,* **Indianapolis, Indiana**

Fig. 10-1 First ad for a Marmon automobile (Horseless Age, Dec. 1904) *the progressive as distinguished from the conventional "good enough" and the obsolete, acquaint yourself with Marmon details . . . Double Three Point Suspension, advanced type of air-cooled motor, perfect automatic lubrication, and many other unique and exclusive features, all thoroughly tested through two years of unsparing road work, summer and winter."*

The ad concluded with a definition of the First Era Slogan: and a guaranty statement which no other manufacturer could approach: *"The Marmon is 'A Mechanical Masterpiece' in whole and in every part. We support this statement with a guaranty that has been respected worldwide for over fifty years."*

Soon after the start of sales promotion, a booklet "Good Words About The Marmon" was made available; it reproduced letters of satisfaction from owners of 1904 Marmons. A product brochure was issued in a format retained through the First Era. It was identified by a code 530 in which the first digit (s) was the model year and the 30 designated a 'Product Catalog'. This publication code would persist for many decades, with '33' being a service manual and '34' an owner's manual. It was characterized by very high quality photography and multiple photos to explain the function of the dual-frame chassis; the text format was farsighted. And the page size was 9 inches wide by 6 to accommodate the automobile's height to length ratio. Amid great controversy in **1968;** *Antique Automobile,* the publication of the Antique Automobile Club of America, wisely reversed its page/binding orientation to 11 wide by 8.5 for the same reason.

By 1907 the secondary slogan 'Easiest Riding Car in the World' had become a part of every Marmon ad.

A March 1906 ad was headlined "IN A CLASS BY ITSELF". The following text began:

True in more respects than one — but one will illustrate.

Ever sit in the tonneau of an automobile going at a good rate of speed over a rough road? Didn't you spend half your time in the air, bouncing eight inches off the seat, not counting the lurches to one side or the other?

Isn't any machine (car) 'In a Class by Itself' if it takes you over any road, at good speed, with the easy gentle sway of a sail boat.

Whether the First Era Marmon ride was best compared with sedan chair or sail boat ride-character, it was unique. A wise passenger in the rear seat of most cars in the 1920's still noted the location of top bows relative his head before it was too late.

Marmon ads ran in *Country Life* (in America), *Harper's Weekly* and *Mc Clure's* in addition to trade publications.

The 1908 models carried a brass script 'Marmon' badge centrally on the grill or radiator . . . an effective medium. The earlier manufacturer identification on the brass front step plates was continued.

A Marmon advertisement during the Second Era, illustrated a Model H (First Era) touring with an accompanying letter from an Indianapolis bank official reporting his purchase of the car at a 50,000 mile odometer reading, his satisfaction and pleasure in the car, and his projection of another 50,000 miles remaining.

## SECOND ERA ADVERTISING 1909-1915

The decision to shift from air-cooling to water-cooling and to abandon the patented dual-frame chassis presented a critical challenge in marketing. Advertising was carefully managed to smooth the transition.

Certain product decisions, improvements and modifications made the task easi-

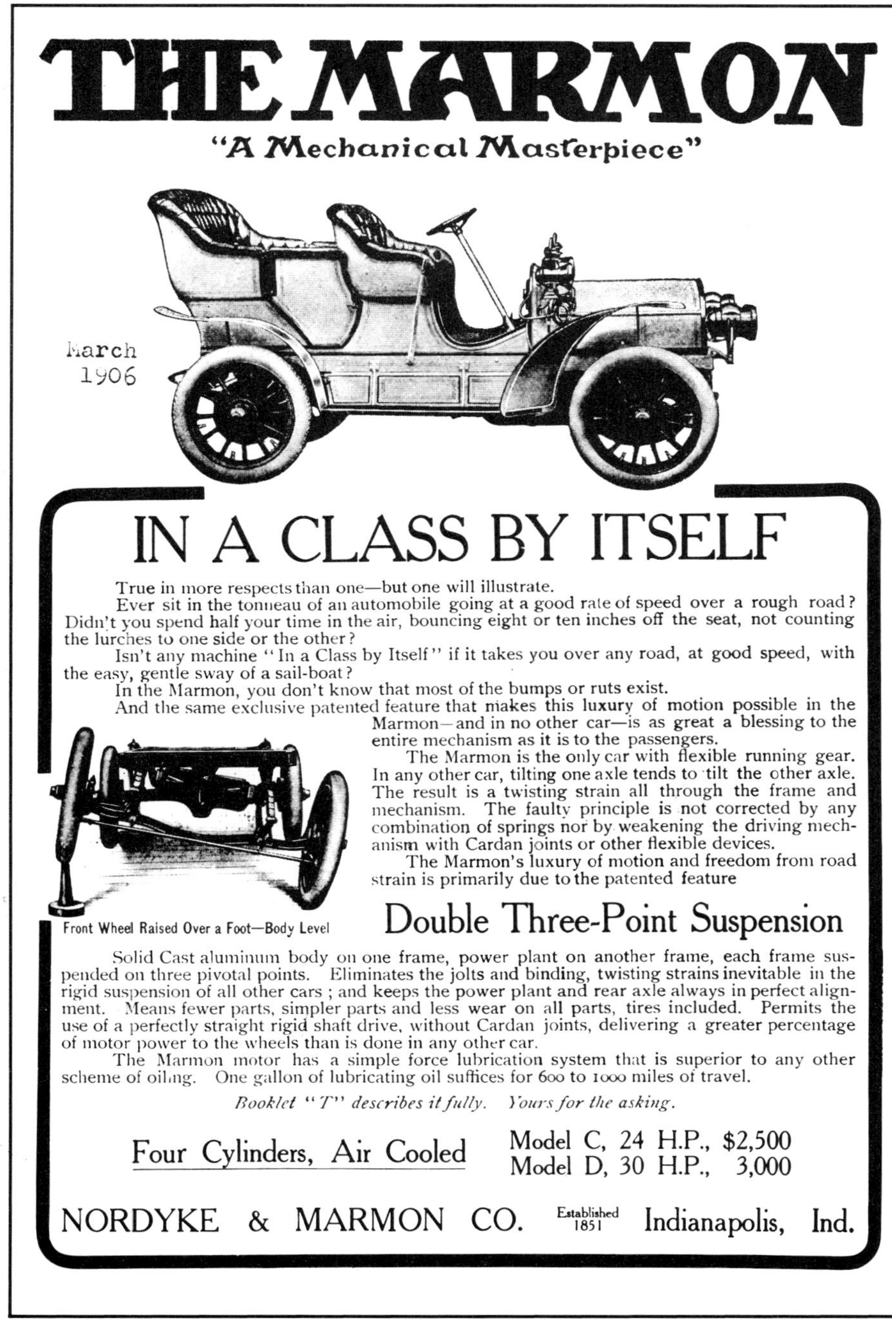


Fig. 10-2 "In a class by itself"

Fig. 10-3 Country Life in America, January 1911.

Fig. 10-4 The Literary Digest, August 24, 1912

er. The 1908 air-cooled Model G had a hood and grill configuration not unlike a water cooled car and was offered with water-cooling as an option. The completely new 1908 Model H, although water-cooled, was offered with the traditional air-cooled engine as optional equipment.

Traditional Marmon customers were not alienated and purchasers of 1908 models were protected from excessive depreciation upon introduction of the new Model 32 for 1909.

Initial Model 32 ads stressed the light weight (2100 pounds) and simplicity of construction of the new car and the firm's commitment to a single model concept with progressive improvement. The Glidden victories of the First Era had brought significant publicity. The new model 32 proved highly reliable and as easy on tires as its complex predecessor. This set the theme for Second Era advertising as a remarkable string of victories provided fresh victory results for each succeeding ad which typically showed the growing array of trophies and listed the imposing list of records. The climax . . . victory in the first Indy 500 . . . provided the Era slogan "Marmon . . . International Champion' and overnight made Marmon a household word. The press accounts constituted 'free' advertising exceeding paid advertising by orders of magnitude equaled and exceeded only by the major pronouncements of Henry Ford.

Whereas Era One cars were illustrated unoccupied, Era Two practice was to show at least one car with seats fully occupied. Closed cars were shown as early as January 1912.

In January 1913 a color ad appeared in *Country Life in America*... Model 32 and 48 limousines passing on a wet city street with snow covered trees in the background. Spokeless wire wheels implied motion. The word Marmon was in bright orange with pale orange in the clouded sky reflected on the black cars for an impression of early evening. The message appeared on a letter format superimposed on the scene.

Every year 1910 through 1913, an "announcement ad" detailed all the improvements or additonal equipment on the Model 32 for the new 'season'.

The entire 1910 production run was sold to dealers and half of that output sold by the dealers to individual buyers before Nov. 1, 1910.[1] This proud claim made after the start of 1911 production shows how different marketing was in 1910.

## THIRD ERA ADVERTISING

### MODEL 34 SUB-ERA

The Marmon pre-announcement advertising program for the Model 34 A in the fall of 1915 was a spectacular promotion, especially so considering the annual production rates.

The 34 A was so new that pictures of the car and of mechanical details could be shown without providing even a hint as to the manufacturer. This situation was advantaged by releasing a 'teaser campaign' . . . a series of weekly sequential ads which conveyed the impression that they were authored by the magazine staff; inquiries were to be directed to the offices of the periodical. Even the badge design was new; it appeared in all the ads but with "Made in USA" substituted for "Marmon".

The first (two page) ad showed the front view of the car, the heading 'Announcing a New Automobile', and a sub-heading 'A Statement by the Publishers of This Paper'. It began: *"In succeeding issues of this publication the attention of its readers will be forcefully directed to a new and very remarkable automobile."* And concluded: *"We are well acquainted with the company manufacturing this car; its financial standing, long and successful record as manufacturers and its demonstrated ability to produce motor cars of the highest type and quality."*

Performance claims were stated to have been verified by the technical expert of the publication. Photographs (album like) showed prototypes in various terrains in southwestern states.

The second ad addressed "The Development of the New Car' and showed the side view. Work was reported started in November 1912, and the engines running in October 1914. The third, 'Unsprung Weight and Balance', showed running gear details. The fourth was a two-page ad with the first addressing 'What Weight-Really Means' and the second 'The Motor of the New Car'. A three-quarter rear view of the car was revealed; the engine was shown and the wet-sleeve aluminum construction revealed. The fifth week addressed 'The Selling Value of the New Car' and promised full details the following week, exhibition at the New York and Chicago shows and deliveries immediately thereafter; a second fifth week ad pleaded 'Wait for the New Car' and suggested prospective customers register, without commitment, their interest as a prospective buyer "before we dispose of our entire 1916 production."

The new car was finally identified as a Marmon and as Model 34 in a four page extravaganza.

Page 1 was a retouched photograph of the touring style. To maximize car size and attendant detail, this page was turned 90 degrees. Page 2, Specifications, was similarly oriented. Two other pages provided a detailed discussion of features.

A remarkable similarity was evident in Henry Ford's pre-introduction ad campaign for the Model A in 1927. This is discussed in detail in Chapter 14 Part 2 'Howard Marmon and Henry Ford — a Parallel'.

To introduce the 'Scientifically Constructed Light-Weight Motor Car' in 1915, Nordyke & Marmon published a remarkable 45 page (8.5 x 11) technical support document. It was republished by Floyd Clymer Publications in 1951 and was successively reprinted into the 1960s. Clymer described it: *Probably the finest and most informative catalog of early days. A collector's item without comparison.*

Every detail of the car's design was discussed and justified. If a little technical knowledge was essential to understanding, it was clearly and concisely provided. Even a little management philosophy and future technology predictions were included.

The two pages addressing the selection of wheels for the car are reproduced in Chapter Four. They illustrate the clever format and general style. A remarkable publication only fifteen years into the automotive century.

On July 29, 1916 at 5:00 PM a Marmon 34 stock car, sans tonneau body section and top, arrived at Third and Market Streets in San Francisco. Millionaire sportsman S.B. Stevens had departed Columbus Circle, New York City just 5 days, 18.5 hours before arriving at his California destination. The driving was done by Mr. Stevens and other amatuers and the trip was made under the auspices of the American Defense Society. The car carried a letter from Major General Leonard Wood, Commander of the Eastern Department, US Army, to his Western Department counterpart in San Francisco.

Mr. Stevens' record transcontinental crossing was quickly exploited by Marmon. A special advertising section in Motor Age and other magazines began with a four-page Nordyke & Marmon ad and was followed by eleven more pages of congratulatory and "me too" ads sponsored by component or equipment manufacturers. Even the Marmon ad was partly sponsored by Lynite Aluminum.

Page one of the ad showed the car at the North Platte, Nebraska, refueling stop and provided trip total statistics. Pages two and three together provided specifics on the individual segments of the trip and are repeated here for the record.

Charles Betts and William C.Binz (d 1964) recounted the latter's experiences in driving the transcontinental Marmon from Cleveland, Ohio, to Merrillville (Chicago area), Illinois in *Antique Automobile* for May-June 1966.

# Across the Continent in 5 Days–18½ Hours

The Car at North Platte, Neb.

## MARMON 34

### Establishes a Transcontinental Record New York to San Francisco–3476 miles

S. B. STEVENS, a millionaire sportsman of Rome, New York, driving a Marmon 34, arrived at Third and Market Streets, San Francisco, at five o'clock, Saturday afternoon, July 29, just 5 days, 18½ hours after the car had left Columbus Circle, New York City, 3476 miles away.

This completed the most remarkable and fastest journey ever made across the United States in a motor car—the average rate of speed being almost equal to the schedule of the fast transcontinental railroad trains.

The car was a Marmon 34 touring car of regular production and was driven by Mr. Stevens and other amateurs under the auspices of the American Defense Society to demonstrate the possible speed and reliability of motor transportation across the continent. This epoch-making run by the Marmon 34 demonstrates again the soundness of its advanced scientific construction—the extensive use of aluminum, the deep section frame, the new rear spring suspension, the self-oiling chassis lubrication system and other features

Fig. 10-5 Nebraska pit stop on 1916 race against the clock.

## FIRST DAY

**Start, New York 1:30 a.m. Monday, July 24. Syracuse, 10:30 a.m. Monday, 295 miles, average 34 m.p.h. Cleveland, 7:55 p.m. Monday, 647 miles, average 34 m.p.h.**

The start was made at 1:30 a.m., Monday, July 24, from Columbus Circle, New York City. Fred Barbour, of New York, drove the first leg of the trip. Two hundred and ninety-five miles from New York to Syracuse were covered in nine hours, the car arriving at 10:30 a.m.

Stevens took the wheel at Syracuse and drove to Cleveland, by way of Batavia, Buffalo, and Erie, arriving in Cleveland, 647 miles from New York, at 7:30 p.m. Monday night. There was some delay in Cleveland on account of the car getting caught in a parade and bad luck at railroad crossings. At one crossing the car was held twelve minutes.

## SECOND DAY

**Merrillville, Ind. (near Chicago) 6:45 a.m. Tuesday, 977 miles, average 33.3 m.p.h. Cedar Rapids, 3:45 p.m. Tuesday, 1235 miles, average 33 m.p.h.**

At Cleveland, William Binz, of Chicago, took the wheel and drove during the night 330 miles to the next station which was Merrillville, Indiana, near Chicago. The time made on this stretch was limited to the time which could be made by Stevens in getting from Cleveland to Merrillville by train. The train stop near Merrillville is Gary, and it was impossible for Stevens to get to Gary until 6:45 a.m. Tuesday morning.

The Marmon car reached Merrillville exactly on time and after taking on supplies another car flashed over the hill, bearing Stevens, the driver, who was to take the car to Omaha. The car left the station as per schedule at 7:00 a.m. On this leg the car passed through Elyria, Norwalk, Fremont, Bryan, and South Bend.

## THIRD DAY

**Omaha, 1:05 a.m. Wednesday, 1536 miles, average 33 m.p.h. Cheyenne, 4:50 p.m. Wednesday, 2113 miles, average 32.3 m.p.h.**

Upon leaving Merrillville the car sped through Aurora, Dixon, Clinton, Cedar Rapids, Marshalltown, Ames, Grand Junction, and Carroll into Omaha arriving on schedule at 1:05 a.m. Wednesday morning. At Omaha the car was turned over to Fred Barbour while Stevens took the train for Cheyenne, Wyo.

From Omaha the tourists dipped down to Kearney, going by way of Fremont, Columbus, and Grand Island. After leaving Kearney, they passed through North Platte, Big Springs, Sidney, Kimball, Pine Bluffs, Wyo., into Cheyenne arriving there at 4:50 p.m. and stopping for thirty minutes. At Cheyenne they were about thirty minutes ahead of their schedule.

## FOURTH DAY

Stevens was driving from Cheyenne to Evanston, a distance of 407 miles. He encountered about 20 miles of muddy roads during the last hundred miles and arrived at Rawlings at 1:10 a.m. Between Cheyenne and Rawlings the car had to be driven up the long Sherman hill, reaching an approximate elevation of 8,000 feet and crossing the Great Continental Divide. At Evanston, Stevens wired:

"Delay on account of much rain. Roads muddy. Car in best of condition."

Heavy rains in Utah reported by Salt lake papers as cloud-bursts, made the trip across Utah dangerous and slow. Walter Bieling drove the car from Evanston, to Ely, Nev., passing through Salt Lake City, County Well, Fish Springs, and over the Shelbourne Pass, crossing through the great American Desert, which was one of the worst parts of the trip. The heavy rains in Utah had put the roads in bad condition and caused numerous washouts. Three hours were lost west of Salt Lake City because the pilot became confused in the route through the desert.

## FIFTH DAY

At Ely, the car was turned over to Robert Creighton, who drove to Reno. Here again road conditions were very bad. The trails over the Nevada desert were in terrible shape. The transcontinental car, however, was in fine condition. Creighton passed through Austin, Alpine Ranch, and Fallon. Aside from being delayed by the rough trails, a short circuit left him without lights. He located the trouble, however, and remedied it so that delay from this source was not considerable. Breaking through a bridge several miles out of Fallon caused a further delay of 3½ hours.

At Reno, Stevens was again waiting to take the car. He left Reno at 5:40 a.m., going by way of Truckee and Colfax, arriving at Sacramento at 12:22 Saturday afternoon, and thence to Vallejo, arriving at 3:10 p.m. Here a special boat was waiting to take the victorious car to San Francisco, and at 5 o'clock p.m. the car was officially checked at Third and Market Streets.

The involvement of the firm in aircraft engine production during 1917-18 and the award of the 'Champion Liberty Motor Builder' pennant was reported in advertisements and recorded on the face of the Motometer[13] for several years thereafter. The attendant increased capability in manufacturing precision was extoled as the underlying reason for the redesigned Model 34 engine . . . the 34 B for 1920.

Marmon used single color illustrations in catalogs and ads for the New Series 34 (A). They featured the illustration of a car, in color, with passengers and background added in the contemporary style of fashion magazine illustration.

A similar format was used for black and white magazine ads in the same period but with the car drawn rather than photographed. The example includes a novel impression of motion by omission of the wire wheel spokes.

During promotion of the first series 34A all advertising had shown only the vehicle . . . no people.

When the surprising decision was made to redesign the Model 34A engine, which was an aluminum unit block/crankcase with wet sleeves, into an iron-blocked Model 34 B, Marmon repeated their incognito multiple-ad extravaganza of 1915.

The ads ran in a number of trade and general periodicals for a period of five weeks. The message of the two-page ads was the impending introduction of a motor car with High Efficiency Motor in a new series. They progressively addressed the features of the new engine, and in the fourth ad the aircraft motor test procedures and inspections used in its manufacture; the new engine would have dimensions held to one-quarter of one-thousandth (.00025) inch, a precision unknown before WWI.

Inquiries consequent the first four ads were to be addressed to 'High Efficiency Motor' care of the magazine. The fifth (3 pages) revealed the 'manufacturer parent' to be Marmon and announced the Model 34B . . . Born of the War, with precision, metallurgy and heat treatement not previously in the realm of practical pro-

Fig. 10-6 A 'TORRENT' of power awaits the 'GO'.

Fig. 10-7 MOTOR June 1918.

Fig. 10-8 Country Life, April 1922.

duction.' As a consequence the new series was to have "an extraordinary smoothness of running, a freedom from vibration, and an immunity to trouble and wear." (It did, indeed.)

Like the 1915 series, each ad provided a performance bulletin. The fourth provided the following data with the 3.75:1 rear axle (4 — passenger open models):

| | |
|---|---|
| 10 to 50 mph, high gear | 18.8 seconds |
| Standing half mile | 69.75 mph |
| 8.16% grade, 1000 feet | 5 to 24 mph |
| 6.14% grade, 1775 feet | 5 to 43 mph |

The slogan for the Model 34 Sub-Era was "A Scientifically Constructed Motor Car . . . weighing 1100 pounds less (than competitors) and giving 14 instead of 11 miles per gallon."

Through the *Sub-Era,* Marmon used a unique ad format. A dark colored page (except margin), a color rendering of one style and a brief message and address in a small frame against the dark background. During the 34A model run, the car was shown relatively large, but during the 34 B model run, the car was shown quite small. The cars were shown unoccupied.

Some automotive historians consider these ads the 'dark ages' of automotive advertising. I rather like them. The concise localized message always had something specific and timely to convey and its size certainly did not discourage the casual page-turner.

The message varied with major developments . . . the French General Staff selection of Marmon, the Permanent Award of the Champion Liberty Engine Builder. When sublte changes, not immediately apparent, were made to enhance the cars appearance, the message was: *The wider seats, the deeper sides and the slanting windshield are attractions that distinquish the New Series Marmon 34.*

The message always concluded in small type with: "1100 pounds lighter" and frequently included "12 to 14 miles per gallon".

Marmon issued a 3 x 5 inch booklet (5 x 12 folded) to commemorate the French General Staff selection of Marmon. That honor was particularly significant when the French origin of the Marmon (t) family is considered. Three photographs were color tinted very effectively and resembled color photographs. They showed French President Poincare en route to dedicate the San Mihiel monument to the American Army, the President visiting the ruins of the City of Lens, and a village liberation celebration . . . all Marmon 34 tourings flying the Tricolor.

Ads had repeatedly stressed the selection of the Marmon for the French Army Staff cars. When Marshall Foch visited Indianapolis during a US tour following the Armistice, he was furnished a touring car in French Gray with the French insignia on the rear door.

## RENEWED MARMON 34 PROGRAM

In the summer of 1920, Nordyke & Marmon ran a series of full-page ads . . . eight or more . . . in the Saturday Evening Post which discussed motor car values. The ad was basically dedicated to the message of the week, with exception that a portion of one (of three) columns included a tiny replica of the classic Marmon ad . . . colored car against a black background.

Number Seven of the Series described the "Marmon Method of Complete Renewal" . . . a new concept made feasible by the Marmon policy of stablized design and advanced engineering. Renewal inside and out was accomplished only by authorized Marmon distributors. The ad was entitled "Compare a Renewed Marmon 34 with any NEW car of like price". It stated: "A Renewed Marmon is all that the name implies. It is a Marmon 34 made new."

The objective of the Renewed Marmon 34 Program was expansion of the market "to include additonal thousands who would otherwise have to remain content with lesser cars".

A series of ads appeared February-October 1923 with the text written by prestigious owners of successive Marmons. The upper half page was typically a signed painting of the currently owned Marmon in an appropriate setting. Each ad was by a different artist.

One owner reported 228,000 miles of service, expressed doubt that his 1910 Model 32 would ever wear out, and acknowledged two overhauls with but $30 repair cost in the last two years.

A second reported 142,756 miles (97 percent off--pavement) on a 1917 Model 34A in seven years with two overhauls.

The third had just acquired his eleventh Marmon and reported hundreds of thousands of troubleless miles including five trips to Florida covering in one instance 180 to 240 miles per day in the rain.

A Mrs. Anna Sage Hatch was painted in front of her home and between facing Marmons . . . her new 1923 Phaeton and her old 1916 34A Sedan: *"My first Marmon gave me 201,000 miles of the most luxurious trouble-free miles I can conceive of. I shall keep it as well, knowing that it will give me complete satisfaction in its performance and pride in its ownership."*

Fig. 10-9 President Poincare en route to dedicate the Battle of the Marne Monument.

Fig. 10-10 When Marshall Foch visited Indianapolis in 1919, he rode in a Marmon duplicate of the French Staff tourings, in style and finish.

One in this series deserves rank among the all-time classic ads. It was written by a Marmon owner who also sat for her portrait in her Marmon. It provided the ultimate assessment of the sophistication of passenger isolation achieved by Marmon in the Model 34 B sedan. Beneath the painting was the caption:

**'I knew we were in the Catskills by the atmosphere.'**
*By Helen Keller*

Miss Keller has attained international fame for her mastery of handicaps. Though blind, deaf, and dumb, she now "sees" and "hears" through touch. She converses fluently. She has devoted her life to the helping of others similarly afflicted.

*Miss Keller touring in the Catskills*
*—A painting by the Countess Elizabeth Zichy*

# "I knew we were in the Catskills by the atmosphere"

## *By Helen Keller*

*"I am delighted with my Marmon. To my touch the workmanshihp seems perfect. Borne along on deep, springy cushions, I find a long drive in it lulling and alluring. I do not have to hold on to keep my balance. In other cars I am keenly conscious of curves, but in my Marmon I hardly know when we turn a sharp corner. Recently we drove over Catskill roads, but I hardly felt any difference from pavements. I knew we were in the mountains by the atmosphere and the odors peculiar to high altitudes. Riding in a Marmon is just like sailing . . . the same smooth, vibrationless motion. I can hardly say enough in praise of this wonderful automobile. It gives me so much enjoyment."*

* * *

Two of the series appeared in July and were each written by a famous race driver. One was a half page painting of Cliff Durant crossing the desert in his 2-passenger Sport Speedster en route to the 1923 Indy 500. He reported having acquired the car at Indianapolis after the 1922 race, and commented: *"The Smoothest Riding, Easiest Handled Car I have ever driven."* The other was a portrait of Tommy Milton in racing dress, beside a small painting of Milton passing a mission church in his 4-passenger phaeton en route to the Indy race. His comments included: *"I loafed across the Continent in seven days in my Marmon, resting up for the big Race."*

The letter was written on May 10th and Milton won the race (for the second

Fig. 10-11 The high-mileage owner endorsement ad series.

time) on Memorial Day. He continued: *"Of all the cars I have ever driven, I have never seen one that stands up under hard driving, that has the punch or that holds the road like the Marmon. Mrs. Milton (like Mrs. Oldfield... author insertion) drives a Marmon closed car and is particularly well pleased with the way it handles, its comfort and sense of control."*

Initially this series of ads carried the slogan 'The Foremost Fine Car' but this did not appear after April 1923. Thereafter, it was replaced by the Marmon badge superimposed on a symbolic pyramid formed by two intersecting lines, representative of the pyramid characteristic of the Marmon Model 34 radiator.

A new series of ads, which the author identifies as 'The Valentine Series' appeared approximately January — July 1924. They were characterized by paintings of scenes involving Marmons framed fully or partially by ornate designs reminiscent of the first generation of valentines with their open paperwork trim. The Marmon badge surmounting the pyramid was similarly surrounded. Typically one third of the page, full length, was devoted to either favorable letters from owners or to a commentary personifying the Marmon's character: *'The first big difference between a Marmon and other truly fine cars is its mechanical manner... its attitude towards its job.' 'You can't get it fussed up or red in the face. It has a great big unused mechanical reserve and can afford to be good natured.' 'And the Marmon is probably the only fine car in the world with the same kind of uncanny, mysterious road adhesiveness... with automatic self-balance. You soon learn that you can give it its head, like an intelligent trained horse.' 'If there is such a thing as a hairy two-fisted man falling in love with a machine, that thing can happen with respect to a Marmon. Strong men have fallen in love with good ships, good horses, good battalions and good cannon. Why should they feel any less strongly towards a good automobile?'*

All ads in the Valentine Series featured only the frontal portion of the car... that which would change least in the Model 74 for 1925. An exception to the Series was an ad entitled "Friendly Enemies" which appeared in July 1924 and is indeed a classic automobile ad. In anticipation of that status, it is the single Marmon ad bearing a copywrite notation.

The following month, August 1924, saw the appearance of the last ad in the Valentine Series. Its portrayal of a woman at the wheel of a Marmon was in a long standing tradition of Marmon advertising. It is noteworthy because a woman was also at the wheel in "Friendly Enemies" the prior month and because of the contemporary perspective of the female character which has since undergone such a revolution.

The 'Valentine' format reappeared briefly in a Model 74 ad appearing in May 1925.

Marmon distributors also sponsored ads with original art work. The example shows a woman at the controls of a 1920 Model 34B coupe.

The Michigan distributor ran a series of excellent original ads in the *Detroit* (Auto Club) *Motor News.* The first example, October 1919, is a photograph of a soldier in a 34A Cloverleaf Roadster at a (still) prestigious Detroit location; the text speaks for itself. The second is from the May-June 1920 issue but in format and illustration it was a decade or more ahead. Note how the car appears to be supported by the letters forming "MARMON 34'. Note also the tasteful selection of a Lowell quotation, and the 'art device' which lead the reader's eye from message to illustration to 'where to buy'.

The traditional Model 34 B ad... small color car on dark background... persisted into the 1924 year.

Fig. 10-12 A rival to "Somewhere — Laramie."

## THIRD ERA
## MODEL 74 — 75 SUB-ERA

The Model 74 was promoted as 'The New Marmon'... It's a Great Automobile". Pins bearing only the latter legend were passed out at the 1925 shows.

For the first time Marmon ads had a catalog format, ie, they showed all the body

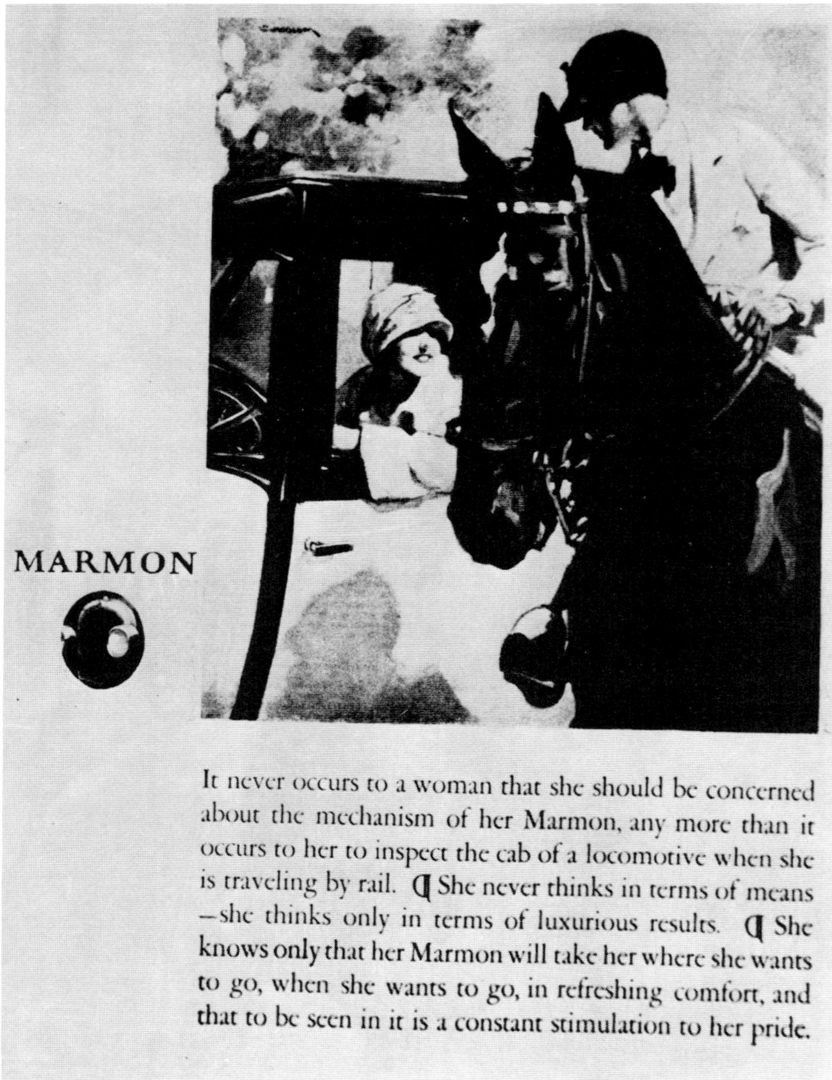


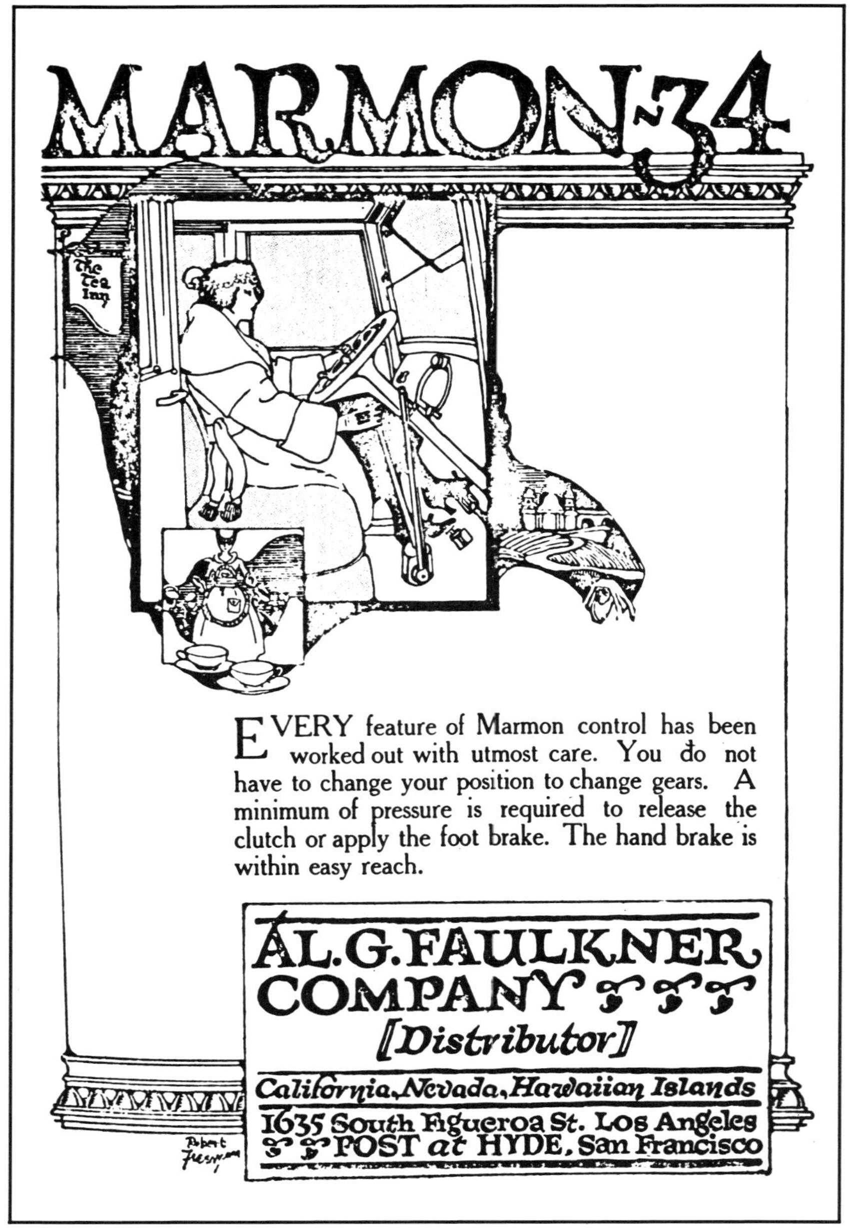


Fig. 10-13 Saturday Evening Post, August 16, 1924.
Fig. 10-14 Ad by Marmon's California distributor.

styles or depending upon the message, several styles. This was often facilitated by the use of two-page ads.

Likewise, the possibility that price could be a consideration in the selection of a Marmon was addressed for the first time in this Sub-Era. In the fall of 1924, the ads bore a headline 'Only $130 more than an open car.' This differential referred only to the 5 passenger standard sedan until January 1926 when the new brougham was included, and to the 7 passenger standard sedan as well when an April ad featured the three price leaders side by side, being surveyed in a showroom with distinguished observers all about.

In the May 1925 *House & Garden,* an ad stressed the luxury of the Marmon with a three-quarter page painting of a formally attired couple boarding their floodlight-

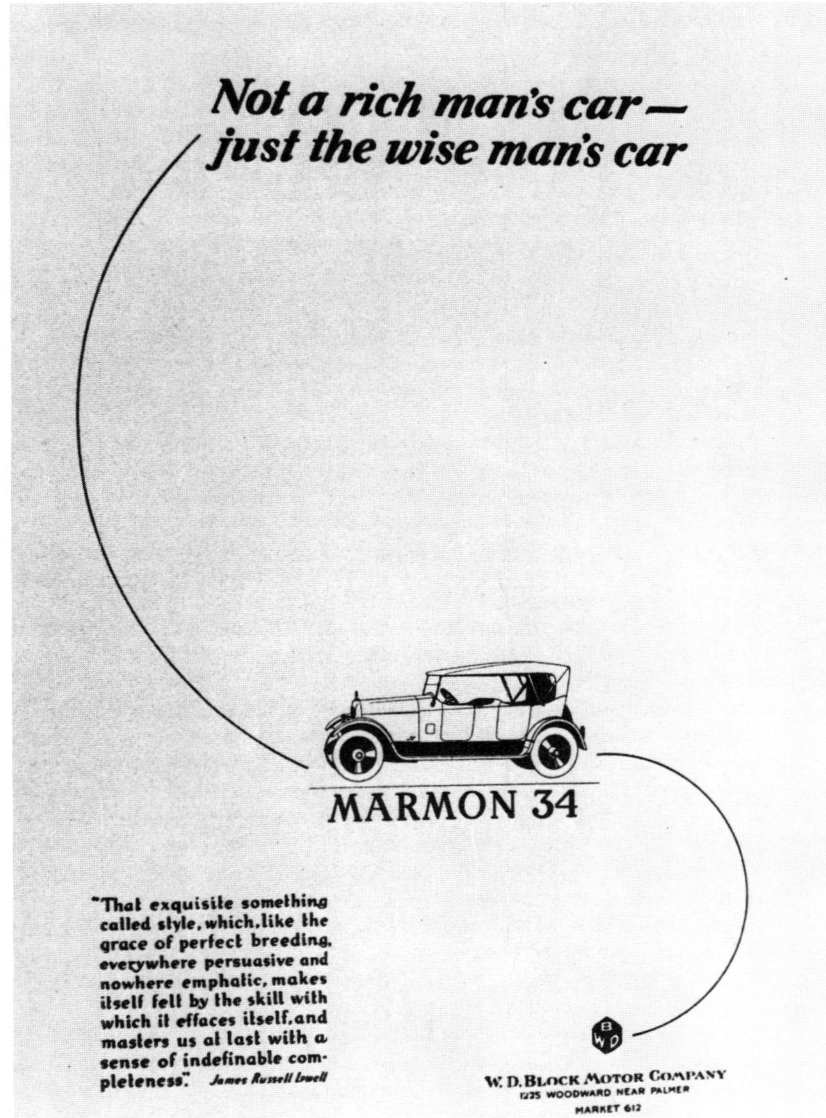


Fig. 10-15 and 16 Advertisements by the Detroit distributor.

ed chauffeur-driven DeLuxe 7-passenger sedan. Beneath it: *"The New Marmon . . . a successful car for successful people . . . Its a Great Automobile."*

Promotion of the 1926 model year began with yet another grouping of body styles. Headed 'The Preferred Fine Car Investment', the ad showed four closed body styles offered at the open car price, and a fifth for a $75. premium.

There followed a series of ads, resembling the earlier distinctive Marmon color ads on a black background but with occupied cars in motion, which typically carried a one sentence message in a white oval. An example: 'The New Marmon Brougham is one of four models selling at open car price.'

During the 1926 Model Year, the following appeared in small print below the Nordyke & Marmon identification: 'Observing its 75th Anniversary . . . Leaders in fine Motor cars since 1902.'

Fig. 10-17 The "here or there" series of ads.

Fig. 10-18 Model 74 'Safety' and Series 75 'Jewel' ads.

A fine series of split or contrast ads ran during the central portion of that Model Year. They portrayed Marmons in opposite extremes of climate or weather or circumstances with the common by-line 'Its All the Same to a Marmon': *'Broadway or Highway' . . . A night portrayal of the same sedan both in Times Square traffic and on a lonely moonlit highway; 'Touring or Detouring' . . . A brougham shown at speed on-pavement and a touring speedster (top down) negotiating a muddy stretch of detour. Marmon's reputation as 'the world's greatest road car' was extoled. Four other examples are illustrated, as they appeared in National Geographic. 'A woman's Intuition/A Man's Decision' . . . Interior views showing a woman at the wheel of a (standard) victoria coupe and a man at the wheel of a deluxe sedan, with both proclaiming 'It's a Great Automobile'. This ad is reproduced as it appeared in ASIA for March 1926. (note dual rear view mirrors).*

That same month, another ad format was introduced which addressed the feminine influence exclusively.

In a circular frame central on an unframed white page appeared a Fred Mizen (black and white) painting of a fashionable lady in the rear compartment (tonneau) of a Marmon deluxe sedan. Beneath it the legend: "With unerring perception they have discovered in its subtle color harmonies and in the 'petite salon' of its tonneau an authoritative and satisfying expression of their own motoring ideals": At the very bottom of the ad appeared a picture of the deluxe sedan, but only about two inches long.

An interesting promotion . . . evidently one of a series in Harper's Bazaar about 1926 . . . showed two models alighting from a Marmon 74 brougham. One paragraph beneath the picture addressed features which made the Marmon especially suitable for women; the other provided details of their attire and the place of purchase.

Marmon was a pioneer in advertising safety features incorporated in the car. A 1926 advertisement showed a painting of a broadside collison: A heavy truck was pushing a Model 74 sideways following collision at an intersection in Washington, D.C. The documentation was impressive: the uninjured driver had been Admiral Grayson, previously personal physician to President Wilson, and the actual occurrence had been recorded on film by an Underwood and Underwood camera crew who chanced to be present. A booklet "The Safest Car in the World" was offered upon request. Other 1926 ads carried a footnote "Only Marmon Provides Maximum Built-in Safety".

The Fourth Era overlapped the last year of the Third Era because the last iteration of the Third Era product, the E — 75, was a 1927 (and 1928) product as was the first offering of the Fourth Era, the Little Marmon.

A single (one page) ad in the Saturday Evening Post portrayed BIG and LITTLE Marmon open models side by side. Thereafter separate ads were used for the two product lines. The Series 75 ads were typically in color on the reverse side of the magazine front cover; representative magazines were: *ASIA, The New Country Life,* House and Garden, Art and Decoration and SPUR, The Little Marmon ads were frequently two-page, black and white and took over the traditional Marmon space in the *Saturday Evening Post* and *National Geographic;* other mass market magazines were also used. The Little Marmon ads ignored big brother; only in May 1927 did a Series 75 ad acknowledge little brother: "Also the little Marmon 8... $1795 and upward." (Note lower case 'little').

The Little Marmon ads stressed performance and economy, claiming a new industry standard in acceleration. This claim evidently referred to rate of change of acceleration or 'jerk', the response rate to a throttle pedal request for acceleration. The dual downdraft manifold was credited with the 'lightning' acceleration.

The Series 75 promotion stressed the Jewel Colors: "Not only the harmonies but the proportionings and the stripings are analogous with those used by nature itself in gem stones and their matrices." An example: 'A Symphony in Grey and Green ... its harmonies directly inspired by the emerald and its matrix, the mica — schist, which shows subdued gray toneings in contrast with the brilliant green of the jewel crystal. (Two-tone gray with green wheels and splash pan, green striping).

In another ad: "Jewel Colors . . . Marmon designers sought colors most suitable to making **an art of painting cars.** Nothing short of inspiration prompted Marmon to seek out sources where Nature works upon hard surfaces. Experimentation led into the realm of precious and semi-precious stone." It should be noted here that the Jewel Colors program was one of the first fruits of the Styling Studios then recently established at the Marmon plant.

However, ads stressing the more sporty models . . . the Locke 4 and 7 passenger speedsters, and (Convertible) Coupe Roadster . . . pictured the models against

Fig. 10-19 The Marmon 74 featured in one of a series of automobile reviews in *Harper's Bazaar* magazine.

less-traditional renderings of specific locales. The text addressed the pleasures of driving to that destination in a Marmon and provided a capsule trip agenda to emphasize the Marmon's capabilities. An example: "Boston to Moosehead Lake, Maine, a distance of 295 miles. Leave Copley Plaza at 8 AM, arrive in Poland Springs for luncheon at 11:45, departing at 1:00 with arrival at Greenville, Maine (on Moosehead) at 6 P.M." A second example: "Los Angeles to San Francisco...451 miles...on the Coast Highway in 11 hours and 50 minutes. Time on the Daylight Limited (train): 12 hours."

Fig. 10-20 Advertisement for the Model 74 by the Parisian agent.

For the first time Marmon elected to offer a price-leader, a model created to allow the advertising of a lower price than otherwise possible. In October 1926, a two-door close-coupled sedan was advertised as a "Town Coupe" at $3195. Ads for this and all standard Series 75 models noted 'cars sold, if desired, on new convenient credit plan'. This promotion certainly did not enhance the image of the large Marmon, particularly at a time when the offering of the Little Marmon at grossly **lower** prices was confusing the traditional clientele.

By 1924, the Marmon account was being handled by The Homer McKee Co., Kahn Building, Indianapolis. That agency also carried the Apperson, Cole and Premier accounts.[2] In 1926 Mr. Homer McKee was elected a member of the Board of Directors of Marmon Motor Car Co.

### *The Marmon Miniatures*
### *1921*

The Marmon engineering model shop received a very unusual assignment sometime in 1920. They were directed to build two sets of eight small-scale replicas[3]. . . two only of each of the eight Marmon 34 body styles. The work was performed by apprentices in the shop. The replicas were one-quarter scale, i.e. one fourth as long, as wide, and as high as the real article.

Each car was mounted on a heavy glass plate supported at adult eye level on two steel tubular legs. The cars first greeted the public at the 1921 New York Auto Show where they made it possible for Marmon to "display" their full product line. Visitors picked up a quality brochure, entitled "Something New At The Show" and sub-titled "Unique Models of the Marmons in Miniature", which showed each miniature in color against a pen and ink background of ornate exhibition hall and the very latest clothing fashions for the whole family; a Gem among automotive literature pieces. It would be an injustice to do other than repeat the text of the introductory page with its added insight into the sophistication of the Marmon design activity so long ago:

### *Something New at the Show*
### *Marmon Motor Cars in Miniature*

As usual, Marmon contributes a sensation. Last year it was the famous motor "tear-down" that attracted the throngs . . . a great educational

Fig. 10-21 The refurbished 2-passenger Speedster Miniature in the Los Angeles County Museum of Natural History.

Fig. 10-22 Craftsmen making the Marmon miniatures in the model shop.

exhibit, staged in an extensive manner. This year brings the unveiling of unique models of the Marmon in miniature.

Nothing of this sort has ever been attempted before. It has meant much preparation . . . but so astounding is the result that it was well worth the time and effort.

By displaying miniature models built to one-quarter scale, it is possible to have a complete exhibit of Marmon models in the limited space at the show.

**The genesis of the idea came from small models built by our engineering department, in order that the designing engineers might better judge the lines.** The Marmon exhibit of miniatures is an artistic event; it is the first time that automobiles have been brought down to small size and yet portray faithfully their actual charm.

These cars are a rare concept in fine craftsmanship . . . well worth the study of all who appreciate the niceties of artistic effort. So faithful is their execution that people almost expect them to be endowed with tiny engines and capable of locomotion.

Their alliance, however, is more with art than mechanics. They are painted and upholstered as in life size. Even the smallest details have had careful attention.

After viewing this artistic treat, study the Marmon 34 from a mechanical standpoint. Learn its advancement. Make comparisons. Then you'll appreciate its combination of distinctive beauty and rare performance.

Credits are in order to The Marmon Advertising Department and Rogers & Company of New York and Chicago. They produced Marmon works of art.

The Marmon Miniatures are remarkable in another way . . . their survivability. Two of the sixteen not only survive, but have been or are in the process of full restoration and are considered treasured works of art by their highly regarded owners.

The 4-passenger (victoria) coupe was located and purchased in the 1970s by none other than John R. Bond who with Mrs. Bond was instrumental in the success of Road & Track Magazine. It is being displayed at the Briggs Cunningham Museum. The 2-passenger Wasp Speedster was restored, and is now displayed, by the Los Angeles Museum of Natural History. It was presented to the Museum in the

Fig. 10-23 The miniatures on display in New York and Milwaukee (lower).

original shipping crate.

The survival rate is even more surprising when one realizes that the models permitted Marmon to display the product in remote domestic and foreign dealerships where it would have been totally unreasonable to commit a complete car or cars. The town car miniature was displayed at the Centennial Exhibition in Lima, Peru by the Marmon dealer of Callao, Peru who also alternated a full size touring and a New Haven Suburban. A Birmingham, Alabama dealership, who dualed the Marmon with the Chandler line displayed a full size Marmon Wasp Speedster beside the Cloverleaf Roadster miniature. The sedan was shown at the first automobile show in Mexico City, April 20 to May 5, 1921; did it ever return?

The little cars were known to have silken window shades, windshields that opened, instruments with dials, radiators made from soda straws, model Goodrich white wall tires, and radiator badges and hubs made by manufacturing jewelers. Kid leather was used for the pleated leather upholstering. A picture survives showing the models in process; a touring in the foreground confirms that the doors were individual parts and that the unbelievably detailed wire wheels mounted on authen-

### The Marmon 34—Roadster

FOR cozy, conversational grouping, this four-passenger Roadster is a favorite. The top is lower than on the Touring Car, the windshield shorter—which add to the striking appearance. This and other models are equipped with wire wheels, inbuilt shock-absorbers, cord tires, and every motor-car necessity. Examine all the conveniences—note the instrument-board, for instance. The instruments are grouped together and located under a common glass—with indirect lighting. And remember that extras don't have to be bought, but are included in the Marmon price. Remember this in price comparisons.

### The Marmon 34—Speedster

THIS two-passenger Speedster is our latest creation, a dressed-up speedway car, particularly designed for the man or woman who wants a racer type, with the resultant smart effort. This car typifies speed, without being freakish. Carrying space is provided in the rear deck for the convenience of those who who wish to tour.

While a car like this is extremely individual, he who drives it is immediately identified as one who picks and chooses with great distinction. Everyone pauses to admire this speedy—yet extremely comfortable—Marmon 34.

### —a Perfectly Balanced Car

A SHORTER-BODIED, four-door Touring Car seating four—a happy medium between a seven-passenger Touring Car and a four-passenger Roadster. In every Marmon 34, whatever the model, you will find true balance, provable on scales. This brings greater riding ease, easier handling. It prevents side-sway and skidding. It saves on tires and gasoline. It precludes undue strain on any one unit. With true balance Marmon also introduced a more extensive use of aluminum, which eliminates needless weight. Aluminum is used, for instance, for hood and fenders—parts which do not carry weight.

### The Marmon 34—Sedan

OF closed cars, the greater call is still for Sedans and Coupes. The Marmon Sedan, seven-passenger, is especially commodious. It is fitted and equipped with that same care shown in the Limousine and Town Car.

For the man who likes to drive himself, midst his family or friends, this model combines all the pleasures of a closed car with that "feel at the wheel" which delights experienced motorists.

The popularity of this model has grown amazingly during the past few years—more and more people seem to prefer it. Possibly it is the type of the future—it certainly affords a family car of many uses and one that can be kept permanently.

### The Marmon 34—Coupe

SHARING equal interest with the Marmon Sedan is the Marmon Coupe. It appeals to the man or woman preferring a lesser seating capacity. It provides for four, without crowding, but is cozy enough to drive alone.

For the business man who needs automobile transportation between home and office, or for the woman who likes to do her own driving, this model is particularly appealing.

Nothing has been left undone to make this Coupe a smaller-scale, personal Limousine. It has the same fine upholstery, the same attention to details, and is mounted, of course, on the standard Marmon chassis. It is a model which will bring constant comfort and service.

### The Marmon 34 Touring Car

A ROOMY seven-passenger Touring Car, embodying many refinements, yet retaining all the basic advancements introduced by Marmon in 1916. The High Efficiency Motor brings the "close-limits" craftsmanship learned in building Liberty Motors during the war. Here is high power, ever dependable—perfect lubrication, long life, low fuel consumption. Ask a Marmon representative to explain the many features of the Marmon engine—the crankshaft of exceptional size and strength, combination iron and aluminum pistons, shimless bearings, etc. The logic will be convincing, comparisons interesting.

tic axle hubs. Interior and exterior lights and door latches were operable originally. The Marmon Miniatures were first mentioned in automotive historical literature by Charles L. Betts, Jr. in his 'American Cars'[4]

In 1979 a member of the Society of Automotive Historians wrote to the Society Newsletter editor inquiring "What ever became of the Marmon miniatures?" and triggered a long string of letters from the membership in subsequent Newsletters.

The victoria coupe did not have an engine and the hood does not open. The front axle is functional including tie rod but there is no functional steering gear.

But the author owns a quarter-scale 1921 Model 34B engine of remarkable exterior detail. The intake manifold and rocker cover are buffed aluminum; individual spark plugs are threaded into the blocks; external oil lines have hex nut fittings. This model engine was never drilled for mounting bolts and is mounted on a mahogany block. It reportedly graced the desk of the manager of the Los Angeles Marmon dealership.

The models were: 2 p Speedster, 4 p Cloverleaf, 4 p touring, 7 p touring, 4 p coupe, 7 p sedan, town car, limousine.

It is the concensus of the interested historians that the original intent was to have fully detailed chassis and accordingly patterns were made for engine parts. But the task was without precedent and presumably time limitations precluded completion of unnecessary detail on the cars. Once the show deadline had been met, attention could be given to the engine models which presumably served as some kind of recognition awards.

The Marmon press release on the models noted that the making of the models was unusual only in the number of styles and in the public showing. It stated that for many years similar models had been made for styling evaluation purposes.

There is another Marmon miniature and it too resides in prestigious quarters. It is one-tenth scale styling pre-prototype of the SIXTEEN produced by Walter Dorwin Teague, Jr. as a design confirmation tool. It was presented by him in the 1970s to the Cooper Hewitt Museum in New York City where it is displayed. "That Museum is dedicated to the history of style and preserves artifacts that best exempllify the period styles of our century. Marvelous examples of nouveau art, deco art, and todays super graphics are displayed." "The Marmon miniature represents the deco art style and was executed when that look was at its highest point."[5] It is illustrated in Chapter Six.

The SIXTEEN model was produced by the prestigious Boucher firm in New York. It is one sixth scale . . . an unusual selection for a very practical reason. According to Darwin Teague, Jr., that was the scale of the Firestone tire replicas which adorned ash trays popular at the time.

## FOURTH ERA ADVERTISING

The Fourth Era began with the introduction of the Little Marmon 8.

Initial advertising stressed the unique character of the new offering . . . the smallest eight cylinder engine on the market with more power per cubic inch than any other engine except in racing and aviation and the consequent performance . . . "fastest acceleration ever known", 50 mph in second and 70-75 in high depending on style. Fuel economy of 18-22 miles per gallon was claimed and all regular models were priced below $2000.

Motor Trend had not yet been conceived, but a January 1927 two-page ad acknowledged that *"Trade experts have proclaimed it not only* ***The Car of the Year*** [6] *but the most interesting development of the decade."*

Marmon's long time claim of "The Easiest Riding Car in the World" was addressed with the longest springs for a given wheelbase of any car in the world. Every Fourth Era Marmon would meet that criteria. The Little Marmon ride was claimed to be the equal of any car in the world, regardless of wheelbase or price.

The long-held Marmon reputation for superior handling was sustained by virtue of the very low chassis made possible by the adoption of Hypoid rear axle gearing. This new development was only found on Packard and the Little Marmon.

A late March 1927 ad showed a Little Marmon sedan on the Indianapolis Speedway under the headlines . . . 'The greatest proving ground in the world . . . THE INDIANAPOLIS MOTOR SPEEDWAY WHERE MARMON CARS OF THE PAST DECADE HAVE BEEN PROVED OUT!' The picture caption proclaimed that Little Marmon 8 test cars had been an almost constant sight on the Speedway during 1926. This ad carried the phrase 'AMERICA'S FIRST TRULY FINE SMALL CAR' beneath 'the little MARMON 8'; all previous ads had the phrase 'COMPANION TO THE LARGE MARMON, Series 75'. It was also significant because of the first appearance of the 'Lightning Flash', symbolic of the performance of the Little Marmon, in an ad; it had graced the cover of the introductory color catalog.

By June, all reference to size had been abandoned and the ads were headed 'the MARMON 8'. First hand experience was encouraged: "It takes just fifteen minutes in the Marmon 8 to find out more than you ever knew existed in any automobile". The car was described as 'the easiest, most modern and thrillful way of getting about so far ever developed'.

A color ad was headed 'Other cars will serve but this is the ideal woman's car'. A cabriolet, top collapsed, was shown being driven from an estate garage by a woman while the chauffeur closes the doors behind her. The ad claimed that 'an almost unheard-of percentage of owners of the Marmon 8 are women . . . for reasons quite obvious!

The 1927 advertising had produced mediocre results. The anticipated market for a small luxury car had not materialized. The Little Marmon had put on considerable weight and become the top-of-the-line Model 78 as the Large Marmon Series 75 was absent from all ads, although listed as a 1928 model. Evidently it had become evident that the Era of the Big Marmon Six had ended . . . ended suddenly much like the concurrent demise of the Ford Model T which like the Model 34/74/75 had been offered for more than a dozen years. During the few weeks between announcements of the Model 78 and of the Model 68, exclusive '78' ads appeared. Initial catalogs portrayed the Model 78 in standard trim compatible with the low advertised price of $1895 . . . wood-spoked wheels and rear mounted spare rim and tire. Only the speedster was shown with wire wheels, and still with a rear spare. But the exclusive '78' ads showed the car with wire wheels and side mounts as did all subsequent catalogs and ads. An example of the exclusive ads showed the fully equipped sedan with luxury home and chauffeur in the background. The Model 68 was shown usually in standard trim but occasionally, depending on the audience, in deluxe trim with side mounted wire wheels. It was shorter than the '78' and styled to look even shorter. It was a price leader and the ads proclaimed: *"Now you can own a Marmon Eight at the price you have been accustomed to pay for the average six . . . $1395." (The latter price in very large type).*

Attractive color ads used highly legible type widely spaced. In one example the car was shown in a level scene but with the picture and text tilted; in another the lines of text and the invisible road beneath the car fan out.

Significantly the 1928 ads stressed 8 cylinders in-line under the hood and provided no other technical information. Introductory ads for the New Series in Sep-

Fig. 10-23 Examples of the contemporary world series of ads.

tember 1928 did note the addition of double chromium plating, a coincidental ignition/transmission lock and a vibration absorbing clutch.

When the Roosevelt was introduced in January 1929, the ads had a distinct format. The handsome Roosevelt was shown in photographs of contemporary scenes, e.g. with a Ford tri-motor transport. By mid year 'Marmon-Built' appeared directly beneath 'Roosevelt' and for 1930 the names were reversed . . . Marmon above Roosevelt . . . and the Marmon pyramided radiator replaced the likeness of

Roosevelt
MARMON-BUILT

Theodore Roosevelt. Various technical details were related but were typical of the contemporary industry rather than individual.

For 1929, the lines of text were conventionally horizontal and the single car was illustrated in a painting. An excellent example is shown.

Nearly all 1929 ads carried logo: 'Eight firing in line', and some ads had that as the main theme.

A limited advertising budget restricted 1930 advertising largely to the BIG EIGHT — the reentry to the luxury market. Sales of the Eight-69 and Eight-79 probably suffered as a result; only 2675 and 1462 units respectively being sold. BIG EIGHT sales totaled only 884.

A series of black and white ads showed the Big Eight but addressed the full line of four straight eights. These were followed by a series of color ads which were distinctive, innovative and probably the best Marmon series of any Era. They are sometimes referred to as 'Neon Ads', because the **black** car was illustrated, e.g., with the body outlined in yellow, the wheels and tires in red, and the chromium trim in green.

The example page was headlined 'MARMON' in green. Text phrases alternated black and red; references to the four product lines were in green, black, red and black.

Interspaced in the text were people shown in the same color treatment as the car and in some related activity: a dancing couple and a resort scene, a couple in English and cowboy riding clothes and a rodeo scene, mother and children with balloons/a circus tent.

A single body style was illustrated in each ad. The colors for outlining, headline and text were different, ie, interchanged, for each of the six (or more) ads in the Series which, with one exception, showed BIG EIGHT styles. The illustrations accurately portrayed the cars with the noteworthy exception that the roofs were "chopped" . . . the window pillars were shown about 60 percent actual height . . . much like 'customized' cars of the 40s and 50s. A somewhat surprising distortion to enhance apparent length; however, for many years most auto show cars were displayed with springs compressed.

Fourth Era advertising ended with a pair of color ads in a new format . . . presumably planned to be compatible with the launch ads for the Fifth Era.

The first ran in October: 'Marmon — Presents a new custom series' and the second in January 1931 'Marmon — Presents two new Eights'. The former showed the Locke Grand National — Sportsman's Convertible on the Second Series 1930 Big Eight — chassis. The latter a (Locke designed) club sedan on the 8 — 88 chassis. For the first time in the Fourth Era, these ads used the black background of the Third Era ads but restricted it to the central page and otherwise had a 'Modern' motif.

The last Fourth Era model, the 8 — 125, shared a two-page ad with the MARMON SIXTEEN in January 1932. This single ad introduced the 8 — 125 and reported the winning of the Stevens Trophy by the SIXTEEN. It evidently expended the advertising budget for the year.

The Homer McKee Co., 320 N. Meridian, Indianapolis, continued as the Marmon agency during the Fourth Era. The 1928 advertising budget was in the $250,000 to $500,000 bracket.

## FIFTH ERA ADVERTISING

The advertising that launched the MARMON SIXTEEN promotion was on a par with the product . . . spectacular and exhilarating, deserving of an historic niche.

The car featured extensive use of aluminum; the ads appeared to be printed on aluminum foil. Only the white border destroyed the illusion. The car . . . only one style per ad . . . was black, detailed in aluminum lines.

The first page of the four-page announcement carried only the legend 'NINETEEN THIRTY ONE' against the Art Deco theme repeated on the following pages. Pages two and three were in effect a double width page, with the coupe on one and the announcement of the new car on the other. The fourth showed the three 5-passenger sedans for 1931, all left side views to the same scale with front bumpers aligned; the size of the SIXTEEN was emphasized and its advanced styling obvious. In *MOTOR,* this ad was on pages 32-35 of the January show issue. But just in

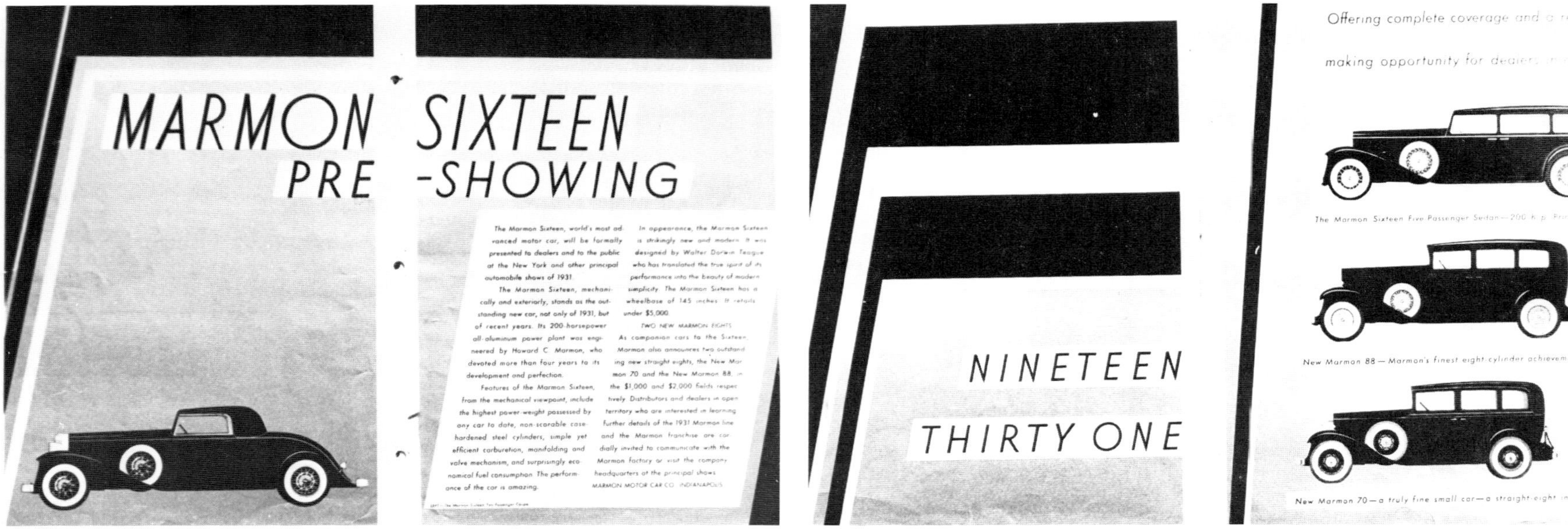


Fig. 24 The SIXTEEN announcement ad: Pages 1 and 4 right, 2 and 3 left.

case the reader missed the insert, another full page ad in the same clear treatment was on page 239.

It was headlined WHAT and asked six questions about 'The world's most advanced motor car'. An example: 'What famous engineer's most notable achievement will soon make its appearance?' The manufacturer was not identified; the reader was referred to page 32 of the issue for answers.

This announcement ad was followed by a two page ad (double-sided insert) in the same treatment. On the first page, the words MARMON and SIXTEEN and each line of text appeared to have been printed in black on strips of white paper affixed to the aluminum background. The 5-passenger sedan was shown. Page 2 repeated the decor and heading and was really an independent ad which discussed and illustrated the car's feature. Car details were black on aluminum background; the aluminum engine details were aluminum on black.

Marmon's last national ad was their first and only in *TIME,* running on October 31, 1932. It announced the Series 1933 of the Marmon Sixteen. An artist's rendering of the sedan was shown in the black and white ad. Beneath the word 'MARMON' was shown the numeral 16 surrounded by a pair of laurel wreaths.

By 1932, the Marmon account was held by the Central Advertising Corp., Indianapolis.[7]

## PART 2 — The Distribution System

*Domestic Marketing*

Direct factory advertising during 1905 resulted in factory sales of only 25 units. A distribution system was obviously necessary.

### FIRST ERA

First Era dealer marketing of the Marmon car began in 1905 with two dealers — agents, then — Frank E. Wing in Boston, Massachusetts and Theodore E. Schulz in New York, New York. By 1907 there were outlets in Chicago. Dealers in Milwaukee and Minneapolis were subsequently added, with representation in Nashville, Tennessee and Waterloo, New York the following year. By the end of the Era there were 13 dealers in cities including St. Louis and Los Angeles. Three were in New York state: New York City, Binghamton and Rochester. Listed below are the pioneer dealers:

| | | |
|---|---|---|
| 1905 | Newton (Boston), Massachusetts | F.E. Wing |
| 1905 | Providence, Rhode Island | |
| 1905 | New York City, New York | Theodore E. Shulz (ex Homan & Shulz) Broadway at 10th. |
| 1905 | Chicago, Illinois | Dan B. Southard, 1409 Michigan Avenue |
| 1905 | Nashville, Tennessee | Tennessee Auto Co. (Nashville Motor Car Company by 1907) |
| 1906 | Philadelphia, Pennsylvania | Brazier Auto Works 28th and Market |
| 1906 | Los Angeles, California | C.S. Anthony |
| 1906 | Minneapolis, Minnesota | A.F. Chase & Co. |
| 1906 | Atlanta, Georgia | Piedmont Motor Co. |
| 1906 | St. Louis, Missouri | Van Automobile Co. |
| 1906 | Baltimore, Maryland | Snodeal Automobile |
| 1906 | East Orange, New Jersey | Rickey Machine Co. |
| 1906 | Wilkinsburg, Pennsylvania | George J. Treadgold |
| 1906 | Brighton, New York | O.S. Heller |

## SECOND ERA

During the Second Era, the dealer group grew to 80.
Marmon factory ads no longer listed distributors or dealers.

## THIRD ERA

During the 34A Production Period, the dealer group grew from 80 to over 200 dealers (220 in Jan. 1920). By the end of the 34B Model run there were 325 dealers.

In 1920, C.F. Kettering was president of the Dayton Engineering Laboratories Co. *(DELCO)* and neither he nor the firm were yet affiliated with General Motors Corp. Nordyke & Marmon held a major dealer convention timed to include the Indy 500 paced by Oldfield in a Model 34B. The day following the race, the man later known as Boss Ket of GM delivered a major address to the assembled Marmon "Round-up". Because of the career stature of the speaker and what he had to say that 31st day of May, the talk is repeated herein as it appeared in *The Marmon Post* (which required two issues to record the convention; #9 and 10 of Vol. 3).

### WHAT THE FUTURE WILL BRING FORTH IN MOTOR CAR MECHANICS

**By C.F. Kettering**

Let us discuss this morning the new angles that there are in the automotive business today. This business was started, as any new business usually is, more or less as a sporting proposition. It was a sport to take an automobile and see if you could go some place on Sunday and get back. You took a chance on whether you would or you would not.

Then, in order to promote the thing, the word pleasure car got hitched onto it, and today, thousands of people think of the automobile as a pleasure vehicle.

Later on, the truck came, and the tractor, but today the automobile business is just as essential a business as is the bread and butter business, because it has so worked itself into the everyday life of folks that it is an essential. It is a form of transportation of which the most imaginative writer of twenty years ago could have hardly hoped. In other words, I do not believe that any of us appreciate to what extent the internal combustion engine has entered into our lives. We think of an automobile, we think of a truck or a tractor, or what not, without ever thinking of one of the reasons for it.

Fundamentals of Automobile Success

There are two fundamental things in the automobile which have made it a success: One has been the internal combustion engine: the other has been pneumatic tires, and yet how passé have we become on tires. We don't think of the thing. Of course, an automobile ought to have tires, everybody knows that, and everybody cusses them out when he has a puncture, and yet the pneumatic tire has made possible the industry. If we had had engines, we could never have had an industry without the tires.

The engine has grown from the great cumbersome thing to the modern engine. It was simply a question of making up so many cylinders, putting on an intake and a carburetor and running in the early days. The question of fuel, etc., did not enter into it for a minute, because there was no fuel problem with gasoline at ten cents a gallon.

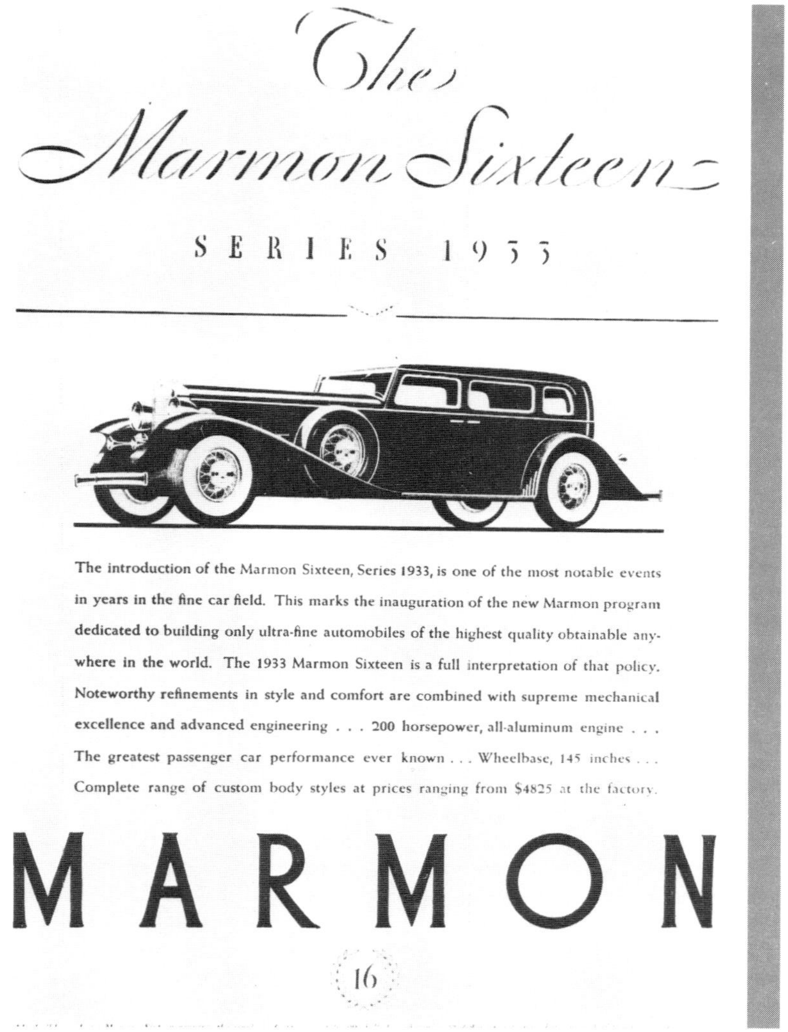


We are at a peculiar point in the industry today. The demand today for cars exceeds the supply, for several reasons. Reason number one is that during the war there was a great shift of finance. A great many people who did not have money acquired money; some who had money lost it; and the newly-acquired gang immediately bought diamonds and high-priced automobiles. So, we have an inflation in that direction. On the other hand, everybody came back from the war with a new idea about how much work they were going to do, and we were called upon for motive power, transportation, to fill that in. So today there are two reasons why the automobile business in general is at a high tide. One is because a lot of people want them because they have got the money to buy them; another reason is that we have got to supplement the deficiency in our ability to produce things.

Now, every city in the world — I say every city in the United States of any considerable size — is short of housing facilities. New factories are being built, and this housing question has gone the rounds. Did you ever stop to think where are the people coming from? Why do we need so many more houses now than we did before the war? Well, if you will get hold of your state agricultural report and study it for a little while, you will get some surprises.

In Ohio, last year, seventeen per cent of the farm labor left the farm and went to the cities. The average age of all farm labor in the state of Ohio is fifty-five years. That is true almost over the United States. New York state lost a greater per cent than that. Now, that is simply meaning that we have got to supplement the work of those men that have come into the cities with machinery and other devices for the farm. It means that many of those people who have come to the cities to work have been attracted by the high wages. It means that the work which they are doing, they are unfitted for. And in every line of human activity there is a demand which none of us can imagine for the supplanting of human effort by machinery. In other words, we are at a point today where mechanical developments, mechanical facilities to aid human activity have never been at such high tide.

### Industry in Excellent Condition

Now, there was never a time in the history of the world when we were more able to handle a situation than now, because the war upset every social condition. This unrest is not an alarming thing, because it is a perfectly normal thing. It ought to be after the upheaval that we have had.

Now, out of this war, while we get the upheaval, we get another sort of thing, because during the war everybody who knew anything pitched into the pot of human information that little thing which he knew. Research laboratories were opened, trade secrets, manufacturing secrets, and everything else was thrown out, and I do not believe that a carpet tack is made the same way now that it was before the war, because we have all learned something.

Have you ever stopped to think just what an important part the internal combustion engine played in the war? The dreaded submarine was possible because of that; the submarine chaser was possible; the tank was possible; all the motor transports, your trucks and your tractors, and almost everything; your airplanes were made possible because of this engine.

The Germans say one of the things that upset most of their plans was the fact that the allies could move great bodies of troops from one point to another more rapidly than their figures indicated that they ought to. That is where we all fall down, you know.

### Service and Adaptation to Economic Conditions

Now, what has that got to do with the particular business that you and I are interested in? Here is the one thing that we have gotten for the first time in the world injected into the automobile business. There are two very important things that we have got to do, and one is, we have got to give more service in the automobile today than we have ever given before, because it has become a utility piece of apparatus. The next thing, we have got to fit into the economic conditions as things go along.

There is only one great big cloud on the horizon of the automobile industry at all, and that is the fuel situation. That is not an alarming thing if you know what the facts are. The thing is only alarming when you don't know what the facts are, because if you know what the facts are, you can begin to readjust yourself to what they are.

An old fellow expressed that very well. He said, "It ain't them things that you don't know that get you into trouble; it is them things that you know for sure that ain't so." So, if you will pick up the magazines and read them, there is a lot known about this fuel subject that "ain't so."

### Fuel Situation

Now, if you will permit me for just a moment, I will briefly touch the fuel situation here, because it is of so much importance to every one of you. This country is divided into some five or six economic districts by the oil industry. Those districts are made up to suit their methods of distribution, their sources of supply, etc. Therefore, today when we discuss gasoline, it is nothing definite at all, because each one of those districts has a different set of conditions to meet, and the gasoline in each one of them is different.

The east coast district, which takes in all the coast around through Texas, has a very peculiar situation on gasoline. Of course, you know all of your fuels are gotten from crude oil. You take a barrel of crude oil, taken over the whole of the United States last year, twenty-one per cent of that barrel of oil goes into gasoline. So, you take that away.

Now, twelve per cent of that goes into kerosene. We have been cutting down into the kerosene just as far as we could, and today ordinary gasoline contains about thirty per cent of what we used to call kerosene. It is perfectly all right. There is no objection to it being in fuel at all. We only damn it when we don't recognize it as a part of our problem. And we have got to recognize that the automobiles of the future have got to forget as an excuse, "Well, the reason you don't get along is because you have poor fuel," because that is one of the things we have got to swallow. You are never going to get any better fuel than you have got now. It is going to get worse. If you recognize that that is what is going to be, you can design your motors and everything else to handle it just as well one way as another. It only becomes a problem when we refuse to move, to progress.

Now, why can't we take all of that twelve per cent of kerosene — and here is some very interesting data: Ninety-five per cent of the world is lighted with kerosene lamps. About thirty per cent of the houses of Indianapolis are wired for electricity.

Now then, we come into quite a section of this barrel of crude oil that is called gas and fuel oil and out here in Indiana they can take that gas and fuel oil and make pretty good gasoline out of it. But over in the east, they don't let them do that, because they need that oil to carburete their gas-making machines for cooking purposes, and so forth. Just recently it was ruled in one of the superior courts over there when they began to take this gas oil out for other purposes, that it was more important to have gas to cook with than it was to have gas to joy ride with.

That shows that we haven't the point quite over to all the people that the automotive industry is an essential industry, and it is just as essential as the cooking industry.

Now, there is another little section comes out of that barrel of crude oil from which we get our lubricating oils, and you can't take anything out of that. Now, we have got the fuel oil, and it is very difficult to get any of that, because the demand for that is equally as great, and it simply becomes a question of how much you can pay, and whether the law will allow these companies to divert it or not.

So, in the east district, the gasoline, the kerosene, the gas oil, the lubricating oil, and the fuel oil are in equal demand; consequently, you can't divert. Out in this particular section there is not such a great demand for the gas and fuel oil and a great deal of that is being converted into gasoline.

## Sectional Variations

Out in the west, the percentage of oil to be converted is still greater. Now, the situation is about this: The increase in the number of automobiles has been a great deal more rapid than the increase in fuel supply; in fact, our crude supplies are falling off today. By the first of August, this year, we will have a refining capacity in this country for a million eight hundred thousand barrels of crude oil per day, and at that time, if the scale holds out as it is now, we will be getting about nine hundred thousand barrels of crude oil per day. In other words, we will have just double the refining capacity that we have crude oil for. The crude oil is falling off at the rate of about fifty thousand barrels every sixty days.

Immediately we will say, where are we going to get some substitutes for that? Now, there aren't any substitutes for it, yet we have got two other sources of fuel — one is alcohol, and the other is benzol. One is coal tar stuff, and the other is vegetation. Today, if we were to raise enough vegetation to make the fuels required for our automotive industry, it would take about four million square miles of the best farming area of the United States. In other words, it would take so much of the area of the United States, we would not have food enough to live on, so we would not need the alcohol. Some might want to drink it, but we are not considering alcohol as a food this morning. We are considering it as a fuel.

Alcohol is just about two-thirds as valuable as fuel. In other words, where you burn two gallons of gasoline, three gallons of alcohol are needed. Today our alcohol production is about five per cent of the total demand of the United States. Benzol is furnishing about three per cent.

## Probable Trend of Motor Industry

Now, when we get that picture of what the thing is, what are we going to do? What is going to be the trend of the motor industry from what we know to-day? In the first place, there are a great many things that the motor car people can do that they haven't done. There are a great many things that the fuel people can do that they haven't done. Anything which you do to lighten the weight of your car, to cut down the mechanical friction of that car, to increase the efficiency of the motor, is all in the right direction.

The size of a car does not necessarily mean its wheelbase or its seating capacity, or anything of the kind. **In the future the size of the car will be a measure of how many miles that car will run on a gallon of fuel.** There are a great many things that enter into fuel consumption on an automobile besides carburetion and engine; there are clutches and gear shifts, etc.

We had a little resume made in Philadelphia, Baltimore, Washington and Chicago on the average mileage of small cars. To our great surprise, we found out that the mileage was about **twelve miles to the gallon on some of our very popular small cars,** yet you could run tests and show that you could get seventeen or eighteen or nineteen miles. So I started to drive one of those cars.

There is a perfectly definite reason for that. Don't say that report isn't true, because we have checked it up a dozen times. There is a reason for it. The reason for it in the case of the Ford automobile is purely and simply its gear shift, because when you get in congested traffic you are not going to take any chance of dropping back and letting that motor choke, especially when you haven't any starter on and have to get out and start it by hand. The ability of a clutch to come in and out, ease with which you can handle a job, all contribute to these.

So it doesn't make any difference what the piston displacement of your motor is, what the weight of your car is, it is, after all, going to be a measure of how far you can go on a given amount of fuel. I don't believe, and I don't think that we want to look at all with any fear and trembling on the fact that **fuel may go to sixty cents a gallon.** It would not mean anything. It would mean that we would all get awake, I think, after this long sleep that we have had.

## Situation Needs Study

I don't think it would have any more effect than the reducing of the piston displacement from three hundred cubic inches to a hundred eighty-three cubic inches had on the mileage here on the Speedway. In fact, you can go faster with those motors now than you ever could with the three hundred cubic inch motor. But you can't do this unless you put a little brains into it, and the brains can't all be put into it at the factory and at the engineering department. You men have to become students of your own situation, and you have to begin to formulate the opinions of the people in your community — and, believe me, that is a hard job. I know because I have been trying to convince a great many people for the last ten years that they were using the **most abominable ignition in the world when they had magnetos** in their car.

We had some difficulty in even persuading some very high officials of the Nordyke & Marmon Company about that, and the only reason we went into this Speedway race out there this year — we have never been in the racing game before — was just for the sole purpose of showing folks that we could run a car around that track just as fast as anybody else could. And, of course, the fellow who ought to have won didn't win because his magneto went bad on him, all of which contributed to our original conclusion on the matter.

Now, I am just mentioning that because you have got to start everlastingly at the thing, and it is not a thing that you can quit for Sunday. It is a thing you have got to talk, you have got to begin to get a point of view on this thing. You don't need to get scared, as I say, on this fuel situation, any more than the fellow who says, "What's the use to go to Indianapolis next year, because they are going to have those little bits of motors?" That hasn't anything to do with the situation.

## Progress Absolutely Necessary

When we once know what the problem is, we have brains enough in this country to solve it. But we have to recognize that we have to move with the world. We can't stand still, and that is the hardest ting we have got in the scientific side of this industry — to get the other side of it to recognize that the worst thing on earth for them to do is to stay put.

Most people, if you would listen to them, would still be riding around in ox carts, because it is safer — it is much safer. I can remember the time when I started to ride a bicycle, some one said, "You had better keep off of that fool thing. You will get your neck broke." But I didn't. Then I got to driving an automobile. Then they were sure I would get my neck broke. If I don't get my neck broke in an airplane, I haven't any chances left, because I don't know any other way to try it.

Now, I want to tell you, you might just as well have them sprinkling a little dirt over you and shovel you under as to have the attitude — "My God, why haven't we got the things as they used to be?" And the trouble with the world to-day is that it is sitting back and wishing that things were as they were before the war.

They won't get into the new financial way of figuring; they won't get into anything. They say, "Oh, if things were just as they were before the war!" Now, they are not that way and they never will be, because we are not back there any more, we are some place else; and if you want to be back where you were before the war, you were dead then, and the chances are you are just as dead now. So go ahead and get buried and get out of the game.

### Outlook Optimistic

I am more enthusiastic to-day over the motor car business than I have ever been in my life before, because of the fact that we are reaching a utility demand for this thing, so important that the business can not be deflected or can not be curtailed. People have to have these automotive devices.

And the next thing, our natural resources have reached the point at which we have got to recognize that we have to put more engineering, more skill, more science, more study, more everything into them, in order to accomplish those results.

Why, it is perfectly ridiculous that we are willing to suppose to-day that we have reached any point at all in perfections, because we haven't. The industry is too new. We make steps along certain lines. We get a commercial demand, and the commercial demand is so great, we say, "Now don't let's change anything, for God's sake don't let's change." That is when you are dying, right then, when you say, "That's all right; let's leave that alone."

Nor, on the other hand, do I believe a man should be so foolish that he wants to change everything every day. But when you can see from the situation on your job and from the conditions that there is a difficulty in the thing that ought to be changed, then change it. But on the other hand, you should so analyze your situation that you know that the change should be made, you have the whole thing worked out ready to slip in the operation when the commercial and economic conditions of your organization get such that you can do it. And there are more companies being ruined by foolish changing of their apparatus and upsetting of their stuff when the business situation was not permissible than there have been by not changing something the other way.

### Analyze Proposed Changes

So, you have to to analyze and weigh those things both ways. But the thing I am going to try to get at is, let us know what our changes are, let us be looking ahead on this thing.

Our only difficulty to-day in the fuel situation is the fact that nobody is ready for it. We talked to the fuel people beginning five years ago, we talked to the motor people. Each one said it was the other fellow's problem, and that is the reason why it is not solved.

Now, we do a lot of foolish hoping. The oil people would say, well, we are not worrying because we know they will strike new wells, so why should we worry. We always have struck new wells when we have needed them. But they didn't strike the new wells in this particular case. They struck a lot of wells that had salt water in, and I suppose some of you gentlemen in this organization have some oil stock — most of them have in almost every organization. But if we had gone back five years ago and began to analyze this economic situation, we could have apprised everybody of what the situation is, and in a couple of years from now the motors would have come along in a normal way.

I don't mean that you pull out of your chassis the motor which you have developed and stick an entirely new thing in; it doesn't mean that at all, but means that we just gradually as we know what the problem is, meet it with this gradual change and go from one condition into the other.

The war contributed the information with which we can do almost any of these things. There was never so great an opportunity for mechanical development as there is to-day because of the information which we have.

The question of steels, the question of all of the materials, which enter into the automotive industry have never been so well understood as they are now. The question of steel production, tire production, all of those things are well under way. I can remember the time when we used to put just as little rubber in tires as we could, because rubber was difficult to get.

### Preparation Solves Problems

I want to tell you one of these problems, because it shows what is the result of preparation, not spectacular preparation, in the sense of going out and showing what a wonderful work you are doing. The rubber industry was in bad shape. They could not get enough rubber to make the tires properly. All right, they started these rubber plantations, and to-day the only trouble with the tire business is they haven't got the cotton, because everybody knew that the cotton situation would be taken care of. By whom? By nobody, because nobody was interested in it outside of the fellow who farmed his individual farm. Now to-day they are starting the cotton thing in the same way they started the rubber situation.

That is just exactly the way with every one of these problems, if we know them in time. If the tire industry required 50 per cent more tires than it does, we have got the rubber, but we haven't got the cotton. You men from a psychological standpoint have got to predict, and do it just as a matter of mental gymnastics — you project what the demands in your territory are going to be, and project what you think the changes in the car ought to be. Do that for five years ahead and just lay it away, and see how near you run true to form. Now, if you don't run true to form, you find out you don't know anything about your business. That is the easiest way of checking it up without letting anybody know it.

In other words, we found some interesting things when we were over in Europe last year. A lot of those export cars, they have got to have just a certain kind and type, and a certain kind of this and a certain kind of that. Well, to our great surprise, we found out that the people in France were paying a hundred fifty dollars more for a standard American top than they were for one with the changes in it. They said, "We don't want these damn khaki tops." They were giving one hundred dollars more for a black top. Why? Because some fellow goes over there and they ask what is the standard kind of top over here. He says everybody uses this yellow top. That is the only thing they can get. All right, there is no use to break precedent, but we found out that if you took the other top over there accidentally, the fellow would say, "Where are you getting that beautiful material?"

They were paying a hundred dollars more for a standard black top than a khaki top. A lot of us ask somebody who doesn't know anything about this subject what his opinion is of it, and then we say that is the opinion of the public. That is the thing you have got to watch out for. The fellow's opinion on a subject he doesn't know anything about does not amount to much. So you have got to watch out for the opinions which you get from your community, whether they mean anything or not.

### Authoritative Opinions Only Are Valuable

An authority on a subject is a fellow who knows about the subject. You don't find any two individuals that want the same motor car, and yet if you tried to build a motor car to suit the individual tastes of every one of your customers you would have some **awful-looking cars, worse than they are now.** But you just build three or four models. If you will go back and trace the history of some of our automobile companies in the early days who had a very strong lead — we had one of them over in our little town of Dayton, Stoddard-Dayton Company, who had a real lead on the motor industry at one time — if you will try to analyze what caused them their trouble, it was trying to design cars to suit their customers.

That is all right to a certain extent, but if you are sick, you don't dictate to the doctor just what kind of medicine he should give you. You don't say, "Well, I don't want typhoid fever, I want appendicitis." The average customer's knowledge of what a motor car should be is mighty crude. It is the most technical combination of things in the world. For that fellow to dictate to your engineer and your sales division what a motor car should be is utter fallacy, and yet we find there is a tendency of people to say, "Well, now, if they would only make their cars this way or that way or the other way.".

The final answer to the automobile business is what is the maxiumum amount of dollar value and service that you can give for the dollar which you spend for that automobile. That is the only reason the Ford Company is in business to-day, because they turn out at the end of their road a greater dollar value for their automobile within the range of the buying price of that car than any other concern in the world, and that is the reason why they are making thirty-five hundred of them a day. Now, we sometimes get this point of view. A fellow says, "I don't see how they sell those things." I have gotten a bunch of engineers who would say when you suggested making a thing a certain way, "But we couldn't make it like that." And when I say, "Look at the Fords that are made that way," the answer is, "I know, but we couldn't get away with that Ford spring."

You people are getting away with it in very good shape, because your springing is the Ford type of springing. There isn't a thing wrong with it. I said to **Mr. Ford** one day, "How have you determined those springs?" He **said, "You know when we built that car we knew it was going to be used for a lot of things. We made it so that one man could ride it twenty-five miles and still be alive, and so it wouldn't strike the axle with ten men in it."** Perfectly scientifically designed for the purpose. There isn't a thing in that car that hasn't been put there for a perfectly definite purpose, and if you want to get a real study of economics, just tear a Ford car down and study it, and I know of no greater example of engineering economics in the world to-day than a Ford automobile. I make the gangs that I have anything to do with tear one of them down every year, and we take a great big blackboard, and we put the parts of a Ford car up there, its price and what it will do, and the car that we are comparing that we think is a world beater right along in under it, and then we stand back and we never need to worry about their getting busy. There is a reason for the whole thing.

Now, all we need to do is to recognize that there is a reason for our business, and that is how much value it gives to the man that has it. It isn't how much he pays for that car in the original at all. We have all made the great mistake — Mr. Marmon and I were discussing this thing here — we have all made the great mistake of determining how much the car cost when you bought it. How much does the car cost you after you have had it one or two or three years? That is the final price you pay for that automobile. And the most encouraging thing to me to-day is the fact that the lower priced cars are putting in better apparatus, better parts, better steel, better this and better that than they ever have done before. We are learning that it is not the selling price of the job, it is the ultimate price of the job that counts.

## Problems of Today

Now, I can say to you gentlemen this, that there is enough known scientifically to-day to take care of the problems which come up. The problems to-day are two-fold. They finally go into one, and that is economic operation. The fuel cost to-day is not the disturbing factor, because the average number of gallons of gasoline used last year for a motor was about **two hundred gallons per car.** At thirty cents a gallon you see that is only sixty dollars a year for fuel. It can't go much higher than that — I mean the number of gallons — we have got about two gallons of gas per day per car right now. With three million more cars going on the market this year, and with the normal increases of cracking that they get in their fuel production next year, it will be a gallon and a half per day per car. That means that all we have to do is that we have got to go farther on a gallon and a half next year than on two gallons this year. Those set things are a good thing.

Some of you may be quite familiar with the typewriter business. We have gotten very wonderful typewriters in this country, the typewriting machines have been developed, and the thing that developed the typewriter more than anything else was when the Typewriter Association got together and fixed the price of the typewriter at a hundred dollars. They said, "You can put anything in that typewriter that you want to, but the standard price will be $100." They tried to leave everything out of the typewriter that wasn't essential so as to get some essential feature into it, and it was the finest thing in the world. We have in the motor car business something we have never had before. We have got a few fixed normal things beyond which we dare not go, and you are going to see a constructive thing in the design and engineering of motor cars that is going to go right ahead. In this country we have thought of the motor car as a very much shorter-lived thing than they do in Europe. A man who buys a motor car here expects to get a new one every year. I think we will see that changed. I think that we will drive motor cars for longer than a year. That means the question of servicing, and I haven't seen anything that impressed me so much as that motor tear-down job that you had out at the dinner last night, for the reason that what you are trying to do is to get a picture across that a motor can be taken down, that the parts can be taken out, that they can be replaced and that motor can be put back together practically as good as it was when it left the shop.

## Service Equipment

Now, you go into the ordinary repair shop — you have all been through them — and you know what their tools consist of: a few banged-up old monkey wrenches and pairs of plyers and screw drivers that never had a point on them or are never ground or anything. It is perfectly ridiculous, and I think that there is nothing in the world that will give you such interest in the business as to go to work and collect from your shops the tools which they are using, and then just get a set of the tools that they ought to have and show them to the boys. We murder more beautiful automobiles with those plyers — if the devil ever contributed any one thing to the automobile industry it was the plyer.

Now, here is another thing you do — I am going to talk from the standpoint of a user, because I have always taken that standpoint. There are a few things on automobiles that we need to fix. The tops on automobiles are rotten. They are the must terrible thing on the automobile. They call them one-man tops — well, they are not.

And then the matter of putting curtains on. Of course I don't think we are going to make very many automobiles in the next ten years with tops on them. I think everything is going to go to closed jobs. The tops are not a very good thing. But that is one of the things that came in from the buggy business. They will stay on there as long as we regard it as a buggy; I think when we regard it as a vehicle they will come off.

Tire change is another thing that is bad. Most of you people have chauffeurs and the chances are that they don't know they have any tires on, or have punctures, but to the fellow who has to change his own tires it is another matter. The only way I ever find anything about an automobile is to drive it myself, and take care of it myself and fix my own tires. Then I find out something about the job. The tires and the wheel change business is something that we need to change.

Now we go to work and take a beautiful set of tools, you may have gone way ahead of this, but the last time I investigated this you put a beautiful set of tools in a motor car and a lot of wrenches. For a little less than nothing you could stamp on that wrench what it should be used for. Do you do that? Well, you ought to do it. Because you know, you say, "Sure the damn fool ought to know that that wrench is for that nut." Why, he has never seen the wrench or has never seen the nut before.

### Explain Uses of Equipment

You say you give an instruction book.

An instruction book is the greatest alibi in the world for not being on the job.

Now, I want to tell you why. I have fought instruction books ever since I have been able to talk, and I started to talk quite young. I will tell you what the difficulty with the instruction book game is. You put an instruction book in there and the owner gets this nice new car, and he is proud of it, and he sits down and reads the instruction book over and he gives it to the chauffeur, and he reads it over. The next week he fires that chauffeur and gets another one. The chauffeur maybe has put this instruction book in his pocket. The second-hand car is sold next year to another man, and I bet, by God, you don't put instruction books in your second-hand cars. They make your reputation more than your new cars where you haven't anything to do with it; that is where your reputation is being made. After you have forgotten about the car there is where your reputation comes in.

A fellow says, "There is a Marmon I have had for ten years, and God, how she does run." That is where your reputation is made.

You say you are furnishing instruction books with second-hand cars, if so, you are one point ahead of most of them.

Now, no matter whether you do furnish instruction books or not, here are these wrenches and these tools. The fellow opens up this tool box and he sees all these picks and punches, and says, "My God, what is this thing for?" Now, you can put the name on the side of that wrench as to what it is to be used for. It does not cost you hardly anything to put it in the dies, and the only thing is that your purchasing department can't go out and get bids on these wrenches, and you will find out that the difference between having them put on and left off is about eighty-five cents a car, as near as I have been able to figure. You can put the name on the wrench and put the name on the tool, telling what it is for. I have found it a very good thing that in the cover of that tool box, or if it is in the door, it doesn't cost very much to put a nice little picture of just where that tool goes on to the thing.

I have seen fellows worry about taking this thing or that thing off because they didn't know how. Some one comes along, opens the door, takes out the proper wrench and says, "See how easy it was done." The only reason those boys got by with that motor change last night was because of the tools they had for doing it. You give them the ordinary garage tools and they would have been out there yet, maybe. That is why I was so impressed with that demonstration, because the cost of the outfit that those boys had to have to do that trick is so insignificant compared with the result that you get when you use it.

And every one, every dealer, every man that sells ten cars ought to have the tools for taking these things apart and putting them back together readily. We go to work and go to all the trouble in the world to facilitate our production of motor cars, and then we go out and try to service them without any tools at all. You go out — not in your fine garages, your dolled-up ones, but some of us have to try to make stuff to go out into all places.

I was talking to an Englishman the other day, and he was saying that our American motor cars were not very good, and he was one of the fellows who had not yet gotten into his mind that in America we didn't take all of the motor cars that were built in a given factory back to that motor car company once a year to have them overhauled, that they didn't come back to the factory, because over there they try to bring them back to the factory.

### Sales Dependent on Service Facilities

I took this particular man out to a small garage in a town of about fifteen hundred people. We went in. I said, "How would you repair your car in here?" And I want to tell you a few little fairy tales that will interest you. I made a canvass last year on why there were not more relatively high-priced cars sold in towns of from fifteen hundred to three thousand. I just picked that out. I just had a boy make a trip around over two or three states finding that thing out. The principal reason was because the boys in the small town knew how to repair the cheaper cars, but if they ever took a high-priced car down they never could get it back together so it would run. I can give you a little town in this state, a particular fellow who has owned a Cadillac, who owned a Stutz, two or three other high-priced cars, and finally sold them, got rid of them and bought a Buick, because he could get the Buick repaired in the town and he couldn't get these other things repaired. Now, that is just a matter of education. The fellow gets scared. He doesn't know how. I was never so delighted in my life as I was to see the work that you had laid out in this room to get you to understand that the machine, after all, is the thing, it is the product that you build your business on, it is the knowledge of that product that makes you an expert man or not.

I don't mean by knowledge of the product that you should be able to go out and take a piston out, but you should know enough about that product to know whether the fellows in your organization are doing properly or not. That is all. This crank-case dilution and all that sort of thing — the boys don't understand it. There is a different theory in every garage about what that is. We have got to get technical information out so that the boys understand it. If you want to know really how thoroughly things are understood, you want to stand around these Speedway jobs out here and hear the boys discuss design. You will find them discussing everything from the statement that the exhaust valve ought to be smaller than the intake up to the point where the exhaust valve ought to be twice as big as the intake. Each one perfectly convinced. They are supposed to be real good mechanics. It is all theory. To-day we know how to measure things. You don't have to guess any more at all. If those boys could have gone to some place and had the measure, they could easily find out. But they don't know; therefore, they get a theory and stand on the theory. You would be surprised to find the number of theories about how things ought to be done in different sections of the country. We have one type of trouble with our ignition apparatus on the Pacific Coast that is just as much different as day is from night to what we might have down in New York or Boston. Why? Because you will find out that there has been some automobile school in that community that has taught the boys a certain way of thinking, and you can't get that out of their minds. We had a very interesting thing the first year we put our apparatus on Packard cars. It was a notorious fact that a Packard car would stall every time you slowed it down. It was another notorious fact that you couldn't get the thing up to speed, and we had never found a car that we couldn't take and drive right down through the congested part of the city, or that we couldn't go out on the speedway with and make maximum speed.

Service-Knowledge in the Field

I went down to New York one time. Colonel Vincent was running some tests out on Sheepshead Bay track and he wired for me to come down there, and had a driver of a Packard car meet me at the depot. The fellow didn't know who I was and we got to talking about things. He said, "This car would be all right if it wasn't for this ignition." I said, "What kind of ignition have you?" He told me. I said, "What seems to be the trouble?" He slowed this car down and she stalled. "Well," I said, "it may be out of adjustment." He said, "No, it isn't out of adjustment at all. It is all right." I said, "Aren't there some little breaker points that you adjust a certain way, and didn't they send you the gauges along with this thing?" He said, "Gauges — what do those gauges amount to? Any damn fool could see they are all wrong." I said, "What did you do with the gauges?" He said, "I just filed them in two, made them half as thick as they were." We finally got out to the speedway. I said, "I would like to see this car on the speedway." So he took it on the speedway. He couldn't get hardly any speed out of it at all. I said, "Now, if you will just let me have this car for a minute I think I can fix it." He said, "What are you going to do?" I said, "I am going to set these spark gaps on this breaker to where they should be." He said, "Anybody knows they can't work that way; what is the use to monkey with it?" I said, "You are a darned sight smarter than any fellow I ever met. I happened to help design this, and I know you don't know what you are talking about." When I had finished, you couldn't run that car so slow that it wouldn't hit. We did all the mileage you could expect of any car. "I will be damned," he said, "there is something wrong; it oughtn't to work that way."

That gave us a cue to our Packard difficulties, and the Packard Company sent two men all over the country, and honestly, if I could show you the bunch of stuff that they got as the theory of that stuff, it would make Puck and Judge look silly. We have got to begin to work that kind of stuff out of the business. It is in transmissions, axles, engines, it is in spark plugs, it is in every old thing. The fellows have more fool notions of what things ought to be, and a lot of your trouble comes from the fact that you have some fellow in your service that knows more about it than the man who made it.

Now, that is the part of your research and development work in which you can help this industry. We have got to have more fellows coming up that know the whys and the wherefores of this thing. When we get our economic problems solved in the field, and our factories producing, this automotive industry is going to go just as positively and honestly as any other business in the world. We are right at a very wonderful time in the history of this country, just at the time when the transitions are coming; we are going right ahead and it seems to me that **every live, red-blooded man to-day ought to just be willing to get down on his knees for once and thank God that he is alive and living in a country like this.**

* * *

By January 2, 1926, a claimed increase of 55 percent[1] over 1924 would indicate 525 dealers and an estimated 550 by the end of the Era.

A 1922 brochure aimed at attracting both domestic and foreign dealers had the following comment: *"For years the production of the Marmon 34 has been small and largely absorbed by a few distributors in the major cities. But gradually the plant has been expanded and enlarged with the aim of increasing the production steadily, consistently, but never abnormally. Now that the plant has reached its state of completeness and the Model 34 has been on the market for so many years, (this) is a logical time to expand the distribution. This opens up what can be termed 'the smaller city field of distribution'. Already the Marmon 34 has nationwide distribution in the larger cities, and in a growing list of smaller cities. The Marmon 34 is being bought by well-to-do men everywhere. It is no longer the exclusive car of 'big city' men."*

## ERA FOUR

When the dealers and distributors met in convention in the first week of January, 1927, they numbered 700. An increase of 200 from January 1, 1927, was reported in mid-April, probably bringing the April total to about 900.

By October 10, 1929, a three-fold increase from 1926 was claimed — based on 550 in the summer of 1926, that would be a total of 1750. Since this was also claimed to be a 50 percent gain from the prior year, there must have been about 875 dealers in 1928.

## ERA FIVE

Era Five saw the collapse of the dealer organization and the closing of the factory branches as well.

Demand for Marmon cars increased dramatically (300%) with the introduction of the Model 34A for 1916. Sales then were constant through the 1926 Model 74, ranging from 2500 to 4500 irregularly. The number of outlets rose steadily from 80 in 1915 to 195 in 1920 (240%) when the Model 34B was introduced; during the next five years, the growth continued — from 195 to 525 outlets in 1925 (275%). The average increase was 44.5 dealers per year.

During 1926, Marmon faced strong competition in their traditional market segment, particularly from the Cadillac 314. Only 25 dealers were recruited that year.

The increased dealer population made the Marmon available to a much larger geographic area; the constant sales level during this expansion marked a declining popularity which was reflected in grossly decreased unit sales per outlet. This sales rate had increased from 5.3 in 1915 to a high of 19.5 units per outlet in 1920 but had fallen all the way back to 5.7 in 1926.

But the introduction of the Little Marmon for 1927 began a steady escalation in sales volume for the Marque, with sales in 1927, 1928, and 1929 being greater than 200, 300 and 400 percent of 1926 sales volume.

This sales spurt sharply reversed the decline in sales per outlet, with a uniform rise from 5.7 in 1926 to a healthy 25.5 in 1929 (based on calendar year registrations) when there were 875 dealers.

A strong advertising promotion of the Marmon franchise was initially slow in effectiveness with 150 and 175 new outlets in 1928 and 1929 but matured in 1929

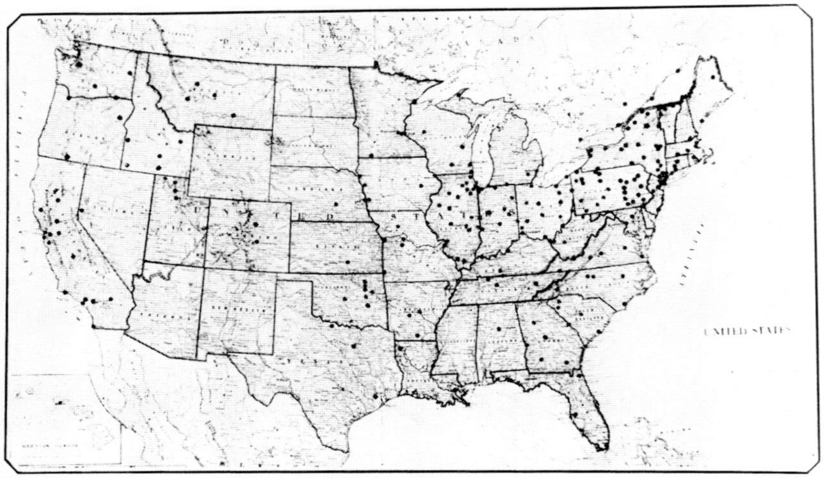


Fig. 10-24 Marmon dealer organization, 1922.

Fig. 10-25 Sea-going Model F in its box.

with a doubling of dealers by early 1930 to a total of 1750!

The collapse of the market following the Wall Street crash dropped Marmon 1930 sales to less than half the volume of the preceding year. With half of the sales

Fig. 10-26 A Model 34 heading overseas.

spread over twice the number of dealers, units per outlet fell back to the 5.3 level of 1915 (assuming all dealers survived the year).

The consequences of this sales level were disastrous to the established dealerships and catastrophic to the 875 new dealerships who had barely opened their doors. The gross contraction of the number of outlets, which followed, compromised proper servicing of the Marmons on the road with very many owners becoming geographically remote from any dealer. Marmon owner satisfaction and car value suffered greatly and disproportionately as a result.

## *Foreign Marketing*

Nordyke & Marmon was known throughout the world before it undertook the manufacture of automobiles. Consequently soon after the start of domestic car production, foreign orders were received.

### FIRST ERA

Export packaging of Marmon automobiles in the First Era was described in the *The Motor World.*[8]

SPECIAL BOX FOR EXPORT CARS
Ocean Traveling Marmons Are Given the Protection Needed on Shipboard . . .
Preparing Cars for Shipments.

While considerable energy was fruitlessly spent in an endeavor to induce the ocean steamship companies to carry automobiles uncrated, there were and are those who would not take advantage of such a concession were it offered them. They are of the class who had rather be sure than sorry and who believe that no protection is too great to apply to such an expensive possession as a motor car. Of this class is the Nordyke & Marmon Co., who for export are not content to merely crate their cars, but who securely box them. As an example of a job well done, their method as shown by the accompanying illustration is of no small interest.

The picture shows a 35-40 Marmon car ready for shipboard save for the placing of the end panel of the box, which fits the car on sides and top with just a little clearance at the ends. The box is made of 1¾ inch poplar matched with loose tongue. It is built up in sections, sides, top and ends with bottom spiked to two 4x6 inch skids, set in far enough to receive the weight of the car. The skids are reinforced outside with 2x3 inch pieces to receive the side battens. The sides and ends are made up by using 2x6 inch battens screwed to the panels, the battens extending down far enough to bolt to the skids. The top is made up by spiking on to 3x3 inch timbers which set it, the ends of the top boards resting on edge of the side panels. When placed in position the top and side panels are fastened together with ⅜x2x24 inch angle irons and bolted through the 3x3 inch pieces. There is also 3x3 inch strips bolted to the bottom and sides.

The box is lined with tarred paper and is practically air tight and moisture proof. The car is placed in the box so there is no side motion. A chock block is placed forward and back of each wheel, bolted to the floor and clamped to the wheels. The axles are blocked up so that the weight is removed from the tires and blocks are placed between the elliptic springs. The car is then bolted to the floor, front and rear, with ½ inch diameter long bolts. A 2x6x18 inch piece is bolted to sides over each wheel hub. Heavy felt doubled is used wherever blocking comes in contact with the car. Lamps, etc., are left attached to the car, spare parts are boxed, and top, extra tires, trunk rack, etc., are secured to the floor under the car. The end panels are bolted on. The car is thus shipped ready to run and the box may be turned on end, on its side or upside down without disturbing the machine inside. Marks "This side up" and the fact that it is on skids make it improbable that the box will be handled any other way than right side up. Oil and grease in the motor, transmission-axle and steering gear is not drawn off before shipping.

* * *

## THIRD ERA

Export packaging practice in the Second Era was described in the Marmon brochure 'Information for Overseas Dealers' (1922): "Every precaution is taken in the crating and boxing of the Marmon car to insure its safe arrival across the water. The illustrations above give an idea of the several steps necessary before the car is finally ready to be sent on its way.

In tight boxing for export shipments, tongued and grooved yellow pine is used on the top and bottom of the box. The car is first run on a platform with its wheels and top removed. This platform is spiked to 4x6 inch timbers. The tool boxes and wheels are crated under the car, the top is carted on the ceiling. The entire car is then crated, braced and boxed, the corners being locked with iron bands.

The dimensions and weights of the boxes are kept to the lowest limits consistent with safety. The weights of the cars boxed ready for shipment are as follows:

| | | |
|---|---|---|
| Chassis crates | 51x78x188 inches | 433 cubic feet |
| Open car crates | 62x78x188 inches | 526 cubic feet |
| Closed car crates | 77x85x188 inches | 712 cubic feet |

The model 34 A Marmon had magneto ignition but Marmon changed to the Delco system of ignition on the Model 34 B, stating that their experience had taught them that the latter was superior to the magneto. However the export market had sufficient importance that the magneto option was engineered into the 34 B and was fitted to export cars when specified in the order.

The model 34 B was the first Marmon model to have right-hand drive available and ready for shipment to those countries desiring it!

The Third Era brought worldwide recognition of the Marmon 34. The remarks of British motor car authority Edgar N. Duffield in the British *Auto* magazine were so complimentary that Marmon repeated them *in a full page advertisement in the Saturday Evening Post.* The ad was headed **What England Thinks of Marmon.**

The text follows: *"Those who have anything but a proper appreciation of the higher-grade American car need to meet the Marmon. It is a relatively high-priced car, certainly, but it is the type which will give a lot of folk quite new conception of American automobile engineering, from no matter what standpoint he looks at it. Lots of Marmon cars have come here, from time to time. One in particular, which arrived at the end of 1918, was one of the busiest machines in the British Isles. It went from factory to factory, up and down the kingdom, and lots of people made little notes of certain of its features. I got a ride in it. I drove it a little over two hundred miles, going northward from Manchester over roads very sorely war worn, and it was this run in the Marmon which gave me my first really useful insight on the matter of modern suspension. That big, fast car traveled as all cars could, and some day will. Its road adhesion, at from sixty to seventy miles an hour, was perfectly wonderful. Springs and frame, gravity-center location and steering, acting in concert, made it the 'safest' fast car I had ever driven. I knew at the time that whole chunks of the design were bound to be stolen. They were. I can identify several of them today. Pass & Joyce, Ltd. of Euston Road, has some rippingly good agencies, Austin, Sunbeam, Fiat, and so forth. They have only one American concession, that for the Marmon Company, an engineering house of three generations, or just on. Doubtless P & J would sell any car, of any nation's production, that one wished to buy; but is fairly significant that they 'run' only one American, the Marmon. When I heard that they wanted me to try a Marmon I leaped at the chance. We started out under murky skies, we met a thin, warm Scottish kind of drizzle early on, and we lived in it for the rest of the outing. But that did not matter a little bit. Having admired a sumptuous body, asked the price of the complete rigout, and stepped into the (passenger's) front seat, I began to absorb impressions. First, of a really noise-free body and a very comfortable seat, as to both cushion and squab. In all that body there was not a sound. Neither floor boards nor . . . when I put a finger to them . . . gear shift or brake lever gave any indication of vibration, or any gear, at any speed. We just oozed over a very narrow, tram-cursed bridge, through a riverside town's High Street that is always congested at 10 a.m., and were called to halt at the foot of a suburban hill which has probably a gradient of about one in eighteen or twenty, but is yet a very trying ascent on top gear, from a standstill. The Marmon sleeked up that hill on top, checking at times to four or five miles an hour, accelerating instantly to twenty-four or twenty-five and then slowing down again, as traffic compelled, but always so silently, so sweetly. The Marmon is a big, husky, powerful car, which might be expected to weigh somewhat on the wrists. It does not. Anybody could handle it, all day, without getting tired or worried. You want to get out in the Marmon. You want to loaf around London, at its busiest, you want then to get into the real outdoors to find hills, to seek long, straight levels, and just play with accelerator and spark lever; play tunes on them. Try to hit an engine speed and spark setting at which you can make the pistons audible. You won't do it but try. Try to produce anything in the way of a skid, without absolutely locking your wheels. Same result, but it will amuse you to try. The Marmon is a very low-price car, as and for what it is and will do, making allowance for the really excellent finish of everything, outside and in, top, sides, and bottom. We have seen very creditable American productions in the past season or two, all real sellers at their prices, but this is a car of another class, a more costly class, per-*

*haps, a smaller-public class; but a car which will salutarily and necessarily modify many British folk's ideas of what America can produce for those who will spend broadmindedly."*

The Marmon which Mr. Duffield drove with such enthusiasm may well have been the car which influenced the Armstrong-Siddeley so extensively. The 'Armstrong-Siddeley Clone' is discussed in the Third Era chapter.

By 1922, there were Marmon dealers in the following fifteen foreign countries:

| | | |
|---|---|---|
| Cuba | Holland | Portugal |
| Denmark | Japan | South Africa |
| England | Java | Spain |
| France | Norway | Sweden |
| Greece | Porto Rico | Switzerland |

In late 1925 a supply depot was established at Antwerp, Belgium to take care of growing demand in Germany, Switzerland, Norway, Sweden and Denmark.

European offices were opened at 63 Champs d 'Elysees, Paris under the direction of W. L. Nicoll. Marmon acceptance in England was such that right hand drive cars had become a factor in production.[9]

In the fall of 1925, the Marmon distributor for Germany, Wender A.G., who also had the Willys-Overland franchise, learned that a popular theatrical production was to open at the Nelson Theater in Berlin. In that show, a comedian act used a backdrop depicting an automobile agency. For the Berlin engagement Mr. Werner saw an opportunity to advertise and arranged to have the Berlin backdrop be an exact likeness of his business establishment. One show window bore the words 'The NEW MARMON' (the Model 74), the other 'Willys Knight' and 'Overland'.

The Puerto Rico Insular Police purchased a fleet of three 7-passenger touring cars for patrol purposes.

The Model 74 introduction was followed by a doubling of exports during 1925. Markets in England, France and Germany were particularly active with Francesco Alberici, Marmon distributor in Spain, reporting a growing demand.

By July 1927 ten percent of Marmon production was being shipped abroad.[10]

## FOURTH ERA

The factory sales organization was expanded in October 1926. The USA was divided into three zones to be under direction of S. A. Sweibel, Otis Lucas, and HWD Brown. The three were then subdivided into nine districts (three each) under R. W. Greulich, Porter Smith, H. C. Edwards, J. K. Gregory, **John Tainsh,** Paul Morford, L. F. Johnson, John Boe and George F. Green.

The foreign zone continued under Frank L. Hambly, as it had since 1923 or earlier, but with a new European district under W. L. Nicoll and a new South American District under Pablo Homs.[11] (A Canadian Division was added In October 1929.)[12]

In 1928, regional sales offices were established in San Francisco, Kansas City, KS, Chicago, Indianapolis, Atlanta and New York City.

In the spring of 1927, five foreign distributors were added including one in **China** and one in Cairo, **Egypt.**.[13]

Three foreign sales representatives were appointed in January 1929. H. Allan Kingsley was sent to open offices in Sydney, Australia, where there would be a Marmon-Roosevelt exhibit at the Melbourne Auto Show.

J. E. O'Brien was appointed special representative for Mexico. Cuba, Central America and northern South America; J. A. Munos for Argentina, Chile, Uruquay and Brazil. Both would open offices in Havana, Cuba.

Of 1970 cars imported into the Canary Islands during 1929, thirteen percent were Marmons. Thomas Dorreste Marrero was the distributor.

When the third segment of the Cuban Central Highway was officially dedicated with much fanfare, the procession over the new road was led by President Machado in his Marmon Eight limousine.

Frank Hambly and President Williams sailed to Europe on the Aquitania for the purpose of driving a Roosevelt to the principle cities of England, France, Germany and Switerzerland. Marmon was an exhibitor at the 1929 Geneva Show which opened March 14th.

In 1931, there was little doubt that in Brazil, THE status symbol was a Marmon. Among prominent owners were the following:

City Magistrate of Sao Paulo (city)
President of State of Sao Paulo
Secretary of the Treasury (**two** Marmons)
Secretary of Agriculture (**two** Marmons)
Secretary of State
Secretary of Public Roads
Secreatry of Public Works
Director of Traffic
Commander-General of the Fire Department
British Consul
German Consul
Mexican Consul
Norwegian Consul
Hungarian Minister

## FIFTH ERA

Little is known of export activity in the Fifth Era except that a Prince Lekow, a nobleman of Japan, ordered a MARMON SIXTEEN, "his third recent order", and a SIXTEEN convertible coupe as well as two other SIXTEENs were shipped to a single customer in the Netherlands, Dr. Rost Omnes.

### Surviving Evidence

In 1984 there were 24 Marmon-built cars owned by 20 foreign members of the Marmon Club. Members from Brazil and New Zealand attended the 1984 Marmon Muster in Indianapolis.

| | | | | | |
|---|---|---|---|---|---|
| Australia | 8 cars | 5 members | New Zealand | 2 cars | 2 members |
| Brazil | 5 | 2 | Norway | 2 | 2 |
| Canada | 5 | 6 | Sweden | 1 | 1 |
| Netherlands | 1 | 0 | West Germany | 1 | 1 |

The SIXTEEN convertible shipped to the Netherlands is today exhibited in the National Car Museum, Driebergen.

## PART 3 — Production Volume and Market Penetration

### *Marmon Car Production Volume*

#### Sources of Data

The available information on production quantities varies with the several Eras. For the First Era, specific data is available for units built in the period 1902-1905

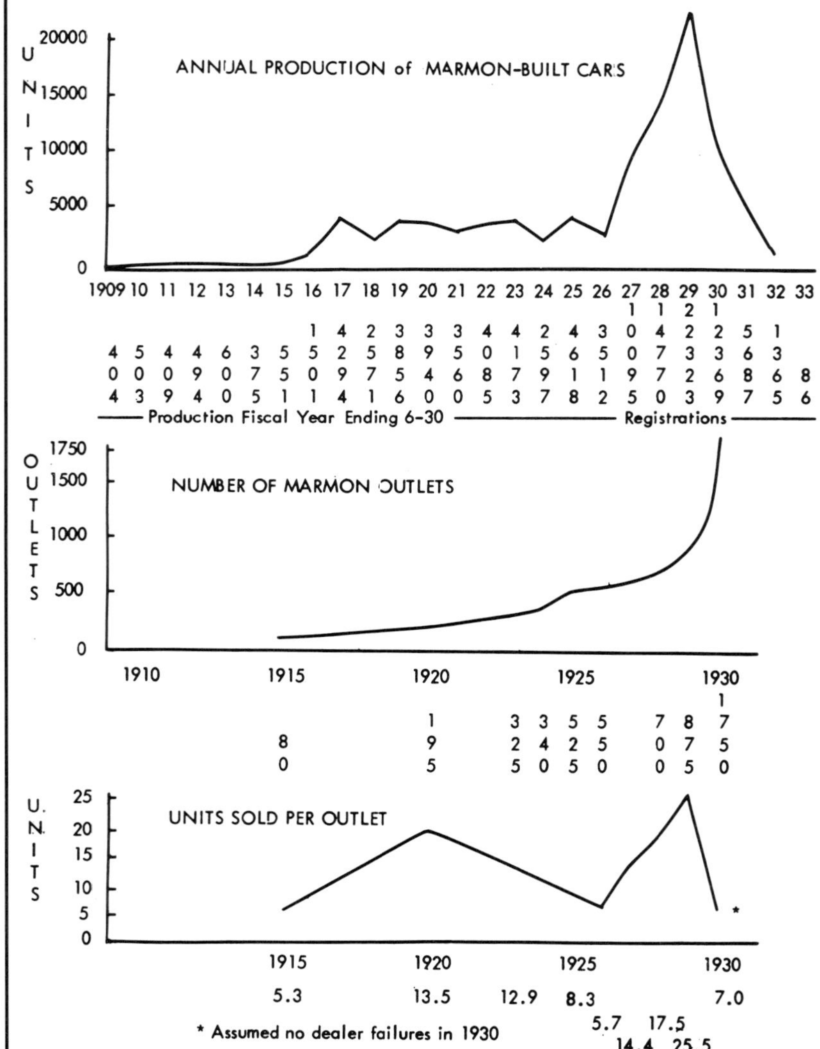


Figure 10-25. GROWTH IN MARMON SALES AND OUTLETS IMPACT ON SALES PER OUTLET

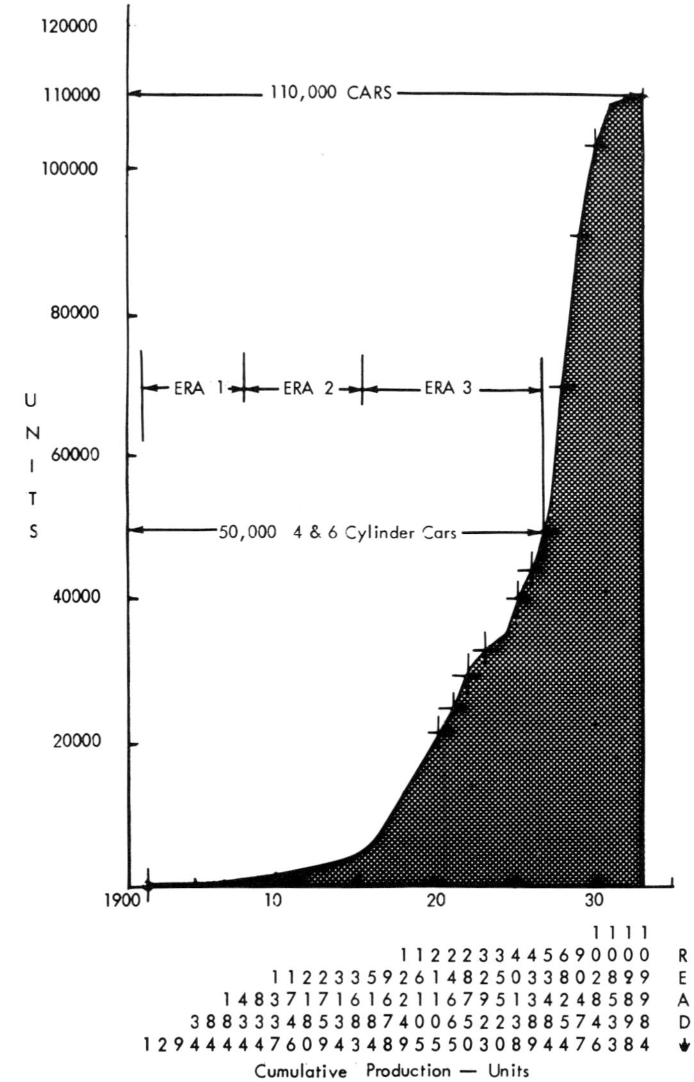


Figure 10-26 CUMULATIVE PRODUCTION OF CARS

| | Manufactured | | Manufactured | | Registered | Manufactured | |
|---|---|---|---|---|---|---|---|
| 1902 | 1 | 1913 | 599 | 1924 | 2597 | 4000 | |
| 1903 | 1 | 1914 | 378 | 1925 | 4618 | 4363 | |
| 1904 | 7 | 1915 | 551 | 1926 | 3512 | 3253 | |
| 1905 | 25 | 1916 | 1753 | 1927 | 10095 | 10489 | |
| 1906 | 50** | 1917 | 4494 | 1928 | 14770 | * | 7562 8 mon. |
| 1907 | 100** | 1918 | 2571 | 1929 | 22323 | 29216 | 16551 12 mon. |
| 1908 | 250** | 1919 | 3856 | 1930 | 12369 | 10115 | |
| 1909 | 425** | 1920 | 3981 | 1931 | 5687 | 3812 | |
| 1910 | 476 | 1921 | 3670 | 1932 | 1365 | | |
| 1911 | 534 | 1922 | 3887 | 1933 | 86 | | |
| 1912 | 486 | 1923 | 3866 | 1934 | 0 | 0 | |

Manufactured data through 1923 is model year serial basis.
16500 owners reported in April 1920. ** Estimated.
* Fiscal year changed, resulting in an 8 month reporting period.

Figure 10-27 MARMON CAR VOLUME — ANNUAL UNITS BASIS

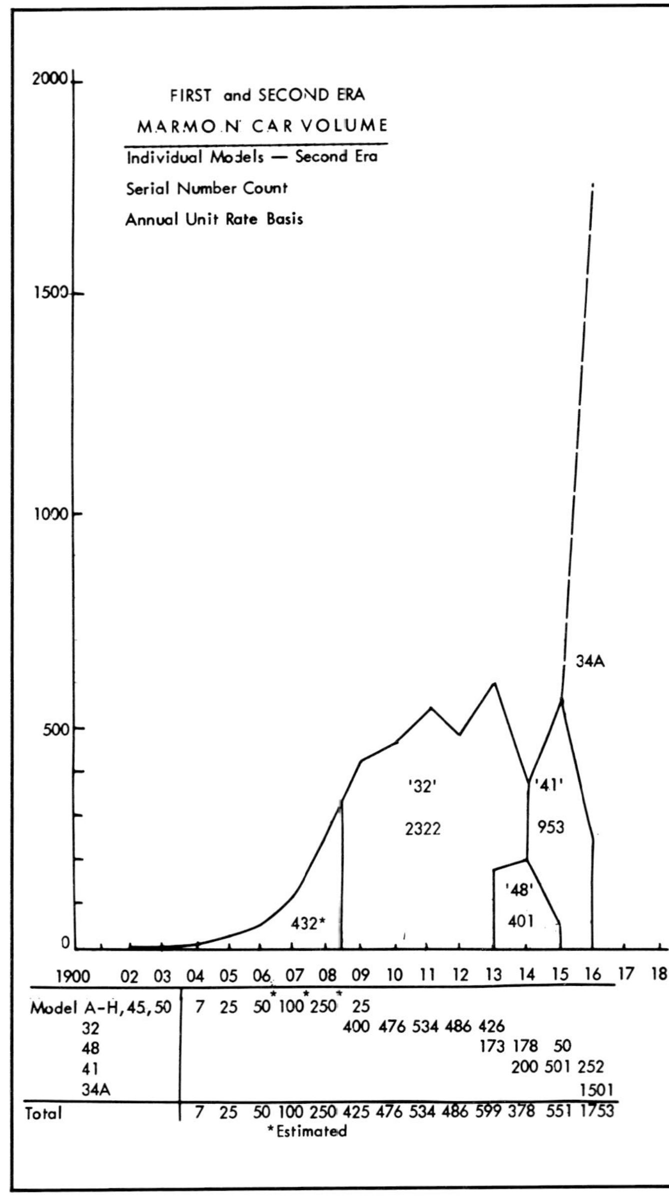


Fig. 10-28 Marmon car volume — First and Second Eras.

only and best extimates must be used for the remainder of the Era. The same is true of the first year of the Second Era. Beginning with 1910, the data which Marmon furnished to the several state Secretaries of State for vehicle license regulations is available through 1931. This includes beginning and ending serial numbers for each body style for each **model** year, which generally began in mid-September; the combined model sum is then the annual production. This source of data provided satisfactory Second Era quantities to the end of the Era (except for the initial year of 1909). Production for 1909 is estimated based partly on the ratio of the highest surviving 1909 serial number to the 1910 production of the same body style.

The registration document also provided good data for the Third Era except: (1) for 1924, when a mid-year (April 1) change in serial numbering practice was not accompanied by sufficient data to identify annual volumes; (2) for 1926 when the number sequence was carried over from 1925 (Beginning 1926 number serve as ending 1925 numbers.); and (3) was then replaced for 1927-28 with the Fedco theft-deterrent serial number plate system. That system is as baffling to the historian as it was intended to be to the thief. An alpha-numeric code etched on both a plate **and** its attaching rivets was fastened to the instrument 'board' with special tools; any attempt to remove it resulted in defacement.

Fortunately Moody's Manual of Investments for 1932 lists Marmon unit sales on a fiscal year basis for 1924 through 1931. These figures are used herein as factory data.

Registration data during the Third Era was six months displaced from the Mar-

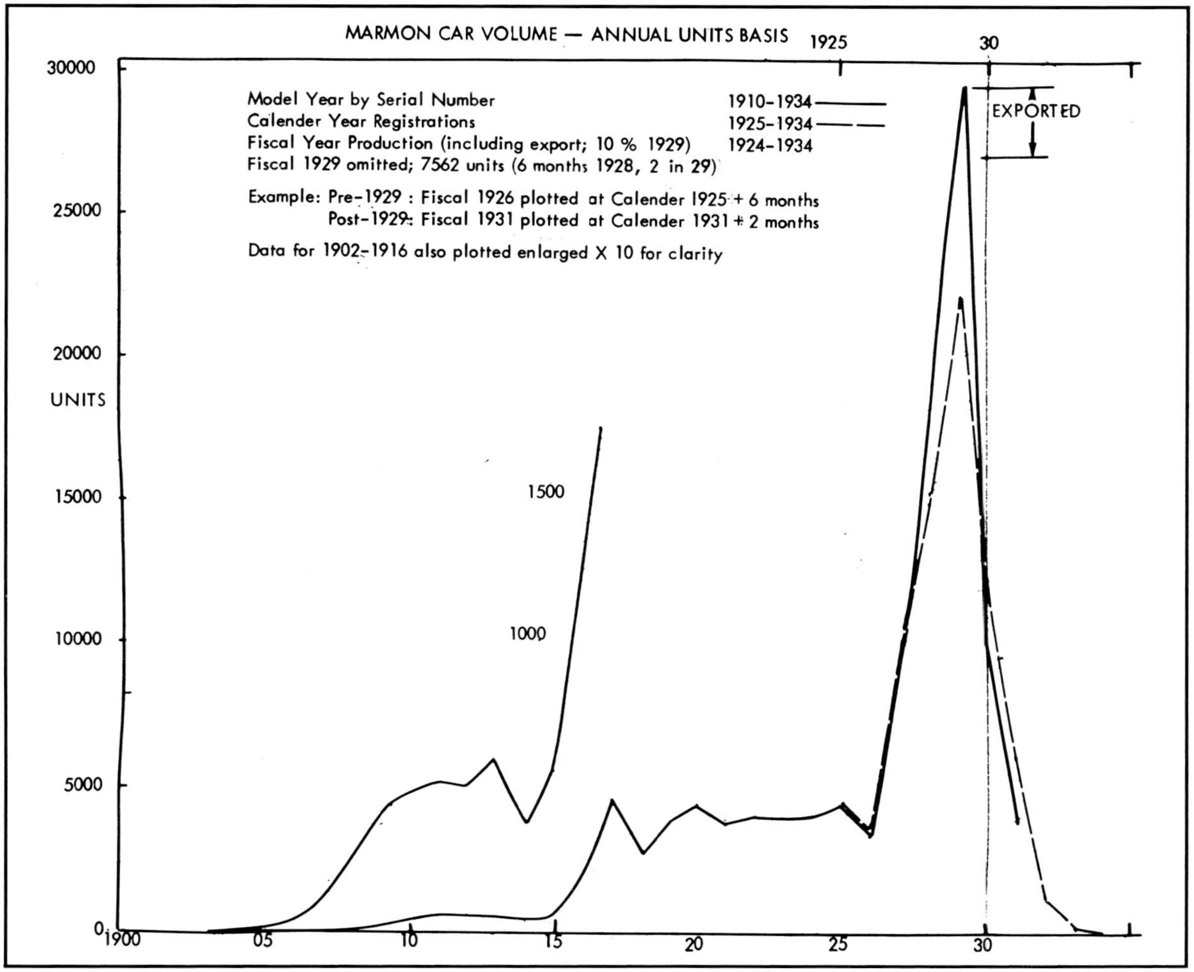


Fig. 10-29 Marmon car volume on an annual units basis. Five Eras.

mon fiscal year and about four months from the model year. Because the Marmon fiscal year coincided with their model year, the fiscal year volumes are used here-in.

A change in the Marmon fiscal year about 1928 brought it within two months (actually 10 months ahead) of the calendar year. This necessitated an eight month fiscal period for '1929' (six months of 1928 and two of 1929); fiscal 1930 began with March, 1929 and ended with February 1930. The 7562 units manufactured in that interval are used in computing total production but are necessarily omitted in figure plots of production on an annual basis. Moody's also provides a production figure of 16,551 for the twelve months ending February 28, 1928, but this double counts 8,989 units in the prior year count.

Consequent to the shifted fiscal year and particularly due to a lack of factory production data after 1930, both registration and manufacturer data for the Fourth Era is recorded. When addressing monthly production (expressed as cars per hour), the plotted data beginning with the start of new model production reflects public announcements on a variety of bases. For the remainder of the year (or in some instances until a cessation of operations) following the onset of an inventory liquidation or major plant rearrangement, the production rates have been progressively reduced to agree with the stated number of units manufactured during the period.

## Cumulative Production

Fig 10-26 shows the cumulative production of Marmon cars from 1901 through 1933. Total production was 110,000 cars (109,984). The midway point of 55,000 cars was reached in mid 1927 (end of the 1927 fiscal year). Therefore, production was approximately evenly divided between the First-Third Eras (25.5 years) and the Fourth-Fifth Eras (6 years), with approximately half the cars being

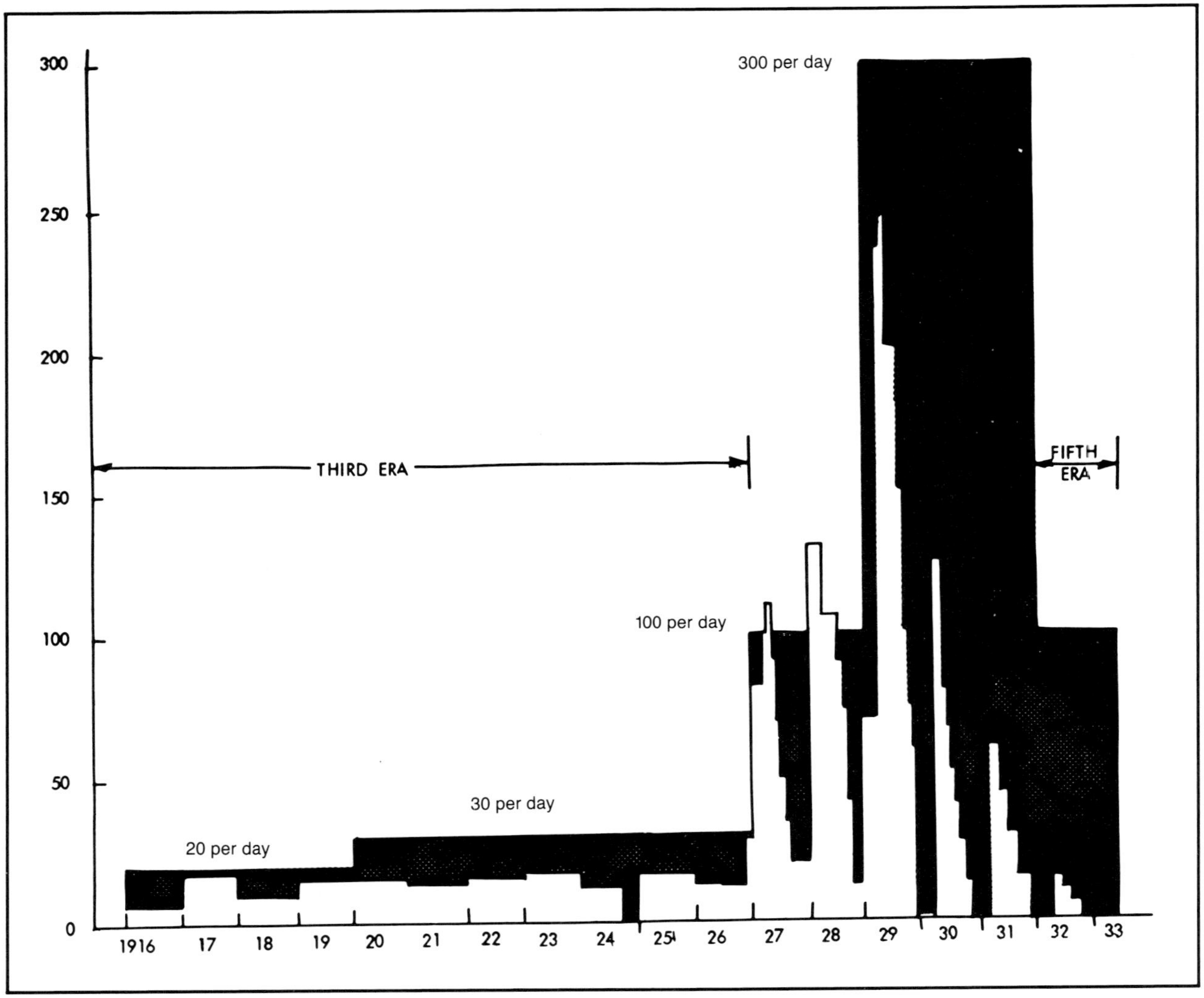


Fig. 10-30 Installed capacity and actual production; Eras Three, Four and Five.

straight eights. Assuming that the 10,489 cars made in 1927 were divided, 6,000 Little Marmons and 4,500 Series 75 cars, 45 percent of Marmons preceded the straight-eight generation. Because the Fifth Era SIXTEENS account for only 0.3 percent of all Marmons, the straight-eights constituted 55 percent of total production.

## Annual Production Volumes

Figure 10-27 presents, in tabular form, production on an annual unit rate basis. Any cumulative tabulation from that data would be compromised by the time offsets between registration data and manufacturers fiscal year data, by a change in the fiscal year, by the one year differential between calendar and fiscal year designations. Prior to 1925 only manufacturer data is shown; due to uniformity and continuity of production, registration data is similar. Figure 10-28 presents, in graphical and tabular form, Era One and Two production with the latter on a model basis. Figure 10-29 graphs production for the Five Eras. Data points for registration and fiscal years are appropriately offset from model year data points, the latter based on serial numbers. The export volume (1929) is identified to give a correct perspective to the difference between production and (domestic) registrations. Era One and Two Production was:

| | | |
|---|---|---|
| First Era | 432 | estimated |

| | |
|---|---|
| Model 32 | 2,322 |
| Model 48 | 401 |
| Model 41 | 953 |
| Total | 4,108 |

Surviving cars have engine numbers as low as 102 (1909 Model 32 could have started at 100) and as high as 30928 (1915 Model 41). However, no surviving car has a vehicle serial number which contradicts the basis for the above tabulated volumes. The engine serial numbers may have incorporated a code for major engineering change points to facilitate service part logistics.

Figure 10-30 shows the installed capacity, in cars per hour, for the Third through the Fifth Eras. It also shows the achieved production rates. The underutilization is shaded to emphasize this dramatic factor in the profitability of a manufacturing operation; the capital commitment and associated loan burdens place a dreadful burden on a firm when expected production levels are not realized. This figure is repeated and discussed in the 'Financial History' chapter; it appears here to illustrate the severe cyclic production rates experienced as, in each of the years of the Fourth Era, a new or revised product and supporting sales promotion was launched, only to be followed by an oversupply of cars and severely curtailed production.

The increased capacity installed for the Model 34B (from 20 to 30 per day) was never utilized. However, the successive economic crises of the early 1920's could not have been anticipated. But the increase to 100 per day capacity installed for 1927 was over-utilized, at premium costs, during 1928 and 1929. The 10,000th Roosevelt came off the assembly line on May 15, 1929. The gross increase of an additional 200 per day Roosevelt capacity appeared to have been a prudent program in 1929 when its full utilization in 1930 appeared probable. But all that changed in the latter half of 1929 as the Great Depression set in.

The contraction of capacity to 25,000 per year (100 per day) by the closing of the entire Plant 1 and 3 complex for 1932 operations still left a fatal overcapacity. Only 247 employees were on the roll on April 30, 1932.

### Production by Model and Style

Figure 10-31, 32 present the number of units, by body style and model, produced in each year of the Third and Fourth Eras. Total production of body style and of model for each Era are also given. The body style code numbers used in the serial number (S/N) in a particular Era are stated as is the digital sequence in the S/N for complete identification of the model, body style, year and body-style based S/N.

This data is particularly important in determining the relative rarity of a particular model and style when new and in assessing the probable rarity in the surviving population today.

Factory serial number data furnished by the manufacturer to the 48 Secretaries of State for car registration purposes provides data on sales of individual styles for the period 1910 through 1923. Because of the change in serial numbering practice which occurred on April 1, 1924, and the lack of information on ending numbers on March 31, 1924, 1924 model year production must be estimated.

A minimum volume, with a high degree of confidence could be arrived at by considering the volume of 1924 styles represented by the highest surviving serial number of each. Such numbers are shown on Figure 10-31 followed by a + sign. Their total is however meaningless because so many styles are not represented among the known survivors.

Instead the author has elected to base his estimate on the several pairs of actual 1923 and surviving 1924 volume. Surviving cars in the limousine and grouped 4-passenger open styles indicate a higher production in 1924 (322 vs. 309 and 428 vs 408 respectively) and about equal production of speedster styles (171 vs 170). Therefore a figure of 4000 (103% of 1923) was adopted.

Total production of Model 34 cars was therefore about 31,889 units, an average of 3543 for each of the nine years. Model 34A production was 12485 units, averaging 3121 per year, and Model 34B/C production was about 19404 units for an average of 3881 per year.

Model 74 production data is complete with exception of the Standard Coupe which was replaced by the Deluxe Coupe; hence no starting number for 1926 was available to establish a final 1925 value. Volume is estimated at 100 units because of less than a full year run and sale of 150 units of the Deluxe Couple in 1926. Model 74 production was therefore about 7616 units, an average of 3808 per year. Model 75 production is simply estimated arbitrarily at one-third of 1927 production of Models L and 75. About 10 percent are assigned to 1928 sales; they were probably manufactured in 1927. Combined production of Models 74 and 75 is then about 11,000 (10916) units.

Recognizing that the D74 and E75 are the D and E versions of the Model 34 A to E series, the totals can be combined: 42800 units in 12 + years of 3567 per year.

Only limited information on body style production is available for the Fourth Era. Therefore the chart generally serves only to identify which styles were offered and which were not. Figure 10-32 lists the letter model code, corresponding to the numerical series designation, which was used for engineering and service parts purposes. For example, the Series (or Model) 68 was a Model T, the Big Eight was a Model H. Engine models used in the several car models are also identified; e.g., the Series 78/Model N had a Model N engine but the Roosevelt/M-R Model S used initially the Model T and alter Model Q engines 'borrowed' from corresponding car models.

## MARKET PENETRATION

During Marmon's year of highest unit sales, 1929, they achieved a sales position of number eleven rank in May (when all badges of Hudson (Essex), Chrysler (Plymouth Dodge and DeSoto), Oakland (Pontiac), Nash (Ajax) and Buick (Marquette) were treated collectively with their parent. Figure 10-45 illustates the distribution of volume among the 18 leading producers.

The May 1929 market of some 562,995 cars was divided into three main segments: [1] **Ford** (36) and **Chevrolet** (28) at **64%;** [2] Chrysler (10), Hudson (8) and Oakland (5) at 23%; and [3] thirteen manufacturers dividing up the remaining 13 percent. Four of these had two percentage points each, Hupp and Marmon had one percent each and the other seven ranged from Packard and Reo at 0.8, Cadillac at 0.6 down to Peerless at 0.15 percent.

Penetrations of 0.1 percent in the Third era and 0.4 percent in the Fourth are representative, with a high of 0.6 in 1929.

Figure 10-34 shows the sales performance of Packard, Cadillac, Pierce-Arrow, Lincoln and Marmon during the period 1926-1933. The industry total is also shown; note the steady growth from 1927 through 1929 and the even more rapid steady decline through 1932.

Packard had climbed from 0.9 to 1.2 to 1.5 percent of the market in 1928; although the industry growth continued, Packard sales fell in number and to 1.1 percent. Interestingly, Packard maintained that 1.1 percent throughout the decline that followed.

Cadillac sales declined with the market in 1927, but failed to recover at all and

| BODY CODE | BODY STYLE | Model 34A | | | | | Model 34B, 34C | | | | | | Model 74 | | | Model 75 | | |
|---|---|---|---|---|---|---|---|---|---|---|---|---|---|---|---|---|---|---|
| | | 1916 | 17 | 18 | 19 | | 20 | 21 | 22 | 23 | 24 | | 25 | 26 | | 27 | 28 | |
| 1 | Speedster 4 pass | - | - | - | - | | - | - | 640 | - | - | 640 | - | - | | | | |
| 2 | Speedster 2 p | - | - | - | - | | 100 | 481 | 238 | 170 | 171+ | 1160+ | - | 104 | 104 | | | |
| | Roadster 2/4 p | - | - | - | - | | - | - | - | - | - | | 201 | - | 201 | | | |
| 3 | Cloverleaf 3 p | 74 | 63 | - | - | 137 | - | - | - | - | - | | - | - | | | | |
| 4 | Cloverleaf 4 p | 321 | 1264 | 674 | 909 | 3168 | 481 | 131 | 116 | 17 | - | 845 | - | - | | | | |
| | Touring 4 p | - | - | - | - | | 1300 | 682 | 601 | 66 | - | 2649 | -- | - | | | | |
| 5 | Touring 5 p | 116 | 115 | 28 | 22 | 281 | - | - | - | - | - | | 103 | 17 | 120 | FEDCO | | |
| | Phaeton 4 p | - | - | - | - | | - | - | - | 325 | 428+ | 753+ | - | - | | Theft | | |
| 6 | | | | | | | | | | | | | | | | Deterent | | |
| 7 | Touring 7 p | 922 | 2315 | 1116 | 2012 | 6365 | 2100 | 1508 | 1296 | 110 | - | 5014 | - | - | | System | | |
| 8 | Closed styles | 68 | 800 | 753 | 913 | 2534 | - | - | - | - | - | | - | - | | | | |
| | Phaeton 7 p | - | - | - | - | | - | - | - | 1229 | 294+ | 1523+ | 171 | 30 | 201 | | | |
| 9 | Cpe-Willoughby | - | | | | | ? | 273 | - | - | - | 273 | - | - | | | | |
| | Speedster 4 p | - | | | | | - | - | - | 531 | 96+ | 627+ | - | - | | | | |
| 10 | Sedan-N H 7 p | - | | | | | - | 473 | - | - | - | 473 | - | - | | | | |
| 11 | Limo-Rubay 7 p | - | | | | | - | 110 | - | - | - | 110 | - | - | | | | |
| 12 | Town Car-Rubay | - | | | | | | | | | | | | | | | | |
| 13 | Limo-Willoughby | - | | | | | ? | 12 | 34 | 24 | ? | 70 | - | - | | | | |
| 14 | Town Car " | - | | | | | ? | ? | 18 | 13 | ? | 31 | - | - | | | | |
| 15 | Coupe-N&M 4p | - | | | | | - | - | 212 | 336 | ? | 548 | - | - | | | | |
| | Coupe DeL 2 p | - | | | | | - | - | - | - | - | | 100* | - | 100* | | | |
| 16 | Limo-N&M 7 p | - | | | | | - | - | 365 | 309 | 322+ | 996+ | - | - | | | | |
| | Sedan DeL 7 p | - | | | | | - | - | - | - | - | | 341 | 126 | 467 | | | |
| 17 | Sedan-N&M 7 p | - | | | | | - | - | 199 | 267 | ? | 466+ | - | - | | | | |
| | Limo DeL 7 p | - | | | | | - | - | - | - | - | | 252 | 77 | 329 | | | |
| 18 | Sedan-N&M 4 p | - | | | | | - | - | 168 | 469 | ? | 637+ | - | - | | | | |
| | Sedan DeL 5 p | - | | | | | - | - | - | - | - | | 222 | 45 | 267 | | | |
| 19 | Limo Std. 5 p | - | | | | | - | - | - | - | - | | 56 | 7 | 63 | | | |
| 20 | Brougham 5 p | - | | | | | | | | | | | 885 | 335 | 1220 | | | |
| 21 | Sedan Std. 5 p | - | | | | | | | | | | | 1108 | 400 | 1508 | | | |
| 22 | Sedan Std. 7 p | - | | | | | | | | | | | 873 | 468 | 1341 | | | |
| 23 | Victoria 4 p | - | | | | | | | | | | | 51 | 1342 | 1393 | | | |
| 24 | Coupe Std. 2 p | - | | | | | | | | | | | - | 151 | 151 | | | |
| | TOTAL | 1501 | 4557 | 2571 | 3856 | 12485 | 3981 | 3670 | 3887 | 3866 | 4000* | 19404+ | 4363* | 3253 | 7616* | 3000* | 300* | 3300* |
| | *Estimated | +Highest surviving serial number | | | | | | | | | | | | | | | | |

Fig. 10-31 Era Three production by body style and model.

| BODY CODE | SERIES-MODEL | | L | 68-T | | 78-N | | 'R'- S | -'MR' | 69-Q | 79-W | B8-H | 70-U | 88-CC | 125-HH | 16-DD |
|---|---|---|---|---|---|---|---|---|---|---|---|---|---|---|---|---|
| | ENGINE MODEL | | L | T | | N | | S | T&Q** | Q | W | H | Q | H | H | DD |
| | BODY STYLE | | | 28 | 29 | 28 | 29 | 29 | 30 | 30 | 30 | 30 | 31 | 32 | 32 | 31-33 |
| 0 | Sedan | 5p | ? | ? | ? | ? | ? | 20000 | 3309 | 1606 | 959 | 474 | ? | 197+ | ? | |
| 1 | Coupe | 2p | ? | ? | ? | ? | ? | 3000 | 713 | 197 | 125 | 124 | ? | 319+ | ? | |
| 2 | Victoria | 4p | ? | ? | ? | ? | ? | 2000 | 172 | - | - | - | ? | - | - | |
| | Touring | 7p | - | - | - | - | - | - | - | - | - | 38 | - | - | - | |
| 3 | Speedster | 2/4p | ? | ? | ? | ? | ? | - | - | - | - | - | - | - | - | |
| 4 | Conv. Coupe | 2/4p | ? | ? | ? | ? | ? | 1500 | 212 | 93 | 102 | 18*** | ? | 218+ | 319+ | See Chapter 6 |
| 5 | Speedster | 6p | ? | ? | ? | ? | ? | - | - | 38 | 41 | 5**** | - | - | - | |
| 6 | Sedan 2 door 5* | 7p | ?* | - | - | - | - | - | - | - | - | 271 | - | 657+ | - | |
| 7 | Limousine | 7p | - | - | - | - | - | - | - | - | - | 128 | - | ? | - | |
| 8 | Brougham | 5p | - | - | - | - | - | - | - | 99 | 81 | 96 | - | ? | - | |
| 9 | Club Sedan | 5p | - | - | - | - | - | - | - | 155 | 152 | 152 | - | - | - | |
| | TOTAL | | | | | | | <26500 | 4404 | 2188 | 1457 | 1306 | | 1391+ | | |
| | Series Prefix to S/N | | | | | | | | | | | | | | | |
| | Open Style 130* | | | 3 | | 4 | | | | | | | | 16* | | |
| | Closed Style 136* | | | 6 | | 5 | | | 7 | 13 | 10 | 11 | 12 | 17* | 18 | 14 |
| | First unit number | | 01 | 001 | | 001 | | | 001 | 501 | 501 | 501 | 501 | 501 | 501 | |

Serial Number VO-AA-01 E2-AA-01 Y1-AA-01 Series Prefix+Body Code+Style Serial Number
Chassis Number Alpha Model+Chassis Serial Number
Body Number Alpha Model+Body Code+Chassis Serial Number

** Tengine prior S-25545
Sengine after S-25544

***Grand National 5p Conv.
****Weymann Fabric Body Sedan

1930 count does not include 1st Series 1931 units which probably are quite small count.

TYPE = Series Prefix + Body Code; eg, Model 68 coupe was Type 61. +Unit count based on highest surviving serial.

Fig. 10-32 Era Four production by body style and model.

continued to decline through 1932; the success of the V-16 and V-12 flattened the rate of decline in the 1930-32 period. By 1932 Packard and Cadillac had comparable sales, consequent to the less rapid decline in sales for the latter during the 1930-32 period.

Lincoln sales declined throughout the 1926-32 period, but much more rapidly after 1929; however, they climbed from 0.23 to 0.29 percent of market.

Pierce Arrow sales escaped the 1927 decline, failed to follow the industry up until 1929, and then declined but at a lesser rate than experienced by the three marques discussed above. Pierce-Arrow outsold Lincoln in 1929, 1930 and 1931 with 0.25 percent of the market in 1932.

| Mfg. | Units | % | Cum.% |
|---|---|---|---|
| Ford * | 200903 | 36 | 36 |
| Chevrolet | 160895 | 28 | 64 |
| Chrysler * | 54595 | 10 | 74 |
| Hudson * | 42086 | 8 | 82 |
| Oakland * | 28444 | 5 | 87 |
| Nash | 12117 | 2 | |
| Buick | 12000 | 2 | |
| Graham-P | 11677 | 2 | |
| Durant | 10094 | 2 | |
| Huppmobile | 5329 | 1 | 13 |
| Marmon | 5221 | 1 | |
| Packard | 4561 | 0.8 | |
| Reo | 4453 | 0.8 | |
| Cadillac | 3511 | 0.6 | |
| Auburn | 3032 | 0.6 | |
| Franklin | 1707 | 0.3 | |
| Pierce-Arrow | 1563 | 0.3 | |
| Peerless | 807 | 0.15 | |
| Total | 562995 | | |

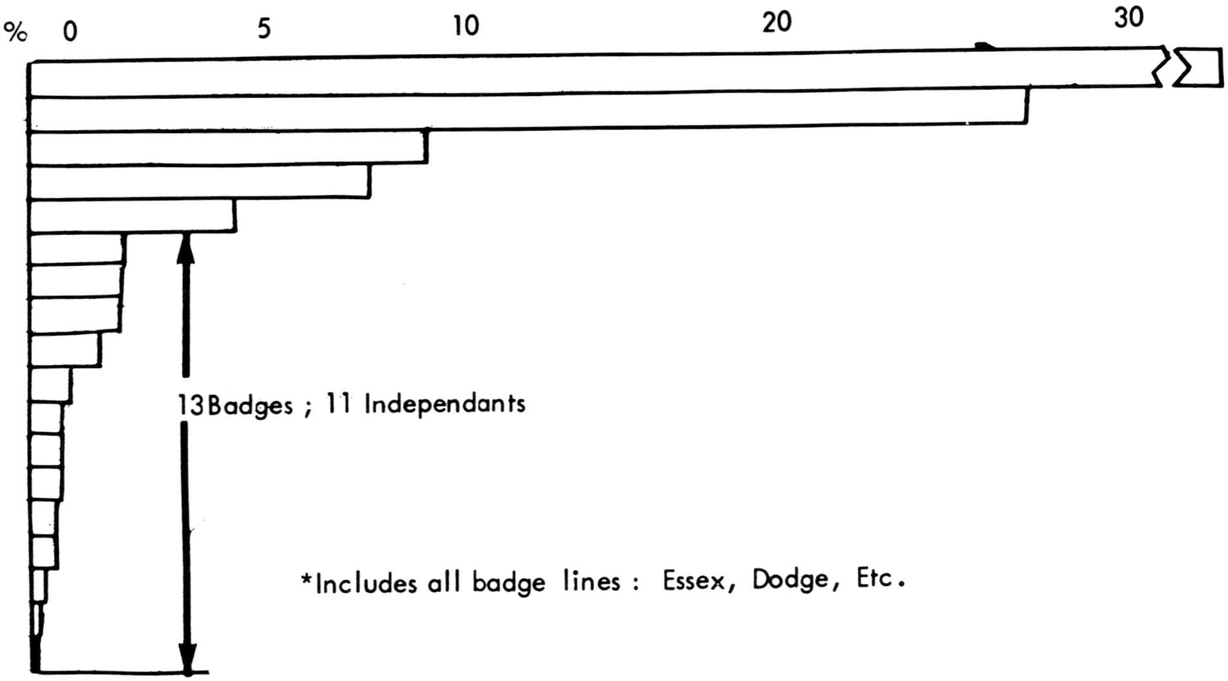


Fig. 10-33 How the market was divided in May 1929 (Lincoln and Oldsmobile missing).

During the '34B' period, there were 6 new Cadillacs for every new Marmon, and 2 new Lincolns. In the first year of the '74' period, there were 4 new Cadillacs and 1.5 new Lincolns per new Marmon, but Marmon sales fell in 1926 while the others surged ahead.

But in 1927 the dual offering of the Series 75 and Little Marmon resulted in sales which equated to 2 new Cadillacs and 1 new Lincoln per new Marmon **nameplate.**

## A MARKET PERSPECTIVE — THEN AND NOW

How often was a motorist likely to encounter a Marmon?

Marmon sales per outlet ranged from 5.3 to 19.5 1915-1928 and 25.5 in 1929. Jaguar dealers (USA) showed a comparable performance over the 1973-1981 period. Marmon sales volume in the Third Era was about that of Jaguar in 1981 and at the peak of the Fourth Era about that of Chrysler Cordoba in 1981.[14]

But in 1929 the domestic industry production was 5,227,087 units (the greatest until 1949),[15] whereas in 1982 it was 5,756,660.[15] Hence the Marmon eights would be as uncommon a sight in 1929 as a Chrysler Cordoba among domestic cars in 1981, about 0.4 percent or one in 250; in terms of total U.S. sales in 1982 (8,521,514), as common as an AMC Spirit.

In 1923, the domestic industry also set an all-time record — 4,034,012 cars. Marmon built about 4,000 of them or only one tenth of one percent (0.10%) — one in a thousand and typical of the Third Era.

There were 16,550 Marmons in use in early 1920. The first year of the Second Era (1909) saw US domestic production of over 125,000 units. The year the first Third Era Marmons were introduced, US domestic production was over 800,000 units — an increase of 550 percent over the duration of the Second Era.

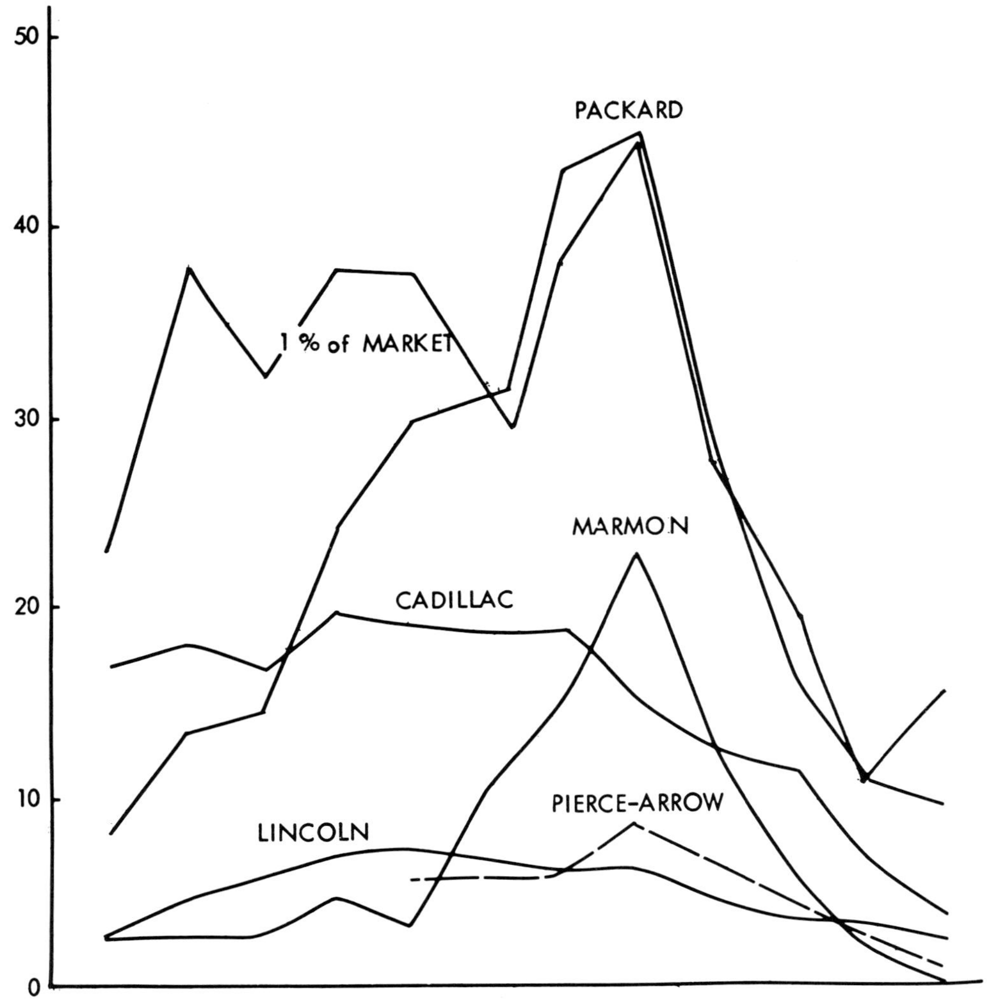


| Cal. Year | 22 | 23 | 24 | 25 | 26 | 27 | 28 | 29 | 30 | 31 | 32 | 33 |
|---|---|---|---|---|---|---|---|---|---|---|---|---|
| Industry/100 | 22742 | 37624 | 31858 | 37351 | 36923 | 29365 | 37754 | 44551 | 27875 | 19481 | 11035 | 15605 |
| Packard | 9050 | 13382 | 14820 | 23948 | 29589 | 31255 | 42890 | 44634 | 28318 | 16256 | 11058 | 9081 |
| Cadillac | 16718 | 17809 | 16740 | 19445 | 25027 | 18413 | 18783 | 14936 | 12078 | 11136 | 6269 | 3903 |
| Lincoln | 2911 | 4348 | 5672 | 6808 | 7211 | 6460 | 6039 | 6151 | 4356 | 3466 | 3129 | 2532 |
| Pierce-Arrow | | | | | 5369 | 5834 | 5736 | 8386 | 6795 | 4523 | 2692 | 2112 |
| Marmon | 2893 | 2830 | 2613 | 4618 | 3512 | 10489 | 14770 | 22323 | 12369 | 5687 | 1365 | 86 |

Figure 10-34 Sales Trends of the Leading Luxury Marques 1922-1933

Reproduction of the Frontispiece and Forward from a Marmon Motor Car Co. Catalog, circa 1924.

## *Foreword*

*Nearly everyone in America today has some business, some profession, some endeavor into which he pours his thought, his energy and his years. Rising above financial gain, it becomes also a matter of sentiment and pride—really a second hearthstone. We, in the second generation of an institution which was making fine machinery in the days when Abraham Lincoln was still a country lawyer, know what Marmon is; know how fine we are daily trying to make it.*

*We cannot expect others to share with us the full measure of our feelings and convictions. We only can ask them to think of Marmon in terms of their own undertakings. Their feelings about Marmon will then approximate our own. That is the only way anyone can understand us, when we say that Marmon is the second hearthstone of thousands of men who approach their daily tasks with the responsibilities and incentives of the present, sharpened by the challenge of the past.*

*Some of the quaint but solid masonry of the seventies is still standing at the Marmon factory. Surely we will not be accused of unwarranted sentiment or idealism when we cherish this brick and mortar and find in it an emblem of the craftsmanship which is our heritage.*

*Towering above and extending vast distances beyond, are the fine new buildings, models of industrial architecture, where Marmon craftsmanship is practiced today.*

*In building the Marmon car we enlist the finest of human skill, the most rigorous mechanical precision and the most uncompromising standards of metallurgical science. Applying these factors in the practical terms of what we know is wanted in a motor car, the Marmon could hardly be other than the splendid car it is; it could hardly perform less brilliantly or give less comfort or render service of a less reliable character.*

*But the insurance of such excellence comes neither from machines nor methods; it is found in the human fibre of a great body of men at the Marmon plant, with each of whom Marmon success is a vital personal matter—men who are literally patriotic in the cause of Marmon quality.*

*That is the thing everyone will understand, because everyone has a second hearthstone—everyone is devoted to the ideals of doing some task supremely well.*

MARMON MOTOR CAR COMPANY

INDIANAPOLIS, IND.

CHAPTER ELEVEN

# HOME OF THE MARMON CAR

*THE MARMON PLANT*
*1851-1933*

The Marmon plant in Indianapolis was a large complex of industrial installations that had begun to grow in the latter part of the nineteenth century and by the dawn of the twentieth was a significant facility. Its size, in fact, was remarkable for its mid-western location. Because the ultimate Marmon plant was an expansion of the nineteenth century facility, the story of The Marmon Plant must begin with the ancestral plants and the relocation to Indianapolis. The development of the plant facilities will be addressed for each of the FIVE ERAS of the Marmon car.

## PART I — NORDYKE & MARMON 1851-1901

### PRE-INDIANAPOLIS

The ancestor of the Marmon plant was a small building on the bank of the White-water River in Richmond, Indiana. It was a small farm, one-story building, containing room enough for only a few work benches and a small amount of material. The purpose of the facility was the manufacture and erection of flouring mills. This building stood in its original state until well into the twentieth century when it was dismantled. This was the plant of E. & A. H. Nordyke from 1851 to 1866; it was adjacent to Ellis Nordyke's residence.

The firm became Nordyke, Marmon & Co. in 1866 and promptly purchased a site in Richmond only a few blocks from the original building. A brick and stone structure was erected; it comprised a foundry, machine shop and millwright shop. Rocks for the building foundations were quarried from the banks of the adjoining river which also provided the motive power for the machinery.

The first structure on the new site was a three story brick and stone building. Later a one-story wood-working shop was added on the east side of the main structure (which therefore must have faced south, the new wing being to the right in the photograph).

The plant was situated high above the river and mill race; it was necessary to transmit the power from the water wheel up to the plant. Daniel Marmon designed a shaft which despite a length of about two hundred feet successfully did the job, and continued to do it well into the twentieth century.

Later, there was additional construction including a lumber shed and a steam generating plant to augment the water power.

As recently as the 1920s the Nordyke, Marmon & Co. plant was occupied and operated by the Richmond Milling Co. The lumber shed became the site of the Cheasapeake & Ohio (C & O) Railroad Depot, but the new owner expanded the facility in other directions.

The plant was visible from the Pennsylvania Railroad bridges as the railroad entered Richmond from Indianapolis.

By 1875 the firm had achieved worldwide recognition and had a significant export business. Because of local transportation logistics and absense of space for expansion on the Richmond site, a move was made to Indianapolis in 1876. But not before efforts to acquire a new Richmond site at a reasonable price had been unfruitful.

### MOVING TO INDIANAPOLIS

An additional and important consideration in the decision to relocate to Indianapolis was the immediate availability of a desirable plant in that city.

It was the "Quaker City Works" owned by Artemus Hadley who had suffered financial reverses in the panic of 1873-74.

The move was greatly hindered by political maneuvers and transporation limitations perhaps difficult to envision today. However some of the circumstances that occurred have their nearly identical contemporary events. Some of them were related in the *Marmon Books of Facts:*[1] *After moving arrangements had been consummated, the banking interests of Richmond began to realize that the city was losing a great and promising industry. Flattering offers were made by prominent Richmond citizens in an endeavor to influence the company to remain. But as Mr. Nordyke related, "It was too late then. We had a hard time to decide upon moving, and once the decision was made we would not change our minds." Then came the moving problem. In 1876 the railroad facilities were limited to a few slow and inefficient systems covering only part of the state of Indiana. The equipment was shipped (from Richmond) to Hamilton, Ohio via the C., H & D Railroad,*[2] *and from that point to Indianapolis, a roundabout route but the only one available at that time. The riddle was made harder to solve by the attitude taken by Richmond bankers, who grasped at every straw in an endeavor to keep the industry. Equality of freight rates was unheard of, and it required but little persuasion on the part of the Richmond bankers to induce the railroad to boost the tolls. The cost of leaving Richmond came high but the officers of the company were determined to establish their business in Indianapolis in spite of opposition that was coming from all quarters. The process of moving was slow, as orders had to be filled in the meantime. Machinery was dismantled as it could be spared and then "hurried" to the new plant, to be set up and again take its place in the* ***line of production.*** *The* ***majority of the skilled men*** *who had been with* ***the Richmond organization moved their families*** *and took up at Indianapolis the unfinished work they had left in the abandoned plant. (Author's Emphasis).*

Fig. 11-1 Original shop of the Nordykes
From the Marmon Book of Facts

Fig. 11-2 Original shop at Richmond, Indiana
From a Marmon catalog.

Fig. 11-3 The Richmond plant prior to the 1876 relocation to Indianapolis.

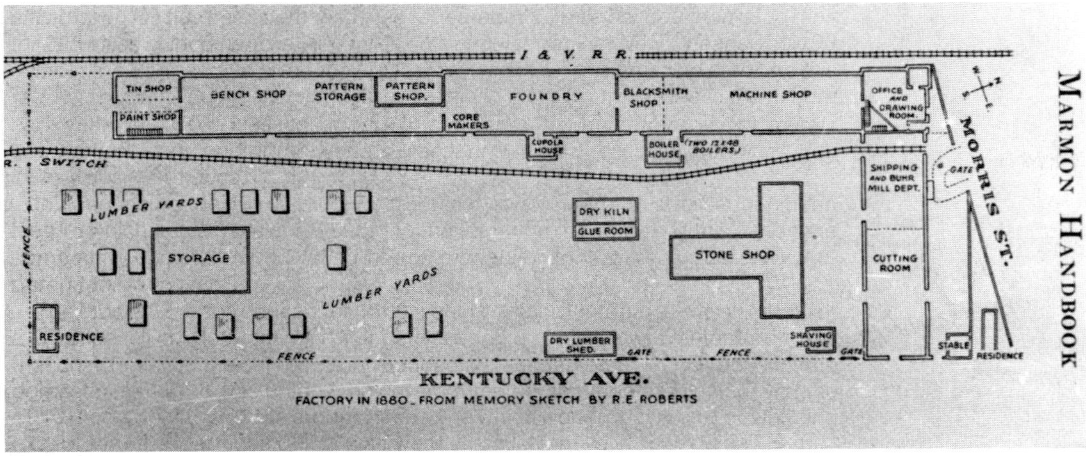


Fig. 11-4 Indianapolis site plan circa 1880 as recalled by R.E. Roberts.

After the crisis of 1873-74 there were few, if any, other firms in a position to undertake a relocation and expansion. Undoubtedly, Nordyke & Marmon's involvement in international marketing was a favorable factor in their expansion in that period.

## INDIANAPOLIS

The Marmon site in modern Indianapolis was initially 8.5 acres of land in West Indianapolis. It was about two miles southwest of the Indianapolis city center of Meridian and Washington but in 1876 was remote from the city. Kentucky, Avenue, although interrupted twice downtown, runs on a line southwest from Monument Circle, the city center. Just after crossing the White River, it intersects Morris Street which has an east-west orientation.

The triangle formed on the southwest side of that intersection was the site of the Marmon industrial complex. Today it is visible slightly to the south, from Interstate Highway 70 just west of the White River bridge. To a traveler on I-70 it would seem practically in the central city. The relationship of the plant to the central city is shown on the map —.

"The triangle" was further divided by the tracks of the Indianapolis & Vincennes ( & Vandalia) Railroad which paralleled Kentucky Avenue about 250 feet northwest.

The "Quaker City Works" acquired by the firm, as it became Nordyke and Marmon Company, was situated on 8.5 acres of land on the north side of Kentucky Avenue, extending some 250 feet back to the I & V railroad and bounded on one end by angular Morris Street while extending 1750 feet (800 fenced) along Kentucky.

The building on the site was a 700 foot long structure approximately 50 feet wide paralleling the railroad. The entrance was from Morris Street and it successively housed the offices and drafting room, machine, blacksmith, foundry, pat-

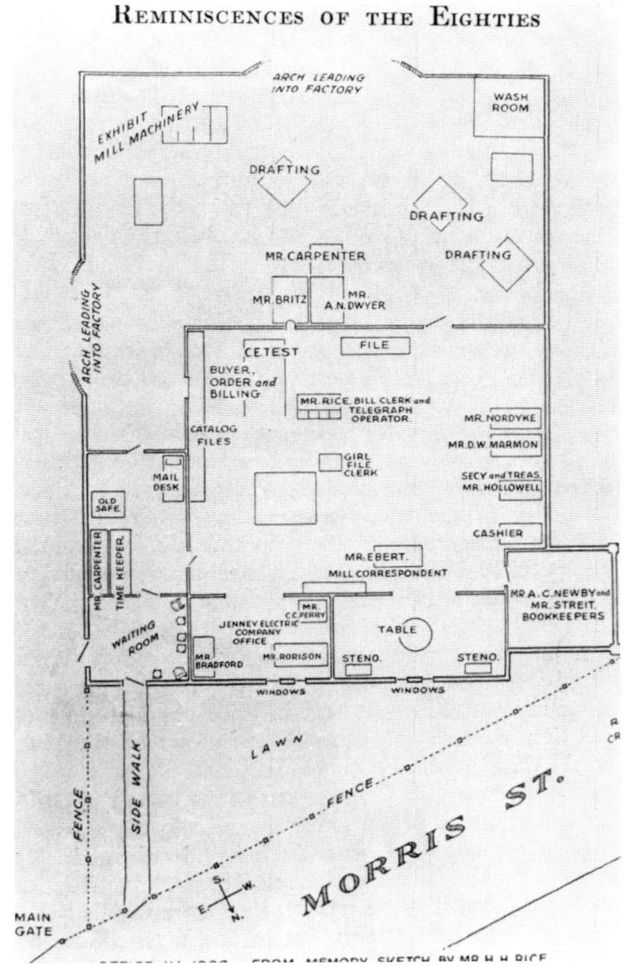


Fig. 11-5 Office plan circa 1880 per Mr. Roberts.

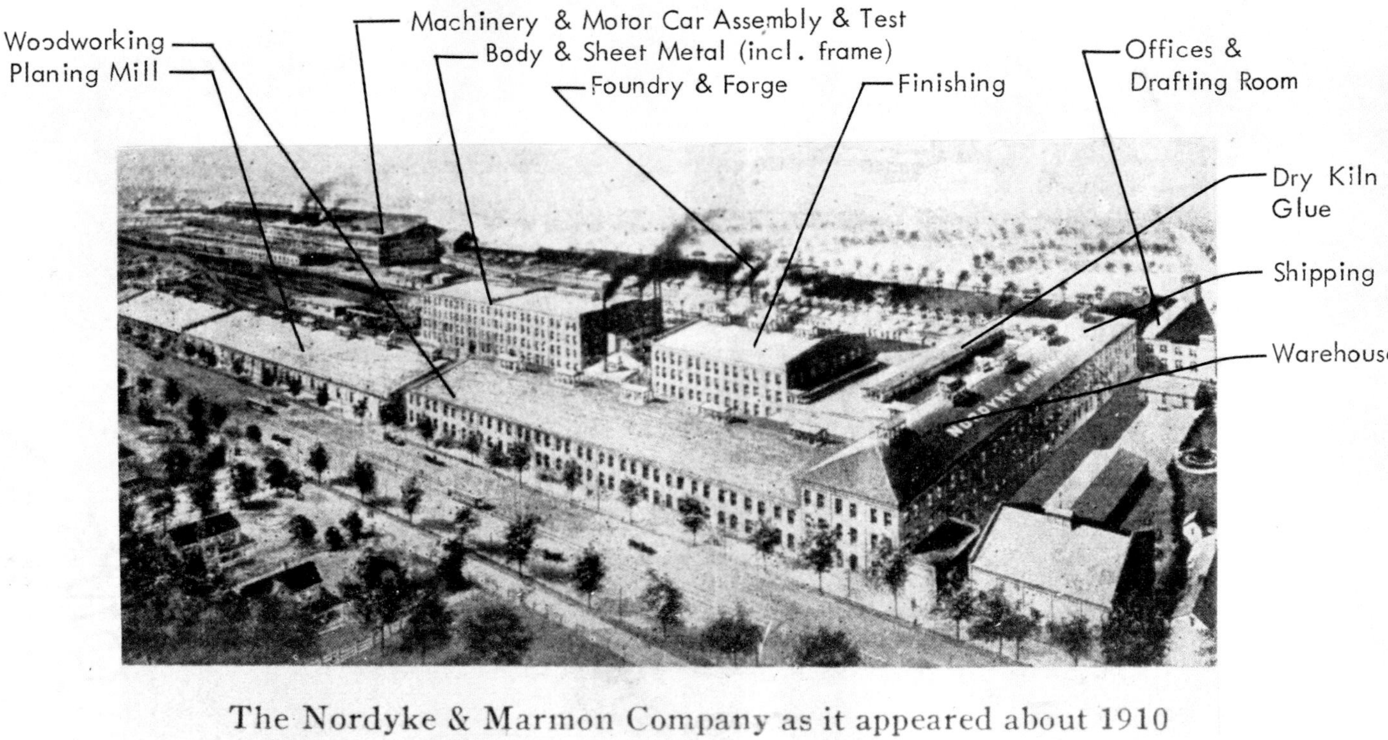


Fig. 11-6 The Nordyke & Marmon complex in 1910 with author's identification of structures.

tern-making shops, a bench-hand shop and a tinsmith shop. It was characterized by a tower of turret structure at the corner by the railroad and street which was always evident throughout the FIVE ERAS. The facilities could accommodate 500 employees. The Vandalia railroad provided rail commuter service initially.

The original structure was enlarged prior to 1880 by an addition of a wing extending from the office section to Kentucky Ave. at a 90 degree angle, i.e., not parallel to Morris. It was approximately 55x190 feet and, based on the single usage noted by Mr. Roberts in his sketch by recollection, only one story high. It housed the shipping and buhr mill departments and a stone cutting room. Within the L shape of the main structure, a T-shaped Stone Shop was added as well as lumber storage facilities and a dry kiln.

In Mr. Roberts' site plan, the railroad appears adjacent the plant. The author's reconstruction of the progressive stages of development on the site led to the conclusion that the 'Quaker City' property extended from the Kentucky Avenue property line to the railroad right-of-way property line which evidently was 90 feet on either side of the track center-line. At some later date Nordyke & Marmon acquired 60 feet from the railroad and extended the original building to 310 feet from the Kentucky property line.

A two story woodworking building (estimated to be 75x300 feet) was constructed along Kentucky Avenue abutting the previous structure and forming a U-shaped composite structure on three sides of a central yard.

The planning mill was constructed also on Kentucky, but an entrance way to the central area was provided. This was a one-story structure approximately 75x300.

The triangular extension was built on the two-story office and drafting room section of the original "Quaker City" structure. This moved the lobby out close to the Morris Street sidewalk.

Artist rendering of the Nordyke & Marmon plant in 1888 and in 1892 are shown in Chapter One. Inspection of them shows no additional construction. The later view appears to be a rework of the earlier view, primarily in the removal of the mature surrounding community which the artist had conceived for the purpose of illustrating the Jenney System of electric community lighting.

Evidently there had been extensive construction in the twelve years following the purchase of the 'Quaker City' property in 1876, which paralleled the railroad.

# PART 2 — MARMON CAR
# 1902-1933

## ERA ONE
## 1902-1908

The initial advertisements by Nordyke & Marmon in the automotive trade press in late 1904 featured the facilities and reputation of the firm and their potential for marketing and supporting a superior product as yet then undefined.

The first brochure, N & M #530 for the 1905 Model B, carried an "aerial view" of the plant and a companion text identifying the facilities associated with the production of the new automotive product.

In the central area, within the U-shaped composite structure consisting on the original purchase and additions connecting thereto, two large multi-storied buildings are shown.

The larger was behind the planing mill; it was a four story flat-roofed building (estimated at 75x180 ft). It was identified as the site for automobile body and automotive sheet metal (including the frame) fabrication.

The smaller was situated behind the woodworking shop and was three stories high (estimated 75x120 feet); this location being between the larger structure and the base of the U-shaped structures. It was identified as the site where the automobile product was finished, ie., painted and possibly upholstered.

The two plants were "joined" by a new steam generating plant, with several steel "smokestacks."

A large new steel and concrete structure, 104x340 feet (35,360 square feet) was evident parallel to the I & V tracks beyond the extremity of the pioneer 700 foot facility. This was a high-bay configuration identified as the shop where automobile engines were machined, assembled and tested; where the frame and chassis and practically every machined part going into the finished car were manufactured, and where final assembly of the complete vehicle product was accomplished. It would appear from inspection of photographs that this structure also incorporated extensive manufacturing offices as well, these being located on the side opposite the railroad.

Most certainly the initial automotive volume did not warrant facilities of the magnitude described . . . 25 vehicles the first year of all-out marketing (1905). Yet this was the support level Nordyke & Marmon considered necessary if the automobile product were to warrant equal respect with the traditional product offerings. It reflected also the magnitude of ambition behind the new venture. The reader may want to assess it in relation to the facilities available to the others either in or entering the market in 1905.

The dedication of facilities was nevertheless timely for in six years the automotive volume would require total facilities equal to those dedicated to the milling machine business!

The period 1880-1900 must have seen nearly continuous construction and expansion considering the number and size of structures erected and fitted-out. The foundry had grown to be the largest in the State of Indiana.

The stone shop was demolished to make room for other construction after roller-milling technology made buhr stones obsolete.

## ERA ONE MANUFACTURING

The cylinders for the V-4 engines had eighteen circular cooling fins and required sophisticated foundry techniques, as is evident from Hugh Dolnar's description as he reported it in *Cycle and Automotive Trade Journal* for October 1905: *The Marmon cylinders are made in dry sand, planed flasks, taper pins replaced by taper bolts before pouring. These flasks and the patterns part in the middle, and the rib (fin) spaces are first rammed up, two makers (men) working together, one each side, then the core ribs are nailed with wire nails "gagers" and the iron flask-half is set on and rammed up, turned over and the pattern drawn, leaving the flask-half with the green core-sand mould in it, ready for baking in the core oven. The cylinder head projections are had by setting a pair of false cores in the dry sand mold thus formed.* ***The whole time on one cylinder is about five hours of skilled labor.*** *(Author's emphasis). The rough cylinders weigh about 32 or 33 pounds. The cylinder bore is finely finished, no chambering (enlargement) at the combustion end, whole combustion chamber finished head and all.*

Mr. Dolnar also described the manufacture and fitting of the piston: "The pistons are rough turned and then heated to a dull cherry red, cooled in open air, then ground, and finally fitted by hand lapping in the cylinder with ground glass and oil."

For the 1908 Model H engine, the cylinder bores and the pistons were ground to size in an air blast which carried away the 'shavings' and cooled the tool and parts.

*Cycle and Automotive Trade Journal* described the manufacture of the 1905 axle: *This is a steel-blank two-piece forging, eye-ends drop-forged,* ***electrowelded*** *in the middle, next milled to H-section and finally finished to form by heating and driving down to fit a full-length forming-block. The 'Mercedes' knuckles have hard pins, .875 diameter, threaded in the axle ends, resting on bronze washers, in the lower fork lugs and pinned with taper pins.*

The rear axle of Era One Marmons had a central differential housing of aluminum sandwiched between the two axle housing halves, each of which was a steel tube on which a conical aluminum inner-end housing was integrally cast. *Cycle and Automotive Trade Journal* reported the manufacturing processes involved in the novel construction. Steel tubes of two diameters were joined by electric welding to a ring spacer, thereby effecting the reduced diameter on which the wheel was bearinged.

Holes were drilled in the larger (3.25 diamter by .187 wall) tube near the ends and the tube was heated red hot before being placed in the foundry mold. Aluminum was then poured in the mold forming the inner bell-mouth and independently the brake attachment structure at the outer end. Hugh Dolnar described the design and associated foundry practice: *The most original construction feature is the manner in which the aluminum casing members are united to the steel tube sleeves of the rear axle casing. It is needful that the aluminum casing sides and the steel tubing should be held together beyond the possibility of separation save by destruction.*

*The tube ends are drilled with six holes, .50 diameter in two circles; 3 holes in a circle* ***nearly through*** *the tube (wall) where (to be) covered by the aluminum 'hubs'; the tube ends are heated red, placed in the molds and the aluminum is poured on them.*

*Marmon gives the casing hubs a long bearing on the steel tube, and pours the casing side on the tube laid in the green sand mold, and so makes this difficult union in the sand, with no finishing cost. The shrink-excess of the aluminum holds the two dissimilar metals together so firmly that the aluminum has to be chipped off with hammer and chisel to remove it. Tests to destruction show more than a 10 to 1 factor of safety for this gear casing.*

## ERA TWO
## 1909-1915

Only two years into the second era, automotive production required facilities equal to those dedicated to the milling machine product.

By 1913:

The two-story woodworking building, which fronted on Kentucky Ave. near Morris Street, was reconstructed to a four story structure.

A large and high masonry chimney had replaced the several steel stacks on the power plant.

Additional office space was obtained by construction of a two-story addition fronting on Morris Street and interfacing with the earlier addition which moved the lobby to the sidewalk. It extended approximately half the distance to Kentucky.

The planing mill along Kentucky Avenue was doubled in length.

Two large multi-story buildings had been built in the central area at the turn of the century, and were associated with the automobile product. Now a third was added. It was a massive three-story structure standing independently and approximately in line with the planing mill on Kentucky and the four-story structure on the railroad side.

Beyond the extended planing mill, a new large one-story building was erected. It featured a high central bay, probably with a crane-way, which was shorter than the full building. It was separated from the planing mill by several hundred feet and the intervening space appears used for vehicle parking. In 1913, employees rode public transportation to work or walked (as did George Freers and later Joseph Felts), so the vehicles in the picture are undoubtedly new vehicles. From this it would be concluded that the building was a test facility; since there is no other structure identified as an engineering garage, the author hypothesizes that this was indeed the engineering test garage.

Opposite the "test garage", on the railroad side, was another large new structure adjacent to the automobile "machining and assembly/test" building. It had a high center bay.

All of the above were in place by 1913.

When Chief Engineer Jeffers and associates arrived at Marmon, they found the drafting room situated in the loft of the foundry building. Smoke, cinders, and noise were very evident. Jeffers promptly arranged for the building, in the fall of 1913, of a new structure which housed the drafting room on the second floor and the purchasing group on the first floor. Roy Milner[3] provided the author with the excellent photograph of the new drafting room which shows the appearance during World War I. Roy is standing at the rear.

The Nordyke & Marmon plant was inundated during the February 1913 flood disasters which struck Indianapolis as well as Johnstown, PA.

The facility had grown to the point that the foundry was the largest in the State of Indiana and the pattern-making shop was operated as a Division of the firm.

## ERA TWO MANUFACTURING

By 1914 and perhaps earlier, Marmon manufacturing practice was to hold piston diameters within 0.0005 (half a thousandth) inch and cylinder bores within 0.001 inch. The clearance between piston and cylinder in the touring cars was 0.004 inch and in the speedster the clearance was increased to 0.008 inch.

Cylinder castings were inspected prior to machining by visual and test hammer, then water tested under 100 psi jacket pressure. After milling, drilling and boring, the water test was repeated; after final cylinder bore grinding two final tests were

Fig. 11-7 The Nordyke & Marmon plant in 1913 — the year facilities were equally divided between flour milling machinery and automobiles.

Fig. 11-8 The N & M drafting (design) room in 1918. Photo furnished by the late Roy Gilmer who is standing in the rear.

Fig. 11-9 The new Plant 2 circa 1918.

made: 100 psi in the water jacket and 300 psi in the cylinder bores. The purpose was not only detection of already existing leaks but also of potential leaks after additional or completed machining.

The Model 32 introduced for 1909 was manufactured with "Perfect fitting interchangeable parts assured by carefully prepared tools and jigs."

## ERA THREE
## 1916-1927

At some date prior t o 1915, a "21 acre site" was acquired for future expansion. This land was across Kentucky Avenue from the main site and paralleled the Belt Railway on the Southwest side.

A new assembly plant was constructed in 1916 on the "21 acre site." It measured 100x275 feet (37,5000 sq. ft). (extended to 857 feet in 1919). This would house the assembly line for the Model 34A.

When the United States entered the World War conflict, a very large one story plant was constructed on an expedited basis for the purpose of undertaking the manufacture of the Liberty V-12 aircraft engine jointly with Ford, General Motors, Leland-Lincoln and Packard. The plant had 146,500 square feet in one "room". That unit of Plant 2 fronted on York Avenue.

This effort was of such magnitude that the production quota for October 1918 was exceeded by 246 percent and resulted in permanent retention by Nordyke & Marmon of a trophy for the Champion Builder of Liberty Engines.

Some measure of the steps taken is evident from the following. N & M chartered and operated, with women motormen, a trolley line to carry employees to and from work. A cafeteria was erected adjacent (towards Kentucky Ave.) to the military engine plant capable of feeding 3000 employees.

A three story addition was made to the (two-story) office building to accommodate military operations. This unit extended along Morris to the Kentucky intersection. It interfaced with the previous addition along Morris, but the rear wall paralleled the center section of the early 'U' shaped structure.

"Of the 3800 persons engaged in this Liberty Motor production, more than 900 are women, and it is the work of these women that is one of the most striking features of the factory."[4]

Mary E. Bostwick continued: "Here are women doing work that was hitherto considered the work of only the most skilled and trained mechanics, and doing it with an accuracy and perfection to detail that could not be improved upon. There is hardly a department where the women are not seen at work, except where heavy parts are handled."[4] The linguistic enthusiasm of the reporter, who was escorted on a plant tour by security guards, warrants perservation: *"It almost reminds one of a packing factory . . . only here the transformation is from rough, unwieldy castings, almost shapeless, rusty and uncouth looking, that emerge at the end into a polished symphony of perfectly synchronized machinery, the triumph of the mechanics art holding in its steel and copper heart the motive power that gives man the means with which to range the clouds."*

Following the cessation of military production, the executive offices were relocated to the second floor of the three story building at the Kentucky-Morris corner. This made them adjacent to the engineering department in the adjoining 1913 addition.

Fig. 11-10 The 1918 office building addition.

Fig. 11-11 The new Plant 3 circa 1922.

Fig. 11-12 Plant 1 circa 1920.

A major expansion, with the object of doubled production of both automobile and milling machine products and doubled employment rolls and necessitating a $2.5 million loan, was announced in June 1919[5] by Walter Marmon, President, who said: *The move was necessitated by our increased facilities which we had to take over from the government after the completion of our war contracts. When we finally took a mental inventory we found that we had large machine shop facilities, but not sufficient body and wood-working facilities. These had to be furnished or we would be in the position of having an unbalanced plant. This, and the fact that the officers of the plant (firm) felt it to be a splendid time to insure our financial position in the automobile industry which we appear destined to fill, impelled us to consider adding to the financial strength of the company.*

Mr. Marmon also noted that 1918 was the best in the history of the N & M milling machine business and that the need to feed the postwar world would greatly increase demand such that output capacity for that product was being doubled as well.

The plant side was enlarged 14 acres by purchase of the wedge of land between Morris Street and the I & V (Pennsylvania) Railroad extending to the Belt Railroad. Marmon now owned the full triangle formed by Kentucky, Morris and the Belt RR.

Principal new building was to be a body plant five stories high of reinforced concrete construction measuring 80x600 feet and scheduled for completion in late 1919. Closed car bodies previously made elsewhere would be built in the plant, probably in addition to relocation of open body production.

Concurrently a second contractor was to erect a 100x800 one story structure of similar construction to be used for car assembly including chassis road test.

Also included in the late June groundbreaking was a new planing mill, dry kiln and power plant.

The assembly plant was built adjacent the Morris Street property line with the body plant parallel to it. The 13.4 acre site was to be known as Plant 3.

The new buildings totaled over 350,000 square feet and brought the total Marmon facilities to 1.5 million square feet. This was the peak floor area. Following sale of the milling machine business, the area was reported to be 1,100,000 square feet; presumably some elderly structures were razed. But additional construction in 1927 increased it to 1,248,000 and by 1930 to a final 1,411,840 square feet.

Marmon employed 5000 people in 1920. The combined floor area of the three plants for cars and milling machines in 1922 was 34.4 acres; the site was 60 acres.

## ERA THREE MANUFACTURING

A Marmon booklet "Information for Overseas Dealers", published in 1922, included an excellent account of the manufacture of the Model 34 in the expanded facilities.

THE MARMON CAR . . . ITS HOME AND HOW IT IS BUILT

*"With the recent completion of the new buildings comprising the third distinct growth of Nordyke & Marmon Company since its establishment on its present site, thirty-four acres are now devoted to the manufacture of Marmon motor cars.*

The feature of the group of buildings, erected during the war and now comprising Plant 2 is the machine shop, with its six acres of floor space, in which the engine and chassis are built and assembled. Practically every part going into the engine and the chassis is made and machined here in whole or in part. This building is 565 feet long and 535 feet wide, its walls are of brick and glass, its roof of a saw-tooth steel construction. It is a model of lightness, airiness and cleanliness. Its equipment is of the latest and most complete type, its facilities, such as rest rooms, first-aid rooms and the like, are of the most modern.

Into this building, then, comes the stock, in the rough or semi-finished, passing through the endless inspections, going through the countless operations to be fitted for the engine or the rear axle or the transmission or for whatever part it is to be used, until it comes out of this same building a chassis complete in every detail, ready to receive its paint and the body.

Accuracy and minute measurements to a degree unheard of before the war in quantity production are the watchword of Marmon production. In this connection it is of interest to follow the production of the engine from the bulk and see how it emerges finally as the High Efficiency Motor that it is. To begin with, once research

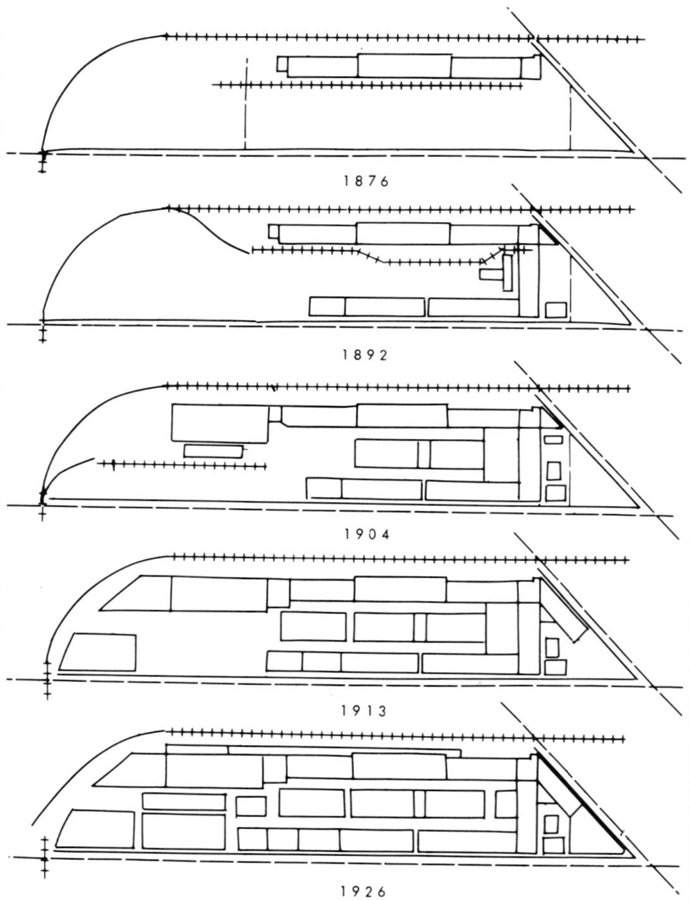


Figure 11-12: NORDYKE & MARMON PLANT ONE
50 YEAR GROWTH 1876 - 1926

and experiment have determined the necessary materials for the individual parts, it becomes essential that there be no variation from these specifications. Experimental and chemical laboratories, elaborately equipped, are responsible for this with the constant checking and rechecking of the inspection department. Samples of each shipment of steel are subjected to chemical anaylsis; each daily run of iron and aluminum in the Marmon foundries is tested. **Chemical and physical tests are applied to every part received from outside whether rough, semi-finished or finished.**

Throughout the actual machining and assembling of parts the same rigid and severe standards are maintained. The basis of the entire inspection system is the famous Johansson blocks. The possibility of human error is at a minimum; the hand is replaced by machines and tools. **Every machine, tool, jig and fixture in the entire plant is inspected thoroughly every half hour.**

Add to this the fact that **every part going into the engine goes through an inspection after every distinct operation on that part, no matter how minor.** With this unceasing vigilance throughout the machining of the parts, with the specialization of operations on the assembly line, the engine reaches the end of the line, as one might suppose, practically ready for placing in the chassis. But in reality it is merely on the threshold of its testing.

It is placed on a dynamometer, where for seven hours and forty minutes, it is thoroughly tested at varying speeds and horsepowers. Throughout this entire period the closest watch is kept over the engine by expert mechanics, who keep a log or test report which accompanies the engine at all times. In addition to noting all defects, knocks and rattles, the tester keeps a complete record giving the oil pressure, the speed at which a certain horsepower is reached, the temperature of the water and the charging of the Delco generator. Particular care is taken in the observation of such parts as the timing gears, valve parts and oiling system. The slightest imperfection is recorded.

This test over, the engine is taken to a **glass-enclosed dust-proof room,** where it is completely disassembled and all the parts thoroughly cleaned with petrol. There follows an exhaustive inspection, following closely the report of the log. All defective parts or parts that seem as if they might become defective are replaced, the inspection is completed and the entire process of painstaking assembling is repeated.

The rebuilt engine is again placed on the dynamometer blocks and put through its paces for another hour. Before the engine reaches the chassis it is subjected to two rigid examinations, interspersed with a complete tear-down inspection and rebuild.

The chassis leaves the machine shop ready for its coats of paint and for the body. Other buildings in the Plant 2 group aid in this process. The aluminum foundry is responsible for the making of the majority of aluminum parts that go into the car.

Opposite the foundry is the sheet metal building, a long, one-story structure, 97x857 feet. Here the work of making and assembling the frame is carried on, as well as the manufacture of aluminum steel parts, such as the shaping of the fenders and parts of the bodies. Auxiliary buildings comprise rooms devoted to heat treat, pickling, sand blast, nickel-plating, brazing and welding and the like, to say nothing of a huge restaurant capable of caring for the more than a thousand employees in this section of the factory.

With the chassis and engine complete, the car is now ready for Plant 3, where it receives the body and final touches. This entire plant reached completion in the spring of 1920. The body shop, final assembly building, lumber storage and plan-

Fig. 11-13 Manufacture of the Model 34 B Engine.

ing shop, dry kilns and power plant comprise, in all, 350,000 square feet of floor space on a thirteen-acre tract adjoining the original plant.

The first step in the body work leads to the dry kilns, where the lumber goes through from three to four weeks of seasoning, which eliminates the moisture and gives the stock a smooth surface which does not shrink or swell. From the kilns the lumber goes to the storage and planing shop, a two-story, concrete, brick and steel affair, where all the motor car woodwork is done.

The next link in the chain is the movement of these wooden parts by elevated runway to the fifth floor of the body shop . . . a thoroughly modern reinforced concrete building, 600 feet long, 80 feet wide. On this floor all the bodies are made, including the building of closed bodies. Descending one floor, the body then passes through the painting operations, fifteen in all, including the final finish varnish. The floor below is used partly for further painting operations, and for the sewing of the curtains and for trimming. On the second floor the body receives its complete trimming, with the highest grade hand-buffed leather, Marshall springs and 100 percent curled cattle hair for the cushions. Here, too, the work is completed on the top and curtains.

Thus the body is now ready to go by overhead runway to the balcony of the final assembly building, a one-story brick structure, 800 feet long by 100 feet wide. The chassis has, in the meantime, entered one end of the building and passed through the painting rooms. It emerges from these to go upon the tracks for final assembling. It receives it body from the balcony above, and as it proceeds down the track it is given its final trimming and upfitting until at the end it is ready to be driven on the road for its final seventy-five-mile test. This completed and the necessary adjustments made, the car again enters the final assembly building, where it is washed and rubbed, given its final finish varnish, and receives the finishing touches of equipment.

* * *

This then is the Marmon motor car of today

Nordyke & Marmon was a pioneer in the development and use of diamond-tipped cutting tools to faciliate manufacture to very close tolerances. The 1927 Edition of the Marmon Book of Facts stated: *"A considerable fortune in uncut diamonds is represented in Marmon vaults. These gems are used in delicate tools for cutting vital Marmon engine bearings. The primary secret of machining bearings with diamond tools is in the method employed for holding the diamond in the end of the tool holder.* ***This is a process developed in the Marmon plant*** *and accounts largely for the* ***exclusiveness that Marmon has enjoyed in the use of these tools. Marmon has used these tools for years*** *and while it has been a matter of knowledge to the Marmon organization, little has been said about it from an advertising standpoint." (Author's emphasis)*

On the Model 75, diamond tools were used to bore the main bearing seats in the crankcase and to finish the surfaces between bearings insert halves. Dual shaper heads in a special machine carried the diamond tools for finishing the bearing insert faces.

Marmon bearings were nickel babbitt molded into the insert cast iron backing under high pressure to eliminate air pockets and insure a closer bond to the backing.

The two-piece piston used in the Model 34B, 1920-24, featured a cast aluminum piston which was bolted at the base to a cast iron skirt, leaving the aluminum free to expand within the iron skirt. The iron skirt controlled piston clearance and slap, and retained the pin as well. The clamped surfaces on the two parts were ground and lapped true to assure that the assembled parts were in line.

Each Model 34B engine was block tested as follows upon completion of assembly:

Driven at 700 rpm for 1 hour to lap wearing surfaces
Fired at 750 rpm, no load, for 5 minutes
Run at 750 rpm, 5 hp, for 5.75 hours to glaze cylinder bores
Run at 6 load and speed pairs for 5 minutes each to 35 hp at 1450 rpm

The engine was then transferred to a glass enclosed room for tear-down, inspection and reassembly. Random 50 hour tests were also run.

The completed chassis was driven 50 miles, of which 25 were over a specific test course. Following final assembly and paint, a 15 mile road test was driven; after clear-coating and a 3-day drying period, a final 5 mile acceptance test was driven on paved surfaces.

The Plant 3 body facility (the 5 story structure) was idled in 1924 when the American Body Company of Buffalo was contracted for body manufacture and trim.

In the spring of 1924 a 2000 square foot space in the lobby of the Nordyke & Marmon administration building was utilized to construct and furnish a product display room. Marmon models would be displayed for visitors in a model dealership decor.

During a two month shutdown for inventory exhaustion in the summer of 1924, chassis assembly operators were transferred to Plant 2, to take advantage of the huge, nearly square, one floor facility for the purpose of implementing a new concept of high efficiency assembly with low inventory levels.

The war-time expansion and subsequent acquisition of those facilities had, for the first time, involved outside banking interests in the business. Despite a high profit rate on sales, these outside interests actively sought escalation of sales to the full utilization of installed facilities.

Mr. G. M. Williams was recruited to be president of Nordyke & Marmon specifically to increase the body and car assembly rates to a level comparable to installed

Fig. 11-14 Model 34B engine assembly.

Fig. 11-15 Twin assembly lines for the Model 34B.

machine shop capacity and overall plant footage. Obviously he also was faced with comparable expansion of the marketing and distribution efforts, but it is the factory reorganization with which we are concerned here.

A two-page Marmon advertisement included a reproduced newsclip dated Aug. 30, 1924 which reported that rearrangement of the factory for more efficient manufacture had been underway for two weeks. Skilled mechanics (note) were reportedly being employed at the rate of 100 per day to expand the work force to 2500 as rapidly as possible. Payroll growth would cost $4 million per year in addition to $5.5 million expenditure for 'new materials.'

In contemporary terminology, Mr. Williams had subscribed to, or perhaps invented, the technique of KANBANNING (Japanese for "just-in-time delivery") and had undertaken the conversion of the Marmon plant and its suppliers into MARMON CITY.

The task was less formidable than it might have been because of the high content of in-house manufacture which was still 80 percent as late as 1927 (Toyota in-house was 32 percent in 1983) [6]. Carburetor sourcing was switched to Schebler of Indianapolis from Stromberg of Chicago, suspension spring sourcing to Ansted of Indianapolis from Sheldon of Wilkes-Barre, Pennsylvania.

While dwarfed by the huge auto assembly plants being erected in the 1980s, the Marmon Plant 2 main-building had the largest single "room" under roof in the state of Indiana as late as 1937: 261,800 square feet; with one exception the largest in the industry.

Mr. Williams found Plant 2 (ex Liberty Engine building) ideal for his concept and the assembly operations were relocated there from Plant 3 and from the Plant 2 elongated structure paralleling the railroad.

Mr. Williams impact was dramatic. *Automobile Topics,* in the October 25, 1924 issue,[2] carried a feature article headed as follows:

NORDYKE & MARMON TRANSFORMED
Old Concern Completely Rebuilt in Harmony with Modern Trends . . .
New Manufacturing System Designed for High Quality Product
Remarkable Economies Effected

The article described both major physical changes in the plant and assembly process as well as significant efforts to involve the hourly employee in the business. Both warrant recording for their historic significance but are equally of renewed interest as the domestic industry struggles to reform itself to counter Japanese advances in both the assembly process and employee involvement.

Mr. Williams' first move was to detach the Tool Engineer, Mr. J. W. Anderson from his assignments and send him on a tour of the Detroit automotive plants. The article continued: Progressive assembly, such as is employed by all the major automobile makers, depends entirely upon an unfailing supply of constituent parts being delivered to the respective stations along the line. To effect this end, it is of course necessary to schedule the various manufacturing departments in such a way that they will all operate in synchronism practically automatic. This scheme is to arrange manufacturing equipment in parallel lines at right angles to the assembly line, so that raw material entering one side of the plant by means of a single delivery aisle, progresses in parallel (paths) to the assembly stations. This has the advantage of eliminating the circuitous movement of material and reduces trucking to a minimum. Notably, it (the scheme) indicates as ideal for such an arrangement a substantially square floor plan with the chassis assembly line bisecting it, (with) manufacturing divisions running at right angles to the line and occupying half the floor while paint, equipment, testing and final assembly departments occupy the other side. Parts removed from one machine, at the conclusion of a stated operation, are placed within reach of the operator who is to perform the next oper-

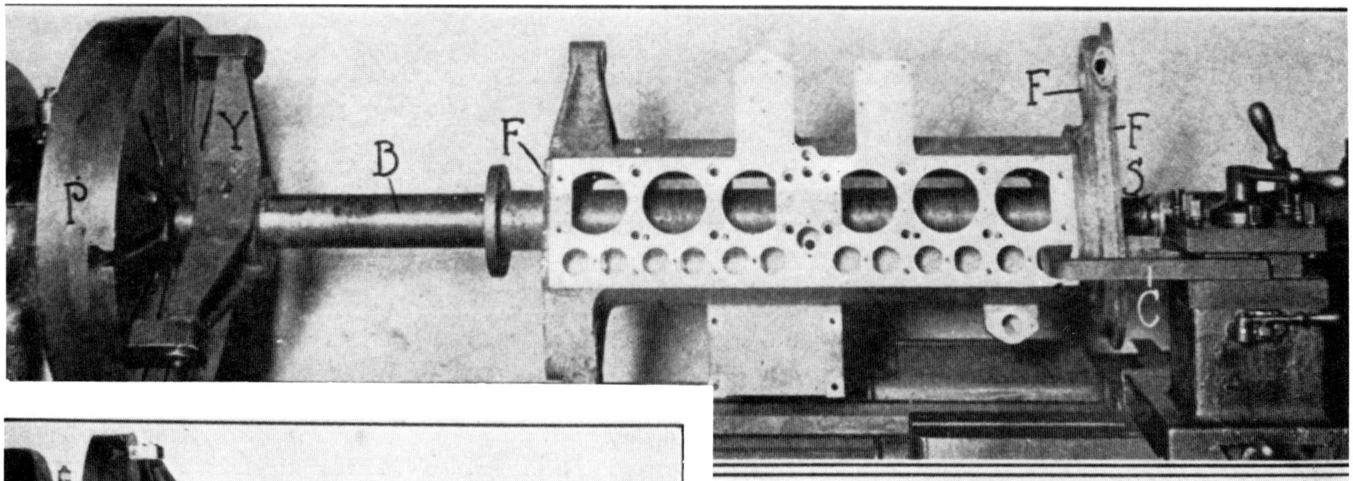


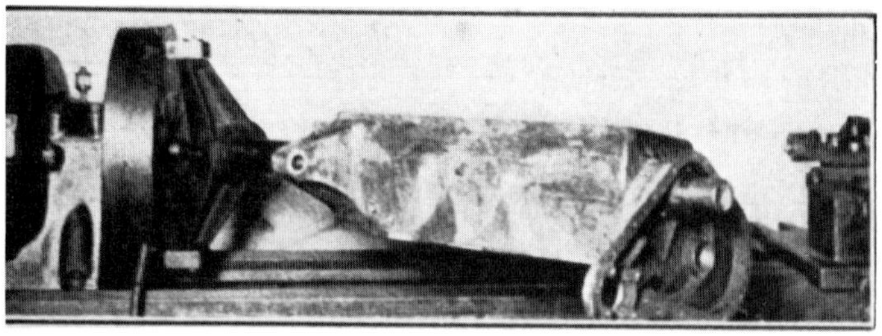

Fig. 11-16 shows (above) lathe turning of crankcase faces from crankshaft axis and (inset) pivoted arbor to facilitate loading of the Model 48 crankcase.

**Marmon spent $40,000 for interactive tooling for their 1914 models to improve precision. The aluminum castings were primed on the inside and baked at 200 degrees for two hours before machining to seal and to relieve casting stresses. Three hours were required to bore and face the crankcase.**

Fig. 11-17 shows (right) drilling of crankcase with jig located by arbor inserted in crankshaft bearing bore and (below) a special Marmon machine for boring the crankshaft, camshaft and accessory shaft bearings at the same time.

ation.

Thus mechanical conveyance from one machine to the next is largely dispensed with. Equipment is so selected and operations so timed that the material progresses down the line at the rate required to deliver it at the assembly (line) at approximately the rate required. . . . Sub-assembly and inspection (operations) are introduced directly into the manufacturing process.

The article reveals an unrelated point of contention between engineering and manufacturing. Engineering decreed a chassis road test which manufacturing considered redundant. The result was a conveyor system for handling the recycling of test bodies . . . in 1924, Engineering still had the last word at Marmon.

The article concluded the discussion of physical changes: . . . the novelty rests not so much in the physical rearrangement of equipment and the sequence of operations as in the system governing them and **the spirit of team play** which it develops **in the working forces.** This system **believed to be brand new in a number of its aspects,** is a development of the "group bonus" payment plan which is highly respected for its resultfulness.

The Marmon "pay-unit" concept held that the efforts of all the factory workers were directed toward accomplishment of definite results, rather than mere repetition of one operation. The AT reporter provided an example: The man milling crankcase castings is really interested in performing his part toward completion of a given number of engines, rather than in (mill) facing the greatest possible number of crankcase castings per day. His work is satisfactorily completed only after several other operations have been performed and (the Part) has been bolted up into an engine.

The sub-pay unit in the example was the crankcase and the mill hand was paid a proportionate part of the cost of the finished full pay-unit. Instead of setting up inspection in opposition to the shop force, the "pay-unit" plan placed the inspector on the side of the workman; he became an ally rather than an enemy, because he could promptly detect a problem and prevent work being rejected.

Inventory at Marmon was reduced to the material stock piles at the heads of the several machine lines. Due to inventory reduction during the last half of 1924 and the associated management controls, the inventory turnover in 1925 was at the rate of 10.3 times per year (ie, once every 36 days). Foremen were reidentified as "group leaders." A group could voluntarily leave early or remain afterhours to balance pay-unit inventories.

Marmon executives were quoted as saying that the system "makes all the difference in the world in the **attitude of the entire shop toward the business as a whole"**, and "It is believed this system satisfies the long-felt want of **contentment with mutual profit in industry."** "This system might not work at some of the other (manufacturer's) plants", Hal L. Purdy, vice-president and assistant General Manager explained: "We have the advantage of an unusually intelligent class of people. Nearly all are Americans born and raised right here. You'd be surprised how many college men we have around the plant."

The number of "stock chasers" was reduced from 160 to 30. The number of factory trucks was reduced by 40. Large reductions in non-productive labor, sweepers and time-office employees were reported and cost accounting was made rapid, simple and positive.

The realization of a MARMON CITY was compromised by developments in the body building industry. Several groups of individual body builders merged and in the process cut off some of the surviving independents from traditional customers. The result was that it became economically advantageous for Marmon to close down their Plant 3 body facility (which was only three years old) in favor of outside procurement of bodies from non-traditional suppliers beginning with the 1925 model production.

Fig. 11-18 Heat treating transmission gears as soon as they are cut — 1925.

The innovation of the Williams era is exactly the concept which was in the process of adoption by General Motors in 1983 for the payment of vendors . . . eg, the supplier of glove-box doors to be paid on the basis of how many cars were completed, not on how many box doors were received for that model.

G. M. Williams reorientation of Marmon Motor Car Co. revealed remarkable perception for the middle 1920s. Not only is the pay-unit concept receiving serious attention at General Motors as noted above, but the automotive giant is concurrently embarking on adoption of a system of warehouseless inventory control much like that implemented at Marmon in 1924. Marmon was 73 years old then; GM was 75 in 1983.

GM is adopting both concepts in a crucial effort to reduce car manufacturing costs to the level attained by the Japanese. Williams had very much in mind the (then) recent successes of Henry Ford in multiplying sales volume by lowering prices to reflect manufacturing economies; this was evident from the out-of-sequence designation of the Marmon 68 as the Model T and the remarkably lowered prices for the top of line offering in 1928 followed by the world's first eight cylinder car under $1000 in 1929 for a Guinness record 'First.'

A Service Department Research Division was established sometime prior to October 1926.

A brochure authored by G. M. Williams and titled "To Those Who Would Know About Marmon" was distributed circa 1925. It began:

We believe here at the factory that the Marmon car
is the most precisely and carefully built American automobile
is the best road car in the country — that it will carry its passengers
more miles per day in genuine comfort than any other car.
is probably the safest car — at least we seem to have every assurance that

Fig. 11-19 Frame manufacture — Model 74.

Fig. 11-20 One of battery of giant presses used to form frame side rails and side-guard running boards.

it is.

It included reproduction of an article from IRON AGE titled "How Marmon Produces Super-Quality Without Excessive Cost" and reporting the "Marmon City" sourcing and pay-unit compensation concepts. Some views of factory operations are reproduced on the following pages.

## ERA FOUR
## 1927-1930

Joe Felts told the author some recollections of working in the Marmon drafting room in the early 1930s. The main entrance to the office building was the 'V' section of the building; it was reserved for executives and visitors. There was a Morris Street entrance for lesser personnel and a time clock to greet them.

The executive offices were in the 'V' on the second floor. The engineering department was partitioned into two rooms: the larger was used by production engineers, clerical help and included the vault for engineering drawings; the smaller by the engineers responsible for advanced design.

The salaried cafeteria was located in a converted residence directly across Morris Street from the offices. Directly across Kentucky Avenue was a bank which Felts recalled was robbed by no less than one John Dillinger. Division Street runs north from Morris Street east of the Belt Railway and near the west end of the Marmon property; The Pathfinder automobile plant was located on the west side of Division a block north of Morris.

The employment office in 1930 was in a building on Kentucky near Morris which

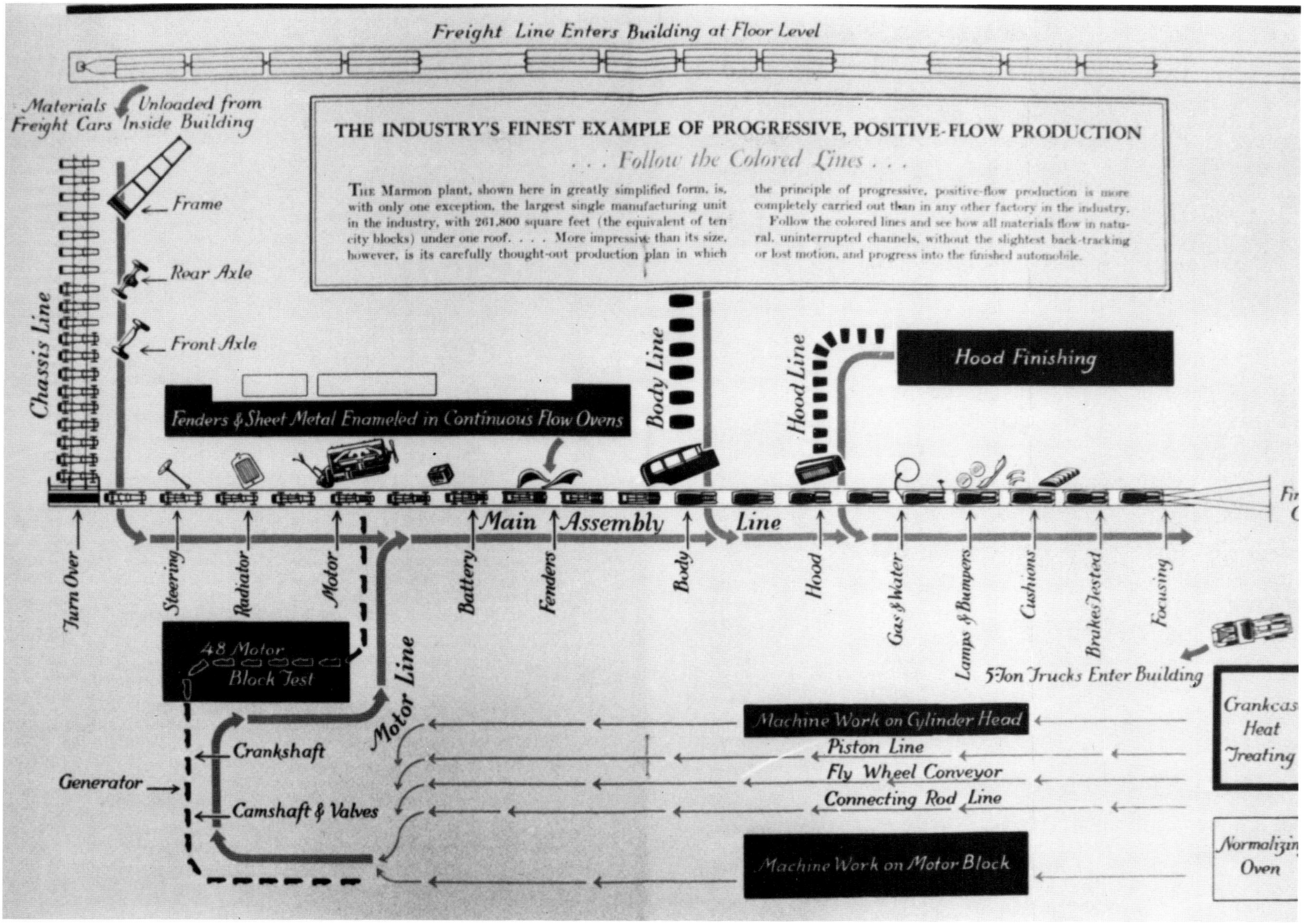


Fig. 11-21 Assembly Flow Chart, Plant 2 — Model 74, 1925.

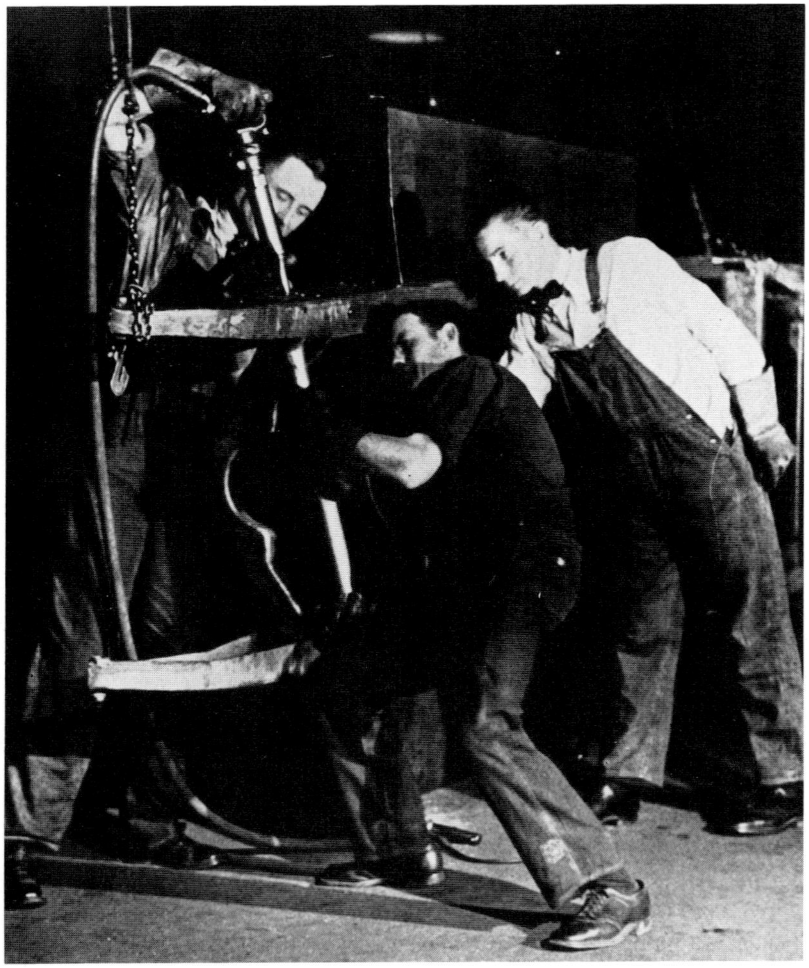

Fig. 11-22 "Assembly of Model 74 frame with hot rivets and human brawn."

Fig. 11-23 The pre-tested power plant meets the Model 74 chassis.

earlier had served as a stable for the Nordyke & Marmon executive horses.

The iron foundry in 1930 was located in the structure where engine and powertrain parts were machined and cars assembled during Era One.

In the 12 month period ending August 1930, the Marmon power plant consumed 455 carloads (22,727 tons) of coal.

By February 1929 Marmon had a Suggestion Award system in operation with a committee assigned to review suggestions by employees. An example from *The Marmon Factory News: "To J.A. Todd, a cash prize of $10.00 for his suggestion: 'That the brackets and links used to assemble the trunk rack be assembled at the machine where they are made. This will save half the time now used to do so and the keyway nut will not need Parkerizing when the bracket and link are assembled. This can be painted with the handling of one piece and the way it is done now, four are handled.*

*'I also suggest that the four trunk rack holes in the rear of the frame be made larger when the frame is made.'*

A lending library was established, stocked with texts and periodicals on manu-

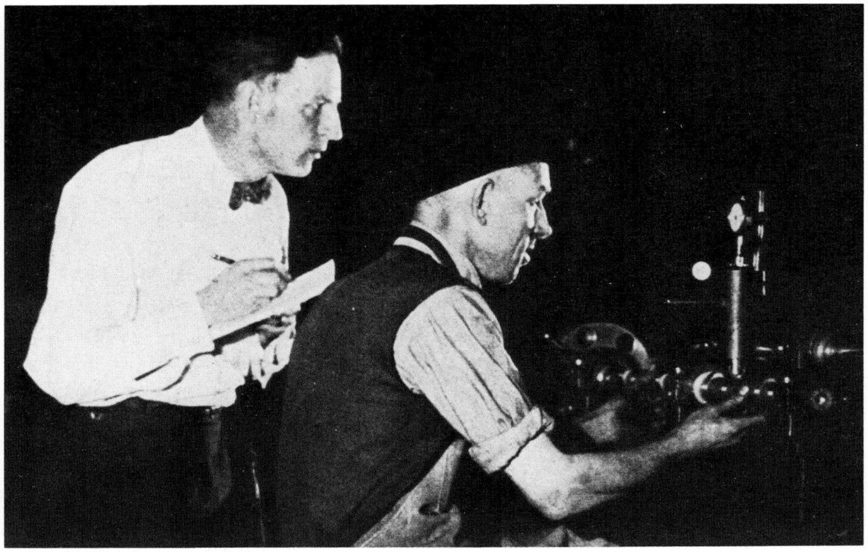

Fig. 11-24 "Super accuracy employed in manufacture and inspection of the camshaft."

facturing processes and made available to the factory employees. New for 1931 was a smoking room for hourly employees which was equipped with games and playing cards.

On June 23, 1926 it was announced that the assembly building on the "21 acre site" built in 1916 would house the Model 75 assembly line following transfer from Plant 3 to make room there for the Little Marmon: Murray was to build the bodies on the upper floors and the car was to be assembled on the first floor.

Fig. 11-26 Frames received from vendor are inverted for start of chassis assembly; Era Four.

Fig. 11-25 First Roosevelt off the line; Plant 3, February 1929.

Two thousand employees of Murray were to produce 75 bodies per day.

Plant 3 was reactivated for body production when it was leased to the Murray Body Corp. of Detroit with commitment for three years production of all Marmon bodies (1927-28-29) but in mid 1928 was sold to Hayes Body Corp. of Grand Rapids, Michigan who assumed control for production of 1930 bodies.[8] Marmon resumed operation for the 1931 models.

In October 1926 construction was reported to be underway on a "third" addition. It was a one-story building 440 feet long, with steel framing and brick walls. It was to be used to increase production capacity.[9] It probably paralleled a structure on Kentucky.

It was July 1927 when the announcement was made that Marmon would rearrange factory facilities to accommodate requirements for increased production of the "Eight." Each of the three plants would be involved in the manufacture of chassis and bodies for a distinct product line. The entire manufacture and assembly of the large Series 75 was to be transferred to the Plant 1 facilities, vacated upon sale of the milling machine business, from the former Plant 2 location. A new 700 foot assembly line was to be installed in Plant 1 surrounded by a unified and compact manufacturing operation. It would share Plant 1 with the largest foundry in the state of Indiana, the pattern shops and the service parts operations.

The Marmon Eight (1928 Model 78) which had been built as the Little Marmon in the new Plant 3 exclusively was to be assembled in Plant 2. Plant 3 was to be committed exclusively to the new low-priced line, the Model 68 (and later to the Roosevelt).

10,366 cars were built in fiscal 1927, ending June 30, whereas only 5701 were built the previous year; 5,600 in the first four months of 1927 versus 1,600 the prior year.

The decision to build both Marmon and Roosevelt cars on the same assembly line was considered revolutionary.

In 1929 the assembly of the Marmon eights was accomplished by unloading frames directly from a railroad siding inside the plant on to a conveyor in an upside down attitude. After assembly of front and rear axles and springs, as well as shock absorbers, a coat of paint was sprayed on the chassis bottom side.

Upon reaching the final assembly line, the chassis was placed upright by a mechanical arm mechanism which set it in a correct position on the moving floor of the final assembly line. Both **the chassis turnover machine and the moving floor were Marmon developments** in the interest of plant efficiency.

The upright chassis was then painted on the top side followed by assembly of steering gear and brake lineage. Following engine and front end sheet metal assembly, the body was delivered and 'dropped' by an overhead conveyor.

The car then rolled across long pits for work beneath the car to the 'clean-up' line where brakes were adjusted and tested and headlights aimed and focused.[11]

The 1930 cars all carried Hayes bodies built in the Plant 3 body plant purchased by Hayes in mid 1928. The business was severely contracted and all four chassis (Marmon Roosevelt, 8-69, 8-79 and Big Eight) were assembled on a single chassis line until reaching the body drop station. Starting with the body drop operation, two lines were utilized: The two smaller lines continued to body drop and completion of assembly on the same line. Chasses for the two larger car lines were transferred by an automatic trolley transfer system to the head of a second assembly line where they received bodies and completed assembly. The hoist was equipped with **automatic grip and release hooks,** two at the rear and one at the front of the chassis.[12]

Note that 1931 car assembly operations were in Plant 2, not Plant 3 which was the site of Little Marmon and Roosevelt assembly in prior years.

Carboloy tooling was being used, in the manufacture of Roosevelt and Marmon engines, for boring operations. Connecting rods were fitted to 0.0015 to 0.00175 clearance without shims.

New milling machines installed in 1928 finished three sides of the Marmon and Roosevelt cylinder blocks simultaneously. All eight bores were honed at one operation. Sixteen valve guide holes were bored simultaneously. In the manufacture of the piston, ten radial oil drain holes were drilled simultaneously.

## BODY BY MURRAY 1927-1929

The April 1927 issue of *Autobody* carried a feature article on the 'Murray Body Plant at Indianapolis' which is repeated here, with tense changed, because it illustrates the challenge and complexity of mass painting automobile bodies with the finishes and equipment available at that time. Granted the problems had been even greater in earlier years but the production rates had been inversely smaller. The Murray plant on the Marmon site is an excellent candidate for this detailed attention because Marmon had only recently pioneered the use of two and three tone combinations on a standard option basis but at a rate less than half that expected of the Murray facility. Surviving examples of 1928-29 Marmons with carefully duplicated original shades are surprisingly pleasing today. The *Autobody* article:

The body plant of the Murray Corporation of America at Indianapolis is interesting as representing a progressive fabrication on conveyor lines of highgrade standard bodies. It is also noteworthy because of the arrangement of its painting department, where bodies are carried along by conveyors for the various operations, and during air-drying periods, are elevated to the ceiling by the same conveyor systems and lowered to the floor again when ready for the next operation thus effecting a great economy of space.

This Murray plant was chiefly contained in a 5-story building, approximately 80 X 600 ft., leased from the Marmon Motor Car Co. for whom the Murray Corporation is building the standard bodies for the "Series 75" chassis and for the "Little Marmon" eight. Possession of the body-plant building was taken by the Murray interests in June 1926, and the company's engineers in Detroit began designing a plant layout for the production of 100 high-grade bodies daily; equipment was installed, and framing was begun on Aug. 20; the first completed body was turned out on Sept. 11. The operation of the Indianapolis plant was under the direct supervision of Division manager, G.J. Connolly, and general superintendent, J.L. Snyder.

Wood body-frame parts and steel stampings were received from other plants of the Murray Corporation so that the Indianapolis plant was mainly an assembly plant although some minor woodworking and metal operations were conducted there. The wood frame members, the steel stampings and other body materials were received and stored in a building adjoining the assembly plant, and as required, they were transferred by electric storage-battery truck and raised by a lift to the fifth floor of the body plant where assembly began. The fourth floor was used as a painting department and the third floor was occupied by the trimming department. Polishing, inspecting and other final operations were conducted on the second floor from which delivery was made by a bridge and crane to the Marmon company's assembly floor in the adjoining building where certain accessories were added before placing the body on the chassis.

The sales department in Detroit received a schedule from the Marmon company for body models desired during the succeeding month. This schedule and formal orders were sent to the corporation's Bay City woodworking plant, to the stamping plant in Detroit, to the works manager at Indianapolis and to such other executives as were concerned. The Detroit office ordered all production material and only sundry special items were purchased direct by the Indianapolis plant.

In initiating a new model the first jobs were put through on day work and then wherever possible the group-bonus plan was used. A standard group time was set after a time study. The standard time or pay was for 100 jobs and no cut was made under ordinary circumstances until three months had elapsed to determine whether the rate was fair. There were two or three "pay points" in each department where the work was finished and inspected. A bonus of 20 percent was paid for 100-per cent efficiency. The "actual efficiency" percentage was obtained by dividing the "standard" hours for certain operations by the actual hours, and was posted each day in the plant where it could be inspected by the groups. The bonus did not cease until the efficiency percentage fell to 75 per cent, but if this point was reached or approached, a special study was made to determine and correct the cause.

Assembly, began on the top floor of the body building and for the most part was conducted on conveyor lines. Body-frame members, metal stampings and other parts, brought to the fifth floor by the electric storage-battery truck, were delivered to stock racks beside the assembly lines. The assembly of the body frames was done in short transverse bays. The body-frame assembly groups might work on different models or on the same models, depending on the demand for a given body style. When the framework was assembled it was bolted by the panelers to a special steel dolly truck and put on the longitudinal conveyor line, traveling along the north side of the building.

Although most of the metal work was received already stamped, there was a considerable amount of welding done on certain parts, especially on the fronts. The company used 10 or more spot welders. Oxy-acetylene welding, was used especially for back panels and sundry other places, and butt welding is employed for the center or the cowl, which is stamped in two pieces. Most of the welding and other metal work was done on the south side of the fifth floor. The door framing was also done on the side of the body-in-white department. Steel jigs were used for the door assembly.

Details of the body-frame assembly did not differ much from those at other plants where high-grade standard bodies were made but one or two features may be mentioned. The Marmon bodies were of the composite type but more steel was used than was customary with the old type of body. To secure good vision the fronts were entirely of steel except that there was a wood filler in the top member. The wood framing of the body was mostly of hard maple with a little oak and ash in special places. The joints were glued and screwed and where special strength was required steel braces were used. The glue was heated in an electric glue boiler and the glue was kept hot at the working places by small electric-heated glue pots. Anti-squeak or white lead was used in all joints. Soft roofs were used, some models having chicken wire and others roof slats. On those models that had roof slats a covering strip of thin steel binding was nailed to the front roof rail as an extra assurance against the roof slats becoming loose.

A spot welder was used in the assembly line for attaching to the steel front the hooded metal visor, which continues the roof line, and for certain other minor operations on the front. Most of the moldings were pressed into the panels when they were stamped but the upper molding around the back from rear-quarter window to rear-quarter window was an applied molding. The rear pillars of the standard closed bodies are all-steel, made in one piece except for the side brackets which are riveted and welded.

A bit of interesting equipment used at the Murray plant is the steel truck to which the body frame is bolted and to which it remains attached throughout all operations until the body is completed and ready for delivery to the Marmon company. This all-metal truck was of sufficient length to protect the bodies from bumping into each other while traveling on either ceiling or floor conveyors. A guide strip was placed in the floor so as to engage guides on the bottom of the transverse members. The trucks operated on conveyors most of the time during passage through the body plant but when necessary could be readily pushed about on the floor on four ball-bearing casters provided with 4-in. wheels. Inter-departmental movement of the trucks from floor to floor was by means of the lift-type elevators.

The painting department of the Indianapolis plant was unusual by reason of the special conveyor system which carried the body during air drying along the ceiling, thereby conserving floor space and increasing the drying efficiency by having the bodies in the warmest part of the room. This arrangement was a matter of practical necessity in order to economize space and provide the necessary capacity on a single floor. Another unusual, although not entirely novel, feature of the painting operations was the piping of the paint materials at the booths. The painting depart-

Fig. 11-27 View of the Murray body-in-paint department showing ceiling and floor conveyors. Note the working space provided by the use of the ceiling conveyor.

Fig. 11-28 Metal finishing line

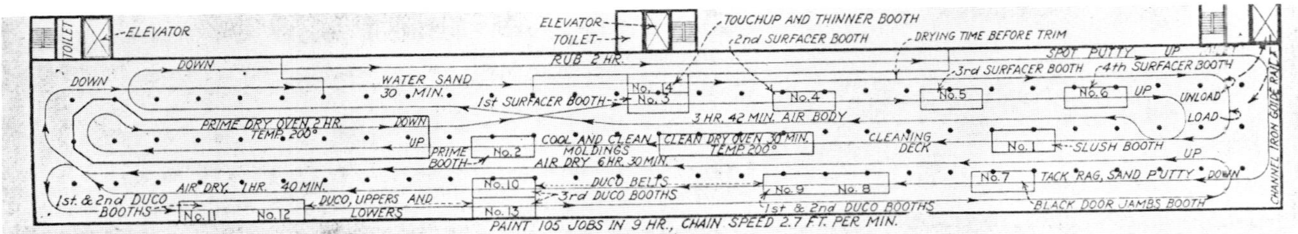


*Layout of the body-in-paint department at Indianapolis plant of the Murray Corporation of America. This layout and the operations shown were for the production of 105 jobs in nine hours with conveyor speed of 2.7 per min. The current schedule of operations is slightly changed, but the layout shows the travel of the interesting conveyor system which carries the bodies during drying time on an elevated chain near the ceiling. This arrangement not only increases the available floor space but also carries the bodies in the warmest part of the room during the air-drying period. In the drying oven the bodies are similarly carried at the top of the oven to take advantage of the greater heat in the upper portion. The paint department is not so congested as would appear from the drawing because the use of the ceiling conveyor leaves ample floor space for passageways and working places. The large dots represent building pillars at 20-ft. centers.*

Fig. 11-29 Layout of the body-in-paint department at Indianapolis plant of the Murray Corp. of America on the Marmon Plant 3 site.

Figure 11-30 Right view shows bodies-in-white entering the paint department and about to be suspended from the overhead conveyor. Left view shows a rumble seat roadster at the Dexodine cleaning deck. The hoods are suspended from the conveyor adjacent the body and both receive the same spraying and drying treatment. There is a floor guide rail to steady the bodies during work. Note bodies in both views traveling on the ceiling conveyor while air drying.

ment occupied the fourth story, approximately 16 ft. high. The conveyor system was about **4600 ft. long and was driven from a single shaft** so that the various lines were synchronized to operate at the predetermined speed, **2.7 ft. per min.** A good idea of the extraordinary effect of this conveyor system, which carries the bodies first on the floor and then up along the ceiling will be obtained from the accompanying illustrations.

All paint materials were delivered to the mixing room at the southeast corner of the building. The mixed materials were drawn from agitators which were continually filled so as to prevent variation in the color. The mixed lacquers were distributed to the various painting booths by hand truck in 2½ gal. and 12 gal. containers which replaced those already at the booth; any material remaining in the old containers was returned to the corresponding agitator. At the end of the shift all mixed lacquer was collected and returned to the agitators, so that there was no possibility of the materials settling out before the new shift started and thus producing a different shade. Another precaution that was taken to assure uniformity of shade on bonnet and body panels was to have them sprayed by the same gun with the same mixture. This was done by hanging the bonnet on the conveyor just ahead of the body. By this system both received the same spraying and drying treatment, and "matching troubles" were obviated.

The layout of the body-in-paint department, as originally planned, is shown in an accompanying line drawing. This layout was slightly altered in practice as the spraying of double pyroxylin coats in some booths reduced the number of booths required, but the general features of the layout remained the same. The most noteworthy feature of the body-in-paint department was the conveyor system to which reference has been made. This system was made necessary to secure the required working space for the desired capacity, but it had the special advantage of carrying the bodies during the air-drying period in the warmest part of the room — along the ceiling.

The spray booths varied in size, some being arranged for a single operation and others being staggered for a double operation, one operator spraying at one end and the other operator beginning as the body reached the second or staggered half of the booth. The sides of the booths were made of sheet steel up to the 4-ft. line, above which glass was set in the steel to provide light. These panes became coated as spraying proceeded and artificial light was also provided from the lower part of the booth by eight 100-watt lights set in pairs, with reflectors, in special sheet-metal boxes having one glass side flush with the inside of the booth. Although these panes also get coated and eventually require cleaning, they stay clean much longer than the upper glass sides of the booth, and the light bulbs themselves are better protected than when suspended openly. The booths have four pressure feed tanks, equipped for air agitation, and mixed lacquers or other paint materials are piped to both the near and far sides of the booths by lines, properly tagged, from each of the corresponding feed tanks. The air pressure on the feed tanks is 10 lb. and on the spray guns approximately 65 lb. per square inch.

Bodies in the white, mounted on the steel dolly trucks were received at the west end of the building and after being attached to the conveyor were given a general cleaning. Exposed wood parts were brushed with a waterproof pyroxylin wood sealer, and on arriving at the next deck were brushed with Deoxidine. They were sanded and wire-brushed with the acid still on them, and then flushed with hot water, dried with rags, and the corners and moldings blown out with compressed air. After being wiped again the bodies were passed through a drying oven at 200 deg. F. for 30 min., given a light sand and rubbed with wood alcohol.

An oil-base primer was sprayed on and the bodies then passed into the long drying oven at the east end of the plant; they were in this oven for 2 hr. at 200 deg. F. and upon emerging were spot puttied and glazed. After a 40-min. air dry, they were lightly sanded. In Booth 3 they received two double coats of a pyroxylin surfacer. Following a 10-min. interval, a double coat of surfacer was again sprayed and the conveyor took the bodies up to the ceiling for air drying. At the end of 4½ hr., they were lowered to the rubbing deck where they were water sanded with 220 and 280 Speed-grits. After reputtying and sanding the spots, the bodies were inspected and if approved were taken to the ceiling again for 3 hr. They were then tacked off and, if for a 3-tone job, a double coat was sprayed on the door jambs and on the belt and saddle panels; at the next booth a double coat was sprayed on the same parts. After a drying interval of 40 min., a masking coat was laid on the belt and saddle. Sanded, tacked off and then to Booth 10 for a double coat of black on the superstructure and of body color below the belt; following an interval of 10 min., these operations were repeated in Booth 11 (and again in Booth 12, if needed).

The bodies were taken to the ceiling again for 10 hr. As the bodies were being lowered at the eastern end of the building, dings were removed and, after an oil sand, the bodies were spot sprayed and, if the job required, window reveals were put in. The bodies were slushed inside and then given a final thinner coat and if passed by the inspector, they were sent to the trim shop on the floor below.

The trimming operations followed the usual course, the dolly trucks being moved at 1.4 ft. per min. by floor conveyors along two separate lines separated by a screened storage space. The trimming department was well equipped with electric cutting tables, pasting and sewing tables distributed along a 36-in. flat belt conveyor that carried the items along progressively to the appropriate sections where they were removed, finished and returned to the conveyor for delivery at the end of the building where they were inspected and taken to the proper storage. A glass department was in the northwest corner on this floor and there all glass was cut, polished, inspected, mounted in windshields or if intended for door windows was edged on a 6-unit edge grinder and polisher.

After trimming operations were completed the bodies were taken to the second floor, lightly sanded, spot sprayed, spot rubbed, polished, moldings striped and windows cleaned. A general re-conditioning of the bodies, where necessary, and a final inspection preceded their delivery to the "bridge" where the Marmon crane picked them up and delivered them to the floor of the adjacent Marmon plant where wiper motors and instruments were installed. Murray built for Hupp, Paige, Jordon and others in 1927.

## BODY BY MARMON AGAIN 1931

After one model year of Hayes ownership and operation of the Plant 3 body facility, it reverted to Marmon operation; whether the property ownership reverted is not known to the author. It would not be surprising if Hayes was unable to complete the purchase. Hayes continued to build the body shell.

*Men of Marmon* for April 1931 reported that some three months earlier, probably January 1, 1931, Marmon Motor Car Co. resumed operations in its own body shop, following a four year lapse during which completed bodies were purchased. Bodies were assembled (probably from stampings shipped from Grand Rapids) by Hayes Body Co. on the first floor of the five story building. Upon completion, the body-in-white (ready for painting, trimming and finishing) was inspected and accepted by Marmon on the first floor. It was then transferred to the fourth floor paint shop and placed on an overhead chain conveyor. The paint process description warrants full quotation: *First in the series of applications is the use of thinner*

Fig. 11-31 The rubbing deck is covered with duckboards; all water drains toward the inner edge where it is taken away by 6 inch drain pipes running just below the ceiling of the third floor. Primer coat drying oven is to the right. Air-dry conveyor at upper right.

*in cleaning the metal. This treatment is followed by a thorough acid cleaning and still again by a hot water bath before entering the drying oven. After emerging from the dryer, successive applications of a prime coat, a surfacer and a yellow guide coat are made. The latter brings out the high spots in the metal to expedite the efforts of the water sanders whose work follows after another drying process. Rigid inspection in connection with the water sanding operations insures an excellent condition on the surface of the metal before the lacquer is applied.*

*Still further processes are required, however, before the actual painting is permitted. A black guide coat is next sprayed onto the bodies to elucidate any defects yet undiscovered. Dinging and spot puttying correct any imperfections at this point, and a spot spray and sanding are the final operations before the several lacquer coats are sprayed on.*

*At this point bodies are oil sanded and inspected, followed by the final mist coats. With these operations completed, units are dispatched to the third floor for trimming.*

*Five groups compose the trim shop: the cutting department, pasting and sewing room, back assembly, cushion assembly, and the trim and upfit lines. The first four are auxiliary units, making the sub-assemblies to be installed on the trim line. Sewing and cutting occupy the southwest section of the third floor, back assembly is located in the southeast corner, and the trim and upfit lines extend the length of the building along the north wall. The stock room, which runs almost the length of the floor, is between the lines. All cloth and leather upholstery, inside and outside hard-*

Fig. 11-32 Trimming at the Murray plant. Model L, 1927.

Fig. 11-33 Body Paint and Trim by Marmom — Inspection after water sanding.

*ware fittings and other minor assemblies are installed in the trim shop before bodies are released to the finishing department on the floor below. Skillful trimmers and upfitters, working under expert supervision, are important factors in the maintenance of Marmon quality standards in automobile interiors.*

*From the trim shop, bodies pass down to the second floor for final finishing and polishing. The finishing line begins in the east end of the building and extends to the*

Fig. 11-34 Body Paint and Trim by Marmon — Oil sanding operation.

Fig. 11-35 Body Paint and Trim by Marmon — Sewing room.

Fig. 11-36 Body Paint and Trim by Marmon — Back assembly department.

The Marmon
FACTORY NEWS

me I—Issue 6 Published at INDIANAPOLIS, INDIANA March,

Fig. 11-37 Plant 2, 1932

Fig. 11-38 Milling block and two heads at one pass.

Fig. 11-39 Rough boring 4 cylinders per pass. Finish boring block one cylinder at a time.

*southwest corner where, following a very thorough inspection, bodies are placed on the ramp, conveyed to the loading dock and loaded on trailers for the assembly division in Plant 2.*

## ERA FIVE
## 1931-33

The SIXTEEN was assembled in Plant 2. Operations for 1932 were concentrated in Plant 2 and the entire Plant 1 and Plant 3 complex was idled, as it turned out, permanently. Undoubtedly the thousand or so Model 8-125s produced, beginning in March 1932, shared the V-16 assembly facility. Company offices were in the office structure at the corner of Drover and York, constructed at some time subsequent to World War 1, probably for Plant 2 Engineering and/or Plant 2 management.

For the first time since 1876, the 1101 W. Morris Street address was changed; changed to 1001 York.

## ERA FIVE MANUFACTURING

Tooling for the SIXTEEN engine was planned as much as possible to produce an engine set of parts per set of operations. For example, two cylinder heads and one cylinder block, all of nearly identical material, were surface milled at the same time. The joint surface was then lapped prior to drilling stud holes and push rod holes. The combustion chambers were rough and finish machined.[13]

The aluminum bronze valve seats were made with a .006 inch interference fit. They were installed by heating the cylinder head to 600 degrees F. on a cast iron plate 2.5 inches thick in a gas-fired furnace. The plate was removed with the head and acted as a heat source to maintain head temperature while the 16 valve seats were put in place.[13]

The cylinder liners (called sleeves then, possibly due to Knight engine influence) were saw cut to rough length from Shelby brand SAE 4615 steel tubing. After a rough bore . . . face one-end-chamfer operation, they were carburized but not quenched. The outer surface was turned and the flange formed, using multiple tools to reduce cycle time. Then the liners were hardened to 75-80 Scleroscope on the inside, with the OD remaining soft because the carbon rich surface had been removed.

The OD was then ground to .001, the bore rough ground, the OD semi-finished ground and a relief ground in the center. Next the flange OD was ground to .001 and the sharp corner under the shoulder relieved. The bore was then finish ground and, locating therefrom, the outside was finish ground.

The sleeves were measured and selected for best fit in the block. Pistons were also selected with the desired fit being an 8 pound pull on a 0.003 inch ribbon.[13]

The repeated plant rearrangements that characterized the Williams regime,

Fig. 11-40 Cutting grooves for packing rings.

Fig. 11-41 Push broach for shaft bearing surfaces.

Fig. 11-42 Fixture for drilling bolt holes in con rods.

Fig. 11-43 Plants 1, 2, 3, 1929.

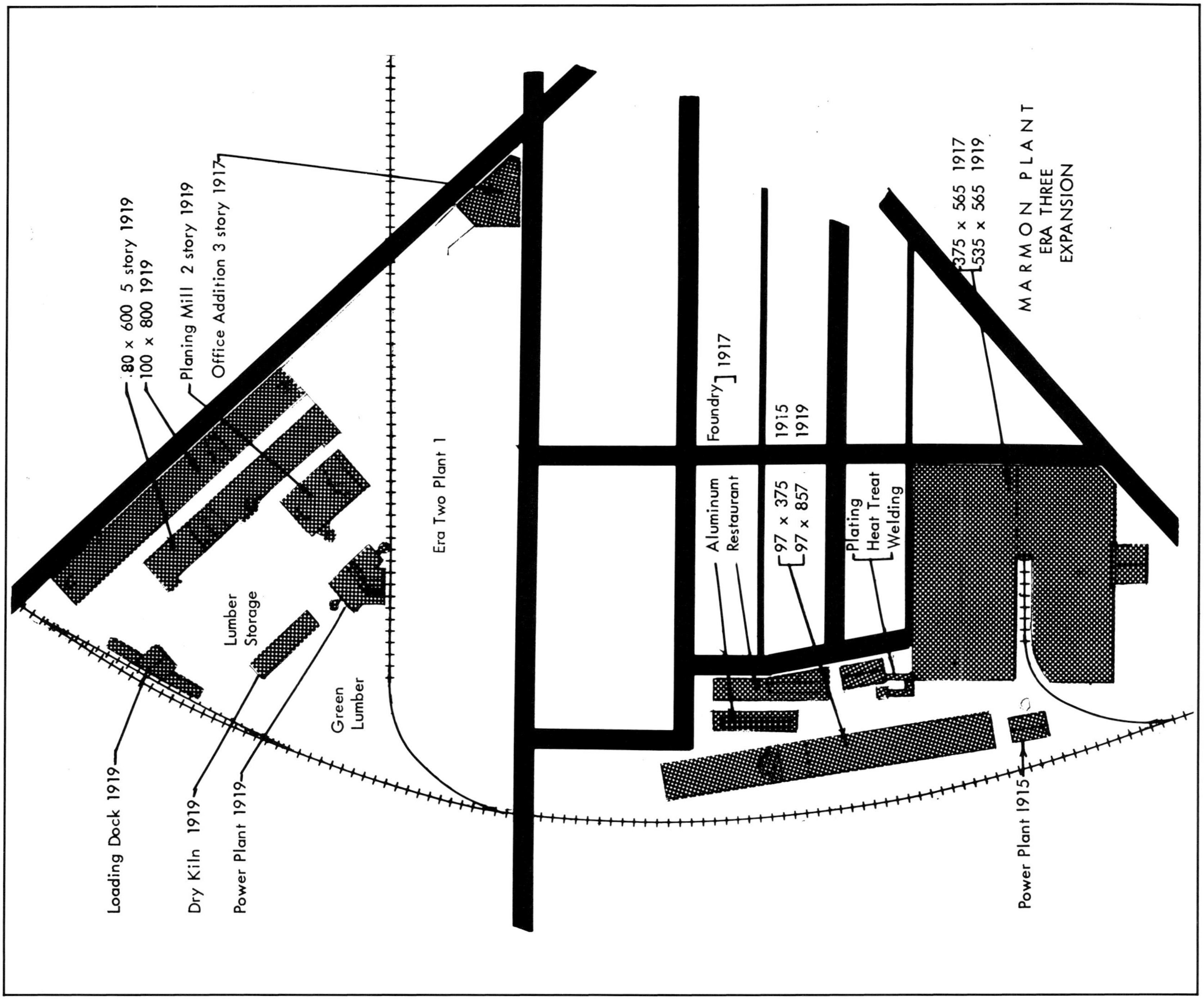
MARMON PLANT
ERA THREE
EXPANSION
375 x 565 1917
535 x 565 1919
80 x 600 5 story 1919
100 x 800 1919
Planing Mill 2 story 1919
Office Addition 3 story 1917
Foundry 1917
1915
1919
Aluminum
Restaurant
97 x 375
97 x 857
Plating
Heat Treat
Welding
Era Two Plant 1
Lumber
Storage
Green
Lumber
Loading Dock 1919
Dry Kiln 1919
Power Plant 1919
Power Plant 1915

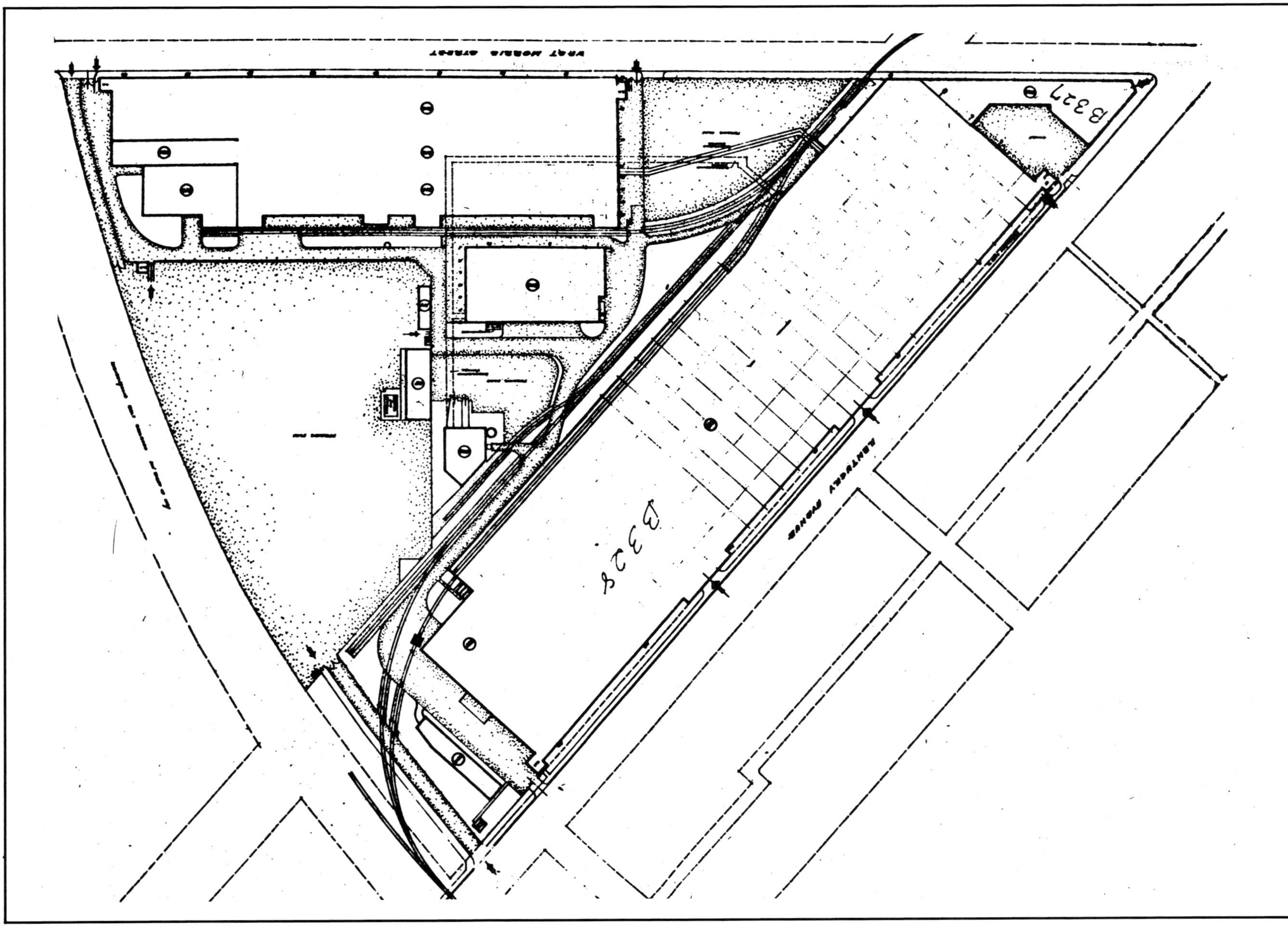

Fig. 11-43 Former Plant 1, 2 site plan circa 1945. Plant 1, except offices, replaced by Curtiss-Wright engine plant. Plant 2 nearly intact.

including Era Four, must have been a serious economic burden on Marmon Motor Car Co.

## THE SITE AND PLANTS TODAY

Today, the triangular site bounded by Kentucky and Morris Avenues, and the Belt Railroad is occupied by the Eli Lilly complex. The original Nordyke and Marmon facility (prior to 1915) has been completely replaced by modern structures fronting on Kentucky Avenue.

The major additions built in 1919 for Model 34B production survive today: The assembly building paralleling Morris Avenue, the five-story body plant behind it (Plant 3) and a third structure (original woodshop) behind that are today well maintained major components of the Lilly complex.

The Stewart Warner Corporation acquired Plant 2 in November 1937. Except for two frame structures removed at that time, the entire complex survives: the assembly building built for model 34A production in 1915 and later extended, the adjacent power plant and the Liberty engine building (still the largest in Indiana at the time of sale).

The property fronting on Morris and Kentucky Avenues, and paralleling the original Marmon complex, continues to be a facility of the Indianapolis Power and Light Co. (IPALCO). This location reflects the Marmon Parentage of IPALCO.

The Marmon properties embracing Plants 1 and 3 were acquired by the Stokely Company in 1938. The Plant 1 site was acquired by Curtiss-Wright Corporation in 1940 and the Marmon facilities razed, with the exception of the office building (at Kentucky and Morris). A large new plant structure B328 encompassing the entire remaining site was erected for the manufacture of military aircraft engines; it abutted the Marmon office structure. Following the end of hostilities, Eli Lilly and Company purchased the C-W plant in February 1946. The following May, Lilly also purchased the Plant 3 site from Stokely who had previously leased it to Curtiss-Wright. The Marmon office building (Structure B327) survived until 1972.

By 1952 Lilly had removed two small structures between the plant proper and the power plant and replaced them with three much larger structures and two small ones distributed around the available site; a large new building, structure 315, (approx 100x225) was erected on Morris Street abutting the Marmon body and assembly buildings on the east.

By 1977 a large structure adjoining the power plant had replaced one of the larger structures erected before 1952 and two others had been erected along the Belt Railway.

Panoramic views of the Marmon plant circa 1929 showed two multi-story buildings not referenced otherwise. One was directly east of, and adjoining, the body and assembly structures on Morris Street; the site was vacant in 1945. The other was on the southwest corner of Kentucky and Stout Streets, extending from Stout to the Belt Railroad, and from Kentucky to the Plant 2 long assembly building which paralleled the tracks. The author has concluded that these may have been planned but never constructed.

The 'Marmon Organization' progressively transitioned to the Marmon-Herrington Company over the following decade. Therefore the story of that organization is told in two parts — in the following Chapter Twelve and in Chapter Six, Part 1 of Book Two 'Organization (and Financial History)' of Marmon-Herrington.

CHAPTER TWELVE

# THE MARMON ORGANIZATION

## PART 1 — Operating Management

"Starting with the work of their own hands, Ellis Nordyke and his son Addison added other help gradually as the business warranted . . . Each year since 1851, new blood, the best that can be had, has been transfused into the institution. No distance is too far to reach for a man, whether he be scientist or painter, engineer or accountant, mechanic or clerk, that will enable the great factory to better serve its patrons. These specialists, coming each year from the colleges and practical training schools of the **world,** with few exceptions elect to serve long periods of time, often devoting their entire lives."[1]

The Marmon business organization will be reviewed as it was during the Nineteenth Century, and separately as the Marmon and Marmon-Herrington firms in the Twentieth Century. The first two will be addressed in this chapter but Marmon-Herrington will be addressed in Book Two, Chapter Six, Part 1. Emphasis in this chapter will be on the organization during the Marmon car period, but the periods 1851-1901 and 1902-1933 will be successively addressed because any distinction is product, not organization, oriented.

Like the preceding product treatment, the organization treatment will be divided into corresponding eras. However, the Organization Eras will be displaced backward in time from the Product Eras for which they were responsible. In addition, some of the Organization Eras will be subdivided to reflect major non-product transitions, e.g., financial crises.

Upper management . . . corporate officers and the board of directors . . . will be discussed and tabulated first, followed by the engineering and sales departments, and then by supporting groups such as advertising.

The first addition to the firm of E. and A.H. Nordyke was young Daniel W. Marmon in 1866; there would be no event of greater significance in the history of the firm. He had graduated from Earlham College in Richmond the prior year, but during his years at Earlham had been a regular visitor to the Nordyke shops. His curiosity and evident aptitude brought him the invitation to partnership. His own meager savings, combined with financial assistance proffered by his uncle (Eli Stubbs who had raised his orphaned nephew), bought his interest in the firm which became Nordyke, Marmon and Company.

The move to Indianapolis in 1876 marked the start of a more formal organization. A.H. Nordyke was president from then until 1898 when he retired, remaining a director until his death in 1904. D.W. Marmon was secretary to 1898 succeeding to the presidency and remaining in that office until his death in May 1909. Amos K. Hollowell sought a better investment than his general store in Fountain City, Indiana (then Newport) and, after a brief period in Indianapolis with a wholesale hardware firm, joined Nordyke & Marmon Co. at Richmond in 1875. Mr. Hollowell was employed to care for the financial end of the business, on one years probation because Mr. Marmon said, "We don't want any disquieting influences and Ad (A.H. Nordyke) and I are a little afraid that another member in the firm might cause friction, something that we have never had in this place since we started."[1] In March 1876, Mr. Marmon said, "Well, A.K., Ad and I are satisfied if you are. Your term of probation is over." Mr. Hollowell bought his stock and became treasurer of the Nordyke & Marmon Co.

Mr. Marmon and Mr. Nordyke used high desks backed against each other and that of Mr. Hollowell was grouped intimately so that the discussion of daily problems was an easy matter. Mr. Hollowell quipped "We had a meeting of the directors a dozen times each day."

Mr. Marmon looked after factory management, Mr. Nordyke handled purchasing and business correspondence while Mr. Hollowell attended to the financial problems and letter writing.

In March 1880 R.E. (Bob) Roberts, a youth of 22, joined the firm and for many years was in charge of the mill machine room . . . the pride of D.W. Marmon . . . and in 1902 supervised the manufacture of the first experimental car. *Men of Marmon* for April 1931 noted that "Uncle Bob" Roberts had reached the age of 74 and marked his fifty-first year with Marmon. He was reported to be quite active and "looking forward to many more pleasant years with the organization."

In 1881 Miss Mamie Kelly joined the 250 Nordyke & Marmon employees and for 42 years directed the department with critical responsibility for cutting the bolting silk and selecting the appropriate grades for specific types of milling machines.

There were no commercial stenographers in Indianapolis and the court stenographer was the only individual with any knowledge of shorthand. Mr. Marmon noted an advertisement by an eastern stenographic school and had Mr. Hollowell telegraph the school to send a stenographer at once. It was Miss Hattie Clock who married two years later and was replaced by Miss Hattie Sperry from the same school.

In 1886 H.H. Rice, just nineteen, was employed as telegraph operator and bill clerk. By then there were 14 office employees, four of them exclusively involved with the Jenney Electric Co. (a separate company within Nordyke & Marmon). Mr. Rice would become a director in 1916 and resign in 1926.

The concurrent Jenney Electric Company officers (beginning in 1885) were D.W. Marmon, President; A.H. Nordyke, Vice-President; A.K. Hollowell, treasurer; Mr. Brainard Rorison, secretary and Mr. Charles Jenney, chief electrician. They retained these positions until the sale of the company to the General Electric Co.

In 1897 Walter Marmon joined the firm and was followed by his brother Howard in 1899. In 1898, Mr. Nordyke retired and Daniel Marmon became president. Howard then assumed the product engineering functions previously attended by his father. Nordyke remained a director until his death in 1904.

And so the stage was set for the motor car epoch! From a management perspective, the Eras are displaced backward from the product eras they fostered:

| | | Management | Product |
|---|---|---|---|
| First | Era | 1898-1908 | 1902-1908 |
| Second | Era | 1908-1914 | 1909-1915 |
| Third | Era | 1915-1923 | 1916-1927 |

| | | | |
|---|---|---|---|
| Fourth | Era | 1924-1931 | 1927-1932 |
| Fifth | Era | 1932-1933 | 1932-1933 |

The Nordyke and Marmon families and their origins are discussed in Chapter One, Part 2.

## FIRST ERA
## 1898-1908

The organization that would bring Nordyke & Marmon into the automobile business was in place in 1898. It was an all-Marmon team, Mr. Nordyke having relinquished the presidency that year; he retained his directorship and died the year of the first Marmon car production. Although the new product would carry the single name, the stature of the badge's reputation warranted retention of the double name for another quarter century.

The team of Daniel Marmon, President, and A.K. Hollowell, Secretary, Walter and Howard Marmon was joined by Harvey G. Shafer in 1899 and by Leslie L. Zeigler in 1900. (Shafer and Zeigler would become vice-presidents by 1920 and 1925 respectively). A.R. Heiskell came aboard in 1905 to complete the team that ran Nordyke & Marmon in the First Era. (He would be a vice-president by 1925).

The illustrated letter, announcing a stockholders' meeting on January 17, 1899 over D.W. Marmon's signature, attests by virtue of Mr. Marmon's notation of copy distribution, to all of the stock being held by himself, Mr. Nordyke and Mr. Hollowell. A companion letter announcing a meeting of the Board of Directors on the same date confirms the directorate to be limited to D.W. and W.C. Marmon and A.H. Nordyke.

As of Jan. 17, 1899, the officers thereafter were to be "President, Secretary and a Treasurer." Previously there had been only President and Secretary.

The author is indebted to the unknown author (s) of the Marmon Book of Facts for the assemblage and preservation of facts and personalities relating to the firm's management, particularly so in the preceding century. The remainder of this First Era Organization section is directly quoted from the Book of Facts to foster extended preservation. Note that in 1929 Marmon had 25 employees with more than 30 years of service, and an esteemed female department head.

Fig. 12-1 Early office employees

View of the machine shop about 1890. R. E. Roberts is shown standing at the end of the building

Fig. 12-2 View of machine shop circa 1890; R. E. Roberts standing in rear.

In 1899 Mr. Harvey G. Shafer was employed as a foreman millwright. This was early in the history of the high roller mills. After two years as a traveling technical expert, he entered the factory to begin a long series of foremanships, which led him into practically every factory department. In quick succession he was placed in charge of buhr stone shops, bench department, planning mills, shipping department, gradually assuming the responsibility for plant maintenance and general factory upkeep.

Mr. Shafer is a native of Indianapolis. He was born July 24, 1869, and received his education in this city. (He would become factory manager and secretary of the Nordyke & Marmon Company in the spring of 1917.)

Mr. A.R. Heiskell, Purchasing Agent, was born in Indianapolis December 27, 1872, and received his education in the public schools of his home city. At the age of thirteen, during summer vacations, he worked for what was then the McCormick Harvester Machine Company of Chicago, and what is now known as the International Harvester Company. Starting as an office boy, he was eventually appointed manager of their branch house in Atlanta, Georgia, from which point he directed the sale of International Harvester products in the states of Georgia and Florida.

In 1902 Mr. Heiskell returned to Indianapolis as collection manager for the district, a position he held up until 1905, when he resigned to become purchasing agent for the Nordyke & Marmon Company. He continued in this capacity until 1917, when he was given a year's leave of absence on account of ill health. (He would become a director, vice-president in 1920).

Leslie I. Ziegler, manager of the milling machinery department, was born in Sterling, Ohio, September 22, 1864. Prior to joining the Nordyke & Marmon Company in 1909 he was associated with the Bradford Mill Company, Cincinnati, Ohio. Mr. Siegler is nationally known as one of the country's greatest engineers in the milling machinery industry. Under his supervision the Nordyke & Marmon Company has designed and built many of the largest mills of the world. Among many may be mentioned a two-thousand-nine-hundred-barrel mill for Valier & Spies Milling Company, St. Louis, Missouri; The Crown Mills, with a capacity of three thousand barrels, for Balfour, Guthrie & Company, Portland, Oregon; Texas Star Flour Mills at Galveston, Texas, with a three-thousand-barrel capacity; a thirty-six-hundred-barrel mill for the Sperry Flour Company, Vallejo, California; an eight-hundred-barrel for S.A. Grandes Moinhos Do Brazil, Pernambuco, Brazil, S.A.; a three thousand-barrel mill for the Dairi Flour Mills, Dairi, Japan; a six-thousand-barrel mill for the Dominion Flour Mills, Ltd., Montreal, Quebec, Canada; a six-thousand-barrel mill for the Larabee Flour Mills Corporation, St. Joseph, Mo., and the Fisher Flouring Mills Company, Seattle, Washington, with a capacity of five thousand barrels. (He would become a vice-president, assistant secretary/treasurer).

The millers of the United States, Japan, China, Belgium, Canada and South American countries laid their problems before Mr. Ziegler and his department.

In 1873 the Nordyke & Marmon Company employed S.S. Griffith as a foreman for their cutting room. Today this man buys every foot of lumber that is used in the great institution; in fact, he has done so since he was first employed. In spite of his age, he is considered one of the keenest judges of lumber in the United States. Forty-six years of faithful and loyal service! Indeed, no record of Nordyke & Marmon Company's achievements would be complete if it did not pay a tribute to this man. November 21, 1844, Smith S. Griffith was born near Winchester, Ky., moving to Indiana when a small boy and settling near Richmond. Shortly before he was employed by the company he was to serve so well, the Nordyke & Marmon Company had completed a large contract for a prominent Texas miller. So pleased was the customer that he came to Richmond to banquet the company's employees. "This event," Mr. Griffith often comments, "made a great impression upon me. I had a great desire to be employed by a company so well known and one that was held in such high esteem by its customers. At that time the company had about fifty employees and was really a great industry, judged by the size of manufactories in those days."

Thus began the career of a man who is but one of many long in the employ of the company they have faithfully served. Twenty-five are today playing their parts in the success of Nordyke & Marmon Company who have devoted over thirty years of their lives to the institution.

West Indianapolis presented transportation difficulties for the employees, and travel to and from work was accomplished by the best means at the disposal of the individuals. At first the Vandalia Railroad added a train that stopped at the factory and made frequent runs to accommodate the employees and visitors. Bus lines and carriages were pressed into service connecting the city with the Nordyke & Marmon Company factory, a practice that was continued until the advent of mule street cars.

In the mill-building business there is no more important department than that in which the bolting cloth is cut and prepared for the machines. The slightest error in the assignment of the various grades of bolting silk for different types of machines would prove costly. And the selection of this material is made more difficult by the similarity of appearance in the different grades of silk; they can only be distinguished by the use of a magnifying glass. For thirty-eight years this department of the milling machinery division has been under the care of Miss Mamie Kelley. When she first took up her work, in 1881, she was but a girl in the twenties. At that time, there were about two hundred and fifty employees. Both Mr. Marmon and Mr. Nordyke then drove to their work in two-wheeled gigs and Miss Kelly smiles as she recalls the days when the few women employed at the plant were invited to ride to the city in the gigs driven by their employers.

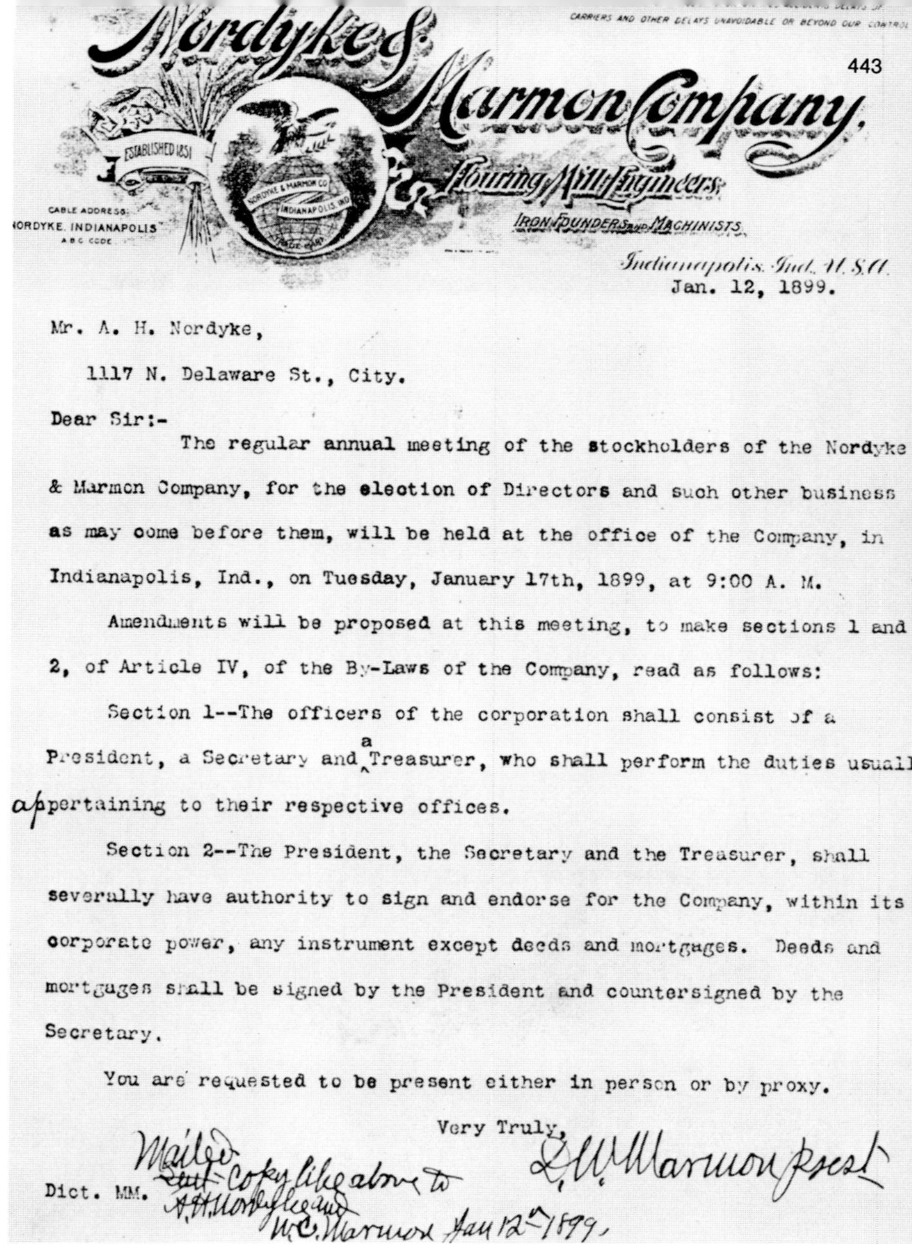

Nordyke & Marmon Company
Flouring Mill Engineers
Iron Founders and Machinists
Established 1851
Cable Address: Nordyke, Indianapolis. A B C Code
Carriers and other delays unavoidable or beyond our control

Indianapolis, Ind. U.S.A.
Jan. 12, 1899.

Mr. A. H. Nordyke,

1117 N. Delaware St., City.

Dear Sir:-

The regular annual meeting of the stockholders of the Nordyke & Marmon Company, for the election of Directors and such other business as may come before them, will be held at the office of the Company, in Indianapolis, Ind., on Tuesday, January 17th, 1899, at 9:00 A. M.

Amendments will be proposed at this meeting, to make sections 1 and 2, of Article IV, of the By-Laws of the Company, read as follows:

Section 1--The officers of the corporation shall consist of a President, a Secretary and a Treasurer, who shall perform the duties usually appertaining to their respective offices.

Section 2--The President, the Secretary and the Treasurer, shall severally have authority to sign and endorse for the Company, within its corporate power, any instrument except deeds and mortgages. Deeds and mortgages shall be signed by the President and countersigned by the Secretary.

You are requested to be present either in person or by proxy.

Very Truly,
D.W. Marmon Prest

Dict. MM.
Mailed copy like above to A.H. Nordyke and W.C. Marmon Jan 12th 1899.

Fig. 12-3 Announcement of the annual meeting of stockholders, 1899.

Fig. 12-4 Pioneer employees in 1929.

## SECOND ERA
## 1909-1914

When Daniel Marmon died in 1909 there was a realignment. Walter Marmon became president; Howard Marmon (joined 1899) became secretary and C.C. Hanch treasurer. The directors were Daniel's wife and heir Elizabeth Carpenter Marmon, Walter and Howard. This organization continued through the Era but was significantly reinforced by the addition of Fred E. Moskovics in 1913 to direct automotive sales and marketing and the entering of a consultant agreement with Alanson P. Brush of Detroit who had earlier designed and manufactured the Brush car and was well respected as an innovative and competent design engineer. W.K. Swigert was the General Factory Superintendent.

About 1913 the widow upon receiving her inheritance elected to share it with her three children.

During 1913 Mr. F.E. Moskovics made a substantial purchase of stock and was elected a director. He and Mr. Heiskell were then the only non-family holders of record; however the board now numbered six directors. Mr. Moskovics was born in Budapest, Hungary in 1879, coming to the United States in 1883. He became affiliated with Nordyke & Marmon in a consulting capacity in July 1912 and made a permanent connection in January 1913. He possessed a combination of engineering skill and commercial talent.

The elevation of A.R. Heiskell (1905) and C.C. Hanch to directorships by 1913 raised the Board to five members, but with the family retaining control.

Nordyke & Marmon Company bought membership in the Motor Car Manufacturers Association in 1907 under threat of litigation by the Selden Patent holders but dropped the affiliation when the Court of Appeals reversed the lower court and declared the patent invalid. The membership was only publicly acknowledged in 1911 advertisements just prior to the termination.

## THIRD ERA
## 1915-1923

Although Daniel Marmon's widow lived to age 90, she evidently relinquished her directorship to her daughter, Caroline Marmon Fesler, at age 70 in 1920 or earlier.

The Third Era was one of great expansion which saw advancement of members of the 'Old Guard' and the advent of many newcomers.

Among the 'Old Guard', H.H. Rice (1886) became a director about 1916 and Sales Manager in 1918, A.R. Heiskell (1905), already a director, became a vice-president about 1920, as did H.G. Shafer. (1897).

F.E. Moskovics who had acquired a large block of stock in 1913 was elected a vice-president and General Manager about 1918. There were now four vice-presidents and five directors.

The next significant change came in 1916 when C.C. Hanch who had been closely associated with the establishment of installment buying for automobiles suddenly went to another car company. A.R. Heiskell who had been Purchasing Agent for some years, assumed the additional post of treasurer and H.H. Rice (1886) took Hanch's seat on the board. The following year Howard Marmon temporarily relinquished his responsibilities to accept a military assignment and there were several other personnel changes, the direct result being promotion of H.G. Shafer (joined 1899), who had been factory superintendent, to the position of secretary. D.S. Ritter became purchasing agent relieving Heiskell of temporary dual responsibility; H.L. Purdy who had assisted Heiskell in purchasing became assistant treasurer. In 1919 E.C. Badger joined Nordyke & Marmon; two years later he was Controller. Mr. Badger was probably the first of the "outsiders" to be brought aboard by financial interests concerned about rates of return on their wartime expansion loans. The following year Heiskell and Shafer were elected vice-presidents while retaining the offices of treasurer and secretary until 1922

Fig. 12-5 The first Marmon photographed in 1914. Walter Marmon is the driver; with him is H. H. Rice (Sales Mgr.). In the tonneau H. G. Shafer (Factory Supt.), C. C. Hanch (Treasurer) and F. E. Moskovics (Commercial Mgr.).

when Hal Purdy (1919) moved up from assistant treasurer and H.H. Price became secretary; Heiskell was no longer a director.

## FOURTH ERA

In late 1923 Fred Moskovics resigned and Col. Edgar S. Gorrell took over as sales vice-president; there were now four vice-presidents and still only five directors. In the spring of 1924 the directors named an outsider, G.M. Williams, as president; Walter Marmon became the first chairman of the board. It appears probable that Williams acquired Moskovic's interest in the company in the interim. With Williams came C.J. Sherer (both from General Motors' Dayton Wright Co. subsidiary) to be vice-president and treasurer, relieving Purdy of that responsibility. C.E. Jeffers resigned as chief engineer and Howard Marmon assumed that position while continuing to be vice-president of engineering. Williams replaced Caroline Marmon Fesler as a director; the others were the Marmon brothers, H.H. Rice and A.R. Heiskell.

Near the end of 1925 L.I. Ziegler (1900) was elected a vice-president and assistant treasurer/assistant secretary. This marked the last advancement of what the author elects to call the 'Old Guard,' who together with the family members had brought Nordyke and Marmon to its pinnacle.

On Feb. 2, 1926 the car and milling machinery businesses were separated and the former became the Marmon Motor Car Company; Col. Gorrell (1919) and H.H. Rice (1886) resigned.

The director's seat of H.H. Rice went to Hal Purdy who also became Chairman of the Executive Committee and General Manager. Homer McKee, president of Homer McKee Co. . . . who held the Marmon advertising account . . . was elected a vice-president and director. Two other outsiders were elected directors . . . Elmer

Fig. 12-6 Executive Committee — 1920

W. Stout of Indianapolis and George A. Ellis of 59 Broadway, NYC.

After all these developments, the organization in 1927, less than a year after the first Fourth Era model went into production, was:

| | | |
|---|---|---|
| W.C. Marmon | 1897 | Chairman of the Board |
| G.M. Williams | 1924 | President and Director |
| C.J. Sherer | 1924 | Treasurer (VP and Director Feb. 1930) |
| A.R. Heiskell | 1905 | Secretary |
| H.L. Purdy | 1918 | General Mgr. and Director |
| H.C. Marmon | 1899 | Vice-President and Director |
| H. McKee | | Vice-President and Director |
| E.C. Badger | 1919 | Asst. Treasurer / Asst. Secretary |
| E.W. Stout | | Director |
| G.A. Ellis | | Director |

The directors had been increased to seven to accommodate Stout and Ellis, representing respectively Indianapolis and New York financial interests.

T.J. Litle Jr. was Chief Engineer
G.H. Freers was Asst. Chief Engineer
P.E. Chamberlain was Sales Manager
H.H. Brooks was Asst. Sales Manager (Sales Manager 1929)
H.H. Rice directed the Export Division

Two years later, with Marmon stock already anticipating the October 1929 stock market crash, Fred B. Johnson and Maurice Newton (of 44 Pine, NYC) were elected directors, increasing the board to nine members, of whom four represented financial interests.

The year 1930 saw the last advancement of an 'Old Guard' member: E.C. Badger (1919) was promoted, from assistant secretary and assistant treasurer, to Vice-president of Purchasing. His former duties were inherited by Helen Ray and J.M. Petersen. The latter shared the assistant treasurer title with A.P. Smith who replaced G.A. Ellis as a director.

Also in mid 1930, P.H. Nolan came aboard to replace H.L. Purdy, as Vice-president and General-Manager, who had resigned earlier in the year.

1932 saw a new generation of financial-interest representatives with their member increased by two chief executives of major supplier-creditors. Fred Barnichol of Indianapolis Drop Forge Co. and W.W. Hoagland, President of the Hayes Body Co., who became directors, taking the seats of departed A.P. Smith and F.B. Johnson. H.H. Davidson representing the First National Bank of Detroit replaced director Maurice Newton of NYC.

Paul A. Crouch and D.A. Powell came aboard as 'Limited' vice-presidents representing the interests of Bankers Trust Co. of New York City.

Also in 1932 P.H. Nolan (1930) Vice-President and General Manager, became Secretary of the Company, probably signaling the departure of A.R. Heiskell (1905).

Bankruptcy came on May 1, 1933 and all officers resigned. The administration of the firm following bankruptcy is addressed in Chapter Thirteen 'Financial History'.

## ENGINEERING

Prior to 1913, about all that is recorded about engineering staff members other than Howard Marmon is that George Freers entered the organization as a draftsman-designer in 1910 (or 11). In 1913 Alanson P. Brush, who had designed the Brush economy car came to Marmon first as a consultant and then as employee. He brought with him Owen Nacker a gifted designer later to be credited with a major contribution to the Cadillac V-16. In early 1913 Roy Milner, later to be one of the foursome which Alfred Sloan would send to revitalize Oakland and be Chief Body Engineer at Pontiac Motor Division for 25 years, joined Marmon as a draftsman and became responsible for coachbuilder liaison. This then was the engineering department that had designed and developed the Marmon car up through the Model 34 A.

Fig. 12-6 H. H. Rice

Fig. 12-7 Executive Committee 1928

Fig. 12-8 Executive Committee 1929

Howard Marmon's acceptance of the call to military service in 1917 brought in a new engineering crew. C.E. Jeffers directed engineering with the title of Chief Engineer while George Freers was in charge of the design room. David E. Gallup, formerly a professor at the Worcester Polytechnic Institute became principal consultant.

It was 1918 when Jerome (Joe) Felts graduated from high school and was hired by his neighbor, George Freers, as a draftsman.

D.E. Gallup resigned at the end of 1922 and Jeffers followed, resigning in August 1924. Howard Marmon had resumed the vice-presidency in 1919 but left the day to day operation of the engineering department to Jeffers. The resignation, evidently unexpected and coming only months after G.M. Williams' arrival, necessitated Howard assuming the dual roles of vice-president and chief engineer and resuming active management of engineering while a search was undertaken for a new chief. Jeffer's departure came just before the start of production of the Model 74; he had been responsible for both the Model 34 B and C and the Model 74. In January 1925 George Freers was elevated to Assistant Chief Engineer reporting to Howard Marmon. In late summer Delmar G. (Barney) Roos who had been chief engineer at Locomobile became Chief Engineer and Mr. Turyman became a consultant. Designers included Paul Bastien and Al Leamy. This was the team that would engineer the Little Marmon; Barney Roos resigned in November 1925 about the time that tooling would have been released and prototype durability testing begun.

Fig. 12-10 C. E. Jeffers

Fig. 12-12 Thomas J. Litle, Jr.

Fig. 12-11 George H. Freers.

T.J. Litle, Jr. (pronounced 'Lie-Tell') was chief engineer of Lincoln in the fall of 1926 when he was singled out to be the 1927 President of the Society of Automotive Engineers, a most distinguished honor then and now. This development was reportedly greeted by Mr. Ford with an ultimatum "I understand you have two jobs . . . which one do you intend to keep?" or words to that effect. The Marmon brothers had a much more professional perspective . . . Litle became Chief Engineer of Marmon, replacing Roos, following a brief interim post as Vice-President of Copeland Products — Detroit, then and now a refrigeration equipment manufacturer.

One day in 1927 when Howard Marmon and George Freers made a routine visit to the "Midwest Aircraft Corporation" to check on progress, they brought along the new chief engineer, Thomas Litle. Joe Felts recalled that when he was again working in the design room at the Marmon plant, Mr. Litle was seldom seen there. He evidently was an advocate of delegating work assignments from his office base — he would fit today's popular concept of engineering management.[2]

In early 1927 E.C. Blaine was appointed chief body engineer and Ralph Chestnut was made head of the Experimental Test Department. He had been chief engineer of the Indianapolis plant of the Murray Corp.

George H. Freers dutifully continued as Asst. Chief Engineer under Litle as well as Roos, but when Litle resigned in October 1929 'to undertake a consulting practice,' Freers became Chief Engineer. S.E. Rowe came from Stutz to fill the Assistant Chief Engineer slot. The Fifth Era engineering organization that would produce the MARMON SIXTEEN was in place; in fact, the Howard Marmon / George Freers' team would go on to produce the prototype HCM . . . the last Marmon.

## SALES ADMINISTRATION

The earliest recorded sales staffing was the employment of Frederick E. Moskovics as Commercial Manager in 1911, early in the Second Era. A.J. Rogers, working in the Advertising Department in 1914, became assistant (sales manager) to Moskovics in 1915. Moskovics was named a Vice-President (in charge of sales) in 1917. In mid 1922 H.L. Peterson[1] replaced Rogers, working first for Moskovics until his resignation in late 1923 and then for Col. Edgar S. Gorrell until Rogers brief return to Nordyke & Marmon for 1925; Gorrell and Rogers resigned when the Marmon Motor Car Co. was created on Feb. 2, 1926 and a major reorganization followed — Purdy became a Vice-President and Assistant General Manager — ultimately (1928) General Manager. During the last years of Nordyke & Marmon in the automobile business, H.H. Rice (1886) directed the Export Division.

The Marmon Motor Car Co. began with new faces in the sales department: P.E.

Fig. 12-9 H. H. Brooks

Chamberlain was Sales Manager assisted by H.H. Brooks (1916). Chamberlain departed in mid 1927 and was replaced by Brooks pending the recruitment of John Tainsh in Dec. 1927, whereupon Brooks was elevated to the new post of General Sales Director; When Brooks retired at the end of 1928, T.E. Jarrard replaced him.

One year later the post of assistant General Sales Director was created and John Tainsh moved up to it, being replaced as Sales Manager by S.V. Harding who in late 1924 had become Manager of the Indianapolis Marmon branch. Nicholas Moore became successively Assistant Sales Manager and Advertising Director. A greatly enlarged sales organization was in place just as the market evaporated; Six months later, Jarrard departed and A.J. Rogers, then Assistant Vice-president, filled the post for the latter half of 1930 until George C. Tenney (1924) took over in June 1930. This was the final sales organization.

### MANUFACTURING

General factory Superintendent W.K. Swigirt was replaced by John W. Anderson in mid 1916. In late 1924 Robert K. Mitchell became Production Superintendent, being replaced by Anderson at the end of 1929 with the new title of Works Manager; C.A. Dalton assumed the new post of Master Mechanic (he had been

Fig. 12-8 Factory executives, 1929. Left to Right, First Row: Frank Castle, R. B. Little, D. R. Meo, Ted Collier. Second Row: Dan Schidler, Max Thoms, C. O. Peck, E. L. Roberts, Frank Updyke, Mark F. Mathews. Third Row: George Freers, W. F. McGlasson, William Marshall, L. M. Engle. Fourth Row: W. D. Guyton, J. W. Anderson, R. S. Wiegand, E. C. Blaine, E. C. Sudhoff.

mechanical supt., a position in which he was preceded by Leroy E. Jolls.).

## PART 2 — Distinguished Alumni

The Marmon engineering department was a highly professional activity decades before work began on the first Marmon Car. Advertisements for the first Marmon model claimed Nordyke & Marmon to be "One of the World's Greatest Engineering Houses." The famous Model 34 was advertised as "A Scientifically Designed Motor Car."

When the editor of "Motor" visited the offices of the major American motor car manufacturers to interview their chief executive officers in 1911, he had an unusual experience at Nordyke and Marmon. He reported meeting an unusually high percentage of men whom he assumed, from their conversations and bearing, were college graduates. This according to Mr. W.A.P. John was a situation quite different from anything he had encountered elsewhere in the industry. Mr. John mentioned his observation to Mr. Walter Marmon and inquired whether it was accidental or intentional?

"Oh, it's intentional," Mr. Marmon explained, "Although we don't, of course, make a college education one of the requirements for a position in the organization. But most of the executives are college men . . . father (Daniel) was an Earlham graduate, you know . . . and taking it all in all, we found that the college trained man after he has absorbed the hard knocks, turns out to be more valuable to us. He's less of an individualist and more of a cooperator. And cooperation is the fundamental of success in any large institution."

Mr. Marmon continued **"We have an executive school here and I believe that every man in the particular class (now) going through it this time is a college man."** It is truly remarkable that an automobile manufacturer was conducting executive training classes in 1911!

The Marmon organization and its executive school was a training ground for many famous automotive industry personalities who left their mark on the industry at many renowned manufacturers in subsequent years . . . they are the Marmon Alumni and this chapter is devoted to their accomplishments at Marmon and elsewhere . . . it is a Roster of Marmon Alumni.

Let the Roster begin with George Freers because of the longevity of his career spent entirely with the "Marmon Organization."

GEORGE FREERS — Marmon to Vice-President Marmon-Herrington Co.

George Freers was born in Terre Haute, Indiana, January 21, 1886 and was a 1908 graduate of Rose Polytech. He joined Nordyke & Marmon Co. in 1912, as a draftsman. He rose through the design organization to Chief Draftsman and was made Chief Engineer in 1929. He directed the design of the Marmon Sixteen engine under the guidance of Howard Marmon; when Marmon returned to Indianapolis, after receiving the prestigious award from the Metropolitan Section of the Society of Automotive Engineers in recognition of his design of the Sixteen engine, he is reported to have tossed the medal on George Freer's desk with the comment "This should have been yours".[2] Freers so respected and regarded his boss that he named his son 'Howard' in his honor. When the Marmon plant was closed by bankruptcy proceedings, Freers went to work for Howard Marmon personally and directed the design of the V-12 HCM prototype, sharing many of the related patents with Marmon and accompanying him on travels seeking a buyer for the design when it was evident Marmon Motor Car Co. was finished.

When Howard Marmon despaired of finding a buyer to produce the V-12, Freers joined Stutz and by 1938 was Chief Engineer when Stutz was building a multi-stop delivery truck. He subsequently was briefly with Holcomb & Hoke (stoker manufacturer).

George Freers joined Marmon-Herrington in 1940 and became Vice-President of Engineering in 1949. He retired from that position in 1963 at the age of 77 and died in 1964.

Joe Felts, who worked for Freers beginning as a teenager, said in 1982 that to him "Freers was always ten feet high; a great and good man full of compassion and loyalty and had a great influence on my life as a person and as an automotive engineer." (From underlined remarks prepared for the Marmon Muster, Sept. 25, 1982).

JEROME (Joe) FELTS — Marmon to Executive Engineer
Lincoln-Mercury Div. Ford

Jerome Joseph Felts went to work at Nordyke & Marmon in September 1918. He was hired by George Freers, Chief Draftsman, to trace drawings for the Liberty Engine. He left N & M in 1921 to join the new Lafayette Motors Co. in Mars Hill (Indpls) where D. McCall White was chief engineer. In May 1926 he returned to Marmon. During an interview on July 18, 1977, he expressed the opinion that Chief Engineer Litle did not make any significant contributions to the product design and observed that Robert Lewis was responsible for the design of the flat-head Marmon engines. Mr. Felts recalled copying Continental Motors drawings on Marmon drawing forms for the Model L engine.

In 1931 Mr. Felts made layout drawings of a transfer case for a four-wheel-drive truck under the direction of Col. Herrington. Again in July 1977, he recalled working with a Rod Hoffman (now deceased) who held patents on a front wheel drive configuration similar to that used on the 1929 Cord, and who was working on a front drive concept car for Marmon.

Joe was transferred to Howard Marmon's personal payroll on August 10, 1931 and remained until August 19, 1934. He then went to work for Auburn where he stayed thru August 1935, when he joined Rod Hoffman in Detroit to work on an engine, reportedly for a Fisher Brothers car, which had four cylinders in an 'X' configuration; the car is preserved in the Brooks-Stevens Museum. He then worked at GMC, going to Reo in 1937 and to Ford Motor Co. in December 1938. By February 1952, he was Asst. Chief Engineer for Lincoln-Mercury.

At the Lincoln Mercury Division, Mr. Felts had the unusual experince of hiring the son of George Freers . . . Mr. Howard Freers is today Chief Engineer Body and Electrical at Ford Motor Co. and was previously Chief Engineer — Engine and Power Train.

Mr. Felts retired as Lincoln-Mercury Executive Engineer in 1966 and was a member of "The Marmon Club" and a contributor to "Marmon News," until his death.

When the Marmon Club held the first Marmon Muster at the Indianapolis Speedway in September 1982 to mark the eightieth anniversary of the first Marmon, Joe Felts was the guest of honor. Following his viewing of the cars on Saturday afternoon, Mr. Felts became ill and was hospitalized. Death came the following week in Methodist Hospital. En route to Indianapolis, Joe had been interviewed at the Auburn Cord Duesenberg Museum, relative to his Auburn experience, and on Friday had served as a tour guide when Marmon Club members visited the several remaining historic automobile plants in Indianapolis. He had moved to within six blocks of the Nordyke & Marmon plant when only eight years old. His willingness to share his recollections with automotive historians, including the author, made a significant contribution to automotive history documentation.

FRED E. MOSKOVICS — Marmon to Vice President of Franklin
and President of Stutz

Mr. Moskovics was born in Budapest, Hungary on May 23, 1879 and came to Nordyke & Marmon Co. in July 1912 in a consultant capacity. He became a N & M Co. employee in Jan. 1913. He obtained his technical education at Armour Institute, Chicago.

Fred Moskovics was listed as the only stockholder other than A.N. Heiskell outside the Marmon family in the Stockholders Report of June 29, 1922. He resigned as sales manager in November 1923 and was succeeded by Colonel Edger S. Correll.

In February 1924 he became vice-president of sales for H.H. Franklin Co. In 1925 he became president of Stutz, a position he resigned in 1928. In 1931 he was instrumental in the creation of the Marmon-Herrington Co. and served as the first Chairman of the Board, resigning in late 1931. His subsequent career was as an executive of the A.O. Smith Co.

In 1950 Frederick E. Moskovics was the recipient of a 'Distinguished Service Award' from the Automotive Hall of Fame (former A.O.T.). His citation reads: For pioneering with the Daimler Company in Germany . . . for designing the first American taxicab . . . for administrative duties for Remy Electric, Nordyke & Marmon, Stutz Motor Car Co., President and Chairman of Marmon-Herrington Co.

A co-inductee was Mr. Charles S. Mott.

### COLONEL EDGAR S. GORRELL — Marmon to President of Stutz

Colonel Edgar S. Gorrell, West Pointer and former USAC General Staff Officer, joined Nordyke & Marmon in January 1920, replacing Fred Moskovics in 1923 as Sales Manager. By July 23, 1925 he was vice-president of sales but left later that year, joining Moskovics at Stutz and succeeding him as president in January 1929.

### JAMES A. BOHANNON — Marmon to President of Peerless

James A. Bohannon joined Nordyke and Marmon in 1922 as Purchasing Agent. In April 1928 he was promoted from Director of Purchases to Vice-President of Manufacturing. He resigned in mid-1930 to become President of Peerless Motor Car Co. He is listed as the donor of the sole Peerless V-16 to the Thompson Museum in Cleveland, Ohio.

### CHARLES C. CHAYNE — Marmon to Chief Engineer of Buick to Vice-President of Engineering Staff — General Motors Corporation

Charles C. Chayne came to Marmon following graduation from Massachusetts Institute of Technology and went to work for George Freers on the drawing board. In the spring of 1927, Joe Felts and Charles Chayne had adjacent drafting tables in the Marmon design room. According to Felts, when George Freers was appointed chief engineer, Chayne became Assistant Chief Engineer for Product Design and S.E. (Ed) Rowe came in from Stutz to be Assistant Chief Engineer for Product Development. Evidently this organization change was not publicized due to Chayne's departure soon after. S.E. Rowe was announced as having joined Marmon in the position of Assistant Chief Engineer (Design and Development). Felts believed that Chayne had acquired a mentor at Dupont during his MIT days. Chayne left in 1930 to join the Buick engineering department where he rose to chief engineer in 1936 and ultimately to Vice-president of the General Motors corporate Engineering Staff in 1955. He retired to California in 1963 and died in December 1980. Mr. Chayne's last (of many) patents assigned to GM covered a hydraulic lash-adjuster for stud-mounted rocker-arms. The patent illustrated the application of the invention to the hollow-stud oil supply (Pontiac), to the hollow pushrod oil supply (Chevrolet) and to the cylindrical-bushed (instead of spherical) design (Marmon 34B and post-patent Ford 351C). During his GM Vice-Presidency he restored a Bugatti Royale Coupe and subsequently presented it to the Henry Ford Museum in Dearborn, Michigan.

### ROBERT P. LEWIS — Marmon to Dana Corporation

Robert P. Lewis was a graduate of Pratt Institute who had joined Nordyke & Marmon in 1918 and had become Advanced Design Engineer by the late 1920s. He was the responsible design engineer on the series of L-head eights and on the SIXTEEN. He eventually became chief engineer of the Dana Corporation.

Marmon dealers appreciated the interesting and instructive discussion of future motor car mechanics by C. F. Kettering, President of the Dayton Engineering Laboratories Company.

As The Marmon Post cartoonist sketched Vice-President F. E. Moskovics at "The Round-up" May 30, 1920.

### AL LEAMY — Marmon to Body Designer — Auburn Cord Dusenberg

Al Leamy, according to Joe Felts who worked with him at Marmon, presented art work with his job application to Mr. Ames of Duesenberg which ultimately became the body, hood and front end design of the Model 'J' prototype. He subsequently became body designer for Auburn and Duesenberg, preceding Gordon Buehrig, and is credited by the latter with the classic styling of the L-29 Cord. David S. Henderson, speaking at the Auburn Cord-Duesenberg Seminar in Auburn, IN, in September 1977[3] noted that Leamy's styling patents included the frontal ensemble and four of the body styles for the L-29 Cord. The National Automotive History Collection in Detroit includes an original example of Mr. Leamy's art work.

### BERT DINGLEY — Marmon to President of Marmon-Herrington

Bert Dingley was national AAA racing champion in 1909. When he was sidelined following an accident in 1914, he became Western Sales Mgr. for Mitchell Motor Car Co. in 1916, and the following year joined the US Bureau of Aircraft, Liberty Motors Div., as Chief Inspector on Liberty Engine production at the Nordyke and Marmon Co. At the end of the war he joined the H.H. Franklin Co., returned to N & M Co. but left in 1925 to join Stutz where he became Vice-President[4] in charge of manufacturing in 1929. The year after Marmon-Herrington was organized in 1931, Dingley joined as Executive Vice-President, advancing to the Presidency in 1942 and retiring in 1948.

OWEN MILTON NACKER — Marmon to Engine Designer at Cadillac

A musical chairs exercise in 1927 brought Owen Nacker to Cadillac from Marmon and he is recognized as the creator of the Cadillac V-16 and its derivative V-12 engines. Maurice Hendry, in his "Cadillac — The Complete History"[5] states: *"In 1927 an important change occurred. Frank Johnson, one of the old Leland 'family group' had left Cadillac in 1917 with the Lelands and remained with them at Lincoln until September 1921 when he returned to Cadillac (Assistant Engineer — engine and transmission design, member of the Cadillac 314 engineering group) . . . until in 1926 Thomas J. Litle, chief at Lincoln, shifted to Marmon. Johnson went back to Lincoln while his place at Cadillac was taken by Owen Milton Nacker who had been a consultant with A.P. Brush."*

Evidently in this latter capacity, Nacker was working at Marmon in 1926-27 when the Marmon Sixteen engine was being designed and built at a secret off-site location. This circumstance had led to much conjecture that Nacker carried knowledge of the Marmon program to Cadillac. However Jerome Felts in his recounting of personal involvement in the Marmon Sixteen program makes no mention of Nacker whatever.

W. ROY MILNER — Nordyke & Marmon to Chief Body Engineer, Pontiac Motor Div., GM

W. Roy Milner came to N & M as a detail draftsman accompanying C.E. Jeffers from Studebaker on Jan. 15, 1913. He was promptly given responsibility, at age 24, for liaison with the custom body builders, who were contracted to provide certain N & M closed body models as well as one-off specials. Milner's responsibilities included ⅛ scale preliminary sketches and styling approval following consultation with the customer. On Sept. 15, 1918 he left to join Cadillac where he was in charge of body engineering at the Jefferson Ave. plant. When the Cadillac management team resigned in 1921 to take control of Peerless, Milner was included in the "syndicate". The abortive Peerless effort was followed in 1925 by the recruitment of Milner, along with Anibal, Schwarze and Holden to reinforce Oakland (introduce Pontiac) under the new Glancy management. Milner held the position of Chief Body Engineer at Pontiac for the next 30 years until retirement in 1955.

HAROLD BLANK — Asst. Service Manager at N & M to Service Manager at M-H

Harold Blank went to work for the Nordyke & Marmon Co. in October 1919. He left in 1923 to go with Duesenberg, remaining through March 1926 when he moved to Stutz. He was with Stutz until 1929.

He joined Marmon-Herrington in 1934, retiring on March 31, 1970 after 36 years at Marmon-Herrington and 51 years after first joining the "Marmon Organization."

H.H. RICE

Mr. Rice spent most of his active career with Marmon, having joined the firm in 1886 at the age of 19 and having resigned in 1926. He is noted in this Distinguished Alumni section because he was not the H.H. Rice who became president of Cadillac, but has been confused with him by some writers.

DELMAR G. ROOS — Chief Engineer of Locomobile, of Marmon, and of Willys

Joe Felts described "Barney" Roos as he remembered him during his somewhat brief period at Marmon: "An active, capable, sometimes abrasive engineer who was fully in charge of engineering."[2]

CHAPTER THIRTEEN

# FINANCIAL HISTORY

*Marmon Motor Car Company*
*and*
*Predecessor Firms*

"In 1851, under the name of Nordyke, Ham & Company, the manufacturer of flour mill machinery was first begun in a shop in Richmond, Indiana, USA. In 1858 Mr. Nordyke's son, Addison H. Nordyke was taken into the business which was carried on as E. and A.H. Nordyke"[1] . . . a simple partnership. All that is known of the initial capitalization is the statement in the 'Marmon Book of Facts' . . . a corporate history document reissued and updated as late as 1929: *"The partnership was somewhat limited in capital but abundantly supplied."*

In 1865 Daniel Marmon supplied the 'Marmon' name to the firm. He became a partner of the Nordykes. The same source continues: *"His small estate of some few thousand dollars, added to the financial assistance offered by his uncle, Eli Stubbs, bought his interest in the company."*

The following year the name was changed to Nordyke, Marmon & Company. Capitalization is unknown until a change in Indiana corporate laws occurred. "An Act for the Incorporation of Manufacturing and Mining Companies and Companies for Mechanical, Chemical or Building Purposes, May 1852 and Amended" required filing of an "Articles of Association" on March 9, 1871. The original hand written document is preserved in the Indiana State Library archives. Sections 3, 7 and 5 provide the following pertinent information:

Sec.3: The Amount of Capital Stock Shall be Eighty Thousand Dollars
Sec.7: The Capital Stock Shall be divided into Shares of fifty dollars each.
Sec.5: The Number of Directors shall be Five:
Addison H. Nordyke (President), Daniel W. Marmon (Secretary), James Speer (Treasurer), Ellis Nordyke, Charles A. Nordyke.

When David Beecroft of "The Automobile" made a written inquiry to Nordyke & Marmon Co., the treasurer A.R. Heiskell replied on Feb. 8, 1916 and confirmed that no records existed at that time for the period 1851-1871. He reported that the 1871 Incorporation of Nordyke, Marmon & Co. with capitalization of $80,000 was superseded by the Incorporation under the name of Nordyke & Marmon Company.

Following relocation to Indianapolis in 1876, changes in the laws of the State of Indiana required a reincorporation on June 27, 1876. The original hand written copy of these Articles of Incorporation of the Nordyke & Marmon Company is preserved in the Marion County, Indiana, Library. It is worthy of note that the word 'and' was erased and the symbol therefore substituted on the original Articles.

The incorporators were:

| | |
|---|---|
| Addison H. Nordyke | Amos K. Hollowell |
| Daniel W. Marmon | Isaac H. Richards |
| Henry C. Wright | |

Article 1 stated: "The object of the formation of said company shall be and is the manufacturing and sale of machinery and machines, including portable and stationary steam engines, flouring mills, agricultural machinery and agricultural implements of every description, shafting, pulleys and gearing for all kinds of machinery, machinery for grain elevators, saw mills and all other machinery for working wood and iron, together with any other kind of machinery not above enumerated." Certainly this was a comprehensive and ambitious set of objectives.

Capitalization was increased to $100,000 (2000 shares of $50 par value).

Although the Nordyke name remained in the corporate marque until 1926, the Nordyke participation in the firm ended before the automobile was even a potential product line and Nordyke interests were acquired by the Marmon family members in 1904 just as the first production of automobiles was launched. Ellis Nordyke, because of his advanced age, early placed the burden of the business upon the shoulders of the two younger members of the firm. He died in 1871. When the Articles of Incorporation were revised in 1876 and the name expanded to include Marmon, Addison H. Nordyke remained President and held the position until he retired from the business in 1898 at age 60. He continued as a stockholder and director until 1904 and survived until death in 1921 at age 93. Beginning in 1904 the ownership was entirely in the Marmon family.

Indiana law did not require annual reporting until the 1908 fiscal year. The Annual Domestic Corporation Reports required by the State of Indiana provide an insight into the corporate structure and finances of the Marmon firm thereafter.

Daniel W. Marmon died on May 9, 1909 leaving his estate to his widow, Elizabeth Carpenter Marmon. Because the stock was privately held, little is known of the financial history but some assessment of the success of the firm can be made by consideration of the estate of Daniel Marmon.

The widow called her three children together when she had received her inheritance. She told Walter C., Howard C. and Caroline that she wished to share it with them . . . approximately $1 million for each. From this it can be concluded the inheritance was either four or six million.

When Mrs. Marmon died at age 90 in February 1940[2] she left an estate of $2,993,720.[3] She presumably lost heavily in the bankruptcy of Marmon Motor Car Co. and the large estate, despite that loss, gives a hint of her inheritance from basically the Nordyke & Marmon milling machine business. Marmon car sales prior to 1909 were not in sufficient volume to be a major contributor to the corporate profit.

No significant changes are shown by the reports through 1912. The entry into the automobile business did not influence Nordyke & Marmon capitalization until midway through Era Two when expansion of facilities for production of the Era Three products was in the planning change.

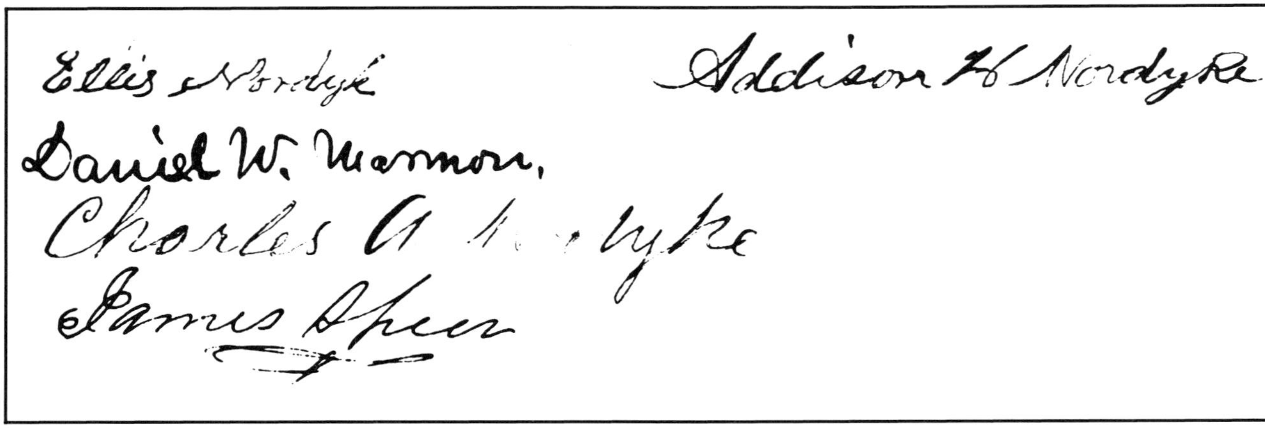


Fig. 13-1 Founders' signatures on the 1871 incorporation document.

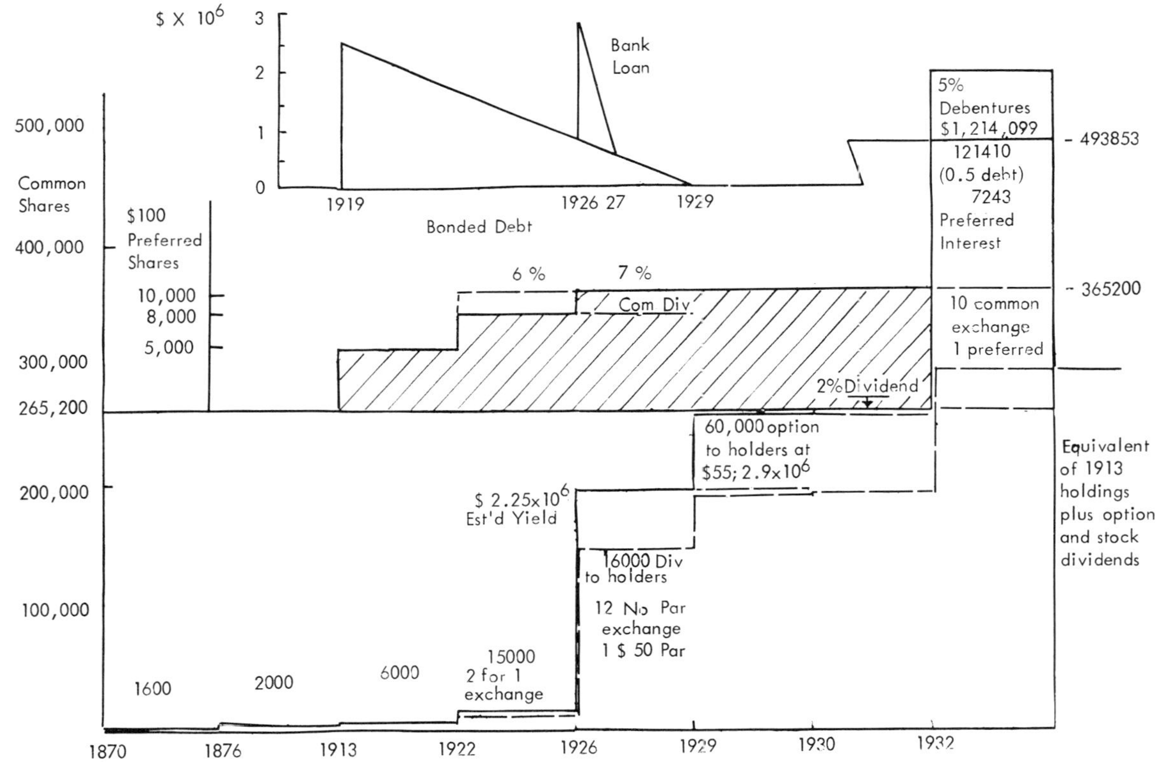


Figure 13-2: CAPITALIZATION of NORDYKE, MARMON & CO., NORDYKE & MARMON CO., MARMON MOTOR CAR CO.

By 1911, facilities devoted to motor car production equaled those for the traditional milling machine products. The need for additional capital was evident. The fact that equal facilities had been achieved without increased capitalization highlights the efficiency and profitability of the firm up to that time.

On June 26, 1913 Nordyke & Marmon recapitalized. In accord with provisions of An Act of the General Assembly of Indiana (Mar. 9, 1907), the firm issued a "Certificate of Corporate Existence, enlargement of corporate objectives, increase of common stock and creation of issue of preferred stock of the Nordyke & Marmon Co". This certificate, over the signatures of Walter C. Marmon, President and Howard C. Marmon, Secretary:

1. Extended the corporate existence to June 26, 1963.
2. Added to Article II of original Articles of Incorporation: "The manufacture, construction and sale of all kinds of mills and parts thereof, including mill supplies, **motor vehicles of all kinds,** their parts and accessories, and to buy and sell merchandise of all kinds similar or incidental to any of the machinery, mills, and parts thereof, including mill supplies, **motor vehicles of all kinds,** their parts and accessories, and to buy and sell merchandise of all kinds similar or incidental to any of the machinery, mills merchandise or other products so to be manufactured".
3. Increased the common stock from $100,000 to $300,000 (4000 additional shares of $50 par value).
4. Created an issue of $500,000 6% cumulative preferred stock (5000 shares of $100 par value).

Following the end of the hostilities of the first World War, the Liberty Engine facility was acquired from the government and was converted to motor car engine and chassis manufacture; new car assembly facilities were constructed for the introduction of the Model 34B.

In 1919, the Board of Directors authorized issuance of $2.5 million of 6% serial gold notes. They were dated July 1, 1919 and were offered by Fletcher American Co. of Indianapolis, Bankers Trust Co. and Montgomery & Co. of New York. They were to yield 6.25 to 6.75 percent according to maturity dates; were callable at 102 on 30 days notice and were due at a rate of $250,000 each anniversary through full retirement in 1929.

This bonded indebtedness was a direct obligation and the company convenanted the bondholders: (1) that none of its properties would be mortgaged; (2) that net total and net quick assets would be maintained, at 2.5 and 2.0 times, respectively, the sum of the outstanding notes; and (3) that no further obligations would be issued unless net earnings for the prior twelve months were 5 times the interest and retirement obligations of the notes.

At the annual meeting on **June 29, 1922,** at the Nordyke & Marmon Co. site, an increase in capital stock from $800,000 to $1,600,000 was authorized (effective June 30). Common stock authorized was doubled by 6000 additional shares of $50 par value and preferred stock was doubled by 5000 additional shares of $100 par value. The report was signed by Walter C. Marmon, President, and H.H. Rice, Secretary.

Stockholders voting and their holdings were:

| | | |
|---|---|---|
| Walter C. Marmon | 1325 shares | 23⅔% |
| Howard C. Marmon | 1325 shares | 23⅔% proxy voted by WCM |
| Mrs. Elizabeth C. Marmon | 1025 shares | 18⅓ proxy voted by WCM |
| Caroline M. Fesler | 1023 shares | 18⅓ proxy voted by WCM |
| J.W. Fesler | 2 shares | |
| F.E. Moskovics | 500 shares | 9% |
| A.R. Heiskell | 400 shares | 7% |
| | ———— | ———— |
| | 5600 stated to be | 100% of the common stock. |

The next insight into the profitability of the combined milling machine and motor car business, when both were at their peak, came when Walter Marmon evidently found it appropriate to assure the distributors and dealers gathered in convention on December 10, 1923, of the viability of the Nordyke & Marmon Company despite the tumultuous financial experience of the nation in the early 1920s. He described the holding of Marmon stock: *"Of course, you all know this is a close corporation. The stock is all held here, in the family, so to speak. Hence we do not publish statements, nor is our stock listed on the Stock Exchange. This has its decided advantage. We are fully in control of our business and stock manipulators cannot wrest it from us."*

Mr. Marmon's statement had been valid for more than seventy years but its validity would end on the seventy fifth birthday of the firm.

Mr. Marmon continued and revealed financial information on operations for fiscal years 1916 (July through June) through 1922: *"On July 1, 1917, we completed the most properous year that we had, up to that time, ever had. Our net worth was: July 1, 1917 — $3,570,473.28; July 1, 1918 — 4,341,159.15; July 1, 1919 — 4,804,090.85. "You all know the reverses that every manufacturer met with in 1920. Despite this fact and notwithstanding the period of bitter competition and price reductions that followed, the Company has grown and prospered. So that its net worth has increased to: July 1, 1923 — $5,911,036.32. "In other words, this Company has weathered the most severe storm that ever shook the financial structure of the Country; has withstood the period of bitter competition that resulted and stands today better off than it did when the storm clouds gathered. Today our indebtedness is absolutely current. Our trade credit is in the best shape. We do not owe a dollar that is due or past due. We have a bank balance of a third of a million dollars and quick assets of $6,827,218."*

On May 17, 1924 came the surprising announcement that Walter Marmon was relinquishing the presidency of Nordyke and Marmon to an outsider. This event was of such significance in the subsequent history of the firm that it is appropriate to reproduce here the exact announcement as it appeared in Automotive Topics for that date:

## G.M. WILLIAMS MADE PRESIDENT OF MARMON

Succeeds Walter C. Marmon — Remainder of Organization Remains Intact — Howard C. Marmon to be Vice-President and Engineer in Charge.

G.M. Williams, whose achievements in the industry have been striking, has been made president of the Nordyke & Marmon, Co., Indianapolis, to succeed Walter C. Marmon. The news comes unexpectedly but naturally follows events which have been developing for some time past. Howard C. Marmon, who has been identified with the mechanical development of the lines from the beginning many years ago, continues as vice-president in charge of engineering and it is understood that the organization as a whole is to remain intact.

Williams, in an almost increditably short time, has made himself an enviable reputation in the industry as an organization builder. Only eight months ago he left Dayton, Ohio, the scene of his earlier work, to take the presidency and active general management of the Wire Wheel Corp. of America, at Buffalo. In that relatively short time he brought the products of the corporation into national prominence and so marked was his success that the corporation in accepting his resignation with regret, successfully prevailed upon him to continue to serve both as a member of the board and executive committee.

For five years Williams was general manager of the Dayton-Wright division of the General Motors Corp. Prior to 1916, when he became affiliated with General Motors he was chief engineer of the H.E. Talbott Co., at Dayton, and vice-president of the Dayton Metal Products Co. By profession he is an engineer and enjoys a wide acquaintance in engineering circles. At one time he was president of the Manufacturers' Aircraft Assn., and a governor of the Aeronautical Chamber of Commerce of America.

* * *

Other sources stated that Williams had made a significant purchase of Nordyke & Marmon Co. stock concurrently with his joining the firm.

When Williams election as Marmon president[4] was followed by appointment on June 1, 1924 of Carl J. Sherer as special assistant to Williams, rumors spread rapidly that a General Motors takeover was imminent. Williams had been general manager and Sherer secretary and treasurer of the Dayton Wright Division of General Motors.

It became necessary for both General Motors and the banking interests, representing the new factor in Marmon management, to deny emphatically any planned affiliation. Actually the opposite was true . . . by the end of 1924 General Motors had ceased operations at the Dayton plant and begun liquidation.

Although operations were profitable, the increased capacity represented by the additional facilities installed for the war effort was not being fully utilized.

Although operations were profitable, the additional facilities constructed in 1919-1921 were not being fully utilized. The huge growth in industry registrations of passenger cars was not being reflected in Model 34 sales which were relatively flat. (Registrations increased 250 percent in the period 1915-1920 and 115 percent in the period 1920-1925). Walter Marmon had consulted his former classmate at MIT, Alfred P. Sloan, Jr., concerning a possible candidate to improve the performance of Nordyke & Marmon. Williams had been selected as the man to reorganize the car manufacturing organization and to seek greater sales and production volume.

A brief resume of Mr. Williams' life and career is to be found in Chapter Four, Part 2, "Introducing G.M. Williams."

The Stockholders Meeting Report for **July 23, 1925** lists the corporate address as 1101-1135 W. Morris Street. Of the authorized capitalization ($600,000 common and $1,000,000 9% preferred), $1,400,000 was reported issued and outstanding. None of the preferred had been redeemed.

Officers and directors were listed as:

| | |
|---|---|
| W.C. Marmon | Chairman of the Board |
| G.M. Williams | President and Director |
| A.R. Heiskell | Vice President, Director and Secretary |
| H. C. Marmon | Vice President, Director |
| H. L. Purdy | Vice President, Director |
| E.S. Gorrell | Vice President |
| C.J. Sherer | Treasurer |

The report was signed by H.L. Purdy and C.J. Sherer, as Acting Secretary.

## MARMON MOTOR CAR CO.

On February 2, 1926, the stockholders at the annual meeting approved a change in the name of the corporation from Nordyke & Marmon Co. to Marmon Motor Car Co.

Holders of 11,118 of the outstanding 12,000 authorized shares were represented. The report was signed by H.L. Purdy, Vice President and A.R. Heiskell, Secretary.

## NORDYKE & MARMON AGAIN!

The Nordyke & Marmon Marque reappeared on February 17, 1926. The address of Nordyke and Marmon, Inc. was stated as 1101 W. Morris Street. A stock issue of 1,200 shares of no par value stock with fifty year life was authorized with a selling price of less than $10 per share.

The Articles of Incorporation show A.R. Heiskell and H.L. Purdy as incorporators "for the purpose of manufacture and sale of machinery for every purpose; manufacture and sale of products of an iron foundry; the manufacture and sale of all kinds of mills and parts thereof, including mill supplies, and said corporation shall have power to acquire real estate necessary to its business and all other acts insiborated thereto."

The annual report for 1926, dated July 7, was signed by A.R. Heiskell; the reports for 1927 and 1928 dated respectively July 28 and July 26 were signed by E.C. Badger. No reports were filed for 1929 through 1949 and on April 15, 1950 the franchise was forfeited by order of the Indiana Attorney General.

The officers and directors of Nordyke & Marmon, Inc. through 1928 were:

| | |
|---|---|
| W. C. Marmon | Chairman of the Board |
| G.M. Williams | President |
| A.R. Heiskell | Vice-President and Secretary |
| C.J. Sherer | Treasurer |
| H.C. Marmon | Vice-President |
| H.L. Purdy | Vice-President |
| L.I. Zeigler | Vice-President (1926 only) |
| E.C. Badger | Asst. Treasurer and Asst. Secretary (not a Director) |

This reconception of Nordyke & Marmon was obviously a tool for the disposal of the original milling machine product line and of the iron foundry. Sale of the milling machine business and the 'Nordyke' marque to Allis-Chalmers Company was completed in December 1926. The amount realized from this transaction is not known to the author. Allis-Chalmers continued manufacture of the 'Nordyke' line of milling machines for many years with service support extending into the 1960s.

It would be most interesting to know to what degree this corporate division protected the wealth of the original Nordyke & Marmon Co. stockholders when Marmon Motor Car Co. failed in 1933.

In anticipation of "going public," financial performance data of the "closely held" (preferred and common) Marmon Motor Car Co. first appeared in Moody's Manual of Investments 1926 edition.

It noted that Marmon Motor Car Co. controlled the Marmon Automobile Company of New York, Inc. and the Marmon Detroit Company.

Net earnings for the year ended Dec. 31, 1925 were reported "in excess of $2,250,000 on gross sales of more than $16,000,000 (14% of sales). Working capital was identified at $2,386,175 and inventory was valued at $2,507,695.

A special meeting of the stockholders on **May 6, 1926** resulted in the following:

1. Change of the common stock to "no par value" in the "best interests of the Company and in accord with the Statutes of Indiana."
2. An increase in the common stock of 56,000 shares (68,000 total) of no par stock.
3. Direction to the Board of Directors to sell these shares at not less than $15 each.
4. Direction for the exchange of all par-value shares for no-par shares.
5. An increase in the preferred stock dividend to 7% (from 6%) maximum from profits prior to a common dividend.
6. Holders of preferred stock to be paid full par value, plus interest due, prior to any liquidation or dissolution. The Company may retire the preferred issue entirely upon 30 days notice at $100 par value plus dividends and should do so by December 31, 1935.

The new stock was to be dispersed as follows: 10,000 shares preferred . . . issued and outstanding. 144,000 shares common, no par value, in exchange for 12,000 shares of $50 par value common (12 for 1 basis) 56,000 shares common, no par value, to be disposed of by the Board of Directors for no less than $15 per share, ($840,000 min.)

The report was signed by G.M. Williams, President and A.R. Heiskell, Secretary.

This action set the stage for a major venture. The potential prize was great but its attainment fraught with hazards. The combined actions of this meeting and the separation of the Nordyke & Marmon milling interests was certainly significant and in fact prophetic. It was to influence the post-depression wealth of the Marmon family and did indeed anticipate the demise of the Marmon Motor Car Co.

Capitalization of the Marmon Motor Car Co. and the predecessor films, 1851-1933, is charted in Figure 13-2.

After seventy-five years as a closely held business, without a requirement for disclosure of financial results, Marmon was going public.

Financial results for the year ending July 3, 1926 were publicly disclosed. "Net income for the year was reported at $1,669,800 after depreciation, interest and expenses for contingencies and Federal taxes. The company had outstanding $1 million preferred stock and 144,000 no-par shares of common stock."

"Current assets in the Company's balance sheet amounted to $3,310,125. and current liabilities $692,090. a ratio of almost 5 to 1. During the year, bank indebtedness which at the start of the year totaled $1,888,000. was retired and the annual installment on serial gold notes amounting to $250,000 was paid."

A block of 40,000 shares of Marmon Motor Car Co. common stock was purchased by Hallgarten & Co. and Green, Ellis & Anderson during the first week of September, 1926 and they offered it publicly at $48.50 a share. After seventy-five years, Marmon had gone public.

The financing was described as placing the company in an extremely good financial position, with cash of $2,154,542 and current assets of $4,714,125 together totaling $6,868,667. Current liabilities were only $536,000, for an asset to liability ratio of 12.8:1. The capital and surplus account was increased to more than $6,800,000.

Marmon common stock was first listed on the New York Stock Exchange (NYSE) on September 4, 1926 when it was reported at 45.[9] At that time, automotive stock performance was measured independently by a 'Twenty Auto Stock' Index which stood at 57 on that date. To better illustrate the fluctuations of Marmon stock, the author has normalized the price on the basis of the Twenty Auto Stock Index and expressed it as a percent of the index. Both the un-normalized price and the Percent Index are regularly given; the latter filters out market perturbations and reflects Marmon-related developments. When Marmon was first listed on the NYSE, the Index was at 57 . . . the Marmon listing at 45 was 79 percent of Index.

Marmon stock is charted for the period 1926-1933 on the NYSE on Figure 13-4.

Beginning with October 23, 1926 Marmon consistently paid $1.00 quarterly dividends on its common stock. This was to continue through August 10, 1929 with the exception of $0.65 dividends on May 15, 1928 and May 15, 1927.

In mid March 1927 Marmon reached the 100 percent Index value of $51. Marmon stock had closely followed a decline in the Index and stood at its initial offering price of $45 in April 1927 but importantly remained at 93 percent Index.

The stockholders met on **May 19, 1927** and the holders of 177,071 of the 200,000 outstanding shares were represented in person or by proxy. Article V of the Articles of Incorporation was amended to provide for seven (rather than five) directors "who would direct and manage affairs of the Marmon Motor Car Co." The report was signed by G.M. Williams, Pres. and A.R. Heiskell, Secy.

Rumors of the coming introduction of a low priced Marmon car drove the stock steadily upward and by mid-November it was at 65.375, 113 percent of the 56.6 Index. With announcement of the Roosevelt only two weeks away, the stock took off, on announcement day reached 76.75 and peaked a week later at 87.5, a healthy 137 percent of Index.

The report of the **July 27, 1928** stockholders meeting showed that J.A. Bohannon had replaced (after two years) Homer McKee as a Vice-President and that E.C. Badger had become Asst. Secretary and Asst. Treasurer. The election of Elmer Stout and George A. Ellis brought the Board to the authorized strength of seven; Eastern US banking interests had made their appearance, reflecting the new public corporate status. Mr. Ellis is presumed to be the partner in the underwriting firm of Green, Ellis and Anderson.

The stockholders met again on November 21, 1928 to double the no par common from 200,000 to 400,000 authorized shares. When that action was taken the stock was at 65 and rising; it peaked at 88 in December and then began to decline.

For the fiscal year ending February 28, 1929 Marmon reported earnings of $6.01 per share of common stock after provision for the preferred dividend ($700,000) and for depreciation of assets ($283,000). Net profit was $1,447,919.

The Roosevelt had been introduced at the 1929 shows and production was underway when the fiscal year began.

A market decline in February had brought the Index down to 60.4 and Marmon to 68.5. By early March 1929 a significant development first evidenced itself: The auto Index had begun a decline which would not be reversed significantly through 1933. Marmon however again displayed its counter-trend performance of late 1928 . . . it recovered ten points in a week and began a rapid climb which carried it to **104 in mid-May 1929 . . . nearly double the Index which stood at 55.5!**

On June 8, 1929, the press reported "Marmon Closes Best Quarter in History . . . Shipments of Marmon and Roosevelt Lines in First Period of Fiscal Year Reach 14,847 as against 6,893 in same Period Last Year" (A gain of 115%). The report by G.M. Williams continued: "Shipments in May this year totaled 5,221 units against 1,904 in May 1928" (a gain of 174').

The **July 24, 1929** report listed outstanding shares as of June 30, 1929 at 260,000 common and 10,000 preferred shares. Of these, 60,000 common had been issued during the preceding fiscal year, selling at $55 and totaling $3,300,000; 140,000 authorized common shares remained unissued.

Fred B. Johnson and Maurice Newton were new directors, bringing the total to nine. Of these, two (Ellis and Newton) were representing New York banking interests. In addition D.P. Kingsley and A.G. Cameron became assistant secretary and assistant treasurer respectively; both had New York City addresses.

Marmon stock was $90⅞ on the close of the fiscal year and at $80½ when the stockholders met. After provision for depreciation of $283,880 and for a $700,000 preferred dividend, Marmon had reported a net profit for fiscal 1928 of $1,447,919. The stock jumped reflecting the favorable report. The presence of banking interests in management reflected only caution and a due share in future decisions.

Following a slight decline to 80 in late June, Marmon rebounded to 92.75 at month end (180 percent Index) when first quarter earnings of $1,238,637 were reported. It then declined through July and August to 81.5. In September H.L. Purdy was credited with the financial rejuvenation of the Marmon Motor Car Co. His glory and his house of cards would both be short lived. Earnings for the quarter ending August 31, 1929 were only $61,281.

The trauma of late October began for Marmon shareholders in early September as the Marmon stock performance provided a portent of what was to come for other investors. Marmon and the Index fell to 61 and 45 respectively by late September but in the next three weeks Marmon continued down to 41 while the index leveled off at 40.25. Marmon was still at 102 percent index...a reasonable position, but no longer a Superstar.

Gray Thursday (October 24th) came early in Indianapolis — in the third week of October Marmon fell from 41 to 28 at a stable Auto Index of 40.3; it had fallen to only a fourth of its value in May!

On October 29, 1929 — Black Tuesday — the entire market joined the Marmon plunge. On that infamous day Marmon held all day and closed unchanged at 28.

## AUTOMOBILE STOCKS IN THE 1929 CRASH

The stock market crash in 1929 came as a double whammy. Black Tuesday, October 29, followed Gray Thursday. To comprehend the significance, some adjustments must be made to 1929 statistics; one analyst projected a 1979 (50th anniversary) equivalent of Gray Tuesday at 450 million shares traded in a single day!!

General Motors common stock declined from a peak of $215 in 1928[9] to a closing on October 29 at $40; Marmon had peaked at $104 in May 1929 and closed at 28. Although Marmon had peaked at only 48 percent of GM's peak, it closed Black Tuesday at 70 percent of GM.

Some automotive stocks showed apparent remarkable stability during trading on October 29. Graham Paige declined from 11 to 10, International Harvester from 85 to 80, and Packard only from 16.75 to 16.25 while Pierce-Arrow and Marmon closed unchanged at 22.25 and 28 respectively. Willys-Overland closed down from 11 to 10, but that statistic masks the true pathos of that historic day. The W-O closing at 10 was little solace to the investor whose shares had been sold by his broker earlier in the day for as little as 5.75 to cover his purchase on a "margin" basis (partial payment on purchase with intent to sell before full payment was required). GM investors had a similar experience: GM dropped from 45.50 to 33.50 and then recovered to 40 leaving most margin purchasers sans their investment. Similarly Federal Motor Truck closed unchanged at 8 despite a low of 5. The most dramatic change among the autmotive stocks was Auburn which plummeted from 175 to close at 130 that memorable day.

Marmon continued to slide to a low of 21 in mid-November but recovered to 27 and ended the year at 24.5.

A favorable report, on February 8, the largest ever January shipment of Marmon cars was followed on the 22nd by announcement of a reduced dividend.

The stock was typically steady at 23.75 through January and February of 1930 but jumped to 29, 97 percent Index, in early March following a declared dividend of $0.50 (versus regular $0.65) to holders of record February 15, but it reverted to the prior value of 23.75, 80 percent of the 29.5 Index.

Results for fiscal year 1929 were very discouraging: a large third quarter loss had reduced the year's profit to $936,251. Sales had decreased 65 percent in terms of units (10,115 vs 29,616) and **58 percent in dollars** ($12,854,366 vs $30,462,261) **while the industry average decline was 39 percent** on dollar sales. Mr. Williams sought to improve the apparent performance by noting that **exclusive of Roosevelt, Marmon Motor Car Co. sales had indeed suffered only the average industry decline** ($6,232,269 vs $10,186,291) while the **Roosevelt sales declined a whopping 74 percent,** ($4,368,682 vs $16,602,817). This impact of rumored mechanical defects in Roosevelt cars on the company's performance is charted in Figure 13-3.

Marmon closed April at 20 and May at 10; the auto industry Index closed at 27 and 20 . . . Marmon had fallen from 74 percent Index to 50 percent in a month!

By the time the shareholders met in June 1930 there had been significant changes in management:

1. H.L. Purdy, the reputed architect of Marmon finance had resigned and been replaced on the Board of Directors by Fred B. Johnson.
2. C.J. Sherer, a VP and Treasurer, was elected a director.
3. Helen Ray and J.M. Paterson were the new assistant Secretary and Assistant Treasurer.

A share dividend was declared to holders of record May 5, 1930; it consisted of 5200 common shares. A surplus of $78000 was transferred to capital account (5200 shares at $15.00 equal to $78,000). Outstanding stock was then 271,200 shares consisting of 265,200 common and 1000 preferred whereas 400,000 and 1000 respectively were authorized. It would appear that the Board of Directors had not been successful in floating more than one third of the 200,000 additional shares authorized.

Production had been continued during the first quarter of the calendar year at a rate which was inconsistent with the declining sales pace. As a consequence Marmon had embarked on a liquidation program in May 1930 which failed to fully clear out the troublesome inventory on a timely basis . . . 200 units still remained unsold in mid January 1930.[1]

Meanwhile there was a brief market improvement in late July, the Index rising to 23.5, Marmon to 14 or 60 percent Index. But August was a repetition of May and a bad portent for the Industry and a worse one for Marmon.

September 1930 ended with the Index at 20, Marmon at 9 (45 percent Index).

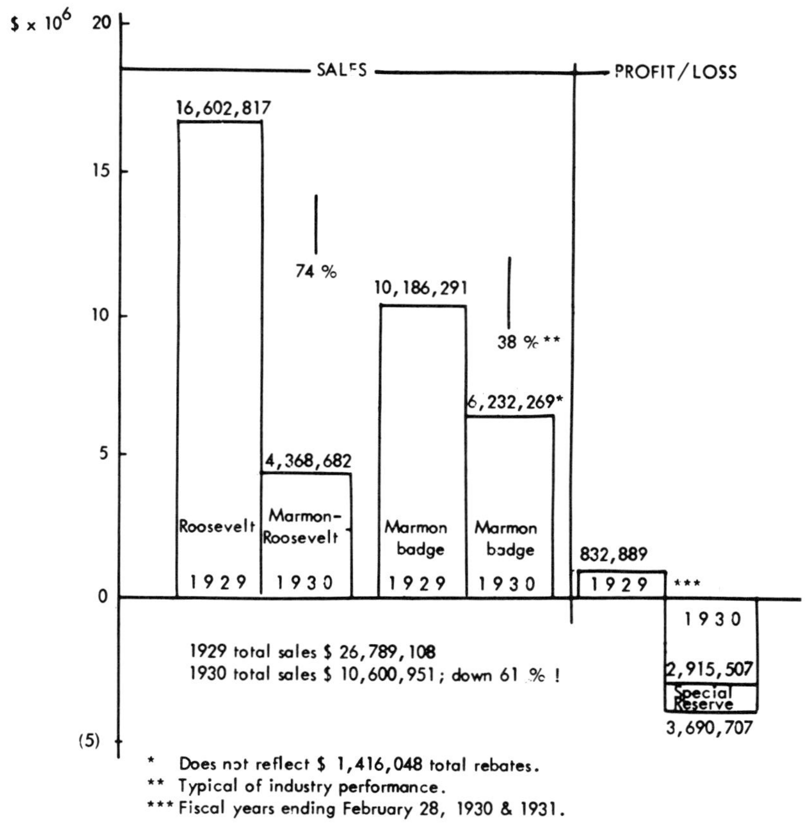


Figure 13 - 3 : MAJOR CONTRIBUTION OF ROOSEVELT PROGRAM TO THE 1930 PERFORMANCE OF MARMON MOTOR CAR COMPANY

October saw the trend continue with the Index down to 16.2 and Marmon at 6 (37 percent Index).

Marmon reported a nine month loss[5] of nearly $2 million ($1,937,508) most of it during the inventory liquidation ($1,791,788) when the factory was presumably closed. Because of the surplus inventory problem, production of new 1931 eight-cylinder models could not be started until February, when indeed Model 88 production did begin.

It was a great misfortune that the new assembly line for the MARMON SIXTEEN was also inactive due to start-up scheduling problems in the new product. There was no inventory problem for the SIXTEEN for it was an entry into a new market segment . . . one which was exhibiting great interest in the new car. The first SIXTEEN did not come off the line until late February and true series production did not begin until late April . . . a full quarter after the enthusiastic reception at the 1931 Shows. This unfortunate development resulted in more than $1 million of Marmon funds being tied up in materials and parts inventories.

The year 1931 had begun with Marmon stock at 5 or 35 percent Index, having held its own with the industry during the last quarter of 1930. It climbed to 6 in late January, buoyed by a general rise in auto stocks rather than by the MARMON SIXTEEN announcement, and again in early May until the fiscal year loss of $3,690,707 was finally revealed on May 9. Prior to the news, Marmon was at 6 and the Index at 13.7 for a relative performance of 43 percent, but Marmon was in serious difficulty and the stock would soon reflect it.

Working capital had declined from $5,407,538 on Feb. 28 1928 to $4,180,622 one year later and to $1,627,731 on that date in 1931. The ratio of assets to liabilities had declined from 2.87 ($8,293,290/$2,886,452) to 1.83 ($3,583,820/$1,956,089) in the same period.[9]

A great hazard of the automobile business is the rate at which capital reserves become unsold product inventory. This is the 'Achilles Heel' of all new automotive ventures. But an even more threatening hazard of the business is to have working capital transformed into parts and material inventories due to a start-up 'snafu.' Assembled vehicles can be merchandised at some price, even less than cost if necessary, to reconvert capital for other needs; no such alternative exists if assembly operations are impossible.

Twenty of the principle creditors of Marmon Motor Car Co. organized a Committee in May 1931 to cooperate with the Marmon management in the working out of plans for payment of debts for materials "which resulted from unfavorable operating conditions during the month of April." The committee was headed by H.H. Hornbrook, an Indianapolis attorney.[9] This situation combined with the gross shrinkage of the multi-cylinder luxury car market to force a financial reorganization of Marmon Motor Car Co.

During the week of May 23 Marmon slumped dramatically from 6 to 3.125, from 44 to 23 percent Index; an effective drop of 50 percent in a week.

The stockholders met on **May 27, 1931**[1] and approved the Articles of Reorganization. Authorized stock was unchanged and G.M. Williams survived as president. Officers representing Eastern financial interests were increased:

Paul A. Crouch —Vice-President, (Limited) Bankers Trust Co. NYC

D.A. Powell — Vice-President, (Limited) Bankers Trust Co. NYC

A.P. Smith, P.H. Noland and E.C. Badger were new directors replacing Elmer Stout, George A. Ellis and Maurice Newton, the latter two having New York addresses.

The Reorganization was promptly followed by the necessary refinancing.

The press release[7] describing the Marmon refinancing follows:

*"MARMON REFINANCED — INDIANAPOLIS JUNE 15, 1931*

*Eastern banking interest are entering into a financing program with the Marmon Motor Car Co. to provide additional working capital, it has been announced by G.M. Williams, President. These new arrangements, with the favorable cooperation of vendors, place the company in an excellent position to take immediate advantage of the unusual demand for the New Marmon Sixteen and to proceed with comprehensive plans for the future, Mr. Williams said. Although the company* *has already produced a number of Model Sixteen cars, it has maintained a satisfactory order position and has on its books unfilled orders for this model with a retail value of $500,000 (100 cars — author). The new financing arrangements follow closely the announcement that the company has reduced overhead expenses to the point where operations are profitable in spite of a curtailed output."*

Prior year operations had resulted in a loss of $3,690,707. Undoubtedly known at the meeting was the subsequent first quarter loss of $580,460. This brought to $4,271,167 the total loss incurred from operations in the last five quarters, not one of which had been profitable.

While the index remained stable, Marmon gradually climbed from the May 1931 closing of 3.125 to a June closing of 5.125, or 37 percent Index, but promptly reverted to the earlier low and held it through July when the Index began a steady decline with Marmon holding relatively at 25 percent Index:

| | Auto | MARMON |
|---|---|---|
| Aug. Closing | 12.0 | 3.0 |
| Sept. Closing | 9.9 | 2.5 |
| Oct. Closing | 7.8 | 2.0 |

The business situation continued to worsen. On November 10, 1931, the Marmon Parts and Service Departments were sold under terms of a contract with the Indianapolis Machinery and Supply Co. which made Marmon their agent to operate the severed departments. In event of receivership of the Marmon Motor Car Co., Indianapolis Machinery and Supply Co. would be authorized to use the name "Marmon Motors, Inc." This contract was subsequently (Nov. 19, 1931) assigned by Indianapolis Machinery and Supply Co. to the Beelef Corporation, of which Jackiel W. Joseph was a principal; three members of the Borenstein family of Indianapolis were also involved.

A plan of reconstruction was submitted by President G.M. Williams on January 23, 1932 and accepted by the creditor committee on April 29 and confirmed by Marmon directors on May 2. It involved:[1] Concentration of operations in the single most modern plant. Drastic write-offs of old assets, including field and factory inventory, cancellation of unfulfilled commitments, to the extent of $2,289,740. Exchange of all preferred stock for common on a 10 for 1 basis. This would increase the equity back of a common share from $5.35 to $6.29. A composition settlement of claims as of May 16, 1931, half by $1,213,919 in five year 5% debenture notes and half by common stock on the basis of $10 a share[9] (121,391 additional common shares).

The debenture notes were to be due February 28, 1937 and were to be redeemable by annual payments of 65% of net profits before depreciation and taxes and by payments equal to 75% of net proceeds from sale of any real estate, buildings or equipment. The holders of preferred stock agreed to a substitute date of retirement (instead of Dec. 31, 1935) subsequent to retirement of the debentures; they also had waived preference rights in relation to common.

By February 1932 Marmon had fallen to 1.25 despite a rebound of the Index to 6.3; Marmon's performance was down to 18% of Index. By mid-April it was down to 0.625 and 14% of Index.

Production of the single eight cylinder offering for 1932 . . . the new Model 8-125 . . . was reported underway in early March 1932. An unfortunate repetition of the delayed start up of the SIXTEEN the prior year; almost certainly this was an inventory clearance problem. Regardless, dealers did not have cars to show prospective buyers brought into the show rooms by the introductory advertising which ran concurrently with the auto shows and pretty well exhausted the advertising budget.

Marmon sales for the fiscal year ending February 28, 1932 were only $5,392,819; in 1930 sales had been $30,462,261.

The stockholders met on May 19, 1932 to consider and accept the 'plan of reconstruction.' Significantly they met at 1001 York Street rather than the traditional 1101 W. Morris; this reflected the contraction of all office and manufacturing facilities to the southeast side of Kentucky Ave. opposite the original Marmon complex that predated the now ill-fated automobile venture.

Production capacity was reduced to 25,000 cars annually from a high estimate of 50-60000 vehicles.[9]

The shareholders voted to increase authorized common shares from 400,000 to 500,000 and directed that 100,000 unissued common shares be exchanged for the outstanding 10,000 preferred shares of $100. par value; this was equivalent to valuing the common at $10 per share. On that basis the total outstanding shares would have a value of $3,652,000." The equity back of a common share was increased to $6.29 from $5.33. Net worth at February 27 was shown as $3,061,217, including $1 million in the 7 percent preferred which is now . . . eliminated."[8]

An issue of five year, five percent debenture notes was approved for payment of half the creditor claims of $2,427,829; the remainder to be paid with common stock. "Reflecting the anticipation of continued improvement, the debentures carried a provision for interim retirement by annual pro rata payments equal to 65 percent of net profits (before depreciation and taxes), and also by pro rata payments equal to 75 percent of the net proceeds of sale of any of the Company's buildings, real estate, etc."[8]

Vice-President Noland had become Secretary of the Board, H.H. Davidson representing First National Bank of Detroit, Fred Barnickol of Indianapolis (1321 N. Meridian) and W.W. Hoagland representing the interest of the Hayes Body Co. (he was president) of Grand Rapids, Michigan were new directors, replacing A.P. Smith and Fred B. Johnson.

Approval was received in June 1932 from the New York Stock Exchange to list 234,800 additional shares of no par common stock "in order to carry out the provisions of the company's financial reconstruction plan. This stock, to be admitted on notice of issuance, will bring the total of Marmon shares authorized up to 500,000."[10] There were 121,408 shares required for issuance to creditors with claims of $2,428,179.2[11].

The exchange rate at which preferred had been exchanged for common had implied a $10 value to a common share. Since the creditors were issued common on a basis of $20 per share, acceptance constituted settlement of claims at 50 cents on the dollar. A situation very much like that in 1980 when Chrysler vendors were required to reduce claims by 50% as a condition to the Federal loan guarantee legislation.

While the industry index continued its tragic decline, Marmon held and mid-June 1932 found the Index at 3.5 and Marmon again at 18 percent.

The author initially referred to the "Twenty Auto Stock Index" and subsequently has referred to it as the "Index." The latter usage was not only a convenience; as the Depression deepened, the number of firms averaged in the Index decreased due to bankruptcies.

The first week of September saw strong resurgence in the market: Marmon at 2.75 and 37.5 percent of the 7.3 Index. However by the end of October Marmon had plummeted to 1.125, back again to 18 percent of the 6.3 Index.

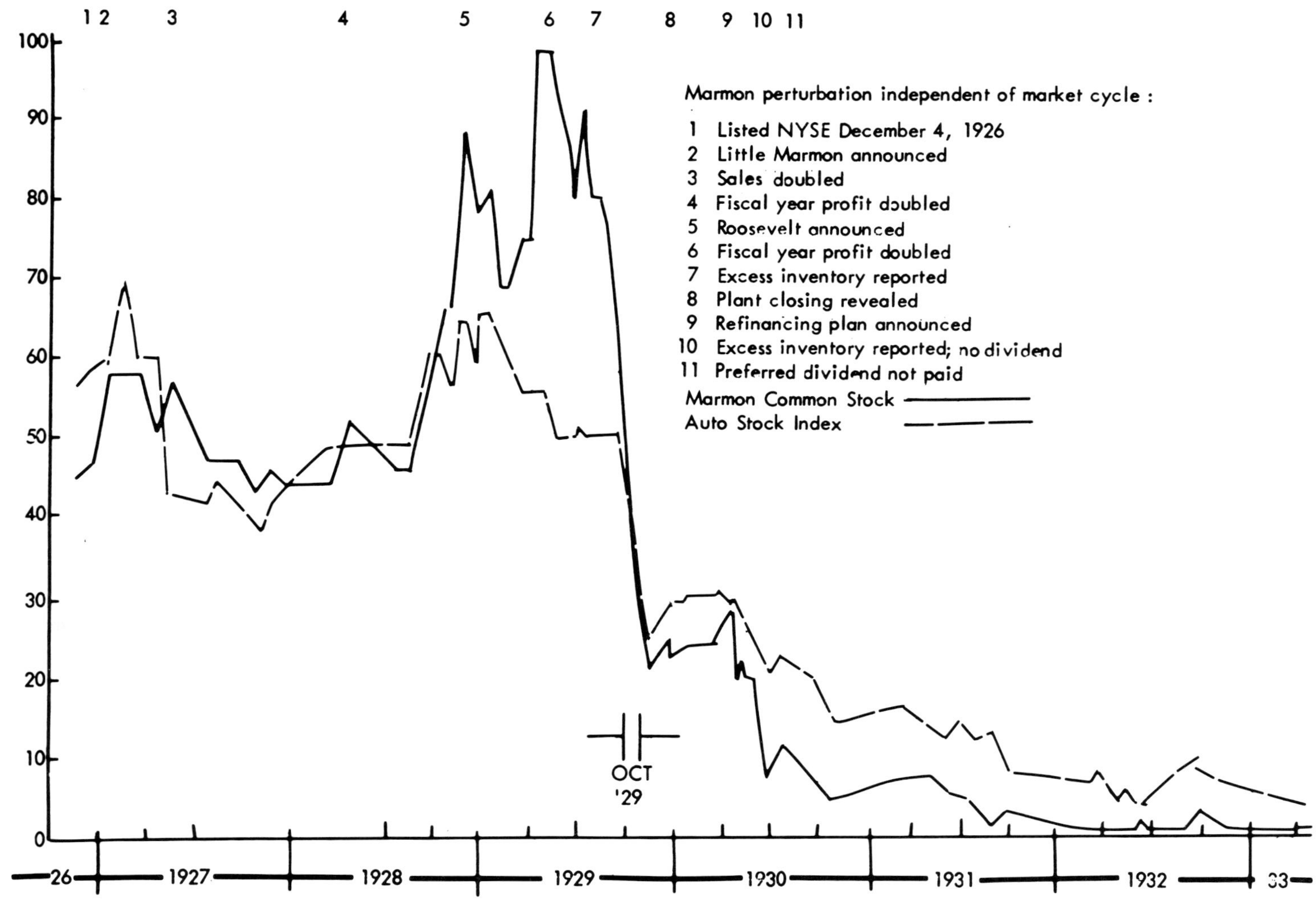


Figure 13-4 : MARMON COMMON STOCK PERFORMANCE — NYSE

By the end of calendar 1932, Marmon Motor Car Co. had suffered seven consecutive quarterly losses totaling $5,478,931 and ten losing quarters out of eleven for a total net loss of $5,608,745 between February 1930 and November 1932. Despite these tremendous losses Marmon could still state in September 1932, when the tenth quarterly loss was reported, that the ratio of assets to liabilities was still 2.4:1. This was only half the ratio value in 1926, just prior to the introduction of the Little Marmon, but a one third improvement from February 1931.

Ratio of Current Assets to Liabilities

| Year Ending | A/L |
|---|---|
| 7-03-1926 | 4.9:1 |
| 2-28-1928 | 2.9:1 |
| 2-28-1931 | 1.8:1 |
| 8-21-1932 (9 Mon.) | 2.4:1 |

## SUMMATION ANNUAL STOCK RANGES

Moody's Manual of Investments reported the annual ranges of Marmon Motor Co. common stock, after the initial offering in September 1926 at $48.50, as follows:

| | | | |
|---|---|---|---|
| 1927 | 62.25-39.50 | 1931 | 10.-1.25 |
| 1928 | 88.875-38.625 | 1932 | 3.5-0.50 |
| 1929 | 104.-19. | 1933 | 2.125-0.25 |
| 1930 | 30.875-4.25 | 1934 | 0.28 asked |

Figure 13-4 illustrates the behavior of Marmon common stock during the period it was listed on the NYSE. The Auto Stock Index[12] (an automotive Dow-Jones) is also shown and market changes explain many perturbations in Marmon common. Marmon perturbations independent of the market are identified and briefly justified. From an overall aspect, Marmon quickly achieved industry index and held it for 1927 and 1928. During the first three quarters of 1929 it grossly exceeded index, but fell to index in October. Thereafter it progressively declined as a percentage of index while following market trends.

## RECEIVERSHIP

On May 1, 1933 in a hearing before the Marion County Superior Court, receivers were appointed for the Marmon Motor Car Co. This action terminated the agency relationship with the Beelof Corp. and automatically authorized Beelof to become Marmon Motors, Inc. This untimely development followed closely upon the declaration of a National Bank Holiday in April 1933 by President Franklin Roosevelt. One of the two receivers was Fred Barnickol who had become a director the prior year and resigned to assume his receivership task. The significance of this event warrants a full quotation of the reporting by Automotive Industries Magazine for May 6, 1933.

RECEIVERS APPOINTED FOR MARMON COMPANY

*Reorganization Plans Under Way*
*Marmon-Herrington Co. Not Affected by Receivership*

*INDIANAPOLIS . . . Receivers were named on Monday of this week for the Marmon Motor Car Co. in superior court here. The suit asking a receiver was filed by the Eaton Spring and Axle Company on an account of $1,100. Fred Barnickol, President of the Indianapolis Drop Forging Company, and Robert A. Adams, attorney, were named receivers for the Marmon Company.*

*Walter C. Marmon, chairman of the Board for the Marmon-Herrington Company, Inc., truck manufacturers, explained that this action does not affect the truck firm in any way as the two firms have been completely independent for more than a year.*

The appointment of the Marmon receiver follows several months of reorganization seeking to conserve the assets of the company. G.M. Williams, president of the Marmon Motor Car Co., said that the national bank holiday caught the company at a critical time and that production could not be brought to profitably high levels. Williams added that work already has started to attempt a plan whereby the business may be kept alive after the release from receivership.

*Auditors have been at work on Marmon books for some time and as yet no report has been made.*

The Marmon Herrington Company subsequently purchased the former Duesenberg plant in Indianapolis and relocated there. This left vacant the entire Marmon complex between Kentucky and Morris Avenues, the Marmon Motor Car Co. having previously retrenched across Kentucky to the York Ave. address.

The Annual Domestic Corporation Report dated June 1933 stated that there were 500,000 outstanding shares, that 100,000 had been exchanged for 10,000 preferred shares of $100 par value, thereby retiring the preferred issue. The report stated: *"Consideration other than cash for 128,653 such shares is $1,286,530 accepted by creditors in lieu of cash for payment of indebtedness." Preferred holders received 7243 shares of common for their $72430 interest claim. Other creditors received 121410 shares valued at $1,214,100 and $1,214,099 in 5% debentures in full settlememt of their claims.*

No officers other than Assistant Secretaries Helen Ray and A.P. Smith were listed. The resignations of G.M. Williams, C.J. Sherer and Fred Barnickol were acknowledged, the latter necessary to accept the receivership appointment.

Surviving directors were the Marmon brothers, A.R. Heiskell, H.H. Davidson, W.W. Hoagland.

The two receivers wrote to the creditors in early December 1933 to report that the Fletcher American National Bank of Indianapolis was expected to file a single claim covering all debentures against Marmon and also to advise that March 1, 1934 was the last date for filing of claims. The Receivers' Inventory and Appraisal of the inoperative plant put the value of equipment and machinery at $149,233. They had disposed of some obsolete machinery for $43,240.[13]

The June 30, 1934 report listed shares outstanding as 493,853 of the authorized 500,000. The report was signed by the Receivers (since May 1, 1933), Fred Barnickol and Robert A. Adams.

Automotive Topics reported in the July 1, 1933 issue that the NYSE had stricken Marmon Motor Car Co. Common from the Board. It was subsequently listed on the New York Produce Exchange.

The June 1935 report simply restated the stock situation of the prior report and noted that there had been no officers since May 1, 1933.

## MARMON MOTORS, INC.

Marmon Motors, Inc. was the name which Beelof Corp. of Indianapolis had contractual authorization to adopt upon receivership of Marmon Motor Car Co. Beelof already held title to the parts and service operations of Marmon and the receivership terminated Marmon's agent position for their operation. The address was 1959 S. Meridian St., Indianapolis.

Annual Domestic Corporation Reports of Marmon Motors Inc., were filed for 1940 through 1946. They give May 12, 1933 as the date of establishment.

## AMERICAN AUTOMOTIVE CORPORATION

Formation of a new company to reactivate the Marmon facilities and product was announced in Indianapolis on Jan. 9, 1934, by A.D. Sterner and S.N. Wheeler, heading a bank syndicate. On Feb. 13, the American Automotive Corporation was incorporated in Delaware to buy from the receivers of the Marmon Motor Car Co. certain real estate, factory buildings and automobile equipment, and to manufacture and sell Marmon automobiles, Miller Special automobiles and racing cars, aviation and marine motors, automotive equipment, etc.

Officers and directors were identified as: H.A. Miller, President; Preston Tucker, Vice-President; R.C. Street, Secy-Treasurer. The official address was familiar . . . 1001 York St., Indianapolis.

Class A and B stock issues were authorized: 200,000 of $10 par Class A with 72,000 to be outstanding; 500,000 of no-par Class B all to be outstanding. Class A to have preference over Class B as to assets and .60 per share annual dividend. Issued for the purpose of acquiring from receivers a portion of the real estate and factory buildings and equipment of Marmon Motor Car Co., for working capital, and for other purposes.

American Automotive Corp. stock (50,000 Class A and 500,000 Class B shares) was offered in April 1934 by Panton & Co., Inc. New York.

Class A were offered at $10.10 cash or for 10 cents cash and the remainder ($10.00) payable by assignment of possible Receiver's dividends which might be declared and/or paid to debenture note holders and/or open-claim creditors of Marmon Motor Car Co.

Class B were offered at 10 cents per share, cash plus assignment of debenture notes and/or open claims against Marmon and/or by exchange of Marmon Motor Car Co. stock.

The Prospectus of American Automotive Corporation was reported[14] received in Indianapolis by former shareholders of Marmon Motor Car Co. in April 1934. According to the Prospectus, A.D. Sterner, representing the new corporation, had acquired from the receivers an option to purchase at a price of $232,500 the principle assets of the Marmon Company.

Moody's Manual of Investments listed American Automotive Corporation in the 1935 edition and noted the development under Marmon Motor Co., but made no mention of it the following year. This was probably the result of the court decision, reported in the following paragraph, to refuse to sell the property at the current valuation. That action sealed the fate of the Marmon car! Preston Tucker would try to enter the business again a decade later.

## TRANSFER OF JURISDICTION

The Indianapolis Star reported that on April 24, 1935 Judge Joseph T. Markey of Superior Court had ordered the immediate sale of machinery and equipment of the Marmon Motor Car Co. His efforts to dispose of the real estate of the company, valued with the machinery at approx. $500,000 had been unsuccessful. The court refused to sell at a sacrifice to creditors and stockholders and the Judge announced he would lease the buildings for two or three years and await a more favorable market. Machinery and equipment valued at $40,000 was ordered sold within thirty days. Part of the machinery had been sold by the former receiver (Fred C. Barnicol); Judge Markey was reported to have recently appointed Wilfred R. Boreinstein (See Marmon Motors, Inc.) to be co-receiver with Robert Adams and had appointed Will C. Kern as attorney for the receivership. Judge Markey had assumed charge of the receivership on January 1, 1935 and had expressed his determination to curtail the operating cost of the receivership, already $90,000.

On May 23, 1935, Marmon Motor Car Company was transferred to Federal Court, from Marion County (Indiana) Superior Court, for reorganization under Section 77B of the Federal Bankruptcy Act.[16]

The Receivers reported to the Federal Court that during the period of the Superior Court receivership, a total of $331,575.76 was realized from the sale of equipment and rental of buildings. Of that, $305,835.59 was spent in operating costs, taxes, preferred claims, attorneys' fees ($9500) and a three percent dividend ($45,361.57) to creditors. Each of the two receivers was paid $17,500.

Reorganization of the Marmon Motor Car Co. was formally approved by a final decree entered in Federal Court by Judge Robert C. Baltzell on December 31, 1936, under Section 77-B of the Corporate Bankruptcy Act in Proceeding No. 8465. The Company was declared insolvent and consent of the 2659 shareholders was not required.

Under the plan, holders of allowed claims were to receive an equal amount of Certificates of Beneficial Interest, bearing 6 percent interest from date of issue. These were exchanged for old 5 percent debentures due February 28, 1937 of which $1,214,099 were outstanding and for other allowed claims.[13]

The Reorganization Plan established a three-person Board of Trustees who would administer all property and who would have broad powers to manage, lease or liquidate the holdings over a long period. Proceeds therefrom were to be paid to creditors until **they would have received full payment on amounts due.**

On January 1, 1937 the value of the holdings was $1,355,530 whereas creditors' claims amounted to only $1,600,000. Should the asset value ultimately exceed the creditors' claims, the additional **proceeds were to be prorated among the stockholders.**[13]

Otto J. Feucht representing banking interests, A.A. Anderson and Homer H. Davidson representing creditors were named trustees.[2] In the event of resignation of Mr. Feucht, his replacement was to be chosen by the Indiana National Bank, Fletcher Trust Co. and the liquidating trustees of the Fletcher American National Bank; of Mr. Anderson, by the Hayes Body Co.; of Mr. Davidson, by Automotive Credit Services Co. On January 12, 1937, legal title to all company property and assets was transferred to the trustees.

**On November 22, 1937,** the trustees sold Plant 2, 22 acres of land with 539,000 square feet of building floor space. The buyer was Stewart-Warner Corp. of Indiana. Payment was to be made over five years and include liquidation of a $25,000 mortgage to the Indianapolis Life Insurance Co. After the sale, the remaining property was the 9 and 14 acre sites designated Plants 1 and 3.

**In July 1938,** 24000 square feet of floor in Plant 3 was leased to Stokely Brothers & Co. Inc., bringing leased space to 100 percent.

On December 10, 1938, a liquidating dividend of 4.5 percent was paid to holders of record on that date of Marmon Motor Car Co. Certificates of Beneficial Interest. On May 1, 1939, outstanding Certificates were $1,651,697; they had been $1,729,525 one year earlier.

During World War II, Plant 1 was sold and a new plant erected for the manufacture of Curtiss-Wright aircraft engines. In 1945, the Eli Lilly Co. acquired both the Plant 1 and Plant 3 sites and buildings.

## MARMON MOTOR COMPANY OF INDIANA

The Office of the Secretary of State, Indiana, recorded the chartering on January 24, 1969 of the Marmon Motor Company of Indiana. The principals were Keith Buckner and Gladys J. Elias of Cumberland, Indiana, and Richard H. Crokin of Indianapolis. The charter was revoked on June 30, 1978.

1920 Marmon Model 34B Four-passenger Touring, Serial 4200679, Engine 2771.

CHAPTER FOURTEEN

# MARMON MUSINGS

*A Potpourri of Vignettes*

## PART 1 —Marmon and Aviation

"On April 7, 1917 the Nordyke & Marmon Co. went to war with the German Empire."[1] All of the 3000 men and women together with the company's entire equipment was offered to the U.S. Government on that day by President Walter Marmon and Vice-President Moskovics.

"Within a few weeks, the company was commissioned to build 1000 Hall-Scott aircraft engines for training aircraft. Immediate construction was begun on a new building on the '21 acre site;' *(This would become Plant 2)*. Surveying began in June 1917 and motors were being produced in November of the same year. The contract was the first motor contract given by the U S Government and the Hall-Scott motors were the first 'turned-out' by any motor car manufacturer."[1]

A survivor of the Hall-Scott production with Nordyke & Marmon plate attached is displayed by the Ford (Edison Institute) Museum in Dearborn, Michigan.

The engine was a four cylinder water-cooled unit having individual cylinders, with sheet metal water jackets, on an aluminum crankcase and having an overhead camshaft with vertical shaft drive at the rear. Rocker arms were exposed.

It had a 5.25 bore and 7.0 inch stroke. It weighed 415 pounds and measured 49.25 inches long over the propeller hub. It featured full pressure lubrication and was rated an undisclosed power at 1450 rpm. Ignition was by dual Dixie magnetos.

No sooner was Hall-Scott production underway than the firm was asked to prepare for the production of the Liberty aircraft engine, a V-12 intended for combat aircraft.

"Nordyke & Marmon Co. was the only motor car manufacturer to produce two distinct and absolutely different types of aircraft engines.'[1]

'Again the factory had to be increased in size, until the aircraft machinery buildings alone contained over a quarter million square feet of floor space (all on one floor).' As late as 1929, this would be the second largest "room" in the automobile industry.

The increased office staff including government employees and Army officers necessitated the enlargement of the office building by erection of a three story structure right on the corner of Kentucky and Morris streets.

'Approximately 5000 (2000 additional) men and women were employed. New cafeterias, one with capacity for 3000 employees, were erected adjacent to Plant 2. The company chartered the street car line running to the (central) city and 'manned' it with feminine operators thus releasing men for more strenuous work.'

In the fall of 1918 a pennant was to fly over the plant which was building and shipping accepted engines fastest in relation to their quota for the preceding month. The Liberty Engine builders were: Ford Motor Co., General Motors Corp, Leland-Lincoln Motor Co., Packard Motor Car Co. and Nordyke & Marmon Co.

On Nov. 1, 1918, the first award of Champion Liberty Engine Builder was made to Nordyke & Marmon for exceeding their quota by shipping test-perfect engines at a rate of 246 percent of quota. A flight of Liberty-engined military aircraft, led by a N & M Liberty-powered plane landed at the Speedway for the presentation ceremony. As H.R. Heiskell, treasurer, accepted the award, the Speedway again resounded to cheers for a great Marmon victory.

A parade of 4300 Nordyke & Marmon employees through downtown Indianapolis followed; company officials and visiting dignitaries rode in a string of Marmon cars. It was a mile and a half long.

Of 20,478 Liberty Engines built, 1000 (Serial 4402-5401 were built by Nordyke & Marmon. A contract with Duesenberg for 1000 engines was transferred to General Motors. Lincoln and Packard built 6500 each, Ford 3950 and General Motors 2528.[2]

The late Roy Milner (Pontiac Chief Engineer for some 30 years) worked on the tooling to produce the Hall-Scott engine. He informed the author in an interview that Nordyke & Marmon also manufactured the wooden propellers to go with the engines they produced. The Marmon pattern shops, so large that they were operated as a division, produced the props.

A full page advertisement in the Aero Digest for February 1931 offered "NEW 100 H.P. AIRPLANE ENGINES" for unlicensed ships, wind wagons, racing cars or ground school use. The Peerless Merchandising Co. of Detroit offered these brand new Hall-Scott 4-cylinder airplane engines **built by the Marmon Motor Car Co.** for the U.S. Government and carefully stored in their original boxes."

William G. Stout had been successful in convincing a number of Detroit and national executives to invest $1000 each in his aircraft manufacturing business in Detroit. They included Henry Ford, Edsel Ford and Edward G. Budd. Stout proposed to design a commercial plane which would outperform the DH-4, then widely used, so notably that it would take over the air mail and other commercial air service. It would be an all-metal aircraft seating 8 passengers and flying twice the speed of the DH-4 with fuel reserve for 2.5 hours flight. It would be powered by a single engine — the Liberty V-12 initially developed during the first World War.

A stockholder in the Stout enterprise, Mr. Newton Skillman, happened to be at the Marmon plant in Indianapolis and noted the nickel-plated Liberty Engine displayed in a glass case in the Marmon lobby; the display bore the Champion Liberty Engine Builder citation. When the display was mentioned, his Marmon host commented that "the war is long over and we plan to junk it — if we get a couple hundred dollars for it we'll be lucky." "Sold" said Mr. Skillman who promptly wrote out a check and directed that the engine be shipped to Stout Aircraft at Dearborn, Michigan, where shareholder Henry Ford had provided a building at nominal cost. The aircraft was constructed, the N & M Liberty installed and a series of flight demonstrations undertaken. Henry Ford was so impressed he bought out the other shareholders and directed work on the Ford tri-motor to begin.[1]

The Marmon organization had been very successful in diversifying into an emerging technology industry with the Jenny Electric Co. venture before the turn

1910 Model 32 Four-passenger Suburban (three views)
*Owner* — Ronald C. Blanding
Winner of the 1911 Indianapolis 500 Mile Race
*Owner* — Indianapolis Motor Speedway Museum

Model 34B Two-passenger Speedster with detachable top. "The Corvette of the Early Twenties."
*Owner* — Craven Foundation
Inset: Eagle and traditional Pyramid on 1927 Little Marmon radiator.

Fig. 14-1 The Liberty Engine that powered the first Stout all-metal transport.

of the century. The second recognition of an emerging technology industry led to preemption of the parent product . . . Nordyke and Marmon gave way to Marmon Motor Car Co. The Jenny Electric Co. had been profitably disposed of to a major competitor and the Nordyke name and milling machine business were sold to Allis-Chalmers.

The budding aviation industry was also being watched intently in Indianapolis.

Marmon had purchased a Stinson cabin bi-plane, in late 1926, for executive use. G.M. Williams had been general manager of the Dayton-Wright Aircraft Division of General Motors prior to joining Marmon. It was to be expected that Marmon Motor Car Co. would be a pioneer in corporate aircraft ownership.

On April 2, 1927 *Automobile Topics* magazine reported that an airplane service for passengers and freight was being planned between Indianapolis and Detroit. The leader of the group of Marmon Motor Car Co. officials responsible was Hal L. Purdy, vice-president and assistant general manager. Associated with him were G.M. Williams, president; A.J. Rogers, assistant vice-president and J.A. Bohannon, Purchasing Agent.

The same magazine reported in its October 29, 1927 edition that H.H. Brooks, the sales manager, was undertaking a 5000 mile swing through the eastern and southern states for the purpose of conducting a series of sales meetings. He would use the Marmon corporate Stinson 'Detroiter' aircraft and accomplish in twelve days a mission otherwise requiring upwards of a month! Stops in 26 cities were scheduled. Captain J. Nelson Kelly, Marmon chief pilot, estimated 50 flying hours for the trip.

Brooks was accompanied by Ray J. Barbin of the engineering department for the purpose of conferring with service managers.

*Automotive Industries* magazine for November 17, 1928 (p743) reported that a Marmon affiliate was to be formed for the manufacture of commercial aircraft and that a highly regarded aircraft designer was to join the new venture.

Specifically *Automotive Industries* reported "Marmon Execs Back Plane Co.; "Will be separate organization but the Marmon officers will aid in development." "H.H. Brooks, retiring General Sales Director of Marmon (12 yrs service) will head a new company in Indianapolis to develop and manufacture, on a large scale, commercial aircraft engines and later the production and nationwide distribution of airplanes." This organization concept was to some degree similar to the Jenny Electric Co. relationship.

The article continued: "It is stated that in all likelihood Col. Howard Marmon, while retaining the Vice-Presidency of Marmon, will be responsible for engineering development of the aircraft engine. Mr. Williams said that the complete plane will

Fig. 14-2 G. M. Williams with the corporate Stinson-Detroiter bi-plane.

be the development of one of the best known aircraft engineers in the world and the culmination of years of experimental work. Was it Eddie Stinson, the Northville, Michigan builder of aircraft so successful that today the crowded Ford Museum displays two Stinson Detroiters, each uniquely famous for its exploits.

Did the tragic death of Stinson in Chicago, during an emergency weather landing, end the Marmon airborne ambitions? More likely the advent of Wall Street representation on the Marmon board of directors brought a new perspective of financial realities. Or had the Midwest Aircraft Company (shelter for the Marmon Sixteen cylinder engine project) trapped the press?

## PART 2 — Howard Marmon & Henry Ford A Parallel

As the Twentieth Century dawned, people of foresight recognized the potential of the automobile as a major transportation media. Hundreds of vehicle "brands" were offered to the public in the first decade. Two men did a superior job of evaluating the problems confronting the operator of a motor car on the roads of the early century. They marketed products which, were engineered to suit the buyers requirements in a highly satisfactory manner. Consequently they both achieved a standardized model run of many years, made automobiles a national rather than locally marketed item, and this era coincided with the foundation of our highway system.

These two men are engineers Howard Marmon and Henry Ford.

The one is universally known, the other perhaps until recently a total stranger to the reader. Their fame is probably in proportion to the volume of cars they produced but both achieved equal pinnacles of professional success; were in fact the pioneers of reliability and maintainability.

Experience with earlier models by both companies was (by each) drawn upon to produce classic models of longevity exceeded only by that of the Volkswagen, *'Beattle,'* (1939-1979).

The Marmon Model 34 appeared in 1916 and survived until 1927. (Eleven years so early in the industry is equivalent to a much longer period today). The Ford T appeared first in 1908 attired in brass radiator and flat "cowl" but achieved its classic form, with open and closed models, in 1917, beginning a run that lasted until 1927. Both "34" and "T" were "face-lifted" in 1926, acquiring all new sheet metal in an attempt, not very successful saleswise, to preserve the original personalities while conforming to current competitive styling.

Reliability was achieved by both firms in this era (1) by virtue of the continued refinement programs, analgous to the VW program, free from extensive redesigns and (2) by manufacture of a high percentage of the total product inplant, at a time when modern procurement and inspection communication was unknown.

The Model AA Ford truck drive shaft between transmission and torque tube was analogous to the Marmon 34 clutch output shaft which was between the clutch and transmission/torque unit. On both vehicles, the function of the shaft was identical — transmission of torque with compensation for assembly variations and frame deflection — and Ford selected the identical configuration[3], the use of a crowned-tooth, internal/external gear coupling at the forward end and the universal joint (for the torque tube pivot) at the rear.

By February 1929, the Model A Ford passenger and commercial car chassis had adopted a "six-brake system" with two additional brake shoes mounted inside the rear brakes and the brake drums correspondingly deepened.[4] that was the system favored by Howard Marmon and used on all Era Two Marmons; the parking brake was truly an emergency brake.

The Marmon 34 achieved additional leg room for an emergency third front seat passenger by the offsetting of the shift and brake lever assembly to the left; a practical solution when the transmission was controlled via linkage. Ford achieved a similar result by the novel placement of the Model A brake forward of, and in line with, the shift lever.

George DeAngelis, in his "Ford Facts — How the V-8 Distributor Was Designed" article for *Antique Automobile* (M-J 1984) recounts Ford designer Emil Zoerlein's frequent amazement "at Henry Ford's uncanny memory for mechanical detail and his ability to recall a mechanical item to fit his specific need." In that instance it was a Wright governor on an Armington-Sims steam engine. Not infrequently it was some detail that Henry remembered from Marmons he had owned or noted in the owner's books or press coverage.

The flat-head Ford V-8 mounted the oil pump on the rear of the camshaft as did the Model 34 Marmon. It also mounted the ignition distributor on the nose of the camshaft as did all V-4 Marmons. The combined concept gave Ford a camshaft free of integral gears, like the Model 34, thereby making remanufacture feasible and timing more precise. The flange mounting of the coil on the distributor of the V-8 paralleled the Model 34B 1920-23 design. The generator and fan assembly of the 1949 V-8 was clamp attached to the intake manifold by a single bolt, as was the Model 34B/C generator. Such a parallel cannot be made with brakes. The exact opposite applied: Marmon brakes were increased in diameter from the prior year in 1905, 1906, 1907 and 1908 and significant mechanical advances were made as well. Ford persisted throughout the Model T Era with notoriously inadequate miniscule brakes; owners in hilly terrain had to resort to retrofitting 'Rocky Mountain' brakes. When Marmon-Herrington began converting Ford cars and trucks to all-wheel drive, the conversion included M-H hydraulic brakes; when Fords did have hydraulic brakes (Lockheed type) in 1939, the conversion still included 'M-H hydraulic brakes.'

When Henry Ford and his firm finally broke the long silence which followed the announced end of Model T production after 20 years, it was accompanied by an introductory advertising campaign which dwarfed any prior effort.[5] For five straight days, Monday through Friday, Nov. 28-Dec. 2, 1927, Ford ran full page advertisements in all of the nation's 2000 English language daily newspapers.

The first ad, signed by Henry Ford, discussed the new car in general terms. The second and third ads promised the vehicle would be low-priced and disclosed mechanical specifications. The fourth provided a photograph of the Model A and revealed the price. The fifth summarized the preceding four.

Twelve years earlier Nordyke & Marmon had launched their Model 34 with a series of multi-page ads, in weekly magazines, which used the same suspenseful promotion sequence (and compounded the suspense by concealing the manufacturer's identity until the fourth ad). Henry Ford must have been impressed by the Marmon introductory series for he bought at least one for his personal use over the next five years and very likely recalled the series in 1927 and brought it to the attention of the N.W. Ayer agency.

## PART 3 — House Organs

The founders of Nordyke & Marmon early recognized the importance of communications . . . both within the organization and with the public in general as well

1924 Model 34 C Four-passenger Sport Speedster
*Owners* — George P. and Stacey P. Hanley

# MARMON 34

1930 BIG EIGHT Five-passenger Sedan
Former owner — Russell E. Stadt
1932 Model 8-125 Convertible Coupe
Owner — Philip W. Belote
1928 Model 78 Deluxe Sedan
Owner — Bruce S. Williams
1929 Roosevelt Deluxe Sedan
Owner — Joe Franko

as their consumers. As in many other matters, where precedent was lacking, they established the recognized state of the art.

The first company periodical, although characterized as their first house organ, was for practical purposes a monthly national magazine. The 'Mill Stone' appeared in 1876 when house organs and national magazines were both rarities, and national advertising was, as Marmon described it in 1929, 'in swaddling clothes.'[1] The editor during many years of publication was David H. Rank who subsequently became a well-known journalist as editor of the 'Ram's Horn.'

The 'Mill Stone' was 'devoted to the advancement and diffusion of practical science and home entertainment' and was published monthly by 'Nordyke & Marmon Co., Proprietors' who installed complete printing facilities in their offices. Its pages contained a variety of items including poetry, short stories, statistical data, weather prophecies, current event discussions, editorials and both financial and commercial reports. It even featured a 'boys and girls' department and, of course, information about the products of the company.

It was a semi-commercial endeavor with advertising space sold at a rate of $4.00 per inch; the annual subscription price (twelve issues including postage) was $1.00. Copy number two of volume four, dated December 1878 contained the announcement: "The largely increased demand upon our space, both in the body of the paper and on our advertising columns, make it appear as if we shall soon have to further increase the size of the "MILL STONE' and make it a 24 page paper at no distant day." In the same issue, the editor noted the extent of the circulation and market: "the MILL STONE has no superior, its circulation extending to every state and territory of the Union, reaching millers, millwrights, capitalists, grain dealers, farmers, stockmen, manufacturers, merchants, etc.".

The publication became successively the official organ of the Miller's Association of Indiana (July 1877, Vol. 2, No.9) and of the Miller's National Association (June 1878, Vol.3, No.8). Reflecting the revolution in milling processes wrought by the Nordyke, Marmon & Co. patent roller mill, the name was changed to THE MILLER'S JOURNAL & HYDRAULIC ENGINEER.

The MARMON POST began publication with the February 15, 1918 issue. It was a factory to dealer media, initially with 24 issues per year and diminishing until August 1926, Vol. 5, No. 10. It was replaced by the 'Marmon WASP' for September 1926, (Vol. 1, No. 1). The WASP recounted significant product and organizational developments and featured numerous letters from owners and photo-documented accounts of foreign or unusual trips undertaken by Marmon owners. It was directed primarily at owners or prospects and was typically of eight pages on large premium gloss-finish paper sheets. It was still being published as of January 1931 when Vol. 5, No. 5 was issued. The POST was issued semi-monthly except for the dealer convention in 1920 when the coverage consisted of the combined June 15/July 1 and July 15/August 1 issues. Mr. C.F. Kettering's address on that occasion is reprinted from the POST in Chapter Ten Part 2. At that time, the POST staff included a very competent, but unidentified, charicaturist, whose work illustrates the Kettering reproduction. The POST usually had eight pages.

The MARMON NEWS was published for Marmon owners beginning in January 1919 or earlier. It was an eight page publication, 10.25 x 13.5 inches. By 1921 it was being mailed to 16,500 owners and distributor/dealer personnel. Publication has been confirmed only through January 1920. It reported management and factory as well as product developments and was probably integrated in the MARMON POST.

THE FIRING LINE appeared with the April 18, 1929 issue. It was a factory to dealer media which explained selected technical features in considerable detail and compared the Marmon specific models with the competition — e.g., the Marmon harmonic rubber-hub crankshaft damper versus the 'front-flywheel' and Lanchester damper types and a listing of the type, if any, used on 52 car makes. It was issued randomly as required until No. 14, April 2, 1930 or later, in an 8.5 x 11 size.

THE MARMON FACTORY NEWS first appeared in October 1928. It was primarily an internal employee-oriented house publication. It described itself as the 'Personal Magazine of Every Employee' and bore the motto 'Of the Employee, For the Employee, By the Employee.' Sometime between March 1929 and August 1930 the name was changed to 'Men of Marmon' but the issue numbering system continued from the 1928 base.

It was a monthly typically consisting of twelve 8.5 x 11 pages. It was edited successively by Nickolas Moore and Howard Dirks. Circulation was about 3000 copies consequent to being mailed to the 1400 dealers as well as to a select list of manufacturers who requested it or with whom house organs were exchanged.

It recognized significant company developments, both at Indianapolis and throughout the field organization, provided the employee familiarization with members of management, and recounted employee personal, recreational and civic activities. Somewhat in the 'Mill Stone' tradition it offered poems, personal and gossip columns, recipes and cross-word puzzles . . . even a classified ad column. Unlike the 'Mill Stone,' it had no subscription fee rates and left other magazine coverage to the publishing industry.

'The Marmon-Herrington NEWS' began publication with the September 1942 issue. It served two important functions: Historical documentation of a rapidly growing firm and wartime employee motivation to both increased work output and increased commitment to War Bond purchases. It appeared monthly through 19 and typically had 16 pages of 8.5 x 11 size; black and red ink were generally used. The firm produced a galaxy of vehicle products and the NEWS documented within wartime constraints their nature and destination.

## PART 4 — First-Choice of Rum-runners

Motor Magazine for January 1924 published a feature article "How Fast Can Your Car Go." This was described as an academic question with possible exceptions of Montana and Nevada where you might be able to drive at top speed. "You may not be permitted to drive at 70 miles an hour (Sound familiar?) but the power capable of driving the machine at that speed may be legally used in quick getaway, hill climbing and comfortable touring." Selected cars and their maximum speeds are listed below:

| **Make** | **Model** | **MPH** |
|---|---|---|
| Auburn | 6-63 | 75 |
| Buick | 6 | 60-70 |
| Cadillac | | 70 |
| Chevrolet | Superior | 55 |
| Cole | | 65 |
| Cunningham | | 70 |
| Ford | T | 45 |
| Franklin | | 50 |
| HCS | 6 | 70 |
| Kissel | Speedster | 65 |

| | | |
|---|---|---|
| MARMON | CLOSED | 60-65 |
| MARMON | SPEEDSTER | 80-85 |
| Mercer | Raceabout | 80 |
| Mercer | Open | 70 |
| Mercer | Closed | 65 |
| Packard | 8 open | 65-70 |
| Packard | 6 open | 60-65 |
| Rolls Royce | open | 73 |
| Star | F | 50 |

The superior speed capability of the Marmon 34 was widely publicized following the 1920 Indianapolis race when Barney Oldfield brought the field to 80 mph before pulling his prototype Marmon Wasp Speedster into the pit ramp.

A full page back cover advertisement by Marmon, in search of additional dealers, appeared on a major trade magazine about 1925. It extoled the success of a Wyandotte, Michigan dealership which had sold six Marmons in six months . . . in a town of perhaps 2500 population. It predicted a similar potential for other dealers who might take on the Marmon line.

The true circumstances on the shore of the Detroit River were not apparently understood in Indianapolis. The River was patrolled by US Internal Revenue Agents in armed speedboats in a not-very-successful effort to stem the flow of illegal alcoholic beverages from Canada.

The village of Ecorse, Michigan was known in that era to be a major base of illicit alcohol operations. The adjoining community was the highly respectable town of Wyandotte with lovely residential property, — and a Marmon dealer.

Were Wyandotte not adjacent to Ecorse, it is likely that there would have been but one or two prospects for Marmon ownership, e.g., the local banker and local business tycoon, and they would most likely trade them only infrequently. There is no doubt as to the service into which the Wyandotte Marmons were impressed.

When the author was asked to pinch-hit as the dinner speaker at a local service club when the intended speaker was hospitalized an hour before dinner, he elected to recount various unusual, and generally humorous, experiences and encounters consequent the Marmon restoration project in which he was involved. He told the preceding Wyandotte story early in the talk. When he concluded, a guest arose and said: "Mr. Hanley — I believe everything you said tonight because when I was a young boy, my parents would take me on a Sunday to visit an uncle in Wyandotte — he **was** a rum runner and he **did** own a Marmon!"

Reagor Motlow, President of the Jack Daniels Distillery in Lynchburg, Tennessee, purchased a 1924 Marmon 4-passenger phaeton with Panorama Top and used it until the repeal of prohibition in 1933.

During the Hanley's expedition to the Maine woods in search of the "Book of the Month Marmon," they encountered an antique shop proprietor, a Mr. Brown who provided the following Marmon anecdote:

He had grown up in Bethel, Maine where in his youth he lived next door to the local sheriff Harry King. He happened to be in Sheriff King's company sometime in 1922-24 when a phone call was received from the sheriff of Upton, Maine which was closer to the Canadian border. The call was to alert Sheriff King to a southbound rum-running Marmon traveling at high speed on the road leading to Bethel. The sheriff not having his gun with him, a guard at the county jail was recruited and the party elected to await the speeding car on the south side of the railroad overpass where they would be hidden from sight. A well informed enforcement officer in the 1920's knew the location of the Marmon fuel tank . . . in the cowl, and one shot was fired as the car sped past.

The abandoned Marmon was found about a mile away on the climb up Paradise Hill. The cargo was identified as 80 gallons of alcohol in cartons of "Maple Syrup" cans.

## PART 5 — Book-of-the-Month Marmon

A Marmon 34C was in the cast of characters when Louise Dickinson Rich wrote of the adventures she and her family experienced when "We Took to the Woods"[1] of northern Maine. It was a 1942 Book-of-the-Month selection which recounted their life alongside the Rapid River upstream from the community of Middle Dam. That community is located on the west shore of Lake Richardson and is accessible only by boat from the community of South Arm, except in winter when travel is over the frozen lake. Both communities were once transfer points on a stage/ship route to a New Hampshire resort.

Mr. Rich planned to make his living as a wilderness guide for hunters and fishermen. During a winter in the mid or late 1930s they drove in over the ice a 1924 Model 34C 7-passenger phaeton and a 1917 Packard Twin-Six. These were to provide routine transportation over the old stage tracks from the homesite several miles to the supply source at Middle Dam and to transport hunters and fishermen upon their arrival. Once each year the Marmon carried the Movie Newsreel photographers and their equipment to cover the annual Whitewater Championship competition. This was a national event held on the Rapid River because of the capability of the (Middle) dam to adjust the water flow for the occasion. Following the 1939 New England hurricane, the Marmon had the task of reopening the road which was blocked by many fallen trees.

Mrs. Rich's comments on the family vehicles are worth repeating here. "The Packard was a 1917 Twin-Six touring car, built on the general lines of a pre-revolutionary four-poster with canopy, and I always felt like Queen Mary when I sat enthroned in it." Its passing is confirmed by the appearance of its front axle assembly as a logging dolly in the post-hurricane photograph. "The Marmon was left-over from the days when Ralph (Mr. Rich) lived on The Outside. I won't regret the passing of the Marmon, known locally as Rich's big green Mormon, at all. It is a 1924 sports touring model at least half a block long. Because of its tremendous power, Ralph uses it for hauling, and I have to ride in back and watch through the rear window that we don't lose our load. I hate the thing — there is no place to brace my feet and the frame of the car is so long that a twig in a rut turns it into a catapult that tosses me helplessly into the air. The Carry-Road is nothing but bumps, so its like riding in a corn-popper. And besides I have a sneaking notion that Ralph is just a little bit fonder of the 'Mormon' than he is of me."

The author did not learn until 1980 that Mrs Rich had authored a sequel to "We Took to the Woods" titled "Happy the Land." Therein she recorded the demise of the Marmon: "The pitch up from Long Pool means to me that we are almost home from Sunday Cove, and that's where the Marmon dropped her transmission, an accident which proved fatal. **She** never went on the road again, thank God, but has spent the rest of her days powering the carriage of the sawmill. I never liked the Marmon, anyway, but Ralph went into full mourning for a week. I always thought, although I have no way of proving it — and far be it from me to pry into the past

THE NEW MARMON PHAETON

THE NEW MARMON COUPE DE LUXE

"It's a Great Automobile"

*. . . From the Model D 74 Catalog for 1925.*

THE NEW MARMON ROADSTER

THE NEW MARMON FOUR-DOOR BROUGHAM-COUPE

# MARMON SIXTEEN

Coupe owned by Robert Atwood reprinted from *Classic Car,* June 1979, by permission CCCA.

Convertible Coupe from *'Marmon Sixteen Impressions'* by Bill Harrah, Sedan and details from *Marmon — A Quest for Perfection* by Maurice Hendrey, from Automobile Quarterly magazine Vol. 7, No. 2 by permission of Automobile Quarterly Publications.

Fig. 14-3 The "Big Green Mormon" at work.

of any husband of mine — that the chief reason for his sorrow was that the car had a great sentimental value to him, from the good old days when he was free and twenty-one. It must have been a swell parking car."

In 1980 the author and his wife, ever on the lookout for elusive body parts required for the restoration of their 1924 Marmon 4-passenger phaeton, sought to learn the fate of the Rich's Marmon. At that time they were unaware of the documentation of its passing from active service and presumed the remains might still be around. A letter was carefully addressed using all references to location in the book, even to a map, on the envelope. The Postal Service returned it, denying the existence of a South Arm, Maine. We then traveled to South Arm, crossed Lake Richardson on the outboard powered mailboat, and continued on foot down the Carry-Road to the Winter House of the Richs by the picturesque white waters of the Rapid River. It was abandoned and the only motor vehicle was a rusting Reo Speed Wagon chassis. Subsequent conversation with some of the principles in the book still in the area led to acquisition of a copy of the sequel book and the knowledge that about 1950 someone had rescued the remains and transported them by raft to The Outside.

About 1970 the author had purchased the forward half of a 1924 Marmon in Romulus, Michigan which had also run a sawmill, but in Mount Clemens, Michigan. In 1982 a friend in Virginia called to say that he had answered an ad offering Marmon parts and the party identified them as having been from Middle Dam. During the raft crossing of the lake, the salvagers had pushed all the body parts overboard, retaining only the chassis.

## PART 6 — The Marmon Sixteen Thesis

Seniors at the University of Detroit College of Engineering are (or were) required to conduct a thesis project during their final (or fifth) year. The author's brother had acquired the rights, jigs and fixtures for the manufacture of the (Marmon) Sixteen engine in 1940. As a result the author, jointly with Robert Gies and Eugene Hemming, undertook a thesis which consisted (1) of a literature search and assessment effort to determine if progress made by the industry, as represented by 26 car models during the ten years following the introduction of the Marmon Sixteen, was complementary (i.e., catch-up), or supplemental to the advanced performance of the engine when designed, and (2) of dynamometer testing of a Marmon Sixteen engine for the purpose of evaluating possible improvements including dual carburetors applicable to the fire apparatus application.

Each modification reported by the 26 car model manufacturers was mathematically evaluated or otherwise assessed to determine the contribution of each to the improvements claimed during the period 1931-1941. About the only item which it appeared could be adopted with benefit was the ducting of cold air intake from forward of the radiator as Cadillac had done briefly before their new engine was available in 1936. Some unexplained improvements appeared to be relatable only to carburetor substitutions; an assumption confirmed when the original Stromberg DDR-3 was replaced by the Stromberg EE series carburetor specifically calibrated

Fig. 14-4 The "Marmon" hauling hurricane pine, 1939.

Fig. 14-5 Eric "Gene" Hemming with the SIXTEEN engine en route to the dynamometer at U. of Detroit. 1938 LaSalle is the author's car.

for the Hanley application. The irregularities of the original torque curve were eliminated and the peak torque increased.

The engine used in the thesis was removed from a used Marmon automobile procured for the purpose. Compression pressures after valve reseating averaged a uniform 100 psi compared to a new engine value of 120 psi. It was decided to use the cylinders in that condition but to replace all bearing shells to prepare the engine for full throttle operation at rated rpm.

On the dynamometer at the University of Detroit, the engine developed 160 horsepower at 3700 rpm with the original carburetor and had a flat spot in the power curve in the 2700-2900 rpm interval. When the new Stromberg EE-3 carburetor was substituted, 177 horsepower at 3400 rpm was developed and there was no evidence of any irregularity. The original carburetor gave a full-throttle brake specific fuel economy ranging from a low of 0.57 at 1500 rpm to highs of 0.63 at 600 rpm and 0.70 at 3400; the EE-3 showed a low 0.515 from 500 to 2600 rpm and a high of 0.56 at 3600 rpm.

The thesis literature study and evaluation led to the conclusion that individual intake ports for each cylinder had resulted in improved breathing and improved uniformity of mixture distribution over all the cylinders. The Marmon SIXTEEN head already had identical passages to every cylinder — four siamesed ports per head; industry improvements had resulted from splitting the two siamesed ports, usually used, with two single ports on in-line sixes. Presumably the SIXTEEN would have benefitted from a redesign of the head which would place a dividing wall in the head and manifold at each siamese port. However costly pattern, core-box and fixturing costs would be incurred.

An independent conclusion of the thesis was that pistons with a concave upper surface would improve breathing and charge mixing. During the 1970s and early 1980s, the concept has found wide acceptance for production automobile engines subject to fuel economy regulations. Procurement and test of pistons on one bank was recommended post-thesis.

Due to time and economic considerations, the remaining thesis efforts were directed toward the design of a dual carburetor installation on the existing cylinder heads.

Ideal carburetor metering occurs under conditions of uniform air flow; the successive individual intake gulps result in less than ideal flow and the flow at the carburetor and in the individual runners must be made as even as possible. The most obvious dual manifold concept was the simple division of the original into identical forward and rearward manifolds, each mounting a dual carburetor. However, the flow from a specific throat would be forward for half of the full 720 degree cycle (sixteen firings) and rearward for the other half. Excessive condensation and inertial flow delays would result. Throttle response would be poor unless masked by otherwise unnecessarily rich mixture.

It must be remembered that the SIXTEEN engine was a very tight fit in the engine compartment; the 45 degree bank angle and the radical downdraft carburetion both increased the height above the crankshaft centerline. Only the first considered dual manifold would fit within the unmodified compartment. However, the objective of the thesis was a design applicable only to motor fire apparatus where additional overhead room was available.

The next configuration investigated was the same as the first with one pair of runners reversed or crossed. Totally unacceptable cessations in flow combined with successive pulses in opposite directions to rule out the concept.

The most favorable configuration used two identical manifolds with the throttle barrels arranged in a fore and aft rather than in a side by side location. This carburetor placement was ideal for linkage and plumbing connections and to some degree retained the best feature of the original carburetor — the throttle axes were parallel to the manifold runners so that part-throttle mixture distribution was not biased to one end of the manifold. This design would also lend itself to the use of two dual horizontal carburetors; such an installation might not exceed the original engine height. The maximum interval between pulses would be one quarter of the 720 degree cycle. Three successive intervals would be 180, 135 and 90 degrees on all barrels. Directional reverses would be limited to about 90 degrees.

Graduation in June 1942 left no time for procurement or test of manifolds and the associated procurement of specially calibrated carburetors. During the course of the thesis Pearl Harbor was bombed and normal commercial activity ceased. There would be no custom fire trucks for the duration. Instead air-raid trailers and firetugs, propelled and steered by water jet reaction forces, became the products

Fig. 14-6 The SIXTEEN engine on the dynomometer at the University of Detroit, College of Engineering in the spring of 1942.

Hanley Triple Combination Fire Apparatus rated 1,000 gallons per minute — 1941. Sixteen cylinder (Marmon) engine manufactured by Hanley Engineering Service. Prospect, Ohio.

The Story of Marmon-Herrington
MARMON-HERRINGTON
All-Wheel-Drive

BALANCED COMPONENTS

MARMON

| | Model | 34A | | | | 34B/C | | | | | D74 | | E75 | L | 68 | 78 | R | 68 | 78 | MR | 69 | 79 | B8 | 70 | 88 | 125 | 16 | | |
|---|---|---|---|---|---|---|---|---|---|---|---|---|---|---|---|---|---|---|---|---|---|---|---|---|---|---|---|---|---|
| | Year | 16 | 17 | 18 | 19 | 20 | 21 | 22 | 23 | 24 | 25 | 26 | 27 | 27 | 28 | 28 | 29 | 29 | 29 | 30 | 30 | 30 | 30 | 31 | 31 | 32 | 3X | | |
| 0 | | | | | | | | | | | | | | 5 | 11 | 13 | 20 | 15 | 12 | 16 | 5 | 4 | 6 | 6 | 3 | 2 | 13 | 0 | Sedan 4,5p,2,4 dr |
| 1 | Speedster 4p | | | | | | | 3 | 0* | | | | | 1 | 4 | 1 | 8 | 2 | 4 | 2 | 1 | 1 | 2 | 2 | 3 | 0 | 6 | 1 | Coupe 2/4p |
| 2 | Speedster 2p | | | | | | 4 | 7 | 6 | 3 | | 6 | 3 | 5 | 2 | 1 | | 1 | 0 | | | | | | | | | 2 | Speedster 2p |
| 2 | Roadster 2/4p | | | | | | | | | | 2 | | | 1 | 1 | 1 | | 0 | 1 | | 0 | 2 | 2 | | | | 2 | 2 | Phaeton # |
| 3 | Cloverleaf 4p | | | | | 1 | 0 | 1 | | | | | | 1 | 1 | 1 | 2 | 1 | 3 | 0 | | | | | | | 10 | 3 | Victoria 4p |
| 4 | Clov/Trg 4p° | 2 | 5 | 0 | 2 | 1 | 0 | 0 | 0 | | | | 0 | 3 | | 2 | 4 | 0 | 4 | 3 | 0 | 0 | 1 | 2 | 1 | 3 | 8 | 4 | ConvCpe 2/4,4 |
| 5 | Phaeton 4p | 0 | 4 | 0 | 0 | | | | 1 | 2 | 1 | 2 | | 0 | | | | | | | | | 0 | | 1 | | 11 | 5 | ConvSed 5p ° |
| 6 | Roadster 2/4p | | | | | | 1 | | | | | | | | | | | | | | | | 3 | | 1 | | 6 | 6 | Sedan 7p |
| 7 | Touring 7p | 2 | 3 | 0 | 0 | 6 | 2 | 3 | 1 | 1 | | | | | | | | | | | | | 1 | | 0 | | 3 | 7 | Sed Lim 7p |
| 8 | Phaeton 7p ° | 0 | 0 | 0 | 0 | | | | 1 | 6 | 0 | 0 | 1 | | | | | | | | | 1 | 1 | | 0 | | | 8 | Brougham 5p |
| 9 | Cpe/Spdr 4p | | | | | 0 | 0 | 0 / | 0 | 3 | | | 4 | 0 | | | | | | | | 1 | T | | | | 12 | 9 | Club Sedan 5p |
| 10 | Sedan 7p NH | | | | | 0 | 0 | 0 | | | | | | | | | | | | | | | | | | | | 10 | |
| 11 | Lim 7p Rubay | | | | | 0 | 0 | 0 | | | | | | | | | | | | | | | | | | | | 11 | |
| 12 | TownCar " | | | | | 0 | 0 | 0 | | | | | | | | | | | | | | | | | | | | 12 | |
| 13 | Lim 7p W'by | | | | | | | 0 | 0 | 0 | | | | | | | | | | | | | | | | | | 13 | |
| 14 | TownCar " | | | | | | | 0 | 0 | 0 | | | | | | | | | | | | | | | | | | 14 | |
| 15 | Coupe 4p/DeL 2p | | | | | | | 1 | 3 | 1 / | 2 | | | | | | | | | | | | | | | | | 15 | |
| 16 | Sedan DeL 7p | | | | | | | 0 | 0 | 3 | 0 | 0 | 1 | | | | | | | | | | | | | | | 16 | |
| 17 | Limou DeL 7p | | | | | | | 0 | 0 | 0 | 0 | 0 | 0 | | | | | | | | | | | | | | | 17 | |
| 18 | Sedan DeL 5p | | | | | | | 0 | 0 | 1 | 1 | 0 | 0 | | | | | | | | | | | | | | | 18 | |
| 19 | Limou DeL 5p | | | | | | | | | | 0 | 0 | | | | | | | | | | | | | | | | 19 | |
| 20 | Brougham 5p | | | | | | | | | | 1 | 0 | 1 | | | | | | | | | | | | | | | 20 | |
| 21 | Sedan Std 5p | | | | | | | | | | 7 | 6 | 4 | | | | | | | | | | | | | | | 21 | |
| 22 | Sedan Std 7p | | | | | | | | | | 0 | 2 | 4 | | | | | | | | | | | | | | | 22 | |
| 23 | Victoria 4p | | | | | | | | | | 1 | 4 | 1 | | | | | | | | | | | | | | | 23 | |
| 24 | Coupe Std 2p | | | | | | | | | | | 0 | 0 | | | | | | | | | | | | | | | 24 | |

* Includes 3 with Hume closed bodies.

o 34A Cloverleaf Roadster; 34B touring 4p.

\# L Speedster 4p, 68/78 Speedsters 4,6p; 69/79 phaeton 5p, B8Touring 7p.

E75, L, 68, 78 used FEDCO Code S/N without style code number.

° Style 8 was used for all closed cars on 34A; Style 5 was Weymann club sedan on L, B8 & 88.

Figure 14-7: NUMBER OF SURVIVING CARS FROM THIRD, FOURTH AND FIFTH ERAS by STYLE CODES

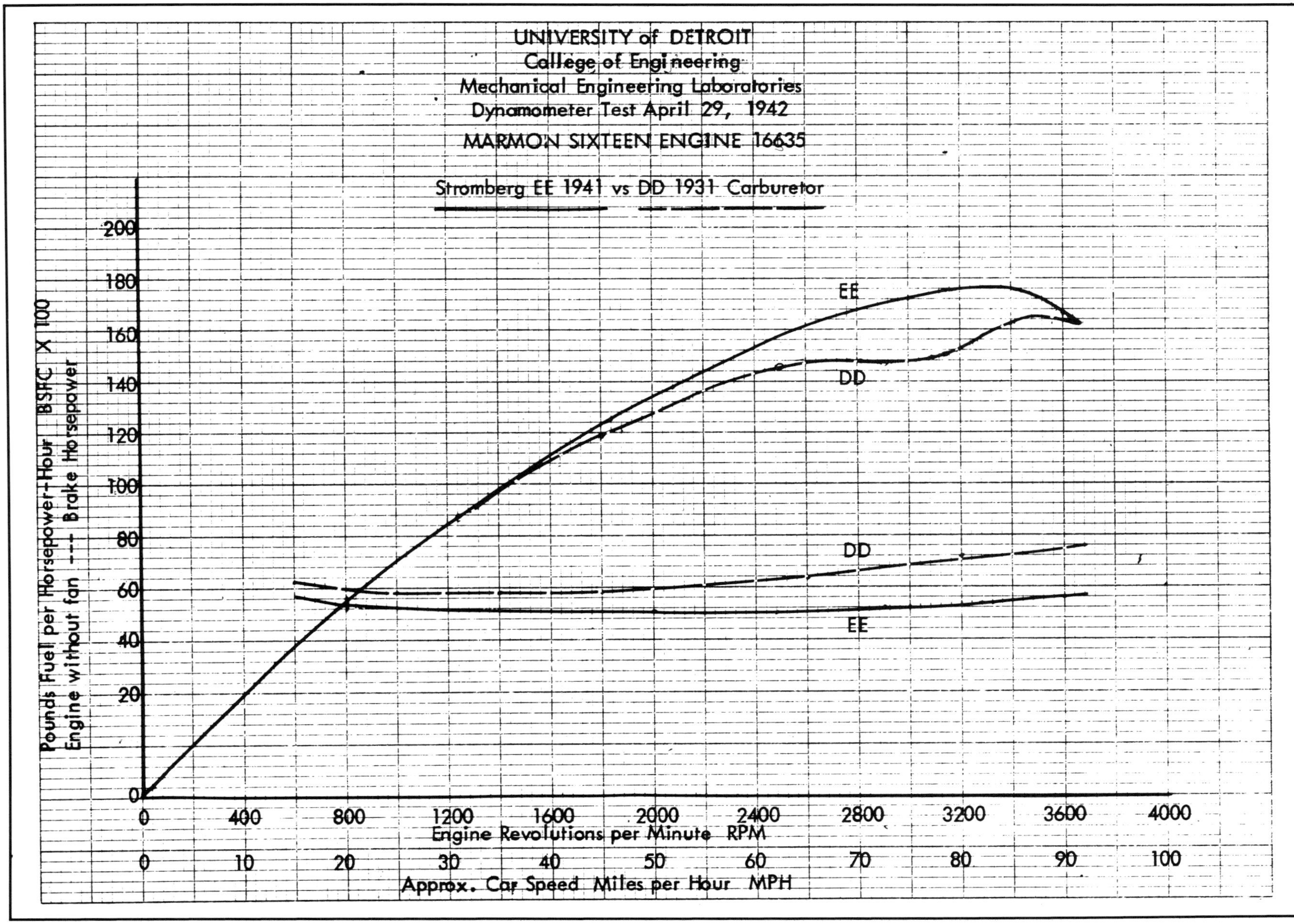


Fig. 14-7 Improvement in SIXTEEN engine with Stromberg EE carburetor. Note that this was not a new engine and had less than "new" compression.

of the Hanley Engineering Service. Postwar, propulsion research for the military and abortive efforts to launch a pleasure craft propulsion revolution replaced the fire apparatus line.

## PART 7 — The Survivors

The author estimates the surviving population of Marmon Cars to be less than 500. About 450 are known to be in the hands of collectors or museums. The estimated distribution of survivors is as follows:

| ERA | QUANTITY |
|---|---|
| 1 | 1 |
| 2 | 32 |
| 3 | 136 |
| 4 | 208 |
| 5 | 73 |
| | 450 |

The known survivors represent 0.4 percent of the some 110,000 Marmon built cars. That means that one out of every 250 cars remains — a surprising survival rate.

But the SIXTEEN has a phenomenal survival rate! The 73 survivors from a production of only 390 cars make a rate of 19 percent — one out of every five built remains!

From the Second Era, only 3 Model 48s exist and possibly only 2 qualify as survivors. There are no examples of closed styles on either First or Second Era Marmons. There are 24 Model 32s and 5 Model 41s.

The Third Era total of 136 includes 82 Model 34s and 54 Models 74 and 75. There are 17 Model 34A and 65 Model 34BC. Nine of the 34As are the then very popular cloverleaf roadster style. Among the 34Bs, the 2-passenger Sport Speedster — the Corvette of its time — accounts for 17 of the 54 Model 34Bs or about 30 percent. They are distributed 4 from 1921, 7 from 1922, 3 from 1923 and 3 from 1924. There are however only six examples of the 4-passenger Sport Speedster. Three are Model 34B and 3 are Model 34C. The two and four passenger versions of the Sport Speedster together account for 35 percent of the surviving population of Models 34B and 34C.

Most common of the other Model 34 surviving styles are the 7-passenger open cars at 25. Five are the Model 34A from the 1916-17 period; 13 are Model 34B and seven are Model 34C.

Only 12 Model 34 closed cars survive and are all Model 34B. There are five 4-passenger coupes, one 5-passenger sedan and three 7-passenger sedans. Three Hume custom bodied cars carry 2-passenger Speedster body codes to identify their speedster chasses; a sedan, coupe and victoria survive.

The 54 survivors from the 74-75 Sub-era are divided 35 for the Model 74 and 19 for the Series 75. Eleven five passenger standard sedans include 7 and 6 respectively, but only 3 of 9 two-passenger speedsters are Series 75. The eight four door open cars are evenly divided; only one of them is a seven passenger — a Locke bodied Series 75 Touring Speedster — and four are Series 75 Locke bodied five passenger Speedsters. There are examples of the very popular victoria — five 74s and one 75 —, and one each of the popular brougham style.

Despite the small percentage of open cars produced — less than ten percent for the 74-75 Sub-era — more than one third of the survivors are open styles.

There are 208 examples of Fourth Era Marmons. The three year Roosevelt, Marmon-Roosevelt and Marmon 70 sequence in the $1000 price class accounts for 31 percent or 14 percent of all survivors. The two-years of Models 68 and 78 accounts for 39 percent (68-19%, 78-21%) or 18 percent of all survivors. Fortunately single examples survive of the Models L, 68, 78 four passenger speedsters, of the Second Series 68 and 78 six-passenger speedsters and of the Big Eight LeBaron touring. Two Model 79 phaetons complete the survival of at least one of all four-door open styles. With the exception of the Second Series 78, all two-passenger speedsters or roadsters are represented. Considering the relatively few 8-79 and Big Eight survivors (9 and 17) there is surprisingly good coverage of the styles.

Figure 14-9 charts the Number of Surviving Cars from the Third, Fourth and Fifth Eras by style code number. The styles assigned to the number series during the Third Era appear with the numbers on the left margin and apply to the left half of the chart. Different styles were assigned to the number series for the Fourth Era and they appear on the right margin and apply to the right half of the chart. In certain cases an inactive number was reassigned; in that case the two styles are shown, separated by a slash; a slash may also appear on the chart matrix to identify the transition year. Every body style offered by the factory appears with either a survivor quantity or a zero entry.

The chart will assist owners of surviving Marmons in assessing the rarity of their specimens. It unfortunately emphasizes the large number of styles for which there are no known surviving examples. The greatest regret is the complete absence of any of the V-4 dual-frame cars of Era One. The sole survivor from that Era is the water cooled transition Model 50 with the semi-dual frame construction.

Only cars which have a serial number identification appear on the chart. This avoids any double counting with ownership changes but probably omits some cars from the count. Previously unknown and uncounted cars surface every year but the number is declining.

The high rate of survival of Marmon cars is partly the result of a surprising number of cars remaining with the original owner from youth to death. This was especially true of the Third Era cars, and of both the two and four passenger Sport Speedsters in particular. The author acquired his four-passenger Sport Speedster from the estate of the professional owner who had traded-in a Marmon and a Cadillac on it in the fall of 1923. Three parties called the author during 1984 seeking information on unrestored Third Era Marmons which they had inherited from relatives who were the original owners!

# BOOK TWO — The Story of MARMON-HERRINGTON

MARMON-HERRINGTON FORD ALL WHEEL DRIVE CONVERSION

# BOOK TWO THE STORY OF MARMON-HERRINGTON

## CHAPTER ONE

*"The Most Useful Trucks on Earth"*

### PROLOGUE *Enter Colonel Herrington*

Early in 1931 Walter Marmon received overtures from an entrepeneur who sought his financial support in the launching of a new enterprise in the automotive field. It is appropriate that the reader be introduced now to this unusual and dynamic man.

Arthur William Sydney Herrington was born in Coddenham, England on March 30, 1891 and came to New Jersey, USA, when he was five years old. He graduated from Stevens Preparatory School and the Stevens Institute of Technology at Hoboken, N.J. He began his professional career with the Harley-Davidson Motorcycle Company and when the United States entered the war he became a motorcycle dispatch rider with the rank of Lieutenant in the American Expeditionary Force in Europe. At the cessation of hostilities, he was discharged from the Motor Transportation Corps with rank of Captain. He subsequently rose to Major in the Army Reserve but never to Colonel.

The title was an unofficial one of respect by his associates in the Motor Transportation Corps during the post war years when he was a consultant to that agency and to several firms. Art Herrington liked that recognition and retained the title throughout his life. How intriguing that the architect of the upcoming Marmon truck products should espouse that title which was also retained for his lifetime by the architect of the past generations of Marmon car products, Colonel Howard C. Marmon.

While serving as a government consultant in Washington, D.C., Herrington met and married Nell Ray Clarke in 1926. She was a writer for several national publications at that time. The Honeymoon began with a trip to the Indianapolis 500 Mile Race . . . the new Mrs. Herrington spent her second night in the Dusenberg factory as she watched her husband and Fred Dusenberg assemble the racer to be driven by Pete DePaolo to fifth place.

Herrington was no stranger to the Classic. He watched the Marmon driven to victory by Ray Harroun in the 1911 race. He became involved first in 1922 when he was chosen as pit manager for the victor, Jimmy Murphy. He had the same assignment in 1925, that time for DePaolo who also made it to Victory Lane, and later served as Chairman of the Technical Committee.

Like the Marmons, the Herringtons were to become active in advancing the cultural life of Indianapolis. Mr. Herrington was a trustee of the Indianapolis Museum of Art (then the Art Association of Indianapolis), was chairman of the Fine Arts Committee, the Horticultural Committee and a member of the Executive Committee. Importantly, he was a member of the development committee responsible for building the new Indianapolis Museum of Art at Oldfields, the former J.K. Lilly estate — Mrs. Anne Marmon Greenleaf was the 1980-81 President of that Museum. He was a past president and trustee for the Indiana State Symphony Society which organized and sponsors the Indianapolis Symphony Orchestra. Mrs. Herrington also served on numerous civic committees concerned with cultural enhancement.

Over the years the Herringtons developed an art collection that was considered one of the best in Indiana and valued at several million dollars.

The Colonel was an avid sailor. His home at 4500 Kessler Boulevard, Indianapolis, contained an extensive library on the history of exploration, particularly that carried out under sail. He sailed frequently in the Bahamas and the Caribbean and attended the major yacht races. He founded the Out Island Squadron, a Bahama Islands yacht club, and was an official of the 1966 America's Cup Races. In 1969 he sailed the Atlantic as Captain of the 55 foot yacht "Integrity," with a crew whose average age was 72, for the 250th Anniversary of the Royal Cork Yacht Club in Ireland. His pride was his own 44 foot auxiliary sloop "Ruevalla" ("wanderer" in Arabic), and Colonel Herrington's Christmas cards often featured his boat. Indianapolis had a Royal visitor when the Persian Shah (Father of the Shah Reva Pavlavi) was a houseguest at the Herrington home prior to WWII.

Mr. and Mrs. Herrington traveled widely and were frequent guests of European royalty. The government of Portugal presented him with a set of five books, containing reproductions of all the old navigational charts which were developed by Portuguese explorers. He once raced with Prince Philip, Duke of Edinburgh in the Bahamas.

During the War, Colonel Herrington twice was the recipient of honorary degrees: on May 1, 1934 a Doctorate in Mechanical Engineering from his alma mater Stevens Institute and on October 2, 1943 a Doctor of Engineering as the Commencement Speaker at Rose Polytechnic Institute, Terra Haute, IN. The Stevens citation noted that Col. Herrington's knowledge of the terrain and soils in North Africa, the Near East and the Far East had contributed significantly to the success of American military transport operations in those war theaters. President Roosevelt appointed him as technical advisor to the American Mission in India during the War; he subsequently served (1949-50) as economic advisor to Pakistan.

The Colonel was President of the Society of Automotive Engineers for 1942 (Howard Marmon was the 1913 SAE President and Thomas J. Litle, Marmon Chief Engineer, was the 1926 president.) He was a member of the prestigious Newman Society and the American Ordnance Association. He was a past president of the Automotive Hall of Fame, Inc. then known as the Automotive Old Timers.

In 1962 Arthur W. Herrington was the recipient of a 'Distinguished Service Citation' from the Automotive Hall of Fame (formerly A.O.T.). Such awards are given annually to recognize the significant contributions of living individuals in a variety of automotive disciplines. Col. Herrington's Citation reads: *For his service to his country in peace and in war . . . an engineer, designer and builder of tanks and mili-*

Fig. 1-1 Mr. and Mrs. Herrington pose, with items of their art collection, for a Marmon-Herrington Christmas Card.

*tary vehicles . . . for the Marmon-Herrington equipment he produced.*

Mrs. Herrington preceded her husband in death in April 1966. The Colonel died in 1970 at 79 in an Indianapolis hospital following an accident in a small foreign sports car in which he suffered fatal head injuries . . . another automotive giant to die on the road. The Herringtons are survived by a son, Doctor Arthur Clarke Herrington, a nuclear physicist at Massachusetts Institute of Technology.

In "75 years of SAE," W.S. Sherman records that "during his (Herrington's) term, the SAE Automotive Technical Advisory Board became the War Engineering Board and greatly expanded its scope of action."

Now that the reader has an appreciation of the professional stature of this automotive great, he is ready to peruse The Story of Marmon-Herrington.

Nordyke died before the advent of the automobile, hence the Nordyke and Marmon Co. entry into the automobile field bore the "Marmon" badge which generally stood for Walter — Chairman, Howard — Vice President of Engineering, an inferred "Marmon-Marmon" badge. Now a new badge was to be born: "Marmon-Herrington" reflecting Walter Marmon — Chairman and Art Herrington — Vice President of Engineering.

Fig. 1-3 The faces behind the new MARMON-HERRINGTON badge.

Fig. 1-2 George Freers, Rose Poly '08, Chief Engineer, congratulates A. W. Herrington on award of honorary degree.

## PART 1 — The Beginnings
## A Precocious Stepchild Is Born

In the very same month that series production of the Marmon Sixteen got underway, a most significant event occurred. The Marmon-Herrington Co. Incorporated was chartered on March 13, 1931, under the laws of the State of Indiana. The full significance was of course unknown and unforseen that March day, but this fortunate development was to perpetuate the organization that began in 1851 past a 125th birthday . . . shared with the Nation's 200th . . . and on into the 1980s. It would in fact provide an unbroken chain of motor vehicle offerings from the first Marmon car of 1902 through the Marmon-Herrington combat vehicles of WW II to the Marmon heavy duty trucks of the 1980s, and the Marmon-Herrington multi-wheel-drive conversions of the trucks of the 1980s.

About the time when a Marmon '34' was driven coast to coast in some 5 days to demonstrate the potential for motorized military operations, a young Lieutenant was commissioned — Arthur W. Herrington, not long out of Stevens Institute of Technology with an engineering degree. By the end of World War One, Captain Herrington of the Motor Transport Section was distraught by the inferior performance of military motor transport in the European combat theater. He vowed then that never again should the US enter an armed conflict with inadequate transport. As a first step in that direction he became a consultant to the US Army and the US Marine Corps. The author has not been able to ascertain whether Herrington was involved in the 1919 evaluation of the Marmon 34 in the rugged terrain of the Mexican border lands. W.S. Sherman in his "75 years of SAE" (1980) notes that Herrington's dedication to development of better off-road vehicles was also based on personal experience with General Pershing's Mexican border forces.

He most certainly was involved in the Military Pageant and Exposition at the Holabird Quartermaster Depot of the US Army at Baltimore, Maryland on June 12-14, 1929. That event provided commercial truck builders an opportunity to compare the performance of their vehicles with military-designed prototype trucks on the same test course. All-wheel-drive trucks were demonstrated by Coleman, FWD (Four Wheel Drive Auto Co.), Freeman and Walter. The Christie Crawler Co. demonstrated 6x4 trucks with both steel and rubber tracks over the rear tandem wheels. The front and forward rear wheels were driven. It was a state-of-the-art exposition which just preceded Colonel Herrington's period of contribution to that art; his front-wheel drive axle patent application was filed on May 15, 1930.

Significant at the Holabird exercise was the Coleman 6x6 truck using single tires on all wheels on one truck and duals on all wheels of another truck. Equally interesting was the use on the FWD 4x4 truck of duals on all wheels with the outer front tires slightly smaller than the inner and consequently out of contact with a paved road surface. This concept had previously appeared, in 1928 on 4x2 fire trucks for small airports built by The Prospect Fire Engine Co. where Keenan Hanley was the Chief Designer. These "Airporter" vehicles were built on Model A Ford commercial chassis units. They successfully negotiated off-runway soil conditions with two-wheel drive by virtue of the dual front tires of differing diameter while providing normal handling on paved roads. The nineteen twenties may have been roaring elsewhere in the country but they were dreadfully peaceful for the armed forces. The world had been "made safe for democracy" and the League of Nations established at Geneva, Switzerland to assure it remained that way. Relatively new naval vessels were being sunk to comply with arms reduction agreements. Congress saw no need for military transport development or procurement . . . after all, the Army had more surplus WWI equipment than it could possibly utilize despite the thou-

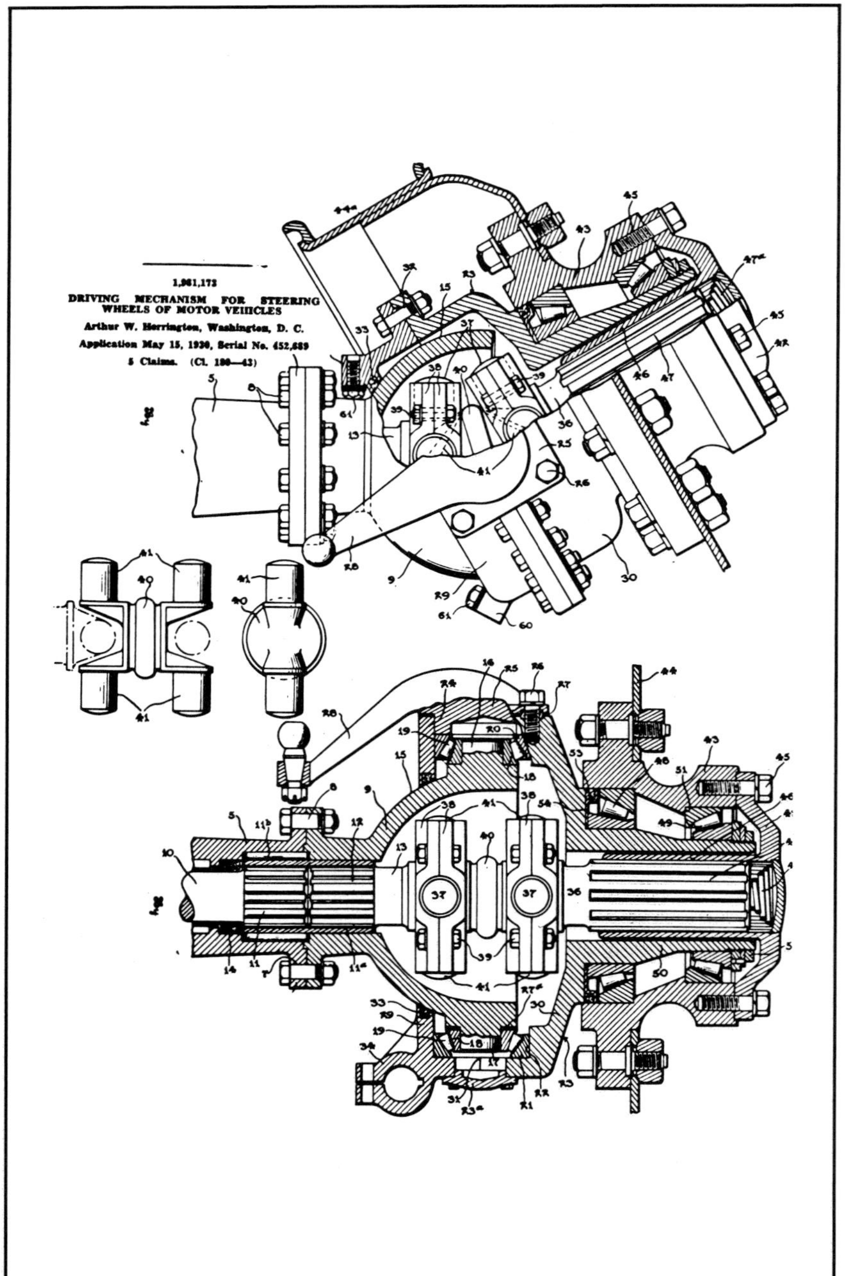


Fig. 1-4 The Herrington patent driving mechanism for steering wheels of vehicles.

sands of vehicles deliberately destroyed or abandoned when the AEF sailed home. And then the stock market crash in the fall of 1929 . . . and the roaring died to a hush.

The Congress did however recognize the rapid strides being made in aircraft technology — dramatically brought home by the Lindbergh transatlantic solo flight in 1927 and the beginning of scheduled air transport service. Hence limited funds were available to the US Army Air Corps for new combat aircraft and ground support equipment. Purchase of a fleet of refueling trucks was proposed and the specifications written around Colonel Herrington's concepts of an advanced design all-wheel drive chassis. But who would bid and manufacture such a vehicle? In 1931, Herrington correctly sensed that political forces were moving inevitably toward another world conflict. With the potential Quartermaster Corps order for refueling trucks in hand, he approached Fred E. Moskovics who had been an associate of the Marmon brothers in the youth of the Marmon car and who, despite his departure from Marmon in 1920, was the only non-family stockholder of the Nordyke and Marmon Co. other than A.R. Heiskell. More recently he had been President of Stutz through 1928.

On January 5, 1931 Herrington and Moskovics entered into an agreement of joint venture to develop and exploit Herrington's invention related to all wheel drive vehicles for which he had applied for patent protection on May 15, 1930.

## THE HERRINGTON PATENT

The usual form of universal joint is the Cardan type — a cross to which are pivoted a pair of yokes. An undesirable characteristic of that construction is that an irregular rotation of the output member results from regular rotation of the input member; the irregularity begins as the shafts depart from a common axis of rotation and becomes increasingly severe as the angle between the axes is increased.

The use of Cardan joints in a front drive application might appear to be vibration-free, and unfelt in the steering-wheel, when in a straight ahead mode. In actual practice, this is not so. Only on a flat horizontal pavement, and with zero wind yaw direction would the wheels be directed straight ahead. Also the wheels must be castered (and cambered) to provide directional stability to the vehicle . . . hence they do not operate in parallel planes.

Herrington's invention claimed avoidance of the irregular rotation, and its consequent unpleasant reflection in steering 'feel,' by the opposing use of a pair of Cardan joints connected by a 'shaft' of negligible length and mass; only the 'shaft' element would suffer the irregularity. The cancellation concept was well known; the innovation was in its application to powered steering knuckles. US Patent 1,981,173 finally issued to A.W. Herrington on November 20, 1934.

The following illustration shows vertical (bottom of page) and horizontal (top of page) sectional views of the design as they appear on the patent. In the vertical section, the wheel is in the straight ahead position; in the horizontal section, the wheel is in the extreme turn position. The other illustration is the small intermediate double yoke element which is the heart of the concept.

Other significant claims for the design were:

1. The axle shafts were independent of the U-joints and the U-joint/pivot joint assembly was separable from the axle housing to allow replacement of the axle shaft, in event of breakage, without other disassembly.
2. Use of a single seal to prevent entry of foreign matter.
3. "Means to adjust the pivot (steering) bearings which eliminates the placing of the weight of the vehicle upon screw threads or the like, ordinarily

employed for such adjustment."

4. General interchangeability of parts between left and right wheels for manufacturing and service logistics convenience.

## THE MARMON-HERRINGTON COMPANY

By the terms of the Moskovics-Herrington contract, each party agreed to use his best endeavors in the development of designs and devices relating to motor truck construction and to give such time as was necessary until the completion of arrangements for the manufacture of trucks using such devices. In that paragraph was the following provision: "It is understood that no definite obligations are assumed under the foregoing provision beyond a period of six months from the date hereof, nor beyond the time that the corporation hereinafter referred to is organized and arrangements made by said corporation to compensate the parties hereto."

By the terms of paragraph two, Herrington sold, transferred and assigned a one-half interest in all patents, patent applications, designs, devices and inventions then owned or controlled, or thereafter owned or controlled by Herrington, relating to motor truck construction and particularly to four-wheel drive motor trucks.

Paragraph three set forth the intention of the parties to form a corporation to which the parties were to transfer all of their rights in and to said patents, patent applications, designs, devices and inventions, and whose capital stock would be divided on the basis of 49 percent to each of the parties and 2 percent to them jointly.

Not surprisingly Moskovics involvement led to discussions with the Marmon Motor Car Co. officers. Negotiations progressed favorably . . . a treatment of the issue was prescribed which was not unlike that invoked when the Jenny Electric Company (ancestor of Indianapolis Power and Light Co.) was established in 1885. An independent firm would be created, sharing many of the Marmon officers and directors, and would be housed in available space within the Marmon complex in Indianapolis.

"Articles of Incorporation of the Marmon-Herrington Company, Incorporated" were drawn on March 10, 1931 and recorded on March 13, 1931.

In a letter dated March 10, 1931 Herrington and Moskovics proposed to turn over to the Marmon-Herrington Company an exclusive and non-assignable license covering any and all patents, applications for patents or patents pending which they then had together, with any improvements thereon and/or reissues thereof and to further assign, transfer and deliver to said company all rights and benefits under the then existing contract with the United States Army for thirty-three four-wheel drive trucks. In the same letter, Moskovics agreed to subscribe to one-fourth of the preferred stock and one-sixth of the common stock of said company on the same basis as the balance of preferred and common stock was subscribed for by the Marmon Motor Car Company. (Moskovics' subscription to 500 shares of common stock at a price of $1 per share and to 500 shares of preferred stock at a price of $50 per share was confirmed in the Opinion of Mr. Justice Eder in Moskovics vs. Herrington).

The letter continued: "In consideration of the above, the company is to cause to be issued, as fully paid up and non-assessable, 500 shares of its common stock to A.W. Herrington and 500 shares of its common stock to F.E. Moskovics out of an authorized issue of 3000 shares . . ." Based on this, the stock ownership would ultimately have been as follows:

| | Preferred | | | Common | Subscription | |
|---|---|---|---|---|---|---|
| Marmon Motor Car Co. | 1500 | 75% | $75,000 | 1500 | 1500 | $76,500 |
| F.E. Moskovics | 500 | 25% | 25,000 | 1000 | 500* | 25,000 |
| A.W. Herrington | - | 0% | | 500 | - | - |
| Total | 2000 | | $100,000 | 3000 | 2000 | $101,500 |

The letter also contained the following condition: The "Company is to employ A.W. Herrington as Chief Engineer at a salary of $600 per month for one year. It is further understood that you will cause Mr. F.E. Moskovics to be elected a director and Chairman of the Board and Mr. A.W. Herrington to be elected a director and vice-president."

The proposals in the letter were accepted and agreed to by a resolution of the stockholders of the Marmon-Herrington Company, unanimously adopted, at their first meeting on March 14, 1931. Moskovics and Herrington thereupon undertook their duties as directors, officers, and Chairman of the Board and Chief Engineer respectively.

The "Incorporators" and initial members of the board of directors were:

*F.E. Moskovics, Chairman, 1 Fifth Ave. N.Y.C., N.Y. — Incorporator
*G.M. Williams, President — Incorporator
*A.W. Herrington, Vice-President, 1661 Crescent Place, W.DC. — Incorporator
*A.R. Heiskell — Incorporator
*P.H. Noland — Incorporator
W.C. Marmon — H.C. Marmon — C.J. Sherer

MARMON SUBSIDIARY ORGANIZED TO MAKE COMMERCIAL VEHICLES

The Marmon Motor Car Co. stock subscription was in the names of G.M. Williams, P.H. Noland and A.R. Heiskell (President, Secretary and Treasurer of Marmon Motor Car Co.), all of 1100 Morris Street, Indianapolis. Walter C. Marmon and C.J. Sherer were not original subscribers (Fact Six, Agreement of Merger of Matteson Metals into Marmon-Herrington Co., December 1963). The initial press coverage as it appeared in the *Indianapolis Star* on March 14, 1931 (Page 1) and *Automotive Industries* March 21, 1931 is repeated below: "Announcement of the organization of the **Marmon-Herrington Co.** was made on March 13, 1931, through filing of the papers of incorporation. G.M. Williams stated that the new company would be a subsidiary unit for the purpose of making multiple-wheel-drive commercial vehicles of a highly specialized type. The new subsidiary will be headed by G.M. Williams, President, F.E. Moskovics, Chairman of the Board, Arthur W.S. Herrington, Vice-President and Chief Engineer. Walter C. Marmon is a Director and other Marmon Motor Co. officers and directors have corresponding capacities.

A segregated portion of the plant at Morris Street and the Belt Railroad was developed for Marmon-Herrington operations." (The address would be "York Street & Kentucky Avenue).

MARMON FORMS TRUCK SUBSIDIARY — WILL MANUFACTURE HERRINGTON DESIGN *(Automotive Ind.)* "G.M. Williams, President and F.E. Moskovics, Chairman of the Board, of newly formed Marmon-Herrington Co. announced that the company will manufacture multiple wheel drive trucks with features new to the commercial transport field in America."

"The truck design was brought to the company by A.W. Herrington who also will serve as Vice-President. Walter C. Marmon will be a Director."

"Operations will start as soon as possible to give delivery of trucks already sold. Orders on hand assure factory activity for some time. One of the plants of the Marmon Motor Car Co. has been taken over."

The announcement of the establishment of the Marmon-Herrington Co. in "Men of Marmon" for April 1931 provided a much more perceptive (but not newsworthy in 1931) assessment of the purpose: *"In making the announcement of the new subsidiary, President Williams said: The new truck was of a type "which the government uses for mobilization purposes" and "President Williams stated that company officials felt that opportunity existed to build up an intensive business in this line."*

The interlocking of the Boards of Directors is evident from the following:

| | Marmon only | Common to both firms | Marmon-Herrington only |
|---|---|---|---|
| Chairman | W.C. Marmon | | F.E. Moskovics |
| President | | G.M. Williams | |
| V-P Eng. | H.C. Marmon | | A.W. Herrington (V-P & Ch. Engr.) |
| V-P/Treas. | | A.R. Heiskell | |
| V-P Sales | | P.H. Noland | |
| Director | | A.R. Heiskell | |
| Director | | W.C. Marmon | |
| Director | | H.C. Marmon | |
| Director | E. Stout | | |
| Director | F.B. Johnson | | |
| Director | G.A. Ellis | | |
| Director | M. Newton | | |
| Director | J.A. Bohannon | | |

Note that the first four of the exclusive Marmon directors are the "uninvited" directors representing New York financial interests. Mr. Bohannon was President of Peerless and formerly a Marmon VP.

Early in 1931, Herrington appeared in the Marmon Motor Car Co. design room and Joe Felts was assigned to do his bidding on the drawing board. The project was design of a transfer case for the prototype truck (a transfer case receives power from the transmission output and directs power to forward and aft driveshafts; it may incorporate a gear-splitter, ie, an alternate set of gears which double the number of ratios in the main transmission).

## FINANCIAL CRISIS

Moskovics and Marmon Motor Car Company made only partial payments on their subscriptions . . . $10,500 and $20,200 respectively. The latter furnished space, tools and manufacturing facilities. Both parties defaulted on the balances of their subscriptions, precipitated by the take-over of Marmon control by a Committee of Creditors. It therefore became necessary for the Marmon-Herrington Company to seek financial support elsewhere.

An approach for financial assistance was made to Walter Marmon as an individual. Most likely he had been intimately involved with Moskovics in obtaining the original Marmon Motor Car Co. commitment. Mr. Marmon was highly regarded as an astute financier of significant personal fortune. In a later letter to Moskovics, dated January 9, 1932, Walter Marmon wrote: "You will remember that when the developments in the Marmon Motor Car situation made it impossible for that company to carry out its understanding with Marmon-Herrington, I felt a definite moral obligation to do all in my power to protect Art Herrington from a default in the performance of the pending Government contract and the consequent loss of his entire proposition. As a result I have made available $25,000 to enable him to secure his vendors and complete his original quota of forty jobs. It became impossible for him to sell the extra seven jobs that he made in addition to the thirty-three for the Government because of not only the possibility but also the definite probability that they would be orphaned, so he struggled along in an effort to develop some other interests who would refinance his proposition and carry-on, at which time he stated that since Marmon Motor could not proceed that you (Moskovics) would like to be relieved from any further responsibility —. The situation again became critical, and I could well see that I would never get the money I had originally advanced out of the proposition unless we could sell the seven trucks left over from the first quota. Accordingly I made a deal with Art to put up some money to enable him to carry on and take additional orders and provide the wherewithal for a small operation which would enable him to dispose of the trucks on hand. This he is now doing and I only trust that Art is now on his way so that I will not be compelled to advance any additional funds."

Fred M. Young, president of the Young Radiator Company (then and now) espoused the cause of Herrington and successfully encouraged several major component vendors to extend credit to the fledgling firm on the basis of the Quartermaster contract.

The "Marmon" in Marmon-Herrington had in infancy undergone a redefinition — initially conceived to represent "Marmon Motor Car Co." It now permanently would stand for "Walter Marmon."

Walter Marmon set certain conditions which had to be met before he would provide financial support in lieu of the original Marmon Motor Car subscription: These required the surrender to the Marmon-Herrington Treasury of the stock issued to Marmon Motor Car Co. and that issued to Mr. Moskovics, and the resignation of Mr. Moskovics and the representatives of the Marmon Motor Car Co. who were officers and/or directors of Marmon-Herrington Co. In return, the stock subscribers were to be released from their liability on their stock subscription agreements.

In late Nov. 1931, in consideration of the Marmon-Herrington Company releasing Mr. Moskovics from his stock subscription liability, he released all his interest in Marmon-Herrington Co. The company was then reorganized in December, 1931 with Walter Marmon in a position to dictate the disposition of the common stock of the company. Mr. Marmon appreciated the earnest and sincere efforts of Art Herrington and others in the development and exploitation of the enterprise and sought to reward them by allocating certain shares of the common stock to them. He allocated 600 shares to Art Herrington which, added to his 500 original shares, gave him a holding of 1100 shares. To Fred Moskovics, with whom he had been associated for many years in the automotive industry, he gave a gratuitous allocation of 100 shares in recognition of the money and time spent on the initial Marmon-Herrington organization. The small size of the allocation probably reflected Mr. Marmon's displeasure with Moskovics failure to go through with his stock subscription contract.

Mr. Moskovics was keenly disappointed at the small allocation and declined to accept the shares. He expressed his displeasure in a letter written to Herrington on Nov. 27, 1931, upon learning of what the allotment was to be: *"In view of the time, money, energy, sweat, effort and labor I have put into this thing, if my share of it is estimated at 100 shares of common stock, I am delighted to know it. I am going to sign any papers you send and in addition I will make you a present of the 100 shares. It is probably just as well that I awoke to this situation when I did, so please do not have the stock issued to me, but give it to someone who will be more appreciative than I."*

Mr. Moskovics then signed a release of all interest in the Marmon-Herrington Co. — a decision he would have reason to seriously regret. He requested Walter Marmon write him a letter to the effect that he was released from liability under his stock subscription agreement and Mr. Marmon complied on Jan. 9, 1932. On Jan. 12, 1932 Moskovics wrote Walter Marmon. *"I acknowledge receipt of your letter of January ninth, which I appreciate very much, and I think it covers the point very well. I understand the situation thoroughly. I appreciate that your interest in this was primarily a rescue party and to regain the money you invested in it. I still think it is going to be a good thing . . . You probably know I told Art I did not want the stock in the Company because if what I had previously done and put in was not worth more than 100 shares, I did not think it was worth anything. I still feel that with proper handling of the patent situation, a nice position could develop, but I am not familiar with what has been done since I dropped out."*

The officers of the newly reorganized Marmon-Herrington Company were:

| | |
|---|---|
| Walter C. Marmon | Chairman |
| A.W. Herrington | President & Chief Engineer |
| Bert Dingley | Executive Vice-President |
| D.I. Glossbrenner | Secretary-Treasurer (as of June '32) |

And this was the cast for the great growth that characterized the Military Eras of Marmon-Herrington.

Some fourteen years later . . . years of spectacular growth for Marmon-Herrington Co. . . . Moskovics sued Herrington, claiming that the Agreement of Joint Venture signed by both parties on January 5, 1931 remained in force and that he, Moskovics, was entitled to a half interest in certain patents owned by Herrington and to assignment of one half of stock in Marmon-Herrington Co. owned by Col. Herrington.

The litigation was eventually resolved at the State Supreme Court level. Mr. Justice Eder, presiding over Special Term, Part V of the . . . Supreme Court in the matter of Moskovic vs Herrington.

Mr. Justice Eder, in reviewing the case to establish the basis for his decision in the matter authored a remarkably fine automotive history document. He provided insight and detail on the tumultous first year of the Marmon-Herrington Co. that is not otherwise documented. The author in particular, and automotive historians in general, are indebted to Justice Eder for his excellent recounting and perceptive assessment of events that transpired some fourteen years earlier. The accounting herein of the Moskovics association with Col. Herrington and the infant Marmon-Herrington Co. is largely drawn therefrom.

The Judge did rule in Moskovics favor with regard to patent ownership, but only for one specific patent, directing Herrington to acknowledge to Moskovics the latter's ownership of a one-half interest in United States Letters Patent No. 1,981,173.

Justice Eder noted that Herrington had never contested the half-ownership rights and his prior "offer of judgement" in this regard had been rejected by Moskovics. He dismissed the complaint upon all other matters and awarded Herrington court costs.

For those who might desire additional legal specifics in the matter and in the interest of preservation herein of the full content of Justice Eder's recounting, the following is provided.

Mr. Moskovics' complaint: "Plaintiff seeks a decree of specific performance compelling the defendent to assign him a one-half interest in patents and inventions then owned or thereafter acquired by defendent . . .; an accounting of all profits received by defendent in trust for the joint venture; that a trust be imposed upon the stock of Marmon-Herrington Co., Inc., owned by defendent and upon proceeds of any stock sold by him, and that he be decreed to execute and deliver to plaintiff an assignment of one-half interest in such stock and directing payment of one-half of proceeds realized by defendent from sale of any such stock."

The significant issue of contention was the duration of the Agreement of Joint Venture: "After due consideration of the oral and documentary evidence, I am of the opinion and find that the agreement of January 5, 1931, established a joint venture and that the object thereof was accomplished by the agreement of March 10, 1931, when it was accepted by the Marmon-Herrington Co., that the joint venture agreement was thereupon terminated and that it did not survive the agreement of January 5; that thereafter the parties owed no duty or obligation of any kind to each other under that agreement and that they acted separately and independently of each other."

An Internal Revenue Service affidavit detracted from the credibility of Mr. Moskovics' claim: "In his 1931 personal income tax return, plaintiff had written off the part payment of his subscription to the company's stock as a deductible loss; he did not want to jeopardize that deduction by retaining ownership of a stock interest (the 100 shares offered by Walter Marmon-author's comment) which he regarded as worthless. In support of his claim for this deduction he made an affidavit on July 6, 1933, filed with Bureau of Internal Revenue, and therein averred, among other matters, 'That the said report disallows the deduction of a loss of $10,000 advanced for the account of Marmon-Herrington Co. Inc., in 1936 . . . That in December 1931; the stock subscription agreements were canceled and the deponent (Moskovics) abandoned all interest in the company and its enterprise, and knew at that time that it had no assets or equities sufficient to meet its liabilities. . . . That some of the persons interested in the company reorganized it and offered the deponent 100 shares of its stock in recognition of his advances and efforts in the original promotion of the company. That the stock so offered was worthless and he absolutely refused to receive delivery of the same or to have the same registered in his name, and he completely abandoned all his interests and rights in the company or its stock or its assets' . . . It seems clear to me that when plaintiff was of the belief that the Marmon-Herrington Co. was about to collapse or would be unable to longer continue he unhesitatingly abandoned all his interest in the company and its stock and in its enterprise and that upon the company later becoming successful he is now endeavoring to capitalize upon a subsequently conceived claim of a continuing joint venture and partnership in an effort to share in the benefits, when he was content to forego all risk when the collapse and extinction of the company and its enterprise appeared imminent and inescapable . . . a course not without precedent."

It is important to establish the continuity of the Marmon Motor Car Company into Marmon-Herrington Company despite the default by the former on its stock subscription agreement.

Colonel Herrington's own assessment is very specific: When the Indianapolis Speedway celebrated the golden anniversary of the "500" race, Marmon-Herrington ran a full page advertisement extending their "congratulations."

Of particular interest in this regard is Herrington's reference to "50 years of . . . between the Speedway and **our company** and his reference to Marmon cars and vehicles produced by the **Marmon organization.** (Author's emphasis).

The "Marmon organization" to which Colonel Herrington was referring was grossly disrupted by the bankruptcy of Marmon Motor Car Co. but progressively

*First in the first 500—Marmon Wasp • Driver—Ray Harroun • Average speed—74.59*

# Congratulations

## on the golden anniversary of the 500

• "The name 'Marmon' has been an influence factor in the automotive industry for more years than most people can remember. The first 500-Mile Race was won in 1911 by the Marmon Wasp, with the legendary and still active Ray Harroun behind the wheel. Even then the present Marmon-Herrington Company was well over 50 years old. It was established as Nordyke & Marmon Company, ten years before the start of the Civil War—1851. The name changed to the Marmon Motor Car Company in 1926, and then became, in 1931, The Marmon-Herrington Company, Inc.

"So, in 1911, we could lend a bit of know-how to the first 500-Mile Race. Since then, through 50 years of friendly cooperation between the Indianapolis Speedway Corporation and our company, we have traveled the path of automotive progress together. The Marmon organization no longer races cars, but all Marmon cars and vehicles have been proof-tested at the Speedway. The lessons learned there in durability, safety and roadability have been built into our specialized vehicles and buses.

"From one old-timer to another—and both still going great—best wishes for the next 50 years . . ."

**THE MARMON-HERRINGTON COMPANY, INC.**

*A. W. HERRINGTON, Chairman of the Board*

reassembled under the Marmon-Herrington banner in the following years as the new business prospered. This is well illustrated by comparing the positions held at Marmon in 1931, when M-H was born, and those held by the same individuals at M-H ten years later:

| | **Marmon Motor Car Co.** | **Marmon-Herrington Co. Inc.** |
|---|---|---|
| | Dec. 1931 | 1941 |
| Walter C. Marmon | Chairman | Chairman |
| Bert Dingley | (Chief Inspector-Liberty Engine; Service Manager 1924) | Vice-President |
| George Freers | Engineer reporting to H.C. Marmon; Subsequently HCM employee | Chief Engineer |
| S.E. Rowe | Asst. Chief Engineer | Asst. Chief Engineer |
| Harold Blank | Service Manager | Service Manager |
| Wayne Kinsey | Parts Manager | Parts Manager |
| Harold Fisher | Test Driver; 1931 record speed runs | Supv. Final Test and Acceptance Engineer |
| George McMath | Engineer; later HCM employee | Engineer |
| Frank Updike | Production Foreman, 25 years | Production |
| D.R. Meo | Purchasing Agent | Purchasing Agent |

The history of the Marmon-Herrington organization is recounted in a separate chapter. It is addressed here only to illustrate the continuity of the "Marmon organization" as staff members moved individually from Kentucky and Morris to Washington and Harding.

Presentation ceremonies for 10 year service awards at Marmon-Herrington provided photographs of several members of 'The Marmon Organization' who had previous Marmon Motor Car Co. service records.

Much later, in 1961, when announcing the entry of Marmon-Herrington into the inter-city truck-tractor market, David M. Klausmeyer would be identified as "President of the 110 year old firm...the company whose existence antedates the automobile itself appears ready to add a new chapter to its distinguished history."[1]

The most significant management additions in the formative stage of the new organization were Bert Dingley and Robert Wallace.

The first of these events was the arrival of Bert Dingley on the scene. He had been a race driver from the infancy of the automobile and was National AAA Champion in 1909 but retired following a serious accident in 1914. By the onset of US participation in WW I, he was Western Sales Manager for Mitchell Motor Car Co. He promptly joined the United States Department of Aviation, Liberty Engine Division, as Chief (Government) Inspector at the Nordyke and Marmon Co. Indianapolis plant for Liberty engine manufacture.

Following the Armistice, he joined the H.H. Franklin Mfg. Co. but soon shifted to Nordyke and Marmon as Western Region sales and service representative. He was next Master of Ceremonies for the tear-down and assemble exhibit demonstration which was a premier attraction at the 1920 auto shows from coast to coast. Following that assignment Dingley was named manager of the new Field Maintenance Department with responsibility for adequate service facilities and expertise at all Marmon dealers. In December 1920 he was transferred to manufacturing and placed in charge of all inspection — parts, assemblies, units and finished cars. By 1924 he was Service Manager but left to join the Stutz organization concurrently with Moskovics joining Stutz as Vice-President of Sales. Moskovics became Stutz president in 1925 and in 1928 Dingley became Vice-President of Manufacturing. By the founding of Marmon-Herrington, Stutz was in serious financial plight.

Fig. 1-5 President Dingley presents 10 year awards to L. N. Bowman (center) and Harold Blank, Service Manager.

Fig. 1-6 President Dingley presenting 10 year awards to Harold Fisher, George McMath and Frank Updike, all from the engineering department.

Fig. 1-7 Robert C. Wallace watches as President Dingley recognizes ten years of service by Charles Ambrose, 'Dutch' Kim and 'Nick' Meo.

BERT DINGLEY • PRESIDENT

Fig. 1-8 Bert Dingley

Fig. 1-9 Robert C. Wallace

As Moskovics departed after overseeing the incorporation of Marmon-Herrington from the Board Chair, Bert Dingley came aboard as Executive Vice-President. He would ultimately be President (1942-48).

Robert C. Wallace was the other significant addition, joining M-H in 1932. He was a 1927 graduate of Massachusetts Institute of Technology and had previously been Factory Service Manager at Stutz where Bert Dingley had become aware of his abilities, and Dingley brought him on board as Shop Engineer at Marmon-Herrington, where he was soon Factory Service Manager. He would ultimately be Director of Engineering for the period of the Dingley Presidency.

Let's look at the infant Marmon-Herrington firm in the words of Bert Dingley some twenty-one years later when he was speaking, as President, of the firm's progress. *"Ten years before Pearl Harbor, in Indianapolis, Indiana, two men 'declared war' against the enemies of Democracy. Col. Herrington was a Captain in the AEF in France in WWI and saw the transport vehicle problems of that war. He decided that America should have better vehicles in the next war and that he would supply them. Even in 1931 the certainly of another great conflict was apparent. Walter C. Marmon had a large and completely equipped plant and ample capital at his disposal . . . He saw the opportunity in Col. Herrington's plans and designs. The Marmon-Herrington Co. can rightly be called the* ***Laboratory of the arsenal of Democracy."***

Through an intriguing series of events, M-H was to provide transport for the building of the Middle-East oil pipelines in the thirties, was to provide armored vehicles for their defense, and hordes of combat and transport vehicles for the British forces of Montgomery in their defeat of the German thrust for control of those oil fields.

## PART 2 — An Overview — "Most Picturesque of all US Truck makers"

I find no better way to introduce the unfamiliar reader to this most significant firm than to repeat the introduction by the Editors of Time to the equally unfamiliar US citizen of 1942.

Time Magazine, issue of January 5, 1942, reported that "the Marmon-Herrington Co. . . . **most picturesque of all U.S. truckmakers** (author's emphasis) . . . last week was a closed corporation no longer. It had almost finished selling 15,000 of its 450,000 shares of common stock to the public at $10 per share."

"Although unknown to most U.S. citizens, M-H nevertheless is the king of its field: the world's most enormous trucks, all-wheel drive, extra-heavy duty, able to haul almost anything almost anywhere. Only ten years old, it has sold thousands of square-headed, locomotive-like behemoths to lumbermen, oilmen, miners, highway departments. In the Louisiana oil fields M-H trucks are called "Mud-Cats" because they slosh through hub-deep mud as though equipped with web wheels, (No pun intended by Time's Editor, but today's Marmon Group, Inc. has a Webb Wheel Division manufacturing very conventional truck wheels.) in western lumber camps they climb rough 40 degree grades so easily they are known as "Mountain Goats". In all, M-H has 33 models, going from five to 35 tons with six driving wheels, twelve forward and four reverse speeds. Special jobs include $130,000, 12-wheel, 65 ton giants used to build pipelines in the Near East." © 1942, TIME, Inc., Reprinted by permission.

After relating the birth of M-H in the Marmon Motor Car Co. plant in 1931 and Colonel Herrington's background as a Motor Transport Corps. Captain in World War I, the Time article continued: *"Because of his World War I stint, Herrington knew that the tremendous traction of his all-wheel drive trucks would be useful in anyone's army. So he prepared for the boom; 30 days after the Germans broke into Poland, Herrington broke ground for a five-fold expansion of plant facilities. Then he expanded again and again. His output is now (Jan. 1942) 20-25 times that of August 1939, still going up. But almost all of this increase is in tanks, not trucks. Equipped to make almost anything big that moves, Herrington soon had orders from Britain, the Netherlands Indies, Greece and China. The M-H equipment has been war-tested aplenty. The British were so pleased with their first M-H trucks (they went over desert sand like camels) that they bought about* ***100,000*** *units, are now using many of them in the Libyan push. In Crete, scores of British-and-Greek-owned M-H light tanks fought mightily before being blasted to bits by Nazi bombers. Dutch-owned M-H tanks and gun carriers now plough through the jungles of Sumatra in pursuit of Jap paratroops. But this week the main concern of Herrington is not where his products fight but how fast he can get them out. To speed production, he has already subcontracted 60-70% of his $35 million backlog, hopes to boost the ratio even higher. His deliveries in 1941's first eight months were $4,703,000, only 2% less than 1940's record-breaking twelve-month total. But December sales alone were roughly $3.5 million, including $2.75 million worth of tanks for the Dutch. For next year, if the expected orders come in, Herrington is shooting at deliveries of $60 million."*

Obviously that Time Editor was excited by what he had learned about the Indianapolis firm. His excitement was well founded for the M-H story is an exhilarating one and the events related above pale when compared with others that followed.

More wartime citizens learned of the trail followed by "Hell-on-wheels Herrington" when Francis Chase, Jr. authored a feature article by that title in the *Saturday Evening Post* for August 8, 1942.

It began: "The only company which was set for a flying start when the Government decided to swamp the Axis with tanks and armored vehicles was a small outfit in Indianapolis named Marmon-Herrington, Inc."

"Unlike the larger automotive plants, which naturally had to close up shop pending conversion of their mass-production tools, this one had been converting itself at a dizzy clip for nine years in order to make strange vehicles for 'impossible' off-the-road tasks in many parts of the world . . . Marmon-Herrington merely slipped into high gear and was off in a cloud of government orders.

*"Today the Indianapolis plant of Arthur W. Herrington is a mecca for allied purchasing missions. He's slated to unleash $60 million worth of rolling death against the Axis this year."* © 1942 Curtis Publishing Co., by permission.

The famous war reporter Ernie Pyle visited Marmon-Herrington in July 1940 when he was authoring a column "Hoosier Vagabond" which appeared in The *Indianapolis Times*. Ernie's assessment of the "Birthplace of the Arsenal" is repeated below as it appeared in the *Marmon-Herrington NEWS* for October, 1942, reprinted from the July 25, 1940 edition of the *Times:*

THE OLD MARMON AUTOMOBILE, one of the many that helped make Indianapolis famous, is practically extinct now. They haven't been made for years.

But the Marmon name is carried on in the new products of the Marmon-Herrington Co., an outgrowth of the old firm. And they have scattered their products far. They have armored cars, tanks, tractors, marsh-wagons and trucks running on every continent.

They have equipment on the front lines in Europe and China. They have great trucks and trailers in the oil fields of Persia. They have grotesque, mammoth-wheeled machines dragging steel pipe through the marshy jungles of Venezuela. They have tanks scattered all over Central America. They have machines in South Africa and the Dutch East Indies.

The company is proudest, as it should be, of its non-war, commercial vehicles. It makes stuff that does the heavy work of the world. To put it in layman's language, its keynote is making wheeled vehicles that won't get stuck.

The backbone of its product is an all-wheel-drive truck. They take a Ford truck, and add their own axles and transmission to transmit the engine's power directly to all the wheels, not just the rear two.

They make trucks with power going to three sets of axles — meaning as high sometimes as a dozen power-driven wheels on one truck. If you put enough power directly into enough wheels, you can pull out of a bottomless pit. That's their theory.

### THE INVASION OF HOLLAND

A.W. Herrington, president of the company, spends much of his time in far-away lands, selling new stuff or getting equipment running. He travels 10 times as much as I do. He got out of Germany just two days before the war started.

Marmon-Herrington has practically mechanized the entire Rumanian Army in the last six years. And today in the factory are row upon row of great wooden boxes, full of truck equipment, ready for shipping to New York. How it gets from New York is Rumania's worry.

Just last week two officers of the Dutch Army were here, talking over more equipment for the Dutch East Indies. These two men went through the German invasion of Holland. They say there never was such awful destruction of life. The Germans just kept coming, and falling. They said it amounted, in the end, to the

Today, the Indianapolis plant of Arthur W. Herrington (standing) is a mecca for Allied purchasing missions. He's slated to unleash $60,000,000 worth of rolling death against the Axis this year.

Hell-on-Wheels Herrington

By FRANCIS CHASE, JR.

**He started planning for war in the early 1920's. No one has done more to mechanize our fighting men for victory.**

Fig. 1-10 Reassuring news from Indianapolis. Reprinted from The Saturday Evening Post. Copyright 1942, Curtis Publishing Co., by permission.

simple fact that Germany had more soldiers than Holland had bullets. But they feel the East Indies are almost impregnable.

The Marmon-Herrington plant somehow has a good personality. You can sense it. It isn't so big that humans lose their identity. I expect that General Manager Alfred Campbell knows every workman by name. Or at least he did up until now.

Things are changing everywhere, now, of course. Marmon-Herrington is swinging rapidly into the defense program. New men are being taken on daily.

Carpenters are partitioning off new offices, and the hammering is so terrific that salesmen, waiting for appointments, think somebody is shooting at them.

GANG WAY FOR ERNIE

Two big new buildings are underway, which will almost triple their production capacity. They will be ready in less than two months, too. And furthermore, they're doing it themselves; not asking help from the Government.

They can't say what Government orders they have, or expect, outside of a new batch of tanks. Probably don't know themselves yet. But they will soon be ready.

When we got through looking at the plant, General Manager Campbell said, "Now if there's anything else you'd like to ask, go right ahead, and we'll answer if we can, and if we can't we'll lie to you." That threw me so badly off base I couldn't think of a thing to ask.

We wound up the afternoon by taking an all-wheel-drive Ford station wagon and going out to the company's sand pile. It's an old gravel pit, several miles from the plant, which the company bought for a test ground. It is such a horrible place they didn't have to do a thing to it to make it worse.

It's rough and rugged and the sand lies several feet deep. We just plowed all over the place. As an old sand-driver from Arizona, I know what you can drive through and what you can't. We drove for half an hour through stuff you can't drive through. Then they turned it over to me, and even I drove through it.

As we left, I put in an order for an all-wheel-drive station wagon. For when the Nazis enter New York I'm going to start driving due west, and I don't want anything to stop me.

* * *

The house organ *Men of Marmon* issue for May 1931 was indeed a significant and prophetic one.

On the front cover was pictured the first production MARMON SIXTEEN car coming off the assembly line at **Plant 2** on April 7, 1931, in the presence of James Anderson, Factory Manager and Vice-President Howard Marmon, and President Williams.

Inside was a picture of Art Herrington standing beside the pre-prototype Marmon-Herrington truck. Under it was the caption. 'First Truck Unit Soon'; the text: All material for the first Marmon-Herrington truck has been shipped into the factory. Assembly operations will go forward at once in the **Plant 3** assembly unit. Material for the first forty jobs is expected by May 1; completion of the 33 military trucks by May 30. Mr. Herrington reported that the first axle, the part which makes the Marmon-Herrington truck **unique,** was placed on the torque stand when completed and functioned perfectly. The first Marmon-Herrington distributor was announced as the Logan Sales Co. of St. Paul; Mr. A.O. Logan, President.

Unknowingly 'Men of Marmon' had concurrently reported the first production of the last car model of the Marmon Car epoch and the impending assembly of the first production truck of the Marmon-Herrington epoch!

The prototype Marmon-Herrington truck when completed in the spring of 1931 exhibited some unusual features. It incorporated prescriptions to correct the serious deficiencies of the four-wheel-drive army transport trucks of World War One. The Nash and Jeffrey "Quads" overturned on soft shouldered roads and buried their overhanging snouts in steep embankments. Herrington took the wheel for the first public demonstrations — an event that was to become a tradition.

The new truck was characterized by a "drop" frame which lowered the load and the vehicle's center-of-gravity. The front wheels were far forward, not unlike contemporary passenger cars, to ride up, or rather to climb up, embankments. A high road clearance was maintained between the front and rear wheels by careful configuration of the transfer gear case.

Importantly, the new truck was configured **without** a center differential. Colonel Herrington's experience at the Indianapolis Speedway included learning that race cars fitted with four wheel drive and a center differential encountered steering control difficulties to an unsatisfactory degree; elimination of the center differential corrected the control problem! In addition, the excessive front tire wear which had characterized all-wheel-drive racers was minimized.

Herrington was also well aware of the requirement, inherent in a vehicle with Ackerman (conventional universal practice) front steering, for the front wheels to travel a greater distance than the rear wheels during other than straight ahead movement. To allow for this requirement, the M-H transfer case incorporated an innovative ratchet clutch in the "positive" drive to the rear axle (s).

At the advice of their consultant, the US Army had incorporated this type of transfer case in the official specifications for all-wheel-drive transport vehicles.

The Herrington philosophy of all wheel drive was recorded in his own words and is reproduced on the following two pages — as it appeared in *M-H NEWS.*

The worldwide acceptance of front wheel drive for passenger cars and four wheel drive for utility vehicles with off-road capability is descended from a very noteworthy and highly innovative invention — the Rzeppa constant velocity universal joint.

## CONSTANT VELOCITY (CV) UNIVERSAL JOINTS

Wide acceptance of front wheel drive awaited the arrival of Hans Rzeppa (pronounced Jeppa) who conceived a design of joint in which the input and output shafts rotated at the same angular velocity regardless of the angularity of the output to the input shaft axes. He brought his invention from his native Hungary to England and interested Thomas Powell in its promotion and manufacture.

Powell organized a firm, Universal Power Drives, Limited, for the purpose of promoting the Rzeppa joints in marine applications. Manufacture of the new universal joint necessitated tooling and machinery almost as innovative on the joint itself. Unable to find sufficient financial support for the tooling expenditures in Great Britain, arrangements were made for manufacture by the Gear Grinding Machine Co. in the United States. Mr. Rzeppa subsequently settled in rural Michigan.

Several important developments promptly occurred in Indianapolis, Indiana. Harry Miller installed Rzeppa joints in one of his race cars which ran in the Indianapolis "500.' Arthur Herrington, always well versed on developments at the "brickyard," saw in the Rzeppa joint the solution to certain problems being experienced with practical applications of his invention and elected to adopt the Rzeppa concept.

### A Word About Universal Joints

To appreciate the significance of this breakthrough, it is necessary that the serious shortcomings of the familiar universal joint, as used in drive shafts for rear drive axles and known as the 'Cardon' (or Hook's) type be understood.

When the input and output of a Cardon joint are in line, the input and output shafts rotate uniformly; if a line were scribed on each, the lines would always align regardless of rotation degree. When the input and output are **not** in line (the reason for using the joint) the lines would align only twice each revolution, with the leading and lagging lines being reversed in the second half revolution. The amount of misalignment of the scribed lines would increase as angularity of the output shaft was increased. In technical terms, the output, resulting from a uniform input rotation is non-uniform rotation or non-uniform 'angular velocity.'

Non-uniform angular velocity can excite vibration and produce gear noise — a special form of vibration — at certain resonant speeds. This characteristic of the Cardon joint favors the open drive shaft with two joints over the torque-tube drive with its single joint. With the former, the engine and axle can rotate uniformly and vibration-free despite the continual irregular rotation of the intervening shaft. This is the incentive for the tubular light weight driveshaft. However, the pair of joints must be correctly releted or magnification, rather than cancellation of the irregularity will result. When, in later years, efforts to reduce the height of rear-drive passenger cars necessitated a two-piece drive shaft, the advantages of the two-joint system were retained by the use of a third joint which was really a close-coupled pair of Cardon joints.

# Some WHYS and WHEREFORES of MARMON-HERRINGTON CONSTRUCTION

*—In which are presented highlights from a discussion by A. W. Herrington before the Annual National Meeting of the Society of Automotive Engineers in Detroit on the subject of various safety factors of all-wheel-drive design.*

THE single point which in my opinion determines the stability or instability of any truck which has a steering and driving front axle is the means of driving the wheels. It has been common practice in the industry up until recently to use the Carden or pin type universal joint for this purpose. It is also commonly known that the fluctuating velocity of the driven end or outboard driven shaft of a Carden joint when it is turned to an angle of 30 degrees will be approximately 35 per cent. If there is any single factor which could create front steering instability in a motor vehicle it is due to this particular cause.

In addition to this, it does not appear to be necessary in any front axle design to have the wheel bearing point at the ground, four and three-quarters inches outboard of the subtended steering pivot center. The combination of this peculiar conception of front axle layout with the Carden type joint makes a reaction through the steering gear which is so pronounced as to be, in my opinion, objectionable. If a steering gear which is sufficiently unreversible is used, failures of the gear will most certainly result. If the gear is moderately or completely reversible, the resulting gyrations of the steering wheel in any turn of 15 degrees or over, seems to be unsatisfactory from the standpoint of some users.

Front wheel driving joints of the uniform velocity type are available, have been used in the industry, and eliminate all of the unsatisfactory phenomenon noted above. The use of such joints combined with the standard practices of normal front wheel steering geometry will result in the production of an all-wheel-driven vehicle which in every respect will handle as easily and safely as the best motor vehicles to be found on our highways today.

(It is important that the construction of Marmon-Herrington trucks is based on the conclusions drawn above.)

I want to take direct issue with any statement that the location of the center of gravity has no effect on the braking performance of motor trucks.

As a homely illustration, may I suggest that a pair of roller skates having relatively short wheelbase, attached to the feet of an individual having relatively high center of gravity and no experience with the use of these appendages, would probably demonstrate very quickly that there is some relationship between center of gravity, wheelbase, and general vehicle stability. The height of the center of gravity in relation to tread and whether single or dual rear tires are fitted, has a very definite effect on vehicle lateral stability against side over side turning forces.

(See Page 7 for facts pertaining to the unusually low center of gravity of Marmon-Herrington all-wheel-drive motor trucks.)

*To give you a comparison of size. This shows a Marmon-Herrington Model TH320-4, four-wheel-drive, capacity 8 to 9 tons, with one of the latest model Ford coupes.*

Experience has shown us that the largest individual factor influencing the effort which it is necessary to exert on the rim of the steering wheel is the caster angle of the front axle. In a four-wheel-drive truck where the power is transmitted through the Hotchkiss drive type of front spring installation, it is necessary to make allowances for this torque reaction in determining the caster angle to be used. To obtain the same steering effects the caster angle of a four-wheel-drive truck must be set at less than common rear-drive truck practice when the vehicle is standing still. When power is applied torque reaction increases the static caster angle, the amount being determined by the stiffness of the front springs.

* * * *

Any inter-axle differential on a four-wheel-drive truck is essentially a power-dividing mechanism. It will always subdivide the power equally to the front and rear axles. This fact probably accounts for the unusual front wheel tire wear that is always observable on trucks possessing center differentials of the "full and free-acting" type. When these vehicles get into a heavy pull with the front axle carrying the lightest load, the front wheels spin on the pavement causing tire wear and power loss.

The facts are that the tread of the front axle on most of the four-wheel-drive trucks which do not possess a center differential approximately lines up with the center of the two dual wheels fitted on the rear. In turning a circle, the track described by the outer front wheel will be approximately the same distance outside of the outside dual rear wheel, as the tracks of the two dual rear wheels are from each other. It is quite obvious in turning a corner with an ordinary dual installation where the wheels are solidly bolted together that there must be some tread slippage. The amount of tread slippage that occurs between the front wheels and the rear wheels of a four-wheel-drive truck not having a center differential, is no greater than the slippage between the two duals, if the above conditions are adhered to. This being true, and taking into consideration the adequate and satisfactory mileage that is being obtained from thousands of rear dual installations throughout the United States, it is apparent that no greater front tire wear could be expected on a four-wheel-drive truck without center differential than would be normal to a dual-tired axle installation.

(See opposite page for further reasons why Marmon-Herrington vehicles have no center differentials.)

* * * *

I would like to add that it has been definitely proven that a satisfactory, stable, dependable, safe truck, driven on all wheels, can be manufactured and sold.

This vehicle will be found to possess elements of front axle geometry which are in complete accord with the practices of the rear-drive truck industry of today. After all, the factors which determine a satisfactory front-end geometry are based upon running experience and the developments of many engineers.

* * * *

We are not living in the Stone Age and we should not be governed by Stone Age engineering.

# We Repeat—MARMON-HERRINGTON All-Wheel-Drive Trucks Have NO CENTER DIFFERENTIALS

BY A. W. HERRINGTON, PRESIDENT.

THERE is no center differential in any Marmon-Herrington truck. It has been left out purposely and for some very good reasons, regardless of the fact that it has received so much prominence in the sales literature of one of our competitors that customers sometimes wonder whether they are buying a four-wheel-drive truck or a center differential.

Practice clearly demonstrates that when no center differential is used the vehicle has better tire wear. Where a center differential is used, front tire wear is very pronounced. The possibilities of designing an all-wheel-drive truck so that when it is loaded the front and rear axles will bear the load equally are very remote. Assuming, however, that a vehicle were so loaded, the minute it arrived on a 20 pct. grade the uniform loading of the axles would no longer exist.

If the truck had a center differential, this differential would divide the power equally. The power, then, would do the natural thing—leak out through the front wheels and cause them to spin on the pavement because the rear wheels would be the more heavily loaded. Perhaps the best indication that this situation gives difficulty is the fact that one manufacturer provides a manual means of locking out the actions of the center differential in such cases. Certainly, this lock would not be provided unless there was a very necessary reason for its existence.

Some companies use center differentials of the automatic locking type. These are peculiar instruments and usually are completely enclosed in a housing so that when they fail the broken parts cannot get into the transmission and cause difficulty. At one time we had an Army design known as the "Militor". It was fitted with a differential of this type. Within the first thousand miles this locking differential invariably would resolve itself into a solid mass of broken parts. Being completely enclosed in a housing, this unit operated as a solidified mass and the vehicle in effect continued to function in the same manner as a truck with no center differential. This condition was rarely, if ever, discovered until the truck was taken into the shop for its overhaul period.

A survey of operators of four-wheel-drive trucks having center differentials brings forth the illuminating information that these differentials are a source of great demand for spare parts for replacement purposes. Inasmuch as the differential is not necessary in the first place, it is only common sense practice to eliminate all of these parts and all of their replacement costs by leaving this troublesome mechanism out of the vehicle entirely.

Experts of the U. S. Army, who are probably the best and the most unbiased judges of whether or not a center differential is necessary, have answered the question in the preparation of their own specifications by saying—"a differential between front and rear axles will not be required."

Let me emphasize again that Marmon-Herrington all-wheel-drive trucks have no center differentials.

TURNING CIRCLE OF ANY TRUCK WITH DUAL REAR TIRES

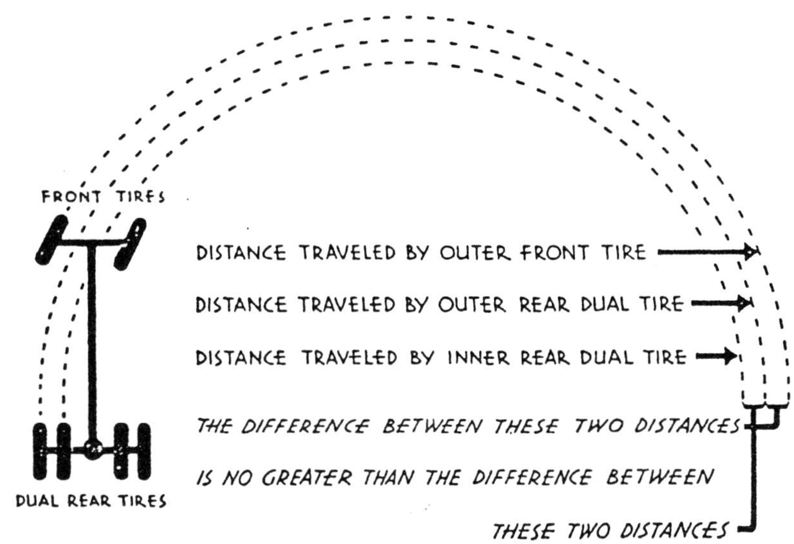


As explained by A. W. HERRINGTON—

"The engineer will point out that the front wheels travel farther than the rear wheels. He forgets that the differential in the axle averages the distance of the front wheels and averages the distance traveled by the rear wheels. Take a truck with a normal wheelbase and a 30-degree turn of the front wheels and lay off the wheel tracks. You will find that the front wheel circle is no farther outside of the outer wheels on the rear dual wheels than the distance between these dual rear wheels themselves. The hundreds of thousands of dual wheel combinations running on rear-drive motor trucks give undisputed satisfactory service. The tire wear on the front wheels of a four-wheel-drive truck with no center differential is no greater, and there is no reason for it to be greater, than the corresponding wear between the two tires of a rear dual wheel installation."

Fig. 1-11 Experimental truck and its designer, A. W. Herrington. *Men of Marmon*, May 1931.

The fundamental requirement for homokinetic (ie, uniform) rotation is that the contacts between driver and driven elements must always occur in a plane which bisects the angle between the axes of the elements. This means that the balls transmitting the motion must always be in that homokinetic plane. In a ball-type joint, the balls ride in grooves milled in drive and driven elements. In the Rzeppa joint, they are restrained in the plane by an intermediate member (similar to a bearing ball or roller separator) having a ball guide on the axes in the plane, and at the end of an appendage therefrom, another ball riding deep in a bore in the driving shaft.

A second form of CV (constant velocity) joint was used on the Cooper (Marmon) Indy cars and on the 1937 Cord automobile. This Bendix-Weiss joint, by virtue of the curvature of the two mating raceways being identical in both elements but of opposite hand, causes the engaging balls to always be in the homokinetic plane without the need for guidance other than a single ball at the intersection of the axes. The Bendix-Weiss joint accommodates end motion by allowing the balls to roll back and forth in the races, thereby eliminating the need for splined slip-joints with their high frictional resistance, and permits a greater (30 degrees) steering angle.

A third CV joint concept is the Tracta Joint which uses mating profiled yokes for torque transmission.

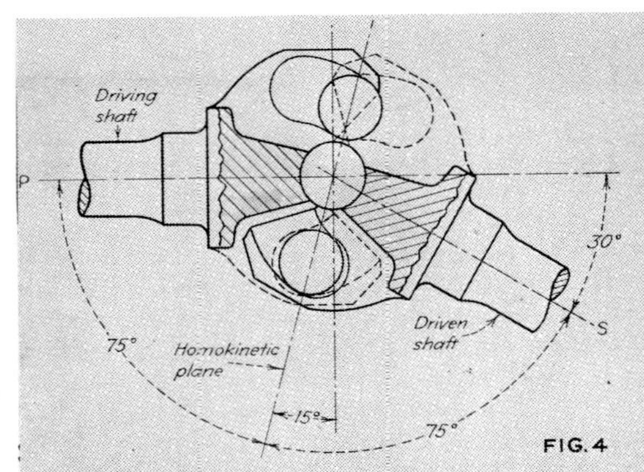


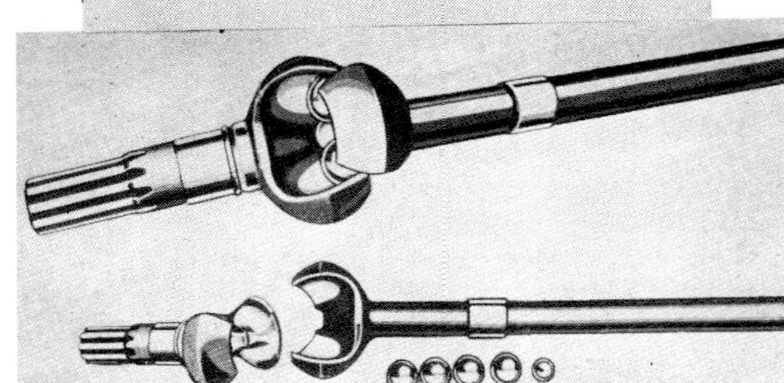

Fig. 1-12 Rzeppa (Gear Grinding Machine Co.) CV Joint (left). Weiss (Bendix) CV Joint (right)

| MODEL | TH-310A-6 | TH-320-6 | TH-330-6 |
|---|---|---|---|
| *Gross Capacity* | 34070 | 43075 | 50130 |
| *Tires—Front* | 9.75 x 22 | 10.50 x 22 | 11.25 x 24 |
| *Tires—Rear* | Quad. 9.75 x 22 | Quad. 10.50 x 22 | Quad. 11.25 x 24 |
| *Engine* | Hercules R X C | Hercules H X B | Hercules H X D |
| *Cylinders* | Six | Six | Six |
| *Bore & Stroke* | 4⅝ x 5¼ | 5 x 6 | 5½ x 6 |
| *Piston Displacement* | 529 | 707 | 855 |
| *Max. B. H. P.* | 115 @ 2200 | 150 @ 2000 | 180 @ 2000 |
| *Main Bearings* | 7 | 7 | 7 |
| *Bearing Diameter* | 3" | 3½" | 3½" |
| *Carburetor Make* | Zenith | Zenith | Zenith |
| *Carburetor Type* | Down-draft | Down-draft | Down-draft |
| *Fuel Feed* | Cam Pump | Cam Pump | Cam Pump |
| *Electrical Units* | Delco | Delco | Delco |
| *Voltage* | 6 | 12 | 12 |
| *Clutch Type* | Multiple Plate | Twin Plate | Twin Plate |
| *Transmission Make* | Fuller | Brown-Lipe | Brown-Lipe |
| *Mounting* | Unit Power | Unit Power | Unit Power |
| *Forward Speeds* | 5 | 4 | 4 |
| *Auxiliary Make* | Own No. 2 | Own No. 3 | Own No. 3A |
| *Aux. Forwd. Speeds* | 2 | 3 | 3 |
| *Total Forwd. Speeds* | 10 | 12 | 12 |
| *Total Reverse Speeds* | 4 | 3 | 3 |
| *Universal Make* | Blood | Blood | Blood |
| *Universal Size* | 5 | 6 | 6 |
| *Front Axle* | Own No. 3 | Own No. 3A | Own No. 4 |
| *Joint Type* | Constant Velocity | Constant Velocity | Constant Velocity |
| *Rear Axle—Make* | Wisconsin | Wisconsin | Wisconsin |
| *Rear Axle—Type* | Tandem Dble. Red. | Tandem Dble. Red. | Tandem Dble. Red. |
| *Shaft Size* | Fr. 2½, Rear 2¼ | Fr. 2½, Rear 2⅜ | Fr. 2$\frac{11}{16}$, Rear 2$\frac{11}{16}$ |
| *Axle Ratio* | 9.11 | 9.11 | 10.266 |
| *Other Ratios* | 7.33—8.14 | 7.33—10.266 | 7.33—9.11 |
| *Tot. Low Gear Red.* | 164.2:1 | 189.3:1 | 189.5:1 |
| *Service Brakes* | West. Air | West. Air | West. Air |
| *Compressor Cap* | 6 Cu. Ft. | 6 Cu. Ft. | 6 Cu. Ft. |
| *Brake Size* | 17¼ x 5 | 17¼ x 5½ | 17¼ x 5½ |
| *Lining Make* | Brakbloc | Brakbloc | Brakbloc |
| *Brake Drum* | Gun Iron | Gun Iron | Gun Iron |
| *Hand Brake* | 14" Disc | Double 14" Disc | Double 16" Disc |
| *Steering Gear Make* | Ross | Ross | Ross |
| *Frame Size* | 8⅛ x 3 x $\frac{5}{16}$ | 10⅛ x 3 x ⅜ | 10⅛ x 3 x ⅜ |
| *Fishplate* | 14 x 3$\frac{5}{16}$ x $\frac{5}{16}$ | 20 x 3⅜ x ⅜ | 20 x 3⅜ x ⅜ |
| *Frame Material* | Chrome Manganese | Chrome Manganese | Chrome Manganese |
| *Gasoline Tanks* | 2 | 2 | 2 |
| *Tank Capacity* | 50 Gals. Each | 75 Gals. Each | 75 Gals. Each |
| *Wheels Driving* | 6 on 10 Tires | 6 on 10 Tires | 6 on 10 Tires |

On heavy commercial vehicles requiring off-road capability and maneuvering capability beyond the 30 degree limitation, single Cardon joints continue to be used today in powered front axles.

Dr. (Doc) B (everley) W. Keese of Wisconsin Axle Co. worked jointly with A.W.S. Herrington to develop a modern front-wheel drive axle incorporating the Rzeppa type constant velocity universal joints. Later Bendix-Weiss type (Mechanics and Spicer) were used in the Ford conversion axles. Wisconsin Axle Company was acquired by the Timken Axle Company in the 1940s. As a result of that and a subsequent merger, the M-H axles were initially referred to as "Wisconsin," later as "Timken" and still later as "Rockwell."

The introduction of a new product into a marketplace lacking in competitive products poses a most difficult marketing challenge. High fixed charges associated with the new manufacturing facilities pose a serious financial burden while the consumer is educated to his need for the product.

Delivery of the first truck to the Quartermaster Corp. was followed promptly by the promised order for thirty three units. This instant prosperity was short lived; no follow-up orders were in sight. The only alternative to a dearth of military orders was development of a commercial market for all-wheel drive equipment to smooth the irregular military procurements. No less a sales team than the President and Vice-President (Bert Dingley) took to the sales trail in a Marmon Sixteen loaned by Walter Marmon. Francis Chase Jr. in his *Saturday Evening Post* article described this odyssey: "They barnstormed the West, the Southwest, the Midwest, appearing before county boards of supervisors and city councils, talking with oil and lumber contractors in a vain effort to sell their trucks for road-building and heavy hauling in tough terrain."

## PART 3 — The Iranian Connection

"One night in Des Moines, Herrington had a long-distance call from his wife, who had come across an interesting item in the "Oil Journal." The Iraq Petroleum Company, which was then building a 1200 mile pipe line from Kirkuk to Haifa and Beyrooth across the Syrian Desert, had been having trouble getting trucks capable of hauling the heavy casing (pipe) across the dry sands. Officials of the company were meeting at that very moment in London to award new truck contracts."

Chase continued: "From Des Moines, Herrington telephoned London, asking them to delay their decision until he could get there and talk with them.

Two days later, he was on his way. When he arrived in London, unbelieving construction engineers listened to him, wide-eyed. He promised them in positive terms that he could build trucks capable of hauling 80,000 pounds of casing across desert sands unpierced by highways, at a speed of twenty miles an hour."

"The oil officials were dubious but told him to build such a truck to his own specifications and promised, if it lived up to his claims, they would buy it and order three more like it. Back in Indianapolis, Marmon, Dingley and others about the plant were inclined toward the oilmen's view. Only Herrington believed that he could design and build a truck that would live up to his claims. On August 18, 1932, the test run was made before a big crowd at the Indianapolis Speedway."

The truck hauled forty tons at an average speed of 25, not 20, miles per hour for one hundred miles. It successfully started, stopped and U-turned on a 25 degree hillside, utilizing its 12 speed forward, 3 speed reverse transmission. After only a week of Syrian Desert operation, the order for three more at $35,000 each was placed.

This truck model became the TH-300-6 Series Six Wheel Drive Truck offered in three gross weight ratings: 34070, 43075, and 50130 pounds. The units used by the Iraq Petroleum Co. to build the 1200 mile pipeline were the middle rating, the Model TH-320-6 used as a truck tractor pulling a semi and a full trailer.

This model was powered by a Hercules six cylinder gasoline engine model HXB of 5 inch bore and 6 inch stroke, 707 CID displacement rated at 150 hp @ 2000 rpm. The Wisconsin axles were double reduction units with **constant velocity joints on the front axle.** Full specifications for the Series are shown in the tabulation.

Prior to shipment of the first unit, this truck-train was operated on the Indianapolis Speedway loaded with 40 tons (80000#) for 100 non stop miles at an average speed of 26.28 miles per hour (30 max.). The test was supervised by the Contest Board of the American Automobile Association who certified the performance.

The M-H brochure for the TH 300-6 series described the appearance of these trucks as attractive and pleasing. The styling was in fact very well done and the distinctive radiator shell, somewhat reminiscent of the Marmon Sixteen was handsome. The V-shell carried square screening panels and rugged but pleasing headlight brackets of unusual shape.

The brochure also noted that the handling ease was remarkable . . . steering and braking being no more difficult than in the ordinary passenger car (pre power steering) regardless of the road or terrain.

The channel front bumper was spring mounted to the main frame. The forward end of the front springs were carried on vertical hydropneumatic shock absorbers which contributed to the distinctive overall appearance.

An advertisement by the International Nickel Co. in *Automotive Industries* magazine, Nov. 4, 1933 reported their product "Strengthened this Modern Ship of the Desert." The caption for a photo of the fully loaded rig describd the equipment as "believed to be the largest of its kind ever produced — overall length 100 feet; weight without load over 20 tons . . . loaded to 60 tons." The ad reported that "in actual service on the desert during a recent month, this (prototype) equipment covered 5000 miles through sand and mud at an average speed of 20 mph."

The remarkable and reliable performance of the M-H trucks building the Iraq Petroleum pipeline did not go unnoticed. It was observed by two brothers of interesting ancestry. Their father was a New Zealand physician who had served as General Allenby's liaison officer to Laurence of Arabia. Norman and Jerry Nairn had conceived the idea of introducing motorized freight and passenger service across the Syrian desert. Because motor vehicles had not been able to negotiate the desert sands, the trip from Baghdad to Damascus was made by sea, a trip lasting 30 days, and the brothers had had little success with the motor crossing despite the potential for a three day trip . . . a full order of magnitude shorter timewise than the sea route.

In 1933, Norman set out for the United States and met Herrington in Washington, D.C. When the requirements were defined and a deal struck, Marmon-Herrington undertook the design and manufacture of a huge double-deck trailer bus — perhaps more like a railroad pullman/dining car for the trackless sand trails than like a precursor of the vehicles which travel the roads today.

The significance and novelty of this super-bus, and its companion erstwhile prototype of today's highway diesel eighteen wheelers, warrant repeating here the excellent reporting of the development by an unidentified staff reporter for *Automobile Topics* in the August 26, 1933 issue:

Fig. 1-12 The Marmon-Herrington TH-320-6.

Fig. 1-13 M-H TH-320 with semi and full trailer during construction of the Iraqi pipeline.

## NEW KIND OF DESERT TRANSPORT
### Marmon-Herrington Building Semi-trailer Diesel-powered Coach and 20-ton Truck

The entry of Marmon-Herrington into the field of large all-wheel drive trucks is signaled by the beginning of construction of two vehicles ordered by the Nairn Transportation Company, Ltd. for service between Damascus and Baghdad.

One is a 31 passenger semi-trailer coach and the other a 20-ton semi-trailer freight unit. Both are built around the M-H TH-315-6 six wheel drive tractors equipped with the newly announced Hercules heavy duty automotive diesel engine.

All metal bodies for both are being constructed by the Bender Body Company of Cleveland. The overall length of the caravan coach is more than 60 feet, and the vehicle is equipped with all modern conveniences having both first-class and Pullman compartments and room for 6400 pounds of baggage. The cab is so constructed that there is a full size bunk running lengthwise back of the driver's seat, so that the relief driver may sleep during the long trip across the desert. The full width driver's seat may be occupied comfortably by three or even four persons, bringing capacity to 35 or 36 people.

The forward Pullman compartment has four rows of double seats and four of single seats with 45 inch back-to-back spacing. The rearward first-class compartment has three rows of double seats on one side and four on the other with a full rear cross seat for five. The rear compartment floor is slightly higher than the forward.

The lower halves of the windows are stationary and the upper halves may be lowered and raised, dust-tight when closed. The rear of the coach has a dust-proof compartment for 2000 pounds of baggage. There is another dust-proof compartment in the front end of the coach and large size luggage shelves throughout. Back of the front baggage compartment is a bulkhead in which are carried two spare wheels and tires, and back of the bulkhead is a complete buffet three feet deep with sliding door extending full width of the coach.

Seats are of the reclining type with card tables and ash trays. There is a telephone system from the buffet compartment to the driver's cab. The entire vehicle is insulated and ventilated for desert temperatures which exceed 120 degrees F.

The buffet compartment has a built-in ice box serviced from outside, equipped with copper cooling coils for drinking water, space for 50 imperial quarts of beverages and shelves for perishables. There is a double-plate burner and a sink with running water from an air-pressurized underfloor tank of 80 gallon capacity.

* * *

The two level passenger deck of the Nairn coach anticipated the configuration of the Greyhound Scenicruisers of the 1950s.

Although the pipeline rigs and the Nairn sleeper coach used tandem axles, the freight hauler was only fitted with a single rear axle. Consequently the trailer's tandem suspension was located quite forward; very likely the requirement to negotiate ancient city streets dictated the configuration. The TDH 315-4 (4x4) tractor with standard cab listed then at $12,500 and the THD 315-6 (6x6) at $15,650.

When Marmon-Herrington selected Hercules diesel-power for the pipe-haulers and the Nairn coach and freighter, only 150 diesel-powered trucks had been built in the United States with diesel engines. They immediately offered both gasoline and diesel power-train options on the TH series of 4x4 and 6x6 chassis.

Fig. 1-14 The Bagdhad-Damascus 14-wheeler — birth of a breed.

**MARMON-HERRINGTON**
***All-Wheel-Drive*** **TRUCKS**

Fig. 1-15 The Damascus-Baghdad desert pullman, impressive both in size and aerodynamics.

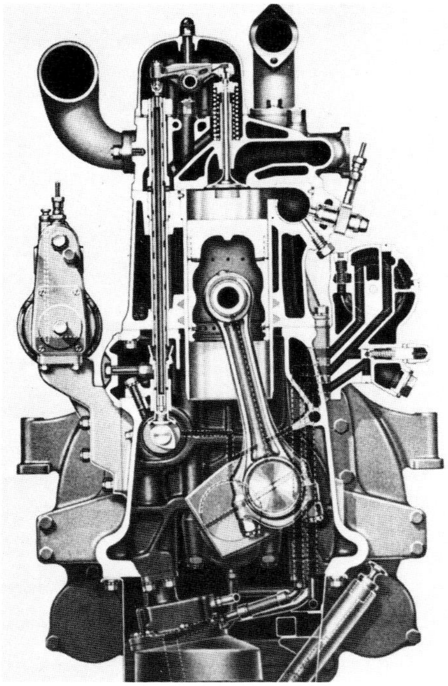

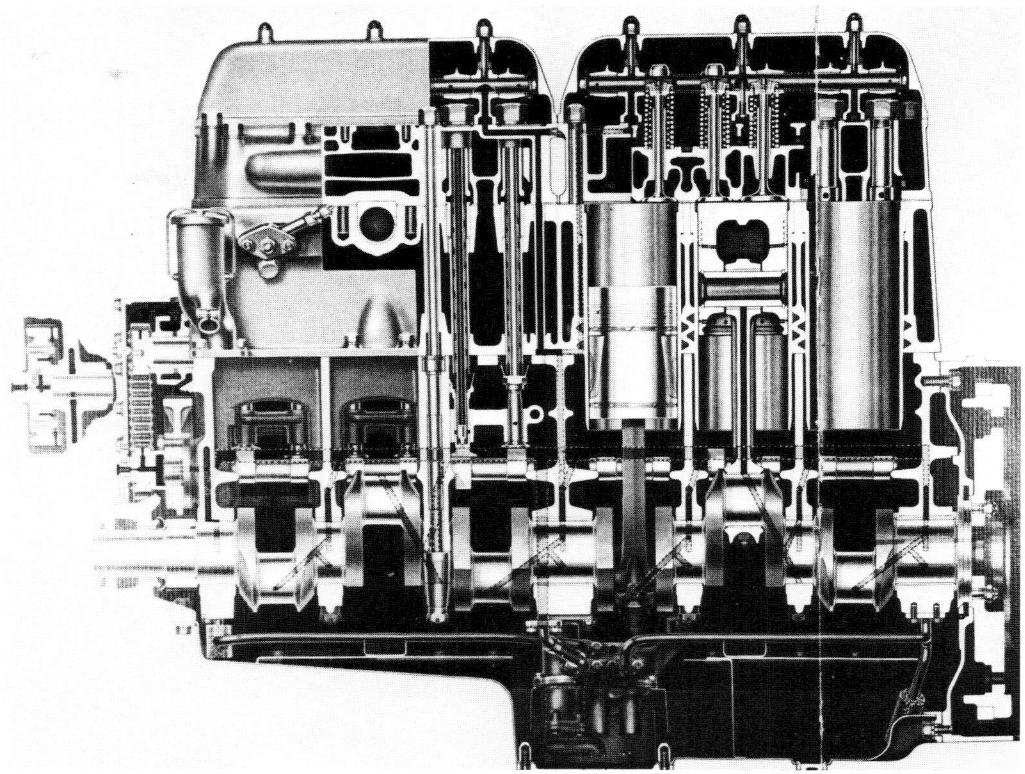

Fig. 1-16 The Hercules DHXB 185 horsepower pre-chamber diesel for the TH-320.

The Hercules automotive diesel engine Model DHXB was a 5 inch bore, 6 inch stroke engine rated 185 horsepower at 2000 rpm. The torque peak was 525 foot-pounds at 1400 rpm; torque exceeded 500 foot-pounds from 800 to 1800 rpm.[4] The Hercules catalog described the engine as a "solid injection, compression ignition type" with the "Hercules progressive increasing turbulence controlled combustion." Brake specific fuel consumption ranged from 0.38 at 1000 rpm (100 hp) to 0.46 pounds per horsepower-hour at 2000 rpm.[4]

Automobile Topics in a feature article on the "sudden and dramatic increase in the use of diesel engines" chose, as stellar examples of pioneer applications, the familiar Burlington Zephyr unit-train and the Marmon-Herrington Baghdad-Damascus coach. That train is preserved in the Chicago Museum of Science and Industry and has been designated a National Mechanical Engineering Landmark. Where is the coach?

The Hercules automotive diesel engines employed American-Bosch in-line multi-plunger injection pumps and pintle type nozzles injecting fuel into "auxiliary turbulence chambers" (now referred to as pre-combustion chambers). The pre-chamber type of engine could not compete with the superior fuel economy of the direct injection type in heavy truck service after the latter type was introduced;

a similar situation is now developing in passenger car and light truck diesels. The Hercules diesel was unusual in having the location of the injection nozzles and pre-chambers in the cylinder block rather than the now common practice of a cylinder-head location.

The coupling device by which a truck tractor is connected to a semi-trailer is known as a "fifth wheel" . . . an expression evidently coined when a truck had four rather than six or ten wheels. The novel and innovative M-H for the Nairn Brothers featured a "real life" fifth wheel coupling. To contend with the desert environment, Bob Johnson replaced the conventional fifth wheel and its large flat greased surface with an additional wheel unit identical with the other eighteen. The tire/wheel was clamped to the tractor chassis by three arms bearing down on the tire sidewall and tread. The hub and sealed bearing assembly was attached to the bottom of the trailer frame. Coupling was accomplished by lowering the trailer and hub into the wheel and installing the usual wheel mounting nuts, exactly in the manner of changing a flat tire on a road wheel.

M-H also designed a companion freight trailer of flat bed type, to the same massive scale, with a second tractor. The pair of vehicles posed a massive transportation problem which Herrington solved in his traditional highly resourceful manner. Because they were too large for the US highways and too long for the US railroad flat cars, they traveled east to port of embarkation on two super-sized circus railroad flatcars borrowed from the Hagenbeck-Wallace Circus.

When Francis Chase, Jr. wrote his "Post" article, on Herrington (Aug. 8, 1942) he reported that these "ships of the desert" were still in highly profitable operation but had been forced to travel by evasive routes, guided by compass, and accompanied by armed guards consequent to Nazi infiltration of Syria. Chase also reported that those same circumstances had resulted in the fourteen year old son of Norman Nairn living in the Herrington home in Indianapolis for safety.

Some years later, Marmon-Herrington furnished the tractor for a "Modern Flying Carpet" built for the King of Saudi-Arabia. The tractor mounted an electric power generating plant forward of the fifth wheel, and towed a semi-trailer designed and built by the Spartan Aircraft Co.; the aluminum trailer weighed 12,000 pounds and cost $50,000.

## PART 4 — Armorer To The Shah

The four trucks which were bought by the Iraq Petroleum Co. led unexpectedly to a major role in the Middle East for Colonel Herrington. The surprising turn of events was related in Francis Chase, Jr.'s. "Hell-on-Wheels Herrington" for the *Saturday Evening Post,* (Aug. 8, 1942) and is quoted here: *"In the winter of 1933, the Shah Reza Pahlavi of Persia (now Iran) decided to mechanize his army, and dispatched a staff officer to London to buy the best equipment available there. The officer, arriving in Baghdad several hours late, missed railroad connections for Beyrouth, and the one slender chance he had left for catching his boat was to try for it in an Iraq Petroleum Co. truck. He made the boat in plenty of time and, impressed by the truck's speed, jotted the manufacturer's name on a memorandum slip. In London, he cabled Marmon-Herrington to have someone meet him in Sweden, where he was next heading. Herrington kept the appointment. The outcome . . . his appointment as consulting engineer to the Persian/Iranian army and carte blanche to design and build all mechanized equipment for the Shah's army, which remains today (1942) one of the most up-to-date mechanized outfits in the world."*

Thus began a relationship that brought urgently needed business to the fledgling firm in Indianapolis and would ultimately see the Shah as a house guest of the Herringtons.

Some of the military vehicles created by M-H for the Iranian service became prototypes for US and allied power vehicles in WW II.

One of the first vehicles designed under the new program was the TH 310-ALF Armored Car. The least that can be said of this novel creation would be that it anticipated James Bond/Agent 007 by more than a quarter century. The styling, despite the armor, can only be classed as handsome for a motor vehicle of any class. It had armored side-mount spares and, what appeared to be a saucily tilted rear spare, was actually a stack of four round armor plates, one-half inch thick and nearly the full diameter of the tires. These discs provided protection from side fire.

Front headlights were shielded by pairs of armored doors, vertically divided. Bob Johnson recounts that the control system was procured from a commercial building supply firm . . . an apartment building remote electric door actuator.

Because many Iranian roads climbed steep mountain sides with "hairpin" turns too sharp for a motor vehicle, the armored car was fitted with an auxiliary driver's station in the rear. Except for the compromise of "rear" end steering in reverse, the vehicle was multi-directional.

The four man crew was protected by the ¼ inch armor hull, and by armored observation ports. A 45 gallon rectangular tank in the left hand corner provided water for emergency purposes. The tank in the right hand corner carried fuel for 375 to 450 miles of operation.

Compliance with contract specifications was assured, prior shipment overseas, by obtaining Certificates of Performance from the Contest Board of the American Automobile Association (AAA). Charles Merz and William H. Russell were the official observers. The Certificates (sanction 3044) were also signed by Eddie Rickenbacker, Chairman of the Contest Board.

For the acceptance tests, June 13-19, 1934 the vehicle was driven by Bob Wallace; Mr. Herrington also drove during the 50 mile runs. The customer, the Imperial Persian Army, was represented by Major Mahmond Sartip and Lt. Habib Naficy.

Acceleration, braking and 50 mile non-stop tests were conducted on the Indianapolis Speedway in both forward and reverse directions. Twelve-hour non-stop runs on Indiana State highways were made in both directions (forward and reverse) on succeeding days. Mobility and stability tests were then conducted on the Mann Farm. On the fifth day, deep sand tests were run at the M-H Proving Grounds and river crossing demonstrated on the White River test course.

The Armored Car accelerated from 0 to 20 mph in 13 seconds, stopped in 2.5 seconds, 35 feet. On the 50 mile run; fuel economy was 5.36 miles per gallon forward, 3.77 miles per gallon in reverse. During the twelve hour highway runs, economy was 5.50 and 4.0 miles per gallon at average speeds of 23.5 mph forward and 21 mph in reverse. The course was south on Indiana 37 and US 150 to New Albany and north on US 31.

It required seven minutes to climb a 1000 ft route rising 154 feet with a maximum grade of 30° or 67%. Water depths of 34 inches were successfully traversed during the White River crossings.

The power train which provided this performance was:

| | |
|---|---|
| Engine: | Hercules gasoline Model RXC<br>6 cylinders 4 5/8 bore, 5 ¼ stroke, 529 CID |
| Axles: | Wisconsin, Double-reduction on 139 inch wheelbase, 6.36:1.<br>Semi-elliptic springs: 48 x 3 front, 60 x 3 rear. |

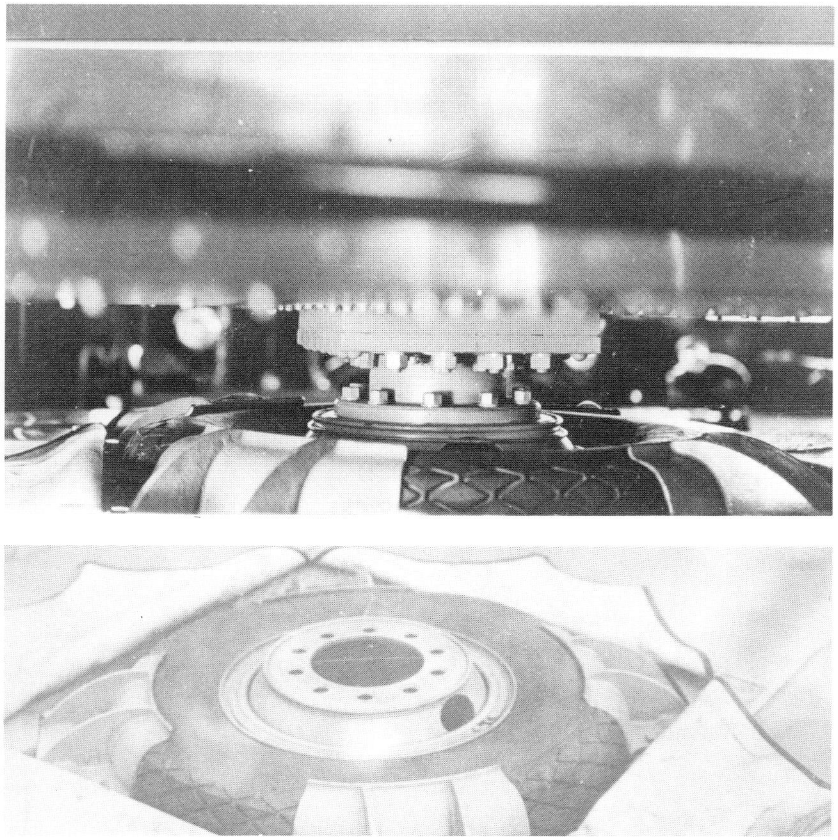

Fig. 1-17 This tractor had a real 'fifth wheel'. Note clamps and dust cover.

Transmission: Fuller with M-H two speed forward/reverse auxiliary, Fuller multiple-disc clutch.
Reduction First gear (of 4) 42.301: Reverse 50.37:1 High Range
Reduction First gear (of 4) 117.85:1 Reverse 168.05:1 Low Range

Mr. Chase's account reported Herrington's first venture in the dictionary business: *"Another Iranian problem gave Herrington more trouble. When he started to prepare an instruction manual and spares lists for the Iranians, he learned that many of the names of motor parts had no counterpart in their language. In order to overcome this deficiency, Herrington and the visiting Iranian officers sat down in his office in Indianapolis and coined the necessary additions to the Iranian vocabulary, cabling them to Iran for government adoption before ordering the manuals printed."*

Years later, Herrington would again be in the dictionary business: The *Marmon-Herrington NEWS* for March 1944 (Vol. 2, #7) reported a Marmon-Herrington Contribution to Chinese-American Relations... *"Recognizing the increasingly important part that transportation will play in the industrial and cultural development of post-war China, the Marmon-Herrington Company is making a substantial contribution to a better understanding between the Chinese and American peoples by underwriting the preparation of an English-Chinese automotive dictionary. Books of technical terminology are few in China. Until now there has been little standardization of Chinese automotive terms, and no organized translation of any great number of American names and terms into the Chinese language. The new book which consists of 252 pages, lists almost 10,000 automotive and related terms and phrases, with their Chinese equivalents. It will prove particularly valuable to the military forces now operating in China and to governmental and*

Fig. 1-18 Marmon-Herrington mobile home for the king of Saudi-Arabia.

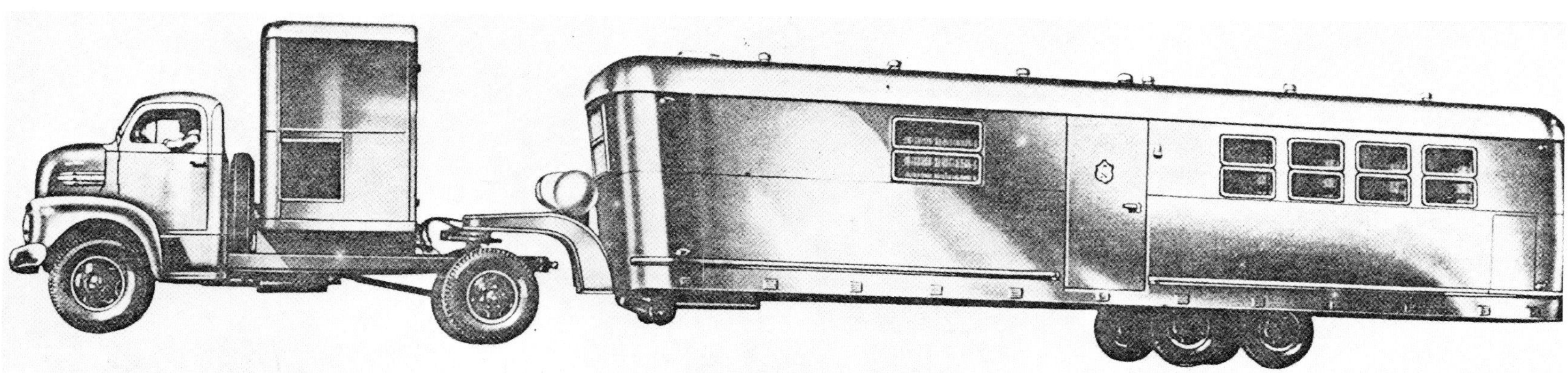

Fig. 1-20 Handsome bi-directional armored car with side-mounted spare tires.

Fig. 1-22 Left to Right: Charles Merz, Walter C. Marmon, Lt. Habib Naficy, Col. Herrington, William H. Russell.

Fig. 1-23 Note armor discs stowed on rear for tire protection in combat.

Fig. 1-22 AAA Sanctioned acceptance test on the Indy Speedway.

Fig. 1-24 The REAR driving station.

Fig. 1-25 The forward driving station. Note Marmon 68 instrument cluster

FOREWORD

The late Dr. Sun Yat-sen, founder of the Republic of China, once said that "This world is benefited by the United States being an industrial and business country." I in turn feel that the future world will be greatly benefited by China becoming industrialized.

We have long admired the wisdom of the Chinese, their good sense of humor, and their philosophic thought. They are both thinkers and realists. They have given us great philosophers like Confucius and Mancius. The world owes much to China, the invention of the mariner's compass, gunpowder, silk, porcelain, paper and ink.

One of the great obstructions to international development is that we do not have a universal language which would enable the masses of our people to communicate readily.

In the absence of such a ready method of communication it is a pleasure on our part to contribute this book to the cause of China's military success and industrial development, and to a more complete mutual understanding between our great countries.

A. W. Herrington
Chairman of the Board of Directors
Marmon-Herrington Company, Inc.

Indianapolis, Indiana
October, 1943

Fig. 1-26 Foreward, in English and Chinese, to *English-Chinese Automotive Nomenclature* sponsored by Marmon-Herrington; author's acknowledgment and first page of translations.

*civilian transportation organizations in the post-war reconstruction period. The new dictionary, just off the press, has been in preparation for over a year. The huge task of research and translation of terms has been done by Mr. Calvin C. Chang, a university educated, young businessman of Shanghai, China, temporarily marooned in this country. In speaking of the difficulties encountered in this work, Mr. Chang explained that this was not merely a matter of translation but in many instances, the creation of Chinese names for automotive terms. After translation, the Chinese words and characters were inscribed by Mr. Chang for photographic reproduction. The work represents many months of painstaking toil. Mr. Chang became associated with the Marmon-Herrington Company early in 1942. He completed a course of training in the Company's tank school, later gained experience in the Service Technical Division and is now connected with the Public Relations Division. Mr. Chang has made numerous speeches and lectured on the situation in China. Chinese authorities and technicians in this country and American experts on Chinese matters have taken a great interest in the development of this book. It has the endorsement of prominent Chinese officials, foremost of whom is His Excellency Dr. Wei Tao-ming, Chinese Ambassador to the United States, who contributed a personally hand lettered Chinese title page for the book. The dictionary will be distributed free by the Marmon-Herrington Company to Chinese engineering students in this country, and to the Chinese government for distribution in China. A quantity will be donated to the United China Relief for sale by that organization in raising funds. It will also be distributed to the libraries of colleges, of states and other cultural institutions in the United States and Canada."*

The revolution of Mao Tse-Tung intervened in the development of free commercial enterprise in China. Following the Ping-Pong diplomatic breakthrough of the Nixon Whitehouse era, the author suspected that there would have been published a more recent English-Chinese automotive dictionary. However a national computer search by the Kreske Library of Oakland (MI) University in September 1984 failed to locate a latter-day counterpart.

The author purchased a copy of Mr. Chang's "English-Chinese AUTOMOTIVE NOMENCLATURE" in 1982; it had been retired by Georgetown University, evidently on an age basis rather than a supercession basis.

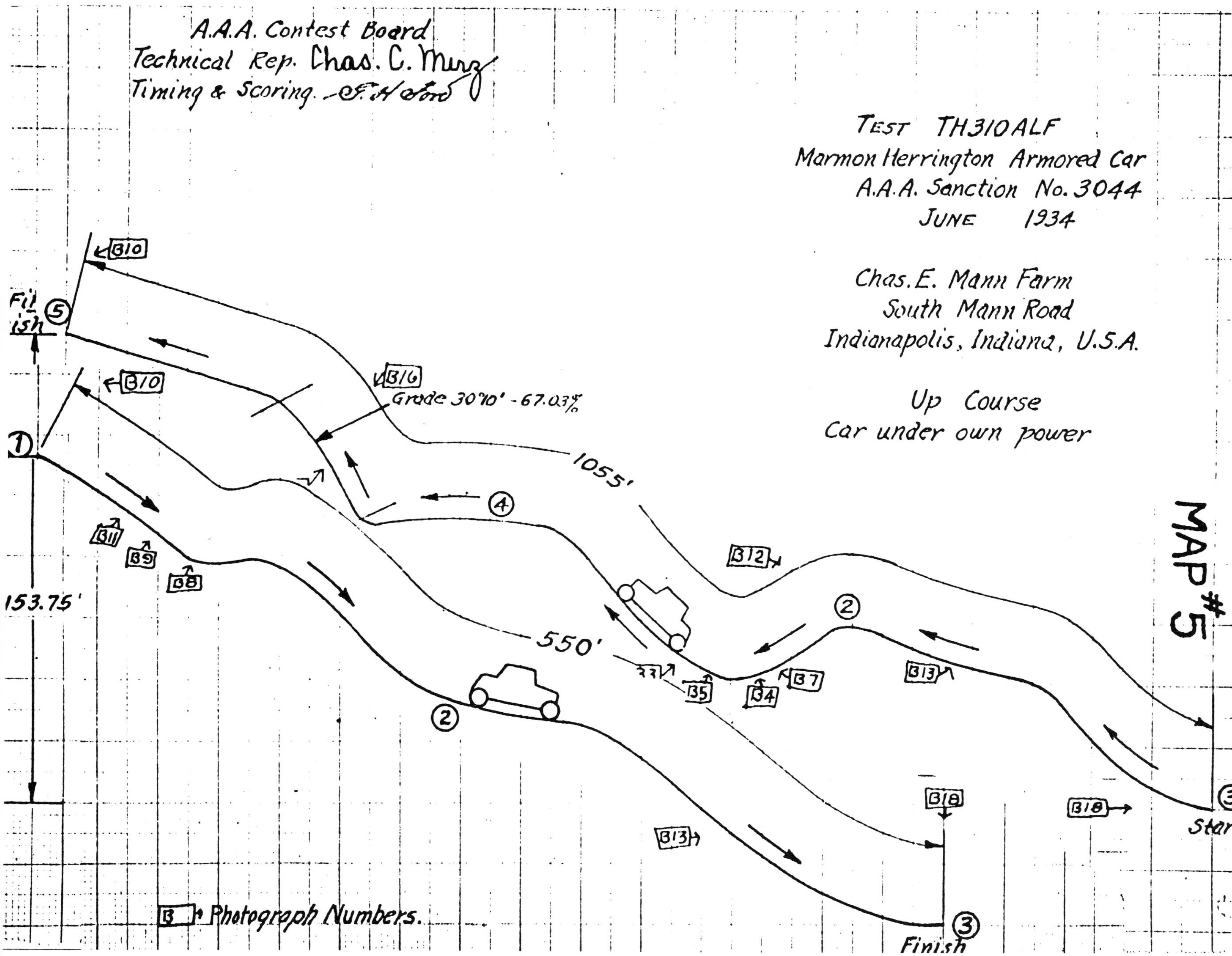
A.A.A. Contest Board
Technical Rep. Chas. C. Merz
Timing & Scoring
Test TH310ALF
Marmon Herrington Armored Car
A.A.A. Sanction No. 3044
June 1934
Chas. E. Mann Farm
South Mann Road
Indianapolis, Indiana, U.S.A.
Up Course
Car under own power
MAP #5
Finish
Grade 30°10' - 67.03%
1055'
550'
153.75'
Start
Finish
Photograph Numbers.

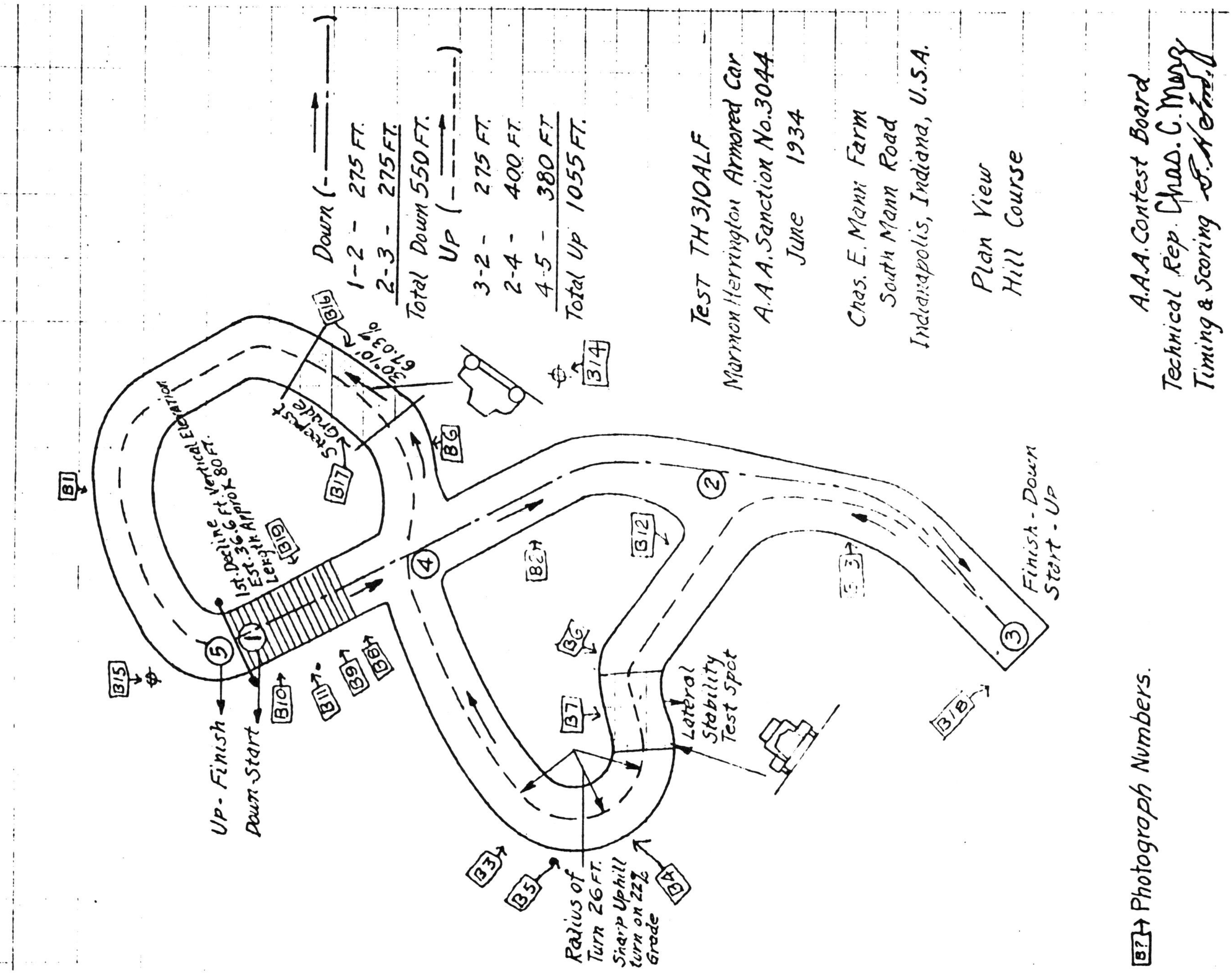
Down (——)
1-2 - 275 FT.
2-3 - 275 FT.
Total Down 550 FT.
Up (- - -)
3-2 - 275 FT.
2-4 - 400 FT.
4-5 - 380 FT.
Total Up 1055 FT.
Test TH 310ALF
Marmon Herrington Armored Car
A.A.A. Sanction No. 3044
June 1934
Chas. E. Mann Farm
South Mann Road
Indianapolis, Indiana, U.S.A.
Plan View
Hill Course
A.A.A. Contest Board
Technical Rep.
Timing & Scoring
Up - Finish
Down Start
Finish - Down
Start - Up
1st. Decline 36.6 Ft. Vertical Elevation
Est. Length Approx. 80 FT.
Steepest Grade
30°10'
67.03%
Lateral Stability Test Spot
Radius of Turn 26 FT.
Sharp Uphill turn on 22% Grade
Photograph Numbers.

MARMON-HERRINGTON
HIGH-SPEED
Track-Laying
Tractor
MARMON-HERRINGTON
Amazing TRACTIVE POWER
MANEUVERABILITY AND ECONOMY

CHAPTER TWO

# "PURVEYORS OF MOTORIZED MILITARY EQUIPMENT"

## PART 1 — The Domestic Marketplace 1931-1941

The chassis for the first M-H fleet, the 33 aircraft refueling trucks built for Holabird Quartermaster Depot in 1931, was designated M-H Model T-1.

This fleet of T-1s was unique. The hood and cowl sported ventilation doors, as did the hoods of the contemporary Marmon 70, 88 and Sixteen. The cabs had stylized windows in the doors with curved tops and there was a short sun-visor built into the roof line.

This was the first "drop frame" all wheel drive vehicle procured by the military. The "drop-frame" concept originated by Herrington retained the normal frame height for the length of the body, keeping a normal height to the center of gravity of the load. The frame then rose higher under the cab to provide the necessary clearance with the "live" axle. The usual "dead" axle of a 4x2 had a drop center for the same purpose, allowing a straight frame rail.

A slightly larger and more rugged 4x4 chassis had been developed and christened TL-30 when the request was received from the Shah's procurement minister for Herrington to meet him in Sweden. In those days, you rushed to Europe by the fastest ship; the TL-30 prototype went along with the Colonel to Bofors, Sweden.

The Bofors demonstration resulted in the first large Iranian government contract. It was for an entire series of a new model A 30 S-4 trucks rated 3 to 3.5 tons, with powered axles on a 139 inch wheelbase. They were used to tow Bofors anti-aircraft guns, carrying ammunition and the gun crews. Some chasses were fitted with a M-H designed reconnaisance car body and provision for carrying a motorcycle.

Satisfaction with the refueling vehicles led to an order for a pilot model truck capable of transporting a portable liquid oxygen generator. This was the TH 315-6 with a 193 inch wheelbase and rated 17 to 20 tons. Other experimental vehicles were built on orders from the Quartermaster Corps and Ordnance Department. These included the TL 29-6, a 6x6 prime mover with large engine in a light chassis and the T 310-6, a 6x6 with the first modern mobile machine shop body — sides and rear hinged at the top and extended as roof members while in use at a fixed site.

A fleet of seven TH-310 A-6 were furnished the Army which designated them T-9 prime movers. This fleet of 6x6's carried the military "covered wagon" type cargo body.

A group of 20 A-10 chassis also with typical military "covered wagon" bodies was built for the US Army. The US Army jobs had roadster cabs with canvas top and side-curtains, and dual side mounted spare tires.

Two similar vehicles were built for the Army on the A-20 chassis with oversize tires to be used for cross-country wire laying missions.

It is interesting that in the same year that the last Marmon automobile was produced — the first Marmon-Herrington passenger carrying vehicle was produced — the Iranian scout car!

The "Marmon-Herrington News" (Vol. 2 #5) proudly announced that delivery of the fleet of approximately fifteen Model A-10 trucks to the US Army made the Army the first purchaser of the A-10 production model. The vehicles as delivered included brush guard (forward of radiator), open cab, cargo body and two spare tires.

Fig. 2-1 First production; Model TL-30-4 for Holobird Quartermaster Depot.
Fig. 2-2 Model TL-30-4.

The 1934 product line is described on the following page reproduced from *M-H NEWS:* The several truck series were priced as follows:

| Series | Price | Series | Price |
|---|---|---|---|
| A | $2250-5940 | THD-4 | $8,995- |
| TH-4 | $6285-12,745 | THD-6 | $11,075-19,450 |
| TH-6 | $10,000-15,165 | | |

16 MARMON-HERRINGTON NEWS

# THE COMPLETE 1934 LINE

## —All Wheels Driving—

### NEW A SERIES

Model A10-4......$1\frac{1}{2}$ to 2 Tons Capacity
Model A20-4......$2\frac{1}{2}$ to 3 Tons Capacity
Model A30-4......$3\frac{1}{2}$ to 4 Tons Capacity
Model A40-4......4 to $4\frac{1}{2}$ Tons Capacity
Model A50-4......4 to $4\frac{1}{2}$ Tons Capacity

*Four Wheels Driving*

### TH SERIES

Model TH300-4.......$4\frac{1}{2}$ Tons Capacity
Model TH310-4...5 to $5\frac{1}{2}$ Tons Capacity
Model TH310A-4........6 Tons Capacity
Model TH315-4.........7 Tons Capacity
Model TH320-4.....8 to 9 Tons Capacity
Model TH330-4....9 to 10 Tons Capacity
Model TH340-4........20 Tons Capacity

*Four Wheels Driving*

### TH SERIES

Model TH310A-6.......10 Tons Capacity
Model TH315-6...12 to 13 Tons Capacity
Model TH320-6...15 to 18 Tons Capacity
Model TH330-6........20 Tons Capacity
Model TH340-6........35 Tons Capacity

*Six Wheels Driving*

### THD SERIES-DIESEL

Model THD310A-6.....10 Tons Capacity
Model THD315-6..12 to 13 Tons Capacity
Model THD320-6..15 to 18 Tons Capacity
Model THD330-6.......20 Tons Capacity
Model THD340-6.......35 Tons Capacity

*Six Wheels Driving*

### THD SERIES-DIESEL

Model THD310-4..5 to $5\frac{1}{2}$ Tons Capacity
Model THD310A-4......6 Tons Capacity
Model THD315-4........7 Tons Capacity
Model THD320-4....8 to 9 Tons Capacity
Model THD330-4...9 to 10 Tons Capacity
Model THD340-4.......20 Tons Capacity

*Four Wheels Driving*

MARMON-HERRINGTON COMPANY, Inc.
Indianapolis, Indiana, U. S. A.

Printed in U. S. A.

Fig. 2-3 Model TH-300-4 Note change to louvered hood, and down-draft carburetor on Hercules engine.

Fig. 2-4 Marmon-Herrington mobile machine shop circa 1932.

Fig. 2-5 Model TH-310-6

Fig. 2-6 Model A-10

Fig. 2-8 Interior of Model A-10 cab.

The A Series styling was different from the T-1. The radiator shell followed the hood line, seemed intended for a flat grill, and had a forward lean at top center, as if the Marmon Pyramid had blended. However, the grill was pointed only slightly at the top, increasing to the bottom — very distinctive and easily identifiable. The hood had vertical louvers and a belt molding extending along the cab as well.

The M-H axle construction was improved and many appearance improvements made. The October 7, 1933 Edition of *Automobile Topics* (page 37) reported a test demonstration of superior tractive efficiency of all-wheel-drive vehicles. The firm of E.C. Harris & Son of Indianapolis operated one of the pioneer inter-city common-carrier operations between that city and Chicago, providing regular overnight service over good roads and relatively flat terrain.

In a joint evaluation program with Marmon-Herrington, a M-H 4x4 tractor was substituted for the regular 4x2 unit. In that era, they were 10 wheelers not 18 wheelers! The Marmon tractor was a Model A-20, rated at 2½ tons, and hauled payloads ranging from 14533 to 15,746 pounds in a Harris trailer estimated to be 27 to 30 feet long.

"The Harris firm . . . has figured the distance between the two cities as 210 miles as traveled. At any rate, every time one of its rear wheel drive trucks made the trip, . . . there were 210 miles of tire wear and parts wear."

"The extraordinary feature of the test was that, in four trips between Indianapolis and Chicago, the wheels of the M-H truck traveled an average of only 181 miles per trip or 13.5 percent less than the distance covered by the wheels of the conventional truck."

"This difference, according to Marmon-Herrington officials, is accounted for by the fact that the positive traction achieved by driving through all wheels prevents wheel slippage ordinarily encountered in rear wheel drive vehicles."

Author's comment . . . When a vehicle body bounces up and down, the weight on the wheels varies, and on a severe bounce may not be sufficient to momentarily sustain non-slip traction. But once slippage is initiated, the lesser dynamic coefficient of friction applies instead of the greater static coefficient; in consequence, the slippage is prolonged. The cumulative slippage is composed of an infinite number of minute slippages.

The walking-beam tandem suspension found on many eighteen wheelers today was initially introduced for inter-city buses in the late 1920s because it reduced the vertical movement (and acceleration) of the passenger compartment. This reduced movement also diminished the variation in weight on the drive wheels and the suspension type is widely used today to improve ride and traction of truck tractors.

It is surprising that the application of hydraulic shock absorbers to heavy trucks has only become common on the front axle and an available option on some rear suspensions in recent years. In fact, the whole issue of truck ride dynamics and driver comfort has become a major issue of the 1980s with the Teamsters Union, the National Highway Traffic Safety Administration, The Highway Traffic Safety Institute (Univ. of Michigan) and the Motor Vehicle Manufacturers Association of the US (MVMA) being deeply involved.

The demonstration example would be more credible had some kind of speedometer calibration test been run at low speed and light load, to confirm that there was no error introduced by tire rolling radius or drive ratio error.

By November, 1934 there were 26 models, 35-ton rated models having been added to the TH-6 and THD-6 lines and 20-ton rated models to the TH-4 and TD-4 lines.

Fig. 2-9 One of six T11 armored cars built for Ordinance Dept.

Fig. 2-10 Model A-7 scout car on the Model A-30-S chassis.

It was 1935 when the A 30-6 Iranian cargo fleet was completed and shipped. Nineteen thirty five was a milestone year for the young firm because two major product markets were entered . . . markets, that would shape the growth of M-H for years to come.

The first new category of vehicle was the armored military vehicle . . . both track laying and rubber-tired. The second was the conversion of high-volume commercial vehicles to all-wheel drive.

The Ordnance Department ordered six T-11 armored cars to a M-H original design, following demonstration of the prototype, which were reported still in use ten years later. These were of a class of vehicle manufactured today by the Cadillac Gage Co.

*The Marmon Herrington NEWS* for March 1944 (Vol. 2 #7) reported that the US Government was not alone in its early interest in the outstanding performance of

M-H trucks. Witnessing the spectacular work done by M-H vehicles in hauling pipe across the trackless Syrian desert in 1932, the Persian (Iranian by 1944) Government gave M-H the job of completely mechanizing their army. Among the many different units of the mechanization were armored cars which were unlike any then in use by the US Army. M-H engineers then designed a scout car, the A-7, with the open "bath-tub" type body so familiar subsequently in WW II. Several suggestions of the US Ordnance Department were incorporated in the A-7 and it was tested at the Aberdeen Proving Ground before being shipped to Iran. That car definitely influenced all later designs of Scout cars. It had a low silhouette yet the sides were high and afforded protection to the occupants; the car was armored; it had high power transmitted to all wheels and was fast.

M-H built other Scout car models for the US Army in the following years, incorporating improvements developed by experience. But basically the US Scout Car of 1944 was much the same as that first A-7 for Iran and it was conceded that the A-7 had a most definite influence on the design of all those that followed.

During 1934 and early 1935 Marmon-Herrington undertook the design and development of a new concept of track for track-laying vehicle applications. This was a continuous rubber belt track reinforced with continuous steel cables. The development was done with the cooperation and assistance of the B.F. Goodrich Rubber Co. The track was a Goodrich development, from a French concept, which was under the direction of Bob Maims. Marmon Herrington engineers made significant contributions to the success of the track concept by research and development of the track interface with the vehicle sprockets and support wheels. Herrington was enthusiastic about the potential of the track and during 1935 it was engineered into two prototype vehicles; a half-track and a light combat tank which are subsequently described.

For the 1938 model year, Marmon-Herrington boasted 'A Truck for Every Need' and introduced a 'B' Series ranging in gross capacity from 8400 pounds (**not** GVW) upward. The B Series was a refinement of the A Series. The cab greenhouse was redone eliminating the curved-top windows and the headlights were made free-standing on the B-10. The regular line then included 28 standard models with capacities as high as 52,000 pounds. Seventeen were 4x4s (12 gasoline and 5 diesel); eleven were 6x6s (7 gasoline and 4 diesel). Each model was available in two or more wheelbases and two tire sizes.

Fig. 2-11 Model B-10-6

Fig. 2-12 Model B-20-4

## PART 2 — The Ford Conversions — 1935

In 1935, while Colonel Herrington was in the Middle East on an extended business trip, Walter Marmon confided to Bob Wallace that he had an idea that a large part of the cost of the M-H trucks was in the chassis and sheet metal rather than in the multi-wheel drive system. Marmon directed Wallace to undertake an immediate study of the potential for conversion of the Ford 1½ ton V-8 powered truck to all wheel drive. Walter Marmon was familiar with the Ford products because he owned a Ford dealership in Indianapolis and both he and Howard were enthusiastic over the Ford V-8 engine design and its performance. When Colonel Herrington returned, he expressed violent opposition to the concept. But Marmon prevailed and the prototype was built. When Herrington drove the prototype vehicle, he reversed himself completely and enthusiastically supported the project thereafter.

Engineering of the conversion of the Ford 1½ ton 4x2 to a 2½ ton 6x6 was undertaken soon after, and the prototype was completed in late 1935.

The M-H conversion was a major chassis modification requiring major disassembly of the Ford truck and removal of the forward frame and transverse spring/axle assembly. The conversion included new frame extensions and parallel semi-elliptic springs. The axle utilized Ford internal elements and a special casting for one differential half housing to provide a spring perch necessitated by the offset of the drive shaft to clear the engine. Outer ends of the axle terminated in a unique M-H design with constant velocity universal joints and provision for steering. Cen-

Fig. 2-13 Marmon-Herrington 6x6 conversion of Ford circa 1936. Note the triangular Marmon-Herrington badge on the hood.

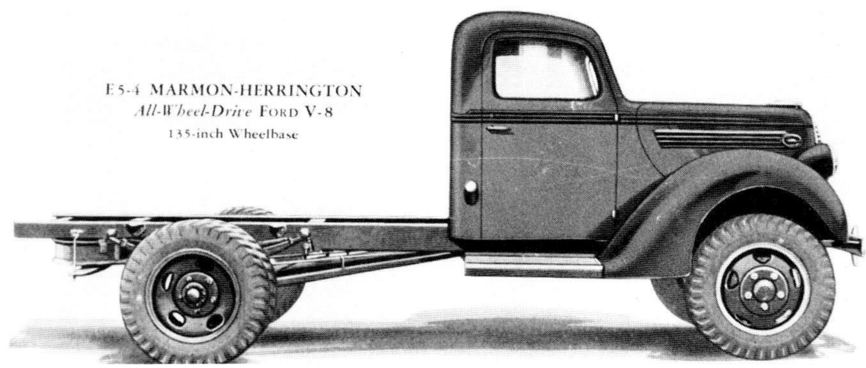


Fig. 2-14 Marmon-Herrington 4x4 conversion of Ford chassis.

Fig. 2-15 Plan view of 4x4 converted long wheelbase chassis circa 1937.

Fig. 2-16 Marmon-Herrington Model E5-4 All-Wheel-Drive Ford V-8; 135 inch chassis.

Fig. 2-17 Original conversion package (inset) and package used after Ford adopted semi-elliptic front springs.

Fig. 2-19 M-H Ford AWD fire truck for wartime Washington National Airport. American LaFrance equipment

Fig. 2-18 Marmon-Herrington Model E5-COE-4 All-Wheel-Drive Ford V-8.

*All-Wheel-Drive Ford Hauling Heavy Draw Works*

Fig. 2-20 All-Wheel Drive Ford hauling heavy draw works.

tral frame cross members were also modified and/or replaced to accomplish mounting of the Marmon-Herrington transfer cars.

The result was an all-wheel-drive vehicle only slightly higher off the ground than the standard rear wheel drive Ford. A casual observer might fail to detect the modification at all.

The Marmon-Herrington Ford conversion provided an all-wheel-drive vehicle at an attractive price, otherwise impossible, with the tremendous advantage of service and parts availability from the thousands of Ford outlets worldwide with parts unique to the all wheel drive system and front suspension the only ones requiring additional logistic support.

Very much to the credit of Marmon-Herrington is the fact that they recognized the realities of the Ford efforts at mechanical braking of motor vehicles. As a consequence of this perception, the consumer could buy a Ford truck in 1935 with hydraulic brakes! Of course, he also had to invest in a Marmon-Herrington all-wheel-drive conversion. And by 1938 the same could be said for the Ford passenger car (Ford adopted hydraulic brakes for the 1939 model year).

The conversion was accomplished by a reverse assembly process until the forward half of the frame was bare. The frame was then modified to accommodate semi-elliptic front springs and to provide clearance and support for the transfer case. Then the Marmon-Herrington Ford was assembled.

## ABOUT THE FORD V-8 4X2 TRUCK . . . "A Great Truck"

Full appreciation of the Marmon-Herrington Ford conversion concept cannot be had until one views the advent, first of the Ford V-8 trucks, and second of the M-H conversion, in the perspective of the early 1930s.

The first trucks were outright substitutions for the dray horse or horse teams, lumbering along at a few miles per hour on solid rubber tires, over cobblestone pavements, unlike the first automobile which quickly outperformed the horse and buggy. As late as 1927, the famous Mack AC truck had a 471 CID (40SAE hp) engine capable of 74 horsepower, but governed to 50 hp for durability reasons, and maximum road speed was 21.5 mph for all models, but only upon special order. Standard maximum speed was 18.5 mph for the 3½ ton rating and 12 mph for the 7.5 ton rating. Tires were still solid rubber, but pneumatics were optional on Mack fire apparatus.

Unlike the Mack, automotive engines were often designed for survival in passenger car service then grossly derated for the truck application. The Ford Model AA truck, although not derated, had only 40hp; it was, like other truck applications of car engines, downgeared to multiply torque to a requisite level for commercial use with a resultant compromise in top speed.

The advent of the 85 hp Ford V-8 (65hp in 1932-33) had major impact on the truck marketplace. It had sufficient power that it could be geared for 50 mph truck service and superior durability that avoided derating for commercial service.

A dual-breaker ignition distributor, with one-set making and one-set breaking the circuit provided exceptional ignition performance and durability at high engine speeds. Unusual floating bearing inserts, with bearing alloy on both surfaces, provided impressive durability.

The performance and durability of the Ford V-8 engine was perhaps best attested by its popularity for repowering the Mack AC trucks.

During the 1934-1941 period, the author's brother Keenan Hanley manufactured fire apparatus . . . The Hanley Engineering Service of Prospect, Ohio . . . the majority of which were built on commercial chassis. Each completed fire truck was required to complete successfully a twelve hour endurance pumping test observed by the National Board of Fire Underwriters. Unless the engine produces the claimed horsepower, the water delivery will not meet specifications and the test will be failed. The experience at Prospect was that the Ford V-8 could be consistently depended upon to produce rated horsepower and to demonstrate the requisite durability. The same could not be said equally of experience with other commercial chassis of similar rating.

The author recalls Roadway Transit operating a fleet of 1934 Ford trucks between Detroit and Chicago hauling 25 foot trailers at speeds up to 50 mph. The Ford V-8 truck provided great impetus to the infant inter-city trucking industry. It made low-cost fire apparatus, rated 500 gallons per minute, available to communities which had been unable to afford custom-chassis apparatus.

Just as Art Herrington had quickly recognized the potential of the Rzeppa U-Joint, Walter Marmon had recognized the potential of the Ford V-8 truck and the "Great Truck Made Even Greater" was the result.

During the initial promotion era for the Marmon-Herrington Ford conversion, the M-H ads stressed the merits of the basic Ford truck and their enhancement by the conversion. The slogan was "A Great Truck Made Even Greater". The ruggedness, power, dependabilty of the Ford V-8 engine when combined with the M-H All-Wheel-Drive traction was claimed "to give a measure of payload capacity never before available in the low-priced field, and to be fast, flexilbe and remarkably safe and economical."[2]

Other ads stressed the theme "Two-fold Leadership Gives You the Greatest All-Wheel-Drive Truck Value in History" . . . a union with "the firm with leadership in volume production and quality, a wide scope of manufacturing operations and world-wide service that always is available."[3]

List prices in 1937 began under $1400 and were described as "less than half the cost of other four or six wheel trucks of similar rating and capacity".[4]

Marmon-Herrington first offered All-Wheel-Drive conversions of Ford passenger cars in 1937. These vehicles carried the usual Ford badges, just as did the pick-ups and trucks. All shared the addition of "Marmon-Herrington All-Wheel-Drive" badges on the hood sides near the cowl.

The 1937 offering was initially restricted to the Ford "woody" station wagon,[5] but beginning in 1938 the full line of passenger cars was available with M-H conversions.

Surprisingly the Marmon-Herrington conversion resulted in a 113.5 inch wheelbase whereas the basic Ford had a 112 inch base.

The M-H conversion slogan was readjusted for 1938: "A Great **Car** Made Even Greater" Conversion of all Ford 112 inch wheelbase models was offered for a charge of $895[6] when the Ford dealer furnished the standard Ford vehicle to M-H; the converted car was a Marmon-Herrington LD2-4. The price list stated that "the installation of this all-wheel-drive equipment can only be made at the Marmon-Herrington factory. This equipment will NOT be sold separately for installation outside the factory".

The conversion price included five 7.50 X 15 six ply tires. LD2-4 conversions were specifically offered for all 85 horsepower passenger car models and for the following models on the Commercial Car chassis: chassis cab, Pick-Up, Driveaway front cab, Panel Delivery, Sedan Delivery, and Station Wagon.

A second M-H price list[7] quoted prices for the complete vehicle whereas the first quoted only the conversion prices. Prices for all-wheel-drive passenger cars ranged from $1560 for the coupe to $1835 for the Deluxe Convertible Sedan. AWD commercial cars ranged from $1546 for the pick-up to $1794 for the Station Wagon.

Fig. 2-21 The ultimate suburban status symbol — 1938.

Fig. 2-22 M-H Ford AWD cars and trucks for 1940.

Fig. 2-23 Ad in Ford Dealer & Service News. Note single tires.

Fig. 2-24 M-H conversion of Ford "One-Tonner." 1941. Note hood badge.

The fitting of 7.50 X 15 tires resulted in a rear tread of 58.25 inches and a front tread of 57 inches. M-H axle #79 with #0 CV steering ends was fitted, as was transfer case #21.

The transverse Ford front spring was replaced by two parallel semi-elliptics 36 X 1.75 inches. The single transverse rear spring 46 X 2.25 was retained.

It is worthy of note that the **M-H** 112 inch conversion had a suspension system closer than any other to the Marmon Model 34, 74, 75 system. The latter system had dual opposed transverse springs joined by a single U-bolt and achieved an essentially deflection free platform for the body mount; the Ford rear spring was rigidly clamped to the frame for a significant length, resulting in effectively two quarter elliptic springs. Marmon-Herrington described the conversion as "three-point mounted."

In mid-April 1938, M-H announced availability of AWD conversions for the 122 inch wheelbase "One-Tonner" chassis and related body styles. Price list 380415-2 showed the same conversion charge applied as quoted in the prior November for the 112 inch conversion ($895). List 380415-1 quoted complete AWD vehicle prices, pre-tax, FOB Indianapolis ranging from $1570 for an Open Driveaway Front End to $1855 for the Panel Body. The 122 inch conversion was M-H Model OT1-4 and had a 123 inch wheelbase.

Whereas the passenger and commercial cars used a 4.44:1 axle reduction, the One-Tonner used a 4.11:1. The transfer case differed in some way, being designated #21a, not #21, for the One-Tonner. Both the 122 and the larger conversions were modified to hydraulic brakes. The 122 inch chassis when modified had a 57 inch tread front and rear; two 45 X 2.25 semi-elliptic springs were used.

It is noteworthy that 90% of the parts in the M-H conversion transfer case were Ford standard parts stocked by dealers. This combined with the use of the standard Ford differential carrier assembly in the front and in rear tandem axles to provide a high level of confidence that an average Ford dealership could put a M-H Ford conversion vehicle back on the road quickly. This clever design tactic also greatly reduced the conversion cost and was a major factor in the unprecedented popularity achieved.

Conversion of the 134 inch wheelbase truck also added one inch to the wheelbase. Evidently there was a clearance problem between engine and axle.

In 1944 several secondary manufacturers were adding windows and two additional seats to the Ford panel delivery truck, thereby creating a heavy duty station wagon. Colonel Herrington foresaw this vehicle as the ideal off-road passenger vehicle and marketed the All-Wheel-Drive conversion as the "Marmon-Herrington Ranger". It anticipated the General Motors GMC/Chevrolet "Suburban" four wheel drive vehicles of the 1970's. Admittedly, Chevrolet had offered a factory station wagon conversion of their panel delivery as early as 1935, but of course only with 'rear' wheel drive'.

Fig. 2-25 The Marmon-Herrington Ranger conversion of the Ford panel truck.

Suprisingly Ford Motor Co. has never offered such a vehicle (except as a Marmon-Herrington Ford Ranger), leaving the market to General Motors. Prophetically the M-H ad writer stated: *Wherever you find men whose work — or play — takes them into the wide open spaces, where the going is rough and tough, there you are sure to find the new All-Wheel-Drive RANGER in increasing numbers.*

He continued: *The RANGER is exactly the kind of passenger-cargo vehicle you would expect Marmon-Herrington to build. With live power and traction on all wheels, it is as much at home off the highway as on — goes through deep mud, sand or snow, up steep hills and mountain grades with sure-footed speed and ease. Powerful and rugged as it is, the RANGER provides passenger-car comfort for seven. Gives you lots of cargo space, too — up to 160 cubic feet with rear seats out. And because the RANGER is built primarily of standard Ford parts, you'll find Ford dealers everywhere ready to give you the kind of fine servie for which they are famous. If you haven't already seen and driven the All-Wheel-Drive RANGER, you've a real treat in store. You'll say it's the nearest thing ever to a miracle on wheels. No hill too steep . . . no mud too deep . . . yet smooth going on highways . . . the flip lever puts the RANGER into conventional (drive) with passenger car speed and economy*''.

## THE MARSH BUGGY

The great grand-daddy of ''BIG FOOT'' was another Marmon-Herrington ''First''. Developed during 1938, the ''Marsh Buggy'' was offered in 1939 as Model F5MB-4 at an additional charge of $1210 over a Model F5-4 or a total AWD conversion cost of $2030. The price, including the Ford basic chassis with cab, was $2975. Four 13.50 X 24 six-ply tires were included. Additional tires (on the road) were only available on the ''Special Marsh Buggy'' MBS. Model F5MBS-4 was a much more expensive vehicle. Conversion was priced at $3630 and included replacement of the rear axle; with the Ford basic chassis-cab included the price was $4550. Options included **rear dual tires** $371, **duals all around** $740, **duals front and triples rear** $1270. The dimensions outline drawing for the Model E5-MB-4 (1938) shows a double reduction rear axle; evidently only the Special duty chassis was offered initially.

Fig. 2-26 A Big-Foot sighted in 1938! The M-H Ford Marsh Buggy.

The following tabulation is provided to help understand the M-H practice of model designation. Only the Ford truck conversions are traced for the period 1937-1942, followed by the passenger One-Tonner conversions for 1937-1939. The eventual addition of a letter T to the 6 X 6 designation probably was necessitated by the limited offering of ''Tag'' axle tandems in which the front and forward rear wheels were powered, the T identified the all-wheel-drive 6 X 6.

| Year | Base Vehicle | M-H Conversion | Model | M-H Conversion | Model |
|---|---|---|---|---|---|
| 1937 | 134″wb 4x2 | 135″wb 4x4 | B5-4 | 157″wb 6x6 | B5-6 |
| | 157″ | 158″ | B6-4 | 180″ | B6-6 |
| 1938 | | | E5-4 | | E5-6 |
| | | | E6-4 | | E6-6 |
| 1939 | | | F5-4 | | F5-6 |
| | | | F6-4 | | F6-6 |
| 1942 | 134″ | 134″ | CCM 5-4 | | CCMT 5-6 |
| | | | CCM6-4 | | CCMT6-6 |
| 1937 | 112″ | 113″ | LD1-4 | | |
| | 122″ | N/A | | | |
| 1938 | 112″ | 113″ | LD2-4 | | |
| | 122″ | 123″ | OTT-4 | | |
| 1939 | 112″ | 113″ | LD3-4 | | |
| | 122″ | 123″ | OT2-4 | | |

The 4 X 4 truck conversion was rated from 1.5 to 3.0 tons payload capacity depending on component selection. The 6 X 6 conversion was rated from 3 to 6 tons.

The complete specifications table for 1942 Ford conversions is reproduced below. Note that the brakes on all models are ''M-H Ford Hydraulic'', not ''Ford Hydraulic''.

# SPECIFICATIONS

| | MODEL | CMM5-4 Four Wheels All Driving | CMM6-4 Four Wheels All Driving | CMMT5-6 Six Wheels All Driving | CMMT6-6 Six Wheels All Driving | LD6-4 Four Wheels All Driving | |
|---|---|---|---|---|---|---|---|
| 1 | Wheelbase........... | 134" | 158" | 156" | 180" | 114" | 1 |
| 2 | Ton Rating........... | 1½-3 | 1½-3 | 3-6 | 3-6 | ½ | 2 |
| 3 | Tires Front........... | 7.50 x 20 | 7.50 x 20 | 7.50 x 20 | 7.50 x 20 | Com'l "15" | 3 |
| 4 | Tires Rear............ | Dual 7.50 x 20 | Dual 7.50 x 20 | Tan. Dual 7.50 x 20 | Tan. Dual 7.50 x 20 | Com'l "15" | 4 |
| 5 | Engine.............. | Ford V-8* | Ford V-8* | Ford V-8* | Ford V-8* | Ford V-8 | 5 |
| 6 | Cylinders............. | Eight | Eight | Eight | Eight | Eight | 6 |
| 7 | Bore & Stroke........ | 3 3/16" x 3¾" | 3 3/16" x 3¾" | 3 3/16" x 3¾" | 3 3/16" x 3¾" | 3 1/16" x 3¾" | 7 |
| 8 | Piston Displ.......... | 239 cu. in. | 239 cu. in. | 239 cu. in. | 239 cu. in. | 221 cu. in. | 8 |
| 9 | Max. B. H. P.......... | 100 at 3600 | 100 at 3600 | 100 at 3600 | 100 at 3600 | 90 at 3800 | 9 |
| 10 | Main Bearings........ | Three | Three | Three | Three | Three | 10 |
| 11 | Bearing Diameter..... | 2.5" | 2.5" | 2.5" | 2.5" | 2.5" | 11 |
| 12 | Carburetor Make..... | Ford | Ford | Ford | Ford | Ford | 12 |
| 13 | Carburetor Type...... | Dual Down Draft | Dual Down Draft | Dual Down Draft | Dual Down Draft | Dual Down Draft | 13 |
| 14 | Fuel Feed............ | Cam Pump | Cam Pump | Cam Pump | Cam Pump | Cam Pump | 14 |
| 15 | Electrical Units....... | Ford | Ford | Ford | Ford | Ford | 15 |
| 16 | Voltage.............. | 6 | 6 | 6 | 6 | 6 | 16 |
| 17 | Clutch Type.......... | 11" Single Disc | 11" Single Disc | 11" Single Disc | 11" Single Disc | 11" Single Disc | 17 |
| 18 | Transmission Make... | Ford | Ford | Ford | Ford | Ford | 18 |
| 19 | Mounting............ | Unit Power | Unit Power | Unit Power | Unit Power | Unit Power | 19 |
| 20 | Forward Speeds....... | 4 | 4 | 4 | 4 | 4 | 20 |
| 21 | Auxiliary Make....... | MH | MH | MH | MH | MH | 21 |
| 22 | Aux. Forward Speeds.. | 2 | 2 | 2 | 2 | 1 | 22 |
| 23 | Total Forward Speeds. | 8 | 8 | 8 | 8 | 4 | 23 |
| 24 | Total Reverse Speeds.. | 2 | 2 | 2 | 2 | 1 | 24 |
| 25 | Universal Make....... | Ford-Spicer | Ford-Spicer | Ford-Spicer | Ford-Spicer | Ford-Spicer | 25 |
| 26 | Universal Size........ | Standard | Standard | Standard | Standard | Standard | 26 |
| 27 | Front Axle........... | MH | MH | MH | MH | MH | 27 |
| 28 | Joint Type........... | Constant Velocity | Constant Velocity | Constant Velocity | Constant Velocity | Constant Velocity | 28 |
| 29 | Rear Axle Type....... | Ford | Ford | MH-Ford | MH-Ford | Ford | 29 |
| 30 | Axle Type........... | Single Reduction | Single Reduction | Single Reduction | Single Reduction | Single Reduction | 30 |
| 31 | Shaft Size............ | 1¾"—16 Spline | 1¾"—16 Spline | 1¾"—16 Spline | 1¾"—16 Spline | 1¼" | 31 |
| 32 | Axle Ratio........... | 6.67:1 | 6.67:1 | 6.67:1 | 6.67:1 | 4.55:1 | 32 |
| 33 | Total Low Gear Red.. | 86.44:1 | 86.44:1 | 86.44:1 | 86.44:1 | 30.12:1 | 33 |
| 34 | Service Brakes........ | MH-Ford Hyd. | MH-Ford Hyd. | MH-Ford Hyd. | MH-Ford Hyd. | MH-Ford Hyd. | 34 |
| 35 | Brake Size........... | 14" x 2"-15" x 3½" | 14" x 2"-15" x 3½" | 14" x 2"-15" x 3½" | 14" x 2"-15" x 3½" | 12" x 1¾" | 35 |
| 36 | Lining Type.......... | Moulded | Moulded | Moulded | Moulded | Moulded-Woven | 36 |
| 37 | Brake Drum.......... | Alloy Iron | Alloy Iron | Alloy Iron | Alloy Iron | Alloy Iron | 37 |
| 38 | Hand Brake.......... | Mechanical | Mechanical | Mechanical | Mechanical | Mechanical | 38 |
| 39 | Hand Brake Type..... | Drive Shaft | Drive Shaft | Drive Shaft | Drive Shaft | Internal | 39 |
| 40 | Frame Size........... | 7" x 2¾" x 7/32" Reinforced | 7" x 2¾" x 7/32" Reinforced | 7"—7½" x 2¾"— 3" x 7/32"—¼" | 7"—7½" x 2¾"— 3" x 7/32"—¼" | 5.92 x 2.5 x .1495 | 40 |
| 41 | Frame Material....... | Medium High Carbon Steel | Medium High Carbon Steel | Medium High Carbon Steel | Medium High Carbon Steel | Medium High Carbon Steel | 41 |
| 42 | Steering Gear Make... | Ford | Ford | Ford | Ford | Ford | 42 |
| 43 | Gasoline Tank........ | One | One | One | One | One | 43 |
| 44 | Tank Capacity........ | 19 Gal. | 19 Gal. | 19 Gal. | 19 Gal. | 19 Gal. | 44 |
| 45 | Wt.—Chassis w/Cab.. | 4850 | 4975 | 6673 | 6690 | 3170 | 45 |

*The Marmon-Herrington Company, Inc., Indianapolis, Indiana, U. S. A., reserves the right at any time to revise, change or modify the above specifications.*

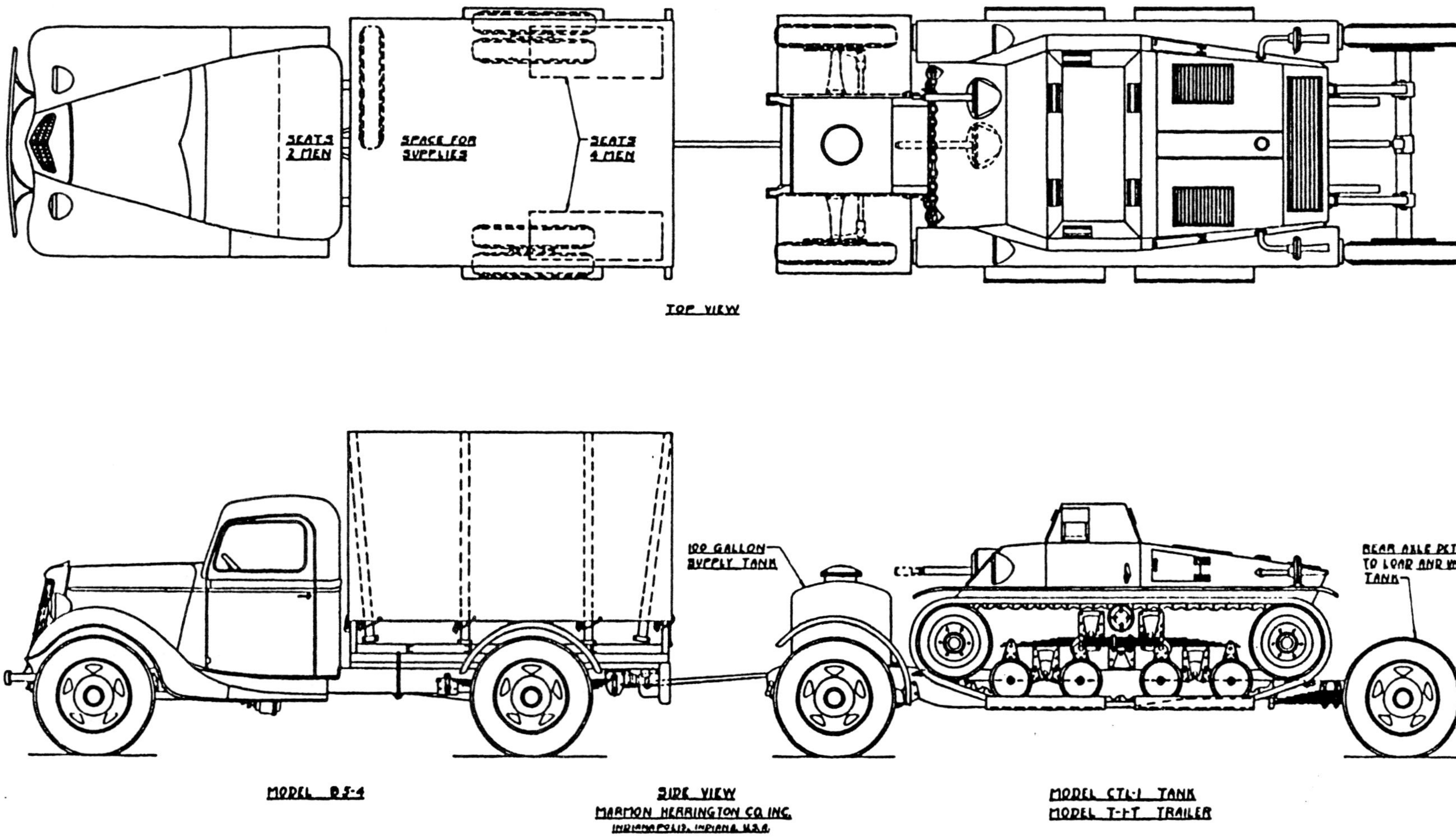


MARMON-HERRINGTON CTL COMBAT TRAIN

## PART 3 — The Combat Train — 1935

Col. Herrington and Walter Marmon did not limit their enthusiasm for the Ford V-8 power train to the conversion of the Ford Truck to AWD but proceeded to create a new vehicle: a combat tank which would also benefit technologicaly and economically from the Ford componetry.

A major demonstration was planned to acquaint U.S Government military procurement personnel with the Marmon-Herrington concept of highly mobile armored combat vehicles derivable to some degree from commercial componetry.

A mobility "Test of the Marmon-Herrington Combat Tank with Auxiliary All-Wheel-Drive Supply Truck and Carrying Equipment" was conducted in Indianapolis in June 1935 under sanction of the Contest Board of the American Automobile Association — Certified Test #3207.

The AAA Technical Representative was Charles C. Merz and the timing and observation was managed by G.H. Ford. Merz would later be President of the M-H subsidiary Merz Engineering Co. and Ford would become Director of the M-H (Armor) Proof Range.

Invited guests included some who later gained renoun:

U.S. Ordnance Dept: Colonel Munroe and Mr. W.F. Beasley

U.S. Marine Corps: Col. N.J. Larsen and Maj. M.J. Kelleher

U.S. Air Corp.-Wright Field: Maj. F. Hackett, Lt. J.F. Early, Messrs G.R. Johnson, J.F. Bolgiano, W.E. Luffman

U.S. Army 38th Nat'l. Guard Div.: Gen. Robert H. Tyndall and Lt. Col. H. Weir Cook

Imperial Iranian Army: Lt. Habib Naficy

Unidentified affiliation: Col. Wm. Guy Wall and Mr. H.E. McCrady

Goodrich Tire & Rubber Co.: Messrs Aspell, Leguillan, Goering, Wayne and Little.

Mr. Beasley was Chief Automotive Engineer of the Ordnance Department. Col. Guy Wall had been Chief Engineer at National (automobile). Indianapolis International Airport is named for Lt. Col. Weir Cook.

Both the truck and tank were powered by the Ford V-8, 85 hp., 221 CID engine/clutch/4 speed transmission unit. The truck was a Ford 4 x 4 conversion with cargo body having seats for four and tarpaulin top. The tank had a two man crew.

The truck axle ratios were 6.6:1 for total low-low ratio of 85.24:1. The tank had a brake-controlled steering axle with ratio of 9.05:1 for total low-low ratio of 113:1.

The trailer featured a removable rear axle assembly attached to Ford Model AA rear springs or a similar design. A jack system then lowered the rear of the trailer to facilitate rapid unloading.

The CTL-1 tank had air assisted hydraulic internal service brakes and manual external drum hand brakes. It measured:

| | | |
|---|---|---|
| Height | 62½ inches | (1587.5 mm) |
| Length | 132 | (3353. ) |
| Width | 76½ | (1943. ) |
| Clearance | 12 | ( 304.8 ) |
| Turning R | 101½ | (2578. ) |

The metric equivalents are repeated as reported in 1935. Like all M-H Ford conversions, the truck had hydraulic brakes. The loaded trailer had a height of 80 inches, width of 87 and a wheelbase of 172.5. It had four-wheel brakes, 34 x 7 tires and carried a 100 gallon fuel reserve.

Fig. 2-27 Invited guests for the Combat Train demonstration with Col. Herrington.

Fig. 2-28 CTL Tank loaded on trailer in 3 minutes 28.6 seconds.

Fig. 2-29 Trailer with rear axle removed, ready for tank to mount.

Fig. 2-30 M-H Special trailer for tank transport using Ford components.

Fig. 2-32 TEST CREW — R. C. Wallace driver, E. C. King.

Fig. 2-31 PIT CREW — #1 R. C. Wallace; #2 E. C. King; #3 H. H. Fisher; #4 L. N. Bowman; #5 C. H. Douglas; #6 J. Humble.

Fig. 2-33 CTL-2 passes famed Indy Pagoda during AAA sanctioned test.

Tests were conducted on the Indianapolis Motor Speedway, under AAA supervision, on June 18, 1935. The AAA Contest Board Certificate, signed by Chairman Eddie Rickenbacker stated that the Combat Train covered a distance of 40 miles at an average speed from standing start of 30.7 mph. Fuel consumed was 7.725 U.S. Gallons (5.2 mpg.). The 21,685 pounds including 8 men was decelerated 20 mph in 3 seconds to stop in 37 feet, 2.3 in.

The CTL-1 Tank was then unloaded and driven ten miles on the bricks at an average speed of 26.3 mph from a standing start. A straightaway speed of 28.8 mph was attained. Fuel consumption was 2.96 mpg.

In the best tradition of Indy Speedway pit crews, the AAA determined the time, from deadstop of the train, for **unloading** and starting the tank to be **2 minutes and 31 seconds,** with a 6 man pit crew. **Reloading** and securing the tank, to initial movement of the train required only **3 minutes and 28.6 seconds.**

The pit crew was composed of (1) R.C. Wallace (who drove the tank), (2) E.C. King, (3) H.L. Fisher, (4) L.N. Bowman, (5) C.H. Douglass, (6) J. Humble.

Following the tests, the train was weighed on the scales of the Indianapolis Power and Light Co. in the presence of the AAA reps.

| | |
|---|---|
| B5-4 Transport Tractor (no men) | 6550 pounds |
| T-I-T Trailer and CTL-1 Combat Tank | 13,720 pounds |
| Crew (6 men) | 1100 pounds |
| Crew & 2 AAA reps (Brake Test) | 1415 pounds |
| Total (6 men) | 21,370 pounds |

Fig. 2-34 CTL-2 climbing two-stage abutment 22 plus 14, 36 inches total.

Fig. 2-35 Combat Train passing ancestors — Models 74 and 70 — at trackside.

Fuel tanks were filled about 50%: Tractor 15 gallons, "Tank" 36 gallons (2 tanks), Trailer reserve 45 gallons.

The combat train was then driven to the Marmon-Herrington Proving Grounds where representative combat obstacles had been constructed. Heavy rain began and resulted in failure to accomplish some obstacle tests due to slippage. They were successfully negotiated on June 25. Obstacles included two depths of "shell holes", two trenches (40 and 50 inches), three abutments (2 cement and one earth), a 4 foot drop off and 36 inches of water in a tank.

On June 19, the pit crew (6 men) removed a track in 4 minutes 30 seconds and replaced it in 7 minutes 30 seconds. A deliberately thrown track was reinstalled in 4 minutes 15 seconds by the two-man crew using on-board tools.

On June 20, a non-stop run of the combat train was made to simulate military transport over hilly terrain on public roads. The 245 mile route was south by west from Indianapolis on Indiana 37, returning on US 31 from slightly north of Louisville, with connection between Paoli and Scottsburg. The truck fuel tank was refilled in motion from the reserve tank on the trailer and lunch was passed from a support vehicle. An average speed of 30.8 mph was made and 5.33 mpg attained. Wallace and E.C. King alternated as drivers.

CHART #1

OFFICIAL TRIALS
ON
INDIANAPOLIS MOTOR SPEEDWAY
UNDER SANCTION OF THE
CONTEST BOARD, AMERICAN AUTOMOBILE ASSOCIATION
SANCTION # 3207 DATE JUNE 18-1935
TEST OF CTL COMBAT TRAIN AND CTL-1 COMBAT TANK
FOR
MARMON-HERRINGTON COMPANY INC.

1. 25 MILE NON-STOP RUN OF TRAIN
2. DISMOUNTING TIME COMBAT TANK
3. 10 MI. NON-STOP RUN CTL TANK 9.05 GEAR RATIO
4. MOUNTING TIME COMBAT TANK
5. 15 MI. NON-STOP RUN OF TRAIN
6a. OIL CONSUMPTION TRANSPORT TRACTOR 40 MI.
6b. FUEL CONSUMPTION TRANSPORT TRACTOR 40 MI.
6c. WATER CONSUMPTION TRANSPORT TRACTOR 40 MI.
6d. FUEL CONSUMPTION COMBAT TANK 10 MI.
6e. WATER CONSUMPTION COMBAT TANK 10 MI.
7. DECELERATION (BRAKE TEST) 20-0 M.P.H.
8. MAXIMUM SPEED OF COMBAT TANK ½ MI. 9.05 GEAR RATIO

TEST #1 25 MILE NON-STOP RUN CTL COMBAT TRAIN STANDING START

| LAP START 0 | 1 | 2 | 3 | 4 | 5 | 6 | 7 | 8 | 9 | 10 |
|---|---|---|---|---|---|---|---|---|---|---|
| MILES | 2.5 | 5.0 | 7.5 | 10 | 12.5 | 15 | 17.5 | 20 | 22.5 | 25 |
| LAP TIME | 6:22.2 | 5:20.0 | 5:05.2 | 4:49.6 | 4:43.4 | 4:35.6 | 4:32.2 | 4:33.2 | 4:30.8 | 4:36.4 |
| TOTAL TIME | 6:22.2 | 11:42.2 | 16:47.4 | 21:37.0 | 26:20.4 | 30:56.0 | 35:28.2 | 40:01.4 | 44:32.2 | 49:08.6 |
| LAP AV. M.P.H. | 23.54 | 28.12 | 29.48 | 31.07 | 31.75 | 32.65 | 33.06 | 32.92 | 33.23 | 32.56 |
| TOTAL AV. M.P.H. | | | | | | | | | | 30.52 |

TEST #2 DISMOUNTING TIME COMBAT TANK 2 MIN: 31 SEC.

TEST #3 10 MILE NON-STOP RUN COMBAT TANK 9.05 GEAR RATIO STANDING START

| LAP | 1 | 2 | 3 | 4 |
|---|---|---|---|---|
| MILES | 2.5 | 5 | 7.5 | 10 |
| LAP TIME | 06 06.0 | 05:34.4 | 05:29.4 | 5:38.0 |
| TOTAL TIME | 06 06.0 | 11:40.4 | 17:09.8 | 22:47.8 |
| LAP AV. M.P.H. | 24.59 | 26.91 | 27.32 | 26.63 |
| TOTAL AV. M.P.H. | | | | 26.3 |

TEST #4 MOUNTING TIME COMBAT TANK 3 MIN: 28.6 SEC.

TEST #5 15 MILE NON-STOP CTL COMBAT TRAIN STANDING START

| LAP | 11 | 12 | 13 | 14 | 15 | 16 |
|---|---|---|---|---|---|---|
| MILES | 27.5 | 30 | 32.5 | 35 | 37.5 | 40 |
| LAP TIME | 5:30.4 | 4:45.6 | 4:44.0 | 4:35.6 | 4:39.0 | 4:48.0 |
| TOTAL TIME | 54:39.0 | 59:24.6 | 1:04:08.6 | 1:08:44.2 | 1:13:23.2 | 1:18:11.2 |
| LAP AV. M.P.H. | 27.24 | 31.51 | 31.25 | 32.65 | 32.35 | 31.27 |
| TOTAL AV. M.P.H. | | | | | | 30.69 |

TEST No.

6a OIL CONSUMPTION B5-4 TRANSPORT TRACTOR 40 MILES. NONE
6b FUEL CONSUMPTION B5-4 TRANSPORT TRACTOR 40 MILES. 7.75 US. GAL. 5.16 MI. PER GAL.
6c WATER CONSUMPTION B5-4 TRANSPORT TRACTOR 40 MILES. NONE
6d FUEL CONSUMPTION CTL-1 COMBAT TANK 10 MILES. 3.375 U.S. GAL. 2.96 MI. PER GAL.
6e WATER CONSUMPTION CTL-1 COMBAT TANK 10 MILES NONE
7 DECELERATION (BRAKE TEST) CTL COMBAT TRAIN 20-0 M.P.H. AV. 37 FT. 2.33 IN. AV. 2.96 SECONDS
WEIGHT OF CTL TRAIN WITH CREW OF 6 MEN & 2 A.A.A. OBSERVERS (BRAKE TEST) 21,685 LBS.
8 MAXIMUM SPEED CTL-1 COMBAT TANK ½ MILE 1 MIN: 02.8 SEC. 28.8 M.P.H.
MECHANICAL FAILURES DURING TESTS NONE
MECHANICAL ADJUSTMENTS DURING TESTS NONE
MECHANICAL REPAIRS DURING TESTS NONE

Chas. C. Merz
A.A.A. REPRESENTATIVE

A.A.A. TIMING & SCORING

Fig. 2-36 AAA Sanction #3207, CTL combat train and CTL-1 combat tank.

## T-2 SCOUT CAR

During 1935 the Ordnance Department procured the first group of Marmon-Herrington Ford military vehicles — the T-2 Scout Car with an interesting seating arrangement for 8 soldiers.

Contest Board of the American Automobile Association

CERTIFICATE OF PERFORMANCE

The undersigned Certify in the name of the

Contest Board, American Automobile Association

that
at the

Indianapolis Motor Speedway, Indianapolis, Indiana, U.S.A.
a
Model CTL-1 Track-Type Two-Man Combat Tank
manufactured by
The Marmon-Herrington Co., Inc.
Was timed at 22 Minutes 47.8 Seconds
Over a certified distance of 10 miles
Averaging 26.3 miles per hour from a standing start
and
Was timed at 1 Minute 02.8 Seconds
Over a certified distance of 1/2 mile
Averaging 28.8 miles per hour from a running start
On July 9, 1935.

Fuel Consumption for 10 miles: 3.375 U. S. Gallons
Average:- 2.96 Miles per Gallon.
Water Consumption for 10 miles: None.
Gear ratio used: 9.05 to 1

Sanction No.
3207

Technical Representative

Chas. C. Merz
Official Representative

The Secretary

Chairman of the Board

Fig. 2-37 AAA Certificate of Performance issued for the combat train.

Contest Board, American Automobile Association

that
on the
Marmon-Herrington Proving Ground Courses
(a. Combat; b. Sand; c. Grades or Inclines)
at
Indianapolis, Indiana, U. S. A.
a
Model CTL-1 Track-Type Two-Man Combat Tank
manufactured by
The Marmon-Herrington Co., Inc.
Was Officially Observed Over the Above Courses
on
June 18, 19 & 25 and July 9, 1935.

CTL-1 Track-Type Two-Man Combat Tank negotiated all Combat Proving Ground Obstacles, Deep, Wet Sand Course and Steep Inclines Without Fault or Failure.
Mechanical adjustments - none. Replacements - none. Breakage - none.

Contest Board, American Automobile Association

that
at the

Indianapolis Motor Speedway, Indianapolis, Indiana, U.S.A.
a
Model B5-4 All-Wheel-Drive Tractor Transport Truck
drawing a
Model T-1-T Transport Trailer
carrying a
Model CTL-1 Track-Type Two-Man Combat Tank
(and crew of six men)

Manufactured by the Marmon-Herrington Co., Inc.
Was timed at 1 Hour 18 Minutes 11.2 Seconds
Over a certified distance of 40 miles
Averaging 30.70 Miles per Hour
From a standing start on June 18, 1935.

Consumption: Fuel - 7.725 U.S. Gallons; Oil - None; Water - None
Deceleration: 20 M.P.H. to stop; Time - 2.96 Seconds; Distance - 37ft.2-1/3 in.
Deceleration Weight - 21,685 pounds.

Fig. 2-38 Marmon-Herrington Ford squad car.

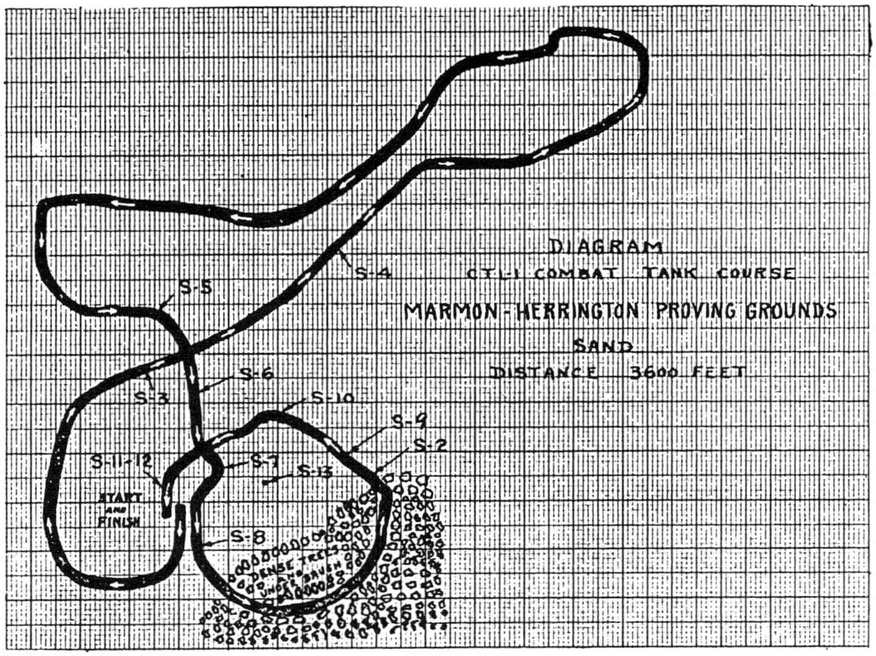


Fig. 2-39 CTL-1 Combat tank test course at M-H Proving Grounds.

## PART 4 — Two Historic Developments — 1936

### JEEP AND HALF-TRUCK

Marmon-Herrington made history in 1936. It is best reported therefore by quoting the development from M-H's own recounting of the first decade . . . "Ten Years Before Pearl Harbor", a 1944 brochure.

"Two outstanding achievements made 1936 a memorable year at Marmon-Herrington. During that year M-H engineers introduced several vehicles which were complete innovations . . . distinct departures from previous design and practice".

"Two of these innovations, or "firsts", vie in importance from the military standpoint. One of these was the first ½ ton passenger-carrying unit with power and traction applied through all four wheels. This was the Ford pick-up truck, converted to all-wheel-drive in July 1936 and tested on the King Ranch in Texas in September, and which Earl Godwin in his Ford Motor Company radio broadcast March 15, 1943, called the "Granddaddy of the Jeep".

During the summer of 1936 Colonel Herrington was in Belgium demonstrating the M-H conversion of the Ford 1½ ton truck when Bob Wallace received a trans-Atlantic cable. Was it feasible to convert a Ford ½ ton V-8 pick-up to four wheel drive? There would be a market for such in Belgium. Wallace headed at once for Walter Marmon's Ford dealership in Indianapolis where a Ford truck was placed on the hoist. Bob Wallace and M-H designer Chet Blair studied it for a while, and then returned to the plant. Wallace cabled the Colonel to say "Yes, it could be done".

Wallace recognized the problem of front axle torque reaction with the relatively light suspension springs necessary for a satisfactory ride. His solution was a secondary quarter elliptic spring paralleling the rear half of the usual M-H half elliptics which were substituted for the transverse factory spring.

The T-9 transmission from the Ford 1½ ton truck with four speeds was substituted for the factory three speed, ahead of the M-H transfer case.

The order from Belgium did materialize. 100 sets of non-Ford conversion parts were shipped to Belgium for assembly there. This production run also included parts for the first Ford conversions sold to U.S. Ordnance. They were bare bones squad cars designated B5-4 SQC.

The Belgium vehicles were rather "workhorse" in appearance, with fixed canavas tops and boxy canvas covered cargo bodies. But their performance enthused Herrington and he envisioned an entirely new concept of military vehicle that would enchance the mobility of the infantryman and artillery man . . . **be a steed rather than a faithful plodding mule** which had been the role of the military motor vehicle.

A bare-bones Ford pick-up conversion was prepared. The cab was replaced by a cowl and folding windshield. Twin bucket seats were fitted as well as top "irons". Wide 15 inch rims were riveted to basic Ford Wheel centers. A massive towing eye was added to the rear of the frame to permit hauling artillery.

In recounting the pioneer conversion of Fords to All-Wheel-Drive, the Marmon-Herrington copy writer prophetically reported:

"The new unit proved to be so practical economically, so sound mechanically and so outstanding in its performance that an entire line of fast, dependable all-terrain vehicles, lighter and smaller than any All-Wheel-Drive trucks previously available, was created. These conversions "blazed the Trail" for the production of multiple-wheel-drive vehicles by the great mass-production automotive plants of America . . . "

The copy writer was referring to the medium and heavy duty conversions and the explosion in volume of 4 x 4 and 6 x 6 military-trucks which occurred during the war. Unknowingly he was also referring to the similar explosion of personal-use 4 x 4's which characterized automobile production in the 1970's. Both the expression "blazed the trail" and the designation "Ranger" for the Marmon-Herrington conversion of the Ford passenger cars, panel-wagons and pick-up trucks, were prophetic of the Chevrolet "Blazer" personal 4 x 4 and the Ford "Ranger" 4 x 4 pick-ups of the 1970s.

In his "The American Automobile — A Brief History" (The University of Chicago History of American Civilization-Daniel J. Boorstein, Editor) John Bell Rae would state (P246): *"The design of the Jeep was the work of Captain Robert G. Howie of the U.S. Army working in cooperation with Colonel A. W. S. Herrington (of Marmon Herrington). The prototype Jeep, built by American Bantam Car Co. was displayed at Fort Benning, Georgia in March 1940".*

On Monday, March 15, 1943, Earl Godwin, noted newscaster, in his broadcast (series) "Watch the World Go By" (Remember "Watch the Fords Go By"?) paid gracious tribute to Marmon-Herrington in the commercial. This broadcast was carried by 126 stations and was sponsored by the Ford Motor Company. From H.C. Doss, General Sales Manager of Ford, came the statement that "this (broadcast commerical) is in accord with our purpose to give credit where credit is due, to industries and others doing an outstanding job in behalf of this country's war effort".

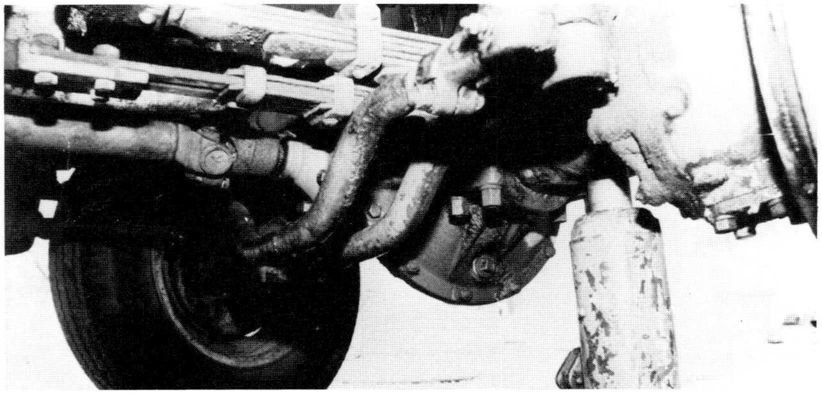

Fig. 2-40 The torque reaction spring crucial to the pick-up conversion. *Special Interest Autos,* David Brownell.

Mr. Godwin's broadcast was as follows: *"A converted Ford was the grand-daddy of the American Jeep now being turned out by Ford and other manufacturers and performing sensationally on all the major battlefronts of this war. The story goes back to World War I, when the principle of all-wheel-drive for military vehicles was proved an outstanding advantage. A young captain of motor transport in the A.E.F., Arthur W. Herrington, now board chairman of the Marmon-Herrington Company, was impressed with the advantages of the multiple-drive military vehicles as they performed in the French mud and decided to do something about it. In 1936, Marmon-Herrington converted the first half-ton passenger-carrying vehicle to all-wheel drive and gave it a tryout on the King Ranch in Texas. The first quantity order came from the Belgian Army. Today hundreds of thousands of Marmon-Herrington converted Fords are in military service all over the world . . . with the British 8th Army in Tunisia, in the Solomons, with MacArthur in Australia, as well as with the builders of the Alcan Highway and the Texas-Illinois pipeline. These all-wheel drives are doing a good job. The jeep merely applies the same principles of design to still smaller and lighter vehicles for a variety of duties over the worst possible terrain. It's another example of the fact that the automobile industry was ready with the "know-how" and experience when Uncle Sam sounded the call to arms. And they've adopted their plants, machinery and men to meet the emergency quickly".*

Time Magazine in an article on contemporary Jeeps, December 18, 1978, noted that "a generation of G.I's depended on the frill free GP's (for General Purpose and hence jeep) that could growl through rivers of mud and over impossible obstacles. General George C. Marshall called the Jeep "America's greatest contribution to modern warfare" and the infantryman developed a love affair with his Jeep that was sketched by Cartoonist Bill Maudlin in his 'Willie and Joe' series".

When Lincoln-Mercury Division of Ford Motor Co. celebrated the thirtieth anniversary of Lincoln in 1952, they noted the wartime production of 145,000 Jeep bodies and 12,785 Jeep amphibian bodies (and 24,929 nacelles for B-24 bombers).

Fig. 2-41 The first ¼ ton military 4x4 is demonstrated at the King Ranch. Col. Herrington at the wheel.

Fig. 2-42 The army exercises its newfound mobility.

The M-H half-ton 4 x 4 become known in the Service as "Darling" and the name persisted until the advent of the G P vehicle with the "Jeep" handle. Ultimately the term "Jeep" appeared in official U.S. Army literature: TM 10-460 dated May 1942 said the "Darling" was dubbed "Granddaddy of the Jeep".[8]

The second of the two great "firsts" of 1936 began when Colonel Herrington won government support for development of a new concept of combat vehicle . . . the half-track with powered wheels. Considering the importance of the half-track concept to the Allied forces, it is appropriate to be precise in the documentation of its origin in Indianapolis. There is no better way to accomplish this then to quote directly the official Marmon-Herrington version.

The *Marmon-Herrington NEWS* (Vol. 2, No. 6 Feb. 1944) Charles H. Green., Editor reported: *"Among the many "firsts" in military automotive design and manufacture which, by the record, must be credited to the Marmon-Herrington Company, is the half-track with driving front-axle which (was subsequently) used by American and Allied armies. In July of 1936, the Ordnance Department of the US Army commissioned Marmon-Herrington to design and build a pilot model of a new half-track vehicle which would have a driving front axle. Half-track vehicles were in use but none had a driving front axle and additional advantages were sensed if such a vehicle could be developed. The order for this experimental vehicle was placed by the Rock Island Arsenal and entered on the Marmon-Herrington books on July 10, 1936. The pilot model was completed and the first photographs made November 1, 1936. Other series of photos were made on December 28 and December 31 1936. The chassis of the vehicle was a Marmon-Herrington All-Wheel-Drive converted Ford B5 4 x 4. It was equipped (alternately) with two types of rear-bogies: a four roller type and a two roller type. It was on this order that special gear ratio combinations, along with automatic compensation, were developed to coordinate the operation of the front driving axle and the half-track bogie, thus eliminating the loss of pulling power of either mechanism if the other should not find effective traction".*

Concurrently M-H were building a fleet of armored scout cars derived from the Ford conversion 4 x 4 chassis; they featured a louvered grill design. These two vehicles developed in 1936-37 together incorporated the concept for the M-3 half-tracks produced during 1941-45 by White Motor and Autocar.

"Tests on the rugged Mann Hill proving ground produced another interesting "first". It was on these units that Marmon-Herrington installed the first front rollers used on half-tracks. This practical feature was to prevent 'rooting' when the vehicle started climbing an acute grade".

"Ordnance Department took delivery and after subjecting it to several months of testing, ordered a **second series** of half-track trucks. This series was to prove out, further, the relative merits of the two or four-wheel bogie designs and to be compared on a different type of track from either of those supplied on the first order. The order was completed in December 1937, (and included the) first half-track with front driving axle **and** first use of continuous band rubber track on a half-track".

The two-wheel bogie trucks had regular front fenders and front bumper; the four-wheel bogie units had "relieved" front fenders and anti-rooting rollers.

"Earlier, in May 1937, M-H designed and built a half-track truck utilizing another type track suspension and continuous band rubber track previously developed for the Marmon-Herrington TA-36 tractor. This unit was ordered by the California-Arabia Standard Oil Company".

"Out of this pioneer engineering work developed one of the most effective military vehicles of World War Two. It can operate at relatively high speeds on the road or plow through mud or sand or extremely rough ground off the road. It is used as a personnel carrier, as a reconnaisance vehicle, ammunition carrier, prime mover, artillery tractor and for other purposes as use and experience has suggested. Most spectacular has been the Tank Destroyer, mounting a heavy field piece"

Fig. 2-43 The grand-daddy of all the US Army half-tracks, with powered front wheels and nose roller.

CHAPTER THREE

# THE BIRTH PLACE OF THE ARSENAL OF DEMOCRACY

## PART 1 — The Marmon-Herrington Proof Range

### The Merz Engineering Co.

Colonel Herrington had begun the enterprise bearing his name with the entreprenuering of a technology breakthrough . . . the Rzeppa constant velocity universal joint. That invention had made fully practical the application of all wheel drive (or front drive) to wheeled vehicles for use on hard surfaces. Similar circumstances marked his undertaking of the design and manufacture of tracked vehicles for both military-tactical and utilitarian applications which would be equally "at home" on paved and rugged terrain. Herrington joined forces with a rubber product firm to develop a practical application of a French invention . . . a steel-rope reinforced rubber track for "crawler" or tracked machines. But vehicles equipped with the new track did not "crawl," could in fact speed along paved surfaces. The year was 1935 and Herrington set about the first demonstration of the potential for "blitzkreig" warfare.

This first vehicle was followed in 1936 by the application of the same track system to a utilitarian tractor capable of a variety of tasks, both military and civilian. Both classes of vehicle came down Indianapolis assembly lines from the time of the first prototypes until the infamy of Pearl Harbor. Thereafter a variety of tracked, armored tanks were series produced until the final victory in 1945.

But the all-wheel-drive truck business prospered concurrently for both civilian and military purposes. In fact, the contribution of Marmon-Herrington wheeled vehicles to the ultimate Allied victory far excelled that of the tracked vehicles. As the world came to know North America as "The Arsenal of Democracy" it would also recognize the Indianapolis firm as "The Birthplace of the Arsenal of Democracy" and owe a tremendous debt of gratitude to the foresight, creativity, energy and enthusiasm of Arthur W.S. Herrington.

In the interest of continuity, the concurrent stories of the Wheeled Military Products and of the Tracked Military Products will be addressed separately, respectively in Parts Two and Three of this Chapter. During World War II, Marmon-Herrington was the sole US manufacturer of rear-drive, all wheel drive, half-tracked and full-tracked vehicles.

Fig. 3-1 Marmon-Herrington Ford half-track conversions with driving front axle.

## THE MARMON-HERRINGTON PROOF RANGE

Before an Ordnance contractor could use a shipment of armor plate, it was required by contract specification that sample plates be "proofed" by subjection to specified projectiles under specified conditions. Strict timing was required to confirm projectile velocities; precise atmospheric data was also essential.

In 1939, upon receipt of the first contract for armored tanks from a country later one of the Allies, Marmon-Herrington found that it was necessary to have all armor plate "proofed" at government ranges. The advent of the "Marmon-Herrington Proof Range" was reported in an article of that title which appeared in the Nov. 1942 issue of the *The Marmon-Herrington NEWS* (Volume 1, No. 3). The author was Ruth Wengi of the M-H Test and Proof Department and the following quotations are exerpted from the article. *"At that time, long before most people believed the US would be involved in the war, Colonel Herrington foresaw that to be a completely equipped plant the company should have a proof range. He believed that conditions would change so rapidly that demand would so far outstrip production of proof facilities that a proof range, if properly equipped and manned could be* ***offered to the War Department.*** *As to the soundness of that analysis, it can be said today that the M-H Proof Range is proof-firing practically all of the armor plate being heat-treated in the Middle West. This relieves the War Department ranges at a time when their facilities and personnel are needed for the development and testing of their ordance and combat supplies. With the cooperation of the Office of Chief of Ordnance and the Aberdeen Grounds the Proof Range was built and completely equipped. A man was found with sufficient ballistic experience to enable him to be trained at Aberdeen Proving Ground. This was G.H. Ford, nationally known in professional ballistic circles, who was made Director of the Range. The balance of the personnel were instructed at the Range. Mr. Ford was chief timer of the Indy 500.*

The M-H Range could fire projectiles up to 75 mm but most frequent shots were 50 caliber and 37 mm.

The same security measures found at any War Department range were imposed. Consequently the Range staff including clerical help, etc., were isolated from other employees.

Fig. 3-1 The Marmon-Herrington Proof Range; Chief Gunner John Schwed at the breech block, Col. Herrington talks with G. H. 'Snappy' Ford.

## MERZ ENGINEERING CO.

The formation of the Merz Engineering Company was reported in Automotive Industries for November 21, 1927. The announcement noted the purchase by the new firm of $14,000 worth of machine tools at the sale conducted by the Receiver for the bankrupt H C S (Harry C. Stutz) firm in Indianapolis.

The Merz Engineering Co. was acquired in 1939 by Marmon-Herrington. The chairman and officers were as follows:

| | |
|---|---|
| Chairman, Board of Directors | Arthur W. Herrington |
| President, | Charles C. Merz |
| Vice-Pres., | Bert Dingley |
| Secy. | Wm. P. Nottingham |
| Treas. | Earl J. Breech |
| General Mgr. | J.C. Thompson |

Charles C. Merz was the AAA official observer of the Sanctioned demonstration of the M-H Combat Train in 1935. Mr. Merz was a participant in the first day of racing at the Indianapolis Speedway; he drove a National in that event. Mr. Switzer (Switzer-Cummins Co. — auto parts and superchargers/turbochargers) won the first race, in 1909.)

Fig. 3-3 Mr. Merz and Mr. Switzer.

Fig. 3-4 Charles Merz in the National No. 5 he drove to seventh place in the 1911 Indianapolis 500, and in a Peugeot in 1916. Indianapolis motor Speedway photos via Jerry E. Gebby.

The reader will recognize the other officers as holding the same positions with Marmon-Herrington.

A new plant was erected on Harding Street directly south of the sixteen acre M-H site and across the New York Central tracks. Ultimately some 200 craftsmen skilled in the precision arts were employed.

Additional insight is provided by an item in the *Marmon-Herrington NEWS:*

"WE PAY A VISIT TO THE MERZ ENGINEERING COMPANY"

It will have to be a short visit, far too short, because there are many things of interest to see at the plant of our neighbors, The Merz Engineering Company.

Most of us know, of course, that it is the business of the Merz Company to design and manufacture the precision measuring instruments and tools, used by other manufacturers to produce industrial parts and products in large quantities.

Only a few of us have had the opportunity to observe first hand, that the Merz plant built in 1940, is one of the most modern and well-equipped institutions of its kind in this part of the United States. Officials of the company realized long before the war that there would be a tremendous demand for gages, tools and fixtures and made sure that the Merz plant should be complete in every facility and item of equipment. This foresight on their part has had a great deal to do with our success as producers of military vehicles, as we have had access to the entire resources of the Merz organizaton for the gages and tools we needed.

In order to insure extreme accuracy in the gages and tools manufactured by Merz, several areas of the plant are operated under temperature and humidity control, and the machines in these divisions are capable of the highest precision in operation. Many of these machines are found only in a few of the finest engineering plants in the country.

Of course the full productive capacity of the Merz plant, like our own, is now devoted to war, but gages, tools and fixtures will be in great demand for the production of civilian goods after the war, so the company has a bright future.

* * *

It should now be evident to the reader that Col. Herrington and his firm had indeed assembled in Indianapolis the birthplace of the Arsenal of Democracy. A major facility for series manufacture of diverse motor vehicles was complemented on the same site by a sophisticated gage manufactory and a fully qualified armor-proofing firing range and all these extraordinary facilities were installed with reserve capacity beyond the on-site support requirements.

Colonel Herrington staffed his engineering and test departments with skilled and innovative engineers and other personnel the majority of whom had previously been associated with the Marmon Motor Car Co. It is therefore not surprising that the products of M-H were as innovative and as highly regarded internationally as had been the pre-automobile products of Nordyke and Marmon Co. and the generations of Marmon automobiles that followed.

The United States military might not have been fully mobolized in 1940 but Marmon-Herrington clearly was.

Truly, if the United States was properly described during World War II as the "Arsenal of Democracy", then Marmon-Herrington equally well deserves the title: "Birthplace of the Arsenal of Democracy".

## PART 2 — The Wheeled Military Products

The first tremors were being felt in **1936** that prefaced the eruption in September 1939 of World War Two. Frantic rearming was beginning in Europe and even the United States Army seemed to find some funding for updated mechanization.

Marmon-Herrington received an order for "a large group" of B 5 Motor Carriages for the 4.2 Mortar. This was a 4 x 4 conversion of the 1½ ton Ford for the Ord-

Fig. 3-5 The T11-E2 armored car with Hercules engine.

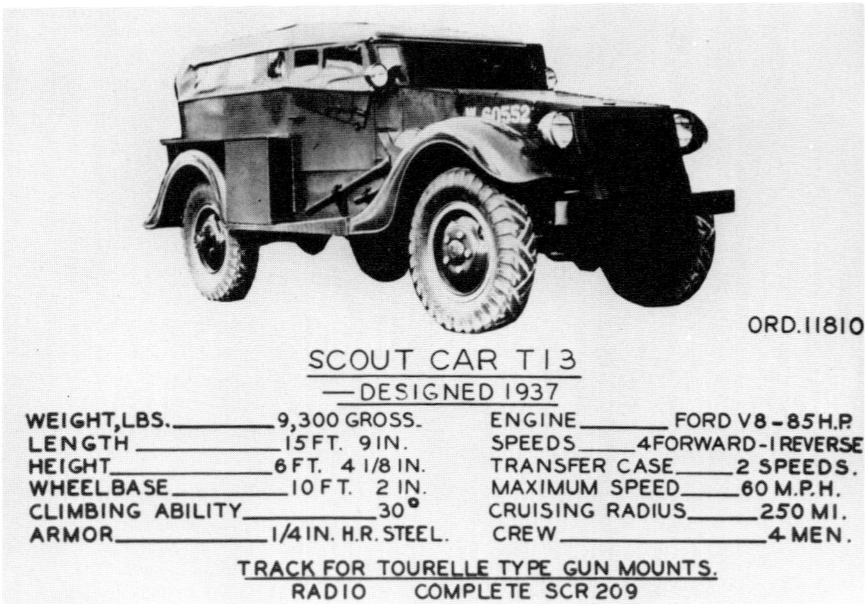


Fig. 3-6 Marmon-Herrington Ford scout car.

Fig. 3-7 M-H Ford half-ton 4x4 close coupled ambulance chassis.

nance Department, similar to the B 5-4x4 S.Q.C. Squad Cars built the previous year. Both had a tandem pair of "church pew" seats, the seats being closer coupled than on the earlier vehicles to allow a larger storage "trunk" and having an Ordnance towing eye and rugged rear fenders, instead of pick-up fenders, over dual rear wheels instead of singles. At least one B 5 4 x 4 was finished into a fire truck for the Aberdeen Proving Grounds using a front mounted centrifugal pump.

T 11-E 2 Armored Cars were built for the Ordnance Department. These were production units based on a Marmon-Herrington prototype development and benefited from experience with the T 1 armored car series for Iran and prototypes delivered to Ordnance in 1934. The T-11 had used a rear mounted Cadillac V-8 engine. The T-11-E2 was reconfigured to accomodate a Hercules line-6 engine and a redesigned turret.

The year **1937** was characterized by orders from the full spectrum of US military agencies, limited in number of units, but significant in prototyping of the WW 11 fleet to come.

"A Series" of T-13 armored cars, based on the B 4 x 4 Ford conversion were built for US Army Ordnance. There were high sided, cloth topped vehicles with single military tires and Ford fenders on an armored hood and body. The sloping armored grill with horizontal louvers set a significant precedent. Thirty eight were built with soft ('armor') plate for use by the National Guard.

To strengthen the National Guard forces, the **US Army** ordered 2½ ton Ford 6 x 6 conversions. The low cargo bodies carried parallel slat seats on both sides. The trailer hitch was sized to handle the 155 mm Howitzer. Rear fenders were fitted, extending over both rear duals much like those beginning to be used today on commercial tractors.

Fred W. Crissom, when discussing this Marmon-Herrington Ford Model C-5, 6 X 6 of 1937 in his "US Military Wheeled Vehicles"[8] stated "This C-5, 6 X6 formed the basic concept from which the 2½ ton 6 X 6 of WW11 fame eventually evolved."

The **Marine Corps.** also placed a "large order" for 2½ ton Ford 6 x 6 conversions fitted with special bodies for covered transport of sky-search lights. The Corps also procured at least ten ½ ton Ford 4 x 4 conversions fitted with canopy enclosure for driver and personnel load and designated "personnel carriers".

The **U.S. Navy** bought " a series" of crash fire trucks for Naval Air Stations. These were medium Ford 4 x 4 conversions with chemical fire fighting equipment.

The **U.S. Coast Guard** bought ½ ton Ford pick-up 4 x 4 conversions with canvas "covered wagon" enclosure and dual aft sidemounts.

The **Army** bought ½ ton Ford 4 x 4 Cab-over-engine conversions. Unusual for Marmon-Herrington military orders, these were sold as cab-chasis for subsequent mounting of ambulance bodies. The cabs, however, were very special and pleasing in appearance. They featured fixed vee windshields configured like those used on a convertible coupe, hardtop type doors and a cloth top lacking the usual tent appearance of such military tops.

This was a Marmon-Herrington original design. The sheet metal work was done by Mike Macella, an Indianapolis builder of racing cars.[2] Because it was built on a typical M-H conversion of the basic Ford chassis, it had parallel semi-elliptic front springs. It was indeed another in the long line of Marmon-Herrington "firsts", for in the spring of 1938 the Ford Motor Company introduced the first factory cataloged low priced cab-over-engine (COE) models on either a 101 or 134 inch wheelbase, the former allowing a 60 inch axle to rear of cab dimension and paralleling the M-H of the prior year. And, notably, these models were the first Ford trucks to be built at Ford without the transverse front spring . . . they, like the predecessor M-H, had parallel semi-elliptics mounting a wide-track front axle despite the two-wheel drive.[1]

The Ford design used a standard cab widened by 7 inches and elvated 7.62 inches.[2]

Most significantly, Chevrolet offered a COE in 1939 and Dodge in 1940, following the M-H of 1937 and Ford of 1938, a remarkable expansion of offerings in a previously unrecognized market.

The years **1938 and 1939** saw the start of M-H business with the customers destined to be the three largest until the United States entry into the War: The Netherlands government (Sept 1938) the Ford Motor Co. of Canada (Nov. 1938) and the Canadian Traction Co. Ltd. (Jan. 1938).

A large order for ½ ton 4 x 4 units for Belgium was completed in 1938 as well as a series of similar units for the **U.S. Marine Corps**.

A prototype C30-6, 6 x 6 Marmon-Herrington 6-8 ton dump truck was shipped to Technoprominport, Moscow, USSR.

A series of light combat tanks were supplied to Mexico and another series to the **U.S. Marine Corp.** a TA-30 track-laying tractor to the Netherlands for Java service, and a larger TA 40 to Egypt. These vehicles are discussed in Part 3 of this chapter.

For the **U.S. Army Air Corps** a fleet of heavy duty AWD Marmon-Herrington Observation balloon Winch Trucks were produced.

In **1938-39,** the **U.S. Coast Guard** procured, from Marmon-Herrington, ½ ton International long wheelbase pick-ups with 4 x 4 conversion. That Service also bought more enclosed Ford pick-ups, but this time with the 1 ton 4 x 4 conversion and oversize tires.

The **U.S. Quartermaster Corps.** acquired a fleet of post hole diggers on 1½ ton Ford 4 x 4 conversions. Front fenders were severely cut-back. All four wheels carried dual tires . . . the front duals consisting of two tire sizes on the same size wheels.

The **U.S. Public Health Service** bought a 1938 Ford coupe 4 x 4 conversion with truck lid omitted and a fifth-wheel on the compartment floor; The companion trailer had a gooseneck nose. Both were in anticipation of today's familiar beverage trailers hauled by pick-ups. The PHS trailer housed a mobile dental office.

The **U.S. Army Air Corps.** procured ½ ton Ford 4 x 2 conversions without windshield, mounting a pair of bucket seats and an overhanging hoist for transporting heavy bombs to airplanes.

1938 was the last year that the full line of Ford cars and trucks were characterized by the transverse front spring and wishbone radius rod, for in May of that year Ford introduced a cab-over-engine (COE) chassis with conventional front suspension . . . parallel semi-elliptic springs also serving as radius arms. This, of course, was necessary because of the increased proportion of the load bourne by the front suspension. But the idea caught on and appeared on the entire truck line for 1940 with the exception of the ½ ton model. Concurrently all trucks except the ½ ton switched from the traditional torque-tube to Hotchkiss Drive, taking the wheel torque reaction through the rear springs which also served as radius arms, being single shackled rather than the previous double-end shackling.

This major change in the Ford chassis construction meant a major revision to the Marmon-Herrington conversion process. The highly-regarded Ford power train was now available is essentially a conventional chassis. It was no longer necessary for M-H to substitute a new forward frame section and parallel heavy duty spring attachments. Thereafter the conversion began with frame reinforcement and the substitution of cross members for the transfer case with optional substitution of heavy duty springs.

Very significant in 1939 was the delivery of the first Ford C.O.E. 101 wheelbase 4 x 4 conversion for **Canada.** This first job used the factory closed cab but mounted a pick-up body with parallel slat-type seats and heavy duty hitch. This vehicle instigated the design by Marmon-Herrington of the CO 1 Q front drive axle subsequently adopted by Canada as a basic unit of the Canadian Composite Vehicle.

During 1940, a fleet of 6 x 6 conversions of the Ford 1½ ton were built into air field crash trucks for the **Royal Canadian Air Force.** Also the **US Corps. of Engineers** procured 4 x 4 conversions with earth boring equipment installed. The Corps. also acquired Ford COE conversions mounting large Sullivan self-powered air compressors. The latter had iron bar lamp/grill guards and front fenders relieved for off-road protection. The **U.S. Marine Corps.** bought ¾ ton 4 x 4 conversions with vertical rigid windshield and semi-open canvas tops for ambulance service.

There was military demand for the rugged Marmon-Herrington chassis as well. An additional large group of C90-BWS-4 Observation Balloon Winch Trucks were furnished the **Army Air Corps.** continuing a procurement sequence begun in 1938. These were the heavy series 4 x 4 short wheelbase chassis and made a very impressive vehicle.

The DSD series was introduced about 1940. The DSD 400-6 rated 6-9 tons was powered by the Hercules WXC3 previously used in the 440, and the DSD

800-6, 10-12 tons, used the RXC of the 555 model. The smaller models DSD 200-6, 300-6 and 400-6 were rated 5-6, 6-8 and 7-9 tons. They used Hercules JXC and JXD engines. The larger models DSD 900-6 rated 14-17 tons and DSD 1000-6, 17-20 tons, used the HXD and HXE engines displacing 855 and 935 cubic inches and rated 180 and 198 horsepower at 2000 rpm.

The year 1940 was characterized by a large increase in military truck procurement by the Iranian Army. The Iranian army ordered mobile machine shops with fold out-and-up side and rear panels. These were mounted on the medium 6 x 6 chassis and designated DSD 800-6. During the year there were two production runs of the 7-9 ton DSD 400-6 and the 10-12 ton DSD 800-6. The former was an ammunition and personnel carrier 6 x 6 with a transverse crew seat behind the cab but integral with the body; A canvas curtain served as windshield. The latter was a gun tractor and ammunition carrier which had the first M-H cab featuring a split windshield.

By the start of **1941** the influence of the European war was evident. Production was confined to large volume runs for the Netherlands and the US Army Air Corps. while experimental work concentrated on a new light truck for the Ordnance Department. Early in the year a fleet of ½ ton 4 x 4 Ford conversions equipped with windshield, doorless body with seats for four and a machine gun on a stanchion on the left running board were built for the Netherlands Procurement Commission and designated LLD MG 5-4. Orders for small and medium track-laying tractors, for service in Java in the Netherlands East Indies, resulted in parallel assembly lines at Marmon-Herrington for the first time.

The **Air Corps** orders were for the production of a large number of barrage balloon winches. Some were for high altitude surveillance and some for low altitude; some were skid mounted and some were mounted on heavy duty Marmon-Herrington trucks, some on M-H 4 x 4 converted Ford COE chassis (BB-2 winches). Later in the year, BB-1 winches were furnished the **U.S. Army Engineers** for mounting on crawler tractors.

Fire truck chassis using Ford COE 4 x 4 conversions were furnished for airport crash (American La France equipment) and ordnance plant protection. Chassis with 6 x 6 conversions were furnished as complete airport crash trucks for Java, Netherlands East Indies service. Ford 4 x 4 COEs were fitted with **6 passenger cabs** for oil field work.

An amphibious Jeep was prototyped for subsequent manufacture by the **Ford Motor Co.** production and is discussed in PROTOTYPES & LICENSES.

A one and one half ton airfield utility and bomb carrier truck was designed by Marmon-Herrington but **produced by the Ford Motor Co.** for both the US Army and Navy.

World War Two developments had significant impact on Marmom-Herrington in **1942.** It was the first year of really big production of tanks . . . these were M-H designed machines, whereas the rest of the domestic motor industry was struggling to retool for production of Ordnance designed tanks. Tanks, originally ordered for shipment to the Netherlands East Indies and China, were produced steadily throughout the year. As their countries were overrun by the Japanese, the Contracting Agencies released the inventory to the US Ordnance Department. Similar action was taken with one of the largest fleet orders M-H had ever produced . . . for the Netherlands but diverted before delivery to U.S. contracting agencies. Many of these were Ford 6 x 6 conversions equipped for searchlight and sound detection duty.

Fig. 3-8 Marmon-Herrington balloon winch on M-H chassis, and on M-H Ford all-wheel-drive chassis.

Fig. 3-9 The new DSD Series.

Fig. 3-10 Six passenger (crew) cab unit for the Argentine oil fields.

Fig. 3-11 Canadian Lynx vehicle.

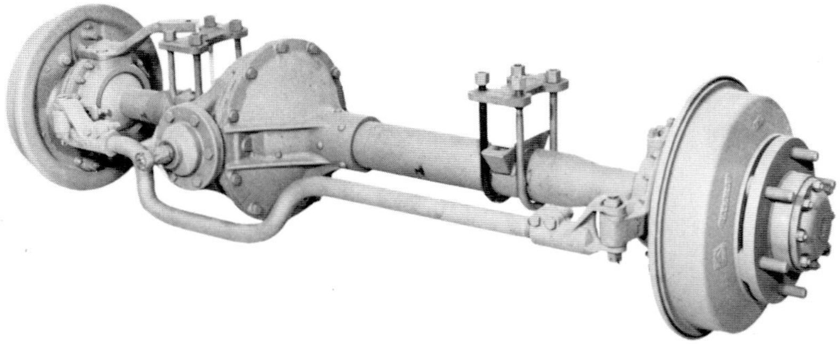

Figure 3-11 The CO1Q axle for the Canadian Military Pattern vehicles.

Figure 3-11 Stateside demonstration of the Marmon-Herrington barage balloon winch mounted on a Marmon-Herrington All-Wheel Drive truck. One of a fleet built in 1943.

FIELD ARTILLERY TRACTOR

Fig. 3-12 All-wheel drive Ford based Canadian field artillery tractor. Lloyd A. Brown

Fig. 3-14 British armored cars — M-H and Ford parts assembled in South Africa and fitted with armored bodies made there.

Fig. 3-13 Ford of Canada CMP production with CO1Q axle.

A tank contract for the **U.S. Marine Corp.** was completed in 1942 and was followed late in the year by the first large order for the new, secret light tanks.

Although most of the production facilities were converted to the building of tanks, truck manufacture continued with another fleet of All-Wheel Drive Barrage Balloon Winch Trucks produced for the **Air Corps** early in 1942. These were followed by a large fleet of Ford 4 x 4 Conversions for use on the Alcan Highway and Canol Projects by the **U.S. Corps of Engineers.** Also a large fleet of Marmon-Herrington TD 600-6 6 x 6 chassis with very heavy duty wrecker equipment was built for Iran.

Part of the tooling program at Marmon-Herrington in 1943, for production of the secret "Locust" tank, was the design and construction of a special duty M-H truck. The hulls for the "Locust" were to be welded together from armor plate and castings at Bedford, Indiana. This was a wartime utilization of granite quarry personnel and organization. Transport of the hulls to Indianapolis required special truck equipment.

A custom 4 x 6 was built with side-by-side twin Ford V-8's each driving one of the tandem rear axles through the regular 4 speed transmissions, both of which were equipped with a variant of the Bendix "Electric Hand" power shift system used on the Hudson and Cord cars. A single electric selector unit allowed the driver to make concurrent shifts on both power trains. Bob Wallace recalls now that the vehicle and the shift system performed very well and accomplished the task for which it was created. The author has not identified the existence of any photographs of this M-H custom job.

## PROTOTYPES AND LICENSES

Marmon-Herrington initially manufactured the Ford conversion axles partially from standard Ford parts. Wartime volume demands required procurement of the axles from Timken Axle Co., to the M-H design, after some 3 to 500 units had been built.

Lloyd A. Brown in his "Canadian Military Vehicles of World War II" (*Vintage Canada,* Vol. 3, #1, Sept. 1976) stated". *"It has been written in history, how Canada was instrumental in the design, production, maintenance and supply of some of the most unique Military Vehicles ever built. . . . the Canadian Military Pattern (CMP) vehicles were designed just prior to the start of World War II. The Canadian CMP vehicle with four wheel drive would pull through a tremendous amount of*

Fig. 3-15 ANZAC (Australia-New Zealand Army Corps) operations.

Fig. 3-16 ANZAC convoy. Note single rear tires.

Fig. 3-18 M-H ad in Fortune magazine showed M-H Ford armored car leading a convoy in the African campaign. "Hundreds of M-H Ford conversions and armored cars kept Rommel on the run for 1400 miles."

Fig. 3-17 "Leading the way to Tunisia." M-H ad in Fortune.

*snow, sand and mud. The engineering of the drive wheels gave these trucks a great advantage over contemporary trucks, because they utilized a large cross-section tire, with direct tracking rear single wheels . . . duals were not employed as testing proved the direct tracking rear singles were more effective. The German Army in North Africa issued a directive in 1942 that they try to capture, for their own use, Canadian vehicles because of their superior performances in the sand. Standardization of design was of utmost importance in the production of the CMP trucks which were built by General Motors and Ford".*

The standard axle for all CMP vehicles was the Marmon-Herrington CO 1 Q design.

The Canadian Traction Co., Ltd., where Bill Duffield was responsible for engineering, was licensed to build the M-H Model CO 1 Q and transfer cases. These were built into Ford 4 X 4, 6 X 6, and 6 X 4 (tag or unpowered rear tandem axle)

Fig. 3-19 Field Marshall Viscount Alexander's Staff car. The Canadian War Museum and Lloyd A. Brown.

trucks for the military. Conversion of the basic 4 X 2 truck occurred at assembly in the Ford of Canada plant. The M-H components were also designed into the line of Canadian Military Pattern vehicles including the Field Artillery Tractor manufactured by Ford of Canada, beginning in 1940. The "Tractor" hauled an 18 pounder 4.5 inch Howitzer, was built on a 101.25 inch wheelbase and weighed 8800 pounds.

During World War II, Ford Motor Co. of Canada produced some **hundreds of thousands** of all-wheel drive military trucks for the British Empire Countries using basic M-H components manufactured under the license agreement with Canadian Traction Co.

Another example of the foreign assembly of Marmon-Herrington components into Ford truck chassis was the very important strategic move by the government of the Union of South Africa who saw the possibility of inter-allied cooperation at an early date, sensed the urgency and had the courage for prompt action.

The Ford Motor Co. of Canada and the Marmon Herrington Co. furnished the chassis and the parts necessary for conversion of the standard Ford truck to All-Wheel-Drive. The chasses were assembled by the Ford Motor Co. (Pty), Ltd. of South Africa at Port Elizabeth. Some 3336 trucks and 1643 cars with all-wheel drive were produced. For the armored cars, the armor plate was fabricated and the superstructures built in South Africa. Then the vehicles were shipped to Egypt and went into action manned by the British and Australians. Similar programs operated in Australia and in New Zealand.

There was even a Marmon-Herrington assembly facility in Burma where 3 ton 4 x 4 chassis were assembled and fitted with native sourced bodies which Bob Wallace recalls were entirely of teakwood.

Two armored vehicles were built in considerable numbers in Australia using locally sourced bodies fitted to Ford truck chasses assembled and converted with Marmon-Herrington components by Ford Motor Co. of Australia (Pty.), Ltd. One was the series of Mark II, III, IV, and IVF armored cars; the latter two had a rear mounted engine and all were 4 X 4s. Another was the "Dingo" armored scout car on a shortened 110 inch wheelbase and fitted with a Bren or Lewis gun; it carried a crew of two and was fitted with 10.58 X 18 run-flat tires. There were also limited quantities of an armored command car on the 156 inch wheelbase converted 6 X 6 Ford truck chassis.

The Mark VIII was an eight wheeled armored car used by the British in North Africa. It was a Marmon-Herrington design.

The staff car used by Field Marshal Viscount Alexander of Tunis during the Mediterranean campaigns of the war is today displayed in The Canadian War Museum, Ottawa, Ontario. It is a Canadian Ford Marmon-Herrington conversion of a 1941 "woody" station wagon subsequently reworked into an open phaeton with folding windshield and tonneau windshield. It still bears the Union Jack flag surmounting the five-star cluster.

The British drive on Bardia in Italian Libya took the Italian defenders by complete surprise because a hundred miles of hot wind-blown desert sands separated the forces. The Italians thought it would take days of hard travel for the British to cross the same ground with the best equipment they had... hence the Italian position was secure from the flank. What the Italians did not know was that hundreds of trucks designed and built especially with power applied to all wheels for desert travel had been convoyed over thousands of miles of oceans by the British Navy and landed in Egypt. Concentrated in an unsuspected direction from Bardia, the British forces plunged across the "impossible" desert barrier in these All-Wheel-Drive trucks hauling troops, guns and supplies. New Zealand field artillery units used M-H 4 x 4 Ford Conversions with hi-flotation single tires to haul British field guns at the Battle of Bardia.

The M-H Fords used by the New Zealanders so successfully were fitted with high-flotation single tires because tests had confirmed that dual rear tires caused considerable resistance by their inability to follow in the groove of the front tires.[1]

"In the later stages of the campaign. General Montgomery lunched an attack from which the Axis army never recovered. General Rommel, the 'Desert Fox', was outfoxed by better fighting and better equipment which enabled Montgomery's famous Eighth Army to keep Rommel on the run for more than 1400 miles of battle-scared desert. Hundreds of M-H converted Ford trucks and armored cars took part in this desert chase."[3]

"The Ford Motor Company publicity staff exercized considerable ingenuity in associating the Model T car with current events. For instance, when troulbe with Mexico was anticipated in 1913, the publicists recommended that every American battleship be equipped with Model Ts to enable the Marines "to charge on Mexico City from Vera Cruz in record time." The idea was hailed in metropolitan newspapers all over the country.[4]

It may have been a fecitious proposal in 1913 using Ford Model Ts. It was not so in Africa in WW II with Marmon-Herrington Ford conversions.

Marmon-Herrington was called upon to perform design and development tasks on several unusual amphibious vehicles which subsequently were produced in vol-

Fig. 3-20 Marmon-Herrington engineers, with the help of Sparkman & Stephens of New York noted yacht designers, experimented and produced the original amphibian. Because our facilities were already taxed with other military orders, the manufacture of the amphibian was placed in the hands of the Ford Motor Co.

ume by other manufacturers with whose marques they are normally associated by the public.

In all these projects M-H joined forces with Rod Sparkman of the prestigious marine architectural firm of Sparkman-Stevens.

The first of these was the amphibious conversion of the Model G.P. Jeep under Ordnance Dept. contract. A single prototype was demonstrated, followed by production by Ford Motor Co.[5]

The second was the famous "Duck" Model DUKW subsequently series produced by GMC Truck and Coach Division of G.M. The single vehicle prototype produced by M-H under Ordnance Dept. contract utilized many GMC components. Another amphibious conversion project involved the M18 Howitzer Carrier built by Buick Motor Division of General Motors Corp.[6]

Marmon-Herrington engineers participated with the US Navy in the initial experiments which led to the amphibious landing of tanks from water-borne craft. Tests conducted with the fleet Marine force led to the development of the familiar front

AMPHIBIAN "JEEP" SERVES INVASION FORCES

Sea-going "jeep," donning steel scow body, boathook, and running lights, and transferring power drive to propeller, operates afloat and ashore as agile reconnaissance and patrol vehicle for landing forces

ramp vessels used in all water-borne combat landings.[6]

Universal Power Drives, Ltd. of Great Britain began importing Marmon-Herrington axles in the mid 1930s and incorporated them in specialty trucks for seismic survey and exploration and for forestry work using integral winches for self-loading. These truck tractors were produced until interruped by the war and production was resumed after the war, continuing under the "Unipower" marque. The prewar Unipowers were powered by the American Hercules J X D; the postwar trucks were redesigned under the direction of George Wildey and used Gardner diesel engines. George had been with Universal Power Drives since 1938 and emigrated to the US becoming Safety and Environmental Engineer for Terex, then a Division of General Motors.

Unipower eventually manufactured the M-H axles under license and marketed the Rzeppa joints in Great Britian until 1949; Post war, the joints were procured from United Kingdom machine shops. In 1949 the manufacturing and marketing rights were acquired by Hardy-Spicer who adopted the name "Con-Vel" joint.

George Wildey provided the author with the information on the Universal Power Drives, Ltd. venture. He stated that he had met Colonel Herrington in England.

Marmon-Herrington had an affiliate, Marmon-Herrington SAF, a French firm which manufactured M-H components and all-wheel-drive conversions for military vehicles beginning in 1946. Later the firm would manufacture M-H trucks. In 1963 the name was changed to Marmon-Bocquet and the interest increased to 24 percent.

## PART 3 — The Tracked Products
## Tanks and Tractors

### LIGHT COMBAT TANKS CTL

Marmon-Herrington unveiled their first tracked, and armored, vehicle in June 1935. It was a component of a 3 unit combat package conceived and developed by M-H for submittal to military customers. The light combat tank was designated CTL-1, the trailer which transported it was model T-1-T and the towing vehicle was a Ford conversion B5-4. The complete package was designated the CTL Combat Train.

The CTL-1 was significant because it was to remain in intermittent production through 1942 both for the U.S. Marine Corp. and for allied forces.

The CTL series was a turretless tracked armored vehicle. Armament was typically a pair of forward facing machine guns. It had a low profile with the superstructure raised only over the crew stations. There were 3 forward, 1 right hand and 1 left hand viewing ports for the crew. An exception was the Model CTLS-4TAC where the S and T refer to **S**teel track and **T**urret. That model was produced in a large quantity under contract to the Netherlands government; delivery was interrupted by the German defeat of the Dutch and the fleet was subsequently delivered to the US Ordnance Department.

Production runs for the US Marine Corps. were made in 1937, Model CTL-3 and 1939, Model CTL-3A. Wartime material priorities probably dictated the change to steel track for 1942 production as Model CTL-3M and Model CTL-6, also for the US Marine Corp.

The CTL-1 Light Combat Tank was a rear longitudinal engined vehicle. Drive from the unit engine/transmission was forward through a short U jointed shaft to the differential and steering/brake unit. Double (close-coupled) U-joint shafts of negligible length coupled with final reduction gear boxes on the hull wall which carried the drive sprockets.

Bob Wallace recalls that the representative of the U-joint vendor, Blood Brothers Co., considered the application to be impractical. Johnson was obviously correct for all modern U.S. tracked vehicles use this type of attachment between the power pack (engine-transmission-differential) and the final drive units on the hull. Another Marmon-Herrington 'first'.

Power was supplied by a Hercules engine of 4¼ inche bore and 4¾ stroke, 404 CID, rated 160 horsepower at 3200 revolutions per minute.

Driver and gunner sat in bucket seats on each side of the transmission whose shift lever was then in a "console-mounted" position.

Fig. 3-22 Marmon-Herrington light tank CTL-3A

Fig. 3-23 First track-laying vehicles built for the US Marines; 1937

## TRACK-LAYING TRACTORS

This same basic lower hull and machinery concept was used for the Marmon-Herrington track-laying tractor (the generically used 'Caterpillar' is a registered trademark of that firm) which was first offered in 1936 as Model TRL-1. The rubber track with steel-cable tension elements was also shared.

The description of the M-H tractor by the M-H Staff Writer in the 1938 catalog cannot be improved upon and is repeated here:

### Whizzes down the road at 30 M.P.H.

*Automotive engineers and manufacturers, dealers in heavy-duty construction and hauling equipment, users of trucks and tractors on all sorts of projects, newsmen, photographers and the general public — all have stood amazed at the remarkable speed and flexibility of this new track-laying tractor. All have been quick to see its advantages over conventional, slow-moving, noisy and cumbersome track-laying tractors for many kinds of work — where speed as well as sure-footed power is a factor. They have seen it "scoot" down the highway at the speed of a fast truck, and then off the road it pulls like a forty-mule team. Hooked on to heavily loaded cargo trailers, it will pull through soft sand, mud or snow, to the highway, and once on the pavement, will deliver the load at truck speed to destination or terminal. Completely engineered, and manufactured in the Marmon-Herrington plant, this new Track-Laying Tractor incorporates many novel departures from conventional track-laying tractor design. Probably the most important of these, and the one making most of the others possible, is the new exclusive track-suspension and rubber track, which so greatly reduce operating vibration, friction and rolling resistance as to permit the high speeds of which this unit is capable. So freely does this tractor move, that brakes are necessary for the first time on a track-laying vehicle. Here again is the "secret" of the Marmon-Herrington extraordinary fuel economy of four to five miles to the gallon, and the astonishing flexibility of the unit. Then too, there is the water-tight hull which permits going through deep water without discomfort to the driver, or damage to the mechanism. Driver and passenger sit well forward for full view ahead. Handles like a passenger car . . . Pulls like a locomotive. Marmon-Herrington Track-Laying Tractors are now being manufactured in three sizes, each size available either with a gasoline or Diesel engine. Each of these models is equally easy to drive. Each operates with the same freedom from noise and vibration. There is no "galloping" even over the roughest terraine, due to the scientifically designed spring suspension. The low center of gravity provides unusual safety on steep grades. Steering is accomplished through new type fully air-cooled steering clutches which operate with fingertip ease. The absence of metal contact with the road permits the operation of the Marmon-Herrington on pavements where tractors with steel lugs are forbidden. Starting, stopping, backing . . . on or off the highway . . . are done with passenger car ease, and the unit turns in its own length from a standing start, or at top speeds. The maximum pull of these units, in comparison with weight and horse-power, is the best evidence of the unusual efficiency of The Marmon-Herrington Track-Laying Tractor. No other automotive vehicle combines such high tractive power, with so great speed, maneuverability and safety. Great pulling power is to be expected of a track-laying tractor operating at one or two miles per hour . . . but when the same vehicle will get up and go down the road at 30 m.p.h. its usefulness is multiplied many times. One of the primary reasons for The Marmon-Herrington's greater tractive ability, of course, is the rubber track. It adheres to wet pavement, packed snow and ice as only rubber will. . . and practically 100 % of the motive power is tranmsitted to the road. A Marmon-Herrington will shorten the job, and cut hauling costs, wherever there is need for speed as well as power*

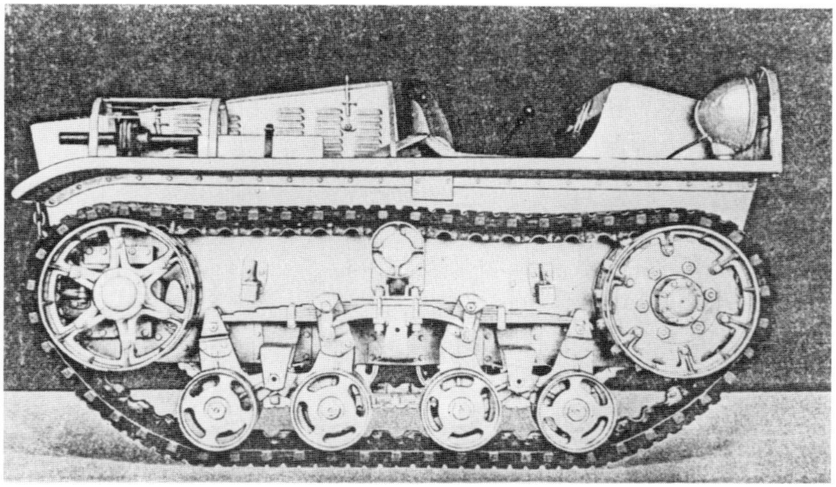

Fig. 3-24 Marmon-Herrington Tractor Model TA-20

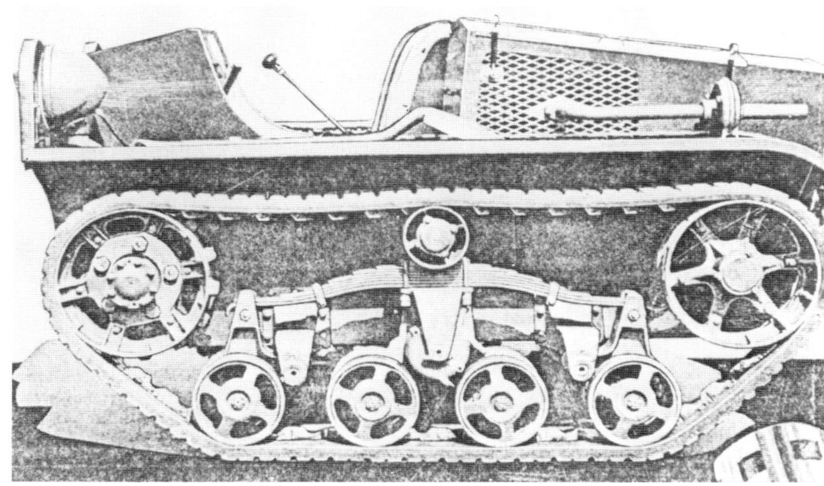

Fig. 3-25 High Speed Tracklaying Tractor Model TA-30

*Hundreds of Uses . . . in industry, large scale agriculture, road construction, oil fields, mining, airports, etc. Merely to enumerate the hundreds of industrial uses for The Marmon-Herrington Track-Laying Tractor would require far more space than is availabe in this literature. The possibilities are almost unlimited. An entirely new horizon is established in the sales field of track-laying vehicles. In railroad and overland freight terminals; on wharves and docks; for rapid transportation of crops from fields to loading platforms or mills; for moving equipment over soft or rough ground in the mining country or oil fields; for pulling heavy airplanes across the landing fields; for excavating and road construction and maintenance; for logging and lumbering; snow removal; in manufacturing establishments; public utilities; replacing industrial railways; for coast guard beach service . . . all of these, and dozens more come to mind. We look to the individual imagination and initiative of management executives and engineers for a further enlargement of its field of usefulness.*

How appropriate . . . the pioneering of a superior suspension system and consequent ride for a new class of road vehicle, just as the first generation Marmon cars were known as the "Best Riding Cars in the World" and their successors as "Scientifically Designed" cars. The TA-30 and TA-40 were equipped with air power steering for fingertip control.

Further extending the M-H espousement of the Ford V-8 powerplant, the M-H tractors were powered by Ford power units: the Ford V8-60hp (136 CID) in the TA-20, the V8-85hp (221 CID) in the TA-30 and the Lincoln Zephyr V-12-110hp (276 CID) in the TA-40. The standard corresponding truck clutches and 4-speed transmissions were used.

The model designations match the rated drawbar horsepower, which was the same for both diesel and gasoline option. Diesel tractor models had a D prefix to the model designation.

Diesel tractors were powered by Hercules engines Models DOOB, DOOC, and DJXC respectively with 62, 70 and 83 hp. but equaling or exceeding the torque rating of the gasoline engines.

The three tractors had lengths and weights of 97″-4310#, 103″-5766#, and 130″-8020#.

The TBS designation identifies a second generation product and reflects the switch to steel track, probably due to wartime material priorities.

Prices for the track-laying tractors, as of January 24, 1938, were as follows, pre-tax at Indianapolis.

| Gasoline | Diesel |
|---|---|
| TA-20 $4500 | DTA-20 $5500 |
| TA-30 $4950 | DTA 30 $5950 |
| TA-40 $5950 | DTA-40 $7250 |

According to the *Star* for December 4, 1938, a Marmon-Herrington tracked tractor had been shipped to Republic Pictures in Hollywood via Railway Express. It was to be made into a tank for use in the film "Army Girl" and was selected because of its high performance despite its basic three ton weight plus an additional two tons due to the tank simulation.

A TA-20 tractor was demonstrated in the Goodrich Rubber Co. display at the New York Worlds FAir in 1939, because the Marmon-Herrington track design was produced by Goodrich.

The Model CTVL tanks produced for the Mexican government in 1938-39 were derivatives of the tractor series. Mexico subsequently acquired a fleet of CTMS-1TBI tanks.

## LIGHT TANK DEVELOPMENTS

Miniature Marmon automobiles are discussed in Chapter 10, Part 2 of the Marmon Story. One miniature of a Marmon-Herrington product is known to have been made. Several years after the Mexican tank fleet went into service, Col. Herrington made an inspection trip and was so impressed with the level of maintenance that he commissioned a silver trophy subsequently presented to Mexican Army Chief of Staff General Cardenas by Vice-president Campbell.

Fig. 3-26 US Marine Corps tractor with modified suspension.

During 1939, the suspension structure was reinforced with the addition of some additional elements and a controlled differential was adopted for steering instead of the former steer-by-braking arrangement.

The earliest Marmon-Herrington combat tanks incorporated a M-H designed transmission driving a truck axle with the usual outboard brakes which were utilized for steering control by one-side actuation. A new metallic brake lining material developed by the S.K. Wellman Co. was used to avoid concurrent loss of steer and deceleration control following submerged brake incidents. Col. Herrington, according to Bob Wallace, resisted adoption of controlled differential steering but finally had to concede.

A typical plan layout of a M-H tank of the early period is that of the CTLS-4TAC.

The tank bogie and support assembly of the MTLS-1G14 is representative of M-H tanks.

The CTL-3M resorted to steel-link track and conventional bogie construction to replace the continuous rubber track and longitudinal leaf spring suspension, for strategic material considerations. Later units had rubber track pads, a development significantly influenced by the experience with compounds for the M-H continuous rubber track beginning in 1937.

In addition to the light combat tank series CTL, Marmon-Herrington also designed and manufactured both a medium tank series CTM and a single model medium heavy tank, the MTLS-1G14.

## MEDIUM TANK CTM

The CTM was a 3-man crewed medium tank equipped with a 360° swing turret. Initial production was in 1942 with the CTMS-1TB1 designation. It is presumed the S had reference to the steel track. All CTMs were of rivited-plate hull construction. The suspension was similar, but not identical, to the steel-tracked CTLs.

When the Netherlands was overrun by the German army, the severed contracting agency turned over the tanks on the Marmon-Herrington site to the US Ordnance Department.

A diesel powered version was also produced for the U.S. Marine Corp. designated Model CTM-3TBD, beginning in early 1942. Power was supplied by a Hercules Model DXRB diesel with six cylinders, 4⅜″ bore and 5¼″ stroke, 474 CID rated 125 horsepower.

Armament on the USMC medium tank was a 37mm gun in the turret and three .30 cal machine guns in the hull facing forward.

## MEDIUM HEAVY TANK MTL

The largest tank designed and manufactured by M-H was the MTLS-1G14 with a four man crew built initially in 1942 for the Netherlands but rededicated to US Ordnance Department and still in production in 1943 for US forces. Riveted plate hull construction was used. The turret was fitted with cal. 30 Browning machine guns installed on both ball and bracket mounts.

The suspension was very similar to the CTM medium series but the two sets of bogie wheels were spaced farther apart and the drive sproket was larger in diameter.

Power was a Hercules L head six cylinder engine of 5¾″ bore and 6″ stroke, 935 CID rated 238 hp at 2100 rpm with dual downdraft carbuetors and equalizer tube reminiscent of the 1941 Buick system.

MARMON-HERRINGTON TRACKED PRODUCTS
COMBAT TANKS

| Year | Model | Type | Customer | Track | Crew |
|---|---|---|---|---|---|
| 1935 | CTL-1 | Light | Speculative | Rubber | 2 |
| 1937 | CTL-3 | Light | USMC | Rubber | 2 |
| 1938 | CTVL | Light | Mexico | Rubber | 2 |
| 1939 | CTL-3A | Light | USMC | Rubber | 2 |
| 1942 | CTL-3M | Light | USMC | Rubber | 2 |
| 1942 | CTL-6 | Light | USMC | Rubber | 2 |
| 1942 | CTLS-4TAC | Light | Netherlands US Ordnance | Steel | 2 |
| 1942 | CTMS-1TB1 | Medium | Netherlands Mexico | Steel | 3 |
| 1942 | CTM-3TBD | Medium | USMC | Steel | 3 |
| 1942 | MTLS-1G14 | Medium Heavy | Netherlands | Steel | 4 |
| 1944-45 | M-22 (T9E1) | Airborn Light | US Ordnance | Steel | 2 |

TRACKED TRACTORS

| Year | Model | Customer | Crew |
|---|---|---|---|
| 1936 | TRL-1 | | 2 |
| 1937 | TVL-1 | U.S. Marine Corps. | 2 |
| 1937 | TA-30 | Russia (Amtorg Corp.) | 2 |
| 1938 | TA-30 | Netherlands (Java) | 2 |
| 1938 | TA-40 | Egypt (Anglo-Egyptian Motors) | 2 |
| 1939 | T-7 | US A-AF | 1 |
| 1939 | TA-30 | US Ordnance | 2 |
| 1941 | TBS-30 | Netherlands Steel track | 2 |
| 1941 | TBS-45 | Netherlands Steel track | 2 |

Marmon-Herrington also undertook the design and development of a unique armed vehicle for the Canadian Army in the latter years of the war. Willys-Overland Corp. was the original contractor but for unknown reasons asked Marmon-Herrington to assume it during 1944. The contract specified a floatable tracked vehicle capable of carrying a crew of two with light armor protection from

Figure 3-27 Marmon-Herrington tanks: Light CTVL (left), Medium CTM-3TB (center), Medium Heavy MTLS-1G14.

ground gunfire. A unique requirement was capability to be loaded, under its own power, through the side cargo door of the specified transport airplane.

Marmon designed a vehicle powered by the Jeep engine and propelled both in and out of water by steel tracks. Bob Wallace, who provided the pictures of this vehicle, told the author that even the **wheels** and the **suspension arms** were **air-tight hollow structures.**

The engine was transversely mounted in the rear with unit transmission and offset gear case whose driving flange faced towards and to the left of the engine. A close-coupled U-joint shaft carried the power to a spur-gear input on the final-reduction unit which incorporated steering clutches and axle housing extending to the two final drive sprockets.

Bob Wallace resolved a problem of track throwing (disengaging the sprocket,) on the rear sprocket during turns. An air cylinder was used to tension the track under normal operation but air was also admitted to a larger diameter of the piston/cylinder whenever a steer mode was initiated. The excess track wear which would have occurred from continious use of the high loading was avoided.

Novel means were resorted to for compliance with the aircraft loading requirement. The right hand track tension unit was removed and the track shortened temporarily to route it downward from the forward support roller to the forward road wheel. When so configured, the armored vehicle could maneuver through the specified cargo door; upon arrival and unloading, the configuration was restored to normal.

The vehicle was designated TJ-1 for "Tracked Jeep-1" and five units were built in early 1945, following successful compliance demonstration including floation without auxiliary aids.

In 1942 Marmon-Herrington had performed concept engineering studies for a secret light tank for the Ordnance Department. General Motors and Walter P. Christie also made concept proposals. In April 1943 M-H received a contract for

Figure 3-28 Tracked Jeep TJ-1 for the Canadian Army. Note transverse Willys Jeep engine.

Figure 3-28 Crew compartment and armament of the TJ-1.

Light tank company maneuvering in an Alaskan mountain pass. T-14 (M-H CTLS-4TAY) left turret and T-16 (M-H CTLS-4TAC) right turret. US Army Signal Corps photo

design, development and production of the airborn tank known experimentally as the T9E1[1] and series produced as the M-22 "Locust". The T9E1 differed from the initial T9 both in the forward hull configuration and in armament.

The design and development of the T9E1 tank was under the direction of R.C. Wallace. Denny Dusenberg (son of Fred D.) worked under Wallace. 1070 units were built with many in process of assembly when VJ Day ended production.[7]

Marmon-Herrington designed the T9E1 tank around the Lycoming air-cooled aircraft Model 0435T which was a six cylinder horizontal-opposed unit (like the GM Corvair of 1960-65) rated 162 horsepower at 2800 rpm. The transmission and final drive units were of M-H design for the application; the controlled steering differential was a Cletrac (Cleveland Tractractor Co.) unit.

The engine was rear mounted with a longitudinal crankshaft position. Drive was forward from a 2:1 step-up drop gear box unit-mounted on the engine. A U-jointed propeller shaft extended forward to the transmission/differential/steer unit. Very close coupled U-jointed shafts joined the out puts to the final drive units on the hull walls. Bob Wallace recalls that the Mechanics Division of Borg-Warner Corp. did not approve the application but subsequently noted the advantages of such use, which is common tank design practice today.

The engine, according to Wallace, had torsional characteristics which required a large escalation from the initial U-joint size determination and the use of a "soft" propellor shaft which featured concentric steel tubes coupled with a rubber tube. The 2:1 step-up engine rpm was necessary to provide the speed required for differential steer control; a resultant problem was the durability of the bearings for the idler in the drop case which was resolved by use of P.R. Mallory Co. silver bearings.

The transmission provided one overdrive and one underdrive ratio plus direct drive.

The rear engine placement resulted in a dust problem with the clutch. Bob Wallace worked with Charley Space of the Lipe-Rollway Co. to develop a dust proof application which was confirmed by pouring bags of foundry parting-sand over the housing.

M-H designed the T9E1 suspension-along the lines of other Ordnance designs. It was found during development that transverse reenforcing rods were required at each of the bogies and at the track tensioner location.

The T9E1 (M22) tank was designed with a removable turret to facilitate transport by air. The turret was reportedly procured from Bofors who shipped complete turrets to Indianapolis for vehicle final assembly. The turret incorporated a 37 mm gun and a 30 caliber machine gun. The vehicle weighed 15,975 pounds (8 tons); in fighting trim, 17000 pounds. It measured 155 inches long, 68.5 wide and 87.75 high.

Bob Wallace recalls that the solid-rubber tired bogie wheels performed satisfactorily on the rubber-block (highway) track but failed by blister-blowout when the steel tread (cross country) was used. He worked with Waldo Stein and Jim Hale, Firestone Indianapolis representative and Chief Engineer respectively and after fifteen rubber compound evaluations eventually achieved 4-5000 mile life.

The T9E1 became the M22 "Locust" as series production was undertaken in 1943. The combat mission of the "Locust" was capability for air transport in British "Hamilcar" gliders towed by heavy bombers and cut loose for landing and deployment behind enemy lines. They proved highly effective in demoralizing enemy troops, who expected the gliders to disgorge lightly armed infantrymen, and performed critical support in securing beach-heads and securing bridges before the enemy could effect destruction. The Locust was used by the British 6th Airborne in securing the Rhine crossing on March 24, 1945. Postwar releases refer to the gliders as "Hamiltons"; "Hamilcar" may have been a code name.

The secret status of the Locust during the period immediately following 'D' Day made reporters subsequently hesitant to risk censorship of dispatches by reference to Locusts. The result was a surprising unfamiliarity with the M22 by both the media and troops who did not experience the Locust in action.

The unquestioned importance of the Locust is best assessed by quotation of the text of a large sign displayed in the lobby of the Chief of Ordnance, Detroit, MI.

Additional official detail is revealed in the War Department press release[8] on the "Locust":

### WAR DEPARTMENT
### OFFICIAL RELEASE ON T9-E1
### (U.S. Model M-22)

A new combat vehicle which has made its appearance on the battlefields in Normandy, is the airborne tank carried by gliders to places which would ordinarily be inaccessible to tanks. The element of surprise gained by the appearance of a tank instead of troops when a glider lands has already had its effect on the successes of the invasion of Hitler's fortress in the west.

The new tank is familiarly known to troops as the "Locust" as a result of initial action scenes of huge gliders landing in battle zones disgorging tanks, which immediately blast positions where Nazis have been waiting to fire on advancing Allied soldiers.

Designed especially for airborne use, the tanks were produced from specifications furnished by the Ordnance Department to the Marmon-Herrington Company, Indianapolis, Indiana.

The Locust is a compact, streamlined tank with a low silhouette. It is a highly mobile vehicle and will cross over rough terrain at 15 to 25 miles an hour and travel on roads at speeds up to 40 miles an hour. It mounts a rapid-fire, hard-hitting 37 mm. cannon, which fires both high explosives and anti-tank shells, and a .50 caliber machinegun. The vehicle combines all the technical skill and manufacturing know-how that has gone into other heavier tanks now in use by the United States Army in successful battles in other theaters of this global war. The tank has many tactical uses, an important one being the advantage of effecting tank concentrations behind enemy lines on short notice. This Firepower is very effective in creating confusion. The tanks and crews can arrive on the scene as a surprise move and the vehicles are in fresh mechanical condition, the crews rested and ready for battle.

* * *

During the latter half of 1944, procedures and fixtures were developed to accomplish air transport of the Locust by both the C-82 Fairchild Packet and C-54 Douglas Skymaster four-engined transport. This development was made to permit trans-oceanic transport of the M22 for deployment in the final onslaught on the Japanese forces.

For transport by the C-54, the tank was parked below the side cargo door and the 2000 pound turret was hoisted and swung aboard using hoist equipment built into the aircraft. Then the tank was maneuvered directly below the belly of the aircraft and between the landing gears. In only six minutes the 7 ton turretless tank was hoisted by cables and secured to the special rack built into the aircraft belly.

The Marmon-Herrington NEWS for December 1944 reported: *"A sensation was a gliderborne weapon of war . . . widely publicized in newspapers and magazines throughout the world, and featured on the front cover of Popular Mechanics magazine for Dec. (1944) — Marmon Herrington's M22 "Locust" Airborne Tank promises to become an even more potent weapon, now that a huge new plane has been developed to transport it to scenes of action". "This mammoth tank-carrying plane was developed by Fairchild and has been approved by the Army. Designated the "Packet" C-82 the plane is powered by two 18-cylinder Pratt & Whitney engines each of 2100 take-off horsepower and has a range in excess of 3500 miles at an unrevealed speed.*

The T9E1 tanks in the European theater were all assigned to British forces and carried the British name **Locust. The subsequent production units intended for the US forces in the Pacific were designated *Stuart* for Confederate General Jeb Stuart.[9]**

**SIGN IN LOBBY OF CHIEF OF ORDNANCE, *Detroit, Michigan***

**T9-E1** The most spectacular of all is the tiny 7-ton airborne "Locust" tank which has proved the most flexible weapon used by General Bradley against the Germans. Its use in the paratroop invasion of South France now permits Ordnance officials to release this closely guarded secret weapon for publication. It is flown to the scene of action in gliders, mounts a 37 mm. cannon and a caliber .50 machine gun and is manufactured for Army Ordnance by Marmon-Herrington Company.

2

Fig. 3-29 The new baby poses with big brother: T9E1/M22 with MTLS-1G14.

Fig. 3-30 Windshield/enclosure for non-combat driving; note armored hatch/periscope/gun port assembly stowed above.

Fig. 3-31 Crew compartment with turret removed: Driver zone and gunner zone.

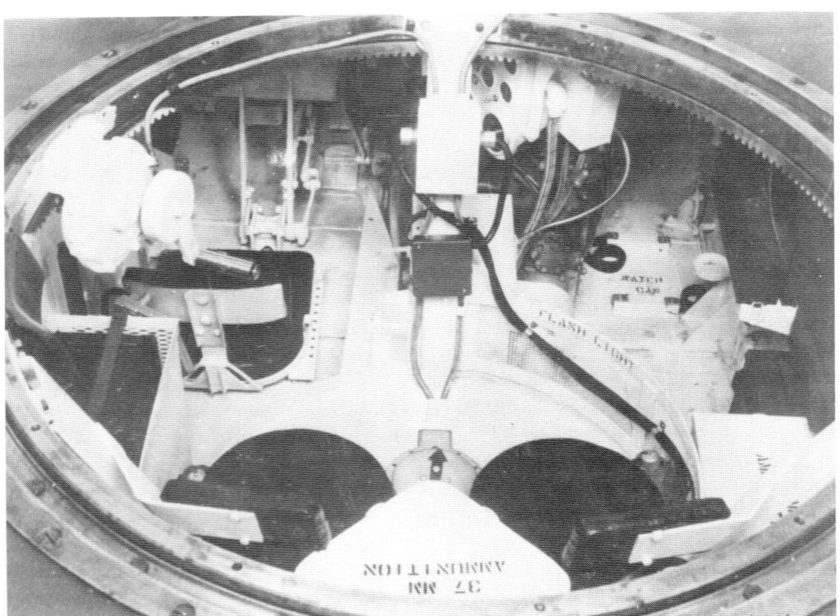


Fig. 4-1 Model 440-4

CHAPTER FOUR

# CIVILIAN TRANSPORTATION PRODUCTS

## PART 1 — All-Wheel-Drive Trucks

In 1944 the Ordnance Department required more truck-tractors than could be produced by the prime manufacturer. Kenworth and Marmon-Herrington were contracted to expand the production rate of the H-542/G 671 4x2 (roadster) tractor of International Harvester origin. These trucks were powered by an I-H engine Model RED-450-D, 6 cylinder, 451 CID producing 124 horsepower at 2600 rpm. They weighed 11660 pounds including the fifth wheel (for trailer interface) and ready for the road, as shipped. The transmission provided 5 forward and one reverse speed. The trucks were intended to haul 20 ton trailers for long distances on supply routes in Europe and were the basis for the European trucking industry in the immediate post-war period.

This production run was followed by a run of a similar 4-5 ton 4x4 tractor initially produced by Autocar and subsequently by Kenworth as well as Marmon-Herrington. This tractor used a Hercules RXC-6 of 529 CID rated 112 horsepower at 2200 rpm and was intended for long haul operation over the difficult roads of the Far East.

There were the trucks which carried the famous and proud "Red Ball" insignia and furnished essential and remarkable support to the rapidly advancing armored divisions.

This development was favorable to ultimate resumption of civilian production when hostilities would cease. It was reported to stockholders in the 1944 Annual Report: *"Our company is just starting its sixth year of war production. During 1944 we have been successful in making considerable progress toward the transition from war to peace, without interfering in any way with our policy of all-out cooperation with the demands of the military service. We have been totally engaged in war production since 1939 and we will continue this policy as long as the necessity for it exists.*

*During 1944, due to a change in war contracts, we were able to shift over from the production of combat tanks to the production of military trucks. This change has been very fortunate for us, because our facilities have been converted back to our normal peacetime business of manufacturing motor trucks, although we are still making them for military use and not for civilian use.*

*As we have now returned to manufacturing trucks instead of combat tanks, our past experience has enabled us safely to set prices for the truck units at a level which left lower gross profit margins. Although we are continuing to set aside reserves for renegotiation on the basis of past experience, we are hoping that our repricing policy will secure more favorable consideration for us in 1944, from the Government's renegotiation authorities."*

The M-H NEWS reported:

HEAVY TRUCKS TOP URGENCY LIST

*A bulletin just received from the automotive Council for War Production rates the need for heavy trucks along with B-29 Super-Fortresses and heavy artillery. Even though we are ahead of schedule, we must redouble our effort now — to help shorten the war!*

The post-war plans of Marmon-Herrington were revealed and detailed in President Dingley's letter to stockholders in the Annual Report for 1944, dated March 23, 1945: *Before the war the normal business of the company was the manufacturing of the type of truck which the Army adopted and which during the war our whole American Automotive industry has been producing. In our introduction of the Marmon-Herrington Ford all-wheel-drive conversion unit, our company pointed the way to the mass production of this type of vehicle for army use.*

*After the war it is our intention to return to our former working relationship with the Ford Motor Company and to continue the production of the all-wheel drive conversion of Ford products.*

*In addition to this, we will continue to produce the larger Marmon-Herrington all-wheel-drive trucks, the performance of which has contributed so much to the good name of our company throughout the world.*

*We feel that the universal acceptance of the all-wheel-drive idea which has been brought about by the essential military use of this type of vehicle, will have so reduced any previously existing sales resistance that there will be a world acceptance of the all-wheel-drive principle as essential to "off the road" operation.*

*In addition to our domestic market, there is a large field in export sales for vehicles of this type in countries which do not possess the long mileage of good roads available to us here in the United States.*

*The increased plant which we have acquired during the war will also enable us to take on other products in the transportation field. We propose to do this to the extent necessary to give us adequate dollar volume to support a plant of this size.*

The resumption of traditional product lines and the "taking-on of other products in the transportation field" to which Dingley referred in March 1945 had become definitive soon after. In May 1946 he wrote, in a special 'Interim Report' on first quarter operations: *The production line for the manufacture of this vehicle has been completed, but at the present time the inventory on hand is not sufficient to assure production. We anticipate deliveries from this line at some time within the succeeding three months.*

*Our electric trolley coach production line is now in the process of being installed. This work will be completed within the next quarter and we are planning for production of trolley coaches in the third quarter of the year.*

*Heavy-duty Marmon-Herrington truck orders have far exceeded our expectation and we have been fortunate in being able to accomplish some production of this item.*

*Orders for Marmon-Herrington Ford conversions have been much greater than our ability to secure the necessary Ford units for conversion.*

*Our All-Wheel-Drive truck orders in all sizes have been much greater than we had hoped for. The expected sales resistance brought about by the disposal of Army surpluses of this type of vehicle has not developed to the extent that we had anticipated.*

Not evident in Bert Dingley's roster was one other ultimate new product line — motor coaches, evident on Colonel Herrington's business card from later years:

Note his position that Marmon-Herrington of Indianapolis had become so well known worldwide that no additional address information was provided.

## POST-WAR TRUCKS

The first of the post war era products to appear was a pair of All-Wheel-Drive truck models. The announcement advertisement stated: *After several years of uninterrupted war production Marmon-Herrington is pleased to announce that it had been granted permission and allotted materials by the War Production Board to build a limited number of light-heavy All-Wheel-Drive trucks. First production, scheduled to start in October (1944), will be concentrated on two models to be known as the model MH 440-4 and MH 555-4. Prices will be announced immediately upon approval by OPA (Office of Price Administration). Into these two new M-H trucks will be incorporated all of the engineering skill and automotive craftsmanship for which this company has long been famous. Both models will be equipped with the latest type of improved air brakes and will be characterized throughout by the sturdiest kind of construction. They will have the abundance of power and built-in ruggedness needed to cope with the most severe on-and off-the-road duty.*

New features included a divided windshield, rounded leading edges around the sides and top thereof, and new front fenders with totally convex surfaces and "skirts." The cast aluminum radiator shell with distinctive vertical center rib and trailing fin was continued. The radiator design used on the lighter "A" and "B" series trucks did not reappear after the war. It, too, was a very distinctive design and immediately identified the vehicle as a Marmon-Herrington.

A production rate of 20 per day was achieved by the end of 1944. A shortage of heavy truck tires was the principle constraint.

The catalog specifications for the two models are reproduced; they provide a concise information package with surprising detail. Certain items deserve additional comment however. The frame rails for the two models differ in depth by 0.125 inches and the frame rail gauges differ by 0.062; the same male die member was being used to form both.

The engine in the 440 was the Hercules WXC3; it had a bore and stroke of 4.25 x 4.5 inches, displaced 383 cubic inches and was governed at 106 horsepower, 2800 rpm. It weighed 1280 pounds. The 555 used the Hercules RXC with 4.625x5.25 cylinders displacing 529 cubic inches; it was governed at 114 hoursepower at 2200 rpm, had 265 foot-pounds torque at 1100 rpm.

The Zenith carburetor on the RXC engine in the 555-4 was a new type which incorporated an air inlet elbow and was claimed to provide air cooling of the float bowl which was suspended from above.

The W.C. Lipe clutch on both models was a single disc (one driven member) or a single plate (one pressure plate spring-loaded against the flywheel) type. It incorporated a single conical spring and pull type release. A feature which made it a popular heavy duty clutch was the provision for adjusting the spring load to compensate for load loss due to facing wear; a threaded ring in the center of the cover could be rotated. A flexible lever system assured a uniform loading of the pressure plate. No provision for torsional damping was made in the driven disc. The transmission was manufactured by the Fuller Mfg. Co. of Kalamazoo, Michigan. For the M-H application, it provided direct drive on 4th, and 5th was an overdrive ratio for use under partial load conditions. Constant mesh helical gears were employed for 3rd and 5th gears.

Fig. 4-1 The 600 Series introduced for 1952.

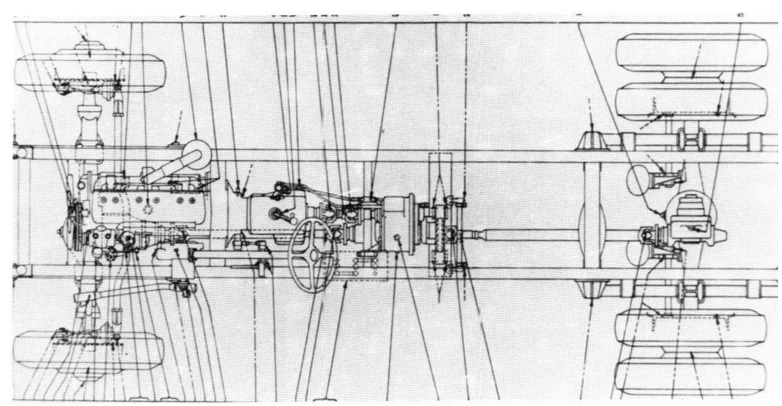

Fig. 4-2 Model 555-4 chassis plan.

Fig. 4-3 Typical 600 Series chassis.

Fig. 4-5 Ratchet-type compensator between front and rear drives and provision for lock-up.

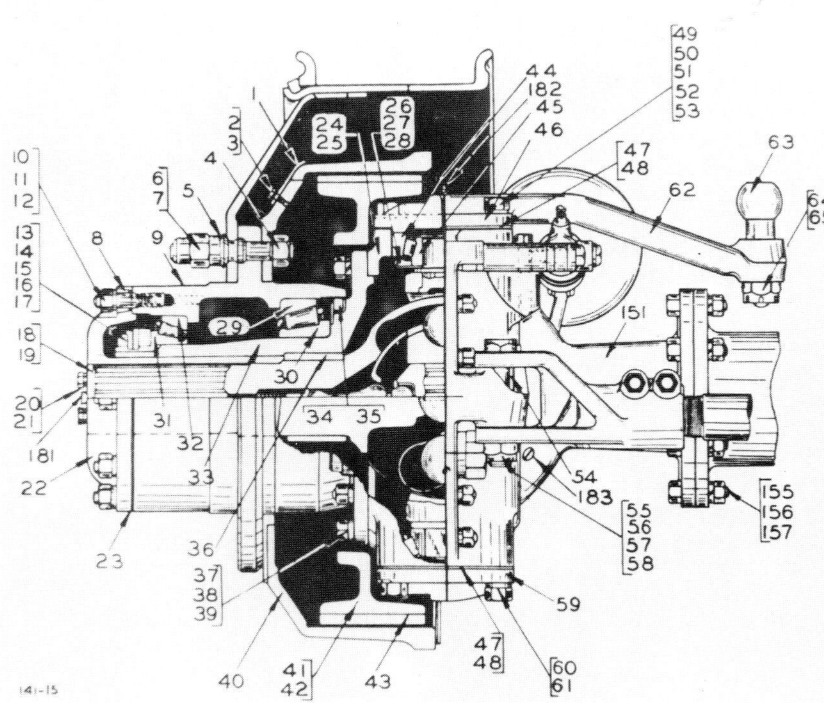


Fig. 4-4 Front spindle; Model 555 with Rzeppa type CV joint.

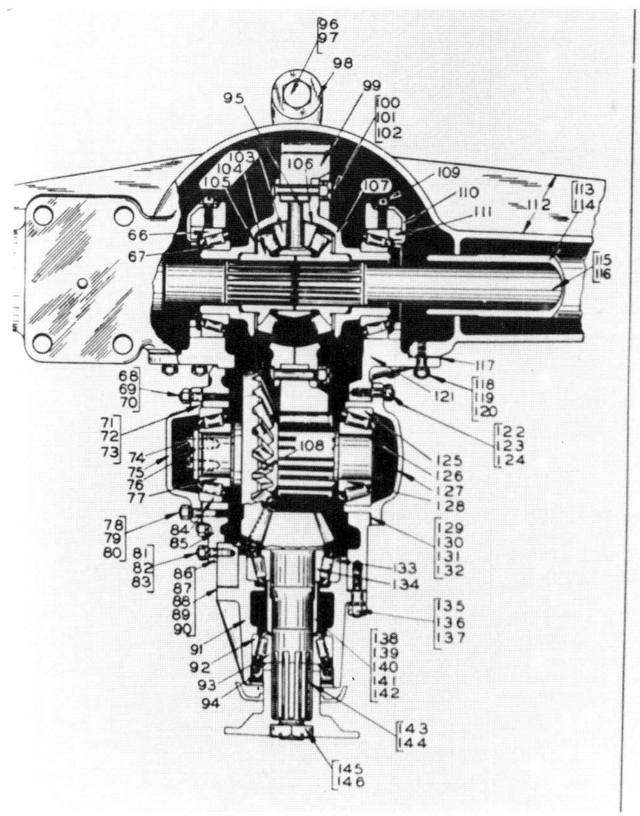

Fig. 4-6 Front differential of Model 440.

The front axle incorporated a double reduction of input speed on the same center-line as a single reduction axle. The rear axle was the more usual configuration.

The steering universal joints were Weiss type with a center ball configuration.

Ross cam and double-lever steering gears were used. Both cam followers were mounted on tapered roller bearings making the system fully anti-friction bearinged. The double lever system overcame a shortcoming of the cam and lever gear — limited angular travel of the output arm. With the double lever system, one of two followers is always in the cam groove at the extremes of travel.

## THE "Q Series" FORD CONVERSIONS

Ford Motor Company expanded their truck product line for the 1948 model year. A new large V-8 engine had been developed for the Lincoln and larger Mercury lines and a commercial version was a logical next step. Rated at 145 horsepower, the engine powered the biggest and most powerful truck ever built by Ford, the F-7 model.

Because Marmon-Herrington had a close working relationship with Ford, an All-Wheel-Drive conversion was promptly available. Marmon-Herrington labeled it the "Q" Series in a press release in April 1948. Six models, based on the Ford F-7 models, were announced. Three were 4-wheel drive with GVW of 21,5000; three were 6-wheel drive with GVW of 35,000 pounds. The wheelbase range was 136.5 to 220 inches.

The release on the 'Q' Series, noted that the Marmon-Herrington line also included twelve other Ford conversions and four heavy duty models of its own manufacture. The twenty-two models covered a GVW range of 4700 to 42,000 pounds.

A subsequent magazine announcement ad listed the frame as 9″ deep with 3″ flanges of ¼″ stock. Hypoid front and rear axles offered a final reduction of 6.275:1 center:, 4 places (high gear) to 47.595:1 in high range; 11.762:1 to 89.215:1 in low range of the two speed auxiliary. Ten forward and two reverse ratios were avilable. Helper or secondary spring leaves were used at the front as well as at the rear. . . . On re-manufactured Fords, M-H added an over-riding frame to the Ford frame. This resulted in an extremely strong, flexible rear tandem assembly.

## THE AIR FORCE CRASH TRUCK 0-10

In late 1951, the U.S. Air Force awarded Marmon-Herrington a contract valued in excess of $10 million "for some hundreds" of a new post-war generation of airport crash trucks for fighting fires at and near airports.

The new truck was a 6x6 vehicle designated Model 0-10 and destined to be a familiar site at airports across the country for many years as they replaced equipment of WW II vintage.

The entire crash truck was designed for operations ranging from the heat of the desert to the Arctic cold. The vehicle would operate in temperatures down to 65° below zero and up to 130° above zero. This requirement necessitated among other things, an automatic generating system so that batteries were kept fully charged at all times and the engines kept warm and ready to go without delay. The new 0-10 was a high speed vehicle designed to reach a crash within seconds after it occured; the truck incorporated All-Wheel-Drive so that it could operate over rough terrain such as ploughed fields, ditches and other uneven ground which might surround an airstrip.

Each vehicle was equipped with a special high velocity large volume air-foam turret nozzle controlled from within the cab. This nozzle was fed with a mixture of air-foam liquid, water and air. Pressure for this very high velocity nozzle was derived from a pump driven by a separate engine carried amidships in the vehicle.

While Marmon-Herrington was to produce the major portion of the new crash truck, many of the special accessories were to be sub-contracted. At the time of contract signing, it was anticipated that a number of additional workers would be required over the next several months as production advanced.

Production of the Model 0-10 at Indianapolis ultimately numbered 726 vehicles. In addition, 134 Model MB-1 crash Trucks were built for the U.S. Navy.

Production of All-Wheel Drive chassis for the MA1 and NC5 ground servicing units for starting and servicing jet aircraft for the U.S. Navy and U.S. Air Force totaled 2246 units.

## THE 600 SERIES "AWD" TRUCKS

Marmon-Herrington announced for 1952 a completely new truck product line with new styling, new cabs and sheet metal, new frames and rear axle.The new trucks were designated the "Anniversary Series" for Marmon-Herrington's twenty years of truck manufacture. The styling was unusual and distinctive, the vehicles massive appearing and the overall impression was reflective of the all-business purpose of the trucks.

The all-new cabs featured a V-windshield 62 inches wide with 870 square inches of glass area and mounted in rubber. Wrap-around corner rear windows and a vertically-opening center rear window were standard equipment. Cab construction was of 14 gauge steel throughout. The front seat was six feet wide and featured a separate driver cushion adjustable 4.25 inches fore and aft. The roof configuration provided 38.25 inches from cushion to ceiling. Instruments were relocated to be in front of the driver.

The new series were designated Models 610, 615, 620, 625, and 630. The three pairs had engines rated 139, 145, and 180 horsepower. The trucks were rated from 24000 to 33000 GVW.

The fender treatment was unique. The two fenders were joined by a central panel, in which there were air intake slots, forming a flat "platform" on which the radiator shell and hood were positioned. An unusual feature was the separate forward fender sections which were removable to facilitate close-coupled mounting of snow plow blades with attendant advantages.

The new chassis lacked the traditional M-H frame kick-up over the front suspension but a low center of gravity of the load was still claimed. The flat-top channels were 12¼ inches deep tapering upward to about 6 inches over the front suspension. M-H described it as "one of the biggest, strongest frames ever built." Cross members were joined to side rails by gussets top and bottom and angles full length. Assembly was by riveting.

Rear spring were located by double wrapped-leave eyes at the forward ends; they had sliding support at the rear. Helper springs were used front and rear. Springs were engineered for an ideal (tire) weight distribution of one-third on the front axle. All models carried five speed transmissions which in combination with the M-H auxiliary transmission/transfer case provided 10 forward and two reverse speeds. The auxiliary transmission included the automatic compensator for the difference in travel between front and rear wheels in turning corners.

The 600 Series would ultimately survive the end of vehicle production at Marmon-Herrington. The UNIC Division of SIMCA was reported assembling the MS600B chassis for Marmon-Bocquet as late as 1982. Marmon-Bocquet is the 1963 successor to Marmon-Herrington SAF (France). The MS designation is Marmon-Simca reflecting the Simca V-6 (or Peugeot 4) engine now used in these military chasses.

## FRONT AXLE COMPENSATION AND DECLUTCHING

Marmon-Herrington vehicles had always incorporated an automatic compensator in the transfer case or auxiliary transmission. The function, in the absence of an inter-axle differential, was to compensate for the difference in travel between the front and rear wheels. Sometimes front wheels must rotate faster than rear wheels. For example, in turning a corner the front wheels travel faster because they must cover a greater distance at the beginning of the turn than do the rear wheels (the rear wheels cut inside the radius of turn of the front and travel on a shorter circumferential segment).

When the front wheels travel faster than the rear, the compensator automatically goes into action. It allows the front wheels to over-run the rear wheels (the front wheels are driven faster by the road surface than by the transfer case; the energy coming from the rear wheels). Without the compensating action, front and rear wheels would be driven at the same speed and sliding would have to compensate for the differential distance between the paths of the wheels. Because even straight line driving is really a continuous left turn/right turn cycle, the compensator makes steering the vehicle on a hard surface much easier and less tiring.

The compensator device used by M-H was a simple rotary ratchet device in the auxiliary transmission output.

By 1955 M-H had made the compensator device an option. An internal/external toothed coupling was made standard equipment. When manually disengaged from the driver's compartment by mechanical linkage, the vehicle performed as a rear wheel drive vehicle; when engaged, positive all wheel drive was available for off-road operation. Addition of the compensator made all-wheel-drive available on paved surfaces for increased control and safety under slippery conditions. For use entirely off-road, a vehicle could be ordered with full-time positive drive to all wheels.

Owners were warned that the front driving axle must never be disconnected when low range had been selected in the auxiliary transmission. To do so would direct the additionally multiplied torque entirely to the rear axle with consequent severe or possibly destructive forces.

## "21st ANNIVERSARY" FORD CONVERSIONS

Marmon-Herrington had introduced "The 21st Anniversary Models" in 1952. This unusual anniversary was again acknowledged in 1956 with the "21st Anniversary Models of Marmon-Herrington All-Wheel-Drive Fords," the first having been offered in 1935. In the intervening years, one hundred thousand conversions had been completed.

The 1952 line-up of Ford conversions included light, medium and heavy duty models with a GVW range of 5300 to 29000 pounds. Power was provided by 5 V-8 engines and an optional six in the light and medium models.

The M-H model designation reflected the Ford model with the digit "four" (for 4 wheel drive) replacing the final zero in the Ford model. For example a M-754 was a M-H conversion of a Ford F-750. Ten conventional and six cab-over truck conversions were offered.

All-Wheel-Drive was also available for all Ford Bus and Parcel Delivery Models.

For the 1955 model year, Ford Motor Co. announced new engines for their truck product line. They were overhead valve V-8s with crankcase walls extended well below the crankshaft centerline for increased rigidity. M-H releases referred to them as Y-8s in recognition of this superior construction.

All Marmon-Herrington Fords now incorporated transmissions with silent helical gearing and synchronized shifting with exception of the lowest gear. The five-speed was a Clark Equipment Co. transmission incorporating the Clark Split-Pin synchronizer.

Marmon-Herrington had been a regular advertiser in Ford Dealer and Ford field, typically appearing in color on inside covers. The ad in the 1956 issue of Ford Field carried the message: "Ford and Marmon-Herrington Teamed together for **21 years**" and the time span "1935-1956." The leader and the underscore were significant — the underlying concern was hinted at in the text:

"Marmon-Herrington is the only All-Wheel-Drive engineered specifically and exclusively for standard Ford chassis. As a consequence, they are the only mass-produced All-Wheel-Drive vehicles of completely matched and balanced components."

It was the **last** ad for the M-H Ford conversions to appear in Ford field. Only a few months later, Ford Field began to carry a regular series of ads for: *"The best 4-Wheel Drive for Ford and the lowest in cost!"* It was manufactured by the Front Wheel Drive Division of Napco Industries, Inc. of Minneapolis, MN, *"the largest supplier of 4-wheel drive assemblies to major truck manufacturers."*

## THE AIR FORCE CRASH TRUCK 0-12

During 1956, the Engineering Department was involved in the development of another generation of US Air Force crash, fire and rescue trucks — the Model 0-12. The Project Engineer was Harold F. Baker, who reported to George H. Freers, Vice-President in charge of engineering.

## THE RHINO

Popular Science Magazine, in the September 1943 issue, devoted half a page to a revolutionary vehicle with hemispherical wheels. The wheel concept and the vehicle utilizing them were credited to a Turkish-born Greek engineer, Elie P. Aghnides of New York city. Mr. Aghnides envisaged a vehicle which he called "Roller" as the successor to the military tank vehicle. Propulsion was to be by the two large hemispherical drive wheels and steering was by a single spherical wheel in the rear. The inventor listed five advantages of his concept over a tracklaying tank:

1. The hemispherical wheels had a small footprint on a hard surfaced road and could therefore travel at high speeds not feasible with track. The wheels had a tire tread at their largest diameter.
2. When soft terrain was encountered, the wheels rapidly reduced ground pressure as they sank in the mud or sand and ribs on the surface provided traction. The radial surface ribs were not as deep as the tires and only came into use on soft ground. A tracked vehicle becomes mired when a certain ground pressure is exceeded.
3. The tricycle steering system would permit turning within the vehicle length "by swiveling as easily as an office chair." Hence the gun would be fixed, dispensing with the turret of the tracked tank and permitting nearly instantaneous retreat.
4. The hemispherical wheels avoided the pivot formed by the track edge when upsetting forces were encountered; when the center of gravity of a tracked

Fig. 4-7 The Aghnides Amphibious Roller or 'Rhino' photographed at the M-H plant in 1957. The vehicle had undergone considerable reconfiguration from the initial design. Access was through the water-tight bubble canopy. All Hanley Hydrojets were characterized by the patented 360° stering jet.

Fig. 4-8 BIG FOOT again; M-H Ford conversion model available in late 1950s.

vehicle passes that pivot, recovery is impossible and roll over follows. Mr. Aghnides vehicle was expected to approach the 90 degree point before it was unable to recover.

5. The wheels were to be light weight, air tight, and of large volume. They were to provide floatation when a water crossing was required and provide propulsion by the paddle-wheel effect of the ribs.

Colonel Herrington was always interested in all-terrain mobility. And, he had built his firm by developing, before the war, new military vehicle concepts at the firm's expense and then demonstrating them to convince the procurement authorities that they indeed had a need and that he had the product to satisfy it. He had also developed and marketed as early as 1938 specialized models of Ford conversions equipped with high flotation tires for use in off-road exploration. Therefore it was not surprising that when Marmon-Herrington had gotten their post-war products underway, they undertook the design and development of Mr. Aghnides concept vehicle. The year was 1948 and the vehicle was dubbed "The RHINO."

The prototype was 19 feet long, 9.5 feet wide and 9 feet high. It weighed 10,000 pounds including the aluminum wheels which were six feet in diameter and weighed 1500 pounds each. Power was provided by an 85 horsepower Ford V-8 engine. There was provision for the driver and one passenger. It differed from the 1938 concept in having two small hemispherical wheels at the rear and in having the axes of the front land propulsion wheels inclined; propulsion afloat was provided by a Hanley-Kermath Hydrojet water propulsion system developed by the Hanley Engineering Service of Prospect, Ohio. When completed in 1954[1] the prototype RHINO was driven down Harding Street to proving grounds north of the Troy-Harding intersection. Vic Maier and George McMath were still actively assigned to the project in 1963!

When the CBS "Sixty Minutes" TV program covered the Inventors show on January 25, 1981, the Aghnides invention was discussed briefly — "better than a crawler track" and "Ordnance was never interested" were comments.

## BIRTH OF BIG FOOT

The RHINO development program may possibly have been responsible for the apearance of BIG FOOT in the late 1950s. When the mind-boggling giant tracks were followed to their source, it turned out to be the Marmon-Herrington Ford granddaddy of all those intimidating elevated 4x4s that travel the pavements these days. M-H had marketed a Ford conversion "Marsh Buggy" in 1938 but this one had 13.5 x 24 tires and an 8 to 10 pound per square inch ground pressure.

The Greyhound Corporation had attracted much attention in 1954 when they introduced the Scenicruiser Model. That introduction had been a particular ambition of Greyhound Chairman Orville S. Caesar who envisioned that the future prosperity of his firm depended upon carriage of both passengers and priority package freight.

The regular passenger Greyhounds were powered by the six cylinder Series 71 Diesel manufactured by Detroit Diesel Division of General Motors. The Scenicruiser required an eight cylinder Series 71 engine which was not then available as a single engine; It was however available in the form of a pair of four cylinder Series 71 engines. The unique kit-like concept of the in-line Detroit Diesels made available a pair of mirror-image fours — essentially a V-eight but with the bores vertical and twin crankshafts.

Each engine was coupled to the gear case, which combined the outputs of the two engines, by fluid couplings. The gear case output drove the single clutch and transmission with center-line drive forward to the rearmost of the tandem rear axles.

Certain advantages were anticipated from the arrangement. A failed engine could be isolated by draining the coupling for that engine and the trip continued to destination at moderately reduced speed. During terminal operation, one engine could be isolated and shut down while the mate provided power for the air conditioning system. Because both had starting motors; either engine could be started electrically and then the other started by filling the hydraulic coupling in event of starting motor failures.

Disadvantages also soon became evident. The gear case and the two fluid couplings absorbed a surprising amount of power and the bus fuel consumption was excessive.

In 1957 a V-eight configuration Detroit Diesel Series 71 engine became available. The Scenicruisers had accumulated about 500,000 miles each and a major overhaul of engines, power train and vehicle was due.

Greyhound invited bids for replacement power modules incorporated the DD 8V-71 engine. (On most transit and inter-city buses, the lower rear section of the "body" is detachable and the entire power system including radiators can be operated, detached, for service purposes.) Marmon-Herrington was the successful bidder to design and furnish the complete modules for the 1000 Scenicruisers which Greyhound would concurrently refurbish. GMC Division of General Motors, who manufactured the buses originally, also bid.

The complete new power train which Marmon-Herrington furnished included a 275 horsepower V-8 Series 71 Detroit Diesel, clutch, 4-speed transmission, differential carrier assembly, radiator and engine doors. This contract was for $10 million of the $13 million total refurbishing cost. The contract specified delivery over an eleven month period beginning in Octobeor 1958 Expansion of the work force, which currently numbered 600 was anticipated in the press announcement.

## PART 2 — DELIVR-ALL

The first of the innovative post-war products from Marmon-Herrington was the DELIVR-ALL house to house delivery truck.

The announcement release on Oct. 2, 1945 stated: *"The ideal vehicle for this type of work should provide for ease of entrance and exit in order to minimize driver fatigue. It should be economical in operation and be unusually roomy in order to carry a large number of parcels or units to be delivered. It should have a low, unobstructed floor and an extra wide body. For efficiency on the route it should be highly maneuverable and capable of making quick stops and starts. For driver convenience and comfort it should be easy to drive and steer from either a standing or sitting position. For servicing and repairs the mechanical units must be readily accessible. With these ideas in mind, the Marmon-Herrington Company of Indianapolis, Indiana, set out to create a home delivery truck that would meet all the demands of these exacting services. The name of the new vehicle, the DELIVR-ALL, is a particularly happy selection — for it indicates the wide range of industries to which it is especially adapted. By utilizing a front driving axle of the company's own proven design, the propellor shaft, rear axle and it's differential are eliminated. The level and unobstructed floor is brought down to a height of only 13 inches from the pavement, all the way from the driving controls clear to the rear of the body.* ***Monocoque*** *(airplane) body type of* ***construction*** *provides the necessary rigidity to support the load without a conventional chassis frame. In addition to providing a low, unobstructed floor and much more headroom, extra width and load space within the truck, front wheel drive makes possible the* ***inclusion of the engine, steering gear and all operating controls in a single unit, which can be easily and quickly detached from the truck body for servicing.***

*Not only can the entire power plant be removed for major mechanical attention, but all the mechanical units are readily accessible for inspection, adjustment or minor repairs while the power unit is in place in the vehicle. Ample access panels are quickly and easily removable. These panels are held in place by quick acting aircraft cowling type fasteners. The DELIVR-ALL has a short wheelbase, is highly maneuverable, has a fast pick-up, and is particularly pleasing in its modern appearance. Wide, fast operating, finger touch opening folding doors on either side and at the rear make for quick, easy loading and ready accessibility of merchandise. The 60 h.p. engine is a sturdy, economical, dependable 4 cylinder unit, capable of handling the loaded vehicle with ease. Controls are arranged for standing or seated driving. Acceleration is controlled by either hand throttle or foot accelerator. The hand throttle operates in combination with the gear shift lever. An auxiliary transmission provides two speed ranges, a low range for frequent stop operation and a higher range for driving to and from the route. This gives a total of*

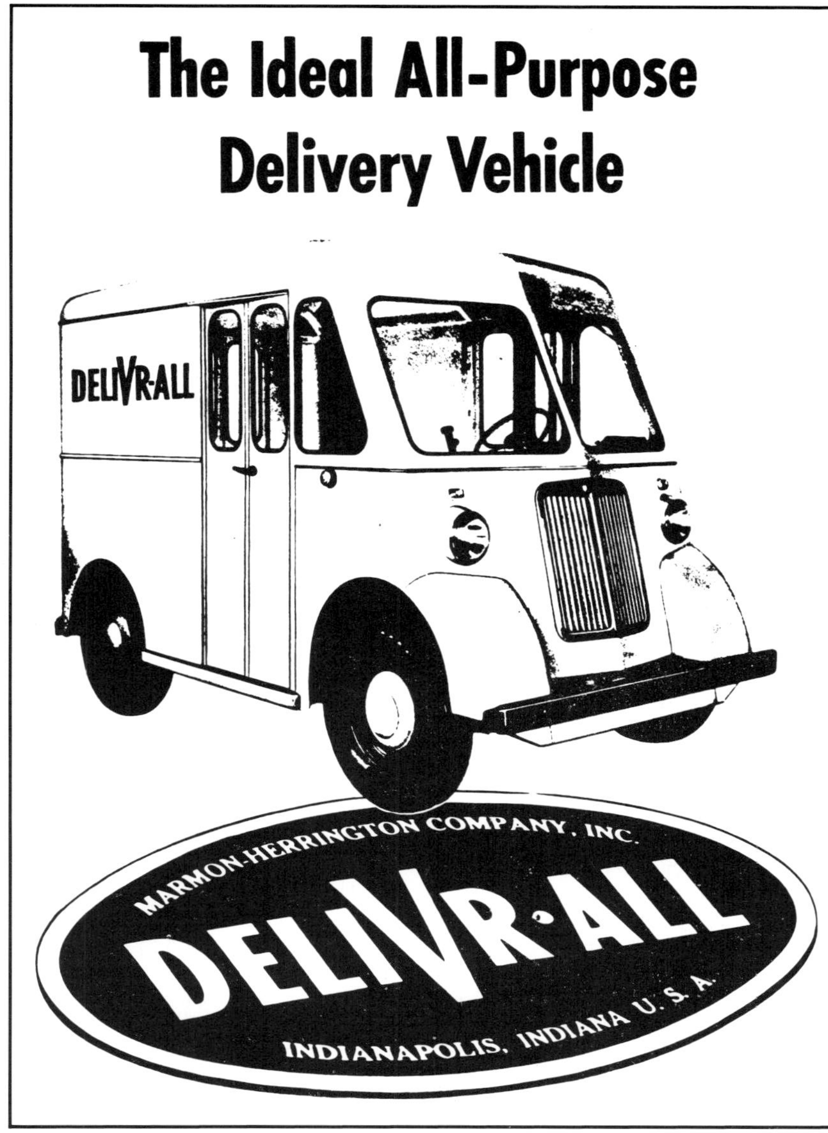

*six forward and two reverse speeds. The vehicle is designed to offer faster, more economical deliveries and more "stops" per day, with less driver fatigue. The introductory sales brochure called DELIVR-ALL "the newest thing on wheels" and proclaimed it "A Revolutionary Advancement in truck Design! To build such a vehicle, it was necessary for Marmon-Herrington engineers to create an entirely new conception in truck design."*

The name of Alec Issigonis is universally known consequent to his design of the innovative Austin and Morris, "Mini" cars for 1959 production. Many believe that advent to have been the pioneer application of the front-mounted transverse engine and front wheel drive combined concept. The Fiat 1100 series came soon after and, following the energy crisis of the early 1970's the concept spread firefast throughout the worldwide industry and became the hallmark of the "World Car" concept. It appeared on North American General Motors cars with the introduction of 1980 'X' body carlines in the spring of 1979. The GM design was planned for integration in the full spectrum of corporation products and was characterized by very compact packaging of larger engines than previously used in the Issigonis concept. Service accessibility was also accommodated by the integration of the full power train — engine to wheels — in a sub-chassis unit attached to the body/chassis unit through four vibration isolators. Major service of engine/transmission/differential/drive shafts is accomplished by engine unit removal.

Few people, however, are aware that Marmon-Herrington pioneered and marketed a vehicle, the DELIVR-ALL, which incorporated the true pioneer application of a transverse mounted 4 cylinder engine driving the front wheels through a helical geared differential and constant velocity universal joints **AND** the pioneer application of packaging the complete power system including the front wheels as a detachable unit for service access. Moreover, this omnibus concept was conceived and introduced concurrently, whereas the Austin-Mini to GM 'X' cycle of maturity encompassed twenty years and began more than ten years after introduction of the DELIVR-ALL.

But Walter Christie was the originator of the concept and his Christie Iron Works at Eleventh Avenue and Twenty-third Street, New York City produced taxi cabs in 1908. They embodied most elements of the now favored design including independent suspension and a four cylinder engine arranged transversely with output to the driver's left and via the transmission countershaft to a spur gear driven differential, and unequal length double (Cardan) universal jointed driveshafts. "This vehicle drives and steers with the front wheels, and is so designed that the uncoupling of the gasoline, water, change-gear (shift linkage) and steering connections and the removal of eight bolts permit the entire front axle housing with the power plant to be removed from the frame," The radiator, fuel tank, steering gear and shift lever remained on the vehicle.

The cab design was a variation of the probably more familiar Christie race cars of the early century with their huge V-4 inclined transverse engines. The cab engine was configured to have all systems accessable from the front of the vehicle and placed the radiator behind the engine in the early Renault position. Rear suspension was by semi-elliptic springs and drop-center axle without a frame kick-up.

Bob Wallace credits Colonel Herrington with the DELIVR-ALL innovation. Wallace and Pierre Schon, the latter formerly with GMC, developed the concept for production. Wallace's chief innovation was the T-bolt attachment scheme for joining the two units into a complete vehicle. He does not recall any effort to seek patent coverage.

The DELIVR-ALL power unit was all-encompassing. It included the fuel tank, battery, and all steering and driving controls as well as instrumentation and head lights. The vehicle was not only intended to appeal to all door to door delivery operations but had a special feature for the small fleets which characterized much of the dairy product, baked goods, and dry cleaning/laundry business operations of the immediate post-war period. That special attraction was the ability to keep all the vehicles in a small fleet in productive operation by ownership, not of a spare truck, but only of an extra power unit. Only the rear suspension/brake units and the tail lights were a part of the body unit. A quick-disconnect hydraulic coupling permitted separation of the hydraulic brake line without fluid loss or air ingestion. An electric plug separated the body wiring and a clevis pin provided for the release of the parking brake cable.

Jacks were provided for the support of the body unit during absence of the power unit. A tubular handle and quick-attachment hardware was provided for handling the severed power unit in the manner of a 'wheel-barrow.' "Tailskids" were integral with the power unit.

Initial production of DELIVR-ALLs differed in styling from later production. The early units carried the forward part of the front fenders across the full width of the vehicle; the grill was consequently divided horizontally with the upper half being Vee-shaped with vertical bars and lower half being curved with horizontal bars. The later units restricted the fender curvature to the fenders themselves and the Vee line of the divided windshield extended all the way, through the full length Vee-grill, to the bumper apron.

The flat load floor, extending all the way aft from the control console, was made possible by the substitution of suspension modules at each rear wheel for the conventional rigid rear axle. With the module concept, there is no interconnection between the rear wheels and nothing to interfere with the floor. The rear suspension module concept, with flat floor, has been a feature of two trucks (or truck-like vehicles) in the past thirty years — the M-H DELIVR-ALL of 1946 and the GMC luxury motor home of the early 1970s.

The Marmon-Herrington rear suspension module consisted of a pair of parallel semi-elliptic springs (37 long by 1.75 wide — 7 leaves) arranged longitudinally directly above each other. The complete independent suspension units were easily removed. The DELIVR-ALL had the suspension characteristic of most passenger cars of the 1934-1980 period — a rigid driving axle and independent suspension of the non-driving wheels. Uniquely, the M-H product was the reverse of the usual configuration, or could perhaps better be characterized as differing in the steering of the powered rather the free-rolling wheels.

The DELIVR-ALL was offered in three lengths and corresponding wheelbases:

| Model | DVL-1 | DVL-1A | DVL-1B |
|---|---|---|---|
| Wheelbase-in. | 90 | 105 | 118 |
| Load capacity | | | |
| cu. ft. aft front door | 182 | 284 | 344 |
| Power unit-lb. | 1900 | 1900 | 1900 |
| Body unit | 2150/2300′ | 2400/2550′ | 2600/2750′ |
| Total vehicle | 4050/4200′ | 4300/4450′ | 4500/4650′ |
| Gross weight-max. | 6750/8200′ | 6750/8200′ | 6750/8200′ |
| Load-lb. | 2700/4000 | 2450/3750′ | 1850/3550′ |
| Tires | (4) 7.50x16 | (2) 7.50x16; | (2) 8.25x16 rear) |

Note that the body length increment identifies the weight of the monocoque body structure at less than 200 pounds per foot of length.

Vehicle width was 78 inches and height 94.

Because of the novelty of the power unit module incorporating the full power train and it's support systems as well as the exterior sheet metal and driving con-

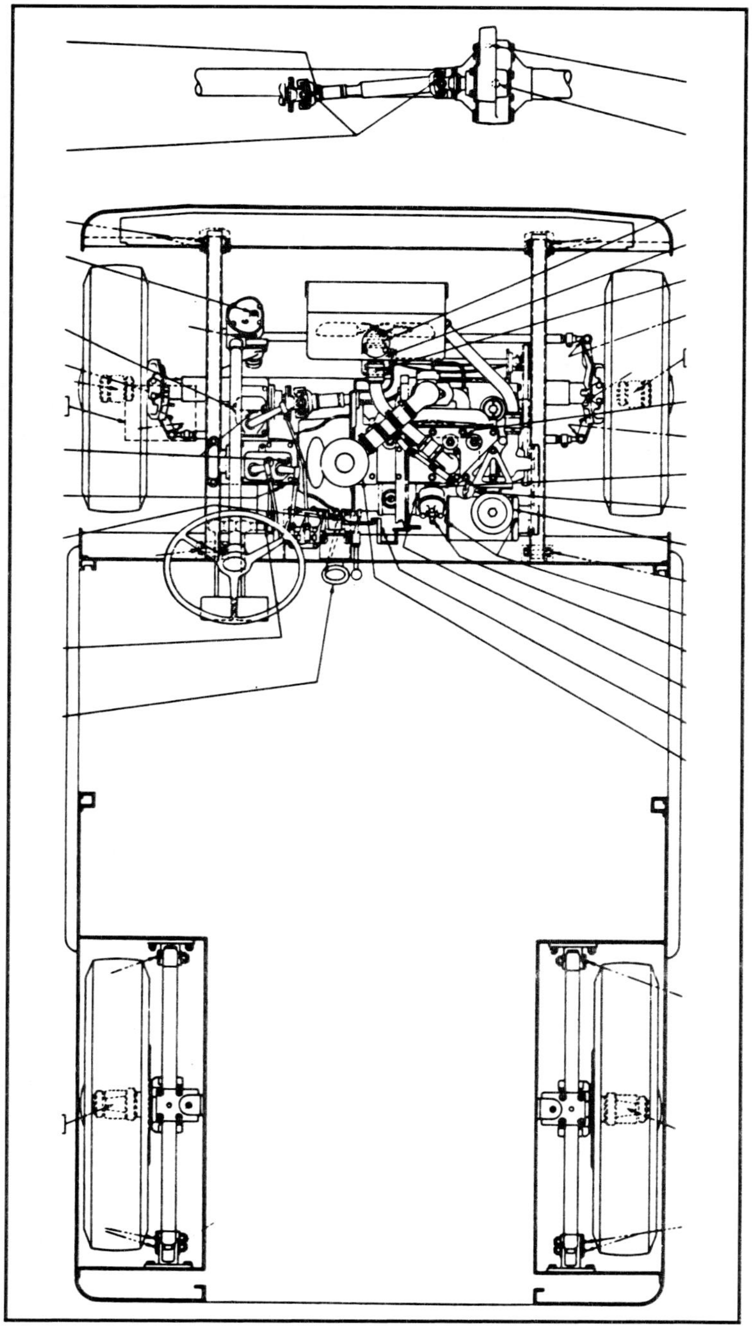

Fig. 4-9 DELIVR-ALL vehicle plan.

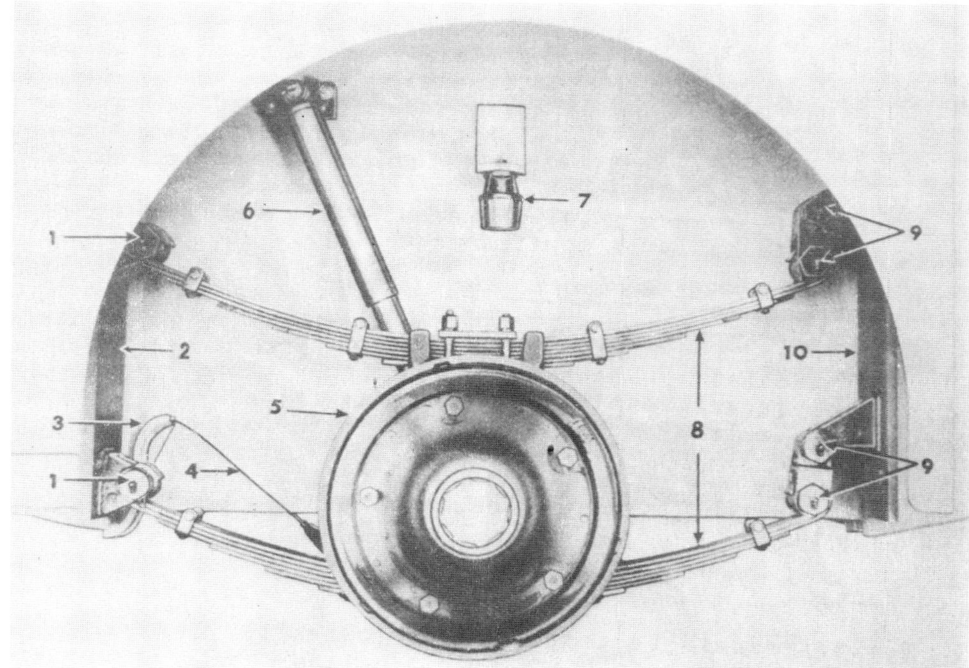


Fig. 4-10 The compact rear suspension responsible for the low cargo floor.

Fig. 4-11 Severed DELIVR-ALL with handle in place on Power Unit.

trols, the complete factory directions for module separation are repeated here as stated in the factory Maintenance Manual:

(1) Disconnect wiper hose at lower left of windshield.
(2) Disconnect wiring harness at instrument cluster.
(3) Push Dolly DVL-1-8516 under power unit and tighten clamp over bumper.
(4) Place a jack under either forward side of the body floor.
(5) Disconnect self-seal brake connection at hole in lower frame cross beam.
(6) Remove four attaching bolts in lower cross beam, two near each end.
(7) Remove 6 nuts, washers clamps — one at each upper corner of unit frame and two at upper front center of frame.
(8) Jack body up three or four inches until unit is loosened from body. Then pull unit forward far enough for body floor to clear pedals. Jack body high enough to permit steering wheel to clear windshield.

## POWER UNIT — POWER TRAIN

The power train concepted and developed for the Canadian Tracked Jeep (TJ-1) incorporated all the basic elements of the packaged transverse-engine/front wheel drive power train of the DELIVR-ALL truck.

However, the DELIVR-ALL power train system required an extension of the prop shaft length to accommodate the relative motion of the sprung axle on the truck. This was accomplished by a severe offset of the differential. Vehicle length constraints also dictated folding the axle beneath the engine, from the beside-engine position of the TJ-1 which was appropriate to the height of the track drive-sprokets.

The power unit used a Willys Jeep engine (134CID, 69bhp) transversely mounted with output to the vehicle left. Drive was through the clutch to a special Marmon-Herrington transmission unit with output parallel to input, in the opposite direction (to the right). In typical M-H fashion, the transmission was derived from a commercial unit both in interest of low initial cost and service logistics. The basic housing was extended on the side of the case to house an auxiliary (splitter) gear set; the normal output shaft extension mounted only the parking brake drum. Including the reverse idler, there were no less than five parallel shafts in the transmission.

An additional helical gear was splined to the modified mainshaft requiring an extended length case, which engaged a two-gear cluster on a second layshaft (in addition to the normal transmission layshaft or countershaft). This cluster gear included a helical (input/output) and spur (output) gear which selectively meshed with corresponding gears on the output shaft.

The high-low range selector lever was located adjacent to the gear shift lever and permitted the operator to select a spur-geared low ratio of 1.97:1 for door-to-door operations, or a 'high' ratio of 1:1 for travel to and from the delivery route. The lower overall ratio of 37.28 to 1, versus 18.92:1, made clutch engagements easier and cut in half the minimum speed with clutch fully engaged. Transmission ratios were typical automotive values of 2.665:1, 1.564:1, and 1:1.

The axle was a special Marmon-Herrington design with helical double reduction of 7.10 to 1. The differential was offset severely to the right, directly adjacent the right spring. Input was above, forward and parallel to the axle shafts and from the left of the vehicle.

Front tread was 66 inches (Rear was 64.5) steering was Ross cam and lever with an 18 inch wheel. Turning radius for the three models was 24.26 and 30 feet. Main and auxiliary shift levers, clutch and brake pedals, accelerator pedal (when in use for seated driver mode), folding seat and parking brake lever all extended from the rear face of the power unit.

Fig. 4-12 DELIVR-ALL power train viewed from below.

A motorcycle seat was standard. An option provided a swivel mounted bucket seat with vertical adjustment of 2.75 inches by means of a lock screw near top of seat pedestal. Cushion and back were integral and could be tilted forward or back and were held in place by wing nuts and spring tension.

The instrument cluster and steering column extended from the top surface of the power unit. The hand throttle was on the instrument cluster.

Rotation of the gear shift knob provided hand throttle control for the driver standing mode. The (optional) accelerator pedal motion was horizontal to safeguard against accidental actuation by a standing driver. Means were provided for locking the clutch and brake pedals together for one-foot actuation.

The top and face of the power unit enclosure incorporated large primary and small secondary (frequent) access panels attached with retained aircraft type half-turn fasteners. Both the TJ-1 and the DELVR-ALL required the radiator to be at the side of the engine with the consequent requirement for an "around the corner" drive.

Over the production life of the DELIVR-ALL, two different drive systems were used to propel the engine cooling fan. With the transverse engine position, because the radiator remains in the normal attitude, it is necessary to drive a fan at 90 degrees to the engine axis. This is universally accomplished today by electric fan drive.

On vehicles DVL-1-4700 to DVL-1-5015 the fan was gear driven, mounted on a fan drive shaft and gear case bracket, on the manifold (front) side of the engine. A gear case at the fan end of the bracket housed the gears, which were driven from a pulley on the opposite end of the drive shaft. A belt from the crankshaft pulley operated the fan drive pulley. A spring loaded idler kept constant tension on the belt, eliminating adjustment.

On vehicles after serial D VL-1-5015, the fan was belt driven without the intermediate gear case. The fan was mounted on the manifold side of the engine using a slotted bracket for adjustment of the continuous belt. The fan bracket incorporated both the fan drive pulley and an idler pulley in an oblique position to turn the belt, in the same manner that the same objective wsa accomplished in the drive of the Corvair direct-cooling fan some fourteen years later.

The constant velocity universal joints used in the DELIVR-ALL had a center ball charcterized by a bore for a spring-pin and a groove. Disassembly was accomplished by driving the spring-pin into the hollow output shaft, whereupon the joint separated sufficiently for fingers to rotate the center ball 90 degrees to an offset position (by virtue of the groove) which allowed the adjacent ball (s) to pass and come out of the assembly when the joint was in an angular attitude.

In the interest of extending the component historical developments related in the Story of the Marmon, the engine, steering system, clutch and brakes of the DELIVR-ALL are briefly addressed in that sequence because the vehicle was of passenger car size and the components, all of which were vendor designed, were representative of the state-of-the-art or contemporary practice.

## ENGINE

The engine was the same Willys passenger car engine which had been used in the military standard Jeep vehicle as well as in the Marmon-Herrington TJ-1 development vehicle for Canada. It was a four cylinder engine Model CJ-3A with bore and stroke of 3.125 and 4.375 inches; it displaced 134.2 cubic inches, had an SAE rating of 15.63 horsepower and produced 60 brake horsepower at 4000 rpm with a compression ratio of 6.48:1. That was 0.447 horsepower per cubic inch.

It was a typical engine of the 1940s in being a side-valve or 'L-head' type and in having two crank throws between main bearings. The crankshaft had four counterweights. It was equally typical in having flat-faced cam followers offset from the lobe centerline to encourage valve rotation during seating.

Intake valves opened 9 degrees before top center and closed 50 degrees after bottom center for a 239 degree duration. Exhaust valves opened 47 degrees before bottom center and closed 12 degrees after top center for a 239 degree duration and 21 degree overlap. Both had a lift of 0.36 inches and a hot clearance of 0.014 inch.

Pistons were aluminum (Lynite alloy) with a T-slotted, oval (cam) ground skirt and a heat dam configuration. The two compression rings were 0.094 wide; the single oil ring was 0.1875 wide. The upper compression ring was installed with an inside beveled edge up. The lower had a 0.005 tapered face. Piston length was 3.75 inches; connecting rod length was 9.1875 inches with the pin clamped in the rod.

All engine bearings were steel-backed and babbitt lined. Main journals were 2.334 diameter by 1.92 front (thrust), 1.825 center and 1.75 rear. Cam journals were 2.188 front, 2.25, 2.188 and 1.75 rear diameter. Thrust control was by plunger/spring loading against a thrust washer. Drive was by silent-chain from the crankshaft.

The oil pump was a typical with a planetary gear pumping system. The location however was universal practice on line engines with the pump external to the engine but internally plumbed, coupled to an inclined distributor shaft skew-gear driven from the camshaft.

A single throat downdraft Carter carbureator of metering-rod type with manual choke was fitted as was a velocity type overspeed governor beneath the carburetor.

Ignition advance was by flyweights only. It was and is common to omit vacuum-diaphragm secondary control on commercial engine applications.

Willys engine Model MB was substituted beginning in 1951.

## STEERING SYSTEM

A Ross cam and lever steering gear was used. Subsequent to the Marmon Sixteen period, Ross introduced a twin-pin lever design which allowed a greater angular travel of the pitman arm; as one pin was disengaging, the second was engaging the cam thereby maintaining engagement. The pins were tapered and operated in a groove correspondingly shaped. At the mid-steering position of each pin, the groove was narrower to minimize end-play in the usual driving position, while the extra clearance in the infrequently used position range could be used to readjust for wear compensation in the mean position.

The Marles type gear made by Gemmer and earlier used on the Marmon Series 75 was likewise improved by the adoption of a two-toothed roller in the 1930s.

The wheel alignment specification is of particular interest because of the front drive application. The king pins were inclined 9 degrees, 5 minutes and tilted 2 degrees at the spring pads. Caster was set at 2 degrees, camber at 0 degrees 50 minutes and toe-in was in the range 0-0.125 inch.

## CLUTCH

A Borg & Beck push-lever type clutch with multiple direct-acting springs was used. It was attached to a flat flywheel using body-fit bolts for location. More powerful engines with single-plate clutches required stiffer clutch springs; pedal forces would have increased proportionately had it not been for the introduction of anti-friction devices in the release lever system. The Borg & Beck solution was a round pivot pin in line contact with a D-hole in the lever, combined with line contact between a strut link, the lever tip and the lug on the pressure plane.

Concurrently the Long (brand) clutch used variable-rate counterweighted levers, a pin in a D-hole, and needle bearings on the pin in the pressure plate.

## BRAKES

Lockheed hydraulic brakes were used. Symmetrically opposed and individually anchored shoes canceled any inherent servo action. Two types of this brake were used on the DELIVR-ALL: the front wheel brakes had shoes floating radially on fixed anchors; the rear had adjustable (by rotation) anchors. All shoes were brought close to the drum by eccentric adjustment cams located near the shoe ends. Rear shoes were cable applied for parking. A self-sealing connection allowed disconnection of the rear brake hydraulic line for separation of the vehicle units. The Lockheed brake was at the peak of its popularity, being used on Ford, Nash and Chrysler product lines. Concurrently the Bendix type was on the senior GM cars and the Hudson lines.

Drums were 12x1.75 front and 14x2 rear; area of lining was 202 sq. inches.

The Interim Report to the Stockholders for the first quarter of calendar 1946 noted the acceptance of the DELIVR-ALL light truck by the trade as "unusual" and orders on hand for approximately 2000 units and evidence of demand increasing daily.

Although the production line for the manufacture of the DELIVR-ALL had been

## SPECIFICATIONS

**ENGINE:**
4 cylinder $3\frac{1}{8}$" x $4\frac{3}{8}$", L. Head, 60 H.P., 134.2 cu. in., 105 ft. lbs. torque at 2000 R.P.M.
Carburetor: Downdraft.
Oil Filter.
Air Cleaner: Oil bath.
Crankcase Capacity: 4 quarts.
Water Pump: Centrifugal type.
Piston: Cast iron. Three rings above wrist pin.
Connecting Rod Bearings: Steel back, babbit lined, (replaceable) size 2.334.
Main Bearings: Front 2.334" x 1.920", center 2.334" x 1.8125", rear 2.334" x 1.75".
Governor: (Approx. 38 miles per hour).
Compression Ratio: 6.48 to 1.
Gasoline Consumption: Idling speed, 4 hr. 20 min. per gal.

**RADIATOR:** Tube and Fin. Capacity 17 quarts.

**CLUTCH:** $9\frac{1}{4}$" disc.

**TRANSMISSION:** Three speed in each of two ranges. (6 speed forward, 2 reverse.)

| *High Range* | *Low Range* |
|---|---|
| 3rd: 1.00 to 1 | 3rd: 1.97 to 1 |
| 2nd: 1.564 to 1 | 2nd: 3.08 to 1 |
| 1st: 2.665 to 1 | 1st: 5.25 to 1 |
| Reverse: 3.554 to 1 | Reverse: 7.00 to 1 |

**AXLE:** Special Marmon-Herrington front driving axle with constant velocity universal joint steering ends. Driven from the engine through double reduction helical gear set. Reduction: 7.10 to 1. Total maximum reduction 37.31 to 1.

**REAR SUSPENSION:** Rear wheels are on independently sprung suspensions which do not obstruct the lowest possible floor. Complete suspension assemblies easily removed. Two semi-elliptic springs on each wheel. Four shackles on each suspension.

**SPRINGS:**
Front: $40\frac{1}{8}$" long, $1\frac{3}{4}$" wide. Number of leaves: 8 each.
Rear: 36 15/16" long, $1\frac{3}{4}$" wide. Number of leaves: 7 each. (2 rear springs on each side.)

**SHOCK ABSORBER:** Airplane type, front.

**TREAD:** Front 66 in., rear $64\frac{1}{2}$".

**OVERALL WIDTH:** Outside 78".

**OVERALL HEIGHT:** $92\frac{7}{8}$".

**FLOOR HEIGHT:** 14" approx.

**BRAKES:** 4-wheel hydraulic 12" x $1\frac{3}{4}$" dia. front. 14" x 2" dia. rear. Lining area 202 sq. in. Drum area 327 sq. in. Hand brake: mechanical on the rear wheels only.

**STEERING:** Cam and lever. 18" steering wheel, easy steering, comfortably located for standing or seated driving position.

**CONTROLS:** All controls arranged for seated or standing driving position. Hand throttle in combination with gear shift lever.

Clutch and brake pedal controls arranged for independent operation or so that both pedals can be operated with one foot.

High-low range gear shift convenient to transmission shift lever.

Low range for frequent stop operation. High range for driving to and from the route.

**POWER UNIT:** Completely removable as a unit assembly. May be disconnected from body and service unit installed very easily as all essential parts are integral with power unit. Emergency brake stays with body. Self sealing brake connection avoids necessity of bleeding brakes when replacing unit.

**FRAME:** No actual frame is employed as body panels and top form the structural members. This provides exceptional low continuous floor height from the driving controls to the rear of the body. Floor and body panels are made of metal covered plywood reinforced with welded steel channel iron. Top all steel.

**EQUIPMENT:** Instrument panel containing speedometer, ammeter, fuel gauge, heat indicator, oil pressure gauge, choke, throttle, starter button, light switches. Windshield wiper. One folding seat. Stop light, headlamps, tail lamp, instrument light, body light, front and rear bumpers. Easy access panels are provided for service operations on power unit. Wide folding side doors; solid panel in back.

**ELECTRICAL:** Battery 100 AMP, 6 volts, generator and voltage regulator. Special heavy-duty generator. Starting motor, distributor and coil.

### MODEL VARIATIONS

#### DVL-1

**WHEELBASE:** 90 inches.

**LOADING CAPACITY:** 182 cu. ft. usable volume behind front door.

**TURNING RADIUS:** 24 feet.

**WEIGHTS:**
Front power unit with 7.50 x 16—8 ply tires.....1900 lbs.
Body complete with 7.50 x 16—8 ply tires*......2150 lbs.
Total empty weight with above tires.............4050 lbs.
Maximum gross vehicle weight with above tires..6750 lbs.

#### DVL-1A

**WHEELBASE:** 105 inches.

**LOADING CAPACITY:** 284 cu. ft. usable volume behind front door.

**TURNING RADIUS:** 26 feet.

**WEIGHTS:**
Front power unit with 7.50 x 16—8 ply tires.....1900 lbs.
Body complete with 7.50 x 16—8 ply tires*......2400 lbs.
Total empty weight with above tires.............4300 lbs.
Maximum gross vehicle weight with above tires..6750 lbs.

#### DVL-1B

**WHEELBASE:** 118 inches.

**LOADING CAPACITY:** 344 cu. ft. usable volume behind front door.

**TURNING RADIUS:** 30 feet.

**WEIGHTS:**
Front power unit with 7.50 x 16—8 ply tires.....1900 lbs.
Body complete with 7.50 x 16—8 ply tires*......2600 lbs.
Total empty weight with above tires.............4500 lbs.
Maximum gross vehicle weight with above tires..6750 lbs.

*Substituting 8.25 x 18—10 ply tires in the rear adds 150 lbs. to the weight of the truck and adds 1450 lbs. to the gross allowable vehicle weight, making a total G.V.W. of 8200 lbs.

**INDIANAPOLIS 7, INDIANA**

Fig. 4-13 DELIV-R-Al Specifications.

completed as of mid-May 1946, the company had been unable to acquire an inventory of parts sufficient to assure production because of strikes and material shortages in supplier plants.

The beginning serial number is believed to be DVL-1-4700. Presumably the delayed start of production made it a 1946½ model designated a 1947 model (ala the 1970 Camaro and 1980 GM 'X' cars), and the model year was reflected in the serial number. After the first one hundred units, the serial number digits had no model year relationship. Product on continued through 1955.

Success of the DELIVR-ALL was restricted by the lack of a sufficiently large dealer organization. The type of customer required local sales, service and parts coverage. In the case of the Ford conversions, the Ford dealer organization provided that requirement. Former owners of small DELIVR-ALL fleets encountered by the author complained of excess ve down time awaiting parts delivery.

## PART 3 — The Trolley Bus 1946-1956

The entry of Marmon-Herrington into the trolley bus market was a classic example of successful entrepreneuring. Starting from recognition by its Board of Directors of a need to utilize expanded (wartime) facilities and perception of a potential market, an engineer with a record of innovation and extensive transit vehicle experience was recruited. The resultant new product (1) reflected state-of-the-art technology which had not previously been applied to the product category and (2) exhibited and demonstrated technical and manufacturing superiority which made Marmon-Herrington the world's largest manufacturer of trolley buses within a three year period, and a multi-international licensor within a decade.

Art Herrington had met Charles Guernsey when both were in the US Army and he now brought him into his company to head the trolley bus venture. An April 3, 1948 News Release announced that Guernsey had been elected a Vice-President of Marmon-Herrington and made Manager of the new Transit System Division.

Guernsey had been Vice-President of the J.G. Brill Company, a transit firm which he had served for 21 years. Since 1936 he had directed Engineering, Sales and Service departments but had resigned from Brill in September 1947. During the war Guernsey had been on leave to serve as Technical Director and later Chief of Production of the Philadelphia Ordnance District. Significantly, he attracted to Marmon-Herrington associates at Brill who were skilled in electric power transmission and application.

The first layout and general assembly drawing was completed in April 1945, and the first prototype was assembled in the fall of 1946.

Guernsey directed the design of a vehicle structure which weighed 4000 pounds less than a 44 passenger contemporary bus — certainly in the best Marmon tradition! The integrity of the completed structure was confirmed by deflection; after being "stress-coated" (painted under closely controlled conditions with a special brittle lacquer finish). This was the largest exercise of this technique yet attempted as the entire body including floor pan and underbody were involved. Tests were conducted at 2 AM during summer weather to accommodate the stress-coat product requirements. Only one area of the coach, as indicated by a crack pattern in the lacquer subsequent to deflection on a severe bump, required a design modification. No field structural problems were encountered, despite ultimate fleet use in New York City, which is considered the most severe test for a transit vehicle. In 1979 negotiations were underway for the rebuilding and updating of a Marmon-Herrington fleet in use in Mexico City for decades and still exhibiting viable, servicable structures.

The weight-reduction was achieved by use of a two-ply skin construction with stringer channels between them. Bob Wallace does not recall any effort to acquire patent coverage on this construction. He states that the basic frame structure drooped about ten inches at the rear under the weight of the electric traction motor, but came into alignment when joined to the monocoque riveted double-skinned body.

A Marmon-Herrington advertisement described this vehicle as *a new concept in trolley coach design and construction — a big, light weight coach that would carry more passengers, faster, more comfortably and at lower cost than any transit vehicle in history.* Another modestly stated: *"the new Trolley Coaches are true symphonies in streamlined styling, comfort and efficiency. Beautiful, rapid, odorless, they embody every conceivable improvement in design and construction. They are the very last word in inexpensive city transportation."*

The initial orders came from Indianapolis Railways, Inc. and the Columbus and Southern Ohio Edison Co.

The *Indianapolis News* for July 23, 1949 reported that in 1948 Marmon-Herrington outsold all other trolley manufacturers with sales three times that of the nearest competitor. 202 orders were received in 1948 and 103 more to the date of reporting. The 'News' reported that the light weight postwar design — 3000 to 5000 pounds less than competitors and capacity for four to eight more passengers — resulted in srong competition for M-H against the pre-war designs of competitors.

Three years from introduction, Marmon-Herrington was the largest producer of trolley coaches in the world. In that short period, one third of the trolley coaches operating in the United States were M-H built.

Over the first nine years, Marmon-Herrington trolley coaches were purchased by 19 major properties and 1500 were in operation and had logged 36 million miles.

A M-H advertisement in October 1950 noted that one half of all the trolley coaches sold by M-H since their 1946 introduction were the 48 passenger model. In that month, Marmon-Herrington received the largest order ever issued by any transit operating company in the U.S. — $5,979.417 for 349 trolley coaches for Chicago with a special configuration to accommodate 49 passengers despite wider front and rear doors than used on the 48 passenger models. The coach was lengthened six inches. The order specified General Electric apparatus and Westinghouse air brakes. In November 1950, M-H reported that 1420 trolley buses had been sold since introduction of the product line.

In February 1952, a Marmon-Herrington trolley bus was shipped to Europe for exhibition at the Brussels, Belgium Auto Show. The unit was borrowed from the Cleveland Transportation System and was subsequently to be used as a pilot model for initiation of European manufacture.

Following a newsbreak by the *New York Times* of Feb. 28, 1952 the announcement came in April that Ateliers Metallurgiques of Nivelles, Belgium had been licensed to produce Marmon-Herrington trolley buses. The firm Ateliers de Construction Electriques de Charleroi, Belgium's largest manufacturer of electric equipment was licensed to manufacture the traction motors and controls by Westinghouse Electric International Company. These were reported to be the first American style trolley buses to be made in Europe.

One large property reported cost of operation to be 3.25 cents less per mile than for motorbuses of the same age and capacity, and 4.85 cents less per mile than diesel buses of same age and capacity. Seven of the properties had made from one to four repeat orders.

Domestic properties operating M-H trolley coaches in 1954 were:

| | |
|---|---|
| Cincinnatti, Cleveland, Columbus and Dayton, Ohio | |
| Philadelphia, PA | Kansas City, MO |
| Newport, KY | Little Rock, AK |
| Indianapolis, IN | New Orleans and Shreveport, LA |
| Milwaukee, WI | Dallas, TX |
| Chicago, IL (20) | San Francisco, CA (40 pass.) |

Foreign Properties were:

| | |
|---|---|
| Toronto, Canada | Mexico City, Mexico |

A total of 1513 units were produced for domestic transit companies according to "Transit's Stepchild: The Trolley Coach."[5] That tabulation does not account for the Mexico City and Toronto, Canada fleets nor Brazilian and Belgian production by licenses. One estimate places total worldwide production near 2500.

Fig. 4-14 The Marmon-Herrington Trolley Coach introduced in 1948.

Conspicuously absent are New York City and Detroit, Michigan who had acquired fleets of Marmon-Herrington trolley coaches in the late 1940s.

*The New York Times* for May 4, 1946 reported Marmon-Herrington to be the low bidder for 200 electric trackless trollies for replacement of antiquated street cars on Brooklyn lines. The M-H bid was $16,652.50 per coach with delivery at the rate of forty coaches per month beginning one year from the date of order. Competitive bidders were Pullman Standard Car Manufacturing Co. and Twin Coach Company of Kent, Ohio.

Six years later, when the *Times* reported the upcoming licensing of Belgian firms for the manufacture of M-H trolley coaches, the reporter commented *Old Hat in Brooklyn — Residents of the more remote fastnesses of Brooklyn have a nodding acquaintance with these conveyances.*

While the reason for the disappearance of the Brooklyn trolley bus fleet is not known to the author, it may very well have been the same as that for the disappearance of the trolley — both tracked and trackless — from the streets of the Motor City.

Following the return to peacetime pursuits in 1945, the Detroit Department of Street Railways (DSR) undertook an ambitious program to refurbish — i.e., re equip — the system that had carried a heavy passenger burden throughout the war years, primarily in Ford buses of about 27 (seated) passenger capacity.

A large fleet of PCC (President's Conference Committee) street cars were procured to replace the aged Peter Witt center entrance/exit cars and the antique rolling stock, wood stoves et al, acquired with purchase of the Detroit United Railway (DUR) in the 1920's.

The DSR had had earlier experience with electric trackless trolleys. A fleet of Twin Coach trolleys had operated on Plymouth Road from Grand River to Rouge Park in the 1930s. Postwar, the idea again looked attractive and Detroit was among the early customers for Marmon-Herrington trolley coaches. The Warren Line north of downtown was converted to trackless operation, and conversion of the arterial Grand River line followed.

Something that had been overlooked by the Detroit Street Railway planners, and which undoubtedly eventually ended trolley bus production in Indianapolis and elsewhere in the USA, was the advent of the arterial express highway. Prewar, a four mile depressed dual two lane "Davison Expressway" was constructed to resolve traffic congestion in Highland Park, MI, an island community encompassed by Detroit; it had no intermediate entrances nor exits and replaced an existing surface street. Housing was demolished to make way for the new surface service roads.

The postwar construction of the John Lodge and Edsel Ford Expressways disrupted all main avenue traffic. The former ran north-south and the latter east-west and together they bisected probably every transit line in the city.

The cost of catenary relocation for trolley bus detours was prohibitive; frequently extensive rerouting was necessary because residential streets could not accommodate the large trolley buses. The cost of catenary relocation and temporary tracklaying for street cars was astronomical (by 1950's standards); the degree of rerouting and disruption of automotive traffic and trolley-boarding safety islands further compounded the problem.

The Detroit trolley fleets were consequently retired in their youth and were subsequently acquired by Mexico City, Mexico. As the construction, or planned construction of arterial expressways spread to smaller cities, the market for trolley buses declined rapidly. Unfortunately for the trolley bus industry, and perhaps for the nation, the new expressways were almost universally designed without provision for mass transit. The pioneer Edsel Ford Expressway was built with bus-loading lanes where it passes beneath Woodward — Detroit's "Main Street"; they have never been used in their thirty years existence. The Kennedy Expressway in Chicago was planned with provision for rapid-transit in the median and is presently having the rail line therein extended to O'Hare Airport.

A long term royalty agreement was negotiated under which Troleibus Villares, S.A. of Sao Paulo, Brazil, manufactured trolley coaches of Marmon-Herrington design for the Brazilian market, including electric motors and control equipment. The New York Times for March 14, 1956 reported that the agreement was between a partnership of three Brazilian firms and three American firms: Westinghouse International Electric, Marmon-Herrington and Ohio Brass. The *TIMES* printed a picture of President Knox of Westinghouse and President Luis Villares of Trolleibus Villares S.A. of Sao Paulo, signing the agreement documents, and a picture of a Marmon-Herrington trolley coach. (New York Times, March 14, 1956) Mr. Villares was also president of one of the participating Brazilian concerns, Elevadores Atlas, the Westinghouse licensee. Senor Villares was quoted by the *TIMES* as saying that the transaction would save Brazil $35 million in foreign exchange over five years and most importantly would save gasoline which was critically short in Brazil. Only about $500,000 (Brazilian) additional capital investment was to be required. The Industrialist was further quoted as noting that existing transit systems were handicapped by a shortage of gasoline and a spare parts problem associated with foreign exchange constraints.

Production was to begin, within one year of the agreement, at a rate of 20 buses a month. Capacity was to be on the basis of a minimum market of 1500 buses within five years. The Marmon-Herrington licensee was Grassi, S.A., a long established coach and car manufacturer. Companhia Paulista de Material Electrico was licensed by Ohio Brass Co. to manufacture overhead equipment.

Subsequently the cities of Belo Horizonte and Recife, Brazil were added to the roster of cities operating Marmon-Herrington trolley buses. Eight others were considered prospects.

In 1970 the Lowell Transport Co. of Chicago was purchased and physically combined with Marmon-Herrington. Lowell marketed trolley wire, truck and bus parts and used trolley coaches on an international basis.

Subsequent to cessation of trolley coach production, Marmon-Herrington rebuilt 124 units to original configuration for the City of Guadalahara, Mexico.

In 1975 Slim Swinford arranged the purchase of a former San Francisco M-H trolley coach. The coach had been purchased by a collector and was stored in the barns of the San Francisco Muni. It was purchased for $300 in 1969; M-H paid $750 but it cost $2700 to transport it to the Marmon Transmotive Division/Marmon-Herrington Operations in Knoxville, TN. That coach was reengineered and modified with windows twice the size of the original, and fitted with flourescent lights, and new seats. It formed the basis of a proposal to remanufacture the entire M-H fleet in operation in Mexico City; negotiations were still active in 1979.

In "Bus Industry" for Aug. 1977, the Bill Johnson column "United States Report" stated: *Marmon-Herrington, a/k/a Marmon-Transmotive of Knoxville, is still active in trolley coach projects. Marmon is hoping for a sizeable order of new trolley coaches from Mexico City in addition to working out service contracts for existing trolley coaches. Work on Marmon's mock-up of a new-look coach was to be continued in August. The coach being used for mock-up purposes was purchased from Harold Perazzo of San Francisco. It was a fully restored 40 passenger coach number 541, originally built in 1948 for San Francisco Muni.*

## PART 4 — Busses — Transit and School

Marmon-Herrington had been in the transit business for two years building exclusively trolley buses with noteworthy success. They had also had a remarkably fine relationship with the Ford Motor Co. associated with the conversion of Ford trucks and cars to All-Wheel-Drive, and the wartime application of Ford components to specialized vehicles.

The Ford Motor Co. had manufactured transit bus chassis and marketed complete buses beginning in 1937 with a forward engine and in 1940 with a rear transverse engine. These chassis were driven to Union City, IN, where they were fitted with Union City bodies to accommodate 25 passengers. (In that same era, a Gar Wood division built an inter-city bus designed by William (Bill) Stout with a longitudinal rear Ford V-8 engine and tubular structure). Production of the Ford chassis and the "Universal Bus" continued throughout the war.

For the post war market, Ford paralleled the M-H entry into the trolley bus market with an entirely new transit bus chassis with transverse rear engine.

The new model was announced March 30, 1948; production began in May 1948. The new line was designated the Ford Metropolitan Coach 8MB — 27 to 29 passengers and the 8M2B — 31 to 35 passengers. The buses were distributed to 12 regional distributorships by Metropolitan Motor Coaches, Inc. whose address was 14300 Woodward Ave., Highland Park, MI — the old Ford Highland Park plant. Prices were in the $9-10000 range, and some 600 units had been produced when two developments changed the direction of the Ford bus program. Henry Ford died in 1948 and Henry II became president of Ford Motor Co. which was in serious financial circumstances.

Henry Ford II undertook massive and successful changes to lead the company back to profitability. Among these were abandonment of small operations in the Village Industries chain of plants built on sites, in southeastern Michigan, having water power available. (The Northville valve plant did survive even the domestic automotive depression of 1980). The transit bus operation was inappropriately small for such a large corporation and, in addition, the new models in the field were experiencing serious structural failures aft of the rear axle. Also, the planned production rate of 20 units per day had not been achieved.

In 'Ford trucks Since 1905,' James K. Wagner records that the merchandising of Ford transit buses was initially by the Bus Department of the Ford Motor Company but was transferred to an independent firm — Transit Buses, Inc. — in April 1941 and again to Metropolitan Motor Coaches, Inc. on October 6, 1947. He further comments that the latter firm was unable to sell buses at the rates achieved by the predecessor firm. This problem was compounded by delayed start-up due to a strike at the Wayne Works which had replaced Union City Body Co. as the body fabricator and second stage vehicle assembler effective with start of 8MB production.

## MARMON-HERRINGTON (Ford) TRANSIT BUS

Henry Ford II wanted 'out' of the transit bus business so that he could concentrate his efforts on the major problems. Early in 1950, Charles Guernsey, Vice-President of Marmon-Herrington in charge of the Transit Division and Seth Klein, Vice-President in charge of production, met with representatives of Henry Ford II in Dearborn, MI to negotiate a sale.

The consummation of a deal for an undisclosed sum (Ford was not yet a public company) was announced in the *Indianapolis News* on March 23, 1950. Marmon-Herrington officials were reported to have said that Ford had $2 million invested in tools and engineering. No changes to the design were planned except for the name change from Ford Metropolitan to Marmon-Herrington.

Roger Budrow, Business Editor of the *News* continued: *From now on the city transit buses which Ford made at Dearborn, Michigan will be manufactured in Marmon-Herrington's factory at West Washington and Harding Streets. Marmon-Herrington's president, David M. Klausmeyer, said he hoped production will reach a total of 500 buses a year. At first, however, the chassis will be built elsewhere and the bodies will be constructed at the Wayne Works in Richmond. The company already has a backlog of 200 orders for buses," Klausmeyer said. Eventually employment will be increased at the Indianapolis plant. Klausmeyer said the number of new jobs depends on whether the entire bus will be manufactured here or some parts subcontracted to plants in other cities. Deliveries will start from the Indianapolis plant by April. Asked about Ford's decision to sell its bus business, a Marmon-Herrington official said: 'They shouldn't have been in it in the first place — they're too big. The business is just the right size for us.* Charles Guernsey said: *the bus is ideal for feeder lines, crosstown service and branch lines in larger cities, and for main line service in smaller cities such as Kalamazoo, Kokomo or South Bend. The Ford bus is used by Indianapolis Transit Co. on its Indianapolis-Danville Line.*

The 8 MB chassis had a 146 inch wheelbase; the 8 M2B had 186. Both had frame rails which were 6.5 inches deep with 2.44 flanges of .21 stock. Springs were semi-elliptic, 54 x 3 front and 58 x 3 rear. Drive from the engine was through a 12″ Long clutch, Ford 3-speed transmission, and a 1.07:1 transfer angle gear set to a 5.83:1 axle reduction. Brakes were Hydrovac operated and featured aluminum brake shoes.

The engine was an exclusive bus model, being an overbored version of the Ford Model 7H six cylinder, with 254 CID rated 104 hp., installed transversely and inclined at a 30° angle to the rear.

Marmon-Herrington took over the furnishing of accessories and replacement parts for the 12,000 buses which Ford Motor estimated to be in use. Like Ford's experience, M-H would not achieve the expected volumes. Some 1500 units would be built in Indianapolis. Additional units were manufactured by Welles-Marmon-Herrington in Canada.

Marmon-Herrington engineers redesigned the bus chassis to reinforce it aft of the rear axle and thereby correct the body flexure which they determined to be responsible for the many in-use failures of the body structure. It is probable that the increased wheelbase of the 8M buses over their predecessors was at the root of the problem. The 8 MB was five inches and the 8M2B eighteen inches over the prior wheelbases of 141 and 168 inches.

Ford Motor Co. sponsored a $250,000 rework/retrofit program by Marmon-Herrington which was accomplished in a leased facility — the old Hicks School-Bus plant equipped as a modification center.

The 8 MB buses marketed in Canada carried a W/MH logo on the nose reflecting the Welles/Marmon-Herrington license arrangement with the Windsor, Ontario firm of Welles Corporation, Limited. Marmon-Herrington 8MBs were in service from Maine to Washington and California, from Alaska and Quebec to Texas and as far away as Australia and Columbia.

## MOTOR COACH CUSTOMERS OF MARMON-HERRINGTON

Dallas Railway & Terminal
Dallas, Texas

Augusta Coach Company
Augusta, Georgia

Cheyenne Motor Bus Co.
Cheyenne, Wyoming

Topeka Transport Company
Topeka, Kansas

San Antonio Transit Co.
San Antonio, Texas

Southern Cities Transit
Los Angeles, California

Worcester St. Railway Co.
Worcester, Massachusetts

Sydney Bus Company, Ltd.
Sydney, Australia

Springfield City Lines
Springfield, Ohio

Hudson Bus Lines
Medford, Massachusetts

City of San Angelo
San Angelo, Texas

Northern Trans. Co.
Fargo, North Dakota

City of St. Petersburg
St. Petersburg, Florida

Transit Investment Company
Columbia, Missouri

Neucus Transport Company
Corpus Christi, Texas

City of Alexandria
Alexandria, Louisiana

Laredo Transport Company
Laredo, Texas

Queen City Coach Company
Charlotte, North Carolina

Sunset Stages
Lawndale, California

Juneau Transit System
Juneau, Alaska

Sioux Transit Company
Sioux Falls, South Dakota

Owensboro Rapid Transit
Owensboro, Kentucky

Waterloo Cedar Falls Trans.
Waterloo, Iowa

Eastern Cities Transit
Whittier, California

Greyhound Corporation
Chicago, Illinois

La Cie de Transp.
Boischatel, Quebec

Boston & Maine Transp. Co.
Boston, Massachusetts

Miami Transit Company
Miami, Florida

Omaha & Council Bluffs St.
Railway Omaha, Nebraska

Paducah Bus Company
Paducah, Kentucky

Bellevue City Hospital
New York, New York

Intercity Coach Line Comps
Lansing, Michigan

City of Bogata
Bogata, Columbia

Lorain City Lines
Lorain, Ohio

Overlake Transit
Bellevue, Washington

## PUSHER BUS CHASSIS

Marmon-Herrington further expanded its coverage of the transit field in 1954. A pusher type school bus chassis was introduced which soon became known as the 'Cadillac of the School Bus Industry.'

The initial offering was a Model R225-D-V8. The D-V8 segment of the model designation referred to the Cummins V8-185 diesel engine, which was rated 177 horsepower at 3300 rpm governed speed, with 312 ft. lb. torque at 1750, 283 at governed speed. The 225 segment of the model designation related to the wheelbase; the chassis overall was 401 inches long. The "R" prefix denoted right hand drive.

The engine was longitudinally mounted in the rear. The Allison automatic transmission (6 forward, one reverse) drove forward to the rear axle. The fan was driven by a 90° gear box with the radiator on the off-curb side and belt drive from the engine.

The pusher chassis was later marketed under the "Safe-Way" badge, as "the world's finest and safest transit bus chassis." The 'Safeway' chassis evidenced much redesign. The 42 inch frame width was retained full length except at the front step well.

A lighter series was subsequently offered, with gasoline power and the Allison automatic as an option. Frame rails were 8x2¼ of 7-16 stock instead of 9x3 of 5-16 stock. An 800 pound weight reduction was accomplished. Mr. Richard Duncan, owner of the Jerome-Duncan Ford dealership in Sterling Heights, Michigan recalls the sale of 3 Marmon-Herrington school buses to a southeastern Michigan school district and the dissatisfaction of the customer. These were powered by the 534 CID Ford V-8. Drivers complained of the "heavy" vehicles and the school board of the fuel consumption.

## ONEIDA DIVISION

Marmon-Herrington acquired the Oneida Bus Division of the Henney Corporation located in Canastota, New York, on February 3, 1960. The intent was entry into the school bus market; The Oneida bus bodies would be mounted on the

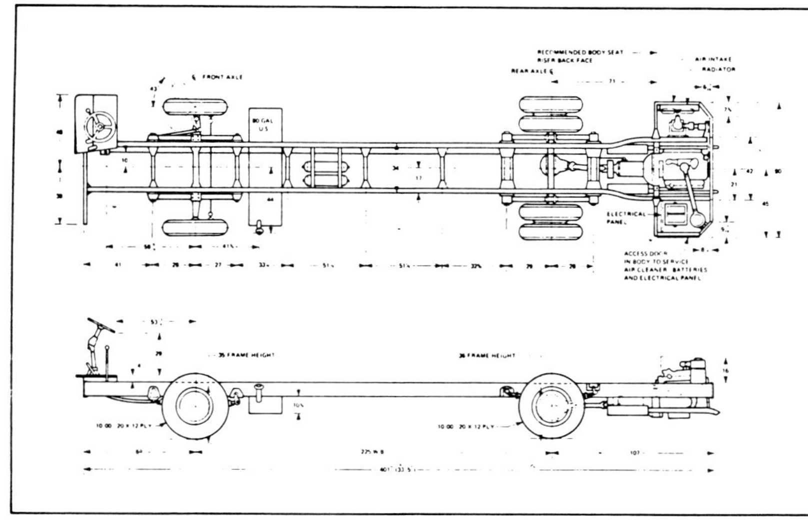

Fig. 4-15 M-H Safe-Way bus chasses

rear-engined Safeway chassis and marketed as a Marmon-Herrington Oneida bus for school or transit applications. The rest of the Oneida product line — bodies for conventional bus chassis would continue to be marketed, under the Oneida name, by the new Oneida Division of Marmon-Herrington Co. Inc. whose address was given as Indianapolis, Indiana.

The chief engineer of M-H's Oneida Division was William De Capua who had previously been chief engineer of a Belgian bus firm. According to Slim Swinford, De Capua was responsible for the undertaking of a major redesign of the Oneida product line which was to have serious consequences. The redesign consisted of incorporation of the Guernsey double-skin body structure into the school bus bodies, similarly to its application in the trolley bus product. The redesign program delayed the start of production at Indianapolis and interrupted the continuity of the traditional Oneida business.

A brochure issued by the Oneida Division in 1961 (#610130) used the "Starline" model designation, but esentially the same brochure appeared subsequently (#610215) with a "Guardian" badge The brochure numbers may be interpreted (1961, Jan. 30) and (1961, Feb. 15). Both carried the message: *"Marmon-Herrington engineers added their thirty years of experience and know-how to the famous Oneida body. Employing the monocoque principle of construction, new Hi-Arch (design) combines greater body strength and rigidity with lighter weight. Riveted skin panels are attached to the beams and cross bars for outer and inner skin stressing. The one-piece construction rub-rail is the widest in the industry."*

Additional insight was provided by the three-column article in the *Indianapolis Star* for February 26, 1961. The Oneida plant was reported to have been enlarged to 300,000 square feet and the most modern equipment to have been installed; the site included 16 acres with adequate chassis storage. The "Starline" was identified as the deluxe version of the base "Guardian" model. The article continued: *Lyle L. Parks, sales manager, said the new models were introduced to meet a*

Fig. 4-15 Safeway Bus — Body and chassis by Marmon-Herrington

Fig. 4-16 Marmon-Herrington Model 8 MB Transit Bus.

## 44-PASSENGER TROLLEY COACH ★ CONDENSED SPECIFICATIONS

DIMENSIONS

| | |
|---|---|
| Length over bumpers | 36′ 5¾″ |
| Height over roof | 9′ 4″ |
| Height floor to ceiling | 6′ 6″ |
| Height floor to top of windows | 63″ |
| Height ground to floor | 33″ |
| Height ground to step | 15″ |
| Step riser height (2) each | 9″ |
| Width overall | 102″ |
| Width over side posts | 101″ |
| Width between side panels at hip height | 100⅜″ |
| Width between posts | 97″ |
| Aisle width | 28″ |
| Aisle width at hip height | 30″ |
| Net open width front door (approx.) | 48″ |
| Net open width center door (approx.) | 30″ |
| Seat length | 34″ |
| Effective seat length (side panel recessed) | 36 3/16″ |
| Seat spacing | 29½″ |
| Post spacing | 29″ |
| Turning Radius over end of body | 42′ |
| Wheelbase | 235″ |
| Tread front | 86¾″ |
| Tread rear | 77⅝″ |
| Tires | 11.00-22 |
| Tire spacing | 12.9″ Approx. |
| Weight (estimated) | 19,000 lbs. |

Note: Height dimensions are given under seated load.

UNDERFRAME—The underframe is made up of steel and aluminum alloys. 7″ steel center sills. Steel cross members or bolsters at spring ends and other highly stressed positions. Aluminum alloys where weight saving can be made without sacrifice of durability. Fully gusseted and hot riveted.

SEATS—Seats for 44 passengers. Transverse two-passenger seats, 34″ long, tubular frame type with stainless steel jacketed handrail across top. Cushions of spring construction 5½″ deep, leather covered. Backs coved and filled with vulcanized curled hair. Other seats designed to harmonize with cross seats. Driver's seat fully adjustable. Arm rests on open ends of longitudinal seats.

HEATING AND VENTILATION—Forced draft ventilation of about 1800 c.f.m. by means of fan on traction motor and resistor blower. Heat by accelerating and braking resistors supplemented when needed by auxiliary resistors. All under thermostat control.

### MECHANICAL UNITS

FRONT AXLE—Timken 86¾″ tread specially designed for trolley coach service.

REAR AXLE—Timken special trolley coach axle 77⅝″ tread, gear ratio 11.42 to 1.

BRAKES—Two shoe air diaphragm operated friction brakes each wheel. Front shoes 16½ x 5, rear 16½ x 7. Meehanite drums.

SPRINGS—Semi-elliptic chrome vanadium steel. Front 4″ wide, 60″ long; rear 5″ wide, 68″ long. Springs flat under seated load. Threaded type shackle bolts and bushings. Alemite hydraulic lubricating fittings.

PROPULSION EQUIPMENT—The power equipment may be either G.E. or Westinghouse type provided with electric braking.

Propulsion motor mounted longitudinally at rear driving forward through propeller shaft to rear axle. Motor mounted on rubber bushings for sound insulation.

Master controller and brake controller operated by organ type pedals at driver's right foot.

BRAKE EQUIPMENT—Dynamic Brakes giving a maximum rate of 2 M.P.H. P.S. Air Brakes may be by either Westinghouse Air Brake Company or General Electric Company as specified.

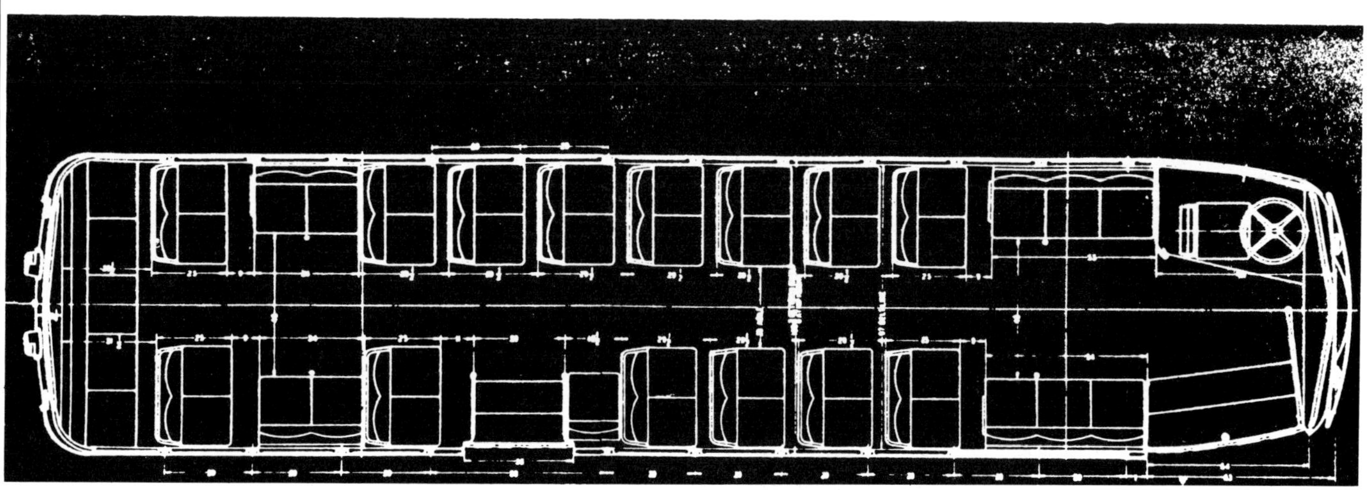

Fig. 4-17 Specifications of the Marmon-Herrington Trolley Bus

Fig. 4-18 Marmon-Herrington 'Oneida' schoolbus. 'Slim' Swinford at the wheel.

Fig. 4-19 The 1956 M-H 'Safeway' school bus chassis and with a M-H 'Oneida' transit type coach body.

1956 **ONEIDA** line features two new transit type coaches, the This model features Oneida-ramic vision with full wrap-around

ONEIDA DIVISION · MARMON-HERRINGTON CO., INC.
INDIANAPOLIS 7, INDIANA

ONEIDA Starline

*heavy demand for school buses and featured increased safety, heightened durability and extended comfort. Both model series have four protective roof beams extending the length of the bus, a skin-stressed roof, a "hi-arch" ceiling, heavy duty 16-gauge steel side panels, a double reinforced rear structure and extra-strong rear bumpers. The four (longitudinal) steel roof beams coupled with roof crossmembers form the base for application of the skin-stressed roof. In designing the roof, Oneida engineers applied the Monocoque principle used extensively in aircraft. In effect it makes (body) frame and covering an integral unit. In addition to increasing the strength, Monocoque makes possible the "hi-arch" ceiling. This provides greater headroom than any other school bus body,* said Parks.

Bill De Capua died suddenly of a heart attack at age 47. Slim Swinford was appointed Chief Engineer of the Oneida Division following De Capuo's untimely death.

When the Marmon-Herrington built Oneida bus appeared, it was obvious that it was targeted to be the Cadillac of school buses.

Both the windshield and the rear window wrapped around the body corners, the former extending to the front door frame. Both glasses were divided on vehicle centerline.

The transverse framing members — "carlines" — were extended downward, below the floor structure, to the bottom of the side skirts. This feature combined with the Guernsey double wall structure, which extended around the top radius on each side to be joined to parallel full length continuous roof beams, to make a very rugged structure, indeed. Only the central roof panel was single layer steel.

The new bus was to be marketed to both the transit and school bus applications. Such is also done today: however the M-H Oneida looked more like a transit bus in a school application than the contemporary inverse; with exception of the California built Crown Coach which I would place in the same category as the Oneida bus.

Only a month after the Oneida acquisition, on March 11, 1960, Marmon-Herrington purchased the Long Co. of Oak Hill, W. Virginia. It was combined with the Airdox Division, which had been acquired in 1955, as the Long-Airdox Division. The event was of great significance for with the acquisition came John B. Long who had been president, and who would later become president of Marmon-Herrington. It also represented the first move toward diversification and ultimate redirection of the company. Then, two months after the Long Co. acquisition, in May 1960 came another diverse move — the purchase of 80 percent of the stock of the Flash Perforating Co. of Abilene, Texas.

By 1963 there were Long-Airdox Division plants in Oak Hill, Uniontown, PA., and Ottumwa, Iowa — and a recently purchased plant in Kentucky for expanded production of coal-mining machinery And there was the Cardair Division plant in Lebanon, IN where four-stage compressors were produced. They were capable of 15,000 pounds per square inch pressure for defense missile applications. This market had declined and during 1963 the manufacturing operations were combined with the mine air compressor production of Long-Airdox Division.

Major losses from the Oneida operation had resulted in corporate losses for three years and management became disillusioned with the bus business. Production was discontinued during 1962. When efforts to find a buyer for any part of the bus business were unrewarding, the entire bus venture was written off; the Oneida Division closed in July, 1963.

Diversification had begun, traditional product lines were being abandoned. The old order was ending; another was beginning. The associated tumultous events are discussed in Chapter Six, Part 1, 'Organization and Financial History.'

## PART 5 — The HDT . . . One For the Road

## THE MARMON-HERRINGTON HDT . . . HIGHWAY TRACTOR

In 1960, the Board of Directors authorized development of a product for entry into the inter-city trucking market. Design and development of a Class VII-Class VIII highway tractor was undertaken by the following organization: G.H. Freers, Vice-President of Engineering; J. Mann, Director of Engineering; M.E. Swinford, Chief Engineer-Trucks

The first public word of the new offering, came in an *Automotive Topics* report in the July 1961 issue: *Marmon to Build 7 ton Truck of own Design*

President Klausmeyer was quoted as making the official disclosure and revealing that production would take place at he Oneida School Bus Division in the main plant in Indianapolis.

*Fleet Owner* for June 1961 carried the first photograph and description. Marmon-Herrington's first over-the-road tractor was designated HDT-1. It weighed 14,000 pounds and was rated up to 43,000 pounds GVW (Gross Vehicle Weight) as a truck and up to 72,000 GCW, (Gross Combined Weight) as a tractor-trailer. The three-axle tractor was a 6 x 4 with the rear tandem suspension rated 34,000 pounds. The rear most axle with 4.1/4.625 optional ratios drove the other rear axle by V-belt driver . . . a Timken Model TK-571 drop center axle.

Standard equipment engines were Cummins NH-195 or NH-220 naturally aspirated models, driving through a Fuller 10 speed transmission.

The steel cab was a 52 inch BBC (Bumper face to Back of Cab) unit of M-H design. It featured a divided windshield with vertical transverse glazing having a full 90° wrap at each end. Individual seats on each side of the engine enclosure provided satisfactory driver comfort despite the 52 inch lengh due to maximum utlization of space. The trucks were to be well equipped, including full instrumentation and ICC air brake and lighting equipment.

By October, 1961, a number of the new trucks had been built and were undergoing road tests in fleets of common carriers, primarily the Ellis Trucking Co. of Indianapolis.

Series production began in 1962. Hardbound, looseleaf, catalog M-H #620129 (presumed to be January 29, 1962) featured "Marmon-Herrington CAT Diesel Powered Trucks . . . Tailor-made to fit ALL Trucking needs". The Caterpillar Tractor Co. sought entry into the OEM (Original Equipment Manufacturer) truck engine market and had a large distributor-dealer organization for their construction machinery products. For whatever reason, Marmon-Herrington had elected to jointly promote the HD-1 and Caterpillar diesel power.

Marmon-Herrington, like any other heavy duty truck manufacturer, then or now, had to be responsive to bid specifications stated in the advertisement by the potential customer as to almost every component.

The tilt cab feature dictated a two piece radiator grille . . . a lower rectangular opening between the headlights and an upper square opening in the cab skin. The latter was surrounded on three sides by a reveal sloping outwards toward the bottom.

Very early in series production, the wrap around windshield was abandoned. It was replaced with a two piece, flat-glass, design in a slight Vee configuration. The cab was unchanged below the bottom of the windshield; a triangular ledge at the base of the windshield resulted from the substitution, the new glass having identi-

Fig. 4-20 Field testing the HDT.

Fig. 4-21 Prototype Model HDT — the first Marmon-Herrington highway truck.

cal extremities to the former curved one and a resultant Vee configuration of the two panes.

The Mayflower field test tractor had the final windshield and grill configuration.

The Marmon-Herrington cab was characterized by a distinctive, innovative, roof configuration seen frequently today. The steel roof panel rolled over on each side to the top of the doors and, extending forward, formed a short visor.

The tractor, with recommended equipment, was offered at a dealer cost of $13,869., FOB Indianapolis, plus Federal Excise Tax of 10%. The vehicle was rated 70,000 pounds GVW and had a road-ready weight of 13,500 pounds (steel cab).

The merchandising program was directed by T.M. Hoban, Fleet Sales Manager, who arranged the joint M-H/Caterpillar promotion with all factory photos evidencing CAT power and Caterpillar Yellow as the standard paint color. Service and Parts Managers were H.B. Blank and W.L. Kinsey respectively.

Following completion of the field test program some 45 to 50 Model HDT-1 tractors had been built at Indianapolis and parts were on hand or in-process for some 15 additional units.

In July 1963, a surprise announcement reported that the assets of the Truck Division of the Marmon-Herrington Co. had been acquired by A.H. Roop, President of Marmon Motor Company. Manufacture was to be transferred to Denton, Texas with resumption of production in May 1964. Slim Swinford recalls that the decision to sell the HDT product and close down the school bus line "was made by Bob Pritzger when he came down to Indianapolis after the purchase".

Production was to be supervised by Louis B. Martin who had seventeen years experience at Marmon-Herrington and was Factory Superintendent. He had set-up similar plants in Pakistan, Burma and Indonesia, and had worked for Ford of Belgium on establishment of production facilities in both Belgium and the Belgian Congo, as well as for Ford in Egypt.

Slim Swinford told the author in 1979 that Adrian Roop had been the Marmon-Herrington distributor, for the Southwest Region, of the HDT tractor. He also said that Arthur LeBon, an engineer and Charles Bartlemay, a line foreman went with the HDT project to Denton.

*Automotive Industries* reported March 15, 1964 that the first truck had been assembled at Denton and that capacity for 1200 trucks per year was to be installed with initial production of one unit per day.

## THE MARMON MOTOR COMPANY

The story of the Marmon (-Herrington) highway truck subsequent to the sale of the Truck Divison of Marmon-Herrington will be told here. The ongoing story of Marmon-Herrington (sans the Truck Division) will then be resumed.

As stated previously, a truck had been assembled on March 15, 1964 in the new plant of Marmon Motor Co. at Denton, Texas and series production was to have begun in May 1964. Mr. A.H. Roop had acquired the assets of the Truck Division of Marmon-Herrington Co. and established a plant in Denton. The need for additional working capital resulted in the acquisition of Marmon Motor Co. by Space Corp., a military contractor with forging and mobile home subsidiaries. Space Corp. had manufacturing facilities available at Garland, a suburb of Dallas, Texas, and Marmon operations were relocated there.

In 1973 Marmon, Space Corp. and its subsidiaries were purchased by Interstate Corp., a Chattanooga, Tennessee insurance company. The purchase was generated by General Transportation Service, a member of the Interstate family. Mr. Robert A. Thornell became president of Marmon Motor Co.; he had sales and manufacturing experience with International and Peterbilt and came directly from General Transportation Service.[7] By 1977, when truck industry sales had plummeted, he had returned to Peterbilt as Fleet Sales Manager. He was replaced as president by Mr. Bill Roth; Mr. Gerald Arnold was sales manager.[8] In 1979, Craig Wilson was sales manager.

During an interview for "Overdrive" magazine early in 1977, Mr. Roth told the reporter that four times in the prior ten years there had been serious disruptions in both the truck (Marmon) and military hardware (Space) markets but they had been out of phase and one had supported the other and assured survival of the Marmon/Space operations.[8]

Mr. Harold Robinson is now president. Constantine M. (Gus) Xeros is vice-president for engineering. Gus is responsible to a large degree for the fine reputation that the Marmon truck has earned.

By 1974 Marmon, Space and its subsidiaries had been grouped as TIC Industries, Dallas, Texas. Financial data is reported collectively for Marmon and Space for the reason discussed above, and separately for W. Pat Crow Forgings, Inc. and Chemical Express Co.

## MARMON HDT HIGHWAY TRACTOR

*Commercial Car Journal,* for December 1964, under the heading "What's New — in Trucks and Tractors" reported that "Diesel-Powered COE Tilt-Cab trucks and tractors are built by Marmon Motor Co., **Garland,** Texas, to meet most heavy duty requirements. The companion photograph showed a new and more conventional roof treatment and dual-headlights . . . definitely a more pleasing and less utilitarian appearance; and one more competitive with the established makes. The new badge was an M, with the word MARMON above the M, both within a square border. The badge appeared three places . . . above the grille on the raised panel, and on both cab skirts forward of the wheel opening.

Both cabs were available in steel or aluminum and both were raised by an independent hydraulic system.

The Cummins NH-220 (220hp @ 2100 rpm; 606 foot pounds torque at 1700 rpm) had joined the CAT 1673 as standard engine offerings, but other powerplants were available.

The two axle model was rated 34,000 GVW and 66,000 GCW (including trailer). The three axle model was rated at 45,000 GVW and 76,800 GCW. Model designations continued as HDT with appropriate suffix numbers for cab selection; -51 for standard and -86 for sleeper. The numbers are the bumper to back of cab dimension. Letter suffixes were also used for axle selection: A for two axle and B for three axle.

## CHDT CONVENTIONAL TRACTOR

Five years after the start of production in Garland, Texas, Marmon Motor Company announced[9] a conventional tractor companion to the COE HDT product. The new offering was designated CHDT and featured a chassis designed for ruggedness. The frame was made of 125,000 psi high tensile steel, but an aluminum frame, for maximum weight saving, was available. Wheelbase was as required by the customer's order. In October, 1969 the Sales Manager was Bill Fortenberry who announced "We can furnish, and are engineered for, all Caterpillar, Cummins and Detroit Diesel engines". Available transmissions were heavy duty models by

Fuller or Spicer, and rear axle choices included all Eaton and Rockwell Standard models in both single and tandem suspensions.

R.G. Myers, Sales Engineer for Marmon, was quoted as saying that the cab was insulated and of all-aluminum construction. He said that it was designed to meet or exceed the structural strength of a conventional steel cab and provided a considerable weight saving. The "firewall" or bulkhead between the engine and cab compartments was insulated with fiberglass and an aluminum facing to provide cooler temperatures within the cab.

Electrical circuit breakers, regulators and terminal boards were located in an enclosed watertight compartment accessed through the 'firewall' with the hood open. Easier maintenance and extended life were anticipated.

The photo accompanying the announcement release carried the caption 'Marmon Model CHDT tractor offers a host of bright trim and accessory options'. Mr. Fortenberry was quoted as saying that two hood designs were available. A butterfly type (triple parallel hinge) hood was offered for construction and other very heavy duty applications.

A tilting hood and fender assembly was offered for unrestricted access to underhood components. The single piece hood and fender construction was manufactured by hand layup of sheets of woven fiberglass cloth. Advantages noted by Mr. Myers were ability to withstand road shocks and vibration and to be easily repaired in event of damage.

## SECOND GENERATION

A half decade after the CHDT bowed, Marmon Motor Co. announced[7] for 1974 a second generation pair of conventional and cab-over-engine models. The press release, made concurrently with the showing of the trucks at the Tenth International Trucking Show in San Francisco, stated that the new trucks to be built in the Dallas, Texas plant would be aimed at the owner-operator market, primarily in the Southern states. The trucks would be priced comparably with Kenworth and Peterbuilt and an expanded sales force would extol the extensive list of former options now made standard; air conditioning, padded and rolled upholstery interior, AM Radio, air-ride driver's seat, twin 75 gallon aluminum fuel tanks and a polished aluminum grille. The standard engine appropriately was a Cummins NTC 350 coupled to a Fuller RTD 9513 transmission and driving Rockwell FF 921 or SQHD axles fitted to the Reyco 101 Suspension. A Farr air cleaner (dry type), a 23 inch Air-Slide Holland fifth wheel, and 22 inch Budd wheels completed the impressive list of standard features.

Both models were characterized by much larger radiator openings and standard 1200 square inch radiators to cool the larger engines, up to 450-600 hp range, projected to be in demand in the mid seventies. The fuel crisis accompanying and following the OPEC embargo of 1973 ended the horsepower growth but the larger radiators were still necessary to accommodate the increased heat rejection of less efficient, but lower emitting, engines and the larger diameter, slower-speed, fans both of which resulted from compliance with air and noise regulations of the US Environmental Protection Agency and the California Air Resources Board.

The massive appearing CHDT hood was one of the first of the now common box (or rectangular prismatic) hoods faced with large area, nearly square grilles. No disconnections were necessary prior to tilting the new cab a full 90°. The Marmon fiberglass hood was of sufficient novelty to warrant a special half-page illustrated article in *Automotive Industries* for Sept. 15, 1974, entitled "Marmon Tilt Hood Made From Fiberglass".

The article on the hood details also provided additional detail information on the new truck models themselves and the facilities for their manufacture. The frames were noted to be of heat treated alloy material assembled with bolts. The doors were hung with piano type hinges. An acoustical padded headliner and tinted glass were other standard equipment items not previously reported. The trucks were assembled from the frame up on two assembly lines, one for the chassis and one for the cab. Production of 800 units at the Garland, TX, facility was forecast for 1974. During the second half of 1973, 300 had been assembled.

The COE cab was described as built with an all-welded aluminum frame and the conventional cab as a rugged all-aluminum welded design.

The Marmon truck found enthusiastic support from *Overdrive,* the magazine of the trucker, during the seventies. "Cadillac of the Industry" was one of the plaudits. When Marmon introduced a sleeper box in 1976 to tandem with the conventional CDHT truck cab, *Overdrive* said it set a standard that the rest of the industry should strive to attain. Tractors equipped with the sleeper box were designated Model CHDT-BC-110, the 110 being the bumper to back of cab dimension and the B denoting a 6 X 4.

The 1979 Marmon color brochure carried the by line "Excellence by Marmon". The inside cover repeated the theme and was blank except for a central paragraph in rather small type entitled 'MARMON — Where Quality is a Tradition'. The paragraph began, "In 1902, Howard C. Marmon..." a capsule history of the Marmon car Eras followed. Then "In 1931, Marmon combined its resources with Herrington...In 1963 the Marmon-Herrington truck was brought to Texas. The dedication to building a quality truck still carries through **and Marmon continues to build horseless carriages for a few of our friends.**"

For 1983 Marmon made a number of significant changes in their conventional tractor. Subtle changes in styling and aerodynamics enhanced the traditional appearance. The cab roof was raised, the corner radii increased and the visor deepened and rounded. A metal tilt hood option to the fiberglass hood was introduced. It reflected much attention paid to small styling details. New fiberglas fenders incorporated housings for twin rectangular headlights; the lights were raised relative the fiberglass hood, to have their protrusion extend horizontally and tangent to the fender radius. Both grill and bumper were formed with a slight V-shape. The grille was extended upward to the top of the hood, encompassing the badge (as on the COE), and around the sides as well... basically becoming a shell rather than a panel.

Like the fiberglass hood, the new hood and grille are characterized by a raised center section which diminshes towards the rear and outlines the upper part of the badge where the word MARMON is located.

## THE ULTIMATE RECOGNITION

In 1982 the Cummins Engine Co. determined to seek out the oldest Cummins Diesel powered truck still in active commercial service. To motivate the search, a highly attractive prize was offered... the owner of the winning truck/engine would receive his (her) choice of any truck-tractor on the market. There was one condition: the order for the new truck must specify Cummins power.

The winner was Johnny Campbell whose 1939 Diamond-T still moved trailers in a terminal in Farmington, N.M. There were over 500 entries. His choice? A Marmon conventional tractor with walk-in sleeper box having twin bunks, color television and a refrigerator. He also selected the optional air-ride suspension. The truck keys were presented to Mr. Campbell in April 1983 by the Cummins distributor in

Los Angeles, California.

The purpose was to suitably acknowledge the 50th anniversary of diesel trucking in the United States. But what could have been a more fitting endorsement of the stature of the MARMON marque after 80 years of traveling America's roads on Marmon Cars, Marmon-Herrington trucks, and Marmon trucks.

## PART 6 — The People Mover Concept

On November 29, 1962, the Indianapolis News carried a front page announcement that the Westinghouse Electric Co. had unveiled details of an automatic crewless train for rapid transit service which would be a joint product venture with the Marmon-Herrington Co.

The system concept, designed by Art LeBon, assistant to the vice-president and M.E. Swinford director of engineering at Marmon-Herrington would use electrically-powered 23 foot light-weight automated cars which would seat 20 passengers each in plastic molded bucket seats. The curved glass windows extended into the roof essentially the same as today's contemporary practice on European buses.

The cars were powered individually. They were to run on rubber-tired wheels along a pre-cast, pre-stressed roadway at elevations ranging up to 14 feet. The cars were to be stabilized on the roadway by side wheels in contact with the curbs cast integral in the roadway slabs. The electrical supply rails were located centrally on the roadbed.

The initial customer was expected to be the Pittsburg (PA) Port Authority . . . probably a demonstration installation. Col. Herrington noted that the west coast smog problem was causing renewed interest in electric mass transportation and Marmon-Herrington officials had been on the coast recently talking with city administrators.

Marmon-Herrington was to manufacture the train and Westinghouse was to supply the electrical power system. Deliveries were considered feasible in two to three years after a go-ahead.

Nothing ever matured on the project. But any reader who rides on the Ford people-mover demonstration installation in Dearborn, MI, adjacent Ford World Headquarters will be amazed at the similarity of that entire system to the Marmon-Herrington concept proposal to Pittsburg in 1962. The Fairlane system was erected in 1977 to connect the Hyatt Regency Hotel with the Fairlane Shopping Center and was a prototype by a division of the Ford Motor Co. established specifically to market the product. The Ford system uses electric power lines on the roadway walls which facilitates switching operations for passing on the two-way single 'track' system.

CHAPTER FIVE

# ENTER THE PRITZKER BROTHERS

## PART 1 — Diversification and Merger

Two months after acquiring the rights to the Ford 8MB/8M2B transit bus product, Marmon-Herrington bought another firm. This acquisition, however, was a first step in the direction of diversification. In May 1955 the Airdox-Cardox Products Co. of Chicago was purchased. Cardox engineered and marketed carbon dioxide, chlorine dioxide and fire equipment utilizing gaseous fire suppressants. Marmon-Herrington had built airport crash trucks during the war which carried Cardox equipment. But in November, 1958 the Company sold all properties, equipment, inventories and investments related to the carbon and chlorine dioxide business. The gain on the transaction was credited to retained earnings.

There were two acquisitions during 1960: the Oneida school bus body business of Henney Motor Company (discussed in detail in Chapter Four) and the Long Company of Oak Hill, a mining equipment manufacturer. The former was an expansion of the traditional vehicle business; the latter was outright diversification. These two events, particularly the latter, were events of greater significance than was evident in 1960; They would completely change the character of Marmon-Herrington within a relatively short time. In the two decades to follow, many firms would follow in the experience of Marmon-Herrington: a diverse acquisition would quickly outgrow the traditional parent product and became the primary source of profit.

By May of 1964, the President would be John B. Long who would report to the shareholders that "The principle sales of the Company had shifted to mining machinery".

The series of tumultous events that preceded that announcement will now be reviewed in detail.

In 1962, control of the Marmon-Herrington Company passed to outside investors, ending thirty one years under the founding interests.

Diversification begun in 1955 had been successful and fortuitous because as the automotive business sagged, partly reflecting the decline in the transit business, the mining equipment business thrived. In addition, there were three promising new ventures in the formative stage in early 1962: (1) a contract to manufacture motor coaches for the Mexican government, (2) a sizeable chassis contract with a bus company, and (3) an exciting dream of an automated high speed urban transit system worked out with Westinghouse Corporation.

Working control of the company was held by Colonel Herrington and his friend Captain Henry Pascale, a director. Because of health and age considerations, Herrington was known to be receptive to sale of his interest. He was not so concerned with the price to be offered for his stock, as he was with assuring the continuity of the firm in its Indianapolis location.

Several Indianapolis firms expressed interest but wanted assurance that they would be able to get 51 per cent of the outstanding stock. That assurance was not to be had.

Two Chicago brothers were the only party willing to buy Herrington's interest and gamble on picking up enough additional shares on the open market to gain control. Robert A. and Jay A. Pritzger had been looking for a promising corporation to purchase. They had been in the negotiations for Marmon-Herrington before the local firms and came back in as the negotiations dragged on. Together they controlled the L.A. Darling Co. and through that intermediary they acquired 150,000 shares of M-H stock from Colonel Herrington in October, 1962. At the same time, the brothers filed a request for tenders (offered to buy) to other holders of the common at $10 per share, and made additional stock acquisitions during 1963.

Captain Pascale did not like the development and began buying additional shares. Pascale was a retired Air Force officer, residing in Rumson, New Jersey.

"The tender by Darling brought in 60,655 shares; bids to two stockholder groups brought 46,947 shares, and in September 1963 John B. Long, M-H president, and members of his family sold 36,249 shares bringing the Darling acquistions to 293,851. In November, 1963 Matteson Metals, another Pritzker company, bought 309,224 shares from Darling (greater than their recent acqusition) and became the majority shareholder in Marmon-Herrington. The proxy statement showed that, through trusts, the Pritzkers own **all** outstanding shares of Matteson Metals and also the plants, machinery and equipment acquired from Darling.

Thirteen months after venturing the purchase of Colonel Herrington's interest in October, 1962, the Pritzkers, as Matteson Metals, had achieved their objective of majority control! Captain Pascale had only been able to acquire slightly more than one third of the stock.

In July 1963, Colonel Herrington's dream of continued Indianapolis operations ended. Less than a year after his sale of interest, the Mexican bus contract did not materialize and the plant was permanently closed.

On July 12, 1963, John B. Long, President of Marmon-Herrington confirmed the shut down of the Indianapolis plant the following Monday. Colonel Herrington was quoted by the *Indianapolis Times* as saying the move "is an election to get out of the automotive business", and indicative that automotive production "is being reduced to zero. Top executives of the firm have elected to concentrate on the manufacture of coal mining equipment". Herrington noted the sale a few days previous of the heavy duty truck division to a Texas firm.

Officials, however, denied the firm was abandoning the automotive business. They noted continuing contracts with Ford Motor Co. and willingness to accept automotive equipment contracts anytime they might be available.

Employment had declined from a peak of 1000 to 525 in 1960; only 40 were on the rolls when the closing was announced. Of these, only nine were expected to move with the operation to Lebanon, according to officials of UAW LocaL 341.

John Long was quoted "Large scale contract work has not been steadily available". The 16 acre, 275,000 sq. ft. facility was completely utilized only sporadically in the preceding several years.

The final obituary for Marmon in Indianapolis is quoted from the 1968 Annual

Report of The Marmon Group. *"Historical Marmon's manufacture and sale of trucks, buses, vehicles and parts therefore has been carried on in its plant at Indianapolis, Indiana. This plant is of brick and steel construction, occupies sixteen acres and has a total floor area of approx. 263,000 square feet. During 1962 Marmon discontinued the production of school bus bodies; inventories and tooling related to such products were written down to established amounts realizable on disposition. In (July) 1963 the operations at the Indianapolis plant were discontinued due to continuing operating losses, the approximate pre tax amounts of which are set forth in the table below; (omitted); the inventory was disposed of, the machinery and equipment was auctioned (July 30, Aug. 1) at a book profit of $65,000 and the real estate has been offered for sale or lease".*

So ended eighty-seven years of Marmon manufacture in the Indiana capital city.

The December 1963 roster of Marmon-Herrington officers was as follows:

Jay A. Pritzker, Chairman; (Member, Pritzker & Pritzker, Chicago)
John B. Long, President; (Owner of 26000 shares of LAD (Darling)
Paul C. Manley, Vice-Pres.
Robert A. Pritzker, Director; (Pres. Matteson Metals, Inc.)
James E. Burgess, Director; (V-P Finance, Matteson Metals, Inc.)
Fred M. Young, Director; (Pres. Young Radiator Co.)
Robert C. Nelson, Vice-Pres.
John C. Ross, Vice-Pres. — Finance
E.G. Larrick, Jr., Treasurer
George H. Shevlin, Jr., Vice-Pres., Sanford-Day
Wm. P. Nottingham, Secretary
A.H. Newbury, Asst. Secy.
J. I. Oglesby, Asst. Secy.

The address of Marmon-Herrington Company, Inc. was given as Oak Hill, West Virginia, the address of the Long-Airdox Co., one of three Operating Divisions; the others were identified as Marmon-Herrington Automotive of 510 Indianapolis Ave., Lebanon, Indiana and the Sanford-Day Company of Knoxville, Tennessee.

On December 13, 1963, stockholders of Marmon-Herrington Company, Inc. received a Notice of Special Meeting of Stockholders to be held on December 30 in Indianapolis for the following purposes:

1. To consider and take action upon an Agreement of Merger providing for the merger of Matteson Metals, Inc., a Delaware corporation, into Marmon-Herrington Company, Inc., and

2. To transact such other business as may properly come before the meeting or any adjournment or adjournments thereof".

The Notice, over signature of William P. Nottingham, Secretary, restricted voting to holders of record of Dec. 11, 1963.

The meeting, and its repercussions, was reported in the *Indianapoles News* for July 13, 1964 by staff writer Charles Vaughan under a four-column heading: "Marmon-Herrington Fuss". The preceding and following discussion of post-sellout developments is essentially re-edited and/or directly quoted from Mr. Vaughan's comprehensive coverage which included personal attendance at the December 30 stockholders meeting.

Colonel Herrington, holder of a single share of M-H common, attended that meeting "because I have friends on both sides". He had hopes of mediating the quarrel but said "the disputants are diametrically opposed". The proposed merger was approved by a majority of the stockholders, but vigorously opposed by Captain Pascale who went first to the Marion County Superior Court and then to the Federal District court in West Virginia (M-H Exec. Offices were in Oak Hill) seeking to set aside the merger.

The Marion County action contended that a two thirds majority was required but the court affirmed the simple majority under which the proposal was approved, and denied an injunction.

The Federal District action sought to set aside the terms of the merger which by mandatory stock exchange provisions effectively preempted the right of M-H holders to continued ownership of shares of the business subsequent the merger. It was determined that the proposed action was not permissable under Indiana law. Instead the matter was settled out of court with the majority holders buying the minority shares at some fifty percent higher value than the original proposal would have returned. Full details of the merger and the establishment of The Marmon Group, Inc. are discussed in Chapter Six, Part One — Organization and Financial History. The Marmon-Herrington story following the departure from Indianapolis is recounted in Part Two of this chapter.

## PART 2 — Unit of The Marmon Group, Inc.

In 1964, the three Marmon-Herrington units: M-H Automotive, Cardox and Sanford-Day were identified as "The Marmon Group", Incorporated". The Lebanon operation became the Marmon-Herrington Division. Then in late 1973 the three units were combined at one site in Knoxville, Tennessee, as the Marmon Transmotive Division; the three product lines continued under their original names.

Marmon-Herrington had set up the firm of Marmon-Herrington SAF at Villiers-le-Bel, France, to convert Ford trucks to all-wheel-drive and to construct special vehicles for the French army . . . 1.5 ton forward control 4 X 4s powered by a Simca V-8 or a Panhard 4 and a 6 X 6, powered by the gasoline Simca V-8 or the diesel Perkins 6.[1]

After Marmon-Herrington control was acquired by the Pritzkers, the interest in M-H SAF was sold in 1963 and became Marmon-Bocquet. The latter developed a new version of the M-H 630 truck designated the MH 600 BS vehicle. As of 1982, the vehicle was still being manufactured by the UNIC Division of Simca Industries for Marmon-Bouquet and was also known as the Simca-Marmon; it was the basis for numerous military vehicles.[1]

The full line of Marmon-Herrington conversion axles as used in the 1964-1984 time frame is illustrated in the reproduced catalog page. The source of the axles is evident from the model prefix: MD for Marmon-Dana, MC for Marmon-Coleman and MR for Marmon-Rockwell.

GMC conversions were first offered in 1964, concurrently with the first offering of the Coleman center-point steer axle. The novel construction of these axles is evident from the exploded view of the MC-9 axle.

The following account of technical progress at M-H is taken directly from "M-H History . . . Specialists in Tomorrow's Transportation" circa 1970:

More and more innovations were introduced in Marmon-Herrington All-Wheel-Drive equipment during the 1960s.

1964 brought "Center-Point" steering to eliminate the wheel-fighting still common to front wheel drives despite the use of constant-velocity joints. The same redesign provided axles with easily replaced axle shafts and drive-flange assemblies. This development brought to powered front axles a basic development pioneered by Nordyke & Marmon on the Model 48 introduced in 1913. The Center-Point steering axle reduced steering effort by 35 percent. It was (and is) procured from the designer/manufacturer, Coleman Motor Co who elected in

Fig. 5-2 Ford 4x4 conversion circa 1960.

Fig. 5-3 Marmon-Herrington AWD Conversion of GMC truck.

1964 to be exclusively an OEM (Original Equipment Manufacturer) and made an exclusive arrangement to furnish Marmon-Herrington with axles for the AWD conversion market.

An entirely new convenience for drivers of All-Wheel-Drive vehicles was introduced in 1965 with the development of a vacuum shift for engaging or disengaging the front driving axle. This innovation permitted the driver to **engage** the front driving axle **at any speed** when leaving the highway or encountering conditions that required additional traction. It also eliminated stopping to disengage the front driving axle when returning to hard surfaced roads.

1969 brought the introduction of the "Shuttle-Wagon", an advanced and unique railroad car "spotter" (for positioning freight cars where needed without a locomotive). This vehicle was capable of highway or track mobility and proved a flexible yard truck with both push and pull capability.

By 1972 M-H was advertising "Yes, you can shift a two-speed axle on a 4 X 4 or 6 X 6." This was accomplished by the use of Marmon-Herrington Fail-Safe Axle Interlocks which synchronized the shifting of the rear axle and the front axle. Also the front driving axle could be engaged or disengaged at any speed.

M-H ads promised Ford truck dealers "One-week flat for conversion of any F-Series Ford truck in the dealer's inventory to M-H All-Wheel-Drive." Another full page ad promised delivery to the customer "two weeks from truck selection from the dealer's inventory."

Ford conversions were made entirely with Rockwell components: front axle F.D.S.-75 (7500 pound spring-pad capacity); auxiliary transmission T-223-C15 (two-speed transfer case). Driveshafts and installation bracketry were provided by M-H. The 40 degree turning angle of the standard vehicle was duplicated. Conversion to a 4 X 4 was priced at $2,327.28 including average freight and Federal Excise Tax.

Following several management changes in an effort to make the Lebanon operation profitable, Marmon-Herrington was combined with the Railway Products Division (new in 1969) and Sanford-Day Divison at Knoxville, Tennessee (the home of Sanford-Day) in a new facility. The three former divisions became operating units of the new Marmon Transmotive Division but product identities were retained. An early Marmon Transmotive ad was headed:

"Marmon Transmotive Makes Products That Won't Come Back for Customers That Will" The ad message continued:

'"The three companies that formed Marmon Transmotive in 1974 brought together an established reputation for quality workmanship and reliable service. Since 1851 Marmon-Herrington has added innovations in the automotive industry. Like designing the winner of the first Indy 500. And pioneering all-wheel-drive capabilities for everything from tanks that invaded Normandy to trucks currently roughing the Alaskan pipeline construction.,"

Marmon-Transmotive introduced a specialized motor truck . . . the MT-Truck . . . in the 1970s. It was a two-axle dual rear-tired vehicle with on-road and on-rail capabilities. It was rated 12000 to 14000 pounds GVW. The power train was an American Motors 360 CID V-8 driving a Chrysler 2400 four-speed automatic transmission. Tread of the front and inside rear tires matched the railroad gauge. A full-width cab with seating for six was installed in a COE configuration on the 138 inch wheelbase chassis; cab to centerline of rear axle was 88 inches.

In the early 1970s, Marmon-Herrington introduced the Model MH-240 Transfer Case for all-wheel-drive vehicles. It featured motorized shifting of range and AWD selection in response to an electric signal.

In 1974 production of a special purpose mechanized Marmon-Herrington trailer was begun for the container shipping industry. It was a side-loading design which was capable of expanding or contracting its length to fit different sizes of ship-

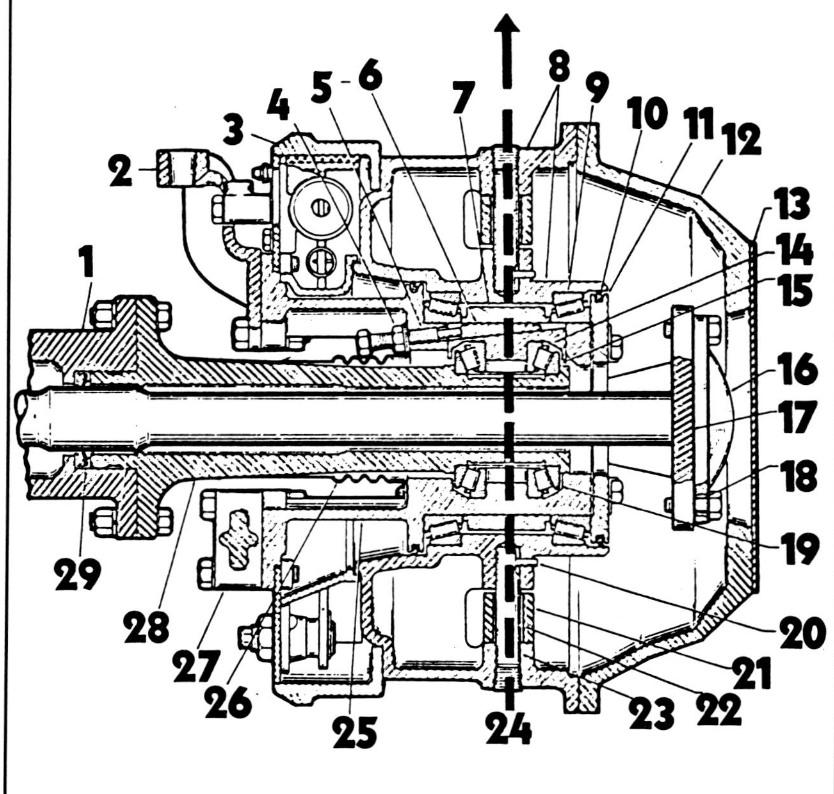


SECTIONAL VIEW OF COLEMAN MODEL 9 FRONT AXLE DRIVE WHEEL

Fig. 5-4 MC-9 Front Driving Axle

board containers. It features one-man operation with capability to stabilize itself and load or unload itself in seconds.

The letterhead of Marmon Transmotive Division states the business as "Prototype and Special Vehicle Production". During the 1970s a special heavy duty chassis with one man cab was designed and developed for use in on-site concrete preparation. Also specially designed recreational vehicle chassis were developed utilizing Marmon-Herrington experience in bus and power train engineering. These special chassis provided a more rigid undercarriage for the vehicle while allowing additional space for equipment, storage and passenger comfort. Customers were not identified.

Although a product of the Railway Products Operation, the Marmon Yard Patrol vehicle is worthy of note, particularly because it has highway capability. It is similar to a 6 X 4 pneumatic tired road grader in general configuration. However it also has flanged guide wheels for on rail operation and the necessary railroad track gauge.

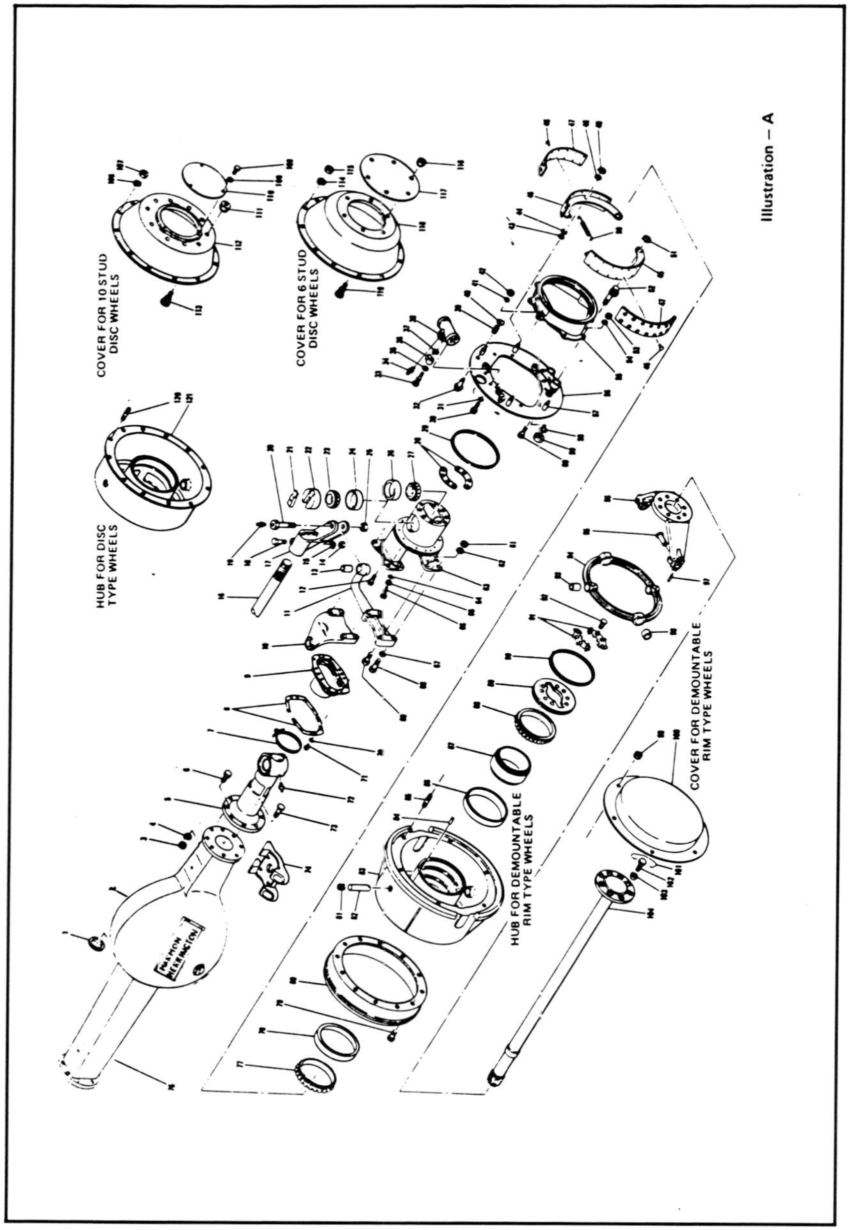


It features a number of independent mechanisms appropriate to the maintenance of railroad yard properties.

## A MARMON ATV

In the early summer of 1973, The Marmon Group, Inc. acquired from the

Cummins Engine Co. the assets related to the COOT all-wheel-drive utility/recreation off-road vehicle. For an undisclosed purchase price, Marmon acquired the inventory, equipment and engineering, financial and sales data from Cummin's Frigiking Division in Dallas. This was transferred to Knoxville.[2]

The press release described the COOT as an all-terrain vehicle capable of moving on both land and water. Maximum speeds were stated to be 20 miles per hour on land and 6 miles per hour in water. Loads of 1000 and 600 pounds could be carried respectively on land and water by the 1000 pound vehicle. The COOT featured forward and aft hull sections which were rotatably pivoted about the drive shaft. All machinery was inboard, with the 12 horsepower engine (Tecumseh 27 CID) under the front seat. A torque converter and 2-speed transmission drove the four 8.85 X 15 tires. Ground pressure was only 4 psi. The COOT could be transported in a pick-up truck. Options included four wheel steer and tires giving 1.5 psi ground pressure.

Marmon Transmotive never seriously promoted the COOT for reasons unknown. In early 1983 a press release was issued by COOT, Inc. of Round Rock, Texas. It announced that the firm was marketing the same vehicle; however, the Tecumseh engine was rated 18 horsepower and the top speed claimed was 30 mph. Load ratings were unchanged and the vehicle was priced at $5995.[2]

Fig. 5-6 AWD transfer case rated 3000 ft-lb input; 2-speed (1:1, 2.05:1) with neutral for PTO drive.

Fig. 5-5 Marmon Transmotive Rail or Road Truck Model MT.

MARMON-HERRINGTON CO. INC

CHAPTER SIX

# THE PEOPLE AND THE BUSINESS

## PART 1 — Organization and Financial History

The beginning of the Marmon-Herrington Co. in 1931 was described by the author as the Birth of a Precocious Stepchild (Chapter One). Some thirty years later, advancing age provoked Colonel Herrington to seek a buyer for his one-third or more interest. In the developments that followed, Marmon-Herrington underwent a kind of metamorphosis and emerged as the The Marmon Group.

When other potential buyers of Herrington's shares despaired of ultimately acquiring a controlling interest, the Pritzger family accepted the challenge in 1963. Under their skillful financial and management guidance, The Marmon Group grew steadily and achieved the exclusive Fortune 200 status by the late 1970s.

As the transition began, president Jay A. Pritzger took the firm out of the production vehicle business which had proven less profitable than the mining machinery business into which Marmon-Herrington had diversified. The HDT (heavy duty truck highway tractor) business including the right to use the name Marmon on trucks was sold; the tooling and inventory for school-bus manufacture was written off. Remaining operations, including a Marmon-Herrington Division continuing the all-wheel-drive conversion business, were restructured as The Marmon Group, Inc. (Indiana) in 1964. By 1967 it was a Michigan Company publicly held but on June 30, 1969 became The Marmon Group, Inc. (Delaware) (1969), a company privately held by the Pritzger interests.

The Marmon Heritage coverage of the Marmon-Herrington story ends with that transition. The story of the Marmon Group, Inc. (Delaware) is still unfolding and will someday be the subject of a book every bit as exciting to write as this book has been. Similarly the story of the Marmon Motor Co. and their Marmon heavy duty trucks is an ongoing one as well.

This chapter will address the management, organization, and the financing of the Marmon-Herrington Co. 1931-1964 and subsequent developments of 1964 and conclude with a description of the manufacturing and other facilities.

The author earlier categorized the production period of the Marmon car into five eras, each quite unique in product, in technical and design philosophy, and in management.[1] And the same basis proved suitable for discussion in this book of all aspects of the Marmon car business.

When assembling the organizational and financial histories of the Marmon-Herrington Co., it is more suitable to categorize on the basis of periods associated with a Chairman/President team. There were five:

**The (W.C.) Marmon/(A.W.S.) Herrington Period 1931-1941**
A ten year period of growth and maturity

**The Herrington/Dingley Period 1942-1946**
Mobilization for Allied War and peacetime prosperity

**The Herrington/Klausmeyer Period 1946-1961**
Fifteen years of decisive acquisition and diversification.

**The Klausmeyer-Pritzger/Long Period 1962-63**
Merger and new Direction

**The J.A. Pritzger/R.A. Pritzger Period 1964-19XX**
Unit of the Marmon Group

Because the story of plant facilities was one of rapid expansion for war production and a post-war struggle to keep the enlarged plant busy and profitable, it does not lend itself to a subdivided treatment and will follow the Organizational and Financial History period sections. Preceding those sections is a brief summary of the overall financial performance 1939-1964.

### FINANCIAL SUMMARY 1939-1964

Prior to becoming a public company in November 1941, financial results for 1939 and 1940 were made public. The public offering did **not** represent new financing.

The number of shareholders varied between 800 and 1000 until 1955 when it climbed to 1300 in two years and crested at 1500 in 1961. It then declined to under 1000 due possibly to poor profitability but more likely to certain parties increasing their holdings. Figure 6-1 shows annual sales.

The authorized stock just before the company went public was 150,000 common shares. It was increased to 600,000 shares to accommodate a 200 percent stock dividend in 1941 prior to the public offering. In 1955 the limit was raised to 1 million as additional stock was issued to accomplish an acquisition. Outstanding shares increased from 150,000 to 450,000 due to the dividend in 1941, did not exceed 500,000 until 1955, 600,000 until 1951 and 750,000 until 1963. Stock dividends of 5 percent in 1957 and 1959 accounted for about a 50,000 share growth, as did the additional issue for the 1955 acquisition.

Sold at $10 in Nov. 1941, the stock reached a high of 14.5 the following year, fell as low as 4 in 1944 as earnings declined, then rebounded to 18.25 in 1946 as new products were announced. It did not reach 10 again until 1955 when an acquisition and increased earnings sent it to an all-time high of 17.25, from which it steadily declined to 9.75 in 1962. The merger activities in 1963-64 sent it back up to 14 as major share holders sought additional leverage.

Figure 6-2A and B shows stockholders equity and stock price range. Figure 6-3A, B and C show annual earnings on a total and per share bases and stock dividends paid. Losses in the 1945-1947 period resulted from tooling and development costs for new product lines and delayed or reduced production due to material shortages. Losses in 1960 and 1962 were consequent delayed utilization of acquired facilities and related development costs. The diversification of 1955 was responsible for the subsequent increase in earnings.

As a condition to merger in 1964, all minority common stock was purchased for cash, thereby transferring full ownership to the majority holders.

Full details are provided in the separate discussions of the Organization and

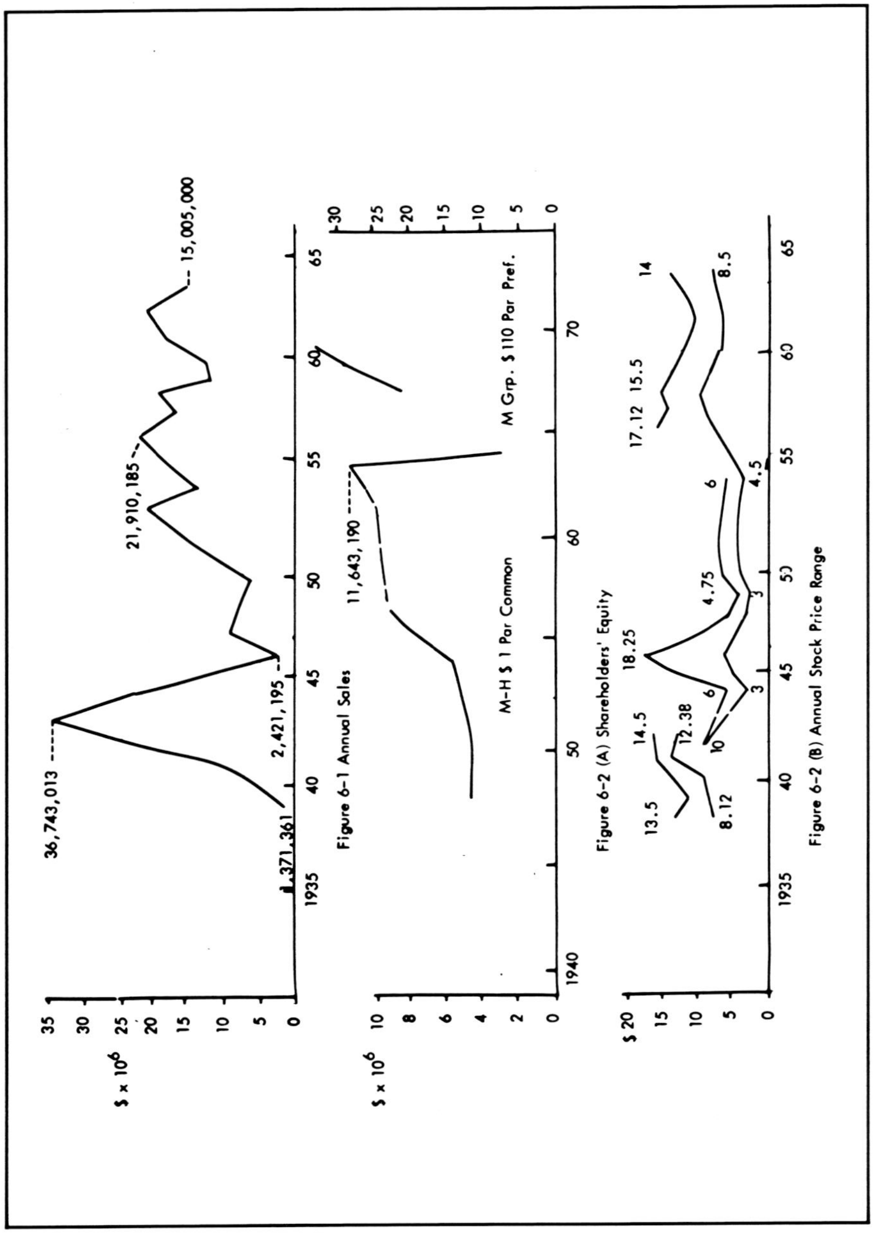


Figure 6-1 Annual Sales

Figure 6-2 (A) Shareholders' Equity

Figure 6-2 (B) Annual Stock Price Range

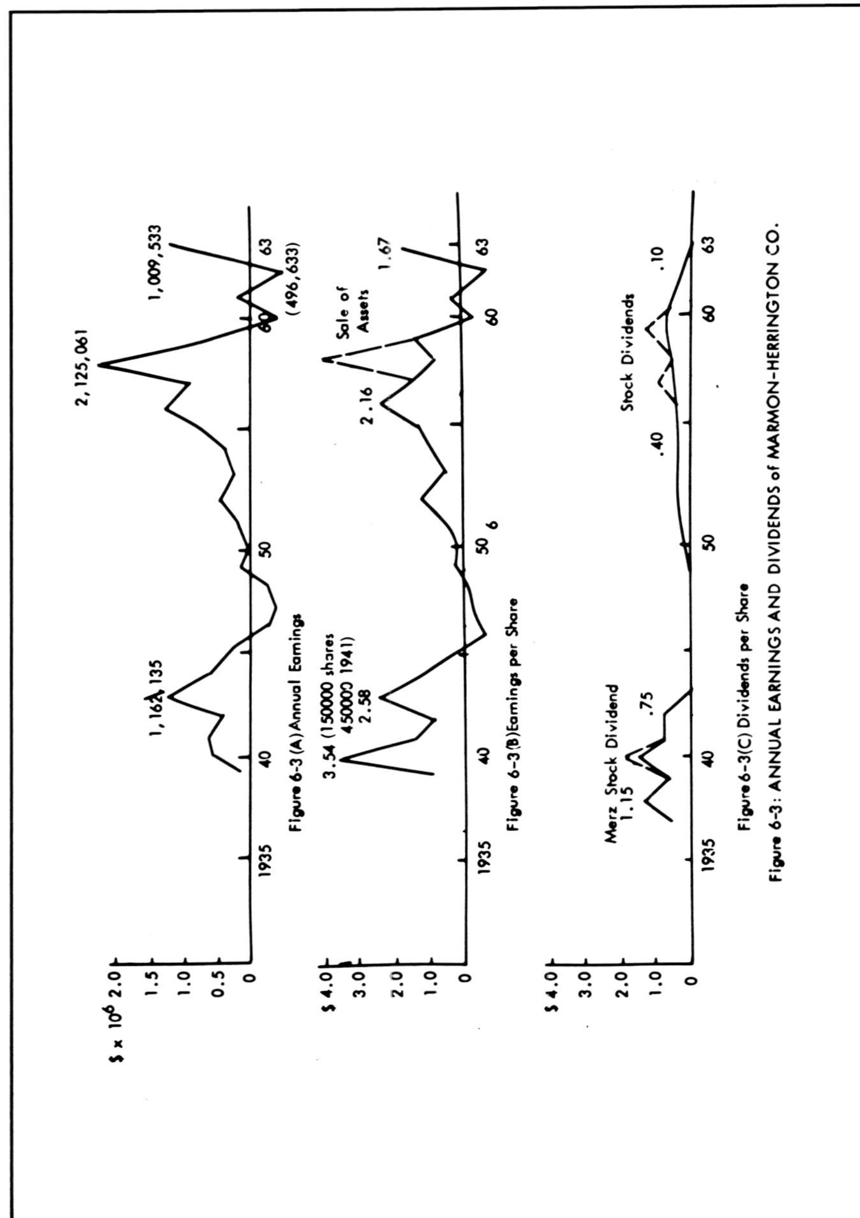


Figure 6-3 (A) Annual Earnings

Figure 6-3 (B) Earnings per Share

Figure 6-3 (C) Dividends per Share

Figure 6-3: ANNUAL EARNINGS AND DIVIDENDS of MARMON-HERRINGTON CO.

Financial History for the several management periods, which follow; The figures in this summary will provide a useful reference.

All stock values are as reported in Moody's Industrial Manual.

### The (W.C.) Marmon/(A.W.S.) Herrington Period
Growth and Maturity
1931-1941

This initial period saw the establishment of the firm, a worldwide search for a market, development of a commercial product from the Herrington concept vehicle, the brilliant Ford conversion concept, innovative military vehicle proposals and limited production of them. During the period, the firm was relocated to the former Duesenberg plant and expansion undertaken to accommodate the burgeoning combat vehicle business following the outbreak of hostilities in September 1939. Walter Marmon died in 1940.

"Marmon-Herrington officials had an unusual opportunity to evaluate conditions in the Old World over a period of more than a decade of commercial contacts with military and industrial leaders of the continent, of the Near and Far East, and of the British Dominions. The first hand observations gained in these contacts enabled (them) to forsee the inevitable conflict, and make preparations to do (their) part toward meeting it".[1]

The military products had been designed, developed and produced in limited numbers for the US and friendly foreign governments. The Merz Engineering Unit provided captive tool and gage capacity and the only private Proof Firing Range allowed in-house qualification of armor shipments.

Only expanded facilities were required to begin large scale production.

The plant at Washington and Harding was not only the 'Birthplace of the Arsenal of Democracy' but it was expanded rapidly and achieved accelerated war production at a rate not even approached by the aircraft industry and triple that of any other industry.

Business Week, in the May 10, 1941 issue, showed a plot of the rates of expanded production achieved by the basic industries and by Marmon-Herrington between the base of August 1939 (100) and March 1941. In nineteen months production was increased 600 percent by M-H, while aircraft and shipbuilding increased 390 and 240 percent respectively, and defense industries only 160 percent.

Exponential growth was not new to Marmon-Herrington in 1939. Production had increased 600 percent between 1932 and 1933. Orders received in January 1934 were already double 1933 annual production.

## ORGANIZATION

The officers of the company during the First Period were: Walter C. Marmon, Chairman of the Board; A.W.S. Herrington, President and Chief Engineer; Bert Dingley, Vice-President; D.I. Glossbrenner, Secretary-Treasurer; and all four were the directors. Mr. Marmon died in 1940; the Chair was temporarily vacant and W.B. Nottingham of Clarksburg, West Virginia, became the fourth director. Mr. Nottingham was a lawyer and member of the Washington, D.C. law firm of Steptoe and Johnson, General Counsel to M-H.

The Second Period began in 1941 with the advancement of Robert C. Wallace to Director of Engineering, thereby relieving Mr. Herrington of that second responsibility. Mr. Wallace had been with the firm since 1932, holding a variety of posi-

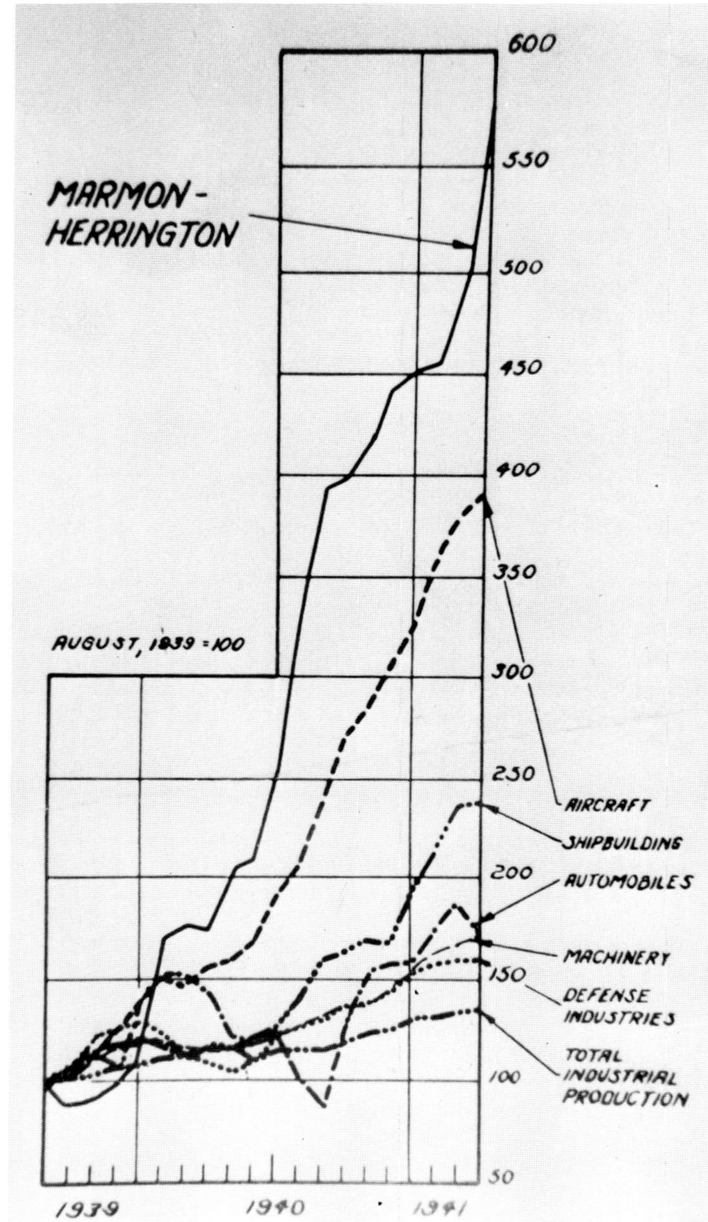


Fig. 6-4 Truly remarkable and unequaled rate at which Marmon-Herrington expanded during WW 2. Curve as it appeared in *Business Week*, May 10, 1941.

tions. S.E. Rowe was assistant Chief Engineer; he had been Asst. Chief at Marmon Motor Car Co.

## FINANCIAL

The Marmon-Herrington Company was established March 13, 1931 as an outgrowth of the Marmon Motor Car Co. The inability of the Marmon Motor Car Co. to meet its stock purchase commitments and the consequent reorganization of December 1931 with Walter Marmon personally involved is reported in Chapter One, Part 2 . . . The Beginnings. This chapter addresses only the reorganized (1931) Marmon-Herrington Co.

Walter Marmon provided limited capital and Fred M. Young, Chairman (then President) of the Young Radiator Co. persuaded several suppliers of major components to join him in extending credit to the infant firm so that it might produce the first group of vehicles for which it held a firm order.

Out of an authorized issue of 2000 shares of $50 par preferred and 3000 shares of no-par common stock, 1100 common (37 percent) were issued to Colonel Herrington for his technical contribution to the new firm.

## PERFORMANCE SUMMARY

The First Period was one of continued and rapid growth with extensive export of trucks and tanks. Figure 6.5 identifies the products of the Period.

Sales escalated beginning in 1939, when they were $0.37 million, to $4.8 million the following year and $9.6 million in 1941.

Pre-tax earnings rose from $1.4 million in 1939 to $5.3 million the following year and to $6.1 million in 1941.

On a per-share basis (adjusted to 450,000 shares), earnings grew from $0.32 in 1939 to $1.18 in 1940 and $1.37 in 1941.

Dividend were paid as follows: 1937, $0.15; '38, $0.38; '39, $0.25; '40, $0.42; '41, $0.75. (Pre 1941 dividends adjusted to 450,000 shares; 150,000 in 1937-39 assumed).

## FINANCIAL DEVELOPMENTS

On August 31, 1937, the Articles of Incorporation were amended to assign par values of $50 to the 2000 preferred shares and $1 to the 1 million common shares. Because the stock was privately held, no information on financial performance is available prior to 1939. Required disclosures to the Security and Exchange Commission prior to a public offering provide some financial results and related information during the period 1939-1941.

The Articles of Incorporation were revised on June 28, 1940 to provide (for a preferred issue of 2000 shares of $50 par stock and) for **600,000** shares of $1 par common stock, instead of 3000 no-par shares; subsequent to August 21, 1937, $1 par had been changed to no-par.

The Merz Engineering Co. was a wholly-owned subsidiary of Marmon-Herrington Co. It had been established in 1939 in anticipation of the unprecedented demand for precision tools and gages which would follow a US entry into the war and became operational in new quarters in 1940. The officers were essentially those of Marmon-Herrrington: Chairman, A.W.S. Herrington; President, Charles C. Merz; Vice-President, Bert Dingley; Secretary, William Nottingham; Treasurer, Earl J. Breech; General Manager, J.C. Thompson.

Mr. Merz had been the AAA Official Observer of the Sanctioned demonstration of the Marmon-Herrington Combat Train in 1935.

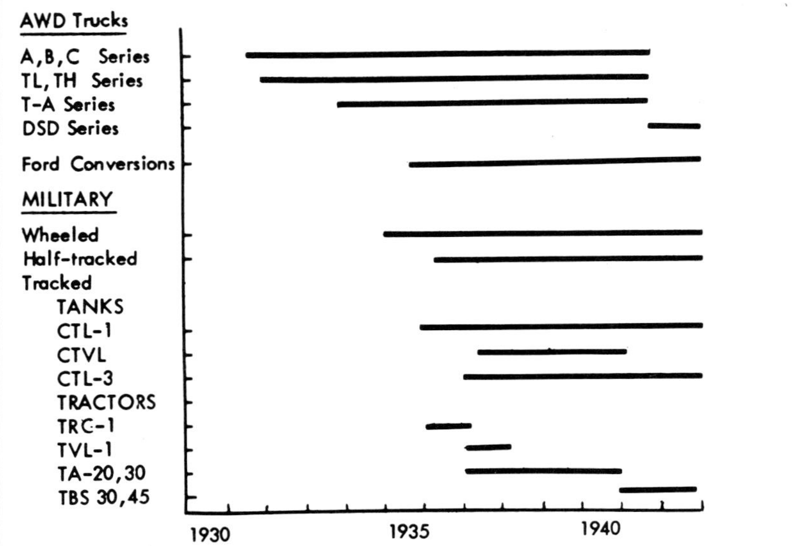


Figure 6-5: PRODUCTS of the (W.C.)MARMON - (A.W.S.)HERRINGTON PERIOD

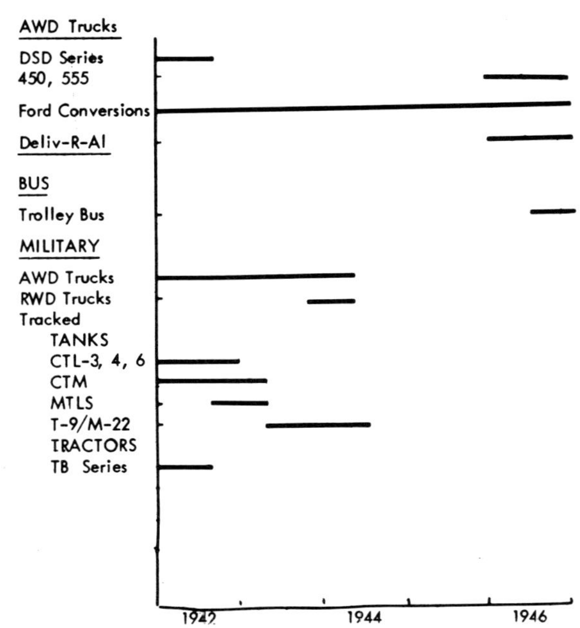


Figure 6-8 PRODUCTS of the HERRINGTON - DINGLEY PERIOD

Fig. 6-6 Management of the Herrington-Dingley Period.

• Ray Harroun (left), winner of first 500-mile Speedway Race and Lee Oldfield of Speedway fame.

# Ray Harroun Visits Plant

On Dec. 17, 1940 all 405 outstanding shares of Merz were distributed as a dividend to holders of 70,000 shares of Marmon-Herrington. The stock was valued at $32,630 or $0.47 per share.

Mr. Breech, after 17 years with the Fairbanks-Morse Co., had briefly been the Asst. Treasurer and a Director of the Indianapolis Motor Speedway Corp. before joining Marmon-Herrington.

During 1940, a premium of $320,000 was paid 'for preferred stock acquired'. Thereafter a preferred authorization of 2000 shares of $50 stock bearing 6% interest was reported as unissued. The transaction may have been related to the death of Walter Marmon.

In preparation for 'going public', several changes were made to Marmon-Herrington capitalization. On October 7, 1941 the company purchased, from the shareholders, the total outstanding stock of the Merz Engineering Company which had been distributed as a dividend the previous year.

A 200 percent stock dividend was issued to holders of the 150,000 common shares outstanding during 1941, bringing the total to 450,000 shares outstanding, of 600,000 authorized. These were changed on November 1, 1941 (meeting of October 17, 1941) from a 'no par' to $1 par value.

Marmon-Herrington common stock was first offered on November 28, 1941. A syndicate headed by Brown, Schlessman, Owen & Co. of Denver, Colorado, and Alexander & Co., Inc. and Straus Securities Co. of Chicago made the offering. Time Magazine reported on January 5, 1942 that: 'The Marmon-Herrington Co. . . . most picturesque of all U.S. truckmakers . . . last week was a closed corporation no longer. It had almost finished selling 150,000 of its 450,000 shares of common stock to the public at $10 a share'.

Colonel Herrington's one third interest in the firm now was in the form of 150,000 shares valued at 1.5 million (1941) dollars (when a top Chevrolet was under $900 and a Cadillac was under $3000); And this just a few months over ten years from the birth of M-H.

### The Herrington/Dingley Period

War and Peace

1942-1946

This period was divided between five years of war production and ten years of civilian production. The preceding period had been one of preparation for the war . . . development of military products. Due to that far-sighted perception, Marmon-Herrington was the first firm to convert to wartime production and expanded at a phenomenal rate unequaled by any other domestic manufacturer. When the war ended, civilian products and markets had to be found which would utilize, and maintain profitable, the expanded facilities. Civilian Truck production was authorized in October 1944. A novel delivery vehicle with front drive by a transverse mounted engine was designed for 1946 introduction. Ford trucks for conversion were not available at the rate required by orders during 1946.

## ORGANIZATION

On August 5, 1942, Bert Dingley was named president as Col. Herrington assumed the Board Chairmanship. Mr. Glossbrenner was replaced as Secretary-Treasurer by William B. Nottingham, (a director) as Secretary and by H. De Braun as treasurer. The new position of assistant secretary/treasurer was filled by L.M. O'Connor.

In 1948 Charles Guernsey came aboard as vice-president to oversee entry into the transit market. He had been vice-president and chief engineer of the A.C.F. Brill Co. manufacturer of buses and trolley cars. Earlier he had been with the Service Motor Truck Co. and most recently was Chief Inspector and then Chief of Production of the Philadelphia Ordnance Distinct. He was 53. That same year Mr. O'Connor, the asst. secretary-treasurer, relinquished the latter responsibility to Mr. J. H. Gant.

Four vice-presidents, effective January 1, 1945 were named: Mr. Wallace, Engineering; C. Alfred Campbell, Public Relations; Seth Klein, Production; and George E. Reynolds, Eastern District (WDC). Public Relations included responsibility for advertising and for post-war sales. Mr. DeBraun was subsequently replaced as Secretary-Treasurer by Mr. Guy C. Dixon, who joined M-H in 1942. Earl J. Breech, Treasurer of Merz Engineering became General Manager as well. Harold Blank and D.R. Meo were Service Manager and Purchasing Agent respectively.

The Dingley organization is shown in portrait form as it appeared on the M-H Christmas greeting card for 1946.

Two events warrant presentation here of resumes of four Marmon-Herrington employees: The promotion of Mr. Campbell to the vice-presidency, and the 1943-1944 presentation of service longevity awards to five former Marmon Motor Car Co. employees who came over to M-H in 1933-34 in similar capacities.

### C. ALFRED CAMPBELL

Mr. Cambell attended the Nordyke & Marmon Executive School about 1921. He held a Bachelor of Civil Engineering from Ohio State University (1920), earlier had studied at the Sorbonne in Paris, France and held a BA degree. After an eleven year career at Stutz where he rose to be Sales Manager and a Director, he followed Dingley and Wallace (and Moskovics briefly) to Marmon-Herrington in 1933.

HAROLD FISHER

Harold Fisher of the Nordyke & Marmon Test Department was a driver on the famous record-breaking coast to coast run in 1916 with millionaire sportsman Sam Stevens. Later he drove most of the 1800 mile, 24 hour run at the Speedway in a stock Marmon 16 sedan. 'A heavy rain slowed him up a bit', he says. A native of Columbus, IN he began with the American Motor Car Co. (underslung) and was then with Marmon, Stutz and Peerless before joining Marmon-Herrington in 1933. He received his Ten Year Award in Sept. 1943, when he was Final Test and Acceptance Suprvisor having been chief of test and chief inspector at M-H

GEORGE McMATH

George McMath came to M-H on Sept. 14, 1933 after several years in the Eng. Dept. of Marmon Motor Car Co. and subsequently in the direct employ of Howard C. Marmon He assisted in the design of the Marmon Sixteen. Born in New Mexico he was raised in Indiana.

FRANK UPDIKE

Frank Updike, native to Indiana (Greensburg), spent his youth watching a blacksmith, worked for Tries Manufacturing Co. in Greenfield who were gas engine builders, but in 1907 began twenty-five years with Nordyke and Marmon. He was, by his own say, at one time or another foreman of almost every production department at Marmon.

Ray Harroun winning driver in the first Indy 500 when he was a Nordyke & Marmon engineer visited Marmon-Herrington in October 1943 to renew old acquaintances with Lee Oldfield (Merz Engineering), Bert Dingley, Vice-president Seth Klein and G.H. (Snappy) Ford. At that time Harroun was General Manager of Saginaw Products Co., Saginaw, Michigan.

HAROLD BLANK

Mr. Blank was Service Manager at Marmon Motor Car Co. and was in the same position at M-H when he received hs ten year longevity pin in January 1944.

LOWELL BURMAN

Mr. Burman was Assistant Superintendent at M-H when he marked his tenth anniversary in 1944. He had previously been with Premier, Stutz and Marmon automobile firms.

## FINANCIAL

PERFORMANCE SUMMARY

The Second Period saw continued rapid growth in sales...28.7 million in 1942, a peak of 36.7 in 1943 and 27.4 in 1944. The end of the war and contract terminations reduced 1945 sales to 11.5 million. Although the resumption of civilian truck production was authorized in October 1944, delayed start-up and parts shortages limited production. Figure 6-8 identifies the products of the Period.

While sales tripled in 1942, from the prior years, pre-tax earnings doubled to 1.3 million but renegotiated war-time excess profit taxes reduced this to 0.45 million; the prior year profit had been 0.6 million. However 1943 earnings were 1.16 million and 1944, after a 5 percent reduction by renegotiation, 0.6 million.

Earnings per share were: 1942, $2.92 reduced to $0.90 by renegotiation of excess profit; 1943, $2.58; 1944, $1.26 (reduced from $1.32); 1945 $0.73.

A dividend of $0.75 was paid in 1942 but none in 1943-1945. A T (US Treasury) Loan arrangement providing a maximum credit of $4 million was obtained in 1944. Although described in the Annual Report as 'Not a present obligation but a reserve', the conditions of the loan agreement prohibited dividends.

The stock which had been sold in November 1941 at $10 declined to $6/$3 range by 1944. The end of the war and the promise of new civilian product lines sent it up to 15/6 in 1945 (and to an all-time high of 18.25/7.25 the following year).

FINANCIAL DEVELOPMENTS

The Articles of Incorporation were revised on November 27 1945 to provide for (600,000 shares of $1 par common stock, each carrying one vote) and **no preferred issue).** The right to issue preferred was withdrawn at the stockholders meeting of January 14 1946.

**The Herrington/Klausmeyer Period**
Acquisition and Diversification
1946-1962

The Herrngton-Klausmeyer Period was characterized by major disruptions in the marketplace for Marmon-Herrington products and by acquisitions and by diversification maneuvers to compensate for the market problems.

The market for trolley buses (and street cars) peaked and promptly deteriorated due to the unfavorable economics of detouring overhead electric power lines during expressway construction in most major cities. Overseas licensees for the trolley bus were contracted in Belgium and Brazil as North American production ended. The post-war period had seen escalating wages for bus drivers and a rapid decline in ridership. Operation of the small transit bus became unprofitable and the 8MB production ended in 1956. Expansion of the school bus business was sought as an alternative. Production and marketing of complete vehicles was undertaken by acquisition of the Oneida bus body business to compliment the pusher-chassis introduced in the prior managment period.

A major move was the entry into the highway tractor business. A cab-over-engine tractor was developed and introduced to the marketplace in 1962. The assembly-line conversion of Ford trucks to all wheel drive begun in 1935, was terminated as was the marketing arrangements with Ford Motor Co. The more conventional construction adopted by Ford for their trucks made the conversion much simpler. The business was continued through the Marmon-Herrington distributor organization.

Earnings plummeted during the latter part of the period following all time high earnings in 1958. Operations were unprofitable in 1960-62. Non-transportation acquisitions were made and proved profitable but were offset by the poor performance of automotive operations.

Fig. 6-7 Management of the Herrington-Klausmeyer Period.

## ORGANIZATION

Mr. Dingley retired as president on June 30, 1946 and was succeeded by an outsider: David M. Klausmeyer, a mechanical engineer who had climaxed a 22 year career with General Motors Corp. by rising to be Chief Engineer (1932) and Plant Manager (1935) of the Chevrolet Commercial Body Plant in Indianapolis. He was then 53; Col. Herrington was 64. Five additional director seats were added. New directors included Russell L. White, President of the Indiana National Bank and of the White Baking Co. (Dayton and St. Louis). Complexity of the changes warrants a tabular display.

| | | |
|---|---|---|
| A.W.S. Herrington* | Chairman | |
| D.M. Klausmeyer* | President | |
| Wm.P.Nottingham* | Secretary | |
| Guy C. Dixon* | Treasurer | |
| J.H. Gant | Asst. Secy./Treasurer | |
| C. Alfred Campbell* | Vice-president-Sales | Russell L. White* |
| Charles Guernsey | Vice-president-Transit | Bert Dingley* |
| Seth Klein | Vice-president-Prod'n | Earl J. Breech (Merz) * |
| Robert C. Wallace* | Vice-president-Engrg. | |

*Directors (9)

Mr. Wallace resigned early the next year and was replaced on the Board of Directors by Mr. Guernsey, and by E.M. Martin, Chief Engineer. Then in 1949 George H. Freers was promoted to the vice-presidency left vacant by Wallace's resignation two years earlier. F.B. Croner was also elected a vice-president.

George H. Freers was promoted to vice-president at age 74. He had held major engineering responsibilities at Nordyke & Marmon and Marmon Motor Car Co. and had worked directly for Howard Marmon thereafter. See "Distinguished (Marmon) Alumni" in Part 2 of Chapter Ten. He had been Chief Engineer of Stutz in 1937-38 and joined M-H in 1940. He had a BS degree in Electrical Engineering.

C.B. Croner, promoted to vice-president, had joined the company in 1947, having been General Superintendent of Material Control for the Chevrolet Commercial Body Division of General Motors in Indianapolis.

Harold Blank, Service Manager and D.R. Meo, Purchasing Agent continued in these positions. Both had held the same positions at Marmon Motor Car Co.

The Klausmeyer organization is shown in portrait form as it appeared on the M-H Christmas card for 1946.

During 1950 Mr. Breech was replaced on the Board by Fred M. Young, the president of Young Radiator Co. The following year R.L. White was replaced by W.P. Flynn, White's successor as president of the Indiana National Bank (one of two M-H bankers).



Seth Klein, a vice-president since 1942 was no longer an officer in 1952. Mr. Campbell a vice-president and a compatriot of Mr. Klein was gone by 1954. The treasurer, G.C. Dixon, was elected a vice-president and director; Mr. Gant became treasurer and asst. secretary. Henry Pascale, a retired military officer, holding a large block of stock, became a director replacing Mr. Campbell.

During 1955, Vice-president Dixon was appointed General Manager as well. Mr. Nottingham acquired the title of General Counsel in addition to that of Secretary. The Board was expanded to twelve with the addition of three members who had been directors of Cardox Corporation prior to Marmon-Herrington purchasing all stock of the firm. They were: J.H. Bell, R.T. Omundson and A.T. Dallstream. The firm of Dallstream, Schiff, Hardin, Waite and Dorschel was General Counsel to M-H from 1959 through 1964. The resultant officer and director roster was as follows:

| | | |
|---|---|---|
| A.W.S. Herrington* | Chairman | B. Dingley* |
| D.M. Klausmeyer* | President | F.M. Young* |

| | | |
|---|---|---|
| W.P. Nottingham* | Secretary and Gen. Counsel | W.P. Flynn* |
| J.H. Gant | Treasurer, Asst. Secy. | H. Pascale* |
| G.C. Dixon* | Vice-president and General Manager | J.H. Bell* |
| C. Guernsey* | Vice-president | R.T. Omundson* |
| G.F. Freers | Vice-president | A.J. Dallstream* |
| C.B. Croner | Vice-president | |

*Directors (12)

Director Omundson replaced Mr. Guernsey as vice-president in 1957-58; the number of directors was reduced to eleven and in 1959 to ten with the departure of J.H. Bell. That same year Cole Younger replaced Mr. Dallstream and J.C. Ross filled the long vacant position of asst. treasurer. It should be noted that the firm of Lybrand. Ross Brothers and Montgomery had been the corporate auditors since 1951 or earlier. Upon the departure of Mr. Guernsey, M.E. Swinford became Chief Engineer of Transit.

### GUY C. DIXON

The arrival of Mr. Klausmeyer and the concurrent elevation of Mr. Dixon to the vice-presidency and a directorship and, in 1955, his additional assumption of the title General Manager makes this an appropriate place to review his resume.

While studying at the Butler University School of Business Administration he worked in the accounting department of Nordyke & Marmon. After graduation he studied business law at Indiana University and the Walton School of Commerce in Chicago. After five years with Ernst & Ernst, he became Supervisor of Factory Accounting at the Chevrolet Commercial Body Plant in Indianapolis. Following a brief stint as Chief Accountant of Hecker Products Co. he joined Marmon-Herrington in 1942 as Secretary-Treasurer.

### ENGINEERING PERSONNEL

The Engineering Department grew from an initial four draftsmen in 1931 to a peak staff of more than fifty.

E.M. Martin, Chief Engineer beginning in 1948 was a University of Missouri graduate in mechanical engineering with some eight prior years experience on specialized vehicles.

S.E. Rowe, Asst. Chief Engineer, Harold F. Baker, Project Engineer (0-12 crash fire truck) and Victor E. Maier, Technical Mathematician supported Mr. Martin in 1950. Mr. Rowe had been Asst. Chief Engineer at Marmon Motor Car Co. beginning in 1927.

William A. Cost, with an ME degree from Purdue, and H.L. Lewis both had extensive prior experience with the J.D. Adams Co. Mr. Cost was associated with the chassis for the NC 5/MA-1 military airport service vehicle.

A.F. LeBon had over twenty years experience with controls and equipment for electrically powered vehicles. He had been at the Philadelphia Ordinance District with Mr. Guernsey.

## FINANCIAL

### PERFORMANCE SUMMARY

The Third Period spanned fifteen years and was the civilian product period. Mr. Klausmeyer inherited the problems of conversion and reestablishment of profitability on a civilian basis. Figure 6-10 identifies the products of the Period.

The year 1946 was an unprofitable one with sales of only 2.4 million. Start-up of the DELIVR-ALL assembly line was delayed for over six months due to parts shortages; a 2000 unit order backlog resulted. Ford standard trucks were not available in anywhere near the volume necessary to meet orders for converted vehicles. Meanwhile tooling was progressing for the transit vehicle debut in 1947.

Sales in the 1947-50 interval were in the 7.5 to 10 million range: 9.4, 8.8, 8.1 and 7.5 million successively. As M-H became the world's leader in trolley bus manufacture, sales escalated in 1951-1953 (12.4, 15.8, and 20.6 million) and declined in 1954 (14.4 million).

Performance in 1955 reflected the June 1955 purchase of all preferred stock of the Cardox Co. for $1,305,083 less than the book value.

Sales continued strong through 1962 remaining between 15 and 15 million except for 1959, 1960 (11.3 and 12.3), while profitability soared through 1958 and then collapsed. Sale of a major asset in 1958 resulted in an extraordinary gain.

Following losses in 1946-48 (0.3, 0.4 and 0.35 million), pre-tax earnings rose exponentially from 0.1 in 1949 to 2.13 million in 1958. After reporting only 0.72 million in 1959, there was a loss in 1960 with a negligable profit in 1961 (-0.41, +0.17).

There were per share losses in 1946-48 (0.18, 0.10, 0.08) and offsetting profits in 1949-51 (0.24, 0.10, 0.23). Earnings were then consistent and averaged about $1.00 through 1959 except for 2.16 in 1956. The asset sale in 1958 resulted in a $0.92 dividend on earnings becoming a $3.86 distribution. A loss of $0.66 in 1960 was followed by a profit of $0.27 per share in 1961. Cash dividends were resumed in 1950 (for the first time since 1942) at $0.10, doubled in 1951 and rose steadily to $0.60 in 1961. In addition stock dividends of five percent were paid in 1957 and 1959. Per share earnings 1951-58 are adjusted for the effect of stock dividends.

The stock ended a two year climb in 1946 at 18.25/7.25, declined to 4.75/3 in 1949 and then averaged about 6.5/4.75 through 1954, uninfluenced by the resumption of dividends. It then rose steadily to 17.25/8 in 1956, and remained at about 15/9 for three years (including the two stock dividend years). A steady decline set in which carried it down to 9.75/6.75 in 1962 (11.50/6.75 in 1961).

Shareholders equity increased, from $4.2 million in 1948-51, steadily to $4.9 million in 1954, and then rapidly grew to $7.4 million in 1955, to $8.4 million in 1956 and to $8.9 million in 1957.

### FINANCIAL DEVELOPMENTS

Marmon-Herrington purchased the entire stock of the Cardox, Corp. of Chicago in May 1955. The purchase for $539,300 was accomplished at $1,305,083 less than the book value of the stock. The acquisition of this manufacturer of low pressure Carbon Dioxide (dry ice), Airdox and Cardox mining systems, drilling equipment, orchard spraying and fire fighting equipment increased stockholders equity from $4,818,425 (1954) to $7,398,567 (1955).

During 1955, 135,753 shares of common stock were sold for $551,959 in excess of par value and that amount credited to capital surplus. Authorized issue was increased on July 19, 1955 from 600,000 to 1,000,000 shares; 574,600 were outstanding.

The Cardox purchase was financed with a $3,000,000 issue of notes bearing 4.5 percent interest: payable on January 1, 1956, $100,000; on Feb. 29, 1956, $300,000; on April 30, 1956, $100,000; and $150,000 quarterly thereafter through 1962. The notes were secured by the pledge of common stock of the Cardox Corp. and by assignment of defense contracts. Net working capital had to be maintained at $1,750,000 and consolidated net at $5,250,000. Dividends could not be paid unless capital remained $250,000 above those minimums. Dividends were paid during the term.

Debentures of the Cardox Corp. in the amount of $400,000 were assumed with minimum payments of $100,000 annually through Mar. 1, 1960, and interest at 4.5 percent. These were paid off during 1958.

Following the collapse of the domestic market for trolley buses in 1955, licenses to manufacture M-H trolley-buses were granted to Ataliers Metallurgiques of Nivelles, Belgium in February 1956 and to Grassi, SA of Sao Paulo, Brazil in March 1956. At least in the latter case, the M-H license was part of a grant of a system license agreement with Trolley bus Villares SA involving manufacture of Westinghouse propulsion motors and controls and trolley-wire in Brazil.

The Brazilian license arrangement is found interesting, when the popularity of Marmon cars in Sao Paulo in the 1920s is noted. (See Chapter 12, Marketing the Marmon-Foreign Marketing). When the merger with Matteson was approved in Dec. 1963, it was noted that: (1) The City of Recife, Brazil was in arrears on payments of $348,115 on Dec. 31, 1962, having stopped making payments in August of that year and (2) that a renegotiation of the loan terms had reduced the balance to $208,869 by Dec. 31, 1963 with prospects for full collection. Circumstances not unlike those of 1983 in international banking!

Only once was Marmon-Herrington struck by the union. In 1956 the local UAW union walked out.

In 1956, wholly-owned Cardox Corp. acquired a controlling interest in Dean-Cardox Chemicals Corp. of Detroit and acquired the CO2 dry ice business of Alkice Corp. of Detroit. Stockholder's equity increased $1,014,565 to $8,413,132.

On March 20, 1957 an issue of 20,000 shares of $100 par preferred stock was reauthorized in addition to the 1 million shares of $1 per common.

A 5 percent stock dividend was paid on January 28, 1957. The issue of 28,745 shares ($1 par) was valued at $373,685 (approx. market price). The former was credited to 'common stock' and the difference to 'paid-in capital'.

In 1957, Cardox acquired control of Well Production Services, Inc. of Denver and the following year control of General Carbonic Corp. of Cleveland. M-H stockholder's equity increased $492,300 to $8,905,432.

During 1958 the (Cardox) CO2, Chlorine dioxide, and fire equipment business, including properties, equipment, inventories and investments was sold. The $344,940 received were not reported as net income but were credited to capital surplus.

In 1959 the school bus body business of the Henney Division of Eureka-Williams Co. of Conastota, NY, was acquired by Marmon-Herrington. This permitted M-H to market complete rear-engined school buses of their own manufacture but proved to be an unwise move.

Then in 1960 a very significant acquisition was made. All assets of the Long Company, a mining equipment manufacturer of Oak Hill, West Virginia, were acquired by Marmon-Herrington.

In March 1960, 50,000 shares were issued as partial payment for certain assets of The Long Co. Of these 17,050 were repurchased in March 1963 at $18 per share ($306,900) following an offer and acceptance. The book value of shares was $16.57 on October 31, 1963; the market price range for 1963 was $11 to $8.5.

An 80 percent interest in the Flash Percolating Co. of Abilene, Texas, was acquired by Cardox in May 1960. The Tung-Set Corp. and a 57 percent interest in Cardox (Great Britain) Limited were also acquired by Cardox in 1960 (or earlier).

**The Klausmeyer-Pritzker/Long Period**
Merger and New Direction
1962-1963

## ORGANIZATION

The 1961 appearance of John B. Long as vice-president and director, following acquisition of the assets of the Long Company of Oak Hill, West Virginia (a manufacturer of coal mining systems and equipment), really marked the beginning of a new era. His board seat was that formerly held by Bert Dingley.

Three new vice-presidents replaced Mr. Dixon who remained as a director. They were R.O. Thomas, School Bus Body Division. J.C. Ross, (former asst. treasurer), and Mr. Long.

The following year, 1962, Colonel Herrington relinquished the chair to former president Klausmeyer while retaining his directorship. Mr. John B. Long became president and was replaced as a vice-president by Mr. P.C. Rice. Mr. Thomas became Executive Vice-president (but was Vice-President of Sales in 1963, replacing Lyn L. Parks). Mr. E.F. Ray was General Sales Director for the Truck Division.

Except for the passing of the chair and the departure of Mr. Dingley, the board was stable.

Very significantly Moody's Industrial Manual noted, beneath the Marmon-Herrington Co. title, 'Affiliated with L.A. Darling Co.' in their 1962 listing. This reflected the sale in October 1962 of Colonel Herrington's 150,000 shares (he retained or reacquired a single share) to the L.A. Darling Co. following negotiations in which the Pritzker brothers of Chicago were the only investors confident of acquiring sufficient additional shares to achieve a 51 percent controlling interest; the Pritzkers controlled L.A. Darling.

The sales and engineering personnel supporting the HDT truck program in 1962 were: G.F. Freers, Vice-President Engineering; J. Mann, Director of Engineering; M.E. Swinford, Chief Engineer; T.M. Hoban, Fleet Sales Manager; H.B. Blank, Service Manager; V.L. Harris, Service Technical Manager.

### CHANGE AT THE HELM

A new organization was fully evident in 1963 when a member of the Pritzker family came aboard as Chairman replacing Klausmeyer, and the only surviving vice-president was Mr. Ross. The secretary, Mr. Nottingham, was the only other surviving officer.

Fred Young and Henry Pascale were the only surviving directors, Colonel Herrington having disposed of all his stock holdings (save one share) and all other directors having been replaced.

The new organization, effective February 18, 1963, was as follows:

| | | |
|---|---|---|
| Chairman | Jay A. Pritzker* | (Director of Matteson Metals) |
| President | John B. Long* | |
| Secretary | W.P. Nottingham | |
| Treasurer | E.G. Larrick, Jr. | |

| | | |
|---|---|---|
| Vice-President | Paul C. Manley | |
| | R.C. Nelson | |
| | J.C. Ross | |
| | G.M. Shevlin, Jr. | (official at Sanford-Day) |
| Directors (7) * plus: | F.M. Young | |
| | R.A. Pritzker | (President of Matteson Metals) |
| | J.E. Burgess | (Vice-President Matteson Metals) |
| | H. Pascal | |

Only Messrs Nottingham (but no longer a director), Ross, Young and Pascal remained from the pre-1961 era. Except for Ross, they too were gone by 1964.

The shareholders remaining after the Herrington sale to L.A. Darling were offered $10 per share to tender their holdings; as a result 60,655 shares were acquired. Additional purchases brought the Darling interest to 309,224 shares. On November 29, 1963, Matteson Metals acquired the Darling holdings at cost. Pritzker family trusts owned all outstanding Matteson shares.

This brought the Matteson holdings to 319,419 or approximately 53 percent of Marmon-Herrington common and the holdings of all board members to 66 percent.

The 1963 board carried over to the expanded firm with the following changes: Jay Pritzker assumed the additional responsibility of Attorney. J.C. Ross, a vice-president assumed the additional office of secretary, replacing Mr. Nottingham (Mr. Ross had been assistant-treasurer in 1960); R.E. Jones replaced G.M. Shevlin as vice-president, and E.W. Smith replaced J.E. Burgess on the board, both after one year. Messrs Young and Pascal were not replaced, thereby reducing the directorate to four, including the two Pritzkers.

When Mr. Long became president on Feb. 18, 1963, the Executive Offices were relocated to the Long-Airdox plant at Oak Hill, West Virginia with the objective of concentration on mining machinery. The automotive operations were relocated to the Cardair Division facilities at Lebanon, Indiana and operated as the Marmon-Herrington Automotive Division of the Marmon-Herrington Company, Inc., 502-510 Indianapolis Avenue, Lebanon, Indiana, 46052. It was under the direction of Mr. M.E. Swinford (who had recently been chief engineer) as General Manager. Harold Blank (originally a Marmon Motor Car Co. employee) became Manager of Sales and Service and second in command to Mr. Swinford. The engineering staff consisted of R. Nothnagel responsible for all-wheel-drive products, Vic Maier and George McMath (ex Marmon Motor Car Co.) assigned to the Aghnides swamp vehicle project and Development Engineer E.V. Brown. Operations at Lebanon began on July 15, 1963 after a thirty day relocation interval during which a $1 million inventory of service parts was moved.

### M.E. (Slim) SWINFORD

The appointment of Slim Swinford as General Manager of the Lebonan Operation noted above is an appropriate time to provide some background information on this individual who made significant contributions to Marmon-Herrington from 1945 to 1980.

When Marion E. Swinford was in his early twenties, his mother ran a furrier shop in Indianapolis; One of her customers was Mrs. A.W. Herrington. The net result of the contact was that young "Slim" Swinford joined Marmon-Herrington in August 1945.

Slim's first industrial experience was with the Emerson-Sharing Tank and Mfg. Co. of Indianapolis, where Frank Sharing provided Slim valuable training. However, the opportunity at Marmon-Herrington offered a salary comparable to Asst. Superintendent at Emerson-Sharing and Slim was quick to take advantage.

Thus began a career with Marmon-Herrington that would extend through 1980 when Slim would retire as Sales and Product Manager at the Marmon-Herrington Automotive operations of the Marmon Transmotive Division of the Marmon Group, Inc.

Slim was hired at M-H by George Freers when Bob Wallace was Director of Engineering. His first assignment was as a sheet metal draftsman, but his career would embrace Chief Engineer of the Transit Equipment Division, Chief Engineer of the Truck Division, Chief Engineer of the School Bus Division, participation in the discussions with Westinghouse and Alcoa on the Pittsburgh People Mover 'DVL", and General Manager of Marmon-Herrington Division of the Marmon Group at Lebanon.

## FINANCIAL

### FINANCIAL SUMMARY

Sales in 1962 of 20 million and in 1963 of 15 million were typical of the preceding decade but a serious delay in resuming production of an acquired bus body product line and failure to find automotive volume to utilize the Indianapolis plant facilities resulted in losses which offset the profits of non-automotive operations.

A loss of $0.50 million in 1962 exceeded that of 1960 but the closing of the Indianapolis plant in mid 1963 resulted in a pre-tax profit of $1 million despite sales 25 percent below the prior year.

Per share earnings were: 1962, ($0.80); 1963, $1.67.

The dividend was reduced from $0.60 in 1960 to $0.15 in 1961 and $0.10 in 1962, none was paid on 1963.

The stock price decline ended in 1962 at 9.75 asked/6.75 bid. Efforts of certain holders to increase holdings, following the sale of Col. Herrington's interest, were probably responsible for a climb to 11/8.75 in 1963 and to 14/8.5 in 1964.

Shareholders equity increased to $10.6 million in 1962 (from $8.9 million in 1957) and to $11.6 million the following year. However it was reduced consequent to discontinued operations to $10.0 million, or $13.83 per each of 723,440 shares outstanding.

As a condition of the merger with Matteson Metals, all common stock would be exchanged on a 10 for 1 basis for a new issue of preferred stock paying $5 until a mandatory surrender option was exercised by the Board of Directors.

### FINANCIAL DEVELOPMENTS

Although the company remained profitable until 1960 and posted small profits in 1961 and 1963, the Indianapolis operations were in serious trouble beginning in 1968.

| Year | Profit/ (Loss) |
|---|---|
| 1958 | (996,000) |
| 1959 | (955,000) |
| 1960 | (2,017,000) |
| 1961 | (1,065,000) |

| | | |
|---|---|---|
| 1962 | | (1,485,000) |
| 1963 | (10 mon.) | (826,000) |
| Total | | (7,344,000) |

Overall net profit as percent of net sales was only 6.95 percent in 1963. Without the Indianapolis net loss of 117.31 percent, the return would have been doubled (875,041 corporate profit vs 826,000 loss from Indy operation) to over 13 percent!

For 10 months of 1963, automotive sales activity was only:

| Vehicle Conversions to AWD | AWD Kits | HDT Trucks |
|---|---|---|
| Factory 33 | 29 | 28 |
| Distributor 43 | 72 | |
| Total 76 | 101 | 28 |

Diversification had saved the company. A story frequently to be reported during the depressed economy of the 1973-1983 period.

During the spring of 1963, when an anticipated major vehicle contract failed to materialize, a decision was made to discontinue Indianapolis operations and effectively to retire from the business of series production vehicles. Except for the highway truck; the inventories and tooling were written down to realizable amounts on disposition; the inventory was disposed of and the machinery and equipment was auctioned at a book profit of $65,000. The real estate was offered for sale.

Holdings of Cardox (Great Britain) Ltd. were increased 57 percent to 77 percent in 1963 at a cost of $57,930 and an obligation assumed to increase the holdings to 85 percent at a cost of $22,600.

In July 1963 the Sanford-Day Corp. of Knoxville, Tennessee, was acquired. Associated M-H indebtedness was $659,045 in the form of non-interest bearing notes payable monthly to August 1968 at $11,983.

During 1964, Marmon-Herrington automotive operations remaining . . . allwheel drive conversions and kits . . . were relocated to the Lebanon, Indiana, facility initially shared with Cardair Division of M-H, primarily a manufacturer of special compressors for military missiles. There was a declining market in 1963 and the following year Cardair was recombined with Long-Airdox, who like Cardox (GB) Ltd. was a manufacturer of special compressors for the coal mining industry. This made the Lebanon operation entirely Marmon-Herrington Automotive Division.

The 1 million dollar inventory of spare parts which was moved to Lebanon was subsequently managed with a fair return on investment. The trolley bus fleet in Santiago, Chile was refurbished with an even more favorable return.

In November-December 1963 as the merger with Matteson Metals was maturing, Marmon-Herrington invested $74,000 in a 'foreign affiliate'. This is presumed to be associated with the transition from Marmon-Herrington SAF (France) to Marmon-Bocquet in which a 24% interest was held. This French affiliate manufactured M-H components and all wheel drive conversions for military trucks (and was reported doing so in 1981, including manufacture, under M-B contract, of the M-H 600 series chassis by Unic Division of Simca.)[1]

The shareholders of Marmon-Herrington approved on Dec. 30, 1963 an Agreement of Merger of Matteson Metals Inc., into Marmon-Herrington. Each holder would receive one share of the new preferred issue of $110 par value for each ten of the old common held.

The preferred stock was to be subject to redemption upon ten days notice, dividends thereon to cease on stated redemption date **"and the respective holders thereof shall have no other right or interest except to receive payment of the redemption price."**

The significant issue in the stock conversion provisions of the merger agreement concerned the ability, of pre-merger holders of Marmon common to control the duration of their continued interest in the post-merger corporation. Because all of the authorized common of the post-merger corporation was dedicated to conversion of Matteson shares, none was available except at the will of the pre-merger Matteson holders. Unable to acquire the new common, a pre-merger holder of common was assured only of ten days continued participation at any time subsequent the merger . . . the ten day notification period before arbitrary redemption by action of the Board of Directors. Only two of the seven directors on the new Board were long-term survivors of the original firm . . . Fred Young and Henry Pascale.

Pre-merger stockholders' equity was $10.0 million (9,998,696) from Marmon-Herrington and $0.9 (.945,996) from Matteson.

Sixty one percent of the 558,359 shares voted were in favor; the 342,080 yes votes included all the Pritzker shares and about nine percent of the remainder. The non-Pritzker holders therefore opposed the merger by nine to one.

The combined common/preferred stock conversion increased the Pritzker control from 57 percent to 71.5 percent. Exercise of the arbitrary redemption provision by the rebalanced board would increase the control to 100 percent.

The post-merger shareholder equity was divided on the basis of preferred and common holdings: $3.1 and $7.8 million.

Captain Pascale went to court in an effort to overturn the Agreement of Merger. Colonel Herrington said in an interview that he had friends on both sides of the issue. Fred Young voted in favor of the merger.

Charles Vaughan, Business Editor of the *Indianapolis News,* reported that the Pascale suit had been settled out of court and a final hearing on approval of the merger would be held on September 28, 1964.

The settlement was based on payment to the minority stockholders of $15 per share held. This arrangement replaced that reported in the *NEWS* on July 13, 1964 wherein preferred stock, subject to redemption, was to be exchanged for common. Captain Pascale owned 173,757 shares to be exchanged for $2,606,355!

Mr. Vaughan, to whom the author is indebted for his excellent coverage of a complex issue, quoted a minority holder: "This is no munificent sum but does work out to be fifty percent better than the initial offer, and a lot better than receiving interest bearing bonds in lieu of stock. Nearly all common shares acquired by the majority holders had been purchased around $10 per share.

Evidently the ambiguity of Section 32 (a) (3) of Indiana Corporation Laws had been clarified as prohibiting involuntary conversion of shares in a surviving corporation. This determination forced a substitute arrangement on the majority holders.

Colonel Pascale told Mr. Vaughan that he would petition the Indiana Legislature to formally clarify the above section and to make amendments to protect minority holders in the future. He sought: a two-thirds rather than majority approval of merger; and determination of 'value' by impartial audit and immediate payment of 75 percent thereof irregardless of appeal for a higher return.

The Board of Directors, with the Pritzker brothers abstaining, voted in favor of the merger agreement. Henry Pascale and Fred Young resigned.

Many years later, (July 27, 1976) Fred Young wrote to M.E. (Slim) Swinford at Marmon Transmotive/Marmon-Herrington Operations and had this to say: *From the very beginning . . . I was a director and always made their radiators and loved to work with Col. Herrington, Bob Wallace and others . . . then you folks (the Pritz-*

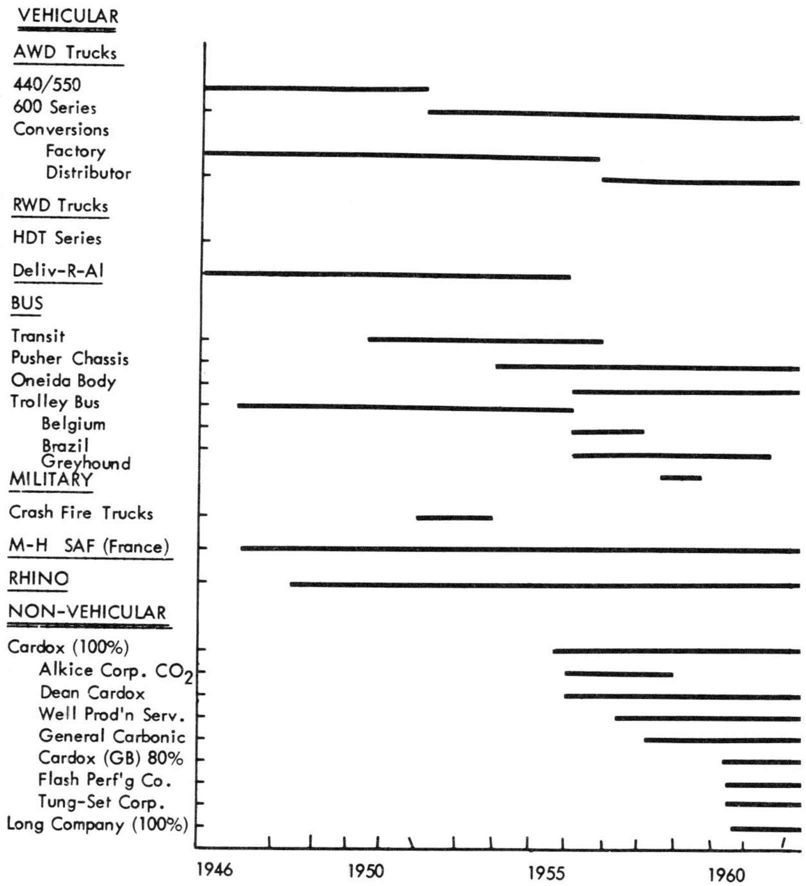


Figure 6-10: PRODUCTS, ACQUISITIONS AND DIVERSIFICATIONS of the HERRINGTON - KLAUSMEYER PERIOD 1948 - 1962

Figure 6-11: GENESIS of the MARMON GROUP, INC.
1953 - 1969

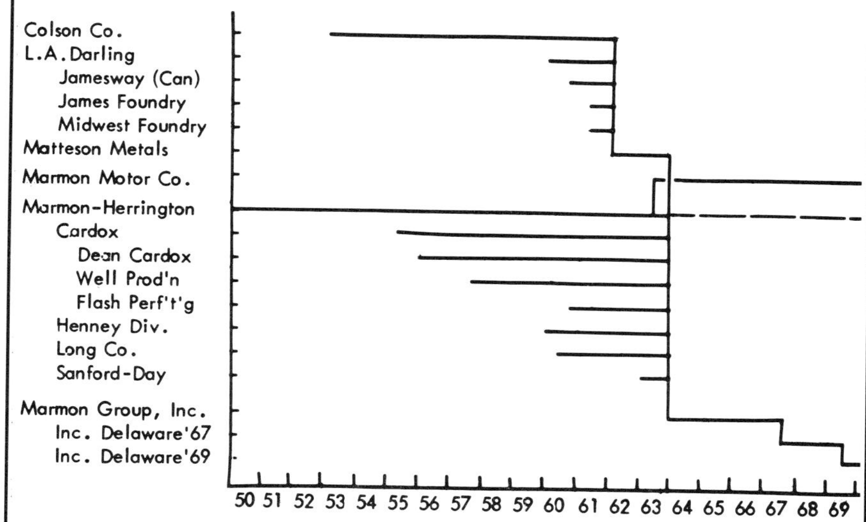


Bars originate on acquisition date and terminate upon merger integration.
Primary indents are subsidiary acquisitions.
Secondary indents are acquisitions by subsidiaries.
Interrupted bar reflects continuation of M-H as a unit of the Marmon Group, Inc.
Marmon Motor Company established mid-1964 to acquire HDT assets of M-H.

*ker team) came along. I was honored to remain on the Board of The Marmon Group and I have been very sorry that I resigned as I should like to be with Bob and Jay Pritzker . . . they are honest, they are brilliant and wonderful fellows.*

The effect of the merger was to increase current assets from $9,345,273 to $15,276,595.

### The (J.A.) Pritzker/(R.A.) Pritzker Period 1964-19XX

Unit of The Marmon Group

THE MARMON GROUP

The three operating units of the pre-merger Marmon-Herrington Co.: Marmon-Herrington Automotive (Lebanon, Indiana), Long-Airdox Co. (Oak Hill, West Virginia) and Sanford-Day Co. (Knoxville, Tennessee) became part of The Marmon Group (Indiana) in 1964.

On September 24, 1965, The Marmon Group, Inc., (Indiana) became a Delaware chartered firm by merger into The Marmon Group Inc., (Delaware) chartered April 30, 1965. There are regulatory advantages to a company chartered in Delaware . . . for example, General Motors is a Delaware company.

Authorized stock of The Marmon Group (Delaware 1964) was:
20,000 shares — $100 Preferred; none issued
1,000,000 shares — Common; 452,520 issued including; 120,081 M Grp (IN) Treasury stock

In 1969 it was briefly merged into Whitehall Enterprises (Delaware 4-23-69), a privately held company holding title to 90 percent of Marmon-Herrington stock. Upon purchase of the outstanding stock remaining at $1650 per share, all Marmon Group, Inc., (Delaware 1967) stock was canceled and retired upon petition of Whitehall Enterprises by Jay A. Pritzker, Chairman and Allen M. Turner, Asst. Secretary. On July 1, 1969 a Certificate of Ownership was filed with the Delaware Secretary of State merging The Marmon Group with Whitehall Industries, relinquishing the Whitehall name and assuming the name "The Marmon Group, Inc." The Marmon Group Inc. (Delaware 1969) had become a privately held company, at least for the immediate future.

The Board of Directors of Whitehall consisted of A.N., Jay A., and Robert A. Pritzker. The officers were generally those of The Marmon Group Inc. (Michigan 1967): Robert A. Pritzker, President and Treasurer; Walter S. Rogowski, Secretary; George A. Jones, Vice-President; Robert C. Gluth, Vice-President and Assistant Secretary; Thomas M. Conley, Asst. Treasurer.

Authorized stock of Whitehall was 1000 shares of preferred no-par and 1000 shares of common no-par. Issued and outstanding were 190.3345 shares[4].

The Officers of The Marmon Group (Delaware 1969) were: Jay A. Pritzker, Chairman; Robert A. Pritzker, President; Walter S. Rogowski, Vice-President; Franklin B. Allen, Secretary and Gen. Counsel; Robert C. Gluth, Controller; Lloyd J. Kurkowski, Treasurer.

Directors were: Jay A. Pritzker, Chairman of the Board; Marshall Bennett; David S. Burnett; Robert H. Hayes; Robert A. Pritzker; Dale W. Wonus.

Of the 6,000,000 common shares authorized, 3,880,000 were reserved for conversion of the preferred stock (on a basis of two common for one preferred ($1 annual dividend) at a rate beginning in 1969 of 100,000 per year and progressively increasing starting in 1972 to full conversion in 1975. Per share earnings were based on the 3,880,000 reserve shares plus the 534,410 outstanding in 1967 and 467,222 in 1968.

### MARMON-HERRINGTON AUTOMOTIVE DIVISION

In 1970, the Marmon Group acquired Lowell Transport Parts and combined it into Marmon-Herrington Automotive Division at Lebanon. Lowell was a Chicago based supplier of trolley wire, truck and bus parts, and used trolley buses on an international basis.

In late 1971 the organization of the Marmon-Herrington Automotive Division of The Marmon Group, Inc. was restructured as follows: F.B. Tuttle, Vice-President and General Manager; M.H. Griggs, Chief Engineer; R.H. Hawk, Sales Manager; B.D. Wiora, Material Control Manager; M.E. Swinford, Production Manager.

Mr. Tuttle had previously been Staff Management Consultant. He was replaced by Colonel L.L. Davis, USMC, Ret'd in mid 1972 with a mandate 'to get things straightened out'; he resigned on March 12, 1973.

At that crucial point, Mr. Swinford and David Pethick (Automotive Manager of the Marmon Group and Detroit based) again took control and evolved the decision to move Marmon-Herrington to Knoxville, Tennessee and integrate it with the operations of the Marmon Group already located there. There was a concurrent decision to liquidate Lowell Transport Parts (acquired the prior year) which shared the Lebanon facility. Marmon-Herrington was in business in Knoxville on May 1, 1973. The Webb Wheel Division of the Marmon Group then relocated to Lebanon from Siloam Springs, Arkansas for logistic considerations.

After relocation to Knoxville, Tennessee in 1973 to share a facility constructed in 1967 for the Sanford-Day (1963) and later Railway Products (1969) Divisions, the Knoxville operations were merged into a single heavy-manufacturing unit designated Marmon Transmotive Division. This occurred in late 1973, officially on January 1, 1974. Thereupon the automotive operations were designated Marmon-Herrington Automotive (operations). Marmon Transmotive employment was about 250.

R.C. Crosby is General Manager of Marmon Transmotive Division including Marmon-Herrington Automotive where Slim Swinford was Sales Manager until his 1980 retirement. Robert E. Jones, with Sanford-Day since 1920 and a Professional Engineer was manager of mining sales. It occupies a 150,000 square foot building on a 40 acre site in Forks of the River Industrial Park in suburban Knox County, ten miles from downtown Knoxville.

Marmon-Herrington Automotive specializes in all-wheel drive conversions, all-wheel drive kits for distributor installation and the design and construction of prototype and special purpose vehicles.

## DISTINGUISHED ALUMNI

Certain Marmon-Herrington employees went on to significant careers subsequent their departure. Among these were the following:

### DENNIS DUESENBERG

Denny Duesenberg, son of Fred, was a project engineer at Marmon-Herrington. After a stint at Perfect Circle Corp., he joined the Ford Motor Company from which he retired in 1982. He specialized at Ford in piston/cylinder development.

### JOSEPH PROSKE

Mr. Proske was one of the original four draftsmen at Marmon Motor Car Co. assigned to the Marmon-Herrington program. He was the first to work on military ordnance vehicles. After leaving Marmon-Herrington he became Chief Engineer at the Detroit Tank Arsenal of the Ordnance Department.

### ROBERT C. WALLACE

Bob Wallace resigned as a Director and Vice-president of Engineering at Marmon-Herrington in 1947, having been with M-H since 1932. He accepted the position of Executive Engineer of the Diamond-T Motor Co. and was promoted to Chief Engineer in 1958. Following the merger of Diamond-T into White Motor Co. in 1961, Mr. Wallace was Director of Engineering at the Lansing Division, builder of Diamond-Reo trucks. Following retirement in 1967, he was a consultant to the Chief Engineer of Brockway Motor Truck Co. for four years.

On September 25, 1980 Bob Wallace was honored, for his contributions to motor truck development, by the American Truck Historical Society. At a dinner meeting of the Illinois Trucking Association he was presented with a Special Citation as 'A Founder of the Trucking Industry'. He was also honored by life membership in the American Ordnance Association and the Automotive Hall of Fame.

His motor truck career began at Marmon-Herrington; he had been associated with Stutz in Indianapolis from 1924 to 1931 where he worked for Bert Dingley. He quit Stutz after five pay cuts of ten percent each. While at Stutz he (1) developed the first standard-equipment power brakes for a production car, the 1929 Stutz; (2) devised brake stopping distance determination firing blank cartridges marking black spots on the red bricks of the Speedway road; and (3) designed, for use in Saudi Arabia, all-wheel drive Stutz chasses for a 4 x 4 oil field crew station wagon and a 6 x 6 drilling chassis for exploratory slim hole work. Born in 1903, Mr. Wallace was a graduate of the Lake Forest (Illinois) Academy, and the Massachusetts Institute of Technology where he earned membership in Tau Beta Pi and Pi Tau Sigma.

## ADVERTISING AGENCIES

The Marmon-Herrington advertising account was placed with the Caldwell-Baker Co. of Indianapolis in 1943. Mr. C.M. Aldred was the account executive and the agency address was Merchants National Bank Building. This relationship

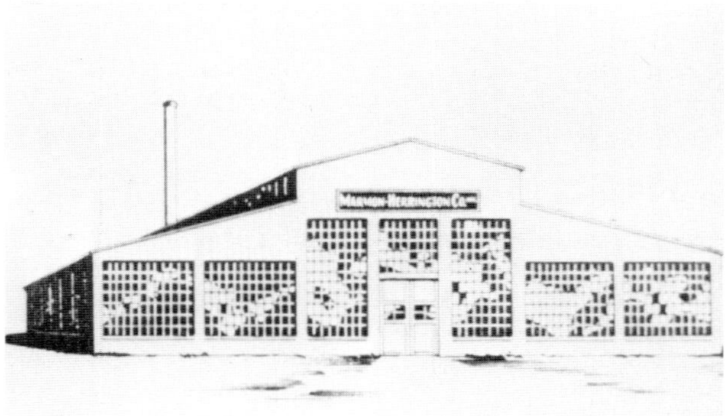

Fig. 6-9 Original Marmon-Herrington Plant within Marmon Plant 2.

Fig. 6-13 Assembly operations on Model A-10 in 1932.

Fig. 6-12 Washington-Harding plant, 1937, after purchase from Duesenberg.

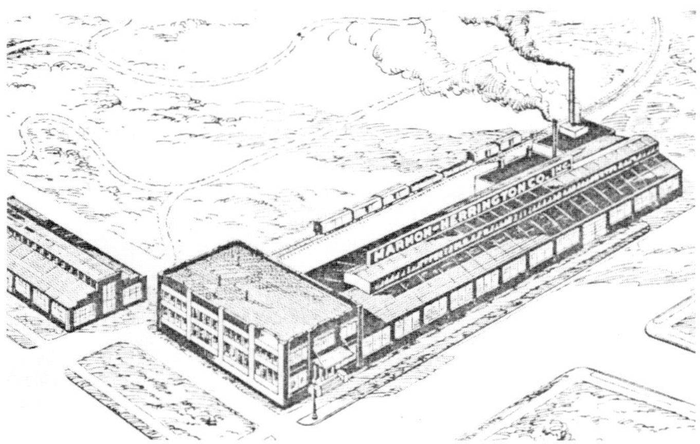


continued through 1962 by which time the agency had become Caldwell, Larkin and Sidener-Van Riper, Inc. and John Merkle was the account executive in 1961-62.

By April 1964 the account had been shifted to Ketchum, MacLeod and Grove, Inc. whose address was Gateway Center, Pittsburgh, PA. It remained there through 1970, after which agencies came and went on an annual basis.

1971 MacGill-Ross of Indianapolis
1972 Marcom of Indianapolis
1973 Quinlan, Keene, Peck and McShay of Indianapolis
1974 Lin Lockhart of Richmond, VA.

## PART 2 — The Marmon-Herrington Factory

During its existence, Marmon-Herrington occupied successively two plants in Indianapolis, one in Lebanon, Indiana, and the present plant in Knoxville, Tennessee.

The Marmon-Herrington plant is of particular interest because of the remarkable rate at which it was expanded during the Allied and US involvement in World War II. Between mid-1939 and mid-1941 production was increased over 300 percent, twice the rate for all defense industries and unequaled even by the critical aircraft industry. Dramatic facility expansion was obviously necessary as was a total management commitment to victory.

The infant company set up operations, following the March 13, 1931 incorporation, in 'a segregated portion of the Marmon Motor Car Co. plant at Morris Street and the Belt Railroad'.[5] This description evidently is a general location of the Marmon site rather than a specific location on that site. The incorporation documents listed the address as 'York Street at Kentucky Avenue'. The specific location was the extreme end of the Plant 2 assembly building, near the powerhouse.

Based on a photograph of parallel assembly lines for 1937 Ford V-8 conversions which appeared in Marmon-Herrington *All Wheel Drive NEWS* (Vol. 6 #2), that operation appears to have been in the Marmon Motor Car Co. Plant 2 machine shop/assembly building on York Street where Marmons were built in 1932. A companion photograph shows an extended rail car loading area characteristic of that plant. The associated text notes recent production of a single order for 225 (6 x 6 Ford) units and states: The two scenes give an idea of the remarkable activity in **the two Marmon-Herrington plants in Indianapolis.**

The impending sale of Marmon Plant 3 to Stewart-Warner Corp. was announced on November 11, 1937.

By October 23, 1937, when the acquisition of a new plant for planned relocation was announced, the press could refer to the **internationally known engineering and manufacturing firm.** The former Duesenberg plant was purchased from the E.L. Cord Corp. and operations of Marmon-Herrington relocated to 1511 West Washington Street, at Harding Avenue on the southwest corner. Harding Avenue

Fig. 6-15 Merz Engineering Co. plant 1940.

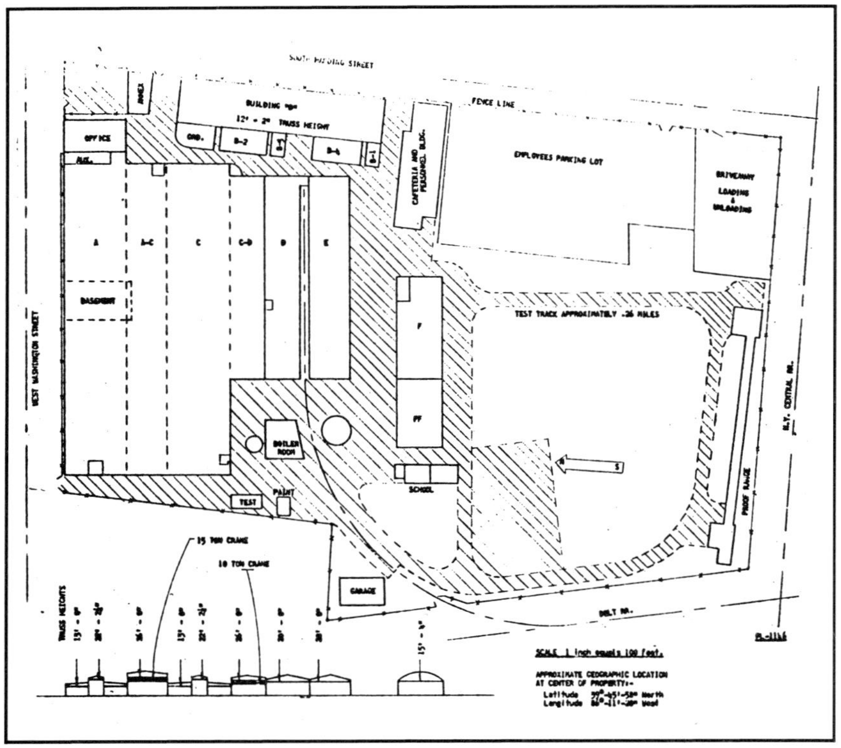

Fig. 6-14 Plant layout circa 1956.

runs north-south just west of the Marmon Motor Car Co. site and Washington Street, the principal east-west artery (US 40) of Indianapolis, is about 1.25 miles north of Morris Avenue (the northern boundary of the Marmon Motor Car Co. plant site).

When the Duesenberg plant was acquired, it consisted of a two-story office building and two factory buildings (one large and one small) extending along the two streets. The site included 16 acres, extending 700 feet along Harding Street to the Big Four (Conrail) Railroad tracks. Construction of the plant had begun in 1920 with the official opening on June 1, 1921 and was little changed when Marmon-Herrington moved there.[6] Colonel Herrington immediately announced the intention to expand the facilities. The site was served by a siding of the Indianapolis Belt Railroad on the west of the site.

It had been the offices and factory of the Duesenberg, Inc. firm. The Cord Corporation was represented in the transaction by the former receiver of Duesenberg, Inc., William T. Rasmussen. The newspaper account quoted George H. Bellstein, purchasing agent for Duesenberg as saying that he had no knowledge of future plans of the Cord Corp. for manufacturing Duesenberg automobiles or for transfer of any equipment. Ray Faulkner, the president of Auburn and R.F. Pruyt, the secretary of the Cord Corp. were in Indianapolis but the press was unable to reach them for comment. The purchase price was not revealed.

The release, carried by both the *News* and the *Star* on October 23, reported that "the Marmon-Herrington Company will take over the machinery and equipment now in the plants besides installing all of its present equipment in the new location".

The move was completed by Jan. 1, 1938, soon after the first demonstrations of the high-speed track-laying tractor.

Buildings on the Washington-Harding site will be identified herein using the official alphabetic designations evident in the 'Building Area Layout — 7-20-56'. Certain buildings were extended in length and the original structures will be designated by, e.g., A East and B North. It shows the final site population of structures and will be a convenient reference during discussion of the plant growth.

The property as purchased included: (1) the 50 x 90 foot 'fireproof' two story (later three) office building; (2) a connecting assembly building of steel and brick construction, designated 'A East', extending along Washington and measuring 88 x 300 feet; and (3) a separate brick and steel building 50 x 150 feet extending along Harding Street and designated 'B North'. The three structures totaled 52,000 square feet. A boiler house adjacent 'A East' added about another 1,000 square feet under roof.

At an unknown date, an area 1,577 square feet in the adjacent factory building was dedicated 'Auxiliary Office' space and included in the floor area of the 'Main Office'.

Subsequently a small site on Harding Street directly across the Pennsylvania tracks was acquired for the construction of the Merz Engineering Co. (wholly-owned) facility of 31,000 square feet. The address was 200 Harding Street. The Plant was equipped with the very latest equipment for the manufacture of precision gages, tools and fixtures. It provided Marmon-Herrington with essentially on-site expedited supply of all its requirements in those categories and was an important source for the domestic mobilization effort as well.

Just north of the tracks and paralleling them M-H constructed the only non-government range for proof testing of armor plate samples. It would serve a large segment of midwest industries during the US involvement and precluded any bottlenecks in the approval of armor for M-H production. It added 5121 square feet to the facilities.

Within 30 days of the invasion of Poland (Sept. 1939), ground was broken for construction to double the plant facilities. A second building paralleling building 'A East' was constructed and building 'B North' was extended to double its length.

The new building 'C was nearly identical to A East, measuring 90 feet 9 inches (versus 87 feet 9 inches) but extending to 456 feet 6 inches. B East became build-

Fig. 6-17 Expanded plant in late 1940.

ing B, 60 x 300 feet. The added area of 52,000 square feet brought the total, without Merz Engineering but including four small structures $B^1$, $B^2$, $B^3$, and $B^4$ abutting the east wall of B, to 116,000 square feet. Building A and C each had a central section 22 feet high under trusses and two side sections 13 feet under trusses; Building B had only a 12 foot clearance, and served as the machine shop.

The additions involved relocation of the power plant.

Just nine months after that groundbreaking, the Indianapolis News reported on June 19, 1940 that Marmon-Herrington would expand their plant in order to build combat tanks for the US War Department. A new building already underway was to double production capacity when completed in September 1940. The present plant on Washington Street was to be extended and serve as main assembly building. The new structure, of brick and steel construction would conform to existing monitor style of buildings. Also the boiler house was to be modernized and a spur track run from Belt Railroad to 300 feet of new loading docks.

Specifically, building D was erected and A East (300 ft.) extended to 475 feet. Building D measured 60 feet 6 inches x 297 feet and 20 feet under the roof trusses. Other than building B which had only a 12 foot truss height, D was the only one with a clear span construction. Buildings A and D together totaled 66,692 square feet; the increase in roofed area was 38,191 square feet.

A third story addition to the main office structure (4089) and erection of a one story personnel building (2116) (facing Harding Street and nearly abutting the main office on a corner) and expansion (500) of the power plant added 6700 square feet of roofed area.

Square footage had been increased to 155,500 or three times the original area.

During the fall of 1940 another 33,550 square feet was constructed as follows: An abutting companion building (19431) to D was erected and designated E (60 x 297) which construction enclosed the rail siding on the south face of D; a 12,000 square foot structure F (80 x 150) was erected south of E to serve as an experimental facility; a second floor (2116) was added to the personnel building.

The Standard Oil gas station on the Washington-Harding corner was omitted on

Fig. 6-16 Expanded plant in mid 1940.

the artist renderings of earlier plant phases. It continued to serve the public throughout the war years and was earlier used to fuel test vehicles.

In the winter of 1940 a new one story personnel building (10185) at the opposite end of B was erected; the former personnel building was redesignated 'Office Annex'. A school for mechanic instruction (2145) was erected east of building F which was extended 100 feet (8000) and designated FF. The original purpose of F had been pre-empted by the need for a spare parts facility. A new two story building for resident Ordnance personnel was erected (4220), the boiler house was modernized and expanded (867), and a 'test' structure erected (1127) near the boiler house.

The project added additional square feet to bring the area under roof to 214500 square feet; the Merz Engineering building added another 31,000 for a total of 245,500 square feet.

Additional space was required and the structures were already encroaching on the area dedicated to a test track, to product and employee parking. By spring 1941, building's A and C and buildings C and D had been joined by roofing the two intervening spaces. Both had high unobstructed bays with 26 foot clearance and crane ways. Building A-C was 62 feet 9 inches wide (29050) with a 15 ton crane; C-D was 49 feet (14587) with a 10 ton crane.

A 2800 square foot garage and a 580 square foot paint storage structure brought the area for this addition phase to 47,000 square feet, bringing the total roofed area to 261,500 and including Merz to 292,500 square feet.

For security reasons, at the time the Locust tank programs got underway, the entire M-H site was fenced with chain-link and barbed wire. On the street sides, the fence line was the curb line. A water tower was erected for fire emergency considerations. An extra 2500 square feet added in 1944 may have been related thereto.

Also in 1944 the Merz plant was enlarged by 2600 square feet for a total M-H area of 264,100, a total Merz area of 33600 for a grand total of 297,700 square feet.

In 1944 the company described[7] their facilities and the financing thereof as follows: *"The manufacturing plants and offices of the Marmon-Herrington Company,*

Fig. 6-18 Twin assembly lines for the M-H tractor — 1941.

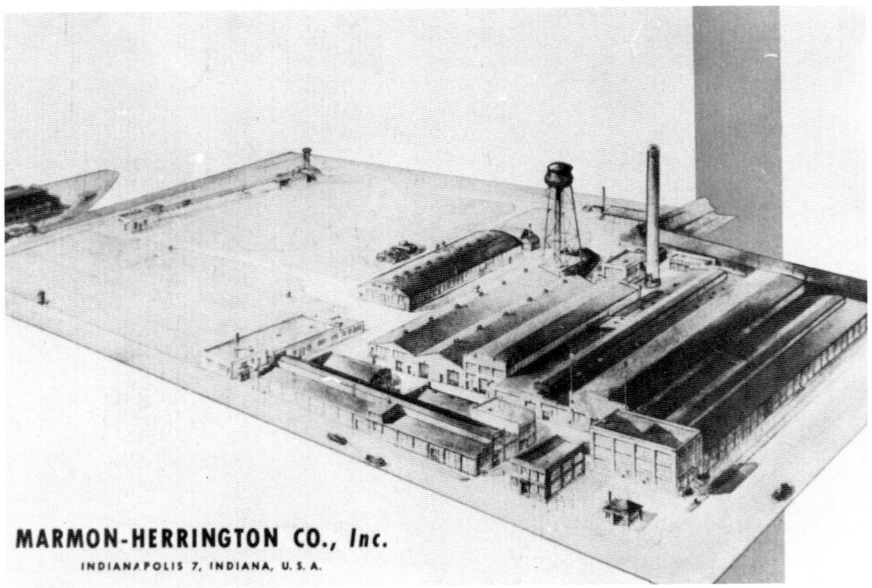


Fig. 6-19 The completed expansion 1944.

Fig. 6-20 The Marmon-Herrington plant in late 1944.

Fig. 6-21 Medium Tank Assembly in Building A-C in 1943; Model CTMS-ITBI.

Fig. 6-22 Marmon-Herrington assembled H542 4x2 tractors — 1944.

Fig. 6-23 M-H Trolley Bus assembly operations in Building A-C — 1950.

*and its wholly owned subsidiary. The Merz Engineering Company are among the most modern in the industry. Plant expansions representing a capital investment of more than three million dollars, without recourse to government or public financing, have been built, almost in their entirety, since early in 1940. The plans for these buildings were drawn in advance of actual need and were ready for immediate use when the first large scale production program required their erection".*

Apparent disposal of Merz in 1948 reduced the area correspondingly.

The Indianapolis plant was closed permanently in mid 1963 and the property sold to the American Fabricated Products Co. in December 1964.[8] As of 1981, the plant was owned by J.W. Hodges Co. and for sale at $1.75 million. Sale to the City of Indianapolis was considered probable, for the construction of a bus garage.[9] Earlier Remco Hydraulics had been a tenant. The Marmon-Herrington sign on the roof was still evident in 1982.

Efforts were initiated in 1985 to have the site marked as a "National Historic Site . . . The Birthplace of the Arsenal of Democracy."

Remaining automotive operations were transferred in 1963 to an existing plant at Lebanon, Indiana, located at 502-510 Indianapolis Ave.

In 1973, The Marmon-Herrington Automotive Division of The Marmon Group was relocated from Lebanon to Knoxville, Tennessee to share a building, erected in 1961, with the Sanford-Day and Railway Products Divisions of The Marmon Group. The integrated operation was identified Marmon Transmotive Division with the three operations retaining product line identity. This plant has 150,000 square feet under roof on a 40 acre site in the Fork of the River Industrial Park ten miles from downtown Knoxville. The address is P.O. Box 1511, Governor John Sevier Highway, Knoxville, TN 37901.



## The Billionaires

| **NAME** Age | | **WEALTH** billions of U.S. dollars | **SOURCES OF WEALTH** Background |
|---|---|---|---|
| **JAY PRITZKER,** 67<br>**ROBERT PRITZKER,** 63<br>**and family** | Chicago<br>ILLINOIS | **$4.5** | **100% of Hyatt hotels; 100% of Marmon Group, a collection of 60 industrial corporations; stockholdings; trust funds.** Jay, a lawyer, and Robert, an engineer, have one of the globe's most diversified portfolios. The brothers hold properties like playing cards; one added, one discarded. Recent play: out of Braniff Airlines, into Ramada Inc., then out again. Not all their dealings are successful, however. Last year they joined forces with First Boston for a 13th-inning grab of RJR Nabisco and missed. Still, the Pritzkers produce: Earnings for Robert-run Marmon Group jumped 41% last year to $204 million. |

-74 **FORTUNE** SEPTEMBER 11, 1989

## SOURCE REFERENCE LIST
## BOOK ONE


### CHAPTER 1

1 Indpls HERALD V1631p1
2 N & M Cat 1893
3 MBF 29
4 Memoirs of Wayne County and City Richmond, IN. Henry Clay Fox, Ed. Western Historical Assn., Madison, WI, 1912 V1p469-70.
5 Indpls NEWS 06JY49p12C1
6 Secy-Treas 1904, Secy 1906
7 AT 22 NO 24 p112
8 MBF 29 Ch VII p 14
9 See Chap. Ten, Part 2
10 MTR NO 22
11 "An Engineer's Story of Marmon History-1918/1934" J. Felts Marmon (Club) News 1981
12 Indpls NEWS 22 MY 79 p8
13 Rob't L. Baker & Marvin Carmody, Ind. Univ. Press 1975
14 First successful gasoline powered car in the USA designed by Chas. E. Duryea and built by T. Frank Duryea in 1893; MVMA Automobiles of America 1961
15 N & M Cat 630p37
16 CTJ OC 05 p145-8
17 Charles Scribner's Sons, NY, NY, 1974
18 Car Collector JA 81

### Chapter 2

1 NAHC
2 MBF29Sec1p24 — "The First Seventy-Five Years"
3 MW MY07p123
4 Oldsmobile Automobile Quarterly Publications 1972.
5 NAHC, letter H.C. Marmon to Editor, Chilton Co.
6 MTR NO22
7 MA04JA06p25
8 H.C. Marmon scrapbook, Speedway Museum, Indpls, IN. (circa 1975).
9 "Marmon-A Quest For Perfection"; M. Hendry, Automobile Quarterly, Vol. 7, No. 2.
10 Copper-cooled Chevrolet of 1922; all were recalled except one purchased by Henry Ford and preserved in the Ford Museum.
11 AE NO81p59-62
12 HA03FE05p188
13 HA25SE07p399-400
14 MW 26SE07p1143
15 CATJO C05p116, 120
16 MA06DE06p31-32
17 AE MA82V90#3p71
18 HA FE05p189
19 CATJ DE08p117
20 CATJ FE08p75,76,119,120

### Chapter 3

1 CATJ01MA09p117-122
2 A02DE08
3 MTR JA12p17
4 N & M ad A14JY10p105 & American Magazine AP10p75
5 N & M Model 32 Introductory Cat.
6 N & M ad National Geographic
7 A24JY13p114; ATJ JY13p6
8 ATJ AU14p87
9 Motor Print JY15p32-33
10 HA 23AU11p280
11 N & M Parts Book 1334-B
12 A 04JA12p52
13 AN 16SE83
14 N & M ad LD JY14
15 MTR 10FE10p 105
16 HA 16SE08p340
17 MTZ (Motortechnische Z) Vol. 38, 1977
18 CATJ AU14p88

### Chapter 4

1 An Automobile Quarterly Marque History Book, 1977
2 Automobile Quarterly, Vol 7, No. 2.
3 Car Collector JA81
4 Car Life AU61p38
5 Autocar 05JU15
6 Pick-up, Van & 4WD AU83p9
7 'Zelda — A Biography', Nancy Milford, Harper-Roe & Co. 1970
8 The Spur 15MY24p154
9 Men of Marmon MY31p5
10 A23DE15p1140
11 HA01FE16p136
12 AT JA21p864
13 AT05MY21p978
14 AT12MY21p1040
15 Marmon POST 15SE21V4#12
16 AT04FE22p
17 Castle Books, NY, NYp56
18 AT19AU22p19
19 AT21AP23p390
20 MA10JA24p21
21 Chas. Scribner's Sons, NY, NY 1943
22 SEP15NO24p76
23 Hugo Pfau, C & P MA74p147
24 AT08NO24p1230-31
25 SEP 06SE24p126-7
26 SEP07FE25p81
27 SEP21MA25p89
28 N & M Certification to State Registration Authorities
29 AT12JU26p429

30 Indpls NEWS 11MA68p30C5
31 Hemmings' Motor News
32 'Automotive Steels — They Still Do It Better' AISlad
33 NAHC
34 Interview with Jerome (Joe) Felts
35 N & M Salesman's Handbook 1920
36 N & M Cat A-23-10
37 AT26SE25p617
38 AT03AP26p718
39 AT08JA20p60
40 N & M Mod34 Introductory Cat. MY1916p41, 33.
41 A119Au26p300
42 N & M 2034 (Owner Info) p92-93
43 MW29SE23p22 & AI27SE23p632-633

### Chapter 5

1 AT05OC29p752
2 AT03NO26p723
3 AT31MA23p621
4 AI23OC26p999
5 AT01JA27p705
6 LM Introductory Ads
7 MTR JA28p328
8 AT13JY29p815
9 MBF29 Sec4p19, Sec3p12, p5, 6
10 AB JU30p225
11 Founder and chairman, Merle Norman Cosmetics, Classic (157 car) Beauty Collection, San Sylmar, CA.
12 NAHC
13 AB MA30p186
14 AT13OC28p831
15 Men of Marmon JA31
16 An Automobile Quarterly Marque History Book, 1977
17 Marmon — A Quest For Perfection; M. Hendrey, Automobile Quarterly Vol. 7., No. 2
18 Marmon L-57Cp23Fig.30
19 AI16DE27p86 (Serv. Man)
20 AI24AP28p668
21 See Chapter 1 (12)
22 'A Brief History of the Continental Engine Co.; G.M. Nau & Jan B. Norbye, SIA J-F77
23 'Custom Body Era'; Hugo Pfau, A.S. Barnes & Co. 1970 p116
24 MTR NO29p124
25 US Pat 1,812,474 George Freers 30JU31
26 AT26NO24p1065
27 Sole Licensee, Willys-Morrow Co., Elmira, NY. Alo7DE29
28 MTR JA30p140
29 Style S-70 @ 22389; S-72 @ 2143; S-74 @ 1658 from start of prod'n
30 AI 04JA30p22

### Chapter 6

1 AA J-A78p32
2 The American Motor Car — A Brief History; (Univ. of) Chicago History of American Civilization.
3 'San Sylmar — A Treasure House of Functional Art'; 1978
4 Heritage House, Arcadia CA, 1979
5 See Chapter 1 (11)
6 MTR JA29p80,81,364
7 'Sixteen Cylinders — A Question of Priority Unresolved'; AA J-A79p33-38
8 'Cadillac — The Complete History'; M. Hendrey, an Automobile Quarterly Marque history book, 1977
9 A115NO30p742
10 AI25AP31p677
11 AT03JA31p682; 2-6
12 AT05SE32p400
13 'Prophetic Design'; W.D. Teague, AT06JA34p563
14 CCJU79p29-30
A "Marmon 16 Revisited." W.S. Snyder, by permission
15 'W.D. Teague-No Gee-Gaws on the 707'; I-N Co. ad Machine Design 1969 p47-49
16 'Housing 200 Horses'; M. Hendrey, Car Collector JA81p12
17 C & P MA68p36,39
18 'Marmon V-16'; E.L. Robinson, R & T JY52p46
19 'not enough people bought Marmons'; E. Janderquist & G. Borgeson TRUE Auto Yearbook #3
20 Marmon (Club) News N-D82; J-F83
21 Mr. Ray Day was honored by Detroit Section SAE in 1982 for 50 years membership.
22 CC Sum55p10 "True Tales of the V-16" Reprinted by permission
23 Automobile Quarterly, Vol. 7, No. 2, p137
24 'American Auto Racing' Penquin Books, The Viking Press, Inc. 1974
25 AI10JA31p74
26 'Custom Body Era'; Hugo Pfau, A.S. Barnes & Co. Inc. 1970
27 "Mechanical Magnificance —", W. Fitzgerald Car Life Classic JU67p35
28 Time Books, NY, NY 1979
29 AI 06MY33p570
30 AT 03JA31p719-21
31 AI NO81p36
32 Tech Ed., R & T JY52p46
33 Marmon Transmission Service Bulletin (Typewritten without illustration)

### Chapter 7

1 Marmon ad 1928
2 See Chapter 1 (12)
3 NAHC
4 TRUE Auto Year Book #3
5 Author of 'Public Image of Henry Ford'; Wayne State Univ Press 1976
6 'Marmon's Last Gamble'; W.F. Robinson, Jr. SIA AU79p, 37
7 Motor (UK) NO31p503
8 'Car Designed and Built by Marmon & Freers'; Promotional Brochure 1933

9 National Geographic JY82p42,58

Chapter 8

1 July 78 Cat 3806
2 'Winging It With The Wasp'; Marion George, Automobile Quarterly, Vol. 7, No. 2
3 Marmon ad
4 MBF29 Sec1p5
5 'Marmon-International Champion'; N & M Co. 1911
6 'The Grand Prix Car'; Lawrence Pomeroy, Motor Racing Publications Ltd., London, Temple Press Ltd 1949, 1950
7 Everybody's Magazine AU16
8 'Serial Numbers of the First Fifty Years'; Grace Brigham, Brigham Press, Atlanta GA 1974
9 A22AU12p396-397
10 AT10NO23p1250
11 AT14AP28p817
12 'American Auto Racing — An Illustrated History', A.R. Bochrach, Viking Press, 1974. Reprinted by permission, Viking Penguin
13 MA31MY28p13
14 'Golden Age of the American Racing Car'; G. Borgeson, W.W. Norton Co. NY, NY 1966
15 'Miller Dynasty'; Mark L. Dees, Barnes Publishing Co. Scarsdale NY, 1981
16 'The 132 Most Unusual Cars That Ever Raced at Indianapolis'; Lyle Kenyon Enger, Arco Publishing Co. NY, NY 1970
17 LA Daily News 1946 & CC JU77p34

Chapter 9

1 See Chapter 12 Part 2
2 MA13JY16
3 Marmon Post V3#5p39
4 The Marmon Post V3#14p189
5 'Custom Body Era'; Hugo Pfau, A.S. Barnes & Co. 1970, p80, 184, 191
6 Marmon WASP house organ
7 See Chapter 5 (17)
8 Country Life JA20p100
9 AB DE26
10 CC, the Ed. AU71p36
11 N & M Cat 2231
12 MA21JA24p27

Chapter 10

1 Marmon ad, Harper's Magazine Advertiser DE08
2 Standard Advertising Register
3 Marmon POST house organ
4 'Modern Sports Car Series'; Sports Car Press, Ltd. 1963
5 Jim Quinlan, TORQUE (MI region CCCA) N-D78p49
6 SEP22JA27p81
7 McKetrick Directory of Advertisers 1932
8 MW 07,p338
9 AT06FE26p1321
10 AT JY27
11 AT02OC26p703
12 AT19OC29p
13 AT19MA27p505 & AT09AP27p793
14 AN Market Data Book 27AP83p22
15 'Automobiles of America'; AMA (MVMA) 3rd ED 1970p89

Chapter 11

1 MBF29S1P8
2 Chicago, Hamilton & Dayton RR
3 See Chapter 12 Part 2
4 Indpls Sunday STAR 27) OC18Pt2p1C4
5 Indpls STAR 24JU19p1C7
6 Ward's Auto World JU83p39
7 AT02JU26p622
8 AT13OC28p831
9 AT09OC26p
10 AT10MY27 AT16JY27p863
11 Marmon Factory NEWS MA29p12
12 AI12AP30p
13 American Machinest 27AU31 p344

Chapter 12

1 MBF29S1p4,7
2 See Chapter 6 (12)
3 Old Cars Newspaper 04OC77p71-72
4 AT02MY29p270
5 AT30JA31

Chapter 13

1 N & M Information For Overseas Dealers 1921p1
2 Indpls STAR 13FE40p6C5
3 Indpls STAR 24 MY41p26C2
4 See Chapter 4, Part 2
5 AT03JA31p682-687
6 Indiana Corp Report date; Moody's states MY 17th
7 AI 20JU31p965
8 AT14MY32p73
9 Moody's Manual of Investments
10 AT25JU32p
11 AT10JU31p965
12 More than 20 firms in 1929; progressively fewer subsequently.
13 AT09DE33p209
14 Indpls STAR 26AP34
15 AI JA34p62
16 Indpls STAR 01JA37p1

## SOURCE REFERENCE LIST
## BOOK TWO

Chapter 1

1 (Fuller) TRANSMISSION (Co) TOPICS OC61
2 AT22NO24p42
3 M-H Anniversary Series Cat 520930 1952 p13
4 AT26AU33p137-142

Chapter 2

1 R.C. Wallace Interview 1980
2 FDSF AP36p10
3 FDSF JA37p31
4 FDSF OC37p29
5 M-H AWD NEWS V6#12p3
6 M-H Price List 371104-1 (decoded 1937 Nov 04 list #1)
7 M-H Price List 371105-1
8 "The Observer's Fighting Vehicle Directory — WWII" Bart Vanderveen, Frank Warne & Co., NY, NY 1972

Chapter 3

1 Crestline Publications, Sarasota, FL.
2 "Ford Trucks Since 1903", James K. Wagner, Crestline Publ. Sarasota, FL.
3 Quoted or exerpted from M-H NEWS V2#7MA44
4 'Public Image of Henry Ford', David L. Lewis, Wayne State University Press, Detroit MI, 1976 p51
5 'Background Study of the M-H Co. Inc., Specialists for Unique and Unusual Military and Civilian Requirements', 1956
6 M-H NEWS V1#9MY43
7 See Chapter 2 (8)
8 M-H NEWS 44
9 Col. Robert. J. Icks (ret'd) letter to W. Frank Robinson, Jr. 06Au83

Chapter 4

1 Ford Field DE51p27
2 Indpls NEWS 26AU54
3 Indpls TIMES 24JU38
4 HA 3#23p790
5 "Transit's Stepchild: The Trolley Coach", Mac Sebree & Paul Ward, 1973
6 Marmon Transmotive Fact Sheet 03AP75
7 AN22AP74p9
8 OVERDRIVE AP77p140
9 FLEET OWNER OC69p210

Chapter 5

1 'Illustrated Encyclopedia of Trucks and Busses', Dennis Miller, A Quattro Book, in US: Mayflower Publishing Co. 1982
2 AI15JU73p54
3 AN18AP83p26

Chapter 6

1 'Five Eras of the Marmon Car', G.P. Hanley, AA J-A78
2 Moody's Industrial Manual
3 M-H Cat.
4 'Certificate of Withdrawal of the Marmon Group (Delaware 1967) from Business in Indiana'.
5 AT21MA31p
6 Indpls STAR C82p4C
7 'Ten Years Before Pearl Harbor', M-H 441001-2 Rear Cover
8 Indpls TIMES 09DE64p24
9 Indpls STAR 03OC82p4C

# ABBREVIATION CODE FOR SOURCE REFERENCE LIST

| | |
|---|---|
| AA | Antique Automobile/Antique Automobile Club of America |
| AB | Autobody |
| AE | Automotive Engineering/SAE (Society of Automotive Engineers, Inc.) |
| AI | Automotive Industries |
| AN | Automotive News |
| AQ | Automotive Quarterly |
| AQP | Automotive Quarterly Publications |
| AT | Automotive Topics |
| ATJ | Automotive Trade Journal |
| CATJ | Cycle and Automotive Trade Journal |
| C & P | Cars and Parts |
| CC | Classic Car |
| FDSF | Ford Dealer and Service Field |
| HA | Horseless Age |
| Indpls | Indianapolis . . . |
| MA | Motor Age |
| MBF | Marmon Book of Facts published annually 1920-29, Education Div, Sales Promotion Dept. |
| MTR | Motor (US) |
| MW | Motor World |
| NAHC | National Automotive History Collection/Detroit Public Library |
| R & T | Road and Track |
| SEP | Saturday Evening Post |
| SIA | Special Interest Autos |

Infrequently referenced sources are fully identified at specific reference location. Certain complex references are referred at specific chapter reference to a prior chapter reference.

The source name, if a periodical, will be typically a 1 to 3 digit letter code, followed by a 6 digit day, month and year, and ending with the page number (s) preceded by the letter p. MA for March, MY for May, JU for June and JY for July.

**DICK NESBITT**
*Marmon Chief of Industrial Design*

In 1989 MARMON MOTOR COMPANY celebrated 25 years of Class VIII truck manufacture. During that time they have built an enviable reputation in the USA and abroad as builders of premier trucks with superior cabs and cabins.

6233 AV

**SUBJECT INDEX**
**BOOKS ONE & TWO**

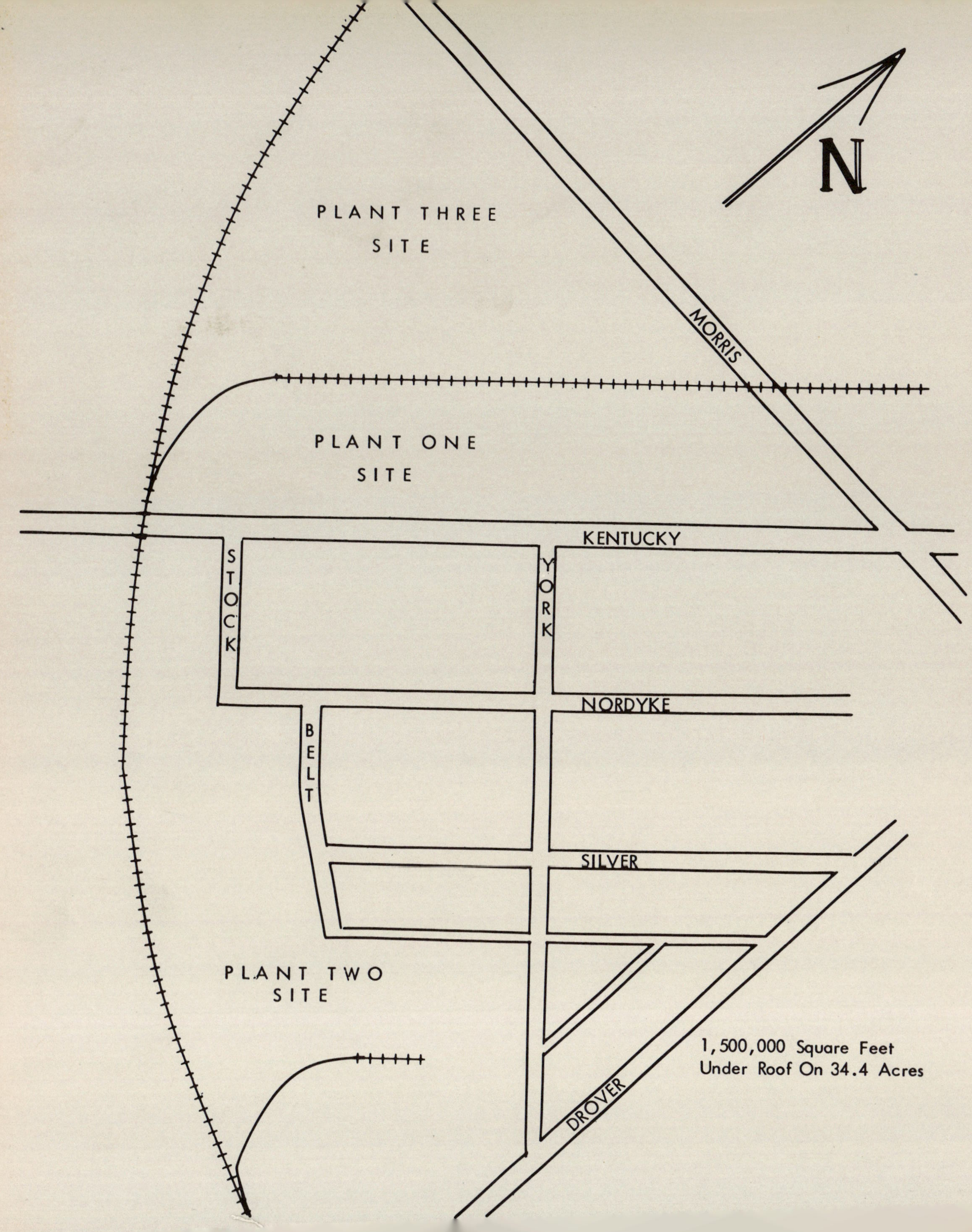
N
PLANT THREE
SITE
MORRIS
PLANT ONE
SITE
KENTUCKY
STOCK
YORK
NORDYKE
BELT
SILVER
PLANT TWO
SITE
DROVER
1,500,000 Square Feet
Under Roof On 34.4 Acres